CONSTITUTIVE LAWS FOR ENGINEERING MATERIALS

RECENT ADVANCES AND INDUSTRIAL AND INFRASTRUCTURE APPLICATIONS

ASME PRESS SERIES ON MATERIALS, MODELLING, AND COMPUTATION

These proceedings constitute Vol. 1 of this new series. This monographs and proceedings series will include publications on such areas as modelling of metals, composites, geomaterials, ceramics, polymers and plastics, materials for electronic and optical devices, space materials, interfaces and joints, and engineered and tailored materials, laboratory testing and parameter identification, and computational procedures.

Constitutive Laws for Engineering Materials: Recent Advances and Industrial and Infrastructure Applications, edited by C. S. Desai, E. Krempl, G. Frantziskonis, and H. Saadatmanesh, 1991

CONSTITUTIVE LAWS FOR ENGINEERING MATERIALS

RECENT ADVANCES AND INDUSTRIAL AND INFRASTRUCTURE APPLICATIONS

Proceedings of the Third International Conference on Constitutive Laws for Engineering Materials: Theory and Applications, held January 7-12, 1991 in Tucson, Arizona, USA

EDITORS

C.S. Desai

Department of Civil Engineering and Engineering Mechanics
University of Arizona
Tucson, Arizona, U.S.A.

E. Krempl

Department of Mechanical Engineering and
Aeronautical Engineering and Mechanics
Rensselaer Polytechnic Institute
Troy, New York, U.S.A.

G. Frantziskonis

Department of Civil Engineering and Engineering Mechanics
University of Arizona
Tucson, Arizona, U.S.A.

H. Saadatmanesh

Department of Civil Engineering and Engineering Mechanics
University of Arizona
Tucson, Arizona, U.S.A.

ASME PRESS NEW YORK 1991

SPONSORS

Center for Material Modelling and Computational Mochanics, CEEM Department, University of Arizona, Tucson, AZ, U.S.A.

Departmenr of Mechanical and Aeronautical Engineering and Mechanics, Rensselaer Polytechnic Institute, Troy, N.Y., U.S.A.

FINANCIAL SUPORT

National Science Foundation (Design and Manufacturing Systems Division), Washington, D.C., U.S.A.

Air Force Office of Scientific Research, Bolling AFB, Washington, D.C., U.S.A.

Allied-Signal Aerospace Company, Garrett Engine Division, Phoenix, Arizona, U.S.A.

Salt River Project, Phoenix, Arizona, U.S.A.

Library of Congress Cataloging-in-Publication Data

International Conference on Constitutive Laws for Engineering Materials : Theory and Applications (3rd : 1991 : Tucson, Ariz.)
Constitutive laws for engineering materials : recent advances and industrial and infrastructure applications : proceedings of the Third International Conference on Constitutive Laws for Engineering Materials—Theory and Applications, held January 7-12, 1991, in Tucson, Arizona, U.S.A. / editors, C.S. Desai . . . [et al.].
p. cm. — (ASME series on materials, modelling, and computation)
Includes index.
ISBN 0-7918-0024-5
1. Materials—Mechanical properties—Congresses. I. Desai , C. S. (Chandrakant S.) , 1936- . II. Title. III. Series.
TA405.I55 1991
620.1' 1292—dc20 90-22578
CIP

TABLE OF CONTENTS

GENERAL PLASTICITY, FINITE STRAINS

CYCLIC PLASTICITY

VISCO-, CREEP, RATE DEPENDENCE

THERMAL EFFECTS

FRACTURE, DAMAGE, LOCALIZATION

MICRO-MACRO CORRELATION

DISCONTINUITIES: INTERFACES, JOINTS

SPECIAL MATERIALS: METAL MATRIX COMPOSITES, CERAMICS, POLYMERS, ENGINEERED MATERIALS

IMPLEMENTATION, EVALUATION

TESTING, PARAMETER IDENTIFICATION

MANUFACTURING ASPECTS; WORKSHOP

WORKSHOP: Participants (Tentative List)

Name/ Affiliation	Topic
Dr. K. Bakhtar Bakhtar Associates Long Beach, CA, U.S.A.	Double-Scale Physical Modeling in Geomechanics
Dr. A.D. Freed NASA-Lewis Research Center Cleveland, Ohio, U.S.A.	Viscoplastic Model with Application to LiF-22% CAF_2 Hypereutectic Salt
Dr. R. W. Griffiths G2 Systems Corporation Pacific Palisades, CA, U.S.A.	Structural Monitoring Systems Using Fiber Optics: Infrastructure Rehabilitation
Dr. G. Harritos Air Force Office of Scientific Research Bolling AFB, Washington, D.C.	Engineered Materials
Professor F. G. Kollmann Technische Hochschule Darmstadt, Germany	State of the Development and Application of Inelastic Constitutive Equations in Germany
Dr. R. L. McKnight General Electric Company Cincinnati, Ohio, U.S.A.	Constitutive Modeling in the Era of Probabilistic Tailored Structures
Dr. L. Matsch Allied-Signal Aerospace Company Garrett Engine Division Phoenix, Arizona, U.S.A.	Application of Composite Materials in Gas Turbine Engines: Goals and Challenges
Dr. J.D. Murff Exxon Production Research Company Houston, TX, U.S.A.	Materials for Offshore Construction
Professor N. Ohno Nagoya University Nagoya, Japan	Evaluation of Inelastic Constitutive Models: FEM Analysis of Notched Bars of 2.1/4CR-1MO Steel at 600°C. *Second Bench Mark* Report of Subcommittee on Inelastic Analysis and Life Prediction
Dr. A. Phillipp MTU-Motoren und Turbinen-Union Munich, Germany	Present and Future: Aeroengine Materials
Dr. M.J. Salkind Ohio Aerospace Institute Cleveland, Ohio, U.S.A.	Integrating Materials Design into System Design
Dr. T. Triffet NASA Center for Utilization of Local Space Materials University of Arizona Tucson, Arizona. U.S.A.	Materials for Space Applications

PREFACE

Our knowledge of the mechanical behavior of engineering materials and its modeling in constitutive equations has lagged behind advances in such other related areas as increasingly sophisticated computers and computational methods for solution of complex problems of today and tomorrow. Proper understanding and characterization of available and new materials and implementation of the models in modern (computational) solution procedures is a vital component for safe, economical and competitive design of industrial and public works (infrastructure) systems. It is believed that unless constitutive models for the materials composing these systems, based on sound scientific principles, laboratory/field testing and verification, are employed in the analysis, the computer results will have only limited validity, if at all. This recognition has spurred significant national and international activities towards research for constitutive modelling and testing for a wide range of available and newly emerging materials. The major thrust of these activities is rather recent, but is occurring at a very rapid pace.

Following the overwhelming reception and success of the two previous conferences, the "Third International Conference on Constitutive Laws for Engineering Materials: Theory and Applications" is held at the University of Arizona, Tucson, Arizona, during the period January 7-12, 1991. A workshop on "Innovative Use of Materials in Industrial and Infrastructure Design and Manufacturing," which is expected to be of interest to industries and academia alike, is also organized during the conference.

A great number of models have been developed for a wide range of materials; however, many of these models involve common and unifying characteristics. In addition to theoretical developments, it is also essential to calibrate the models based on appropriate laboratory tests and identify and evaluate significant parameters. Once the models are developed, it is important to implement them in solution (computational) procedures for modern engineering problems for industrial and infrastructure design and manufacturing. Here, numerical characteristics of computer algorithms must be established so as to lead to reliable and robust solutions. Furthermore, in addition to the conventional areas and material systems, it is important to consider recent developments for special areas such as new materials, engineered and tailored materials, and electronic devices.

The foregoing issues are covered in the conference proceedings through presentation and discussion of 195 papers, and in the workshop sessions. A tentative list of workshop participants is provided in these proceedings. It is believed that the conference and workshop will provide significant and meaningful opportunities for deliberations on recent international developments in the subject areas and toward establishment of future directions.

A symposium on "Problems, Theories and Solutions in Plasticity" will be held during the conference to honor Professor E.H. Lee, Department of Mechanical Engineering and Aeronautical Engineering and Mechanics, Rensselaer Polytechnic Institute, Troy, N.Y., U.S.A., on his retirement after a long and distinguished academic career. A separate volume will be published to include about 15 papers in the Symposium Proceedings, which will be edited by Prof. W. H. Yang, University of Michigan, Ann Arbor, Michigan, U.S.A.

A new series, "Materials, Modelling, and Computation" that will include publication of monographs and proceedings is being initiated with this volume. The series will be published by ASME Press, an imprint of The American Society of Mechanical Engineers .

We take this opportunity to honor Professor D. C. Drucker for his significant and substantive contributions, over a long and distinguished career, to the fields covered in this conference and proceedings.

We wish to thank the authors for their cooperation and timely completion of the manuscripts. We also wish to gratefully acknowledge the financial support by the sponsors listed herein.

September, 1990

C. S. Desai
E. Krempl
G. Frantziskonis
H. Saadatmanesh

SPECIAL INVITED PAPER

We take this opportunity to honor Professor Daniel C. Drucker for his significant and substantial contributions to the fields of this conference.

CONSTITUTIVE RELATIONS FOR SOLIDS RETROSPECT AND PROSPECT

Daniel C. Drucker
University of Florida

ABSTRACT

A perspective is offered on the continued intensive attention to constitutive relations. An assessment is presented of the value of more accurate representations of the behavior of materials across the spectrum from ductile metals and alloys to composites, ceramics, concretes, and geomaterials. A number of open questions and uncertainties are exhibited in the hope that the significant will be resolved and the others will not remain a distraction.

INTRODUCTION AND RETROSPECT

There is no space here for either an extensive discussion of most of the points and questions that will be raised or for proper citation of the literature. Only some highlights can be offered. Reference is made to a few previous papers that go into greater depth on the personal assessment that is presented. They do contain the references to the great contributions that have been made by so many. The proceedings of this and the two preceding international conferences also are full of relevant views and information and of references to the world literature.

Elasticity, plasticity, viscoelasticity, viscoplasticity and their many combinations have been under development for a long time. The materials of interest have included ductile metals and alloys, ceramics, rock and concrete, granular media, and polymeric materials individually and in combination. Sometimes driven primarily by the needs of engineering practice, sometimes primarily by curiosity, experimental results and mathematical developments have been synthesized through the years with an increasing understanding of the behavior of materials on the microscale to achieve both a clearer picture of the constitutive relations of materials and useful idealizations of value in analysis and design.

Great debates have raged for decades on issues that seemed very important at the time but which in retrospect are seen to have been ill-conceived or fundamentally in error. Yet others have shaped our understanding today. They have led to idealizations such as time independence, initial and subsequent isotropy of response, and a simple dependence or independence of the effect of hydrostatic pressure on flow. These and many other idealizations have proved enormously helpful as stepping stones on the path of progress even when they are far from realistic descriptions of behavior of a material of interest.

Drastic simplifications, such as perfect plasticity or isotropic hardening or linear elasticity or particularly simple forms of nonlinear viscosity have served and continue to serve us well. Until 1950 [1] the community dealt almost solely with elementary forms of these and other idealizations or simplifications. Despite the availability of more realistic idealizations based upon our great increase in understanding, and despite our far greater ability to compute, they still are in extensive practical use today.

In the development of a complex subject that has a long history, both the basic concepts and the procedures for their application are reviewed periodically as new information and ideas become available and old ones become obsolete or are forgotten. Once more we are in a period of great resurgence of interest in constitutive relations for materials. Although not exclusively, it is primarily and surprisingly for time-independent materials. There is a growing recognition in diverse design communities of the need to translate qualitative understanding of the behavior of materials on the macroscale to constitutive relations for the solution of problems. Materials of interest now include brittle ceramics and concretes as well as ductile metals, fiber-reinforced and other composites as well as quasi-homogeneous materials. Designers are under ever greater pressure to produce very reliable designs and to predict their performance with high accuracy. Consequently they call for more and more accurate representations of behavior in three dimensions in the traditional time-independent regime and for reasonably good estimates in the important but less explored domain of time and temperature effects.

The ever-growing availability of increasingly powerful computers at decreasing cost certainly is another of the key driving forces. It must seem to the less fully initiated that it no longer is necessary to employ drastic simplifications of behavior in order to obtain solutions. Certainly the most general elastic response, anisotropic and nonlinear with finite displacement gradients (large strains and large rotations) can now be taken into account as needed. That complete success in the elastic domain creates an expectation to do similarly well in the inelastic.

THE INFINITE COMPLEXITY OF INELASTIC RESPONSE

However, it should always be kept in mind that the inelastic response of materials, even when time effects are negligibly small and temperature is held constant, is essentially infinitely complex in its detail. This is in stark contrast to the elastic response which is fully determined by a small number of tests rather simple to perform. The reality of inelastic response is that no finite number of tests is able to predict uniquely the response to yet another test on identical material. Also, materials are not that reproducible. Elastic moduli will not vary much, but the details of inelastic stress-strain behavior will vary considerably from one batch to another of what is nominally the same material. Furthermore, it is not possible to perform most of the tests on a strong material that would be needed to determine its behavior in three dimensions.

Consequently, drastic simplification or over-simplification of inelastic behavior is not to be viewed as an undesirable option but as a necessary approach in the development of constitutive relations for strong materials. Even with geomaterials, to which variable loads can be applied on each of the faces of a specimen in the shape of a cube, the appreciable inhomogeneity of the resulting state of stress makes the determination of stress-strain relations quite uncertain in detail.

BEHAVIOR TO BE MODELED DEPENDS UPON THE PROBLEM

The appropriate simplifications to be employed in design or analysis depend fully as much on the problem to be solved as the actual behavior of the material in all its detail. Identification of the most essential and relevant features of material behavior for the problem or class of problems at hand is a necessary first step in the development of a useful constitutive relation or the choice of an appropriate existing one. Micromechanics is always relevant but micromechanics on the atomic or dislocation level is more a guide to understanding and the design of materials than a method of calculation for design use. Relations among the possible experimental observations of the response of material on the macroscopic level, basic understanding on the microscopic level, and design criteria are often far from obvious. Most often they are more easily clarified when one or more drastically simplified constitutive relations are employed than when an attempt is made to use the most comprehensive relation available.

GOOD DESIGN IS BETTER THAN SUPERB ANALYSIS [2]

The analysis of a given design with constitutive relations known to be satisfactory for the material selected by the designer is both interesting and important. However, any improvement in the design that results will normally lie within the domain of the initial design. It would be far better to help shape that preliminary design with the selection of a desirable class of constitutive relations and properties that will lead to a good design satisfying all requirements with close to optimum configuration and cost. Then a suitable cost-effective existing material would be selected or, if economically feasible, a new one designed for which the relations chosen provide an adequate representation.

The proper selection of an existing material or the design of an appropriate new one to provide desired macroscopic properties is a much more valuable contribution than the most complete analysis of behavior with a less suitable material. In the general sense, constitutive relations include flow strength, ductility, fracture toughness, rate of crack growth in fatigue, rate of wear, etc. as well as conventional and unconventional stress-strain-time-temperature relations.

It is not enough to consider just the initial response of a material. The response to be taken into account must include the behavior in any subsequent damaged or altered state that can reasonably be expected over the design lifetime. A satisfactory estimate of the safety of a damaged structure and its ability to function adequately will be more significant than an accurate prediction of the response of a perfect structure to working loads. This is especially true when injury or loss of life may result or when the structure is out in space and inaccessible or otherwise is not easily accessible. Satisfactory (adequate) performance is essential over the entire spectrum of requirements. Excellent performance in some respects cannot compensate for inadequate performance in others.

ASSEMBLAGES OF MATERIALS OR MODEL ELEMENTS [3]

A structure or a machine or a geotechnical mass will often be an assemblage of a variety of materials, each of which may in itself be an inhomogeneous assemblage. In the modeling of the behavior of a single material it often is helpful to view the material as an assemblage of simple and well-understood model elements. Consequently the behavior of assemblages and of assemblages of assemblages is of some fundamental importance as are the associated constitutive relations.

When thinking about the behavior of materials, it is always well to keep in mind that combinations of isotropic constituents can produce anisotropic material; combinations of anisotropic constituents can produce statistically isotropic material; no plastic volume change on the microscale can lead to macroscopic "plastic" volume change; perfectly plastic components combine to give workhardening; weakly workhardening components may combine to a strongly workhardening material; components with high creep rates may produce a composite with little creep; elastic-perfectly brittle constituents may simulate elastic-plastic behavior and vice versa.

An idealized, well-defined material system will have known or computable state variables. The laws of thermodynamics or special postulates of a thermodynamic nature then will lead to laws of behavior and thermodynamic statements. Such a thermodynamic (or mechanics) statement can be tested for its validity by examining whether or not it applies to a system that combines two or more materials or bodies which individually obey the rule. If the combination is a member of the original class, the rule must apply. If it is not, and the rule does not apply, the conclusion is that the thermodynamic statement is at best of limited validity. It cannot be counted upon as a principle of behavior in any but the special case for which it was developed.

A proper but limited thermodynamic approach is consistent and may be of value for particular problems or particular materials. It is its generality and its broader implications that cannot be depended upon. The use of plastic strain as a state variable is one such example. Principles of maximum entropy production appear to be another, but more study of this question would be worth while.

DISTRACTIONS

There are a number of topics in the plastic behavior of ductile metals and alloys to which considerable attention has been devoted in the past and which now and in the future should be viewed as mere distractions. Among them are: Do yield surfaces really have corners? What is the exact size of each current yield surface? What is the real shape of each surface?

The more sensitive the measurements that are made, the smaller will be the diameter of each yield surface. When the motion of a modest number of dislocations are detected as macroscopic plastic deformation, the observed yield surface will shrink to zero size.

Corners in the yield surface at the current stress point and elsewhere will appear or disappear with the method employed for their detection and the sensitivity of measurement, whether they are there or not. The yield surface shape obtained will vary similarly.

The appropriate question to ask is not whether corners really exist or do not but rather whether the behavior of the material for the problem at hand is better or more conveniently described with adequate accuracy by a yield surface with corners or without corners. Similarly for the size of the yield surface. The question is simply the suitability of the size chosen for the reasoning or the analysis or the computation being done.

How about time effects? Are they real? Of course they are and do in fact govern in many problems, especially those involving elevated temperatures and either very short or long times of loading. They are present even in the most time-independent material. Should they be studied, therefore, and always taken into account in design and analysis. Studied - yes!; taken into account - not necessarily! If a time-independent idealization is a sufficiently accurate description of behavior for the problem at hand, it is entirely appropriate to use it as is the practice in most structural analysis and design. This does not mean that it is inappropriate to employ a time-dependent "unified theory" in a computer program to obtain much greater speed of computation in what is basically a time-independent problem. If the contribution of the time effects to various aspects of the solution are small so that the stresses, strains, displacements, etc. will be close to results for a time-independent formulation all is well.

Of course, when time effects are significant, it is essential to use an appropriate time-dependent constitutive relation.

A FEW OF THE OPEN QUESTIONS [4]

Enormously more attention has been paid and continues to be paid to isothermal time-independent behavior that to time and temperature dependent constitutive relations for general states of stress. Many questions have been raised that have not been answered; many more questions have yet to be asked. Far more attention should be devoted to elucidating the behavior of materials in the time and temperature dependent regime and codifying the results in realistic classes of constitutive relations.

Also, many materials at working temperatures are chemically active over times of interest. Yet little has been done to evaluate the effect of such activity on the constitutive relations for metals and alloys, ceramics, composites, and geomaterials.

At the high density of dislocations and other defects in metals and alloys subjected to high strain, a significant fraction of the volume (of the order of 10 or 20%) is in far from regular atomic array. It may well be closer to an amorphous than a crystalline state. If so, then such regions may have constitutive relations reminiscent of granular rather than crystalline materials. The constitutive relations on the macroscale should then should reflect this behavior properly.

The constitutive relations for granular media appropriately are given a great deal of attention these days for a variety of reasons ranging from the possibly devastating effect of earthquakes to the safety of small earth dams, from the design of blast resistant structures to the load carrying capacity of footings. Much experimental work is done and analysis and computation undertaken. However, for the most part, the constitutive relations employed are at the primitive stage of invariants of the stress tensor as were the constitutive relations for metals and alloys some 40 years ago. Much remains to be done, therefore, to achieve comparable generality and realism.

Yet it should always be kept in mind that the objective of constitutive relations is to obtain a reasonably satisfactory result for the problem at hand. Therefore, the inability of one simple form to cover other experimental facts does not preclude the use of that simple form in its range of approximate validity.

Ceramic materials are assuming greater and greater importance in the industrial world as their ability to carry appreciable tensile stress is enhanced. Modification of the material to ensure adequate fracture toughness is far more important than determining the details of the stress-strain-damage response of an inadequate material. A very modest number of dislocations are generated and move through a ceramic as contrasted with a ductile metal or alloy. Therefore a study of single crystal and bicrystal dislocation interactions and other effects should be translatable to the macroscale in a direct and useful manner that is not possible for metals.

Many ceramics, concretes, and mortars may be idealized as either elastic-perfectly brittle with cracks that initially propagate in a stable manner or as elastic-plastic with a high local ductility in a binder phase joining the brittle bulk constituents. Sintered carbides, such as tungsten carbide with a cobalt binder phase that is ductile, clearly fit into the latter category. When a glassy phase provides some or all of the binding, as may happen in silicon carbides or nitrides, the picture is less clear. However, even for concretes and mortars where the evidence of crack initiation and growth is so clear, it turns out that a treatment of the cement phase as plastic provides a consistent approach that often correlates well the tensile and the bending strengths of mortars and concretes of different aggregate sizes and shapes. Despite the enormous difference between the idealizations, both can explain the appreciably higher bending strength than tensile strength obtained from direct or splitting tests. Both are consistent with the stable growth of microcracks and the applicability of a Griffith type of criterion for unstable fracture. Both are worth much further exploration.

A COMMENT ON INSTABILITY

The question has arisen repeatedly over the years about the necessity of a link between a non-associated flow rule (non-normality of the plastic increment of strain to the current yield surface) and instability as conventionally defined. In recent years, for example, experiments on sands and on concrete in the so-called wedge region have not shown the instability predicted for materials that tests indicate obey so strong a non-associated flow rule that the product of the increment or rate of stress and the increment or rate of <u>total</u> strain is negative. The same question of instability arises with the use of such a rule in static or dynamic computer calculations.

For a time-independent material undergoing small strains and rotations, stability in the small in the forward sense or positiveness of the scalar product of the increments (rates) of stress and total strain for all possible stress and total strain increments or rates (elastic plus plastic) at each state of stress and strain, ensures uniqueness or stability of configuration [4]. The postulate of stability over any path in stress space involving small plastic deformation includes stability in the small in the forward sense and so is more restrictive still. It leads to an associated flow rule or normality of the increment or rate of plastic strain to the current yield or loading surface at the current stress point. It is sufficient but clearly is not necessary for stability of configuration. If the positive contribution from the elastic response is greater than any possible negative contribution from the plastic component of response, stability in the small is preserved and no instability will result.

What therefore is to be made of the results reported by very careful and knowledgeable experimenters of stability of material and apparatus while traversing the "unstable" wedge region? The large pressure sensitivity of sand and concrete along with their frictional character does argue for their modeling by a non-associated flow rule. However, a similar instability question arises for any time-independent material, pressure sensitive or not, that is modelled with a non-associated flow rule over any appreciable domain of stress space.

Surely, if it is at all possible, it is safer to avoid computations with non-associated flow rules. If they are called for, and the elastic response is negligible in comparison with the plastic, local and global instabilities must be anticipated. However, the basic issue needs further careful study.

PRESSURE SENSITIVITY OF METALS

It is most unfortunate that the pioneering work at US Steel on high strength martensitic steels that do show a small but not insignificant pressure sensitivity, or SD effect, had to be abandoned. Much more could have been learned. The side remark I made at the first of these international conferences [5] may be worth repeating. When some interesting and unexpected primary or secondary effect is discovered in a scientific study, considerable money and manpower is devoted to its elucidation and properly so. Scientific curiosity provides ample justification, as indeed it should. Strangely, when an effect of possible practical as well as scientific interest is uncovered in the course of engineering research, attention fades rapidly unless the practical effect is found to be large in current applications. Would it not be reasonable to expect that every highly competent materials engineering group would have a facility in which constitutive relations could be determined under high hydrostatic pressure and other extreme environments? It is obvious that with time there will be more and more important engineering applications of materials under what we now consider to be extreme conditions. Such a facility will be expensive but not in comparison with a variety of routine tools available and used in many fields of basic science. Crucial tests are called for, but no one instead of everyone is in a position to do them.

Fortunately in soil mechanics such facilities are available in several laboratories and many curious effects of importance can be and are being explored.

CONCLUSION

Much is known about the time and temperature behavior of materials under such states of stress as simple tension or compression or shear with added hydrostatic pressure in a few instances. An enormous amount of experimental and analytical work remains to be done in more than one dimension just to begin to obtain the best framework for the inelastic constitutive relations of ductile metals and alloys in the moderate to extreme range. Similar effort is needed for geotechnical materials in their time-independent regime and for ceramics and polymeric materials in all regimes.

The hope that an (almost) all inclusive constitutive relation can be found for each class of materials, misguided though it is, is likely to provide more than enough impetus to guarantee a high level of analytic and computational effect. However, only a very small fraction of the extensive experimental work required across the entire spectrum of materials is now in progress. Until far more research is begun and completed, the adequacy of all proposed constitutive relations of far greater breadth will remain unknown.

ACKNOWLEDGEMENT

It is a pleasure to thank the Office of Naval Research, Solid Mechanics Program, Mechanics Division, Dr. R.S. Barsoum, for support under Grant Number N00014-87-J-1193.

REFERENCES

1. D.C.Drucker "Stress-strain relations in the plastic range - a survey of theory and experiment" Brown University Report A-11, S-1 to the Office of Naval Research (1950).
2. D.C.Drucker "Some classes of inelastic materials-related problems basic to future technologies" Nuc Eng Des 57, 309-322 (1980).
3. D.C.Drucker "On the continuum as an assemblage of homogeneous elements or states" in Irreversible Aspects of Continuum Mechanics and Transfer of Physical Characteristics in Moving Fluids edited by H.Parkus and L.I.Sedov, 77-93 (Springer, Vienna 1968).
4. D.C.Drucker "Conventional and unconventional plastic response and representation" App Mech Rev 41, 151-167 (Apr 1988).
5 . D.C.Drucker "From limited experimental information to appropriately idealized stress-strain relations" in Mechanics of Engineering Materials edited by C.S.Desai and R.H.Gallagher, 231-251 (Wiley, Chichester 1984).

UNIFIED AND COUPLED THEORIES

EULERIAN VERSUS LAGRANGIAN DESCRIPTIONS OF RATE-TYPE CONSTITUTIVE THEORIES

J. CASEY and P.M. NAGHDI
Department of Mechanical Engineering
University of California, Berkeley

ABSTRACT

A number of issues connected with the relationship between Eulerian and Lagrangian descriptions of rate-type constitutive theories, including hypoelasticity and finite rigid plasticity, are addressed. Included among these are the role of invariance requirements under superposed rigid body motions, the question of objective rates and the transformation of Lagrangian and Eulerian variables into one another.

INTRODUCTION

The controversy regarding the superiority of the Eulerian over the Lagrangian formulation of finite plasticity strikes at the very foundations of the subject. Its implications are widespread, and have surfaced most recently in the debate concerning the choice of objective rate in the constitutive equations for elastic-plastic and rigid-plastic materials. On the one hand, a large number of researchers[1] have expressed a preference for one or other of the infinitely many available rates. On the other hand, we find it difficult to disagree in principle with Truesdell and Noll's statement [3, p. 97] that the properties of a material should be independent of the objective rate that is employed to describe them. Central to the entire controversy is the role played by invariance requirements under superposed rigid body motions (s.r.b.m.). In particular, the invariance requirements ensure that material response is unaffected by superposed rotations and spins. They constitute a powerful tool for the construction of physically meaningful constitutive theories, and become doubly important when rates are involved.

In the present discussion, we take a class of rate-type constitutive equations that occur in many theories of rate-independent behavior; and, by appropriate use of invariance requirements under s.r.b.m., show that (i) entirely equivalent Eulerian and Lagrangian forms of the constitutive theory can be developed and (ii) the choice of objective rate in the constitutive theory is immaterial. When the Lagrangian and Eulerian forms are viewed in conjunction with one another, much light is shed on the procedure for formulating satisfactory rate-type theories. Moreover,

---

[1] See the papers by Dienes, Lee *et al.*, Dafalias and others, cited in Sects. 4G and 4H of the review article [1] and in [2].

the method employed automatically renders the structure of the constitutive equations form-invariant under arbitrary transformations of objective rates, so that within this framework *all* objective rates are acceptable.

RATE-TYPE CONSTITUTIVE EQUATIONS

For a three-dimensional continuum undergoing an arbitrary deformation, let **F**,**E** and **L** denote the deformation gradient, the Lagrangian finite strain tensor and the spatial velocity gradient, respectively, and recall the formulae

$$\dot{\mathbf{E}} = \mathbf{F}^T\mathbf{D}\mathbf{F}, \quad \mathbf{D} = \frac{1}{2}(\mathbf{L} + \mathbf{L}^T), \tag{1}$$

where the superposed dot signifies material time-differentiation. Also, let **T** and **S**, respectively, be the Cauchy and the symmetric Piola-Kirchhoff stress tensors, and note that

$$J\mathbf{T} = \mathbf{F}\mathbf{S}\mathbf{F}^T, \quad J = \det \mathbf{F}. \tag{2}$$

Under s.r.b.m. the fields **F**, J, **E**, $\dot{\mathbf{E}}$, **D** are carried into

$$\mathbf{F}^+ = \mathbf{Q}\mathbf{F}, \quad J^+ = J, \quad \mathbf{E}^+ = \mathbf{E}, \quad \dot{\mathbf{E}}^+ = \dot{\mathbf{E}}, \quad \mathbf{D}^+ = \mathbf{Q}\mathbf{D}\mathbf{Q}^T, \tag{3}$$

where the proper orthogonal tensor **Q** represents the rotation in the superposed motion. Similarly,

$$\mathbf{T}^+ = \mathbf{Q}\mathbf{T}\mathbf{Q}^T, \quad \mathbf{S}^+ = \mathbf{S}, \quad \dot{\mathbf{S}}^+ = \dot{\mathbf{S}}. \tag{4}$$

It is useful to consider a class of kinetical fields for which the same relationship exists between their Lagrangian and Eulerian representations as exists between **S** and **T**. Thus, let Σ_R be a second-order tensor field defined on the reference configuration of the continuum which under s.r.b.m. transforms as $\Sigma_R^+ = \Sigma_R$. Then, it is always possible to define a field Σ on the present configuration by

$$\Sigma = \frac{1}{J}\mathbf{F}\Sigma_R\mathbf{F}^T = \pi\{\Sigma_R\}, \tag{5}$$

where the operation π is called a *Piola transformation*. Conversely, $\Sigma_R = \pi^{-1}\{\Sigma\}$. In view of $(3)_1$, under s.r.b.m., Σ transforms into $\Sigma^+ = \mathbf{Q}\Sigma\mathbf{Q}^T$. The material time derivative of Σ_R transforms like $\dot{\mathbf{S}}$ (see $(4)_3$), and consequently is objective. We can construct an objective rate of Σ by simply applying a Piola transformation to $\dot{\Sigma}_R$:

$$\overset{+}{\Sigma} = \pi\{\dot{\Sigma}_R\} = \dot{\Sigma} - \mathbf{L}\Sigma - \Sigma\mathbf{L}^T + (\mathrm{tr}\mathbf{D})\Sigma, \tag{6}$$

where tr stands for trace. Under s.r.b.m., $\overset{+}{\Sigma}$ (known as the *Truesdell* rate) transforms like Σ.

Suppose now that in our constitutive theory the field Σ_R were specified through an evolution equation of the type

$$\dot{\Sigma}_R = \mathcal{A}_R[\dot{\mathbf{E}}], \tag{7}$$

where the fourth-order tensor $\mathcal{A}_R$ may depend on **E**,**S**,Σ_R, but not on their rates, and where the square brackets denote linear action (*i.e.*, in component form

$(\mathcal{A}_R[\dot{\mathbf{E}}])_{KL} = (\mathcal{A}_R)_{KLMN}\,\dot{E}_{MN})$. A constitutive equation of the type (7) appears in many theories which describe rate-independent material behavior. An Eulerian form of (7) can be obtained by applying the operation π to both sides of (7) so that

$$\overset{t}{\Sigma} = \mathcal{A}[\mathbf{D}]\,, \tag{8}$$

where the components of the tensors $\mathcal{A}_R$ and $\mathcal{A}$ are related by $J\,\mathcal{A}_{ijkl} = F_{iK}\,F_{jL}\,F_{kM}\,F_{lN}\,(\mathcal{A}_R)_{KLMN}$.

In the Lagrangian description of a constitutive theory, the objective rate $\dot{\Sigma}_R$ arises naturally. By $(6)_1$, the corresponding Eulerian objective rate is the Truesdell rate. The latter, however, might well appear to be peculiar. Other choices, such as the corotational (or Jaumann) rate, or the convected rate might seem to be more natural. As a means of exploring the significance of the choice of objective rate, Casey and Naghdi [2] considered a general class of objective rates and examined the effects of interchanging these rates in the constitutive theory. Thus, let $\overset{a}{\Sigma}$ be an objective rate of Σ and suppose that[2]

$$\overset{a}{\Sigma} = \overset{t}{\Sigma} + \mathcal{B}^{a,t}\,(\Sigma)[\mathbf{D}]\,, \tag{9}$$

where $\mathcal{B}^{a,t}$ is a fourth-order tensor-valued function of the variables $\mathbf{T}$ and $\mathbf{F}$ as well as Σ. For two arbitrary objective rates $\overset{b}{\Sigma}$, $\overset{a}{\Sigma}$ in the class defined by (9), we have

$$\overset{b}{\Sigma} = \overset{a}{\Sigma} + \mathcal{B}^{b,a}\,(\Sigma)\,[\mathbf{D}]\,, \tag{10}$$

where $\mathcal{B}^{b,a}\,(\Sigma) = \mathcal{B}^{b,t}(\Sigma) - \mathcal{B}^{a,t}\,(\Sigma)$. Invariance requirements under s.r.b.m. imply the following restrictions on the functions $\mathcal{B}^{b,a}$: $\mathbf{Q}\mathcal{B}^{b,a}\,(\Sigma)[\mathbf{D}]\mathbf{Q}^T = \mathcal{B}^{b,a}\,(\mathbf{Q}\Sigma\mathbf{Q}^T)[\mathbf{Q}\mathbf{D}\mathbf{Q}^T]$. (For some examples, see [2, Appendix C].)

For argument's sake, let us assume that for some choice of objective rate $\overset{z}{\Sigma}$ the evolution of Σ is specified by a rate-independent constitutive equation

$$\overset{z}{\Sigma} = c\,\mathbf{D}, \tag{11}$$

where c is a constant. While this choice may appear plausible, there are compelling reasons to doubt its appropriateness. Let us reflect momentarily on the process by which one ordinarily arrives at a constitutive relation between, say, stress and strain for an elastic material. First, an observation -- usually made in the context of a simple experiment -- suggests a proportionality between a component of stress and a component of strain (*i.e.*, Hooke's law). Second, one attempts to extend this relationship to more general circumstances; and normally at this stage, a number of different theoretical directions may be pursued, but ultimately one must generate critical testable predictions. Third, experiments (over and above those required to measure material properties) must be performed to evaluate the status of the theory. Revision can then take place if necessary. For our present purpose, it suffices to focus on the second stage of the above process, especially on the question of what constitutes a proper generalization. Returning to Hooke's law for definitiveness,

---

[2] A more primitive form of (9) was given in [2], but it reduces to (9) when invariance requirements are invoked.

suppose one postulates that for every homogeneous isotropic elastic material undergoing small deformations, the relationship between stress and strain is given by

$$\mathbf{S} = 2\mu\mathbf{E} + \mu(\mathrm{tr}\,\mathbf{E})\mathbf{I}\,, \qquad \mu = \text{shear modulus}\,. \tag{12}$$

To be sure, $(12)_1$ is linear, and a theory can be built upon it. In fact, such a constitutive theory *was* developed (by Navier and Poisson) and corresponds to a Poisson's ratio $\nu = 1/4$ (or equal Lamé constants), and it took a great deal of effort to dislodge it.[3] The mistake incurred by asserting $(12)_1$ is this: it is only a generalization from one special linear response (Hooke's law) to another special linear response $(12)_1$. Instead, one should proceed from Hooke's law to the *most general* type of linear response, namely

$$\mathbf{S} = \mathcal{L}[\mathbf{E}]\,, \tag{13}$$

where $\mathcal{L}$ is a constant fourth-order tensor. Using well-known mathematical arguments, a correct two-constant constitutive theory can be derived from (13).

In light of the foregoing discussion, we proceed to generalize (11). Under a change of objective rate,

$$\overset{a}{\Sigma} = \{c\, I + \mathcal{B}^{a,z}(\Sigma)\}[\mathbf{D}]\,, \tag{14}$$

where I is a fourth-order tensor with components $I_{ijkl} = \frac{1}{2}(\delta_{ik}\,\delta_{jl} + \delta_{il}\,\delta_{jk})$, δ_{ij} being the Kronecker delta, and where (10) has been invoked. One immediately notices that the coefficient of $\mathbf{D}$ in (14) depends explicitly upon the choice of objective rate. For a general material of the type being considered, one should therefore write an evolution equation of the form

$$\overset{a}{\Sigma} = \mathcal{A}^{a}[\mathbf{D}], \tag{15}$$

where the fourth-order tensor $\mathcal{A}^{a}$ depends not only possibly on $\mathbf{T}$, $\mathbf{F}$, Σ *but also on the choice of objective rate*. (The latter dependence is indicated in the superscript a.) Then, under a change of objective rate, (15) becomes

$$\overset{b}{\Sigma} = \mathcal{A}^{b}[\mathbf{D}], \tag{16}$$

with $\mathcal{A}^{b} = \mathcal{A}^{a} + \mathcal{B}^{b,a}(\Sigma)$. Consequently, the constitutive equation (15) is form-invariant under change of objective rate. Equation (15) is the desired generalization: it is the most general linear form of (13) and, in addition, it accommodates all objective rates equally.

The procedure which we advocate contrasts with that generally pursued in the literature, where (11) is retained and a variety of alternative objective rates are utilized.[4] We emphasize that even though (11) is a member of the class (15), and consequently may possibly be satisfied for some special materials, it is just as wrong to base a constitutive theory on (11) as it is to base a theory of elasticity on the

---

3 See Love's *Elasticity*, pp. 12-14.

4 See the discussion on pp. 365-366 of [2].

supposition that $\nu = 1/4$. There may in fact be *no* real material for which (11) is a good approximation. Likewise, there may be no real material for which (7) holds if $\mathcal{A}_R$ is restricted to be a constant tensor.

Hypo-Elasticity

The concept of a hypo-elastic material, developed mainly by Truesdell, embodies a notion of elasticity that partially overlaps that of finite elasticity.[5] The constitutive equation for a hypo-elastic material can be written in the general form

$$\overset{a}{\mathbf{T}} = \mathcal{A}^a(\mathbf{T})[\mathbf{D}] \,. \tag{17}$$

Thus, (17) is a member of the class defined by (15). To obtain a Lagrangian form of the constitutive equation one may apply a Piola transformation to (17) so that

$$\overset{A}{\mathbf{S}} = \pi^{-1}\{\mathcal{A}^t(\mathbf{T})[\mathbf{D}]\} = \mathcal{A}_R{}^A(\mathbf{S},\mathbf{F})[\dot{\mathbf{E}}]. \tag{18}$$

It therefore becomes clear that the material time-derivative is only one of infinitely many objective rates that can arise in the Lagrangian form of a constitutive theory. Thus, corresponding to the Eulerian rates in (9), we have the Lagrangian rates

$$\overset{A}{\Sigma}_R = \overset{T}{\Sigma}_R + \mathcal{B}^{A,T}(\Sigma_R)[\dot{\mathbf{E}}] \,, \tag{19}$$

the T being used here to refer to the material derivative (since it is the Lagrangian analog of the Truesdell rate). In retrospect, we see that the coefficient $\mathcal{A}_R$ in constitutive equations of the type (7) may depend explicitly on our original choice of the material derivative as objective rate.

Rigid Plasticity

In the context of a strain-space formulation of plasticity, equivalent Lagrangian and Eulerian descriptions of rigid plasticity have been recent developed (see [4] and [2]). The elements of the theory are as follows. In the Lagrangian description, a yield function f depending on **S**,**E**, a scalar work-hardening parameter κ, and a shift (or back-stress) tensor α_R, is assumed to exist. The Eulerian form is given by a function f_* such that

$$f_*(\mathbf{T},\mathbf{F},\alpha,\kappa) = f(\mathbf{S},\mathbf{E},\alpha_R,\kappa) \tag{20}$$

with $\alpha = \pi\{\alpha_R\}$. (The condition that f_* be unaltered under s.r.b.m. ensures that f can be recovered from f_*.) Loading criteria are defined by the Lagrangian strain-space conditions

$$\text{(a) } \dot{\mathbf{E}} = \mathbf{0}: \quad \text{Non–loading}\,; \quad \text{(b) } \dot{\mathbf{E}} \neq \mathbf{0}: \quad \text{Loading}\,, \tag{21}$$

or equivalently by the Eulerian conditions

$$\text{(a}'\text{) } \mathbf{D} = \mathbf{0}: \quad \text{Non–loading}\,; \quad \text{(b}'\text{) } \mathbf{D} \neq \mathbf{0}: \quad \text{Loading}\,. \tag{22}$$

---

[5] A comprehensive discussion of hypo-elasticity can be found in [3, Sects. 99-103].

During loading, the Lagrangian form of the flow-rule can be written as

$$\dot{\mathbf{E}} = \gamma \rho \,, \tag{23}$$

where ρ is a symmetric tensor of *unit* magnitude that depends on the same variables as f, and[6] $\gamma = \|\dot{\mathbf{E}}\| > 0$. Equivalently, in Eulerian form

$$\mathbf{D} = \gamma_* \,\xi, \tag{24}$$

$$\xi = (\mathbf{F}^{-T}\rho\mathbf{F}^{-1}) \Big/ \|\mathbf{F}^{-T}\rho\mathbf{F}^{-1}\| \,, \gamma_* = \|\mathbf{D}\| = \gamma \, \|\mathbf{F}^{-T}\rho\mathbf{F}^{-1}\| \,.$$

The tensor ξ has unit magnitude and depends on the same variables as f_*. The constitutive equation which κ satisfies during loading can be written in the forms

$$\dot{\kappa} = \gamma \lambda = \gamma_* \, \lambda_* \,, \tag{25}$$

where the scalar function λ and λ_* depends on the same variables as f and f_*, respectively. Regarding an evolution equation for α_R, let us suppose that during loading

$$\dot{\alpha}_R = \mathcal{H}_R^{\,T} [\dot{\mathbf{E}}] \,, \tag{26}$$

where $\mathcal{H}_R^{\,T}$ depends on **S**, **E**, α_R,κ *and on the choice of objective rate*, which is the present case is the material derivative itself. The Eulerian form of (26) is

$$\overset{t}{\alpha} = \mathcal{H}^t[\mathbf{D}] = \pi\{\mathcal{H}_R^{\,T} [\dot{\mathbf{E}}]\} \,. \tag{27}$$

It follows from (23) and (26) that

$$\dot{\alpha}_R = \gamma \, \beta_R^{\,T} \,, \ \ \beta_R = \mathcal{H}_R^{\,T} [\rho] \,. \tag{28}$$

Similarly, from (24) and $(27)_1$, we have

$$\overset{t}{\alpha} = \gamma_* \, \beta^t \,, \ \ \beta^t = \mathcal{H}^t[\xi] \,. \tag{29}$$

Other matters, especially the characterization of strain-hardening behavior in terms of an arbitrary rate, are also discussed in [2].

REFERENCES

1. P.M. Naghdi, "A critical review of the state of finite plasticity", J. Appl. Math. and Physics (ZAMP), 41, 315-393 (1990).
2. J. Casey and P.M. Naghdi, "On the relationship between the Eulerian and Lagrangian descriptions of finite rigid plasticity", Arch. Rational Mech. Anal. 102, 351-375 (1988).
3. C. Truesdell and W. Noll, "The non-linear fleld theories of mechanics", Handbuch der Physik, edited by S. Flügge, Vol. III/3 (Springer-Verlag, 1965).
4. J. Casey, "On finitely deforming rigid-plastic materials", Int. J. Plasticity, 2, 247-277 (1986).

---

[6] In [2, p. 358], it was pointed out that the scalar-coefficient in the flow-rule does not require a constitutive equation since only the *direction* of strain-rate is specified in rigid plasticity. Here, we improve slightly upon the presentation in [4,2] by the explicit identification of γ and also by introducing its Eulerian counterpart γ_*.

MOLECULAR DYNAMICS AND CONSTITUTIVE MODELING

Brice N. Cassenti
United Technologies Research Center

ABSTRACT

As materials are pushed towards their limits it becomes more important to provide accurate descriptions of their mechanical response. Recently viscoplastic constitutive models have been used to extend the range of predictions to extremely high temperatures. Viscoplastic models though are primarily empirical and cannot be readily extended beyond the range for which experimental data exists. One method for extending the range is to develop computational models based on physical principles that simulate materials in regimes outside the range of available data. Molecular dynamics is being applied to such simulations and results indicate it can provide insight useful in the development of constitutive models.

INTRODUCTION

As materials are developed to operate under extreme environmental conditions it becomes crucial to provide realistic descriptions of their thermomechanical response. Several approaches can be taken to provide more accurate descriptions; these include: thermodynamics[1], statistical mechanics[2], solid state physics, and micromechanical simulations. Each of these can provide a better understanding of the thermomechanical response of materials. Micromechanical simulations provide not only a better qualitative description of the thermomechanical response but can also provide a means for performing numerical experiments to determine the nature of the response where physical experiments are difficult to perform. For example, the multiaxial stress-strain response of materials is an extremely difficult experimental task. Numerical experiments furnish a means for obtaining the qualitative nature of the multiaxial thermomechanical response of materials.

Several levels of micromechanical simulations exist for the response of monolithic materials. At the largest level the simulation of crack (or void) nucleation and growth can provide significant insight into the failure of materials. At a somewhat smaller level the motion and interactions of dislocations, and crystalline grains, helps in the understanding of yielding in materials. The actual strength of materials is controlled by interactions at the atomic level. This level controls dislocation growth and motion and hence can be used to increase our understanding of the strength of materials. The lowest level concerns the motion of the outer electrons about their atomic ion cores and is responsible for the actual binding in solid materials.

Molecular dynamics studies concentrate on the interactions between atoms, and are used to study the properties of materials [3]. The interactions between atoms are usually modeled using two body, central force, (Lennard-Jones) potentials. But two body central force potentials predict moduli that satisfy the Cauchy relations resulting in only fifteen independent moduli in a generally anisotropic material, see Refs. [2] or [4]. Recently the Embedded Atom Method (EAM) has been used to describe the interactions between atoms (Refs. [5) or [6]) which results in twenty-one independent moduli for a generally anisotropic material.

FORMULATION

The simulation is based on evaluating the forces acting on each of the atoms and then integrating the equations of motion for each of the atoms. The force, F, acting on each atom is found by taking the gradient of a potential, V,

$$\vec{F}_i = -\vec{\nabla} V_i$$

The potential between atoms is commonly found by a Lennard-Jones approximation

$$V_i = 1/2 \sum_j V_{ij}$$

where $V_{ij} = -\frac{A}{n r_{ij}^n} + \frac{B}{m r_{ij}^m}$ is the potential of atom i with respect to atom j,

r_{ij} is the distance from atom i to atom j,

and the sum occurs over all the atom pairs when i,j are not equal. This formulation for the interaction between atoms can be shown to result in the Cauchy relations, [2]. If

$$\sigma_{ij} = C_{ijkl}\, \varepsilon_{kl}$$

then the Cauchy relations for the moduli are:

$$C_{ijkl} = C_{jkli} = C_{klij} = C_{lijk} = C_{jikl} = C_{ijlk} = \ldots$$

where σ_{ij} is the stress, and

ε_{ij} is the strain.

Although Lennard-Jones potentials accurately reproduce the equilibrium positions of atoms in a crystalline lattice, and the motion of dislocations through the lattice, they cannot predict the correct moduli in real materials. The Embedded Atom Method (EAM) uses an additional summation to remove the Cauchy relations [5]. In the EAM the potential for atom i is given by [6]

$$V_i = 1/2 \sum_j V_{ij}(r_{ij}) + f(\rho_i)$$

where f is the embedding function, and

ρ is the electron density given as a function of the distance between atoms, i.e.

$$\rho_i = \sum_j \Phi(r_{ij})$$

The electron density function, Φ, the embedding function, f, and the pair potential function, V, are chosen to represent the particular chemical elements in the material. Examples can be found in Ref.[3].

Once the forces are known the equations of motion can be integrated numerically using an explicit integration algorithm.

The effects of temperature are included by adding random velocities to the atoms according to a Boltzman distribution. This procedure usually introduces a small rigid body motion which has to be removed. A small amount of additional random kinetic energy can be added to bring the material to the correct temperature.

IMPLEMENTATION

The above formulation has been implemented on the Connection Machine at United Technologies Research Center (UTRC). The Connection Machine is a single instruction multiple data computer containing up to 64k processors. At UTRC the Connection Machine has 16k processors each containing 32k bytes of random access memory. The data for an atom was stored on a single virtual processor, such that up to four million atoms could be stored in a 16k processors. Simulations are typically performed with 256k atoms.

Several crystalline structures can be represented including: hexagonal close packed, face centered cubic, body centered cubic, simple (or ionic) cubic, and diamond cubic. The crystalline structure is initially simple cubic to specify atomic neighbors and is then distorted to other crystalline shapes. A cutoff distance is included in order to retain the symmetry present in the crystalline configuration.

RESULTS

Several test cases have been run and these indicated that the system of particles will find the correct equilibrium positions. Crack propagation and dislocation motion in single crystals where simulated and the results where in qualitative agreement with experimental results. Currently only high strain rate response can be simulated. Monte Carlo algorithms are being examined to simulate slower diffusion controlled effects.

REFERENCES

1. B.N. Cassenti, and B.S. Annigeri, "Thermodynamic constraints on stress rate formulations in constitutive models", Computational Mechanics 4, 429-432 (1989).

2. B.N. Cassenti, and B.G.J.P.T. Murray, "A physical basis for the development of constitutive models", in Advances in Inelastic Analysis, edited by S. Nakasawa K. William, and N. Rebelo, AMD-Vol. 88,(ASME, New York 1987).

3. J. Tersoff, D. Vanderbilt, and V. Vitek (editors),Atomic Scale Calculations in Materials Science, Materials Research Society Symposium Proceedings Vol. 141 (MRS, Pittsburgh 1989).

4. A.E.H. Love, A Treatise on the Mathematical Theory of Elasticity, p. 100 (Dover Publications, New York, 1944).

5. M.S. Daws, and M.I. Baskes, "Embedded-atom method: Derivation and application to impurities, surfaces, and other defects in metals", Physical Review B, 29(12), 6443-6453 (15 June 1984).

6. J.M. Eridon and S. Rao,"Derivation of many-body potentials for examining defect behavior in BCC Niobium", in Atomic Scale Calculations in Materials Science edited by J. Tersoff, D. Vanderbilt, and V. Vitek, Materials Research Society Symposium Proceedings Vol. 141 (MRS, Pittsburgh 1989).

FLOW OF COHESIONLESS GRANULAR MEDIA; A UNIFIED APPROACH

N.C. Consoli and H.B. Poorooshasb
Concordia University, Montreal, Quebec, H3G 1M8

ABSTRACT

A unified approach based on the concept of state parameters is developed to describe the flow of cohesionless granular media. The model is based on the hypothesis regarding the existence of the State Boundary Surface as well as the Ultimate State Surface. Comparison between the experimental results and the model predictions are encouraging.

INTRODUCTION

Although the concept of the "state parameters" , defined as a set of quantities describing the state of a sample, was introduced nearly three decades ago [1] it has been of little use in the development of constitutive laws. Recently the second author [2] has pointed out the merit of the concept as a unifying agent in describing the flow of a cohesionless granular medium. It is the objective of the present paper to quantify this idea. The formulation is based on two hypotheses postulating the existence of a "State Boundary Surface" and an "Ultimate State Surface" which will be defined later on.

STATE OF A SAMPLE

The state of a sample of a cohesionless granular medium may be represented by the set of quantities (p,q,θ,e) where the first three quantities are related to the invariants of stress (or stress deviation) tensor and e is the void ratio:

$p=I_1/\sqrt{3}$ where $I_1=\sigma_{ii}$

$q=\sqrt{2J_2}$ where $J_2=s_{ij}s_{ij}/2;\; s_{ij}=\sigma_{ij}-I_1\delta_{ij}/3$

$\theta=\sin^{-1}[-3\sqrt{3}J_3/(2J_2^{3/2})]/3$ where $J_3=s_{ij}\,s_{jk}\,s_{ki}/3.$

The space of (p,q,θ,e) is called the state space and the use of parameter e, the void ratio, as a state quantity is justified in view of the isotropy of sample assumption.

THE STATE BOUNDARY SURFACE

Not all the points of the state space are accessible

by a sample. The surface enclosing all the state points which may be assumed by a sample is called the State Boundary Surface or simply SBS. Thus during a loading process the sample may follow a path on, or within the volume bounded by the SBS. This surface is very useful in describing the strain softening behavior of the element.

THE ULTIMATE STATE SURFACE

It is postulated that there exists a surface of the state space such that for all the points on this surface the sample may distort continuously without any change of its state. This surface is referred to as the Ultimate State Surface or simply USS. The trace of the USS in a particular θ=constant sub-space of the state space has been known as the Critical State Line [1] which had previously been called by Roscoe et al [3] as the Critical Void Ratio (CVR) Line. The concept upon which the USS is based is also equivalent to the concept of Steady State of Deformation (Castro and Poulos, [4]) although a general agreement has not as yet, been reached in this respect.

The USS divides the region bounded by the SBS into two regions separating the "loose" from the "dense" samples. The USS in conjunction with the SBS provides a framework within which a unified constitutive law may be formulated.

FORMULATION OF THE CONSTITUTIVE LAW

The yield function F is given by the equation;

$$F=q-\eta(\varepsilon^p)g(\theta)p(\mu-\delta e')=0 \qquad (1)$$

where $g(\theta)$ is a function defining the shape of the yield function in the π plane and $e'=e-e_0+\lambda\ln(I_1/3)$ is a derived state parameter judging the position of the state of the sample with respect to the USS. Material constants e_0 and λ are associated with Casagrande's Critical Void Ratio Line. The parameter $\eta(\varepsilon^p)$ records the history of plastic distortion ε^p and is given by the hyperbolic relation $\eta(\varepsilon^p)=\varepsilon^p/(A+\varepsilon^p)$. Parameters μ and δ are two further material constants, μ is the slope of the trace of the USS in the (p,q) stress space associated with the θ=0 (i.e. triaxial compression) condition and δ is the slope of the trace of the SBS in the (p,e') space of the same sub-space, (i.e., θ=0 condition.)

The plastic potential function is given by;

$$\psi=p\psi_0(\eta) \qquad (2)$$

where the parameter η (not to be confused with $\eta(\varepsilon^p)$ which records the distortion history) is given by $\eta=q/pg(\theta)$. In the present paper the specific form of the function $\psi_0(\eta)$ is taken to be $\psi_0(\eta)=\exp(\eta/\mu)$. With the aid of the consistency condition the flow rule may now be derived:

(a) Formulation relating the strain increment tensor to the stress increment tensor;

$$d\varepsilon_{ij}=[C_{ijkl}{}^e+\lambda_{kl}(\partial\psi/\partial\sigma_{ij})/H_p]d\sigma_{kl} \qquad (3)$$

where $\lambda_{kl}=-\partial F/\partial\sigma_{kl}+[(1+e)C_{iikl}{}^e-\lambda\delta_{kl}/I_1]\partial F/\partial e'$ and $H_p=-(1+e)\partial F/\partial e'.\partial\psi/\partial\sigma_{ii}+\partial F/\partial\eta.d\eta/d\varepsilon^p.[dev(\partial\psi/\partial\sigma_{ij})dev(\partial\psi/\partial\sigma_{ij})]^{1/2}$.

(b) Formulation relating the stress increment tensor to the strain increment tensor;

$$d\sigma_{ij}=[D_{ijkl}{}^e+(\Delta_{ijkl}+\xi_{ijkl}+\chi_{ijkl})/(H_p+\varpi+\nu+\beta)]d\varepsilon_{kl} \qquad (4)$$

where $\Delta_{ijkl}=D_{ijpq}{}^e(\partial F/\partial\sigma_{pq})(\partial\psi/\partial\sigma_{rs})D_{rskl}{}^e$, $\xi_{ijkl}=-(1+e)\partial F/\partial e'.D_{ijpq}{}^eC_{ttpq}{}^e(\partial\psi/\partial\sigma_{rs})D_{rskl}{}^e$, $\chi_{ijkl}=(\lambda/I_1)\partial F/\partial e' D_{ijpq}{}^e\delta_{pq}(\partial\psi/\partial\sigma_{rs}).D_{rskl}{}^e$, $\varpi=(\partial F/\partial\sigma_{pq})D_{pqrs}{}^e(\partial\psi/\partial\sigma_{rs})$, $\nu=-(1+e)\partial F/\partial e' C_{ttpq}{}^e C_{pqrs}{}^e(\partial\psi/\partial\sigma_{rs})$ and $\beta=(\lambda/I_1)(\partial F/\partial e').D_{pqrs}{}^e\delta_{pq}.(\partial\psi/\partial\sigma_{rs})$.

COMPARISON WITH TEST RESULTS

A series tests performed and reported by Seed and Lee [5] are used here. The material parameters employed in the analysis were obtained by a process of trial and error and had the following values; $\mu=.65$, $\delta=.79$, $e_0=.78$, $\lambda=.088$, $A=(-.0038+.008e)(p/p_a)^{.69}$, $E=19000$ MPa and Poisson's ratio=.3. (p_a = atmospheric pressure.)

Figure (1) shows the predicted and the experimental results obtained for a dense sample tested under conventional drained triaxial compression condition. A loose sample of the same sand tested under similar conditions yielded results shown in Fig.(2.) The tendency of the dense sample to expand and that of the loose sample to contract is clearly demonstrated by both theory and experiment. In Fig.(3) is shown the undrained stress path followed by three samples at different initial void ratios and consolidation pressures. Again the influence of the initial state of the sample on its subsequent flow behavior is very clear.

ACKNOWLEDGEMENT

The financial assistance of CAPES (Brazil) and NSERC (Canada) is gratefully acknowledged.

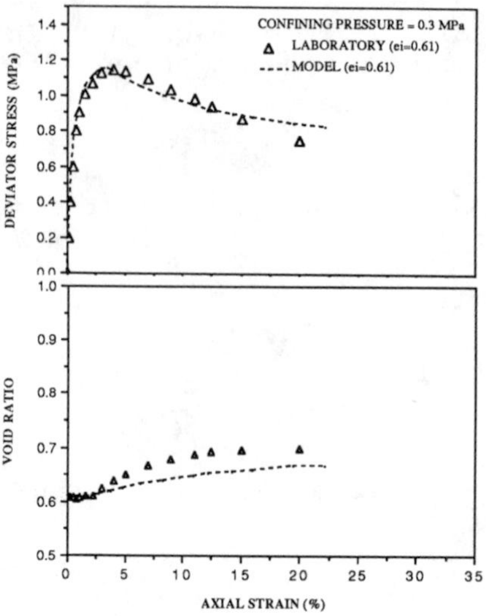

FIGURE 1

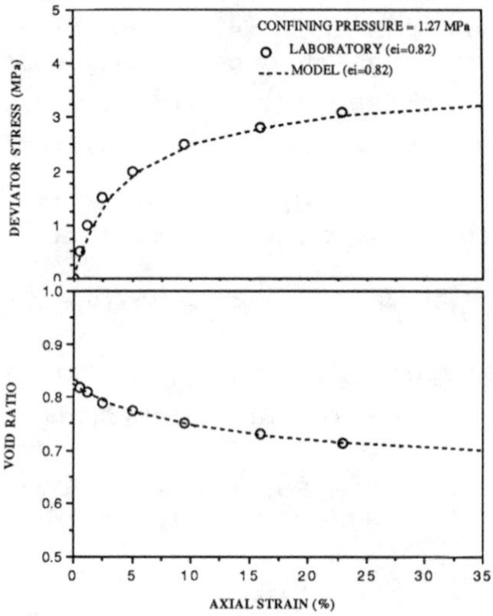

FIGURE 2

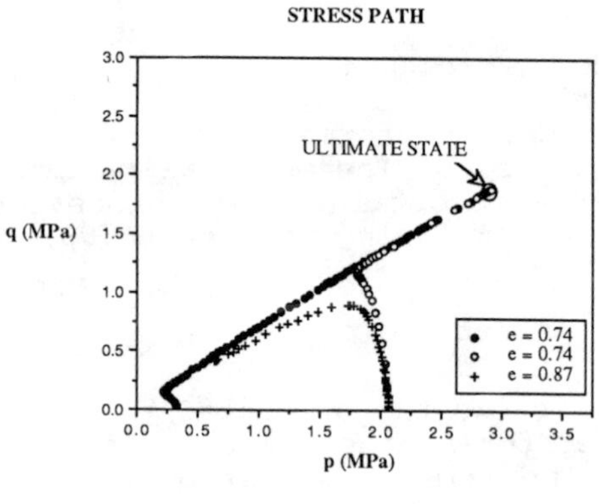

FIGURE 3

REFERENCES

1. Poorooshasb, H.B., Ph.D.Thesis, Cambridge,1961.

2. Poorooshasb, H.B.,"Description of Flow of Sand Using State Parameters," Comp. & Geotech. 8(3),195-218,1989.

3. Roscoe, K.H., Schofield, A.N. and Wroth, C.P."On the Yielding of Soils," Geotechnique,8,22-53, 1958.

4. Castro G. and Poulos, S.J. "Liquefaction and Cyclic Mobility of saturated Sands" ASCE, 103,GT6,501-516, 1977.

5. Seed, H.B. and Lee, K.L.,"Liquefaction of Saturated Sands During Cyclic Loading," ASCE,92,SM6,105-134,1966.

MODELLING OF SOLIDS AND CONTACTS USING DISTURBED STATE CONCEPT

C.S. Desai, Regents' Professor and Head
S.H. Armaleh, Assistant Professor
D.R. Katti, Graduate Associate
Department of Civil Engg. and Engg. Mech.
University of Arizona, Tucson, AZ 85721

and

Y. Ma
Geomatrix Consultants
San Francisco, CA 94105-1001

ABSTRACT

A general mathematical framework, referred to as the Disturbed State Concept (DSC) is proposed for constitutive modelling of solids and contacts. It provides for incorporation of factors such as friction, induced anisotropy, and damage and disturbances with reference to two basic states, intact and ultimate. The concept is verified with respect to laboratory tests for a sand and rock joints.

INTRODUCTION

The Disturbed State Concept (DSC) involves modelling of the behavior of solids and contacts (interfaces and joints) through incorporation of modifications treated as disturbances with respect to certain basic responses of idealized states of a solid material or a contact (interface or joint) [1-6]. The latter can include states such as (hardening) response of an intact or continuous material that remains intact, and the ultimate state at which the material will not carry any further shear loading or will deform at constant volume under shear, the so-called critical state. Disturbance function D is defined to allow representation of the observed behavior of the material with respect to the idealized states.

DSC Concept for Solids

The disturbed state concept is based on the idea that the observed behavior of the material under external influences (mechanical and environmental loads) can be interpolated with respect to responses of two reference (asymptotic) states of the material, the intact and the ultimate. Under external loads (mechanical and/or environmental), the disturbance increases asymptotically to a maximum value at which the material tends toward its isotropic state. A main attribute of the DSC is that factors such as friction (nonassociativeness), induced anisotropy, and damage (softening) can be introduced as general disturbances. Hence, the conventional damage approach [7] can be considered as a subset of the DSC. Figure 1 shows schematic of the material at disturbed state as a mixture of the intact and ultimate zones.

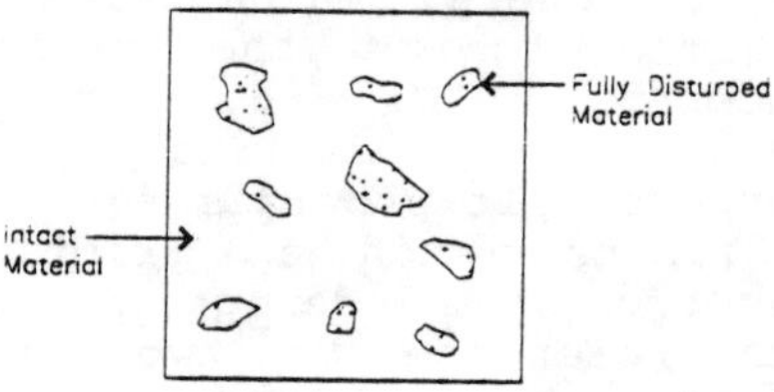

FIG. 1. Schematic of an element of a disturbed material

The average or observed response, Fig. 2, is then expressed with respect to behavior for the intact and critical states through the disturbance function D. The disturbance function is defined as the ratio of the mass of the solids at the fully disturbed or ultimate state to the total mass of solids.

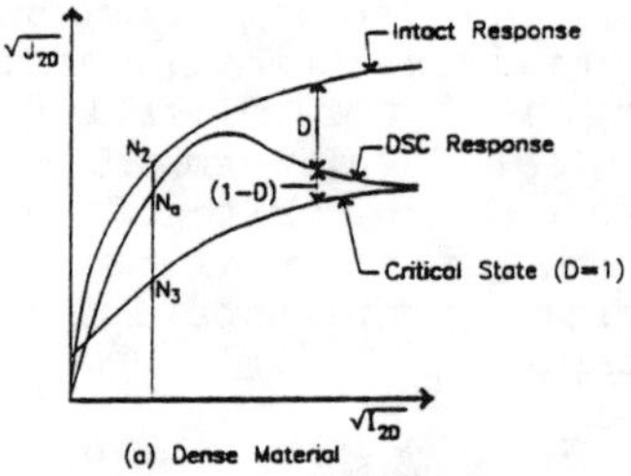

FIG. 2. Schematic diagram of DSC concept

The disturbance function is given as [3]

$$D = D_u \, [1 - \exp(-A \, \xi_D^Z)] \qquad \textbf{(1)}$$

where D_u = the ultimate value for disturbance and it is assumed to be equal to unity, A and Z = the disturbance parameters, and ξ_D = the trajectory of the deviatoric plastic strains. Figure 3 shows the plot of the disturbance function D.

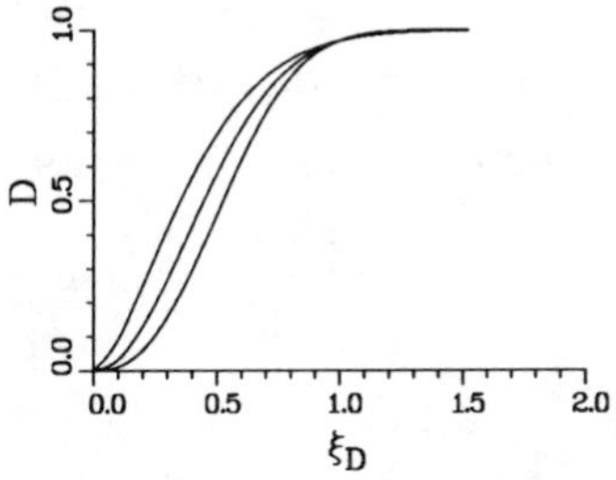

FIG. 3. Schematic diagram of disturbance function D

The intact state is assumed to be initially isotropic and to follow associative plasticity with isotropic hardening. It is modelled using the basic model δ_o of the hierarchical single surface (HISS) modelling approach developed by Desai and co-workers [8] where the yield function is given by

$$F = \frac{J_{2D}}{p_a^2} - \alpha \left(\frac{J_1}{p_a}\right)^n + \gamma \left(\frac{J_1}{p_a}\right)^2 (1-\beta S_r)^m = 0 \qquad \textbf{(2)}$$

where J_1 = first invariant of the stress tensor; p_a = atmospheric pressure; S_r = stress ratio = $\frac{\sqrt{27}}{2} \frac{J_{3D}}{J_{2D}^{3/2}}$; J_{2D}, J_{3D} = second and third invariants of deviatoric stress tensor, respectively; γ, β, n, m = material parameters found from experiments; α = hardening function

$= \dfrac{a_1}{\xi^{\eta_1}}$, ξ = trajectory of the plastic strain increment $d\epsilon^p_{ij} = \int (d\epsilon^p_{ij}\, d\epsilon^p_{ij})^{1/2}$; a_1, η_1 = hardening parameters found from experiments.

By incorporating the two reference states through the disturbance function, the average or observed response is obtained. The resulting incremental constitutive relations are then obtained as

$$\begin{aligned} d\sigma^a_{ij} = {} & [mD\eta + (1-D)]\, C^{ep}_{ijk\ell}\, d\epsilon^i_{k\ell} \\ & + (D - mD\eta)\, C^{ep}_{nnk\ell}\, d\epsilon^i_{k\ell}\, \frac{\delta_{ij}}{3} \\ & + [dD\,(m\eta - 1) + mD d\eta]\, S^i_{ij} \end{aligned} \qquad (3)$$

and

$$\begin{aligned} d\epsilon^a_{ij} = {} & dE^a_{ij} + \frac{1}{3}\,\delta_{ij}\, [D\,(\frac{-\lambda}{1+e_o})\,\frac{dJ_1}{J_1} + (1-D)\, d\epsilon^i_v \\ & + dD\, \frac{(e^c_o - \lambda\, \ell n\,(J_1/3\,pa) - e^i)}{1+e_o}] \end{aligned} \qquad (4)$$

where the superscripts (a), (i) and (c) denote the observed, intact and critical, respectively; $d\sigma_{ij}$, $d\epsilon_{ij}$ denote the incremental stress and incremental strain tensors, respectively; $[C]^{ep}$ is the elastoplastic constitutive matrix for the intact response; $\eta = J_1/\sqrt{J_{2D}}^{\,i}$ S_{ij} = deviatoric stress tensor; m, λ, e_o^c = the critical state parameters; $d\epsilon_{ij}$ is the tensor or deviatoric strain increment; S_{ij} = Kronecker delta, e_o = initial void ratio.

Testing and Verification

Comprehensive laboratory tests have been performed on samples of Leighton Buzzard Sand under different stress paths and with various initial densities and confinements [3]. In this way, a wide range of loose to dense states and confinements are included in the model. Verification of the model is obtained by comparing back prediction with laboratory test data. Figures 4 and 5 show satisfactory prediction for the observed behavior of loose and dense LB sand.

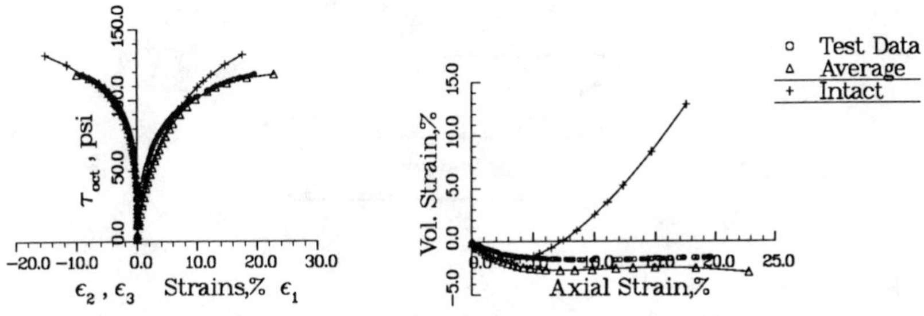

(a) Stress strain response (b) Volumetric response

FIG. 4. Comparison of stress strain and volumetric responses of CTC test on LB sand for 10% relative density and 120 psi (8273.37 kPa) confining pressure

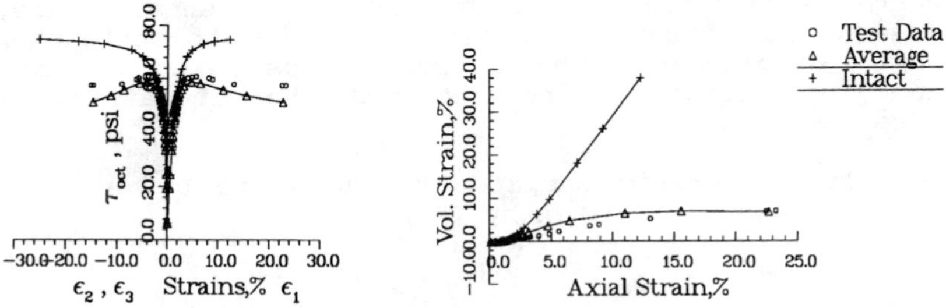

(a) Stress strain response (b) Volumetric response

FIG. 5. Comparison of stress strain and volumetric responses of CTC test on LB sand for 95% relative density and 40 psi (275.79 kPa) confining pressure

DSC FOR JOINTS

A schematic of the joint zone consisting of (a) intact and (b) ultimate or critical parts is shown in Figure 6. The intact zone is that part which has not reached the ultimate state and has not experienced damage of asperities. The ultimate zone is that for which the disturbance has caused a certain level of damage and (shear) stress and the normal (dilative) displacement have reached the asymptotic invariant values.

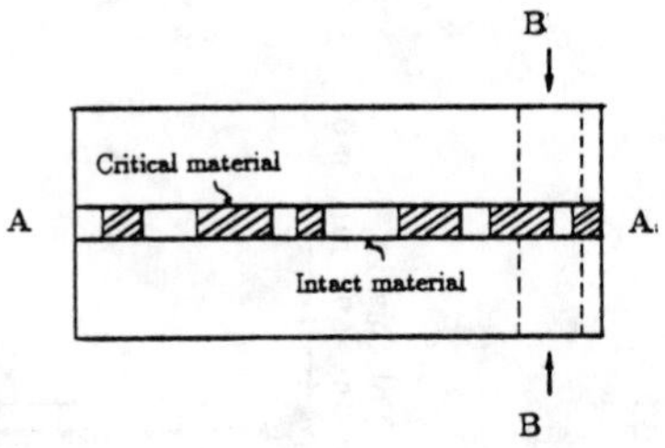

FIG. 6. A joint at disturbed state

The intact response is modelled by using the specialized form of the δ_o-model in which the yield function F is given by

$$F = \tau^2 + \alpha\ \sigma_n^n - \gamma\sigma_n^2 = 0 \qquad (5)$$

where α = hardening or growth function, τ = shear stress, σ_n = normal stress, n = parameter related to state when normal displacements change from compression to dilation and γ = parameter associated with the ultimate shear stresses.

At the critical state, the ultimate or critical shear stress τ^c is given by [9]

$$\tau^c = c_o\ \sigma_n^m \qquad (6)$$

where σ_n = normal stress and c_o and n = parameters, and the ultimate or critical normal displacement, v^u, is given by [10]

$$v^u = v^o \exp\ (-k\ \sigma_n) \qquad (7)$$

where v^o = maximum dilation when σ_n = 0 and k = material parameter.

The disturbance function D represents ratio of length of ultimate zone to the total length, Figure 6. It is expressed as

$$D = 1 - \exp\ (-k\xi^R) \qquad (8)$$

where ξ = trajectory of plastic shear and normal displacements and k and R = disturbance parameters.

The incremental relation is then obtained as

$$\begin{Bmatrix} d\tau^{a} \\ d\sigma_{n}^{a} \end{Bmatrix} = [k] \begin{Bmatrix} du^{a} \\ dv^{a} \end{Bmatrix} \tag{9}$$

where superscript (a) denotes observed behavior, [k] = constitutive matrix is function of intact part $[k^{ep}]$ corresponding to the δ_0 model, critical state and disturbance parameters.

The DSC for joints was verified with respect to series of tests conducted by four different researchers [4]. Figure 7 shows the prediction of tests performed by Schneider [10]. As can be seen, the model provides satisfactory predictions.

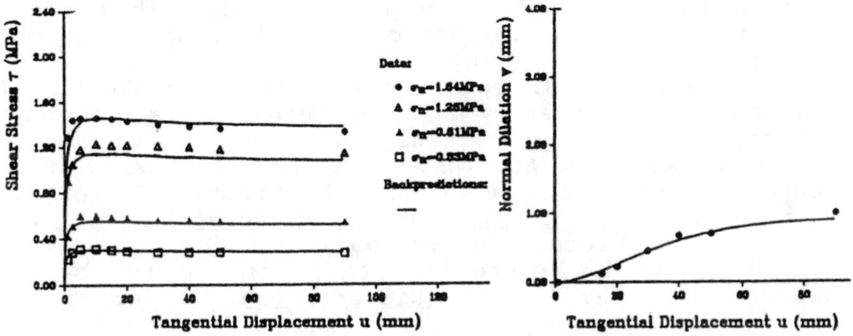

(a) Shear responses

(b) Dilatant response, normal stress σ_n = .69 MPa

FIG. 7. Shear and dilatant responses for type A joint

CONCLUSIONS

From the above results, it is concluded that the DSC provides an alternative and general approach in conjunction with the hierarchical single surface (HISS) approach for the modelling of solids and contacts. It allows for the incorporation of a wide spectrum of loose to dense states of the material. Furthermore, its general framework provides for modelling of both solids and contacts.

ACKNOWLEDGMENTS

Parts of the research herein were supported by Grant Nos. MSM 8618901/914 and CES 8711764, from the National Science Foundation, Washington, D.C. and Grant No. AFOSR-830256 from the Air Force Office of Scientific Research, Bolling AFB, D.C.

REFERENCES

1. C.S. Desai, "Further on unified hierarchical models based on alternative correction of damage approach," Report, Department of Civil Engineering and Engineering Mechanics, University of Arizona, Tucson, Arizona (1987).
2. G.W. Wathugala and C.S. Desai, "'Damage' based constitutive model for soils," Proc., 12th Canadian Congress of Applied Mechanics, Ottawa, Canada (1987).
3. S.H. Armaleh and C.S. Desai, "Modelling including testing of cohesionless soils using disturbed state concept," Report, University of Arizona, Tucson, Arizona (1990).
4. Y. Ma and C.S. Desai, "Constitutive modeling of joints and interfaces by using disturbed state concept," Report, University of Arizona, Tucson, Arizona (1990).
5. C.S. Desai, S. Armaleh, and Y. Ma, "Disturbed state concept for modelling of solids and contacts," Conference on the Mechanics and Physics and Structure of Materials, Thessaloniki, Greece (August 1990).
6. D.R. Katti, "Disturbed state concept for cyclic behavior of clay," Ph.D. Dissertation, University of Arizona (in progress).
7. G.Frantziskonis and C.S. Desai, "Constitutive model with strain softening," Int. J. of Solids and Struct. 23(6), 751-767 (1987).
8. C.S. Desai, S. Somasundaram, and G. Frantziskonis, "A hierarchical approach for constitutive modelling of geologic materials," Int. J. Num. Analyt. Meth. in Geomech., 10, 225-257 (1986).
9. J.F. Archard, "Elastic deformation and the laws of friction," Proc. Roy. Soc. London, A243, 190-205 (1958).
10. H.J. Schneider, "Rock friction - a laboratory investigation," Proc. 3rd Cong. Int. Soc. Rock Mech., Denver, CO, 2 (Part A), 311-315 (1974).

MODELING POROUS AND JOINTED ROCK AS A CONTINUUM

B. C. HAIMSON, M. E. PLESHA AND T. F. CHO
College of Engineering, University of Wisconsin, Madison, Wisconsin 53706

ABSTRACT

We have developed three-dimensional constitutive equations for porous rock, possessing an arbitrary number of joint sets of specified orientations, by using continuum representations of mechanical and hydraulic properties of both rock and joints. Linear elasticity and isotropic permeability were assumed for the intact rock; linear elastic stiffness and isotropic in-plane permeability were assumed for the joint sets. The continuum representation was accomplished by combining the compliances and the fluid diffusions in the intact rock with those in each of the joint sets. The resulting continuum theory allows for fluid diffusion through and between joint sets and intact-rock pores, and also accounts for the anisotropy of mechanical properties due to different joint-set elastic stiffnesses, spacings and orientations.

INTRODUCTION

Wilson and Aifantis [1] extended Biot's consolidation theory [2] (coupling rock deformation with fluid diffusion) to a double porosity medium (a uniformly fissured porous rock) by proposing equilibrium between rock deformation and two Darcian flows. They derived constitutive equations for the solid deformation and fluid diffusions of a double porosity medium by assuming linearity and isotropy for both the strain field of the solid and the volumetric strain of the fluid. Their coupled equations were developed, however, for fissures that are randomly distributed throughout the solid rock. Such a distribution cannot account for the more commonly encountered one or more sets of joints, each characterized by nearly identical features such as strike and dip, spacing, roughness, width, and hence directional permeability. Realistically, therefore, joint deformation will lead to anisotropic rock mass behavior affected by the normal and shear joint set stiffnesses and spacing. Similarly, fluid flow along the joints belonging to a set will result in hydraulic anisotropy affected by the joint set permeability and orientation. Aifantis did develop later [3] an advanced double porosity model which accounts for the hydraulic anisotropy but is still limited to isotropic mechanical properties.

We have developed a model which considers the mechanical as well as the hydraulic properties of both the joint sets and the host porous rock. The model is based on continuum characterization of fully saturated jointed rocks in which the number, the orientation, and the mechanical and hydraulic properties of the joint sets are specifiable. In this paper we summarize the main characteristics of the model (for more detailed description and derivations refer to [4 and 5]), and illustrate its capability with examples.

JOINTED POROUS-ROCK MODEL

We consider a representative volume of a fully saturated porous rock containing a specified number of joint sets of known orientation (not necessarily orthogonal to each other). The intact porous rock is assumed to possess isotropic, linearly elastic properties and isotropic permeability. Joint behavior is also assumed to be linearly elastic. Adding the strains due to the deformation of each joint set ($[F_I]\{\sigma\}$) the total joint strain is obtained. Superposing the latter to the strain of the intact rock ($[E]^{-1}\{\sigma\}$), the overall total strain is:

$$\{\epsilon\} = [\sum_{I=1}^{N}[F_I] + [E]^{-1}]\ \{\sigma\} = [C]\ \{\sigma\} \qquad (1)$$

where I denotes the joint set number, out of a total of N sets.

Pore fluid flow in the intact rock is assumed isotropic and governed by Darcy's law:

$$q_i^R = -K^R\ \partial P^R/\partial i\ , \quad i = x,y,z \qquad (2)$$

where q_i^R is the flow rate in the intact rock (R) in the ith direction, P^R is the pore fluid pressure, and K^R is the permeability of the intact rock.

Fluid flow within each joint set is along the joint plane (y'_I and z'_I directions) as well as in the joint-normal (x'_I) direction. In-plane flow in joint set J_I is given by

$$q_j^I = -1/S_I\ K_I\ \partial P^I/\partial j\ , \quad j = y'_I\ ,\ \ z'_I \qquad (3)$$

where q_j^I is joint in-plane fluid flow in the j direction, K_I is the permeability of joint set J_I for isotropic fluid flow in the plane of the joint, S_I is the mean joint spacing, and P^I is the fluid pressure in the joint set J_I. Use of coordinate transformation and continuity of fluid flow enable the rewriting of equation (3) in terms of the x, y, z global coordinates. The total fluid flow in the joint sets of a given rock mass is the summation of all the in-plane flows in each of the sets.

There can also be fluid flow from the joints to the rock pores. The net flow from joint set J_I to the pores is obtained by vectorially summing the flow in the negative x'-direction with that in the positive x'-direction

$$q_{x'}^{I\to R} = 2K^R\ [(P^I - P^R)/\ 0.5S_I] \qquad (4)$$

A transformation similar to that applied to equation (3) enables $q_{x'}^{I\to R}$ to be expressed in terms of the x,y,z global coordinates. The total fluid flow from the joint systems into the rock pores is the summation of the fluid transports for each joint set.

In jointed and porous rock with compressible solid and fluid constituents, rock deformation is coupled with fluid diffusion yielding constitutive equations that can be expressed as follows [4]:

$$[K]d_{i,jj} = -\alpha^J(P^J_{,j}\cdot\vec{n})\vec{n} - \alpha^R P^R_{,j} \qquad (5)$$

$$-\alpha^j\dot{d}_{k,k} + \dot{P}^J/M^J = -K_{iJ}P_{,ii}{}^J - g(P^J - P^R) \qquad (6)$$

$$-\alpha^R\dot{d}_{k,k} + \dot{p}^R/M^R = -K^R P_{,ii}{}^R - g(P^R - P^J) \qquad (7)$$

where [K] is the stiffness tensor, J and R represent values for the joint set and porous rock, respectively, α and M are the material properties defining fluid and solid compressibilities [6], $g(P^J - P^R)$ denotes the fluid flow from the joint sets to the rock pores, and $d_{k,k}$ is volumetric strain. The time-dependent displacements and fluid pressures of a jointed and porous rock mass can be evaluated from equations (5-7) if boundary and initial conditions are known.

EXAMPLES

Several numerical examples have been solved using a finite element program based on equations (5-7) [4].

Example 1. In this first example we attempted to compare our numerical model to the most advanced analytic solution available for a multiple porosity medium, namely the one-dimensional consolidation of an isotropically fissured and porous rock layer 100 m thick under a uniformly distributed load. In this example we restricted the numerical solution to a medium in which the joint set flow properties were isotropic so that a direct comparison could be made. Comparison between the numerical and analytical solutions for the surface deformation history, and for the pore and joint fluid pressure history at 40 m depth shows excellent agreement between the two approaches.

Example 2. We have also calculated the vertical

deflection of a porous half-space surface subjected to uniform downward load in the case of one joint set of specifiable dip. Using the same finite element mesh as in example 1, and assuming that $K^J = 10^3 K^R$, and that joint displacements were insignificant, the effects of diffusion-dependent rock deformation were more clearly observed. In the case of horizontally jointed rock, joint fluid transport was confined to the horizontal plane, and settlement time response was relatively slow; as the joint-set dip increased, diffusion-dependent consolidation time became shorter due to enhanced vertical transport through joints, and reached a minimum at 90°. Since only the porous rock deformation was considered, the initial and steady-state conditions for each dip angle were the same; the steady-state settlement was also in close agreement with the analytically obtained value. The transient behavior as a function of joint dip met intuitive expectations.

More detailed description of this and other examples is given in references [4 and 5].

ACKNOWLEDGEMENT

This project was supported by the Applied Superconductivity Center, Univ. of Wisconsin-Madison.

REFERENCES

1. R.K. Wilson and E.C. Aifantis, "On the theory of consolidation with double porosity", Int. J. Engng. Sci. 20(9), 1009-1035 (1982).
2. M.A. Biot, "General theory of three-dimensional consolidation", J. Appl. Phys. 12, 155-164 (1941).
3. E.C. Aifantis, "On the response of fissured rock", Developments in Mechanics, 10, 249-253 (1979).
4. T.F. Cho, "Continuum and discrete modelings of porous and jointed rock: application to the design of near surface annular excavation", Ph.D. Thesis, University of Wisconsin-Madison (1988).
5. T.F. Cho, M.E. Plesha, and B.C. Haimson, "Continuum modeling of jointed porous rock", Int. J. Num. & Anal. Meth.in Geomech. (1990, in press).
6. J. Ghaboussi and E.L. Wilson, "Flow of compressible fluid in porous elastic media",Int. J. Num. Meth. Eng. 5, 419-442 (1973).

PREDICTION OF NONPROPORTIONAL LOADING OF METALS BY SUBLOADING SURFACE MODEL

K. Hashiguchi
Dept. Agr. Kyushu Univ., Fukuoka, Japan.

ABSTRACT

The subloading surface model has been developed to describe the cyclic loading behavior. It is applied to the prediction of nonproportional loading behavior of metals and is compared with test data of mild steel.

INTRODUCTION

The interior of the normal-yield surface is assumed not to be an elastic domain in order to describe the plastic deformation induced by the stress change within this surface. This is indispensable for the description of plastic deformation under the cyclic loading, while the conventional plasticity is incapable of describing the plastic deformation for the cyclic loading in which a stress changes within the normal-yield surface. The plasticity theory in which the interior of the normal-yield surface is assumed to be the elastic domain and the one in which the interior is assumed to be the elastoplastic domain are called a "conventional plasticity" and an "unconventional plasticity", respectively, by Drucker [1]. For the unconventional plasticity, various constitutive models have been proposed in the last quarter century. Among them the multi, the two, the infinite, the single, the initial subloading [2] and the subloading surface models [3, 4] are well-known. It was concluded that only the subloading surface model is capable of describing the cyclic loading behavior [5].

In this paper, the outline of this model is described and it is applied to the nonproportional loading behavior of metals.

THE OUTLINE OF THE SUBLOADING SURFACE MODEL

The subloading surface model is the extension of the conventional isotropic and kinematic hardening model so as to describe the plastic deformation caused by the stress change within the normal-yield surface. The salient features of this model are as follows:

i) The "subloading surface" is introduced, which expands/contracts, passing always through the current stress point in not only loading but also unloading states and keeping a similarity to the normal-yield surface.

ii) The ratio of the size of the subloading surface to that of the normal-yield surface continuously increases with a plastic deformation so that the continuous field of plastic moduli is formulated as a monotonic function of the ratio.

iii) The center of similarity (similarity-center) of the subloading and the normal-yield surfaces translates with a plastic deformation so that a closed histeresis loop is described. The similarity-center is physically interpreted to be the most elastic stress, i.e., the state of stress in which a material behaves most elastically.

iv) Three plastic internal state variables, i.e., the isotropic and the kinematic hardening parameters of the normal-yield surface, and the similarity-center of the normal-yield and the subloading surfaces are included. While the former two are the conventional variables, the last one is a new variable for the unconventional plasticity.

Let the stretching $\boldsymbol{D}$ be decomposed into the elastic part $\boldsymbol{D}^e$ and the plastic part $\boldsymbol{D}^p$, respectively:

$$\boldsymbol{D} = \boldsymbol{D}^e + \boldsymbol{D}^p \tag{1}$$

where $\boldsymbol{D}^e$ is given as

$$\boldsymbol{D}^e = \boldsymbol{E}^{-1}\mathring{\boldsymbol{\sigma}} \tag{2}$$

The fourth-order tensor $\boldsymbol{E}$ is the elastic modulus. In what follows let a constitutive equation for $\boldsymbol{D}^p$ be formulated.

The normal-yield surface is described as

$$f(\hat{\boldsymbol{\sigma}}) - F(H) = 0 \tag{3}$$

where

$$\hat{\boldsymbol{\sigma}} \equiv \boldsymbol{\sigma} - \hat{\boldsymbol{\alpha}} \tag{4}$$

The scalar H and the second-order tensor $\hat{\boldsymbol{\alpha}}$ are variables for describing the expansion/contraction and the translation, respectively, of the surface. This surface expands/contracts retaining a geometrical similarity in a stress space, and thus the function f is a homogeneous function the order of which is denoted by n.

The subloading surface is described as

$$f(\bar{\boldsymbol{\sigma}}) = R^n F \tag{5}$$

where

$$\bar{\boldsymbol{\sigma}} \equiv \boldsymbol{\sigma} - \bar{\boldsymbol{\alpha}} \tag{6}$$

$\bar{\boldsymbol{\alpha}}$ is the center of the subloading surface, while $\hat{\boldsymbol{\alpha}}$ is the center of the normal-yield surface. The function $f(\bar{\boldsymbol{\sigma}})$ has the same form as the homogeneous function $f(\hat{\boldsymbol{\sigma}})$ in (3). R is the ratio of the size of the subloading surface to that of the normal-yield surface, which is described by current values of $\boldsymbol{\sigma}$, $\bar{\boldsymbol{\alpha}}$ and F as

$$R \equiv \{f(\bar{\boldsymbol{\sigma}})/F\}^{1/n} \quad (0 \leq R \leq 1) \tag{7}$$

Let the evolution equation of the center of the normal-yield surface be given by the nonlinear kinematic hardening rule:

$$\mathring{\hat{\boldsymbol{\alpha}}} = \dot{A}\frac{\bar{\boldsymbol{\sigma}}}{\| \bar{\boldsymbol{\sigma}} \|} - \dot{B}\hat{\boldsymbol{\alpha}} \tag{8}$$

where $\dot{A}$ and $\dot{B}$, a superposed dot designating a material-time derivative, are scalar functions of the plastic stretching $\boldsymbol{D}^p$ in homogeneity of order one and some plastic internal state variables, and the notation $\| \; \|$ represents a norm (magnitude).

The evolution equation of the similarity-center $\boldsymbol{S}$ of the subloading and the normal-yield surfaces is given as

$$\mathring{\boldsymbol{S}} = \mathring{\hat{\boldsymbol{\alpha}}} + \frac{\dot{F}}{nF}\hat{\boldsymbol{S}} + C \| \boldsymbol{D}^p \| \left(\frac{\bar{\boldsymbol{\sigma}}}{R} - \frac{\hat{\boldsymbol{S}}}{\chi} \right) \tag{9}$$

where

$$\hat{\boldsymbol{S}} \equiv \boldsymbol{S} - \hat{\boldsymbol{\alpha}} \tag{10}$$

$C\,(> 0)$ and $\chi\,(0 < \chi < 1)$ are material constants.

It is assumed that R increases when a plastic deformation proceeds:

$$\dot{R} = U \| \boldsymbol{D}^p \| \tag{11}$$

where U is a monotonically decreasing function of R, satisfying the conditions

$$U = +\infty \text{ for } R = 0 \text{ and } U = 0 \text{ for } R = 1 \tag{12}$$

Differentiating (5) and substituting (8), (9) and the associated flow rule, the plastic strain rate is given by

$$\boldsymbol{D}^p = \dot{\lambda}\bar{\boldsymbol{N}} \tag{13}$$

where

$$\dot{\lambda} = \mathrm{tr}(\bar{\boldsymbol{N}}\mathring{\boldsymbol{\sigma}})/\bar{D} \quad (> 0 \text{ for } \boldsymbol{D}^p \neq \boldsymbol{O}) \tag{14}$$

$$\bar{D} \equiv \mathrm{tr}\left[\bar{\boldsymbol{N}}\left\{\frac{F'}{nF}\bar{h}\hat{\boldsymbol{\sigma}} + \bar{\boldsymbol{a}} + C(1-R)\left(\frac{\bar{\boldsymbol{\sigma}}}{R} - \frac{\hat{\boldsymbol{S}}}{\chi}\right) + \frac{U}{R}\tilde{\boldsymbol{\sigma}}\right\}\right] \tag{15}$$

$$\bar{\boldsymbol{N}} \equiv \frac{\partial f(\bar{\boldsymbol{\sigma}})}{\partial \bar{\boldsymbol{\sigma}}} \Big/ \left\| \frac{\partial f(\bar{\boldsymbol{\sigma}})}{\partial \bar{\boldsymbol{\sigma}}} \right\|, \quad \tilde{\boldsymbol{\sigma}} \equiv \boldsymbol{\sigma} - \boldsymbol{S} \tag{16}$$

$$F' \equiv dF/dH \tag{17}$$

$\dot{H}$ and $\mathring{\boldsymbol{\alpha}}$ involve $\boldsymbol{D}^p$ in homogeneity of order one so that one can write as

$$\dot{H} = \dot{\lambda}\bar{h}, \quad \mathring{\boldsymbol{\alpha}} = \dot{\lambda}\bar{\boldsymbol{a}} \tag{18}$$

$\bar{h}$ and $\bar{\boldsymbol{a}}$ are the scalar and the second-order tensor functions of stress and some plastic internal state variables.

CONSTITUTIVE EQUATION OF METALS

Let the normal-yield surface be given as

$$f(\hat{\boldsymbol{\sigma}}) = \sqrt{3/2}\, \| \hat{\boldsymbol{\sigma}}' \| \tag{19}$$

$$F(H) = F_0[1 + h_1\{1 - exp(-h_2 H\}] \tag{20}$$

$$\mathring{\boldsymbol{\alpha}} = \| \boldsymbol{D}^p \| \left(k_1 \frac{\bar{\boldsymbol{\sigma}}}{\| \bar{\boldsymbol{\sigma}} \|} - k_2 \hat{\boldsymbol{\alpha}}\right) \tag{21}$$

$$\dot{H} = \sqrt{2/3}\, \| \boldsymbol{D}^p \| \tag{22}$$

where F_0 is an initial value of F, and h_1, h_2, k_1, k_2 are material constants, and

$$\hat{\boldsymbol{\sigma}}' \equiv \hat{\boldsymbol{\sigma}} - \frac{1}{3}\mathrm{tr}(\hat{\boldsymbol{\sigma}})\boldsymbol{I} \tag{23}$$

The function U is given as

$$U = u_1(1 - R^{m_1})/\tilde{R}^{n_1} \tag{24}$$

where u_1, m_1 and n_1 are material constants and

$$\tilde{R} \equiv \{f(\boldsymbol{\sigma} - \boldsymbol{S})/F\}^{1/n} \tag{25}$$

The elastic constitutive equation is given by the Hooke's law:

$$\boldsymbol{D}^e = \frac{1}{2G}\mathring{\boldsymbol{\sigma}} - \left(\frac{1}{2G} - \frac{1}{E}\right)(\mathrm{tr}\mathring{\boldsymbol{\sigma}})\boldsymbol{I} \tag{26}$$

where E and G are Young's modulus and a shear modulus, respectively.

The comparison of the calculated result by the present constitutive equation with the experimental result mesured by Ohashi et al. [6] is depicted in Fig. 1, in which plastic deformations of thin-walled tubular specimens (20 mm diameter, 1 mm thickness and 50 mm length) of mild steel are shown for various bilinear stress trajectries by applying combined loadings of axial load and torque after the application of axial stress 24 kgf/mm^2 to the specimens. Material constants and initial values for the calculation are selected as follows:

$h_1 = 1.3,\ h_2 = 30,\ k_1 = 100\ k_2 = 20,$
$C = 10,\ \chi = .99,\ u_1 = 300,\ m_1 = 5,\ n_1 = .001$
$E = 21800\,\mathrm{kgf/mm}^2,\ G = 8040\,\mathrm{kgf/mm}^2$
$F_0 = 14\,\mathrm{kgf/mm}^2,\ \hat{\boldsymbol{\alpha}}_0 = \boldsymbol{S}_0 = \boldsymbol{O}\,\mathrm{kgf/mm}^2$

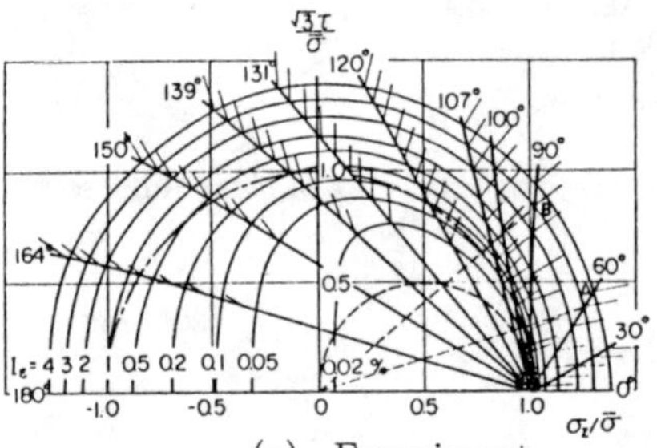

(a) Experiment.

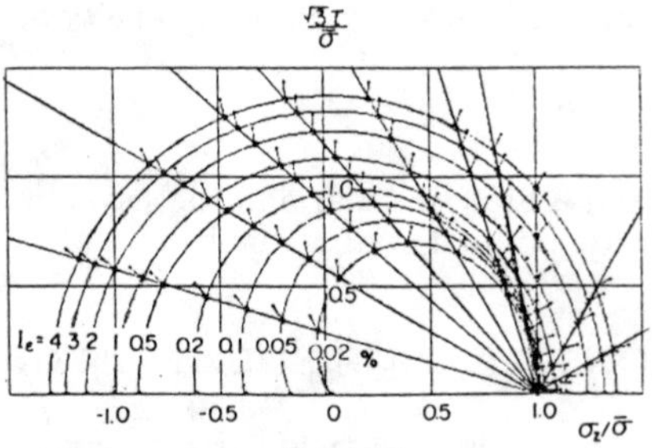

(b) Theory.

Fig. 1 Comparison of theory and experiment: Equi-Equivalent plastic strain curves. σ_z: axial stress, τ: shear stress, $\bar{\sigma} = 24$ kgf/mm^2.

REFERENCES

1. D.C. Drucker, "Conventional and unconventional plastic response and representation, Appl. Mech. Rev. (ASME), **41**, 151-167 (1988).
2. K.hashiguchi, and M. Ueno, "Elastoplastic constitutive laws of granular materials", Constutitive Equtions of Soils (Proc. Int. Conf. Soil Mech. Found. Eng., Spec. Ses. 9), 73-82 (1977).
3. K. Hashiguchi, "A mathematical description of elastoplastic deformation in normal-yield and sub-yield states", Proc. 2nd Int. Symp. Numerical Models in Geomech., Gehnt, 17-24 (1986).
4. K. Hashiguchi, "Subloading surface model of plasticity", Int. J. Solids Structures, **25**, 917-945 (1989).
5. K. Hashiguchi, "Theoretical assessments on basic structures of elastoplastic constitutive models (Panel report), Proc. Int. Workshop on Constitutive Equations for Granular Non-Cohesive Soils, Cleveland, 699-715 (1987).
6. Y. Ohashi, K. Kawashima, and T. Yokouchi, "Anisotropy due to plastic deformation of initially isotropic mild steel and its analytical formulation", J. Mech. Phys. Solids, **23**, 277-294.

Constitutive Equations for Large Dynamic Deformations

W. Herrmann
Sandia National Laboratories[1]
Albuquerque NM 87185

Abstract

Numerous constitutive equations have been proposed which are intended to be generalizations of infinitesimal theories to large deformations. The generalization can be accomplished in a number of different ways that produce very different and sometimes surprising results at large deformation. This point is illustrated by reviewing an example of simple shear of an elastic material. The example shows that generalizations must take explicit account of the finite deformation kinematics, if reasonable results are to be obtained, and points the way to the most convenient means of doing so.

Introduction

Numerical methods, in terms of finite elements or finite differences, are capable of modeling complex dynamic mechanics problems with considerable fidelity, and have become an indispensable tool in mechanical design and performance assessment. There has been a great deal of activity to develop ever more complex constitutive equations describing the behavior of various materials, often based on micromechanical models of material deformation, with the expectation that they would improve predictions of these methods. The models are then often used in calculations of large deformation problems.

Most numerical methods use constitutive equations in incremental form, which specify the stress rate when the rate of deformation is given. It is expected that constitutive equations be objective, that is, that material response not change if the underlying frame, with respect to which stresses and deformations are measured, undergoes a rigid translation or rotation. The ordinary time derivative of the stress is not objective, and a variety of objective stress rates have been formulated. It has repeatedly been stated, and elegantly shown by Truesdell and Noll [1], that any objective stress rate may be used equally well in the constuction of constitutive equations, but that, once constructed, particular forms of the constitutive functions are not invariant under change of stress rate. While this is well understood, its consequences do not seem to have been universally appreciated.

It is the intent of this review to give a brief summary of the geometrical concepts involved in large deformation, sufficient to discuss the above issue. Space limitations permit the discussion only of elastic constitutive equations along with an example of simple shear which has been widely treated in the past. Nevertheless, the example provides an excellent illustration of the kinematical effects

[1]This work performed at Sandia National Laboratories supported by the U. S. Dept. of Energy under contract number DE-AC04-76DP00789

encountered in large deformations, and suggests a convenient methodology for generalizing other constitutive equations.

Kinematics

Since different authors use different names and symbols, an abbreviated synopsis of the kinematics will be given to establish the notation. The equation of motion, which specifies the position $\boldsymbol{x}$ at time t of a material particle P, which occupies the position $\boldsymbol{X}$ in an arbirarily chosen reference configuration, is the mapping $\boldsymbol{x} = \hat{\boldsymbol{x}}(\boldsymbol{X}, t)$. The response of a simple material at P is assumed to depend on the motion only through the material velocity $\boldsymbol{u} = \partial\hat{\boldsymbol{x}}/\partial t$ and deformation gradient $\boldsymbol{F} = \partial\hat{\boldsymbol{x}}/\partial\boldsymbol{X}$ at P. The polar decompositions $\boldsymbol{F} = \boldsymbol{R}\boldsymbol{U} = \boldsymbol{V}\boldsymbol{R}$ describe the local deformation at P as a pure stretch $\boldsymbol{U}$ to an intermediate unrotated configuration, followed by a pure rotation $\boldsymbol{R}$, or as a pure rotation $\boldsymbol{R}$ to an intermediate unstretched configuration, followed by a pure stretch $\boldsymbol{V}$. Clearly $\boldsymbol{V} = \boldsymbol{R}\boldsymbol{U}\boldsymbol{R}^{\mathrm{T}}$. Some strain measures that vanish in the absence of stretch are $\boldsymbol{E} = (1/2)(\boldsymbol{U}^2 - \boldsymbol{I}) = (1/2)(\boldsymbol{F}^{\mathrm{T}}\boldsymbol{F} - \boldsymbol{I})$ and $\boldsymbol{e} = (1/2)(\boldsymbol{I} - (\boldsymbol{V}^{-1})^2) = (1/2)(\boldsymbol{I} - \boldsymbol{F}^{-1}(\boldsymbol{F}^{-1})^{\mathrm{T}})$. It is clear that $\boldsymbol{E} = \boldsymbol{F}^{\mathrm{T}}\boldsymbol{e}\boldsymbol{F}$.

An expression for deformation rate follows from the fact that the second partial derivatives of the motion commute, $\dot{\boldsymbol{F}} = \boldsymbol{L}\boldsymbol{F}$ where $\boldsymbol{L} \equiv \mathbf{grad}\,\boldsymbol{u}$ is the spatial velocity gradient. The rate of rigid body rotation is given by $\boldsymbol{\Omega} = \dot{\boldsymbol{R}}\boldsymbol{R}^{\mathrm{T}}$. Differentiation of the polar decomposition of $\boldsymbol{F}$ provides $\boldsymbol{L} = \boldsymbol{R}\dot{\boldsymbol{U}}\boldsymbol{U}^{-1}\boldsymbol{R}^{\mathrm{T}} + \boldsymbol{\Omega}$. It is then easy to show that the symmetric and antisymmetric parts of $\boldsymbol{L}$ are

$$\boldsymbol{D} = \boldsymbol{R}\overline{\boldsymbol{D}}\boldsymbol{R}^{\mathrm{T}} \qquad \boldsymbol{W} = \boldsymbol{R}\overline{\boldsymbol{W}}\boldsymbol{R}^{\mathrm{T}} + \boldsymbol{\Omega} \tag{1}$$

where $\overline{\boldsymbol{D}}$ might be called the unrotated stretching and is the symmetric part of $\overline{\boldsymbol{L}} \equiv \dot{\boldsymbol{U}}\boldsymbol{U}^{-1}$ and where $\overline{\boldsymbol{W}}$ could, somewhat incongruously, be called the unrotated spin and is the antisymmetric part of $\overline{\boldsymbol{L}} = \dot{\boldsymbol{U}}\boldsymbol{U}^{-1}$. It is also easily shown that

$$\dot{\boldsymbol{E}} = \boldsymbol{U}\overline{\boldsymbol{D}}\boldsymbol{U} = \boldsymbol{F}^{\mathrm{T}}\boldsymbol{D}\boldsymbol{F}. \tag{2}$$

The tensors $\dot{\boldsymbol{E}}$, $\overline{\boldsymbol{D}}$, and $\boldsymbol{D}$ may be viewed as describing the same rate of stretching, referred to corresponding metrics in the reference, unrotated, and current configurations, respectively. Similarly, $\overline{\boldsymbol{W}}$ and $\boldsymbol{W}$ can be viewed as describing the spin referred to metrics in the unrotated and current configurations respectively.

The Cauchy stress tensor $\boldsymbol{\sigma}$ may be related to the second Piola-Kirchhoff stress tensor Σ, and a tensor which might be termed the unrotated stress $\overline{\boldsymbol{\sigma}}$ by

$$\boldsymbol{\sigma} = \frac{1}{J}\boldsymbol{F}\,\Sigma\,\boldsymbol{F}^{\mathrm{T}} = \boldsymbol{R}\overline{\boldsymbol{\sigma}}\,\boldsymbol{R}^{\mathrm{T}}. \tag{3}$$

Note that Σ, $\overline{\boldsymbol{\sigma}}$, and $\boldsymbol{\sigma}$ may be viewed as describing the same stress relative to metrics in the reference, unrotated, and current configurations, respectively. Differentiating Eq. 3 leads to stress rates which may be shown to be objective

$$\overset{\circ}{\boldsymbol{\sigma}} \equiv \dot{\boldsymbol{\sigma}} - \boldsymbol{\Omega}\boldsymbol{\sigma} - \boldsymbol{\sigma}\boldsymbol{\Omega}^{\mathrm{T}} = \boldsymbol{R}\dot{\overline{\boldsymbol{\sigma}}}\,\boldsymbol{R}^{\mathrm{T}} \tag{4}$$

$$\check{\boldsymbol{\sigma}} \equiv \dot{\boldsymbol{\sigma}} - \boldsymbol{L}\boldsymbol{\sigma} - \boldsymbol{\sigma}\boldsymbol{L}^{\mathrm{T}} + \mathrm{tr}(\boldsymbol{D})\,\boldsymbol{\sigma} = \frac{1}{J}\boldsymbol{F}\dot{\Sigma}\boldsymbol{F}^{\mathrm{T}} \tag{5}$$

where the first has been attributed to Green and Naghdi, and the second to Truesdell. Another objective stress rate is that of Zaremba or Jaumann

$$\breve{\sigma} = \dot{\sigma} - W\sigma - \sigma W^{\mathrm{T}}. \tag{6}$$

Still other objective stress rates have been examined by Atluri [4].

Finite Deformation Elasticity

It is instructive to review some concepts in finite deformation elasticity. A simple material is said to be elastic if its stress response at a material point P and time t depends only on the local configuration at P through the gradient of the deformation at P, that is, $\sigma = \mathrm{g}(F)$. Considerations of objectivity [1] reduce this constitutive equation to alternate forms

$$\overline{\sigma} = \mathrm{g}(U) \qquad \Sigma = \mathrm{f}(E) \tag{7}$$

where $\mathrm{f}(\cdot)$ and $\mathrm{g}(\cdot)$ are related explicitly through U. Other objective forms may be obtained in terms of stress measures related to $\overline{\sigma}$ only through U, or strain measures which are functions only of U. Introduction of σ or e explicitly introduces the rotation R. This can be illustrated by considering the linear form

$$\Sigma = \mathsf{C}\,[E] \tag{8}$$

where $\mathsf{C}\,[\cdot]$ is a fourth order linear tensor mapping of E into Σ whose components are the elasticity coefficients referred to the metric in the reference configuration. Introducing e and σ via their defining equations

$$\sigma = \frac{1}{J} F\,\mathsf{C}\left[F^{\mathrm{T}} e F\right] F^{\mathrm{T}} \equiv \mathsf{c}(F)\,[e] \tag{9}$$

where the last relation defines the fourth-order tensor mapping $\mathsf{c}(F)\,[\cdot]$. Its dependence on F has been explicitly indicated. The tensor $\mathsf{c}(F)\,[\cdot]$ has components which are the elasticity coefficients referred to the metric in the current configuration. Note that F cannot enter Eq. 9 arbitrarily, or be left out, if the material response is to be unaltered from that in Eq. 8. In the unrotated configuration it is convenient to define an unrotated strain by

$$\overline{e} = R^{\mathrm{T}} e R = U^{-1} E U^{-1}. \tag{10}$$

Using Equations 3 and 10 in 8 results in

$$\overline{\sigma} = R^{\mathrm{T}}\,\mathsf{c}\left[R\overline{e}R^{\mathrm{T}}\right] R = \frac{1}{J} U\,\mathsf{C}\,[U\overline{e}U]\,U \equiv \overline{\mathsf{c}}(U)\,[\overline{e}] \tag{11}$$

where the last relation defines the fourth-order mapping $\overline{\mathsf{c}}(U)\,[\cdot]$ which has components referred to the unrotated metric.

Since $\mathsf{C}\,[\cdot]$ was assumed to be a constant fourth-order tensor mapping, clearly $\overline{\mathsf{c}}(U)\,[\cdot]$ and $\mathsf{c}(F)\,[\cdot]$ are not, in large deformations. The former depends on the strain, while the latter depends on the rotation as well. If we had assumed that $\mathsf{C}[\cdot]$ was an isotropic fourth order tensor mapping, it is clear that $\bar{\mathsf{c}}(U)[\cdot]$ and $\mathsf{c}(F)[\cdot]$ would not be, in large deformations. Linearity and isotropy of the elasticity tensor apply only when referred to the metric in a particular configuration, if at all. If a configuration exists in which the elasticity tensor has linearity or isotropy properties, it is usually convenient to choose it as the reference configuration. The elasticity tensor referred to a deformed configuration then exhibits kinematical effects often erroneously called strain-induced hardening and anisotropy. Of course, the material does change. The apparent hardening and anisotropy merely reflect changes in the components of the elasticity tensor when referred to metrics in deformed configurations.

Returning to the nonlinear case, a rate equation may be obtained by differentiating the general objective elastic constitutive equation Eq. 7_2

$$\dot{\Sigma} = \mathsf{C}(E)\left[\dot{E}\right] \tag{12}$$

where $\mathsf{C}(\boldsymbol{E})[\cdot]$ is the derivative of $\mathbf{f}(\boldsymbol{E})$. Note that $\mathsf{C}(\boldsymbol{E})[\cdot]$ must satisfy certain integrability conditions if the elastic function $\mathbf{f}(\cdot)$ in Eq. 8_2 is to be recovered by integration. This can be written in terms of stress and strain rates in the current configuration, using Equations 2, 5 and 9

$$\check{\check{\sigma}} = \frac{1}{J} F\,\mathsf{C}(E)\left[F^{\mathrm{T}} D F\right] F^{\mathrm{T}} = \mathsf{c}(F,E)\left[D\right] \tag{13}$$

where the dependence on F, which enters the elasticity tensor $\mathsf{c}(\boldsymbol{F},\boldsymbol{E})[\cdot]$ in a special way, has been kept separate from its dependence on E, which may arise from material nonlinearities. If $\mathsf{c}[\cdot]$ were assumed to be independent of $\boldsymbol{F}$, integration would not reproduce Eq. 7_2, and this constitutive equation would be expected to describe very different material behavior than that described by Eq. 12 at large deformations. In terms of $\breve{\sigma}$, from Eqs. 5 and 6

$$\breve{\sigma} = \mathsf{c}(F,E)\left[D\right] + D\sigma + \sigma D - (\mathrm{tr}D)\sigma \equiv \mathsf{c}'(\sigma, D, F, E)\left[D\right] \tag{14}$$

where the last relation defines the fourth order mapping $\mathsf{c}'(\boldsymbol{\sigma},\boldsymbol{D},\boldsymbol{F},\boldsymbol{E})[\cdot]$. However, this obscures the special way in which $\boldsymbol{\sigma}$, $\boldsymbol{D}$, and $\boldsymbol{F}$ enter $\mathsf{c}'[\cdot]$ to retain the same behavior as that in Eq. 12. If $\mathsf{c}'[\cdot]$ were to be assumed to depend only on $\boldsymbol{E}$, neglecting its dependence on $\boldsymbol{F}$ and omitting the extra terms in Eq. 14, this constitutive equation would be expected to describe very different material behavior than that described by either Eq. 12 or 13.

To examine the form of the elastic rate equation referred to the metric in the unrotated configuration, Eq. 13 could be written as

$$R^{\mathrm{T}}\check{\check{\sigma}}R = R^{\mathrm{T}}\mathsf{c}(F,E)\left[R\overline{D}R^{\mathrm{T}}\right]R.$$

Using Eqs. 4, 5, and 11 this becomes

$$\dot{\overline{\sigma}} = \bar{\mathsf{c}}(U,E)\left[\overline{D}\right] + \overline{L}\overline{\sigma} + \overline{\sigma}\overline{L}^{\mathrm{T}} - (\mathrm{tr}\overline{D})\overline{\sigma} \equiv \mathsf{c}''(\overline{\sigma},\overline{L},U,E)\left[\overline{D}\right] \tag{15}$$

where the last relation defines the fourth order mapping $\mathsf{c}''(\overline{\sigma},\overline{L},U,E)[\cdot]$, but this once again obscures the special way in which $\overline{\sigma}$, $\overline{L}$, and U enter $\mathsf{c}''[\cdot]$ to retain the same behavior as that given by Eq. 12.

Simple Shear

With the background of the foregoing discussion, we can now review solutions to the problem of simple shear for a linear isotropic elastic material. The constitutive equation for infinitesimal elasticity is

$$\dot{\boldsymbol{\sigma}} = \lambda(\mathrm{tr}\boldsymbol{D})\boldsymbol{I} + 2\mu\boldsymbol{D} \tag{16}$$

where λ and μ are elastic constants. The problem to be solved is to find the stress corresponding to the specified homogeneous deformation $x_1 = X_1 + \kappa X_2$, $x_2 = X_2$, $x_3 = X_3$, where κ increases linearly in time. Generalizations of this problem to large deformations have been addressed variously by Truesdell and Noll [1], Eringen [2], Dienes [3], Atluri [4], and others.

The most common generalization simply uses the Jaumann stress rate $\breve{\boldsymbol{\sigma}}$ in place of $\dot{\boldsymbol{\sigma}}$ in Eq. 16. When the specified deformation is inserted, the resulting differential equations have the solution [2]

$$\sigma_{12} = \mu \sin\kappa \qquad \sigma_{11} = -\sigma_{22} = \mu(1 - \cos\kappa). \tag{17}$$

Use of the Truesdell rate $\check{\dot{\boldsymbol{\sigma}}}$ results in the solution [2]

$$\sigma_{12} = \mu\kappa \qquad \sigma_{11} = \mu\kappa^2 \qquad \sigma_{22} = 0. \tag{18}$$

When the Green-Naghdi rate $\breve{\boldsymbol{\sigma}}$ is used, the solution is [3]

$$\begin{aligned} \sigma_{12} &= 2\mu\cos 2\beta(2\beta - 2\tan 2\beta\,\ln\cos\beta - \tan\beta) \\ \sigma_{11} &= -\sigma_{22} = 4\mu(\cos 2\beta\,\ln\cos\beta + \beta\sin 2\beta - \sin^2\beta) \end{aligned} \tag{19}$$

where $\tan\beta = \kappa/2$. Use of $\breve{\boldsymbol{\sigma}}$ in Eq. 16 is equivalent to

$$\dot{\overline{\boldsymbol{\sigma}}} = \lambda(\mathrm{tr}\overline{\boldsymbol{D}})\boldsymbol{I} + 2\mu\overline{\boldsymbol{D}}. \tag{20}$$

This may be integrated immediately, since λ and μ are constants. Substituting for $\overline{\boldsymbol{\sigma}}$ and $\overline{\boldsymbol{D}}$ in the simple shear problem leads immediately to Eq. 20, as verified by Flanagan and Taylor [6].

Motivated by the discussion of elasticity in the last section, one might add the isotropic form of Eq. 12

$$\dot{\boldsymbol{\Sigma}} = \lambda(\mathrm{tr}\dot{\boldsymbol{E}})\boldsymbol{I} + 2\mu\dot{\boldsymbol{E}}. \tag{21}$$

This may also be integrated immediately. Substituting for $\boldsymbol{\Sigma}$ and $\dot{\boldsymbol{E}}$ in the simple shear problem yields the solution

$$\begin{aligned} \sigma_{12} &= \mu\kappa + \frac{1}{2}(\lambda + 2\mu)\kappa^3 & \sigma_{11} &= \frac{1}{2}(\lambda + 4\mu)\kappa^2 + (\lambda + 2\mu)\kappa^3 \\ \sigma_{22} &= \frac{1}{2}(\lambda + 2\mu)\kappa^2 & \sigma_{33} &= \frac{1}{2}\lambda\kappa^2. \end{aligned} \tag{22}$$

If Poisson's ratio is taken to be 1/3, then $\lambda = 2\mu$ and $\sigma_{12} = \mu\kappa(1 + 2\kappa^2)$.

The four solutions for the shear stress are compared in Fig. 1. The first three, labelled Jaumann, Green-Naghdi, and Truesdell, have been discussed in

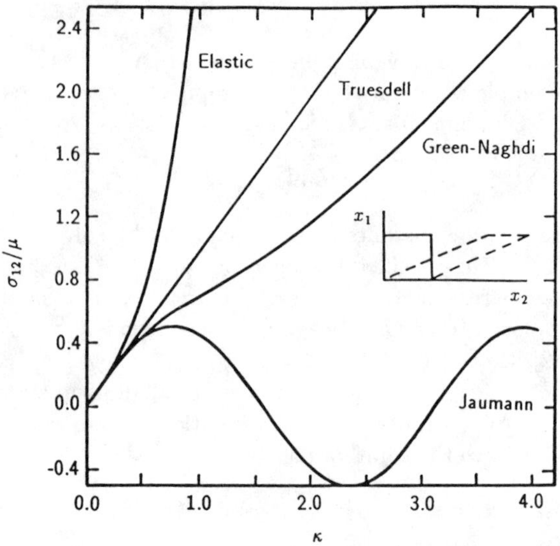

Figure 1: Shear Stress in Simple Shear

detail before, often with a view to arguing that one or another stress rate is a "more accurate" one. We take a very different view here. Each of the four constitutive equations considered above is a perfectly viable one, satisfying all kinematical constraints, but each describes different material behavior.

A hypoelastic material is one in which the current stress rate is assumed to be a function only of the current values of the Cauchy stress and the rate of deformation. Considerations of objectivity require that the stress rate be objective, that the deformation rate reduce to $\boldsymbol{D}$, and that the function be linear in $\boldsymbol{D}$ and jointly isotropic in $\boldsymbol{\sigma}$ and $\boldsymbol{D}$ [1]. Both the Jaumann rate and the Truesdell rate used in Eq. 16 satisfy these requirements.

When the Truesdell rate is used, the response in simple shear shows gratifyingly linear response, Fig. 1. Of course, if a real material remained elastic to large deformations, one might expect on physical grounds that it would develop material nonlinearities and anisotropies as the deformation became large. If experiments could be conducted in pure shear to large deformations, then Eq. 18 could be fitted to the data quite easily. A serious shortcoming of the description can be seen easily from Eq. 18. Once μ has been constructed as a function of κ, the responses of σ_{11} and σ_{22} are fixed. It is unlikely that real material nonlinearities and anisotropies will be described so simply. In fact, it is far from clear how a hypoelastic equation can be extended to describe real material anisotropy.

When the Jaumann rate is used, the shear stress in simple shear exhibits periodic behavior that has been noted previously eg. [3]. While previous authors have taken this to mean that this description is inaccurate, we interpret it to mean that this constitutive description is not particularly useful for modeling elastic response. Truesdell and Noll [1] solved the simple shear problem for a similar hypoelastic equation, and interpreted the maximum in the shear response to correspond to some sort of "hypoelastic yield". It may well have uses for the description of material behavior other than elastic. Of course, the Jaumann rate may be used to give identical results to those using the Truesdell rate if the extra terms in Eq. 14 are retained, but there would seem to be no point in doing so in the present context.

The remaining two constitutive descriptions which have been considered are not hypoelastic. They include dependence on the total deformation from a reference configuration, which is expressly excluded from hypoelasticity. Eq. 21 clearly is one possible elastic description, although there are infinitely many others as noted in connection with Eq. 7. Its solution to the simple shear problem exhibits the apparent kinematical hardening and anisotropy which one would expect in a large deformation elasticity theory. It would, in principle, be relatively easy to fit λ and μ to exerimental data, or if material anisotropies where present, to fit appropriate components of $\mathsf{C}(\boldsymbol{E})\,[\cdot]$ to the data, although this might be a formidable task in practice.

Of course, use of Eq. 12 requires the explicit calculation of the deformation gradient $\boldsymbol{F}$, which, by definition, enters the constitutive equation of an elastic material. This is relatively easily done in the numerical methods by integrating $\dot{\boldsymbol{F}} = \boldsymbol{L}\boldsymbol{F}$, at the expense of a few extra calculations and nine stored variables per element. In the calculation, $\dot{\boldsymbol{E}}$ can be calculated directly from $\boldsymbol{D}$ and $\boldsymbol{F}$ by Eq. 2, and, after integration of the constitutive equation, $\boldsymbol{\sigma}$ can be calculated directly from $\boldsymbol{\Sigma}$ and $\boldsymbol{F}$ by Eq. 3, without disturbing the finite element or finite difference algorithms. No costly matrix inversions, polar decompositions, or transcendental function evaluations are needed.

Use of Equation 20 has been espoused by Dienes [3], Johson and Bammann [5], and implemented in a finite element method by Flanagan and Taylor [6]. It corresponds closely to an elastic relation, in the sense that $\bar{\mathsf{c}}\,[\cdot]$ is related to $\mathsf{C}\,[\cdot]$ only through $\boldsymbol{U}$ as seen from Eq. 11. However, the extra terms in Eq. 15 have been omitted, and the time integral of $\overline{\boldsymbol{D}}$ is not related simply to any of the total strain measures we have introduced. It may be viewed as a constitutive description in its own right. In the present context, it has the disadvantage that it leads to complicated forms which, when fitted to real material data can be expected to involve trancendental functions. The use of Eq. 15 requires integration of $\boldsymbol{U}$ or $\boldsymbol{V}$ and $\boldsymbol{R}$ in the numerical methods, which involves more calculation and the same extra storage as the integration of $\boldsymbol{F}$. In the present context, working in the unrotated configuration would seem to offer no advantages over working in the reference configuration.

Closure

Numerical finite element and finite difference methods perform their calculations using Cauchy stress and material velocity in the current configuration. When velocity gradients have been calculated for an element at a given time step, the constitutive equation is expected to provide new values of the Cauchy stress for use in the next time step. When strain and rotations are small, there is no difficulty. The various measures of stress and strain are indistinguishable, because the reference and current configurations nearly coincide. When strains or rotations become large, then these measures progressively diverge. The different measures represent the same stress and strain referred to metrics in the different configurations, and none are preferred.

It has often been argued that a real material cannot have a memory of a reference configuration which it may never have actually occupied, and that its response should be describable entirely in terms of the current configuration. This is the basic premise of hypoelasticity. Classical theories of elasticity and plasticity, which inherently use a reference configuration in their formulation, are excluded. However, hypoelasticity cannot escape the complications of large deformations because an objective stress rate is needed. Once a stress rate is chosen, the constitutive equation for a specific real material can be constructed and fitted to experimental observations covering the deformation range of interest. At this stage, other stress rates may be used only if they are inserted through their defining equations, otherwise the results may be drastically different than intended, as the present example illustrates.

In the present elastic example, the Truesdell rate provides a simple and useful description, but it may not have enough latitude to describe real material nonlinearities and anisotropies. The elastic theory Eq. 21, or more generally Eq. 12, requires more calculations and computer storage, but is relatively convenient to use and to fit to experimental data. It separates the effects of material nonlinearities and anisotropies from kinematic effects in a relatively transparent way, and extends in a natural way to include real material anisotropy. The Jaumann and unrotated descriptions seem to have little value in the present context.

Of course, few materials are expected to remain elastic at large deformations. There have been attempts to describe elastic-plastic and damaging effects within the theory of hypoelasticity. Classical plasticity theory, however, inherently involves a reference configuration in its formulation. If it is desired to use the classical theory of plasticity, then the description in the reference configuration provides a relatively economical and transparent means of doing so. At large strains, strain hardening and damage are expected to be anisotropic on physical grounds. If details of strain hardening or damage are modeled, then material anisotropy should be modeled with corresponding detail, if the accuracy of solutions really is to be improved. Again, the referential description offers a convenient starting point.

Acknowledgement: Thanks to Karen Mulholland for bibliographic research.

References

[1] C. Truesdell and W. Noll, in *Handbuch der Physik*, ed. S. Flugge, Springer-Verlag, Berlin, 1965.

[2] A. C. Eringen, *Nonlinear Theory of Continuous Media*, McGraw-Hill, New York, 1962.

[3] J. K. Dienes, *Acta. Mech.* **32**, 217, 1979.

[4] S. N. Atluri, *Comput. Meths. Appl. Mech. Engrg.* **43**, 137, 1984.

[5] G. C. Johnson and D. J. Bammann, *Int. J. Solids Structures* **20**, 725, 1984.

[6] D. P. Flanagan and L. M. Taylor, *Comput. Meths. Appl. Mech. Engrg.* **62**, 305, 1987.

PLASTICITY OF SATURATED CLAYS UNDER COMBINED MECHANICAL AND ENVIRONMENTAL LOADS

T.A. HUECKEL and C.M. MA
Department of Civil and Environmental Engineering,
Duke University, Durham, NC 27706

ABSTRACT

Balance and constitutive equations are described for saturated clays subjected to a combined mechanical and thermal or chemical load in elastoplastic range. Mass transfer from solid to liquid fraction describes the adsorbed water degradation due to temperature or chemistry variations leading to changes in permeability.

INTRODUCTION

Dense clay liners up to 4m thick are used in environmental geotechnology to retain domestic and industrial wastes. Clay is also studied for heat generating nuclear waste disposal. The fundamental property of clay in this context is the stability of its low hydraulic conductivity. However, an increase in hydraulic conductivity up to two and four orders of magnitude may be caused respectively by heating to 100°C [1], and by changes in permeant chemistry [2,3]. Most significant chemical variables are: ionic concentration of the permeant c, its dielectric "constant" D, and valency of ions v. Stress acting on clay simultaneously with the above agents or temperature, may induce consolidation [2,4-7] and reduce hydraulic conductivity up to two orders of magnitude [8]. Thus, hydraulic conductivity is coupled to permeant chemistry, temperature and stress. A plasticity theory and equations for inter-constituent mass transfer are proposed to describe the above properties.

A PHENOMENOLOGICAL MODEL

Clay clusters are made of platelets possibly separated with two to four monomolecular layers of internal adsorbed water. Outside, clusters are enveloped by external adsorbed water. The remaining pore space is filled with free water. This water is able to flow under enough high pressure gradient [1]. The adsorbed water is critically sensitive to environmental loads. The range of adsorption forces may change due to heat [9], while at certain temperature the adsorbed water may degenerate into free water [9-11]. Also the replacement of water by a (soluble) liquid with a lower dielectric constant, e.g. benzene, may produce shrinking of the adsorbed water [2,7]. The voids left by the adsorbed water are filled by free water [12], and thus under stress clay may consolidate. Diffuse double layer (DDL) theory can describe some of these phenomena in large clay pores, but it fails for pores smaller than 5 nm [10,11,13,15]. Thus a phenomenological approach is attempted.

BALANCE EQUATIONS

Saturated clay is considered as a two constituent medium composed of a solid and a single liquid. The solid comprises clay mineral and adsorbed water at room temperature. The liquid consists of a mixture of flowing liquids only. The constituents bulk mass densities are

$$\Gamma^S = m^S\phi_1 + m^A\phi_2 \ ; \ \Gamma^L = m^L\phi_3 \ ; \ \sum_1^3 \phi_i = 1 \tag{1}$$

in which m 's are specific masses, and ϕ's are variable total volume fractions.

The degradation of adsorbed water into free water is described as a reversible, inter-constituent mass transfer toward liquid, μ^L or solid μ^S, [14]. Assuming that ϕ_1 = const., and that m^S = const., the mass conservation equation for constituents yields

$$\overline{\dot{m^A\phi_2}} + \Gamma^S v^S_{i,i} + \Gamma^S_{,i} v^S_i = \mu^S \tag{2}$$

$$\overline{\dot{m^L\phi_3}} + \Gamma^L v^L_{i,i} + \Gamma^L_{,i} v^L_i = \mu^L \tag{3}$$

$$\mu^S + \mu^L = 0 \tag{4}$$

where a dot denotes a partial time derivative and v^S_i and v^L_i are respective particle velocities. Because of (4), the mass transfer variables are not externally measurable. Either microscopic experiments [12] or a back analysis of changes in hydraulic conductivity may be adopted instead.

The mechanisms of mass transfer are little known. Following the results for model materials [10,11] and microscopic studies [12], separate, linear equations for mass transfer due to T and D may be assumed as follows

$$\mu_T^S = \frac{\dot{T}m^L}{T_d - T_f}(\phi_2^0 - m^S S_s \delta) \ ; \ T_f \le T \le T_d \ ; \ \dot{D} = 0 \tag{5}$$

$$\mu_D^S = \frac{-\dot{D}m^L}{D\ (T)}(\phi_2^0 - m^S S_s \delta) \ ; \ D \le D^W \ ; \ \dot{T} = 0 \tag{6}$$

where T_f and T_d are respectively the temperature of the onset of failure of adsorbed water (70°C following [11]) and of completion of reversible dehydration (ab. 150°C, [12]. D^W is dielectric constant of water. S_s is total specific surface area of clay, and δ = 0.258 nm is the thickness of a monomolecular sheet of irremovable interlamellar water.

Linear momentum balance for solid, in the absence of inertia effects, with partial stress in solid t^S_{ij} is

$$t^S_{ij,i} + \Gamma^S F_j + \mu^S v^S_j + p^S_j = 0 \tag{7}$$

where F_j is the body force, while the last two terms describe the momentum transfer due to the mass transfer and due to diffusive forces, at permeability k,

$$p^S_j = k(\phi_3)\ (v^S_i - v^L_i)\ \delta_{ij}$$

$$p^S_i + p^L_i = 0 \tag{8}$$

The linear momentum balance for liquid is

$$(\phi_3 u)_{,i}\ \delta_{ij} + \Gamma^L G_j + \mu^L v^L_j + p^L_j = 0 \tag{9}$$

where u is the pressure excess in the free water and G_j is the body force. Total linear momentum balance must be satisfied as well. Following the definition of the solid phase, there is no distinction between contact forces and forces of repulsion and attraction due to the adsorbed water, i.e. they are assumed as equal, as opposed to [7]. Angular momentum is neglected. Energy balance, and diffusion equations are decoupled from mechanics, so that current values of T,D,c and v are considered as given.

CONSTITUTIVE EQUATIONS

The elastic strain is defined via a free enthalpy function Φ^S, depending on variables relative to solid and adsorbed water only

$$\epsilon^e_{ij} = \Gamma^S \partial\Phi^S/\partial t^S_{ij} = A_{ijkl}(M)t^S_{kl} + \alpha_{ij}(t^S_{kl}, \Delta T, \Delta D, \Delta c) \tag{10}$$

where the tensor A_{ijkl} is the elastic compliance tensor depending on total accumulated mass transfer $M=\int\mu\ dT, dD, dc$. Tensor α_{ij} groups the coefficients of reversible thermal expansion and flocculation due to $\dot{D}<0$, $\dot{c}>0$ and $\dot{v}>0$.

Plastic consolidation strain may be induced at an initially elastic stress by an increase in temperature, ionic concentration or valence, or by a decrease in dielectric constant. It is assumed that there is a common mechanism in these types of consolidation, which is degradation of adsorbed water. This leads to an increase in free porosity, inducing shrinking of the elastic domain and yield surface. This is referred to as thermal or chemical softening and is independent of plastic yielding. Since the mass transfer is reversible, the yield surface undergoes a reverse growth at cooling, or rehydration or at $\dot{c} < 0$. A Cam-clay type yield surface is used to describe plastic consolidation

$$f = f(t^S_{ij},\ \epsilon^p_{kk},\ M) = 0 \qquad (11)$$

The plastic strain rule due to changes in environmental loads is non-associated [5], so that

$$\dot{\epsilon}^p_{ij} = \frac{1}{3}\,\dot{\Omega}_v \frac{\partial g}{\partial t^S_{kk}}\,\delta_{ij} + \frac{3}{2}\,\dot{\Omega}_q \frac{\partial g}{\partial I_2'}\,\frac{s^S_{ij}}{I_2'} \qquad (12)$$

The volumetric plastic multiplier Ω_v is determined from the Prager's consistency equation,

$$\dot{\Omega}_v = \frac{1}{H}\left[\!\left[\frac{\partial f}{\partial t^S_{ij}}\,\dot{t}^S_{ij} + \frac{\partial f}{\partial M}\left[\frac{\partial M}{\partial T}\dot{T} + \frac{\partial M}{\partial D}\dot{D} + \frac{\partial M}{\partial c}\dot{c} \right] \right]\!\right] \qquad (13)$$

where H is the usual plastic hardening modulus. The deviatoric plastic multiplier Ω_q may be any function of t^S_{ij}, T, D, c, if the yield surface is not dependent on deviatoric plastic strain. Combining (10-13), the total strain rate is written in matrix notation as

$$\underline{\dot{\epsilon}}^S = \underline{E}^{ep}\,\underline{\dot{t}}^S + \underline{A}\,\underline{\dot{C}} + \underline{M}\,\underline{\dot{C}}\ ;\quad \underline{C}^T = \{\ T,\ D,\ c\ \} \qquad (14)$$

where $\underline{E}^{ep}$ is tangential non-symmetric elastoplastic matrix, $\underline{A}$ is a matrix of thermal and flocculation reversible expansion coefficients and $\underline{M}$ is a stress dependent matrix of irreversible thermochemical moduli.

CONCLUDING REMARK

The full system of balance and constitutive equations may be reduced to 7 equations for 7 unknowns v^L_i, v^S_i and ϕ_3, given temperature and chemical variables.

REFERENCES

1. R. Pusch, Sci. Basis Nucl. Waste Manag. MRS,84,791,1987
2. F.Fernandez and R.M. Quigley, Can. Geotech. J.22,205,1985
3. J.M.E. Storey and J.J.Peirce, Can. Geotech. J.26,57,1989
4. R.G. Campanella and J.K. Mitchell, J. Soil Mech.,ASCE,94,709,1968
5. T. Hueckel and G. Baldi, J.Geotech.E..ASCE,116,1990(in print)
6. S.L.Barbour and D.G.Fredlund, Can.Geotech.J.26, 551,1989
7. A.Sridharan and G.V. Rao, Geotechnique, 23,383,1973
8. J.J.Bowders and D.E.Daniel,J.Geotech.Eng.,ASCE,113,1432,1987
9. H.F. Winterkorn and L.D. Baver, Soil Sci.,40,5, 1935
10. P.M. Claesson et al.,J.Chem.Soc.Faraday Tr.,I,82,2735,1986
11. B.V. Derjaguin et al., J.Coll.and Interf.Sci.,586, 1986
12. R. Pusch and N. Guven, Eng. Geol.28, 303, 1990
13. J.N.Israelachvili and R.M. Pashlay, Nature,5940,1983
14. C.Truesdell,Rational Thermodynamics,Springer, N.Y.1986
15. R. Pusch and H. Hokmark, Eng. Geol.,28, 3-4, 379-389,1990

CONSTITUTIVE EQUATIONS ON A NONLOCAL POROUS, LIQUID–SATURATED CONTINUA

D.KUZMANOVIĆ, Z.GOLUBOVIĆ
Faculty of Mining and Geology, Faculty of Mechanical Engineering
P.CVETKOVIĆ
Faculty of Transport and Trafic Engineering

ABSTRACT

Starting with the current theory of continuum mechanics, this paper presents a derivation of the corresponding constitutive equatins of porous, liquid–saturated materials.

1. INTRODUCTION

As early as in 1983, Duhem noticed the necessity of taking into account the effect of stresses and of the body forces at one point upon the condition of the whole body, i.e., the necessity of including the effects of non–locality in the continuum theories.

The idea which was subsequently elaboreted by C.Eringen and D.Edelen [1] has the aim to examine classical continuum mechanics somewhat more closely in regard to the inclusion of long range effects.

Taking into accaount the nonlocal effects in this papers a corresponding constitutive equations of nonlocal porous liquid–saturated media have been derivated.

2. DEFINITIONS AND NOTATIONS

We have introduced the following definitions, for the volume and mass distributions, respectively:

$$v^\alpha = \int_v n^\alpha dv, \qquad m^\alpha = \int_v \varrho_*^\alpha dv, \tag{2.1}$$

where v – the volume of the body, and m^α, v^α, n^α, and ϱ_*^α denote the mass, the volumes, the volume fraction and the effective mass density of the α-th constituent of the mixture, respectively.

The index α refers to a fluid of a solid (skelet). If the quantity refers to a fluid it is designated by the symbol "f", and to a solid, by "s".

In a special case, when α refers to a fluid, we shall call it:

n^f–the volume porosity and denote it by n.

We also define partial densities as:

$$\varrho^\alpha = \frac{dm^\alpha}{dv} = \varrho^\alpha_* n^\alpha. \tag{2.2}$$

For any quantity $f^\alpha(\mathbf{x}^\alpha, t)$ we define:

$$\frac{D^\alpha f^\alpha}{Dt} = \frac{\partial f^\alpha}{\partial t} + f^\alpha_{,i} v^\alpha_i \equiv \acute{f}^\alpha, \tag{2.3}$$

which are the material derivatives following the α – constituent, where the comma denotes a partial differentiation.

3. BALANCE LAWS

The equation of motion of a material particle $\mathbf{X}^\alpha$ at time t is designated by

$$\mathbf{x}^\alpha = \mathbf{x}^\alpha(\mathbf{X}^\alpha; t). \tag{3.1}$$

We can write all the balance laws in the common form:

$$\frac{D^\alpha}{Dt} \int_v \Phi^\alpha dv = \oint_{\partial v} \varphi^\alpha ds \int_v s^\alpha dv \tag{3.2}$$

or in local form [2]:

$$\frac{\partial \Phi^\alpha}{\partial t} + \operatorname{div}(\Phi^\alpha \mathbf{v}^\alpha) - \operatorname{div}\varphi^\alpha - s^\alpha = \hat{s}^\alpha, \tag{3.3}$$

and $\hat{s}^\alpha$ is subject to the condition: $\int_v \hat{s}^\alpha dv = 0$.

The quantities $\hat{s}^\alpha$ so introduced are called nonlocal volume effects or residuals.

All the non–local effects throughout the paper will be designated by "^".

Thus, the first difference between the local and the nonlocal continuum is the existence of volume residuals.

The coresponding balance equations of the α-th constituent are obtaind when the arbitrary additive quantity, defined in (3.2) is identified with the mass, the momentum, the moment of momentum and the energy:

$$\frac{\partial \varrho^\alpha}{\partial t} + \operatorname{div}(\varrho^\alpha \mathbf{v}^\alpha) = \hat{\varrho}^\alpha, \qquad \int_v \hat{\varrho}^\alpha dv = 0, \tag{3.4}$$

$$\varrho^\alpha \acute{\mathbf{v}}^\alpha = \varrho^\alpha \mathbf{f}^\alpha + \mathbf{r}^\alpha + \operatorname{div}\hat{\mathbf{t}}^\alpha - \hat{\varrho}^\alpha \mathbf{v}^\alpha + \varrho^\alpha \mathbf{f}^\alpha, \qquad \int_v \varrho^\alpha \hat{\mathbf{f}}^\alpha dv = 0, \tag{3.5}$$

$$\varrho^\alpha \mathbf{x} \times \mathbf{f}^\alpha - \mathbf{x}_{,k} \times \mathbf{t}^\alpha_{,k} = \varrho^\alpha \hat{\mathbf{l}}^\alpha, \qquad \int_v \varrho^\alpha \hat{\mathbf{l}}^\alpha dv = 0, \tag{3.6}$$

$$\varrho^\alpha \acute{\varepsilon}^\alpha + \hat{\varrho}^\alpha(\varepsilon^\alpha - \frac{1}{2}\mathbf{v}^\alpha \mathbf{v}^\alpha) - \mathbf{t}^\alpha_k \mathbf{v}^\alpha_{,k} - \operatorname{div}\mathbf{q}^\alpha - \varrho^\alpha h^\alpha = \varrho^\alpha \hat{h}^\alpha - \varrho^\alpha \mathbf{v}^\alpha \hat{\mathbf{f}}^\alpha, \tag{3.7}$$

$$\int_v \varrho^\alpha \hat{h}^\alpha dv = 0,$$

where ϱ^α – partial densities, $\mathbf{v}^\alpha$ – velocity vector, $\mathbf{t}^\alpha$ – partial stress vector, $\mathbf{f}^\alpha$ – body force density, $\mathbf{r}^\alpha$ – interaction force density between the constitutents, $\mathbf{q}^\alpha$ – heat vector, ε^α – internal energy density, h^α – energy source density, and $\hat{\varrho}^\alpha$, $\hat{\varrho}^\alpha \mathbf{f}^\alpha$, $\varrho^\alpha \hat{\mathbf{l}}^\alpha$, $\varrho^\alpha \hat{h}^\alpha$ – are the nonlocal volume residuals for mass, body force, body moment and energy, respectively.

Also, we postulate the entropy inequality for the mixture as a whole:

$$\sum_\alpha \Big\{ -\varrho^\alpha(\acute{\psi}^\alpha + \eta^\alpha \acute{T}) - T[\varrho^\alpha \eta^\alpha (\mathbf{v}^\alpha - \mathbf{v})]_{,i} + (\mathbf{q}^\alpha - T\Phi^\alpha)_{,i} + T_{,i}\Phi^\alpha + $$
$$+ \mathbf{t}^\alpha \nabla \mathbf{v}^\alpha + \varrho^\alpha \hat{h}^\alpha - \hat{\varrho}^\alpha \left(\psi^\alpha - \frac{1}{2} \mathbf{v}^\alpha \mathbf{v}^\alpha \right) - \mathbf{v}^\alpha \mathbf{f}^\alpha \varrho^\alpha - \varrho^\alpha \hat{\eta}^\alpha \Big\} \geq 0. \quad (3.8)$$

where Φ^α – entropy flux.

4. CONSTITUTIVE RELATIONS

We assume that the response of a fluid–saturated porous solid is essentially determined by the following set of variables

$$\mathcal{A}^\alpha = \{T;\, \mathbf{x}^\alpha_{,k};\, \acute{\mathbf{x}}^\alpha_{,k};\, n;\, \varrho^f;\, \overline{T};\, \overline{\mathbf{x}}^\alpha_{,k};\, \acute{\overline{\mathbf{x}}}^\alpha_{,k};\, \overline{n};\, \overline{\varrho}^f\} = \{A^\alpha;\, \overline{A}^\alpha\}, \qquad (4.1)$$

where a "–" on the functions of the cause set indicates their dependence on $\mathbf{Z}$, e.g. $\mathbf{z} = \mathbf{x}(\mathbf{Z}, t), \quad \mathbf{x} = \mathbf{x}(\mathbf{X}, t); \quad \mathbf{X}, \mathbf{Z} \in v$.

In accordance with the principle of equipresence, the constitutive functions $\mathbf{t}^\alpha$; $\mathbf{q}^\alpha$; ψ^α; η^α; Φ^α are assumed to be dependent on the same list of variables.

After introducing the constitutive functions in the entropy inequality and carrying out the indicated diferentiations, we obtain the following explicit form of the entropy inequality:

$$\sum_\alpha \Big\{ -\varrho^\alpha \left[\frac{\partial \psi^\alpha}{\partial A^\alpha} + \int_v \left(\frac{\delta \psi}{\delta \overline{A}^\alpha} \right)^* dv \right] \acute{A}^\alpha - \varrho^\alpha \mathcal{D}^\alpha - \varrho^\alpha \eta^\alpha \acute{T} + [\mathbf{q}^\alpha - T\Phi^\alpha - $$
$$- T\varrho^\alpha \eta^\alpha (\mathbf{v}^\alpha - \mathbf{v})]_{,i} + T_{,i}[\Phi^\alpha + \varrho^\alpha \eta^\alpha (\mathbf{v}^\alpha - \mathbf{v})] + \mathbf{t}^\alpha \nabla \mathbf{v}^\alpha + \varrho^\alpha \hat{h}^\alpha$$
$$- \hat{\varrho}^\alpha \left(\psi^\alpha - \frac{1}{2} \mathbf{v}^\alpha \mathbf{v}^\alpha \right) - \mathbf{v}^\alpha \hat{\mathbf{f}}^\alpha \varrho^\alpha - \varrho^\alpha \hat{\eta}^\alpha \Big\} \geq 0, \qquad (4.2)$$

where

$$\mathcal{D}^\alpha \equiv \int_v \left[\varrho^\alpha \frac{\delta \psi^\alpha}{\delta \overline{A}^\alpha} \acute{\overline{A}}^\alpha - \left(\varrho^\alpha \frac{\delta \psi^\alpha}{\delta \overline{A}^\alpha} \right)^* \acute{A}^\alpha \right] d\overline{v}, \qquad [f(\mathbf{Z}, \mathbf{X})]^* \equiv f(\mathbf{X}, \mathbf{Z})$$

and $\frac{\delta}{\delta}$ denote Frèchet partial derivative.

From thermodynamic restrictions, the folowing relations are obtained:

$$ {}_e t_{ij}^{\alpha} = \varrho^{\alpha}\left\{ \frac{\partial\psi^{\alpha}}{\partial x_{i,k}} x_{i,k} + \int\limits_{v} \left(\frac{\partial\psi^{\alpha}}{\partial\overline{x}_{i,k}} \overline{x}_{i,k} \right)^{*} dv - \left[\frac{\delta\psi^{\alpha}}{\delta\varrho^{f}} \varrho^{f} + \int\limits_{v} \left(\frac{\delta\psi^{\alpha}}{\delta\overline{\varrho}^{f}} \overline{\varrho}^{f} \right)^{*} d\overline{v} \right] \delta_{ij}\delta_{\alpha f} \right\} \tag{4.3} $$

$$ \eta^{\alpha} = -\frac{\partial\psi^{\alpha}}{\partial T} - \int\limits_{v} \left(\frac{\delta\psi}{\delta\overline{T}} \right)^{*} d\overline{v}, \qquad q_i^{\alpha} = 0. $$

There is an additional set of thermodynamic restrictions which are not of an interest at the moment.

The rest of inequality is:

$$ \sum_{\alpha} \left[{}_{\alpha}\mathbf{t}^{\alpha}\nabla\mathbf{v}^{\alpha} + \varrho^{\alpha}\hat{h}^{\alpha} - \hat{\varrho}^{\alpha}\left(\psi^{\alpha} - \frac{1}{2}\mathbf{v}^{\alpha}\mathbf{v}^{\alpha} \right) - \mathbf{v}^{\alpha}\hat{f}^{\alpha}\varrho^{\alpha} - \varrho^{\alpha}\hat{\eta}^{\alpha} - \mathcal{D}^{\alpha} \right] \equiv $$

$$ \equiv \sum_{\alpha} [{}_d\mathbf{t}\nabla\mathbf{v}^{\alpha} + \hat{s}^{\alpha}\hat{v}^{\alpha}] \geq 0. \tag{4.4} $$

Introducing the dissipation function $D = D(d_{ij}^{\alpha};\, \hat{v}^{\alpha})$, from (4.4) we get

$$ {}_d t_{ij}^{\alpha} = \Lambda \frac{\partial D}{\partial d_{ij}^{\alpha}}; \qquad \hat{s} = \Lambda \frac{\partial D}{\partial \hat{v}^{\alpha}}, \tag{4.5} $$

where $d_{i,j}^{\alpha} = v_{(i,j)}^{\alpha}$.

Relations (4.3) and (4.5) give the total stress:

$$ t_{ij}^{\alpha} = \varrho^{\alpha}\left\{ \frac{\partial\psi^{\alpha}}{\partial x_{j,k}} x_{i,k} + \int\limits_{v} \left(\frac{\delta\psi^{\alpha}}{\delta\overline{x}_{i,k}} \overline{x}_{i,k} \right)^{*} dv - \left[\frac{\delta\psi^{\alpha}}{\delta\varrho^{f}} \varrho^{f} + \right.\right. $$

$$ \left.\left. + \int\limits_{v} \left(\frac{\delta\psi^{\alpha}}{\delta\overline{\varrho}^{f}} \overline{\varrho}^{f} \right)^{*} d\overline{v} \right] \delta_{ij}\delta_{\alpha f} \right\} + \Lambda \frac{\partial D}{\partial d_{ij}^{\alpha}}. \tag{4.6} $$

The results so derived will be used to investigate the problem of the consolidation in a nonlocal porous, liquid–saturated and nonsaturated media. This means that the present paper is the first part of the problem in consideration. The equation of consolidation is the subject of the second part of the paper which is forthcoming.

REFERENCE

1. A.C.Eringen, D.B.Edelen, Int. J. Engng. Sci., Vol. 10, 1972.
2. D.B.Edelen, Continuum Phisics, Vol. 4, 1976 (Academic Press).
3. A.C.Eringen, Continuum Phisics, Vol. 4, 1976 (Academic Press).
4. D.Kuzmanović, Theoretical and Appl. Mech., Vol. 15, 1989.

THE ELECTROMECHANICAL EFFECTS IN ELASTIC DIELECTRICS WITHIN MINDLIN'S GRADIENT THEORY

K. MAJORKOWSKA-KNAP and T. MURA
Department of Civil Engineering, Northwestern University
Evanston, IL 60208, USA

ABSTRACT

This contribution deals with the electromechanical effects in elastic dielectrics within Mindlin's linear gradient theory. This theory, as the other coupled theories, violates the classical (purely elastic) constitutive assumptions. The analysis of the constitutive equations shows a linear electromechanical interaction in non-centrosymmetric crystals as well as in centrosymmetric (including isotropic) ones. As an example, the coupled electromechanical effects are shown for the problem of interface wave propagation in a layered dielectric medium.

1.INTRODUCTION

Nowadays the actual technical and natural sciences problems, the appearance of new materials and techniques, the rapidly accelerating demands for higher performance and cost-efficiency in engineering design, the increasing possibilities to produce tailor-made materials emerge the increasing efforts in theoretical as well as experimental research devoted to the mechanical modelling of real materials. The mathematical models have been developed to describe the physical (also novel) properties and phenomena recorded in experiments, also the coupled phenomena as: piezoelectricity, electro- and magneto-striction, piroelectricity, gyro-magnetic phenomena, magneto-acoustics, thermo-electric phenomena, magneto-hydrodynamics etc. The peripheral studies are continually being added to the central subjects of elasticity, plasticity, visco-elasticity and these are developing themselves and growing together within the coupled theories of continua.

This contribution deals with the electromagnetomechanical coupling in elastic dielectrics, basing on Mindlin's linear gradient theory. The quasistatic approximation for the electromagnetic field is used, which corresponds to the zeroth order terms in an iterative expansion of the Maxwell equations, when only electric and material motion phenomena are considered.

The influence of coupled effects, following from the coupling of mechanical and electromagnetic fields in a dielectric continuum, on the interface wave propagation in an elastic dielectric centrosymmetric medium, particularly in layered structures, is shown as an example.

2.MINDLIN'S LINEAR MODEL FOR ELASTIC DIELECTRICS

Mindlin's gradient theory of elastic dielectrics, as the other coupled theories, violates the classical (purely elastic) constitutive assumptions. As distinct from the classical theory of piezoelectricity (Voigt [1], subsequently Toupin [2], Eringen [3]) Mindlin's modified model [4], [5], by incorporating into the stored energy density -

beside strain and the electric polarization - also the gradient of the electric polarization, takes into account more precisely the aspects of microscopic structure and interatomic interactions of matter, and accommodates some remarkable and measured physical phenomena, not presented by the classical theory of piezoelectricity e.g. electromechanical effect in centrosymmetric (including isotropic) crystals. Also a new concept of surface energy of deformation and polarization has been introduced.

Mindlin's theory constitutes an extension of the classical theory of piezoelectricity. The equations of the lattice of the corresponding dynamical theories of crystal lattices have as a continuum approximation the equations of the gradient theory. Mindlin has indicated the relation between the polarization and the shell-shell and core-shell interactions of the theory of lattice dynamics. The gradient theory of elastic dielectrics may be also regarded as a first-order approach to Eringen's theory [6], [7] of nonlocal electromagnetic elastic solids (or as a special case in this theory) with respect to the electromagnetic properties of dielectrics.

Mindlin's variational principle (extending the linear version of Toupin's [2] one), using the assumption $W = W(\varepsilon_{ij}, P_i, P_{j,i})$ for the stored energy density, where ε_{ij} are the components of the symmetric strain tensor: $\varepsilon_{ij} = \frac{1}{2}\left(u_{j,i} + u_{i,j}\right)$, u_i are the components of the displacement vector, P_i are the components of the polarization vector and $P_{j,i}$ are the components of the polarization gradient tensor, yields the following field equations (written within the rectangular Cartesian coordinates):

$$\begin{aligned} &\sigma_{ji,j} + f_i = \rho\, \ddot{u}_i\,, \\ &E^L_{ji,j} + E^L_i - \varphi_{,i} + E^o_i = O\,, \\ &-\varepsilon_o \varphi_{,ii} + P_{i,i} = O\,, \qquad in\ V \\ &\varphi_{,ii} = O\,, \qquad in\ V' \end{aligned} \tag{1}$$

and the boundary conditions

$$\sigma_{ji} n_j = t_i\,,\ \ E^L_{ji} n_j = O\,,\ \ (-\varepsilon_o\left[\varphi_{,i}\right] + P_i) n_i = O \qquad on\ S \tag{2}$$

where V denotes the volume of a body, bounded by a surface S, separating V from an outer vacuum V', **n** is the outer unit normal vector, and $[\ \ ]$ designates the jump across S, φ is the potential of the Maxwell self field $\mathbf{E}^{MS}$: $E^{MS}_i = -\varphi_{,i}$, ρ is the mass density, f_i, E^o_i *and* t_i are the components of the external body force, the external electric field and surface tractions, respectively.

The stress tensor σ_{ij} , the effective local electric field $\mathbf{E}^L$ and the local electric tensor E^L_{ij} are given by

$$\sigma_{ij}=\frac{\partial W}{\partial \varepsilon_{ij}}=\sigma_{ji}\,,\quad E_i^L=-\frac{\partial W}{\partial P_i},\quad E_{ij}^L=\frac{\partial W}{\partial P_{j,i}} \tag{3}$$

3.CONSTITUTIVE EQUATIONS FOR NON-CENTROSYMMETRIC AND FOR CENTROSYMMETRIC MATERIALS

The energy density of deformation and polarization is taken to be

$$W=b_{ij}^{O}P_{j,i}+\frac{1}{2}a_{ij}^{SG}P_iP_j+\frac{1}{2}b_{ijkl}^{SP}P_{j,i}P_{l,k}+\frac{1}{2}c_{ijkl}^{PG}\varepsilon_{ij}\varepsilon_{kl}+d_{ijkl}^{P}P_{j,i}\varepsilon_{kl}+f_{ijk}^{G}P_i\varepsilon_{jk}+g_{ijk}^{S}P_iP_{k,j} \tag{4}$$

(the superscripts S,P,G designate fixed: strain, polarization and polarization gradient, respectively, and will thereafter be omitted), where the linear term $b_{ij}^{O}P_{j,i}$ has been added to include the surface energy effects.

The following coupled constitutive equations are found

$$\begin{aligned}
\sigma_{ij}&=c_{ijkl}\varepsilon_{kl}+f_{kij}P_k+d_{klij}P_{l,k}\,,\\
-E_j^L&=f_{jkl}\varepsilon_{kl}+a_{jk}P_k+g_{jkl}P_{l,k}\,,\\
E_{ij}^L&=d_{ijkl}\varepsilon_{kl}+g_{klj}P_k+b_{ijkl}P_{l,k}+b_{ij}^{O}\,,
\end{aligned} \tag{5}$$

in which the properties of an anisotropic material are introduced. The elastic behavior is dominated by the tensor c_{ijkl} (elastic stiffness), and the electric behavior is represented by the tensors: a_{ij}, g_{jkl} and b_{ijkl}. The piezoelectric tensors f_{jkl} and d_{ijkl} describe the electromechanical interactions in dielectrics. The material coefficients can be subjected to further restrictions by the symmetry transformations of the point groups. As a result of it the internal energy density and the constitutive equations simplify. The relations between $a_{ij}, f_{ijk}, c_{ijkl}$ and the IEEE Standard [8] ones have been done by Mindlin [5]. It may be seen [5] by inspection of the energy density W (4) and the constitutive equations (5), that Mindlin's model accommodates an electromechanical interaction even for materials with centrosymmetric (including isotropic) physical properties through material coefficients d_{ijkl} (for centrosymmetry $f_{ijk}=g_{ijk}=O,$ as tensors of odd rank).

The matrix representation of the quadratic expression for the internal energy density W (we assume $b_O=O$ i.e. initial state is independent of polarization gradient effects) of a centrosymmetric cubic material has a form:

$$
\begin{array}{c|cccccccccccccccccc}
 & \varepsilon_{11} & \varepsilon_{22} & \varepsilon_{33} & 2\varepsilon_{23} & 2\varepsilon_{31} & 2\varepsilon_{12} & P_{1,1} & P_{2,2} & P_{3,3} & P_{1,2} & P_{1,3} & P_{2,1} & P_{2,3} & P_{3,1} & P_{3,2} & P_{1} & P_{2} & P_{3} \\
\varepsilon_{11} & c_{11} & c_{12} & c_{12} & \cdot & \cdot & \cdot & d_{11} & d_{12} & d_{12} & \cdot & \cdot & \cdot & \cdot & \cdot & \cdot & \cdot & \cdot & \cdot \\
\varepsilon_{22} & c_{12} & c_{11} & c_{12} & \cdot & \cdot & \cdot & d_{12} & d_{11} & d_{12} & \cdot & \cdot & \cdot & \cdot & \cdot & \cdot & \cdot & \cdot & \cdot \\
\varepsilon_{33} & c_{12} & c_{12} & c_{11} & \cdot & \cdot & \cdot & d_{12} & d_{12} & d_{11} & \cdot & \cdot & \cdot & \cdot & \cdot & \cdot & \cdot & \cdot & \cdot \\
2\varepsilon_{23} & \cdot & \cdot & \cdot & c_{44} & \cdot & \cdot & \cdot & \cdot & \cdot & \cdot & \cdot & \cdot & d_{44} & \cdot & d_{44} & \cdot & \cdot & \cdot \\
2\varepsilon_{31} & \cdot & \cdot & \cdot & \cdot & c_{44} & \cdot & \cdot & \cdot & \cdot & \cdot & d_{44} & \cdot & \cdot & d_{44} & \cdot & \cdot & \cdot & \cdot \\
2\varepsilon_{12} & \cdot & \cdot & \cdot & \cdot & \cdot & c_{44} & \cdot & \cdot & \cdot & d_{44} & \cdot & d_{44} & \cdot & \cdot & \cdot & \cdot & \cdot & \cdot \\
P_{1,1} & d_{11} & d_{12} & d_{12} & \cdot & \cdot & \cdot & b_{11} & b_{12} & b_{12} & \cdot & \cdot & \cdot & \cdot & \cdot & \cdot & \cdot & \cdot & \cdot \\
P_{2,2} & d_{12} & d_{11} & d_{12} & \cdot & \cdot & \cdot & b_{12} & b_{11} & b_{12} & \cdot & \cdot & \cdot & \cdot & \cdot & \cdot & \cdot & \cdot & \cdot \\
P_{3,3} & d_{12} & d_{12} & d_{11} & \cdot & \cdot & \cdot & b_{12} & b_{12} & b_{11} & \cdot & \cdot & \cdot & \cdot & \cdot & \cdot & \cdot & \cdot & \cdot \\
P_{1,2} & \cdot & \cdot & \cdot & \cdot & \cdot & d_{44} & \cdot & \cdot & \cdot & b_{44}^{\wedge} & \cdot & b_{44}^{*} & \cdot & \cdot & \cdot & \cdot & \cdot & \cdot \\
P_{1,3} & \cdot & \cdot & \cdot & \cdot & d_{44} & \cdot & \cdot & \cdot & \cdot & \cdot & b_{44}^{\wedge} & \cdot & \cdot & b_{44}^{*} & \cdot & \cdot & \cdot & \cdot \\
P_{2,1} & \cdot & \cdot & \cdot & \cdot & \cdot & d_{44} & \cdot & \cdot & \cdot & b_{44}^{*} & \cdot & b_{44}^{\wedge} & \cdot & \cdot & \cdot & \cdot & \cdot & \cdot \\
P_{2,3} & \cdot & \cdot & \cdot & d_{44} & \cdot & \cdot & \cdot & \cdot & \cdot & \cdot & \cdot & \cdot & b_{44}^{\wedge} & \cdot & b_{44}^{*} & \cdot & \cdot & \cdot \\
P_{3,1} & \cdot & \cdot & \cdot & \cdot & d_{44} & \cdot & \cdot & \cdot & \cdot & \cdot & b_{44}^{*} & \cdot & \cdot & b_{44}^{\wedge} & \cdot & \cdot & \cdot & \cdot \\
P_{3,2} & \cdot & \cdot & \cdot & d_{44} & \cdot & \cdot & \cdot & \cdot & \cdot & \cdot & \cdot & \cdot & b_{44}^{*} & \cdot & b_{44}^{\wedge} & \cdot & \cdot & \cdot \\
P_{1} & \cdot & \cdot & \cdot & \cdot & \cdot & \cdot & \cdot & \cdot & \cdot & \cdot & \cdot & \cdot & \cdot & \cdot & \cdot & a & \cdot & \cdot \\
P_{2} & \cdot & \cdot & \cdot & \cdot & \cdot & \cdot & \cdot & \cdot & \cdot & \cdot & \cdot & \cdot & \cdot & \cdot & \cdot & \cdot & a & \cdot \\
P_{3} & \cdot & \cdot & \cdot & \cdot & \cdot & \cdot & \cdot & \cdot & \cdot & \cdot & \cdot & \cdot & \cdot & \cdot & \cdot & \cdot & \cdot & a
\end{array}
\qquad (6)
$$

where the customary contraction of suffices has been introduced, moreover the dots denote zero components and $b_{44}^{\wedge} = b_{44} + b_{77}$, $b_{44}^{*} = b_{44} - b_{77}$.

For a centrosymmetric isotropic material $c_{11} = c_{12} + 2c_{44}$, $b_{11} = b_{12} + 2b_{44}$ $d_{11} = d_{12} + 2d_{44}$,the constitutive equations are as follows:

$$
\begin{aligned}
&\sigma_{ij} = c_{12}\, u_{k,k}\delta_{ij} + c_{44}(u_{i,j} + u_{j,i}) + d_{12}P_{k,k}\delta_{ij} + d_{44}(P_{i,j} + P_{j,i}) \\
&E_i^L = -a\, P_i, \\
&E_{ij}^L = d_{12}\, u_{k,k}\delta_{ij} + d_{44}(u_{i,j} + u_{j,i}) + b_{12}P_{k,k}\delta_{ij} + b_{44}(P_{j,i} + P_{i,j}) + b_{77}(P_{j,i} - P_{i,j}) + b^{O}\delta_{ij}
\end{aligned}
\qquad (7)
$$

and the field equations (1) yield the following equations (we assume $\mathbf{E}^O = O,\ \mathbf{f} = O$) for the displacements, polarization and electric potential:

$$c_{44}\nabla^2\mathbf{u}+(c_{12}+c_{44})grad\ div\mathbf{u}+d_{44}\nabla^2\mathbf{P}+(d_{12}+d_{44})grad\ div\mathbf{P}=\rho\ \ddot{\mathbf{u}}\ ,$$
$$d_{44}\nabla^2\mathbf{u}+(d_{12}+d_{44})grad\ div\mathbf{u}+(b_{44}+b_{77})\nabla^2\mathbf{P}+$$
$$(b_{12}+b_{44}-b_{77})grad\ div\mathbf{P}-a\mathbf{P}-grad\varphi=O,$$
$$-\varepsilon_O\nabla^2\varphi+div\mathbf{P}=O, \qquad in\ V$$
$$\nabla^2\varphi=O, \qquad in\ V' \tag{8}$$

which are completed by the boundary conditions (2).

4.PROPAGATION OF ELECTROMECHANICAL INTERFACE WAVES

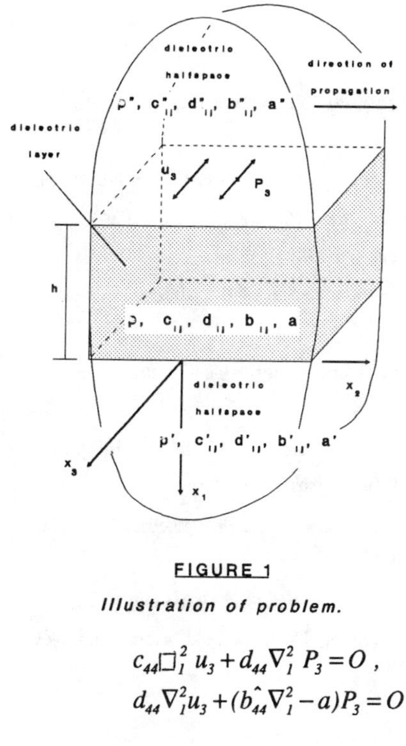

FIGURE 1

Illustration of problem.

The small amplitude waves of SH-type [9] are studied, propagating with frequencies far below those of optical ones [10],[11] in structure (Fig.1), which is idealization of a possible real wave delay line, where a thin isotropic dielectric layer is attached to two comparatively thick dielectric substrates (modelling as halfspaces) materially distinct from the layer and, in general, from each other [12],[13]. Here wave motion is characterized by

$$\mathbf{u}=(O,O,u_3), \qquad \mathbf{P}=(O,O,P_3),$$
$$u_3=u_3(x_1,x_2,t),\ \ P_3=P_3(x_1,x_2,t),$$

To find solution we use the reduced field equations (8)

$$c_{44}\Box_1^2\ u_3+d_{44}\nabla_1^2\ P_3=O\ ,$$
$$d_{44}\nabla_1^2u_3+(\hat{b}_{44}\nabla_1^2-a)P_3=O\ , \quad for\ \ -h\leq x_1\leq O \tag{9}$$

and the similar ones written for the halfspace $x_1\geq O$ with the material constants $c_{44}^{'}$, ...etc., and for the halfspace $x_1\leq -h$ with the material constants $c_{44}^{"}$ ....etc., the continuity conditions

$$u_3(O,x_2,t)=u_3^{'}(O,x_2,t)\ ,\ \ u_3(-h,x_2,t)=u_3^{"}(-h,x_2,t)\ \ ,\ \ P_3(O,x_2,t)=P_3^{'}(O,x_2,t)\ ,$$
$$P_3(-h,x_2,t)=P_3^{"}(-h,x_2,t),\ \sigma_{13}(O,x_2,t)=\sigma_{13}^{'}(O,x_2,t)\ ,\ \ \sigma_{13}(-h,x_2,t)=\sigma_{13}^{"}(-h,x_2,t),$$
$$E_{13}^{L}(O,x_2,t)=E_{13}^{L'}(O,x_2,t),\ \ E_{13}^{L}(-h,x_2,t)=E_{13}^{L"}(-h,x_2,t)\ . \tag{10}$$

and regularity conditions at infinity.

This work was supported by the U.S. Army Research Office DAA L03-89-K-0019.

REFERENCES

1.W.Voigt,Lehrbuch der Kristallphysik, Jonson Reprint Co., New York and Teubner, Stuttgart, 1966.

2.R.A.Toupin, The elastic dielectrics, J. Rat. Mech. Anal.5, 849-915(1956).

3.A.C.Eringen, On the foundation of electroelastostatics, Int. J. Engng. Sci.1, 127-153(1963).

4.R.D.Mindlin, Polarization gradient in elastic dielectrics, Int. J.Solids Struct.4, 637-642(1968).

5.R.D.Mindlin, Elasticity, piezoelectricity and crystal lattice dynamics, J. Elasticity,2, 217-282,(1972).

6.A.C.Eringen, Theory of nonlocal electromagnetic solids, J.Math. Phys., 14,733-740(1973).

7.A.C.Eringen, Theory of nonlocal piezoelectricity, J. Math. Phys.,25, 717-727(1984).

8. IEEE Standart on Piezoelectricity, ANSI/IEEE Std 176-1987.

9.K.Majorkowska-Knap, Surface waves in piezoelectric materials of the class $\bar{4}2m$, Bull. Pol. Ac.: Tech.28, 417-424(1980).

10.K.Majorkowska-Knap, Coupled mechano-electric fields in deformable dielectrics, I, Bull. Pol. Ac. :Tech., 27, 377-382(1979).

11.K.Majorkowska-Knap, Coupled mechano-thermo-electric waves in a solid piezoelectric continuum, Mech.Teor. i Stos.,22, 509-523(1984), (in Polish).

12.K.Majorkowska-Knap, Piezoelectric surface waves in layered elastic dielectrics, Z. angew. Math. Mech.,66, T66-67(1986).

13.K.Majorkowska-Knap, J.Lenz, Propagation of SH-surface waves in elastic layered dielectrics ,in: M.F.McCarthy, M.A.Hayes(eds.),Elastic wave propagation, North-Holland, Amsterdam, New York, Oxford, 1989, pp. 81-87.

A CONSTITUTIVE LAW FOR GRANULAR MATERIALS TO METALS

HAJIME MATSUOKA and DE'AN SUN
Dept. of Civil Eng., Nagoya Institute of Technology, Nagoya, Japan

ABSTRACT

A unified failure criterion and constitutive law for frictional and cohesive materials is presented on the basis of the Matsuoka-Nakai criterion for frictional materials and the Mises criterion for cohesive materials, and is checked by both triaxial compression and extension tests on a cement-treated sand as an intermediate material with friction and cohesion.

INTRODUCTION

It is an interesting problem to present a unified failure criterion and constitutive law for a wide range of engineering materials from frictional materials such as granular materials without bond between grains, to cohesive materials such as metals with strong bond due to crystalline structure. One of the authors has already proposed the Matsuoka-Nakai criterion as a failure criterion for granular materials such as soils on the basis of "spatially mobilized plane (SMP)" [1, 2 and 3]. In order to extend the Matsuoka-Nakai criterion and constitutive relation on the "SMP" for frictional materials to those for frictional and cohesive materials, the idea of "extended spatially mobilized plane (extended SMP)" with a parameter of "bonding stress σ_o" is introduced. To check the constitutive law based on the "extended SMP", both triaxial compression and extension tests are carried out on a cement-treated sand. By the arrangement of test results of the cement-treated sand on the "extended SMP", unique stress-strain relations are obtained in both triaxial compression and extension.

A UNIFIED CONSTITUTIVE LAW BASED ON "EXTENDED SMP"

In Fig.1, three tangent lines from a negative point on the normal stress axis, whose absolute value is named "bonding stress σ_o", to three Mohr's stress circles are considered following [4 and 5]. Then σ_o is expressed by the following equation.

$$\sigma_0 = c \cdot \cot\phi \qquad (1)$$

where c is the cohesion and ϕ is the internal friction angle. According to the new $\hat{\tau}$ vs. $\hat{\sigma}$ coordinates, the principal stresses $\hat{\sigma}_i$ (i=1,2 and 3) and the stress invariants $\hat{J}_1$, $\hat{J}_2$ and $\hat{J}_3$ are expressed as follows:

$$\hat{\sigma}_i = \sigma_i + \sigma_0 \qquad (i=1,\ 2 \text{ and } 3) \qquad (2)$$

$$\left.\begin{aligned} \hat{J}_1 &= \hat{\sigma}_1 + \hat{\sigma}_2 + \hat{\sigma}_3 \\ &= (\sigma_1+\sigma_0) + (\sigma_2+\sigma_0) + (\sigma_3+\sigma_0) \\ \hat{J}_2 &= \hat{\sigma}_1\hat{\sigma}_2 + \hat{\sigma}_2\hat{\sigma}_3 + \hat{\sigma}_3\hat{\sigma}_1 \\ &= (\sigma_1+\sigma_0)(\sigma_2+\sigma_0) + (\sigma_2+\sigma_0) \\ &\quad (\sigma_3+\sigma_0) + (\sigma_3+\sigma_0)(\sigma_1+\sigma_0) \\ \hat{J}_3 &= \hat{\sigma}_1\hat{\sigma}_2\hat{\sigma}_3 \\ &= (\sigma_1+\sigma_0)(\sigma_2+\sigma_0)(\sigma_3+\sigma_0) \end{aligned}\right\} \qquad (3)$$

Fig.2 shows the "extended SMP". The direction cosines $\hat{a}_i$ of the normal of the "extended SMP" are written as follows:

$$\hat{a}_i = \sqrt{\frac{\hat{J}_3}{\hat{\sigma}_i \hat{J}_2}} \qquad (i=1,\ 2 \text{ and } 3) \qquad (4)$$

It should be noted in Eq.(4) that when $\sigma_o=0$, $\hat{a}_i = a_i = \sqrt{J_3/(\sigma_i J_2)}$, the direction cosines of the normal of the SMP, which is successfully applicable to granular materials, and when $\sigma_o \to \infty$, $\hat{a}_i = 1/\sqrt{3}$, the direction cosines of the normal of the octahedral plane, which is successfully applicable to metals. The normal stress $\hat{\sigma}_{SMP}$ and the shear stress $\hat{\tau}_{SMP}$, the normal strain $d\hat{\varepsilon}_{SMP}$ and the shear strain $d\hat{\gamma}_{SMP}$ on the "extended SMP" are represented as follows:

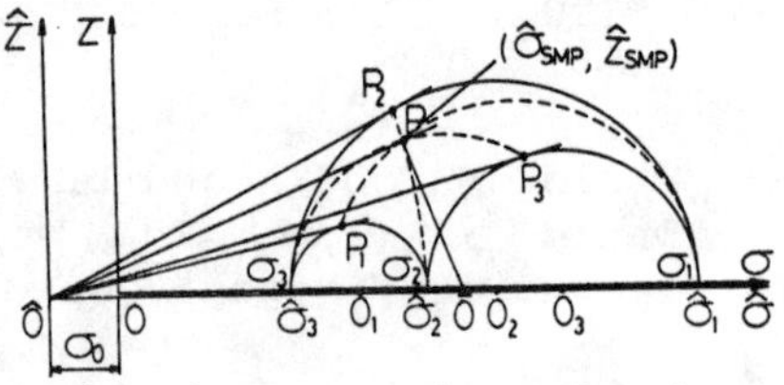

Fig.1 Normal and shear stresses on extended SMP for frictional and cohesive materials

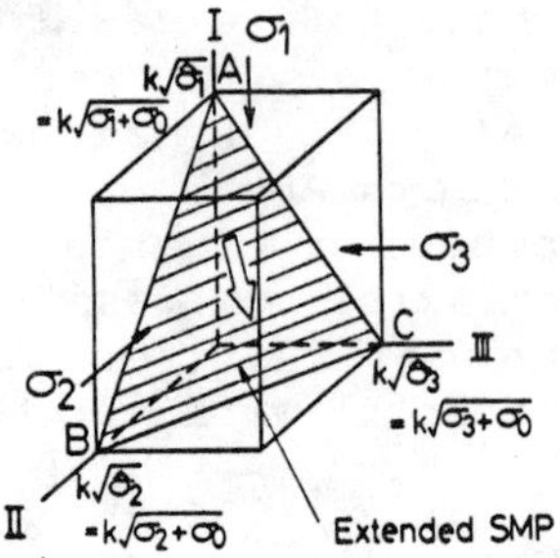

Fig.2 Extended SMP for frictional and cohesive materials

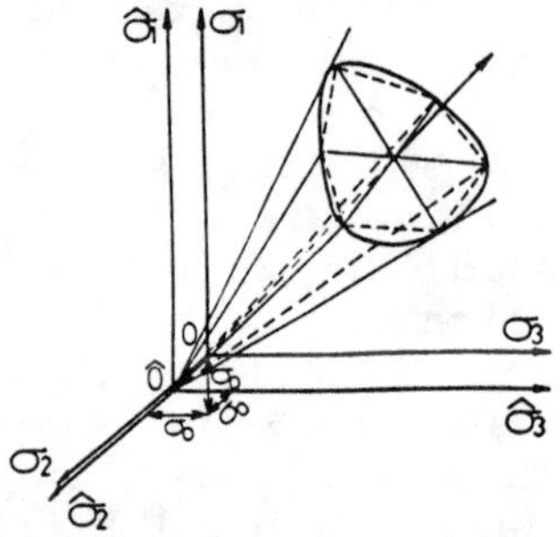

Fig.3 Proposed failure criterion for frictional and cohesive materials based on extended SMP

$$\hat{\sigma}_{SMP} = \hat{\sigma}_1\hat{a}_1^2 + \hat{\sigma}_2\hat{a}_2^2 + \hat{\sigma}_3\hat{a}_3^2 \qquad (5)$$

$$\hat{\tau}_{SMP} = \sqrt{(\hat{\sigma}_1-\hat{\sigma}_2)^2\hat{a}_1^2\hat{a}_2^2 + (\hat{\sigma}_2-\hat{\sigma}_3)^2\hat{a}_2^2\hat{a}_3^2 + (\hat{\sigma}_3-\hat{\sigma}_1)^2\hat{a}_3^2\hat{a}_1^2} \qquad (6)$$

$$d\hat{\varepsilon}_{SMP} = d\varepsilon_1\hat{a}_1^2 + d\varepsilon_2\hat{a}_2^2 + d\varepsilon_3\hat{a}_3^2 \qquad (7)$$

$$\frac{d\hat{\gamma}_{SMP}}{2} = \sqrt{(d\varepsilon_1-d\varepsilon_2)^2\hat{a}_1^2\hat{a}_2^2 + (d\varepsilon_2-d\varepsilon_3)^2\hat{a}_2^2\hat{a}_3^2 + (d\varepsilon_3-d\varepsilon_1)^2\hat{a}_3^2\hat{a}_1^2} \qquad (8)$$

If frictional and cohesive materials fail when $\hat{\tau}_{SMP}/\hat{\sigma}_{SMP}$ on the "extended SMP" reaches a constant value, a new failure criterion as shown in Fig.3 can be obtained from

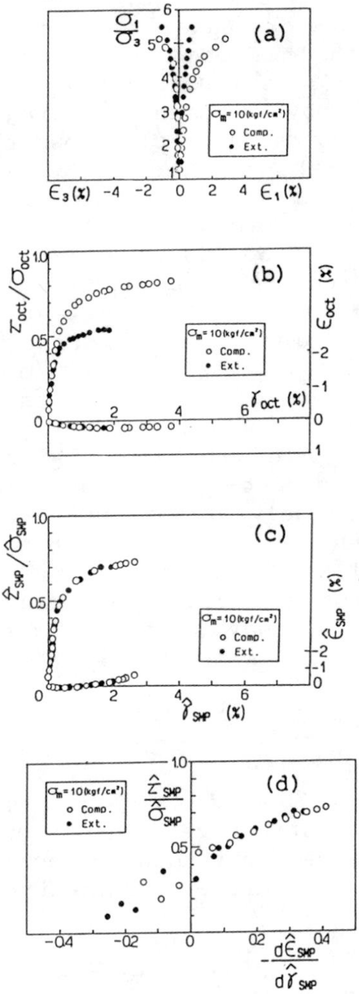

Fig.4 Various kinds of arrangement of a cement-treated sand under triaxial compression and extension (σ_m=10kgf/cm^2)

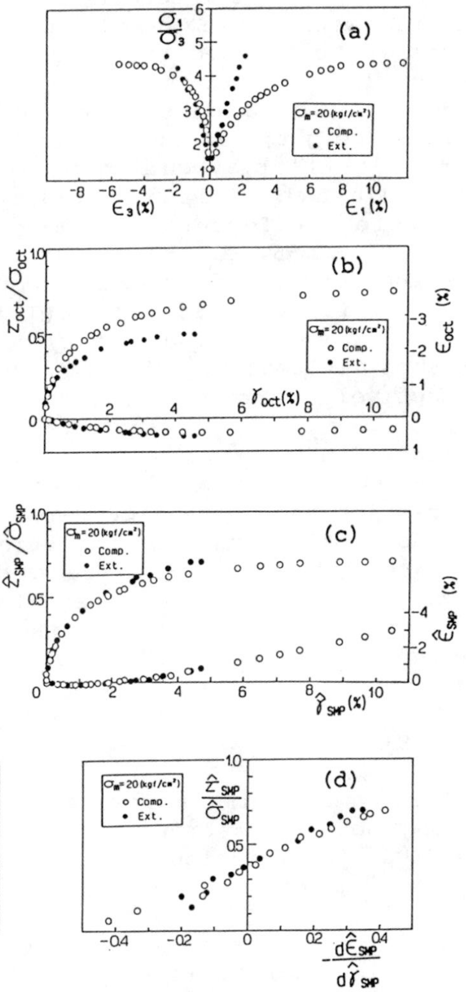

Fig.5 Various kinds of arrangement of a cement-treated sand under triaxial compression and extension (σ_m=20kgf/cm^2)

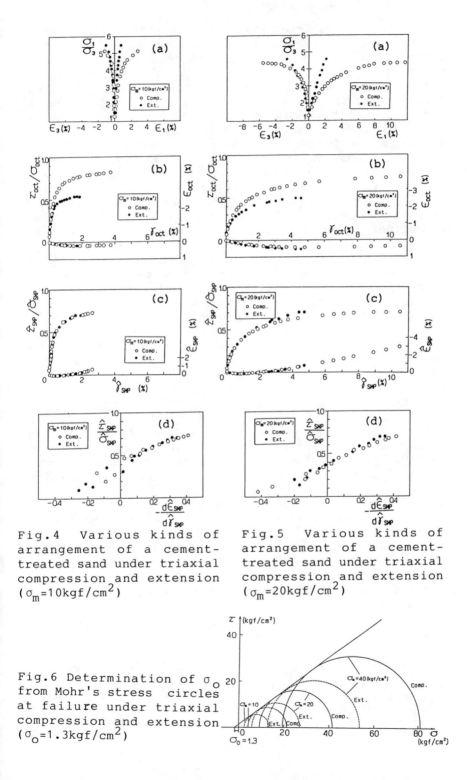

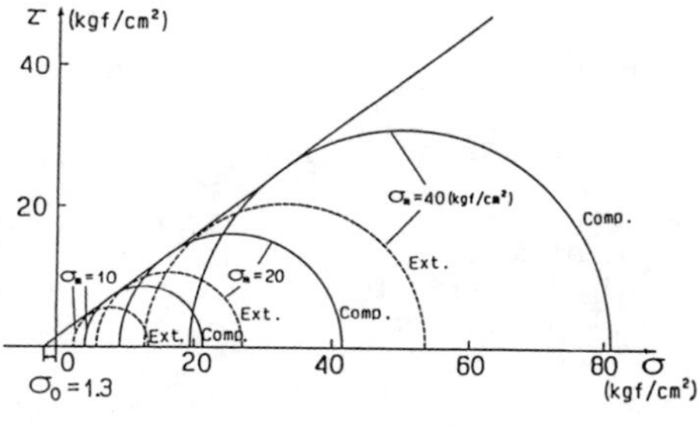

Fig.6 Determination of σ_o from Mohr's stress circles at failure under triaxial compression and extension (σ_o=1.3kgf/cm^2)

Eqs.(2)-(6). It is interesting to know that Figs. 2 and 3 coincide with the SMP and the Matsuoka-Nakai criterion when $\sigma_o=0$ (c=0), and coincide with the octahedral plane and the Mises criterion when $\sigma_o\to\infty$ ($\phi=0$), respectively.

We select a cement-treated Toyoura sand (mixing ratio by weight; water: cement: Toyoura sand: clay powder =4.8:1.0:11.7:1.5, curing period=about 3 months) as a material with friction and cohesion. The physical properties of Toyoura sand are as follows: D_{50}=0.2mm, U_c=1.3, G_s=2.65, e_{max}=0.95 and e_{min}=0.58. Figs. 4 and 5 show the results of triaxial compression test ($\sigma_1>\sigma_2=\sigma_3$) and triaxial extension test($\sigma_1=\sigma_2>\sigma_3$) on the cement-treated Toyoura sand under constant mean principal stresses $\sigma_m=10kgf/cm^2=980kN/m^2$ and $\sigma_m=20kgf/cm^2=1960kN/m^2$ respectively, arranged by four kinds of relationships. The value of σ_o is determined from Fig.6 as $\sigma_o=1.3kgf/cm^2=127kN/m^2$. It is seen from (c) and (d) in Figs. 4 and 5 that the test results in triaxial compression and extension are uniquely arranged on the "extended SMP". Based on the unique arrangement of these test results on the "extended SMP", we can easily formulate a constitutive equation for frictional and cohesive materials, just in the same way as the unique arrangement on the "SMP" for frictional materials such as sands.

CONCLUSIONS

(1) By introducing a parameter of "bonding stress σ_o", the Matsuoka-Nakai criterion and constitutive relation on the "SMP" for frictional materials were extended to those on the "extended SMP" for frictional and cohesive materials.

(2) A unique stress-strain relation in both triaxial compression and triaxial extension was obtained by the arrangement on the "extended SMP" of test results of a cement-treated Toyoura sand as a material with friction and cohesion. This suggests that the stress-strain behavior and failure under three-dimensional stresses of frictional and cohesive materials are uniquely defined on the "extended SMP".

REFERENCES

1. H.Matsuoka et al., Proc. Japan Soci. of Civil Eng., 232, 59-70 (1974).
2. H.Matsuoka, Soils and Foundations, 16, 1, 91-100(1976).
3. H.Matsuoka et al., Soils and Foundations, 25, 4, 123-128(1985).
4. K.Hashiguchi, Dr. Eng. Thesis, Tokyo Inst. of Tech., 164-194(1975)
5. S.Ohmaki, Proc. 3rd Int. Conf. on Numer. Meth. in Geomech., 465-474(1979).

PLANE STRESS IN FIBER-REINFORCED COMPOSITES

R. S. RIVLIN
Lehigh University, Bethlehem, PA, U.S.A.

ABSTRACT

Constitutive equations are developed for plane stress deformations in a lamina consisting of a soft, elastic matrix reinforced by a system of parallel, high-modulus, elastic fibers. Some particular loading conditions are considered in the case when the fibers are initially straight. These results are applied to the calculation of the load-extension relation for simple extension in the case when the fibers follow sinusoidal paths with ambient direction parallel to the tensile force.

1. INTRODUCTION

The continuum theory of fiber-reinforced materials has undergone a considerable development since the work of Adkins and Rivlin early in the 1950's. This is due mainly to Pipkin, Rogers, and Spencer. A good picture of this development can be obtained from the book by Spencer [1] and a volume edited by Spencer [2]. In this paper I outline the main features of a theory of plane stress in a lamina consisting of a soft, elastic matrix reinforced by a system of parallel, high-modulus, elastic fibers.

In the case when the fibers are straight, the material of the lamina is characterized by the strain-energy, per unit initial area, which is a function of the two-dimensional Lagrange strain, and in terms of which a constitutive equation for the two-dimensional Piola-Kirchhoff stress is obtained. This is applied to particular loading conditions: (i) tensile loading with arbitrary inclination to the fiber direction, (ii) simple shear parallel to the fiber direction, (iii) tensile loading with transverse shear. The motivation for the calculations (i) and (ii) is to obtain relations between the load and deformation from which, by comparison with experimental results, the dependence of the strain-energy function on the Lagrange strain can be determined. The motivation for the calculation (iii) is to provide a relation which can be used as a basis for the calculation in Sec. 9 of the extension vs. load dependence for tensile loading of a lamina in which the fibers follow parallel, sinusoidal paths about an ambient direction parallel to the direction of the tensile load.

My attention was drawn to these problems in 1986, by S.-Y. Luo, a graduate student at the University of Delaware, who was conducting experiments on laminae of the type considered here. At that time I sketched the theory presented in the present paper. However, the development of this by Luo and his thesis advisor, T.-W. Chou, which was published by them in [3], is fallacious, and was published against my advice.

A fuller account of the theory outlined in the present paper will be published elsewhere.

2. KINEMATICS

We are concerned with plane stress deformations of a thin rectangular lamina ABCD which consists of a soft, elastic matrix, in the mid-plane of which is embedded a layer of straight, parallel, high-modulus, elastic fibers, initially inclined at an angle θ to the edges AB, DC of the lamina. Alternatively, two similar layers, equidistant from the mid-plane may be embedded.

We idealize the lamina as a highly anisotropic, two-dimensional, elastic continuum, whose mechanical properties can be characterized by the dependence of the strain-energy W, per unit initial area, on the deformation. In the description of the deformation we employ two two-dimensional rectangular cartesian coordinate systems x and $\bar{x}$, both of which have their origins at the center of the lamina. x has its 1-axis parallel to the edges AB, DC of the lamina and $\bar{x}$ has its 1-axis in the initial fiber-direction.

Let $\mathbf{X}$ and $\bar{\mathbf{X}}$ be the vector positions of a particle of the lamina in the undeformed configuration, referred to x and $\bar{x}$ respectively, and $\mathbf{x}$ and $\bar{\mathbf{x}}$ be its corresponding vector positions in the deformed configuration. We define the deformation gradient matrices $\mathbf{g}$ and $\bar{\mathbf{g}}$ by

$$\mathbf{g} = \partial\mathbf{x}/\partial\mathbf{X}, \quad \bar{\mathbf{g}} = \partial\bar{\mathbf{x}}/\partial\bar{\mathbf{X}}. \tag{2.1}$$

The corresponding Lagrange strain matrices are defined by

$$\mathbf{E} = \tfrac{1}{2}(\mathbf{g}^T\mathbf{g} - \mathbf{I}), \quad \bar{\mathbf{E}} = \tfrac{1}{2}(\bar{\mathbf{g}}^T\bar{\mathbf{g}} - \mathbf{I}). \tag{2.2}$$

We adopt the notation

$$c = \cos\theta, \quad s = \sin\theta. \tag{2.3}$$

Then $\mathbf{E}$ and $\bar{\mathbf{E}}$ are related by

$$\bar{E}_{11} = E_{11}c^2 + 2E_{12}cs + E_{22}s^2,$$

$$\bar{E}_{22} = E_{11}s^2 - 2E_{12}cs + E_{22}c^2, \tag{2.4}$$

$$\overline{E}_{12} = \overline{E}_{21} = (E_{22} - E_{11})\, cs + E_{12}(c^2 - s^2).$$

3. THE PIOLA-KIRCHHOFF STRESS

Since the lamina is elastic, W is a function of **g** or $\overline{\mathbf{g}}$. Rotation invariance implies that it must be expressible as a function of **E** or $\overline{\mathbf{E}}$. We choose the latter description. Since the lamina is symmetric under reflection in the $\overline{x}_1$ and $\overline{x}_2$ axes, W must be an even function of $\overline{E}_{12}$. We accordingly write

$$W = W(\overline{E}_{11}, \overline{E}_{22}, \overline{E}_{12}^2). \tag{3.1}$$

Let $\Pi = ||\Pi_{Ai}||$ and $\overline{\Pi} = ||\overline{\Pi}_{Ai}||$ be the two-dimensional Piola-Kirchhoff stress matrices referred to x and $\overline{x}$ respectively. Then $\overline{\Pi}$ is given by

$$\overline{\Pi}_{Ai} = \tfrac{1}{2}\overline{g}_{iB}(W_{AB} + W_{BA}), \quad W_{AB} = \partial W/\partial \overline{E}_{AB}. \tag{3.2}$$

Π can be obtained from $\overline{\Pi}$ by transformation from $\overline{x}$ to x. We find that

$$\begin{aligned} \Pi_{11} &= \Phi_1 g_{11} + \Phi_3 g_{12}, \quad \Pi_{22} = \Phi_2 g_{22} + \Phi_3 g_{21}, \\ \Pi_{12} &= \Phi_1 g_{21} + \Phi_3 g_{22}, \quad \Pi_{21} = \Phi_2 g_{12} + \Phi_3 g_{11}, \end{aligned} \tag{3.3}$$

where

$$\begin{aligned} \Phi_1 &= c^2 W_{11} + s^2 W_{22} - csW_{12}, \\ \Phi_2 &= s^2 W_{11} + c^2 W_{22} + csW_{12}, \\ \Phi_3 &= cs(W_{11} - W_{22}) + \tfrac{1}{2}(c^2 - s^2)W_{12}. \end{aligned} \tag{3.4}$$

If we idealize the problem by assuming the fibers to be inextensible, then $\overline{E}_{11} = 0$ and, from $(2.4)_1$,

$$E_{11}c^2 + 2E_{12}cs + E_{22}s^2 = 0. \tag{3.5}$$

We must now replace (3.1) by

$$W = W(\overline{E}_{22}, \overline{E}_{12}^2) \tag{3.6}$$

and in (3.4) we must replace W_{11} by T, a quantity which is undetermined if the deformation is specified.

In the following sections we discuss a number of particular cases in which the lamina is deformed by specified dead loads (i.e. for specified values of the Piola-Kirchhoff stress) applied to its edges and uniformly distributed over them.

4. TENSILE LOAD

In this section we suppose that the lamina is subjected to a pure, homogeneous deformation by dead loads applied to the edges BC, AD, the other edges being force-free. Then,

$$\Pi_{22} = \Pi_{21} = \Pi_{12} = 0, \ \Pi_{11} \neq 0. \tag{4.1}$$

From (3.3) it follows that $g_{21} = 0$. The deformation may therefore be described by

$$x_1 = \lambda_1 X_1 + K\lambda_2 X_2, \ x_2 = \lambda_2 X_2. \tag{4.2}$$

This describes a pure, homogeneous deformation with principal extension ratios λ_1, λ_2 and principal directions parallel to the axes of x, followed by a simple shear in the x_1 direction of amount K.

For the deformation (4.2), we obtain from (3.3), (3.4), and (4.1)

$$\Pi_{11}/\lambda_1 = W_{11}/c^2 = W_{22}/s^2 = -W_{12}/(2cs), \tag{4.3}$$

and $\overline{\mathbf{E}}$ in the expression for W is given by

$$\overline{E}_{11} = \tfrac{1}{2}\{(\lambda_1^2-1)c^2 + [\lambda_2^2(1+K^2)-1]s^2 + 2K\lambda_1\lambda_2 cs\},$$

$$\overline{E}_{22} = \tfrac{1}{2}\{(\lambda_1^2-1)s^2 + [\lambda_2^2(1+K^2)-1]c^2 - 2K\lambda_1\lambda_2 cs\}, \tag{4.4}$$

$$\overline{E}_{12} = \tfrac{1}{2}\{[\lambda_2^2(1+K^2)-\lambda_1^2]cs + K\lambda_1\lambda_2(c^2 - s^2)\}.$$

If the value of λ_1 corresponding to specified values of Π_{11} and θ is measured, then (4.3) enables us to determine the corresponding values of W_{11}, W_{22}, W_{12}. If further the corresponding values of λ_2 and K are also measured, we can calculate from (4.4) the corresponding values of $\overline{E}_{11}, \overline{E}_{22}, \overline{E}_{12}$. By repeating the measurements for various values of Π_{11} and θ, we can, at any rate in principle, determine the dependence of W_{11}, W_{22}, W_{12} on $\overline{E}_{11}, \overline{E}_{22}, \overline{E}_{12}$ over some surface in the $\overline{E}_{11}, \overline{E}_{22}, \overline{E}_{12}$ space. Of course, if we *assume* some specific form for W, say a polynomial of sufficiently low degree, then the envisaged measurements may enable us to determine the coefficients in the polynomial. The general validity of the expression for W so determined can be checked by examining its effectiveness in predicting the results of other experiments corresponding to values of $\overline{E}_{11}, \overline{E}_{22}, \overline{E}_{12}$ which lie outside the surface corresponding to the tensile measurements.

This procedure becomes very much simpler if the fibers are of sufficiently high modulus so that they may be idealized as inextensible. In that case $(4.4)_1$ yields

$$(\lambda_1^2-1)c^2 + [\lambda_2^2(1+K^2)-1]s^2 + 2K\lambda_1\lambda_2 cs = 0. \tag{4.5}$$

With (4.5), we obtain from (4.4)

$$\begin{aligned}\overline{E}_{22} &= -(1/2s^2)\{(\lambda_1^2-1)(c^2-s^2) + 2K\lambda_1\lambda_2 cs\},\\ \overline{E}_{12} &= -\tfrac{1}{2}\{(\lambda_1^2-1)(c/s) + K\lambda_1\lambda_2\}.\end{aligned} \tag{4.6}$$

In this case, if λ_1 and λ_2 are measured, then K can be calculated from (4.5). Alternatively, if λ_1 and K are measured, λ_2 can be calculated from (4.5). More importantly, W is now a function of two variables instead of three. The tensile measurements determine the values of W_{22}, W_{12} (and T, which replaces W_{11} in (4.3)) over a domain in the $\overline{E}_{22},\overline{E}_{12}$ plane.

5. SIMPLE SHEAR

For a simple shear of amount K in the x_1–direction, the deformation is described by

$$x_1 = X_1 + KX_2, \quad x_2 = X_2. \tag{5.1}$$

Then from $(2.1)_1$ and (3.3)

$$\Pi_{11} = \Phi_1 + K\Phi_3, \ \Pi_{22} = \Phi_2, \ \Pi_{12} = \Phi_3, \ \Phi_{21} = K\Phi_2 + \Phi_3 \tag{5.2}$$

with Φ_1,Φ_2,Φ_3 given by (3.4), and from (2.1) and (2.4)

$$\begin{aligned}\overline{E}_{11} &= \tfrac{1}{2}K^2s^2 + Kcs, \quad \overline{E}_2 = \tfrac{1}{2}K^2c^2 - Kcs,\\ \overline{E}_{12} &= \tfrac{1}{2}K^2cs + \tfrac{1}{2}K(c^2 - s^2).\end{aligned} \tag{5.2}$$

We note that the deformation (5.1) is not possible if the fibers are inextensible unless s=0, or K=–2c/s, since otherwise $\overline{E}_{11} \neq 0$

6. TENSILE LOADING WITH TRANSVERSE SHEAR

We now suppose that the lamina ABCD undergoes a pure, homogeneous deformation described by

$$x_1 = \lambda_1 X_1, \quad x_2 = \lambda_2 X_2 + K\lambda_1 X_1. \tag{6.1}$$

(We note that this is (4.1) with the 1- and 2-directions interchanged.) It is easily calculated from (3.3) that

$$\Pi_{11} = \lambda_1\Phi_1, \quad \Pi_{22} = \lambda_2\Phi_2 + K\lambda_1\Phi_3,$$
$$\Pi_{12} = K\lambda_1\Phi_1 + \lambda_2\Phi_3, \quad \Pi_{21} = \lambda_1\Phi_3, \tag{6.2}$$

where Φ_1,Φ_2,Φ_3 are given by (3.4) and, from (2.4),

$$\overline{E}_{11} = \tfrac{1}{2}\{[\lambda_1^2(1+K^2)-1]c^2 + (\lambda_2^2-1)s^2 + 2K\lambda_1\lambda_2 cs\},$$
$$\overline{E}_{22} = \tfrac{1}{2}\{[\lambda_1^2(1+K^2)-1]s^2 + (\lambda_2^2-1)c^2 - 2K\lambda_1\lambda_2 cs\}, \tag{6.3}$$
$$\overline{E}_{12} = \tfrac{1}{2}\{[\lambda_2^2-\lambda_1^2(1+K^2)]cs + K\lambda_1\lambda_2(c^2-s^2)\}.$$

We now suppose that the deformation (6.1) is produced by tensile forces $\pm\Pi_{11}$, per unit initial edge length, applied to the edges BC, AD, and such forces as may be necessary applied to the edges AB, DC; i.e. we take $\Pi_{12}= 0$. For reasons which will appear later in Sec. 9, we develop the particular case in which $\Pi_{22}= 0$. It then follows from (6.2) and (3.3) that

$$\Phi_2 = -(K\lambda_1/\lambda_2)\Phi_3 = (K\lambda_1/\lambda_2)^2\Phi_1 \tag{6.4}$$

$$\Pi_{11} = \lambda_1\Phi_1, \quad \Pi_{21} = -(K\lambda_1/\lambda_2)\Pi_{11}. \tag{6.5}$$

With these relations (3.4) yield

$$W_{11} = (c\lambda_2-K\lambda_1 s)^2(\lambda_1\lambda_2^2)^{-1}\Pi_{11},$$
$$W_{22} = (s\lambda_2+K\lambda_1 c)^2(\lambda_1\lambda_2^2)^{-1}\Pi_{11}, \tag{6.6}$$
$$W_{12} = -2(c\lambda_2-K\lambda_1 s)(s\lambda_2+K\lambda_1 c)(\lambda_1\lambda_2^2)^{-1}\Pi_{11}.$$

If W is a known function of $\overline{E}_{11},\overline{E}_{22},\overline{E}_{12}$, and hence, from (6.3), of λ_1,λ_2,K, (6.6) are three equations from which the latter can be determined for specified values of Π_{11} and θ. Π_{21} can then be determined from (6.5). We note, for future use, that, from the symmetry of the problem, K and hence Π_{21} must be odd functions of θ.

If the fibers are inextensible, equations (6.5) are still valid, but (6.6) can be replaced by

$$T = (\lambda_2-|s|B)^2(c^2\lambda_1\lambda_2^2)^{-1}\Pi_{11},$$

$$W_{22} = B^2(\lambda_1\lambda_2^2)^{-1}\Pi_{11}, \tag{6.7}$$

$$W_{12} = 2B(|s|B-\lambda_2)(c\lambda_1\lambda_2^2)^{-1}\Pi_{11},$$

where

$$B = (1 - \lambda_1^2c^2)^{½}, \tag{6.8}$$

and the expressions $(6.3)_{2,3}$ may be written as

$$\begin{aligned} \overline{E}_{22} &= (2c^2)^{-1}(\lambda_2^2+s^2-c^2-2\lambda_2|s|B), \\ \overline{E}_{11} &= (2c)^{-1}(-|s| + \lambda_2 B). \end{aligned} \tag{6.9}$$

If W is a known function of $\overline{E}_{22},\overline{E}_{12}$, then for specified Π_{11} and θ, $(6.7)_{2,3}$ provide *two* equations for the determination of λ_1 and λ_2. The corresponding value of K can then be obtained by taking $\overline{E}_{11}= 0$ in $(6.3)_1$. We obtain

$$K = -\lambda_2|s| + B. \tag{6.10}$$

7. SMALL DEFORMATIONS

We now consider the simplifications in the results of Sec.6 which follow from the assumption that the deformation is small, restricting our discussion for the moment to the case in which the fibers are inextensible, so that W is given by (3.6).

We suppose that W is a sufficiently smooth function of $\overline{E}_{22},\overline{E}_{12}^2$, so that it can be expressed as a Taylor series. Then if the extensions undergone in the deformation by line elements of the lamina are sufficiently small, we may write

$$W = ½\alpha\overline{E}_{22}^2 + ½\beta\overline{E}_{12}^2, \tag{7.1}$$

where α and β are constants. Here we assume that the stress is zero when the lamina is undeformed and T=0.

We write

$$\lambda_1 = 1 + e_1, \quad \lambda_2 = 1 + e_2 \tag{7.2}$$

and assume $e_1,e_2,K \ll 1$. Then equations (6.7) yield, with (6.9),

$$\begin{aligned} &e_1 = (2c^2)^{-1}(s^2 - B^2), \quad e_2 = \Pi_{11}\alpha^{-1}c^2B^2 - s^2 + |s|B, \\ &K = -c^{-1}(|s| - B), \quad T = \Pi_{11}c^2(1 + \Pi_{11}\alpha^{-1}B^2)^2, \end{aligned} \tag{7.3}$$

where

$$B = \frac{4\Pi_{11} + \beta}{8|s|\Pi_{11}}\left\{1 - \left[1 - \frac{16\beta s^2\, \Pi_{11}}{(4\Pi_{11}+\beta)^2}\right]^{1/2}\right\}. \tag{7.4}$$

The expression (7.4) can be simplified if $s \ll 1$. Then

$$B = \beta|s|/(4\Pi_{11}+\beta). \tag{7.5}$$

It can also be simplified if $\Pi_{11}/\beta \gg 1$. Then

$$B = \beta|s|/4\Pi_{11}. \tag{7.6}$$

8. NEARLY INEXTENSIBLE FIBERS

If the elastic modulus of the fibers is very large compared with that of the matrix, we can in many problems treat them as inextensible and apply a correction for the departure from ideal inextensibility. The strain-energy W can be approximated by an expression of the form

$$W = \tfrac{1}{2}\mu\overline{E}_{11}^2 + W^*(\overline{E}_{22},\overline{E}_{12}^2). \tag{8.1}$$

For example, in the problems discussed in Secs 4 and 6 we calculate the value of T obtained with the assumption that the fibers are inextensible. The correction to λ_1 is then given approximately by Tc^2/μ.

9. SINUSOIDAL FIBERS

We now suppose that the fibers embedded in the lamina ABCD, instead of being straight, follow parallel, sinusoidal paths described in the undeformed configuration by

$$X_2 = A\sin(2\pi X_2/a) + c, \tag{9.1}$$

where A and a are constants and c is constant on a fiber but varies from fiber to fiber.

We suppose that the lamina undergoes a deformation described by (6.1) in which λ_1,λ_2,K are now functions of X_1 (but not of X_2). We shall determine the dependence of λ_1,λ_2,K on X_1 if no body forces are applied to the lamina and

$$\Pi_{11} = \text{constant},\ \Pi_{12} = \Pi_{22} = 0. \tag{9.2}$$

It follows from the equations of equilibrium that Π_{21} is a function of X_1 only.

Let A'B'C'D' be an elementary strip of the lamina parallel to the edges BC, AD. Let θ be the angle at which the fibers in the strip are initially inclined to the x_1-direction. Then, from (9.1),

$$\tan\theta = 2\pi\bar{A}\cos(2\pi X_1/a), \quad \bar{A} = A/a. \tag{9.3}$$

The strain-energy in A'B'C'D', per unit initial area, will, in general, depend on θ:

$$W = W(\bar{E}_{11}, \bar{E}_{22}, \bar{E}_{12}^2; \theta). \tag{9.4}$$

$\bar{E}_{11}, \bar{E}_{22}, \bar{E}_{12}$ are given in terms of λ_1, λ_2, K by (6.3), and (6.6) are three equations from which the latter can be calculated if Π_{11} is given. The average value of the extension ratio in the x_1-direction, denoted Λ, is given by

$$\Lambda = \frac{1}{a}\int_0^a \lambda_1 dX_1. \tag{9.5}$$

The calculations become simpler when the fibers are inextensible. Then (9.4) is replaced by

$$W = W(\bar{E}_{22}, \bar{E}_{12}^2; \theta) \tag{9.6}$$

and λ_1 and λ_2 can be calculated from $(6.7)_{2,3}$ with $\bar{E}_{22}$ and $\bar{E}_{12}$ given by (6.9).

While W depends, in general, on θ and hence varies with X_1 in accordance with (9.3), it can be argued heuristically that this variation can perhaps be neglected if $\bar{A}$ is sufficiently small and the volume fraction of fiber is small. However, the validity of this contention should be verified experimentally.

We note that the force system required to maintain the deformation consists not only of the tensile loads $\pm\Pi_{11}$ on the edges BC, AD, but also of loads $\pm\Pi_{21}$ applied to the edges AB, DC. The latter depend on X_1. However, as has been seen in Sec.6, Π_{21} changes sign with θ and consequently in successive half wave-lengths of the fibers. We can now invoke Saint Venant's principle to conclude that the effect of not applying the forces $\pm\Pi_{21}$ (i.e. of maintaining the edges AB, DC force-free) will be restricted to narrow regions close to the edges. This is particularly the case in view of the high modulus of the fibers.

REFERENCES

1. A.J.M. Spencer, *Deformations in Fibre-Reinforced Materials,* Clarendon Press 1972.
2. A.J.M. Spencer (ed.), *Continuum Theory of the Mechanics of Fibre-Reinforced Composites*, Springer 1984.
3. S.-Y. Luo and T.-W.Chou, *Finite deformation of flexible composites,* Proc. Roy. Soc. Lond. A (in the press).

CONSTITUTIVE RELATIONS FOR FORCE RESULTANTS IN LAMINATED PLATES

R. S. SANDHU and M. MOAZZAMI
The Ohio State University, Columbus, Ohio 43210

ABSTRACT

A mixed variational principle is used for systematic development of constitutive relationships for force resultants in a laminated plate for a higher order theory.

INTRODUCTION

In this paper we present a generalization of Reissner's method [1-4] to the discrete laminate theory and derive the constitutive relations for force resultants in the laminate. The behavior of various layers is seen to be coupled, the coupling depending upon the lay-up and the material properties. We first present a summary of the variational procedure and then apply it to a higher order discrete laminate theory.

A VARIATIONAL PRINCIPLE

The Hellinger-Reissner functional for linear elasticity is

$$\Omega(u_i,\sigma_{ij}) = {}_R\int(\frac{1}{2}\sigma_{ij}C_{ijkl}\sigma_{kl} - u_{i,j}\sigma_{ij} + u_i f_i)\, dR + {}_{S_1}\int \sigma_{ij}n_j(u_i - \hat{u}_i)\, ds + {}_{S_2}\int u_i\hat{t}_i ds \tag{1}$$

Here, R is the open, connected, spatial region of interest with boundary S consisting of complementary subsets S_1 and S_2, $\hat{u}_i$ and $\hat{t}_i$ are the components of the displacement vector specified on S_1 and those of the traction vector specified on S_2, respectively. A rectangular cartesian reference frame and the standard index notation are used. Quantities $u_i, f_i, \sigma_{ij}, C_{ijkl}$ have the usual meaning. If there are no body forces and the displacements identically equal the specified values on S_1, the functional in (1) specializes to:

$$\Omega\ (u_i,\sigma_{ij}) \doteq {}_R\int(\frac{1}{2}\sigma_{ij}C_{ijkl}\sigma_{kl} - u_{i,j}\sigma_{ij})\, dR + {}_{S_2}\int u_i\hat{t}_i ds \tag{2}$$

Vanishing of the Gateaux differential of Ω_1 along the path $\{v_i,\tau_{ij}\}$, upon use of the divergence theorem leads to the variational equality:

$$_R\int[\tau_{ij}(C_{ijkl}\sigma_{kl} - u_{i,j}) + v_i\sigma_{ij,j}]\, dR - {}_{S_1}\int v_i(t_i - \hat{t}_i)\, ds = 0 \tag{3}$$

If stresses are required to satisfy equilibrium over R and on S_2 the variational equality, for the case of a laminate, becomes:

$$\int_A \left[\sum_{k=1}^{N} \int_{-\frac{t_k}{2}}^{\frac{t_k}{2}} \tau_{ij}^{(k)} \left(C_{ijkl}^{(k)} \sigma_{kl}^{(k)} - u_{i,j}^{(k)} \right) dx_3^{(k)} \right] dA = 0 \qquad (4)$$

where t_k is the thickness of the kth layer, N is the number of layers, and A is the surface area of the plate.

CONSTITUTIVE EQUATIONS FOR FORCE RESULTANTS

In order to set up constitutive equations for force resultants in laminated plates, it is necessary to assume a stress distribution satisfying equilibrium, on the plate as well as on its boundary, in terms of force resultants and interlayer tractions. The variational equality for arbitrary paths in the space of each of these resultants furnishes the constitutive equations. We illustrate this approach by applying it to a higher order theory.

A Higher Order Theory

Assuming the inplane stress components as well as displacements to be quadratic and the transverse displacement to be linear in $x_3^{(k)}$,

$$u_\alpha^{(k)} = v_\alpha^{(k)} (x_\beta^{(k)}) + x_3^{(k)} \phi_\alpha^{(k)} (x_\beta^{(k)}) + \frac{1}{2} (x_3^{(k)})^2 \Psi_\alpha^{(k)} (x_\beta^{(k)}) \qquad (5)$$

$$u_3^{(k)} = v_3^{(k)} (x_\beta^{(k)}) + x_3^{(k)} \phi_3^{(k)} (x_\beta^{(k)}) \qquad (6)$$

$$\sigma_{\alpha\beta}^{(k)} = \eta_1^{(k)} N_{\alpha\beta}^{(k)} + \eta_2^{(k)} M_{\alpha\beta}^{(k)} + \eta_3^{(k)} P_{\alpha\beta}^{(k)} \qquad (7)$$

where η_i, i = 1, 2, 3 are quadratic expressions in $x_3^{(k)}$. To satisfy the equilibrium equations [5]:

$$\sigma_{\alpha 3}^{(k)} = \zeta_1^{(k)} Q_\alpha^{(k)} + \zeta_2^{(k)} M_{\alpha 3}^{(k)} + \zeta_3^{(k)} T_\alpha^{(k-1)} + \zeta_4^{(k)} T_\alpha^{(k)} \qquad (8)$$

$$\sigma_{33}^{(k)} = \xi_1^{(k)} N_{33}^{(k)} + \xi_2^{(k)} T_3^{(k-1)} + \xi_3^{(k)} T_3^{(k)} + \xi_4^{(k)} T_{\rho,\rho}^{(k-1)} + \xi_5^{(k)} T_{\rho,\rho}^{(k)} \qquad (9)$$

Here $\zeta_i^{(k)}$, i = 1,2,3,4 and $\xi^{(k)}$, k = 1,2,3,4,5 are, respectively, cubic and fourth degree polynomials in $x_3^{(k)}$,

$$(N_{33}^{(k)}, M_{\alpha 3}^{(k)}, P_{\alpha\beta}^{(k)}) = \int_{-\frac{t_k}{2}}^{\frac{t_k}{2}} (\sigma_{33}^{(k)}, x_3^{(k)} \sigma_{\alpha 3}^{(k)}, (x_3^{(k)})^2 \sigma_{\alpha\beta}^{(k)}) \, dx_3^{(k)} \qquad (10)$$

and $T_3^{(k)}$ is the value of $\sigma_{33}^{(k)}$ at the top of the kth layer. Equation (4) implies the following constitutive relationships:

$$v_{(\alpha,\beta)}^{(k)} = C_{\alpha\beta\gamma\delta}^{(k)} \left[\frac{1}{4t_k} N_{\gamma\delta}^{(k)} - \frac{15}{t_k^3} P_{\gamma\delta}^{(k)}\right] \tag{11}$$

$$+ C_{\sigma\beta 33}^{(k)}\left[\frac{60}{t_k} N_{33}^{(k)} - \frac{5}{2} (T_3^{(k-1)} + T_3^{(k)})\right.$$

$$\left. + \frac{t_k}{56} (T_{\rho,\rho}^{(k-1)} - T_{\rho,\rho}^{(k)})\right]$$

$$\phi_{(\alpha,\beta)}^{(k)} = C_{\alpha\beta\gamma\delta}^{(k)} \frac{12}{t_k^3} M_{\gamma\delta}^{(k)} + C_{\alpha\beta 33}^{(k)} \left[- \frac{6}{5t_k} (T_3^{(k-1)} - T_3^{(k)})\right. \tag{12}$$

$$\left. + \frac{1}{10} (T_{\rho,\rho}^{(k-1)} + T_{\rho,\rho}^{(k)})\right]$$

$$\Psi_{(\alpha,\beta)}^{(k)} = C_{\alpha\beta\gamma\delta}^{(k)}\left[- \frac{15}{t_k} N_{\gamma\delta}^{(k)} + \frac{180}{t_k^3} P_{\gamma\delta}^{(k)}\right] \tag{13}$$

$$+ C_{\alpha\beta 33}^{(k)}\left[- \frac{60}{7t_k^3} N_{33}^{(k)} + \frac{30}{7t_k} (T_3^{(k-1)} + T_3^{(k)})\right.$$

$$\left. - \frac{15}{2t_k} (T_{\rho,\rho}^{(k-1)} - T_{\rho,\rho}^{(k)})\right]$$

$$\phi_{\alpha}^{(k)} + v_{3,\alpha}^{(k)} = \frac{2}{5} C_{\alpha 3\gamma 3}^{(k)}\left[\frac{12}{t_k} Q_{\gamma}^{(k)} - (T_{\gamma}^{(k-1)} + T_{\gamma}^{(k)})\right] \tag{14}$$

$$2\Psi_{\alpha}^{(k)} + \phi_{3,\alpha}^{(k)} = \frac{4}{7t_k} C_{\alpha 3\gamma 3}^{(k)} \left[\frac{120}{t_k} M_{\gamma 3}^{(k)} + \frac{3}{2} (T_{\gamma}^{(k-1)} - T_{\gamma}^{(k)})\right] \tag{15}$$

$$\phi_3^{(k)} = C_{33\gamma\delta}^{(k)}\left[\frac{60}{t_k} N_{\gamma\delta}^{(k)} - \frac{60}{7t_k^3} P_{\gamma\delta}^{(k)}\right] \tag{16}$$

These are fourteen constitutive equations viz. (11) to (16) in as many field variables. These correspond to 'variations' in $N^{(k)}_{\alpha\beta}$, $M^{(k)}_{\alpha\beta}$, $P^{(k)}_{\alpha\beta}$, $Q^{(k)}_{\alpha}$, $M^{(k)}_{\alpha 3}$ and $N^{(k)}_{33}$, respectively. Three additional equations arise corresponding to variations in $T^{(k)}_{\alpha}$ and $T^{(k)}_{3}$ [5]. These have not been listed here in the interest of brevity.

Discussion

To allow properly for effects of shear deformation, shear correction factors have often been used. However, these are not fixed constants [6,7]. The theory outlined above provides a systematic approach to construction of self-consistent constitutive relationship for force resultants in laminated plates.

ACKNOWLEDGEMENT

The work reported herein is part of a research program supported by the Wright Research and Development Center, U.S. Air Force, under Grant No. F33615-85-C-3213. Dr. George P. Sendeckyj is the Program Manager. The U.S. Government is authorized to reproduce and distribute reprints for governmental purposes not withstanding any copyright notation herein.

REFERENCES

1. E. Reissner, "On Bending of Elastic Plates," Quar. App. Math, 5, 55-68, 1947.
2. E. Reissner, "A Consistent Treatment of Transverse Shear Deformation in Laminated Anisotropic Plates," AIAA J. 10, 716-718, 1972.
3. E. Reissner, "Note on the Effect of Transverse Shear Deformation in Laminated Anisotropic Plates," Comp. Math. App. Mech. Engrg., 20, 203-209.
4. E. Reissner, "On a Certain Mixed Variational Theorem and a Proposed Application," Int. J. Numer. Meth. Engrg., 20, 1366-1368, 1984.
5. M. Moazzami, A Higher Order Layerwise Theory of Laminated Plates, Ph.D. Dissertation, The Ohio State Univ., Columbus, Ohio, 1990.
6. J.M. Whitney and N.J. Pagano, "Shear Deformation in Heterogeneous Anisotropic Plates," J. App. Mech., 37, 1031-1036, 1970.
7. S.V. Kulkarni and N.J. Pagano, "Dynamic Characteristics of Composite Laminates," J. Sound vib., 23, 127-143, 1972.

APPLICATION OF A UNIFIED CONSTITUTIVE THEORY TO THE NONPROPORTIONAL MULTIAXIAL STRAIN DEFORMATION OF 1045 STEEL

James A. Sherwood and Endicott M. Fay
Department of Mechanical Engineering
University of New Hampshire

N. Jayaraman
Department of Materials Science and Engineering
University of Cincinnati

ABSTRACT

A unified constitutive model based upon the original Bodner inelastic flow equation and using drag stress, back stress and damage state variables has been developed and applied to simulate the multiaxial strain deformation of 1045 steel at room temperature. An automated computer code has been written to determine the constitutive equation coefficients, run single material point simulations and interface with the ADINA finite element code.

INTRODUCTION

Many efforts have been made to model nonlinear material behaviors such as plasticity (time-independent), creep (time-dependent) and strain-rate sensitivity. While these efforts have proved to be worthy of simulating uniaxial behavior, the true evaluations of these methods are how well they extend to and can capture multiaxial material deformation. Such evaluations are necessary since most components in field service are subject to loads much more complex than pure uniaxial loads.

THE UNIFIED CONSTITUTIVE MODEL

The constitutive model selected for this work is an extension of the original model developed by Bodner [1] which had only one state variable, Z, the drag stress. The intent of the drag stress is to describe the isotropic hardening of the material during deformation. Since the development of the Bodner model, Ramaswamy and Stouffer [2] have incorporated a back-stress state variable, Ω_{ij} to account for the kinematic hardening. Sherwood and Stouffer [3] have included a damage state variable, ω, to capture the reduction in the material's load carrying capability due to microcracks and microvoids.

The constitutive model is composed of an inelastic-flow equation and associated evolution equations for the state variables. The inelastic flow equation is written as

$$\dot{\epsilon}^{I}_{ij} = D_0 \exp\left[-\frac{A}{2}\left(\frac{Z^2(1-\omega)^2}{3K_2}\right)^{n}\right]\frac{S_{ij}-\Omega_{ij}}{\sqrt{3K_2}} \tag{1}$$

where

$$K_2 = \frac{1}{2}(S_{ij}-\Omega_{ij})(S_{ij}-\Omega_{ij}) \tag{2}$$

and S_{ij} is the deviatoric stress. In Equation (1), D_0 is a material constant denoting the limiting strain rate of the material and A and n are temperature-dependent material constants to capture strain rate sensitivity. The associated evolution equation for the drag stress is given by

$$\dot{Z} = m(Z_1-Z)\dot{W}^{I}$$

where the initial value of the drag stress, $Z(0)=Z_0$, Z_1 is the maximum value of the drag stress, m is a material constant which controls the rate of hardening and $\dot{W}^I$ is the inelastic-work rate.
The back-stress evolution equation is

$$\dot{\Omega}_{ij} = f_1\dot{\epsilon}^{I}_{ij} - \frac{3}{2}f_1\frac{\Omega_{ij}}{\Omega_s}\dot{\epsilon}^{I}_{eff} + f_2 S_{ij} \tag{4}$$

where f_1 and f_2 are temperature-dependent material constants, Ω_s is the maximum attainable value of the back stress as determined from uniaxial tensile tests. The evolution of damage is given by

$$\omega = a \exp\left[b\sum W^{I}\right] \tag{5}$$

where a and b are temperature-dependent material constants.

MATERIAL DATA

The data used in this report were provided by Jayaraman [4]. The data consist of a series of fatigue tests run under a plastic-strain controlled system. The loading was biaxial with a cyclic axial loading and cyclic torsional loading. The strains were fully reversed and all tests were run to failure. Failure is defined as the point at which the tensile load dropped off significantly.

THE COMPUTER PROGRAM *UNIPRO*

The optimization technique used to determine the material constants is a modified conjugate direction search. The direction of the search is controlled through the bounds of the material parameter being searched. The conjugate search is a multidimensional search of up to four dimensions in this application. The multidimensional search is composed of a group of one-dimensional searches for the minimum of a unimodal function. The one-dimensional search used here is known as the golden section search. The routine evaluates the function, F without using any derivatives and the function need not be continuous in the range of interest. The routine minimizes a function which results in a convergence of the flow equation to a tensoral strain rate or an integration of the flow equation over a discrete time step.

The range of the search is from x_{min} to x_{max} and the resulting error of the search is approximately one hundredth of one percent of the range of the search. In the multidimensional searches, the range of each parameter remains constant, and the search is conducted five times through each parameter. The algorithm samples two points in the parameter range. The routine then discards the segment of the range beyond the sample with the highest value of "F". The routine adds another sample in the direction toward the lowest "F" and repeats. The method is sequential in that it will travel to the minimum until it passes this extremum and then will reverse direction. The key to the success of the optimization technique is to provide the program with a good function for which to find a minimum and the range of each parameter.

The program will use the present guess for each parameter and the new guess to calculate the inelastic strain over the specified time increment and integrate over the full time span of one-sixth of a cycle. The calculated value of the inelastic strain is then compared to the actual value. This minimization is accomplished through five complete cycles thereby allowing the range to move up or down in value as the program requires to reach the actual value of inelastic strain.

The first optimization is concerned with finding n, Z_0 and the saturated value of Ω_{ij}. The convergence function is the computation of the axial inelastic strain over the last one-sixth of the first cycle. The same optimization is run with the torsional inelastic strain as the convergence function to verify the calculation. The values of the four constants for each convergence routine are within one percent of each other.

RESULTS OF NUMERICAL SIMULATIONS

The simulations are predictions of material behavior based on the constants derived from a single in-phase biaxial fatigue test. The first simulation, which is shown in Figures 1 and 2, is plotted with the data which was used to determine the material constants. In all the figures, the test data is given by the dashed lines and the simulation corresponds to the solid lines. The nonproportional test are given in Figures 3 through 6.

CONCLUSIONS

An automated procedure for determining the optimum material parameters for the constitutive model under consideration has been developed. Simulations of proportional and nonproportional test data have demonstrated that the model and constants can capture the change in the shape of the hysteresis loop as the phase angle changes. Sherwood and Boyle [5] have incorporated this constitutive model in the ADINA finite element program. Thus, the multiaxial mechanical response of field service components can be investigated.

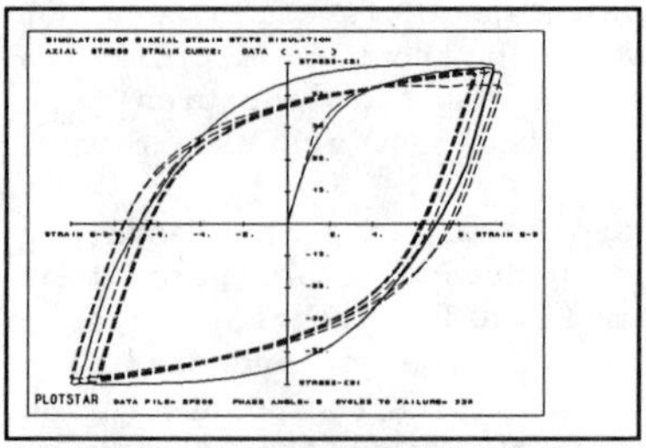

FIGURE 1. AXIAL RESPONSE OF AN IN-PHASE TEST

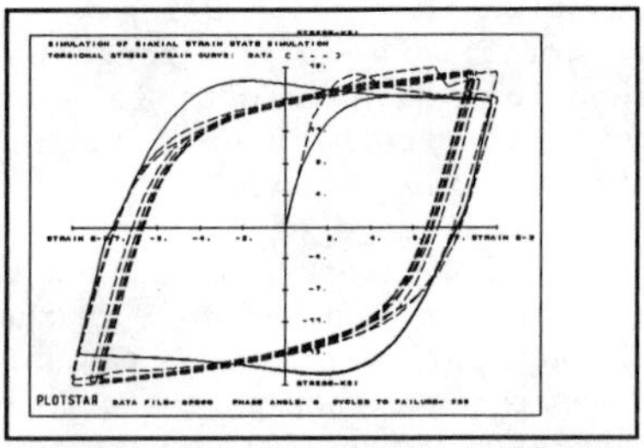

FIGURE 2. TORSIONAL RESPONSE OF AN IN-PHASE TEST

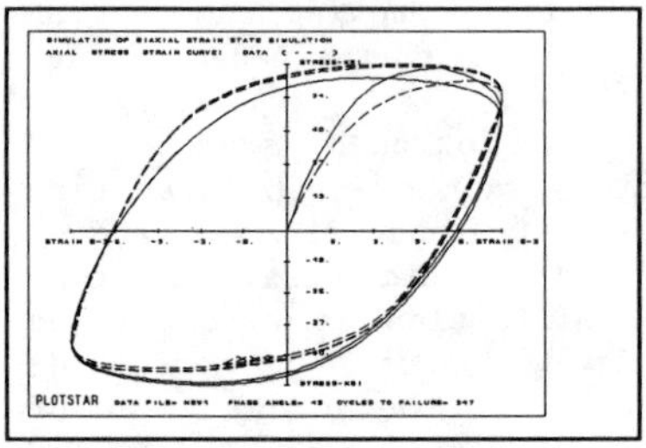

FIGURE 3 AXIAL RESPONSE OF A 45 DEGREE OUT OF PHASE TEST

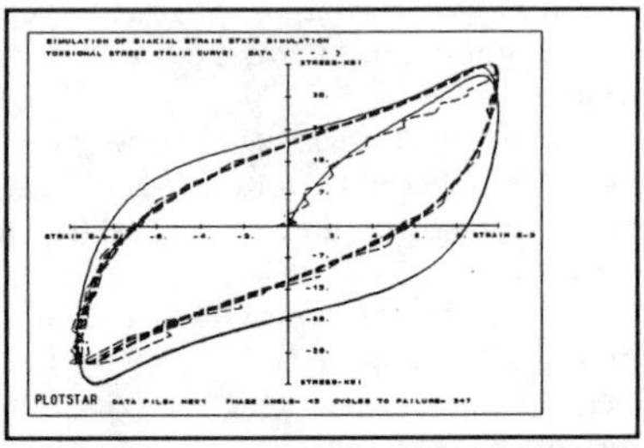

FIGURE 4. TORSIONAL RESPONSE OF A 45 DEGREE OUT OF PHASE TEST

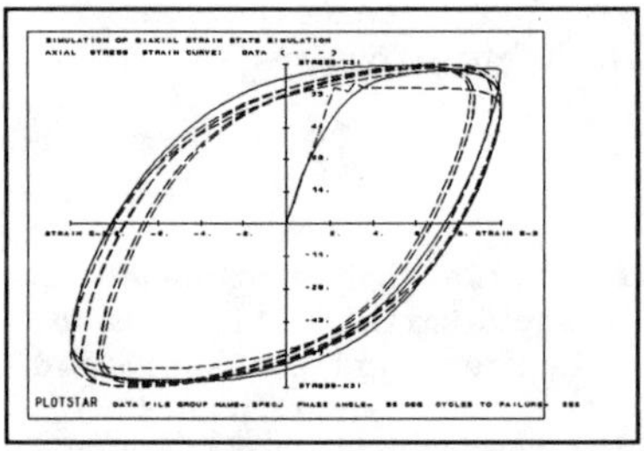

FIGURE 5. AXIAL RESPONSE OF A 90 DEGREE OUT OF PHASE TEST

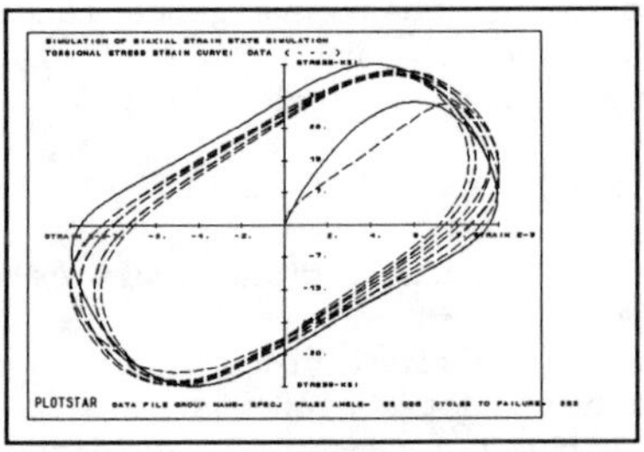

FIGURE 6. TORSIONAL RESPONSE OF A 90 DEGREE OUT OF PHASE TEST

REFERENCES

1. Bodner, S. R., and Partom, Y., "Constitutive Equations for Elastic-Viscoplastic Strain Hardening Materials," ASME Journal of Applied Mechanics, 42, p 385, 1975.

2. Ramaswamy, V.G., Stouffer, D./C., and Laflen, J.H., "A Constitutive Model for the Inelastic Response of René 80 at Temperatures between 538C and 982C", ASME J. of Engineering Materials and Technology, 112, pp. 280-286, 1990.

3. Sherwood, J.A., and , Stouffer, D.C., "A Constitutive Model with Damage for High-Temperature Superalloys", Third Symposium on Nonlinear Constitutive Relations for High Temperature Applications, Akron, OH, June 11-13, 1986.

4. Appel, M. & Jayaraman, N., Senior Project Thesis, University of Cincinnati, 1986.

5. Sherwood, J., and Boyle, "Investigation of the Thermomechanical Response of a Titanium-Aluminide/Silicon Carbide Composite Using a Unified State Variable Model and the Finite Element Method," 1990 ASME WAM, Dallas, TX, November 2-30, 1990.

ENDOCHRONIC PLASTICITY THEORY WITH SHEAR-VOLUMETRIC COUPLING

by
K. C. Valanis
ENDOCHRONICS, Inc.
Vancouver, Washington 98665
and
John F. Peters
US Army Waterways Experiment Station
Vicksburg, Mississippi, 39180

ABSTRACT
In this paper, the asymptotic behavior of a kinematic hardening plasticity model with shear-volumetric coupling is addressed. The simple one-internal variable model of Valanis and Peters [1] describes the behavior at the asymptotic state and in turn provides the basis for comparison to the traditional critical state theory [2,3].

Introduction

In a recent paper by Valanis and Peters [1] an endochronic plasticity theory was developed for soils based on "linear" endochronic plasticity originally described by Valanis [4,5] and applied to concrete by Valanis and Read [6,7]. Reference [1] described the general considerations for introducing shear-volumetric coupling into a plasticity model with kinematic hardening. Although the asymptotic state was briefly discussed the analysis provided was incomplete. In this paper we discuss the relationship between the asymptotic state and the ultimate constant volume state (critical state) and the role of the critical state in endochronic plasticity.

Review of Coupled Model

The theory relates the stress state to the plastic strain and N internal variables through N rate equations which together define the evolution of the plastic state of the material. The constitutive response for shear can be defined more compactly by the resulting history integral given as:

$$\mathbf{s} = \int_0^{z_d} \phi_d(z_d - z') \frac{d\mathbf{e}^p}{dz'} dz' \tag{1}$$

where

$$dz^2 = \|d\mathbf{e}^p\|^2 + k^2 |d\epsilon^p|^2. \tag{2}$$

As a special case of the general hydrostatic response, a stress-dilatancy relationship was derived to describe the volumetric plastic strain for triaxial compression tests which is given by Equation (3).

$$k\frac{d\epsilon^p}{de^p} = \frac{1 - \Gamma_o^2 \frac{s^2}{\sigma^2}}{\Gamma_o \frac{F_h s}{\sigma^2} + \sqrt{\frac{F_h^2}{\sigma^2} + \Gamma_o^2 \frac{s^2}{\sigma^2} - 1}}. \tag{3}$$

The shear and hydrostatic time scales are respectively given as $dz_d = dz/F_d$ and $dz_h = dz/kF_h$ where F_d is the shear hardening function and F_h is the hydrostatic hardening function. F_d and F_h control the isotropic hardening and both depend on hydrostatic stress σ, specific volume v and may depend on z. The kinematic hardening behavior is controlled by ϕ_d and ϕ_h which are rapidly decaying response functions. Note that as for traditional critical state theory, plastic strain with constant volume occurs at a characteristic ratio $s/\sigma = 1/\Gamma_o$.

Shear Response at the Critical State

The critical state generally is viewed as an ultimate constant volume state reached by monotonic shearing and occurs at a unique point in a space defined by s, σ, and v. To compare the critical state theory to the response given by Equation (1) it is convenient to consider two conditions obtained when evaluating the response integral, the fully saturated state and the asymptotic state. The fully saturated state is related to the rapidly decaying nature of ϕ_d such that, for monotonic loading histories, the following is true for some $\Delta z_d < z_d$:

$$\int_0^{z_d} \phi_d(z_d - z')\frac{de^p}{dz'}dz' \approx \int_{z_d-\Delta z_d}^{z_d} \phi_d(z_d - z')\frac{de^p}{dz'}dz'. \tag{4}$$

When condition (4) is satisfied the constitutive response is controlled by the variations in $F_d de^p/dz$. The asymptotic state is one in which condition (4) is true *and* both $F_d de^p/dz$ and $F_h d\epsilon^p/dz$ are constant. As shown by Valanis and Peters [1], a material in the asymptotic state can be modeled by a single internal variable. Accordingly, when ϕ_d in Equation 1 is approximated as a delta function, the surface defining the asymptotic state coincides with a yield surface of a density-controlled isotropic hardening material. When more than one internal variable is used the physical meaning of the asymptotic surface is obscure because $F_d de^p/dz$ and $F_h d\epsilon^p/dz$ are not constant except at the critical state. Evidently, the critical state represents the point on the asymptotic surface where it is intersected by the line defined by $s/\sigma = 1/\Gamma_o$.

Behavior Near the Critical State

The example given by Valanis and Peters [1] will be used to illustrate the interaction between the kinematic and isotropic aspects of response to produce the ultimate critical state condition. In that example, ϕ_d is approximated as the sum of a delta function and an exponential function which leads to a yield surface with isotropic/kinematic hardening. The parameters used in the model remain the same as in reference [1] except that F_d is a function of F_h as well as σ based on the the failure surface derived for a one-internal variable model using Equation (5).

$$F_d = \left[1.4\left(1 - \frac{\sigma}{F_h}\right) + \frac{1}{2\Gamma_o}\left(\frac{\sigma}{F_h}\right)\right]\sigma. \tag{5}$$

The computed response for the example is shown in Figure 1 where the shear stress s is plotted as a functions of shear strain, e, along with the value of $M(z_s)$ given by

$$M(z_s) = \int_0^{z_d} \phi_d(z_d - z')dz'. \tag{6}$$

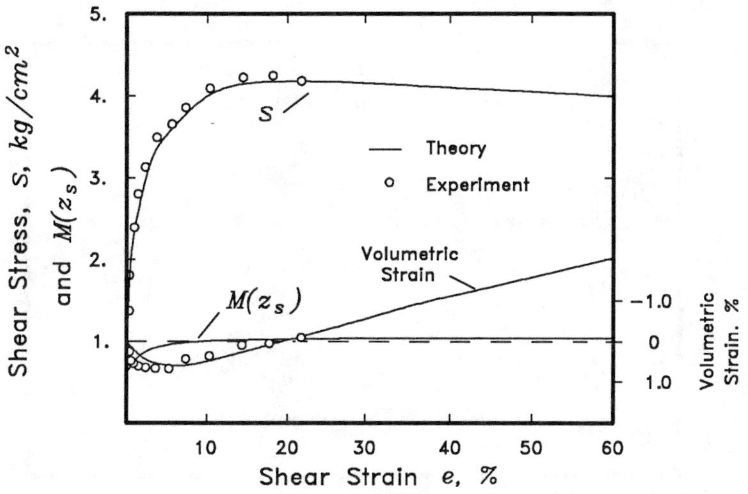

Figure 1: Predicted versus actual response for Sacramento River sand.

The function $M(z_s)$ is seen to flatten by 10 percent strain indicating full saturation of the shear response. However, the stress-strain curve continues to rise and fall as a result of the shear-induced volume change.

On Figure 2, the path taken by the test in a plot of s/F_h versus σ/F_h displays the tendency of the soil to reach the critical state as an asymptotic condition. The stress path is a straight line on a plot of s versus σ but is a curve in the normalized plot because of changes in F_d caused by dilation. Thus as the asymptotic state is reached, dilation reduces F_h, and the normalized stress path bends to the right. Ultimately the path ends where the asymptotic surface intersects the critical state line.

Concluding Remarks

In the previous paper [1] the critical state theory was used as a bench mark in the interpretation of shear-volumetric coupling predicted by the linear theory of endochronic plasticity. The analysis briefly outlined here completes the comparison by demonstrating the role of the critical state in a theory that includes both isotropic and kinematic hardening. At this point it is a demonstration in principle because the problem of predicting the response for a wide range of σ and density is still under investigation. As is apparent from the example, defining the asymptotic state for a wide range of initial conditions is critical to development of a general model.

Acknowledgement

The work described in this paper was developed from research conducted under the Civil Works Research and Development program of the United States Army Corps of Engineers. Permission was granted by the Chief of Engineers to publish this information.

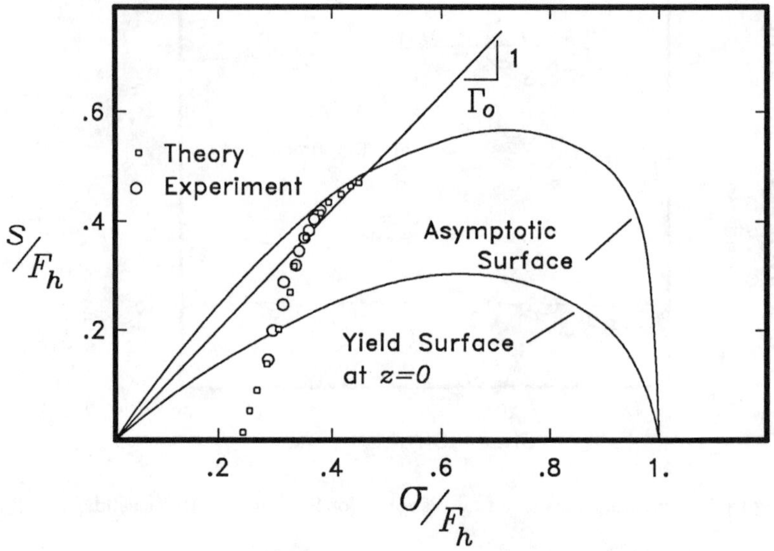

Figure 2: Effects of volume change on stress path.

References

1. K. C. Valanis and J. F. Peters, "An Endochronic Plasticity Theory with Shear-Volumetric Coupling," to appear in the *International Journal for Numerical and Analytical Methods in Geomechanics*, 1990.

2. K. H. Roscoe, A. N. Schofield, and C. P. Wroth, 1958. "On the Yielding of Soils", *Geotechnique*, **8** (1), 25-53, 1958.

3. A. N. Schofield and C. P. Wroth, *Critical State Soil Mechanics*, McGraw Hill London, 1968.

4. K. C. Valanis, "A Theory of Viscoplasticity without a Yield Surface, Part I, General Theory", *Archives of Mechanics*, **23** 517-533, 1971.

5. K. C. Valanis, "Fundamental Consequences of a New Intrinsic Time Measure. Plasticity as a Limit of the Endochronic Theory", *Archives of Mechanics*, **32**, 171-191, 1980.

6. K. C. Valanis and H. E. Read, "An Endochronic Plasticity Theory for Concrete", S-CUBED Report SSS-R-85-7172, 1985.

7. K. C. Valanis and H. E. Read, 1986. "An Endochronic Plasticity Theory for Concrete", *Mechanics of Materials*, **5**, 227-295, 1986.

ELASTICITY

MODELING BONE AS AN ANISOTROPIC CYLINDER OF CLASS mmm USING CYLINDRICAL FINITE ELEMENTS

G. R. BUCHANAN
Department of Civil Engineering

J. PEDDIESON, JR.
Department of Mechanical Engineering
Tennessee Technological University
Cookeville, TN 38505

ABSTRACT

A nine node finite element is formulated in cylindrical coordinates and used to model the cross section of an infinite cylinder. Bone is an anisotropic material and in this instance is assumed to follow the mmm crystal class. The transformed material properties are discussed graphically. Natural frequencies for a cylinder having the properties of bone are computed.

INTRODUCTION

A review article by Natali and Meroi [1] compares the findings of several researchers concerning the anisotropic properties of bone. A review of the work of Van Buskirk *et al.* [2] and Ashman *et al.* [3] leads to the conclusion that bone can be modeled as an orthotropic material of crystal class mmm. In this paper we shall discuss the computation of natural frequencies of vibration for bone and assume that the bone can be analyzed as an elastic cylinder. All elastic constants vary with respect to a rotation about the longitudinal axis of the bone with the exception of constants that correspond to the longitudinal axis. In fact, there are additional nonzero elastic constants for orientations other than zero degrees. A nine node finite element is formulated in cylindrical r, θ coordinates and used to model the entire cross section of the cylinder. A cylindrical finite element offers an advantage over the traditional isoparametric element because the material constants are easily transformed in the same coordinate system as the element. Also, the final results for mode shapes are in the cylindrical coordinate system. The free vibration problem is formulated as an eigenvalue problem and standard methods of analysis are used to compute the natural frequencies.

MATERIAL CONSTANTS

The matrix of stiffness constants for crystal class mmm is given as a symmetric matrix containing 12 constants of which 9 are independent. A general representation for rotation about the 3 axis (cylinder axis) would be a matrix of 20 constants as follows.

$$C_{ij} = \begin{bmatrix} C_{11} & C_{12} & C_{13} & 0 & 0 & C_{16} \\ C_{21} & C_{22} & C_{23} & 0 & 0 & C_{26} \\ C_{31} & C_{32} & C_{33} & 0 & 0 & C_{36} \\ 0 & 0 & 0 & C_{44} & C_{45} & 0 \\ 0 & 0 & 0 & C_{54} & C_{55} & 0 \\ C_{61} & C_{62} & C_{63} & 0 & 0 & C_{66} \end{bmatrix} \tag{1}$$

For rotation about the r or θ axis additional constants are nonzero but they are not of interest in this application. In the standard orientation, $\theta = 0$, $C_{16} = C_{26} = C_{36} = C_{45} = 0$. The same is true for $\theta = \pi/2$, π and $3\pi/2$. In this analysis we wish to model the r, θ plane and include the effect of the variable material constants.

In this study the elastic stiffness constants that are used for computation were given in [3] and are reproduced in [1]. The magnitude of the stiffness constants are given below for the standard orientation ($\theta = 0$) in GPa ($N/m^2\ 10^9$).

$$C_{ij} = \begin{bmatrix} 18.00 & 9.98 & 10.10 & 0.00 & 0.00 & 0.00 \\ 9.98 & 20.20 & 10,70 & 0.00 & 0.00 & 0.00 \\ 10.10 & 10.70 & 27.60 & 0.00 & 0.00 & 0.00 \\ 0.00 & 0.00 & 0.00 & 6.23 & 0.00 & 0.00 \\ 0.00 & 0.00 & 0.00 & 0.00 & 5.61 & 0.00 \\ 0.00 & 0.00 & 0.00 & 0.00 & 0.00 & 4.52 \end{bmatrix} \tag{2}$$

The matrix array of elastic constants can be classified as crystal class mmm according to Nye [4]. The transformation is accomplished using the standard tensor transformation

$$C'_{pqrs} = a_{ip}a_{jq}a_{kr}a_{ls}C_{ijkl} \tag{3}$$

where a_{ij} is the transformation matrix

The variation of the material constants with respect to the θ coordinate is illustrated in Figs. 1 through 3. C_{11}, C_{22}, C_{12}, C_{13} and C_{23} are shown in Fig. 1. These material constants show only a slight variation. At $\theta = \pi/2$ the magnitudes of C_{11} and C_{22} are interchanged and similarly for C_{13} and C_{23}. However, C_{12} has the same value at $\theta = 0$ and $\pi/2$ but is maximum of 10.02 GPa at $\pi/4$. The behavior of C_{44}, C_{55} and C_{66} is shown in Fig. 2. Note that C_{66} is practically constant increasing to 4.56 GPa at $\pi/4$. The magnitudes of C_{44} and C_{55} are interchanged at $\pi/2$. The remaining nonzero elastic constants are shown in Fig. 3. All are positive in the first and third quadrants and negative in the second and forth quadrants. The maximum absolute magnitude of the constants occur at the center of each quadrant where $C_{16} = C_{26} = 0.55$ GPa, $C_{45} = 0.31$ GPa and $C_{36} = 0.30$ GPa.

CYLINDRICAL FINITE ELEMENT

A nine node finite element was formulated in cylindrical r, θ coordinates. The area integration was 0 to 2a in the r direction and 0 to 2α in the θ direction. The shape functions can be derived as Lagrange polynomials and are as follows

$$N_1 = (r-a)(r-2a)(\theta-\alpha)(\theta-2\alpha)/4a^2\alpha^2, \quad N_2 = -r(r-2a)(\theta-\alpha)(\theta-2\alpha)/2a^2\alpha^2,$$

$$N_3 = r(r-a)(\theta-\alpha)(\theta-2\alpha)/4a^2\alpha^2, \quad N_4 = -r\theta(r-a)(\theta-2\alpha)/2a^2\alpha^2,$$

$$N_5 = r\theta(r-a)(\theta-\alpha)/4a^2\alpha^2, \quad N_6 = -r\theta(r-2a)(\theta-\alpha)/2a^2\alpha^2,$$

$$N_7 = \theta(r-a)(r-2a)(\theta-2\alpha)/4a^2\alpha^2, \quad N_8 = -\theta(r-a)(r-2a)(\theta-2\alpha)/2a^2\alpha^2,$$

$$N_9 = r\theta(r-2a)(\theta-2\alpha)/a^2\alpha^2. \tag{4}$$

The area integration for shape functions and their derivatives was carried out using Gaussian quadratures.

A similar four node element was developed for cylindrical coordinates and used to solve Laplace's equation (heat transfer) for a cylindrical cross-section. Solutions from that analysis were compared to results using an isoparametric element formulation and the results were identical. However, that analysis was for isotropic material constants and merely served as a check on the formulation. The nine node element described by Eq. (4) was used to obtain natural frequencies for an infinite circular rod and the results were more accurate than the corresponding (equal number of nodal points) problem that was solved using four node elements. For anisotropic materials, the element, formulated in cylindrical coordinates, has an advantage over a standard nine node isoparametric element because the material constants can be easily transformed in the cylindrical coordinate system. Also, final results, such as mode shapes, are obtained directly in the cylindrical coordinate system.

The three dimensional equations of dynamic elasticity were used to formulate the finite element model. Displacement functions were assumed following the work of Achenbach and Fang [5] as

$$u_i = U_i \, cos(kz - \omega t), \quad v_i = V_i \, cos(kz - \omega t), \quad w_i = W_i \, sin(kz - \omega t)$$

thereby reducing the three dimensional problem to two dimensions.

Natural frequencies were obtained by solving the standard eigenvalue problem and comparison with analytical solutions and other numerical solutions are given by Buchanan and Peddieson [6]. In this study 24 elements were used, eight in the θ direction and three in the radial direction. The elastic constants were computed for each element using Eq. (3) and a value of θ corresponding to the center of the element. Hence, eight sets of elastic constants were used. Frequencies for a solid cylinder with k = 0 are given in Table I and correspond to uncoupled longitudinal or coupled radial and tangential mode shapes. Frequencies are given in terms of the nondimensional parameter

$$\Omega = \omega/(c_{44}/\rho b^2)^{1/2}$$

REFERENCES

1. Natali, A. N. and Meroi, E. A., "A review of the biomechanical properties of bone as a material", J. Biomed. Eng., <u>11</u>, 266-276 (1989).
2. Van Buskirk, W. C., Cowin, S. C. and Ward, R. N., "Ultrasonic measurement of orthotropic elastic constants of bovine femoral bone", J. Biomech. Engrg., <u>103</u>, 67-72 (1981).
3. Ashman, R. B., Cowin, S. C., Van Buskirk, W. C. and Rice, J. C., "A continuous wave technique for the measurement of the elastic properties of cortical bone", J. Biomech., <u>17</u>, 349-361 (1984).
4. Nye, J. F., <u>Physical Properties of Crystals</u>, Oxford, University Press, (1969).
5. Achenbach, J. D. and Fang, S. J., "Asymptotic analysis of the modes of wave propagation in a solid cylinder", J. Acoust. Soc. Am., <u>47</u>, 1282-1289 (1970).
6. Buchanan, G. R. and Peddieson, J., "Vibration of infinite piezoelectric cylinders with anisotropic properties using cylindrical finite elements", IEEE Trans. Ultra., Ferroel. Freq. Cont., submitted.

TABLE I. NATURAL FREQUENCIES FOR A SOLID CYLINDER, k=0.

1.794	long.
2.033	r and t
2.451	r and t
2.981	long.
3.220	r and t
3.769	t only
3.850	r and t

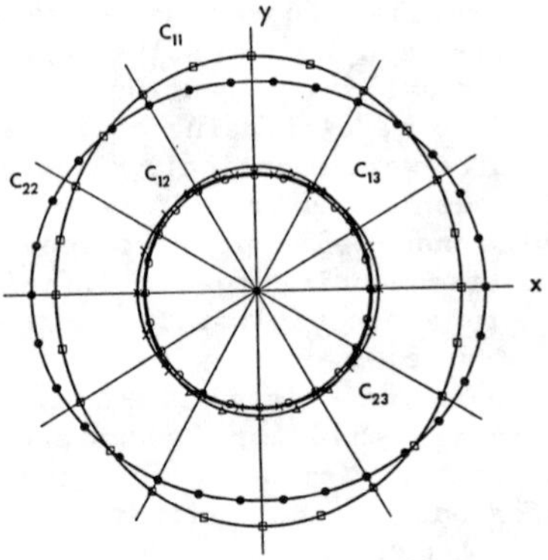

Figure 1.

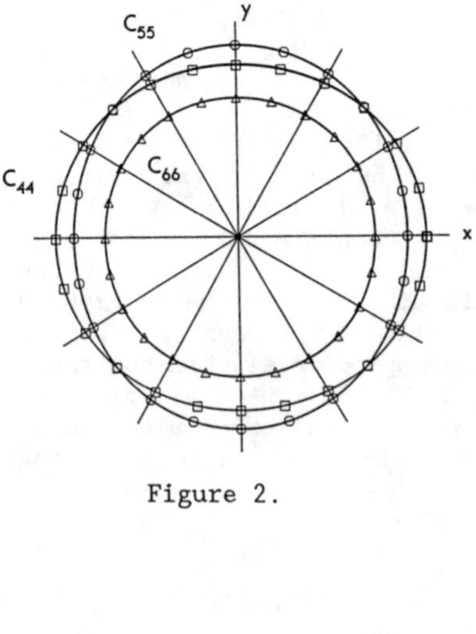

Figure 2.

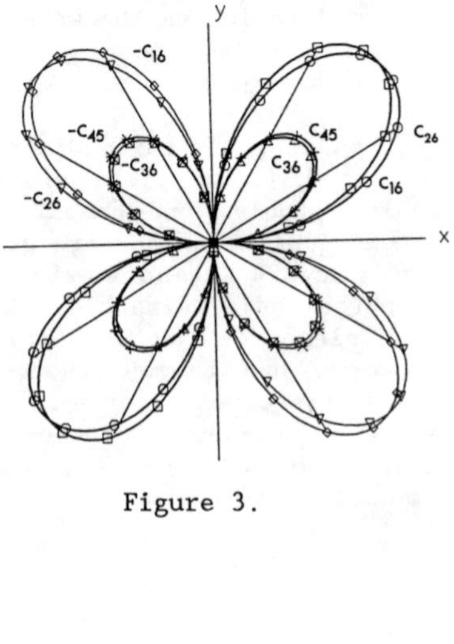

Figure 3.

THE USE OF BUBBLE FUNCTIONS FOR THE SOLUTION OF PROBLEMS WITH INCOMPRESSIBILITY

M.S. Gadala, Director of Research, EMRC
1607 E. Big Beaver, Troy, Mich., 48083
(Also Adjunct. Assoc. Prof. at University of Michigan-Dearborn)

ABSTRACT

A new approach for analyzing incompressible hyperelastic and rubber-like material behavior is discussed. The approach is based on extending the linear analysis bubble function concept into nonlinear analysis. A unified numerical treatment for handling various constitutive relations of hyperelastic materials is presented.

INTRODUCTION

In this paper we consider the formulation and solution of problems involving incompressible hyperelastic material. For such material a strain energy density function W exists and this function may be considered dependent on I_1, I_2, and I_3 which are the invariants of the Cauchy-Green deformation tensor **C**, and where I_3 will be unity for incompressible material [1-4]. For the problem statement, the total potential of the body may be expressed as:

$$\Pi = \int_V W(I_1, I_2)\, dV - \int_V \boldsymbol{b} \cdot \boldsymbol{u}\, dV - \int_{A_\sigma} \boldsymbol{T} \cdot \boldsymbol{u}\, dA \tag{1}$$

where $\boldsymbol{b}$ is the prescribed body forces per unit undeformed volume, $\boldsymbol{T}$ is the prescribed surface traction per unit undeformed surface area and $\boldsymbol{u}$ is the deformation field. The resulting deformation field $\boldsymbol{u}$ is a stationary point of the functional Π, satisfying the incompressibility constraint, $|\boldsymbol{F}| - 1 = 0$, where $|\boldsymbol{F}|$ is the determinant of the deformation gradient tensor, $\boldsymbol{F}$.

We concentrate on two methods of formulation; the average constraint and the new bubble function approach. A more complete review article may be found in reference [4].

FORMULATION METHODS

Average Constraint Approach

In this approach [2,3], the constraint function is *weakened* on the element level. This amounts to finding a stationary point $\boldsymbol{u}$ for the functional:

$$W_a = W + \frac{1}{2\varepsilon}\int_V \left[\frac{1}{V}\int_V (I_3 - 1)dV\right]^2 dV \tag{2}$$

where the constraint equation has been first averaged on the element level through the first volume integral. The variational incremental form of the second term in Eqn. (2), the dilatational energy term U_d, gives

$$\Delta(\delta U_d) = \frac{1}{\varepsilon}\int_V \frac{1}{V^2}\left[\int_V (I_3 - 1)\, dV\right]\left[\int_V \delta E_{ij}(4Q_{ijkl})\, \Delta E_{kl} dV\right] dV$$
$$+ \frac{1}{\varepsilon}\int_V \frac{1}{V^2}\left[\int_V (\delta E_{ij}(2H_{ij})\, dV\right]\left[\int_V \Delta E_{ij}(2H_{ij})\, dV\right] dV$$
$$+ \frac{1}{\varepsilon}\int_V \frac{1}{V^2}\left[\int_V (I_3 - 1)\, dV\right]\left[2H_{ij}\, \delta(\Delta\eta_{ij})\, dV\right] dV \qquad (3)$$

In reference [2,3] averaging of the constraint equation is dealt by projecting the constraint function over the basis of the constraint space. The method is directly applicable to low- as well as high-order elements and seems to be independent of element distortion or the amount of strain the element is undergoing.

Bubble Function Approach

In a multifield or mixed approach, let m and n stand for the number of degrees of freedom, DOF, of the original and the auxiliary variables $\boldsymbol{u}$ and p, respectively. Zienkiewicz et al. [7] define a 'freedom number β'; where $\beta = \Delta m/\Delta n$, by which element efficiency and performance may be examined. For optimum element performance in two-dimensional problems, the freedom number, β, is required to be greater than 2. This is similar to the weakening of the constraint equation in the above approach.

The idea of using bubble functions to handle the incompressibility behavior seems to be logical extension of the above discussion. The bubble function will have the effect of increasing the degrees of freedom for the original variable $\boldsymbol{u}$ over those for the auxiliary variable p and hence increasing the freedom number β. This extension is simple and straightforward in linear analysis. In nonlinear analysis, however, the bubble function or nodeless degrees of freedom have to be condensed and recovered in each iteration. Moreover, these nodeless degrees of freedom should not be associated with any external or residual forces. To achieve this: Starting with the final equilibrium equations in the partitioned form,

$$\begin{bmatrix} \boldsymbol{K}_{uu} & \boldsymbol{K}_{ua} \\ \boldsymbol{K}_{au} & \boldsymbol{K}_{aa} \end{bmatrix} \begin{bmatrix} \Delta \boldsymbol{u} \\ \Delta \boldsymbol{a} \end{bmatrix} = \begin{Bmatrix} \Delta \boldsymbol{f}_u \\ \Delta \boldsymbol{f}_a \end{Bmatrix} \qquad (4)$$

where $\Delta\boldsymbol{u}$ is the incremental displacement vector, $\Delta\boldsymbol{a}$ is the incremental displacement vector corresponding to the bubble function or nodeless DOF, and $\Delta\boldsymbol{f}_u$ and $\Delta\boldsymbol{f}_a$ are the corresponding incremental force vectors. A special loop has to be carried out in the stress calculations routine in which the following iterations are performed:

- Recover the condensed DOF from Eqn. (4) by assuming $\Delta f_a^{(i)} = 0$, where (i) is a local iteration number for this loop in the stress calculation routine.

- Calculate new element strain-displacement $\boldsymbol{B}$ matrix and update the element stiffness to obtain $\boldsymbol{K}_{uu}^{(i+1)}$, $\boldsymbol{K}_{aa}^{(i+1)}$ and $\boldsymbol{K}_{ua}^{(i+1)}$

- Calculate new equivalent nodal forces, $\Delta \boldsymbol{f}_a^{(i+1)}$, and corrected displacements

$$\Delta \boldsymbol{a}^{(i+1)} = \left(\boldsymbol{K}_{aa}^{-1}\right)^{(i+1)} \left(\Delta \boldsymbol{f}_a^{(i+1)} - \boldsymbol{K}_{au}^{(i+1)}\, \Delta \boldsymbol{u}^{(i)}\right) \tag{5}$$

- Continue the process until an appropriate measure for the convergence of $\Delta \boldsymbol{a}$ and/or Δf_a is reached.

An alternative method may be also used in which the unbalanced forces are used to calculate corrective displacements that will be used in the next step, hence eliminating the need for the local iterative loop.

UNIFIED NUMERICAL TREATMENT FOR STRAIN ENERGY FUNCTIONS

The constitutive relation may be given by:

$$S_{ij} = \frac{\partial W}{\partial E_{ij}} \quad , \quad D_{ijkl} = \frac{\partial^2 W}{\partial E_{ij}\, \partial E_{kl}} \tag{6}$$

where S_{ij} are the components of the second Piola-Kirchhoff stress tensor, E_{ij} are the components of the Green-Langrange strain tensor, and D_{ijkl} are the components of the fourth order stress-strain constitutive tensor, and $\Delta S_{ij} = D_{ijkl}\, \Delta E_{kl}$

We consider the generalized Mooney-Rivlin model [6] for strain energy function as an example (other forms of W may be easily casted in the same procedure):

$$W = \sum_{r=0}^{\infty} \sum_{s=0}^{\infty} A_{rs}\, (I_1 - 3)^r\, (I_2 - 3)^s \tag{7}$$

where A_{rs} are material constants.

To obtain the stresses, we define the following tensors:

$$\frac{\partial I_3}{\partial C_{ij}} = \frac{1}{2}\, e_{imp}\, e_{jnq}\, C_{mn}\, C_{pq} = H_{ij} \tag{8}$$

$$\hat{C}_{ij} = I_l\, \delta_{ij} - C_{ij} \tag{9}$$

where e_{ijk} is the permutation tensor and δ_{ij} is the kronecker delta. It is easily verified that:

$$S_{ij} = \alpha_1\, \delta_{ij} + \alpha_2\, C_{ij} + \alpha_3\, H_{ij} \tag{10}$$

where

$$\left.\begin{aligned}\alpha_1 &= \sum_{r=0}^{\infty}\sum_{x=0}^{\infty} 2r\,C_{rs}\left(I_1-3\right)^{r-1}\left(I_2-3\right)^{s}\\ \alpha_2 &= \sum_{r=0}^{\infty}\sum_{x=0}^{\infty} 2s\,A_{rs}\left(I_1-3\right)^{r}\left(I_2-3\right)^{s-1} \quad , \quad \alpha_3 = 0\end{aligned}\right\} \tag{11}$$

To obtain the constitutive tensor, we define the following additional tensors:

$$\frac{\partial \hat{C}_{ij}}{\partial C_{mn}} = \frac{\partial}{\partial C_{mn}}\left[C_{kk}\,\delta_{ij} - C_{ij}\right] = \delta_{ij}\,\delta_{im} - \delta_{im}\,\delta_{jn} = \hat{Q}_{ijmn} \ , \tag{12}$$

$$\frac{\partial H_{ij}}{\partial C_{mn}} = e_{imp}\,e_{jnq}\,C_{pq} = Q_{ijmn} \ , \ \text{and} \tag{13}$$

$$\mathcal{T}_{ijkl} = \frac{1}{2}\left[\hat{C}_{ij}\,\delta_{kl} + \delta_{ij}\,\hat{C}_{kl}\right] \tag{14}$$

With the above definitions, the constitutive relation tensor may be written in the following general form

$$D_{ijkl} = \beta_1\,\delta_{ij}\,\delta_{kl} + \beta_2\,\hat{C}_{ij}\,\hat{C}_{kl} + \beta_3\,H_{ij}\,H_{kl} + \beta_4\,Q_{ijkl} + \beta_5\,\hat{Q}_{ijkl} + \beta_6\,\mathcal{T}_{ijkl} \tag{15}$$

where

$$\left.\begin{aligned}&\beta_3 = \beta_4 = 0 \ , \ \beta_5 = \alpha_2 \ (\text{of Eqn. 11})\\ &\beta_1 = \sum_{r=0}^{\infty}\sum_{s=0}^{\infty} 4\,r\,(r-1)\,A_{rs}\,(I_1-3)^{r-2}\,(I_2-3)^{s}\\ &\beta_2 = \sum_{r=0}^{\infty}\sum_{s=0}^{\infty} 4\,s\,(s-1)\,A_{rs}\,(I_1-3)^{r}\,(I_2-3)^{s-2}\\ &\beta_6 = \sum_{r=0}^{\infty}\sum_{s=0}^{\infty} 8\,rs\,A_{rs}\,(I_1-3)^{r-1}\,(I_2-3)^{s-1}\end{aligned}\right\} \tag{16}$$

NUMERICAL EXAMPLE

Formulations mentioned above have been incorporated into the finite element code NISA [7]. In the following section a numerical example is discussed.

A simply supported flat circular plate subjected to a piecewise linear varying external follower pressure of up to 45 psi at its bottom surface is analyzed. The material is of the Mooney-Rivlin type with C1 = 80 psi, C2 = 20 psi and ν = 0.499. The problem is highly nonlinear and the response departs quickly from linear analysis. Different cases condsidered are: (i) Twenty 8-node axisymmetric solid with reduced integration, (ii) Forty 4-node axisymmetric solid with average constraint, (iii) Twenty 4-node axisymmetric solid with average constraint, and (iv) Twenty 4-node axisymmetric solid with bubble function. Preliminary analyses showed extreme sensitivity of the predicted response to the choice of increment size. It was not possible to predict the response of the plate beyond 43 psi where the cross-sectional areas of the elements shrink to nearly zero (Fig. (2)). The load-deflection curves based on above various formulations are given in Fig. (1) and are compared with those given by Oden [6]. Considering that the results of [6] are obtained by digitizing the curve available in literature, very good agreement is indicated. It is noted that, in this example, and in the high end of the pressure range, the bubble function approach tends to consume more iterations than the average constraint method. Convergence enhancement for the new approach is under investigation.

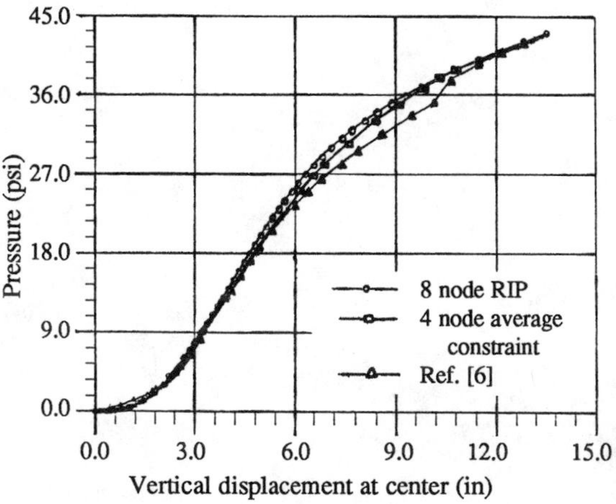

FIG. 1. Pressure-displacement response

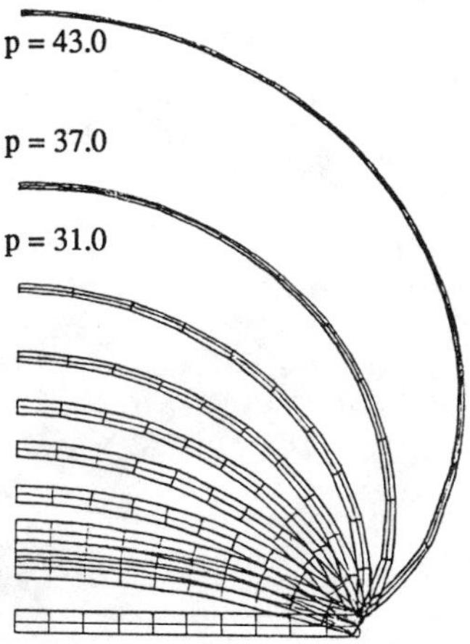

FIG. 2. Bending and infl-ation of a circular plate

REFERENCES:

1. R. S. Rivlin, Large elastic deformations of isotropic materials; IV: Further developments of the general theory, Philos. Trans. Roy. Soc. London Ser., A241 (1948) 379-397.

2. S. Cescotto and G. Fonder, A finite element approach for large strains of nearly incompressible rubber-like materials, Int. J. Solids & Structures, 15(1979) 589-605.

3. M. Bercovier, Y. Hasbani, Y. Gilon and K. J. Bathe, On a finite element procedure for nonlinear incompressible elasticity, In. Hybrid & Mixed Finite Element Methods, Ed. S. N. Atluri et. al., John Wiley & Sons, (1983) 497-517.

4. M. S. Gadala, Numerical solution of nonlinear problems of continua- II: Survey of incompressibility constraints and software aspects, Computers & Structures, 22(1986) 841-855.

5. O. C. Zienkiewicz, R. L. Taylor and J. M. W. Baynham, Mixed and irreducible formulations in finite element analysis, In: Hybrid & Mixed Finite Element Methods, E. S. N. Atluri et al., John-Wiley & Sons, (1983) 405-431.

6. J. T. Oden, Finite elements of nonlinear continua, McGraw Hill, New York, (1972).

7. "NISA User's Manual", Engineering Mechanics Research Corporation-EMRC, Troy, Michigan, (1990).

DETERMINATION OF ELASTIC CONSTANTS IN THE COUPLE-STRESS THEORY OF FIBER-REINFORCED COMPOSITES*

JINGMING MA+ and **MEIFENG LE**#
+Dept. of Civil Engineering; #Dept. of Engineering Mechanics, Xi'an Jiaotong University, Xi'an 710049, The People's Republic of China

ABSTRACT

The fiber-reinforced composite is treated as a continuous elastic medium which is capable of suffering the couple-stress. For a plane stress state in a composite lamina, the couple-stress is introduced by taking account of the bending deformation of fibers.

INTRODUCTION

The classical elasticity takes no account of the influence of the micro-structure of materials, but it is outstanding as the characteristic length of the physical problem approaches that of the material. The direct investigation at the micro-structure lever is very difficult to a certain extent, so it is necessary to establish the continuum theory taking account of the influence of the micro-structure.

The couple-stress theory was originated by Cosserat in 1909 and often bears his name. In the Cosserat continuum each point possesses six degrees of freedom like a rigid body and the stress tensor is unsymmetrical. More developments in theoretical studies may be found in works on the micro-structure theory[1] and the micropolar elasticity[2]. Unfortunately, the investigations on the couple-stress theory for engineering materials have been scarce.

The present paper deals with the lamina of fiber-reinforced composites under the plane stress state by the couple-stress theory. The couple-stress is introduced by taking account of the bending deformation of fibers. The stress-strain relations can be obtained by the classical elasticity theory of composite materials and the Timoshenko theory of beams, and elastic constants are also determined. Finally, a numerical example of the couple-stress theory will be found by the finite element method.

COUPLE-STRESS THEORY

For a composite lamina with unidirectional fibers, it is assumed that in the principal material directions the couple-stress M_{13} exists(see Fig.1). Then,the motion of each point of the lamina will be specified not only by the displacement components(U_1,U_2) but also the rotation component(ω_3).According to the

* Supported by National Natural Science Foundation of China

couple-stress theory in linear elasticity[3], we obtain the strain energy density of the lamina as

$$W = (1/2)T_{ij}\,u_{ij} + (1/2)M_{13}\,\omega_{13} \quad (1)$$

The strain components are defined as

$$u_{ij} = E_{ij} + \varepsilon_{kij}\,(\vartheta_k - \omega_k) \quad (2)$$

$$\omega_{13} = \partial\omega_3/\partial X_1 \quad (3)$$

where

$$E_{ij} = (1/2)(\partial U_i/\partial X_j + \partial U_j/\partial X_i) \quad (4)$$

$$\varepsilon_{kij}\,\vartheta_k = (1/2)(\partial U_j/\partial X_i - \partial U_i/\partial X_j) \quad (5)$$

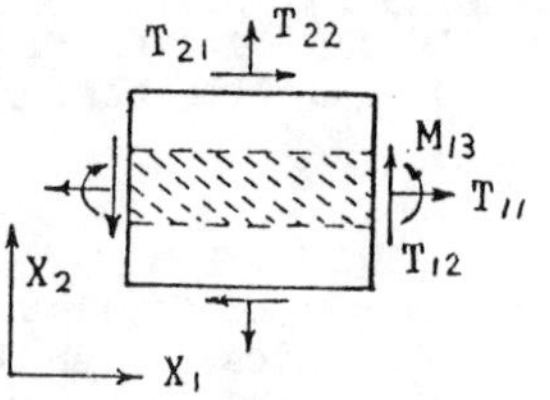

Fig.1 Components of stress and couple-stress

The quantities E_{ij} are the strain components in the classical elasticity theory and ϑ_k the rigid rotation components.

The presence of the couple-stress M_{13} renders the stress tensor unsymmetrical. We express the unsymmetrical real stress tensor T_{ij} in terms of the symmetrical tensor $\mathring{T}_{ij}$ and another unsymmetrical tensor $\bar{T}_{ij}$, i.e.

$$T_{ij} = \mathring{T}_{ij} + \bar{T}_{ij} \quad (6)$$

where $\bar{T}_{ij}$ are defined as

$$\bar{T}_{12} = T_{12} - T_{21} \quad , \quad \text{others} \quad \bar{T}_{ij} = 0 \quad (7)$$

Introducing Eqs.(6-7) into Eq.(1) and using Eqs.(2-5), we obtain

$$W = (1/2)\mathring{T}_{ij}E_{ij} + (1/2)\bar{T}_{12}(\partial U_2/\partial X_1 - \omega_3) + (1/2)M_{13}(\partial\omega_3/\partial X_1) \quad (8)$$

EFFECTIVE STIFFNESS THEORY

The effective stiffness theory of fiber-reinforced composites may be found in the papers by Bartholomew, Torvik[4] and Aboudi[5]. In this paper we make use of the approximate theory presented by Achenbach and Herrmann[6] who allow the interaction of the fiber and the matrix deformations only through the displacement of the center line of the fiber and the displacement of the matrix, consider the matrix as a fictitious matrix, and take account of deformation of the fibers in flexure, torsion and extension. However, we consider the fictitious matrix as a effective orthotropic material whose material constants are the same as those of the composite material in classical elasticity, and only take account of the bending deformation of fibers in the X_1-X_2 plane(see Fig.2). According to the effective stiffness theory[6], we obtain the strain energy density as

$$V = (1/2)\sigma_{ij}E_{ij} + (1/2)(Q_2/D^2)(\partial U_2/\partial X_1 - \omega_3) + (1/2)(M_3/D^2)(\partial\omega_3/\partial X_1) \quad (9)$$

where σ_{ij} are the stresses in the effective matrix, and those stresses are related to the strains E_{ij} through the formula in the classical elasticity of composite materials[7], i.e.

$$\begin{Bmatrix}\sigma_{11}\\ \sigma_{22}\\ \sigma_{12}\end{Bmatrix} = \begin{bmatrix}Q_{11} & Q_{12} & 0\\ Q_{12} & Q_{22} & 0\\ 0 & 0 & Q_{66}\end{bmatrix}\begin{Bmatrix}E_{11}\\ E_{22}\\ 2E_{12}\end{Bmatrix} \quad (10)$$

The bending moment M_3 and the shear force Q_2 can be obtained through the Timoshenko theory of beams, i.e.

$$M_3 = E_f I(\partial\omega_3/\partial X_1) \quad (11)$$

$$Q_2 = \mu G_f A(\partial U_2/\partial X_1 - \omega_3) \quad (12)$$

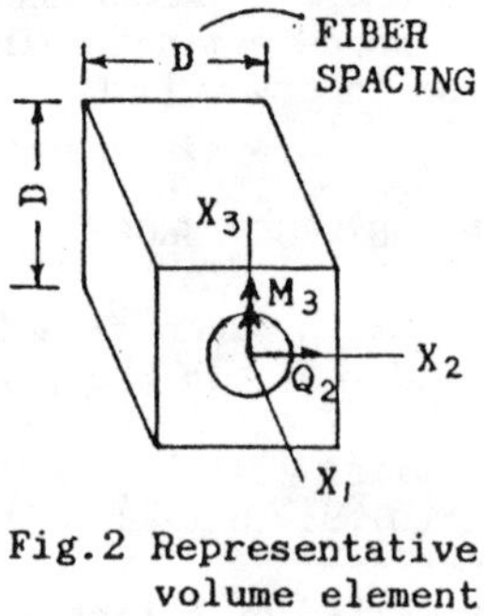

Fig.2 Representative volume element

where

$$I = \pi d^4/64 \tag{13}$$

In Eqs.(11-12), E_f, G_f and A are Young's modulus, the shear modulus and the cross-sectional area (circular area with diameter d) of the fiber, respectively. The coefficient μ is the Timoshenko shear coefficient.

STRESS-STRAIN RELATIONS

It is noted that the strain energy density of the couple-stress theory is corresponding to that of the effective stiffness theory(see Eqs.(8-9)). Then we can established the stress-strain relations of the couple-stress theory in the principal material directions through the effective stiffness theory, i.e.

$$\{\mathring{T}_{11} \quad \mathring{T}_{22} \quad \mathring{T}_{12}\}^T = [Q]\{E \quad E \quad E\}^T \tag{14}$$

$$M_{13} = Q_b\, \omega_{13} \tag{15}$$

$$\bar{T}_{12} = Q_c\, \gamma_c \tag{16}$$

where [Q] is the matrix in Eq.(10), and

$$\omega_{13} = \partial\omega_3/\partial X_1 \tag{17}$$

$$\gamma_c = \partial U_2/\partial X_1 - \omega_3 \tag{18}$$

In Eqs.(15-16), elastic constants Q_b and Q_c are determined as

$$Q_b = V_f E_f d^2/16 \tag{19}$$

$$Q_c = \mu V_f G_f \tag{20}$$

where V_f is the fiber volume fraction. Note, the constant Q_b has something to do with the fiber diameter d or the fiber spacing D (since $d^2=4V_f D^2/\pi$), and the fiber spacing is the characteristic length of composite materials.

Moreover, we can also establish the stress-strain relations of the couple-stress theory in an arbitrary coordinate system.

NUMERICAL EXAMPLE

The finite element method can be applied to the numerical solution of the couple-stress theory above mentioned, and the

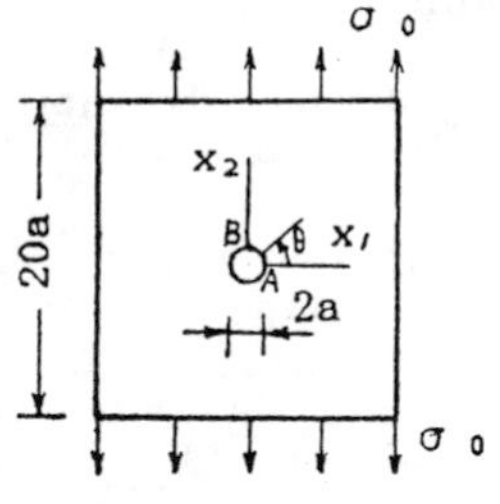

Fig.3 Circular hole in a field of simple tension

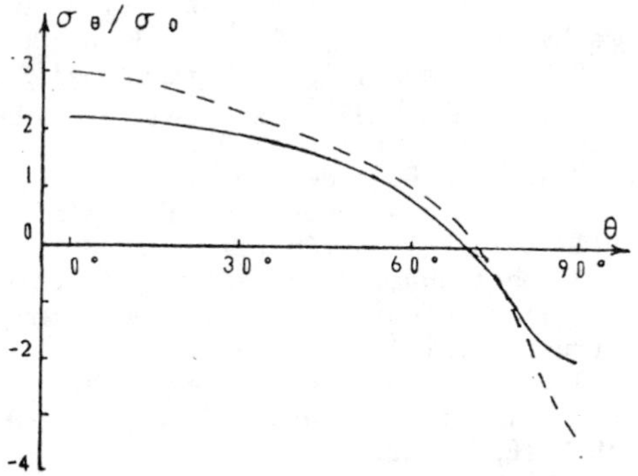

Fig.4 σ_θ-distribution at the surface of the hole (- - - - the analytical solution by classical elasticity[8])

primary variables are the displacement components(u_1, u_2) and the rotation component(ω_3).

Consider a cylindrical hole in a field of simple tension (see Fig.3). The 8-node isoparametric element is used in the numerical analysis. In the carbon/epoxy composite lamina the fiber direction is the x_1-direction. The material constants of the lamina are

$$E_1=10.2*10^5 \text{ kg/cm}^2, \quad E=0.927*10^5 \text{ kg/cm}^2,$$
$$G_{12}=0.454*10^5 \text{ kg/cm}^2, \quad \nu_{12}=0.331$$

and the material constants of the fiber are

$$E_f=2.13*10^6 \text{ kg/cm}^2, \quad G_f=8.07*10^5 \text{ kg/cm}, \quad \nu_f=0.32$$

Fig.4 displays the distribution of the hoop stress σ_θ at the surface of the hole as a/d=1. With the couple-stress taken into account, the stress concentration factor diminishes. For example, if a/d=1, the reductions of the ratio σ_θ / σ_0 at points A and B are 26 and 40 percent, respectively. The numerical example also shows that as a/d becomes large the results of the couple-stress theory approach those of the classical elasticity theory.

CONCLUDING REMARKS

By taking account of the bending deformation of fibers, the fiber-reinforced composites are treated as the elastic continuum which is capable of suffering the couple-stress. By means of the special method of resolving unsymmetrical stress tensor and the effective stiffness theory, the stress-strain relations of the couple-stress theory can be obtained and the elastic constants are also determined. That couple-stress theory of fiber-reinforced composites has a clear conception and is easy to obtain the numerical solution by the finite element method.

REFERENCES

1. R.D. Mindlin, "Micro-structure in linear elasticity", Arch. Rational Mech. Anal. 16(1), 51-78(Mar. 1964).
2. A.C. Eringen, "Linear theory of micropolar elasticity", J. Math. Mech. 15(6), 909-923 (June 1966).
3. V.D. Kupradze, Three-Dimensional Problems of the Mathematical Theory of Elasticity and Thermoelasticity (North-Holland Publishing Company, 1979).
4. R.A. Bartholomew and P.J. Torvik, "Elastic wave propagation in filamentary composite materials", Int. J. Solids Structures 8(12), 1389-1405 (Dec. 1972).
5. Jacob Aboudi, "Generalized effective stiffness theory for the modeling of fiber-reinforced composites", Int. J. Solids Structure 17(10), 1005-1018 (Oct. 1981).
6. J.D. Achenbach and G. Herrmann, "Dispersion of free harmonic waves in fiber-reinforced composites", AIAA Journal 6(10), 1832-1836 (Oct. 1968).
7. R.M. Jones, Mechanics of Composite Materials (Scripta Book Company, Washington 1975).
8. С.Г. Лехницкий, Анизотропные Пластинки (Гостехиздат 1957).

SHEAR BAND FORMATION IN HYPOPLASTICITY

Z. Sikora and W. Wu

Institute of Soil Mechanics and Rock Mechanics, Karlsruhe University, FRG

ABSTRACT

Localized bifurcation for hypoplastic constitutive equations is presented. Both the deformation gradient and the incremental constitutive stiffness can be discontinuous across the shear band. For different incremental constitutive stiffness inside and outside the shear band a necessary condition for bifurcation in form of algebraic equations is obtained. The bifurcation condition is included in a FE-code and a numerical example is presented.

STATICS AND KINEMATICS OF SHEAR BAND

In the present paper, a general approach for shear band analysis with a fairly general form of rate independent hypoplastic constitutive equations is presented. For the mathematical definition of hypoplasticity the reader is referred to [5]. Emphasis is placed on the incremental nonlinearity of the constitutive equations and shear band kinematics.

Consider a solid body subjected to uniform Cauchy stress $\mathbf{T}_o$. For continued uniform deformation, the deformation velocity and velocity gradient fields are denoted by $\mathbf{v}_o$ and $(\partial\mathbf{v}/\partial\mathbf{x})_o$ respectively.

We try to check whether the field equations for homogeneous deformation, *trivial deformation mode*, admit an alternative velocity gradient field

$$(\partial\mathbf{v}/\partial\mathbf{x})_i = (\partial\mathbf{v}/\partial\mathbf{x})_o - [\partial\mathbf{v}/\partial\mathbf{x}]. \tag{1}$$

The subscripts o and i signify *outside* and *inside* of the shear band. The brackets denote the jumps across a shear band.

At the onset of the shear band, two conditions, the kinematic compatibility and static equilibrium conditions, must be satisfied (s. [5]). For the shear band kinematics we define after Hadamard [1] the following discontinuity field of order one

$$[\mathbf{v}] = \mathbf{0}, \quad \text{at least one} \quad \left[\frac{\partial v_i}{\partial x_j}\right] \neq 0 \quad \text{then} \quad [\partial\mathbf{v}/\partial\mathbf{x}] = \mathbf{g}\otimes\mathbf{n}. \tag{2}$$

After Hill [2] we assumed that all components of $[\mathbf{v}]$ vanish, thus the resulting jump of the velocity gradient and consequently the jump of the stretching $[\mathbf{D}]$ and spin $[\mathbf{W}]$ tensors can be written out in the following matrix form

$$\left[\frac{\partial\mathbf{v}}{\partial\mathbf{x}}\right] = \begin{pmatrix} 0 & g_1 & 0 \\ 0 & g_2 & 0 \\ 0 & g_3 & 0 \end{pmatrix} [\mathbf{D}] = \frac{1}{2}\begin{pmatrix} 0 & g_1 & 0 \\ g_1 & 2g_2 & g_3 \\ 0 & g_3 & 0 \end{pmatrix} [\mathbf{W}] = \frac{1}{2}\begin{pmatrix} 0 & g_1 & 0 \\ -g_1 & 0 & -g_3 \\ 0 & g_3 & 0 \end{pmatrix}. \tag{3}$$

The stress and stress rate tensors in the coordinate systems $\mathbf{x}$ and $\bar{\mathbf{x}}$ are related by

$$\mathbf{T} = \mathbf{Q}\bar{\mathbf{T}}\mathbf{Q}^{\mathrm{T}}, \quad \dot{\mathbf{T}} = \mathbf{Q}\dot{\bar{\mathbf{T}}}\mathbf{Q}^{\mathrm{T}} \quad \text{and} \quad \mathbf{Q} = \begin{pmatrix} \cos\theta & \sin\theta & 0 \\ -\sin\theta & \cos\theta & 0 \\ 0 & 0 & 1 \end{pmatrix}. \tag{4}$$

θ in equation (4) denotes the rotation angle from the reference coordinate system $(\bar{x}_1, \bar{x}_2, \bar{x}_3)$ to the band coordinate system $(x_1, x_2, \bar{x}_3)$.

The equilibrium condition across the shear band can be written as follows, see [5]:

$$[\dot{\mathbf{t}}] = [\dot{\mathbf{T}}]\mathbf{n} = \mathbf{0}. \tag{5}$$

CONSTITUTIVE EQUATION

In the present paper, the following form of the hypoplastic constitutive equations is used for the bifurcation analysis, s. [5]:

$$\mathring{\mathbf{T}} = \mathbf{L}(\mathbf{T}, \mathbf{D}) + \mathbf{N}(\mathbf{T})\|\mathbf{D}\| \tag{6}$$

with the explicit form as

$$\mathring{\mathbf{T}} = \frac{\mathrm{C}_1}{2}(\mathbf{TD} + \mathbf{DT}) + \mathrm{C}_2\mathrm{tr}(\mathbf{T})\mathbf{D} + \mathrm{C}_3\mathbf{T}\|\mathbf{D}\| + \mathrm{C}_4\mathrm{tr}(\mathbf{T})\|\mathbf{D}\|\mathbf{I} \tag{7}$$

with the material parameters $C_1 = -66.3$, $C_2 = -44.6$, $C_3 = -149.6$, $C_4 = 49.9$ calibrated according to triaxial tests on dense Karlsruhe medium sand. $\mathbf{I}$ in equation (7) denotes the unit tensor. $\|\cdot\|$ stands for the euclidean norm. $\mathbf{L}(\mathbf{T}, \mathbf{D})$ and $\mathbf{N}(\mathbf{T})$ in equation (6) are isotropic tensor–valued functions. $\mathbf{L}(\cdot, \cdot)$ and $\mathbf{N}(\cdot)$ are bilinear and linear or nonlinear operators respectively. Because of rate independence, $\mathbf{L}(\mathbf{T},\mathbf{D})$ is necessarily homogeneous of degree one in $\mathbf{D}$. Concrete forms of $\mathbf{L}$ and $\mathbf{N}$ of (7) can be obtained by invoking the representation theorems for isotropic tensor–valued functions [4]. Equation (6) defines a class of incrementally nonlinear constitutive equations. It was shown in [3] that equation (7) captures many of the salient features pertinent to granular materials.

SHEAR BAND ANALYSIS

In this analysis, the localization is regarded as an instability in the constitutive equation. The jump of the Jaumann stress rate in the direction $\mathbf{n}$ can be obtained as follows

$$[\mathring{\mathbf{T}}]\mathbf{n} = [\dot{\mathbf{T}}]\mathbf{n} + \mathbf{T}[\mathbf{W}]\mathbf{n} - [\mathbf{W}]\mathbf{T}\mathbf{n}. \tag{8}$$

The jump of the Jaumann stress rate based on the constitutive equation can be written in the following expression

$$[\mathring{\mathbf{T}}] = \mathbf{L}(\mathbf{T}, [\mathbf{D}]) + \mathbf{N}(\mathbf{T})(\|\mathbf{D}_o\| - \|\mathbf{D}_i\|). \tag{9}$$

The quantity $\|\mathbf{D}_o\| - \|\mathbf{D}_i\|$ represents the magnitude jump of the stretching and is designated as $\lambda_{[D]}$ in the sequel. Combining (8) and (9) we obtain a fundamental equation for the emergence of a shear band:

$$\mathbf{T}[\mathbf{W}]\mathbf{n} - [\mathbf{W}]\mathbf{T}\mathbf{n} - \mathbf{L}(\mathbf{T}, [\mathbf{D}])\mathbf{n} - \lambda_{[D]}\mathbf{N}(\mathbf{T})\mathbf{n} = \mathbf{0}. \tag{10}$$

A perusal of equation (10) reveals that the trivial solution of $\mathbf{g}$ is always met. From mathematical point of view, the shear band analysis consists in searching for such a stress tensor $\mathbf{T}$ for which the non-trivial solution exists. In view of (3) equation (10) can be written out in the following matrix form:

$$(\mathbf{R}(\theta) - \mathbf{A}(\theta))\mathbf{g} = \lambda_{[D]}\mathbf{N}(\mathbf{T})\mathbf{n} \tag{11}$$

where $\mathbf{T}[\mathbf{W}]\mathbf{n} - [\mathbf{W}]\mathbf{T}\mathbf{n} = \mathbf{R}(\theta)\mathbf{g}$ and $\mathbf{L}(\mathbf{T}, [\mathbf{D}])\mathbf{n} = \mathbf{A}(\theta)\mathbf{g}$.

Since $\|\mathbf{N}(\mathbf{T})\mathbf{n}\| \neq 0$ for all values of θ, it is necessary that for a non-trivial solution, the matrix $(\mathbf{R}(\theta) - \mathbf{A}(\theta))$ is not singular for all θ. This requirement places a further restriction on the suitable forms of the constitutive equation (7). It can be easily shown that the hypoplastic constitutive equation used in this paper fulfils this requirement.

In solving equation (11), an additional equation is needed, since this equation contains four unknowns g_1, g_2, g_3 and θ, see (3). The additional equation can be obtained from normalization process of this problem according to the velocity vektor or streching tensor respectively in the following form

$$\|\mathbf{D}\| = 1 \qquad \text{or} \qquad \|\mathbf{g}\| = 1. \tag{12}$$

A dilatancy angle ν for the jump of stretching can be calculated according to the following formula:

$$-\frac{g_2}{g_1} = \tan\nu. \tag{13}$$

FEM IMPLEMENTATION

FEM implementation of hypoplastic constitutive equations without localization was described in [6]. In what follows a concise account for the FEM implementation with localization is given.

Let us assume that at N_L Gauss-points localization becomes possible at location ξ in some element of the FE-mesh. As mentioned the localization analysis provides the direction $\mathbf{n}$ of the incipient strain rate discontinuity plane, as well as the remaining characteristic rate direction $\mathbf{m} = \frac{\mathbf{g}}{\|\mathbf{g}\|}$ which completes the description of the kinematics of the localized mode.

We then seek for suplementary shape functions $M(\mathbf{x})_{j;j=1,\ldots,N_L}$ within this element in which the localization condition (10) is met. We enrich an isoparametric element description with suplementary effect of the velocity gradient jump, i.e.

$$v_{i,j}(\mathbf{x}) = \sum_k \frac{\partial N_k(\mathbf{x})}{\partial x_j}\hat{v}_k^i + [v_{i,j}]. \tag{14}$$

Combining (2) and (14) we obtain a patrial differential equation system for the suplementary shape functions in the following form

$$m_i^l n_j^l = \frac{\partial M_l^i}{\partial x_j}(\mathbf{x}) \tag{15}$$

with a particulary solution at point ξ:

$$M_l^i(\mathbf{x}) = m_i \cdot (\mathbf{n}, \mathbf{x} - \xi). \tag{16}$$

In (16) the symbol $(\cdot, \cdot)$ stands for a scalar product. We consider a generic isoparametric element for which the local velocity field in the enriched version are expressible as follows:

$$v^i(\mathbf{x}) = \sum_j \hat{v}_j^i N_j(\mathbf{x}) + \sum_{j=1}^{N_L^e} r_j M_j^i(\mathbf{x}) \tag{17}$$

where $r_j = \|\mathbf{g}(\xi_j)\|$ is the amplitude of the localized mode and $\mathbf{M}(\cdot)$ the specialized shape function exhibiting a jump of the strain rate. For the elimination of the suplementary modes $\mathbf{r}$ in the element level the idea of a static condensation process (i.d. [7]) was adopted.

AN EXAMPLE

We sumarized the results of the bifurcation analysis for the constitutive equation (7) in Table 1. In Fig. 1 a numerical example of FE-calculation for the case of plane strain is presented. The bifurcation condition used in the numerical analysis is underlined in Table 1.

TABLE I. Results of the bifurcation analysis

	$\|\mathbf{D}\|=1$			
$\lambda_{[D]}$	ϕ_{mob}	θ	ϵ_1	ν
0.95	*no solution*			
1.00	55.81	59.06	4.70	22.26
1.05	52.28	59.81	3.20	21.69
1.10	49.26	58.12	2.60	18.70
1.15	46.83	57.95	2.25	17.69
1.20	44.70	57.80	2.00	16.82
1.25	42.72	56.91	1.80	15.21
1.30	41.05	57.20	1.65	14.98
1.35	39.21	53.31	1.50	10.28

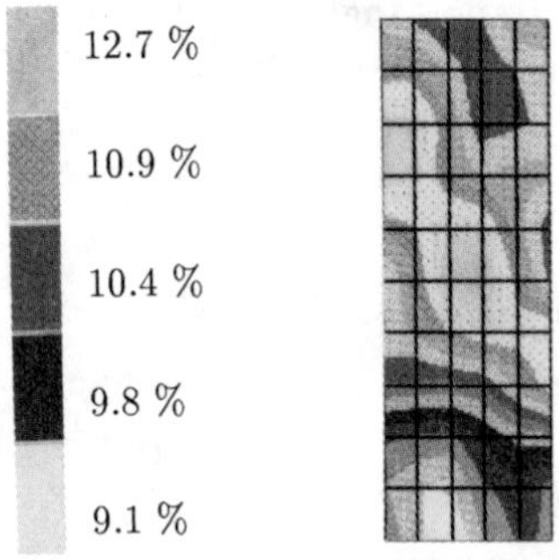

FIG. 1. Normalized strain distribution

CONCLUSIONS

A necesary condition for localized bifurcation within hypoplasticity is obtained without placing any further restrictions either on the kinematics or on the statics. It accounts even for the situation where both $\mathbf{D}_o$ and $\mathbf{D}_i$ are different from the homogeneous deformation, since solely the jumps are concerned. Our results show that localized bifurcation depends on the magnitude jump of the strain rate and the localized mode can be well captured by the numerical scheme in the present paper.

Acknowledgement —— The work was partially supported by Volkswagenwerk Foundation under Grant Az: I/63 374.

REFERENCES

[1] Hadamard, J., *Lecons sur la propagation des ondes et les équations de l'hydrodynamique*, Ch. 6. Librairie Scientifique A. Herrmann, Paris (1903)

[2] R. Hill, Acceleration waves in solids, *J. Mech. Phys. Solids*, 10(1) (1962)

[3] W. Wu, and D. Kolymbas, Numerical testing of the stability criterion for hypoplastic constitutive equations, *Mech. Mater.*, (1990), in print

[4] C. C. Wang, A new representation theorem for isotropic functions, parts I and II, *J. Rational Mech. Anal.*, 36, 166 (1970)

[5] W. Wu, and Z. Sikora, Localized bifurcation in hypoplasticity, *Int. J. Eng. Sci.*, (1990), in print

[6] Z. Sikora, and G. Gudehus, Numerical simulation of penetration in sand based on FEM, *Computers and Geotechnics*, 9, 73-86 (1990)

[7] Y. Leroy, and M. Ortiz, Finite element analysis of strain localization in frictional materials, *Int. J. Num. Anal. Methods Geomech.*, 13, 53-74 (1989)

GENERAL PLASTICITY, FINITE STRAINS

ELASTO-PLASTIC MODELLING OF SILT

K. AXELSSON and Y. YU
Dept. of Civil Engineering, Luleå University of Technology, Sweden

K. RUNESSON
Dept. of Structural Mechanics, Chalmers University of Technology, Sweden

ABSTRACT

Constitutive characteristics of silt - an intermediate type of soil - are discussed. A generalized Cam Clay model seems to predict, within sufficient accuracy, the drained as well as undrained response when the deformations are homogeneous. Localization of deformations at an overconsolidated (dense) state can also be predicted. Calibration via optimization is discussed. The incremental analysis is based on either explicit or implicit integration techniques. For undrained response, the incompressibility condition may be incorporated directly in an explicit algorithm.

CONSTITUTIVE BEHAVIOUR OF SILT

Since silt is a transitional type of soil, it seems to feature properties that are pertinent to both clay and sand. For instance, silt sometimes exhibits swelling and sensitivity to chemicals, salinity, etc.(typical for clay). On the other hand, when sheared, silt (eventually) dilates and may even liquefy under cyclic loading (typical for sand). Further, silt might show characteristics not observed either for clay or sand [1,2]. Some of these characteristics observed in triaxial tests on silt are specially commented upon here.

A variety of tests, such as conventional triaxial compression and extension tests (CTC, CTE) as well as active and passive plane strain tests (PSP, PSA) have been performed on Swedish silt with a silt content of 90%, a clay content of 5%, and an average water content of 25%. This type of silt is prevalent in northern Sweden. For illustration purposes, results from drained and undrained CTC-tests on normally consolidated samples are shown in Fig. 1. In the undrained case it is seen that the peak stress is not approached for strains up to 15%, which indicate the development of shear-induced hardening. In the p-q plane, the drained as well as undrained behaviour can be related to two characteristic lines, a critical state line (CSL) and, above this, a failure line (FL).

Tests on normally consolidated samples can be carried out while the deformations remain homogeneous, i.e. no tendency for deformation localization (shear banding) is evident. On the other hand, test results for heavily overconsolidated (dense) samples indicate tendency for such localization.

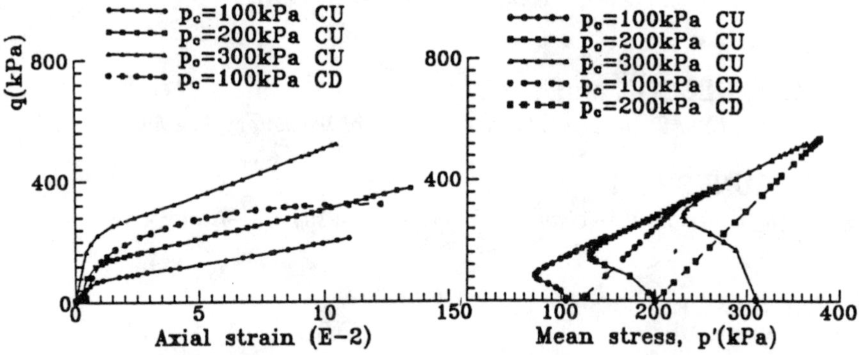

(a) axial strain vs. deviatoric stress (b) mean stress vs. deviatoric stress

Fig. 1 Triaxial experimental results on silt

CONSTITUTIVE MODELLING OF SILT

In light of these experimental results, a few constitutive models that are based on conventional plasticity theory, have been evaluated. One is a three-dimensional generalization of the model proposed by Nova and Wood [3] that is based on the Cam-Clay concept but includes combined volumetric and deviatoric hardening. The modification includes dependence on all stress invariants in a quite general fashion and volumetric nonassociativity. The second model that has been investigated is a modified version of the single-surface model for sand that was recently developed by Lade and Kim [4]. This model bears resemblance with the third model, MRS-Lade model suggested by Sture et al. [5], which is a cone-cap model and features the unfortunate corner irregularities or "grey regions" that cause problems at the numerical integration.

Either of the investigated models seems to describe the development of homogeneous deformations for normally consolidated samples. Likewise, each one can be used to predict localization that is sometimes obtained in plane strain tests of overconsolidated samples. The prediction capability of the Generalized Cam-Clay model for drained and undrained tests on normally consolidated samples is shown in Figs. 2 and 3 with the constitutive parameters of $G=3\cdot10^4$kPa, $\kappa=0.0038$, $\lambda=0.014$, $M=1.215$, $\eta_f=1.43$ (or $D=0.792$) and $e=0.82$.

MODEL CALIBRATION

An efficient optimization method to objectively determine the material parameters from the entire set of available experimental response functions should be used. This optimal fitting process defines a constrained optimization problem, since physical restrictions may be placed on the model parameters. The objective function Π: $R^n \rightarrow R$, is simply the sum-of-norm error function defined by

$$\Pi(\mathbf{x}) = \sum_{I=1}^{N} d_I(\mathbf{x})$$

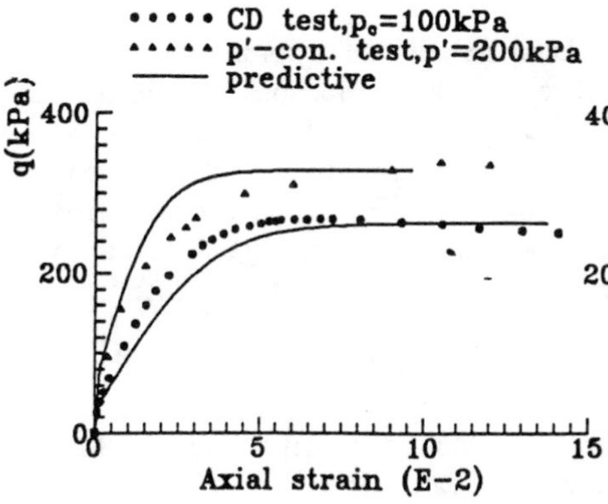

Fig. 2 Predicted stress-strain relation under drained case

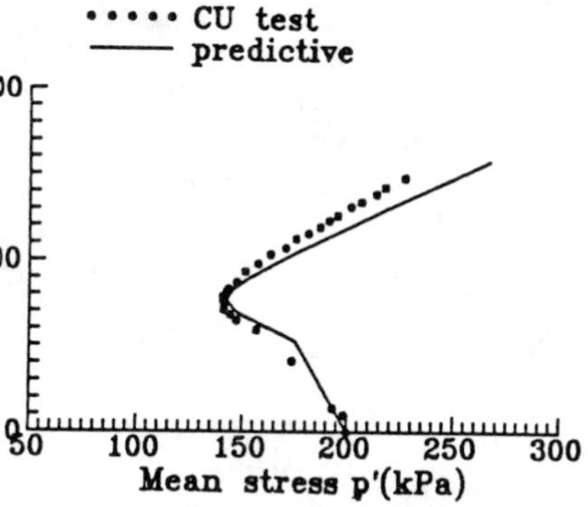

Fig. 3 Predicted stress path under undrained case

where N is the number of observations, **x** the vector of material parameters (in R^n), and d the chosen norm that measures the distance between the predicted and the experimental results.

A variety of algorithms, that are regarded as either direct search or gradient-based methods, exist for solving the optimization problem, Fletcher [6]; however, Rosenbrock's direct search algorithm has proved to be efficient and reliable. A common feature of existing techniques is that only a local minimum is obtained in general, i.e. the quality of the initial guess is crucial for the success in convering to the 'optimal' result. It is thus necessary to have a good understanding of the used constitutive models before performing the optimization.

INCREMENTAL ANALYSIS - INTEGRATION

The constitutive relations may be integrated using (classical or non-classical) explicit algorithms or (more recent) implicit algorithms. In developing a versatile Constitutive Driver it is important to deal with mixed control variables as well as the incompressibility condition that is pertinent to undrained behaviour. For the implementation in "standard" displacement-based finite element codes only strain control is needed, whereas mixed methods require (in-plane) stress control.

Explicit methods are normally based on the pertinent tangent relationship between the control and response variables. In conjunction with undrained conditions, such a relationship is not quite trivial; however, it is possible to incorporate in an exact fashion the constraint of incompressibility in the elastic/plastic loading criterion as well as in the final tangent relations. The response variables include the pore pressure, Runesson et al. [7].

Implicit algorithms have recently gained popularity in conjunction with soil behaviour, e.g. Runesson [8], Borja and Lee [9]. For mixed control, it is convinient to use a core algorithm that is based on the strain-controlled format and iterates on the prescribed stress value. In the undrained situation, incompressibility is satisfied in an iterative fashion as well, see Axelsson et al. [10] and Runesson et al. [11].

LOCALIZATION - BIFURCATIONS

Localized plastic deformation, e.g. related to slope stability problems, is assumed to be connected with the possibility for discontinuous bifurcations, for which a necessary condition is that the acoustic (or characteristic) tangent operator becomes singular. Recent analytical results, [11,12], indicate that plane strain conditions are more viable to bifurcation than axisymmetric states (CTC, CTE). Using actual data for the adopted constitutive models, it has been demonstrated that loading paths pertinent to plane strain tests on dense silt do, indeed, imply such susceptibility for localization.

REFERENCES

1. K. Axelsson, Y. Yu and K. Runesson, "Constitutive properties and modelling of silty soils", Proc. XII ICSMFE, Rio de Janeiro (August 1989).
2. Y. Yu, "Characteristics and modelling of silt", Thesis under preparation, Department of Civil Engineering, Luleå University of Technology (1990).
3. R. Nova and D. M. Wood, "A constitutive model for sand in triaxial compression", Int. J. Num. Analy. Meth. Geomech. 3(3), 255-278 (1979).
4. P.V. Lade and M.K. Kim, "Single hardening constitutive model for frictional materials II. Yield criterion and plastic work contours", Computers and Geotechnics 6, 31-47 (1988).
5. S. Sture, K. Runesson and E. Macari, "Analysis and calibration of a three-invariant plasticity model for granular materials", Ingenieur-Archiv 59, 253-266 (1989).
6. R. Fletcher, Practical Methods of Optimization,(John Wiley & Sons, Chichester 1980).
7. K. Runesson, M. Klisinski and K. Axelsson, "Characteristics of constitutive relations in soil plasticity of undrained behavior", submitted to Int. J. Solid Struc..
8. K. Runesson, "Implicit integration of elasto-plastic relations with reference to soils", Int. J. Num. Analy. Meth. Geomech. 11, 315-321 (1987).
9. R.I. Borja and S.R. Lee, "Cam-Clay plasticity, Part I: implicit integration of elasto-plastic constitutive relations", Comp. Meth. Apply. Mech. Engng. 78 , 49-72 (1990).
10. K. Axelsson, K. Runesson, S. Sture, Y. Yu and H. Alawaji, "Characteristics and integration of undrained response of silty soils", Proc. NUMOGIII, Niagara Falls, (May, 1989).
11. K. Runesson, N. S. Ottosson and D. Peric, "Plane stress and strain discontinous bifurcations in elasto-plastic materials", To appear in Int. J. of Plasticity.
12. D. Peric, "Localized deformation and failure analysis of pressure sensitive granular materials", Doctoral Thesis, Department of Civil, Enviromental, and Architectural Engineering, Univerisity of Colorado (July, 1990).

CONSTITUTIVE MODEL FOR COHESIONLESS SOIL

M.O. FARUQUE, M. ZAMAN and A. ABDULRAHEEM
University of Oklahoma, Norman, OK

ABSTRACT

A plasticity theory based constitutive model is proposed to describe the shear-volume coupling of cohesionless soil in an accurate manner. The model is based on the proposition that two distinct characteristic states exist for a cohesionless soil undergoing shearing. The proposed model is used to predict stress-strain and axial-strain-volumetric responses of a fine sand and excellent agreement is achieved.

INTRODUCTION

Coupling between shearing and volumetric deformation is an important characteristic of cohesionless soil. Two factors that primarily control the nature of such coupled response are the relative density (or density) and the confining pressure. In general, a relatively dense soil, subjected to low confining pressure, exhibits predominantly dilating response during shearing. Increase in confining pressure usually decreases the degree of dilatancy exhibited by a given cohesionless soil. By applying a large enough confining pressure, it is even possible to force a soil to experience only compressive volume change during shearing up to failure.

Based on the coupled shear-volumetric response of cohesionless soil as observed experimentally, two distinct states, termed as characteristic states, can be identified [3]. One such characteristic state is identified with the state of failure when a soil experiences progressive shear deformation at constant volume (i.e. volumetric strain rate is zero). The second characteristic state is identified with the onset of dilatancy at which the rate of volumetric strain momentarily vanishes as the soil passes from the compressive mode of deformation to the dilatant mode of deformation.

Constitutive modeling of cohesionless soil must incorporate both characteristic states in the mathematical formulation to properly account for the volumetric response of cohesionless soil during shearing. Most existing plasticity theory-based constitutive models include only the first characteristic state explicitly in the formulation. As a result, volumetric response is often poorly predicted by these models.

This paper presents a general formulation of a plasticity-based constitutive model for cohesionless soil that explicitly incorporates both characteristic states. The characteristic states are represented mathematically as surfaces in the stress space. Equations for the loading surface and the plastic potential are proposed. Stress-strain and volumetric-axial strain responses of a fine sand are predicted using the proposed model and good correlations are observed with experimental data.

MATHEMATICAL FORMULATION

Equations of the Characteristic State Surfaces

The first characteristic state surface is identified with the failure envelope. For a cohesionless soil, failure envelope is usually a straight line in the J_1-$\sqrt{J_{2D}}$ space. The slope of the straight line, however, varies with the orientation perimeter θ. In the proposed model, the equation of the first characteristic surface is adopted as

$$\sqrt{J_{2D}} = M \cdot J_1 \cdot g(\theta) \tag{1}$$

where J_{2D} is the second invariant of the deviatoric stress tensor, S_{ij}, J_1 is the first invariant of the total stress tensor, σ_{ij}, M is a material constant and $g(\theta)$ is a function that accounts for the change in the slope of the first characteristic line with orientation parameter θ. Following the works of Faruque and Chang [2] and Podgorski [4], $g(\theta)$ is expressed as

$$g(\theta) = \left\{\cos\left[\frac{1}{3}\cos^{-1}(-\cos 3\theta)\right]\right\}^{-1} \tag{2}$$

Unlike the first characteristic state surface, experimental investigation of the second characteristic state surface is non-existent. As a result, the precise form of this surface is unknown. From experimental evidences it is known that the second characteristic state tends to merge with the first characteristic state as the confining pressure becomes very high. Therefore, the second characteristic state surface should approach the first characteristic state surface with increasing confining pressure. An equation of the second characteristic state surface (Fig. 1) is proposed as

$$\sqrt{J_{2D}} = M\left[1 - N \cdot \exp(-\mu J_1/p_a)\right] g(\theta) \cdot J_1 \tag{3}$$

where N and μ are two additional material constants and p_a is the atmospheric pressure. For high values of J_1, the exponential term in Eq. (3) becomes negligible indicating gradual merging of the two characteristic state surfaces.

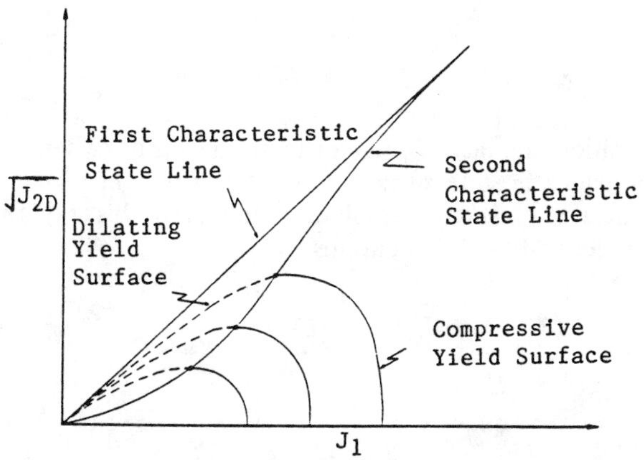

FIG. 1. Elements of the proposed constitutive model.

Yield Surfaces, Plastic Potentials and Hardening Rule

As observed from Fig. 1, the second characteristic state surface separates two stress regions according to the nature of volumetric deformation. Stress-states lying below the second characteristic state surface induce only compressive volume change. As such, an elliptical compressive yield surface, $F_c = 0$, is defined in this region as

$$F_c = \sqrt{J_{2D}} - \sqrt{a^2 - \frac{1}{R^2}(J_1 - C)^2\, g(\theta)} = 0 \tag{4}$$

where R is a material constant and signifies the ratio of the axes of the ellipse (see Fig. 1) and a and c are, respectively, the values of $\sqrt{J_{2D}}$ and J_1 at the point of intersection of the yield surface with the second characteristic state surface. A plastic potential function Q_c is defined in this region by perturbing the yield function F_c as

$$Q_c = \sqrt{J_{2D}} - \bar{A}\sqrt{a^2 \frac{1}{R^2}(J_1 - C)^2\, g(\theta)} \tag{5}$$

where $\bar{A}$ is a material constant lying in the range $0 \leq \bar{A} \leq 1$. It is seen that an associated formulation can be recovered by setting $\bar{A} = 1$ in Eq. (5).

Stress-states lying between the two characteristic state surfaces induce dilation. A dilating yield surface, $F_d = 0$, is introduced to model such response. As evident from Fig. 1, the dilating yield surface $F_d=0$ lies entirely between the two characteristic state surfaces and approaches the first characteristic state surface as $C \rightarrow \infty$. In addition, the dilating yield surface is assumed to satisfy slope compatibilities at both ends (see Fig. 1). The following mathematical form is used to define $F_d=0$.

$$F_d = \sqrt{J_{2D}} - a\left[1 - \left(1 - \frac{J_1}{C}\right)^{\frac{MC}{a}}\right] g(\theta) = 0 \tag{6}$$

By definition, the rate of volumetric strain vanishes when the stress state is on the characteristic state surfaces. This requirement cannot be satisfied within the associated plasticity formulation. In view of this, a dilating potential function Q_d, is defined by the equations

$$\frac{\partial Q_d}{\partial J_1} = q(J_1, J_{2D}) \cdot \frac{\partial F_d}{\partial J_1} \; ; \; \frac{\partial Q_d}{\partial J_{2D}} = \frac{\partial F_d}{\partial J_{2D}} \; ; \; \frac{\partial Q_d}{\partial \theta} = \frac{\partial F_d}{\partial \theta} \tag{7}$$

where

$$q(J_1, J_{2D}) = \alpha \frac{(\beta + \gamma)^{\beta+\gamma}}{\beta^{\beta}\gamma^{\gamma}} \cdot (1 - G)^{\beta} G^{\gamma} \; ; \; G = \frac{MJ_1 - \sqrt{J_{2D}}/g(\theta)}{MN \cdot \exp(-\mu J_1/P_a) \cdot J_1} \tag{8}$$

In Eq. (8), α, β and γ are additional material constants. It may be noted that the explicit equation of the plastic potential Q_d is not defined in this formulation. This is because only the derivatives of Q_d with respect to J_1, J_{2D} and θ appear in the explicit form of the constitutive equations. However, Eqs. (7) and (8) may be integrated numerically to obtain the plastic potential Q_d.

Isotropic hardening rule is used in the formulation to account for responses under monotonically increasing loading. Both F_c and F_d are allowed to expand in the stress space by treating C as an inelastic deformation history dependent function. Referring to Fig. 1, the following relationship between X and C can be obtained:

$$X(\xi, r_D) = C + Ra \tag{9}$$

where ξ and r_D are history dependent parameters [3]. Here $X(\xi, r_D)$ is assumed in the form

$$X = \beta_1 \xi^{\eta} \tag{10}$$

where

$$\eta = \eta_1 \left[1 + \beta_2 r_D^{\eta_2}\right] \tag{11}$$

In Eqs. (10) and (11), β_1, β_2, η_1, and η_2 are material constants associated with hardening during inelastic deformation.

APPLICATION AND FINAL REMARKS

Drained shear test data for a uniform fine sand [1] is used to evaluate the fourteen material parameters associated with the proposed model. The following optimized set of parameters are obtained: $M = 0.27$; $N = 0.442$; $\mu = 0.06$; $R = 6.134$; $\alpha = 0.318$; $\beta = 0.57$; $\gamma = 0.599$; $\beta_1 = 1.61 \times 10^6$; $\eta_1 = 1.17$; $\bar{A} = 0.7275$; $\beta_2 = 0.828$; and $\eta_2 = 3.37$. The bulk modulus K and the shear modulus G are given as [1] $K = 27,369 + 407.7P(KPa)$; $G = 6,598 + 317.6P(KPa)$ where P is the confining pressure.

Stress-strain and volumetric-axial strain predictions for a CTC (conventional triaxial compression) test and their comparison with the experimental data are shown in Figs. 2(a) and 2(b), respectively. As evident, an excellent agreement is achieved in both cases.

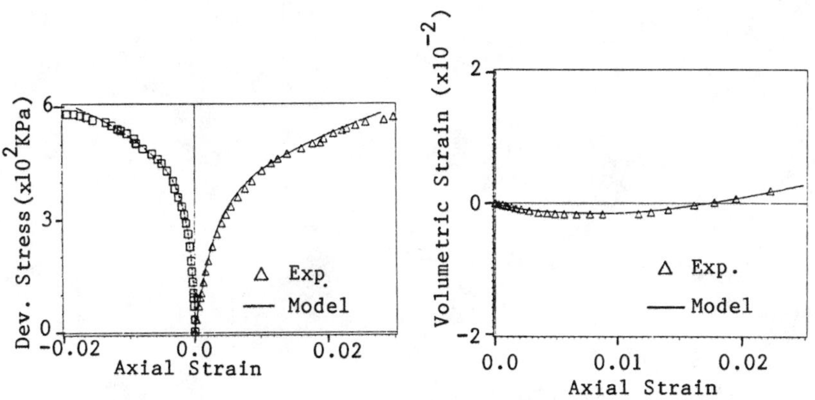

(a) Stress-strain response (b) Volumetric-axial strain response

FIG. 2 Conventional Triaxial Compression Test at 30 psi Confining Pressure

REFERENCES

1. M. Azeemuddin, "Constitutive Modelling of Dhahran Dune Sand Using Cap Model", M.S. Thesis, Univ. of Pet. & Min., Dhahran, Saudi Arabia (1988).
2. M.O. Faruque and C.J. Chang, "New cap model for failure and yielding of pressure sensitive materials", J. Eng. Mech. ASCE 112(10), 1041-1053 (1986).
3. M. O. Faruque and M. Zaman, "On modeling stress-strain and dilatant behavior of cohesionless soil", submitted to Computers and Geotechnics (1990).
4. J. Podgorski, "General failure criterion for isotropic media", J. Eng. Mech. ASCE 111, 188 (1985).

MODELLING THE COMBINED EFFECTS OF THE INTERMEDIATE PRINCIPAL STRESS AND INITIAL ANISOTROPY ON THE STRENGTH OF SAND

Marte Gutierrez
Norwegian Geotechnical Institute
P.O. Box 40, Tåsen, N-0801 Oslo 8
NORWAY

ABSTRACT

A model for the combined effects of the intermediate principal stress and initial anisotropy on the strength of sand is presented. Comparisons with three sets of experimental results demonstrate the validity of the model.

INTRODUCTION

The effects of the intermediate principal stress and initial anisotropy on the strength of sand has been the subject of several studies in recent years. Experimental studies have shown the significance of these two parameters. However, most studies have concentrated on only either one of these factors and there is very little study on their combined effects on the strength of sand. It appears that the possible interaction between these two parameters has so far not been addressed to in constitutive models for soils.

THE MODEL

The model is expressed in terms of the stress invariants giving the mean stress p, shear stress q and Lode's angle θ, and the angle the major principal stress makes with the vertical axis, α. These parameters are defined as:

$$p = \frac{1}{3}(\sigma_1 + \sigma_2 + \sigma_3) \tag{1}$$

$$q = \sqrt{\frac{1}{2}\{(\sigma_1 - \sigma_2)^2 + (\sigma_2 - \sigma_3)^2 + (\sigma_3 - \sigma_1)^2\}} \tag{2}$$

$$\tan\theta = \frac{2\sigma_2 - \sigma_1 - \sigma_3}{\sqrt{3}(\sigma_1 - \sigma_3)} \tag{3}$$

$$\tan 2\alpha = \frac{2\sigma_{xy}}{\sigma_y - \sigma_x} \tag{4}$$

The parameters p, q and θ can be related to the more commonly used parameters in soil mechanics: b -value and angle of friction, ϕ, viz,

$$b = \frac{\sigma_2 - \sigma_3}{\sigma_1 - \sigma_3} = \frac{1}{2}(\sqrt{3}\tan\theta + 1) \tag{5}$$

$$\sin\phi = \frac{\sigma_1 - \sigma_3}{\sigma_1 + \sigma_3} = \frac{\sqrt{3}(q/p)\cos\theta}{3 + (q/p)\sin\theta} \tag{6}$$

The parameter θ, which is a measure of the magnitude of the intermediate principal stress σ_2 relative to the major principal stress σ_1 and the minor principal stress σ_3, has been shown by many experiments to affect the strength of sand. Also, numerous experiments have demonstrated the initially anisotropic strength of sand as manifested by different strengths for different orientations of the major principal stress as given by α.

The combined effects of the intermediate principal stress and initial anisotropy is modelled by a failure criteria consisting of a product of two functions in the form

$$\frac{q}{p} = r_f r_1(\theta) r_2(\alpha) \tag{7}$$

$r_1(\theta)$ and $r_2(\alpha)$ model the effects of σ_2 and the direction of loading on the stress ratio q/p at failure of sand, respectively. r_f is a model parameter giving the stress ratio q/p at failure in the conventional triaxial compression test (i.e., $\theta = -30°$ and $\alpha = 0°$). The functions $r_1(\theta)$ and $r_2(\alpha)$ must be chosen such that $r_1(-30°) = 1$ and $r_2(0°) = 1$.

The effect of the intermediate principal stress is modelled by using the Argyris-Gudehus [1] failure criteria, while the effect of initial anisotropy is modelled using the failure criteria of Gutierrez, et al. [2]. The functions $r_1(\theta)$ and $r_2(\alpha)$ are then defined as

$$r_1(\theta) = \frac{2k_1}{(1+k_1)+(1-k_1)\sin(3\theta)} \tag{8}$$

$$r_2(\alpha) = \frac{2k_2}{(1+k_2)-(1-k_2)\cos(2\alpha)} \tag{9}$$

where $k_1 = r_1(30°)$ is the ratio of q/p at failure for $\theta = 30°$ and $\theta = -30°$, while $k_2 = r_2(90°)$ is the ratio of q/p at failure for $\alpha = 90°$ and $\alpha = 0°$. Note the similarity of the functions $r_1(\theta)$ and $r_2(\alpha)$. The failure criteria, Eq. (7), may be expressed in terms of friction angle and b-value via Eqs. (5) and (6) if necessary.

The formulation of Eq. (7) is based on the postulate that there are no interactions in the effects of the intermediate principal stress and initial anisotropy on the strength of sand. However, experimental results (Refs. [2], [3] and [4]) show possible interaction of these two parameters as manifested by the non-dependency of the angle of friction at failure on θ when the principal stress direction α is oriented 60-90° from the normal to the bedding plane.

COMPARISONS WITH EXPERIMENTAL RESULTS

The failure criteria, Eq. (7), was compared to three sets of experimental results reported in Refs. [2], [3] and [4]. The results in Refs. [2] and [3] were obtained from hollow cylindrical tests while the results in Ref. [4] were obtained from a combination of conventional triaxial and plane strain tests. The comparisons are shown in Fig. 1 in three-dimensional plots of q/p, θ and α. The three-dimensional failure surface Eq. (7) fitted through the experimental points, the surface obtained by kriging of the experimental points, and the experimental data are shown in the figures. Also shown are the model parameters for each set of data.

As can be seen in Fig. 1, the proposed failure criteria adequately models, albeit in somewhat gross manner, the dependency of the stress ratio at failure on θ and α for the three sets of data. This is despite the fact that the simplifying postulate of non-interaction in the effects of the two parameters was assumed in the formulation.

The model should be useful not only in representing the effects of the intermediate principal stress and initial anisotropy on the strength of sand, but also in averaging of results of tests from different apparatus such as those reported in Ref. [4].

REFERENCES

1. Gudehus, G. (1973), "Elastoplastiche Stoffgleichungen fur Trockenen Sand," Ingeniur Archiv, vol. 42, pp. 151-169.
2. Gutierrez, M., Ishihara, K., and Towhata, I. (1988), "Experimental Study and Modelling of the Effects of b-value and Initial Anisotropy on the Strength of Sand," Proc. 43rd Conf. of JSCE, vol. 3, pp. 570-571.
3. Hight, D.W. (1988), Discussion on "The Engineering Application of Direct and Simple Shear Testing," Geotechnique, Vol. 38, pp. 139-140.
4. Lam, W.K. and Tatsuoka, F. (1986), "The Strength Surface of Sand," Proc. 21st Conf. of JSSMFE, vol. 1, pp. 315-318.

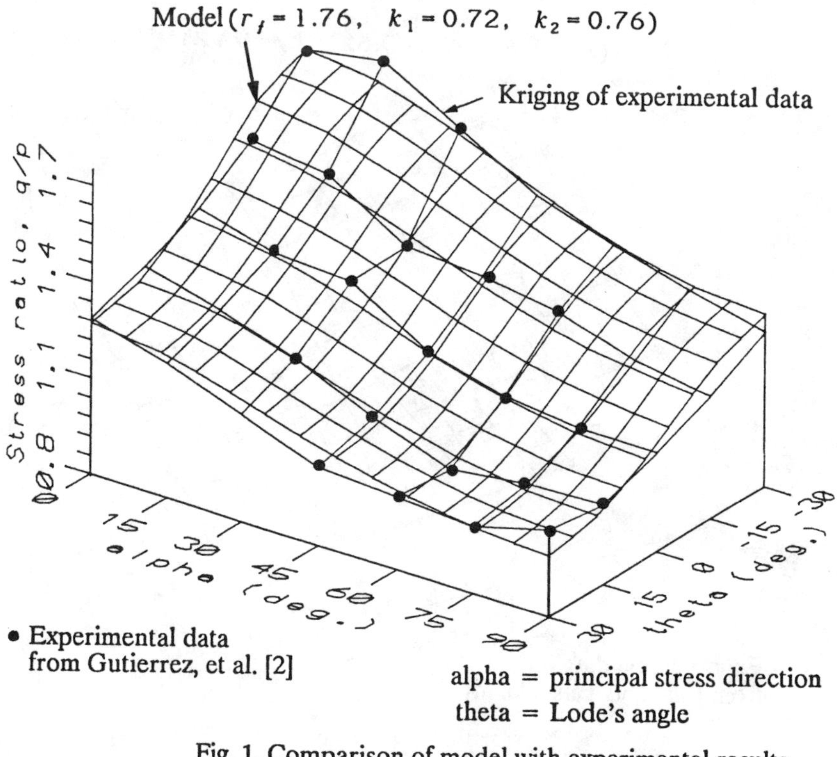

Fig. 1. Comparison of model with experimental results.

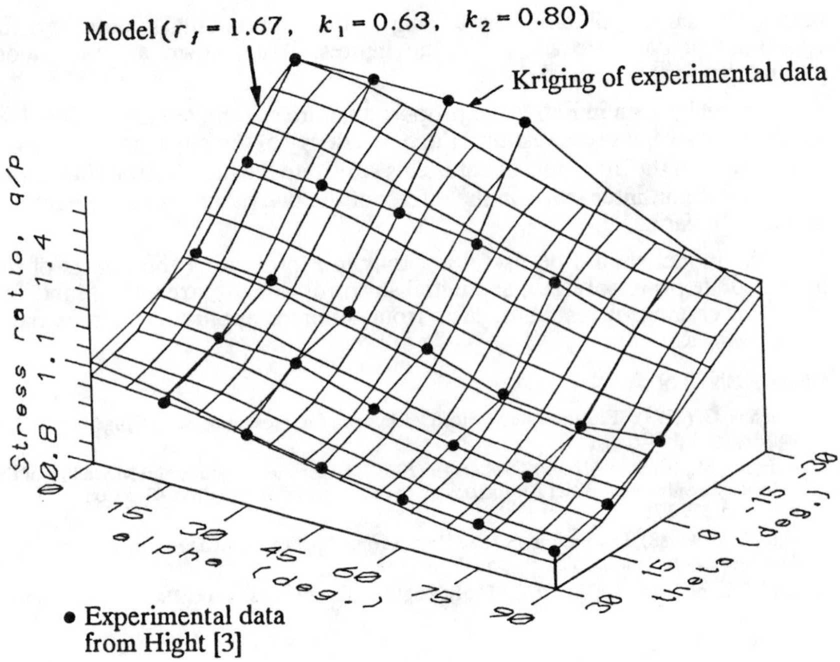
Model(r_f = 1.67, k_1 = 0.63, k_2 = 0.80)
Kriging of experimental data
Stress ratio, q/p
0.8
1.1
1.4
15
30
45
60
75
90
alpha (deg.)
30
15
0
-15
-30
theta (deg.)
● Experimental data from Hight [3]

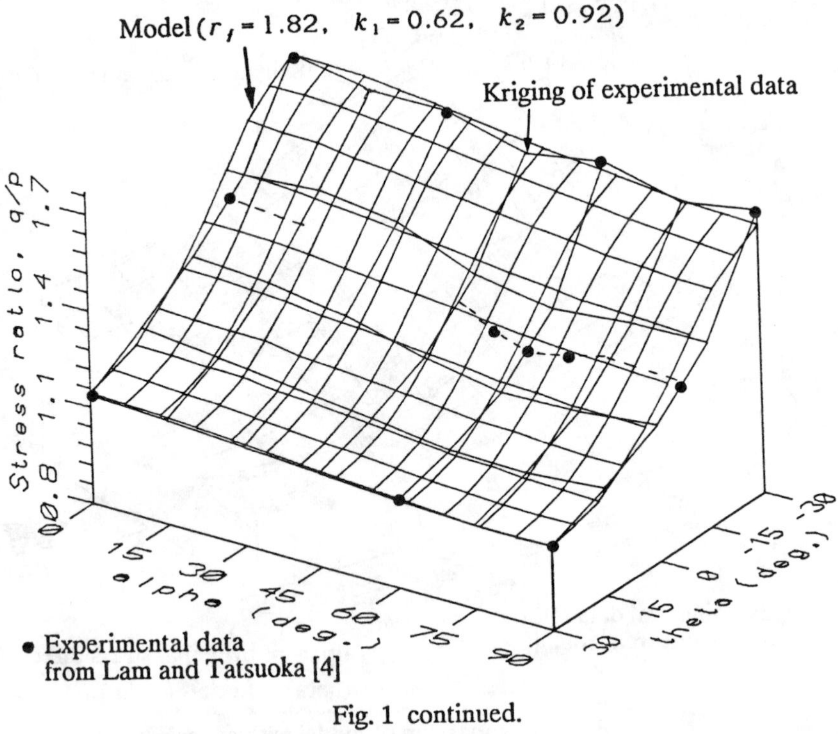
Model(r_f = 1.82, k_1 = 0.62, k_2 = 0.92)
Kriging of experimental data
Stress ratio, q/p
0.8
1.1
1.4
1.7
15
30
45
60
75
90
alpha (deg.)
30
15
0
-15
-30
theta (deg.)
● Experimental data from Lam and Tatsuoka [4]

Fig. 1 continued.

Induced Anisotropy in Normally Consolidated Clay

Pierre-Yves Hicher and Venancio Trueba Lopez
GRECO Geomaterials, Soil-Structure Laboratory
CNRS URA 850, Ecole Centrale de Paris
92295 Chatenay-Malabry Cedex, France

Abstract

Three dimensional triaxial tests have been performed to study the influence of anisotropy on stress-strain relationships in normally consolidated clay. The orientation of major principal stress with respect to the axes of anisotropy was found to play an important effect on the initial modulus. For further large strains the effect was less noticeable due to the increased predominance of induced anisotropy on initial anisotropy.

Introduction

The purpose of the present study is to quantify the influence of induced anisotropy on the mechanical properties of a normally consolidated Kaolinite (Wl = 40, Wp = 20), isotropically consolidated and then subjected to a drained axisymetric compression of a maximum major principal strain of 5% in order to create a well controlled anisotropy of the samples. Three dimensional triaxial tests were then performed on these samples and their mechanical behavior was compared to that of isotropic samples subjected to the same tridimensional loadings.

Testing Programm

Equipment.-The cubical triaxial apparatus was similar to the one conceived by Lade [1,2] with four rigid and two flexible plates. The initial size lengths of the cubical samples were 7 cm each. The tests were controlled by a micro computer linked to hydraulic actuators (GDS), volume change indicator, displacement indicators and load cells [3].

Stress and strain paths.-Drained cubical triaxial tests were performed such that $b = (\sigma_2 - \sigma_3)/(\sigma_1 - \sigma_3)$ and σ_3 remained constant during each test. Selected values of b were: 0, .5, 1. For anisotropic samples the axis of major principal stress σ_1 was identical to the axis of the previous compression or perpendicular to this axis.

Stress-Strain Behavior of Isotropic Samples

Results for initially isotropic samples are presented in figure 1, using the diagramm $q/p' - \bar{\varepsilon}$ where q is the shear stress defined as:

$$q = (3/2 \cdot s_{ij} \cdot s_{ij})^{.5} \qquad (1)$$

in which $s_{ij} = \sigma'_{ij} - p'\delta_{ij}$ and $p' = 1/3(\sigma'_1 + \sigma'_2 + \sigma'_3)$
$\bar{\varepsilon}$ is the shear strain defined as:

$$\bar{\varepsilon} = (2/3 \cdot e_{ij} \cdot e_{ij})^{.5} \tag{2}$$

in which $e_{ij} = \varepsilon_{ij} - \varepsilon_v/3 \cdot \delta_{ij}$ and $\varepsilon_v = \varepsilon_1 + \varepsilon_2 + \varepsilon_3$

The relation between q/p' and $\bar{\varepsilon}$ is practically independant of b for a shear strain up to 5%. For larger shear strains the curves become dependant of b with smaller values of q/p' at a given $\bar{\varepsilon}$ when b increases.

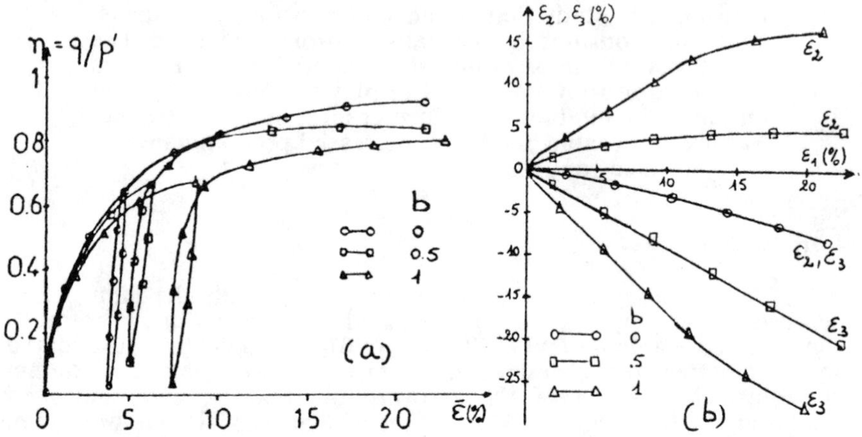

FIG. 1-Cubical Triaxial Tests on Isotropic Clay Samples. (a)-Stress-Strain Relations. (b)-Relations between Principal Strains.

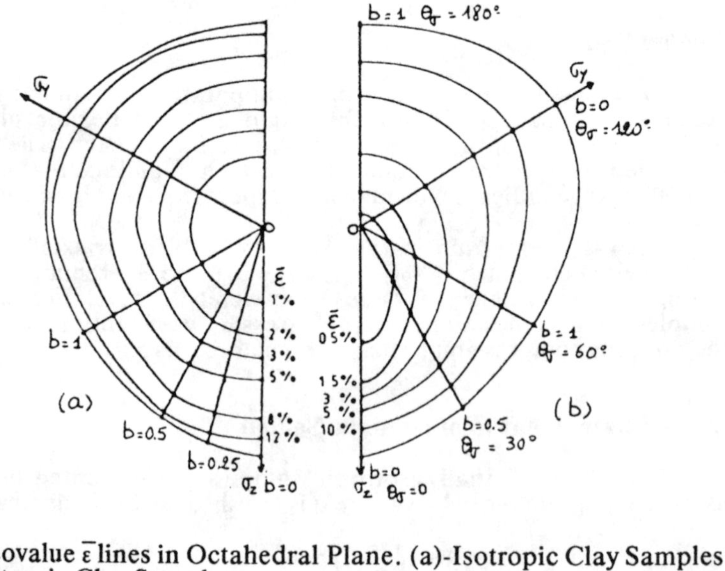

FIG. 2-Isovalue $\bar{\varepsilon}$ lines in Octahedral Plane. (a)-Isotropic Clay Samples. (b)-Anisotropic Clay Samples

In Fig. 2a we have plotted the stress paths in an octahedral plane (p′ constant). Since the samples were isotropic, only 1/6 of the plane was enough to represent the mechanical behavior along paths at constant b values. Along each of these paths shear strain was noted. We observe that the first isovalue $\bar{\varepsilon}$ lines (for $\bar{\varepsilon}<5\%$) are quite circular, which indicates that the q/p'-$\bar{\varepsilon}$ relationship is independant of b. At large deformation the curves tend to take the shape of the failure surface.

The failure surface corresponds to the maximum values of the stress ratio q/p' for the different values of b. It corresponds also to the development of shear bands inside the samples, whose occurence and direction are dependant on b (not discussed here). The failure surface can be represented by Lade's criterion.

The relations between principal strains are presented in Fig. 1b. Calculation of $b_\varepsilon=(\varepsilon_2-\varepsilon_3)/(\varepsilon_1-\varepsilon_3)$ shows differences between b_ε and b_σ when b_σ is different from 0 or 1

Stress-Strain Behavior of Anisotropic Samples

Initial anisotropy was created on isotropically consolidated samples by an axisymetric drained compression (b=0) up to a major principal strain of 5%. Fig. 3 presents the results of 5 three dimensional tests: 2 tests at b=0, one with σ_1 in the same direction as the previous loading, one with σ_1 perpendicular to that direction; 2 tests at b=1 with σ_1 parallel and perpendicular to the direction of previous loading; 1 test at b=.5 with σ_1 in the direction of previous loading. Θ_σ represents the angle between the initial loading and the following stress path in the octahedral plane.

The strength development occurs much faster when σ_1 is applied in the same direction as the initial compression (Θ_σ=0°, 30°, 60°). The isovalue $\bar{\varepsilon}$ lines express this change in initial stiffness with the orientation of major principal stress (Fig. 2b). In comparison with the ones obtained from isotropic samples, they present a translation along the axis of initial loading which suggests a kinematic hardening created by plastic deformations during the axisymetric compression.

The influence of initial anisotropy can also be seen on the evolution of principals strains. The tests at b=0 and b=1 are axisymetric for stress paths and not axisymetric for strain paths. Fig. 4 presents the evolution of ε_2 with ε_1 for tests at b=.5 on isotropic and anisotropic samples. At the beginning of the loading the slopes of the two curves are different, which results in a different orientation of strain increment vectors.

Progressively the two curves become parallel leading to the same orientation of strain increment vectors. The influence of initial anisotropy is erased by further large plastic deformations. This can also be seen in the octahedral plane where the isovalue $\bar{\varepsilon}$ curves become identical for isotropic and anisotropic samples. As a consequence the maximum shear strength and therefore the failure criterion appear to be independant of initial anisotropy.

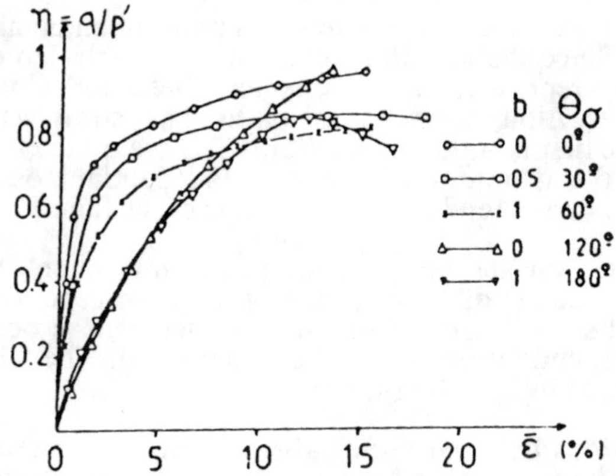

FIG. 3-Stress-Strain Relations in Cubical Triaxial Tests on Anisotropic Clay Samples.

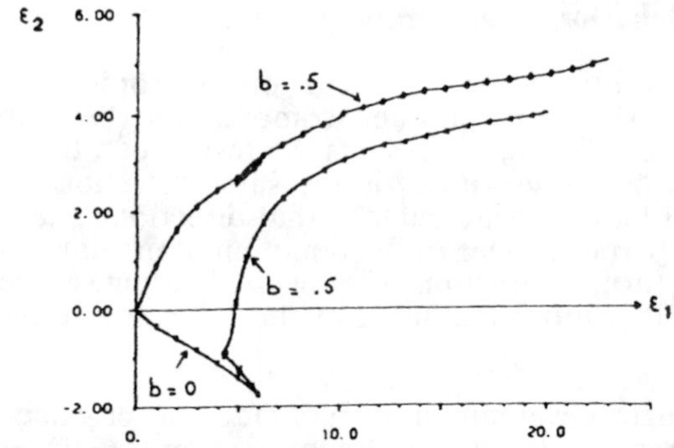

FIG. 4-Relations between Principal Strains in Cubical Triaxial Tests on Isotropic and Anisotropic Clay Samples

Conclusion

The influence of anisotropy induced by an axisymetric loading on a normally consolidated clay has been demonstrated along three dimensional stress paths. The main result is a decrease of initial slope of octahedral stress-octahedral strain curves when the direction of the major principal stress is perpendicular to the direction of initial loading. The effect of initial anisotropy is progressively erased by further large plastic strain and the maximum strength is identical for initially isotropic and anisotropic samples.

References

1. P.Y. Hicher and P.V. Lade, "Rotation of principal directions in K_0-consolidated clay", J. Geo. Eng., ASCE, vol. 113, No. 7, 774-788, July, 1987.

2. P.V. Lade, "Cubical triaxial apparatus for soil testing", Geo. Test. J., GTJODJ, vol. 1, No. 2, 93-101, June, 1978.

3. V. Trueba Lopez, "Etude du comportement mécanique des argiles saturées sous sollicitations tridimensionnelles", Thèse de Doctorat, Ecole Centrale de Paris, 1988.

CAP MODEL WITH TRANSVERSELY ISOTROPIC ELASTIC PARAMETERS

Dana N. Humphrey and Rajendra S. Gondhalekar, Assistant Professor and Research Assistant, respectively, Department of Civil Engineering, University of Maine, Orono, Maine 04469.

ABSTRACT

To better model the behavior of overconsolidated naturally sedimented soil, transversely isotropic elastic parameters have been incorporated into an existing cap model. This results in the first invariant of the stress tensor and the loading parameter being dependent on the strain parallel to the axis of symmetry. The modified model can better represent observed soil behavior.

INTRODUCTION

Elastic plastic constitutive models for soils are generally formulated with isotropic elastic parameters. Yet, it is well known that the elastic behavior of naturally sedimented soils is anisotropic [1]. This discrepancy is particularly important for modeling the behavior of overconsolidated soils. To overcome this discrepancy, transversely isotropic elastic parameters were incorporated into an elastic plastic work hardening cap model [2,3,4,5,6]. The model has a Drucker-Prager ultimate failure surface and an elliptical end cap. The advantage of using this model is that even after introduction of transversely isotropic elastic parameters, the model is still relatively simple [7].

TRANSVERSELY ISOTROPIC ELASTICITY

The elastic stiffness matrix for a transversely isotropic material with the z axis as the axis of symmetry is:

$$\begin{Bmatrix} d\sigma_x \\ d\sigma_y \\ d\sigma_z \\ d\tau_{yz} \end{Bmatrix} = \begin{bmatrix} A_{11} & A_{12} & A_{13} & 0 \\ A_{12} & A_{11} & A_{13} & 0 \\ A_{13} & A_{13} & A_{22} & 0 \\ 0 & 0 & 0 & A_{44} \end{bmatrix} \begin{Bmatrix} d\epsilon_x \\ d\epsilon_y \\ d\epsilon_z \\ d\gamma_{yz} \end{Bmatrix} \quad (1)$$

Where:

$$A_{44} = G$$

$$A_{11} = \frac{ab - d^2}{t}$$

$$A_{13} = \frac{de - ad}{t}$$

$$A_{22} = \frac{at + (de - ad)^2}{t\,(ab - d^2)}$$

$$t = a^2b - 2ad^2 - be^2 + 2d^2e$$

$$a = \frac{1}{E} \qquad b = \frac{1}{E'} \qquad d = \frac{-\nu'}{E'} \qquad e = \frac{-\nu}{E}$$

In the above equations E' is the Young's modulus in z direction, E is the Young's modulus in x and y directions, ν' is the Poisson's ratio in xz and yz planes, and ν is the Poisson's ratio in xy plane. G is the shear modulus in yz and xz planes, and cannot be written explicitly in terms of the other moduli.

This leads to the relationship between the first invariant of the stress tensor and the elastic strains given below:

$$dI_1 = K_1\, d\varepsilon^e_{kk} + K_2\, d\varepsilon^e_{33} \tag{2}$$

where: dI_1 = incremental change in the first invariant of the stress tensor

$d\varepsilon^e_{kk}$ = increment of elastic volumetric strain

$d\varepsilon^e_{33}$ = increment of elastic strain in the direction of axis of symmetry (z-axis)

$K_1 = A_{11} + A_{12} + A_{13}$

$K_2 = A_{22} + A_{13} - A_{11} - A_{12}$

From this equation it is seen that $d\varepsilon^e_{33}$ must be known to compute dI_1.

CAP MODEL WITH TRANSVERSELY ISOTROPIC ELASTIC PARAMETERS

The existing cap model was modified to incorporate transversely isotropic parameters. The most important equations are given below. Further details are given in Gondhalekar [7].

The increment of plastic strain is determined from

$$d\varepsilon^p_{ij} = \begin{cases} d\lambda \dfrac{\partial f}{\partial \sigma_{ij}} & \text{if} \quad f = 0 \text{ and } \dfrac{\partial f}{\partial \sigma_{ij}} d\sigma_{ij} > 0 \\ 0 & \text{if} \quad f < 0 \text{ and } \dfrac{\partial f}{\partial \sigma_{ij}} d\sigma_{ij} \le 0 \end{cases} \tag{3}$$

where f is the loading function and $d\lambda$ is the loading parameter. It was shown that $d\lambda$ is given by

$$d\lambda = \frac{\frac{\partial f}{\partial I_1}[K_1 d\epsilon_{kk} + K_2 d\epsilon_{33}] + \frac{G}{\sqrt{J_2}} \frac{\partial f}{\partial \sqrt{J_2}} s_{ij}\, de_{ij}}{3K_1\left(\frac{\partial f}{\partial I_1}\right)^2 + K_2\left(\frac{\partial f}{\partial I_1}\right)^2 + \frac{K_2}{2\sqrt{J_2}} \frac{\partial f}{\partial \sqrt{J_2}} \frac{\partial f}{\partial I_1} s_{33} + G\left(\frac{\partial f}{\partial \sqrt{J_2}}\right)^2 - 3\frac{\partial f}{\partial I_1} \frac{\partial I_1}{\partial \beta} \frac{\partial \beta}{\partial \epsilon^P_{kk}}} \tag{4}$$

where J_2 is the second invariant of the stress tensor, s_{ij} is the deviatoric stress, s_{33} is the deviatoric stress in the direction of symmetry, de_{ij} is the increment of deviatoric strain, and β is a hardening parameter which is a function of the plastic volumetric strain $d\epsilon^P_v$. Equation 4 shows that $d\lambda$ and thus the elastic plastic stiffness matrix are functions of $d\epsilon_{33}$.

COMPARISON OF PREDICTED AND OBSERVED RESULTS

The predictive ability of the modified cap model was checked using drained triaxial tests results for lightly overconsolidated Lake Agassiz plastic clay from Winnipeg, Canada [7]. Graham and Houlsby [1] determined the transversely isotropic parameters for this soil by assuming that the square root of the ratio between the vertical and horizontal Youngs moduli was the same as the ratio between the two transversely isotropic Poisson's ratios as shown below

$$\frac{E'}{E} = \left(\frac{\nu'}{\nu}\right)^2 = \theta^2 \tag{5}$$

They also gave the isotropic parameters which resulted in the best fit to the observed stress-strain data. These were used to determine the model parameters for the conventional isotropic version of the cap model. A comparison between the observed and predicted shear stress q $(=\sigma_1 - \sigma_3)$ vs. shear strain ϵ $[=2(\epsilon_1 - \epsilon_3)/3]$ for a triaxial test in which the ratio of the increase in vertical to horizontal stress was 1:0.6305 is shown in Fig. 1. It is seen that the modified model with transversely isotropic elastic parameters gives a much better prediction than the model with isotropic elastic parameters.

EFFECT OF TRANSVERSE ISOTROPY ON PREDICTED RESPONSE

The effect of transverse isotropy on the predicted response of soils in drained triaxial compression was investigated by varying θ^2 for a hypothetical lightly overconsolidated soil with properties based on Lake Agassiz clay [7]. In this test the ratio of the increase in vertical to horizontal stress was also 1:0.6305. Values of θ^2 equal to 1.92 (stiffer vertical), 1.00 (isotropic), and 0.5 (stiffer horizontal) were investigated. The effect on the predicted stress strain response is significant as shown in Fig. 2. For θ^2 = 1.92 the strain in the vertical direction is less than the horizontal direction which causes the stress strain curve to lean to the left in the elastic region.

CONCLUSIONS

A cap model with transversely isotropic elastic parameters was developed. The modified model can better represent the stress strain behavior of naturally sedimented soils.

REFERENCES

1. J. Graham and G.T.Houlsby, "Anisotropic elasticity of a natural clay," Geotechnique, 33(2), 165-180 (1983).
2. I.S. Sandler, F.L. DiMaggio, and G.Y. Baladi, "Generalized CAP model for geological materials," J. Geot. Eng. Div., ASCE, 102(7), 683-699 (July, 1976).
3. W.F. Chen and G.Y. Baladi, Soil Plasticity, Theory and Implementation (Elsevier, New York, 1985).
4. W.O. McCarron, "Soil plasticity and finite element applications," Ph.D. Thesis, School of Civil Eng., Purdue University, West Lafayette, IN (1985).
5. D.N. Humphrey, "Design of Reinforced Embankments," Ph.D. Thesis, School of Civil Eng., Purdue University, West Lafayette, IN (1986).
6. D.N. Humphrey and R.D. Holtz, "A Procedure to Determine Cap Model Parameters," Proc. Second Int. Conf. on Constitutive Laws for Eng. Materials, C.S. Desai, et al., ed., Vol. II, 1225-1232 (Elsevier, New York, 1987).
7. R.S. Gondhalekar, "Development of Transversely Isotropic Cap and Bounding Surface Constitutive Models," M.S. Thesis, Dept. of Civil Eng., University of Maine, Orono, ME, (August, 1990).

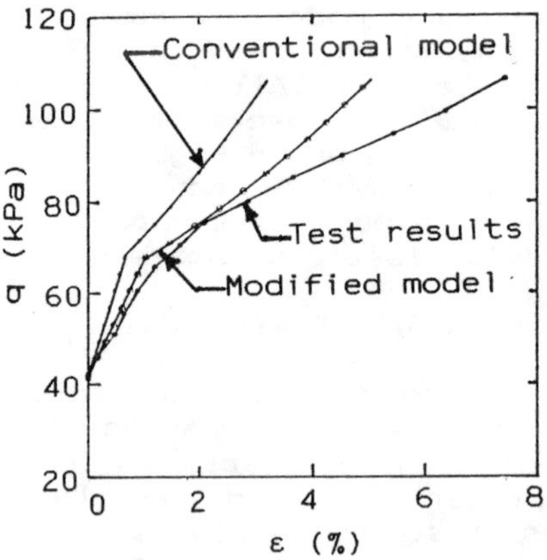

FIG. 1. Comparison of transversely isotropic and isotropic models with test results.

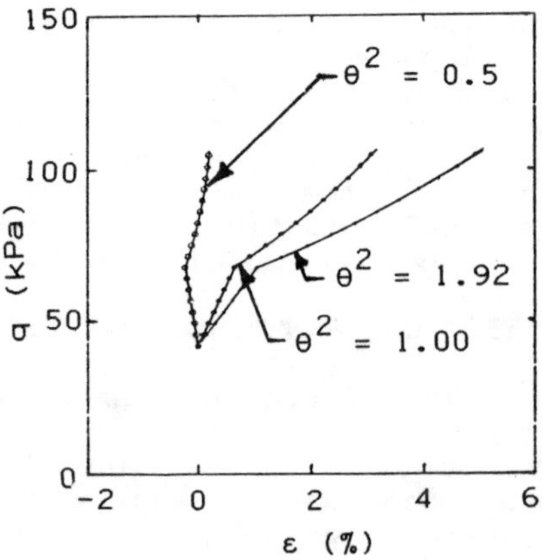

FIG. 2 Calculated response with the modified model for simulated drained triaxial test.

SOME BASIC ASPECTS OF ELASTIC-PLASTIC THEORY INVOLVING FINITE STRAIN

ERASTUS H. LEE
Department of Mechanical Engineering,
Aeronautical Engineering & Mechanics
Rensselaer Polytechnic Institute
Troy, NY 12180-3590

ABSTRACT

This paper discusses some basic aspects of elastic-plastic theory, particularly with respect to the finite-deformation-valid formulation. Such a formulation is often necessary because the elastic-plastic tangent modulus can be of the same order of magnitude as the stress, which brings nonlinear convected terms into the analysis. Because of the incremental or flow type characteristic of plasticity theory, which emphasizes the current configuration, finite strain from the initial state does not play a role in the formulation.

INTRODUCTION

When a ductile metal is stressed beyond the yield limit, elastic and plastic increments of deformation occur simultaneously. Plasticity is a very complex physical phenomenon which is modeled mathematically by a functional law through which the stress can be expressed in terms of the history of the plastic deformation. By contrast the elastic law expresses the current stress as a function of the current elastic deformation. If the stress is reduced to zero, the elastic strain is thus also reduced to zero, so that the residual strain is the plastic strain. On initially loading a test specimen, the gradient of the stress-strain curve is the elastic modulus. When plastic flow sets in, this gradient decreases sharply to the strain-hardening modulus which can be of the same order as the yield stress. In this circumstance, as pointed out by Hill [1], p.54, convected terms in the equations of equilibrium become significant and finite-deformation-valid theory must be utilized. This involves nonlinear kinematics in the coupling of elastic and plastic deformation to determine the total deformation for inclusion in the equation of motion. During the last conference in this series, in the discussion at a plasticity session, a graduate student asked whether plastic strain was, or was not, a state variable. There were several prominent members of the plasticity fraternity present, but no general agreement on the answer to the question was forthcoming. Some related questions concerning the formulation of elastic-plastic theory are considered in this paper.

The incremental or flow type characteristic of plastic flow directs attention to the current state in the formulation of plasticity theory. The migration of dislocations and other defects through the body, driven by the yield stress to which it is subjected, generate the plastic straining. It is thus the rate of deformation or velocity strain associated with the current configuration which contribute the basic foundation of the theory.

THE STRUCTURE OF ELASTIC-PLASTIC THEORY

This paper concerns plastic deformation of structural metals at ambient temperatures. This implies the appropriateness of classical plasticity theory in which the plastic properties, such as the tensile stress-strain relation, are not influenced by the testing speed. Elastic-plastic theory is generally needed for stress analysis problems, even when plastic strain dominates the elastic strain, in order to determine the elastic-plastic boundaries which delineate the regions in which plastic flow is taking place.

Plasticity theory was initially developed on the basis of infinitesimal displacement analysis for which convected terms are neglected. The elastic-plastic law for ideal-plasticity gives an increment of total strain according to:

$$d\epsilon_{ij} = ds_{ij}/2G + d\lambda\ s_{ij} \tag{1}$$

where s_{ij} is the stress deviator, G the elastic shear modulus and $d\lambda$ a parameter which prescribes the extent of plastic flow. In rate form this becomes

$$\dot{\epsilon}_{ij} = \dot{s}_{ij}/2G + \dot{\lambda}\ s_{ij} \tag{2}$$

where the superposed dot denotes the "time" derivative which can be any time-like variable which expresses the forward procedure of the process, such as the displacement of the driving piston in an extrusion process.

The first term on the right-hand side of both (1) and (2) corresponds to the elastic strain change, and the second term to the plastic straining. Because they refer to ideal plasticity, plastic flow could in principle continue indefinitely and the scalar parameter λ is introduced to prescribe the amount of plastic straining. Because infinitesimal displacement theory is adopted, which neglects distortion of a coordinate net considered to be embedded in the deforming material, the rate of strain does not distinguish between changes at a fixed point in space or at a particular material particle. Since, as already stated, convection effects need to be included in elastic-plastic analysis, nonlinear, finite-deformation-valid kinematics must be included.

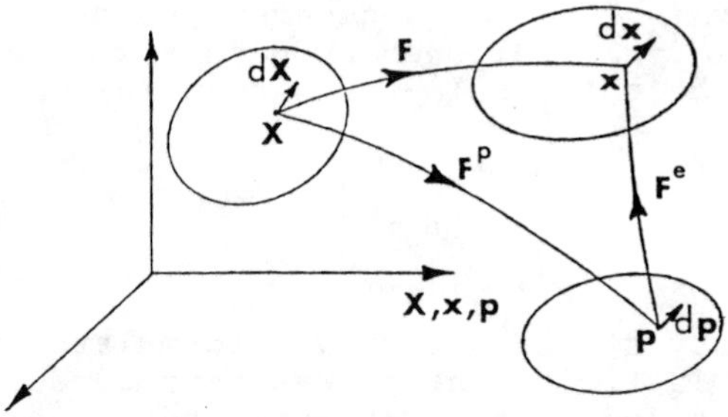

Figure 1 Elastic-Plastic Deformation Kinematics

Figure 1 depicts the elastic-plastic deformation of a body. X are the particle coordinates of the body in its initial undeformed configuration. The particle defined by X has been moved to the position x. It would occupy the position p in an envisaged corresponding unstressed body subjected therefore to only plastic deformation. The change in deformation from p to x thus constitutes the elastic deformation, so that the resultant deformation from X to p, followed by p to x yields the total elastic-plastic deformation. The deformation of the particle considered is expressed by the deformation gradient matrix F:

$$F = \partial x / \partial X \tag{3}$$

The small arrows dX, dp and dx emanating from X, p and x indicate how a small material element is transported and deformed. Their lengths are respectively dS, ds^p and ds. The total Lagrange strain E is given by:

$$ds^2 - dS^2 = dX^T(F^T F - I)\, dX = dX^T(2E)\, dX \tag{4}$$

where the superscript T indicates the matrix transpose. Thus:

$$2E = (F^T F - I) \tag{5}$$

Similarly the elastic strain E^e and plastic strain E^p are given by

$$2E^e = (F^{e^T} F^e - I) \text{ and } 2E^p = (F^{p^T} F^p - I) \tag{6}$$

Since the deformation gradient F of the mapping from X to x is equivalent to the sequence X to p with deformation gradient $\partial p/\partial X = F^p$ followed by p to x with gradient $\partial x/\partial p = F^e$, the chain rule for sequential mappings gives [2]

$$F = F^e F^p \tag{7}$$

The theory of elasticity was formulated and perfected long before plasticity theory was introduced, and many similarities aided in the development of the latter. However, there are basic differences which have not always been fully understood. For example, in the 1940's and 50's the so-called deformation theory of plasticity received some credence. It replaced the plastic strain rate in the flow law by the plastic strain, presumably extrapolated from familiarity with the established elasticity theory. That it was clearly at variance with physical reality was documented by Handelman, Lin and Prager [3]. The need for finite-deformation-valid non-linear kinematics in plasticity theory, already alluded to, calls for care in formulating the flow law. Infinitesimal displacement theory envisaged as the basis for Eq.(2) does not distinguish between the time derivative of the strain and the symmetric part of the velocity gradient - the rate of deformation or velocity strain. When finite deformation is involved these can be quite disparate and care must be exercised in formulating a correct form for the plastic flow law.

Plastic flow is caused by the migration of dislocations and other crystal structure defects through the crystal lattice. They are maintained in motion by the stresses of yield-stress magnitude currently acting throughout the medium. The associated strain-rate-type variable is thus the rate of strain about the current geometry. The velocity strain or rate of deformation, the symmetric part of the velocity gradient, expresses the rate of strain defined in terms of the current configuration, i.e., the rate of strain at the instant the strain is zero. This is clearly the appropriate variable for the flow law, just as it is for viscous flow. In contrast, the rate of Lagrange type plastic strain $\dot{E}^p$ involves the change in configuration from the initial state to the current one and so incorporates information about both configurations.

The velocity gradient, L, in the current configuration is given by [4]

$$L = \frac{\partial v}{\partial x} = \frac{\partial v}{\partial X}\frac{\partial X}{\partial x} = \dot{F}F^{-1} \; ; \; D = L|_s \tag{8}$$

where v is the particle velocity, and D, the symmetric part, gives the rate of deformation. By differentiating the second equation in (6)

$$\dot{E}^p = F^{p^T} D^p F^p \tag{9}$$

due to Green and Naghdi [5]. Since plastic strains can be large, F^p can be far removed from the unit matrix. For a tensile stretch of two, F^p would be a diagonal matrix with diagonal terms $(2, 1/\sqrt{2}, 1/\sqrt{2})$, so that $\dot{E}^p$ and D^p would be quite disparate.

Use of the rate of deformation D in elastic-plastic theory leads to a convenient coupling of elastic and plastic terms. The unstressed configuration p is not unique since changing its orientation leaves it unstressed. According to the polar decomposition theorem p can therefore be selected to prescribe elastic deformation without rotation so that $F^e = V^e = V^{e^T}$, a symmetric matrix. Since elastic strains are of the order 10^{-3} (yield stress/elastic modulus) V^e would differ from the unit matrix only by terms of the order 10^{-3}. This simplifies the expression for the total rate of deformation D in terms of elastic and plastic components D^e and D^p, obtained from (7) and (8), see [4], giving:

$$D = D^e + V^e D^p V^{e^{-1}}\Big|_s + V^e W^p V^{e^{-1}}\Big|_s \tag{10}$$

where W^p is the plastic spin, a skew-symmetric matrix. Since $V^e \sim I$, (10) is accurately approximated by

$$D = D^e + D^p \tag{11}$$

Thus, in terms of the rate of deformation, the elastic-plastic coupling is simple.

This is not the case if the finite plastic strain or its derivative is used. Differentiation of the equations (6) generate expressions analogous to (9):

$$\dot{E}^e = V^e D^e V^e, \; \dot{E} = F^T D F \tag{12}$$

While $V^e \sim I$, with large plastic flow F^p and F will be far removed from the unit matrix. Taking into account (9) and (12), and the additivity expressed in (11), it is clear that additivity of strain rates will be strongly violated. This is also true of Lagrange strain; relation (7) implies that

$$E = F^{p^T} E^e F^p + E^p \tag{13}$$

It is clear that the plastic-strain variable is not an appropriate choice for formulating elastic-plastic theory at finite strain. It is natural for elasticity since it couples the unstressed reference configuration and the current configuration, the stretching between which generates the stress directly. For plastic analysis the initial state is not recoverable. In engineering stress-analysis problems, a virgin state which has not been subjected to plastic flow is usually not known since most engineering components have been formed plastically prior to being presented for stress-analysis, but this presents no difficulty. The circumstance is consistent with the inappropriateness of terming plastic strain to be a state variable.

REFERENCES

1. R. Hill, The Mathematical Theory of Plasticity (The Clarendon Press, Oxford, 1950).

2. E.H. Lee, "Elastic Plastic Deformation at Finite Strains," J. Appl. Mech. 36, 1-6 (March 1969).

3. G.H. Handelman, C.C. Lin and W. Prager, "On the Mechanical Behavior of Metals in the Strain-Hardening Range," Quart. Appl. Math. 4, 397-407 (1947).

4. E.H. Lee, "Some Comments on Elastic-Plastic Analysis," Int. J. Solids Structures," 17, 859-872 (1981).

5. A.E. Green and P.M. Naghdi, "Some Remarks on Elastic-Plastic Deformation at Finite Strain," Int. J. Engng. Sc. 9, 1219-1229 (1971).

ACKNOWLEDGEMENT: Supported in part by the U. S. Army Research Office.

ON INCREMENTAL AND HEREDITARY DESCRIPTIONS OF HARDENING EVOLUTIONS OF PLASTICITY

ZENON MROZ
Institute of Fundamental Technological Research
Warsaw, Poland

S. K. JAIN
Department of Engineering Science & Mechanics
Virginia Polytechnic Institute & State University
Blacksburg, Virginia 24061 U.S.A.

ABSTRACT

A hereditary hardening rule is incorporated into the classical plasticity formalism. A multi-surface hardening rule is discussed. The effectiveness of hereditary rule is shown, and an attempt is made to establish a link between the two rules.

INTRODUCTION

In mathematical modeling of plastic deformation processes, the state of material is usually represented by a yield surface and the associated hardening variables. The evolution of variables in rate or incremental forms provides a relation between the state variables and the plastic strain tensor.

The hardening process is inseparably linked with the anisotropic material structure, and therefore, anisotropy is decomposed into the Bauchinger (or back stress) anisotropy and the texture (or fabric) anisotropy. This decomposition was treated in Ref. [1] by providing rules of texture and back stress evolutions utilizing the multi-surface plasticity concepts where the effect of maximal prestress on subsequent plastic response is naturally incorporated by introducing switching points between specific loading events and by formulating rules for back stress evolution accounting for prestress. An alternative way to describe back stress evolution is to formulate hereditary rule in a form first discussed by Backhaus [2]:

$$\alpha_{ij} = \int_0^{\zeta} H(\zeta - \zeta') \frac{d\varepsilon^p_{ij}}{d\zeta'} d\zeta' \tag{1}$$

where the yield surface is specified by

$$F(\sigma_{ij}, \alpha_{ij}, \sigma_p) = f(\sigma_{ij} - \alpha_{ij}) - \sigma_p = 0 \tag{2}$$

and the flow rule is

$$\dot{\varepsilon}^{p}_{ij} = \dot{\lambda}\,\frac{\partial F}{\partial \sigma_{ij}} \quad , \quad \dot{\lambda} F = 0 \quad , \quad \dot{\lambda} > 0 \tag{3}$$

and

$$d\zeta = (d\varepsilon^{p}_{ij}\; d\varepsilon^{p}_{ij})^{\frac{1}{2}} \tag{4}$$

H(z) is called a hereditary function. Eq. 1 describes an evolution of back stress similar to one associated with multi-surface plasticity. Also, an assumption of vanishing elastic domain provides transition of Eqs. (1-4) to the endochronic theory of plasticity, cf. Valanis [3].

INCREMENTAL EVOLUTION RULES

Consider a multi-surface hardening model constructed by a set of nesting surfaces:

$$f^{(m)}(\sigma_{ij} - \alpha^{(m)}_{ij}) - \sigma^{(m)} = 0 \tag{5}$$

where, $f^{(o)} - \sigma^{(o)} = 0$, is the yield surface enclosing elastic domain, and $f^{(m)} - \sigma^{(m)} = 0$, m = 1,2,... are consecutive loading surfaces. Consider an evolution rule of the form,

$$\dot{\alpha}^{(m)}_{ij} = A^{(m)}\; m^{(m)}_{ij}\; \dot{\zeta} \tag{6}$$

where,

$$m^{(m)}_{ij} = \frac{\sigma^{(m+1)}_{ij} - \sigma^{(m)}_{ij}}{|\sigma^{(m+1)}_{k\ell} - \sigma^{(m)}_{k\ell}|} \tag{7}$$

in which $\sigma^{(m)}_{ij}$ is the associated stress on a consecutive yield surface corresponding to the same direction of the normal vector as for the stress point on the yield surface, as shown in Fig. 1. Thus, for a given stress state σ_{ij}, there exist ℓ, ($\ell \leq m$), associated stress states $\sigma^{(m)}_{ij}$ on consecutive loading surfaces. For the active loading surface, there is $\sigma^{(m)} = \sigma_{ij}$, that is the associated stress equals the actual stress. The functions $A^{(m)}$ depend on the absolute value of the back stress $\alpha^{(m)} = (\alpha^{(m)}_{ij}\, \alpha^{(m)}_{ij})^{\frac{1}{2}}$. For monotonic functions $A^{(m)}(\alpha^{(m)})$, the translation of consecutive surfaces will diminish with their sizes.

In such a formulation the whole field of hardening moduli is modified during plastic deformations. The evolution rule provides a set of back stress tensors $\alpha^{(m)}_{ij}$ at each instant. The complex loading history is thus naturally incorporated in this evolution. In particular cyclic loading effects such as ratchetting, memory of maximal prestress, etc. can be quantitatively simulated.

INTEGRAL HARDENING RULES

Consider a yield surface which is a function of deviatoric stress tensor, s_{ij},

$$f(s_{ij} - \alpha_{ij}) - \sigma_p(\zeta) = 0 \tag{8}$$

and express the back stress evolution (1) in terms of deviatoric plastic strain e^p_{ij} as follows.

$$\alpha_{ij} = \int_0^z H(z - z') \frac{de^p_{ij}}{dz'} dz' \tag{9}$$

$$\text{where,} \qquad dz = \frac{d\zeta}{g(\zeta)} , \quad d\zeta = (de^p_{ij}\, de^p_{ij})^{\frac{1}{2}} \tag{10}$$

in which, $g(\zeta)$ is an isotropic hardening function. The flow rule for such a material takes the form

$$de^p_{ij} = d\lambda \frac{\partial f}{\partial s_{ij}} , \quad f - \sigma_p = 0 , \quad d\lambda > 0 \tag{11}$$

The incremental form of Eq. (9) is as follows.

$$d\alpha_{ij} = H(0)\, de^p_{ij} - h_{ij}\, dz \tag{12}$$

$$\text{where, } h_{ij} = - \int_0^z \frac{\partial H(z - z')}{\partial z} \frac{de^p_{ij}}{dz'} dz' \tag{13}$$

The consistency condition,

$$\frac{\partial f}{\partial s_{ij}} ds_{ij} - \frac{\partial f}{\partial s_{ij}} d\alpha_{ij} - \frac{d\sigma_p}{dz} dz = 0 \tag{14}$$

provides the expression for $d\lambda$,

$$d\lambda = \frac{g(\zeta) \left(\dfrac{\partial f}{\partial s_{ij}} ds_{ij} \right)}{H(0)\, g(\zeta) \left(\dfrac{\partial f}{\partial s_{k\ell}} \dfrac{\partial f}{\partial s_{k\ell}} \right) - \left(\dfrac{\partial f}{\partial s_{mn}} h_{mn} + \dfrac{d\sigma_p}{dz} \right) \left(\dfrac{\partial f}{\partial s_{pq}} \dfrac{\partial f}{\partial s_{pq}} \right)^{\frac{1}{2}}}$$

$$= \frac{1}{K} \frac{\partial f}{\partial s_{ij}} ds_{ij} \tag{15}$$

where K is the plastic hardening modulus. Thus,

$$de_{ij} = de^e_{ij} + de^p_{ij} = \frac{ds_{ij}}{2G} + d\lambda \frac{\partial f}{\partial s_{ij}} \tag{16}$$

For strain-controlled processes, the inverse relations to (16) can be easily obtained.

It is seen that the incorporation of evolution rule (9) into the classical plasticity formalism is straightforward. Moreover, the incremental form (12) can be correlated with the evolution rule (6) by a proper selection of the function H(z).

Assuming the function H(z) to be of the form,

$$H(z) = \sum_{r=1}^{n} A_r e^{-\beta_r z} \tag{17}$$

we proceeded to test the validity of the integral hardening rule by the biaxial experiments on OFHC stabilized copper [4]. The objective was to determine the material parameters from a uniaxial experiment, and then predict the behavior under nonproportional biaxial loading. The following material parameters [5] produced a uniaxial curve shown in Fig. 2.

A_1 = 16040 ksi, A_2 = 1206 ksi, β_1 = 1380, β_2 = 205,
E = 15300 ksi, ν = 0.33, σ_p = 9.8 ksi = const., g = 1

For the random nonproportional loading history of Fig. 3a, the predictions are shown in Fig. 3b. The predictions are similar to those obtained from the endochronic theory [6]. However, the plasticity formalism used here seems to possess two distinct advantages. First, the resulting constitutive matrix is incrementally linear. As shown in the appendix, the endochronic theory leads to an incrementally nonlinear constitutive matrix. Second, a very stiff initial response of the endochronic theory, used in generating an apparent elastic domain, is unpleasant from the viewpoint of numerical analysis.

REFERENCES

1. Z. Mroz and A. Niemunis, "On the description of deformation anisotropy of materials", Proc. IUTAM Symp. Yielding, Damage, and Failure of Anisotropic Solids, edited by J. P. Boehler (Grenbole, 1987).
2. G. Backhaus, "Flresspannungen und Flressbedingung ber Zyklischen Verformungen", ZAMM, 56, 337-348 (1976).
3. K.C. Valanis, "Fundamental consequences of a new intrinsic time measure: Plasticity as a limit of the endochronic theory", Archives of Mechanics, 32, 171-191 (1980).
4. H.S. Lamba and O.M. Sidebottom, "Cyclic plasticity for non-proportional paths", J. Eng. Mat. Tech., ASME, 100, 104-111 (1978).
5. Z. Mroz and S.K. Jain, "Local directional properties of some incrementally linear and nonlinear constitutive equations in plasticity", Tech. Rep. VPI-E-90-12, Coll. of Engrg., VPI&SU, Blacksburg, Virginia (May 1990).

6. S.Y. Hsu, S.K. Jain, and O.H. Griffin, "A procedure for determining endochronic material functions and verification of endochronic theory for nonproportional loading paths", Report No. VPI-E-89-25, Coll. of Eng., VPI&SU, Blacksburg, Virginia (1989). (To appear in J. Eng. Mech., ASCE.)

APPENDIX: ENDOCHRONIC CONSTITUTIVE LAW

Consider the endochronic equation providing an integral relationship between stress and plastic strain, cf. Valanis [3]

$$s_{ij} = \int_0^z H(z - z') \frac{de^p_{ij}}{dz'} dz' \tag{18}$$

The incremental form associated with Eq. (18) is

$$ds_{ij} = H(0)\, de^p_{ij} - h_{ij}(z)\, dz \tag{19}$$

where h_{ij} is defined by Eq. (13). The plastic constitutive matrix can now be established as follows.

$$C^p_{ijk\ell} = \frac{ds_{ij}}{de^p_{k\ell}} = H(0)\ \delta_{ij}\ \delta_{k\ell} - h_{ij} \frac{dz}{de^p_{k\ell}} \tag{20}$$

or,

$$C^p_{ijk\ell} = H(0)\ \delta_{ij}\ \delta_{k\ell} - \frac{h_{ij}}{g} \frac{de^p_{k\ell}}{d\zeta} \tag{21}$$

The matrix is incrementally nonlinear. A similar technique was used earlier [6] for establishing a constitutive matrix for the endochronic theory.

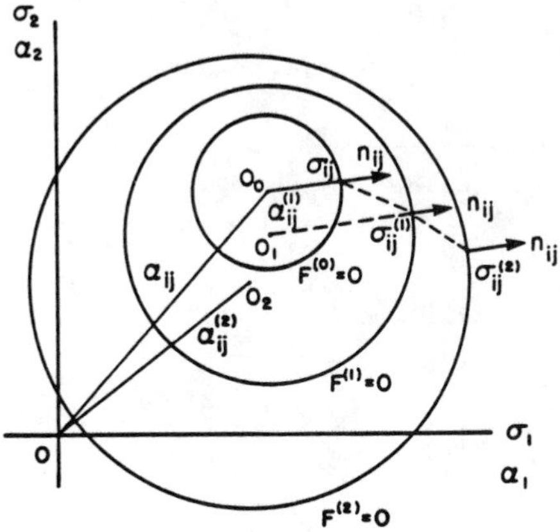

Fig. 1. A Multi-surface Hardening Model

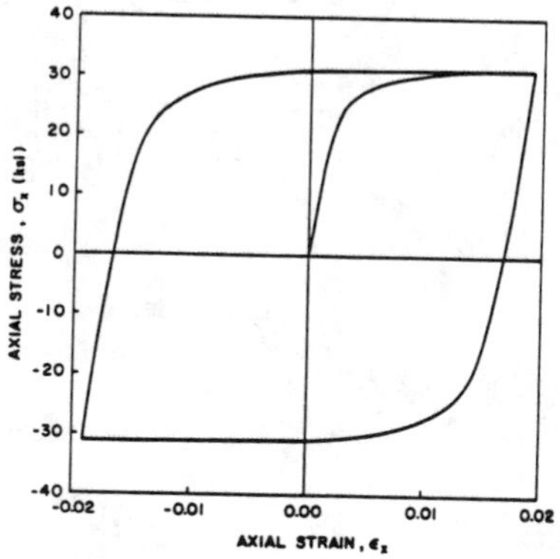

Fig. 2. Response of Plasticity Model with von-Mises Yield Function and Integral Translation Rule

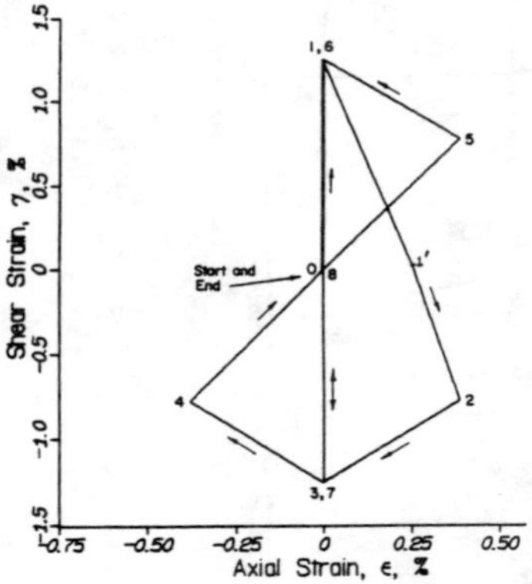

Fig. 3a. Random Non-proportional Strain-controlled Loading Path [4].

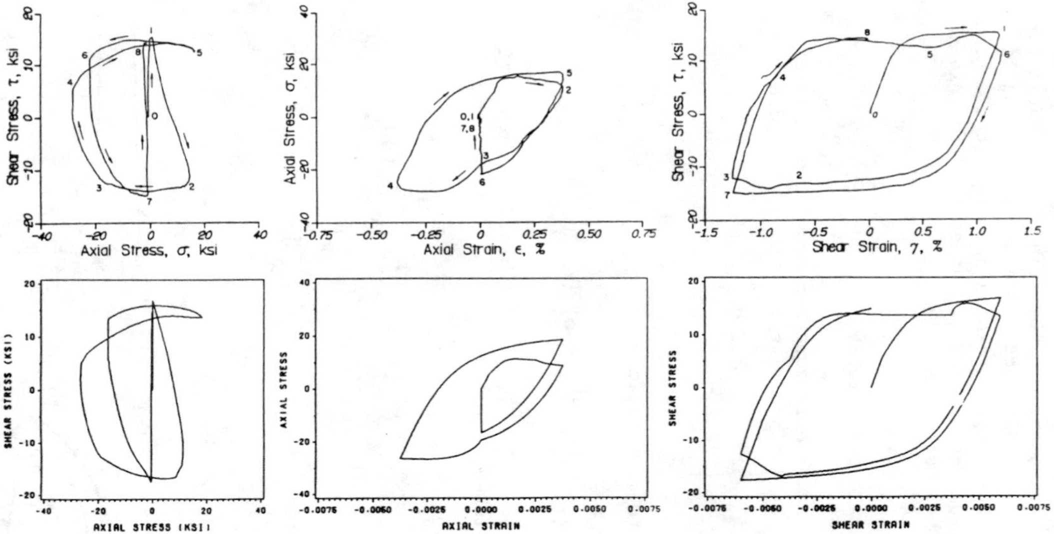

Fig. 3b. Top Row: Experimental Response [4]
Bottom Row: Model Response

A NOTE ON SAND LIQUEFACTION AND SOIL STABILITY

ROBERTO NOVA
Milan University of Technology (Politecnico)

ABSTRACT

It is shown that ,to describe static liquefaction in the framework of strainhardening plasticity with a single yield function, the flow rule must be non-associated.By assuming further that the positive definiteness of the second order work is a convenient criterion for stability, it will be proven that unstable behaviour may occur before liquefaction can take place.

STATIC LIQUEFACTION

Figure 1 shows typical experimental results obtained by Castro [1] in drained and undrained triaxial tests on a loose saturated sand. At the stress level for which a peak in the stress-strain curve occurs in the undrained test, the drained sample appears at first sight quite stable. Indeed any non-linearity can hardly be detected from the measured stress-strain curve. Also, Castro found that the mobilised friction angle at failure in the drained test is the same as the mobilised friction angle in the undrained test calculated at the end of the test, after that liquefaction fully developed.

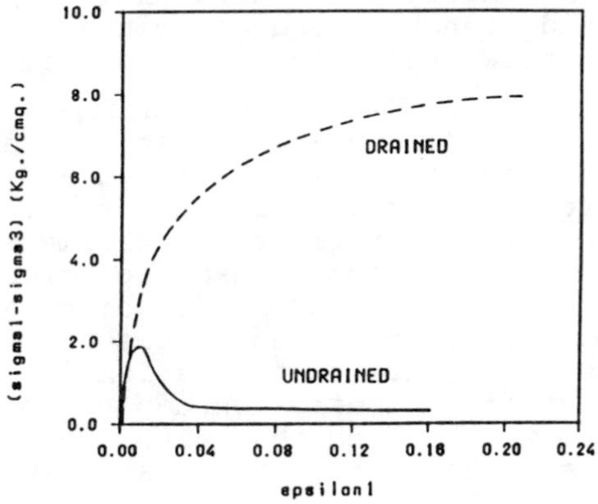

FIG. 1

Fig. 1. Experimental results of drained and undrained triaxial compression tests on a loose sand, isotropically consolidated to a cell pressure of 4 Kg/cm^2 - modified from Castro [1].

This type of behaviour may look rather odd and indeed 'ad hoc'explanations were suggested, as for instance the 'flow structure theory'proposed by Casagrande [2]. Strainhardening plasticity however does not need any to 'explain' why and when it occurs.

To show that, consider a general expression for an elastoplastic strainhardening law, characterized by a single yield function and a single plastic potential, not necessarily coincident with it:

$$\dot{\varepsilon} = (C_e + 1/H\ m\ .\ \tilde{n}\ )\ \dot{\sigma} = C\ \dot{\sigma} \quad (1)$$

where $\dot{\varepsilon}$, $\dot{\sigma}$ are the strain and stress rates, respectively, listed in vectors of six components, C is the compliance matrix while C_e is its elastic part, m and n are the unit vectors normal to the plastic potential, g, and yield function, f, respectively, and H is the hardening modulus. A superposed tilde means transposed.

In an undrained test the volumetric strain rate is zero. Denoting by δ the vector:

$$\tilde{\delta} = \{\ 1\ 1\ 1\ 0\ 0\ 0\ \} \quad (2)$$

we have:

$$\dot{v} = \tilde{\delta}\ C\ \dot{\sigma} = 0 \quad (3)$$

Since at peak $\dot{\sigma} = \dot{a}\delta \neq 0$ it follows necessarily that

$$\tilde{\delta}\ C\ \delta = \tilde{\delta}\ (C_e + 1/H\ m.\tilde{n}\ )\ \delta = 0 \quad (4)$$

Because of the positive definiteness of the elastic matrix, if the flow rule is associated, n = m and Equation 4 can not be fulfilled for any positive value of the hardening modulus. Therefore liquefaction, which starts after a peak in the stress strain relation in a region where hardening is still positive, can not be modelled by an elastoplastic strainhardening model with a single yield function and associated flow rule. If, however, the flow rule is non-associated, then the scalar product $\tilde{\delta}\ m\ \tilde{n}\ \delta$ may be negative and for a convenient value of H and of the elastic parameters it is possible that Equation 4 is fulfilled and liquefaction modelled. Note that this occurs for a certain scalar combination of the constitutive parameters. Elastic parameters as well as those linked to hardening play an important role. For a denser sand, which is stiffer and more dilatant than that tested by Castro, the constitutive parameters are such that Equation (4) can never be fulfilled and static liquefaction can not occur.

It is also worthnoting that although a peak appears in the stress-strain law, the hardening modulus is positive. This situation should therefore not be confused with softening.

STABILITY

When Equation (4) is fulfilled, the compliance matrix C can not be positive definite. This can be easily verified by choosing as a special stress rate the vector $\dot{\sigma} = \dot{a}\,\delta$. We have:

$$\dot{a}^2\, \tilde{\delta}\, C\, \delta = 0 \qquad (5)$$

Equation (5) implies that for the particular stress path chosen the second order work $d^2W = \tilde{\dot{\sigma}}.\dot{\varepsilon}$ is also zero.

Hill [3] gave the condition $d^2W > 0$ as a sufficient condition for material stability. If we assume now that such a condition is also necessary, as suggested by Maier and Hueckel [4], it turns out that on the verge of liquefaction the soil sample is unstable.

Material instability may occur even before liquefaction can take place, however. In fact C will be positive definite if so is its symmetric part C_s, For a symmetric matrix a necessary and sufficient condition for positive definiteness is that all the leading principal minors are positive. In this case, however, the matrix C_s is given by the sum of two components:the elastic compliance, which is positive definite, and the symmetric part of the plastic compliance which depends continuously on the hardening modulus H. For a high value of the hardening modulus the sign of d^2W is controlled by C_e and is necessarily positive. For H tending to zero, however, if $m \neq n$, it certainly exists a stress rate $\dot{\sigma}$ such that the second order work is negative. In fact, in that case elastic strains are small compared to plastic ones and the sign of d^2W is controlled by the latter ones. There will be then a critical value of H for which d^2W is positive semidefinite, see Langhaar [5]. For that value there will exists one and only one $\dot{\sigma}^* \neq 0$, such that $d^2W = 0$, while for all other possible $\dot{\sigma}$, d^2W will be positive.

If $\dot{\sigma}^*$ is such that

$$\tilde{\dot{\sigma}}^*\, C_s\, \dot{\sigma}^* = 0 \qquad (6)$$

and C_s is positive semidefinite and symmetric, a necessary condition is

$$C_s\, \dot{\sigma}^* = 0 \qquad (7)$$

what implies

$$\det C_s = 0 \qquad (8)$$

Therefore the leading principal minor that becomes non-positive first, i. e. for a higher value of H, is the determinant. Thus instability, in the sense of non-positiveness of the second order work occurs before liquefaction may occur. The stress path which causes instability is that fulfilling Equation (7).

Note that although the material compliance depends on the orientation of the stress rate, since plastic strains may occur only if $\tilde{n}\, \dot{\sigma} > 0$, the strain rate direction for which d^2W may become zero always obeys this condition.

The loss of stability may be studied also in terms of the stiffness matrix. With a similar path of reasoning it is easy to show that instability occurs when

$$\det D_s = 0 \qquad (9)$$

where D_s is the symmetric part of the stiffness matrix D.

Note also that uniqueness of solution of the constitutive equation:

$$\dot{\sigma} = D\,\dot{\varepsilon} \qquad (10)$$

is lost when:

$$\det D = 0 \qquad (11)$$

what, for an elastoplastic strainhardening material, occurs when the hardening modulus is zero. The stress level coresponding to that value of H is normally considered as a failure value for a homogeneous specimen.

Uniqueness of solution of Equation (10) does not implies stability. In fact Ostrowski and Taussky [6] have shown that if D_s is positive definite:

$$\det D \geq \det D_s \qquad (12)$$

so that when D_s becomes positive semidefinite and stability is lost, the determinant of D is still nonnegative and the solution of Equation (10) may still be unique. Note that if the determinant of D is positive, the hardening modulus is also positive so that loss of stability occurs in the hardening regime.

REFERENCES

[1] Castro G. "Liquefaction and cyclic mobility of saturated sand" J. GED-ASCE, 101, GT6, (1975), 551-569.

[2] Casagrande A. "Liquefaction and cyclic deformation of sands. A critical review". Harvard SM Series n. 88 presented at the 5th Panam. C. SMFE. , Buenos Aires, (1975).

[3] Hill R. "A general theory of uniqueness and stability in elastic-plastic solids" J. Mech. Phys. Solids 6, 236-249.

[4] Maier G. and Hueckel T. "Non associated and coupled flow rules of elastoplasticity for geotechnical media" Spec Sess. Constitutive Relations for Soils 9th ICSMFE, (1977), 129-142.

[5] Langhaar H. L. "Energy methods in applied Mechanics" New York John Wiley and Sons (1962).

[6] Ostrowsky A. and Taussky O.'On the variation of the determinant of a positive definite matrix'Nederl. Akad. Wet. Proc. (A)54, (1951), 383-351.

Constitutive modeling of a solid exhibiting transformation induced plasticity

H. Okada, N. Ramakrishnan[1] and S. N. Atluri
Computational Mechanics Center,
Georgia Institute of Technology,
Atlanta, Georgia 30332-0356, U. S. A.

ABSTRACT

In this paper, the basic behavior of a solid with the second phase particles undergoing dilatational stress induced phase transformation, is analyzed. A rate (incremental) form constitutive model is developed, using a self-consistent method. Some illustrations on the dilatational stress-strain response are presented, based on the constitutive equation derived. A finite element numerical simulation is also conducted to study the constitutive behavior. Both the results, of the self-consistent method and of the finite element numerical simulation, agree very well.

1 Introduction

In this paper, a constitutive model for describing the macroscopic behavior of a solid exhibiting transformation plasticity is presented. Essentially, this paper aims at developing a constitutive equation which can be implemented in computational methods such as finite element method or boundary element method to analyze crack problems.

Transformation toughening has been found to be a fracture toughness enhancement mechanism in certain ceramic materials, which contain tetragonal zirconia phase. The essential feature of this phenomenon is the allotropic transformation of the second phase (zirconia) from tetragonal to monoclinic phase. This phase transformation is triggered by an applied stress, and results in a nonlinear stress-strain behavior (McMeeking and Evans [5] and Budiansky, Hutchinson, and Lambropoulos [1]). The phase transformation is induced by high stress condition around the crack tip, and produces a significant amount of irreversible strain, thereby reducing the stress intensity at the crack tip. Partially stabilized zirconia (PSZ) and Zirconia toughened Alumina (ZTA) are typical materials exhibiting this phenomenon. A considerable effort has been made in developing constitutive models for the transformation induced plasticity. The behavior of the transformation plasticity solid (TPS) was assumed to be dilatationally nonlinear and distortionally linear, and a phenomenological macroscopic constitutive equation was presented by Budiansky, Hutchinson, and Lambropoulos [1]. Lambropoulos [3] obtained an incremental constitutive model which was governed by a flow rule and included the effect of the distortional stress component also. In Lambropoulos [4], a strain localization criterion was discussed based on the constitutive equation developed in [3], and the material properties of the second phase were assumed to be the same as those of the matrix material. McMeeking [6] presented a brief discussion on the irreversible strain due to the phase transformation of the second phase particles whose elastic properties differed from those of the matrix material. Suresh [10] described a flow rule based on a yield criterion presented by Chen

[1]on leave from DMRL,Kanchanbagh, Hyderabad, India - 500258

and Reys-Morel [2]. The yield condition presented by Chen and Reys-Morel [2] was composed of the first invariant of stress and the second invariant of deviatoric stress.

In the present study, a constitutive model is developed based on micromechanics. The TPS is considered to be a two phase composite material, consisting of the matrix and the second phase particles. The second phase particles undergo phase transformation driven by the applied stress. In order to characterize the amount of second phase material which is transformed, the hydrostatic tensile stress inside the second phase particles is chosen to be the effective stress. The second phase particles are assumed to be spherical in shape, to make the constitutive modeling tractable. The self-consistent method (Mura [7]) is appropriately employed to obtain both the average elastic moduli and the transformation induced nonlinear response of the composite. In addition to it, a numerical simulation using finite element method is conducted to study the constitutive behavior of TPS, and the results of this numerical experiment are compared to the constitutive equation derived using the self-consistent method.

2 Constitutive equation for the transformation plasticity solid (TPS).

In this section, a rate (incremental) form constitutive equation is presented for the solid with the second phase particles undergoing the stress induced phase transformation. The self-consistent method [7] is employed considering three material constituents, the matrix, the transformed and untransformed second phase particles, as shown in Fig. 1. The phase transformation of the second phase particles is assumed to be induced by the hydrostatic tensile stress. When the stress in the second phase particle reaches a critical value, the phase transformation is assumed to take place. The critical stress may not be the same for all the second phase particles, because of slight difference of the chemical constituents. Therefore, it is appropriate to assume a variation of the critical stress among the second phase particles. We assume that the amount of second phase material that is transformed, is related to the stress in the untransformed second phase particles, as shown in Fig. 2.

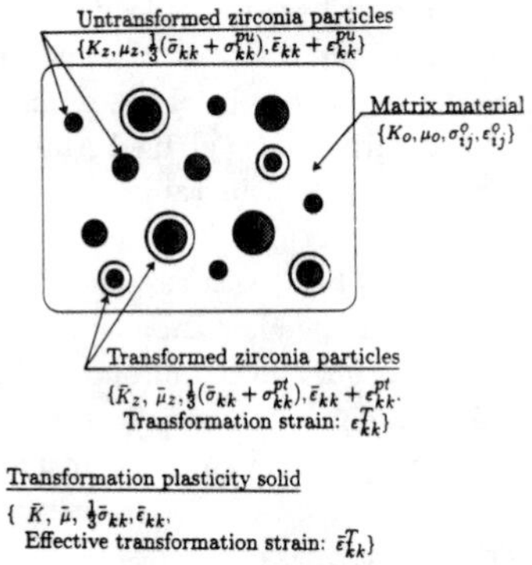

Fig. 1 A solid consisting of the matrix, the transformed and untransformed second phase particles.

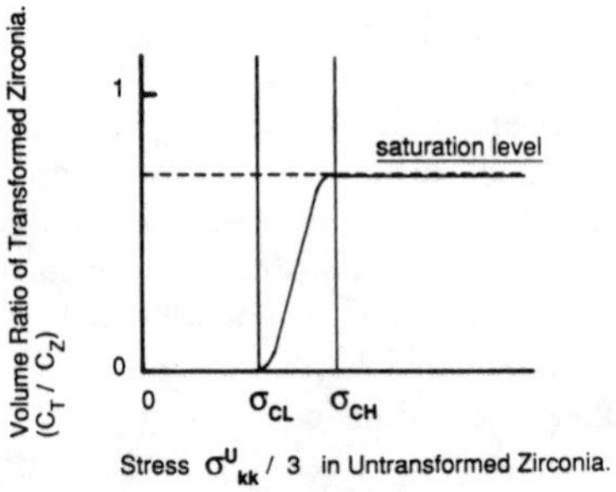

Fig. 2 The volume fraction of transformed zirconia as a function of the hydrostatic tensile stress inside the untransformed zirconia particles.

The phase transformation begins when the stress in the second phase particles reaches some critical value σ_{LC}, and ends when the stress in the untransformed particles reaches σ_{UC}. σ_{LC} and σ_{UC} are called the lower and the upper critical stresses in this study. As a limiting case, the upper and the lower critical stresses are equal to each other. All the second phase particles thus transform at the same stress level.

The constitutive equation is summarized as follows[2]:

$\bar{K}$:	effective bulk modulus.
K_o:	bulk modulus of the matrix material.
K_z:	bulk modulus of the second phase particle.
$\bar{\sigma}_{ij}$:	effective stress.
$\bar{\varepsilon}_{ij}$:	effective strain.
$\bar{\varepsilon}^T_{kk}$:	effective dilatational irreversible strain due to the transformation.
$\frac{1}{3}(\bar{\sigma}_{kk} + \sigma^{pu}_{kk})$:	hydrostatic tensile component of stress in the untransformed second phase particles.
S_{ijkl}:	Eshelby's tensor.
C_z:	volume fraction of the second phase material(Zirconia).
C_T:	volume fraction of the transformed second phase particles.

The effective elastic moduli can be shown [7], as:

$$\bar{K} = K_o + \frac{C_z(K_z - K_o)\bar{K}}{\bar{K} + \frac{1}{3}(K_z - \bar{K})S_{mmnn}}; \quad \bar{\mu} = \mu_o + \frac{C_z(\mu_z - \mu_o)\bar{\mu}}{\bar{\mu} + 2S_{1212}(\mu_z - \bar{\mu})} \tag{1}$$

and,

$$S_{mmnn} = \frac{1+\bar{\nu}}{1-\bar{\nu}}; \quad S_{1212} = \frac{4-5\bar{\nu}}{15(1-\bar{\nu})}$$

When the hydrostatic tensile stress in the untransformed particles is increasing (loading) [$\frac{1}{3}(\dot{\bar{\sigma}}_{kk} + \dot{\sigma}^{pu}_{kk}) > 0$]:

$$\dot{C}_T = h \times \frac{1}{3}(\dot{\bar{\sigma}}_{kk} + \dot{\sigma}^{pu}_{kk}) \tag{2}$$

and,

$$h = h(\frac{1}{3}(\bar{\sigma}_{kk} + \sigma^{pu}_{kk})) \tag{3}$$

The evolution equation for the effective transformation strain:

$$\dot{\bar{\varepsilon}}^T_{kk} = \frac{BhK_z\bar{K}\dot{\bar{\varepsilon}}_{kk}}{\bar{K} - \frac{1}{3}(\bar{K} - K_z)S_{iijj} + \frac{1}{3}K_z\bar{K}hBS_{iijj}} \tag{4}$$

where,

$$B = \frac{\{K_z\bar{K} + \frac{1}{3}K_z(K_o - \bar{K})S_{iijj}\}\varepsilon^T_{kk}}{\bar{K}\{\bar{K} - \frac{1}{3}(\bar{K} - K_z)S_{iijj} - \frac{1}{3}C_z(K_z - K_o)S_{iijj}\}}$$

The constitutive equation is written, as:

$$\dot{\bar{\sigma}}_{ij} = \delta_{ij}\bar{K}\{\dot{\bar{\varepsilon}}_{kk} - \frac{BhK_z\bar{K}\dot{\bar{\varepsilon}}_{kk}}{\bar{K} - \frac{1}{3}(\bar{K} - K_z)S_{iijj} + \frac{1}{3}K_z\bar{K}hBS_{iijj}}\} + 2\bar{\mu}\dot{\bar{\varepsilon}}'_{ij} \tag{5}$$

[2]See Okada [8] for the further details of the derivation.

When the stress inside the untransformed second phase particles is decreasing (unloading) [$\frac{1}{3}(\dot{\bar{\sigma}}_{kk} + \dot{\sigma}^{pu}_{kk}) \leq 0$]:

$$\dot{C}_T = 0 \tag{6}$$

The material is governed by Hooke's law:

$$\dot{\bar{\sigma}}_{ij} = \delta_{ij}\bar{K}\dot{\bar{\varepsilon}}_{kk} + 2\bar{\mu}\dot{\bar{\varepsilon}}'_{ij} \tag{7}$$

Some characteristic behavior of the TPS is presented in the following examples. The material is assumed to be alumina-zirconia type ceramic composite, and no other mechanisms of nonlinear deformation are assumed to take place. The ratio of the bulk modus of the second phase material to that of the matrix (K_z / K_o) is assumed to be 0.4, and the Poisson's ratios of both the materials are taken as 0.2. The value of the unconstrained dilatational phase transformation strain of the second phase particles(Zirconia) is 0.05 which is a typical value reported in the literature. The volume ratio of the transformed second phase material to the volume of the second phase material contained in the composite is assumed to have a linear variation with respect to the level of hydrostatic tensile stress in the untransformed second phase particles, as shown in Figs. 3 and 4.

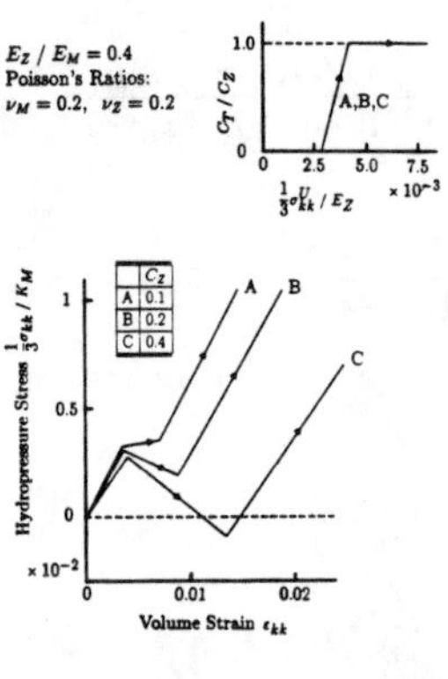

Fig. 3 Dilatational stress-strain curve for different volume fractions of zirconia.

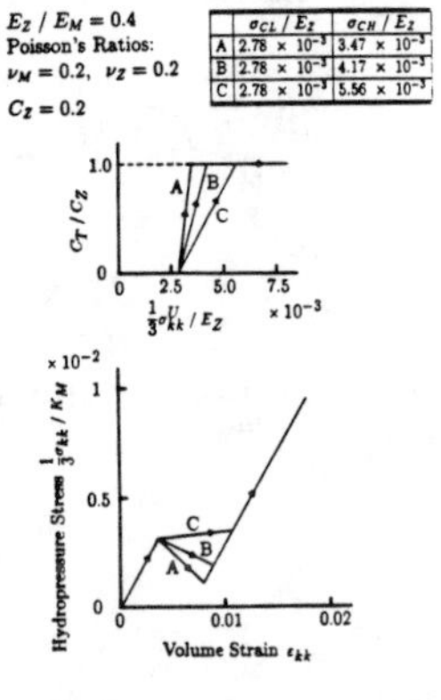

Fig. 4 Dilatational stress-strain curve for different transformation criteria. (The upper critical stress σ_{UC} is varied with the constant lower critical stress σ_{LC})

In order to show the influence of the volume fraction of the second phase material on the dilatational constitutive relation, the volume fraction of the second phase material is varied in Fig. 3. Due to the reduction of the stiffness and the increase of transforming material, the stress-strain curve exhibits stronger strain softening behavior as the volume fraction of the second phase material increases. Fig.4 shows hydrostatic tensile stress-dilatational strain curves for different values of the upper critical stress σ_{UC}. It is seen that the material exhibits a stronger volumetric strain softening when the difference between the upper and the lower critical stresses is small, and the amount of irreversible strain produced is unchanged.

3 Numerical experiment using finite element method

The constitutive behavior of the TPS is simulated, using the finite element procedure outlined by Ramakrishnan et al. [9]. A numerical experiment is carried out, varying the volume fraction and the shape of the second phase particles. Volume fraction is varied from 5 to 30 %. Both angular and spherical shapes of particles were considered. Alumina-Zirconia two phase composite material is considered here. Young's moduli of Al_2O_3 and ZrO_2 are taken as 390 and 210 GPa, respectively, and the Poisson's ratios of both the materials are taken as 0.25. The transformation volume strain in ZrO_2 is assumed to be 5 %. We have assumed the chemistry of the second phase particles to be the same, and consequently the critical stress at which they transform also should be the same. The critical stress for transformation is treated as a material constant.[3]

The results are presented in Figs. 5 and 6. In Fig. 5, a comparison is made to examine the difference between two materials having angular and spherical shapes of ZrO_2 keeping the volume fractions of the second phase to be the same. It is found that there is very little or no difference in the overall behavior. But there is a difference in the local serrations which is attributed to the difference in the particle size, shape and distribution. The second comparison is made, to observe the difference in the effect of varying the volume fraction of the second phase, and is shown in Fig. 6. The strain softening reduces with the reduction in the volume fraction. An important observation made in this work is that the slope of the stress-strain curve at the beginning of the transformation, appears to be equal to $-\frac{4}{3}\bar{\mu}$.

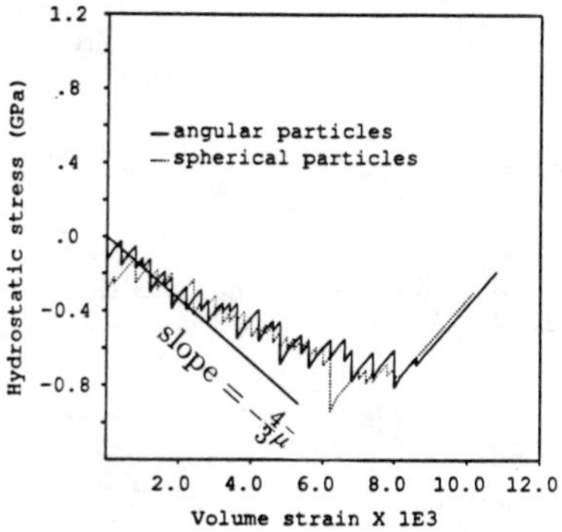

Fig. 5 Effect of shape of the second phase particles on the constitutive behavior of $Al_2O_3 - 30\%\ ZrO_2$.

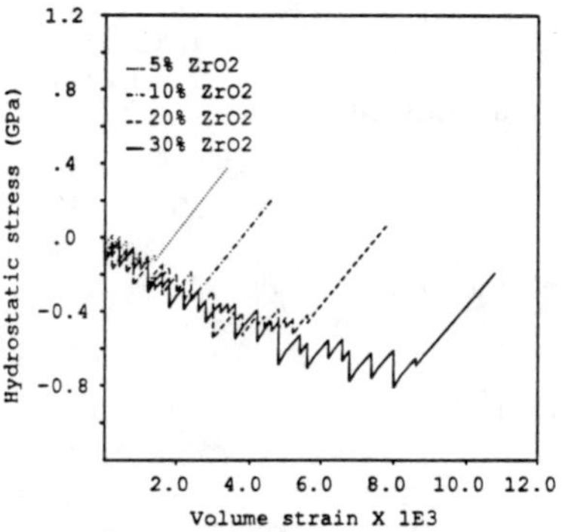

Fig. 6 Effect of the volume fraction of the second phase particles on the constitutive behavior of $Al_2O_3 - 30\%\ ZrO_2$.

4 Concluding Remarks

In this paper, the analysis of the behavior of a solid with its second phase particles undergoing the stress induced phase transformation is presented. The dilatational response of the transformation plasticity solid is such that, when the stress in the second phase particle reaches to a critical value, the phase transformation starts occurring, and the strain softening behavior follows. After the phase transformation reaches a

[3]This corresponds to $\sigma_{UC} = \sigma_{LC}$ ($h \equiv \infty$) in the constitutive equation, derived using the self-consistent method.

completion, the material behaves elastically. Moreover, one can show from Eq. (5), that in a limiting case where the upper and the lower critical stresses are equal to each other, the slope of the stress-strain curve, while the phase transformation is taking place, is $-\frac{4}{3}\bar{\mu}$, which corresponds to the *critical* transformation (as per the definition in [1] [4]) . It is observed in the finite element simulation also that the slope of the stress-strain curve appears to be equal to $-\frac{4}{3}\bar{\mu}$ at the beginning of the transformation, which implies that the transformation is always *sub-critical* and the limiting critical condition exists only at the beginning of the transformation. The results of the present study suggest that the transformation be always *sub-critical*, except for the limiting cases. Another important observation, made in the finite element simulation is that the difference in the stress-strain curve between the angular and the spherical second phase particle cases, is very small. It can be concluded that the constitutive equation, derived using the self-consistent method with the assumption of the spherical second phase particles, may be applicable for the other cases also.

Acknowledgements

The results reported herein were obtained during the course of investigation supported by the U.S. Army Research office. This support is gratefully acknowledged.

References

[1] B. Budiansky, J. W. Hutchinson and J. C. Lambropoulos, "Continuum theory of dilatant transformation in ceramics", *Int. J. Solids Structures*, vol. 19(1983), pp. 337-355

[2] I-W. Chen and P. E. Reyes Morel, "Implications of transformation plasticity in ZrO_2-containing ceramics: I, Shear and dilatational effects", *J. Am. Ceram. Soc.*, vol. 69, no 3(1986), pp. 181-189

[3] J. C. Lambropoulos, "Shear, shape and orientation effects in transformation toughening", *Int. J. Solids Structures*, vol. 22, no. 10 (1986), pp. 1083-1106

[4] J. C. Lambropoulos, "Macroscopic and microscopic localization of deformation in transformed-toughened ceramics", in "Constitutive modeling for nontraditional materials", ed. by V. Stokes and D. Krajcinovic,pp. 217-232, ASME winter annual meeting, Boston Massachusetts Dec. 13-19 (1987)

[5] R. M. McMeeking and A. G. Evans, "Mechanics of transformation-toughening in brittle materials", *J. Am. Cera. Soc.*, vol. 65, no. 5(1982), pp 242-246

[6] R. M. McMeeking, "Effective transformation strain in binary elastic composite", *J. Am. Cera. Soc.*, vol. 69, no. 12 (1986), pp. C-301-C-302

[7] T. Mura, "Micromechanics of defects in solids", Martinus Nijhoff Publishers(1982)

[8] H. Okada, Ph. D. Thesis, Georgia Institute of Technology (1990)

[9] N. Ramakrishnan, H. Okada and S. N. Atluri, "Computer simulation of transformation induced plasticity using finite element method", submitted to *Acta. Metall.* (1990)

[10] S. Suresh, "Constitutive behavior of ceramics: Implications for fracture under cyclic compressive loads", in "Constitutive modeling for nontraditional materials", ed. by V. Stokes and D. Krajcinovic,pp. 233-247, ASME winter annual meeting, Boston Massachusetts Dec. 13-19 (1987)

[4] In the *sub-critical* transformation case, the stress and strain fields are continuous throughout the material, whereas the discontinuities exist in the *critical* and *super-critical* transformation cases. See [1] for further detail.

A Non-associative Multi-surface Models for Geomaterials

Yii-Wen Pan[1]

1. Introduction

Various multi-surface model [e.g. Dafalias 1986 Mroz, et. al. 1981; Pietruszczak and Poorooshasb 1987; Prevost 1978] have demonstrated their effectiveness and capability in modelling the inelastic behavior of geomaterials under both monotonic and cyclic loadings. Among many multi-surface models, the transitional yielding model proposed by Pan and Banerjee (1987), which makes use of a yielding ratio and a transformed stress state. In this paper a generalized transitional yielding approach, with any desired number of limit-surfaces, is introduced. These limit surfaces represent various degree of the hierachical material memory due to stress history. This model includes a new approach for describing non-associative behavior of granular materials without using different functions for yield surface and plastic potential surface.

2. The Transitional Yielding Approach

The fundamental assumption of the transitional yielding model is the existence of a bounding surface and an instantaneous yield surface [Pan and Banerjee, 1987]. The instantaneous yield surface is defined by a transformed stress state and a yielding ratio, without the need to define a conjugate point. A brief outline of the transitional yielding approach is presented here.

Suppose the bounding surface can be expressed as

$$F(\{\sigma\}, a_c) = 0 \tag{1}$$

in which a_c = the size of the bounding surface. The instantaneous yield surface is assumed to satisfy

$$f = f(\{\sigma^*\}, a) = 0 \tag{2}$$

where

$$\{\sigma^*\} = \{\sigma\} - (1-s)\{\sigma^R\} \tag{3}$$

in which $\{\sigma^*\}$ = the transformed stress state; $\{\sigma^R\}$ = the stress tensor at the instance of the most recent stress reversal; s = the yielding ratio, which is the ratio of the size of the instantaneous yield surface to the size of the bounding surface, $0 < s < 1$; and a = the size of current yield surface, which satisfies $a = sa_c$

A stress reversal condition occurs whenever

$$f(\{\sigma^*\} + \{\Delta\sigma\}, a) < 0 \tag{4}$$

in which $\{\Delta\sigma\}$ is a change in the stress state. When a stress reversal condition occurs, the stress state $\{\sigma\}$ at the instance of stress reversal becomes the new $\{\sigma^R\}$, and the yield surface collapses to a point. The yielding ratio, s, simultaneously drops to zero. From then on, the instantaneous yield surface starts to evolve (expand and/or translate) along with the change in stress state.

The yielding ratio s can be updated using an incremental approach. A change in the stress tensor, $\{\Delta\sigma\}$, may result in a variation of the instantaneous yield function, δf. A change in the yielding ratio, δs, has to required to satisfy

$$-\delta f = \left(\frac{\partial f}{\partial\{\sigma^*\}}\right)\left(\frac{\partial\{\sigma^*\}}{\partial s}\right)\delta s + \left(\frac{\partial f}{\partial a}\right)\left(\frac{\partial a}{\partial s}\right)\delta s \tag{5}$$

[1]Assoc. Prof. of Civ. Engrg., Natl. Chiao-Tung Univ., Hsinchu, Taiwan 30049

Note that $(\frac{\partial a}{\partial s}) = \{\sigma^R\}$ and $(\frac{\partial a}{\partial s}) = a_c$, δs can be evaluated.

$$\delta s = \frac{-\delta f}{(\frac{\partial f}{\partial\{\sigma^*\}})\{\sigma^R\} + (\frac{\partial f}{\partial a})a_c} \tag{6}$$

The concept of the transitional yielding approach can be extended to include any desired number of nesting limit surfaces for the representation of a higher degree material memory [Pan 1990]. For a n-surface model, (n-2) limit surfaces can be considered; and the transformed stress state is defined as

$$\{\sigma^*\} = (\{\sigma\} - \{\sigma_{n-2}\}) - (1 - s_{n-1})(\{\sigma^R_{n-1}\} - \{\sigma_{n-2}\}) \tag{7}$$

in which

$$\{\sigma_{n-2}\} = (1 - s_{n-2})(\{\sigma^R_{n-2}\} - \{\sigma_{n-3}\}) + \{\sigma_{n-3}\} \tag{8}$$

$$\vdots$$

$$\{\sigma_i\} = (1 - s_i)(\{\sigma^R_i\} - \{\sigma_{i-1}\}) + \{\sigma_{i-1}\} \tag{9}$$

$$\vdots$$

$$\{\sigma_0\} = \{0\} \tag{10}$$

where

$$s_1 = \frac{a_1}{a_c};\ s_i = \frac{a_i}{a_{i-1}},\ (i = 2, n-2)$$

$$s_{n-1} = \frac{a}{a_{n-2}},\ s = \frac{a}{a_c} = \prod_{i=1}^{n-1} s_i$$

in which a_c, a_i, and a are the sizes of the bounding surface, the $i-th$ limit surface, and the instantaneous yield surface, respectively. The ratios s_i and s are real numbers on the interval [0,1]. $\{\sigma^R_{n-1}\}$ is the stress state when the most recent stress reversal occurs. $\{\sigma^R_i\}$ is the stress state when the $i-th$ limit surface formed (when a stress reversal occured).

The limit surfaces in the proposed approach represent various degree of the hierachical material memory due to stress history. As n approaches infinity, the generalized approach is equivalent to an infinite-yield-surface model. When $n = 2$, this approach reduces to a two-surface model. It should be noted that the generalized multi-surface approach does not require a particular form of the yield surface.

3. The Non-associated Model

A new approach [Pan 1990] for modelling the non-associated behavior of granular materials is introduced subsequently. In this approach, the yield surface and the plastic potential surface are always identical. An associated model can be treated as a special case of which no additional irrecoverable volumetric strain exists.

Assume that the total strain increment can be decomposed into three parts, namely, (i) elastic strain increment, $\{d\varepsilon^e\}$, (ii) associated plastic strain increment, $\{d\varepsilon^p\}$, and (iii) additional irrecoverable volumetric strain, $\{d\varepsilon^a\}$

$$\{d\varepsilon\} = \{d\varepsilon^e\} + \{d\varepsilon^p\} + \{d\varepsilon^a\} \tag{11}$$

The elastic strain increment, $\{d\varepsilon^e\}$, due to a stress increment, $\{d\sigma\}$, is assumed to obey the generalized Hooke's Law as follows

$$\{d\sigma\} = [D]\{d\varepsilon^e\} \tag{12}$$

where $[D]$ = elastic stiffness matrix

The associated plastic strain increment, $\{d\varepsilon^p\}$, can be derived from an associated flow rule.

$$\{d\varepsilon^p\} = \frac{1}{H}\{n\}(\{n\}^T\{d\sigma\}) \tag{13}$$

where $\{n\}$ = the unit normal vector on the instantaneous yield surface; and $H = f(H_c, s)$ = hardening modulus. The function $f(H_c, s)$ should satisfy two necessary conditions: (1) $H = \infty$ when $s = 0$, and (2) $H = H_c$ when $s = 1$.

An additional irrecoverable volumetric strain change is assumed responsible for describing the shear-densification effect in addition to the associated plastic flow. Assume a relationship between the additional irrecoverable volumetric strain change, $\{\delta\varepsilon_v^a\}$, and the increment of plastic shear strain length, δL^p.

$$\delta\varepsilon_v^a = W\delta L^p \tag{14}$$

where $\delta L^p = |\delta\gamma_{oct}^p|$ and $L^p = \int \delta L^p$, in which the term $|\delta\gamma_{oct}^p|$ represents the absolute change in plastic octahedral shear strain. Further assume that the additional irrecoverable volumetric strain increment, $\delta\varepsilon_v^a$ equally contributes to three normal components. As a result,

$$\{d\varepsilon^a\} = [Z]\{d\varepsilon^p\} \tag{15}$$

where

$$[Z] = \frac{4W}{9\gamma_{oct}} \begin{bmatrix} e_{11} & e_{22} & e_{33} & e_{12} & e_{23} & e_{13} \\ e_{11} & e_{22} & e_{33} & e_{12} & e_{23} & e_{13} \\ e_{11} & e_{22} & e_{33} & e_{12} & e_{23} & e_{13} \\ 0 & 0 & 0 & 0 & 0 & 0 \\ 0 & 0 & 0 & 0 & 0 & 0 \\ 0 & 0 & 0 & 0 & 0 & 0 \end{bmatrix} \tag{16}$$

4. the Overall Stress-strain Relation

From Eqn. (11), (12), (13), and (15), the total strain is

$$\{d\varepsilon\} = [D]^{-1}\{d\sigma\} + ([Z] + [I])\{n\}(\{n\}^T\{d\sigma\}) \tag{17}$$

where $[I]$ = unit matrix, (i.e. $I_{ij} = 1$ if $i = j$; $I_{ij} = 0$ if $i \neq j$) The overall stress-strain relation can be found as follows.

$$\{d\sigma\} = [D^{epa}]\{d\varepsilon\} \tag{18}$$

where

$$[D^{epa}] = [D] - \frac{[D]([Z] + [I])\{n\}(\{n\}^T[D])}{H + \{n\}^T[D]([Z] + [I])\{n\}} \tag{19}$$

These relations are readily applicable for numerical methods on the elasto-plastic analyses of various boundary-value problems.

5. Calibration of Material Parameters

The required parameters in the proposed model, in addition to all necessary parameters for the associated flow model, shall also include the parameters that define the function W. Those parameters which are relevant to an associated flow model can be calibrated based on the same procedure for the selected associated flow model. The material parameters defining W can be calibrated based on the difference between the measured volumetric strain and the volumetric strain calculated from the selected associated model.

6. Summary and Conclusions

A generalized transitional yielding approach, with any desired number of limit surfaces representing various degree of the hierachical material memory, is proposed. A new method for modelling non-associative soil behavior is also proposed. In this method, the inclusion of an additional volumetric strain and an associated plastic strain is assumed. The advantage of this method is that the yield surface and the plastic potential surface are always identical. This approach is readily applicable for the implementaion to a numerical method for solving boundary-value problems.

References

Dafalias, Y. F. (1986). "Bounding Surface Plasticity. I: Mathematical Foundation and Hypoplasticity," J. Engrg. Mech., ASCE, 112(9), 966-987.

Mroz, Z., Norris, V. A. and Zienkiewicz, O. C. (1981). "An Anisotropic, Critical State Model for Soils Subject to Cyclic Loading," Geotechnique, 31(4), 451-469.

Pan, Y. W. and Banerjee, S. (1987) "Transitional Yielding Approach for Soils under General Loading," J. of Engrg. Mech., ASCE, 113(2), 153-169.

Pan, Y. W. (1990) "A generalized Non-associative Multi-surface Approach for Granular Materials," Submitted for Possible Publication in J. of Geotechnical Engrg., ASCE.

Pietruszczak, S. and Poorooshasb, H. B. (1987) "On Yielding and Flow of Sand: a Generalized Two-surface Model," Computers and Geomechanics, I, pp. 33-58.

Prevost, J. H. (1978). "Anisotropic Undrained Stress- strain Behavior of Clays," J. Geotech. Engrg., ASCE, 104(8), 1075- 1090.

CRITICAL STATE, STEADY STATE OF DEFORMATION AND THE ULTIMATE STATE OF COHESIONLESS MEDIA

Hormoz B. Poorooshasb
Concordia University, Montreal,Canada, H3G 1M8

ABSTRACT

The concept of the *"Critical State"* is examined in some detail and compared with the concept of the *"Steady State of Deformation."* It is demonstrated that the two concepts are essentially equivalent and that a better terminology to represent this singular state of a sample of a cohesionless medium is the *"Ultimate State."* Under certain conditions it is possible to represent the isometric view of the Ultimate State Surface of the four dimensional state Space. These views are presented and discussed.

INTRODUCTION

An element of cohesionless granular medium may reach a state at which unbounded distortion of the element may take place without any change in its state. This state, which is usually obtained at very large sample distortions, is referred to as the "Ultimate State." It is postulated the the set of points of the state space which represent this singular state are located on a surface which shall be referred to as the Ultimate State Surface or simply USS. The USS is an extension of the Critical State Line (CSL),[1] which in turn was developed from the Critical Void Ratio (CVR) line,[2]. It describes events which are essentially equivalent to the events described through the concept of the Steady State of Deformation, [3]. The present paper is, perhaps, worthy of attention on two counts: it attempts to clarify certain misunderstandings which currently exists in geomechanics and it provides a simple framework within which a unified flow law for cohesionless granular media may be established.

STATE OF A SAMPLE OF A COHESIONLESS GRANULAR MEDIUM

The state of a sample of a cohesionless granular medium is defined by the entire set of its pertinent state parameters. Any quantity that is directly measurable at the moment of examination is a state parameter and if it is judged to be a player in the particular constitutive law under investigation, then it is called a pertinent state parameter. Stated otherwise, a state parameter is any quantity associated with the sample that can be measured without reference to the previous history of the sample. As an example the void ratio of a sand sample is a quantity that can be evaluated without any reference to its history of loading. It is also judged to be a pertinent state parameter. Note that the use of void ratio e, a scalar, is justified if the sample is assumed to be isotropic. The other quantity which is a state parameter and is likely to play a paramount role in the flow behavior of granular media is the state of stress which shall be represented through parameters derived from its invariants. These are p, the mean stress, q, the deviator stress and θ, a quantity very similar to Lode's angle. Thus in its simplest form the state space would be a four dimensional space with its axes representing the set of quantities (p,q,θ,e).

THE CONCEPTS

The independent experimental investigations of Hvorslev and Rendulic conducted more than half a century ago showed that under certain circumstances a unique relationship existed between the void ratio of a clay sample and the state of stress. Roscoe, Schofield and Wroth [2], using a three dimensional space of (p,q,e) provided a graphical representation of these relations, which, consisted of two distinct surfaces, the Hvorslev Surface and the now widely known as the Rendulic-Roscoe Surface. These surfaces intersect along a space curve which was referred to as the Critical Void Ratio (CVR) Line. The use of this terminology was, no doubt, influenced by the terminology used by the late Professor Casagrande who had established certain criteria regarding the density of sands and which shall be discussed further on in this paper.

It was left to the present author [1] to point out that the (p,q,e) space used in [2] was in fact a State Space and hence the Critical Void Ratio Line must legitimately be referred to as the the Critical State Line: it is, after all, a curve of the State Space.

In the four dimensional State Space of (p,q,θ,e) the "Line" would conceivably transform in to a "Surface" which will be referred to as the "Ultimate State Surface". It is emphasized that the Ultimate State Surface is nothing but the generalization of the Critical State Line as defined originally.The main reason for the change of the name of this particular surface from "Critical State" to "Ultimate State" is that, as mentioned before, the use of the term "Critical" was, more than any thing else, a historical accident. There is absolutely nothing "Critical" about it!

Before proceeding further with the developments in this area it is appropriate to provide a formal definition of the the Ultimate State Surface:

It is postulated that there exists a surface in the

State Space for which the relation;

$$\partial p/\partial\varepsilon=\partial q/\partial\varepsilon=\partial\theta/\partial\varepsilon=\partial e/\partial\varepsilon=0 \quad (1)$$

holds true at all and every point of this surface of the State Space. This surface will be called the Ultimate State Surface. The parameter ε, which is not a state parameter, is derived from the second invariant of the strain deviation tensor and is a measure of sample distortion.

Perhaps it was because of the misnomer associated with the word "Critical" that Poulos [3] introduced the term "Steady State of Deformation". It is worth emphasizing, once again, that Poulos uses the term "state" in a completely different context than the one used above: state as used by Poulos indicates a state of affairs (steady deformation) as distinct from a "state of the sample" which indicates the condition in which the sample is found. Poulos asserts that "....The term steady state of deformation was selected because; (1) the term "steady state" is used in analogy to steady-state (non-accelerative) flow in liquids; and (2) the term "deformation" is included to emphasize that the steady state does not exists unless deformation is ongoing. If the velocity of deformation is zero or changing, the specimen is not in a steady state of deformation..." In spite of qualifications imposed by Poulos the "state of the sample" at the instant it is undergoing a "steady state of deformation" must be such that unbounded distortions result without any change in the state parameters; i.e. Eq.(1) must be satisfied. Thus it is clear that the "Steady State of Deformation" and the "Critical State" as intended originally are identical.

The objection to the use of the term "Steady State of Deformation" (versus the Ultimate State) is that if the sample is in fact in a steady state of flow, with constant velocity, then the velocity of flow must be specified. Now if for each constant velocity a particular set of values of (p,q,θ,e) is envisaged then the flow velocity itself must be admitted as a state parameter. There are no philosophical objections to this proposal. It would, however, reduce the usefulness of the concept as a practical tool considerably. The alternative is to postulate that regardless of the magnitude of the flow velocity, *as long as it remains constant,* the sample experiences the same state; i.e. the same values of (p,q,θ,e). This is a very useful postulate but it poses a new problem: zero velocity is also a constant velocity and can not be excluded from consideration. Thus Poulos's condition regarding zero velocity can not be met with.

Thus to avoid all confusion, misnomer and misunderstanding it is decided to use the term the "Ultimate State" for the very simple reason that if all samples are distorted far enough they would Ultimately tend to reach this State.

THE ULTIMATE STATE SURFACE

Certain properties of the Ultimate State Surface will be discussed in this section. By the basic postulate, let the equation of the Ultimate State Surface be;

$$f(p,q,\theta,e)=0 \tag{2}$$

expressing a unique relation between one of the state parameters and the other three. For example let this relation be expressed for the void ratio e and in the form;

$$e=e(p,q,\theta) \tag{3}$$

Thus it is possible to construct contours of equal e in the stress space as a graphical representation of the Ultimate state Surface. This mode of presentation is precisely the same as the one adopted by Rendulic in his classical paper where contours of equal e were plotted in the particular stress space associated with the triaxial testing condition.

An alternative (and from a practical point of view a much more useful) approach would be to adopt a second postulate which relates two of the state parameters of the Ultimate State on a one to one basis.

It is postulated that at the Ultimate State a unique relation exists between the parameter void ratio, e, and the mean normal stress, p, and that this relation is governed by the Casagrande's Equation.

Casagrande's Equation is;

$$e_{Cas}=e_0-\lambda\ln(p/p_0) \tag{4}$$

where λ, e_0 and p_0 are three fundamental soil constants which between them fix the slope and the position of the "Critical Void Ratio" Line (as Casagrande called it) in the e-p domain, see Fig.(1) which shows the test results on the Sacramento River Sand. The tests were performed and reported by Seed and Lee [4].

Having established a relation between two of the state parameters of Eq.(2) the space reduces to three dimensions and now it is possible to produce an isometric view of the the Ultimate State Surface. This is done in Fig.(2) for the aforementioned sand. It is also noted that in preparation of Fig.(2) it has bee tacitly assumed that the Ultimate State Surface and the State Boundary Surface [1] have a set of points in common. The State Boundary Surface is defined as the surface enclosing all the possible states a sample may acquire. Thus a sample

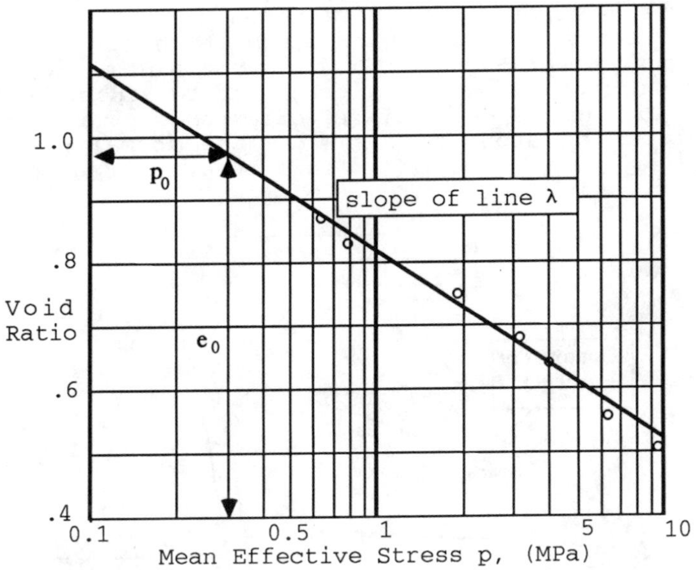

Fig.(1)- Casagrande's Critical Void Ratio Line

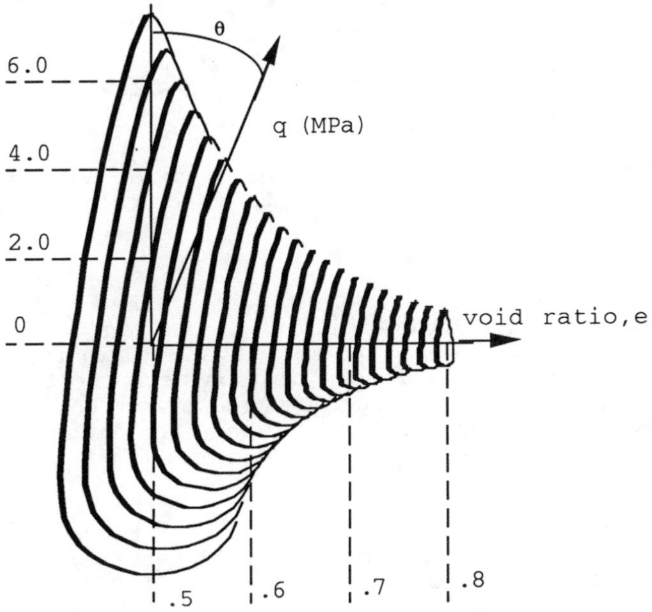

Fig.(2)- The Isometric View of the Ultimate State Surface for the Sacramento River Sand.

may follow a path within the State Boundary Surface or ride on it. It may not, however, cross it. The surface can be represented graphically say by taking e=constant, which is similar to the technique used by Rendulic. Fig.(3) shows one such a representation for a "dense" Sacramento River Sand sample having a void ratio of .5. The use of the term "dense" is justified since the trace of the Ultimate State Surface corresponds to a mean normal stress of about 10 MPa; in the normal practical range (say .1 to .5 MPa) the sand behaves as a dense and dilatant material.

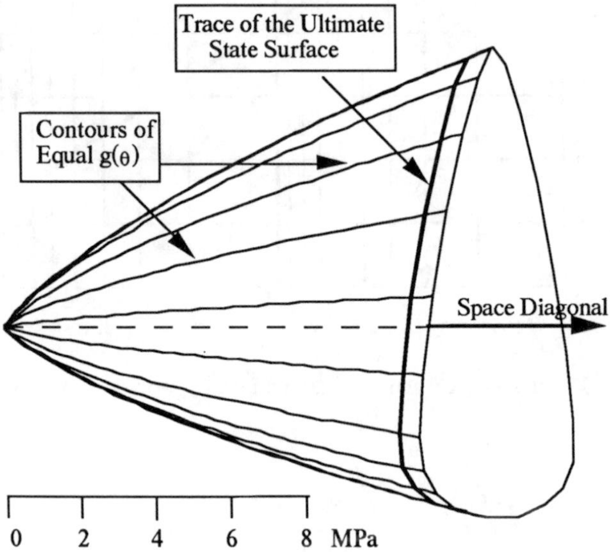

Fig.(3)- The State Boundary Surface for the Sacramento River Sand, void ratio =.5.

CONCLUDING REMARK

The *"Critical State"* and the *"Steady State of Deformation"* are substantially identical and represent a *State* of a sample which it may attain *Ultimately*. The Ultimate State Concept in conjunction with the Concept of State Boundary can provide a framework within which a unified constitutive relation may be established.

REFERENCES

1. Poorooshasb, H.B., Ph.D.Thesis, Cambridge,1961.
2. Roscoe, K.H., Schofield, A.N. and Wroth, C.P."On the Yielding of Soils," Geotechnique,8,22-53, 1958.
3. Castro G. and Poulos, S.J. "Liquefaction and Cyclic Mobility of saturated Sands" ASCE, 103,GT6,501-516, 1977.
4. Seed, H.B. and Lee, K.L.,"Liquefaction of Saturated Sands During Cyclic Loading," ASCE,92,SM6,105-134,1966.

YIELDING AND PLASTIC FLOW OF ANISOTROPIC CONSOLIDATED POROUS MEDIA

J.RACLIN and A.OUDJEHANE
L.T.T.M.M.IUT de MONTLUCON
Université Blaise Pascal,CLERMONT II,France.

ABSTRACT

The aim of this work is to develop a consistent theoretical and experimental study of the constitutive relations for anisotropic consolidated porous media such as oak wood after drying process within Non Linear Continuum Mechanics.

KEYWORDS

Consolidated porous media,initial anisotropy,induced anisotropy,anisotropic criterion, anisotropic flow law,HILL's criterion.

INTRODUCTION

It is well known that during drying process,the internal structure of wood becomes strongly oriented,thus anisotropy [1] .On the macroscopic level,the oriented internal structure results in the anisotropy of the mechanical behavior.For wood,the internal structure presents three orthogonal planes of symmetry generally:the macroscopicbehavior is thus orthotropic.When wood is subjected to further irreversible processes,the anisotropy of the internal structure is modified.The modifications affect not only the degree of initial orthotropy,but also the type of anisotropy,according to the orientation of the principal directions of the irreversible deformation with respect to the mechano-sorptive processes during drying: this is the phenomena of anisotropic consolidation [2] .

This work will attempt to bring together the presentation of constitutive anisotropic relations concerning mechanical behavior and experimental method,intended to disclose and evaluate the various mechanical anisotropies in order to reach a unified approach to the Mechanics of Anisotropic Solids and to develop suitable methods allowing to deal with the problems of elasticity,plasticity, hardening,damage,porosity and failure of consolidated anisotropic porous media.

CONSTITUTIVE RELATIONS

We consider the wood as an homogeneous orthotropic solid in its initial non deformed configuration, when the rate moisture corresponds to the saturation point of the grains. The privileged directions ($\underset{\sim}{v}_1, \underset{\sim}{v}_2, \underset{\sim}{v}_3$) correspond respectively to the radial, tangential and longitudinal directions (Fig. 1).

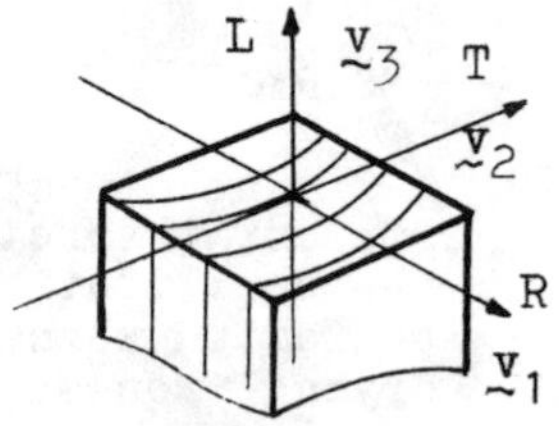

FIG. 1. Privileged directions.

We suppose that during drying of oakwood, the wood is subjected to an irreversible mechano-sorptive prestrain $\underset{\sim}{C}$. This second order tensor C, called "consolidation tensor", results of the humidity mass free and non forced transfer.

The tensor $\underset{\sim}{C}$ refers to the initial configuration and depends of the thermo-hygroscopic shrinkage deformation tensor $\underset{\sim}{E}$ as :

$$\underset{\sim}{C} = \underset{\sim}{f}(\underset{\sim}{E}) \qquad (1)$$

where $\underset{\sim}{E}$ is a second order tensor. Its principal directions coïncide with the privileged directions ($\underset{\sim}{v}_1, \underset{\sim}{v}_2, \underset{\sim}{v}_3$).

Defined to represent the initial anisotropy - orthopic - of the material, the structural tensors $\underset{\sim}{M}_1, \underset{\sim}{M}_2, \underset{\sim}{M}_3$ are introduced with respect to the initial configuration.

The constitutive law of a consolidate material, such as dry oakwood, whose mechanical behavior depends on the deformation history through the consolidation tensor $\underset{\sim}{C}$, by the application of the theorems of representation for anisotropic tensor functions for anisotropic materials takes the form :

$$\underset{\sim}{T} = \underset{\sim}{F}(\underset{\sim}{D}, \underset{\sim}{C}, \underset{\sim}{M}_1, \underset{\sim}{M}_2, \underset{\sim}{M}_3) \qquad (2)$$

$$\underset{\sim}{C} = \underset{\sim}{G}(\underset{\sim}{E}, \underset{\sim}{M}_1, \underset{\sim}{M}_2, \underset{\sim}{M}_3) \qquad (3)$$

where $\underset{\sim}{D}$ is a kinematic second order tensor.

According to the "Principle of Isotropy of Space", $\underset{\sim}{F}$ and $\underset{\sim}{G}$ are respectively isotropic functions with respect to their arguments ($\underset{\sim}{D}, \underset{\sim}{C}, \underset{\sim}{M}_1, \underset{\sim}{M}_2, \underset{\sim}{M}_3$) and ($\underset{\sim}{E}, \underset{\sim}{M}_1, \underset{\sim}{M}_2, \underset{\sim}{M}_3$), but orthotropic functions with respect to the two mechanical arguments $\underset{\sim}{D}$ and $\underset{\sim}{C}$ for $\underset{\sim}{F}$, and to the one arguments $\underset{\sim}{E}$ for $\underset{\sim}{G}$ [3].

An irreducible representation of the constitutive relation (2) is given by :

$$\begin{cases} \underset{\sim}{T} = a_i \underset{\sim}{G}_i \\ a_i = a_i(I_1, \dots, I_p) \end{cases} \qquad (4)$$

The representation (4) includes twelve generating tensors:

$$\left.\begin{array}{l} \underset{\sim}{M}_1 \quad ,\underset{\sim}{M}_2 \quad ,\underset{\sim}{M}_3 \\ \underset{\sim}{M}_1\underset{\sim}{D} + \underset{\sim}{D}\underset{\sim}{M}_1, \underset{\sim}{M}_2\underset{\sim}{D} + \underset{\sim}{D}\underset{\sim}{M}_2, \underset{\sim}{M}_3\underset{\sim}{D} + \underset{\sim}{D}\underset{\sim}{M}_3, \underset{\sim}{D}^2 \end{array}\right\} \text{initial anisotropy}$$

$\underset{\sim}{C}^2 \quad ,\underset{\sim}{D}\underset{\sim}{C} + \underset{\sim}{C}\underset{\sim}{D}$: induced " by $\underset{\sim}{C}$

$\underset{\sim}{M}_1\underset{\sim}{C} + \underset{\sim}{C}\underset{\sim}{M}_1, \underset{\sim}{M}_2\underset{\sim}{C} + \underset{\sim}{C}\underset{\sim}{M}_2, \underset{\sim}{M}_3\underset{\sim}{C} + \underset{\sim}{C}\underset{\sim}{M}_3$: coupling between this two anisotropies

where the coefficients a_i are arbitrary scalar-valued functions of nineteen invariants :

$\text{tr}\, \underset{\sim}{M}_k\underset{\sim}{D}, \text{tr}\, \underset{\sim}{M}_k\underset{\sim}{D}^2, \text{tr}\, \underset{\sim}{D}^3$: initial orthotropy

$\text{tr}\, \underset{\sim}{C}^3, \text{tr}\, \underset{\sim}{C}^2\underset{\sim}{D}, \text{tr}\, \underset{\sim}{C}\underset{\sim}{D}^2$: induced anisotropy by $\underset{\sim}{C}$

$\text{tr}\, \underset{\sim}{M}_k\underset{\sim}{C}, \text{tr}\, \underset{\sim}{M}_k\underset{\sim}{C}^2, \text{tr}\, \underset{\sim}{M}_k\underset{\sim}{C}\underset{\sim}{D}$: coupling between this two anisotropies

$k = 1,2,3$

We have appointed the different parts played by the component units of the representation (4). The canonical invariant form (4) constitutes the most general form of the mechanical behavior law for consolidate materials.

PLASTIC BEHAVIOR OF OAKWOOD

In condition that $\underset{\sim}{D}$ stands the rate of deformation and that $\underset{\sim}{T}$ is homogeneous of order zero with respect to time the representation (4) can describe plastic behavior [4].

We obtain the most general form of both the flow law :

$$\begin{aligned} \frac{\underset{\sim}{D}}{\sqrt{\text{tr}\underset{\sim}{D}^2}} = {} & b_1\underset{\sim}{M}_1 + b_2\underset{\sim}{M}_2 + b_3\underset{\sim}{M}_3 + b_4(\underset{\sim}{M}_1\underset{\sim}{T} + \underset{\sim}{T}\underset{\sim}{M}_1) + b_5(\underset{\sim}{M}_2\underset{\sim}{T} + \underset{\sim}{T}\underset{\sim}{M}_2) + \\ & b_6(\underset{\sim}{M}_3\underset{\sim}{T} + \underset{\sim}{T}\underset{\sim}{M}_3) + b_7\underset{\sim}{T}^2 + b_8\underset{\sim}{C}^2 + b_9(\underset{\sim}{T}\underset{\sim}{C} + \underset{\sim}{C}\underset{\sim}{T}) + \\ & b_{10}(\underset{\sim}{M}_1\underset{\sim}{C} + \underset{\sim}{C}\underset{\sim}{M}_1) + b_{11}(\underset{\sim}{M}_2\underset{\sim}{C} + \underset{\sim}{C}\underset{\sim}{M}_2) + b_{12}(\underset{\sim}{M}_3\underset{\sim}{C} + \underset{\sim}{C}\underset{\sim}{M}_3) \end{aligned} \qquad (5)$$

$$b_i = b_i(\text{tr}\,\underset{\sim}{M}_k\underset{\sim}{C}, \text{tr}\,\underset{\sim}{M}_k\underset{\sim}{C}^2, \text{tr}\,\underset{\sim}{C}^3, t_\ell) \; ; \; k = 1,2,3 \; ; \; \ell = 1,2,\dots 11.$$

and the yield criterion :

$$\begin{array}{ll} f(\text{tr}\,\underset{\sim}{M}_k\underset{\sim}{T}, \text{tr}\,\underset{\sim}{M}_k\underset{\sim}{T}^2, \text{tr}\,\underset{\sim}{T}^3, & \text{: initial orthotropy} \\ \quad \text{tr}\,\underset{\sim}{T}\underset{\sim}{C}^2, \text{tr}\,\underset{\sim}{T}^2\underset{\sim}{C}, \text{tr}\,\underset{\sim}{C}^3 & \text{: induced anisotropy by } \underset{\sim}{C} \\ \quad \text{tr}\,\underset{\sim}{M}_k\underset{\sim}{T}\underset{\sim}{C}, \text{tr}\,\underset{\sim}{M}_k\underset{\sim}{C}, \text{tr}\,\underset{\sim}{M}_k\underset{\sim}{C}^2) = 0 & \text{: coupling anisotropies} \end{array} \qquad (6)$$

Most of the presently proposed criteria for anisotropic solids result from appropriate generalization of the available criteria for isotropic media. They are commonly homogeneous in stress, in particular the case of Von MISES' criterion. Regarding to HILL'scriterion, that's a generalization of Von MISES'criterion too, we suppose that the failure criterion (6) is an homogeneous form of order two in stress. Hence (6) becomes :

$$f(\mathrm{tr}\underset{\sim}{M}_1\underset{\sim}{T}, \mathrm{tr}\underset{\sim}{M}_2\underset{\sim}{T}, \mathrm{tr}\underset{\sim}{M}_3\underset{\sim}{T}, \mathrm{tr}\underset{\sim}{M}_1\underset{\sim}{T}^2, \mathrm{tr}\underset{\sim}{M}_2\underset{\sim}{T}^2, \mathrm{tr}\underset{\sim}{M}_3\underset{\sim}{T}^2) = 0 \quad (7)$$

in the case $\underset{\sim}{C} = \underset{\sim}{0}$ for non consolidate oakwood.

Finally the knowlegde of only six invariants let one define the yield criterion for a non consolidate media.

HILL's CRITERION

Developed for orthotropic metals, HILL's criterion is according to Von MISES distorsional energy theory. It's assumed that the strengths under compression and tension are the same and that the material is insensitive to hydrostatic stress.

Since non consolidate oakwood presents nearly the same strengths under tension and compression, we applied HILL'criterion ascertain if it's suitable for oakwood, tested first under compression. In the privileged directions $(\underset{\sim}{v}_1, \underset{\sim}{v}_2, \underset{\sim}{v}_3)$ reference and in the case of orthotropic materials, the criterion is expressed by :

$$\left\{\begin{array}{l} a_1(\mathrm{tr}\underset{\sim}{M}_1\underset{\sim}{T})^2 + a_2(\mathrm{tr}\underset{\sim}{M}_2\underset{\sim}{T})^2 + a_3(\mathrm{tr}\underset{\sim}{M}_3\underset{\sim}{T})^2 + \\ \quad a_4\mathrm{tr}\underset{\sim}{M}_1\underset{\sim}{T}^2 + a_5\mathrm{tr}\underset{\sim}{M}_2\underset{\sim}{T}^2 + a_6\mathrm{tr}\underset{\sim}{M}_3\underset{\sim}{T}^2 + \\ a_7\mathrm{tr}\underset{\sim}{M}_1\underset{\sim}{T}.\mathrm{tr}\underset{\sim}{M}_3\underset{\sim}{T} + a_8\mathrm{tr}\underset{\sim}{M}_1\underset{\sim}{T}.\mathrm{tr}\underset{\sim}{M}_2\underset{\sim}{T} + a_9\mathrm{tr}\underset{\sim}{M}_3\underset{\sim}{T}.\mathrm{tr}\underset{\sim}{M}_2\underset{\sim}{T} = 1 \end{array}\right. \quad (8)$$

where the a_i ($i = 1$ to 9) are nine constants related to the six material constants of HILL.

Hence only six compression tests are necessary to define the invariant form of the failure (or yield) criterion for non consolidate oakwood.

THEORETICAL SETTING IN UNIAXIAL COMPRESSION TESTS

We denote by σ_n the axial stress, by α and β the angles between the privileged directions and the stress (Fig.2).

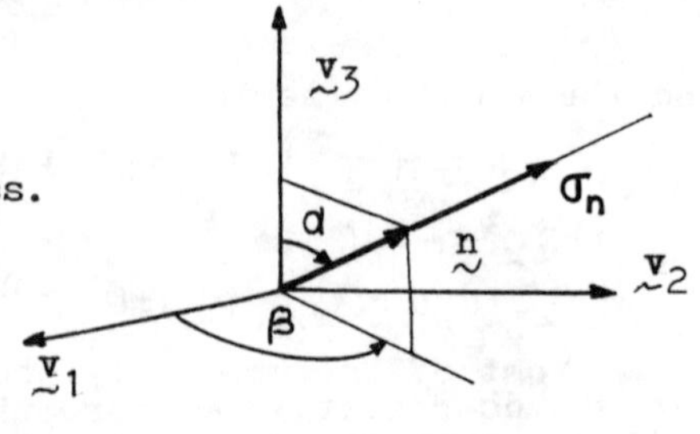

FIG. 2. Orientation of the axial stress.

The yield criterion (8) is reduced to a function of only three parameters α , β and $\underset{\sim}{T}$ under the form :

$$(\sigma_n.(\sin^2\beta + \cos^2\alpha))^2.\left[a_1\sin^4\alpha\sin^4\beta + a_2\sin^4\alpha\cos^4\beta + a_3\cos^4\alpha + a_4\sin^2\alpha\sin^2\beta + a_5\sin^2\alpha\cos^2\beta + a_6\cos^2\alpha + a_7\sin^2\alpha\cos^2\alpha\sin^2\beta + a_8\sin^4\alpha\sin^2\beta\cos^2\beta + a_9\sin^2\alpha\cos^2\alpha\cos^2\beta\right] = 1 \qquad (9)$$

EXPERIMENTAL RESULTS

The specimens of oakwood were cut to the shape of parallelepipeds with 20.20.40 mm3-dimensions. They were prepared in seven orientations, corresponding to the following couples of angles (α , β) :

(0,0), (90,90), (90,0), (45,45), (90,45), (45,90), (45,0).

We give in the three following tables :
- -the results of compression tests Table I
- -the values of the nine coefficients a_i obtained from six compression tests Table II
- -the compressive strengths according to HILL's criterion Table III.

TABLE I. Experimental compressive strengths (MPa)

α,β	0,0	90,90	90,0	45,45	90,45	45,90	45,0
σ_n	269	118	98	132	121,25	86,9	143

TABLE II. Material constants a_k $(1/MPa^2)$

a_1 = 12,4.E-3	a_2 = 2,7.E-3	a_3 = -7,7.E-3	a_4 = -2.E-3
a_5 = 4,4.E-3	a_6 = 8,8.E-3	a_7 = -2,2.E-3	a_8 = 4,4E-3
a_9 = 16,4.E-3			

TABLE III. HILL's criterion compressive strengths (MPa)

α,β	0,0	90,90	90,0	45,45	90,45	45,90	45,0
σ_n	301,5	118,7	98,1	132	128,3	109,4	139,7

CONCLUSIONS

The representation theorems for anisotropic tensor functions allow for a unified description of the mechanical behavior of anisotropic consolidated porous media.In this work,attention was focused most particularly on the evolution of the anisotropic yield criterion during irreversible consolidation process and a generalization of the HILL's criterion was proposed for consolidated oakwood.

The theoretical predictions of the proposed criterion are in good agreement with the test data obtained in a specific experimental program on oakwood specimens.In order to develop further the presented theory and,in particular,to fully determine the anisotropic flow law and its evolution during irreversible consolidation,it is necessary to perform a number of well-organized experiments, consisting of multiaxial mechano-sorptive processes on oriented specimens.

REFERENCES

1. D.Guitard et J.M.Génevaux, "Vers un Matériau Technologique : le Bois",Arch.Mech.,40(5-6), 665-676 (1988).
2. J.Kubik and A.Sawczuk, "A Theory of Anisotropic Consolidation",Ingenieur-Archiv, 53, 133-143 (1983).
3. J.P.Boehler et J.Raclin, "Ecrouissage des Materiaux Orthotropes Prédéformés", J.M.T.A.,nº spécial,23-44 (1982).
4. J.Raclin, "Contributions Théoriques et Expérimentales à l'Etude de la Plasticité,de l'Ecrouissage et de la Rupture des Solides Anisotropes",Thèse de Doctorat-ès-Sciences Physiques,Univ.Grenoble-F,nº 84-89 (1984).

MODELING OF INELASTIC BEHAVIOR OF [AISI 316L] STAINLESS STEEL

J.Schwertel[1], B.Schinke[2], U.Spermann[1], D.Munz[1]

[1]*Universität Karlsruhe, Institut für Zuverlässigkeit und Schadenskunde im Maschinenbau, Kaiserstr. 12, D-7500 Karlsuhe 1*
[2]*Kernforschungszentrum Karlsruhe, Institut für Material- und Festkörperforschung IV, Postfach 3640, D-7500 Karlsruhe 1*

Abstract

The Unified Constitutive Model of Robinson has been applied to describe the inelastic behavior of [AISI 316L] stainless steel (1.4404) at 700°C. After presenting the model and the experimental technique the method for parameter determination is described. Satisfactory simultaneous modeling of both creep and strain-controlled tests − both were used for parameter determination − was not possible. A slight modification improved the situation. The agreement with verification experiments is demonstrated.

1. Introduction

In the last years a number of so-called "unified models" have been developed to describe the inelastic behavior of materials for any kind of loading history. However there is little to find in literature about the accuracy. Therefore this work had the objective to find out to which degree the models are able to describe experimental results and what kind of modifications could improve theoretical modeling.

The results of Robinson's model [1] are shown. Material constants are determined from creep and strain-controlled cyclic tests. The predictions of the model for further loadings are compared with experimental results. Although the results are presented for one material and one temperature only most of the conclusions can be generalized.

2. The Viscoplastic Model of Robinson

This model decribes initially isotropic materials on the basis of external and internal variables. It seperates the strain rate into a linear elastic and an inelastic part.

The growth laws for the inelastic strain rate $\dot{\varepsilon}^{in}$ and the rates for the internal variables $\dot{\alpha}$ and $\dot{K}$ have the following forms (isothermal, uniaxial formulation):

$$\dot{\varepsilon}^{in} = \begin{cases} A\left[\dfrac{(\sigma-\alpha)^2}{K} - 1\right]^n \operatorname{sgn}(\sigma-\alpha) & \text{for } \sigma\alpha > 0,\ (\sigma-\alpha)^2 > K \\ 0 & \text{else} \end{cases} \quad (1)$$

$$\dot{\alpha} = \begin{cases} H|\alpha|^{-\beta}\dot{\varepsilon}^{in} - R|\alpha|^{m-\beta}\mathrm{sgn}(\alpha) & \text{for } \sigma\alpha > 0,\ |\alpha| > \alpha_0 \\ H|\alpha_0|^{-\beta}\dot{\varepsilon}^{in} - R|\alpha_0|^{m-\beta}\mathrm{sgn}(\alpha) & \text{else} \end{cases} \tag{2}$$

$$\dot{K} = (K_s - K)d\,\dot{W}^p \tag{3}$$

$\dot{W}^p$ is the rate of the cumulated inelastic work: $W^p = \sigma\varepsilon^{in}$.

The model contains nine parameters: A,n,H,R,β,m,d K_s and K_i which is the initial value of K. They have to be determined by experiments. The isotropic variable K has been introduced to describe cyclic hardening, while α is intended to account for directional hardening. α_0 has a small value to avoid the occurrence of a singularity at $\alpha = 0$ and causes large changes of α on reversed loading $(\sigma\alpha \leq 0)$. For the following calculations α_0 has been fixed at .01 MPa.

3. Experimental Technique and Results

Different uniaxial tests were performed on stainless steel (type AISI 316L) in order to determine the parameters of a viscoplastic model and to verify it. The cylindrical specimens (s. Fig. 1) were heated in a furnace with three seperate heating zones up to a nearly constant temperature (±1 K). The specimens were loaded by a servohydraulic uniaxial tension machine. The elongation of the cylindric part of the specimen was measured by an extensiometer. It grasps with two ceramic holding units into the furnace. The load was registrated by a dynamometer outside the furnace. All tests were performed with load or elongation control. A computer allows to program almost any arbitrary history of loading. The measured values were registrated by another computer.

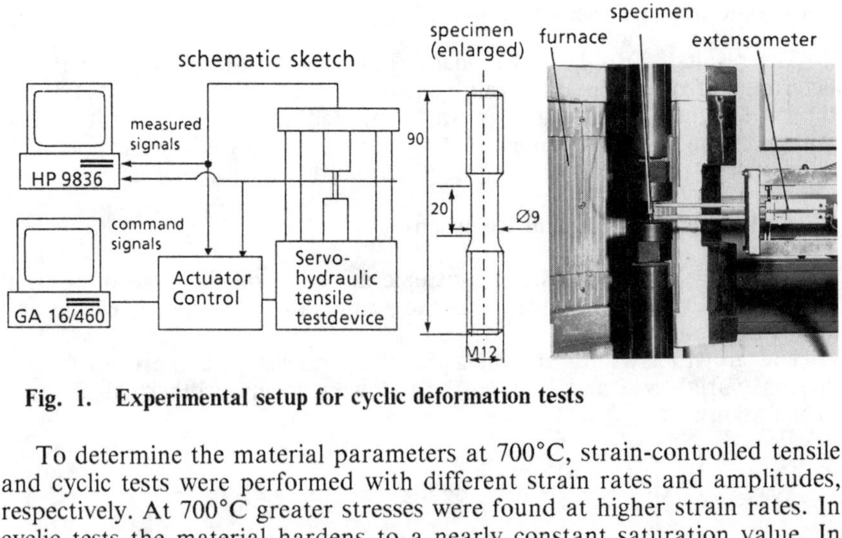

Fig. 1. Experimental setup for cyclic deformation tests

To determine the material parameters at 700°C, strain-controlled tensile and cyclic tests were performed with different strain rates and amplitudes, respectively. At 700°C greater stresses were found at higher strain rates. In cyclic tests the material hardens to a nearly constant saturation value. In addition short-time creep tests were performed up to 2000 h. Some were done with help of a dead weight, others with a servohydraulic device to load the specimen within 100 ms and to measure primary creep very early. Primary creep was found to be clearly visible. The secondary creep could be described by Norton's law. To verify theoretical predictions a lot of different types of experiments were performed. Besides creep tests and relaxation tests after monotonic or cyclic predeformation, cyclic tests with changing amplitudes or with predeformation were performed since these loadings are typical for many structures.

4. Adjustment and Verification

The parameters of the model have been determined from experiments in three steps: First, cyclic saturation hysteresis loops were used to determine the parameters A,n,H,β and K_s. For this vanishing recovery was assumed ($\dot{R} = 0$) to be valid for large strain rates ($\dot{\varepsilon} \gtrsim 10^{-4}\,s^{-1}$). Secondly, from the cyclic hardening curves (peak stresses vs. cycle number) the values responsible for the development of K have been calculated. In a third step, the remaining recovery parameters R and m have been determined from the creep curves. For all steps the sum of the squared difference between theoretical and experimental inelastic strains has been minimized using the results of the preceding steps.

Fig. 2 shows the results for the saturation loops for different strain rates and strain amplitudes. For creep curves (s. Fig. 3) the agreement is unsatisfactory: the initial strains are too high and there is no steady state creep. For that reason the model has been modified slightly. The creep experiments together with the strain-controlled tests indicate that the development of K does not only depend on the inelastic work but also on time. For that reason a static term in the evolution law of K has been added. It increases the growth of K for low strain rates. The introduction of the kinematic hardening variable α along with the inelastic work reduces the increase of the peak stresses for strain controlled cyclic tests:

$$\dot{K} = \left(K_s - c_1 |\alpha| e^{-c_2 W^p} - K\right) d\,\dot{W}^p - r \left| K - \frac{h}{d} \right|^{m_K} \operatorname{sgn}\left(K - \frac{h}{d}\right) \quad (4)$$

In the limiting case for large plastic works this evolution law is reduced to the original one. There are some other models that suggest or contain such a time hardening, see e.g. [2] or [3]. As can be seen in Fig. 3 there is a signifcant improvement of the creep curve description. Fig. 4 demonstrates the results for the cyclic hardening curves.

Some predictive capabilities are evident from Fig. 5 and 6. The stress rate for the relaxation curve in Fig. 5 is too large just after relaxation starts. This is the same effect as for the creep curves in Fig. 6: this plot shows creep behavior starting from the peak stresses of two steady state hysteresis loops.

5. Conclusions

The constitutive viscoplastic model of Robinson has been applied to describe the inelastic behavior of [AISI 316L] stainless steel at 700°C. Cyclic steady state hysteresis loops in addition with cyclic hardening curves and creep curves have been used to determine material parameters. The results were not satisfactory. A modification of the drag stress evolution law resulted in a significant improvement. The agreement with verification experiments – these were not used to calculate the parameters – is similar to that of the characterization experiments.

Acknowledgement: *We are grateful for the financial and personnel support of fast breeder project PSB and Sonderforschungsbereich 167.*

n	4.18	β	0.37
m_k	2	m	8
d	0.111	c_2	0.5
A	1.15×10^{-11} s^{-1}		
H	4.7×10^{4} MPa		
R	8×10^{-19} $MPas^{-1}$		
K_s	420 MPa^2		
K_i	141 MPa^2		
c_1	2 MPa		
c_3	1.5×10^{-5} MPa^2		

Table 1. Parameters

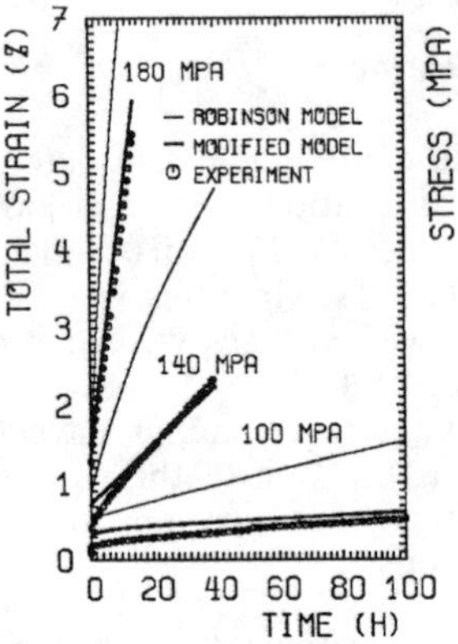

Fig. 3. Creep curves

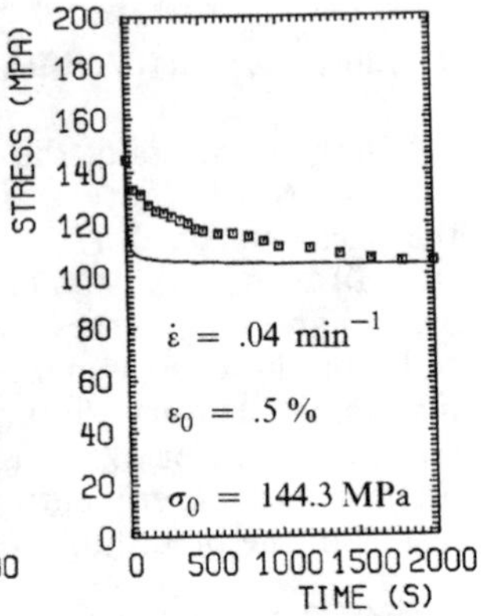

Fig. 5. Relaxation

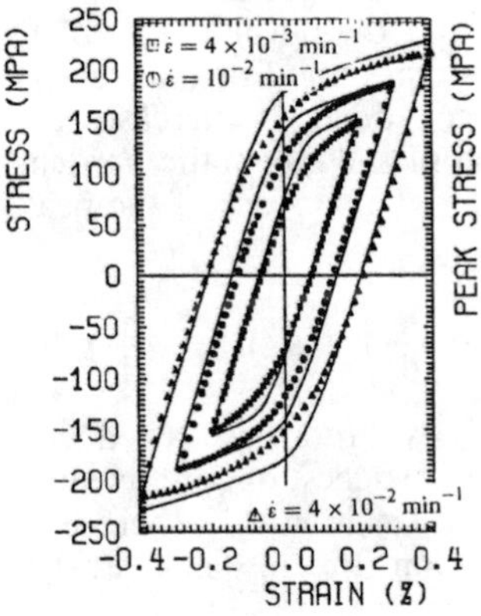

Fig. 2. Saturation loops

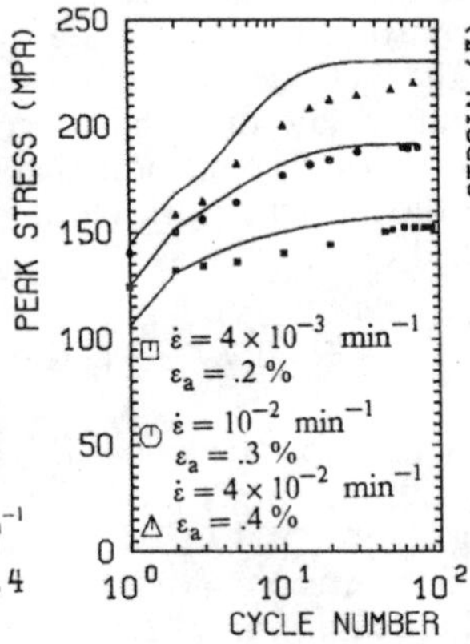

Fig. 4. Hardening curves

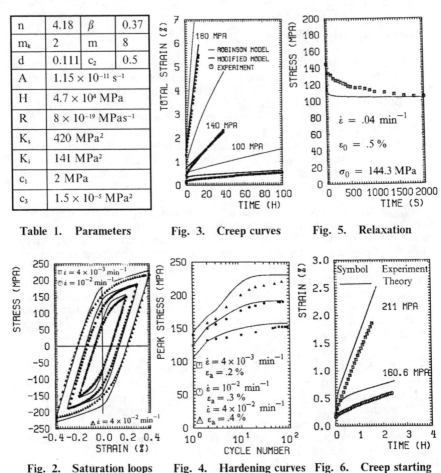

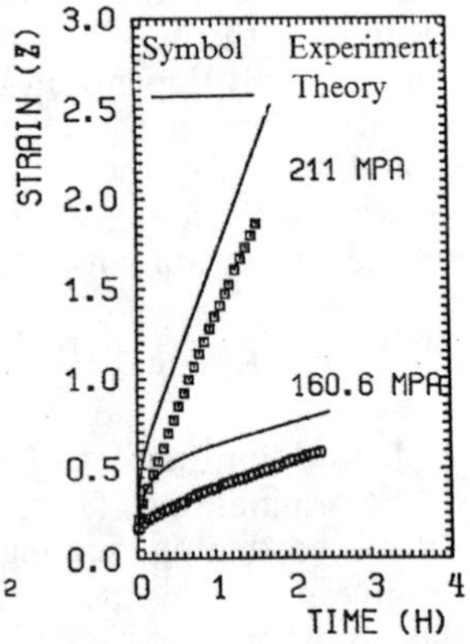

Fig. 6. Creep starting from steady state loops

6. References

[1] D.N.Robinson,P.A.Bartolotta: Viscoplasic Constitutive Relationships with Dependance on Thermomechanical History, NASA CR-174836, University of Akron, Akron, Ohio (1985)

[2] K.P. Walker: Research and Development Program for Nonlinear Structural Modeling with Advanced Time-Temperature Dependent Costitutive Relationships, NASA CR-165533, NASA Lewis Research Center, 1981

[3] J.L. Chaboche, G.Rousselier, Journal of Pressure Vessel Technology, 105 (1983) 159-164

An extended H-B criterion as yield surface for rocks

Lingli Wei & John A Hudson
Imperial College of Science, Technology & Medicine, London, U.K.

ABSTRACT

The Hoek-Brown failure criterion (H-B criterion) is widely used in rock engineering. It has several advantages over other failure criteria for rocks. However, this criterion has its own drawback caused by the singularities in the octahedral plane in the principal stress space, so it is not convenient to use it as a yield function in 3D elasto-plastic analysis. In this paper, an extended H-B criterion is developed to remove the singularities, while the original advantages are maintained. Finally, the convenient form of this criterion for incorporation into computer programs is derived.

INTRODUCTION

Up to now, more than 15 strength/failure criteria for rocks have been developed, but only a few of the criteria are widely used in rock engineering, the H-B criterion being among them.

The H-B criterion possesses the following advantages ([1], [2], [3], [4]): (a) it is a non-linear criterion, and is based on examining comprehensive data; (b) it is valid both for intact rock and for rock masses; (c) it is suitable both for compressive and for extensile cases; and (d) only two empirical parameters are needed, i.e. m and s.

It is a common practice in rock engineering to employ a failure/strength criterion as a yield function when we conduct an elasto-plastic analysis to numerically study the mechanical response of rock masses to the ambient loading and/or unloading. When the H-B criterion is used as a yield function for three dimensional elasto-plastic analysis, difficulties will arise with the singularities in the octahedral plane in the principal stress space. In this paper, an extended H-B criterion is developed to remove the singularities and at the same time retain its original advantages.

The ideal shape of yield surfaces for rocks has been well accepted as a curved, pointed bullet with curved section in the octahedral planes similar to those of Mohr-Coulomb's, but without sharp intersection points, tending to be triangular at low stress levels and towards circular at high stress levels. The extended H-B criterion will achieve this ideal shape without introducing any new parameters.

THE H-B FAILURE CRITERION AND THE YIELD FUNCTION BASED ON IT

In developing their criterion, Hoek and Brown assumed that the intermediate principal stress has a negligible influence on rock failure conditions and a rock or rock mass failure is dominated by brittle behaviour because the parameters used in the criterion are derived from the test data under relatively low confining pressure. If a ductile failure occurs, more attempts should be made to determine the required empirical parameters. Some other assumptions, such as the influence of pore fluid, loading rate, scale, etc. were discussed in detail by Hoek [2].

The criterion assumes the following relation between the principal stresses at failure:

$$\sigma_1 = \sigma_3 + \sqrt{m\sigma_c\sigma_3 + s\sigma_c^2} \qquad (1)$$

in which, σ_1 is the major principal stress at failure, σ_3 is the minor one at failure; σ_c

is the uniaxial compressive strength of the intact rock material. m and s are empirical parameters that depend on the properties and on the extent to which it had been broken before being subjected to the failure stresses σ_1 and σ_3. m varies from 0.008 (for weak rock) up to 25 (for hard rock). s changes between 0 (for heavily jointed rock mass) and 1 (for intact rock).

For plastic analysis in rock mechanics, a yield function/criterion F should be defined. Generally, F is defined as identical to the strength criterion. So, the yield function based on the H-B criterion can be expressed as:

$$F = \sigma_1 - \sigma_3 - \sqrt{m\sigma_c\sigma_3 + s\sigma_c^2} \qquad (2)$$

In equations (1) and (2), the sign of stresses is coincided with rock mechanics convention, i.e., a compressive stress is taken as positive and a tensile stress negative.

In order to use a convenient form of the yield surface equation, σ_1 and σ_3 are replaced by p, q and Lode angle θ_σ as: (see Fig. 2 for the definition of p, q and θ_σ)

$$F = \frac{2}{\sqrt{3}}q\cos\theta_\sigma - [-\frac{2}{3}m\sigma_c q\sin(\theta_\sigma + \frac{2}{3}\pi) + m\sigma_c p + s\sigma_c^2]^{1/2} = 0 \qquad (3)$$

The yield surface of equation (3) is a curved, pointed bullet with curved sections in octahedral planes similar to those of Mohr-Coulomb's and with intersection points which cause the singularity of this yield function.

a. s=0.082, m=2.4, σ_c=30 b. s=1.0, m=20.0, σ_c=50

Fig. 1 The H-B yield surfaces

THE EXTENDED H-B YIELD FUNCTION

In order to remove the singularities on the H-B yield surface, Pan [5] simplified equation (3) by ignoring the influence of Lode angle θ_σ in a similar way to Drucker-Prager yield surface. It is the mean surface between the inner apices and the outer apices surface (see Fig. 2). Obviously, this is a great simplification and it is only valid for some weak rocks. Hence, some advantages of the original H-B criterion will be defected.

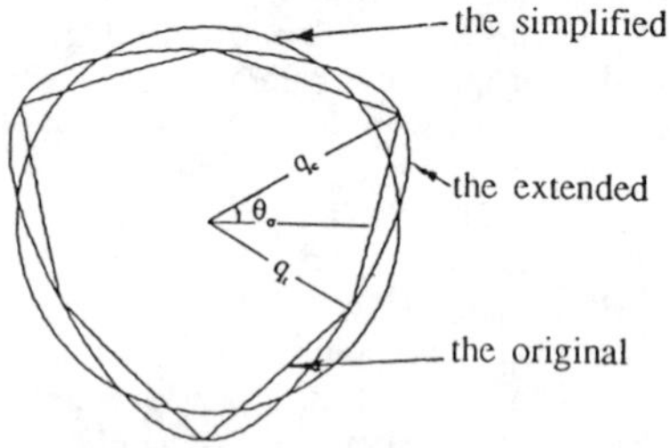

(s=0.1 m=5.0 σ_c=30.0)

Fig. 2 The original, simplified and extended H-B yield surfaces

The proposed extended H-B yield function will take full advantages of the H-B criterion by assuming that it is identical to the H-B yield surface in the compressive meridian section and most part of the extensile meridian section, i.e. it passes all of the outer apices and almost all of the inner apices in the octahedral section.

THE GENERAL FORM

The general form of the extended H-B yield function is adopted as

$$\frac{q}{g(\theta_\sigma)} = f(p) \tag{4}$$

in which, f(p) will be determined by the condition that the extended surface passes through all of the outer apices; $g(\theta_\sigma)$ is the shape function of the octahedral section, which will be discussed in detail in the next section.

Let $\theta_\sigma=\pi/6$, equation (3) becomes

$$q^2 + \frac{1}{3}m\sigma_c q - (m\sigma_c p + s\sigma_c^2) = 0 \tag{5}$$

Solving equation (5), we have

$$q_c = \frac{1}{6}[\sqrt{m^2\sigma_c^2 + 36(m\sigma_c p + s\sigma_c^2)} - m\sigma_c] \tag{6}$$

Similarly, if $\theta_\sigma=-\pi/6$, we have

$$q_t = \frac{1}{3}[\sqrt{m^2\sigma_c^2 + 9(m\sigma_c p + s\sigma_c^2)} - m\sigma_c] \tag{7}$$

q_c and q_t are illustrated in Fig.2.

Since equation (4) involves all of the outer apices, when $g(\theta_\sigma)$ equals unit under $\theta_\sigma=\pi/6$, q must be equal to q_c, i.e. $f(p)=q_c$. So, we have

$$\frac{q}{g(\theta_\sigma)} = \sqrt{\left(\frac{1}{36}m^2\sigma_c^2 + s\sigma_c^2\right) + m\sigma_c p} - \frac{1}{6}m\sigma_c \tag{8}$$

THE OCTAHEDRAL SECTION

According to plasticity theory, the shape function $g(\theta_\sigma)$ should be: (a) a closed convex curve; (b) symmetrical to the three principal stresses in the principal stress space; (c) differentiable within $0\leq\theta_\sigma\leq\pi$; (d) satisfying the following boundary conditions:

$$g(\theta_\sigma) = \begin{cases} 1 & \theta_\sigma = \pi/6 \\ k & \theta_\sigma = -\pi/6 \end{cases} \tag{9}$$

and

$$\frac{\partial g(\theta_\sigma)}{\partial \theta_\sigma} = 0 \qquad when \quad \theta_\sigma = \pm\pi/6 \tag{10}$$

where $k=q_t/q_c$, which will guarantee the yield surface passes through all of the outer and inner apices.

So, the following shape function is adopted,

$$g(\theta_\sigma) = \left[\frac{2k^n}{(1+k^n) + (1-k^n)\sin 3\theta_\sigma}\right]^{\frac{1}{n}} \tag{11}$$

where

$$k = \frac{2\sqrt{1+9p_p}}{\sqrt{1+36p_p}} \qquad \text{and} \qquad p_p = \frac{s}{m^2} - \frac{p}{m\sigma_c} \tag{12}$$

As can be seen, k is a function of the average stress p, and as p increases, k increases, which lets the yield surface tend to be circular; otherwise, when p decreases, k will decrease too, which lets the yield surface towards triangular.

Equation (11) satisfies the above conditions (b), (c) and (d) for $0<k\leq1.0$. While for condition (a), only when n=4.2, k has the largest range, i.e. $0.609\leq k\leq1.0$ to keep the yield surface convex. Therefore, n=4.2 is adopted in the extended H-B surface. Since k must be larger than 0.609 to meet the condition (a), the average stress p should satisfy

$$p \geq (0.05184332 - \frac{s}{m^2})m\sigma_c \tag{13}$$

That is to say, when the average stress is larger than the value defined by equation (13), the yield surface will pass through all the outer and inner apices. If equation (13) cannot be satisfied, k will be kept as a constant, i.e. 0.609.

Equations (8), (11) and (12) form the extended H-B surface, two examples corresponding to Fig.1 are shown in Fig.3.

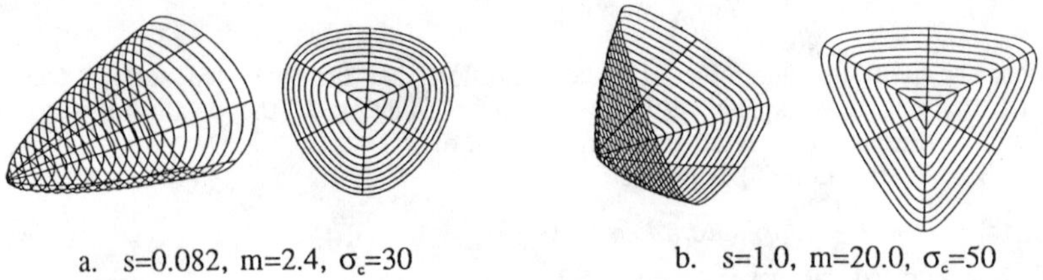

a. s=0.082, m=2.4, σ_c=30 b. s=1.0, m=20.0, σ_c=50

Fig.3 The extended H-B surfaces

THE CONVENIENT FORM

According to Nayak & Zienkiewicz [6], for an associated flow, the plastic flow vector can uniquely be expressed by C_1, C_2 and C_3 for each different yield function. For the extended H-B criterion, the corresponding C_1, C_2 and C_3 are

$$C_1 = -\frac{\sqrt{3}\bar{\sigma}}{4m\sigma_c} \bullet \frac{18\frac{\sqrt{1+9p_p}-1}{\sqrt{1+36p_p}} - 4.5\frac{\sqrt{1+36p_p}-1}{\sqrt{1+9p_p}}}{\left(\sqrt{1+9p_p}-1\right)^2} \bullet q_q^{\frac{1}{n}-1} \bullet \left[1-\frac{3\sqrt{3}}{2}\frac{J_3}{\bar{\sigma}^3}\right] \bullet \frac{1}{k^{n-1}} - m\sigma_c\left(\frac{m^2\sigma_c^2}{36}+s\sigma_c^2+m\sigma_c\sigma_m\right)^{-\frac{1}{2}}$$

$$C_2 = \sqrt{3}\cdot q_q^{\frac{1}{n}} + \frac{27}{4n}\cdot q_q^{\frac{1}{n}-1}\cdot\left(\frac{1}{k^n}-1\right)\cdot\frac{J_3}{\bar{\sigma}^3} \qquad C_3 = -\frac{9}{4n\bar{\sigma}^2}\left(\frac{1}{k^n}-1\right)\cdot q_q^{\frac{1}{n}-1}$$

where

$$q_q = \frac{1}{2}\left(\frac{1}{k^n}+1\right) - \frac{3\sqrt{3}}{4}\frac{J_3}{\bar{\sigma}^3}\left(\frac{1}{k_n}-1\right)$$

and

$$\sigma_m = p$$

$$\bar{\sigma} = \frac{1}{\sqrt{3}}q$$

$$J_3 = -\frac{2}{3\sqrt{3}}\bar{\sigma}^3 \sin 3\theta_\sigma$$

The extended criterion has been implemented into a 3D FE program [7] and it works satisfactorily.

CONCLUSIONS

An extended H-B criterion has been developed in this paper. It keeps all of the advantages of the original criterion while it eliminates the disadvantages.

REFERENCES

1. E. Hoek & E.T. Brown, "Empirical strength for rock masses", J. Geotech. Engng. Div. ASCE 106, 1013-1035 (1980).
2. E. Hoek, 23rd Rankine lecture, "Strength of jointed rock masses", Geotechnique 33, 187-223 (1983).
3. E. Hoek & E.T. Brown, "The Hoek-Brown failure criterion - a 1988 update", Proc. 15th Can. Rock Mech. Symp. University of Toronto, 31-38 (1988).
4. R.K. Srivastava et al. "Finite element analysis of tunnels using different yield criteria", 2nd Int. Symp. on Numerical Methods in Geomech. Ghent, 381-389 (1986).
5. X.D. Pan & J.A. Hudson "A simplified three dimensional Hoek and Brown yield criterion", in Rock Mechanics and Power Plants, M.Romana (ed.), ISRM Symp., 95-103 (1988).
6. G.C. Nayak & O.C. Zienkiewicz, "Convenient form of stress invariant for plasticity", J. of Structural Div. ASCE 98(ST4), 949-954 (1972).
7. L. Wei & J.A. Hudson, "Permeability variation around underground openings in jointed rocks: a numerical study", Proc. Int. Conf. on Rock Joints. N.Barton & O. Stephansson (ed.), 565-569 (1990).

A Fundamental Inequality for Plastic Constitutive Relations

Wei H. Yang
The University of Michigan
Ann Arbor, Michigan 48109

Abstract: Using convex functions to model the yield behavior of crystaline materials began more than a century ago. Such models remain valid even in light of new materials and experimental results. Without loss of generality, a yield function being convex can be represented by a norm (or seminorm) on the 3×3 symmetric stress matrix. Such matrix norm repesentation not only conveys a sense of bound on the stress as a measure of strength limit of a material, it leads to a fundamental inequality which encompasses the Cauchy-Schwarz and Hölder inequalities as special cases. All these inequalities pertain to an inner product of two vectors, matrices or functions with an upper bound in terms of the vector, matrix or function norms. The inner product of stress and strain rate matrices is fundamental in the field of mechanics and, in particular, the theory of plasticity. The norms that model yield functions are generally non-Euclidean. The generalized Hölder inequality is valid for all norms that appear in primal-dual pairs. The convexity and normality relations in plasticity become natural consequence of this inequality form of constitutive law which provides an added feature called duality. It leads naturally to many minimax theorems and mature formulations of plasticity problems and computational methods of solutions.

Introduction

Plastic behavior of crystaline materials lacks one-to-one relation between stress and strain. It is also incorrect to assume an one-to-one relation between stress and strain rate as a flow rule [1] may suggest. A case in point is that, if a yield function has a corner [2] (being a C^0 function), the strain rate can not be uniquely determined there. A single stress state may relate to a set of strain rates. This point-to-set map is obviously not a function in the usual sense. We shall model materials of this type by constitutive relations instead of equations. Only under restricted settings, does an equation apply. Such is the case of an incremental stress-strain equation under the condition of small increment sizes, strict monotone hardening [3] and a smooth yield function. A general model of plastic bahavior must be in the form of a relation. Equations are subsets of relations. A one-to-one relation is an equation.

In this paper, we shall establish a fundamental inequality which must be satisfied by all plastic constitutive relations. This inequality takes the form,

$$|\sigma : \epsilon| \leq \|\sigma\|_{(p)} \|\epsilon\|_{(d)} \tag{1}$$

where $\sigma \in R^{3\times3}$ and $\epsilon \in R^{3\times3}$ are stress and strain rate at a point in a deformed or deforming body in the forms of 3×3 symmetric matrices, the symbol ":" denotes the inner product operator between two matrices, the matrix norms (or seminorms) have subscripts (p) for primal and (d) for dual respectively.

A class of inequalities concerning inner products of vectors and functions began with Cauchy and Schwarz. We shall restrict our discussion to finite dimensional spaces. The Cauchy-Schwarz inequality in the Euclidean vector space has the form, $|\mathbf{x}^t\mathbf{y}| \leq \|\mathbf{x}\|_2\|\mathbf{y}\|_2$, where $\mathbf{x}$, $\mathbf{y}$ are vectors in R^n, t transposes a vector and $\|\cdot\|_2$ denotes the Euclidean norm. The equality holds if $\mathbf{y} = \alpha\mathbf{x}$, $\alpha \in R$, or the two vectors are co-linear.

Inequalities are often used in the upper bounding process of a mathematical analysis. A "sharp" upper bound that includes the equality case is vitally important in the field of functional analysis [4] and its applications. Obviously, a function that is bounded above has a finite supremum. The supremum of a function is contained in the range of a sharp upper bound function. Therefore a search for the least upper bound (supremum) will recover the maximum of the original function. This indirect method of finding the maximum of a function will fail if its upper bound function is not sharp.

The Cauchy-Schwarz inequality does not hold in a general non-Euclidean space. Hölder introduced an extension so that a modified inequality is valid in a family of non-Euclidean spaces with paired norms defined by Minkowski [5]. The sharpness condition is also modified for the Hölder inequality. A further extension of the Hölder inequality will take the form of the inequality (1) for general paired non-Euclidean spaces.

The family of Minkowski norms [5] for a vector $\mathbf{x} \in R^n$ is defined by $\|\mathbf{x}\|_p = (\sum_{i=1}^n |x_i|^p)^{\frac{1}{p}}$, $1 \leq p \leq \infty$ which, called a p-norm, includes Euclidean norm $(p = 2)$ as a special case. A pair of norms, $\|\cdot\|_p$ and $\|\cdot\|_q$, in this family are said to have a dual relation [5] if

$$\frac{1}{p} + \frac{1}{q} = 1, \qquad 1 \leq p, q \leq \infty. \tag{2}$$

The Hölder inequality for any pair of such dual norms is

$$|\mathbf{x}^t\mathbf{y}| \leq \|\mathbf{x}\|_p\|\mathbf{y}\|_q \tag{3}$$

and the equality case holds for $1 < p < \infty$ if either $y_i = \alpha|x_i|^{p-1}sign(x_i)$ or $x_i = \alpha|y_i|^{q-1}sign(y_i)$, $\alpha \in R$, $i = 1, 2, ..., n$. It can be easily shown that this sharpness result can be simply obtained by requiring that

$$either \quad \mathbf{y} = \alpha\nabla\|\mathbf{x}\|_p \quad or \quad \mathbf{x} = \alpha\nabla\|\mathbf{y}\|_q \tag{4}$$

where ∇ is the gradient operator. When operated on a C^1 function of n variables, ∇ produces a gradient vector in R^n. A gradient vector is normal to the level sets (contours) of the norm function. Thus, the condition for sharpness given in (4) can also be called the normality relation. The vectors satisfying (4) belong to a dual pair. To avoid the issue of lack of differentiability of certain norms, we postpone the discussion on the limiting cases $(p, q = 1, \infty)$ in which the norms are C^0 functions.

Generalized Hölder Inequality for Non-Euclidean Vector Spaces

All norms share a common property as convex, non-negative and homogeneous functions of degree one. We shall consider a pair of norms, $\|\cdot\|_{(p)}$ and $\|\cdot\|_{(d)}$, which are said to have a primal-dual relation implied by their subscripts. We use parentheses on the subscripts to emphasize generality and to prevent confusion with the Minkowski norms. One may arbitrarily define a primal norm as long as it satisfies the conditions: (i) $\|\mathbf{x}\| \geq 0$, (i') $\|\mathbf{x}\| = 0$ if and only if $\mathbf{x} = 0$, (ii) $\|\alpha\mathbf{x}\| = |\alpha|\|\mathbf{x}\|$ and (iii) $\|\mathbf{x}+\mathbf{y}\| \leq \|\mathbf{x}\|+\|\mathbf{y}\|$ for all $\alpha \in R$ and $\mathbf{x}, \mathbf{y} \in R^n$. The part (i') is omitted for the definition of a seminorm.

Usually, the definition of a primal norm (or seminorm) arises naturally from a specific application. A dual norm must be matched to the primal norm so that a theorem for the generalized Hölder inequality may be stated below.

Theorem: *For any two vectors* $\mathbf{x}, \mathbf{y} \in R^n$ *where* $\mathbf{x}$ *is measured by a properly defined primal norm (or seminorm), there exists a dual norm (or seminorm) such that the inequality* $|\mathbf{x}^t\mathbf{y}| \leq \|\mathbf{x}\|_{(p)}\|\mathbf{y}\|_{(d)}$ *holds. The case of equality is attained when the normality relation*

$$\mathbf{y} = \|\mathbf{y}\|_{(d)} \nabla \|\mathbf{x}\|_{(p)}. \tag{5}$$

is satisfied.

The proof can be found in [11] for smooth as well as C^0 norms. This theorem states the vector version of the inequality (1). The theorem, of course, covers the original Hölder inequality (3) as a special case.

Constructing the Dual Norm

The specific form of $\|\mathbf{y}\|_{(d)}$ can be determined from a given $\|\mathbf{x}\|_{(p)}$. If the primal norm is given as numerical data of a level set of a convex yield function as it may be the case in an engineering problem, then a local gradient can be computed approximately at each point of the set. These gradient vectors will trace a level set of the dual norm. This numerical method, although practical, may not be a satisfactory demonstration of the dual norm construction.

We shall assume that the primal norm is a close-form C^1 function such that its first derivatives can be computed everywhere. Dividing (5) by $\|\mathbf{y}\|_{(d)}$, then the equation represents a map from $\mathbf{x}$ to $\mathbf{y}$. If the primal norm is strictly convex, the inverse map exists and the equation (5) can be solved for the components of $\mathbf{x}$ in terms of that of $\mathbf{y}$. By taking the (p)-norm of $\mathbf{x}$ from its components and setting it to unity (or a constant), an expression for $\|\mathbf{y}\|_{(d)}$ is obtained in terms of the components of $\mathbf{y}$. If $\|\mathbf{x}\|_{(p)}$ contains a linear portion, the points on the linear portion map to a single point on $\|\mathbf{y}\|_{(d)}$. The map is no longer one-to-one but the construction of the dual norm can still proceed.

We have stated the Hölder inequality (3) without questioning the origin of the dual relation (2) of the Minkowski family of norms. For a given primal norm $\|\mathbf{x}\|_p$ $(1 < p < \infty)$, we shall now derive the form of its dual norm $\|\mathbf{y}\|_{(d)}$ without resorting to (2). Rewriting first equation of (4) in component form and letting $\|\mathbf{x}\|_p = 1$ and $\alpha = \|\mathbf{y}\|_{(d)}$ after differentiation, we have $y_i = \|\mathbf{y}\|_{(d)}|x_i|^{p-1}sign(x_i)$ from which we may solve for $|x_i|$,

$$|x_i| = \left(\frac{|y_i|}{\|\mathbf{y}\|_{(d)}}\right)^{\frac{1}{p-1}}.$$

It is an easy matter to form $\|\mathbf{x}\|_p$ from the components x_i and to obtain

$$1 = \|\mathbf{x}\|_p = \|\mathbf{y}\|_{(d)}^{-\frac{1}{p-1}}\left(\sum |y_i|^{\frac{p}{p-1}}\right)^{\frac{1}{p}}.$$

From the above equation,

$$\|\mathbf{y}\|_{(d)} = \left(\sum |y_i|^{\frac{p}{p-1}}\right)^{\frac{p-1}{p}}$$

is obtained explicitly in terms of y_i and the information of the primal norm. Now letting $p/(p-1) = q$, we confirm the dual relation (2) and conclude $\|\mathbf{y}\|_{(d)} = \|\mathbf{y}\|_q$ as the dual norm for the primal norm $\|\mathbf{x}\|_p$. Construction of the dual norms for other non-Euclidean C^1 primal norms are analogous.

The cases $p = 1, \infty$ correspond to C^0 norms which are not differentiable in the usual sense at certain places of the independent variable. They can be regarded as the limiting cases and the inequality $|\mathbf{x}^t\mathbf{y}| \leq \|\mathbf{x}\|_\infty \|\mathbf{y}\|_1$ is confirmed from the above analysis with the values $q = \infty, 1$ obtained for the corresponding dual norms by taking limits from the p-norms in C^1. Although there is no mathematical difficulty to demonstrate the generalized Hölder inequality and its normality relation for C^0 norms, it is desirable to offer an interpretation of the theorem at those points where the norm function is not differentiable. We shall call those points the vertices of a norm.

Consider the ∞-norm in R^2 as the primal norm. The gradient of $\|\mathbf{x}\|_\infty$ has only four distinctive values (1,0), (0,1), (-1,0) and (0,-1) each evaluated along a respective edge DA, AB, BC and CD on the "unit circle" of the primal norm as shown in Fig. 1a.

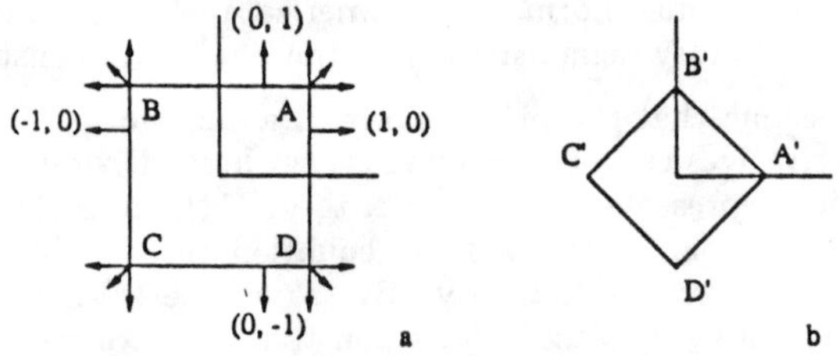

Fig. 1 Duality and Normality of 1-norm and ∞-norm

These four values are shown as four points A', B', C' and D' in Fig. 1b on the "unit circle" of the dual norm, $\|\mathbf{y}\|_{(d)} = 1$, yet to be constructed. Since

These four values are shown as four points A', B', C' and D' in Fig. 1b on the "unit circle" of the dual norm, $\|\mathbf{y}\|_{(d)} = 1$, yet to be constructed. Since the dual norm must be a convex and continuous function, these four points must be connected. Let us connect the four points by straight lines to form a diamond $A'B'C'D'$. Then no points of the "unit circle" should fall inside the diamond or the dual norm function would be non-convex. There exist, of course, convex functions passing through these four points but outside the diamond. We shall show that they can not be the dual norm function either. Since duality is symmetric, the dual norm of the dual norm is the primal norm and the symmetric normality conditions are given in (4). The gradient of the dual norm on the "unit circle" $\|\mathbf{y}\|_{(d)} = 1$ must fall on the original unit circle (the square $ABCD$) of the primal norm. The diamond, satisfying this condition, is the correct dual unit circle.

The gradient of a C^0 norm function at the vertices is not unique and is interpreted as a point-to-set map. This generalized gradient [6] takes the form of triangular fans as shown in Fig. 1a in which the length of all gradient vectors are proportionally drawn. In computational approach to the solutions of applied problems, C^0 norms are less popular for the obvious reason of numerical difficulty. A C^0 norm is often replaced by a smooth norm in applications.

Application to Plasticity

The concept of a norm is not restricted to vectors. There are matrix norms [7], function norms and operator norms [8]. We shall limit our discussion to finite dimensional spaces therefore only vector and matrix norms are discussed. A matrix norm has the same properties as the vector norms. It can be either induced by a vector norm [7] or independently defined. The norms defined on stress matrices in the theory of plasticity are mathematical models of yield behavior based on experimental results and certain physical principles [12].

In plasticity, the natural primal variable is the stress denoted by $\sigma \in R^{3\times 3}$, a matrix representing the state of force per unit area at a point in the material. Since there exists no real material that is infinitely strong, the strength of a material can be modeled by a bound

$$\|\sigma\|_{(p)} \leq \sigma_0 \tag{6}$$

on the matrix σ where σ_0 is a material constant. The specific form of the norm is called the yield function derived from experimental data. The best known yield function is the von Mises yield function [8]. To shorten the discussion and facilitate graphical presentation, but still keep all the essentials relevant to the intended exposition in this paper, the von Mises yield function is presented in a subspace of plane stress where the stress is a 2×2 symmetric matrix

$$\sigma = \begin{bmatrix} \sigma_{xx} & \sigma_{xy} \\ \sigma_{yx} & \sigma_{yy} \end{bmatrix}, \qquad \sigma_{xy} = \sigma_{yx}, \tag{7}$$

which represents a state of stress in a sheet of material located in the (x, y)-plane of a Cartesian coordinate. The von Mises yield function can be stated in two equivalent forms,

$$\|\sigma\|_{\mathrm{V}} = \sqrt{\sigma_{xx}^2 - \sigma_{xx}\sigma_{yy} + \sigma_{yy}^2 + 3\sigma_{xy}^2} = \sqrt{\sigma_1^2 - \sigma_1\sigma_2 + \sigma_2^2} \tag{8}$$

where σ_1 and σ_2 are the eigenvalues of the plane stress matrix in (7) and $\|\cdot\|_V$ denotes the von Mises norm.

Another yield function named after Tresca [8] is physically more precise but is less used because of its non-smoothness. When applied to the plane stress matrix, it takes the form,

$$\|\sigma\|_T = max\{|\sigma_1|,\ |\sigma_2|,\ |\sigma_1 - \sigma_2|\}, \tag{9}$$

where the Tresca norm $\|\sigma\|_T$ can also be expressed in terms of stress components [8]. One can not find these norms in a mathematics book since they are derived from physical considerations. Nevertheless, they are valid norms or seminorms.

The dual variable in the context of plasticity is the strain rate ϵ, also a 2×2 symmetric matrix for the plane stress case. The inner product $\sigma : \epsilon$ of the two matrices appears frequently in the field of mechanics. From a physical principle of non-negative dissipation, Drucker [9] had reached a conclusion that the plastic strain rate matrix is a constant multiple of the gradient matrix (a gradient vector arranged in the form of a matrix) of the yield function such that

$$\epsilon = k\nabla\|\sigma\| \tag{10}$$

where the gradient is taken with respect to each of the stress components in (7). This is known as the normality relation in plasticity between strain rate and stress matrices. Using the mathematical argument of a sharp upper bound, we reach the same conclusion as that in [9].

It was not well understood however in the theory of plasticity that a plastic strain rate matrix and the stress matrix must also satisfy the inequality (1). We restate this fundamental inequality,

$$|\sigma : \epsilon| \leq \|\sigma\|_{(p)}\|\epsilon\|_{(d)} \tag{11}$$

as the matrix version of the theorem presented in this paper. Equality holds if the normality relation (10) is satisfied.

For the von Mises and Tresca yield functions, the corresponding dual norms on the plastic strain rate ϵ in terms of its eigenvalues ϵ_1 and ϵ_2 are

$$\begin{aligned} \|\epsilon\|_\wedge &= \frac{2}{\sqrt{3}}\sqrt{\epsilon_1^2 + \epsilon_1\epsilon_2 + \epsilon_2^2}, \\ \|\epsilon\|_\perp &= max\{|\epsilon_1|, |\epsilon_2|, |\epsilon_1 + \epsilon_2|\} \end{aligned} \tag{12}$$

respectively. They are derived from the corresponding primal norms in (8) and (9) by the constructive process described earlier. Since $\|\cdot\|_T$ is a C^0 norm, its gradient needs be interpreted in a generalized sense for the construction of its dual $\|\cdot\|_\perp$.

For a graphical presentation of the results, the eigenvalues of the plane stress matrix (σ_1, σ_2) and that of the plastic strain rate matrix (ϵ_1, ϵ_2) may be arranged as vectors in R^2. The unit circles for the four norms are shown in

Fig. 2 where the primal norms are shown in solid lines and the dual norms in broken lines.

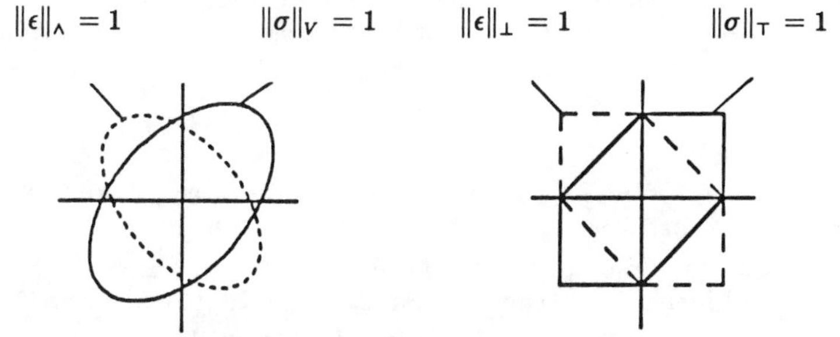

Fig. 2 Geometric Representation of Primal and Dual Norms for von Mises and Tresca Yield Functions

The model of plastic behavior, using a convex yield function in terms of a primal norm (norm functions are naturally convex) in the form of inequality (6) and its normality relation (10) as the associated flow rule (the strain rate stress relation), is still incomplete without using the duality inequality (1) as a fundamental constitutive relation for the theory of plasticity. Using special forms of this inequality, we have established several minimax (duality) theorems for certain structural and manufacturing problems. The results can be found in [13].

Conclusion

In the highly specialized research environment today, similar ideas may be developed independently in different fields. A case in point is the normality relation. Engineers deduced it from their believe that a material can not produce energy during plastic deformation (a pure physical argument). The condition of a sharp upper bound on the inequality (1) produces the same normality relation (a pure mathematical argument). An ultimate truth seems to have many faces.

Science of mechanics in the three centuries from 1600 to 1900 had been a rich source of mathematical ideas and a fertile ground for their cultivation. In fact, a physicist and a mathematician were often found in the same person in that era. Now, mathematics seems to have evolved into a purified art form that is both general and abstract. To apply abstract theorems to concrete engineering problems has become more challenging. But modern mathematics and computational methods have solved more difficult problems ever considered feasible.

Material modeling, neither in the mainstream of modern physics nor in the interest of pure mathematics, is important in many engineering branches. The abstract theorems in non-Euclidean and non-Hilbert spaces [10] are likely to elude the attention of engineers. But these theorems may benefit the mathematical modeling of complex materials. The theorem of dual non-Euclidean norms and the generalized Hölder inequality cited in this paper were motivated by the theory of plasticity. Now this general theorem of mathematical nature can help to provide deeper understanding of the existing theories and to promote further progress in new models of plastic constitutive relations and their applications.

References

1. (a) Prandtl, L. Proc. First Int. Congress Appl. Mech., Delft, (1924); (b) Reuss, A. Zeits. Ang. Math. Mech. 10, 266, (1930).
2. Tresca, H. (a) Comptes Rendus Acad. Sci. Paris, 59 and 64, (1864 and 1867); (b) Mem. Sav. Acad. Sci. Paris, 18 and 20, (1868 and 1872).
3. Owen, D. R. J. and Hinton, E. Finite Element in Plasticity: Theory and Practice, Pineridge Press, UK (1980).
4. Goffman, C. and Pedrick, G. First Course in Functional Analysis, Prentice-Hall, (1965).
5. Birkhoff, G. and MacLane, S. A Survey of Modern Algebra, MacMillan, (1953).
6. Clarke, F. H. Generalized gradients and applications, Trans. Amer. Math. Soc. 205, (1975) 247-262.
7. Householder, A. S. On Norms of Vectors and Matrices, ORNL Report 1759, (1954).
8. Hill, R. The Mathematical Theory of Plasticity, Oxford at Clarendon Press, (1950).
9. Drucker, D. C. A definition of stable inelastic material, J. Appl. Mech. 81 (E101), (1959).
10. Adams, R. A. Solobev Spaces, Academic Press, (1975).
11. Yang, W. H. On generalized Hölder Inequality, to appear in Nonlinear Analysis, (1990).
12. Yang, W. H. (a) A useful theorem for constructing convex yield functions, J. Appl. Mech. 47, (1980) 301-303; (b) A generalized von Mises criterion for yield and fracture, J. Appl. Mech. 47, (1980) 297-300.
13. Yang, W. H. (a) A duality theorem for plastic plates, ACTA Mechanica, 69, (1987) 177-193; (b) Pipe flow of plastic materials, J. Appl. Mech. 47, (1980) 496-498.

CYCLIC PLASTICITY

UNDRAINED CYCLIC LOADING WITH REORIENTATION OF PRINCIPAL AXES : EXPERIMENTS AND MODELLING

B. CAMBOU - PH. DUBUJET
Ecole Centrale de Lyon,
Ecully, France

T. DOANH
Ecole Nationale des Travaux Publics de l'Etat,
Vaulx-en-Velin, France

ABSTRACT

This paper presents experimental results and simulations for cyclic torsional shear stress, under undrained conditions, on normally consolidated clay. Predictions are made using an elastoplastic model by taking into account a kinematic hardening.

1 - INTRODUCTION

Several constitutive equations have been proposed in the literature. Many of them give good results for the simulation of simple loadings. Very few results are available for the accuracy analysis of the models in complex loadings (cyclic loadings, undrained conditions, loadings with reorientation of principal stresses). Nevertheless these kinds of loadings are not uncommon, for example in the foundations of buildings under seismic stress or in the foundations of off shore platforms shaken by the movement of waves.

The aim of this paper is to present experimental results and simulations for loadings taken into account with all the difficulties previously mentioned.

2 - CAMBOU-JAFFARI'S MODEL

The proposed model has been built within a general framework of elastoplasticity. Two types (mechanisms) of plastic strain are defined : the strain ε_{ij}^{dp} depending on the variation of mean stress ($I_1/3$), and the strain ε_{ij}^{ip} depending on the variation of s_{ij}/I_1 (s_{ij} =deviatoric stress).

The total strain is defined as the sum of three terms :

$$\varepsilon_{ij} = \varepsilon_{ij}^{e} + \varepsilon_{ij}^{ip} + \varepsilon_{ij}^{dp}$$

where ε_{ij}^{e} is a nonlinear elastic strain. The evolution of the first mechanism of plasticity is characterized by an isotropic hardening, and the second one by an isotropic and a non-linear kinematic hardening. The latter hardening allows us to take into account the induced anisotropy. This fact is very important for the description of cyclic loadings. The variation of the volumetric strain is characterized by a kinematic condition, representing the characteristic state concept.

For a normally consolidated clay this model takes into account ten constants which can be identified from triaxial tests.

3 - EXPERIMENTS AND MODELLING

3 1 - DESCRIPTION OF EXPERIMENT

Our experiments were performed with a hollow cylinder device, which allows a continuous rotation of the principal stress axes. Our samples were prepared from a slurry of commercial kaolinite powder P300. The saturated specimens were consolidated in a large consolidometer at a vertical effective stress of 200 kPa. Then they were trimmed and consolidated again, isotropically in the apparatus at an initial cell pressure of 500 kPa.

Monotonic and cyclic torsion shear were performed under static loading conditions. A constant shear strain rate $\dot{\gamma}$ was applied to the specimen, and the shear stress $\tau_{\theta z}$ was measured, while the stress difference $\sigma_z - \sigma_r$ and the cell pressure σ_r were kept constant respectively at 50 kPa and 500 kPa. In the two-way cyclic test, the cyclic stress ratio τ_{cyc} / τ_{max} was 0.42 and the angle between the vertical and the direction of the major principal stress was ± 32°.

3 2 - NUMERICAL SIMULATION

The constants of the model had been defined from monotonic drained and undrained triaxial tests. The simulation of the two cyclic torsional tests globally tally with experimental results as we can see in the following different curves: stress-strain curves (Fig.a), pore pressure evolution curve (Fig.b),and stress paths curve (Fig.c). In particular the increasing evolution of pore pressure along the cyclic loadings is accurately simulated (Fig.2.b).

Nevertheless we must point out two observations which are not satisfactorily characterized in the simulation :

- after the initial cyclic loadings the pore pressure reaches great values similar to those obtained in the monotonic test;
- when the pore pressure attains such great values, the material shows a dilatance behaviour due to a great over-consolidation.

The model initially built to characterize granular material has to be improved on these two points in order for it to be useful for clay.

We must emphasize that these simulations are very sensitive to two aspects of modelling : first, the relative importance between the isotropic and the kinematic hardening mechanisms and second, the value of the constant, characterising the plastic volumetric strain evolution.

4 - CONCLUSIONS

Experimental results showed principally that, for large strains the behaviour seems to be independant of the initial stress path (in particular, initial cyclic loadings with a limited number of cycles).

Predictions showed that the model can reproduce the most important aspects of clay behaviour. These kinds of simulations emphasised the necessity of considering kinematic hardening in the modelling. This present work should be considered as a first step towards modelling of clay in complex loadings.

REFERENCES

1. B. Cambou and K. Jafari, "Modèle de comportement des sols non cohérents", Revue Française de Géotechnique 44 (1988) 43-55.

2. T. Doanh, "Contribution à l'étude du comportement de la kaolinite", thesis,Lyon E.N.T.P.E (1986).

3. P.Y. Hicher, "Comportement mécanique des argiles saturées sur divers chemins de sollicitations monotones et cycliques", these de Doctorat d'Etat (1985), Université de Paris VI.

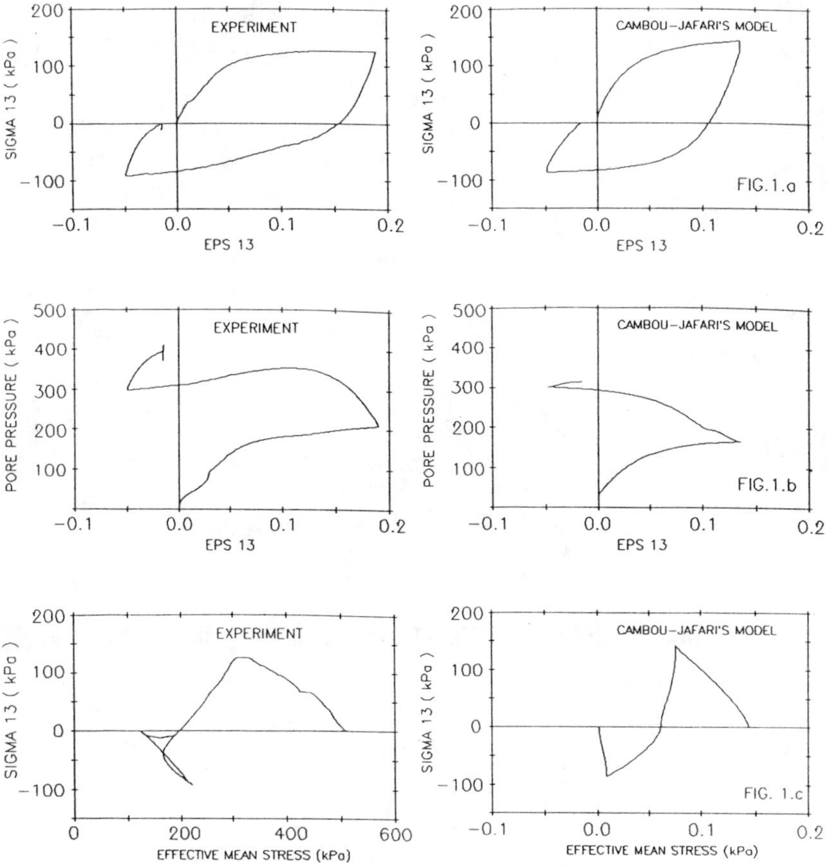

FIG. 1. a,b,c. Test AK06 - Experiments and simulations.

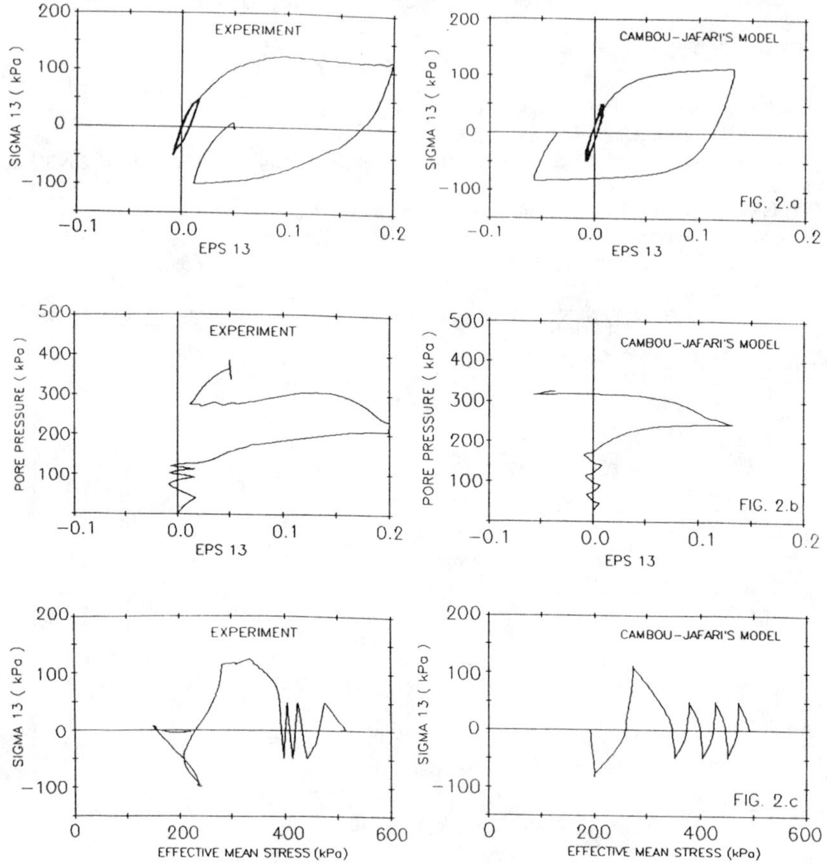

FIG. 2. a,b,c. Test AK05 - Experiments and simulations.

CYCLIC MATERIAL BEHAVIOUR OF TUBES WITH RATCHETTING STRAINS

S.J. HARVEY AND A.S. MANNING
COVENTRY POLYTECHNIC U.K.

P. ADKIN
LAND ROVER LTD. U.K.

ABSTRACT

The paper describes a test program which examines the cyclic material properties of tubes subjected to cyclic plastic torsion with sustained axial stresses and cyclic plastic axial strains with sustained internal pressures. The tests involved loading and unloading from transient and steady state cyclic conditions in order to establish a hardening rule. The materials investigated were a low carbon steel EN32, and an austenitic stainless steel AISI 321. The data acquisition system used enabled the cyclic stress-strain curves to be developed accurately using a Fourier analysis with a data sampling rate of twice the highest harmonic frequency. The results of unloadng and probing into the yield surface established that a kinematic hardening model was appropriate for cyclic plastic push-pull with and without sustained internal pressure.

INTRODUCTION

The cyclic response of materials to inelastic deformation has received considerable attention in recent years. The cyclic material characteristics may be adequately described for simple uniaxial states by using the models of Iwan[1] and Mroz[2] The concept of a bounding surface has been examined by Popov[3] and Kreig[4] For multi-axial states the models become complex and are not easily applied by the design engineer. Appropriate yield criteria and flow rules are now well understood. However, the selection of an appropriate hardening rule for components which are subjected to multi-axial cyclic plastic strains would appear to still pose many problems, and additional experimental data is required in order to help resolve uncertainties. In the experimental program presented data was obtained from tests on tubular specimens subjected to cyclic plastic torsion and cyclic plastic push-pull with sustained steady loads which resulted in ratchetting strains. The cyclic material properties were monitored and hardening rules investigated. The hardening rule is established by identifying the yield surface, defined by departure from linearity, when unloading from various points within a cycle or plastic strain.

EXPERIMENTAL PROGRAM

The materials used were a low carbon steel EN32 and a stainless steel AISI 321.

The nominal dimensions of the tubular specimens were: Gauge Length 50mm, Gauge Section Outer Diameter 18.21mm, Wall Thickness 1.12mm. The test rig was capable of applying any combination of 3 loading modes, a) Push-Pull, b) Cyclic Torsion, c) Internal Pressure, and is fully described elsewhere[5].

Tests were performed at particular loading rates in which loads and displacements were measured, via strain gauge bridges and RVDT's and LVDT's. Fine control of strains and displacement was obtained using stepper motors. Stepper motor control and data acquisition systems were based around a Gemini 64K micro-computer. The system was designed to be capable of sampling sufficient data within each cycle to allow stress-strain curves to be defined and modelled accurately.

A Fourier analysis was used to estimate the required number of discrete data points necessary to describe the cyclic load - deflection curve with sufficient accuracy. Too little data would lead to inaccuracies in subsequent curve fitting and too much data would be unnecessary and could lead to an overly complex data acquisition system. Typically linear displacements needed to be measured to ± 1μm and angular displacements to ± 1' when defining plastic strains by depature from linearity.

EXPERIMENTAL RESULTS

The Fourier analysis was first applied to a previously measured[6] steady state hysteresis loop for AISI 321, in cyclic torsion. The Fourier series was derived as a special case of the Fourier integral using distribution theory. Existing software was used to calculate the coefficients of the truncated Fourier series which approximate to a set of data points when the number of harmonics allowed is specified. The original data was digitised to provide 121 data points per cycle and the result for several harmonics are shown in Fig.(1) and Fig.(2). The choice of sampling rate is determined to give a balance between the amount of data required and noise rejection. The Nyquist sampling theory indicates the sampling rate should be at least twice the highest harmonic frequency. A sampling rate of 12.5 Hz was chosen since it gave good noise rejection qualities and allowed higher strain rates if required. Software limitations prohibited evaluation of harmonics greater than 60th, but sampling at this speed satisfies a performance criteria of accuracies better than 0.1%.

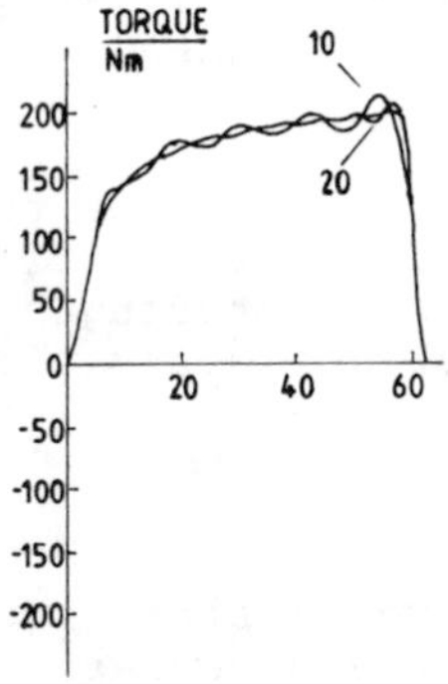

FIG.1. FOURIER ANALYSIS HARMONICS 10 & 20

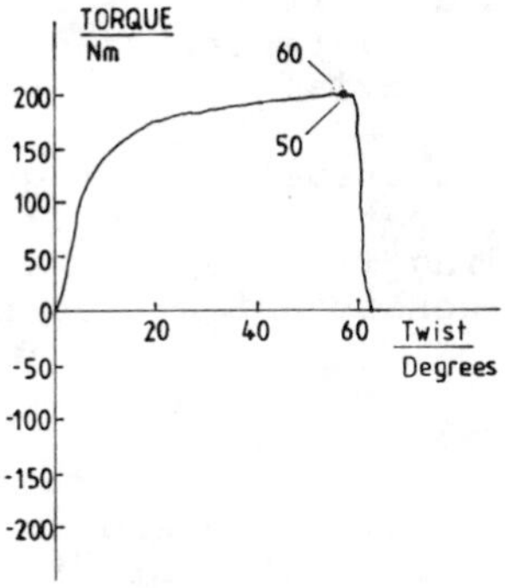

FIG. 2. FOURIER ANALYSIS HARMONICS 50 & 60

In order to apply yield surface models it is necessary to know the current values of yield stress in several directions. At steady state, the hoop yield stress could be determined by internal pressure tests at any time within a strain cycle by increasing the internal pressure until yield is observed. Also, the determination of the yield point on reverse loading is required.

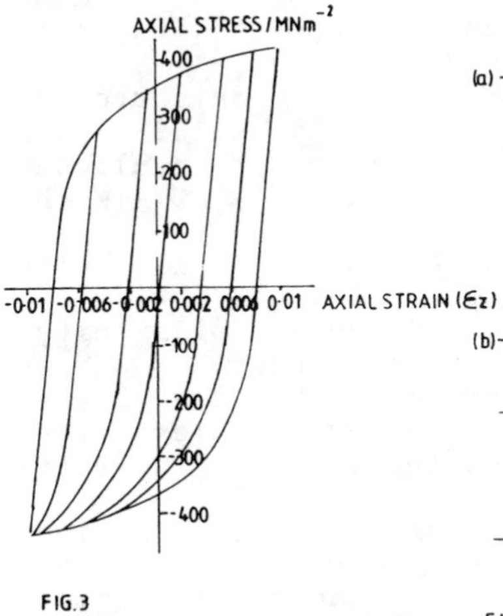

FIG.3
CURRENT REVERSED YIELD POINT DATA

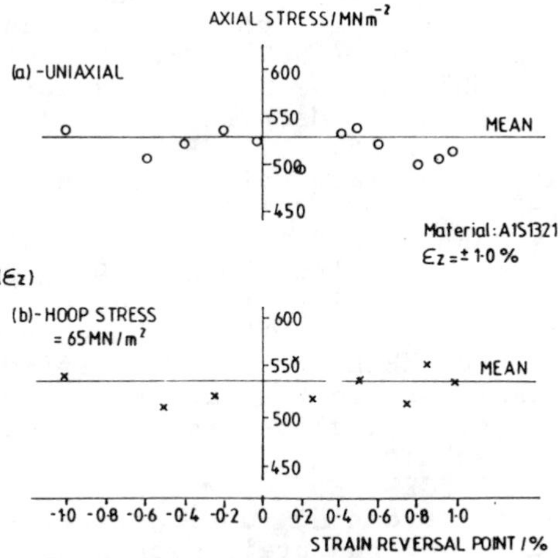

FIG. 4. COMPARISON OF 'CURRENT' DIFFERENCE BETWEEN FORWARD AND REVERSE YIELD STRESSES

The results of tests using AISI 321 and typical hysteresis curves are shown in Fig.(3) where the current reversed yield limit at various points with a cycle are identified. The differences between forward and reverse yielding point were then determined and the results are shown in Fig.(4) for zero internal pressure and 1700 psi respectively. To a first approximation it can be seen that the difference between forward and reverse yield remains largely constant throughout a cycle and is unaffected by an internal pressure (within experimental error). A recent paper[7] has confirmed similar characteristics for an austenitic stainless steel in uniaxial cyclic loading.

These results are significant since they tend to indicate that the chord length of the yield surface on the loading line remains constant throughout a cycle, which is consistent with a form of kinematic hardening.

CONCLUSIONS

The data acquisition and control system used has proved itself capable of achieving the necessary accuracy and reliability required of such relatively long time-scale tests. The use of a Fourier series to model material behaviour was successful using this data acquisition system. The identification of a kinematic hardening rule and its implications in controlling the size, shape and movement of the yield surface have been examined[5] and used to predict ratchetting strains with reasonable accuracy.

REFERENCES

1. IWAN, W.D. On a class of models for the yielding behaviour of continuous and composite systems J.Appl.Mech. (1967) 34 pp.612-617.
2. MROZ, Z. On the description of anisotropic hardening. J.Mech.Phys. Solids (1967) 15 pp.163-175.
3. DAFILIAS, Y., POPOV, E. A model of nonlinear hardening materials for complex loading. ACTA.Mech. (1975) 21 pp.173-192.
4. KREIG, R.D. A practical and surface plasticity theory. Inl.Appl.Mech. (1975) 26 pp.641-645.
5. ADKIN, P. Yield surfaces in cyclic plasticity. PhD Thesis. June 1986. Coventry Polytechnic.
6. BRECKELL, T.H. The effect of martensite transformations on the fatigue behaviour of stainless steel in cyclic plastic torsion. PhD Thesis. 1976. Coventry Polytechnic.
7. TUEGEL, E.J. Measurements of isotropic and kinematic hardening during inelastic cyclic straining. Trans. Structural Mechanics in Reactor Technology (1985) Vol 1. pp.79-84.

RATCHETING IN CYCLIC PLASTICITY

T. Hassan, E. Corona, and S. Kyriakides
Engineering Mechanics
The University of Texas at Austin
Austin, Texas 78712

EXTENDED SUMMARY

The paper is concerned with the phenomenon of cyclic creep or ratcheting of metals cyclically loaded in the plastic range (ratcheting here describes cyclic accumulation of deformation). Time independent ratcheting under uniaxial loading as well as ratcheting under multiaxial loading were examined in a combined experimental and analytical effort. Carbon steel 1020 and heat treated 1026 thin-walled tubes were used in the experiments. In an effort to reduce interaction between the mild cyclic softening exhibited by the materials and the measured ratcheting, the material was cyclically stabilized through an initial axial strain symmetric cyclic prehistory.

Uniaxial ratcheting was studied through a systematic set of stress controlled uniaxial experiments. A typical set of stress-strain results, from CS-1020, is shown in Fig. 1a ($\bar{\sigma}_x = \sigma_x/\sigma_o$, σ_o is the initial yield stress). The effect of the prescribed mean stress and amplitude of the stress cycle on the rate of ratcheting, were quantified experimentally. Figure 2 shows a plot of the maximum strain in each cycle (ε_{xp}) as a function of the number of loading cycles applied (N) from a set of experiments in which the mean stress was kept constant ($\bar{\sigma}_{xm} = 0.13$) . Clearly, increase in the amplitude of the stress cycle leads to an increase in the rate at which ratcheting occurs. From the results presented, it can be observed that, following an initial transient, the material ratchets at a nearly constant rate.

The Drucker-Palgen [1], Dafalias-Popov [2] and Tseng-Lee [3] models were used to simulate the experiments. The first was found to predict a racheting rate which is much higher than that measured experimentally. This is a direct result of the very simple nature of the flow rule adopted. The second model was found to yield good predictions of the rate of ratcheting at the earlier part of the cyclic history. However, the model always leads to eventual arrest of ratcheting if a linear bound is used. The third model predicts a constant ratcheting rate for a cyclically stable material. The rate of ratcheting predicted by this model can be quite good for stress cycle parameters which are close to those of the experiment fitted. However, the predictions were not satisfactory for the whole range of experimental parameters considered (see [4] for details).

During the experimental program, it was observed that large amplitude strain controlled loops induced after significant ratcheting, have the same shape as the stable hysteresis loops (see, for example, Fig. 1a). Thus, the initial stable hysteresis loop can be said to translate at the rate of ratcheting in the direction of the accumulated strain. In view of this, a simple scheme for shifting the linear bound in the Dafalias-Popov model was devised. In this modification, the bound is allowed to translate in the direction of ratcheting at the rate of ratcheting. The modified model was used to simulate the whole set of ratcheting experiments conduced. An example of a stress-strain response predicted is shown in Fig. 1b. All major features of the predicted results are seen to be in good agreement with the corresponding experimental results, shown in Fig. 1a. Ratcheting strains, as a function of N, predicted for other experiments, using the same material parameters, are included in Fig. 2. The predicted rates of ratcheting are seen to be in very good agreement with the experimental ones for all cases. The same quality of predictions was found to occur in many other uniaxial ratcheting experiments reported in [4].

Biaxial experiments involving cyclic, strain-symmetric axial loading and constant internal pressure were performed on heat treated CS-1026 tubes. Under this type of loading, the material exhibits ratcheting of the circumferential strain. A plot of measured circumferential strain (ε_θ) versus the prescribed cyclic axial strain ($\varepsilon_{xc} = 0.5\%$) for a constant circumferential stress of $\bar{\sigma}_\theta = 0.245$, is shown in Fig. 3a. Similar results were obtained for various values of internal pressure and axial strain cycle amplitudes. Results obtained from experiments with constant strain cycle amplitude of $\varepsilon_{xc} = 0.5\%$ and different values of σ_θ are summarized in Fig. 4. The peak values of circumferential strain ($\varepsilon_{\theta p}$) are plotted against the number of cycles prescribed. Clearly, higher σ_θ results in a faster rate of ratcheting.

The three cyclic plasticity models mentioned above were used to simulate the experiments. The models were found to reproduce the major features of the material response, but yielded wrong rates of ratcheting. The rate of ratcheting was found to be very sensitive to the hardening rule adopted in the models. For example, the Prager-Ziegler hardening rule leads to arrest of ratcheting in the first couple of cycles. The Mroz hardening rule yields rates of ratcheting which are much higher than those recorded in the experiments. These discrepancies are, at least, partly due to the simplicity built into the models adopted by requiring that the yield surface retain its shape and size throughout the loading history. Detailed examination of the reasons which cause failure of these hardening rules in this loading history can be found in [5].

A hardening rule proposed by Armstrong and Frederick [6] incorporated into the Drucker-Palgen and Dafalias-Popov models was found to produce the most satisfactory predictions. Figure 3b shows the response predicted with the latter model for the experiment shown in Fig. 3a. All features of the response, including the rate of ratcheting, are seen to be well reproduced in the predictions. Predictions of the ratcheting strains as a function of N obtained for other experiments using the same set of material parameters are shown in Fig. 4. The rate of ratcheting is seen to be quite well reproduced for all experiments shown. Equally good performance of the model was found in other similar biaxial experiments reported in [5].

REFERENCES

1. Drucker, D. C. and Palgen, L., "On Stress-Strain Relations Suitable for Cyclic and other Loading", *ASME Journal of*

Applied Mechanics, **48**, 479-485, 1981.
2. Dafalias, Y. F. and Popov, E. P., "Plastic Internal Variables Formalism of Cyclic Plasticity", *ASME Journal of Applied Mechanics*, December, **43**, 645-651, 1976.
3. Tseng, N. T. and Lee, G. C., "Simple Plasticity Model of Two-Surface Type", *ASCE Journal of Engineering Mechanics*, **109**, 795-810, 1983.
4. Hassan, T. and Kyriakides, S. "Ratcheting in Cyclic Plasticity. Part I: Uniaxial Behavior," *International Journal of Plasticity* (to appear).
5. Hassan, T., Corona, E. and Kyriakides, S., "Ratcheting in Cyclic Plasticity. Part II: Multiaxial Behavior," *International Journal of Plasticity* (to appear).
6. Armstrong, P. J. and Frederick, C. O., "A Mathematical Representation of the Multiaxial Bauschinger Effect", Berkeley Nuclear Laboratories, R & D Department, Report No. **RD/B/N/ 731**, 1966.

ACKNOWLEDGEMENTS

The test facility used to conduct the experiments was developed in part with the financial support of the Office of Naval Research under the equipment grant No. N00014-86-G-0155. The work was conducted with the support of the National Science Foundation under grant MSM-8352370.

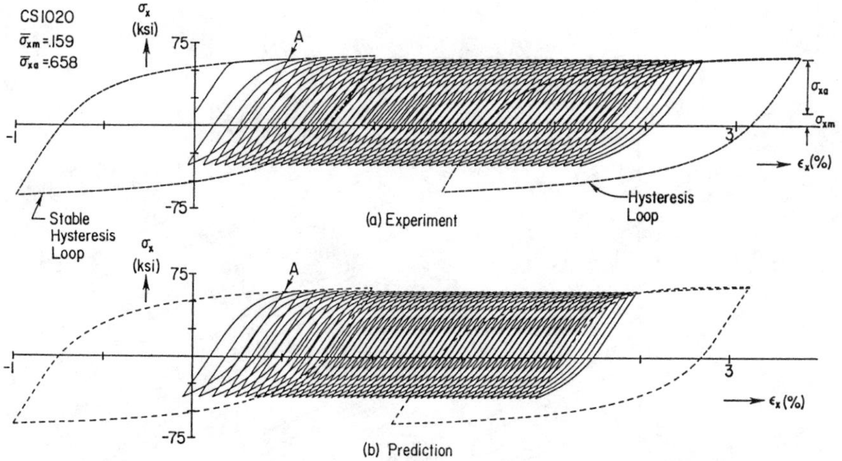

Fig. 1 Axial strain ratcheting of CS 1020. Comparison between experiment and prediction

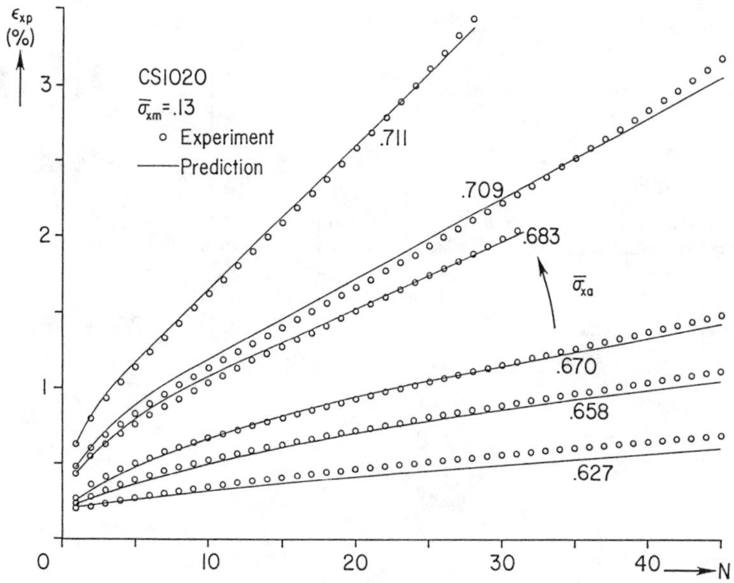

Fig. 2 Maximum strain per cycle as a function of number of cycles in uniaxial ratcheting experiments. Comparison between experiments and predictions

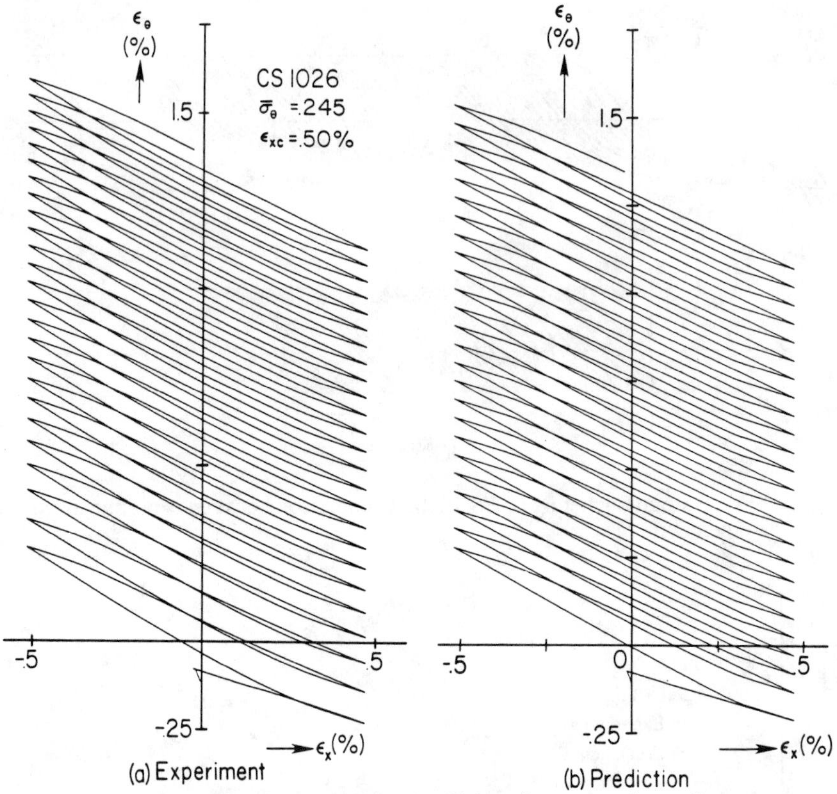

Fig. 3 Circumferential strain ratcheting in axial strain symmetric cycling at a constant internal pressure. Comparison between experiment and prediction

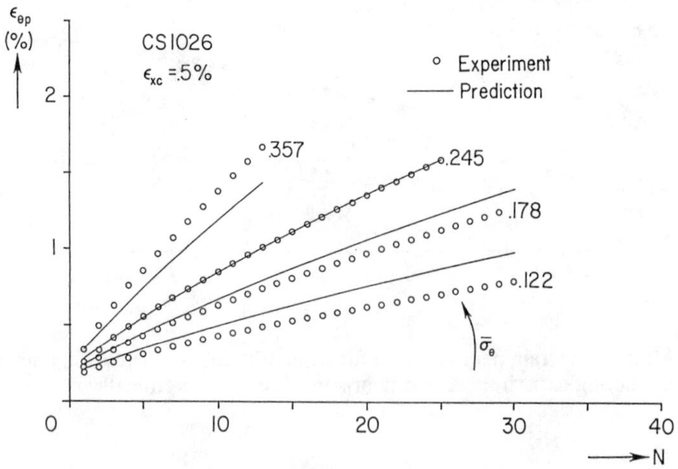

Fig. 4 Maximum circumferential strain as a function of number of cycles. Comparison between experiments and predictions

ANISOTROPIC CONSTITUTIVE MODELING FOR SINGLE-CRYSTAL SUPERALLOYS USING A CONTINUUM PHENOMENOLOGICAL APPROACH

D. NOUAILHAS and J.L. CHABOCHE
Office National d'Etudes et de Recherches Aérospatiales
BP 72. F - 92322 CHATILLON cedex (France)

ABSTRACT

This paper presents the developments made in a phenomenological viscoplastic theory in order to describe the cyclic behavior of single-crystal superalloys. The anisotropy is introduced in a very simple manner, with a very few number of additional material constants with regard to the isotropic case. An illustration of the possibilities of this kind of models is given on the example of the CMSX2 single-crystal alloy at elevated temperatures.

INTRODUCTION

The nickel-base single-crystal superalloys considered in this paper were developped to increase their fatigue and creep properties with regard to polycrystalline superalloys by eliminating the grain boundaries. Various alloys were then elaborated by different companies like the PWA 1480, René N4, CMSX-2 or AM1 alloys, for the same kind of high temperature applications in gas turbine engines. All these single-crystals have the same microstructure: they are two-phases alloys with a large volume fraction of γ' precipitates (around 65%).The precipitates have the L_{12} crystal structure and are interspersed in a coherent face center cubic (f.c.c) solid solution matrix. Appropriated thermal treatment is done to set γ' particule size at about 0.5μm which give optimum properties at the alloys. The final microstructure is then a well-ordered two-phases material with cuboidal γ' precipitates having their edges aligned along the <001> directions of γ and γ' phases. Their elaboration at industrial level now requires the development of new constitutive equations, usable in finite element codes in view of three-dimensional structural computations and life prediction analyses. In addition to the classical features of a viscoplastic behavior, such as strain path history, strain-rate and temperature dependencies, plastic anisotropy must be properly described by a constitutive model.

From a modeling point of view, several works were conducted for single-crystals using crystallographic approaches, starting with the modeling of creep behavior and then the modeling of time independent plasticity. More recently, models for cyclic viscoplasticity have been developed. These works are described in a recent revue made by JORDAN [1] on the modeling of single-crystal behavior. These viscoplastic models introduce octahedral and cubic slips, necessary for single-crystal superalloys. They also take into account latent hardening for octahedral slip and cube cross-slip responsible for tension-compression asymmetry at low temperatures. With this kind of approach, taking into account the main physical mechanisms leads to complex models, requiring an important number of material parameters to determine at each temperature, that are not easy to identify through mechanical testings. Another way is to introduced the anisotropy in a purely phenomenological approach [2,3]. This is the purpose of the present paper which is shared into two main parts:

- The first part gives a short presentation of the anisotropic constitutive model, built up in the context of a continuum phenomenological approach, and using the notion of macroscopical internal state variables.

- In the second part an application of this model to the case of cubic symmetry is done, on the basis of experimental results obtained on the single-crystal superalloy CMSX-2 at 950° C.

ANISOTROPIC VISCOPLASTIC MODEL

This model is a generalization to the case of anisotropic materials of the cyclic viscoplastic model developped at ONERA and mainly based on non-linear kinematic hardening (NLKH) rule [4]. In the cristallographic axes, the constitutive equations are expressed by:

Viscoplastic potential : $\Omega = \frac{K}{n+1} < \frac{f}{K} >^{n+1}$

where f is the yield surface equation : $f = \sqrt{3/2\,(\sigma'_{ij} - X'_{ij}) M_{ijkl} (\sigma'_{kl} - X'_{kl})} - k$

(σ'_{ij} et X'_{ij} are the components of the deviatoric tensors, and $<u> = uH(u)$; $H(u) = 1$ si $u>0$; $H(u) = 0$ si $u \leq 0$).

Viscoplastic strain rate : $\dot{\varepsilon}^p_{ij} = \frac{\partial \Omega}{\partial \sigma_{ij}} = \dot{p} \frac{\partial f}{\partial \sigma_{ij}}$ with $\dot{p} = \sqrt{2/3\, \dot{\varepsilon}^p_{ij} M^{-1}_{ijkl} \dot{\varepsilon}^p_{kl}} = < \frac{f}{K} >^n$

Non-linear kinematic hardening : $\dot{X}^q_{ij} = \frac{2}{3} N^q_{ijkl} \dot{\varepsilon}^p_{kl} - Q^q_{ijkl} X^q_{kl} \dot{p}$ with $X_{ij} = \sum_q X^q_{ij}$

n and K represent the effects of viscosity, both of which are strongly dependent on temperature. k establishes the size of the yield surface. Two NLKH variables are superimposed (q=1,2), as for isotropic materials, to well reproduce both the high non-linearity at the begining of the plastic flow and the correct tangent modulus at higher plastic strains.

The fourth-order tensor M_{ijkl} describes the initial anisotropy and eventually some plastic strain induced anisotropy according to the expression of its components. In each NLKH rule, two fourth-order tensors N_{ijkl} and Q_{ijkl} are introduced in place of the two constants of the classical model, to describe plastic strain induced anisotropy. No isotropic hardening is introduced in the present formulation, this kind of hardening being negligible in the following application to CMSX2 superalloy. Nevertheless, it could be easily taken into account if necessary.

APPLICATION TO SINGLE-CRYSTAL SUPERALLOY

For a cubic single-crystal, the material-constant fourth-rank tensors have only three independent components. Using Voigt vectorial notation for strains and stresses, and in the crystallographic axes, all these tensors (*i.e.* the elastic compliance S_{ijkl} and the inelastic quantities M_{ijkl}, N_{ijkl} and Q_{ijkl}) are represented by 6×6 matrices having the same following common form :

$$M = \begin{bmatrix} M_{11} & M_{12} & M_{12} & 0 & 0 & 0 \\ M_{12} & M_{11} & M_{12} & 0 & 0 & 0 \\ M_{12} & M_{12} & M_{11} & 0 & 0 & 0 \\ 0 & 0 & 0 & M_{44} & 0 & 0 \\ 0 & 0 & 0 & 0 & M_{44} & 0 \\ 0 & 0 & 0 & 0 & 0 & M_{44} \end{bmatrix}$$

Using the hypothesis of plastic incompressibility, the number of independent material constants is reduced to two for the matrices M, N and Q. Then, in this model, only five additional material parameters are introduced for two NLKH variables, with regard to the isotropic case.

In practice, only five parameters have to be determined, the couple (M_{11} , M_{12}) being set to return the isotropic equations in the crystallographic axes, that is to say $M_{11} - M_{12} = 1$.

In addition of the reduced number of parameters as compared with crystallographic models, the interest of this approach lies in the fact that tests in only two crystallographic orientations (<001> and <111>) are needed to completely identify the model. The coefficients of type (C_{11} , C_{12}) and (C_{44}) appear independently for these two directions, as shown by the equations below, established for the case of tension-compression loadings (only the non-zero components are written).

-Tension-compression in the <001> direction :

$$\dot{\varepsilon}_1^p = \left\langle \frac{|\sigma - X| - k}{K} \right\rangle^n \text{sign}(\sigma - X) = -2\,\dot{\varepsilon}_2^p = -2\,\dot{\varepsilon}_3^p$$

$$\dot{X}_1^q = \frac{2}{3}(N_{11}^q - N_{12}^q)\,\dot{\varepsilon}_1^p - (Q_{11}^q - Q_{12}^q)\,X_1^q\,\dot{p} = -2\,\dot{X}_2^q = -2\,\dot{X}_3^q \quad \text{with } \dot{p} = |\dot{\varepsilon}_1^p|$$

-Tension-compression in the <111> direction :

$$\dot{\varepsilon}_1^p = \left\langle \frac{|\sigma - X|\sqrt{\tfrac{1}{2} M_{44}} - k}{K} \right\rangle^n \sqrt{2\,M_{44}}\ \text{sign}(\sigma - X) = -2\,\dot{\varepsilon}_2^p = -2\,\dot{\varepsilon}_3^p$$

$$\dot{X}_1^q = \frac{2}{3} N_{44}^q\,\dot{\varepsilon}_1^p - Q_{44}^q\,X_1^q\,\dot{p} = -2\,\dot{X}_2^q = -2\,\dot{X}_3^q \quad \text{with } \dot{p} = \frac{|\dot{\varepsilon}_1^p|}{\sqrt{2M_{44}}}$$

This uncoupling gives the possibility to model drastically different cyclic behavior for these two directions, as it is observed experimentally [5]. At the microscale, these differences are explained by the different activated slip systems at high temperatures: octahedral slip near <001> and cubic slip in the vicinity of <111>. A rough correlation can be made between the type (C_{11} , C_{12}) parameters and octahedral slip, while the type (C_{44}) parameters can be associated with primary cube slip and cube cross-slip [6,7].

Another particularly interesting loading is the one of shear on a <001> oriented specimen. In that case too, only the type (C_{44}) parameters appear in the equations. It can be demonstrated [8] that there is an equivalence, in the sense of the von MISES criterion, between this shear loading and a tension loading on a <111> oriented specimen. Therefore, provided there is no influence of tension-compression on torsion (*i.e.* no interactions between octahedral and cubic slip systems), then one tension-torsion specimen grown in the <001> orientation is sufficient to entirely determine the model. A second property to be mentioned, given by the cubic symmetry, is the identity between <011> and <123> orientations (coherent with the identity of the elastic moduli $E_{<011>} = E_{<123>}$). These properties seem reasonable compared with the experimental results. Nevertheless, it must be mentioned the large scattering observed near <123> due to the disorientation of the specimens.

This model has been established for the CMSX2 alloy at 650, 950 and 1100°C, but illustrations for the answer of the model are limited here to some examples at 950°C. The corresponding material constants are given in table I. Let us precise that no freedom is given for the M_{44} parameter. It is set by using the answer of the SCHMID's law in tension, when octahedral and cubic slips are considered. That imposed the ratio $\sigma_{<111>}/\sigma_{<001>} = \sqrt{3}/2$ and then the constant $M_{44} = 8/3$.

Figures 1 and 2 show the predictions for cyclic tests at steady state for various strain rates, following <001> and <111> directions. Viscosity effects, as well as the shape of the hysteresis loops are well described for both orientations. Predictions for <011> specimen is shown on figure 3 on the example of a particular cyclic loading where small cycles are inserted in a large one. Contrarily to the observations on various polycristalline materials [9], single-crystal exhibit a non-closure of the small cycle, which is correctly reproduced by the NLKH rule. On the figure 4 are plotted the creep predictions for various stress levels following four crystallographic orientations. Figure 5 presents the differences in term of cyclic curves for two tests performed near <123> orientation at the same strain rate. It is interesting to note that the model returned this effect in the right order when considering the real orientation to perform computations. Many other calculations were made for various uniaxial and tension-torsion loadings, showing good comparisons between the model and the data [8].

TABLE I. Material constants for CMSX-2 at 950° C.

Parameter	Numerical Value	Unit
k	134.	*Mpa*
n	4.65	-
K	1273.	*Mpa.s*$^{1/n}$
$N^1_{11} - N^1_{12}$	154440.	*Mpa*
N^1_{44}	60738.	*Mpa*
$Q^1_{11} - Q^1_{12}$	429.	-
Q^1_{44}	573.	-
$N^2_{11} - N^2_{12}$	2628.	*Mpa*
N^2_{44}	4830.	*Mpa*
$Q^2_{11} - Q^2_{12}$	18.	-
Q^2_{44}	30.	-

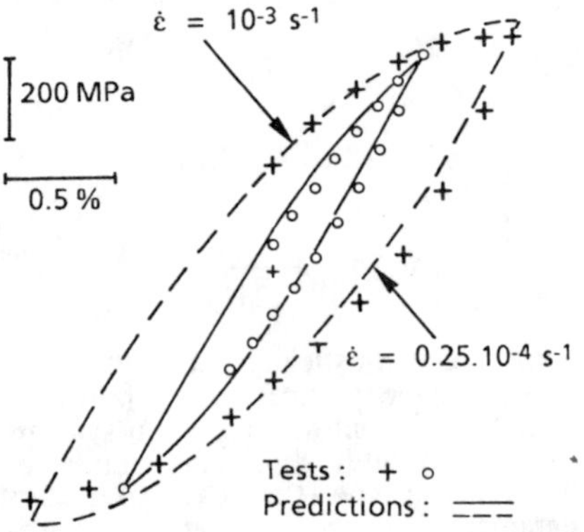

FIG. 1. Stabilized behavior and strain-rate sensitivity for <001> oriented specimens.

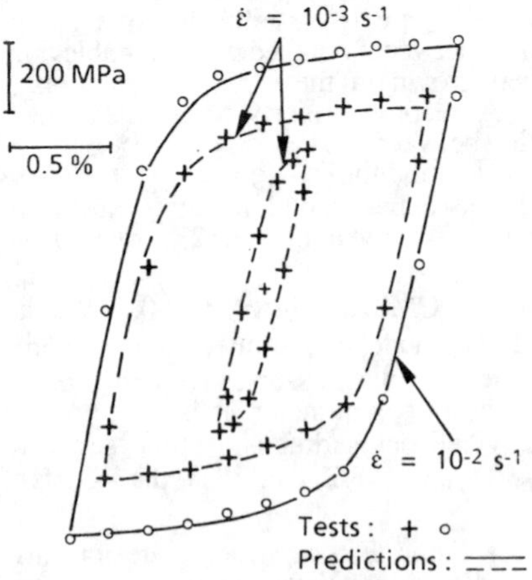

FIG. 2. Stabilized behavior and strain-rate sensitivity for <111> oriented specimens.

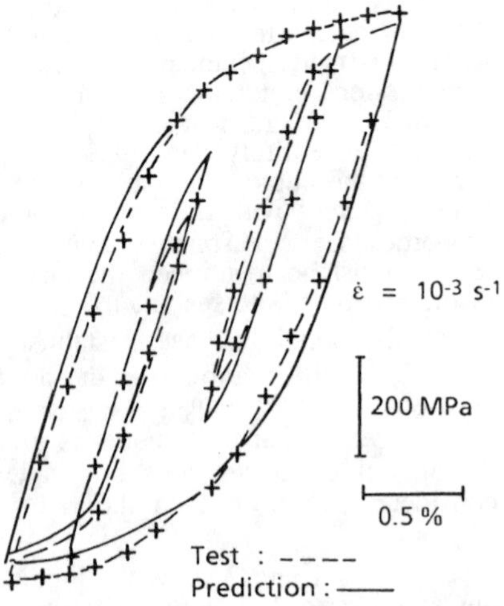

FIG. 3. Cyclic behavior in <011> orientation. Non-closure of the small cycle.

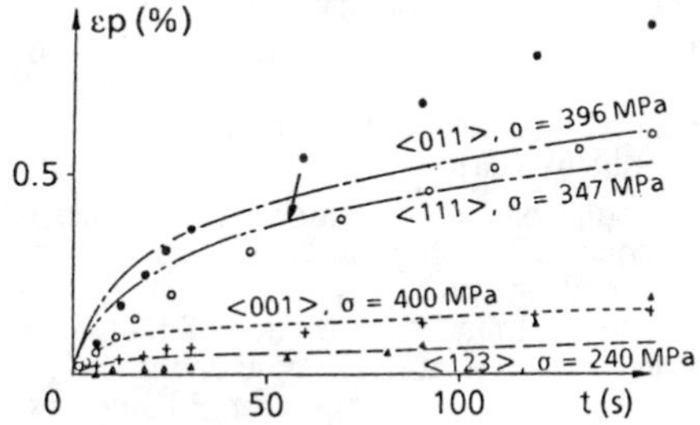

FIG. 4. Creep test predictions for four crystallographic orientations.

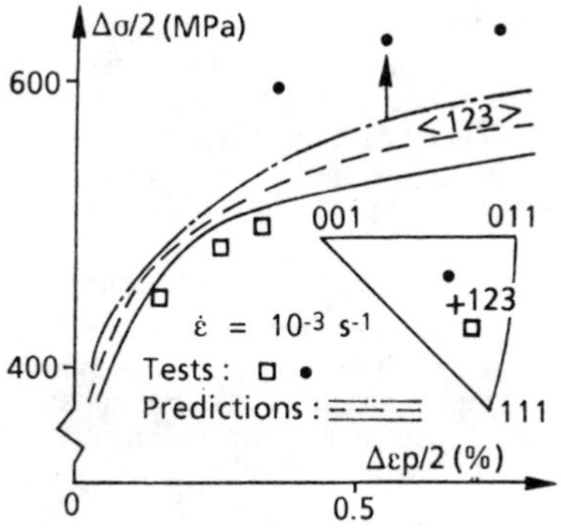

FIG. 5. Influence of disorientation near <123>.

Additional calculations were made using data from literature on other single-crystals. In that case, material constants for CMSX2 were used and the comparison is only qualitative. An example is given in figure 6 concerning recovery effects. Tests are from [10] and were run on RENE N4 at 982°C, on <001> and <111> oriented specimens. Samples are loaded up to ε=1.5% at a strain rate of $10^{-4}s^{-1}$, then unloaded to zero stress and reloaded at $\dot{\varepsilon}=6.10^{-4}s^{-1}$ after a period of 120 seconds at σ=0. The results clearly demonstrate, as mentioned by the authors, the existence of back stress (NLKH) and its orientation dependency. The test results and the corresponding predictions show: *i*) the correct answer of the macroscopical model for all the sequencies of the tests (plotted here only for <001> orientation); *ii*) this answer is very similar to the one given by the model developped by SHEH and STOUFFER based on crystallographic slip theory (continuous curves superimposed to experimental data).

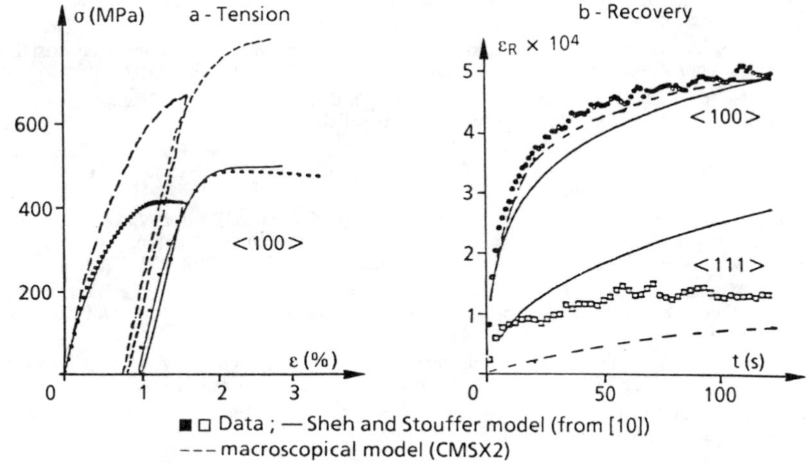

FIG. 6. Double tensile test with hold time on RENE N4 at 982°C.

CONCLUSION

The challenge in this work was to establish a phenomenological model able to reproduce the main features of the cyclic behavior of single-crystal superalloys. That has been done by using a unified viscoplastic theory, based on a type HILL criterion and using anisotropic NLKH rules. The application at 950°C corresponds to two simplifications in its expression: *i*)no isotropic hardening;*ii*)no tension-compression asymmetry, mainly observed at lower temperatures. For this last point, this effect could be introduced as did by WALKER [11].

It has been shown that this simple formulation gives a fairly good description of the uniaxial cyclic behavior and viscous effects for the four studied crystallographic orientations. For tension-torsion loadings, predictions are correct, but in the sense of a mean value of the plastic shear strain. As a matter of fact, it is observed experimentally a non-homogeneous plastic strain along the circumference of the specimens. Shear strains are drastically differents for <100> and <110> areas. The activated slip systems are not the same in both cases and it appears that the <100> areas are less deformed than the <110> ones [12]. The macroscopical model does not distinguish these two type of behaviors and predict a uniform plastic shear strain along the circumference. To eliminate this insufficiency it is necessary to modify the expression of the yield criterion. The theory of tensorial functions offers a general framework for determining the yield function. For FCC crystal, any function of the nine invariants constituing the integrity basis will ensure the symmetry properties of the crystal [13].

REFERENCES

1. E. Jordan, "A review of single crystal plastic and viscoplastic behavior", Proc. of Int. Conf. on The Inelastic Behaviour of solids: Models and utilization, Besancon, France, Sept. 1988.

2. S.H. Choi and E. Krempl, "Viscoplasticity Theory Based on Overstress Applied to the Modeling of Cubic Single Crystal", R.P.I. Report MML 88-4, August, 1988.

3. D. Nouailhas, "Un Modèle de Viscoplasticité Cyclique pour Matériaux Anisotropes à Symétrie Cubique", Note aux C. R. Acad. Sci. Paris, t 310, Série II, 1990.

4. J.L. Chaboche, "Viscoplastic Constitutive Equations for the Description of Cyclic and Anisotropic Behaviour of Metals", 17th Polish Conf. on Mechanicsof Solids, Szczyrk. Bul. de l'Acad. Polonaise des Sciences, Série Sc. et Techn., 25,1977.

5. D. Nouailhas, P. Poubanne, H. Policella, D. Pacou and P. Paulmier, Modélisation du Comportement Cyclique de Superalliages Monocristallins pour Aubes de Turbine", ONERA Technical Report 75/1765RY180R, Avril, 1989.

6. W.W. Milligan and S. Antolovich, "Deformation Modeling and Constitutive Modeling for Anisotropic Superalloys", NASA Contractor Report 4215, February, 1989.

7. P. Poubanne, "Etude et Modélisation du Comportement Mécanique d'un Monocristal en Superalliage pour Aubes de Turbine", Nouvelle Thèse, Ecole Centrale, Mai, 1989.

8. D. Nouailhas, "Lois de Comportement en Viscoplasticité Cyclique : Application au Cas des Matériaux à Symétrie Cubique", La Recherche Aérospatiale, *n*° 1990-3,1990.

9. J.L. Chaboche, D. Nouailhas, P. Paulmier and H. Policella, "Sur les Problèmes Posés par la Description des Effets de Rochet en Plasticité et Viscoplasticité Cyclique", La Recherche Aérospatiale, *n*° 1989-1, 1989.

10. M.Y. Sheh and D.C. Stouffer, "Anisotropic Constitutive Modeling for Nickel Base Single Crystal Superalloys Using a Crystallographic Approach", Constitutive Modelling for Engineering Materials with Applications, Eds. D. Hui and T. J. Kozik, ASME PVP-Vol. 153, 1988.

11. K.P. Walker, "Research and development program for non-linear structural modeling with advanced time-temperature dependant constitutive relationships", Report PWA 5700-50, NASA CR 165533, 1981.

12. L. Méric and G. Cailletaud, "Finite element computation in anisotropic viscoplasticity for single crystals", Proc of 2nd Conf. on Computational Plasticity, Barcelona, Spain, Sept. 1989.

13. G.F. Smith and E. Kiral, "Integrity bases for N symmetric second order tensors- The crystal classes", Rend. Circ. Mat., Palermo II, Ser. 18, 1969.

MULTISURFACE AND MULTICOMPONENT FORMS OF NONLINEAR KINEMATIC HARDENING: APPLICATION TO NONISOTHERMAL PLASTICITY

N. OHNO and J.-D. WANG
Dept. of Mech. Eng., Nagoya Univ.
Chikusa-ku, Nagoya 464-01, Japan

ABSTRACT

Employing two alternative forms of nonlinear kinematic hardening (i.e., multisurface and multi-component forms), effects of a temperature-rate term and a relative translation term involved in them are studied. Examples of stress-strain relations under monotonic and cyclic thermomechanical loads are presented. It is demonstrated that for the multisurface form both the temperature-rate term and the relative translation term are necessary to avoid intersection of surfaces, and that neglect of the temperature-rate term may induce anomalous shift of stress-strain hysteresis loops under thermomechanical cyclic loading.

INTRODUCTION

Temperature-rate terms in nonisothermal kinematic hardening rules have been studied in several works so far [1-4]. The present authors [5] were also concerned with such a term; i.e., discussing two alternative forms of nonlinear kinematic hardening (i.e., multicomponent and multisurface forms), they showed the following:

- When the kinematic hardening variable is decomposed into components, it can be represented in terms of nonintersecting multi-surfaces.
- They remain nonintersecting under nonisothermal conditions, if a temperature-rate term and a relative translation term are considered in their translation equation.

It is worthwhile to study intersection conditions of surfaces for multisurface theories of plasticity and viscoplasticity. Importance of such studies was recognized by McDowell [6]. He examined in detail the intersection conditions for two surface plasticity theories under isothermal conditions.

In the present work, employing the multisurface and multicomponent forms of nonlinear kinematic hardening, effects of the temperature-rate term and the relative translation term are discussed. Stress-strain relations under monotonic and cyclic thermomechanical loads are calculated to demonstrate that these two terms are indeed necessary to guarantee the nonintersection of surfaces. Besides, it is shown that neglect of the temperature-rate term may induce anomalous shift of stress-strain hysteresis loops under thermomechanical cyclic loading if the kinematic hardening rule employed includes a linear or nearly linear component.

TWO EQUIVALENT FORMS OF NONLINEAR KINEMATIC HARDENING

We assume that back stress $\boldsymbol{\alpha}$ is decomposed into components,

$$\boldsymbol{\alpha} = \sum_{j=1}^{N} \boldsymbol{\alpha}_j \ , \tag{1}$$

and that evolution of each component is bounded,

$$(3/2)^{1/2}\|\boldsymbol{\alpha}_j\| \leq r_j \ , \tag{2}$$

where $\|\boldsymbol{\alpha}_j\| = (\mathrm{tr}\, \boldsymbol{\alpha}_j^2)^{1/2}$.

Using $\boldsymbol{\alpha}_j$ and r_j, we introduce

$$\mathbf{A}_0 = \mathbf{0}, \quad \mathbf{A}_k = \sum_{j=1}^{k} \boldsymbol{\alpha}_j \quad (k = 1,2,\ldots,N) \ , \tag{3}$$

$$R_k = \rho + \sum_{j=k+1}^{N} r_j \quad (k = 0,1,\ldots,N-1), \quad R_N = \rho \ , \tag{4}$$

where ρ is an arbitrary scalar. Then, we can show that the following N+1 surfaces are nested, as shown in Fig. 1:

$$F_k = (3/2)^{1/2}\|\mathbf{A}_k^* - \mathbf{A}_k\| - R_k = 0 \quad (k = 0,1,\ldots,N) \ , \tag{5}$$

where $\mathbf{A}_k$ and R_k are the center and radius of the surface $F_k = 0$, respectively, and $\mathbf{A}_k^*$ denotes a point on it.

When we consider $\mathbf{A}_k$ and R_k as well as $\boldsymbol{\alpha}_j$ and r_j, we can express nonlinear kinematic hardening in two equivalent forms.

Let us consider the following nonisothermal evolution rule of $\boldsymbol{\alpha}_j$, which is obtained by imposing the temperature-history independence of $\boldsymbol{\alpha}_j$ on the rule derived by Chaboche [2]:

$$\dot{\boldsymbol{\alpha}}_j = \zeta_j(\tfrac{2}{3}r_j\dot{\boldsymbol{\varepsilon}}^p - \boldsymbol{\alpha}_j\dot{q}) + (\boldsymbol{\alpha}_j/r_j)(\partial r_j/\partial T)\dot{T} \ , \tag{6}$$

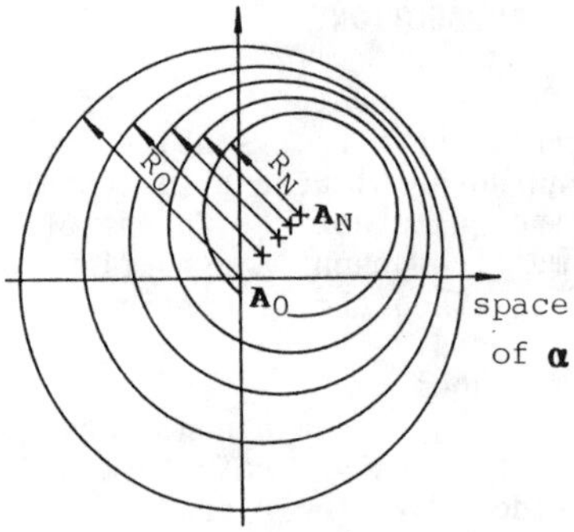

Fig. 1 Nested multisurfaces

where $\boldsymbol{\varepsilon}^p$ denotes plastic strain, and $\dot{q} = (2/3)^{1/2}\|\dot{\boldsymbol{\varepsilon}}^p\|$. It is shown that eq. (6) satisfies the condition (2).

Equations (3) and (4) enable us to rewrite eq. (6) as

$$\dot{\mathbf{A}}_k = \dot{\mathbf{A}}_{k-1} + \zeta_k(\mathbf{A}_{k-1}^* - \mathbf{A}_k^*)\dot{q} + \frac{\mathbf{A}_k - \mathbf{A}_{k-1}}{R_{k-1} - R_k}\left[\frac{\partial R_{k-1}}{\partial T} - \frac{\partial R_k}{\partial T}\right]\dot{T} \ , \tag{7}$$

where $\mathbf{A}_k^*$ is taken to be $\mathbf{A}_k^* = \mathbf{A}_k + (2/3)^{1/2}R_k\boldsymbol{\nu}$, $\boldsymbol{\nu} = \dot{\boldsymbol{\varepsilon}}^p/\|\dot{\boldsymbol{\varepsilon}}^p\|$.

Equation (7), the multisurface form of eq. (6), has the following features:

- The multisurfaces $F_k = 0$ $(k = 0,\ldots,N)$ do not intersect even under nonisothermal conditions.

- To guarantee the nonintersection of surfaces, however, it is necessary to consider the three terms in the right-hand side; i.e., the first term referred to as the $\dot{A}_{k-1}$-term or the relative translation term, the second term representing the classical translation rule of Mroz, and the third term called the $\dot{T}$-term.

NUMERICAL EXAMPLES

The material parameters were determined from isothermal tensile tests of 304 stainless steel. N = 3 was chosen. For simplicity, isotropic hardening was ignored, and rate-independent plasticity was assumed.

Figures 3 and 4 are results of calculation for the history of temperature T and mechanical strain ε^m shown in Fig. 2. It is seen that neglect of the $\dot{T}$-term or the $\dot{A}_{k-1}$-term results in temperature-history dependence of stress response (Fig. 3) and intersection of surfaces (Fig. 4). The intersection is not desirable for multi-surface theories, as discussed by McDowell [6].

Figures 5(a) and 5(b) show stress versus mechanical strain hysteresis loops calculated for in-phase cyclic change of T and ε^m. It

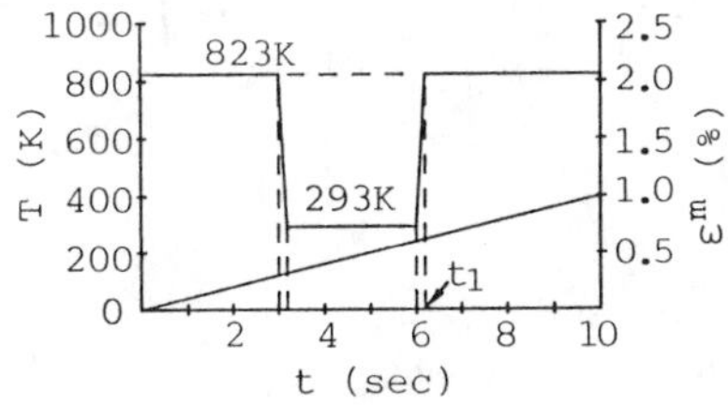

Fig. 2 Nonisothermal tensile deformation.

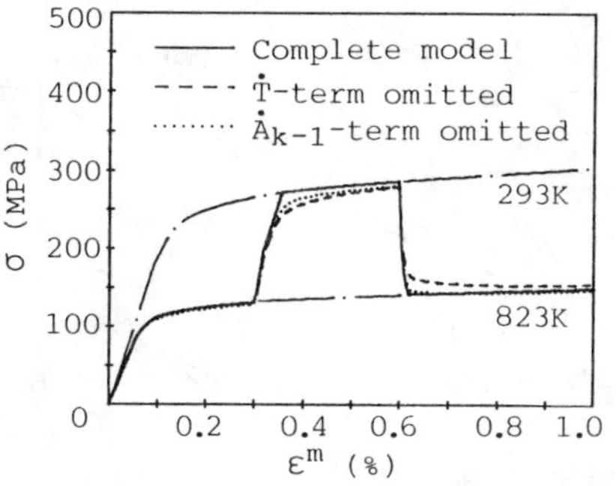

Fig. 3 Effects of $\dot{T}$-term and $\dot{A}_{k-1}$-term on nonisothermal tensile deformation.

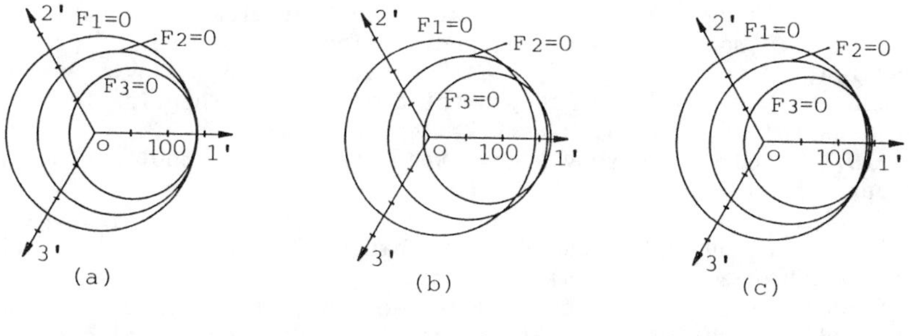

Fig. 4 Configuration of multisurfaces at the end of the second temperature variation (i.e., $t=t_1$ in Fig. 2); (a) complete model, (b) $\dot{T}$-term omitted, (c) $\dot{A}_{k-1}$-term omitted.

is seen that neglect of the $\dot{T}$-term induces anomalous shift of the hysteresis loop. This shift is significant exclusively in the response of the linear component α_1 (= A_1) (Fig. 6(a)).

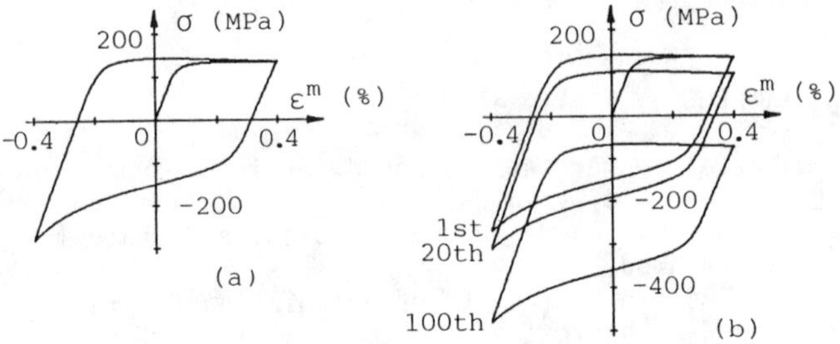

Fig. 5 Hysteresis loops of stress and mechanical strain under in-phase cyclic change of T and ε^m (Tmax = 823K, Tmin = 293K, $\Delta\varepsilon^m$ = 0.8%); (a) complete model, (b) $\dot{T}$-term omitted.

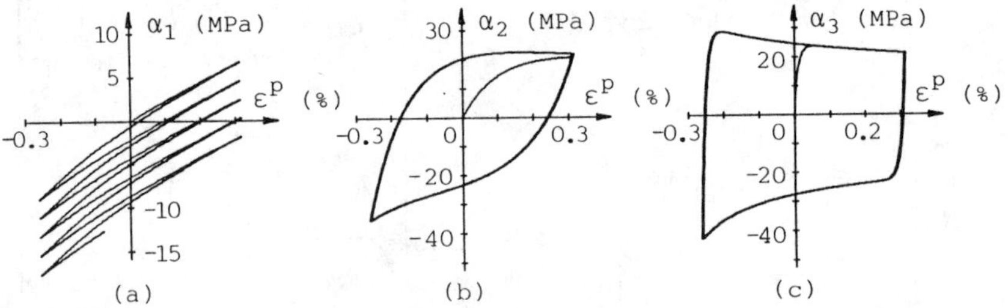

Fig. 6 Response of α_i under in-phase cyclic change of T and ε^m; (a) α_1 (linear), (b) α_2 (nonlinear), (c) α_3 (strongly nonlinear).

REFERENCES

1. K.P. Walker, "Research and development program for nonlinear structural modeling with advanced time-temperature dependent constitutive relationships", NASA Report, CR-165533, (Nov. 1988).
2. J.L. Chaboche, "Time-independent constitutive theories for cyclic plasticity", Int. J. Plast. 2, 149-188 (1986).
3. A.D. Freed, "Thermoviscoplastic model with application to copper", NASA Report, TP-2845, (Dec. 1988).
4. K.-D. Lee, and E. Krempl, "An orthotropic theory of viscoplasticity based on overstress for thermomechanical deformation", RPI Report, MML88-2 (July 1988).
5. N. Ohno, and J.-D. Wang, "Transformation of a nonlinear kinematic hardening rule to a multisurface form under isothermal and nonisothermal conditions", Int. J. Plast. (in print).
6. D.L. McDowell, "Evaluation of intersection conditions for two-surface plasticity theory", Int. J. Plast. 5, 29-50 (1989).

ON A NON-PROPORTIONALITY PARAMETER IN CONSTITUTIVE MODELLING OF CYCLIC PLASTICITY

Eiichi TANAKA and Sumio MURAKAMI
Department of Mechanical Engineering
Nagoya University
Chikusa-ku, Nagoya, 464-01 Japan

ABSTRACT

A non-proportionality parameter plays an essential role in formulating a constitutive model of non-proportional cyclic plasticity. In the present paper, a rational parameter expressing the non-proportionality of a deformation process and the relevant evolution equation are proposed by examining the cross-hardening phenomena induced by the torsional cycles after tension-compression cycles. Then the parameter is incorporated into a viscoplastic model, and the experimental results by Benallal et al. are simulated by the model. The results show that the proposed model can describe the qualitative properties of non-proportional cyclic hardening.

INTRODUCTION

Non-proportional cycles induce more significant hardening than proportional ones. In the framework of the advanced flow theories of plasticity or viscoplasticity, this marked hardening is usually described by the isotropic hardening variable and the relevant evolution equations. In other words, the parameter expressing the non-proportionality of a deformation process is introduced, and the saturated value of the isotropic hardening variable is specified as a function of the parameter.

So far, the parameter is defined as a simple relation between observable or internal variables such as stress, strain, inelastic strain, a kinematic hardening variable, and their rates [1-2]. These variables, however, do not describe rationally the state of internal dislocation structures causing such hardening. Hence the values of non-proportionality parameters defined by them are changed significantly by a small modification of inelastic deformation path. This means that a new internal variable reflecting the dislocation structures and the relevant non-proportionality parameter should be introduced.

In the present paper, these are established by examining the cross-hardening phenomena induced by the torsional cycles after tension-compression cycles. Then the accuracy and the adequacy of the proposed model are evaluated by incorporating them into Chaboche model [3] and by simulating the experimental results by Benallal et al. [4].

FORMULATION OF A MODEL

Macroscopic Variable Describing Internal Dislocation Structures

Let us consider the following experiment: (1) at the first stage tension-compression cycles are applied to a thin-walled tubular specimen, and (2) torsional cycles with the same inelastic strain amplitude are then loaded, and (3) finally the tension-compression cycles at the first stage are again given. Figure 1 shows the schematic figure of the stress amplitude ($\Delta\sigma/2$) versus the accumulated inelastic strain (P) relation. We see that the hardening phenomena (we call it cross-hardening hereafter) occur just after the path changes. The information obtained from this experiment is summarized as follows:

(1) The cross-hardening at the first path change is induced by a certain dislocation structure formed in the preceding tension-compression cycles.

(2) This structure does not resist the tension-compression cycles, but resists the torsional cycles. Hence the structure has the directional character as to the resistance to dislocation movement.

(3) The magnitude h_c of the cross-hardening increases with the accumulated inelastic strain P_0 in the preceding tension-compression cycles, and approaches a constant value depending on the inelastic strain amplitude. Thus the structure has the stabilized state depending on the amplitude.

(4) In the subsequent torsional cycles, the stress amplitude decreases to that for the virgin material. But the cross-hardening is induced again by the subsequent tension-compression cycles. This means that the structure formed in the preceding cycles is destroyed by the subsequent cycles, and that the new structure corresponding to the latter is established.

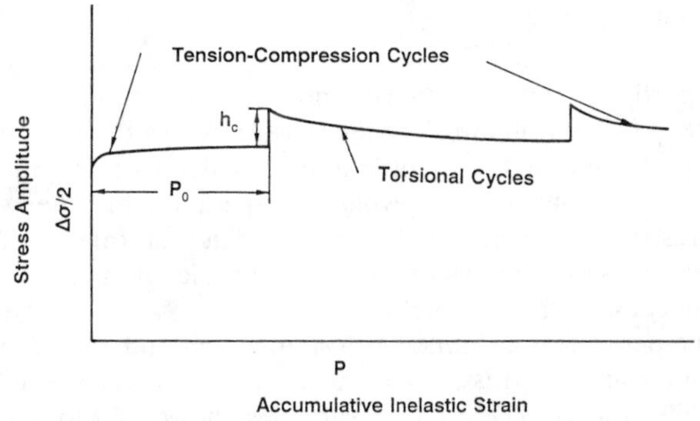

FIG. 1. Cross-hardening

On the basis of this information we represent the internal structure by a second rank tensor **C** in the five-dimensional inelastic strain vector space of von Mises type[5], and expresses the rate $\dot{\mathbf{C}}$ as

$$\dot{\mathbf{C}} = c_c\,(\boldsymbol{u} \otimes \boldsymbol{u} - \mathbf{C})\dot{P} \tag{1}$$

where $\dot{P}$ is the magnitude of an inelastic strain rate vector $\dot{\boldsymbol{P}}$, c_c is a material constant, the symbol $\otimes$ denotes the tensor product, and $\boldsymbol{u}$ is the normalized inelastic strain rate vector ($= \dot{\boldsymbol{P}}/\dot{P}$).

In order to examine the characteristic features of the equation (1), let us discuss the behavior of **C** under typical loading conditions. We denote the base vectors in the vector space by $\boldsymbol{n}_\alpha$ ($\alpha = 1, \cdots, 5$). In the following, the values of $C_{\alpha\beta}$ not specified are identically zero. We first consider simple tension-compression cycles in which $\boldsymbol{u} = \pm \boldsymbol{n}_1$. Equation (1) gives the solution

$$C_{11} = 1 - \exp(-c_c P) \tag{2}$$

This result shows that C_{11} approaches unity from zero with the increase of P.

Let us next consider the case that the torsional cycles of $\boldsymbol{u} = \pm \boldsymbol{n}_3$ is loaded following the preceding tension-compression cycles. The solution of Eq. (1) in the stage of the torsional cycles is

$$\begin{aligned} C_{11} &= [\exp(c_c P_0) - 1]\exp(-c_c P) \\ C_{33} &= 1 - \exp[-c_c (P - P_0)] \end{aligned} \tag{3}$$

where P_0 is the value of P at the end of the preceding cycles. Hence C_{11} decreases to zero, while C_{33} approaches unity.

The final case is the circular cycles with the inelastic strain amplitude of $\Delta P/2$. In this case the vectors $\boldsymbol{P}$ and $\boldsymbol{u}$ are expressed by

$$\boldsymbol{P} = (\Delta P/2)(\cos\theta\, \boldsymbol{n}_1 + \sin\theta\, \boldsymbol{n}_3) \tag{4}$$

$$\boldsymbol{u} = (-\sin\theta\, \boldsymbol{n}_1 + \cos\theta\, \boldsymbol{n}_3) \tag{5}$$

where $\theta = P/(\Delta P/2)$. The solution at the stabilized state is

$$\begin{aligned} C_{11} &= [D^2 \sin^2\theta - D\sin 2\theta + 2]/(4 + D^2) \\ C_{33} &= [D^2 \cos^2\theta + D\sin 2\theta + 2]/(4 + D^2) \\ C_{13} &= [D\cos 2\theta - (D^2/2)\sin 2\theta]/(4 + D^2) \end{aligned} \tag{6}$$

where $D = c_c(\Delta P/2)$.

Non-Proportionality Parameter

Now we define the parameter describing the non-proportionality of the deformation process by the relation

$$A = [\{\mathrm{tr}(\mathbf{C}^T\mathbf{C}) - \boldsymbol{u}\cdot \mathbf{C}^T\mathbf{C}\boldsymbol{u})/\mathrm{tr}(\mathbf{C}^T\mathbf{C})]^{1/2} \quad (0 \le A \le 1) \tag{7}$$

where the symbol (tr) and the superscript (T) denote the trace and the transpose of a tensor, and the symbol ($\cdot$) expresses the inner product of vectors. This variable always takes the value zero in the case of proportional cycles. In the case that the subsequent proportional cycles are loaded after the saturation of cyclic hardening of the preceding proportional cycles, the parameter A takes the value $|\sin\phi|$ just after the path change, and decreases to zero, in which ϕ is the angle between the rectilinear paths in the inelastic strain vector space corresponding to the two proportional cycles. In the case of the circular cycles mentioned above, on the other hand, A takes the value

$$A = 1/[\{2 + (c_c \, \Delta P/2)^2\}]^{1/2} \tag{8}$$

For the value of c_c identified, A is approximately $1/\sqrt{2}$. This value corresponds to the case that the rectilinear path change of $\phi = 45°$ is continuously given.

Evaluation of the Proposed Model

The above model was incorporated into Chaboche model [3], and the experimental result by Benallal et al. [4] was simulated. The experiment is performed by applying the combined loadings of axial force and torque to a tubular specimen of the type 316 stainless steel. The repeated loadings of proportional path in the strain vector space of von Mises type are first applied, and 8 sequences of stairs path joining the ends of the proportional path are then loaded. These sequences are composed of 128, 64, 32, 16, 8, 4, 2 and 1 step, respectively. Following these paths, the square and circular paths are repeated. In Figs. 2 and 3, the numbers denote these path sequences. We see that the variations of the equivalent stress σ_e for the experimental and simulative results agree qualitatively well.

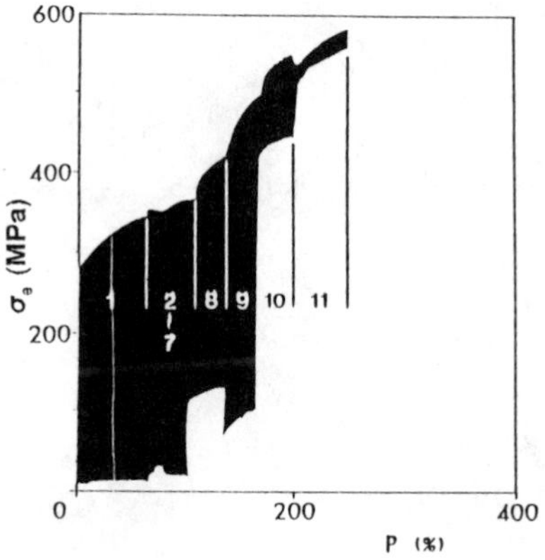

FIG. 2. Experimental results [4]

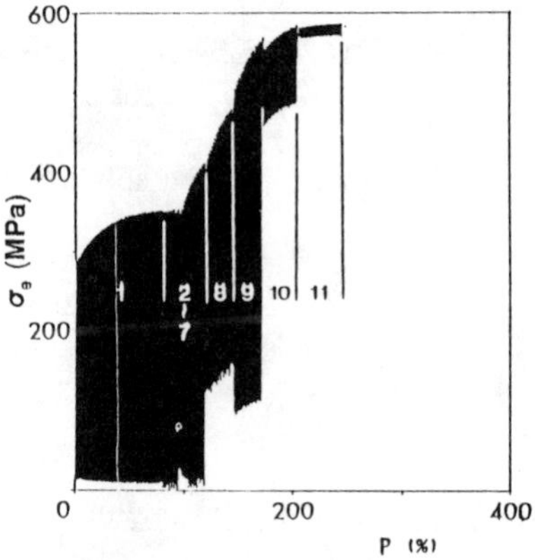

FIG. 3. Simulative results

REFERENCES

1. D. L. McDowell, "A two surface model for transient nonproportional cyclic plasticity", J. Appl. Mech. 52(2), 298-302 (June 1985).
2. A. Benallal, and D. Marquis, "Constitutive equations for nonproportional cyclic elasto-viscoplasticity", J. Eng. Mater. Technol. 109(4), 326-336 (October 1987).
3. J. L. Chaboche, and G. Rousselier, "On the plastic and viscoplastic constitutive equations", J. Press. Vessel Technol. 105(2), 153-158 (May 1983).
4. A. Benallal, P. Le Gallo, and D. Marquis, "An experimental investigation of cyclic hardening of 316 stainless steel and of 2024 Aluminum alloy under multiaxial loadings", Nucl. Eng. Des, 114, 345-353 (1989).
5. E. Tanaka, "Hypothesis of local determinability for five-dimensional strain trajectories", Acta Mechanica 52(1-2), 63-76 (1984).

A NEW KINEMATIC HARDENING LAW BASED ON BOUNDING SURFACE PLASTICITY

GEORGE Z. VOYIADJIS

Professor of Civil Engineering, Louisiana State University, Baton Rouge, LA 70803

PETER I. KATTAN

Research Associate, Dept. of Civil Engineering, Louisiana State University, Baton Rouge, LA 70803

ABSTRACT

A cyclic theory of plasticity is formulated for finite deformation in the Eulerian reference system. A new kinematic hardening rule is proposed based on the experimental observations made by Phillips and Lee [1]. The Tseng-Lee model [2] is also obtained as a special case of the proposed model. The behavior of the model is shown for an arbitrary nonproportional loading path.

GEOMETRY OF MOTION IN THE DEVIATORIC STRESS SPACE

The proposed kinematic hardening rule consists of a combination of a novel suggested rule governing the motion of the yield surface in most of the interior of the bonding surface along with the Tseng-Lee model [2] which is used just before contact of the two surfaces. The suggested model is based on experimental observations made by Phillips and Lee [1]. In this model, the yield surface moves in the interior of the bounding surface along a curved path. The yield surface approaches the bounding surface according to the suggested model. When the yield surface is close enough to the bounding surface, we resort to the Tseng Lee [2] model in order to satisfy the tangency condition at the stress point when the two surfaces meet. The minimum distance between the two surfaces is taken as the criterion to determine how close they are. Switching to the Tseng-Lee rule [2] is achieved when this distance reaches a critical value.

Referring to Figure 1, the center of the yield surface f moves along the curved path 0_1, 0_2, ..., 0_{N+1} (discretized as line segments for numerical implementation purposes). The proposed kinematic hardening rule is implemented when the center of the yield surface moves from the initial center 0_1 to the location 0_N. At any intermediate location 0_j, $1 \leq j \leq N\text{-}1$, the direction of movement of the center is given by

The location of the center 0_{j+1} is then obtained from

$$\nu^{(j)} = \frac{an^{(j)} + bl^{(j)}}{\|an^{(j)} + bl^{(j)}\|}, \; j = 1, 2, ..., N-1 \qquad (1)$$

$$\alpha^{(j+1)} = \alpha^{(j)} + \Delta\alpha^{(j)} \qquad (2)$$

where $\Delta\alpha$ is obtained from $\dot{\alpha}$. The magnitude of $\dot{\alpha}$ at the j-th step is obtained by invoking the consistency condition $\dot{f} = 0$. Therefore, one has

$$\|\dot{\alpha}\|^{(j)} = [\frac{3(\tau - \alpha):\dot{\tau} - 2\,k\dot{k}}{3(\tau - \alpha):\nu}] \qquad (3)$$

As will be discussed later, the yield surface will not change its shape or size and therefore k remains constant in equation (3).

At every step j, the minimum distance $\Delta^{(j)}$ between the yield surface and the bounding surface is computed as follows:

$$\Delta^{(j)} = [\sqrt{2/3}\;K - -(\|\alpha - \beta\| = \sqrt{2/3}\;k]^{(j)} \qquad (4)$$

This minimum distance $\Delta^{(j)}$ is checked against $\|\dot{\alpha}\|$ at the j-th step such that if

$$\Delta^{(j)} > \|\dot{\alpha}\|^{(j)} \qquad (5)$$

We use equations (1) and (2) to determine the next location of the yield surface. In the case when

$$\Delta^{(j)} \leq \|\dot{\alpha}\|^{(j)} \qquad (6)$$

the kinematic hardening rule of Tseng and Lee [2] is used to ensure the tangency of the yield and the limit surfaces at the stress point when the center of the yield surface moves from 0_N to 0_{N+1}.

When the center is located at 0_N, the direction of motion of the yield surface is given by:

$$\nu^{(N)} = [\frac{\sqrt{2/3}\;(K - k)\;\lambda - (\alpha - \beta)}{\|\sqrt{2/3}\;(K - k)\;\lambda - (\alpha - \beta)\|}]^{(N)} \qquad (7)$$

where $\lambda^{(N)}$ is a unit tensor as defined by the Tseng-Lee model [2].

It is clear that the proposed model coincides with the Tseng-Lee model when the yield surface is very close to the limit surface. In the last step, the motion of the center of the yield surface is along a straight line and the two surfaces make contact at the stress point. However, it is important to note that Tseng and Lee [2] originally derived their model for small strains. This model is incorporated in the presented general model for finite strains in the Eulerian coordinate system.

BEHAVIOR OF PROPOSED MODEL UNDER AN ARBITRARY NONPROPORTIONAL LOADING PATH

The example presented here displays the behavior of the model qualitatively. In this example, a and b are taken to be unity for simplicity. In Figure 2, an arbitrary non-proportional loading path is shown. The direction of motion of the center of the yield surface is clearly shown to be a curved path for small increments of load. The dotted line joining 0_1 and 0_{N+1} represents the direction of translation of the center of the yield surface. It is clear that the proposed formulation predicts a curved path $0_1 \ldots 0_N\, 0_{N+1}$ with only the last segment $0_N\, 0_{N+1}$ being a straight line. This motion is thought to represent the material behavior more realistically and to conform with experimental results as given by Phillips and Lee [1]. This motion is not restrictive to a straight stress path and is more flexible dependent on the choice of the parameters a and b. A special case can be obtained by assuming a = 0, in which case the yield surface moves parallel to the stress path.

SUMMARY AND CONCLUSION

A new kinematic hardening rule is proposed based on the theory of bounding surface plasticity. The formulation is given in the Eulerian coordinate system primarily for finite deformation cyclic plasticity (small elastic strains). The proposed model, which includes the Tseng-Lee model [2] as a special case has all the properties of the Tseng-Lee model [2] which has been proven to give good results that conform with experimental observations. However, the proposed model predicts a curved path for the motion of the center of the yield surface in the interior of the bounding surface. On the other hand, the Tseng-Lee model [2] assumes that the center of the yield surface moves in a straight line. The authors do not see any physical justification for this assumption which is considered restrictive to the motion of the yield surface.

REFERENCES

1. Phillips, A. and Lee, C.-W., "Yield Surfaces and Loading Surfaces. Experiments and Recommendations," Int. J. of Solids and Structures, Vol. 15, pp. 715-729 (1979).

2. Tseng, N. T. and Lee, G. C., "Simple Plasticity Model of the Two-Surface Type," ASCE J. Engineering Mechanics, Vol. 109, No. 3, pp. 795-810 (1983).

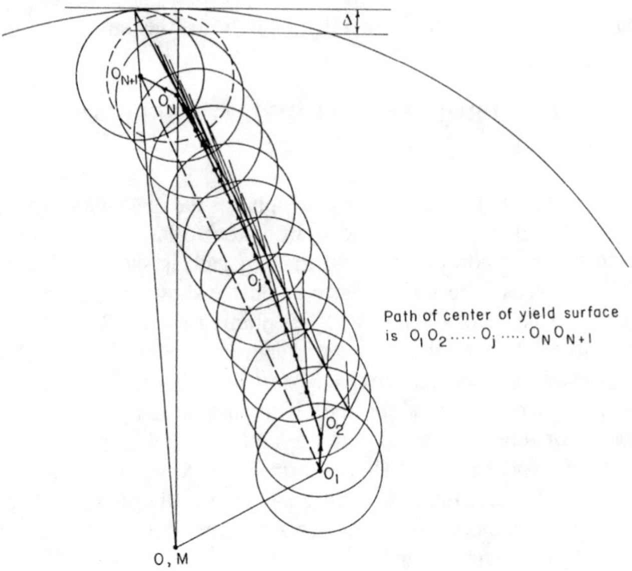

FIG. 1. Generalized kinematic hardening model.

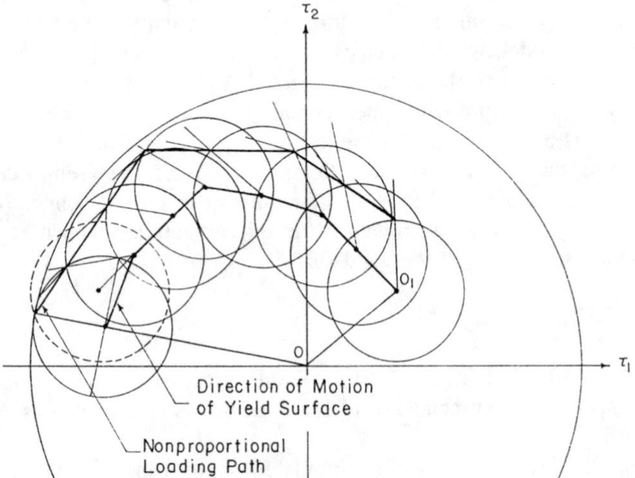

FIG. 2. Geometry of motion under an arbitrary non-proportional loading path.

Hierarchical single surface model for isotropic hardening cohesive soils

G.W. Wathugala

Academic Computing Specialist, User Support, Center for Computing and Information Technology, University of Arizona, Tucson, AZ85721

C.S. Desai

Regents' Professor and Head, Dept. of Civil Engineering and Engineering Mechanics, University of Arizona, Tucson, AZ 85721

ABSTRACT

A constitutive model for isotropic hardening cohesive soils is presented. The model is developed as an extension to the hierarchical single surface (HiSS) models, so that it retains all the previous capabilities of HiSS models. The model is verified by back predicting laboratory cyclic behavior of Sabine clay. It has been further verified by back predicting the field behavior of two instrumented pile segments in saturated clay.

INTRODUCTION

The hierarchical single surface (HiSS) models[1] have been developed and verified for many geologic materials including dense sands, soft rocks and concrete. However, these models need modifications to capture the behavior of cohesive soils. Therefore, a new series of models termed as δ^* series is developed. These models capture the behavior of cohesive soils in addition to all the current capabilities of HiSS models. In this paper, the simplest model in the series, δ_0^* is presented.

BASIC FORMULATION

General incremental stress-strain relationship for any loading may be given by

$$d\sigma_{ij}=\left(C^{e}_{ijkl}-\frac{C^{e}_{ijnm}n^{Q}_{nm}n^{R}_{op}C^{e}_{opkl}}{H^{*}+n^{F}_{rs}C^{e}_{rstu}n^{Q}_{tu}}\right)d\varepsilon_{kl} \tag{1}$$

where C^{e}_{ijlk} is the elastic constitutive tensor. n^{R}_{ij} and n^{Q}_{ij} are defined as unit normal tensors to reference (loading) surfaces R and potential surface Q. For δ_0^*, $R\equiv Q$. H^* is the plastic modulus. In this paper the superscript * can be VL, RL or UL, depending on virgin loading, reloading and unloading, respectively. One of the task of a constitutive model is to define n^{R}_{ij}, n^{Q}_{ij} and H^* for all significant loading situations.

REFERENCE SURFACE, R

This model treats virgin loading (VL), unloading (UL) and reloading (RL) differently. They are identified by using a convex reference surface (R), passing through the current stress point in the stress space as shown in Fig. 1. R is defined in terms of stress invariants J_1 (first invariant of the stress tensor, σ_{ij}), J_{2D} (second invariant of the deviatoric stress tensor), and J_{3D} (third invariant of the deviatoric stress tensor) as

$$R \equiv \left(J_{2D}/P_a^2\right) - F_{br} F_s = 0 \tag{2}$$

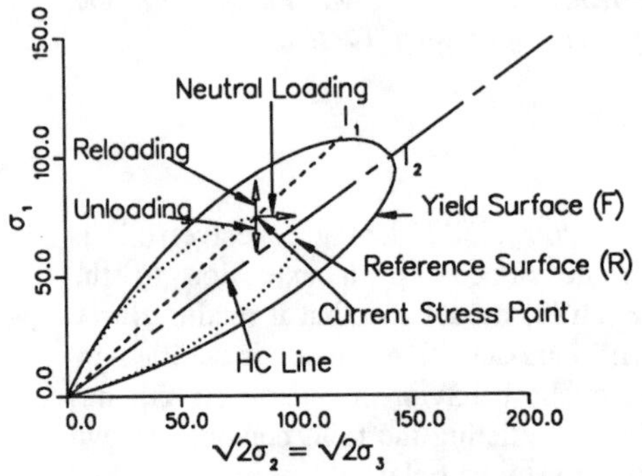

Figure 1 Yield Surface (F) and Reference Surface (R) in Triaxial Plane

where p_a is the atmospheric pressure and $F_{br} \equiv -\alpha_r (J_1/P_a)^n + \gamma (J_1/P_a)^2$. The value α_r is obtained by equating R = 0 and substituting current stresses into the Eq. (2). The function F_{br} describes the shape of R in the J_1-$\sqrt{J_{2D}}$ space. γ and n are material parameters. The function F_s describes the shape of R in the octahedral plane and is given by $F_s \equiv (1-\beta S_r)^m$. S_r is defined as a stress ratio and here $S_r \equiv (\sqrt{27}/2) J_{3D} J_{2D}^{-1.5}$ is used. β and m (= -0.5 used) are also material parameters.

The direction of loading is identified with respect to the surface R as shown in Fig. 1. When a material is yielding, the stress point lies on the yield surface or prestress surface (F) and R coincides with F. (i.e., $R \equiv F$). At this state, loading is defined as the virgin loading, and at all the other states, loading is defined as reloading. The yield function F is defined as

$$F \equiv \left(J_{2D}/P_a^2\right) - F_b F_s = 0 \tag{3}$$

where $F_b = -\alpha_{ps} (J_1/P_a)^n + \gamma (J_1/P_a)^2$ and α_{ps} is the hardening or growth function.

HARDENING FUNCTION

All the hardening functions used in earlier HiSS models were functions of the trajectory of total plastic strains, ξ. Therefore, they usually predict dilative behavior before failure. Since normally consolidated (NC) clays do not dilate, a new hardening function given below is used in this study.

$$\alpha_{ps} = h_1 / \left(\xi_V + h_3 \xi_D^{h_4}\right)^{h_2} \tag{4}$$

where h_1, h_2, h_3 and h_4 are material parameters. The increments of ξ_V and ξ_D are defined as

$$d\xi_D=\left(de^p_{ij}de^p_{ij}\right)^{\frac{1}{2}} \text{ and } d\xi_v=\begin{cases}(1/\sqrt{3})d\varepsilon^p_v & for\ d\varepsilon^p_v>0\\ 0 & for\ d\varepsilon^p_v\leq 0\end{cases} \tag{5}$$

where de^p_{ij} $(= d\varepsilon^p_{ij}-\delta_{ij}d\varepsilon^p_v/3)$ is the incremental deviatoric plastic strain tensor and $d\varepsilon^p_v$ is the incremental volumetric plastic strain due to virgin loadings. It should be noted that the plastic strains developed during non-virgin loadings do not contribute to these trajectories, thus avoiding the expansion of the prestress (or yield) surface during non-virgin loadings. According to Eq. (5), $d\xi_v$ is always non-negative and therefore avoids the possibility of shrinking yield surfaces. Models with shrinking yield surfaces could produce non-positive definite stiffness matrices in finite element procedures and could cause numerical problems.

Virgin plastic modulus, H^{VL}, can be calculated using the consistency condition in the theory of plasticity[2]. Wathugala[2] had carried out a parametric study of the effect of h_3 and had concluded that this hardening function can capture the behavior of both clays and sands. For most clays h_3=0 can be adopted. For this case h_4 is not applicable.

INTERPOLATION FUNCTIONS

The following simple interpolation function is used to evaluate reloading plastic modulus, H^{RL}, in this study.

$$H^{RL}=H^{VL}_{I_1}+H^{VL}_{I_2}r_1(1-\alpha_{ps}/\alpha_r)^{r_2} \tag{6}$$

where r_1 and r_2 are material parameters, and $H^{VL}_{I_1}$ and $H^{VL}_{I_2}$ are virgin plastic moduli at points I_1 and I_2 (Fig. 1) on the prestress surface. The image point I_1 is the intersection between the radial line passing through the current stress and the prestress surface. The point I_2 is located at the intersection of the hydrostatic compression line and the prestress surface. The effect of the distance from the current stress point to the prestress surface is included in Eq. (6) through the ratio (α_{ps}/α_r). When stress point reaches the prestress surface, $\alpha_r\rightarrow\alpha_{ps}$ and $H^{RL}\rightarrow H^{VL}_{I_1}$. This assures smooth transition from reloading to virgin loading. Since $H^{VL}_{I_2}\geq 0$ and $H^{VL}_{I_2}>0$, H^{RL} is greater than $H^{VL}_{I_1}$ and decreases when moving toward the prestress surface. Introduction of $H^{VL}_{I_2}$ into the Eq. (6) assures $H^{RL}>0$. This is necessary since $H^*=0$ represents perfect plasticity and should not occur inside the yield surface.

Most of the plastic strains developed during a cyclic loading occur during the reloading part of the cycle. Therefore elastic unloading behavior is assumed in this study. (i.e. $H^{UL}\rightarrow\infty$)

LABORATORY VERIFICATION

Material parameters for Sabine Clay have been determined from laboratory tests performed by Katti[3] on 'undisturbed' samples obtained from the field test site for the instrumented pile segments. They are Young's Modulus, E=4147kPa; Poisson's Ratio, ν=0.42; γ=0.047; β=0; m=-0.5; n=2.4; h_1=0.0034; h_2=0.78; h_3=0; h_4=na;

r_1=500; and r_2=2.4. The algorithms for the determination of material parameters are given in [2]. Back predictions are compared with the observed test data from the compression test in Fig. 2. Figure 2(a) shows a typical comparison that indicate the model back predicts the observed stress-strain curve very well. Predicted pore pressures are in close agreement with the measured values for the initial and final part of the test (Fig. 2(b)). For the reloading part of the test, model had predicted higher pore pressures than observed. However, overall prediction is considered good.

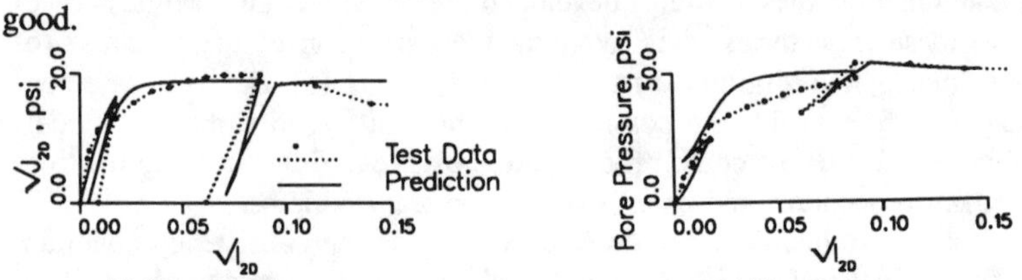

(a) Stress Strain (b) Pore Pressure

Figure 2 Comparison of predicted and observed behavior for undrained triaxial compression test (σ_3=110psi, OCR=1)

FIELD VERIFICATION

The proposed model has been implemented in a finite element procedure based on the theory of dynamics of porous media. This program has been used to simulate the complete test procedure of two instrumented pile segment tests carried out at Sabine Pass, Texas. Details of these simulations are given in [2].

CONCLUSIONS

A HiSS model for isotropic hardening cohesive soils has been developed and verified. The model is simple but captures most of the behavior of clays, in addition to previous capabilities of HiSS models. The model was verified by back predicting laboratory behavior of Sabine clay. Model is further verified by back predicting the field behavior of two instrumented pile segments in saturated clay.

ACKNOWLEDGEMENTS

The research results herein were supported partially by grants No. CEE 8320256 and CES 8711764 from the National Science Foundation, Washington, D.C.

REFERENCES

1. Desai, C.S., Somasundaram, S., and Frantziskonis, G.N., "A Hierarchical Approach for Constitutive Modelling of Geologic Materials," Int. J. Num. Analyt. Meth. Geomechanics, 10, 225-257, (1986).
2. Wathugala, G.W. and Desai, C.S. Dynamic Analysis of Nonlinear Porous Media with anisotropic hardening constitutive model and application to field tests on piles in saturated clays," Report to NSF, U. of Arizona, Tucson, Arizona, (1990).
3. Katti, D.R., "Constitutive Modelling and Testing of Saturated Marine Clay," Ph.D. Dissertation, (Under Preparation) Dept. of Civil Engineering, Univ. of Arizona, Tucson. (1990).

VISCO-, CREEP, RATE DEPENDENCE

CONSTITUTIVE MODELING OF THE CREEP BEHAVIOUR OF SINGLE CRYSTALS WITH APPLICATIONS TO NOTCHED SPECIMENS

A. BERTRAM, J. OLSCHEWSKI, R. SIEVERT, M. ZELEWSKI
Federal Institute for Materials Research and Testing (BAM), Berlin, FRG

ABSTRACT

Tensile loadings of notched bars of single crystal superalloys under monotonous creep conditions are described by means of linear viscoelastic constitutive equations. The anisotropy of the three-dimensional behaviour is represented by linear tensor functions of cubic anisotropy. The complete form of the constitutive model is determined by twelve (temperature dependent) material constants. The orientation dependence of stress redistributions near the notch root due to viscous effects are shown by means of finite element calculations.

INTRODUCTION

During the last years, single crystal superalloys have been developed for high temperature applications in hot section turbine blading. Their use requires design methods which take into account the highly anisotropic inelastic behaviour the material shows at operating temperatures.

The starting point here is a four-parameter model from linear viscoelasticity. The constitutive equations can be brought into the form of two evolution equations by means of introducing an internal variable. The complete three-dimensional generalization is appropriate for implementation into FEM-codes. The model ist capable to describe the material under primary and secondary creep (monotonous and cyclic). Experimental creep data are used to identify the material constants.

CONSTITUTIVE EQUATIONS

A four-parameter model with two elasticities and two viscosities leads to the ordinary differential equation

$$\ddot{\sigma} + a_1 \dot{\sigma} + a_2 \sigma = a_3 \ddot{\varepsilon} + a_4 \dot{\varepsilon} \tag{1}$$

in the stress σ and the strain ε . a_i are (positive) material constants. By introducing the internal variable τ by

$$\tau(t) := \int_{o}^{t} \{b_1\, \varepsilon(s)^{\cdot} - b_2\, \sigma(s)\}\; ds \tag{2}$$

we obtain the evolution equations

$$\tau^{\cdot} = c_1\, \sigma^{\cdot} + c_2\, (\sigma - \tau) \tag{3.1}$$

$$\varepsilon^{\cdot} = c_3\, \sigma^{\cdot} + c_4\, \sigma + c_5\, \tau \tag{3.2}$$

so that all the constants b_i and c_i are determined by $a_1, \ldots, a_4$ (for details see Ref. [1]).

The three-dimensional cubic generalization of equation (1) is

$$\mathbf{S}^{\cdot\cdot} + \mathbf{A}_1[\mathbf{S}^{\cdot}] + \mathbf{A}_2[\mathbf{S}] = \mathbf{A}_3[\mathbf{E}^{\cdot\cdot}] + \mathbf{A}_4[\mathbf{E}^{\cdot}] \tag{4}$$

where $\mathbf{S}$ is the stress tensor, $\mathbf{E}$ the linear deformation tensor and $\mathbf{A}_i$ fourth-order material tensors, which represent the anisotropy of the material. By introducing the fourth-order projection tensors (summation over repeated indices)

$$\mathbf{P}_1 = \tfrac{1}{3}\, \mathbf{I}\otimes\mathbf{I} \tag{4.1}$$

$$\mathbf{P}_2 = \mathbf{I} - (\mathbf{e}_j\otimes\mathbf{e}_j\otimes\mathbf{e}_j\otimes\mathbf{e}_j) \tag{4.2}$$

$$\mathbf{P}_3 = \mathbf{I} - \mathbf{P}_1 - \mathbf{P}_2 \tag{4.3}$$

the material tensors can be brought into the form

$$\mathbf{A}_i := d_{i1}\, \mathbf{P}_1 + d_{i2}\, \mathbf{P}_2 + d_{i3}\, \mathbf{P}_3 \tag{5}$$

with 4 x 3 = 12 real constants d_{ij}. Here $\mathbf{I}$ is the second order identity, $\mathbf{I}$ the fourth-order identity, $\mathbf{e}_j$ the vectors of the crystal directions, and $\otimes$ the tensor product. Instead of τ we define the tensor of inner variables

$$\mathbf{T}(t) := \mathbf{P}_j \Big[\int_{o}^{t} \{k_{1j}\, \mathbf{E}(s)^{\cdot} - k_{2j}\, \mathbf{S}(s)\}\; ds\Big]. \tag{6}$$

In analogy to (3) we obtain the evolution equations

$$\mathbf{T}^{\cdot} = \mathbf{P}_j\, [f_j\, \mathbf{S}^{\cdot} + g_j\, (\mathbf{S} - \mathbf{T})] \tag{7.1}$$

$$\mathbf{E}^{\cdot} = \mathbf{P}_j\, [h_j\, \mathbf{S}^{\cdot} + p_j\, \mathbf{S} + q_j\, \mathbf{T}] \tag{7.2}$$

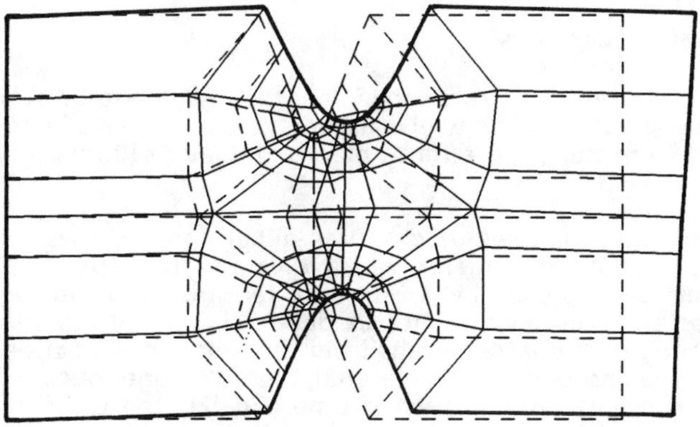

FIG. 1. Longitudinal section of the three-dimensional FE-mesh in the undeformed state and within the primay creep stage.

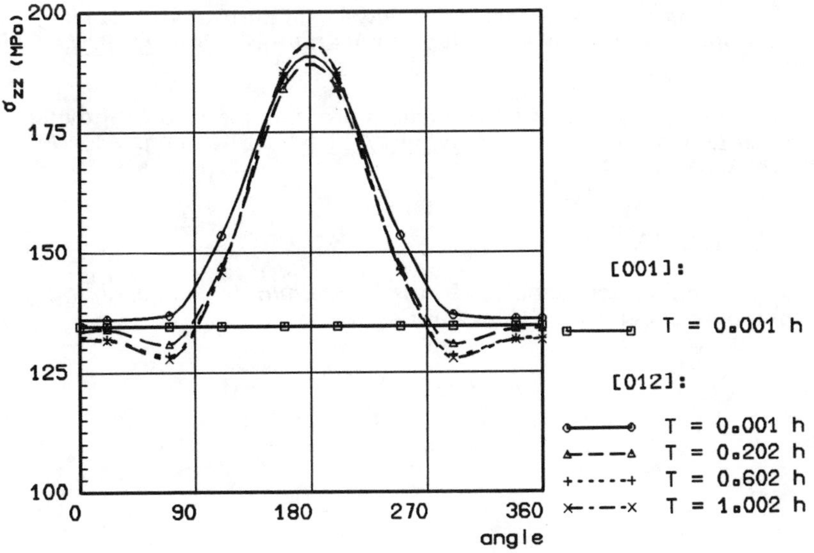

FIG. 2. Normal stress distribution in the net cross section near the notch root for two orientations and different time steps.

wherein the real constants k_{ij}, f_j, g_j, h_j , p_j , and q_j are determined by the 12 constants d_{ij}.

FINITE ELEMENT ANALYSIS

The equations (7.1, 2) are convenient for implementation into a FEM code such as ADINA [2] . The twelve material constants have been roughly determined from creep test data by means of a least square optimization technique.

As an application, we used a three-dimensional FE-mesh with 184 isoparametric 20-node-elements which give rise to 2493 d. o. f. The notched bar was loaded by a constant tension force. In Fig. 1 the undeformed and deformed meshes are depicted in the [0 1 2] orientation. The asymmetry of the latter results from the anisotropy of the material. Due to the creep property of the material, the inhomogenous stresses near the notch root vary with respect to time (Fig. 2). The creep behaviour depends on the orientation of the crystal in the specimen. Anisotropic effects are less pronounced in the [0 0 1] orientation then in the [0 1 2] one, naturally.

More detailed results will be given in a forthcoming article by the authors.

REFERENCES

1. A. Bertram, J. Olschewski, M. Zelewski, "Constitutive modeling of primary and secondary creep at high temperatures". Proc. EUROMAT 89 (1990).

2. K.-J. Bathe, ADINA, "A finite element program for automatic dynamic incremental nonlinear analysis". Report AE 84-1. ADINA Eng. Inc., Watertown MA, USA (1984).

ACKNOWLEDGEMENT

This report was supported by the *Bundesminister für Forschung und Technologie* under grant O3 M 3005 C4.

SOME NEW RESULTS CONCERNING THE NONLINEAR VISCOELASTIC CHARACTERIZATION OF POLYMERIC MATERIALS

O.S. BRÜLLER
Technical University of Munich

ABSTRACT

An often used approach for the nonlinear characterization of viscoelastic materials under isothermal loading conditions is provided by the use of finite exponential series (which describe the linear behavior) multiplied by some load-dependent functions (which describe the nonlinearity of the material). It is shown that for both, creep and stress-relaxation, the same type of constitutive equation can be used. Moreover, for a large class of materials only one nonlinearity function is necessary. As a result, the experimental effort needed for the characterization of the material can be considerably reduced.

INTRODUCTION

Under constant environmental conditions, after a sudden loading step, the long term behavior of polymeric materials can be described as follows [1,2]:

a) under creep (constant stress) by using the creep compliance:

$$J_n(t,\sigma_0) = g_0(\sigma_0)\, J_0 + g_t(\sigma_0)\Delta J(t) \tag{1}$$

with:

$$\Delta J(t) = \sum_{i=1}^{m} J_i[1-\exp(\frac{-t}{\tau_i})] \tag{2}$$

b) under stress-relaxation (constant strain) by using the stress-relaxation modulus:

$$E_n(t,\epsilon_0) = h_\infty(\epsilon_0)\, E_\infty + h_t(\sigma_0)\Delta E(t) \tag{3}$$

with:

$$\Delta E(t) = \sum_{i=1}^{m} E_i \exp(\frac{-t}{\tau_i}) \tag{4}$$

where:

ϵ_0 = constant strain
σ_0 = constant stress
t = time
g_0, g_t = stress-dependent nonlinearity factors
h_∞, h_t = strain-dependent nonlinearity factors

and:

J_0 = instantaneous value of the linear creep compliance
$\Delta J(t)$ = transient term of the linear creep compliance
E_∞ = equilibrium value of the linear stress-relaxation modulus
$\Delta E(t)$ = transient term of the linear stress-relaxation modulus
J_i, E_i = linear material parameters
τ_i = retardation or relaxation-times

In order to describe the creep or stress-relaxation of a nonlinear viscoelastic material it is necessary to know the numerical values of all the parameters of the Eqs. (1) - (4). A procedure for the determination of this data on the basis of a series of long-term experiments is presented in [2].

It should be mentioned that the structure of the mathematical expressions of the creep compliance and that of the relaxation modulus are different: while in Eq. (1) the constant parameter J_0 represents the instantaneous compliance (i.e. the response of the material to the sudden loading step), E_∞ in Eq. (3) reflects the equilibrium (infinite) value of the relaxation modulus. It is clear that only the instantaneous compliance (or its inverse, the initial modulus) can be determined experimentally.

It should also be pointed out here that in the linear viscoelastic range the following identities hold:

$$g_0 = g_t = 1 \qquad h_\infty = h_t = 1 \tag{5}$$

AN EXAMPLE

Results obtained under long-term loading conditions with several thermoplastic polymers, such as PMMA, PVC, PS, ABS, POM, ABS reinforced with chopped glass fibers, under uniaxial loading and pure shear (torsion of thin walled cylindrical specimens) have shown that the above mentioned characterization leads to an accurate description of the nonlinear viscoelastic behavior of those materials. Moreover, more complicated loading histories can be predicted on this basis, as for example sudden periodic loading and unloading [3] or stress-relaxation on the basis of creep data [4].

As an example the creep behavior of PMMA in uniaxial loading is depicted in Fig. 1. A very good agreement between experiments and computed data can be seen. Fig. 2 shows the course of the nonlinearity function vs. applied stress.

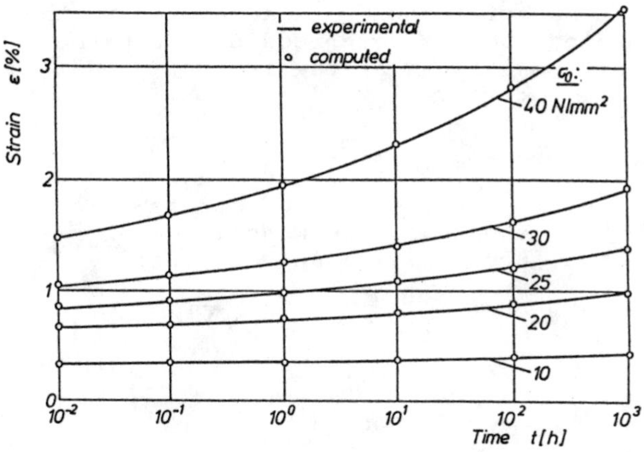

FIG. 1. Creep curves of PMMA

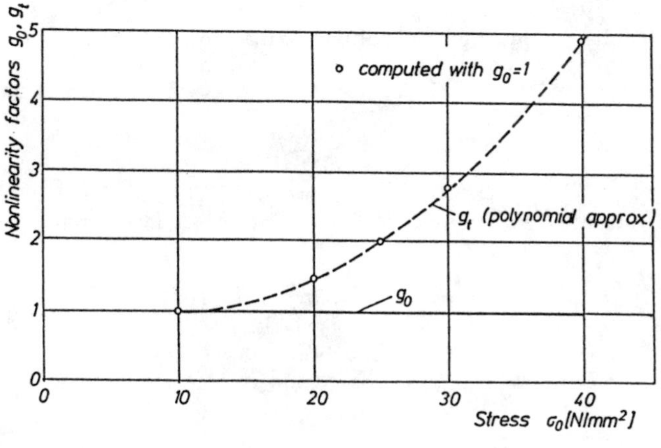

FIG. 2. Nonlinearity factors of PMMA in creep

In this graph one can observe that the nonlinearity function of the instantaneous response of the material is practically equal to one in the entire range of applied stresses, i.e. even in the nonlinear range. This has been confirmed with all the tested materials.

SIMPLIFICATION OF THE CHARACTERIZATION PROCEDURE

The fact that the instantaneous material response to a loading step is linear elastic has major consequences for the material characterization; in other words only one nonlinearity function is needed to describe the material behavior. Accordingly, the linear elastic loading step is described by the parameter which can now be defined as the instantaneous elastic compliance, i.e.:

$$J_0 \equiv J_{el} \tag{6}$$

Consequently, in the case of stress-relaxation the instantaneous response of the material must also be linear elastic. From Eq. (3) one obtains:

$$h_\infty E_\infty + h_t \Delta E(0) = h_\infty E_\infty + h_t \sum_{i=1}^{m} E_i = E_{el} \tag{7}$$

with:

$$E_{el} = \frac{1}{J_{el}} \tag{8}$$

This last equation means that between the two nonlinearity functions of the stress-relaxation representation a linear relationship must exist. This was confirmed with all the tested materials.

With this important result some additional relations can be obtained. If the instantaneous part of the relaxation modulus, which can be derived easily from Eq. (3):

$$h_\infty E_\infty + h_t \sum_{i=1}^{m} E_i \tag{9}$$

is added to and subtracted from the modulus itself:

$$E_n(t,\epsilon) = h_\infty E_\infty + h_t \sum_{i=1}^{m} E_i \exp(\frac{-t}{\tau_i}) \tag{10}$$

one obtains:

$$\begin{aligned} E_n(t,\epsilon) = {} & h_\infty E_\infty + h_t \sum_{i=1}^{m} E_i - h_\infty E_\infty - h_t \sum_{i=1}^{m} E_i \\ & + h_\infty E_\infty + h_t \sum_{i=1}^{m} E_i \exp(\frac{-t}{\tau_i}) \end{aligned} \tag{11}$$

which can be rewritten as:

$$E_n(t,\epsilon) = E_{el} + h_t \sum_{i=1}^{m} E_i \, [1 - \exp(\frac{-t}{\tau_i})] \tag{12}$$

This new form of the stress-relaxation modulus has exactly the same structure as the expression for the creep compliance, with the exception of the sign in front of the time dependent term. It is based, like the expression for the creep compliance, on the knowledge of the initial modulus. Subsequently, the nonlinear behavior under stress-relaxation loading can be described (like the creep compliance) by only one nonlinearity function, namely the nonlinearity function which defines the transient, time-dependent reliance of the relaxation modulus upon the applied strain.

CONSEQUENCES FOR THE EXPERIMENTAL EXPENDITURE

Till now, in order to describe the long-term behavior of a nonlinear polymeric materials accurately, it was necessary to perform series of experiments over the entire range of applied loads (stresses or strains) and full loading time. From these experimental data, the linear material parameters and the nonlinearity functions were determined by numerical procedures, as presented in detail in [2].

On the basis of the simplified relations presented here, it is necessary to conduct only a reduced number of experiments over the whole time-range for which the characterization is needed. From these tests, the linear parameters could be computed, if the nonlinearity function of the time dependent term would be known.

Since this nonlinearity function is valid for the entire considered time-range of the experiments, it is sufficient to perform a series of tests at different loading levels in a very limited time-interval. They allow the numerical determination of the needed nonlinearity function, and subsequently, that of the linear parameters.

As an example, the stress-relaxation of ABS is considered. The experimental data is illustrated by five relaxation curves at strain values between 0.3% and 1.2% (Fig. 3).

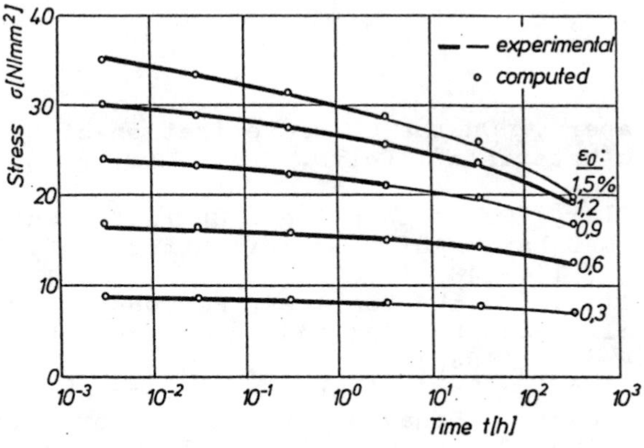

FIG. 3. Stress-relaxation curves of ABS

The experimental data used for the material characterization were taken only from the bulk lines represented in Fig. 3. For the determination of the nonlinearity factors (shown in Fig. 4), all the curves up to 3 hours were used; for the computation of the linear parameters, the curves at 0.6% and 1.2% strain up to 300 hours were considered. In this way about 60% of the experimental time have been saved.

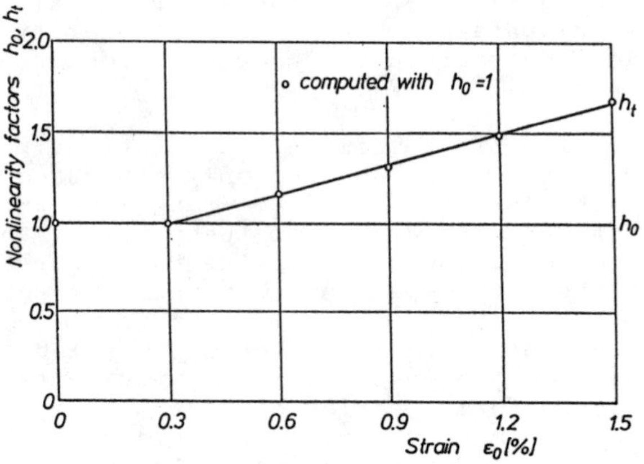

FIG. 4. Nonlinearity factors of ABS in stress-relaxation

The very good agreement between experimental curves and computed data over the whole experimental range can be observed in Fig. 3.

REFERENCES

1. R.A. Schapery, "On the Characterization of Nonlinear Viscoelastic Materials", Polym. Eng. Sci., 9, 295-310 (1969).
2. O.S. Brüller, "On the Nonlinear Characterization of the Long-Term Behavior of Polymeric Materials", Polym. Eng. Sci., 27, 144-148, (1987).
3. O.S. Brüller, "On the Nonlinear Response of Polymers to Periodical Sudden Loading and Unloading", Polym. Eng. Sci. 25, 604-607, (1985).
4. O.S. Brüller and H. Steiner, "Nonlinear Characterization of Relaxation Behavior of Plastics by Using Creep Data", Proc. Ann. Techn. Conf. Soc. Plast. Eng., Dallas, 473-479, (1990).

A Study of the Visco-Elastic Behaviour of Bituminous Mixture at Very Small Strains.

K. CHARIF, E. BARD, P.Y. HICHER.
ECOLE CENTRALE DES ARTS ET MANUFACTURES.
Laboratoire de Mécanique des Sols, Structures et Matériaux
URA 850 CNRS
92295 Châtenay-Malabry CEDEX, FRANCE.

ABSTRACT

By measuring stress and strain directly in bituminous mixtures samples in a triaxial cell, we were able to study the infuence of temperature, frequency and confining pressure on stiffness modulus, Poisson's ratio and phase angle between stress and strain in a range of strain from 10^{-6} to 10^{-2}. At very small strains, we isolated a viscoelastic domain where the equivalence between time and temperature effect was demostrated.

INTRODUCTION

This paper describes the results obtained from a new triaxial apparatus developed at the laboratory of the Ecole Centrale de Paris especially for measuring small strains (10^{-6} to 10^{-3}) at high accuracies. Force and displacements are measured directly on the sample by a load cell placed at the top of the sample and by two non-contacting proximity transducers mounted in the central section of the sample to measure the axial and radial deformations [1].

A study was conducted to characterize the mechanical properties of a 0/10 bituminous concrete composed exclusivelly of crushed aggregates with a bitumen content of 6 per cent by weigth of aggregate. The bitumen used was 40-50 penetration. The cylindrical specimens were sampled in plates obtained by compacting the mixture at 160°C.

The parameters studied were:
- axis of sampling (**V**ertical and **H**orizontal),
- frequency (**f**),
- temperature (θ),
- confining pressure (σ_3).

The following notation is used throughout this paper:
- Stiffness modulus ($\mathbf{S_m}$) : is equal to the absolute value of the complex modulus,
- $\mathbf{E_1}$ and $\mathbf{E_2}$: are the components of the complex modulus,
- phase angle (ϕ) : the phase between stress and strain.

EXPERIMENTAL RESULTS

In order to study the mechanical properties of a bituminous mixture, it is important to know the limits of the linear behaviour and the subsequent degradation of the stiffness modulus in the non linear range.

Experimental results show that the limit of the linear elastic range is between 10^{-5} and 10^{-4} strain, which is larger than that of the dry aggregate (3.10^{-5}) and appears not to be affected by temperature.

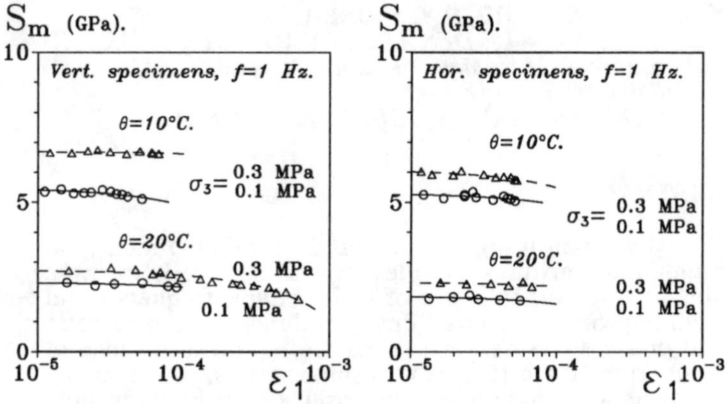

Fig. 1. Influence of **anisotropy, θ, and** σ_3.

The modulus of the specimen sampled vertically is greater than that wich was sampled horizontally. This highlights the anisotropic fabrication of the material. This influence is more marked at low frequencies and at high temperatures (Fig. 1).

The modulus increases with frequency and decreases with temperature. This increase in the modulus with frequency is more marked at high temperatures. At low temperatures the slope of the (S_m,f) curve is shallower than that at high temperature (Fig. 2). We observe also that these curves have a tendency to draw closer together and to become parallel at low temperatures. In the plane (log S_m,log σ_3), the modulus increases linearly with the confining pressure (Fig. 3).

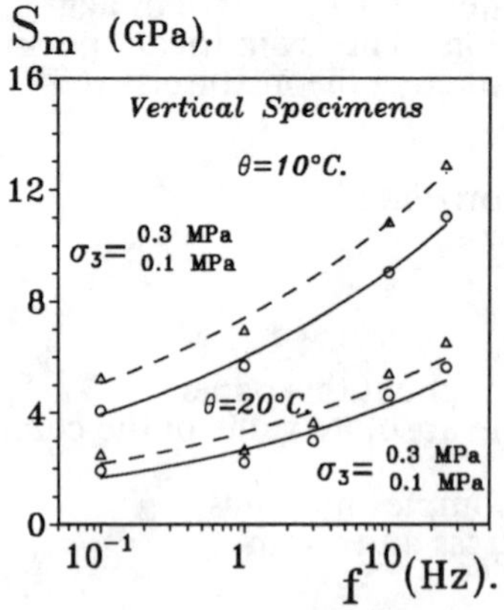

Fig. 2. Influence of **f**, θ, **and** σ_3.

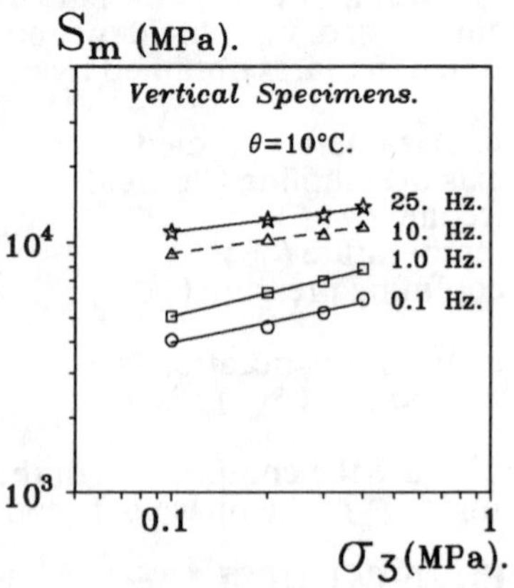

Fig. 3. Influence of σ_3 and **f**.

Poisson's ratio of the bituminous mixture decreases at high

temperatures and frequencies (Fig. 4). It appears that Poisson's ratio is not influenced by the confining pressure. Poisson's ratio of the specimen sampled horizontally is slightly greater than that sampled vertically. Since we did not notice any phase difference between vertical and horizontal deformation, Poisson's ratio always has a real value.

RHEOLOGICAL PROPERTIES OF THE STIFFNESS MODULUS:

Frequency-Temperature Equivalence Law :

One of the most important properties of a visco-elastic material is the influence of **f** and θ on its mechanical properties. It is possible to superimpose by simple translation the curves (S_m,**f**) as a function of the temperature [2,3]. The translation factor (at) is obtained by using a relation of ARRHENIUS type. In choosing a reference temperature, we can obtain one single curve called the master curve (Fig 5).

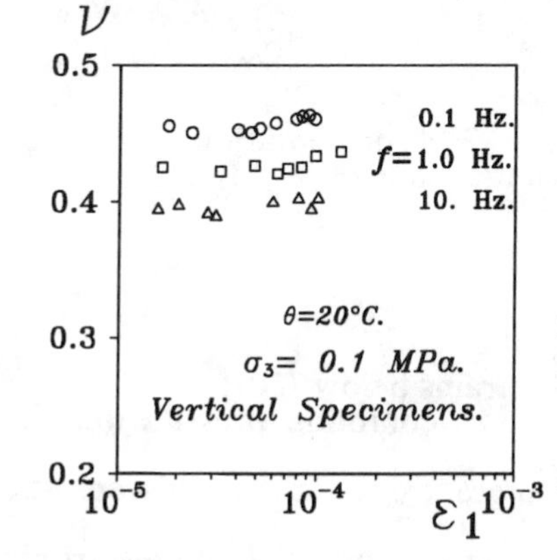

Fig. 4. Evolution of the Poisson's Ratio(ν).

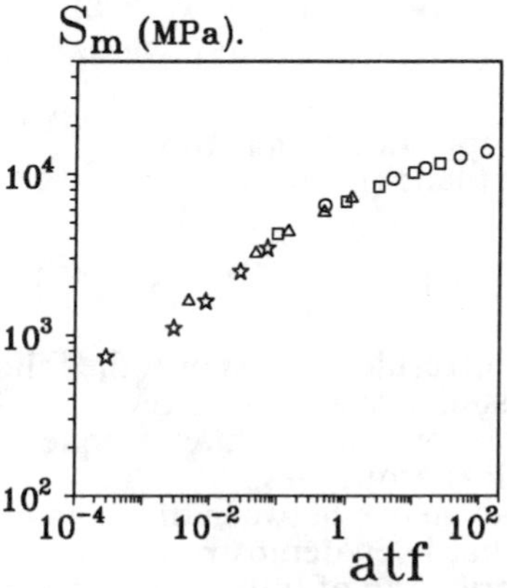

Fig. 5. Master Curve

Representation in the complex plane (E_1,E_2) :

The relation between E_1 and E_2 forms one single curve at different θs and fs for a given value of the confining pressure. At large θ values and small **f** values the curve reachs the X axis; therefore S_m approachs to a real value different from 0, indicating an elastic behaviour that can be compare to the one of the sole aggregate at the effective confining pressure corresponding to the part of the total pressure applied on the grains, the other part being applied on the bitumen (Fig. 6).

The relation between E_1 and E_2 depends on the confinig pressure, leading to higher values of S_m as σ_3 increases.

Representation in the plane (ϕ, S_m) :

The relation between ϕ and S_m forms also a unique curve (Fig. 7). As θ decreases and **f** increases, ϕ approches 0, implying perfect elastic behaviour. More results are needed to obtain the behaviour at high temperature and low frecuency, which theoretically should be also elastic : as a consequence the curve, after reaching a peak, should draw to smaller values of ϕ down to 0.

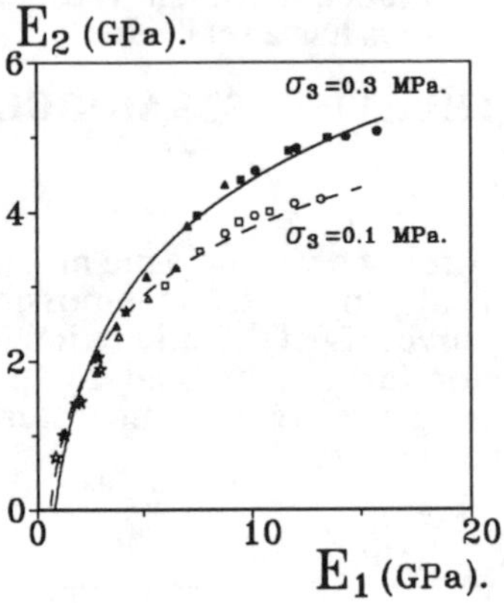

Fig. 6. Representation in the complex plane (E_1, E_2).

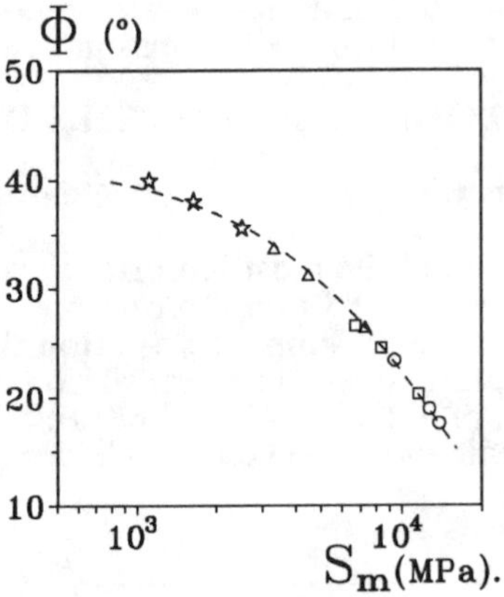

Fig. 7. Representation in the plane (ϕ, S_m)

CONCLUSION:

The results have shown the following:
- the existence of a range of recoverable strains below 7.10^{-5}.
- the role of temperature, frequency and confining pressure on the mechanical properties.
- the equivalence between time and temperature effect in the viscoelastic domain has been demostrated
- the fabrication of bituminous mixture samples induced an anisotropic behaviour, wich is represented by a value of the vertical stiffness higher than the one of the horizontal stiffness

ACKNOWLEDGEMENTS:

This work has been supported by Shell Recherche S.A.. We acknowledge permission to present these results.

REFERENCES :

1. M. S. El Hosri, "Contribution à l'étude des propriétés mécaniques des matériaux", Thèse Docteur ès-sciences, Ecole Centrale de Paris, 1984.
2. C.Huet, "Etude par une méthode d'indépendance du comportement viscoélastique des matériaux hydrocarbonés", Thèse doct.-ing., Fac. sci. Paris, 1963.
3. G. Sayegh, "Contibution à l'étude des propriétés viscoélastiques des bitumes purs et des bétons bitumineux", Thèse doct.-ing., Fac. sci. Paris, 1965.

VISCOPLASTICITY THEORY BASED ON OVERSTRESS APPLIED TO THE MODELING OF CUBIC SINGLE CRYSTALS

S. H. CHOI and E. KREMPL
Mechanics of Materials Laboratory
Rensselaer Polytechnic Institute
Troy, N.Y. 12180-3590, U.S.A.

ABSTRACT

A previously developed phenomenological orthotropic theory of viscoplasticity based on overstress is specialized for the case of cubic symmetry. Rate sensitivity, creep and relaxation are modeled in a unified way. The present version models cyclic neutral behavior. The three dimensional theory is specialized for uniaxial tests in the <100>-, <110>- and <111>-directions. Only specimens tested in the <110>-directions exhibit a distortion of the cross section. This continuum theory needs only two material functions and, at most, twelve constants to characterize the elastic and inelastic rate-dependent behavior of single crystals of cubic symmetry for small deformations.

INTRODUCTION

Within the last decade, single crystal superalloys were developed for high temperature service [1, 2]. Their use in jet engine applications requires that their performance be analyzed in the design stage. Of special interest are the creep performance, the behavior under cyclic temperature changes, as well as the low-cycle fatigue and the creep rupture lives.

In the past, the mechanical behavior of single crystals has attracted the interest of physicists and material scientists. Consequently, modeling of the time-dependent behavior of the superalloys started with the "crystallographic" approach. It was assumed that the behavior on each slip plane is governed by a "unified" constitutive equation and that the contributions of each slip plane are summed-up, using established procedures of time-independent crystal plasticity theory [3-5].

The crystallographic analysis incorporates six cube and twelve octahedral slip systems. The mechanical behavior is different whether the cube slip planes are activated or not [4, 6]. To reproduce the experimental outcomes it is necessary to either include or to exclude the cube slip systems in the calculations. A comparison of the analytical results with experiments is needed so that the micromechanics analysis agrees with macroscopic experiments.

In the present paper, a phenomenological approach is developed for the modeling of the time-dependent behavior of limited ductility single crystals of cubic symmetry intended for elevated temperature service. The orthotropic, unified theory of viscoplasticity based on overstress (VBO) developed previously [7] is specialized for cubic symmetry and applied to the simulation of tension/compression tests in the [100], $[\bar{1}10]$ and [111] directions.

PREDICTIONS FOR UNIAXIAL TESTS IN [100], [$\bar{1}$10] AND [111] DIRECTIONS

The theory presented in detail in [8] is specialized for tensile and compressive tests with loading in cube, or face diagonal, or body diagonal directions. For uniaxial tests in the [100] and [111] directions the theory predicts that circular cross sections deform into circles. The relevant equations are given in [8].

The full equations for a uniaxial test oriented and loaded in the [$\bar{1}$10] direction are given in the following. In this case the 2′-axis corresponds to the face diagonal ([$\bar{1}$10]) direction. The components $\sigma_2' = \sigma'$, $g_2' = g'$ and $f_2' = f'$ are the only nonzero components of stress like quantities.

$$\dot{\varepsilon}_1' = \left(\frac{1-2\nu}{3E} - \frac{1}{6G}\right)\dot{\sigma}' + \frac{1}{k[\Gamma]}\left(\frac{1-2\eta}{3|E_t|} - \frac{1}{6|G_t|}\right)(\sigma' - g') \tag{1}$$

$$\dot{\varepsilon}_2' = \left(\frac{1-\nu}{2E} + \frac{1}{4G}\right)\dot{\sigma}' + \frac{1}{k[\Gamma]}\left(\frac{1-\eta}{2|E_t|} + \frac{1}{4|G_t|}\right)(\sigma' - g') \tag{2}$$

$$\dot{\varepsilon}_3' = \left(\frac{1-5\nu}{6E} - \frac{1}{12G}\right)\dot{\sigma}' + \frac{1}{k[\Gamma]}\left(\frac{1-5\eta}{6|E_t|} - \frac{1}{12|G_t|}\right)(\sigma' - g') \tag{3}$$

$$\dot{\varepsilon}_5' = \left(\frac{-\sqrt{2}(1+\nu)}{3E} + \frac{\sqrt{2}}{6G}\right)\dot{\sigma}' + \frac{1}{k[\Gamma]}\left(\frac{-\sqrt{2}(1+\eta)}{3|E_t|} + \frac{\sqrt{2}}{6|G_t|}\right)(\sigma' - g') \tag{4}$$

$$\dot{\varepsilon}_4' = \dot{\varepsilon}_6' = 0 \tag{5}$$

$$\dot{g}' = \frac{\psi[\Gamma]}{E}\dot{\sigma}' + \left[\phi - \theta\{\phi - E_t(1 - \frac{|E_t|}{E}\psi)\}\right]\frac{\sigma' - g'}{|E_t|\,k[\Gamma]} \tag{6}$$

$$\dot{f}' = E_t\frac{\sigma' - g'}{|E_t|\,k[\Gamma]} \tag{7}$$

$$\Gamma = \left\{\left|\frac{1}{4}(2H_1 + 2H_2 + H_3)(\sigma' - g')^2 + a(\sigma' - g')\right|\right\}^{1/2} \tag{8}$$

$$\theta = \left\{\left|\frac{1}{4}(2P_1 + 2P_2 + P_3)(g' - f')^2 + b(g' - f')\right|\right\}^{1/2} \tag{9}$$

E and G are the elastic moduli in axial and shear, respectively, for cube direction. E_t and G_t are the tangent moduli at the maximum strain of interest in axial and shear, respectively, for cube direction. ν and η are the constant elastic and inelastic Poisson's ratios, respectively. The total Poisson's ratio can be calculated by integrating (1-3) first and then determining their negative ratios at every point in time. $k[\Gamma]$ is the viscosity function and $\psi[\Gamma]$ and ϕ are called shape functions. Both k and ψ are positive decreasing functions of invariant Γ. g and f are state variables and called equilibrium stress and kinematic stress, respectively. H_1-H_3, P_1-P_3 and a, b are suitable constants, see [7, 8].

It should be noticed that the ratios $\varepsilon_1'/\varepsilon_2'$ and $\varepsilon_3'/\varepsilon_2'$ are not equal. It is also interesting that, in this particular case, the shear strain ε_5' develops in the plane perpendicular to the axis of tensile loading, the [$\bar{1}$10] and 2′ direction. The 1′ and 3′ axes are in the directions [111] and [$\bar{1}\bar{1}$2], respectively.

EXPERIMENTAL DETERMINATION OF MATERIAL CONSTANTS

To represent inelastic incompressibility, the inelastic Poisson's ratio is assumed to be equal to 0.5. From the uniaxial tension and torsion tests on thin walled tubes of specimens oriented in the <100>-directions E, E_t, the ψ and k functions, together with the shear moduli G and G_t, and constants H_1, H_3, P_1 and P_3 appearing in the invariants Γ and θ can be determined. Either a uniaxial test in a <110>- or <111>-direction is necessary to determine Poisson's ratio ν, H_2 and P_2.

The independent tests proposed above may also provide evidence which may conflict and is irreconcilable with the predictions of the theory. If this happens, the simple theory will have to be modified.

CROSS SECTIONAL CHANGES

The theory predicts that the circular cross section of tensile bar oriented and loaded in the $[\bar{1}10]$-direction will not deform into circles. Upon loading to the inelastic range and subsequent unloading to zero stress, a permanent cross section change is predicted. For the uniaxial test in the $[\bar{1}10]$-direction, the transverse strain can be calculated as a function of angle α, the angle of rotation around the $2'$-axis measured from the $1'$-direction.

Although no reports were found in the literature, the ratio of the maximum radius r_{max} and the minimum radius r_{min} can be determined experimentally. (Due to the limited ductility of some alloys, this task may not be easy.) The ratio $\delta = r_{max}/r_{min}$ can also be calculated provided the loading/unloading history is known. Recognizing that the elastic strains do not contribute to the permanent cross sectional changes, we obtain

$$\delta = \frac{1 + \left(\frac{1}{4|E_t|} - \frac{1}{4|G_t|}\right)\int_0^{t_0} \frac{\sigma' - g'}{k[\Gamma]}\, d\tau}{1 - \frac{1}{2|E_t|}\int_0^{t_0} \frac{\sigma' - g'}{k[\Gamma]}\, d\tau} \tag{10}$$

where 0 and t_0 denote the time at the start of the test and the time at which zero load is reached again, respectively. For a given loading/unloading history, this expression can be evaluated numerically in conjunction with (6-7). The predicted value of δ can be compared with experiments. The comparison can serve as a check on the predictive capability of the theory.

The above analysis applies to uniaxial tests of <110> oriented specimens. The theory predicts no shape changes for specimens exactly oriented and loaded in the <100>- and <111>- directions.

DISCUSSION

This phenomenological theory for cubic symmetry and isothermal deformations comprises the three independent elastic constants, two tangent moduli, two functions and eight constants appearing in the invariants Γ and θ (only six are required for tension/compression symmetry) for the case of inelastic incompressibility. This is indeed a small number if this theory is compared to the presently used crystallographic theories. In addition the present theory is, compared to existing crystallographic theories, simple in structure. It is only necessary to solve a coupled set of three nonlinear, vector valued differential equations. In crystallographic approaches a unified equation for each of 18 slip systems must be available. Solution requires integration and the summing-up over all the slip systems. The constitutive equation for each slip system is comparable in complexity to the present theory since a flow law and growth laws for the state variables are necessary.

ACKNOWLEDGEMENT

The support of Textron Lycoming Stratford Division and of DARPA/ONR under contract N00014-86-K0700 is gratefully acknowledged.

REFERENCES

1 A. G. Dodd, "Mechanical design of gas turbine blading in cast superalloys," Mater. Sci. Tech., **2**, 476-485, 1986.

2 M. Gell, D. N. Duhl, D. K. Gupta and K. D. Sheffler, "Advanced superalloy airfoils," J. Met., **39**, 11-15, 1987.

3 L. T. Dame, "Anisotropic constitutive model for Nickel base single crystal alloys: Development and finite element implementation," Ph.D. Thesis, University of Cincinnati, Cincinnati, OH, 1986.

4 K. P. Walker and E. H. Jordan, "Biaxial constitutive modeling of a single crystal superalloy at elevated temperature," *Biaxial and Multiaxial Fatigue*, M. W. Brown and K. J. Miller, editors, Mechanical Engineering Publications LTD, London, 145-170, 1989.

5 G. Cailletaud, "Une approche micromécanique du comportement plastique des polycristaux," Rev. Phys. Appl., **23**, 353-365, 1988.

6 G. Cailletaud, Discussion at Symp. on "Constitutive Equations and Life Prediction Models for High Temperature Applications," Berkeley, CA, 1988.

7 K. D. Lee and E. Krempl, "An orthotropic theory of viscoplasticity based on overstress for thermomechanical deformations," Rensselaer Polytechnic Institute, Report MML 88-2, Troy, NY, to appear in Int. J. Solids and Structures, 1990.

8 S. H. Choi and E. Krempl, "Viscoplasticity theory based on overstress applied to the modeling of cubic single crystals," Eur. J. Mech., A/Solids, **8**, 3, 219-233, 1989.

CREEP DEFORMATION MODELLING OF A NICKEL-BASE SUPERALLOY

T.S. Cook, R.L. McKnight, and D.C. Slavik
General Electric Company
Cincinnati, Ohio

ABSTRACT

Classical and unified constitutive models are used to examine the creep behavior of the nickel-base superalloy Rene' 80 at 982C. Tensile, compressive, and step load creep tests are used to evaluate model predictions during load changes. The unified model underpredicts the inelastic strain at high stresses. The creep strain was well predicted by the classical model when no hardening rule was used.

INTRODUCTION

The demand for higher performance in aircraft gas turbine engines has led to higher temperatures and mechanical loading on many hot path components. Some parts occasionally undergo inelastic deformation, so it is vitally important that analysis tools are available to describe this deformation. The available models of inelastic deformation have the ability to handle constant amplitude isothermal fatigue and creep but when the load and/or temperature changes, the results are less successful. The goal of the present program is to evaluate the ability of a classical and a unified model to predict creep deformation.

Constitutive Models

Classical constitutive models separate the deformation into elastic, plastic, and creep components according to

$$\epsilon = \epsilon_e + \epsilon_p + \epsilon_c \qquad (1)$$

Concentrating on the creep term, there are a number of creep models currently available in nonlinear structural analysis codes. One of these describes the primary and secondary creep according to the relation

$$\epsilon_c = K\,\sigma^m\,t^n + Q\,\sigma^r\,t \qquad (2)$$

where the five constants K, m, n, Q, r are found using various optimization and regression analyses.

In addition to classical approaches to creep deformation, a mechanistic based unified model has also been developed. The details of this model are given in [1] but a brief summary will be given here. The mechanical strain is separated into an inelastic component (creep + plasticity) and an elastic component, or

$$\dot{\epsilon} = \dot{\epsilon}_i + \dot{\epsilon}_e \tag{3}$$

The elastic strain rate is calculated from Hooke's Law as

$$\dot{\epsilon}_e = \dot{\sigma}/E - (\sigma/E^2)\ (\partial E/\partial T)\dot{T} \tag{4}$$

where $\dot{\sigma}$ is the stress rate, E is the elastic modulus, and $\dot{T}$ is the temperature rate. The inelastic strain is calculated with the flow rule. The general form of the flow rule is taken as

$$\dot{\epsilon}_i = \text{sgn}(\sigma-\alpha) * f[|\sigma-\alpha|/k] \tag{5}$$

where sgn $(\sigma-\alpha) = \pm 1$, α is the internal back stress state variable, and k is the drag stress state variable. The flow function f( ) reflects deformation mechanisms which need to be determined from experiments. The evolution expression for the back and drag stress are given in [1].

Experimental Data and Model Predictions

The material used in this investigation was Rene' 80, a nickel-base blade alloy. The chemistry, heat treatment, and a brief description of the microstructure are given in Reference 2. All creep tests were conducted in servohydraulic testing machines using axial extensometers. Induction heating was employed; the temperature was controlled by calibrated thermocouples mounted on the specimen shoulders.

The baseline creep data at 982C was obtained at three stress levels; the test data is shown in Figure 1, along with curves depicting the fit of equation (1) to the data. The curves were obtained using a nonlinear regression analysis. The creep rates do not reach a steady state value, but this was not considered a problem as the test times are short.

The data used in the determination of the constants for the unified model was generated by

Ramaswamy [3]. The data analysis procedure is explained in [1]. The resulting model is compared to the strain data in Figure 1. The model fits the low stress data exactly but becomes increasingly nonconservative as the stress is raised.

In order to check the symmetry of the creep deformation in Rene' 80, two compressive creep tests were run at 128.0 and 177.1 MPa. The results of these tests are shown in Figure 2, along with the predictions of the unified and classical models. The unified model is again slightly nonconservative while the classical model brackets the test results. It appears that the creep deformation is symmetric and that the two models adequately predict the compressive creep behavior.

Several tests were run in an interrupted mode, where the loads were changed after a couple of hours. Figure 3 shows a test consisting of 6 creep legs, three in tension and three in compression. Note that when the load is reversed, the tensile inelastic deformation is removed and the compressive deformation starts from approximately zero strain. When the measured strain is compared to the prediction of the unified model, the Rene' 80 is showing larger levels of strain than is predicted. This is true in both tension and compression where the increased loads produce an initially high inelastic strain rate. Figure 4 compares the tensile creep straining with the classical model with no hardening rule. That is, each load increment assumes virgin material and merely sums the total creep strain. The figure demonstrates that at this temperature, Rene' 80 displays a sizeable portion of the primary creep at each load step. This causes constitutive models employing hardening rules to underpredict the observed inelastic deformation.

SUMMARY

At 982C, the short time creep behavior of Rene' 80 can be modelled using either classical or unified constitutive theory, although at higher loads, the unified model underpredicted the amount of strain. This trend continues when the loads are changed; the unified model underpredicts the strain as the load is raised. The classical model with no hardening rule does a fairly good job of predicting this strain. Tensile and compressive creep at this temperature were similar.

REFERENCES

1. D.C. Slavik and T.S. Cook, "A Unified Constitutive Model For Superalloys," International Journal of Plasticity, in press.

2. T.S. Cook, K.S. Kim, and R.L. McKnight, "Thermal Mechanical Fatigue of Cast Rene' 80," Low Cycle Fatigue, ASTM STP 942, H.D. Solomon, et. al., Eds. (ASTM, Philadelphia, 1988), pp. 692-708.

3. V.G. Ramaswamy, R.H. Van Stone, L.T. Dame, and J.H. Laflen, "Constitutive Modeling for Isotropic Materials", NASA CR 175004, October 1985.

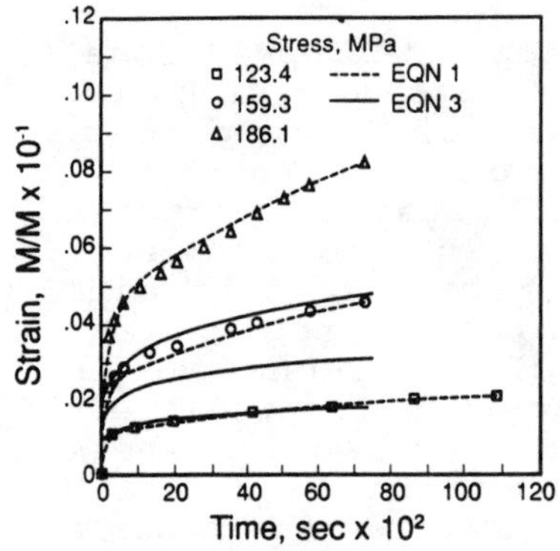

FIG. 1 Tensile Creep Results

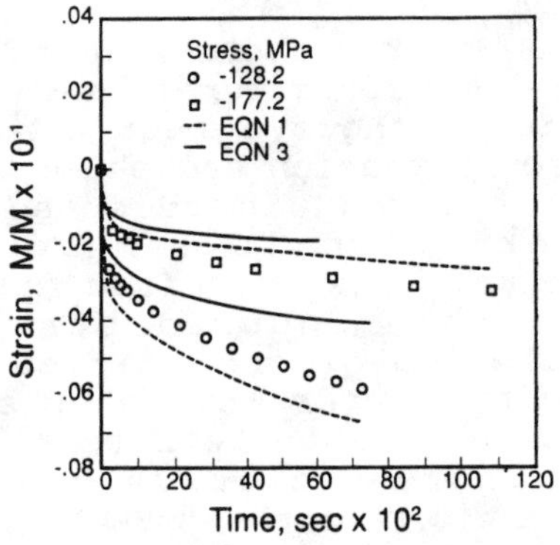

FIG. 2 Compressive Creep Results

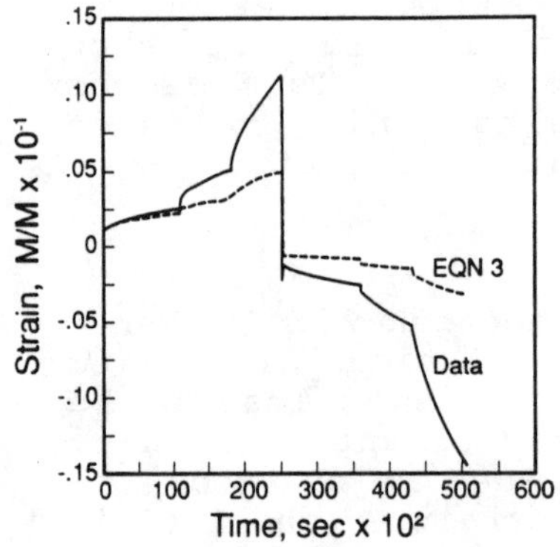

FIG. 3 Unified Model vs. Interrupted Creep Test

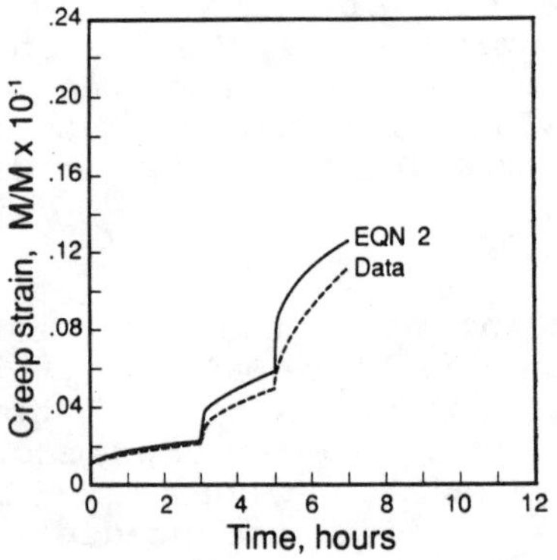

FIG. 4 Classical Model vs. Interrupted Creep Test

BEHAVIOR AND MODELIZATION OF AN AUSTENITIC STAINLESS STEEL IN CYCLIC VISCOPLASTICITY, FROM 20 TO 700°C

P. DELOBELLE
Laboratoire de Mécanique Appliquée associé au CNRS - LA 04
Faculté des Sciences et des Techniques
La Bouloie - Route de Gray
25030 BESANCON CEDEX (France)

ABSTRACT

The objective of the present paper is, on the one hand, to set forth the diversity of mechanical properties of an austenitic stainless steel over a large temperature domain ($20 \leq T \leq 700°C$), including its monotonic, cyclic (uni-directional and bi-directional tension-torsion tests, both in and out of phase) and viscous characteristics, and, on the other and, to present the possibilities of a unified model having internal variables (developed during the past 4 years) with regard to the presented experimental observations.

INTRODUCTION

The industrial materials used in certain components of modern installations (the nuclear industry in the present case) are often submitted to mechanical and thermal loadings and the knowledge of the behavior and the development of the anisothermic laws proves to be indispensable for the prediction of the life duration of the installations. However, before performing and modelizing real anisothermic tests, where the thermal and the mechanical loadings evolve cyclicaly and simultaneously, the identification and the phenomenological modelization for different isotherms of the physical mechanisms taking part in the strain, is necessary. This article presents the results relative to this first step, in the case of an austenitic stainless steel.

EXPERIMENTAL METHODS

The specimen of stainless steel are obtained form slabs cut out of 30 mm thick plates and hyper-quenched from 1200°C. The weight composition of this low carbon steel, with controlled nitrogen content is given in Table I.

TABLE I. Weight composition of the steel.

C	P	Si	Mn	Ni	Cr	Mo	N	B	Co	Cu
≤0.03	≤0.021	0.44	1.084	12.3	17.54	2.47	0.075	0.001	0.15	0.17

Differents machines have been used to perform the tests, notably : hydraulic and electrodynamic tensile-torsion machines controlled by computer.

EXPERIMENTAL RESULTS

Figure 1 shows the evolution of the flow stress $\tilde{\sigma}_{ec}$ for a constant strain rate, $\dot{\varepsilon}^{T}_{zz} = 6.6 \ 10^{-4} \ s^{-1}$, and different strain levels, as a function of the temperature. A plateau appears between 300 and 550°C. Figure 2, curve b, represents, for a constant shear strain rate, $\dot{\varepsilon}^{T}_{z\theta} \simeq 5\ 10^{-5}\ s^{-1}$, and for an V. Mises equivalent plastic strain $(\overline{\Delta\varepsilon}^{p}/2) \simeq 2.2\ 10^{-3}$, the evolution of the equivalent stabilized stress $(\overline{\Delta\sigma}/2)$ as a function of the temperature. The behavior is very different from that observed in Fig. 1 and a significant peak is seen, having a maximum around 550°C and whose width is contained between 200 and 600°C, that is, approximately the zone corresponding to the plateau of the preceeding figure. Note that such a maximum has already been observed for a 304 and a 316 stainless steels [1-2].

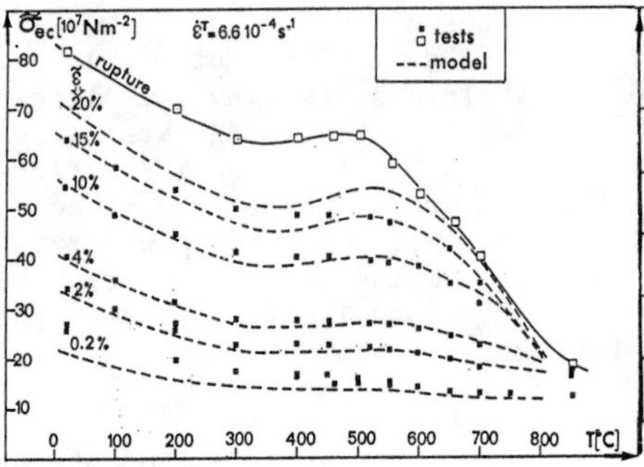

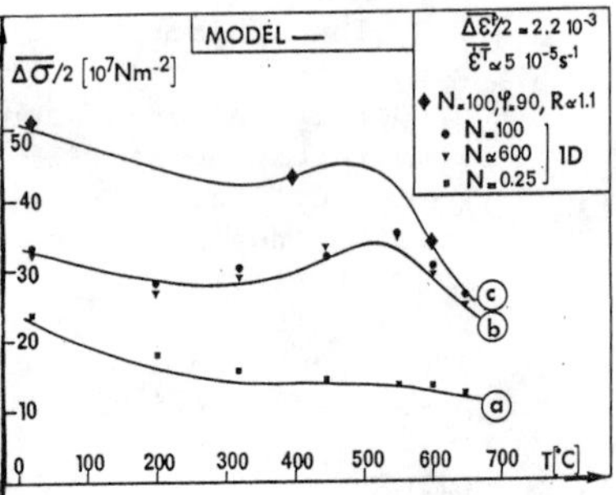

FIG. 1 Variation of the flow stress $\widetilde{\sigma}_{ec}$ with the temperature. Tests and simulations.

FIG. 2 Evolution of the equivalent cyclic stresses, a) first cycle, b) stabilized cycle (1D), c) 90° out of phase stabilized cycles, as a function of the temperature. Tests and modelization.

In terms of strain rate effects (in a ratio of 10^{3}), Figure 3 shows that the sensitivity coefficient n of the sta bilized stress to the strain rate $(\Delta\sigma^{stab}_{z\theta}(1.5\ 10^{-3}\ s^{-1}))/(\Delta\sigma^{stab}_{z\theta}(1.5\ 10^{-6}\ s^{-1}))$ can be positive or negative depending of the temperature of the test. The negative value is observed in the intermediate temperature domain ($300 \leq T \leq 500$°C) and corresponds exactly to that of the peak mentioned earlier. At low temperatures, this coefficient is positive, which is due to a cold viscosity effect. For higher temperatures, $T \geq 600$°C, the sensitivity is naturally positive since the thermally activated recovery effects lead to an increase of the viscosity with the temperature.

For the 2D cyclic loadings, it is shown [2-4] that a supplementary hardening appears, which is mostly a function of the degree of phase difference φ between the strain components and the ratio $R = \varepsilon^{m}_{z\theta}/\varepsilon^{m}_{zz}$ between the maximum amplitudes of these components. This hardening is maximum when $\varphi = 90°$, $R = 2/\sqrt{3}$, and is a strongly decreasing function of the temperature (Fig. 2, curve c, for $\overline{\Delta\varepsilon}^{p}_{Max} \simeq 2.2.\ 10^{-3}$).

As for the viscous properties, the viscous component $\widetilde{\sigma}_{v}$ ($\widetilde{\sigma}_{v} = \widetilde{\sigma}-\widetilde{\alpha}$, where $\widetilde{\sigma}$ is the applied stress and $\widetilde{\alpha}$ the sum of the contributions of the different internal stresses) is measured as a function of the temperature by the method of inverse relaxation [5] (strain dip test technique), during a test conducted at a constant strain rate $\dot{\varepsilon}^{T} = 6.6\ 10^{-4}\ s^{-1}$. Figure 4 shows, for different strains $\tilde{\varepsilon}$, that $\tilde{\sigma}_v$ increases with $\tilde{\varepsilon}$ and passes through a minimum around 300°C.

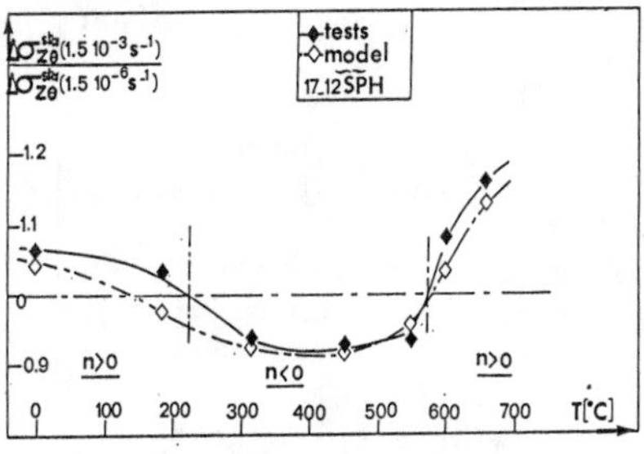

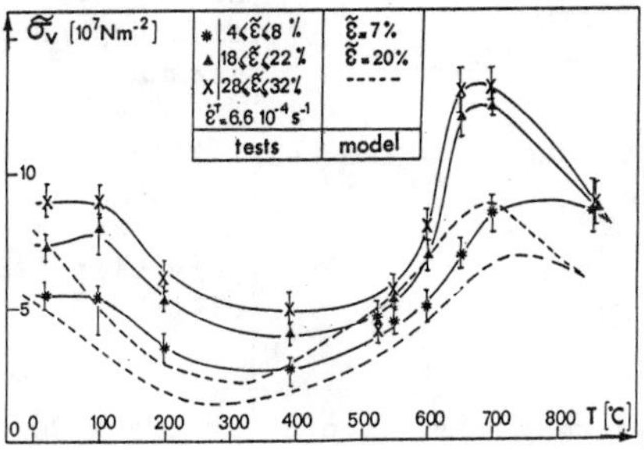

FIG. 3 Evolution of the stabilized stress ratio for two fixed strain rates in a ratio of 10^3. Experiments and modelization.

FIG. 4 Evolution of the viscous stress $\tilde{\sigma}_v$ with the temperature for three levels of strain. Tests and simulations.

Figure 5 presents the results of several creep tests for different temperatures and stresses and corroborates the results presented in fig. 4 : cold and hot viscosity effects, loss of viscosity at intermediate temperature. For the tensile-torsion ratchet [6], the test parameters are such that : $\sigma_{zz} = 50$ Mpa, $\Delta\varepsilon^T_{z\theta}/2 = \pm\ 0.35$ % and $\dot{\varepsilon}^T_{z\theta} = \pm\ 4.6\ 10^{-5}\ s^{-1}$. The response to this type of loading consists in the appearance of a progressive axial strain ε^r_{zz} (fig. 6) due to the axial and shear stresses and in a cyclic hardening in the direction of the shear component. In the limits of experimental accuracy the stress for the stabilized cycle corresponds to the stress obtained in pure torsion for the same strain amplitude (fig. 2, curve b).

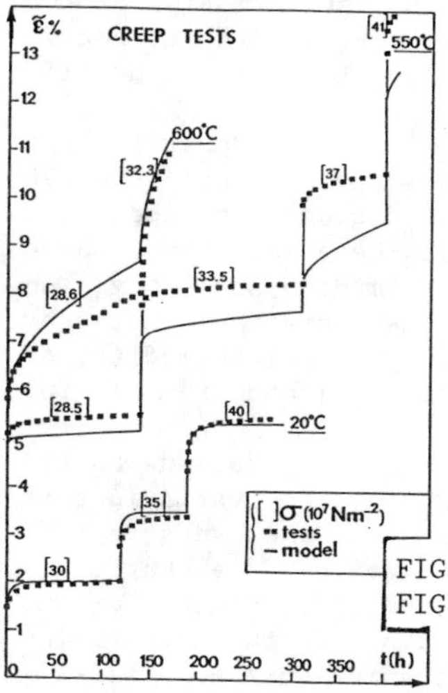

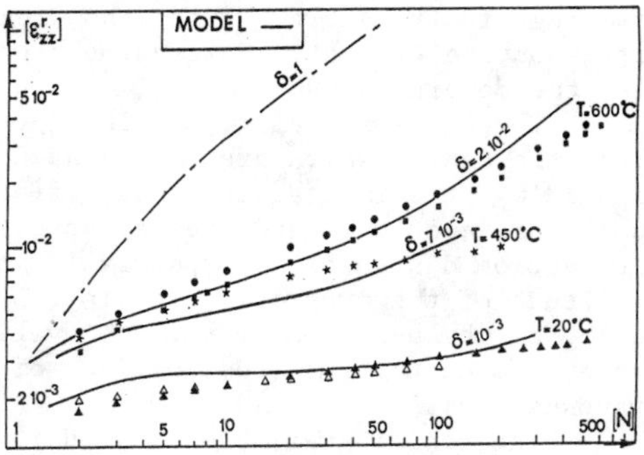

FIG. 5 Simulation of different creep tests.

FIG. 6 Evolution of the progressive strain ε^r_{zz} with the number of cycles N. Tests and modelizations.

In conclusion, the mechanical properties of this alloy can be summarized in table II.

TABLE II. Mechanical properties of the 17-12 SPH steel.

cold creep	no or very little time dependent strain	thermally activated high temperature creep
ratchet (1D and 2D)	little ratchet (1D and 2D)	thermally activated ratchet (1D and 2D)
little cyclic hardening (1D)	very significant cyclic hardening (1D)	fairly important cyclic hardening with time dependence
strong supplementary hardening (2D) due to the phase difference of the stresses	progressive diminution of the supplementary 2D hardening	weak supplementary (2D) hardening due to the phase difference of the stresses with time dependence
positive sensitivity of the stress to the strain rate(monotonic and cyclic)	negative sensitivity of the stress to the strain rate (mo notonic and cyclic)	positive sensitivity of the stress to the strain rate (monotonic and cyclic)
$20 \leq T \leq 200°C$	$200 < T < 550°C$	$T \geq 550°C$

MODELIZATION

Without entering into the details of the formulation of this model, its different components are reviewed together with their associated microstructural parameters [3] [7-8]. A synoptic diagram is presented in Table III.

Different physical properties appear in the viscoplastic state equation on which depend the strain rate, namely : the critical velocities of the Lüders bands, the variation of the stacking fault energy with the temperature and the effective diffusion coefficient. This adimensional equation is identified over the domain 20-700°C.

Three tensorial variables $\underline{\alpha}$, $\underline{\alpha}^{(1)}$, $\underline{\alpha}^{(2)}$ appear in this kinematic model which are associated with the internal stresses induced by the interactions at different distances between the mobile dislocations and the substructure. There are thus three interaction distances, respectively short, medium and long. The amplitude of the progresive strain (2D ratchet) can be adjusted by acting on the position of the $\underline{\alpha}^{(2)}$ variable by the intermediary of the evasnescent terms (combination of normal [9] and radial [10] evanescences).

The scalar variables Y and Y^+ essentially depend on the accumulated plastic deformation and represent the variations of the density of the free dislocations and the substructure dislocations with the strain and temperature. Note that these variables condition the increase of the asymptotic states of the kinematic variables. The partial memorization of the pre-strains is realized by the introduction of non-hardening surfaces [11-12] and reflects the stability, in relation to the strain, of some developed substructures such as ; twinned zones, subgrain

TABLE III.   Synoptic diagram of the model

Viscoplastic state equation
$$\dot{\varepsilon}_{ij}^{v} = f\left(\dfrac{\overline{\sigma - \alpha}}{N(Y^{*})}\right)\dfrac{\sigma'_{ij} - \alpha'_{ij}}{\overline{\sigma - \alpha}}$$

Viscous Properties $N(Y^{*})=..$

Secondary kinematic variables $\alpha_{ij}^{(n)}$, allowing the description of :
- the form of the cyclic, loops
- the 2D ratchet
- the 1D ratchet

Primary kinematic variable, α_{ij} :
$$\dot{\alpha}_{ij} = g(\alpha_{ij}, \alpha_{ij}^{(1)}, \alpha_{ij}^{(2)}, \ldots, Y^{*}, Y^{+}, T)$$

Scalar variable Y^{*} describing the asymptotic state of the kinematic variables α_{ij}, $\alpha_{ij}^{(n)}$;
$Y^{*}=h(Y, Y_{e}, Y^{(i)}, Y^{pp})$
$Y^{+} = \ldots$

Cyclic and monotonic hardenings
$$\dot{Y} = k(Y, Y^{sat}, \overline{\dot{\varepsilon}}, T) \Rightarrow Y$$
$$\dot{Y}^{+} = k^{+}(Y^{+}, Y^{sat+}, \overline{\dot{\varepsilon}}, T) \Rightarrow Y^{+}$$

Memory effect of the pre-strains
$$\dot{Y}^{sat} = \ldots \Rightarrow Y^{sat}$$

Recovery effect directly due to time

Model Responses :
- monotonic loadings (tensile, creep, relaxation, time effects...)
- cyclic loadings (cyclic hardening, memorization of the prestrains...)
- strain rate effects
- bi-directional cyclic loadings
- 2D ratchet
- 1D ratchet (overestima-) ted)

Hardening effect due to the phase difference of the stresses
$$\dot{Y}_{e} = \ldots \Rightarrow Y_{e}$$

Dislocations-point defects interactions
$$\dot{Y}^{(i)} = \ldots \Rightarrow \sum_{(i)} Y^{(i)}$$

Effect of the microstructural instability due to the heterogenous precipitation
$$\dot{Y}^{pr} = \ldots \Rightarrow Y^{pr}$$

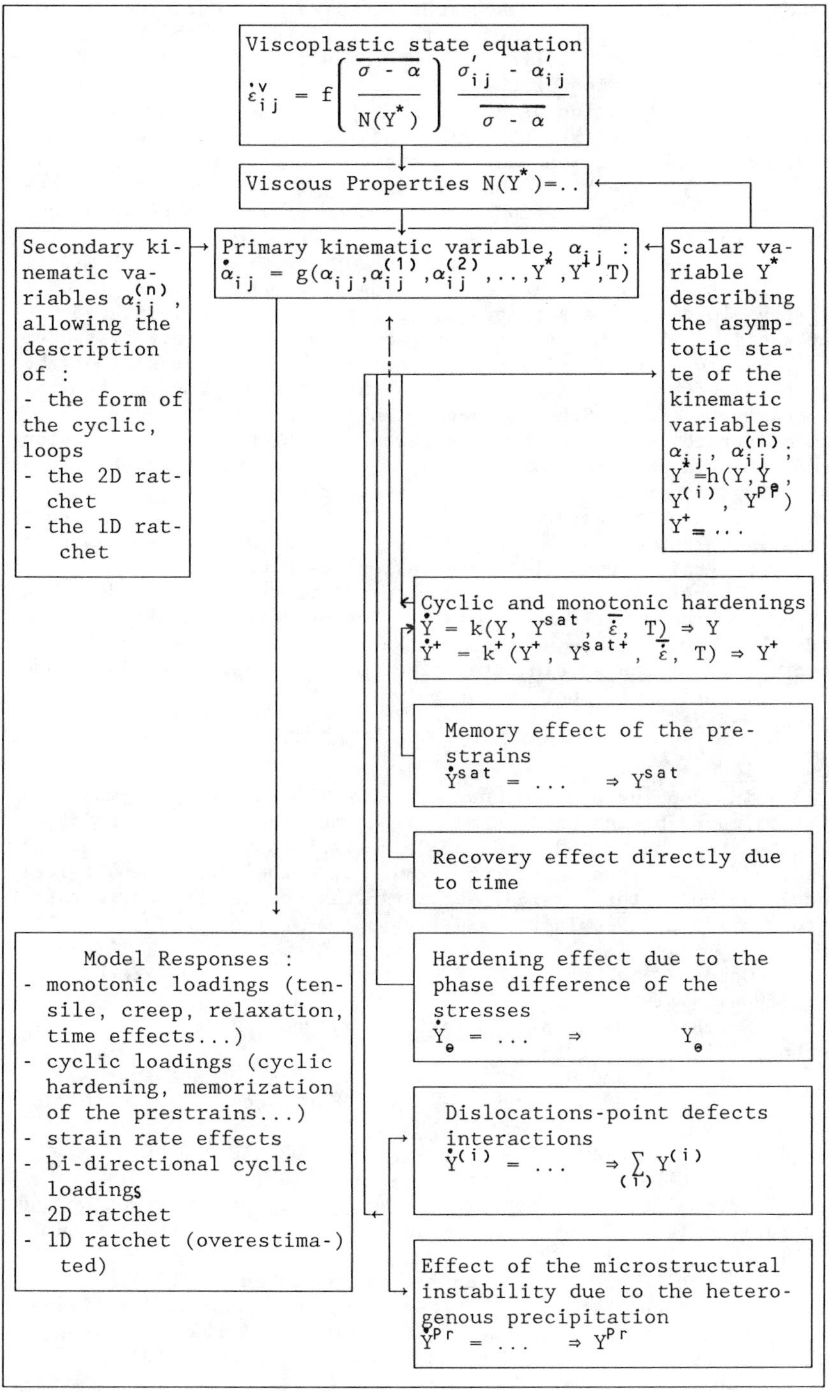

boundaries, tilt boundaries,... The loss of stability, with respect to time, of these substructures is described by the presence of the recovery term in the variables $\underline{\alpha}$, Y and Y^+.

For the 2D cyclic loadings, the supplementary hardening caused by the rotation of the stresses is certainly due on the one hand, to the activation of new secondary gliding systems favorably oriented with respect to the stress components and, on the other hand, to the interactions between the different activated systems. The scalar variable Y_θ is thus introduced, describing the increase in density of dislocations due to the activation of secondary systems [3].

In the intermediate temperature domain, there exists significant agreement in attributing the physical origin of a negative loading rate sensitivity to the dislocation-point defects interactions. For this alloy, the reorientation at short distances has been proposed [7-8] : reorientation in a FCC crystal of the point defects configurations in the stress field of the dislocations (Snoek-Schoeck mechanism). A scalar component $Y^{(i)}$ can be attributed to each interaction (i) which kinetic equation predicts, in agreement with experience, a first order time hardening, exponentially evanescent with the strain.

For higher temperature, T > 600°C, the interstitial atoms migrate to long distances to precipitate in a heterogeneous manner along the grain boundary. A formulation has been proposed.

Figure 1 to 6 compare the predictions of the model with respect to the experimental results. The agreement is satisfactory and the new notable possibilities of the model are : the description of the strain rate effect on the stabilized cycle and the description of the 2D loadings : ratchet and out of phase experiments.

CONCLUSION

In conclusion, except for the 1D ratchetting (which is overestimated by the present model) the mechanical characteristics of this steel between 20 and 700°C, summarized in table II, can be described by the incorporation of the phenomenological modelization of the physical mechanisms taking part in the strain, into a unified viscoplastic model.

REFERENCES

1. N. Ohno, Y. Takahashi and K. Kuwabara, J. Eng. Mat. Techn., 111, 106-114 (1989).
2. S. Murakami, M. Kawai, K. Aoki and Y. Ohmi, J. Eng. Mat. Techn., 111, 32-39 (1989).
3. P. Delobelle and R. Lachat, submitted to Rech. Aerosp. (1990).
4. A. Benallal and D. Marquis, J. Eng. Mat. Techn., 109, 326-336 (1987).
5. A.A. Solomon, Rev. of Sci. Inst., 40, 1025-1030 (1969).
6. P. Delobelle, J. Nucl. Mat., 166, 364-378 (1989).
7. P. Delobelle, in "Constitutive Laws of Plastic Deformation and Fracture", edited by A.S. Krausz and al., Kluwer academic publishers, 253-261 (1990).
8. P. Delobelle, Rev. Phys. Appl., to appear oct. (1990).
9. J.L. Chaboche, Int. J. of Plasticity, 2, n° 2, 149-188 (1986).
10. H. Burlet and G. Cailletaud, Eng. Compt., 3, 143-150 (1986).
11. JL. Chaboche, K. Dang Van and G. Cordier, SMIRT V, L1/13 (1979).
12. N. Ohno, J. of Appl. Mech., 49, 721-727 (1982).

A Viscoplastic Model With Application to LiF–22%CaF_2 Hypereutectic Salt

A. D. Freed
NASA–Lewis Research Center
Cleveland, Ohio

K. P. Walker
Engineering Science Software, Inc.
Smithfield, Rhode Island

July 16, 1990

Abstract

A viscoplastic model for class M (metal-like behavior) materials is presented. A novel feature of this model is its use of internal variables to change the stress exponent of creep (where $n \approx 5$) to that of 'natural' creep (where $n = 3$), in accordance with experimental observations. We apply our model to a LiF–22mol.%CaF_2 hypereutectic salt, which is being considered as a thermal energy storage material for space-based solar dynamic power systems.

1 Stress-Strain Relations

Stress σ_{ij} is assumed to be related to infinitesimal strain ϵ_{ij} through the isotropic *constitutive equations*

$$\sigma_{ii} = 3\kappa(\epsilon_{ii} - \alpha\,\Delta T\,\delta_{ii}) \qquad \text{and} \qquad S_{ij} = 2\mu(e_{ij} - \varepsilon^p_{ij})\,, \tag{1}$$

such that $\varepsilon^p_{ii} = 0$, with bulk κ and shear μ elastic moduli, and where

$$S_{ij} = \sigma_{ij} - \sigma_{kk}\delta_{ij}/3 \qquad \text{and} \qquad e_{ij} = \epsilon_{ij} - \epsilon_{kk}\delta_{ij}/3 \tag{2}$$

denote the deviatoric stress and strain. The inelastic ε^p_{ij} and thermal $\alpha\,\Delta T\,\delta_{ij}$ strains in Eqn. 1 are eigenstrains that represent deviations from deviatoric and hydrostatic elastic behaviors, respectively. The constant α is the mean coefficient of thermal expansion, while $\Delta T \equiv T - T_0$ represents a difference between the current temperature T and some reference temperature T_0. The quantity δ_{ij} is the Kronecker delta.

2 Viscoplastic Theory

The *flow equation* for inelastic strain is given by

$$\dot{\varepsilon}^p_{ij} = \frac{3}{2}\|\dot{\varepsilon}^p\|\frac{\Sigma_{ij}}{\|\Sigma\|}\,, \tag{3}$$

with the effective stress

$$\Sigma_{ij} = S_{ij} - B_{ij} \tag{4}$$

establishing the direction of inelastic strain rate, where B_{ij} is the back stress. A dot '$\dot{}$' placed over a variable denotes its time rate-of-change.

The norms (or magnitudes) of this theory are defined as

$$\|I\| = \sqrt{2/3\, I_{ij}I_{ij}} \qquad \text{and} \qquad \|J\| = \sqrt{3/2\, J_{ij}J_{ij}}\,, \tag{5}$$

where I_{ij} is any deviatoric strain-like tensor, *viz.* $\dot{\varepsilon}^p_{ij}$, and where J_{ij} is any deviatoric stress-like tensor, *viz.* S_{ij}, B_{ij} and Σ_{ij}. These norms are of the von Mises type, where the coefficients are chosen to scale the theory for tension.

The *evolution equations* characterizing the internal state of the material are given by

$$\dot{B}_{ij} = \hat{H}\left(\frac{2}{3}\dot{\varepsilon}^p_{ij} - \hat{D}\,\|\dot{\varepsilon}^p\|\,B_{ij} - \hat{R}\,\frac{B_{ij}}{\|B\|}\right) \tag{6}$$

for the back stress, and by

$$\dot{Y} = \hat{h}\left(\|\dot{\varepsilon}^p\| - \hat{d}\,\|\dot{\varepsilon}^p\|\,Y - \hat{r}\right) \tag{7}$$

for the yield strength, where $\hat{H} > 0$ and $\hat{h} > 0$ are the strain hardening parameters, $\hat{D} \geq 0$ and $\hat{d} \geq 0$ are the dynamic recovery parameters, and $\hat{R} \geq 0$ and $\hat{r} \geq 0$ are the thermal recovery parameters.

Defining functional forms for the six parameters $\hat{H}$, $\hat{h}$, $\hat{D}$, $\hat{d}$, $\hat{R}$ and $\hat{r}$, along with a kinetic equation for $\|\dot{\varepsilon}^p\|$, results in a specific viscoplastic model. Such a model[1] is presented in the next section for class M (metal-like behavior) materials.

3 Viscoplastic Model

A Zener-type *kinetic equation* for the evolution of inelastic strain is considered, *i.e.*

$$\|\dot{\varepsilon}^p\| = \vartheta\, Z\,, \tag{8}$$

where $\vartheta(T) > 0$ acts as a thermal diffusivity, and $Z(S_{ij}, B_{ij}, Y) \geq 0$ is referred to as the Zener parameter. The ZENER-HOLLOMON [1] hypothesis assumes that the kinetic equation can be represented as a product of two functions ϑ and Z; the first is dependent only on temperature, while the latter is dependent only on stress and the internal state.

[1]The derivation of this model, and a more detailed discussion of it, is presented in *NASA TM-103181* (same title and authors as this) from which this conference paper is a synopsis.

For the thermal diffusivity, we use the relationship

$$\vartheta = \begin{cases} \exp\left(\dfrac{-Q}{kT}\right) & T_t \leq T < T_m \\ \exp\left\{\dfrac{-Q}{kT_t}\left[\ln\left(\dfrac{T_t}{T}\right) + 1\right]\right\} & 0 < T \leq T_t \,, \end{cases} \tag{9}$$

which was derived by MILLER [2]. Here k is the universal gas constant ($k = 8.314$ J/mol.-K), Q is the activation energy, T_m is the absolute melting temperature, and T_t is the absolute transition temperature.

For the Zener parameter, we adopt the relationship

$$Z = A \sinh^3\left(\frac{\langle\|\Sigma\| - Y\rangle}{D}\right) , \tag{10}$$

with frequency coefficient A and drag strength D. The Macauley bracket operator $\langle\|\Sigma\| - Y\rangle$ has either a value of 0 whenever $\|\Sigma\| \leq Y$ (defining the elastic domain), or a value of $\|\Sigma\| - Y$ whenever $\|\Sigma\| \geq Y$ (defining the viscoplastic domain), with $\|\Sigma\| = Y$ establishing the yield surface.

The three material parameters associated with the evolution of back stress are taken to be

$$\hat{H} = H \,, \qquad \hat{D} = \frac{y}{bY} \qquad \text{and} \qquad \hat{R} = 0 \,, \tag{11}$$

where H and y are positive-valued material constants, and b is a function of state defined in Eqn. 13.

The three material parameters associated with the evolution of yield strength are taken to be

$$\hat{h} = h \sinh^{3-n}\left(\frac{Y}{yC}\right) , \qquad \hat{d} = 0 \qquad \text{and} \qquad \hat{r} = A\,\vartheta \sinh^n\left(\frac{Y}{yC}\right) , \tag{12}$$

where C, h and n are positive-valued material constants.

The maximum fraction of applied stress that can be associated with the back stress is given by the function

$$b = 1 - y - \frac{yD}{Y} \sinh^{-1}\left[\sinh^{n/3}\left(\frac{Y}{yC}\right)\right] , \tag{13}$$

where the material constant y represents the maximum fraction of applied stress that can be associated with the yield strength.

The material constants for LiF–22%CaF_2 are given in Table 1.

4 Discussion

An important consequence of the viscoplastic model just presented is that the Zener parameter defined in Eqn. 10 reduces, in closed form, to the Zener parameter associated with GAROFALO'S [3] creep equation under steady state conditions, *i.e.*

$$Z_{ss} = A \sinh^n\left(\frac{\|S\|}{C}\right) , \tag{14}$$

Constant	Units	Value
α	10^6 K^{-1}	$18 + 0.03T$
μ	MPa	$52,000 - 29T$
ν	–	$0.65 - 0.001T$
A	s^{-1}	5×10^{15}
C	MPa	27
D	MPa	8
h	MPa	50,000
H	MPa	μ
n	–	5
Q	J/mol.	320,000
T_m	K	~ 1042
T_t	K	850
y	–	0.1
Y_0	MPa	0

Table 1: Constants for LiF–22%CaF_2.

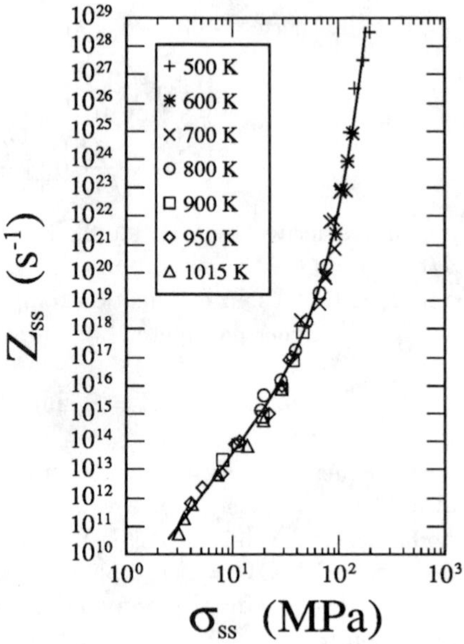

Figure 1: Steady-state kinetic response of LiF–22%CaF_2. Data are from RAJ & WHITTENBERGER [4].

where the subscript 'ss' denotes steady state, and C is the power-law breakdown stress. The capability of Zener's kinetic equation (8) to correlate experimental creep data using the relationships for ϑ and Z_{ss} given in Eqns. 9 and 14 is demonstrated in Fig. 1.

There is no 'explicit' representation for the thermal recovery of back stress in this model, but note that the surface which bounds the state of back stress — characterized by $\hat{D}$ — is a function of the yield strength *via* Eqn. 11. Consequently, since the yield strength thermally recovers, thermal recovery of the back stress is 'implicit'. Unfortunately, inelastic strain must evolve in order for this implicit recovery to take place. Even though this is not a physically accurate representation of thermal recovery, it is a very useful approximation to it. It is useful because it allows the effect of thermal recovery to be modeled, while still enabling the viscoplastic model to reduce to a steady-state creep model (in particular, Garofalo's) in closed form. This allows one to use creep data directly for the purpose of characterizing material constants — a huge benefit.

In the domain of power-law creep, the hardening and thermal recovery parameters, $\hat{h}$ and $\hat{r}$, given in Eqn. 12 are similar in form and function to the hardening/recovery relationships proposed and experimentally verified by MITRA & MCLEAN [5]. Here the stress dependence of recovery, as defined by the product $\hat{h}\,\hat{r}$, has an exponent of 3. This is in accordance with the 'natural' theory

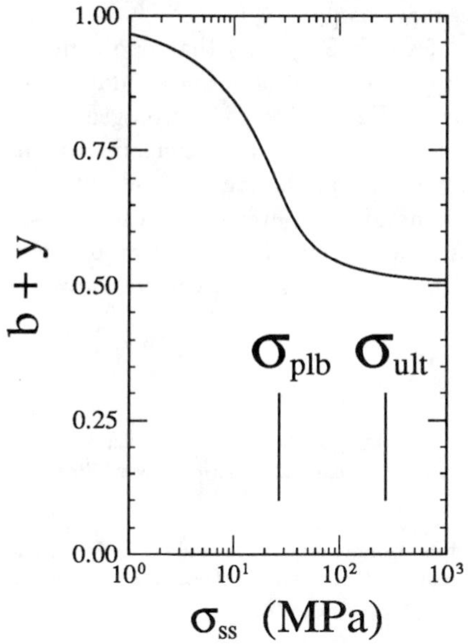

Figure 2: Response of Eqn. 13, where values for the power-law breakdown stress $\sigma_{\rm plb} \equiv C$, the ultimate strength $\sigma_{\rm ult}$, the drag strength D and the creep exponent n pertain to LiF–22%CaF_2.

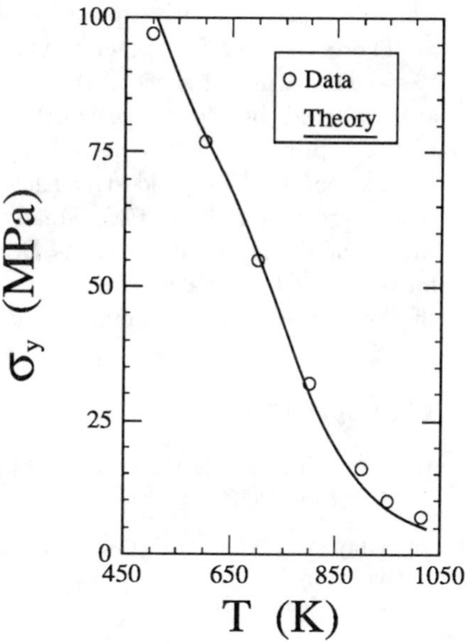

Figure 3: Yield strength (0.2% offset) of LiF–22%CaF_2. $\dot{\epsilon} = 2 \times 10^{-4}\text{s}^{-1}$. Data are from RAJ & WHITTENBERGER [4].

of creep for thermally-assisted dislocation climb [6]. Our particular hardening and thermal recovery parameters extrapolate the Mitra-McLean results into the domain of exponential behavior.

The expression for b given in Eqn. 13 is a direct consequence of the following hypothesis: it is the internal state of stress that accounts for the difference between 'natural' (with a stress exponent of $n = 3$, *cf.* Eqn. 10) and observed (where $n \approx 5$, *cf.* Eqn. 14) power-law creep behavior in class M materials. This internal stress arises from the heterogeneous, honeycomb-like, dislocation microstructure, which is composed of hard cell walls of high dislocation density that surround soft cell interiors of relatively low dislocation density [7]. A graphic representation of Eqn. 13 is presented in Fig. 2. In the domain of power-law creep, this equation reduces to a relationship that ČADEK [8] proposed and experimentally verified, but where b and y were not distinguished one from the other. In the domain of exponential creep, it implies that the fraction of internal stress, $(b + y)\|S\|$, to applied stress, $\|S\|$, at steady state is nearly a constant — note that $Y_{ss} = y\|S\|$ and $\|B\|_{ss} = b\|S\|$ by assumption. The ratio of D/C establishes this plateau. A value of $D/C \approx 1/3$ leads to an internal stress plateau that is between 40 and 50% of the applied stress at steady state, which is consistent with experimental observations [9].

There are very few experimental data for LiF–22%CaF_2 that one can use to validate a model such as the one given above. The only known data that are available, besides those which appear in Fig. 1, are yield strength data. Figure 3 presents the predictive capability of our model, as it pertains to the yield strength. The yield strength σ_y at 500 K, and the fact that about 15% strain was required to attain steady-state in this test, were facts used to set values for the material constants D and h, respectively. The model predictions for all other temperatures are just that, predictions. The agreement between theory and experiment is quite good.

References

[1] C Zener and J H Hollomon, "Effect of Strain Rate Upon Plastic Flow of Steel," *J Appl Phys*, **15**, 22–32 (1944).

[2] A Miller, "An Inelastic Constitutive Model for Monotonic, Cyclic, and Creep Deformation: Part I—Equations Development and Analytical Procedures," *J Eng Mater Technol*, **98**, 97–105 (1976).

[3] F Garofalo, "An Empirical Relation Defining the Stress Dependence of Minimum Creep Rate in Metals," *Trans Metall Soc AIME*, **227**, 351–356 (1963).

[4] S V Raj and J D Whittenberger, "Deformation of As-cast LiF–22mol%CaF_2 Hypereutectic Salt Between 500 and 1015 K," *Mater Sci Eng*, **A124**, 113–123 (1990).

[5] S K Mitra and D McLean, "Work Hardening and Recovery in Creep," *Proc Royal Soc, London*, **A295**, 288–299 (1966).

[6] J Weertman, "High Temperature Creep Produced by Dislocation Motion," *Rate Processes in Plastic Deformation of Materials*, 315–336, eds J C M Li and A K Mukherjee (ASM, Cleveland 1975).

[7] W D Nix and B Ilschner, "Mechanisms Controlling Creep of Single Phase Metals and Alloys," *Strength of Metals and Alloys*, **3**, 1503–1530, eds P Haasen, V Gerold and G Kostorz (Pergamon Press, New York 1980).

[8] J Čadek, "The Back Stress Concept in Power Law Creep of Metals: A Review," *Mater Sci Eng*, **94**, 79–92 (1987).

[9] T C Lowe and A K Miller, "The Nature of Directional Strain Softening," *Scr Metall*, **17**, 1177–1182 (1983).

Spring-Dashpot Models for Viscoelastic Material Damping

S. N. Ganeriwala & H. A. Hartung
Philip Morris Research Center
P. O. Box 26583
Richmond, VA 23261-6583

ABSTRACT

Although complex modulus provides an adequate representation of dynamic mechanical properties of viscoelastic materials it is not suitable for vibration analyses. Spring-dashpot models provide alternate representation of mechanical properties and are good for vibration analyses. Some interesting features were observed in a mathematical analysis of the standard three parameter model for solids (a spring in parallel with a Maxwell unit). With computer simulation it was possible to represent a wide range of complex moduli data by a three parameter model and simple sequential updating of Maxwell unit constants. An initial test of the new formalism was carried out with NBR and polyisobutylene. This model is internally consistent and can be used to obtain transient as well as non-harmonic input responses. It is adaptable to most computer systems.

INTRODUCTION

Polymers are useful for isolation and control of vibrations because of their ability to dissipate energy. Dynamic mechanical properties are generally characterized by complex moduli specified over a range of temperature and frequency. This information is often used to select and even to develop new materials for damping of sound and vibrations. Frequency dependent damping characteristics are not suitable in many engineering analyses such as transient response, inharmonic oscillations, and non-stationary random vibrations [1]. Curve fitting techniques do not work well for medium or heavily damped systems [2]. Fractional derivative representation obscures the physical realities of problems [3].

Viscoelastic properties can also be characterized by series-parallel combinations of springs and dashpots [4,5]. Since a large number of terms are required to represent behavior, such models haven't looked promising. A reanalysis, however, showed that a simple three parameter model, a spring in parallel with a Maxwell unit, can represent the entire range of frequency by sequentially updating the parameters.

SPRING-DASHPOT MODEL ANALYSIS

Expressions for the storage and loss moduli for a three parameter model, also known as Kelvin model are [4-6]:

$$G'_1(\omega) = E_0 + \frac{E_1 \mu_1 \omega^2}{E_1^2 + \mu_1^2 \omega^2} \; ; \qquad G''_1 = \frac{E_1 \mu_1^2 \omega^2}{E_1^2 + \mu_1^2 \omega^2} \tag{1}$$

where G'_1 and G''_1 are storage and loss moduli and other parameters are spring and dashpot constants as shown in Figure 1. Qualitatively, this simple model represents the important features of viscoelastic materials. At high frequencies G''_1 goes to zero while G'_1 extrapolates to $E_0 + E_1$. As the frequency approaches zero G''_1 again vanishes and G'_1 approaches to the spring constant of series spring. A mathematical analysis of the first and second derivatives of the storage modulus showed that it attains constant values at high and low frequencies asymptotically and that it is symmetrical about $\omega = \mu_1 / E_1$ [6]. Also, the limitations of this representation is that it only describes frequency dependent behavior over about a one decade of frequency. This is illustrated in Figure 1. Additional Maxwell units are needed to extend the frequency range.

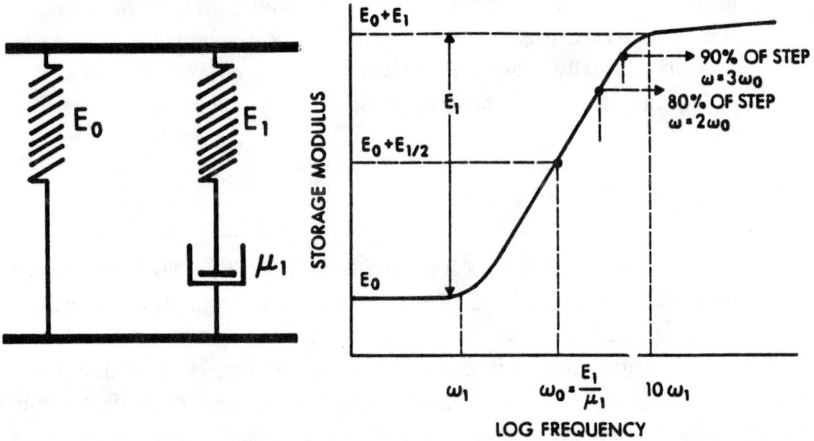

Figure 1. Three parameter model and its analysis.

A close look at the model reveals an interesting pattern. The Kelvin model represents frequency dependence roughly over a one decade. At the lower end, the Maxwell unit is inoperative because dashpot is so weak that it respond without affecting the spring. At the high frequency the dashpot acts like a rigid body and the response is the sum of two springs. This shows that it is possible to select an independent Maxwell unit with its center frequency a decade away from the center frequency of the previous model. The combination of two Maxwell units will have viscoelastic response for two decades which otherwise will not be possible. To cover even further range of frequency, more Maxwell units can be added. It is interesting to note that at a given frequency only one Maxwell unit is effective and the model appears like a standard three parameter model. The general expression for storage and loss moduli are:

$$G' = G'_1 + G'_2 + G'_3 + \text{---} \; ; \; G'' = G''_1 + G''_2 + G''_3 + \text{---} \tag{2}$$

The general forms of the added Maxwell units are:

$$G'_i(\omega) = \frac{E_i \mu_i \omega^2}{E_i^2 + \mu_i^2 \omega^2} \; ; \qquad G''_i = \frac{E_i \mu_i^2 \omega^2}{E_i^2 + \mu_i^2 \omega^2} \tag{3}$$

where i may be 2, 3 or higher. Figure 2 shows model with two Maxwell units and its simulation.

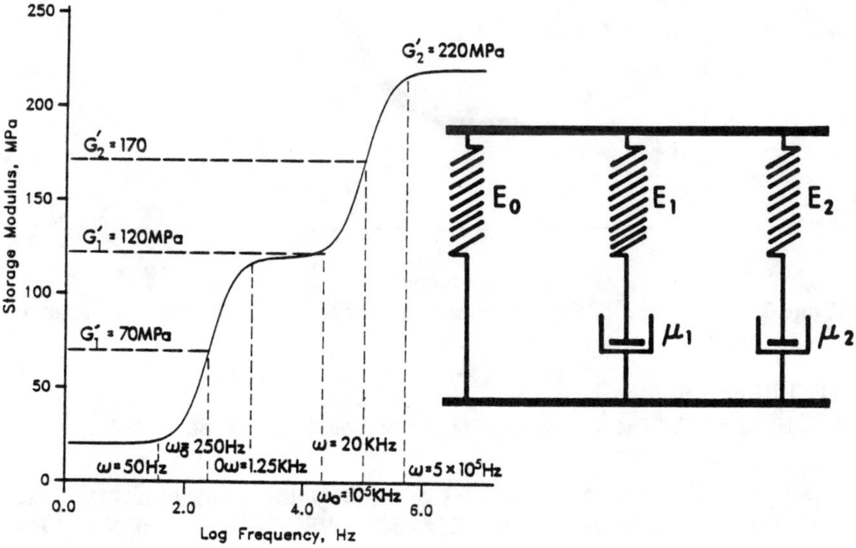

Figure 2. Three parameter model with two Maxwell units and its simulation.

MODEL SIMULATION AND RESULTS

A master curve for the storage modulus of NBR at 20°C. was obtained using our constitutive modeling approach [7]. Three parameter models were generated with one, two, and three Maxwell units. Figures 1 & 2 show simulations with one and two terms. Using nonlinear least square technique G' data were then modeled. Three Maxwell units were needed to model the whole region of viscoelastic response of NBR. Figure 3 shows master curve and model line. To check the internal consistency of the model, G" was predicted using same values of E_i & μ_i's. Polyisobutylene was also studied the same way. Comparisons were good in all cases.

The procedure described above provides a simple spring dashpot model to represent the entire range of viscoelastic response. This can be used for structural analyses. The resulting equation of motion is third order and can be solved. Note that response of a viscoelastic system often depends upon how the motion was started. Kinematically this is equivalent to specifying the acceleration. This can be easily done with our model. Mechanical system response to a non-standard excitation can also be obtained. The model is can be adapted on any computer system.

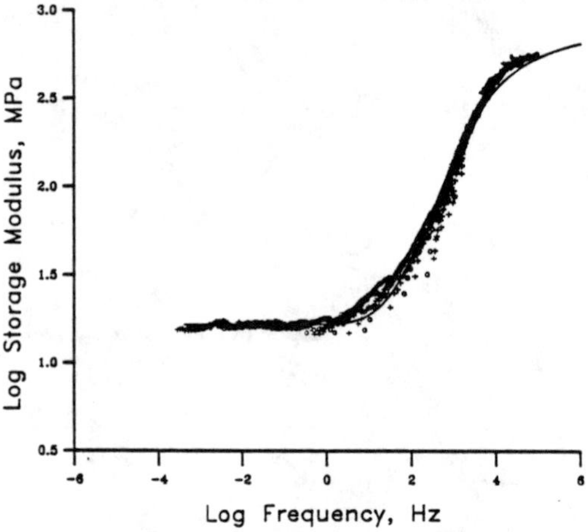

Figure 3. Comparison of three Maxwell units three parameter model with master curve of NBR.

REFERENCES

1. S. H. Crandall, "The roll of damping in vibration theory," J. Sound Vib., 11, (1), 1970, pp 3-18.
2. L. Rogers, V. J. Johnson, and L. G. Kelly, Editors, Proceeding of Damping' 89, Wright-Patterson Air Force Base, Ohio 45433, WRDC-TR-89-3116, Nov. 1989, Sections-CB & CC.
3. See Reference 2, Section-DA.
4. R. M. Christensen, Theory of Viscoelasticity an Introduction, 2nd. ed., Academic Press Inc.: New York; 1982; Ch. 1.
5. W. Flugge, Viscoelasticity, Blaisdell, Waltham, Mass. (1967).
6. S. N. Ganeriwala and H. A. Hartung, J. Sound Vib., a paper in progress.
7. S. N. Ganeriwala and H. A. Hartung, in Sound and Vibration Damping with Polymers, Edited by R. D. Corsaro and L. H. Sperling, ACS Symposium Series 424, American Chemical Society, Washington D. C., 1990; pp.92-110.

ON ONE DYNAMIC MODEL FOR COMPOSITE MATERIALS

Sergej HAZANOV

Ph. D., Senior Scientist in the Swiss Federal Institute of Technology, Department of Materials, Laboratory for Building Materials,
Ecublens MX-G
CH - 1015 Lausanne / Switzerland

1. One of the main modern tendencies in modelling stress-strain behaviour of heterogeneous engineering materials is the utilization of "equivalent" geometrically-homogeneous models, but of more complicated physical nature. For instance, wave propagation in laminated elastic composites can be qualitatively described by classic viscoelastic Maxwell model [1], kinetics of wood - by Boltzmann equation [2], etc..

But what concerns quantitative description, practical use of classic rheologic (viscoelastic) models arises a number of complications (impossibility of predicting geometric dispersion (Fig 1) and oscillating wave shapes (Fig. 2) in composites, multitransition effect in wood [2], etc.).

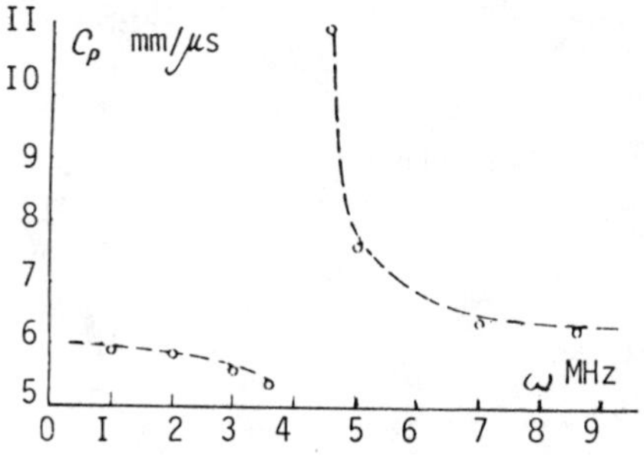

Fig.I. O - experiment;
---- theory.

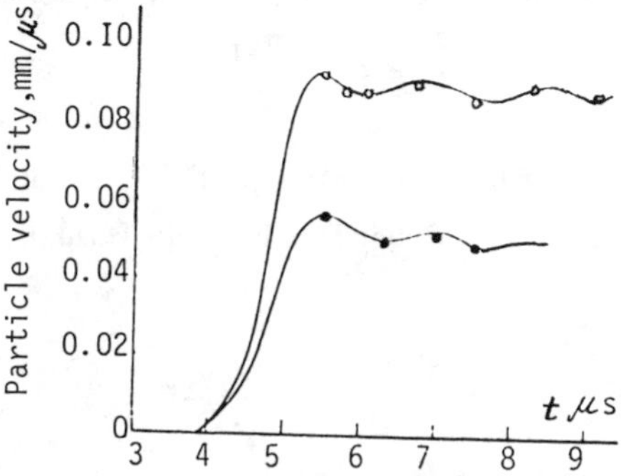

Fig.2. —— experiment;
●,O - theory.

2. Indeed, harmonic vibrations

$$U = U_o \,.\, \exp\,(k_2 x) \,.\, \exp i\,(\omega t - k_1\, x) \qquad (1)$$

of viscoelastic Volterra body :

$$E\,\varepsilon = \sigma + \int_{-\infty}^{t} K(t-\tau)\,\sigma(\tau)\,d\tau \tag{2}$$

are described by

$$C_p = \frac{\omega}{k_1} = \frac{c\sqrt{2}}{\sqrt{E\left(J_1 + \sqrt{J_1^2 + J_2^2}\right)}} \tag{3}$$

where C_p is phase velocity, $C = \sqrt{\frac{E}{g}}$ - elastic velocity, $J_1 = \frac{1}{E}(1 + K^c)$, $J_2 = \frac{K^s}{E}$, K^c and K^s - cos- F and sin-F transformations of creep kernel K.

(3) clearly shows that variations of $C_p(\omega)$ of the type shown in Fig. 1 are possible only for K(t) "non traditional". For instance, taking $K(t) = e^{-\alpha t}$, we can conclude that factor α must be a complex number.

For non stationary wave propagation in viscoelastic body (2) with Heaviside boundary condition, we have in the Laplace space of images :

$$\bar{\sigma} = \frac{\sigma_o}{p}\, e^{-\frac{px}{c}\left(\sqrt{1+\bar{K}} - 1\right)} \tag{4}$$

From where it follows that wave profile $(\sigma|_{x = const})$ can be oscillating as in Fig. 2 only for oscillatiny K(t). Indeed for small $(t - \frac{x}{c})$ we have :

$$\bar{\sigma} \cong \frac{\sigma_o}{p}\, e^{-\frac{px}{2c}\bar{K}} \tag{5}$$

and for $K(t) = e^{-\alpha t}$ one has :

$$\bar{\sigma} \cong \frac{\sigma_o}{p}\, e^{-\frac{px}{2c}}\, e^{-\frac{\alpha x}{2cp}} \tag{6}$$

with the same conclusion that α must be a complex one.

According to [2] thermo-mechanical response of wood involves several transitions in the complex modulus plane, that can also be easily treated in terms of one hereditary model with non-traditional kernel.

3. Trying to explain all these paradoxes and difficulties in term of classic theory, various autors take usually large sets of traditional models, complicating thus greatly the calculation process.

An alternative is, as we have seen, in widening the set of admissible creep (relaxation) kernels, which usually are chosen from the class of positive-definite, monotone decreasing functions (exponential, Abel (parabolic) type, etc.).

But according to Breuer-Onat theorem [3], the only necessary and sufficient condition for them (following from positivity of work during deformation) is the positivity of sin-F transformation of relaxation kernel (resolvent of K), or (equally) positivity of cos-F transformation of relaxation function when constitutive law is taken in the Stielties convolution form.

Surely the class of functions for which $K^S > 0$ is much more wider than traditional one, including for instance non monotone $\frac{t}{a^2+t^2}$, $t^{-\frac{1}{2}}e^{-at}$ and even non-positive $\frac{\sin at}{t^2}$, $e^{-a\sqrt{x}}\sin(a\sqrt{x})$, $J_0(ax)$, . .

The "reality" of such"unusual" kernels can be illustrated by inserting inertial terms (masses) into classic viscoelasting models with springs and dashpots - for several values of parameters the form of the kernels in the constitutive laws can be just as mentionned.

4. The validity of the proposed approach was tested on experimental processing. For this purpose a kernel of the form :

$$K(t) = A\,e^{-\alpha t} \,.\, \sin \beta t \qquad (7)$$

was chosen. Surely, it satisfies the thermodynamic restrictions :

$$K^S(\omega) = \frac{2\,\alpha\,\beta\,\omega}{\left((\alpha-\beta)^2+\omega^2\right)\left((\alpha+\beta)^2+\omega^2\right)} > 0 \qquad (8)$$

For geometric dispersion of harmonic vibrations in laminated composite AL - W [4] (Fig 1) the kernel constants are

$A = 1,835 \qquad \alpha = 0,5 \qquad \beta = 4,17 \,.\, 10^6\ s^{-1}$

Here A has the meaning of amplitude, α – attenuation factor, β – resonance frequency ($\frac{c}{\beta}$- the size of composite lattice).

For wave profiles in shock experiments on epoxide-steel composite [5] (Fig. 2), the processing results are :

$$A = 8,486; \quad \beta = 4,69 \cdot 10^6\ s^{-1}; \quad \alpha = 2,127$$

Coinsidence between theoretical and experimental points proves the effectiveness of the model proposed.

5. Conclusion. Even the simpliest linear viscoelastic model with "widened" class of admissible kernels gives a lot of possibilities for describing several specific heterogeneous effects. Such model can be taken as a basic one for universal dynamic model of solids, including as particular cases homogeneous elasticity (zero kernel), viscoelasicity (classic kernels), and heterogeneous bodies (kernels of oscillating and non-positive type).

Relation between geometry and physics established by this approach (relation between geometric and physical dispersions) resembles the classic ergodic principle in theoretical physics.

Surely, utilisation of introduced here "new" kernels for equation (2) must be interpreted in purely statistic sense - like in hydromechanics, while treating large turbulent flow as a laminar one but with "negative" (statistically) viscosity.

References

[1] Barker, L.M. *J. Comp. Mater.*, 1971, no 4, p. 140-164

[2] Huet, C. A*nnales de ITBTP*, 1988, no 469, p. 35-53

[3] Rabotnov, Y. N. *Elements of hereditary mechanics of solids*, M, 1977, Nauka, p 106.

[4] Sunderland, H.J., Lingle, R. *J. Comp. Mat.*, 1972, no 10, p.490-504

[5] Lundergan, C.D., Drumheller D.S., *J. Appl. Phys.*, 1971, vol 42, no 2, p. 669-675.

The Elastoplastic-Viscoplastic Bounding Surface Model for Cohesive Soils : Recent Developments

VICTOR N. KALIAKIN
Assistant Professor, Department of Civil Engineering, University of Delaware, Newark, DE 19716

YANNIS F. DAFALIAS
Professor of Engineering Science, University of California, Davis, CA 95616

ABSTRACT

A generalized three-dimensional constitutive model for isotropic cohesive soils, based on the concept of the bounding surface in stress space and developed within the framework of coupled elastoplasticity-viscoplasticity and critical state soil mechanics is presented. The consideration of viscoplastic effects sets the current model apart from previous bounding surface formulations for soils and introduces rate and time effects.

INTRODUCTION

The notion of a generalized constitutive model based on the concept of the bounding surface in stress space, and developed within the framework of coupled elastoplasticity-viscoplasticity and critical state soil mechanics was introduced by Dafalias [1]. The prominent feature of the bounding surface concept is the fact that inelastic deformations can occur for stress points located within the surface. The coupling between plasticity and viscoplasticity for stress states within the surface differentiates the present work from the classical formulation of pure viscoplasticity (no coupling), or from formulations involving plasticity and viscoplasticity with a yield surface (coupling only for states on the yield surface). Furthermore, unlike some other time dependent formulations [2,3], this model is not restricted to normally consolidated cohesive soils. The model thus represents a novel approach for simulating the time related behavior of soils. Further details regarding the bounding surface concept and its application to rate independent plasticity for cohesive soils are given in [4] and [5], respectively. A microscopic basis for the present elastoplastic–viscoplastic formulation is presented in [1,6].

FORMULATION FOR ISOTROPIC COHESIVE SOILS

Although the present bounding surface formulation is completely general in nature [7], only a brief overview of the formulation – specialized for isotropic cohesive soils – is given herein. In the subsequent development tensors are presented in indicial form following the summation convention over repeated indices.

Compressive stresses and strains are positive, and only infinitesimal deformations and rotations are considered. The material is defined in terms of the effective stress tensor σ_{ij} and a single internal variable which accounts for the nonconservative nature of soil by keeping track of the past loading history. The dependence of the bounding surface on σ_{ij} is expressed in terms of three stress invariants; namely the first effective stress invariant I, the square root of the second deviatoric stress invariant, J, and the "Lode" angle α ($-\pi/6 \leq \alpha \leq \pi/6$). A section of the surface (for a given value of α) is shown in Fig. 1.

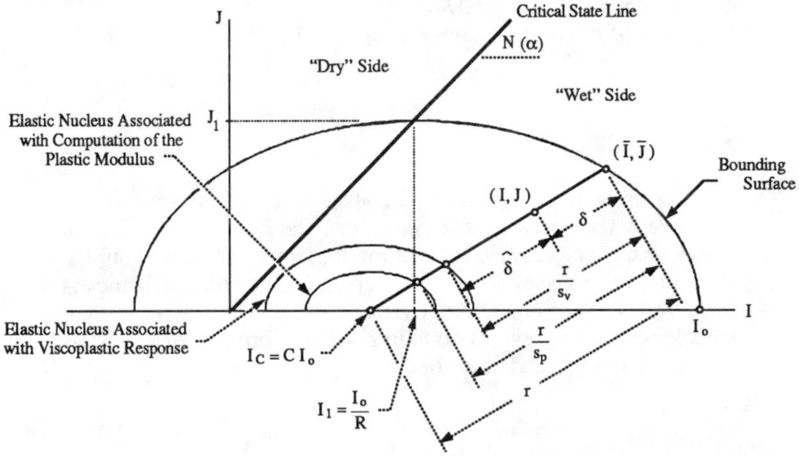

FIG. 1. **Schematic Illustration of the Radial Mapping Rule and of the Bounding Surface in Stress Invariants Space**

The actual stress point (I, J) is related to its "image" value $(\bar{I}, \bar{J})$ on the bounding surface itself through a "radial mapping" rule [5]. In Fig. 1 the quantity C represents a model parameter ($0 \leq C < 1$) and I_o represents the intersection of the bounding surface with the positive I-axis. Using CI_o as the projection center, the image stress is obtained by the radial projection of the actual stress onto the bounding surface (hence the name "radial mapping"). The bounding surface is assumed to undergo isotropic hardening. The hardening is controlled by a single scalar internal variable – the inelastic rate of the total void ratio – which measures the inelastic change in volumetric strain.

The strain rate is additively decomposed into an elastic and inelastic part, the latter consisting of a delayed (viscoplastic) and an instantaneous (plastic) part. Denoting the infinitesimal strain tensor by ε_{ij}, it is assumed that $\dot{\varepsilon}_{ij} = \dot{\varepsilon}^e_{ij} + \dot{\varepsilon}^v_{ij} + \dot{\varepsilon}^p_{ij}$ where the superscripts e, v, and p denote the elastic, viscoplastic and plastic components, respectively, of ε_{ij}. A superposed dot indicates a material time derivative or rate. The response associated with the elastic strain rate is defined in the usual manner. The rate equation for $\dot{\varepsilon}^v_{ij}$ is based upon a generalization of the viscoplastic theory of Perzyna [8]. According to this theory, the viscoplastic strain rate is a function of the "distance" in stress space of the current stress from the stress on the boundary of a quasi-static elastic domain. In the present development, an "elastic nucleus" – a region of purely elastic response around the projection center CI_o (Fig. 1) – constitutes this elastic domain. The "distance" of the actual stress point from the elastic nucleus is represented by the "normalized overstress" $\Delta\hat{\sigma}$:

$$\Delta\hat{\sigma} = \frac{\hat{\delta}}{r - \frac{r}{s_v}} \tag{1}$$

where s_v (> 1) denotes a model parameter which defines the size of the elastic nucleus, and $\hat{\delta}$ and r represent Euclidean distances (Fig. 1). The exact role of $\Delta\hat{\sigma}$ in the rate equations is reflected in a proper continuous scalar function ϕ of the overstress in such a way that $\phi > 0$ when $\Delta\hat{\sigma} > 0$ and $\phi \leq 0$ when $\Delta\hat{\sigma} \leq 0$. On the basis of the above discussion and assuming an associated flow rule, it follows that

$$\dot{\varepsilon}_{ij}^{v} = \langle \phi \rangle \frac{\partial F}{\partial \bar{\sigma}_{ij}} \tag{2}$$

where the symbols $\langle \ \rangle$ denote the Macaulay brackets. An explicit expression for $\partial F/\partial\bar{\sigma}_{ij}$ is obtained using explicit equations for the stress invariants and for the bounding surface F = 0. The plastic response is given by an expression similar to Eq. (2) only with $\dot{\varepsilon}_{ij}^{p}$ replacing $\dot{\varepsilon}_{ij}^{v}$ and ϕ replaced by the scalar loading index L; loading, neutral loading, and unloading occur when $L > 0$, $L = 0$, and $L < 0$, respectively. Within the framework of the radial mapping model, L is defined by [7]:

$$L = \frac{1}{K_p}\left\{ F_{,\bar{I}}\dot{I} + F_{,\bar{J}}\dot{J} + \frac{1}{b}F_{,\alpha}\dot{\alpha} - \langle \phi \rangle \bar{K}_p \left[\frac{1}{b} - C\left(1 - \frac{1}{b}\right)\frac{F_{,\bar{I}}}{F_{,I_o}} \right] \right\} \tag{3}$$

where a comma indicates partial differentiation with respect to the index which follows. Since it accounts for the coupling of plastic-viscoplastic hardening for states on and within the bounding surface, Eq. (3) represents a key step in the development of the present model [7]. The quantities K_p and $\bar{K}_p$ represent the scalar plastic moduli associated with the actual and "image" (i.e., for b = 1) stress states respectively, and allow prediction of loading not only in conjunction with stable (hardening) response, but also for unstable (softening) response as well [4]. An explicit expression for $\bar{K}_p$ is obtained from the consistency condition $\dot{F} = 0$. The feature which distinguishes this bounding surface formulation from classical elastoviscoplasticity formulations is that K_p is obtained not from a consistency condition, but from the following relation which depends upon $\bar{K}_p$ and upon the distances δ and r (Fig. 1):

$$K_p = \bar{K}_p + H\frac{\delta}{\langle r - s_p\delta \rangle} \tag{4}$$

The quantity s_p represents a second elastic nucleus parameter (Fig. 1), and H represents a hardening function which defines the shape of stress-strain curves for points within the bounding surface. Further details regarding H are given in [10].

Explicit expressions for the constitutive relations, obtained in a straightforward manner, are given in [9].

IDENTIFICATION OF MODEL PARAMETERS

Associated with the most general form of the model are fifteen parameters, the values of which fall within fairly narrow ranges and which are determined using a well-defined calibration procedure [7,10]. Twelve of the parameters are determined by matching the results of standard laboratory experiments of short enough duration to ensure negligible viscoplastic effects. This group includes: the elastic shear modulus G (or, alternately, Poisson's ratio ν); the critical state parameters λ, κ, M_c and M_e (the parameters M_c and M_e are associated with triaxial compression and extension, respectively; M is related to N in Fig. 1 through $M = 3\sqrt{3}\ N$); the surface configuration parameters R, C and s_p (Fig. 1); and the hardening parameters h_c, h_e, a, and w which enter the explicit expression for H [10]. The remaining three parameters characterize the viscoplastic effects. These parameters include the elastic nucleus parameter s_v which enters the expression for $\Delta\widehat{\sigma}$ through Eq. (1), and the parameters V and n which enter an explicit expression for $\phi = \phi(\Delta\widehat{\sigma})$ [7].

REFERENCES

1. Y. F. Dafalias, "Bounding Surface Elastoplasticity - Viscoplasticity for Particulate Cohesive Media", in Deformation and Failure of Granular Materials, P. A. Vermeer and H. J. Luger, editors, Balkema, pub., Rotterdam, 1982, 97-107.
2. T. Adachi, and F. Oka, "Constitutive Equations for Normally Consolidated Clay Based on Elasto-Viscoplasticity", Soils and Foundations 22 (4), 1982, 57-70.
3. R. Nova, "A Viscoplastic Constitutive Model for Normally Consolidated Clay", Deformation and Failure of Granular Materials, P. A. Vermeer and H. J. Luger, editors, Balkema, pub., Rotterdam, 1982, 287-295.
4. Y. F. Dafalias, "Bounding Surface Plasticity. I Mathematical Foundation and Hypoplasticity", J. Eng. Mech., ASCE 112 (9), 1986, 966-987.
5. Y. F. Dafalias and L. R. Herrmann, "Bounding Surface Plasticity II: Application to Isotropic Cohesive Soils," J. Eng. Mech., ASCE 112 (12), 1986, 1263-1291.
6. V. N. Kaliakin, "Bounding Surface Elastoplasticity-Viscoplasticity for Clays", Ph. D. dissertation presented to the University of California, Davis, December 1985.
7. V. N. Kaliakin, and Y. F. Dafalias, "Theoretical Aspects of the Elastoplastic-Viscoplastic Bounding Surface Model for Cohesive Soils," Soils and Foundations, 30 (3).
8. P. Perzyna, "Fundamental Problems in Viscoplasticity", Adv. in Ap. Mech. 9, 1966, 243-377.
9. V. N. Kaliakin, and Y. F. Dafalias, "Verification of the Elastoplastic-Viscoplastic Bounding Surface Model for Cohesive Soils," Soils and Foundations, 30 (3).
10. V. N. Kaliakin, and Y. F. Dafalias, "Simplifications to the Bounding Surface Model for Cohesive Soils," Int. J. Num. Anal. Meth. Geom., 13 (1), 1989, 91-100.

A DESCRIPTION OF CREEP ANISOTROPY CAUSED BY PLASTIC DEFORMATION

M. KAWAI
Institute of Engineering Mechanics, University of Tsukuba, Tsukuba 305, Japan

ABSTRACT

A model to describe an anisotropic creep behavior is discussed based on a mapped stress defined by a linear transformation of rank four. The stress mapped by the anisotropic fourth rank tensor proposed by Baltov-Sawczuk is first examined, and then a modified tensor of fourth rank is proposed so as to incorporate both aspects of isotropic and anisotropic hadenings. Finally, demonstrated is a predictability of the Bailey-Norton creep equation generalized by the structural variable defined by the proposed transformation.

INTRODUCTION

An *anisotropy* in strain-hardening develops due to inelastic straining which has a salient influence on subsequent inelastic flow behaviors. The non-collinearity between stress and inelastic strain increment tensors at sudden changes in the principal stress directions may be attributed to a microscopic anisotropy induced by the prior deformation. For the non-proportional cyclic strain hardening, which becomes much larger than the proportional one, the induced anisotropy is intrinsically responsible. Therefore, an appropriate description of anisotropic *state* and its *evolution* becomes essential for a precise formulation of multiaxial non-proportional material behaviors.

In the present study, a description of inelastic strain induced anisotropy is discussed. The formulation is based on an application of the *effective stress* concept. A structural variable tensor is defined in terms of a mapped Cauchy stress by a fourth rank linear transformation [1]. The mapped stress defined by the anisotropic tensor of Baltov-Sawczuk [2] is first examined. Taking it into consideration, a new fourth rank tensor is proposed; it derives a structural variable which accounts for both isotropic and anisotropic hardenings. Finally, a form of creep constitutive equation is formulated and the applicability is discussed based on the experimental results [3].

AN ANISOTROPIC CONSTITUTIVE EQUATION

Rate Equation

Consider an inelastic strain rate $\mathbf{d}$ described in terms of an isotropic function $\mathbf{f}$:

$$\mathbf{d} = \mathbf{f}(\mathbf{s}, \boldsymbol{\Phi}) \tag{1}$$

where $\mathbf{s}$ is the deviatoric Cauchy stress tensor, and $\boldsymbol{\Phi}$ is an internal variable tensor of fourth rank which is characterizing anisotropic properties of materials considered.

Applying the polynomial representation theorem and selecting only linear terms, a reduced expression

$$\mathbf{d} = a_1[\mathrm{tr}(\mathbf{s}_{eff})^2]\mathbf{s}_{eff}, \quad \mathbf{s}_{eff} = \mathbf{s} + k\boldsymbol{\xi}, \quad \boldsymbol{\xi} = \boldsymbol{\Phi}\mathbf{s} \tag{2}$$

can be obtained, where s_{eff} is an effective stress and Φ defines a structural variable ξ characterizing an anisotropy due to inelastic deformation. This expression apparently coincides with the conventional formulation of kinematic hardening when $k = -1$.

Structural Variable ξ

Let us consider a transformation tensor of fourth rank

$$\Phi = \rho \mathbf{J} + \mathrm{A}, \quad \mathrm{A} = A\mathbf{e}\otimes\mathbf{e} \tag{3}$$

where A and ρ denote scalar variables, $\mathbf{e}$ is inelastic strain, and $\mathbf{J}$ is the isotropic fourth rank tensor mapping any second rank tensor to its deviator. When $\rho = 1$, eq. (3) coincides with the expression of Baltov-Sawczuk [2, 4]. Operating Φ on the deviatoric stress $\mathbf{s}$, we can define a structural variable ξ

$$\xi = \Phi\mathbf{s} = \rho\mathbf{s} + A\mathrm{tr}(\mathbf{es})\mathbf{e} \tag{4}$$

The structural variable ξ gives rise to the same magnitude of s_{eff} in the same and opposite directions of plastic prestrains; the Bauschinger effect is not descriptive. Accounting for such anisotropic effect, we consider a modification of eq. (4)

$$\xi = \rho\mathbf{s} + A\Gamma\mathbf{e} \tag{5}$$

where $\Gamma = |\mathrm{tr}(\mathbf{es})|$, denoting the absolute value of tr(**es**). Introduced unit tensors, $\mathbf{m} = \mathbf{e}/[\mathrm{tr}\mathbf{e}^2]^{1/2} = \mathbf{e}/\|\mathbf{e}\|$, $\mathbf{n} = \mathbf{s}/[\mathrm{tr}\mathbf{s}^2]^{1/2} = \mathbf{s}/\|\mathbf{s}\|$, the tensors Φ and ξ are replaced by

$$\Phi = A \ll \mathbf{m}\otimes\mathbf{m} \gg + \rho\mathbf{n}\otimes\mathbf{n} \tag{6}$$

$$\xi = \Phi\mathbf{n} = A\Gamma\mathbf{m} + \rho\mathbf{n} \tag{7}$$

where $\ll \cdot \gg$ defines an operation: $\ll \mathbf{m}\otimes\mathbf{m} \gg \mathbf{x} = |\,\mathrm{tr}(\mathbf{m}\cdot\mathbf{x})\,|\mathbf{m}$. It is noted that the coefficients A and ρ are newly defined in these expressions.

Total hardening developed under monotonic proportional loadings involves both isotropic and anisotropic hardenings. Hence, we assume a decomposition of the structural variable into two parts which play roles as kinematic and isotropic variables:

$$\xi = \alpha + \rho\mathbf{n}, \quad \alpha = cH\Gamma\mathbf{m}, \quad \rho = (1 - c)H \tag{8}$$

where H denotes the whole hardening and c is a parameter of mixed hardening. The above expression results in an effective stress $s_{eff} = \mathbf{s} - \xi = (\|\mathbf{s}\| - \rho)\mathbf{n} - \alpha$. When the effective stress is used, the flow direction is influenced by both isotropic and kinematic variables.

In order to demonstrate the applicability of the present model, a numerical calculation was performed by using a modified version of the Bailey-Norton creep law

$$\mathbf{d} = mA^{1/m} q^{(m-1)/m} \sigma_e^{*\,n/m} \mathbf{n}^*, \quad \dot{q} = [(2/3)\mathrm{tr}\mathbf{d}^2]^{1/2} \tag{9}$$

$$\sigma_e^* = [(3/2)\mathrm{tr}(\mathbf{s} - \xi)^2]^{1/2}, \quad \mathbf{n}^* = (\mathbf{s} - \xi)/[\mathrm{tr}(\mathbf{s} - \xi)^2]^{1/2} \tag{10}$$

VERIFICATION OF THE MODEL

Experimental results on multiaxial creep behaviors after plastic strainings, which were obtained by Ohashi et al. [3], was used to verify the present model. Following the usual procedure, the material constants involved in eq. (9) were determined: $A = 7.42 \times 10^{-16}$ $MPa^{-5.8}$ $h^{-0.31}$; $m = 0.31$; $n = 5.8$. The hardening H was expressed by a linear function $H = H_1(|\mathbf{e}| - H_2)$, where $H_1 = 2188.11$ MPa, $H_2 = 0.28 \times 10^{-2}$. The mixed hardening parameter c was 0.4 in the present study.

Figure 1 shows the predictions in solid lines for torsional creep curves preceded by tensile plastic strains, together with the experimental results in symbols. Figure 2 shows the comparison of the subsequent creep curves at five different stress states, respectively after tensile plastic straining of the amount $\varepsilon^p = 2.0$ %, where the crucial symbols included are the torsional creep data of virgin material. In Figure 3, creep strain paths are shown. The results for the model ($\Gamma = 1$) are also shown in these figures in dashed lines, some of which are overlapping on the solid lines.

A quantitative agreement is not enough in the comparisons between prediction and experiments. However, the model shown in solid lines reproduces as a whole the trend of plastic strain induced creep anisotropy observed in the experiments. The difference between two models $\Gamma = \Gamma$ and $\Gamma = 1$ appears only for $\theta = 45$, 90 and 135 deg; the model $\Gamma = \Gamma$ seems to be preferable to the other.

CONCLUSION

In the present study, an anisotropic constitutive model was discussed with an emphasis on a description of anisotropic *state*. The model was based on the effective stress concept. An anisotropic fourth rank linear transformation was proposed in order to define a structural variable which could incorporate both isotropic and anisotropic hardenings. It was applied to generalize the isotropic Bailey-Norton creep equation. The proposed anisotropic model reproduced fairly well the aspects of creep isotropy and creep anisotropy observed in the experiments referred.

A generalization should be made to describe an *evolution* of the anisotropic state. The irreversible thermodynamic formalism could be applied for the purpose where two kind of potential functions are assumed: free-energy potential and dissipation potential.

ACKNOWLEDGEMENT

The author is grateful to Professor S. Murakami of Nagoya University for his useful discussions. The financial support provided in part by the Ministry of Education and the University of Tsukuba is gratefully acknowledged.

REFERENCES

1. J. Betten, in Applications of Tensor Functions in Solid Mechanics, edited by J. P. Boehler, (Springer-Verlag 1987).

2. A. Baltov and A. Sawczuk, "A rule of anisotropic hardening", Acta Mech. 1/2, 81-92 (1965).

3. Y. Ohashi, M. Kawai, and T. Momose, "Effects of prior plasticity on subsequent creep of type 316 stainless steel at elevated temperature", ASME J. Eng. Mater. Tech. 108(1), 68-74 (1986).

4. M. Waniewski, "A simple law of steady-state creep for material with anisotropy introduced by plastic prestraining", Ing. Archiv 55, 368-375 (1985).

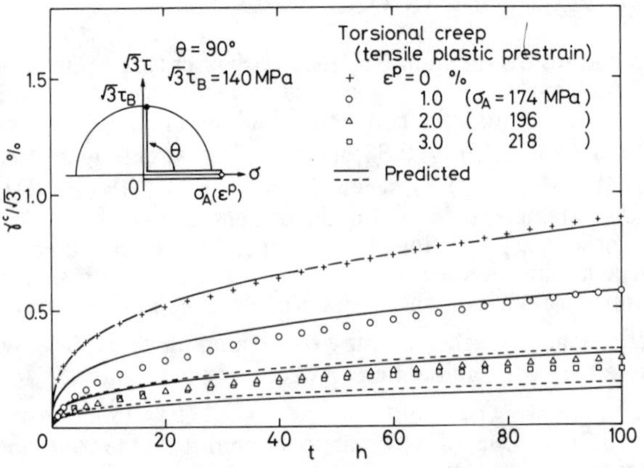

FIG. 1. Torsional creep curves after tensile plastic straining.

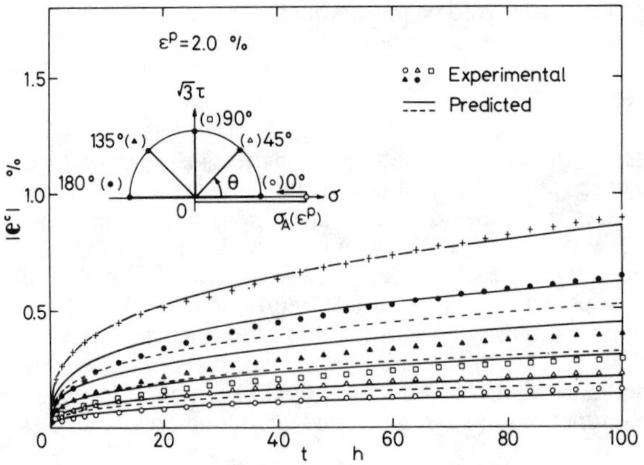

FIG. 2. Creep curves at different stress states after tensile plastic straining.

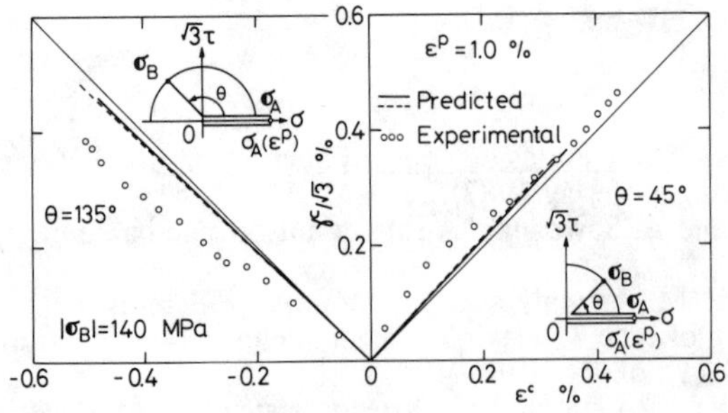

FIG. 3. Creep strain trajectories after tensile plastic straining.

A VISCOPLASTIC MODEL OF THE MECHANICAL RESPONSE OF MASONRY

J. X. LIANG[†], G. N. PANDE, and J. MIDDLETON
Department of Civil Engineering, University College of Swansea, Swansea, SA2 8PP, U. K.

[†]*On study leave from Lanzhou Railway Institute, P. R. China.*

ABSTRACT

This paper presents a constitutive model for masonry in the framework of pseudo viscoplasticity. Using equivalent elastic properties obtained from a homogenization technique the stresses in each constituent viz brick, head joint and bed joint are checked for yielding/failure. A set of strain rates consistent with static and kinematic constraints are computed to simulate the mechanical response of brick masonry.

INTRODUCTION

The viscoplastic framework has been chosen for reasons of its simplicity, although it is possible to formulate the constitutive law as a plasticity model[1]. The equivalent elastic properties of masonry and the structural transformation relations of stresses and strains, from the equivalent to the constituent materials, obtained in the authors' earlier studies [2][3] have been employed. The numerical simulation of stress- strain response of masonry in some simple loading situations is presented.

FORMULATION

Basic assumptions of the model

Considering the anisotropy and composite nature of masonry, the authors [2] have derived the equivalent elastic parameters for brick masonry. It has been shown that the elastic stress-strain relations for the equivalent material can be written as

$$\underline{\bar{\varepsilon}} = [\bar{C}]\ \underline{\bar{\sigma}} \qquad (1)$$

where $\underline{\bar{\varepsilon}}$ and $\underline{\bar{\sigma}}$ are strain and stress vectors and $[\bar{C}]$ is the compliance matrix of the equivalent material. It is assumed that Hill's postulate[4] is acceptable, ie the linear and mutually unique relations between the stresses (strains) in constituents and those in the equivalent material are assumed to be valid. Therefore

$$\underline{\sigma}_i = [\alpha_i]\ \underline{\bar{\sigma}} \qquad (2) \qquad\qquad \underline{\varepsilon}_i = [\beta_i]\ \underline{\bar{\varepsilon}} \qquad (3)$$

where $[\alpha_i]$ and $[\beta_i]$ are non-singular structural matrices of the ith constituent for stress and strain respectively. By using these two transformation matrices, the stresses and strains at any point of the equivalent material can be converted to those of the constituents, or vice versa. In (2) and (3) 'ı' (ı=1, 2, or 3) represents brick, head joint and bed joint respectively. Explicit forms for these structural matrices are given in reference[2] and [3].

Basic equations of the proposed model

It is a basic assumption that each of the constituents exhibits elasto/viscoplastic behaviour.This means that the brick, head joint and bed joint are assumed to respond elastically to instantaneous loading. Adopting the additivity postulate, the total strains in the ıth constituent and the equivalent material may be divided into elastic and viscoplastic components

$$\Delta\underline{\varepsilon}_i = \Delta\underline{\varepsilon}_i^e + \Delta\underline{\varepsilon}_i^{vp} \tag{5}$$

$$\Delta\underline{\bar{\varepsilon}} = \Delta\underline{\bar{\varepsilon}}^e + \Delta\underline{\bar{\varepsilon}}^{vp} \tag{6}$$

Combining (2), (5) and (6), leads to

$$\Delta\underline{\bar{\varepsilon}} = \Sigma\ [\beta_i]^{-1}(\Delta\underline{\varepsilon}_i - \Delta\underline{\varepsilon}_i^{vp}) + \Delta\underline{\bar{\varepsilon}}^{vp} \tag{7}$$

The stress increments for the equivalent material are

$$\Delta\underline{\bar{\sigma}} = [\bar{D}]\ (\Delta\underline{\bar{\varepsilon}} - \Delta\underline{\bar{\varepsilon}}^{vp}) \tag{8}$$

By substituting (8) into (2), the stress increments in the ıth constituent are found to be

$$\Delta\underline{\bar{\sigma}}_i = [\alpha_i]\ [\bar{D}]\ (\Delta\underline{\bar{\varepsilon}} - \Delta\underline{\bar{\varepsilon}}^{vp}) \tag{9}$$

The viscoplastic strain rates are then obtained for the constituents by

$$\dot{\underline{\varepsilon}}_i^{vp} = \gamma_i <F_i> \frac{\partial Q_i}{\partial \underline{\sigma}_i} \tag{10}$$

where γ_i is the fluidity parameter which is set to unity. F_i=0 and Q_i=constant are yield and plastic potential functions respectively for the ıth constituent. Using an explicit Eulerian time stepping scheme, viscoplastic strain increments can be calculated.

Failure criteria

The Mohr-Coulomb yield function in conjunction with a limited tension cutoff is employed to describe the behaviour of the joints. The criterion proposed by Khoo et al [4] is employed for modelling the failure of bricks in a state of biaxial tensile-compressive stress.

$$\frac{\sigma_c}{f_{cb}} + \left(\frac{\sigma_t}{f_{tb}}\right)^{0.546} = 1 \qquad (11)$$

where σ_c and σ_t are the compressive and tensile stresses at failure and f_{cb} and f_{tb} are the corresponding uniaxial strength of brick.

NUMERICAL EXAMPLES

Numerical examples of stress-strain response of the proposed model to some simple loading conditions are presented in this section. In these examples brick and mortar are assumed to be isotropic. The following experimental data is assumed.

Constituents	Physical parameters	
Brick	Young's modulus (E_b)	11000 *MPa*
	Poisson's ratio (υ)	0.20
	Compressive strength (f_{cb})	52 *MPa*
	Tensile strength (f_{tb})	2.6 *MPa*
	Length of brick	0.225 *m*
	Height of brick	0.075 *m*
Mortar	Young's modulus (E_m)	2200 *MPa*
	Poisson's ratio (υ_m)	0.25
	Compressive strength (f_{cm})	14 *MPa*
	Tensile strength (f_{tm})	1.4 *MPa*
	Thickness of mortar	0.01 *m*

These values are generally comparable with experimental results reported in [5]. In figure 1, a uniaxial compressive stress is applied to a masonry panel. In this particular example yielding is caused by the development of vertical tension cracks in the brick unit. Figure 1 also shows the stress-strain relationship in uniaxial compression for the equivalent material, brick, head joint and bed joint. Similar results are shown in figure 3 for the response of a masonry panel in pure shear.

CONCLUSION

A model for the mechanical response of brick masonry has been presented. The stress state in the constituents, which can be found by their corresponding transformation matrices, in conjunction with an appropriate failure criteria are used to judge whether yielding of individual constituents occurs. A set of viscoplastic strain rates consistent with static and kinematic constraints are evaluated. The model has been incorporated in a finite element code and papers reporting the solution of problems of a more practical nature are in preparation.

ACKNOWLEDGEMENTS

The authors wish to thank British Council, the Chinese Government and BDA for financial support.

REFERENCES

1. PIETRUSZCZAK, S., PANDE., G. N. "On the mechanical response of media intersected by two sets of joints", to be published in Int. J. Num. & Analyt. Meths. Geomech.,
2. PANDE, G. N., LIANG, J. X. and MIDDLETON, J., "Equivalent elastic moduli for brick masonry", Computers and Geotechnics, 8, 243-265,(1989).
3. LIANG, J. X., PANDE, G. N. and MIDDLETON, J., "Conversions of stress and strain from an equivalent medium to constituent material", To be submitted to "Computers and Structures" 1990.
4. Hill, R., "Elastic properties of reinforced solids, some theoretical principles", J. Mech. and Phys. Solids, 11, 357-372, (1963).
5.KHOO, C. L. and HENDRY, A. W., "A failure criterion for brickwork in axial compression", Proc. the 3rd Int. Brick and Masonry Conf., Essen, 139-145, (1973).

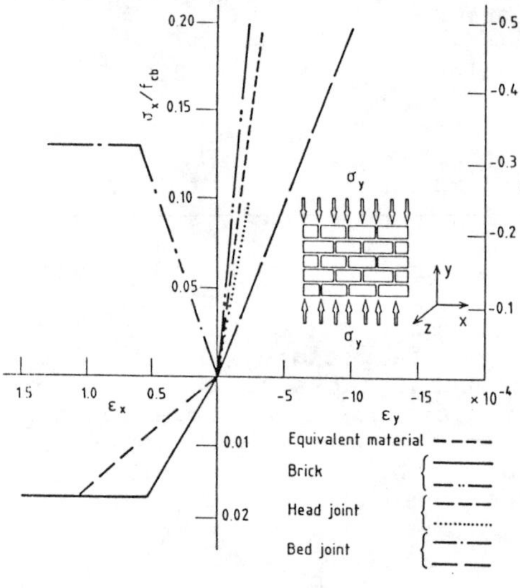

FIGURE 1 STRESSES AND STRAINS IN VARIOUS CONSTITUENTS OF MASONRY UNDER COMPRESSION

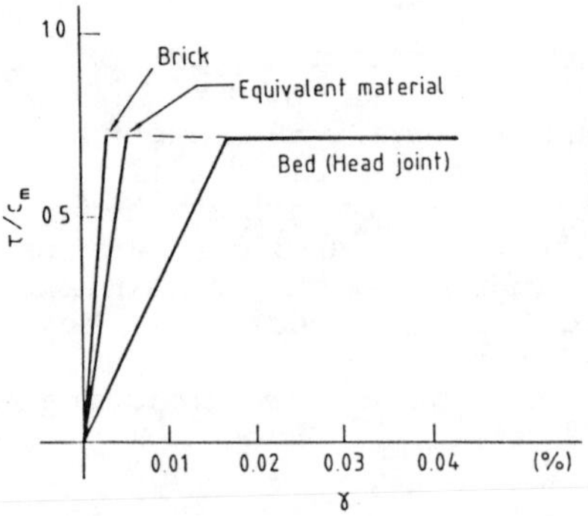

FIGURE 2 STRESSES AND STRAINS IN VARIOUS CONSTITUENTS OF MASONRY UNDER PURE SHEAR

The Nonuniform Flow and Failure of a Creep Cavitating Bar

C.H. Lu
Assistant Professor
Dept. of Mech. Engr.
Memphis State University

A.J. Levy
Associate Professor
Mech. & Aero. Engr. Dept.
Syracuse University

ABSTRACT

In this paper the nonuniform flow and failure of a creep-cavitating bar under constant load, and constant load with superimposed hydrostatic pressure, is studied by means of the finite element method. The constitutive equation used in the analysis is based on net section arguments together with Dyson's micromechanical mechanism of constrained cavity growth. The results indicate (i) a transition from catastrophic necking and brittle fracture to diffuse necking and ductile rapture under constant load with superimposed pressure, (ii) the effects of imperfections on fracture times, (iii) the locations for the initiation of failure, and (iv) the sensitivity of necking behavior to geometrical imperfection and material imperfection.

INTRODUCTION

In studying the creep deformation of polycrystalline metals at elevated temperature a simple power-law constitutive model (power-law strain rate equation plus incompressibility condition) is often used. The model can successfully predict the steady (secondary) creep deformation in a constant stress test and the accelerated (tertiary) creep which arises from the geometric effect of uniform area reduction at constant applied force provided true stress and natural strain are used. This model fails to account for material damage commonly observed in creep tests performed on polycrystalline metal bars. Experimental evidence indicates that cavities nucleate on grain boundaries which are approximately perpendicular to the maximum principle tensile stress axis and that apparently the gross deformation of the material is affected by the interaction of the diffusional growth of cavities in the grain boundaries with dislocation creep occurring within the grain [1]. Accelerated creep (and perhaps an enhanced steady creep) may arise from material damage in a constant stress test, or it may arise in a constant force test from the interaction of material damage and the geometric effect of uniform area reduction. As with the purely viscous case a geometric and/or material imperfection may precipitate neck growth however, unlike the viscous bar which fails at a point, the creep damaging bar will fail at a finite area at the cross section where the radius was initially a minimum [2]. The purpose of the present effort is to obtain the detailed response of a nonuniform creep cavitating bar under a uniaxial tensile load and a uniaxial tensile load with superimposed hydrostatic pressure.

THE CONSTITUTIVE MODEL

The constitutive equation employed here is a physically based model incorporating the constrained cavity growth mechanism [1] together with net section arguments [2,3]. It characterizes elastic, viscous creep and damaging response but neglects transient creep since its effect on long term creep is negligible. The model is suitable for polycrystalline metals at elevated temperatures ($0.3T_m < T, < 0.9T_m$, T_m = melting temperature in degrees Kelvin) and moderate stresses ($10^{-5} < \sigma_m/E < 10^{-3}$, σ_m maximum principle stress, E elastic modulus). Employing a convected spatial description, we first assume that the total spatial strain rate tensor **d** is the sum of an inelastic part ${}^c\mathbf{d}$ and an elastic part ${}^E\mathbf{d}$ which is governed by a hypoelastic law of grade zero,

$$\frac{D}{D\tau} P^{\alpha}{}_{\beta} = \lambda\, {}^E d^{\gamma}{}_{\gamma} g^{\alpha}{}_{\beta} + 2\mu\, {}^E d^{\alpha}{}_{\beta}$$

where $g^{\alpha}{}_{\beta}$ are the mixed components of the convected metric, **P** is the Cauchy stress tensor, λ and μ are the Lame constants, and τ is the dimensionless time defined by $\tau = n\dot{e}_o t = t/t_R$. Here t is real time and t_R is the critical time for ductile rupture in uniaxial tension. The components of the creep strain rate tensor may be written as

$${}^c d^{\alpha}{}_{\beta} = (3/2n)(\bar{J}_2'/P_o)^{n-1}[(P'^{\alpha}{}_{\beta}/P_o) + (2/3)\frac{\omega}{1-\omega}\frac{P_I}{P_o} m^{\alpha}{}_{\beta}]$$

with

$$\bar{J}_2' = \{3[J_2' + P_I[P_I - (J_1/3)]\frac{\omega}{1-\omega} + \frac{P^2{}_I}{3}(\frac{\omega}{1-\omega})^2]\}^{1/2}$$

where **P'** and J_2' are the deviatoric Cauchy stress tensor and its second invariant, respectively; P_o is a reference Cauchy stress; J_1 is the first invariant of Cauchy stress tensor **P**; n is the power law exponent; P_I is the maximum principle tensile stress; and ω is the damage variable (area fraction of cavitated boundaries). The tensor **m** characterizes the orientation of the maximum principle stress axis (I) and its components in convected coordinates have the form

$$m^{\alpha}{}_{\beta} = g^{\alpha}{}_{I} g^{I}{}_{\beta} \quad \text{(no sum on I)}$$

Also, for complex loading in finite deformation, a generalization of the equation governing the rate of the evolution of the area fraction which is consistent with experiment may be postulated to be

$$\frac{D\omega}{D\tau} = \frac{1-\omega}{nC}({}^c d_I)$$

where ${}^c d_I$ is the maximum principle creep strain rate and C is the Monkman-Grant constant which correlates creep fracture data at constant stress. This equation assumes the form given in [2,3] for the case of uniform, uniaxial deformation.

Because of the structure of the hypoelastic and creep rate equations, the following initial conditions for the Cauchy stress tensor, creep strain tensor (${}^c\mathbf{h}$) and material damage must be prescribed:

$$P^{\alpha}{}_{\beta}(\tau=0) = {}_oP^{\alpha}{}_{\beta}, \quad {}^c h^{\alpha}{}_{\beta}(\tau=0) = 0, \quad \omega(\tau=0) = 0$$

The initial "elastic" Cauchy stress ${}_o\mathbf{P}$ may be chosen to be the solution to the related pure hypoelastic problem.

The constitutive model just introduced has been used to study the uniform deformation of a perfect cylindrical bar under a constant axial force, or constant axial force with superimposed hydrostatic pressure. The details of the solution can be found in Reference [4].

IMPERFECTION ANALYSIS

Here we consider both geometrical imperfections characterized by an axial variation of cylindrical radius, and material imperfections characterized by an axial variation of Monkman-Grant constant. The finite element method employed is based on the principle of virtual work in the convected Lagrangian view together with the Newton-Raphson method. In the calculations the creep exponent is taken to be n=5 and Poisson's ratio is taken to be ν=0.33. Also, in order to insure reasonable deformation histories consistent with model limitation, the ratio of initial stress to elastic modulus (P_o/E) is chosen to be 0.0003.

In Fig. 1 the elongation history of the perfect and imperfect purely viscous (non-cavitating) bars subjected to a tensile force is compared. The results indicate that an initial geometrical imperfection does not significantly affect the fracture times. However unlike the uniform thinning of the cross section of the perfect bar, the initial imperfection gives rise to diffuse necking behavior throughout its deformation history (Fig. 2).

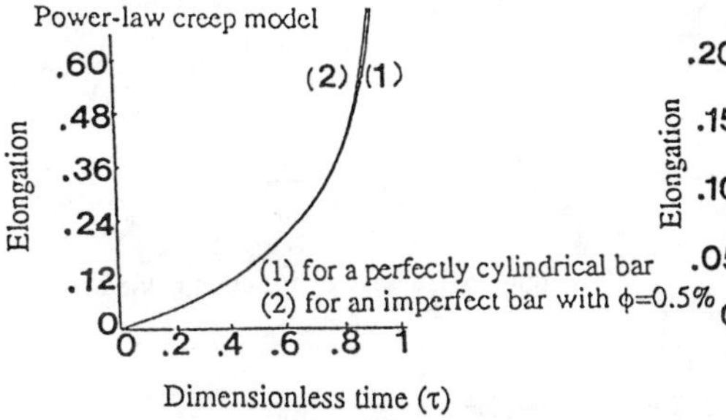

Figure 1. Elongation history of a bar under a uniaxial tensile force (P_o/E=0.0003).

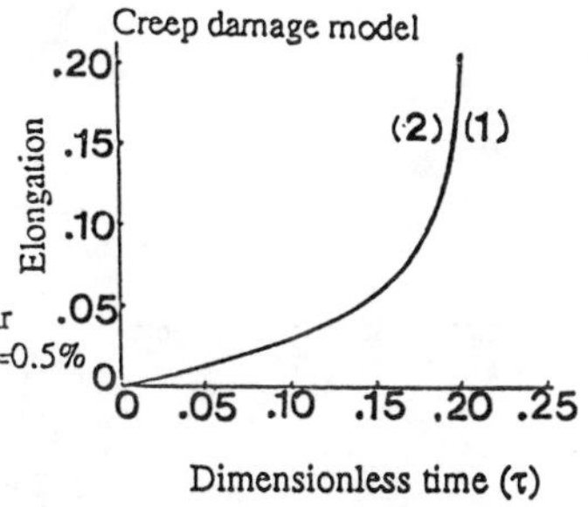

Figure 3. Elongation history of a bar under a uniaxial tensile force (P_o/E=0.0003).

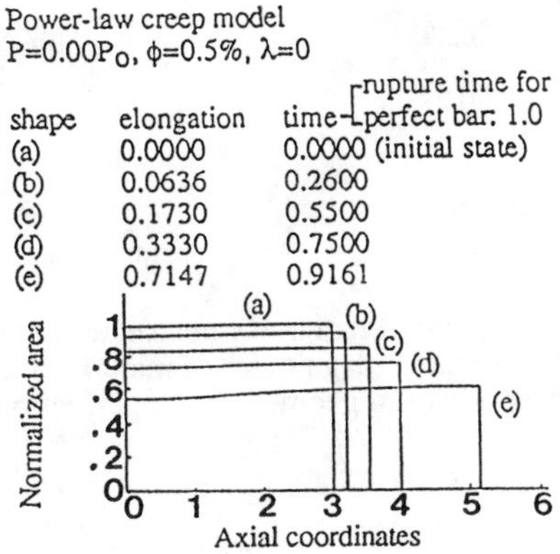
Power-law creep model
P=0.00Po, φ=0.5%, λ=0
rupture time for perfect bar: 1.0
shape elongation time
(a) 0.0000 0.0000 (initial state)
(b) 0.0636 0.2600
(c) 0.1730 0.5500
(d) 0.3330 0.7500
(e) 0.7147 0.9161
Normalized area
1
.8
.6
.4
.2
0
(a)
(b)
(c)
(d)
(e)
0 1 2 3 4 5 6
Axial coordinates

Figure 2. Geometry evolution of a geometrically imperfect bar.

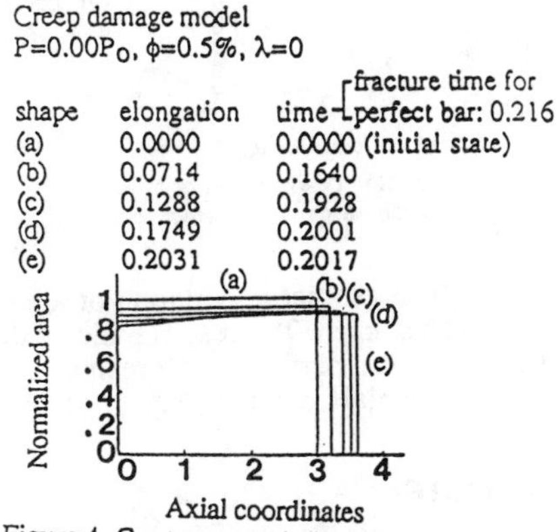
Creep damage model
P=0.00Po, φ=0.5%, λ=0
fracture time for perfect bar: 0.216
shape elongation time
(a) 0.0000 0.0000 (initial state)
(b) 0.0714 0.1640
(c) 0.1288 0.1928
(d) 0.1749 0.2001
(e) 0.2031 0.2017
Normalized area
1
.8
.6
.4
.2
0
(a)
(b)(c)
(d)
(e)
0 1 2 3 4
Axial coordinates

Figure 4. Geometry evolution of a geometrically imperfect bar.

For the creep damage model, Fig. 3 again shows insignificant effect of geometrical imperfection on tertiary creep response and fracture times. The shapes of the imperfect bar are shown in Fig. 4. In contrast to the pure creep model, more significant neck development is observed at smaller elongations just prior to failure at a finite area. The damage distribution, at an elongation of 20.31%, at the neck is depicted in Fig. 5 while the damage distribution along the axis at the center of the bar (r=0) is shown in Fig. 6 in which z and U_z denote the current axial coordinates and current axial displacements of material points, respectively. These figures show that the damage is a maximum at the bar center, and because of this we might expect failure to initiate there. Also, results indicate an almost uniform axial stress through the neck just prior to failure. This implies the validity of the long wavelength approximation used in approximation analyses.

The response of a geometrically imperfect bar subjected to a uniaxial tensile load (P_o) with superimposed hydrostatic pressure (p) of magnitude $0.5P_o$ is shown in Fig. 7. By comparing this plot with Fig. 2 and Fig. 4, it is apparent that there is a transition in failure mode from the tendency for catastrophic nacking and more or less brittle fracture (Fig. 4) to diffuse necking and ductile rupture (Fig. 2)

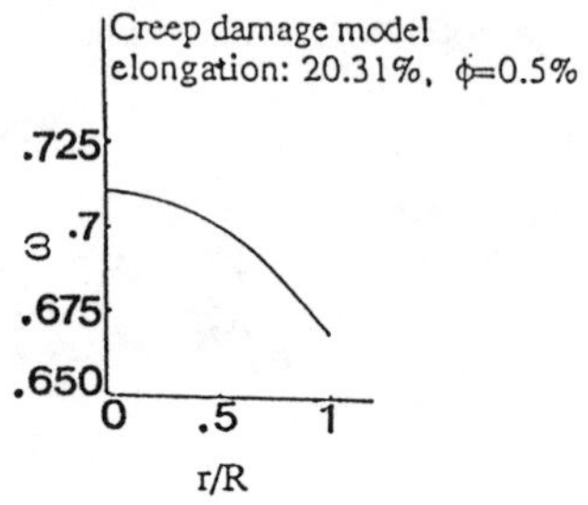

Figure 5. Damage distribution at neck.

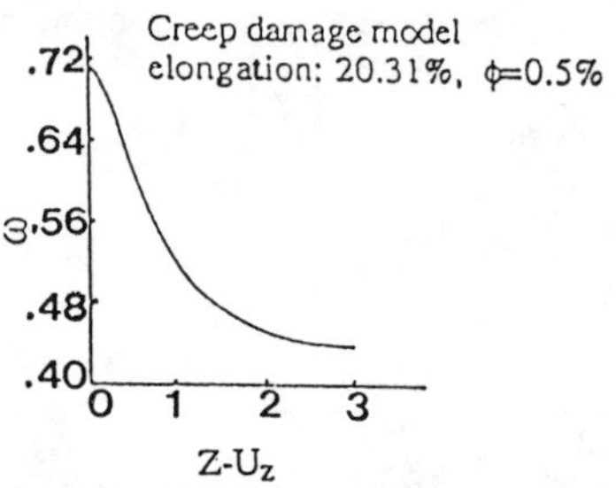

Figure 6. Damage distribution at the bar center (r=0).

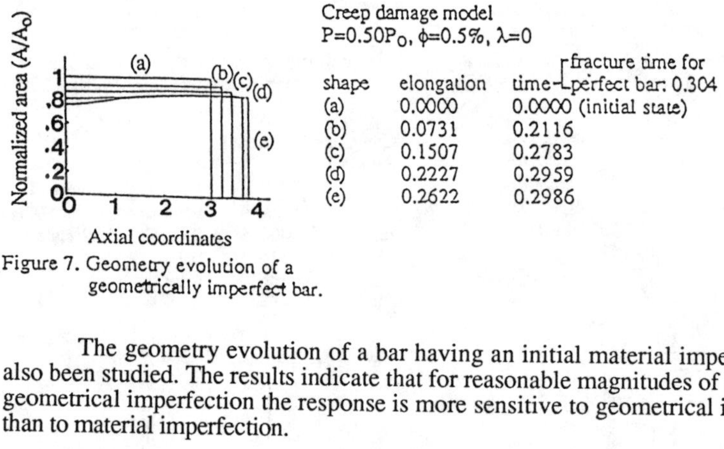
Creep damage model
P=0.50Po, ϕ=0.5%, λ=0
Normalized area (A/Ao)
1
.8
.6
.4
.2
0
0
1
2
3
4
Axial coordinates
(a)
(b)
(c)
(d)
(e)
shape elongation time
fracture time for perfect bar: 0.304
(a) 0.0000 0.0000 (initial state)
(b) 0.0731 0.2116
(c) 0.1507 0.2783
(d) 0.2227 0.2959
(e) 0.2622 0.2986

Figure 7. Geometry evolution of a geometrically imperfect bar.

The geometry evolution of a bar having an initial material imperfection has also been studied. The results indicate that for reasonable magnitudes of material and geometrical imperfection the response is more sensitive to geometrical imperfection than to material imperfection.

REFERENCES

1. B.F. Dyson, "Constraints on Diffusional Cavity Growth Rate," Metal Sci., 10, 349 (1976).
2. A.J. Levy, "The Tertiary Creep and Necking of Creep Damaging Solids," Acta Met., 34, 1991 (1986).
3. A.J. Levy, "A Physically Based Constitutive Equation for Creep Damaging Solids," J. Appl. Mech., 52, 615 (1985).
4. C.H. Lu, "An Analysis of the Flow and Failure of a Creep Damaging Bar," Ph.D. Dissertation, Syracuse University, 1988.

SOME RHEOLOGICAL BEHAVIOURAL FEATURES OF HIGH-STRENGTH CONCRETE

J.C. MASO, V. LUMBROSO, G. PONS
Laboratoire Matériaux et Durabilité des Constructions, I.N.S.A. - U.P.S. Génie Civil
31077 TOULOUSE CEDEX France

ABSTRACT

The purpose of this study is to state the rheological evolution (creep and shrinkage) of High Performance Concretes according to load intensity, composition (cement, superplasticizer, Condensed Silica Fume) and hygrometry.

INTRODUCTION

Increasing the number of main constituents in High-Strength Concrete as compared with traditional structural concrete allows for composition weight ratios to be varied in increasingly larger proportions. Accordingly, consideration must be given to the following question. Do the properties of HSC remain intact or not when different compositional weights are used ? By investigating in particular the delayed responses such as creep and shrinkage, we hope to provide an answer. Among the composition parameters liable to vary, we have chosen to keep constant the nature of the basic constituents and the workability of fresh concrete. Only concrete of compressive strength σ_r=85 MPa at 28 days has been studied. As the behaviour of concrete had to be linear, a stationary load of 0.4 σ_r was chosen. Some additional results were obtained by testing at 0.75 σ_r. In order to obtain as much general information as possible, we have used testing environments ranging from water saturation to severe desiccation at constant temperature.

EXPERIMENTAL METHOD

The various tests were carried out on 28 day old concrete cylinders. The compositions of the 5 concrete samples are given in table 1.

After 24 hours, the test pieces were stripped and stored in air-tight bags to prevent contact with the external environment. Those test-pieces intended for analysis of spontaneous size variations were fitted with two marking studs placed 236 mm ± 2 mm apart. Measurements were taken using a comparator.

Table I. Characteristics of concrete.

concr. type n°	binder (C+CSF) kg/m³	C kg/m³	CSF kg/m³	F kg/m³	F/B %	W kg/m³	W/B	A kg/m³	occlud air %	paste volume %
1	580(10)	522	58	5.4	0.9	216	0.37	1618	1.3	52.6
2	500(10)	450	50	6.5	1.3	195	0.39	1753	1.4	45.5
3	500(5)	475	25	8.4	2.9	170	0.34	1784	1.2	43.8
4	550(0)	550	0	21.5	3.9	167	0.30	1760	1.7	46.8
5	350(10)	315	35	8.7	2.5	166	0.48	1917	1.3	37.6

B = Binder (C+CSF), C = CPA 55R Cement, CSF = Condensed Silica Fume, S = Naphatalene Sulphonate Superplasticizer, A = Dolomitic Limestone Aggregate.

Creep tests were carried out by means of 1000 KN hydraulic jacks . Strain data were collected by means of a differential transducer placed in the centre of the test-piece.

TEST RESULTS

Pure compressive loading: At 28 days, all the concrete types tested had an average compressive strength of 85 MPa. The secant modulus of elasticity varies from 41 to 49 GPa. Although strength at 28 days is practically identical, concrete types 3 and 4 have a higher modulus of elasticity. This must be due to a lower, or even non-existent CSF content.

Spontaneous size variations (shrinkage): After 28 days curing without contact with the external environment, three types of curing conditions were then applied :

Type A - an atmosphere of 20°C and 50 % relative hygrometry,
Type B - a virtual saturation atmosphere (≈ 100 %),
Type C - curing without contact with the external environment.

When cured in a virtually saturated atmosphere such as water vapour, the test piece expands. Strain observations obtained under the three types of conditions are similar for concrete types 1, 2, 3 and 5. Only n° 4 gave greater strain readings. This must be due to the absence of CSF.

Strain under stationary load (creep): At 200 days, creep strain in saturated conditions has in fact ceased. Total creep is noticeably greater than basic creep, whatever the concrete. There is little difference in creep strain in concrete types 1, 2, 3 and 5. Only n° 4 shows different characteristics.

Compressive strength and the modulus of elasticity after creep and recovery: Some of the concrete test pieces are left to recover after a period of creep, and then subjected to pure compressive loading. Compressive strength and secant modulus of elasticity readings taken from cylinder of 0.4 σ_r creep, and those from the reference samples kept in similar conditions are not significantly different. A creep test at 0.4 σ_r does not damage the material. In contrast, at 0.75 σ_r creep, the secant modulus of elasticity is noticeably lower in the creep samples than in the reference pieces. The material is therefore damaged.

DISCUSSION

Modelling of creep law: As the formulae suggested by the CEB-FIP for creep strain are unsuitable for high strength concrete, we propose the following law of creep:

$$\epsilon_c(t) = a \sqrt{(t-t_0)} \,/\, (b+\sqrt{(t-t_0)})$$

where a and b are constants , and t_o is equal to 28 days (age at which concrete is first loaded).

Effect of curing conditions:

Spontaneous size variations: Curing in airtight conditions between 1 and 28 days has emptied certain pores of their water content. In contact with a saturated atmposphere on day 28, the open pores of the concrete became subject to a thermo-dynamic imbalance, sucking in water, which expands the material. At 50 % relative hygrometry there is shrinking as water still present in the concrete continues to be taken out by the surrounding air. What is more, hydration is incomplete at this age. This hypothesis receives additional confirmation in the case of airtight conditions where shrinkage occurs after 28 days.

Creep: The results relating to curing in a saturated atmosphere show that the initial rate of creep is higher than that which can be observed in other environments, although this eventually stabilises. It seems that as soon as there is a hygrometric balance between the concrete and the surrounding saturated atmosphere, creep ceases. A parallel progression can be observed in basic creep and total creep readings, and the difference between the two sets of figures remains constant after a more or less significant lapse of time.

The importance of CSF content: Shrinkage strain is diminished as soon as CSF is added, and there is shown to be little difference between amounts of 5 % and 10 %.

Adding CSF significantly diminishes basic or total creep, whatever the curing conditions. Traditional structural concrete without CSF has a considerably higher creep than when CSF is added, or than high-strength concrete .

The importance of binder and superplasticizer content and paste volume: Creep and shrinkage increase in relation to the amount of superplasticizer. As we also know that the source of creep and shrinkage is to be found in the binder matrix, we can therefore expect creep and shrinkage to be that much higher, when the amount of binder is higher. Examination of compositions in table 1 reveals the simultaneous effects of the amount of superplasticizer and the volume of binder. The two amounts vary in inverse proportion and their effects cancel each other out. The different creep and shrinkage curves confirm this, and thus strain is not affected by binder content.

The effect of increased loading: Test pieces loaded at 0.75 σr reveal a greater amount of creep than those loaded at 0.4 σ_r. At 0.40 σ_r, the loading is still linear. At 0.75 σ_r the non-linear characteristics are clearly distinguishable. At more than 600 days, test pieces subjected to 0.75 σ_r creep failed during the course of a loading. These test pieces had reached the stage of tertiary creep, the material is damaged. A study of microcracking under high loading was carried out concurrently and corroborates our findings.

CONCLUSION

The main conclusions from our research into the delayed responses of unloaded HSC test pieces and those loaded at 0.40 σ_r 0.75 σ_r, are as follows.

1 - As soon as CSF is added, creep and shrinkage strain becomes independent of the composition weight ratio.

2 - Curing conditions affect the delayed responses in proportion to scale effect, and strain is greater in desiccated conditions than when there is no contact with the outside environment.

3 - Delayed responses for HSC without CSF are greater than for concrete with CSF.

4 - Under loading at 0.40 σ_r the effects remain linear. At 0.75 this is no longer the case and the material suffers significant damage.

Multiple Nonlinear Kinematic Hardening Rules for Cyclic Viscoplasticity Models

J.C. Moosbrugger, Clarkson University, Potsdam, NY 13676

Abstract

A general framework for nonlinear kinematic hardening rules with additive decomposition is discussed. Partitioning of rate sensitivity and isotropic hardening between the flow rule and the saturation level of kinematic hardening within this framework is presented. Experimental approaches for determinng the number of kinematic hardening subvariables required and the proper means for accounting for cyclic, isotropic hardening within this model structure is discussed. Determination of model parameters for the kinematic hardening rules is summarized for Type 304 stainless steel at room temperature and Waspaloy at $650°C$.

1 Introduction

Nonlinear kinematic hardening rules of the Armstrong-Frederick [1] type with Chaboche and Rousselier's [2] superposition of multiple nonlinear kinematic hardening rules provides a convenient and accurate representation of the multiaxial Bausinger effect. Constitutive equations for cyclic plasticity and viscoplasticity which incorporate such rules are becoming well established and exceedingly refined [3]. Within the framework of such rules, with some modifications, many possibilities exist for modeling cyclic hardening effects and rate-sensitive material hardening [4-5]. In this paper, some of these possibilities are discussed. Some experimental approaches for characterizing kinematic and isotropic hardening within the framework of multiple nonlinear kinematic hardening rules are presented.

2 Multiple Nonlinear Kinematic Hardening Rules

For moderate strain rate cycling under small strain isothermal conditions, the following convenient framework is offered as an adaptation of approaches previously proposed [3]:

$$\dot{\mathbf{E}}^n = \sqrt{\frac{3}{2}} F\left(\frac{\langle\sqrt{\frac{3}{2}}(\| \mathbf{S}-\mathbf{A} \| - \kappa)\rangle}{1-\eta}\right)\mathbf{N} \quad ; \quad \mathbf{N} = \frac{\mathbf{S}-\mathbf{A}}{\| \mathbf{S}-\mathbf{A} \|} \tag{1}$$

$$\dot{\mathbf{A}}_i = C_i[b_i\mathbf{N} - \mathbf{A}_i] \parallel \dot{\mathbf{E}}^n \parallel \quad , \quad \mathbf{A} = \sum_{i=1}^{m} \mathbf{A}_i \tag{2}$$

$$b_i = b_i^o + \frac{\eta_i}{1-\eta}(\| \mathbf{S}-\mathbf{A} \| - \kappa) \quad , \quad \eta = \sum_{i=1}^{m} \eta_i \tag{3}$$

$$\dot{b}_i^o = \omega_i \dot{\chi} \quad ; \quad \dot{\kappa} = (1-\omega)\dot{\chi} \tag{4}$$

$$\dot{\chi} = \mu[\chi_s - \chi] \parallel \dot{\mathbf{E}}^n \parallel \quad ; \quad \omega = \sum_{i=1}^{m} \omega_i \tag{5}$$

In equations (1)-(5), $\mathbf{S} = \mathbf{T} - \frac{1}{3}\sigma_{kk}\mathbf{I}$ where $\mathbf{T}$ is the Cauchy stress, $\mathbf{A}$ is the deviatoric kinematic hardening variable or backstress, and χ is a scalar isotropic hardening variable. The brackets indicate that $\langle J \rangle = J$ if $J \geq 0$ and $\langle J \rangle = 0$ if $J < 0$; also $\| \mathbf{S}-\mathbf{A} \| = [(\mathbf{S}-\mathbf{A}) : (\mathbf{S}-\mathbf{A})]^{1/2} = [(S_{ij}-A_{ij})(S_{ij}-A_{ij})]^{1/2}$. Function F correlates rate sensitivity; parameters η_i allows partitioning of the rate sensitivity between the effective overstress $\| \mathbf{S}-\mathbf{A}-\kappa \|$ and the saturattion level of $\mathbf{A}_i$ (via the b_i). Likewise, parameters ω_i provide a means for partitioning isotropic hardening between κ and the b_i. Details of the role of the various parameters and a more complete explanation of the framework offered in equations (1)-(5) may be found elsewhere [5].

3 Characterizing the Kinematic Hardening Behavior

Backstress levels may be estimated for cyclic loading conditions using uninterrupted, cyclic nonproportional, axial-torsional tests involving abrupt changes in straining direction. This is accomplished by back-extrapolation of experimentally determined inelastic strain rate directions before and after abrupt straining direction changes [5-6]. Figure 1 shows the behavior of experimentally estimated $|\mathbf{s}-\mathbf{a}|$ versus cycle number from constant strain rate, cyclic nonproportional tests of Type 304 stainless steel at room temperature and Waspaloy at 650^oC; $\mathbf{s}$ and $\mathbf{a}$ are axial-torsional subspace vector counterparts to $\mathbf{S}$ and $\mathbf{A}$ [5]. Also shown in these plots is the maximum effective stress in a cycle versus cycle number. As can be seen in Figure 1, though the maximum effective stress $\bar{s}_{max}$ increases with cycle number (or accumulated inelastic strain), $|\mathbf{s}-\mathbf{a}|$ does not. Thus, for these materials and test conditions, cyclic isotropic hardening is best accounted for as an increase in backstress amplitude (i.e. $\omega = 1$ in the context of equations (1)-(5)).

In the axial-torsional subspace, equation (2) may be written as

$$\dot{\mathbf{a}}_i = c_i \mathbf{d}_i |\dot{\mathbf{e}}^n| \quad ; \quad \dot{\mathbf{a}} = \sum_{i=1}^{m} \dot{\mathbf{a}}_i \tag{6}$$

$$\dot{\mathbf{a}}_i \cdot \mathbf{n} = c_i \mathbf{d}_i \cdot \mathbf{n} |\dot{\mathbf{e}}^n| \quad ; \quad \dot{\mathbf{a}} = \sum_{i=1}^{m} c_i \mathbf{d}_i \cdot \mathbf{n} |\dot{\mathbf{e}}^n| \tag{7}$$

where $\mathbf{d}_i = \sqrt{\frac{3}{2}} b_i \mathbf{n} - \mathbf{a}_i$, $c_i = \sqrt{\frac{3}{2}} C_i$ and $\mathbf{n} = (\mathbf{s} - \mathbf{a})/|\mathbf{s} - \mathbf{a}|$; $|\dot{\mathbf{e}}^n| = \sqrt{\frac{2}{3}} \parallel \dot{\mathbf{E}}^n \parallel$. Using proportional tests, $\dot{\mathbf{a}}$ was computed by differentiating backstress histories (i.e. time sequences of $\mathbf{a} = \mathbf{s} - |\mathbf{s} - \mathbf{a}|$) determined using the mean values of $|\mathbf{s} - \mathbf{a}|$ from the nonproportional tests. Shown in Figure 2 are plots of $\dot{\mathbf{a}} \cdot \mathbf{n} / |\dot{\mathbf{e}}^n|$ versus $\mathbf{d} \cdot \mathbf{n} = \sum_{i=1}^{m} = R^* - \mathbf{s} \cdot \mathbf{n}$ where R^* is an estimate of the saturation effective stress or bounding surface radius, taken as the maximum effective stress in a cycle. For the stainless steel data, $\dot{\mathbf{a}}^s = \dot{\mathbf{a}} - h^* \dot{\mathbf{e}}^n$ where h^* is the asymptotic plastic hardening modulus and $\mathbf{d} \cdot \mathbf{n} = R^* + h^* \mathbf{e}^n \cdot \mathbf{n} - \mathbf{s} \cdot \mathbf{n}$ [5-6]. As can be seen from these plots, both high slope and low slope, approximately linear regions are observed. This would be expected if, for example, $m = 3$ and $c_1 >> c_2 >> c_3$. The $\mathbf{d} \cdot \mathbf{n}$ intercepts on these graphs provide a means for estimating the ω_i and the change in c_1 with cycle number can be used to characterize c_1 as a function of accumulated inelastic strain in the case of Waspaloy at $650^o C$.

4 Acknowledgements

The experimental work cited in this paper was supported by the U.S. National Science Foundation under grant MSM-8601889 (principal investigator - D.L. McDowell, Georgia Institute of Technology, Atlanta, GA., 30332).

5 References

1. Armstrong, P.J., and Frederick, C.O., "A Mathematical Representation of the Multiaxial Bauschinger Effect", C.E.G.B. RD/B/N 731.

2. Chaboche, J.-L. and Rousselier, G., "On the Plastic and Viscoplastic Constitutive Equations-Part I: Rules Developed with Internal Variable Concept", *ASME J, Press. Ves. Tech.*, **105**, 153-158, (1983).

3. Chaboche, J.-L., "Constitutive Equations for Cyclic Plasticity and Cyclic Viscoplasticity", *Int. J. Plasticity*, 5, 247-302, (1989)

4. Moosbrugger, J.C. and McDowell, D.L., "A Rate-Dependent Bounding Surface Model with a Generalized Image Point for Cyclic Nonproportional Viscoplasticity", *J. Mech. Phys. Sol.*, in press, (1990).

5. Moosbrugger, J.C., "Some Developments in the Characterization of Material Hardening and Rate-Sensitivity for Cyclic Viscoplasticity Models", *Int. J. Plasticity*, in press, (1990).

6. Moosbrugger, J.C., "Characterization of the Kinematic Hardening Behavior of Waspaloy at 650°C", *Proc. ASM Conf. on Life Assessment and Repair Tech. for Gas Turbine Engines*, Phoenix, AZ, (April, 1990).

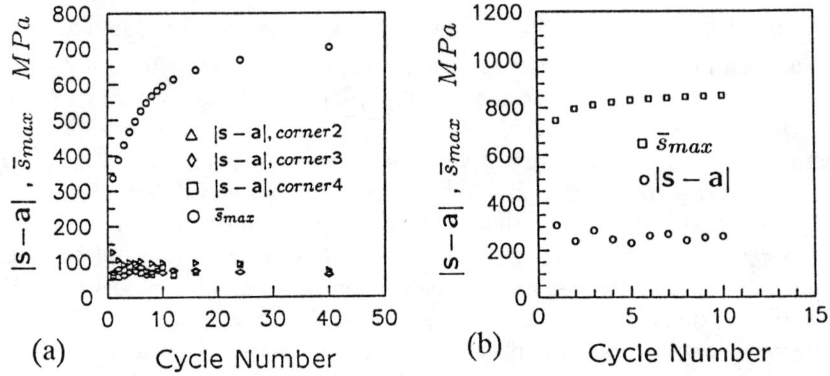

Figure 1: Estimated overstress $|\mathbf{s}-\mathbf{a}|$ and maximum effective stress $\bar{s}_{max}$ versus cycle number for (a) Type 304 stainless steel at room temperature and (b) Waspaloy at $650^{o}C$.

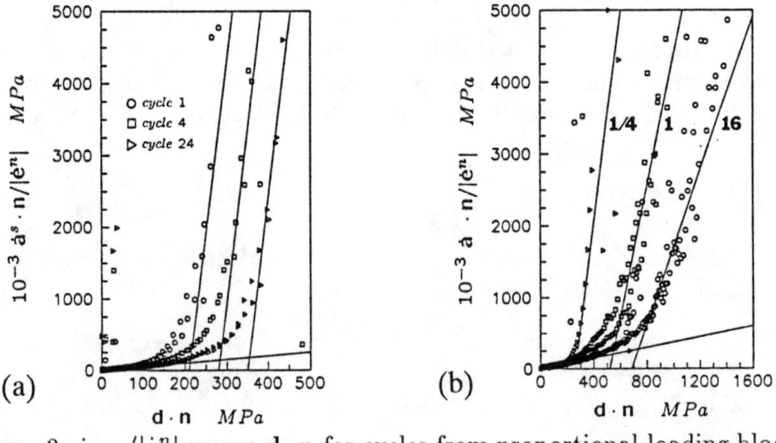

Figure 2: $\dot{a}\cdot\mathbf{n}/|\dot{\mathbf{e}}^{n}|$ versus $\mathbf{d}\cdot\mathbf{n}$ for cycles from proportional loading blocks imposed on (a) Type 304 stainless steel at room temperature and (b) Waspaloy at $650^{o}C$. The solid lines are straight lines drawn through the high and low slope regions of the data for each cycle.

CODE-TYPE MODELS FOR CREEP COMPLIANCE OF CONCRETE

Harald S. Müller

Federal Institute for Materials Research and Testing (BAM)
Berlin, Federal Republic of Germany

ABSTRACT

For code-type prediction of creep compliance of concrete either a product type or a summation type linear constitutive approach is available. Both approaches exhibit typical deficiencies which may not be overcome as long as linearity is assumed for concrete behavior. However, a considerable improvement of the prediction accuracy may be obtained by an adequate modeling and optimization. Such improved prediction models had been derived and will be presented.

INTRODUCTION

Critical investigations of creep compliance models presented in standards and recommendations showed a considerable lack of accuracy of the models primarily due to insufficient optimizations from test data, but also due to inconsistencies and deficiencies of the given models. Many of these weaknesses may be overcome or considerably reduced by an adequate modeling even considering the needs of practicing engineers who prefer concise and operational constitutive models. From the designer's point of view code-type models for creep compliance of concrete should at least contain the following 3 essential features:

- The model should be formulated to allow a simple estimate of the mean creep behavior of a concrete cross-section.
- The model should be based upon the assumption of linearity between stress and strain to assure the applicability of linear superposition at all times.
- In the prediction of the creep behavior of concrete only such parameters should be taken into account which are usually known to the practicing engineer at the stage of design.

These requirements itself restrict the attainable accuracy of corresponding constitutive models considerably, as the real behavior of concrete members is by far more complex, and creep of concrete is affected by a multitude of parameters. However, adequate models may be derived which represent a reasonable compromise between simplicity and accuracy [1]. For these models it is particularly important, to provide the designer with informations on their limits of applicability and on the prediction accuracy by means of a coefficient of variation which may be taken into account in a probabilistic approach.

LINEAR MODELS FOR CREEP COMPLIANCE

The creep compliance $\Phi(t,t_o)$ of concrete may be written as

$$\Phi(t,t_o) = \frac{1}{E_c(t_o)} + \frac{\phi(t,t_o)}{E_c} \tag{1}$$

where $\Phi(t,t_o)$ represents the total load dependent strain per unit stress at time t when concrete is loaded at time t_o; $E_c(t_o)$ is the modulus of elasticity at time at loading t_o, and E_c is the corresponding value at a concrete age of 28 days. The parameter $\phi(t,t_o)$ is the creep coefficient which may be modeled either as product or summation type approach. The product type model is characterized by a product of two different functions to describe the effects of age at loading and duration of loading upon creep of concrete. In the linear summation model creep is separated into the additive components delayed elastic strain (viscoelastic strain) and flow (viscous strain). Due to the underlying assumption of stress linearity the principle of superposition is valid and the constitutive equation for concrete may be given as:

$$\varepsilon_{c\sigma}(t,t_o) = \sigma_c(t_o)\cdot\Phi(t,t_o) + \int_{t_o}^{t} \Phi(t,\tau)\cdot\frac{\partial\sigma_c}{\partial\tau}\cdot d\tau \tag{2}$$

In eq. 2 $\varepsilon_{c\sigma}$ is the stress dependent strain of concrete and σ_c is the stress applied to concrete.

TABLE I. Comparison of the product and the summation type model for creep prediction of concrete

load history	criterion	accuracy of prediction	
		product model	summation model
constant stress	effect of age at loading	+	o
	time development	+	o
increasing stress	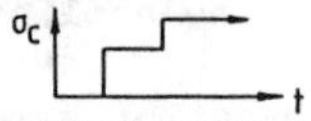	o	o
decreasing stress	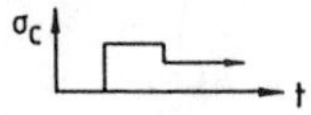	o	o
	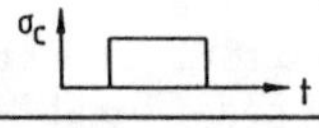	-	o
relaxation	very young age at loading and thick member or high relative humidity	-	o
	other cases	o	o

legend: (+) good; (o) acceptable; (-) poor

The differences between the summation type and the product type model become particularly evident when the models are used to predict the effects of variable stresses or strains on the basis of eq. 2. In Table I the attempt is made to summarize the basic differences concerning the prediction accuracy of the models. The prediction errors of both models mainly result from the fact that creep of concrete is a nonlinear process which requires nonlinear constitutive modeling. However, adequate modeling reduces the errors of linear models. This is particularly true for an improved linear summation model which will be presented below.

NEW PRODUCT TYPE MODEL FOR CREEP PREDICTION

Product type models have been often used as the basis of the creep prediction for concrete in the past [2]. The prediction model presented below belongs to the group of stress linear models without separation into time dependent deformation components. The creep coefficient $\phi(t,t_o)$ may be estimated from the following general relation

$$\phi(t,t_o) = \phi_{RH} \cdot \beta(f_{cm}) \cdot \beta(t_o) \cdot \left[\frac{t-t_o}{\beta_H + t-t_o}\right]^{0.3} \tag{3}$$

where ϕ_{RH} = coefficient taking into account the effects of ambient relative humidity and the size of a structural member;

$\beta(f_{cm})$ = coefficient which depends on the compressive strength of concrete f_{cm};

$\beta(t_o)$ = coefficient taking into account the effect of age at loading considering the type of cement and curing temperature;

β_H = coefficient taking into account the effects of member size and relative humidity on the time development ($t-t_o$= duration of loading) of creep.

Analytical expressions are given for all coefficients in eq. 3 in [1]. The model is primarily valid for ordinary normal weight structural concrete moist cured at normal temperature and subjected to stresses normally not exceeding 40 % of the strength at age at loading and exposed to normal ambient climate conditions. However, some extensions of the model have been developed to allow an estimate of the effects of stresses up to 60 % of the strength and the effects of temperatures up to 80 °C. The model which was optimized on the basis of an extensive data bank on creep by means of a nonlinear optimization algorithm is now proposed for CEB-FIP Model Code 1990 [3]. The coefficient of variation for the prediction of the creep compliance from the model has been found to be V = 20.4 %. Further details on the model and its accuracy may be found in [1].

IMPROVED SUMMATION TYPE MODEL FOR CREEP PREDICTION

The improved summation type model which has been developed for the prediction of creep of concrete under normal loading and climate conditions is based on the following characteristic approach:

$$\phi(t,t_o,\tau) = \phi_d(t,\tau) + \phi_{fb}(t,t_o,\tau) + \phi_{fd}(t,t_o,\tau) \quad (4)$$

Here, creep is separated into the additive strain components delayed elasticity ϕ_d, basic flow ϕ_{fb}, and drying flow ϕ_{fd}. The delayed elasticity coefficient ϕ_d represents an aging viscoelastic strain component (creep recovery) which is to a large extent recoverable upon unloading. As a simplification, ϕ_d is assumed to be independent of the ambient climate. The flow coefficients ϕ_{fb} and ϕ_{fd} represent viscous strain components being irreversible upon unloading. Both flow components following different time functions depend on the age t_o defined as age at the beginning of the load history. The time variable τ represents both the age at loading and the age at a change of stress, respectively. The model takes into account the effect of the same parameters on creep as the product type model presented above; details are given in [1].

As far as the prediction accuracy of the model for various load histories shown in Table I is concerned, it overcomes some of the weaknesses of other summation models. For the prediction of creep under a constant load, its accuracy corresponds to that of the product models, because the introduction of the additional time variable allows to distinguish between the deformation behavior of virgin and of loaded concrete. For decreasing stresses and strains, it is superior to other summation models because the effect of aging of delayed elasticity is included in this model. The coefficient of variation for the prediction of the creep compliance from the model which had been optimized using the same data bank as for the product type model, has been found to be V = 20.6 % [1].

CONCLUSIONS

1. Code-type linear summation and product type models for creep prediction of concrete show different characteristic deficiencies.
2. In particular the summation type model may be improved that it overcomes some major weaknesses of these models.
3. For simple code-type models the coefficient of variation for the prediction of the creep compliance has been found to be approximately 20 %. This prediction error is independent of the type of model.

REFERENCES

1. H.S. Müller, H.K. Hilsdorf, "Evaluation of the time dependent behavior of concrete", CEB Bulletin d'Information No. 199, Comité Euro-International du Béton (Lausanne, 1990).
2. Z.P. Bazant, "Material models for structural creep analysis", Fourth RILEM Intern. Symp. on Creep and Shrinkage of Concrete. Mathematical Modeling, Preprints (Evanston, U.S.A., 1986).
3. CEB-FIP Model Code 1990, First Draft, CEB Bulletin d'Information No. 195 and No. 196, Comité Euro-International du Béton (Lausanne, 1990).

A CONSTITUTIVE EQUATION OF IRRADIATION CREEP AND SWELLING UNDER NEUTRON IRRADIATION

S. MURAKAMI and M. MIZUNO
Department of Mechanical Engineering, Nagoya University, Chikusa-ku, Nagoya 464-01, JAPAN

ABSTRACT

A constitutive equation of irradiation creep applicable to structural analyses in multiaxial state of stress was developed by assuming that irradiation creep can be decomposed into irradiation-enhanced creep and irradiation-induced creep. By taking account of the SIPA (Stress-Induced Preferential Absorption) mechanism, the irradiation-induced creep was represented by an isotropic tensor function of order one and zero with respect to stress, which is, at the same time, the function of neutron flux and neutron fluence. The volumetric part of the irradiation-induced creep was identified with swelling. The irradiation-enhanced creep was described by modifying Kachanov-Rabotnov creep damage theory by incorporating the effect of irradiation.

1. INTRODUCTION

Nuclear reactor components often operate at elevated temperature and under significant flux of high energy fission or fusion production particles. Fuel claddings in fast breeder reactors, for example, are exposed to fast neutron flux (neutron energy E > 0.1 MeV) of the fluence of the order of 10^{23} n/cm^2 at the temperature 450 ° through 750 °C, and show salient irradiation creep and swelling [1,2]. First walls of the fusion reactors, on the other hand, are exposed to neutron flux of E = 14 MeV at high temperature during plasma burn, and the structural materials are subject to severe irradiation damage and irradiation-induced inelastic deformation [3]. Thus, the neutron irradiation in these components has crucial effects on the process of inelastic deformation, damage and final fracture.

The present paper is concerned with the mechanical modeling of irradiation creep and irradiation-induced swelling of polycrystalline metals applicable to structural analyses of nuclear reactor components.

2. MECHANICAL MODELING OF IRRADIATION CREEP AND SWELLING

2.1 Formulation of Irradiation-Induced Creep and Swelling

The creep in nuclear reactor components operating at elevated temperature under neutron irradiation may be divided into two components, i.e., *the irradiation-*

enhanced creep ε^{IEC} and *the irradiation-induced creep* ε^{IIC}. The former is the ordinary thermal creep but accelerated by irradiation, while the latter is caused exclusively by the irradiation and may occur even under vanishing stress.

If we formulate the continuum model of irradiation-induced creep ε^{IIC}_{ij} on the basis of SIPA mechanism, $\dot{\varepsilon}^{IIC}_{ij}$ may be given as a function of stress σ_{ij}, neutron flux ϕ and neutron fluence $\Phi=\int\phi dt$ [4]:

$$\dot{\varepsilon}^{IIC}_{ij} = F_{ij}(\sigma_{kl},\phi,\Phi) \tag{1a}$$

where $(\dot{\ })$ denotes differentiation with respect to time t.

In comparison with the irradiation-enhanced creep, the stress-dependence of the irradiation-induced creep and swelling is small [1,2], and the anisotropy of material properties caused by the irradiation may be insignificant. Then, by taking account of the stress dependence of creep strain rate due to SIPA mechanism [5], the irradiation-induced creep $\dot{\varepsilon}^{IIC}_{ij}$ of equation (1a) may be expressed as an isotropic function of stress σ_{ij} of order zero and order one [4]:

$$\dot{\varepsilon}^{IIC}_{ij} = \eta(\phi,\Phi)\delta_{ij} + L_{ijkl}(\phi,\Phi)\sigma_{kl} \tag{1b}$$

$$L_{ijkl}(\phi,\Phi) = \xi(\phi,\Phi)\delta_{ij}\delta_{kl} + \zeta(\phi,\Phi)(\delta_{ik}\delta_{jl} + \delta_{il}\delta_{jk}) \tag{1c}$$

where L_{ijkl} and δ_{ij} are the fourth rank isotropic tensor and the Kronecker delta, respectively. The symbols η, ξ, and ζ, furthermore, are material functions of ϕ and Φ. Substitution of (1c) into (1b) furnishes

$$\dot{\varepsilon}^{IIC}_{ij} = \eta\delta_{ij} + \xi\sigma_{kk}\delta_{ij} + 2\zeta\sigma_{ij} \tag{1d}$$

We will decompose the stress and strain into isotropic (or volumetric) and deviatoric parts. Then, the volumetric part of strain rate corresponds exactly to the swelling rate $\dot{S}$. Therefore, in view of the incubation period of swelling [6] and the linear dependence of creep and swelling on neutron fluence Φ [7] (as observed in Fig.1), the constitutive equation (1) of irradiation-induced creep may be written in an alternative form:

$$\dot{\varepsilon}^{IIC}_{ij} = (1/3)\dot{S}\delta_{ij} + (3/2)P\sigma_{Dij} \tag{2a}$$

$$\dot{S} = S_0(1 + Q\sigma_{kk})\phi\kappa(\Phi - \chi) \tag{2b}$$

$$P = P_0\phi \tag{2c}$$

where σ_{Dij} and $\kappa(\Phi - \chi)$ stand for the deviatoric part of σ_{ij} and Heaviside step function, while the coefficient $P = (4/3)\zeta(\phi,\Phi)$ represents the anisotropy of the creep rate due to SIPA mechanism. Symbols $S_0=3\eta(\phi,\Phi)/\phi$ and $Q=[3\xi(\phi,\Phi)+2\zeta(\phi,\Phi)]/S_0\phi$ in equation (2b) are material functions of swelling and express the swelling rate under vanishing stress and the coefficient of stress dependence of swelling rate respectively, while χ denotes the incubation fluence for swelling. Though P_0, S_0 and Q are functions of ϕ and Φ in general, we will take them constants by assuming that their dependence on ϕ and Φ is insignificant.

Equation (2) has been derived in the framework of continuum mechanics by taking account of physical mechanisms of irradiation creep and swelling, and is the alternative of physical theories [1,5] proposed so far for uniaxial stress.

2.2 Formulation of Irradiation-Enhanced Creep

The thermal creep of polycrystalline metals at elevated temperature is induced by the diffusion controlled dislocation motion activated by stress, and is usually accompanied by creep damage. Thus, we will assume that the effects of irradiation on the irradiation-enhanced creep ε^{IEC} can be described mainly through the coefficients of the equation for unirradiated conditions.

If we assume the von Mises creep law and the Kachanov-Rabotnov creep damage theory, the constitutive equation of creep and the evolution equation of creep damage under unirradiated condition may be given as follows [8]:

$$\dot{\varepsilon}^{C}_{ij} = (3/2)A_0[\sigma_{EQ}/(1 - D)]^{n_0-1}[\sigma_{Dij}/(1 - D)] \quad (3a)$$

$$\dot{D} = B_0[\sigma^{(1)}/(1 - D)]^{k_0} \quad (3b)$$

where A_0, B_0, n_0 and k_0 are material constants, while $\sigma_{EQ} = [(3/2)\sigma_{Dij}\sigma_{Dij}]^{1/2}$ and $\sigma^{(1)}$ denote equivalent stress and the maximum principal stress, respectively. The symbol D in equation (3), furthermore, is an internal state variable describing damage state (damage variable), and defined as follows:

$$D = 0 \quad \text{at } t = 0 \quad \text{(initial undamaged state)}$$

$$D = 1 \quad \text{at } t = t_R \quad \text{(final rupture state)} \quad (4)$$

According to the preceding argument, equation (3) may be readily extended to the irradiation–enhanced creep as follows:

$$\dot{\varepsilon}^{IEC}_{ij} = (3/2)A(\phi,\Phi)[\sigma_{EQ}/(1 - D)]^{n(\phi,\Phi)-1}[\sigma_{Dij}/(1 - D)] \quad (5a)$$

$$\dot{D} = B(\phi,\Phi)[\sigma^{(1)}/(1 - D)]^{k(\phi,\Phi)} \quad (5b)$$

where $A(\phi,\Phi)$, $B(\phi,\Phi)$, $n(\phi,\Phi)$ and $k(\phi,\Phi)$ are material functions.

In view of the experimental results and the microstructural mechanisms reported so far, we will assume the following functions for A, B, n and k [4]:

$$A(\phi,\Phi) = A_0[1 + a_1(1 - e^{-a_2\phi})][1 + a_3(1 - e^{-a_4\Phi})]$$

$$B(\phi,\Phi) = B_0[1 + b_1(1 - e^{-b_2\phi})][1 + b_3(1 - e^{-b_4\Phi})] \quad (6)$$

$$n(\phi,\Phi) = n_0$$

$$k(\phi,\Phi) = k_0$$

where a_1, a_2, a_3, a_4, b_1, b_2, b_3 and b_4 are constants that can be identified by comparing the creep curves of unirradiated creep, creep under irradiation and of post-irradiation creep under constant stress.

2.3 Constitutive Equations of Irradiation Creep and Swelling

Summarizing the results of the preceding discussions, the constitutive equations of irradiation creep and swelling are given by equations (2), (5), and (6) as follows:

$$\dot{\varepsilon}^{C}_{ij} = \dot{\varepsilon}^{IEC}_{ij} + \dot{\varepsilon}^{IIC}_{ij}$$

$$= (3/2)A_0[1 + a_1(1 - e^{-a_2\phi})][1 + a_3(1 - e^{-a_4\Phi})][\sigma_{EQ}/(1 - D)]^{n_0-1}[\sigma_{Dij}/(1 - D)]$$

$$+ (1/3)\dot{S}\delta_{ij} + (3/2)P_0\phi\sigma_{Dij} \quad (7a)$$

$$\dot{D} = B_0[1 + b_1(1 - e^{-b_2\phi})][1 + b_3(1 - e^{-b_4\Phi})][\sigma^{(1)}/(1 - D)]^{k_0} \quad (7b)$$

$$\dot{S} = S_0(1 + Q\sigma_{kk})\phi\kappa(\Phi - \chi) \quad (7c)$$

In the above equations, swelling has been represented as the dilatational part of the irradiation–induced creep, and hence equation (7) describes the irradiation creep and swelling in multiaxial states of stress consistently in the framework of continuum mechanics. In the case of $\phi = \Phi = 0$, in particular, equation (7) leads to the constitutive equations of ordinary thermal creep, while the case of $\phi = 0$ and Φ = const. corresponds to the post–irradiation creep of materials subjected to prior irradiation of Φ = const.; i.e., equation (7) furnishes a unified constitutive equations of the unirradiated creep, post–irradiation creep and creep under irradiation.

3. ELABORATED CONSTITUTIVE EQUATIONS OF IRRADIATED CREEP AND SWELLING

3.1 Elaboration of Constitutive Equation of Swelling

It has been observed that the magnitude of swelling is usually insignificant up to a specific fluence χ, and beyond this fluence, after a continuous transition from the incubation period, the swelling tends to increase almost linearly with the fluence [6,7]. Thus, in order to avoid the discontinuous increase of the swelling rate described by equation (7c) and to describe the continuous transition from the incubation to the steady state period, Bates and Korenko [7] developed empirical equations of stress–free swelling at different temperatures.

By employing the transient expression of Bates and Korenko, equation (7c) can be revised to give an elaborate constitutive equation of the stress–dependent swelling:

$$\dot{S} = S_0[1 - e^{R(\chi-\Phi)}/\{1 + e^{R(\chi-\Phi)}\}](1 + Q\sigma_{kk})\phi \tag{8}$$

where R denotes the curvature parameter of the swelling curve.

3.2 Constitutive Equation of Irradiation Creep Incorporating Transient creep

Equation (5a), or (7a) has been formulated by assuming the steady–state creep law of Norton. However, the transient creep in metals is often important not only after the application of stress, but also after the change in the magnitude or in direction of stress. In particular, in the case of 20% cold worked type 316 stainless steel at 650 °C discussed in this paper, salient transient creep is observed also in post–irradiation creep and creep under irradiation [2].

Transient creep observed under neutron irradiation is brought about mainly as a part of the stress–controlled irradiation–enhanced creep. However, since transient creep is essential only in a limited duration, we assume that neutron irradiation has not significant influence on the transient part of the stress–controlled thermal creep. Thus, if we employ McVetty's creep law which has separate expressions for transient creep and steady–state creep, the constitutive equation of irradiation–enhanced creep incorporating transient creep may be expressed as follows:

$$\begin{aligned}\dot{\varepsilon}^{IEC}_{ij} &= (3/2)A^* a^* e^{-a^* t}\sigma_{EQ}^{n^*-1}\sigma_{Dij} \\ &\quad + (3/2)A(\phi,\Phi)[\sigma_{EQ}/(1-D)]^{n(\phi,\Phi)-1}[\sigma_{Dij}/(1-D)]\ ,\end{aligned} \tag{9}$$

where t denotes the time, and A^*, a^* and n^* are material constants of transient creep. The first term of the right hand side of equation (9) is based on the time–hardening theory for variable stress creep. For further elaboration, we may employ more accurate constitutive equations of creep for general history of stress [9].

3.3 Elaborated Constitutive Equations of Irradiation Creep and Swelling

To recapitulate the above discussion, the constitutive equations of irradiation creep and swelling incorporating the continuous transition of incubation swelling and the effect of transient creep may be summarized as follows:

$$\begin{aligned}\dot{\varepsilon}^{C}_{ij} &= \dot{\varepsilon}^{IEC}_{ij} + \dot{\varepsilon}^{IIC}_{ij} \\ &= (3/2)A^* a^* e^{-a^* t}\sigma_{EQ}^{n^*-1}\sigma_{Dij} \\ &\quad + (3/2)A_0[1 + a_1(1 - e^{-a_2\phi})][1 + a_3(1 - e^{-a_4\Phi})][\sigma_{EQ}/(1-D)]^{n_0-1}[\sigma_{Dij}/(1-D)] \\ &\quad + (1/3)\dot{S}\delta_{ij} + (3/2)P_0\phi\sigma_{Dij}\ ,\end{aligned} \tag{10a}$$

$$\dot{D} = B_0[1 + b_1(1 - e^{-b_2\phi})][1 + b_3(1 - e^{-b_4\Phi})][\sigma^{(1)}/(1-D)]^{k_0}\ , \tag{10b}$$

$$\dot{S} = S_0[1 - e^{R(\chi-\Phi)}/\{1 + e^{R(\chi-\Phi)}\}](1 + Q\sigma_{kk})\phi\ . \tag{10c}$$

4. ANALYSIS OF IRRADIATION CREEP

To evaluate the validity and the applicability of equations (10), we will now perform the following analyses for 20% cold worked type 316 stainless steel at 650 °C. Material constants in equations (10) can be identified by the procedure discussed in the preceding paper [4], and determined are as follow [10]:

$$\chi = 5.00\times10^{22}\ \text{n/cm}^2, \qquad S_0 = 4.00\times10^{-25}\ (\text{n/cm}^2\cdot\text{hr})^{-1}\ \text{hr}^{-1}, \qquad \text{(11a)}$$
$$Q = 4.75\times10^{-3}\ \text{MPa}^{-1}, \qquad R = 1.25\times10^{-22}\ (\text{n/cm}^2)^{-1}.$$
$$A^* = 1.30\times10^{-7}\ \text{MPa}^{-n^*}, \qquad a^* = 8.50\times10^{-3}\ \text{hr}^{-1}, \qquad \text{(11b)}$$
$$n^* = 2.00.$$

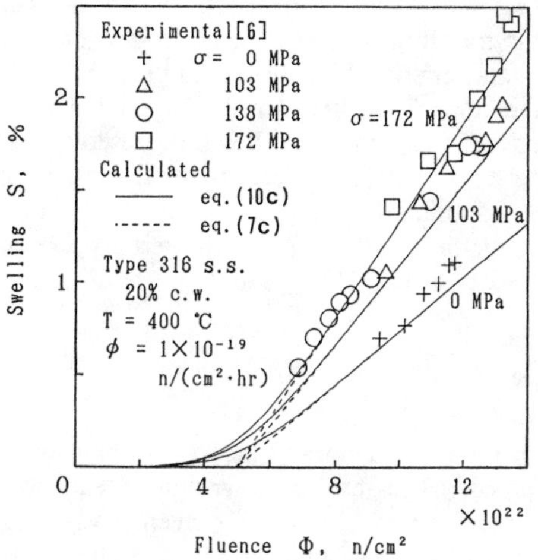

Fig.1 Swelling of 20% cold worked type 316 stainless steel at 400 °C.

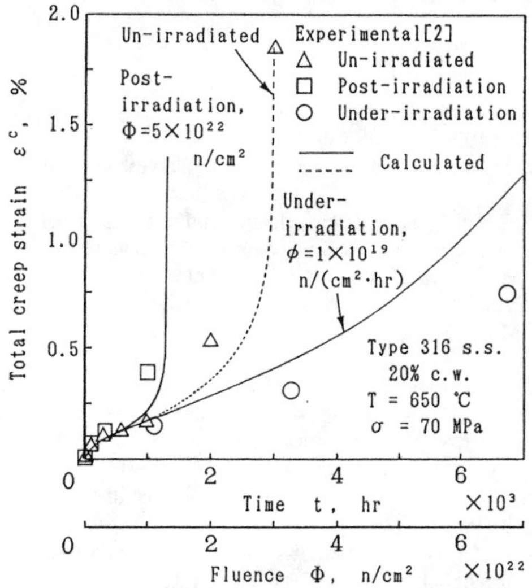

Fig.2 Creep curves of unirradiated, post–irradiation and under–irradiation creep for 20% cold worked type 316 stainless steel at 650 °C.

$$A_0 = 3.20\times10^{-13}\ \mathrm{MPa}^{-n_0}\ \mathrm{hr}^{-1}, \quad a_1 = 5.00\times10^{-2},$$
$$a_2 = 2.60\times10^{-19}\ (\mathrm{n/cm^2\cdot hr})^{-1}, \quad a_3 = -9.00\times10^{-2}, \qquad (11c)$$
$$a_4 = 2.60\times10^{-21}\ (\mathrm{n/cm^2})^{-1}, \quad n_0 = 3.50.$$
$$B_0 = 6.00\times10^{-10}\ \mathrm{MPa}^{-k_0}\ \mathrm{hr}^{-1}, \quad b_1 = -0.95,$$
$$b_2 = 4.5\times10^{-19}\ (\mathrm{n/cm^2\cdot hr})^{-1}, \quad b_3 = 1.30, \qquad (11d)$$
$$b_4 = 2.60\times10^{-21}\ (\mathrm{n/cm^2})^{-1}, \quad k_0 = 2.80,$$
$$P_0 = 3.75\times10^{-28}\ (\mathrm{MPa\cdot n/cm^2\cdot hr})^{-1}.$$

Fig.1 shows the results of prediction of swelling for 20% cold worked type 316 stainless steel at 400 °C. The solid lines and symbols are the prediction of equation (10c) and the experimental results of Porter et al [6]. The results of equation (7c) are also entered by dotted lines. As shown in Fig.1, equation (10c) represent the continuous transition from the incubation to the steady–state period without changing the characteristic of incubation and steady–state periods.

Figs.2 shows the comparison between the experimental results [2] and the corresponding predictions of equations (10). It will be observed that the constitutive equations (10) describe accurately the effects of irradiation on creep; i.e., creep curves under irradiation including transient creep stage, embrittlement due to prior irradiation, and the retardation of the start of the tertiary creep and increase of tensile ductility due to neutron flux.

5. CONCLUSION

Modeling of constitutive equations of irradiation creep and swelling and their elaboration were discussed. The resulting constitutive equations were found to facilitate the unified description of unirradiated creep, post–irradiation creep and the creep under irradiation, for whole process of creep ranging from the transient creep to the tertiary creep and the final rupture. The predictions of the constitutive equations were compared with the results of the corresponding experiments.

REFERENCES

1. K. Ehrlich, "Irradiation creep and interrelation with swelling in austenitic stainless steels", J. Nuclear Materials 100, 149–166 (1981).
2. E. P. Gilbert and B. A. Chin, "In–reactor creep measurements", Nuclear Technology 52, 273–283 (1981).
3. D. R. Harris and E. Z. Zolti, "Structural mechanics and material aspects of the next European Torus", Nuclear Engineering and Design/Fusion 3, 331–344 (1986).
4. S. Murakami, M. Mizuno and T. Okamoto, "Mechanical modeling of irradiation creep and swelling of polycrystalline metals", Nuclear Engineering and Design (contributed).
5. P. T. Heald and M. V. Speight, "Steady–state irradiation creep", Phil. Mag. 29, 1075–1080 (1974).
6. D. L. Porter, M. L. Takata and E. L. Wood, "Direct evidence for stress–enhanced swelling in type 316 stainless steel", J. Nuclear Materials, 116, 272–276 (1983).
7. J. F. Bates and M. K. Korenko, "Empirical development of irradiation–induced swelling design equations", Nuclear Technology, 48, 303–314 (1980).
8. F. A. Leckie and D. R. Hayhurst, "Constitutive equations for creep rupture", Acta Metallurgica, 25, 1059–1070 (1977).
9. S. Murakami and N. Ohno, "A constitutive equation of creep based on the concept of a creep–hardening surface", Int. J. Solids and Struct., 18, 597– (1982).
10. S. Murakami and M. Mizuno, "Constitutive equations of irradiation creep and swelling incorporating their transient behavior", Nuclear Technology, (contributed).

A GRADIENT DEPENDENT VISCOPLASTIC MODEL FOR CLAY

Fusao OKA and Atsushi YASHIMA
Dept. of Civil Engr., Gifu University, Gifu, Japan

Toshihisa ADACHI
School of Transportation Engr., Kyoto University, Kyoto, Japan

Elias C. AIFANTIS, Dept. of Mech. Engr. and Engr. Mech.
Michigan Technological University, Houghton, USA

ABSTRACT

A gradient dependent viscoplastic constitutive model for water saturated clay is proposed. The linear perturbation stability analysis is applied to this model. We have disccused the instability of this model and the problem in the application of the gradient dependent viscoplastic model to the boundary value problem, the consolidation analysis by finite element method.

GRADIENT DEPENDENT VISCOPLASTIC MODEL FOR CLAY

Relating to the strain localization problem, new approach has been proposed where strain gradient is introduced into constitutive model[1]. In this section, we have introduced a second gradient of viscoplastic strain into constitutive model for clay[2]. Adachi and Oka [2] developed an elasto-viscoplastic constitutive model for clay based on the overstress type viscoplastic theory.

The viscoplastic flow rule is given by

$$\dot{\epsilon}_{ij}^{vp} =< \Phi(F) > \frac{\partial f}{\partial \sigma_{ij}} \tag{1}$$

where $\dot{\epsilon}_{ij}^{vp}$ is the viscoplastic strain rate, σ_{ij} is the stress tensor, f is the dynamic yield function, Φ is the material function for strain rate effect and $F = 0$ denotes the static yield function.

It has been found that the shear strength and deformation characteristics of clay depends on the volumetric strain. The volumetric strain is a measure of the deterioration of the granular materials. We have, therefore, introduced the second order gradient of volumetric strain into a constitutive model.

The yield function which includes a gradient term is expressed as follows:

$$f = \frac{\sqrt{2J_2}}{M^*\sigma'_m} + ln\frac{\sigma'_m}{\sigma'_{my}} + X = 0 \tag{2}$$

where X is the gradient term which consists of second order gradient of viscoplastic volumetric strain, J_2 is the second invariant of deviatoric stress tensor S_{ij} , σ'_m is the mean effective stress and σ'_{my} is the strain hardening parameter. In the following, we will use Terzaghi's effective stress σ'_{ij}.

Material function $\Phi(F)$ is given by

$$\Phi(F) = c \cdot exp\{m'(\frac{\sqrt{2J_2}}{M^*\sigma'_m} + ln\frac{\sigma'_m}{\sigma'_{me}} - \frac{1+e}{\lambda - \kappa}v^p + X)\} \tag{3}$$

$$c = c_0 exp\{m' ln \frac{\sigma'_{me}}{\sigma'_{my0}}\} \tag{4}$$

where we assume M^*, m' and c are material constants. In $Eq.(3)$, σ'_{me} is the initial value of σ'_m, λ is the consolidation index, κ is the swelling index, e is the void ratio, and v^p is the volumetric plastic strain.

Elastic strain rate is assumed to be given by the isotropic Hook's law.

TWO-DIMENSIONAL INSATBILITY ANALYSIS

In this section, we consider two dimensional motion under plane strain undrained condition with axial stresses in order to determine the preferred orientation of the growth of the fluctuation (the orientation of shear band)[3].

The plane strain undrained condition ($\dot{\epsilon}_{33} = 0$, $\dot{\epsilon}_{11} + \dot{\epsilon}_{22} = 0$) implies

$$S_{33} = 0, \quad \sigma'_3 = \frac{\sigma'_1 + \sigma'_2}{2} \tag{5}$$

$$\sigma'_m = \frac{\sigma'_1 + \sigma'_2}{2}, \quad \sqrt{2J_2} = \frac{|\sigma'_1 - \sigma'_2|}{2} \equiv \frac{q}{2}, \qquad \frac{q}{2} = \frac{\sigma'_1 - \sigma'_2}{2} \qquad (\sigma'_1 \geq \sigma'_2) \tag{6}$$

From the undrained condition and the initial condition ($v^p = 0$), $Eq.(3)$ becomes

$$\Phi(F) = c \cdot exp\{m'(\frac{\sqrt{2J_2}}{M^* p'} - \frac{\lambda(1+e)}{\kappa(\lambda - \kappa)} v^p - a_3 X')\} \tag{7}$$

where $p' = \sigma'_m$ and X' is the gradient term.

Next we consider the equilibrium equation, undrained condition and constitutive equation in the perturbed configuration. The perturbations of stresses σ_i, velocities v_i, volumetric viscoplastic strain v^p are assumed to be of the periodic form as:

$$\begin{aligned}
\tilde{\sigma} &= \left\{ \begin{matrix} \tilde{\sigma}_1 \\ \tilde{\sigma}_2 \end{matrix} \right\} &&= \sigma^* exp(iq(n_i x_i) + \omega t) \\
\sigma^* &= \left\{ \begin{matrix} \sigma^*_1 \\ \sigma^*_2 \end{matrix} \right\} && \\
\tilde{v}_i & &&= v^* exp(iq(n_i x_i) + \omega t) \\
v^p & &&= v^p exp(iq(n_i x_i) + \omega t)
\end{aligned} \tag{8}$$

where n_i is the component of the unit vector $\tilde{n}$ which is normal to shear band .

$$\tilde{n} = (-sin\theta, \ cos\theta) = (n_1, \ n_2) \tag{9}$$

From equilibrium equations, undarained condition and constitutive equations, following relation are obtained.

$$iq\sigma^*_1 n_1 \ = \ 0, \ iq\sigma^*_2 n_2 \ = \ 0, \ v^*_1 n_1 + v^*_2 n_2 = 0 \tag{10}$$

$$A_1 \sigma^*_1 + A_2 \sigma^*_2 - H_2 v^{p*} - H_3 X' - iqn_1 v^*_1 = 0 \tag{11}$$

$$\omega v^{p*} = H'_0 \sigma^*_1 + H'_1 \sigma^*_2 - H'_2 v^{p*} - H'_3 X' \tag{12}$$

We will consider two kinds of gradient term X'. One is related to isotropic one and the other to directional second gradient.

1) Isotropic gradient[5]

When we assume isotropy, the gradient term can be expressed by Laplacian as:

$$X' = \nabla^2 v^p. \tag{13}$$

Upon substitution of $Eq.(8)_3$, we obtain

$$X' = -q^2 v^{p*}. \tag{14}$$

In this case, the eigen value equation is obtained as:

$$\omega = -H_2' + H_3' q^2 \tag{15}$$

From $\omega = 0$, critical wave number q_c is obtained as:

$$q_c = \sqrt{\frac{a_1}{a_3}},\quad a_1 = \frac{\lambda(1+e)}{\kappa(\lambda-\kappa)},\quad l_c(\text{critical wave length}) = \frac{2\pi}{q_c}. \tag{16}$$

From $Eq.(15)$, it is seen that there is no preferred orientation of the band. Consequently, the perturbation with the wave length l which is larger than l_c never grows. In other words, it is seen that for the stiff clay with the larger value of a_1, the inhomogeneous solution with small wave length grows rapidly compared with soft clay.

2) Directional second gradient

We consider a directional differential operator expressed by operator DO as:

$$DO(1) = n_1' \frac{\partial}{\partial x_1} + n_2' \frac{\partial}{\partial x_2} \tag{17}$$

$$DO(2) = n_1'^2 \frac{\partial^2}{\partial x_1^2} + 2n_1' n_2' \frac{\partial^2}{\partial x_1 \partial x_2} + n_2'^2 \frac{\partial^2}{\partial x_2^2} \tag{18}$$

in which $n' = (n_1', n_2') \equiv (-sin\xi, cos\xi)$ is a arbitrary unit vector.

In order to obtain the directional gradient term X', we used this operator $DO(2)$ for volumetric viscoplatic stain. The eigenvlaue equation similar to $Eq.(15)$ becomes

$$\omega = -H_2 + Bq^2 H_3 \tag{19}$$

$$B = n_1'^2 sin^2\theta + n_2'^2 cos^2\theta - 2n_1' n_2' cos\theta sin\theta \tag{20}$$

The preferred orientation is obtained by maximazing the positive value of ω as:

$$tan2\theta = (2n_1' n_2')/(n_1'^2 - n_2'^2) = tan2\xi \tag{21}$$

in which ξ is the angle measuring the orientation of vector $n' = (-sin\xi, cos\xi)$. Vector n' is normal to the special plane in which direction we apply the differential operator $DO(1)$. We can assume that this special plane can be assumed to be paralell or related to the Mohr-Coulomb mobilized plane.

$$\xi = 45° + \frac{\phi_m}{2},\quad \sqrt{\frac{1+sin\phi_m}{1-sin\phi_m}} = \sqrt{\frac{\sigma_i}{\sigma_j}} \quad (i, j = 1, 2\ i < j) \tag{22}$$

The maximum value of growth speed ω is obtained as:

$$\text{Max}(\omega) = -H_2 + q - 2h_3,\quad when\ \theta = \xi. \tag{23}$$

FINITE ELEMENT ANALYSIS USING A GRADIENT DEPENDENT VISCOPLASTICITY THEORY

Higher order gradient term has been introduced into constitutive equation to simulate a localization phenomenon. However, there is still difficult problem to be solved when we apply to a constitutive equation with gradient term to the boundary value problem defined in the finite domain. Recently, Mühlhaus and Aifantis[6] pointed out the imporatnce of this and proposed the variational principle. In the present study, we assume the weak form of dynamic yield surface and related boundary condition for the first gradient of viscoplastic strain For the first gradient of volumetric viscoplastic strain. For the first gradient of v^p, we consider the flux $\tilde{Q}$ which associates with internal structure change.

$$\int_V (f - a_1 v^p - a_3 \nabla^2 v^p)\delta v^p dv + \int_{S_1} (\tilde{Q} - \nabla_n v^p)\delta v^p \tilde{n} ds = 0 \qquad (24)$$

$\tilde{Q}$ is the microstructure change flux, S_1 is the boundary where $\tilde{Q}$ is defined. δv^p is the virtual volumetric viscoplastic strain associated with virtual displacement δu during the deformation process. In the psresent study, viscoplastic strain rate is assumed to always occurs. From $Eq.$(24), we obtain the relation between the following equation of microstructure change flux Q.

$$\tilde{Q} = \nabla_n v^p \qquad (25)$$

where ∇_n is the normal gradient to the boundary.

For the total equilibirum equation, we use the well known virtual work theorem. When we implement the proposed model into FEM code, we need the boundary value of $\tilde{Q}$. In this respect, we postulate that the flux $\tilde{Q}$ is zero at the boundary between rigid or elastic material and viscoplastic one. From a physical point of view, we could say that from the material that will never deform plastically the flux Q never flows into the material, and also the flux Q never outflows from the viscoplastic material into rigid or elastic materials.

At the boundary where the stresses are specified like a free surface of the ground, the flux $\tilde{Q}$ also assumed to be zero. For the deformation analysis with consolidation, we could use the same method as used by Oka, Adachi and Okano(1986). They combined the finite element method with finite difference method. The equation of motion of fluid phase is discretized by backward finite difference scheme. The finite difference scheme is also used to calculate the second gradient of viscoplastic strain.

REFERENCES

1) E.C. Aifantis, ASME, J. Engr. Materials and Tech., 106, 326-330(1984).
2) T.Adachi, and F.Oka, Soils and Foundations, 22, 4,57-70(1982).
3) L.Anand, K.H. Kim, and T.G.Shawki, J. Mech. Phys. Solids, 35,407-429(1987).
4) F.Oka, T.Adachi, and Y.Okano, Int. J. Num. Anal. Methods in Geomechanics, 10, 1-16(1986).
5)H.M. Zbib, and E.C. Aifantis, Res. Mech., 23,261-305(1988).
6)Unpublished paper,E.C.Aifantis, private communication 1989.

A VISCOELASTIC–PLASTIC–DAMAGE MODEL FOR CONCRETE

V. P. PANOSKALTSIS, J. LUBLINER, and P. J. M. MONTEIRO
Department of Civil Engineering, University of California, Berkeley, CA 94720, USA

ABSTRACT

A modification of the Kuhn model of linear viscoelasticity (originally formulated for rubber), extended to include nonlinearity, is combined with the plastic–damage model due to Lubliner et al. in order to describe the nonlinear behavior of concrete at various rates of loading or deformation. An efficient algorithm is developed for the integration of the rate equations, and the results of some calculations for uniaxial compression are shown.

1. INTRODUCTION

Models for concrete usually fall in one of two categories: a) viscoelastic models, which describe the behavior of the material when it is loaded well below its compressive strength (f'_c), and b) elastic–plastic models describing the behavior of the material around and after it reaches its compressive strength. This paper presents a model combining the two, resulting in a representation of the rate effect on the entire stress-strain curve. The viscoelastic model is discussed in Section 2 and the plastic–damage model is discussed in Section 3. Some numerical results are described in Section 4.

2. THE MODIFIED KUHN MODEL OF LINEAR VISCOELASTICITY

In 1947 Kuhn et al. [1] published a viscoelastic model for rubber to account for the fact that the creep function $J(t)$ had been observed over many logarithmic cycles of time to take the form $J(t) = A + B\ln t$, with the time derivative $\dot{J}(t) = B/t$. Clearly, this creep function cannot account for short-term behavior. To remedy this, Kuhn et al. proposed to replace it by one whose derivative is

$$\dot{J}(t) = \frac{B}{t}(1 - e^{-Ct}), \tag{1}$$

which has the same behavior for $t \gg 1/C$, but has the finite limit BC as $t \to 0$, The creep function derived by Kuhn et al. by integrating Equation (1) is 0 at $t = 0$, accounting for the fact that the instantaneous deformation of rubber is negligible compared with later deformation. The creep of concrete has also been fitted fairly well with a logarithmic creep equation.

In order to account for instantaneous elasticity, a non-vanishing constant of integration A is included in the integral of Equation (1), with the resulting creep function may be written in the relaxation-time superposition form

$$J(t) = A + B\int_{1/C}^{\infty}(1 - e^{-t/\tau})\frac{d\tau}{\tau}.$$

This representation, with its continuous distribution of relaxation times, may be approximated by a discrete model composed of a spring with spring constant $1/A$ and $N+1$ Kelvin elements in series, in which the spring constant of each one is $G_m = 1/(B\ln r)$, and the relaxation times are distributed as $\tau_m = r^m/C$, where $r > 1$ is a constant and $m = 0, \ldots\ldots, N$. The creep function of this model is

$$J_M(t) = A + B\ln r\sum_{m=0}^{N}(1 - e^{-Ct/r^m})$$

It can be shown that $J_M(t) \to J(t)$ as $N \to \infty$ and $r \to 1+$ (with $r^N \to \infty$). This creep function, including the limiting case, will be said to represent the **modified Kuhn model.**

The model may be applied to describe the dependence of the loss tangent on the frequency in dynamic tests. With a suitable choice of the parameters A, B, C, r and N (note that the *number* of parameters is unaffected by the number of elements), the model predicts a nearly constant loss tangent over a wide frequency range, equivalent to what is usually called *structural damping*. Alternatively, the parameters may be chosen to fit experimental results. Such a fit was performed by means of the Levenberg–Marquardt algorithm for data obtained on concrete specimens from the Cypress structure destroyed in the Loma Prieta earthquake; the results are shown in Figure 1.

An extension of the modified Kuhn model to account for nonlinear viscoelasticity is based on the "viscous" strain represented as $\varepsilon^v = \sum_{m=0}^{N} q_m$, where q_m is the partial strain due to the mth Kelvin element and is assumed according to the Eyring rate-process theory to be governed by

$$\dot{q}_m + \frac{C}{r^m}q_m = \frac{BC\ln r}{r^m}\Gamma\sinh\left(\frac{\sigma}{\Gamma}\right), \tag{2}$$

where Γ may be defined as a stress at which nonlinear creep becomes noticeable. A further extension to three-dimensional deformation may be based on the assumption that the viscous strain is purely deviatoric and is governed by a Mises-type potential.

3. THE VISCOELASTIC–PLASTIC–DAMAGE MODEL

In order to describe nonlinear behavior, which includes plasticity and damage, the preceding model is combined in series with the plastic–damage model due to Lubliner et al. [2]. The strain tensor $\boldsymbol{\varepsilon}$ is then decomposed as $\boldsymbol{\varepsilon} = \boldsymbol{\varepsilon}^e + \boldsymbol{\varepsilon}^v + \boldsymbol{\varepsilon}^p$.

In the plastic–damage model [2], the yield surface governing plastic deformation is given by $F(\boldsymbol{\sigma}) - c = 0$, where $F(\boldsymbol{\sigma})$ is a homogeneous function of the first degree in the stresses, and c is the *cohesion*, a scalar variable that may be scaled so that its initial value is the initial yield strength in uniaxial compression, f_{c0}. A form of $F(\boldsymbol{\sigma})$ that has been found to be suitable for concrete is

$$F(\boldsymbol{\sigma}) = \frac{1}{1-\alpha}[\sqrt{3J_2} + \alpha I_1 + \beta\sigma_{\max} - \gamma < -\sigma_{\max} >], \tag{3}$$

where α, β, γ are dimensionless constants, I_1 is the first invariant of stress, and $\sigma_{\max}$ is the algebraically largest principal stress. The internal variables are the plastic strain tensor $\boldsymbol{\varepsilon}^p$, the plastic-damage variable (or "hardening parameter") κ and the cohesion c. Their respective rate equations are

$$\dot{\boldsymbol{\varepsilon}}^p = \dot{\lambda}\,\frac{\partial G}{\partial \boldsymbol{\sigma}}, \qquad \dot{\kappa} = \mathbf{h}^T(\boldsymbol{\sigma}, c, \kappa)\,\dot{\boldsymbol{\varepsilon}}^p, \qquad \dot{c} = k(\boldsymbol{\sigma}, c, \kappa)\,\dot{\kappa}, \tag{4}$$

where $\dot{\lambda}$ is the plastic loading factor (determined by the consistency condition) and G is a plastic potential which may be identified with the yield function for associated plasticity (see [3], for example), while $\mathbf{h}$ and k are respectively matrix- and scalar-valued functions of the indicated arguments. Under certain circumstances (for example, if no two principal stresses have opposite sign), a direct relation between c and κ, $c = c(\kappa)$, may be assumed instead of the rate equation $(4)_3$. The evolution of c is such that after it reaches a maximum value (for which the yield surface coincides with the failure surface) it decreases (softening range) to zero, a condition representing total damage of the material. The plastic-damage variable κ never decreases, and increases only when plastic deformation takes place, up to a limiting value corresponding to the vanishing of c. It is nondimensionalized so that its maximum value is 1; thus $c = f_{c0}$ when $\kappa = 0$ and $c = 0$ when $\kappa = 1$. A form of $c(\kappa)$ that can describe a variety of stress-strain curves for concrete, in both tension and compression, is

$$c(\kappa) = \frac{f_{c0}}{a}[(1+a)\sqrt{\phi(\kappa)} - \phi(\kappa)] \tag{5}$$

where $\phi(\kappa) = 1 + a(2+a)\kappa$, a being a dimensionless constant.

In addition to the plastic damage described by the loss of cohesion, the model accounts for stiffness degradation, i.e. change in the elastic stiffness $\mathbf{D}$ during deformation, which is described by additional internal variables and can take place both before and after plastic deformation begins. Here it is assumed that degradation takes place only in the softening range of plastic deformation and is defined by the single damage variable δ so that the degraded stiffness $\mathbf{D}$ is related to the initial stiffness $\mathbf{D}_0$ by $\mathbf{D} = (1-\delta)\mathbf{D}_0$, the rate equation of δ being

$$\dot{\delta} = \frac{1-\delta}{c} < -\dot{c} >;$$

hence

$$\boldsymbol{\sigma} = \frac{c}{c_{\max}}\mathbf{D}\boldsymbol{\varepsilon}^e. \tag{6}$$

4. EXAMPLE: UNIAXIAL COMPRESSION

As an application of the model described we consider the case of uniaxial compression under strain control. Let $-\sigma$, with $\sigma > 0$, denote the only nonvanishing principal stress, so that $\sigma_{\max} = 0$ and therefore the yield criterion is just $\sigma - c = 0$. Furthermore let $-\varepsilon$ be the principal strain conjugate to $-\sigma$, while the transverse principal strains are ε_t. Thus, with the aforementioned decomposition of ε, we have $\varepsilon^e = [c_{\max}/c(\kappa)]\sigma/E$ (where E is the initial Young's modulus) and $\varepsilon_t^e = \nu\varepsilon^e$ (where ν is the Poisson's ratio, constant in view of Equation (6)); $\varepsilon^v = \sum_m q_m$ and $\varepsilon_t^v = -\frac{1}{2}\varepsilon^v$, where the q_m are governed by (2), while ε^p and ε_t^p are governed according to $(4)_1$ by

$$\left\{ \begin{array}{c} \dot{\varepsilon}^p \\ \dot{\varepsilon}_t^p \end{array} \right\} = \dot{\gamma}\sigma \left\{ \begin{array}{c} 2(1-\alpha) \\ 1+2\alpha \end{array} \right\}, \qquad \text{where } \dot{\gamma} = \dot{\lambda}\frac{1}{1-\alpha},$$

where α is that appearing in (3); and κ is governed by

$$\dot{\kappa} = 2(1-\alpha)\dot{\gamma}\frac{c(\kappa)}{g_c},$$

with $c(\kappa)$ given by Equation (5). In the one-dimensional case g_c can be defined as G_c/l, where G_c is an assumed material property with dimension of energy/area and l is a characteristic length related to the softening behavior ([2]).

The integration of the above rate equations is accomplished by discretizing them by means of the backward Euler method, with the yield condition $(F - c)_{n+1} = 0$ enforced incrementally at every time step. In order to solve the resulting system of six equations the predictor-corrector algorithm is used [4]. A fundamental difference with the algorithm commonly used in elastoplastic analysis is the fact that in the predictor phase some of the internal variables will change (namely the viscous strains). During the corrector phase the total strain ε_{n+1} remains constant, while the plastic internal variables as well as the viscous strains evolve. The algebraic system of the discretized equations is solved by linearization (Newton's method) with respect to the variables σ, $q_0, q_1, \ldots q_N$ in the predictor phase and σ, $q_0, q_1, \ldots q_N, \gamma, \varepsilon^p, \kappa$ in the corrector. Results showing the effect of strain rate on the stress-strain curve are shown in Figure 2.

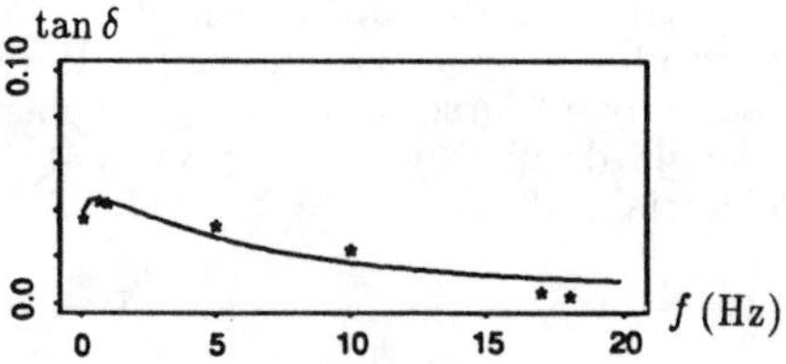

Figure 1

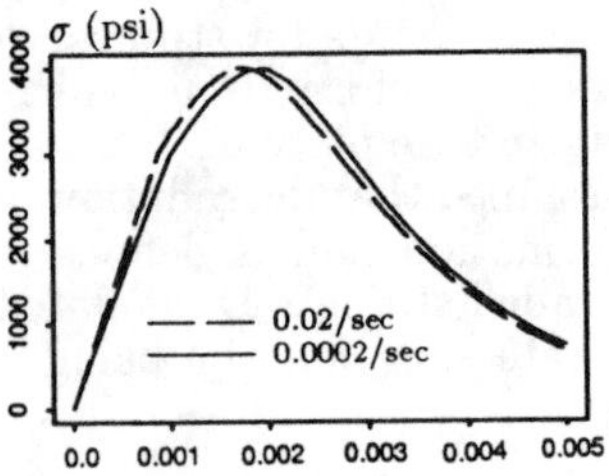

Figure 2

REFERENCES

1. Kuhn, W., O. Künzle and A. Preissmann, *Helv. Chim. Acta*, Vol. 30, pp. 307-328 and 464-486 (1947).

2. Lubliner, J., J. Oliver, S. Oller and E. Oñate, *Int. J. Solids Str.*, Vol. 25, pp. 299-326 (1989).

3. Lubliner, J., *Plasticity Theory*, Macmillan (1990).

4. Simo, J.C. and T.J.R. Hughes, *Elastoplasticity and Viscoplasticity, Computational Aspects*, Stanford Univ., Dept. of Appl. Mech. (1988).

A CONCEPT FOR THE EXTENSION OF ELASTIC-PLASTIC CONSTITUTIVE EQUATIONS INTO RATE-DEPENDENT MODELS

U. ROTT, O. T. BRUHNS
Ruhr-Universität Bochum, Lehrstuhl für Mechanik I
4630 Bochum 1, West Germany

ABSTRACT

A concept is presented which enables the extension of elastic-plastic constitutive equations into rate-dependent ones. Thus, two classes of material functions can be determined successively and the elastic-plastic limit is obtained accurately.

INTRODUCTION

Many technically important alloys exhibit rate-dependence which cannot be neglected, if they are loaded at increased temperature. Constitutive equations developed for this purpose should therefore describe quantitatively not only such effects as hardening and the achievement of a saturated curve due to cyclic loading, but also creep, relaxation, and the increase in stress at elevated rates of loading. Thus, the basic constitutive equations become rather complex and the number of material functions is usually fairly high. Therefore, problems arise in determining these functions.
A concept is presented here which enables the development of rate-dependent models on the basis of approved elastic-plastic constitutive equations with the aim to separate rate-dependent from rate-independent functions. This allows both the application of an appropriate strategy in order to evaluate the material functions and the attainment of an exact limiting value in the case of quasistatic processes. Ordinary viscoplastic models of overstress-type show purely elastic response in the limiting case whereas elastic-plastic behaviour should be predicted.

BASIC IDEAS

The most essential idea of this concept consists of a division of the actual stress tensor σ into an athermal part $\bar{\sigma}$ that is related to the quasi-static state of balance and in a viscous (thermally activated) overstress $\bar{\bar{\sigma}}$ which only arises in case of a non-vanishing rate of loading [1]. Here it is assumed that $\bar{\sigma}$ may be obtained by projecting the actual stress σ onto the yield-surface. Therefore extending the uniaxial evolution equation

$$\dot{\bar{\sigma}} = \langle A \, \dot{\sigma} \rangle - \langle\!\langle B(\sigma - \bar{\sigma}) \rangle\!\rangle \tag{1}$$

to multiaxial case leads to only one additional differential equation for Λ which serves as a measure for the distance between deviators σ' and $\bar{\sigma}'$. Equation (1) contains two material functions $A(\Lambda, v)$ and $B(||\sigma'||, ||\sigma' - \xi||)$ and a variable v describing the rate of loading

$$v = \dot{\sigma} \cdot \mathbf{n} \tag{2}$$

with the unit-tensor $\mathbf{n}$ normal to the yield-surface. Analogously to the separation of stress the inelastic deformation rate is divided into an elastic part $\dot{\varepsilon}_e$, a part $\dot{\varepsilon}_p$ which is equal to the plastic deformation rate in case of quasistatic processes, and an additional viscous part $\dot{\varepsilon}_v$.

APPLICATIONS

As an example for the application of this basic concept inelastic deformations of austenitic stainless steel AISI 316L mod at 550 °C are considered [2]. Quasistatic behaviour of such materials can be accurately described by the INTERATOM-model using two internal variables $\varkappa$ and ξ in order to predict both hardening and saturation due to cyclic loading . Employing the concept leads to the following set of constitutive equations (for details see [3])

$$\dot{\varepsilon} = \dot{\varepsilon}_e + \dot{\varepsilon}_p + \dot{\varepsilon}_v = \frac{1}{2G}\left[\dot{\sigma} - \frac{\nu}{1+\nu}\,\mathrm{tr}(\dot{\sigma})\,\mathbf{1}\right] + \lambda\,\mathbf{n} + «\Phi»\,\mathbf{n}\,, \tag{3}$$

$$\dot{\xi} = c(\varkappa)\,\lambda\,\mathbf{n}\,, \tag{4}$$

$$\dot{\varkappa} = \lambda\,(\bar{\sigma}' - \xi)\cdot\mathbf{n} \tag{5}$$

in connection with (1), v. Mises yield-condition

$$f = (\bar{\sigma}' - \xi)\cdot(\bar{\sigma}' - \xi) - g(\varkappa) = 0\,, \tag{6}$$

the constraints

$$< x > = \begin{cases} x, \text{ if } f(\bar{\sigma}) = 0 \text{ and } v > 0 \\ 0 \text{ otherwise} \end{cases}, \tag{7}$$

$$« x » = \begin{cases} x, \text{ if } f(\sigma) \geq 0 \\ 0 \text{ otherwise} \end{cases} \tag{8}$$

and

$$\Lambda = « \sqrt{f} - \sqrt{g} » \quad . \tag{9}$$

Factor λ can be evaluated from consistency-condition $\dot{f} = 0$ as

$$\lambda = \frac{\dot{\bar{\sigma}} \cdot \mathbf{n}}{c + \frac{1}{2}\frac{\partial g}{\partial \varkappa}} = \frac{\dot{\bar{\sigma}} \cdot \mathbf{n}}{K} = \langle \frac{A}{K} \dot{\sigma} \cdot \mathbf{n} \rangle + \langle\langle \frac{B}{K} \Lambda \rangle\rangle . \tag{10}$$

It may be noted that the loading condition only effects one part of λ which is important for the proof of uniqueness. This has to be shown, because inverting (3) results in a transcendent equation in $\dot{\sigma}$ which possesses a unique solution provided that for

$$A := a_1(\Lambda)\ a_2(v) \tag{11}$$

the relation

$$\frac{d a_2}{d(\frac{1}{v})} \geq 0 \tag{12}$$

holds. During quasistatic processes the rate of loading v vanishes and so does Λ, if A is chosen properly. Therefore (3) - (5) reduce to the limiting case of the underlying elastic-plastic model.
This property enables an exact separation of rate-independent material functions (e. g. $c(\varkappa)$ and $g(\varkappa)$) from those which influence the rate-dependency. Suitable equations for $c(\varkappa)$ and $g(\varkappa)$ are presented in [3], their parameters can be determined in well-known manner by fitting to quasistatic uniaxial tension and compression tests. Thus, the identification remains of only two functions A and B for the evolution of $\bar{\sigma}$ and a third one for evaluation of inelastic deformation-rate.
For their determination a strategy can be employed which is based on the reduction of the material functions into experimentally observable quantities. By means of numerical integration the desired functions can be expressed by discrete values of arbitrary density. This set is approximated by analytical functions using an optimization algorithm.
To demonstrate the practicability of this procedure, fig. 1 and 2 display data acquired by uniaxial tension and creep tests [2] in comparison to those predicted by the constitutive model (denoted by circles).

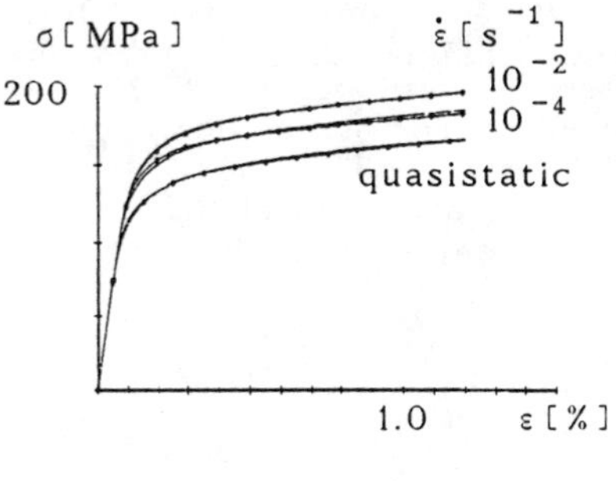

FIG. 1. Tension tests

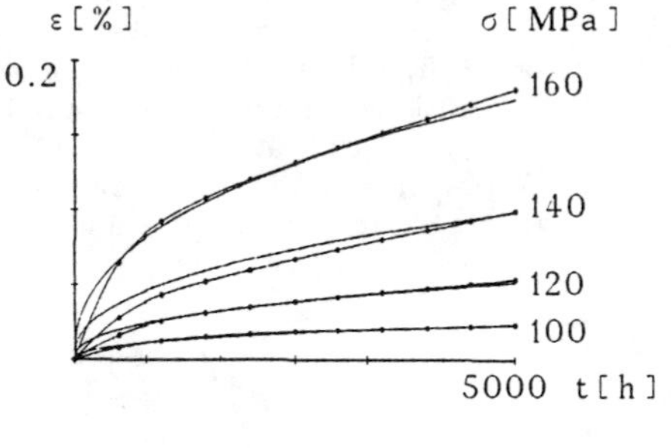

FIG. 2. Creep tests

In both cases the results are very satisfying. Above all, the agreement of not only stationary but also primary creep is excellent.
Fig. 3 and 4 show the results of a FE calculation of a thick-walled tube subjected to internal pressure according to the loading history displayed in the upper left-hand corner of fig. 3. As observed in [3] rather high residual stresses remain at the inner wall after unloading.

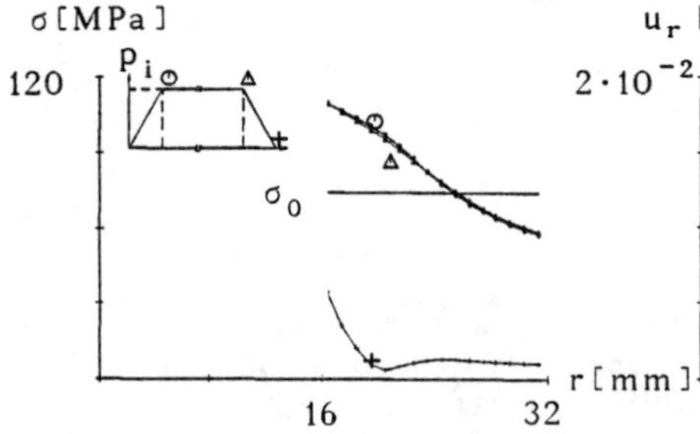

FIG. 3. Equivalent stresses

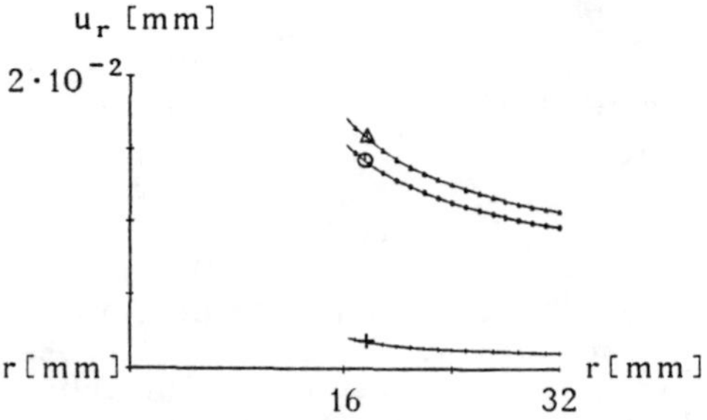

FIG. 4. Radial displacements

CONCLUSIONS

Decomposition of actual stress into an athermal and a thermally activated part results in separating the rate-independent material functions from the rate-dependent ones provided that overstress depends on the rate of loading. Thus, those groups of material functions can be treated successively by which both quasistatic and rate-dependent behaviour are modelled accurately.

References

1. O. T. Bruhns, Th. Lehmann, "Optimum deformation rate in large inelastic deformation" in Metal Forming Plasticity , IUTAM Symp. Tutzing 1978, edited by H. Lippmann (Springer-Verlag, 1979)
2. RCC-MR Design and Construction Rules for Mechanical Components of FBR Nuclear Islands, AFCEN (June 1985)
3. O. T. Bruhns et al., "The constitutive relations of elastic-inelastic materials at small strains", Nucl. Engng. Design 83, 325 - 331 (1984)

VISCOPLASTIC MODEL OF A CLAY USING HIERARCHICAL APPROACH

N. C. Samtani
Graduate Student,
Dept. of Civil Engrg. & Engrg. Mech., University of Arizona, Tucson, USA

C. S. Desai
Regents Prof. and Head,
Dept. of Civil Engrg. & Engrg. Mech., University of Arizona, Tucson, USA

L. Vulliet
De Cerenville Geotechnique SA, Ecublens, Switzerland

ABSTRACT

Soil from an active landslide at Villarbeney, Switzerland, is modeled using Perzyna's viscoplastic theory. Perzyna's theory is similar to the usual theory of plasticity and permits the use of any plasticity based model. Herein, the plasticity based generalized hierarchical single surface (HiSS) model proposed by Desai and co-workers is used. It is shown that the combination of the various merits of HiSS model and the simplicity of Perzyna's theory can provide satisfactory model for viscoplastic response of the soil.

INTRODUCTION

An extensive and detailed experimental program was devised to test Villarbeney Clay from Switzerland in connection to a project involving time-dependent elasto-plastic (i.e viscoplastic) analysis of a slowly moving (creeping) landslide occuring since generations at Villarbeney. The present paper presents the viscoplastic model and verification using the HiSS approach.

VILLARBENEY CLAY

Twenty kilos of clay, collected from the central most active part of the Villarbeney landslide, was obtained. According to the Unified Soil Classification System, the Villarbeney clay is classified as *CL* which means inorganic clays of low or medium plasticity. Further details of soil can be found in [1].

HIERARCHICAL VISCOPLASTIC MATERIAL MODEL

A variety of viscoplastic theories are available. Perzyna's theory of viscoplasticity is very similar to the theory of usual plasticity and affords the use of the concept of yield function. Herein, the single surface yield function as proposed by Desai and co-workers [2-4] in the Hierarchical model is adopted in conjunction with the Perzyna's theory. In the Hierarchical viscoplastic material model, small strain elasto–viscoplasticity is adopted and the strain tensor is additively decomposed into an elastic and viscoplastic part as follows

$$\epsilon_{ij} = \epsilon^{e}_{ij} + \epsilon^{vp}_{ij} \tag{1}$$

$$\text{where} \quad \epsilon^{e}_{ij} = C^{e^{-1}}_{ijkl}\sigma_{kl} \qquad \dot{\epsilon}^{vp}_{ij} = \Gamma < \phi > \frac{\partial F}{\partial \sigma_{ij}} \tag{2-a,b}$$

$C^{e^{-1}}_{ijkl}$ is the elastic tensor, $\dot{\epsilon}^{vp}_{ij}$ is the inelastic strain rate tensor, F is the yield function, Γ is a material parameter (also known as the *fluidity parameter*) and ϕ is a function of F. The angle bracket $<>$ has the meaning of switch-on- switch-off operator as follows (F_o is a normalizing constant):

$$< \phi(F) >= \begin{cases} < \phi\left(\frac{F}{F_o}\right) >, & \text{if } \frac{F}{F_o} > 0; \\ 0, & \text{if } \frac{F}{F_o} \leq 0. \end{cases} \tag{3}$$

The following dimensionless form of the hierarchical single surface (HiSS) model is used [2-4]

$$\frac{F}{p_a} = \frac{J_{2D}}{p_a^2} - \left[-\alpha\left(\frac{J_1}{p_a}\right)^n + \gamma\left(\frac{J_1}{p_a}\right)^2\right](1 - \beta S_r)^m \tag{4}$$

where

$$S_r = \frac{\sqrt{27}}{2} J_{3D} J_{2D}^{-1.5} \tag{4 - a}$$

is the stress ratio, J_1 is the first invariant of the stress tensor, J_{2D} and J_{3D} are the second and third invariants of the deviatoric stress tensor, α, β, γ, m and n are the material response functions and p_a is the atmospheric pressure.

CALIBRATION OF HiSS MODEL FOR VILLARBENEY CLAY

The proposed model can be calibrated for Villarbeney Clay based on careful consideration of the elastic stage (E, ν), plastic stage accompanied by hardening (α, n, γ, β, m) and the transient phase from a given stress state to the ultimate (defined by $\Gamma < \phi >$).

For evaluating these material constants, a number of laboratory tests along different stress paths are required. Tests along four different stress paths, viz., Oedometer test (K_o test), isotropic consolidation (HC test), conventional triaxial compression (CTC test) and reduced triaxial compression (RTE test) were performed using the samples formed from a slurry. Material from the main supply or batch of remolded clay from Villarbeney was again remolded (from slurry state) for triaxial tests using a cylindrical mold. For determination of viscous constants, creep tests are required. Creep tests reported by Vulliet [5] on Villarbeney clay are adopted herein. The plasticity constants need to be determined before the viscous constants can be determined because the viscous constants (are shown to) depend upon the plastic hardening process.

Hardening

In the case of normally consolidated clay, hardening is mainly conditioned by the volumetric plastic strain [4]. Accordingly, a simple hardening law is adopted as follows,

$$\alpha = \frac{a_1}{\xi_v^{\eta_1}} \tag{5}$$

where ξ_v is the trajectory of the volumetric plastic strains. The parameters a_1 and η_1 can be determined from the HC test wherein only volumetric strains occur.

Viscoplastic Flow (Transient) Process

The flow process in a viscoplastic problem is simulated by prescribing a functional form for $< \phi(F) >$. Equation 2-b can be rewritten [1], as

$$\dot{I}_2^{vp} = [\Gamma < \phi >]^2 \frac{1}{2} \frac{\partial F}{\partial \sigma_{ij}} \frac{\partial F}{\partial \sigma_{ij}} \; (= [\Gamma < \phi >]^2 a) \tag{6}$$

where $\dot{I}_2^{vp}$ the second invariant of the inelastic strain-rate tensor. Equation (6) is a general equation for creep experiments using Perzyna's theory. In a creep test with constant stress, $\dot{I}_2^{vp}$ is found by measuring the viscoplastic strain rate (see Eq. (2-b)). Although σ_{ij} is constant, the value of $\partial F/\partial \sigma_{ij}$ will change as F changes with increasing viscoplastic strain (hardening). Thus, in this method the hardening process is made an integral part of the flow process.

The form used for the flow function $< \phi >$ depends upon the experimental data. Since $< \phi >$ involves some function of F/F_o, a plot of F/F_o versus a is required. Using undrained simple shear creep tests [1,3,5], the following function was obtained

$$< \phi >= (F/p_a)^N \tag{7}$$

where N is a new parameter which together with Γ specifies the nature and rate of the transient process. Since a dimensionless form of yield function is adopted the value of F_o is adopted as one. Further details about this method can be found in [1].

The list of material parameters for the Villarbeney clay is given in Table 1. Details for determination of other parameters can be found in [2],[4].

VERIFICATION AND BACK PREDICTION OF TESTS

A reliable constitutive model should be able to backpredict laboratory test used in evaluating its material constants as well as other tests which are not used for finding the constants. The backprediction of the stress-strain response relationship is obtained by integrating the incremental stress-strain relation

$$\{d\sigma\} = [C^{ep}]\{d\epsilon\} \tag{8-a}$$

or its inverse relation

$$\{d\epsilon\} = [D^{ep}]\{d\sigma\} \tag{8-b}$$

where $\{d\sigma\}$ and $\{d\epsilon\}$ are the incremental stress and strain vectors, respectively, $[C^{ep}]$ is the elastoplastic constitutive matrix and $[D^{ep}]$ is the elastoplastic compliance matrix (inverse of $[C^{ep}]$). Equation (8-a) is used for strain-controlled test and Eq. (8-b) for stress controlled tests.

Figure 1 a,b,c shows the verification of the model with respect to a typical CTC test conducted at a confining pressure of 103 kPa (15 psi). Note that this test was not used to determine the hardening parameters. Figure 2 shows the verification with respect to a creep test in which σ=200 kPa and τ=58.80 kPa were applied. Since, herein, creep test is of the simple shear type, the shear strain is plotted against time. It can be seen that the model predicts/backpredicts the laboratory tests satisfactorily.

CONCLUSIONS

A viscoplastic model for clays has been developed using the Hierarchical approach. The model is in a generalized form. The parameters for the model are determined from representative tests. The method of determination of viscous parameters is quite general and infact can be used with any plasticity model based on the hardening concept.

ACKNOWLEDGEMENTS

Part of the research herein was supported by grant No. MSM 8618901/14 from the National Science Foundation, Washington D.C. Dr. L. Vulliet participated in the research, and was supported during his stay, at the University of Arizona, Tucson, by grant No. 82.480.0.87 from the Swiss Science Foundation, Switzerland. The Villarbeney clay was provided by Prof. E. Recordon, Soil Mechanincs laboratory of the Swiss Federal Institute of Technology, Lausanne for testing at the University of Arizona.

TABLE 1.

Material Parameters for Villarbeney Clay

Description	Symbol	Value
Elastic modulus	E	10335 kPa
Poisson's ratio	ν	0.353
Shape parameter	m	-0.5
Ultimate parameters	γ	1.93
	β	0.64
Phase change parameter	n	2.04
Hardening parameters	a_1	1.47
	η_1	0.06
Fluidity parameter	Γ	1.47E-04 /min
Viscoplastic exponent	N	2.58

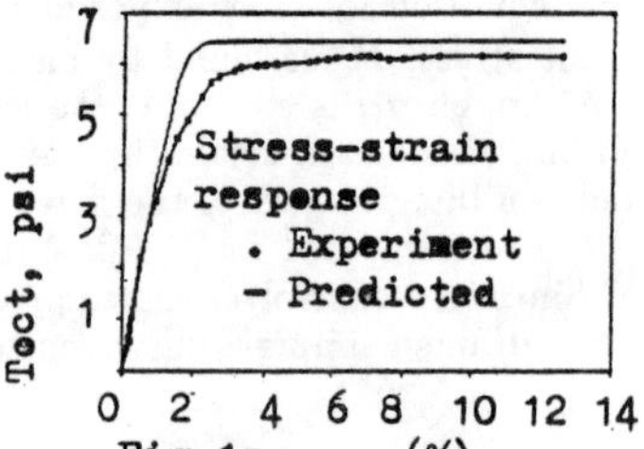

Fig 1a

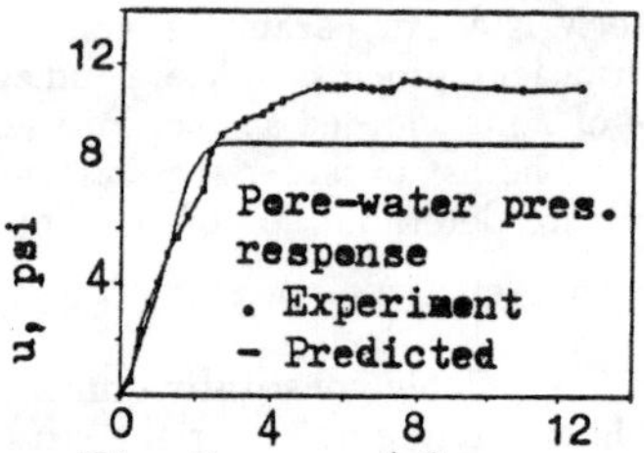

Fig 1b

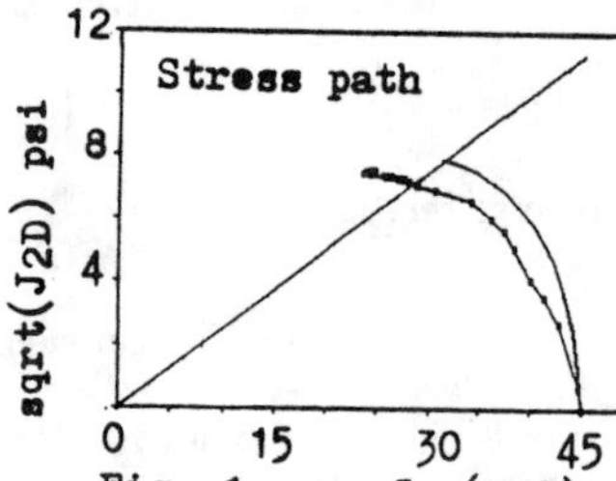

Fig. 1c

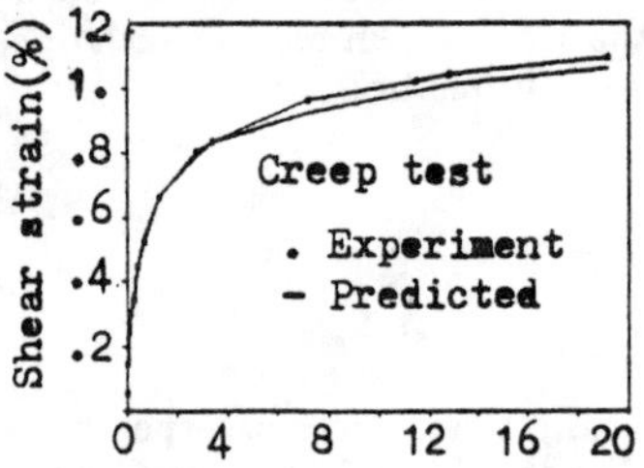

Fig 2

REFERENCES

1. Samtani, N. C., "Constitutive Modeling and Finite Element Analysis of Slowly Moving Landslides using Hierarchical Viscoplastic Material Model," Ph.D. Dissertation, Under Preparation, University of Arizona, Tucson, AZ, 1990
2. Desai, C. S., Somasundaram, S. and Frantziskonis G., "A Hierarchical Approach for Constitutive Modeling of Geologic Materials," Int. J. Num. Analyt. Meth. in Geomech., (10), 225-257, 1986.
3. Desai, C. S., and Zhang, ''Viscoplastic Model for Geologic Materials with Generalized Flow Rule," Int. J. Num. Analyt. Meth. in Geomech., (11), 603-620, 1987.
4. Wathugula, G. W., and Desai, C. S."Finite Element Dynamic Analysis of Nonlinear Porous Media with Applications to Piles in Saturated Clays," Report, University of Arizona, Tucson, AZ, 1990.
5. Vulliet, L., Samtani, N., and Desai, C. S. "Material Parameters for an Elasto-Viscoplastic Law", to be published, *Xth* European Conf. Soil Mech. and Found. Eng., Firenze, Italy, May 14-21, 1991.
6. Vulliet, L., "Modelisation des pentes Naturelles en mouvement", Ph.D. Dissertation, These No 635, EPFL, Lausanne, Switzerland, 1986.

RHEOLOGICAL PROPERTIES OF SOFT SOILS

LUDVIK TRAUNER
Professor, University of Maribor, Yugoslavia

ABSTRACT

A theoretical basis for the rheological properties of soft soils is presented in the form of an elastoviscoplastic model. Solutions are given for a nonlinear Kelvin model taking into account nonlinear and time dependent stress-strain relations in coherent soils but neglecting primary consolidation as an hydrodynamic effect. The results of tests conducted over a period of several years under conditions of slow, drained and monotonic change of triaxial stresses are presented. Rheological properties of three soil specimens are determined.

INTRODUCTION

Dependencies in soils are nonlinear and in general highly complicated to describe. They are time dependant and vary with nonhomogenity, anisotropy, permeability and soil viscosity. They depend upon stress and strain history as well as on the loading rate. Diffusion and dissipation of pore pressure within the soils must be taken into account together with other consolidation processes. It is neither efficient nor practical to consider these complicated factors in applications. For this reason some simplifications must be considered, still giving satisfying results.

The paper presents an analytical interpretation of long term experimental research on rheological properties of soft soils. The results of slow, drained and monotonically changing triaxial tests are shown. Rheological properties for three soils specimen were derived: normally consolidated clay (CI), lake clay (CH) and unconsolidated organic sea clay (OI-CI).

ANALYTIC FORM OF RHEOLOGICAL DEPENDENCIES

In the analytic interpretation relations between isotropic compresion $\varepsilon^{(o)}$, dilatation $\varepsilon^{(d)}$ and secondary consolidation $\varepsilon^{(v)}$, which were determined separately for volumetric (ε_v) and axial (ε_a) deformation from test results on slow, drained and monotonically loaded specimens, were considered.

The results can be written in the following form:

$$\varepsilon_k = \varepsilon_k^{(o)} + \varepsilon_k^{(d)} + \varepsilon_k^{(v)}, \quad k = v \text{ or } a \tag{1}$$

were symbols (o), (d) and (v) represent isotropic compression, dilatation and secondary consolidation and are given by:

$$\varepsilon_k^{(o)} = \{a_{k1} + a_{k2}(p/p_o)^{m_{k1}}\} \ln(p/p_o)^{n_{k1}} \tag{2}$$

$$\varepsilon_k^{(d)} = \{a_{k3} + a_{k4}(p/p_o)^{m_{k2}}\}(q/p - \eta_o)^{n_{k2}} \tag{3}$$

$$\varepsilon_k^{(v)} = \alpha_k \ln(t^*/t_o) = -\alpha_k \ln(\dot{\varepsilon}_k/\dot{\varepsilon}_{ko}) \tag{4}$$

$$\alpha_k = a_{k5} + a_{k6} \ln(p/p_o)^{m_{k3}} + a_{k7}(q/p - \eta_o)^{n_{k3}} \tag{5}$$

Terms p and q are stress invariants defind as:

$$p = \sigma'_{ii}/3 \quad , \quad q = \{(3/2)S'_{ij}S'_{ij}\}^{1/2} \tag{6),(7}$$

$$S'_{ij} = \sigma'_{ij} - \delta_{ij}\sigma'_{rr} = S_{ij} \tag{8}$$

were t* means relative time; p_o, q_o and $\eta_o = q_o/p_o$ are values of p, q and $\eta = q/p$ at the time $t^* = t_o$; δ_{ij} is Kronecker delta and σ'_{ij} the component of the effective stress tensor. The deformations ε_k (1) can be further expressed by elastic $\varepsilon_k^{(e)}$ and plastic $\varepsilon_k^{(p)}$ components.

The elastic (reversible) component is given by similar relations as in Egs. (1) to (8) but without the viscous components (4) and (5)

$$\varepsilon_k^{(e)} = (\varepsilon_k^{(o)})^{(e)} + (\varepsilon_k^{(d)})^{(e)} \tag{9}$$

where

$$(\varepsilon_k^{(o)}) = \{b_{k1} + b_{k2}(p/p_o)^{m_{k1}}\} \ln(p/p_o)^{n_{k1}} \tag{10}$$

$$(\varepsilon_k^{(d)}) = \{b_{k3} + b_{k4}(p/p_o)^{m_{k2}}\}(q/p - \eta_o)^{n_{k2}} \tag{11}$$

Subtracting Egs. (9) and (4) from (1) one obtains plastic strain components $\varepsilon_k^{(p)}$:

$$\varepsilon_k^{(p)} = \varepsilon_k - \varepsilon_k^{(e)} - \varepsilon_k^{(v)} \tag{12}$$

represented by the following components $\varepsilon_k^{(p)}$:

$$\varepsilon_k^{(p)} = (\varepsilon_k^{(o)})^{(p)} + (\varepsilon_k^{(d)})^{(p)} \tag{13}$$

EXPERIMENTAL DETERMINATION OF RHEOLOGICAL PROPETIES

Cylindrical specimens of three soils each with a different consistency index I_c were tested:

(i) Intermediate compressible clay (CI), $I_c = 0.57$,

(ii) High compressible clay (CH), $I_c = 0$,

(iii) Intermediate compressible organic clay (OI-CI), $I_c = -0.52$.

Tests were carried out in a triaxial cell in such a way, that the loading and unloading path was monotonic and prescribed. At each stress level two deformations were measured: volumetric (ε_v) and axial (ε_a) on a logaritmic time scale. The test results were graphical interpreted, and secondary consolidation was either determined or predicted.

The rheological cofficients a_k and b_k, given in Table I, were determined on the basis of consolidation curves evaluating Egs. (1) - (13) to obtain graphs of deformational components in the form of the interrelated families of $\varepsilon_k^{(o)}$, $\varepsilon_k^{(d)}$ and $\varepsilon_k^{(v)}$.

CONCLUSION

The analytical expressions presented in this paper can be used to determine interrelated rheological stress-strain properties in coherent soils. The results were used in several structure-soils interaction applications where the dependencies were given as a function of noncontinuous structure loading and nonlinear, time dependant soil properties.

Analysis is performed by the computer program INTOTAL (Trauner 1986) which enables fast and economic evaluation in practical three dimensional cases. The solutions can be used to directly determine deformational moduli (in hypoelastic form), flexibility matrices and for the calculation of stress increments and displacements in soils.

Soils with rheological properties, such that coefficients a_{k2}, a_{k4}, a_{k6} and a_{k7} (Eqs.(2) to (5)) become negligibly small with the assumption that $n_{k1} = n_{k2} = 1$, reduce to the Sekiguchi (1977) model and furhter to Shibata's (1963) or Roscoe's (1963) model with $a_{k5} = 0$.

Table I. The Rheological Coefficients.

	Soils Symbol CI ($\times 10^{-2}$)	CH ($\times 10^{-2}$)	OI-CI ($\times 10^{-2}$)
a_{v1} / a_{a1}	1.85/0.59	6.83/0.79	7.47/2.13
a_{v2} / a_{a2}	0.11/0.04	0.08/0.36	0.02/0.19
a_{v3} / a_{a3}	0.44/0.92	1.96/3.55	1.58/1.57
a_{v4} / a_{a4}	0.19/0.02	0.31/0.36	0.01/0.02
a_{v5} / a_{a5}	0.19/0.	0.10/0.02	-0.33/-0.34
a_{v6} / a_{a6}	0.06/0.02	0.10/0.07	0.42/0.38
a_{v7} / a_{a7}	-0.15/0.05	0.43/0.43	0./0.
b_{v1} / b_{a1}	0.47/0.26	0.25/0.	0./0.
b_{v2} / b_{a2}	0.03/0.01	0.08/0.	0./0.
b_{v3} / b_{a3}	0./0.11	0./0.	0./0.
b_{v4} / b_{a4}	0./0.	0./0.	0./0.
m_{v1} / m_{a1}	100./100.	100./100.	100./100.
m_{v2} / m_{a2}	100./200.	100./200.	200./200.
m_{v3} / m_{a3}	100./100.	100./100.	100./100.
n_{v1} / n_{a1}	100./100.	100./100.	100./100
n_{v2} / n_{a2}	100./200.	200./300.	150./150.
n_{v3} / n_{a3}	50./75.	200./200.	100./100.

REFERENCES

1. K.H. Roscoe, A.N. Schofield and A. Thurairajah, "Yielding of Clays in States Wetter than Critical", Geotechnique 13(3), 211-240 (1963).
2. T. Shibata, "On the Volume Changes of Normally--Consolidated Clays", Annuals (6), 128-134 (DPRI, Kyoto 1963).
3. H. Sekiguchi, "Rheological Characteristics of Clays", Proc. 9th ICSMFE (1), 289-292 (Tokyo 1977).
4. L. Trauner, "The Flexibility Matrices for Nonlinear Viscous Soils", Proc. 16th JKTPM (C3-7), 401-408 (Bečiči 1984).
5. L. Trauner, "Computer Aided Design of Time Depending Stucture - Soil Interaction", Proc. 10th TCICBRSD-CIB.86 (1), 215-222 (Washington 1986).
6. L. Trauner, "Interaction between Elastic Stucture and Nonlinear Viscous Soils", Proc. 68th PAMM, BAM 398/86 (XLII), 231-247 (Budapest 1986).

APPLICATION OF A VISCOPLASTIC CONSTITUTIVE LAW TO LEAD IN THE IMPACT ANALYSIS OF RADIO–ACTIVE MATERIAL SHIPPING CASKS

ZHIBI WANG, PETER TURULA, and GLENN F. POPPER

Materials and Components Technology Division
Argonne National Laboratory

ABSTRACT

Perzyna's viscoplastic material model is selected to consider the strain rate effect of lead used in radioactive material shipping packages. The model is checked using data from two scale–model tests and the deformations are found to be within 10 percent.

INTRODUCTION

Radioactive material shipping casks are commonly constructed with lead contained between two concentric stainless steel shells. For "Type B" shipping casks for radioactive material, federal regulations require survival of a 9-m drop onto an unyielding surface. Strain rate effects must be considered when dynamic nonlinear analysis is performed, because material test data indicate that lead tends to resist deformation under dynamic conditions more than under equivalent static conditions.

To account for the strain rate effect of lead, Perzyna's constitutive model was selected and incorporated into the ANSYS multipurpose finite-element code as a user-programmable function by invoking the ANSYS USERPL subroutine.

VISCOPLASTIC MATERIAL MODEL FOR LEAD

Perzyna's viscoplastic relation takes the form

$$\dot{e}_{ij} = \frac{1}{2G}\dot{s}_{ij} + \gamma\,\Phi(F)\frac{\partial F}{\partial \sigma_{ij}} \qquad F > 0$$

$$\dot{e}_{ij} = \frac{1}{2G}\dot{s}_{ij} \qquad F \le 0$$

$$\dot{\varepsilon}_{ii} = \frac{1}{3K}\dot{\sigma}_{ii}\ .$$

Here F is the yield function, g is a fluid parameter, $\dot{e}_{ij}$ and $\dot{s}_{ij}$ are the strain and stress deviator rate tensors, ε_{ij} and σ_{ij} are the strain and stress tensors, and G and

K are material constants. Two forms of the function F are selected as follows:

$$\Phi(F) = \exp\,[M(F/F_0 - 1)] \quad \text{case } 1$$

$$\Phi(F) = (F/F_0 - 1)^N \quad \text{case } 2\,.$$

F_0 is the initial value of F when it reaches the yield plane. The parameters M, N, and γ can be determined by appropriate one–dimensional dynamic experiments.

The ANSYS code accommodates a user–defined plasticity constitutive law through a programmable function. The above Perzyna's model of viscoplasticity has been incorporated into subroutine USERPL to account for the strain rate effect, specifically for lead materials. This subroutine deals with the viscoplastic strain increment during one time increment in which the yield criterion is the Mises yield conditions with bilinear kinematic hardening. The viscoplastic strain rate and viscoplastic strain at this time–step are calculated in the subroutine.

It is necessary to choose one set of the parameters to reflect the dynamic behavior of lead from the experimental data with same strain rate range as in the hypothetical accident cases. For this reason, selected experimental data for lead under high strain rates have been studied to develop parameters for Perzyna's material model from a variety of references. Yield function form case 2 was selected and the parameters used were N = 10 and γ = 25000.

APPLICATION TO A RADIOACTIVE SHIPPING PACKAGE

To check the model and corresponding analysis technique, numerical results were compared with published theoretical and numerical results. To apply the viscoplastic model for lead material, experimental results from two tests of scale models of real radioactive material shipping casks [3] were compared with the numerical results.

Experiment 1

The first experiment was carried out at Franklin Institute Research Laboratories on a 0.08–scale cask model (Fig.1). The test model had a lead displacement of 0.409 in. after the 9–m end drop onto an unyielding surface.

In the numerical analysis, the axisymmetric finite–element model consisted of 135 elements for nonlinear dynamic analysis, in which large deformation was considered. The parameters used for lead were σ_y = 1100 psi, E = 45000 psi, and E_t = 0 psi. Here, σ_y is the yield stress, E is the Young's modulus, and E_t is the tangent modulus. Fig. 2 is the deformation time history at different points along the axial length, in which points 73 and 113 correspond to the middle and top of the lead column. Fig. 3 depicts deformation of the lead column at the instant of rebound of the cask after the impact. The calculated settlement of the lead after rebound of the structure is 0.375 in., close to the experimental result of 0.409 in. The result obtained in Ref.[3] was 0.335 in.

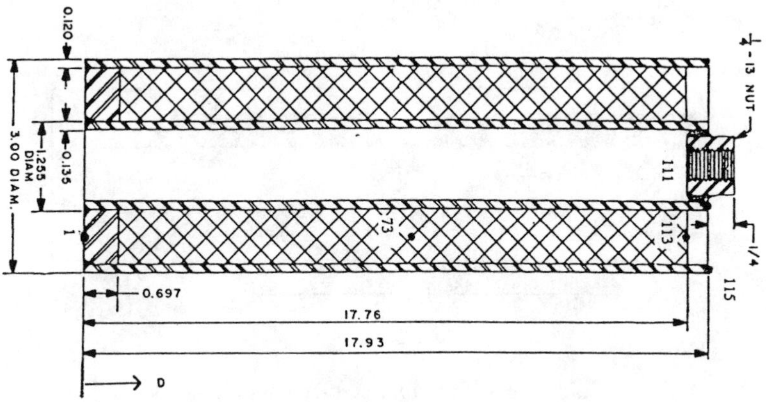

FIG. 1. Model tested at Franklin Institute Research Laboratories (Dimensions are in inches).

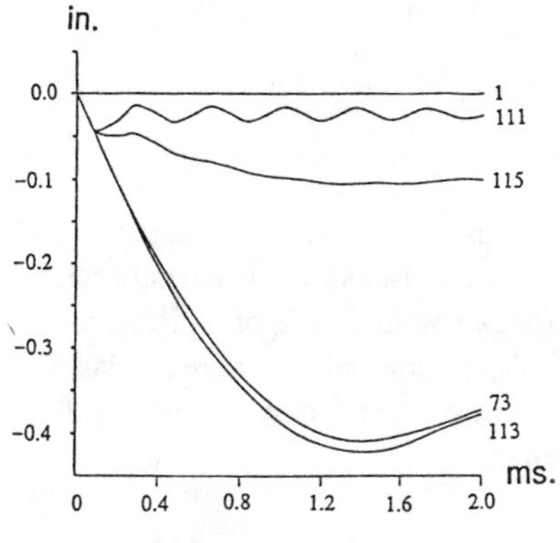

FIG. 2. Deformation vs. Time.

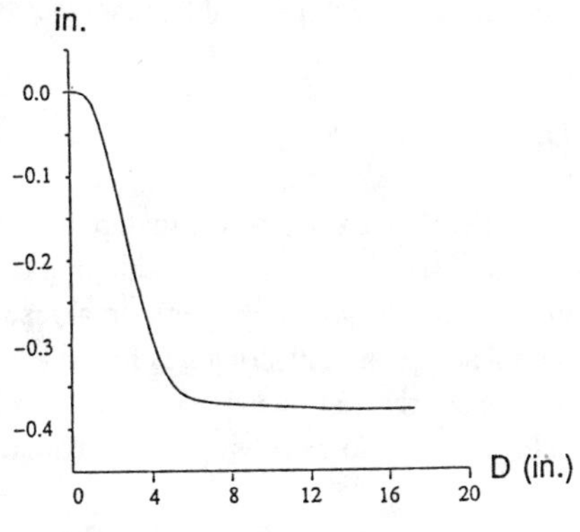

FIG. 3. Lead Column Deformation.

Experiment 2

An experiment with a Hallam steel and lead cask model at 1: 7.5 scale simulated the settling of lead in a lead–filled radioactive material shipping cask due to a 9–m drop impact [3]. A cutaway of the test model is shown in Fig. 4. Lead displacement after the 9–m end drop was 0.7 in.

Dynamic analysis was performed until the structure began to rebound at 3.5 ms. Material parameters were $\sigma_y = 1100$ psi, E = 45000 psi, and $E_t = 0$ psi. Computed settlement of the lead after rebound of the model is 0.670 in., which is consistent with the test result of 0.7 in. The approximate approach in Ref.[3] gave 0.59 in.

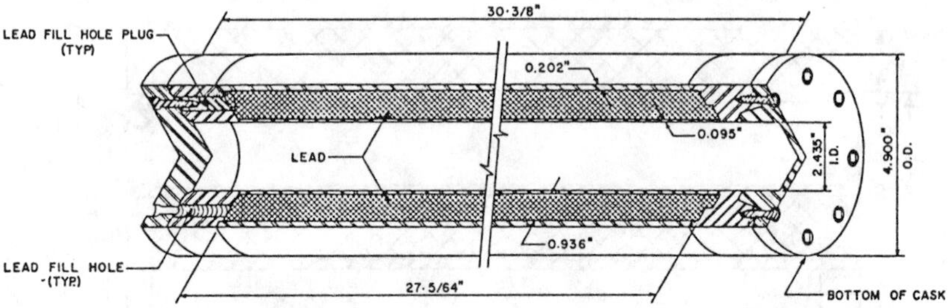

FIG. 4. 1:7.5 Model of Hallam Six–element Shipping Cask
(Dimensions are in inches).

When the numerical results are compared with the experimental results, the lateral deformations of both the inside and the outside shells are almost identical. Maximum radial deformation at the outside shell was about 0.1 in. for the experimental model, while the numerical results gave 0.11 in.

CONCLUSIONS

Perzyna's viscoplastic material model was used for nonlinear dynamic analysis of radioactive material shipping casks. Comparisons were made between the numerical results and the experimental results for two scale models of radioactive material shipping cask structures for which actual experimental data were available. It has been shown that this viscoplastic model works well for the cases investigated. It can be expected to work well for similar structures.

REFERENCES

1. Xiuxian Yang, Guitoung Yang, and Binye Xiu, Introductions to Viscoplastic Mechanics (Chinese Railway Press, Beijing 1985).
2. D. R. J. Owen, and E. Hinton, Finite Elements in Plasticity: Theory and Practice (Pineridge Press Ltd., Swansea 1980).
3. B. B. Klima, L. B. Shappert, and W. C. T. Stoddart, Structural Analysis of Shipping Casks, Vol. 6 - Impact Testing of a Long Cylindrical Lead-shielded Cask Mode, (Oak Ridge National Laboratory Report TM-1312, Mar. 1968).

AXISYMMETRIC ELASTIC VISCO-PLASTIC MODELLING OF COMPACTED SAND-BENTONITE

Jianhua Yin
Jacques, Whitford and Associates, 3 Spectacle Lake Drive, Dartmouth, Nova Scotia, Canada B3B 1W8

James Graham
Department of Civil Engineering, University of Manitoba, Winnipeg, Manitoba, Canada, R3T 2N2

ABSTRACT

This paper presents a new Elastic Visco-Plastic (EVP) model using a new scaling method. The simulation of a CI$\bar{\text{U}}$ test using the new model is compared to the calculation from the Modified Cam-Clay model as viscosity ψ/V approaches zero. The new model is calibrated and verified using triaxial test data of a compacted sand-bentonite buffer material.

INTRODUCTION

Many models have been suggested to describe the time-dependent stress-strain behaviour of soils [1]. The principal difference among the existing elastic visco-plastic models is the methods used for finding *Scaling Function* ϕ (S in this paper) that controls the magnitude of visco-plastic strain rates.

In this paper, the Authors present a new method for determining the scaling function using ideas from Critical State Soil Mechanics and an important new concept called *Equivalent Time* [2],[3]. A new EVP model is then developed by using this new scaling method.

FORMULATION OF THE EVP MODEL

The EVP model employs the commonly used division:

$$\dot{\varepsilon}_{ij} = \dot{\varepsilon}^{e}_{ij} + \dot{\varepsilon}^{vp}_{ij} \tag{1}$$

where $\dot{\varepsilon}^{e}_{ij}$ are time-independent, may or may not be elastic. The visco-plastic strain rates $\dot{\varepsilon}^{vp}_{ij}$ in Eqn.(1) follow the *flow rule*:

$$\dot{\varepsilon}^{vp}_{ij} = S \frac{\partial F}{\partial \sigma'_{ij}} \tag{2}$$

where S is a *scaling function* which controls the magnitude of visco-plastic strain rates, and S ≥ 0. The

F in Eqn.(2) is the *visco-plastic potential function.* The partial differentiation $\partial F/\partial\sigma'_{ij}$ controls the direction of visco-plastic strain rates, see Fig.1.

The flow rule in (2) for calculating visco-plastic strain rates is similar to the flow rule for plastic strain increments in elastic plastic theory. The geometry represented by function F in stress space is called the *flow surface* (or *load surface*) which can go forward and backward in the corresponding visco-plastic region (Fig.1). The low limit boundary of flow surfaces is a *yield surface* which is the boundary of time-independent strain rates only and time-dependent strain rates. The position of a yield surface is controlled by the *limit creep line* in $p'-\varepsilon_v$ space (Fig.1).

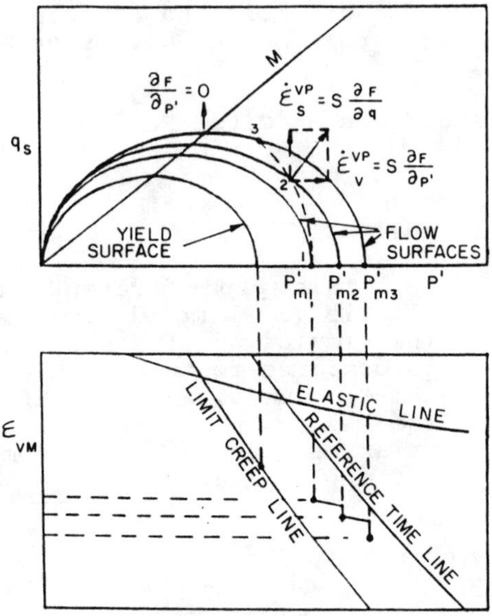

Fig.1 Illustration of flow rule and scaling method

The limit creep line is defined as the extreme points at which creeping stops when equivalent times are infinite [1],[3]. The *reference time line* in Fig.1 is chosen as a reference for counting equivalent times. On the reference time line, the equivalent times are chosen to be zero [3]. When viscosity is zero, the limit creep line and reference time line are combined. And then the flow surfaces and the *yield surface* are the same. Thus the elastic visco-plastic model is reduced to an elastic-plastic model. As in the Modified Cam-Clay, the flow surface F is assumed to be elliptic (Fig.1):

$$F = p'^2 - p'_m p' + q^2/M^2 = 0 \qquad (3)$$

where the p'_m "scales" the various flow surfaces. Using Eqns.(1)-(3), the constitutive relationship for triaxial stress states (axisymmetric) is:

$$\begin{cases} \dot{\varepsilon}_v = \dot{\varepsilon}_v^e + S\ (2p' - p'_m) \\ \dot{\varepsilon}_s = \dot{\varepsilon}_s^e + S\ 2q/M^2 \end{cases} \qquad (4)$$

where the scaling function S is to be determined.

NEW SCALING METHOD

The scaling function in (4) has been related here to a general constitutive equation for isotropic stressing. Yin and Graham ([2],[3]) described a new concept equivalent time which takes account of the differing strain rates observed in both the over-consolidated and normally consolidated ranges of soil behaviour. Using this concept a new 1-D EVP model has been developed for modelling strain rate and time effects in 1-D straining of clays. Here the framework of the 1-D EVP model is used to develop a general constitutive relationship for isotropic stressing. If logarithmic functions are used, the general constitutive equation is:

$$\dot{\varepsilon}_{vm} = \frac{\kappa}{V}\,\dot{p}'_m/p'_m + \frac{\psi}{Vt_o}\,e^{-\varepsilon_{vm}V/\psi}\left(\frac{p'_m}{p'_{mo}}\right)^{\lambda/\psi} \qquad (5)$$

where 'm' denotes strains and stresses in p'-axis.

The scaling condition is that the visco-plastic volumetric strain rates on a flow surface are constant. Using this condition, from Eqns.(4),(5):

$$\dot{\varepsilon}_v^{vp} = S\,(2p' - p'_m) = \dot{\varepsilon}_{vm}^{vp} = \frac{\psi}{Vt_o}\,e^{-\varepsilon_{vm}V/\psi}\left(\frac{p'_m}{p'_{mo}}\right)^{\lambda/\psi} \qquad (6)$$

From (6) and considering $S \geq 0$, the scaling function is:

$$S = \frac{\psi}{Vt_o}\,e^{-\varepsilon_v V/\psi}\left(\frac{p'_m}{p'_{mo}}\right)^{\lambda/\psi} / |2p' - p'_m| \geq 0 \qquad (7)$$

where ε_v is used instead of ε_{vm} to simplify the scaling function. For a given stress state (p',q), p'_m is found from Eqn.(3). This calculation for p'_m implies a *consistency condition* that any stress state must be on a flow surface. The second term in (5) acts as an *evolution law* which is similar to the *hardening law* in the Modified Cam-Clay model.

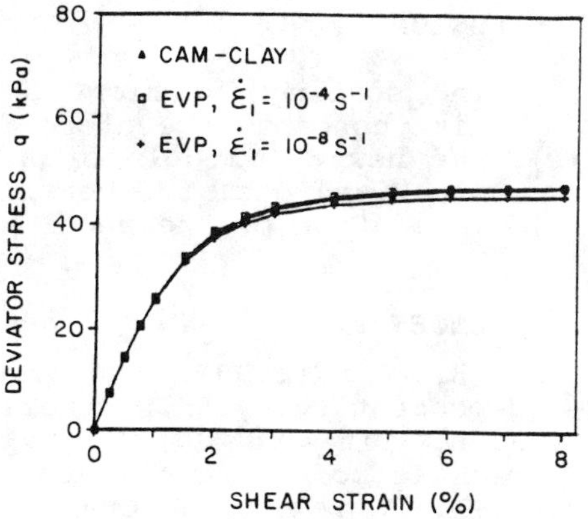

Fig.2 Simulation of a CĪU test, q,ε_s-relationship

SIMULATION OF A CĪU TEST

The parameters used in the modelling were: $\kappa/V = 0.025$; $\lambda/V = 0.1$, $p'_{mo} = 47$ kPa; $\psi/V = 0.004$, 0.0004,

t_o = 1 hour; M = 0.8, G = 1000 kPa. The initial point used was: p'_i = 100 kPa, $q_i = 0$, ε_{vi} = 0.0755, t_i = 0 hour. Fig.2 shows only the simulated q,ε_s-relationship of the CI$\bar{U}$ test using the Modified Cam-Clay model and the new EVP model when the creep parameter ψ/V = 0.0004. It is seen in Fig.2 that when viscosity is very small, the curves from the two models are close.

MODELLING OF SAND-BENTONITE BUFFER

The framework of the new EVP model has been applied to a sand-bentonite buffer material that has been proposed for use in the Canadian Fuel Waste Management Program. Two main improvements were made: (1) using hypoelastic KGJ model for the time-independent strains in Eqn.(1); (2) using a curved strength envelope [1]. The model was calibrated using isotropic consolidation and shear data from the previously-published work. Predictions were made on (1) a creep test using multi-stage q in undrained condition; (2) a step-changed constant strain rate test in undrained condition with unloading/reloading and relaxation. Fig.3 shows only the measured and predicted $q\text{-}\varepsilon_s$ results in the step-changed constant strain rate test.

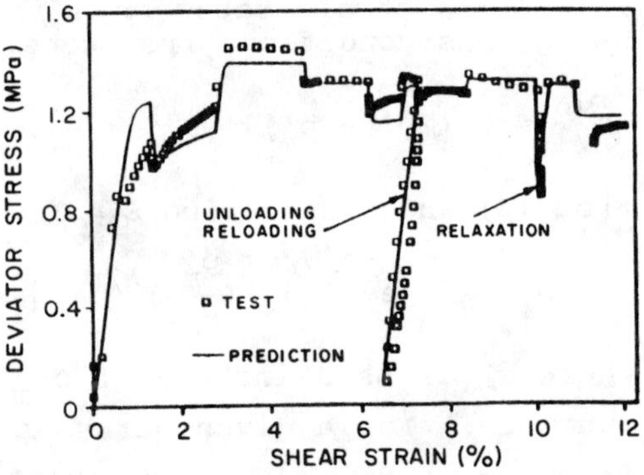

Fig.3 Measured and predicted q,ε_s-relationship

CONCLUSIONS

(1). The scaling function is related to the equivalent time concept, therefore is more meaningful.
(2). The new EVP model is an extension of the Modified Cam-Clay model for considering viscous effects.
(3). Validation of the model needs further examination.

REFERENCES

1. J.-H. Yin (1990), "Constitutive modelling of time-dependent ...", Ph.D. thesis, U. of Manitoba, Canada.
2. J.-H. Yin and J. Graham (1989), "Viscous elastic plastic modelling of ...", CGJ, 26, 199-209.
3. J.-H. Yin and J. Graham (1989), "General elastic viscous plastic constitutive ...", NUMOG III, 108-117.

THERMAL EFFECTS

A Constitutive Model for the Thermomechanical Fatigue Response of René 80

V.S. Bhattachar and D.C. Stouffer

Department of Aerospace Engineering and Engineering Mechanics
University of Cincinnati, Cincinnati, OH 45221, U.S.A.

Abstract

This article describes the study of nonisothermal deformation and thermomechanical fatigue (TMF) response of René 80. The observed TMF response suggests the presence of two deformation mechanisms which interact to produce extra hardening that is not present in isothermal fatigue. A method has been developed to account for the interaction between planar slip (at low temperatures) and dislocation climb (at high temperatures) during a TMF test. In addition, the flow equation from an existing model has been re-written in the form of an Arrhenius equation with explicit temperature dependence. The inclusion of extra hardness and static thermal recovery effects has resulted in a greatly improved TMF response prediction.

Introduction

The constitutive model proposed by Ramaswamy *et al.* [1] and Stouffer *et al.* [2] for the high temperature alloy René 80 is based on a unified strain approach. The inelastic flow equation depends on three state variables, back stress, drag stress and static thermal recovery, which result from the interaction of dislocations with the microstructure. The back stress evolution equation was developed to model strain hardening and dynamic recovery, and static thermal recovery variable allows the equilibrium between strain hardening and dynamic recovery to change with creep loading. Drag stress was formulated for cyclic hardening or softening using accumulated inelastic work. The model can reproduce the isothermal tensile, creep and multiaxial fatigue responses as a function of temperature using material parameters evaluated from the test data at each temperature. It can also predict the isothermal tensile response at 649C using interpolated material parameters. The out-of-phase thermomechanical fatigue (TMF) response was approximated only in the first cycle and the saturated TMF response was not predicted correctly.

The objective of the present study was to understand the mechanisms associated with the nonisothermal deformation of René 80 and various factors that affect these mechanisms. This understanding was incorporated into the constitutive model in order to predict the initial and saturated TMF response accurately. The predicted TMF responses for two tests have been included in this article.

Developing the Nonisothermal Model

First, explicit temperature dependence was introduced in the inelastic flow equation. This was motivated by the Arrhenius equation and the deformation maps for Nickel by Frost and Ashby [3]. This is consistent with the model since the flow law has features of exponential glide and power law creep in the higher stress and temperature regions of the deformation maps. Using the modified flow equation and the existing back stress equation, material parameters were reevaluated at four temperatures. In order to verify the parameters, the isothermal tensile response was calculated at several temperatures using interpolated values (Fig.1). The variation of parameters suggests two deformation mechanisms for René 80 at high and low temperatures. Further, calculations with a ramp change in the temperature during tensile straining suggests an unstable response around 760C (Fig.2). This temperature corresponds to a transition region between the two deformation mechanisms [4]. A similar response has been observed by Chan *et al.* [5] in the nonisothermal modeling of B1900+Hf alloy.

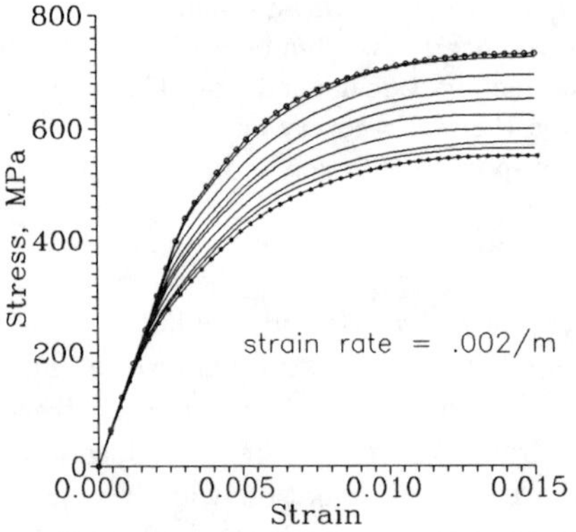

Fig.1 Tensile Response, 760–871C

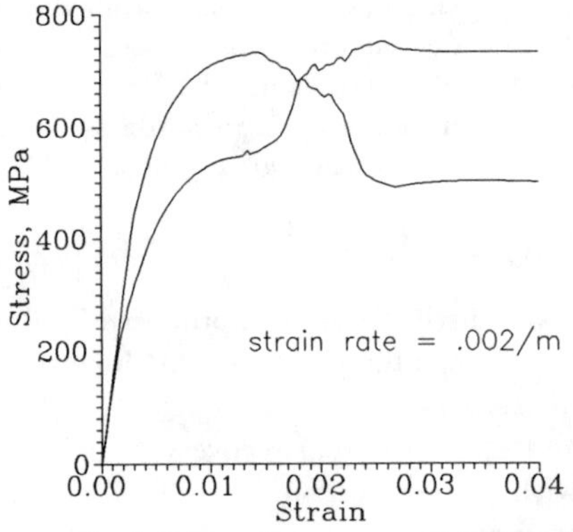

Fig.2 Ramp Response, 760–871C

Second, a thermal recovery equation was used with the parameters calculated from the available slow strain rate test data. Interpolated parameters were used with the recovery equation at all temperatures above 700C, where thermal recovery plays an important role. The extent of recovery is controlled automatically by the existing stress and temperature levels, and the duration of exposure. The revised flow equation and the recovery equation was used to predict the TMF response during the first cycle. The above revisions significantly improved correlation between the model and the experimental results.

Effects of Temperature History

During a TMF test the material is exposed to alternating high and low temperatures. This can lead to interactions between planar slip at low temperatures and dislocation climb at high temperatures. In this section an attempt is made to determine the presence of such an interaction and include it in the constitutive model.

Examination of the data showed that the TMF response of the 649-1093C test saturated after the first cycle. The low temperature response after the first cycle was predicted well. But the stresses in the high temperature portion of the second cycle were greater than the stress predicted by the model by about 30 MPa. The higher stresses were attributed to an increased hardness of the material resulting from alternating exposure to very high and low temperatures. A similar effect has also been observed by Murakami *et al.* in the nonisothermal tests on Type 316 Stainless steel [6]. Deformation at low temperature (below 760C) occurs primarily by planar slip, whereas increasing the temperature increases the opportunity for deformation by dislocation climb. A possible explanation of the extra hardening phenomenon is the interference of dislocation climb with planar slip in the low temperature portion of the cycle, and the interference of planar slip with climb at the higher temperatures. In both cases the resistance to micro-deformation would increase, resulting in the observed increase in hardness.

The above effect was modeled using two parameters, β, which represents the extent of interaction between the two mechanisms, and ω, which represents extra hardness in the material. As a first attempt, evolution equations were chosen for the variables in the form:

$$\dot{\beta} = m_2(\beta^* - \beta)\dot{W}^I \tag{1}$$

and

$$\omega = Z_2(1 - e^{-\beta W^I}) \tag{2}$$

where, β^* and Z_2 are the maximum values of β and ω respectively, at a given temperature. The extra hardness ω is added to the drag stress obtained from the isothermal evolution equation. Depending on the thermal loading sequence and the temperature level, Eqns.(1) and (2) are activated by inelastic work and the values of β^* and Z_2.

Results

The equations described above were used in the model to predict the TMF response of René 80. The effect of extra hardness is seen above 982C from the second cycle onwards in the out-of-phase 649-1093C test (Fig.3). In this test, the temperature range and level were high enough to saturate the response in two cycles. The predicted response shows good agreement with the experimental data.

The response for the out-of-phase 760-982C test is shown in Fig.4. The model has captured the overall trend in behavior. Extra hardness has little effect in this test, and it is postulated that dislocation climb was not significant. The predicted response also captures the increase in mean stress which is observed in the experimental data.

Conclusions

This study has demonstrated the versatility of the unified approach to constitutive modeling of materials and has attempted to account for more of the major deformation mechanisms observed in TMF. Thermal recovery process is important at high temperatures and it cannot be ignored. Also, interactions between deformation mechanisms resulting from an alternating high and low temperature exposure can have a significant effect on the material behavior. Such interactions should be included in a constitutive model.

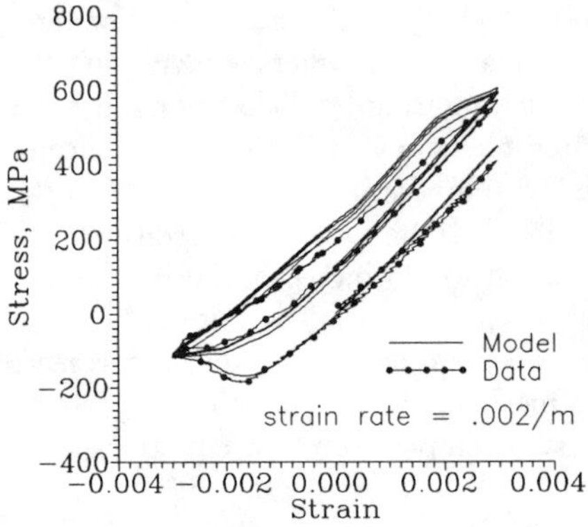

Fig.3 TMF Response, 649−1093C

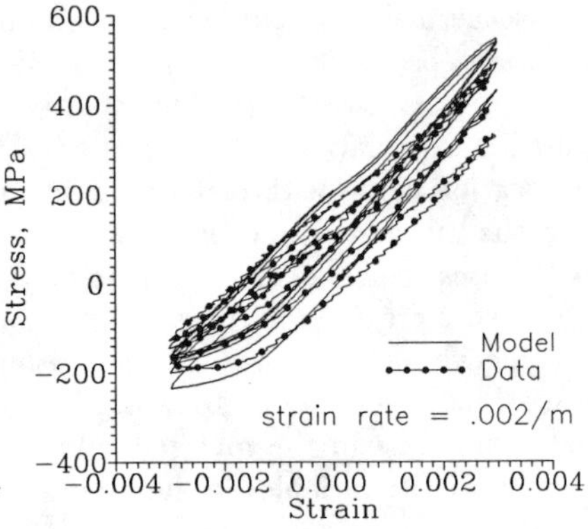

Fig.4 TMF Response, 760−982C

References

1. V.G. Ramaswamy, D.C. Stouffer and J.H. Laflen, 'A Unified Constitutive Model for the Inelastic Uniaxial Response of René 80 at Temperatures Between 538C and 982C', *ASME Journal of Engineering Materials and Tech.*, 112, pp 280-286, July 1990.

2. D.C. Stouffer, V.G. Ramaswamy, J.H. Laflen, R.H. VanStone and R. Williams, 'A Constitutive Model for the Inelastic Multiaxial Response of René 80 at 871C and 982C', *ASME Journal of Engineering Materials and Tech.*, 112, pp 241-246, April 1990.

3. H.J. Frost and M.F. Ashby, 'Deformation Mechanism Maps, The Plasticity and Creep of Metals and Ceramics', Pergamon Press, 1982.

4. V.S. Bhattachar and D.C. Stouffer, 'Progress in Predicting TMF Response of René 80', The Fifth Thermomechanical Fatigue Workshop, NASA Lewis Research Center, Cleveland, OH, October 10-11, 1989.

5. K.S. Chan and R.A. Page, 'Inelastic Deformation and Dislocation Structure of a Nickel Alloy: Effects of Deformation and Thermal Histories', *Metallurgical Transactions*, 19A, pp 2477-2486, October 1988.

6. S. Murakami, M. Kawai and Y. Ohmi, 'Effects of Amplitude History and Temperature History on Multiaxial Cyclic Behavior of Type 316 Stainless Steel', *ASME Journal of Engineering Materials and Tech.*, 111, pp 278-285, July 1989.

Developing a Constitutive Equation for Linear Thermoviscoelastic Materials

S. N. Ganeriwala and H. A. Hartung
Philip Morris Research Center
P. O. Box 26583
Richmond, VA 23261-6583

ABSTRACT

Polymeric materials are used in many mechanical systems and structures. Their utility has been restricted due to a lack of appropriate constitutive equations. The principle of time-temperature superposition has been widely used to obtain linear thermoviscoelastic properties over a wide range of temperature and frequency (or time). The procedure is difficult and does not provide a constitutive equation. This paper presents a new scheme that yields a complete constitutive equation suitable for analysis of complex problems. An initial test of the new approach was carried out with two different classes of polymers. The model is based on postulates of generally observed temperature dependent behavior of polymeric materials and our recent development of a new model of the glass transition phenomenon. It is adaptable to most computer systems.

INTRODUCTION

The viscoelastic properties of polymers make them valuable for control of sound and vibration. A key feature of polymers is strong dependence of mechanical properties on temperature and frequency (or time). Engineering analysis and design requires consititutive equations. Theories developed to date provide only a general form of material functions [1]. The exact constitutive equation has to be derived from studies of mechanical properties over wide ranges of frequency (or time) and temperature. Many dynamic mechanical property studies have been reported but appropriate equations have been elusive [2, 3].

A new technique, Fourier Transform Mechanical Analysis (FTMA), measures the complex moduli over a range of frequencies in one test by exciting the sample by a random signal (band limited white noise) [4, 5]. The primary feature of FTMA is that a complete isotherm is obtained in just a few seconds with minimum heating and other structural changes normally associated with other techniques. FTMA also, readily provides direct assessment of sample inertia and geometry effects. It is virtually impossible to cover the broad ranges of practical interest with a single test. The principle of time-temperature superposition has been widely utilized to obtain master curves that presumably approximate a mechanical response isotherm for a wide range of time (or frequency) [2,3].

A semi-empirical phenomenological model for polymeric materials provides the basis for a constitutive equation. The model is based on physical postulates derived from observed material behavior. An important part of the derivation is a new description of the glass transition (Tg) behavior of polymers [5]. The time-temperature principle is an integral part of this model.

CONSTITUTIVE MODEL DEVELOPMENT

We begin by examining a typical temperature dependent mechanical behavior of most polymers. Figure 1 shows such a plot of the storage modulus G' as a function of temperature at a given frequency. The entire response can be thought of as a sum of three distinct parts without the loss of any generality. First, all materials attain the highest modulus at the absolute zero temperature. Second, the modulus of all materials decreases with increase in temperature. We assumed that the log of G' decreases linearly with temperature. The third part is unique to polymers (or amorphous materials in general). There exists a small region of temperature, known as the glass transition region, in which the material modulus changes by several orders of magnitude. The material changes from a brittle glassy state to a soft rubbery state. It is characterized by three parameters: first, the height of transition; second, the width of transition; and third, the temperature of Tg. We found that a modified form of the hyperbolic tangent function represented the data well. Thus, a complete temperature dependent behavior of a viscoelastic material at a given frequency (or time) is characterized by five parameters: two common to all materials and the other three unique to amorphous materials. Each of them has a definite physical meaning [4,5]. Mathematically the function is given as:

$$\mathrm{Log}G'(\omega,T) = B_0 + B_e T + h\left[1 + \tanh\left(\frac{T - T_m}{S_p}\right)\right] \qquad (1)$$

where T = temperature in Kelvin, and B_0, B_e, h, T_m, and S_p are parameters shown in Figure 1. The model can be applied at many frequencies, and the five parameters can be studied as functions of frequency.

To complete the model development we needed a new description of glass transition phenomenon [5]. We argued, on the basis of thermodynamics, that all isofrequency responses would converge at a point at low temperature. This point is the second-order thermodynamic glass transition temperature. Since glass transition is a nonlinear process we postulated the spread to be a power law function of frequency. Also, we found that the mid-point of transition was linearly related to the spread and Tg. Thus:

$$S_P = S_0 \omega^{K_s} \qquad \text{and} \qquad T_M = T_g + N_s S_p \qquad (2),(3)$$

By combining equations 1, 2, and 3 we obtain a complete constitutive equation:

$$\text{Log}G'(\omega,T) = B_0 + B_e(T - T_g) + h\left[1 + \tanh\left[\frac{T - (T_g + S_0 N_s \omega^{K_s})}{S_0\ \omega^{K_s}}\right]\right] \qquad (4)$$

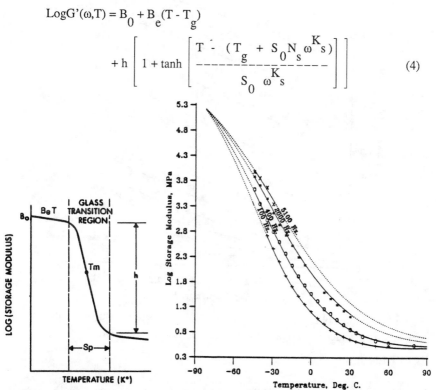

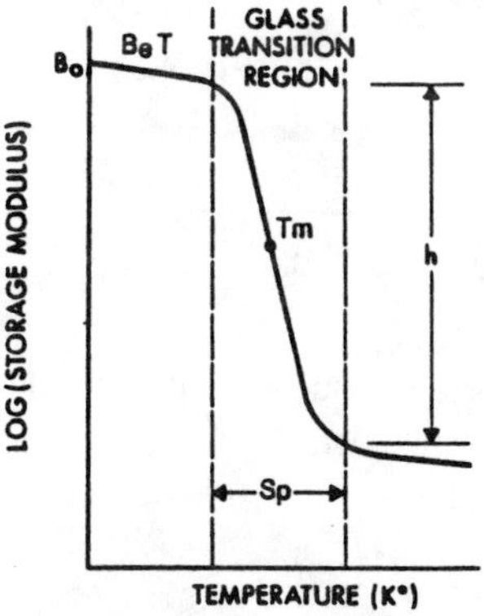

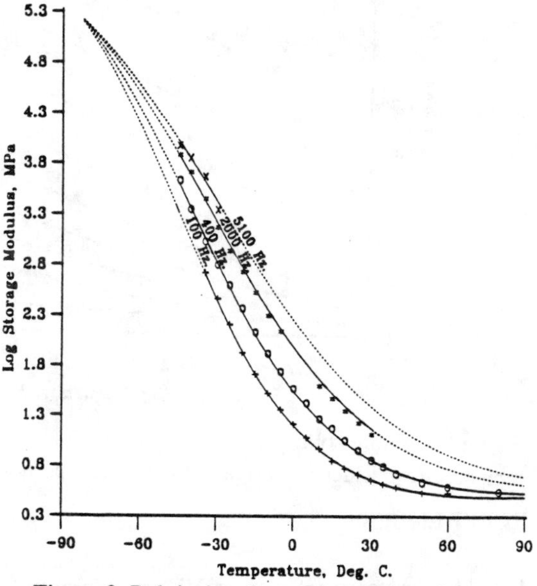

Figure 1. Typical temperature dependent behavior of a polymer.

Figure 2. Polyisobutylene G' at various frequencies. Lines show model and point of intersection is T_g.

RESULTS AND DISCUSSION

Figure 2 shows polyisobutylene storage modulus versus temperature at various frequencies. The lines are model contours and the point of convergence is the thermodynamic second order glass transition temperature. The model parameters were derived by non-linear least squares regression to several hundred data points. The spread of the transition region was consistent with a power law dependence on frequency. It was found that B_0 , B_e and h were essentially invariant over the range of frequencies. To verify the model experimental data were compared with predictions. Figure 3 shows polyisobutylene G' versue frequency data and model lines at various temperatures. Similar comparisons were also made for the loss modulus and loss factor. Another class of polymer, NBR (nitrile butediene rubber), was aslo studied. In all cases, the model predictions agreed well with measurements. Our model provides a simple procedure to study the glass transition phenomenon and construct master curves.

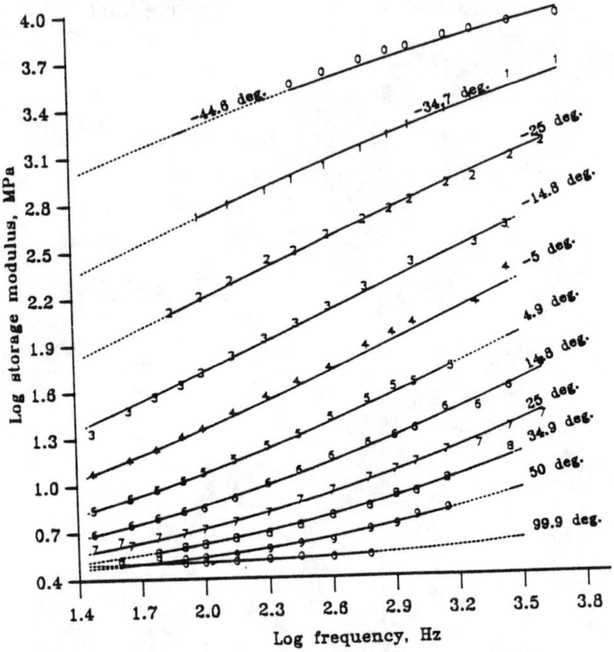

Figure 3. Polyisobutylene storage modulus versus frequency at various temperatures. Lines represent our model and point are actual data.

REFERENCES

1. R. M. Christensen, Theory of Viscoelasticity an Introduction, 2nd. ed., Academic Press Inc.: New York; 1982; Chs. 1,3.
2. J. D. Ferry, Viscoelastic Properties of Polymers; 3rd. ed., John-Wiley & Sons Inc.: New York, 1980; Chs. 6, 7.
3. L. Rogers, V. J. Johnson, and L. G. Kelly, Editors, Proceeding of Damping' 89, Wright-Patterson Air Force Base, Ohio 45433, WRDC-TR-89-3116, Nov. 1989, Section-FA.
4. S. N. Ganeriwala and H. A. Hartung, in Sound and Vibration Damping with Polymers, Edited by R. D. Corsaro and L. H. Sperling, ACS Symposium Series 424, American Chemical Society, Washington D. C., 1990; pp.92-110.
5. S. N. Ganeriwala and H. A. Hartung, J. of Chem. Phys., a paper in progress.1.

EFFECT OF TEMPERATURE VARIATION ON NON-PROPORTIONAL CYCLIC PLASTICITY OF 304 STAINLESS STEEL

Kozo IKEGAMI, Yasushi NIITSU and Seiei FUJIWARA
Tokyo Institute of Technology
Nagatsuta, Midoriku, Yokohama, Japan

ABSTRACT

Plastic deformation at high temperature is experimentally investigated by subjecting thin-walled tubular specimens to non-proportional cyclic strain paths under various temperature conditions. The non-proportional loading is realized by combined axial and shear strain cycling of out-of-phase. The effect of temperature condition on the stress-strain relations for out-of-phase cycling is examined by comparing with those for in-phase cycling.

INTRODUCTION

Structural components for high temperature state is subjected to both cyclic straining condition and high temperature condition. The component materials is deformed in complicated manner by the coupled conditions of temperature and load. The knowledge of material deformation under the complex states is necessary to realize a reliable strength design of high temperature components.

The authors have already reported high temperature plastic deformation under combined loads as well as simple loads[1]. This paper presents some characteristic properties of plastic deformation by cyclic non-proportional loading at high temperature.

EXPERIMENTAL PROCEDURES

Experiments are conducted by subjecting thin-walled tubular specimens to combined axial and shear loads. The material is stainless steel SUS 304. The shape of the specimen and the material composition is previously reported [1]. The machined specimen is heat-treated at 1050 °C for half an hour. The non-proportional cyclic straining is realized by axial-shear strain cycling with out-of-phase of 90 dgrees. The strain amplitudes are chosen to be 0.3% and 0.5%. The locus of the non-proportional cycle becomes circles and the radius corresponds to the strain amplitude. The strain rate in the experiment is 0.2%. The testing temperatures are chosen as 20°C(room temperature), 200°C, 400°C and 600°C.

EXPERIMENTAL RESULTS

Cyclic straining at constant temperatures

The stress-strain relations for cyclic straining are obtained at constant temperature of 20°C, 200° C, 400° C and 600°C. The stress paths during cyclic straining at 600°C are shown in Fig. 1. The stress path is saturated with increasing the cyclic numbers. The cyclically saturated stress values represented by equivalent form are compared between two cases of out-of-phase straining and in-phase straining. Figure 2 shows the saturated stress values for the testing temperature and the strain amplitude. The notations R_t and $\Delta\varepsilon/2$ are the strain amplitudes for out-of-phase straining and in-phase straining, respectively. The stress values in out-of-

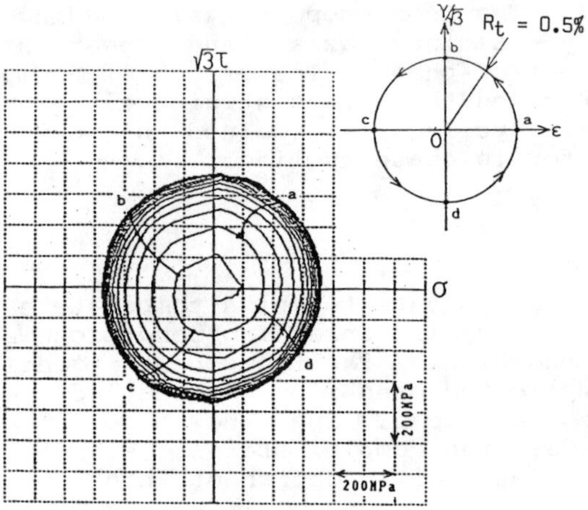

Fig. 1 Stress path for cyclic straining with out-of-phase

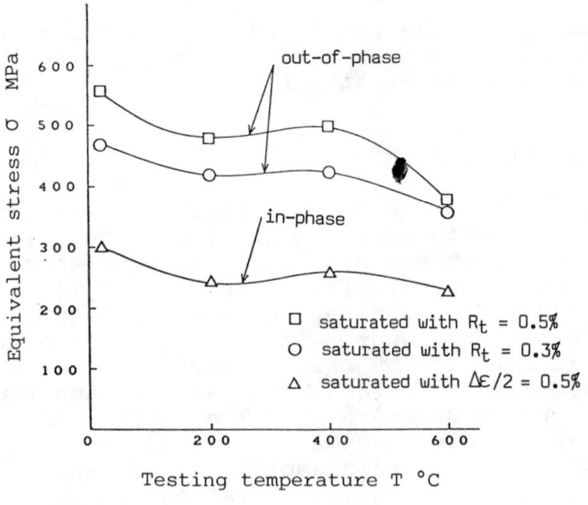

Fig. 2 Cyclically saturated stress values at various temperatures

phase cyclic straining at 0.5 % saturate at the stress magnitude of about two times as much as those of in-phase cyclic straining at 0.5%. The saturated stress values for out-of-phase cycling at 0.3% become also larger than those for in-phase cycling at 0.5%.In out-of-phase cyclic straining, the small strain amplitude gives the low saturated stress values. The saturated stress level decreases generally with rising testing temperature, but the values at 400°C is slightly higher than those at 200° C.

Cyclic straining at variable temperatures

The non-proportional loading is conducted under the temperature condition of step change. The specimen is cyclically loaded at a certain temperature, then the temperature state of the specimen is changed and again cyclically loaded at the changed temperature condition. The same cyclic testing procedures are repeated between a defined temperature interval. The equivalent stress-strain curves are examined at the different temperature condition.

Figure 3 shows the equivalent stress strain relations under the changing temperature conditions between 20°C and 600°C at the strain amplitude 0.5%. The stress-strain curves are obtained under the temperature condition changing from 600°C to 20°C. The maximum stress of the first cycling period at 20°C is larger than the saturated stress in the cyclic straining at the constant temperature of 20° C. The stress values for cyclic straining at 20°C subsequent to cyclic straining at 600°C increases gradually from the initial cyclic straining.

The non-proportional cyclic straining with the strain amplitude of 0.5% is conducted at the four temperature conditions of 600°C, 400°C, 200°C and 20° C, successively. The stress-strain curves are illustrated in

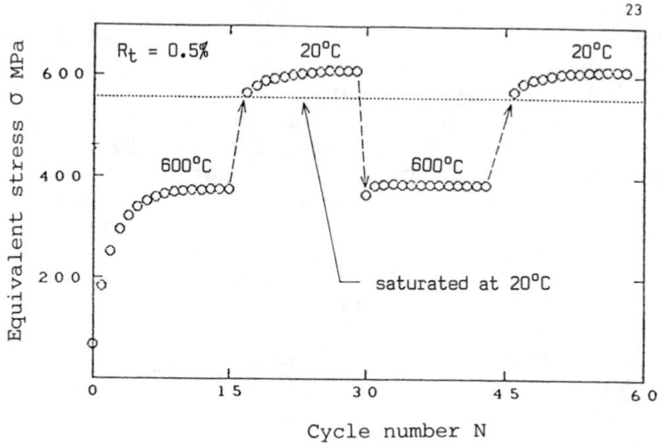

Fig. 3 Cyclic stress-strain relations for two different temperature levels

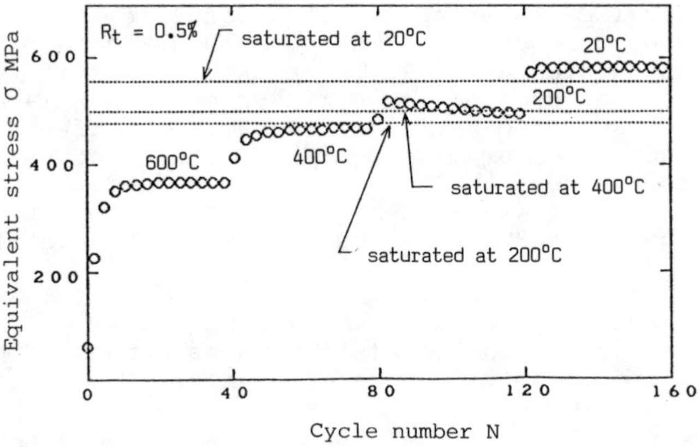

Fig. 4 Cyclic stress-strain relations for four different temperature levels

Fig. 4. The dashed lines in the figure indicate the stress level for saturated state of cyclic stress path at the constant temperature for 400°C, 200°C and 20°C. The saturated stress values at 400°C from 600°C are smaller than those at the constant temperature 400°C, but the saturated stress values at 20°C from 200°C are larger than those at the constant temperature of 200 ° C. The cyclic softening is observed at 200° C after cyclic straining at 400°C. The stress level approaches gradually to the saturated stress values at the constant temperature of 200°C. The saturated property of cyclic straining exhibits complicated behavior with related to both the temperature level and the sequence of temperature change.

CONCLUSIONS

The stress-strain relations of stainless steel SUS 304 for non-proportional straining are experimentally investigated at high temperature by subjecting thin-walled tubular specimens to combined axial and torsional loads. The non-proportional straining is realized by out-of-phase loading of combined tension and torsion. The stress-strain curves for out-of-phase cyclic straining saturate at higher stress levels than those for in-phase cyclic straining. The saturated stress levels for out-of-phase cyclic straining under variable temperature conditions depend on the sequence of temperature change as well as the level of temperature. The coupled effects between temperature condition and cyclic straining condition are observed.

REFERENCES

1. K. Ikegami and Y. Niitsu, "Effect of creep prestrain on subsequent plastic deformation", Int. J. Plasticity 1(4), 331-345 (1985).

CONSTITUTIVE LAWS FOR NONLINEAR THERMO-VISCO-ELASTIC SPRING-AND-DASH-POT MODELS

TOMASZ JĘKOT
University of Zululand
P Bag X 1001, Kwadlangezwa 3886
South Africa

ABSTRACT

The aim of the paper is to introduce nonlinear thermoelastic constitutive equations for a spring in spring-and-dash-pot models. Nonlinear constitutive equations for thermo-visco-elasticity for these models are obtained. The comparison of linear and nonlinear strain in the Voigt model and stress in the Maxwell model are illustrated in the graphs.

INTRODUCTION

An impact of nonlinear constitutive equations of thermoelasticity is worth underlining. The difference between linear and nonlinear solutions for static and dynamic cases is considerable, cf. [1], [2]. This implies a need to introduce nonlinear constitutive equations in theories involving thermoelsticity such as thermo-visco-elasticity. Considerations in the paper are confined to spring-and -dash-pot models. The general approach as well as nonlinear constitutive relations for the Voigt and the Maxwell models are presented. Assuming that the tested material is D-54-S aluminum alloy, the differences between linear and nonlinear strains and stresses in these models are considerable.

CONSTITUTIVE RELATIONS

The following constitutive relations are assumed: for a spring, CF. [1];

$$\sigma = \gamma(\theta) + E\varepsilon + L\varepsilon^2, \tag{1}$$

for a dash-pot;

$$\sigma = \eta(\theta)\ \dot{\varepsilon}, \tag{2}$$

where $\sigma=\sigma(t)$ - stress, $\gamma=\gamma(\theta)$ - stress-temperature coefficient, θ - increment of temperature, $\varepsilon=\varepsilon(t)$

strain, E - Young modulus, $L = l(1-2\nu) - 2m(\nu^2-1) + n\nu^2$, l,m,n - Murnaghan constants, ν - Poissons ratio, (we assume L=0 for linear case), η - coefficient of viscosity, $(\dot{\ }) = \partial/\partial t$. Assuming linear thermal elongation of a spring one can write

$$\varepsilon = \alpha_T \theta \tag{3}$$

where α_T is the linear thermal expansion.
Coefficient γ is expressed by, cf. [1],

$$\begin{aligned} \gamma(\theta) &= \gamma_1(\theta) - L\,\alpha_T^2\theta^2 \\ \gamma_1(\theta) &= -E\,\alpha_T\theta \end{aligned} \tag{4}$$

Inserting:

$$\begin{aligned} \varepsilon_s &= (-E + \Delta^{1/2})/2L, \\ \dot{\varepsilon}_s &= [2\Delta^{-1/2}(\dot{\sigma} - \dot{\gamma}(\theta)]/2L, \end{aligned} \tag{5}$$

...etc.

$$\Delta = E^2 + 4L\,[\sigma - \gamma(\theta)], \quad \dot{\gamma}(\theta) = (d/d\theta)\gamma(\theta)\ (\partial/\partial t)\theta$$

for the elongation of the spring in the Maxwell models, we obtain a constitutive equation

$$P(\sigma, \varepsilon, \theta) = 0 \tag{6}$$

which follows from the construction of the spring-and-dash-pot net. Notice that

$$\varepsilon_s = [\sigma - \gamma_1(\theta)]/E, \quad \dot{\varepsilon}_s = [\dot{\sigma} - \dot{\gamma}_1(\theta)]/E, \ ...\text{etc.}$$

for linear cases and then constitutive equations can be written in the form as follows

$$P(\sigma) = Q(\theta) + R(\varepsilon) \tag{7}$$

(cf. eg. [3] for isothermal case).

MAXWELL AND VOIGT MODELS

Applying relations (1), (2) and (5) constitutive relations are as follows, for the Maxwell model:

$$\sigma/\eta + 2L\,\Delta^{-1/2}(\dot{\theta} - \dot{\gamma}) = \dot{\varepsilon} \tag{8}$$

for the Voigt model:

$$\sigma = \gamma + E\varepsilon + L\varepsilon^{2} + \eta\dot{\varepsilon} \tag{9}$$

Let us assume material constants for D-54-S aluminum alloy, TABLE I, cf. [4], [5]. To compare linear and nonlinear solutions let us suppose that the Maxwell model is clamped (no displacements on ends) and heated with temperature $\theta(t)= 200t$. Stresses in linear and non-linear cases are illustrated in Fig. 1. When the Voigt model is stress free on ends and heated with $\theta(t)= 200t$ then linear and nonlinear strains are shown in Fig. 2.

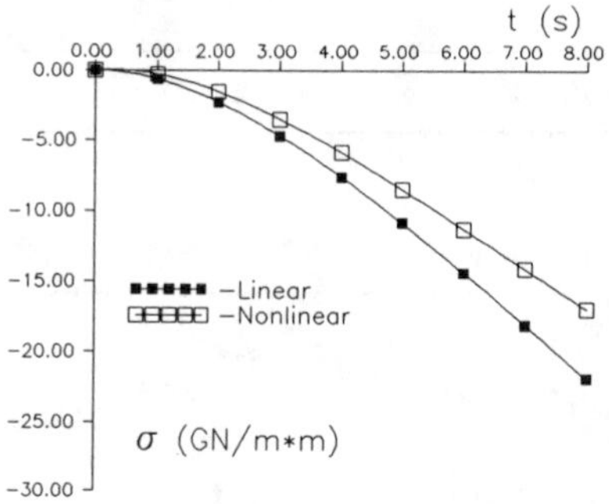

FIG. 1. Comparison of linear and nonlinear stresses for the Maxwell model (ε=0, $\theta(t)$=200t).

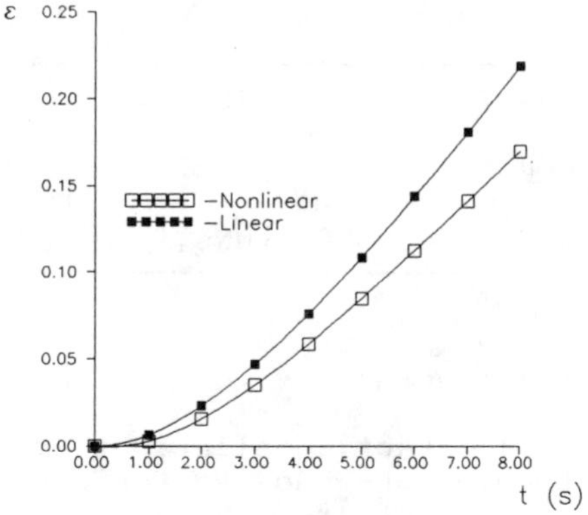

FIG. 2. Comparison of linear and nonlinear strains for the Voigt model (σ=0, $\theta(t)$=200t).

CONCLUSIONS AND REMARKS

1. For the constitutive relations proposed there is a need to know the experimental data including second and third order elastic constants valid for visco-

elastic range as well as coefficient of viscosity and its dependence on temperature.
2 For material data accepted the differences between linear and nonlinear solutions are considerably large.
3. The introduction of the term $L\varepsilon^2$ in Eq. (1) makes considerations of constitutive relations (6) complicated. It is caused by derivation of nonlinear expresion (5)1 . One can suggest a simplified constitutive relation for the spring as follows;

$$\sigma = \gamma(\theta) + E\varepsilon \,, \tag{10}$$

where $\gamma(\theta)$ remains

$$\gamma(\theta) = \gamma_1(\theta) - L\,\alpha^2_T\theta^2 \tag{11}$$

The proposed simplification does not agree with the thermodynamics, but from a qualitative point of view it makes no practical differences for the solutions shown in Figs. 1 and 2.

TABLE I. Material coefficients of D-54-S aluminum alloy

E	l	m	n
	GN/m^2		
34.7	-388.0	358.0	-320.0
ν	η	α_T	
0.31	100 [GNs]	24.0 E-6 [1/K]	

References

1. T. Jękot, Nonlinear thermoelstic problems of homogeneous and isotropic media under great temperature gradients, Arch. Mech., 36,1, 1984.
2. T. Jękot, Spectral Methods for Nonlinear, Coupled, Thermoelstic Rapidly Heated Rod, J. Thermal Stresses, 13, 1, 1990.
3. F. J. Lockett, Nonlinear Viscoelastic Solids, Academic Press 1972.
4. R.T. Smith, R. Stern, P. W. Stephens, Third order elastic moduli of polycrystalline metals from ultrasonic measurements, J. Acoust. Soc. Am. 40, N5,1966.
5. Landolt-Bornstain, Zahlenwerte und Funktionen aus Physik-Chemie-Astronomie-Geophysic und Technik, 1967.

THERMOPLASTIC AND THERMOVISCOPLASTIC FORMS OF BOUNDING SURFACE THEORIES

D.L. McDowell
Georgia Institute of Technology

ABSTRACT

Bounding surface plasticity theories for metals and nonmetals have received much recent attention. We focus here on variations of the models for metallic behavior (c.f. Dafalias and Popov [1]; Krieg [2]; McDowell [3]; Chaboche [4]) under cyclic, multiaxial loading at small strains.

BASIC FRAMEWORK

We adopt the yield condition $f = 3/2||\tilde{s}-\tilde{\alpha}||^2 - R^2$ where $\tilde{s} = \tilde{\sigma} - (\sigma_{kk}/3)\tilde{I}$ is the deviatoric stress and $\tilde{\alpha}$ is the deviatoric backstress. In the multiple surface thermoplasticity theory presented here, we decompose $\tilde{\alpha}$ into two components [4–5], i.e. $\tilde{\alpha} = \tilde{\alpha}^s + \tilde{\alpha}^*$. The yield surface radius (yield strength) R is composed of a temperature dependent component R^o and a component R^{iso} dependent on the degree of isotropic hardening, i.e. $R = R^o(T) + R^{iso}$. We define $\dot{p} = ||\dot{\tilde{\epsilon}}^p|| = (\dot{\epsilon}_{ij}^p\dot{\epsilon}_{ji}^p)^{1/2}$ and T as the absolute temperature. The short range backstress amplitude b is decomposed according to $b = b^o(T) + b^{iso}$, where $b^o(T)$ depends only on temperature and b^{iso} is dependent on the degree of isotropic hardening. Isotropic hardening rules may be partitioned according to

$$\dot{R}^{iso} = \sqrt{\frac{3}{2}}\,(1 - \omega)\,\dot{\chi} \quad , \qquad \dot{b}^{iso} = \omega\,\dot{\chi} \tag{1}$$

where $\dot{\chi} = \mu[\bar{\chi}(\phi,T)-\chi]\dot{p} + (\partial\chi/\partial T)\dot{T}$ is the evolution rate of an isotropic hardening variable representative of the increase of dislocation density, dislocation substructure development, etc. Parameter ϕ denotes dependence on nonproportionality of loading. The factor ω $(0 \leq \Sigma\omega \leq 1)$ partitions the isotropic hardening between the b^{iso} and R^{iso}. Parameter $\bar{\chi}$ represents the saturation level of χ for isotropic hardening. Cmponents $R^o(T)$ and $b_i^o(T)$ are independent of isotropic hardening.

The incompressible flow rule is given by

$$\dot{\tilde{\epsilon}}^p = \frac{1}{H}\langle\dot{\tilde{s}}:\tilde{n} - \Lambda\dot{T}\rangle\tilde{n} = \frac{1}{H}\langle\psi\rangle\tilde{n} \quad \text{(thermoplasticity)} \tag{2}$$

$$\dot{\tilde{\epsilon}}^p = \sqrt{\frac{3}{2}}\,\xi^n\exp(B_0\xi^{n+1})\,\theta\,\tilde{n} \quad \text{(thermoviscoplasticity)} \tag{3}$$

where < > denote MacAuley brackets and $\tilde{n}$ is the unit vector in the plastic rule. Parameter ξ is defined by

$$\xi = \left(\frac{1}{1-\eta}\right)<\sqrt{\frac{3}{2}}\,||\underset{\sim}{s} - \underset{\sim}{\alpha}|| - R>/D = \Sigma_v/D \qquad (4)$$

where Σ_v is the viscous stress and D is the drag stress. To complete the stress-strain relations, we assume linearized, isotropic, decoupled thermohypoelasticity, i.e. $\dot{\underset{\sim}{\epsilon}} = \dot{\underset{\sim}{\epsilon}}^e + \dot{\underset{\sim}{\epsilon}}^p + \dot{\underset{\sim}{\epsilon}}^{th}$; here, the thermal strain rate $\dot{\underset{\sim}{\epsilon}}^{th} = (\alpha \underset{\sim}{I} + \underset{\sim}{\Delta})\dot{T}$ where the latter term enforces path independence of the thermoelastic strain in $(\underset{\sim}{\sigma},T)$ space.

EVOLUTION OF INTERNAL STATE

A convenient assumption is that of temperature history independent evolution of internal variable $\underset{\sim}{\alpha}$ and χ. We neglect static thermal recovery. Let us assume invariance of the ratios $m^s = \underset{\sim}{\alpha}^s/(Cb)$ and $M = \chi/(\mu\bar{\chi})$ with a temperature change, holding all other internal variables fixed [6]. This leads to the requirements that C and μ must be temperature independent and that the sum of the rate independent components of b and $\bar{\chi}$ must be additively separable in T and p, i.e.

$$\Sigma(T)\,\hat{b}(p) = b^o + b^{iso} \qquad (5)$$

$$\bar{\chi} = \Sigma(T)\,\hat{\chi}(p) \qquad (6)$$

Hence, we may select b^o to be of the form $b^o = B\Sigma(T)$ where B is at most a function of p. The complete theory for temperature path history independent evolution of internal variables is summarized as follows, assuming ω is temperature independent:

$$\Lambda = \frac{\underset{\sim}{\alpha}^*:\underset{\sim}{n}}{H^*}\frac{\partial H^*}{\partial T} + \frac{\chi}{\bar{\chi}}\frac{\partial \bar{\chi}}{\partial T}\left\{\frac{\underset{\sim}{\alpha}^s:\underset{\sim}{n}}{b}\,\omega + (1-\omega)\right\} + \sqrt{\frac{2}{3}}\frac{\partial R^o}{\partial T} + \frac{\underset{\sim}{\alpha}^s:\underset{\sim}{n}}{b}\frac{\partial b^o}{\partial T} \qquad (7)$$

$$H = C\left[b - \underset{\sim}{\alpha}^s:\underset{\sim}{n}\right] + (1-\omega)\mu\left[\bar{\chi} - \chi\right] \qquad (8)$$

$$\dot{\underset{\sim}{\alpha}}^s = C(\delta/b,T)\left[b\underset{\sim}{n} - \underset{\sim}{\alpha}^s\right]\dot{p} + \left[\omega\frac{\underset{\sim}{\alpha}^s}{b}\frac{\chi}{\bar{\chi}}\frac{\partial\bar{\chi}}{\partial T} + \frac{\underset{\sim}{\alpha}^s}{b}\frac{\partial b^o}{\partial T}\right]\dot{T} \qquad (9)$$

$$\dot{\underset{\sim}{\alpha}}^* = H^*\dot{p}\underset{\sim}{n} + \frac{\underset{\sim}{\alpha}^*}{H^*}\frac{\partial H^*}{\partial T}\,T \qquad (10)$$

$$\dot{\chi} = \mu\left[\bar{\chi} - \chi\right]\dot{p} + \frac{\chi}{\bar{\chi}}\frac{\partial\bar{\chi}}{\partial T}\,\dot{T} \qquad (11)$$

SOME EXPERIMENTAL CORRELATIONS

It is useful to investigate the performance of the thermoplastic two surface model outlined in the last section. The experiments to be considered herein were conducted on initially annealed OFHC Copper as reported by Freed [7]. The temperature dependent elastic moduli and thermal expansion coefficient are given in [7].

Other material constants and parameters for the temperature path history independent thermoplasticity model include $R^o = 15 + 18\exp(-.002T)$ MPa, $b^o = 0$ MPa, $k_1 = 500$, $k_2 = 30000$, $k_3 = 1$, $k_4 = 3$, $\omega = 0.7$, $\mu = 30$, $\chi = 210$, $H^* = 138 + 2000\exp(-.0105T)$ MPa, $\Sigma(T) = -0.2 - 1.2x10^{-4}T + 1.5\exp(-1.6x10^{-3}T)$ and the initial condition $\chi(0) = 8.16$ MPa. Figure 1 shows the fit of the model to isothermal, cyclically stable hysteresis loops at several temperatures. Figure 2 compares the predictions of the model with the in-phase TMF experimental results. The agreement is qualitatively reasonable. Figure 2 also compares the prediction based on elimination of temperature rate terms with experimental results. Clearly, the model is seriously deficient when the temperature rate terms are neglected, displaying a lack of temperature-induced softening or hardening behavior in the TMF cycle.

CONCLUSIONS & ACKNOWLEDGEMENTS

Thermoplastic and thermoviscoplastic bounding surface theories have been presented. A thermomechanical experiment on OFHC Copper was predicted with the thermoplastic theory. Inclusion of temperature rate terms is essential if the thermoplastic idealization is employed to model high temperature, thermomechanical cyclic loading. For the viscoplastic theory, parallel research indicates that this is not the case.

The author is grateful to Texas Instruments, Inc. and the National Science Foundation (ENG MSM 860-1889) for support of this and related research.

REFERENCES

1. Dafalias, Y.F., and Popov, E.P., 1975, "A Model of Nonlinearly Hardening Materials for Complex Loading," Acta Mechanica, Vol. 21, pp. 173-192.
2. Krieg, R.D., 1975, "A Practical Two Surface Plasticity Theory," ASME Journal of Applied Mechanics, Vol. 42, pp. 641-646.
3. McDowell, D.L., 1985, "A Two Surface Model for Transient Nonproportional Cyclic Plasticity: Part I - Development of Appropriate Equations," ASME Journal of Applied Mechanics, Volume 52, pp. 298-302.
4. Chaboche, J.L., "Constitutive Equations for Cyclic Plasticity and Cyclic Viscoplasticity," Int. J. Plasticity, Vol. 5, No. 3, 1989, pp. 247-302.
5. McDowell, D.L. and Moosbrugger, J.S., "A Generalized Rate-Dependent Bounding Surface Model," ASME PVP-Vol. 129, 1987, pp. 1-11.

6. Ohno, N. and Wang, J., "Nonisothermal Constitutive Modeling of Inelasticity Based on Bounding Surface," Proc. Seventh International Seminar on Inelastic Analysis, Fracture and Life Prediction, SMiRT 10 Post-Conference, University of California, Santa Barbara, August 21-22, 1989, pp. B.1/1-B.1/24.
7. Freed, A.D., "Thermoviscoplastic Model with Application to Copper," NASA Tech. Paper 2845, December 1988.

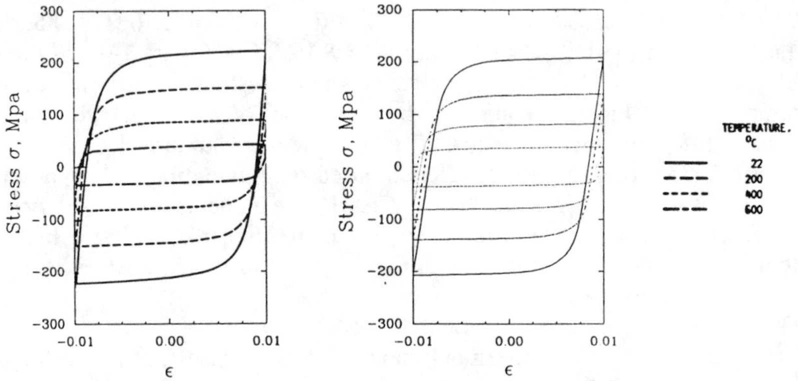

Figure 1. Baseline experimental uniaxial isothermal hysteresis loops at several temperatures (left) and two surface thermoplasticity model correlations (right) for $\dot{\epsilon}$ = 0.001 sec^{-1}.

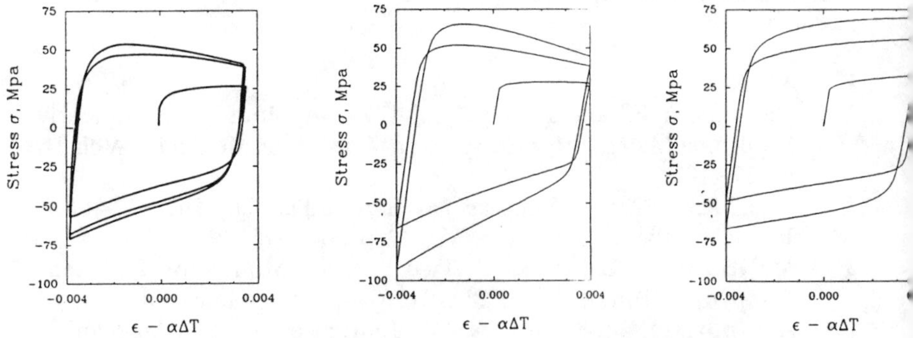

Figure 2. Experimental results (left) and prediction of two surface thermoplasticity model based on temperature path history independence (center) and neglect of temperature rate terms (right) for an in-phase TMF test with a mechanical strain rate of 1.5x10^{-5} sec^{-1} and a temperature range from 200°C to 500°C.

MODELLING OF THERMOMECHANICAL BEHAVIOUR OF SEMI-SOLID METAL

T.G.NGUYEN, D.FAVIER, M.SUERY
Institut National Polytechnique de Grenoble,
Génie Physique et Mécanique des Matériaux,
URA CNRS 793, ENSPG, BP.46,
38402 ST MARTIN D'HERES, FRANCE.

D.BOUVARD
Institut de Mécanique de Grenoble, UMR CNRS 101
BP.53 X, 38041 GRENOBLE CEDEX, FRANCE.

ABSTRACT

This article presents a general non isothermal constitutive model based on the concepts of mechanics of continuous media applied to semi-solid metal. For a given temperature, three types of experiments are used to identify the model.

INTRODUCTION

Semi-solid state processing of metals offers a number of advantages over conventional shaping processes provided that the structure of the solid phase is constituted of quasi-globular particles [1,2] ; these advantages are significant lower operating temperatures, reduced porosity, solidification shrinkage and energy consumption. Owing to the complex rheological behaviour of metal slurries, with history dependency, the prediction of the performance of forming processes is very difficult. After a short presentation of a general non isothermal model, this paper deals with experiments carried out to identify the model at a given temperature.

THEORETICAL ASPECTS

Depending on the volume fraction of solid, semi-solid slurries can behave like a fluid or a viscous solid. The proposed model is restricted to the case for which both solid matrix and liquid of the partially molten system are fully connected ; if the solid matrix is constituted of quasi globular particles of nearly equal size, this assumption restricts our study to solid volume fraction f_s ranging between $\simeq 0.45$ and $\simeq 0.85$.

In these conditions, the deformation of the semi-solid is analogous to that of a porous material saturated with a liquid medium ; the densification of the solid matrix drives the fluid flow behaviour and the resulting pressure distribution of the liquid affects the stresses and densification in the solid. If p_l (M) is the pressure of the liquid, $\underset{\sim}{\sigma}$(M) the total stress applied to the multiphase system at point M, the field of effective stress $\underset{\sim}{\sigma}^*$ (M) is introduced as follows :

$$\underset{\sim}{\sigma}^* (M) = \underset{\sim}{\sigma}(M) + \xi \; p_l (M) \; \underset{\sim}{I} \quad \text{with} \quad \xi = 1 \tag{1}$$

The total stress field must satisfy the equilibrium equation (2).

The behaviour of the solid matrix is assumed to be isotropic viscoplastic and determined by a potential Ω function of $\underset{\sim}{\sigma}^*$ through its invariants. Introducing the deviatoric tensor $\underset{\sim}{S}^*$, a simple form for Ω which takes into account the isotropic-deviatoric coupling is :

$$\Omega = \Omega(\sigma_{eq}) \; ; \; \sigma_{eq}^2 = \frac{A}{2} \, tr\left(\underset{\sim}{S}^* \underset{\sim}{S}^*\right) + B \left(tr \; \underset{\sim}{\sigma}^*\right)^2 \tag{3}$$

with :

$$\frac{\partial \Omega}{\partial \sigma_{eq}} = \beta \; [sh(\alpha \sigma_{eq})]^n \; \exp\left(\frac{-Q}{RT}\right) = D_{eq} \tag{4}$$

This form of the creep flow equation was first proposed by Garofalo for creep and applied to hot working by Sellars and Tegart [3] ; it is selected to express both the power law dependence at small stresses and the exponential dependence at higher stresses of the equivalent strain rate D_{eq} on the equivalent stress σ_{eq}.

The equations (1) to (4) must be completed by equations of continuity expressing the mass conservation, of heat transfer and of liquid flow which can be taken as the Darcy's law in the laminar case. The identification of the general model requires the determination of the coefficients of the creep law (4) together with the A and B parameters. This determination will now be presented using experimental tests carried out both in the solid and semi-solid states.

EXPERIMENTAL RESULTS

The studied material is an Al-base alloy containing 7% Si and 0.3% Mg (A356) for which the semi-solid range extends from 577°C to 610°C. The alloy was subjected to isothermal tensile tests in the solid state at various temperatures (T < 530°C) to identify the creep law (4) and to compression and filtration (drained die pressing) tests in the semi-solid state at 584°C corresponding to an initial solid volume fraction of 0.44. For these tests, the material, initially solidified under a 100MPa pressure, was partially remelted and maintained at 584°C for 30min before deformation ; this generates a globular semi-solid structure with solid globules of 120μm average diameter (fig.1a). The tests in the semi-solid state were carried out at constant strain rates in the range 10^{-3} to $10^{-1} s^{-1}$. For these conditions, it was shown that the interstitial liquid pressure is negligible compared to the material strength [4]. During filtration, homogeneous deformation of the material takes place up to a solid volume fraction of 0.85. On the contrary, the compression tests become rapidly inhomogeneous so that they are exploitable only for small strains. Consequently, compression specimens with various initial solid fractions but with similar size of globules (fig.1b) were obtained by machining the filtration samples after tests interrupted at various strains.

The tensile tests in the solid state allow determination of the coefficients of the creep flow law (fig.2). It is assumed here that these coefficients are also valid at higher temperature including the semi-solid domain. The parameters A and B in equation (3) were determined using the results of both compression and filtration tests in the semi-solid state (fig.3 and 4). The large value of A at relatively low liquid volume fraction confirms the highly deformable character of semi-solid mixtures with globular solid particles and then the advantages of using this kind of material for metal forming. Moreover, A and B are strongly dependent on the solid volume fraction becoming equal to 3 and 0 respectively as the material approaches the fully solid state. In this case obviously, there is no longer isotropic-deviatoric coupling.

CONCLUDING REMARKS

In this paper a model for deformation of semi-solid mixtures has been proposed and identified at a given temperature by experiments carried out both in the solid and semi-solid states. The results show that the behaviour of these materials depends on the isotropic part of the stress tensor, this dependence being however weak compared to that on the deviatoric part and decreasing as the solid volume fraction increases. The morphology of the solid phase influences also probably this dependence ; other experiments dealing with various initial structures are needed to completely determine this dependence.

REFERENCES

1. L. BERNARD, R. MOSCHINI and G. RONCHIATO, Met. Sci. & Tech., 7(1), 84-91 (1989).
2. J.F. SECONDE and M. SUERY, J. Met. Sci., 19, 3995-4006 (1984).
3. C.M. SELLARS and McG. TEGART, Internat. Met. Rev., 17, 1,(1972)
4. P.O. CHARREYRON, Thesis Ecole Nat. Sup. des Mines de Paris, (PARIS 1984).

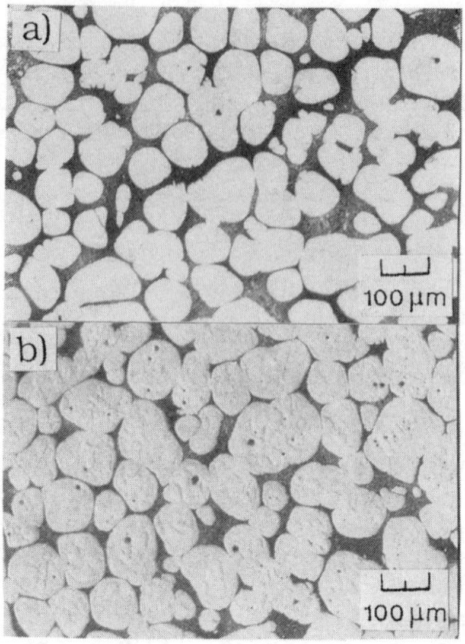

FIG.1 Microstructures of the material :
a) initial (f_s =0.44)
b) after filtration (f_s =0.67)
(The dark phase was liquid at 584°C)

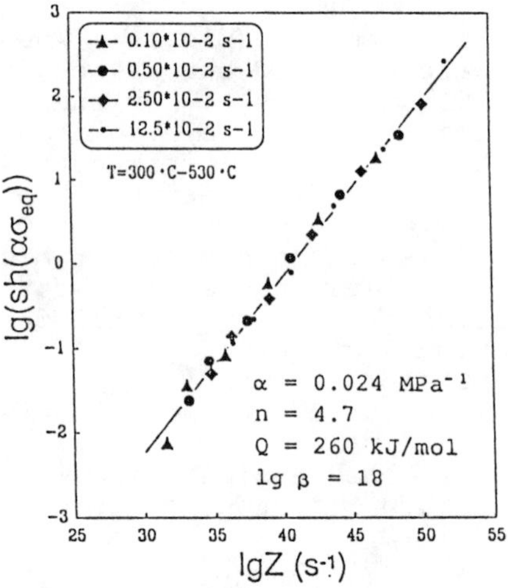

FIG.2 Linear relationship between $\lg(\mathrm{sh}(\alpha\sigma_{eq}))$ and lg Z
Z is the Zener-Hollomon parameter defined as :
$Z = D_{eq} \exp(Q/RT)$

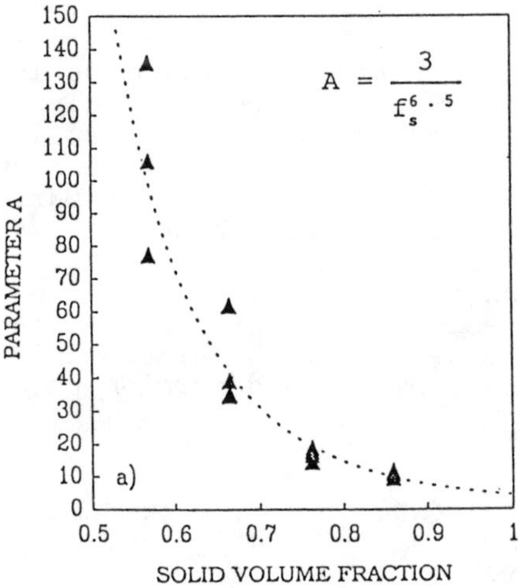

FIG.3 Parameter A versus solid volume fraction.

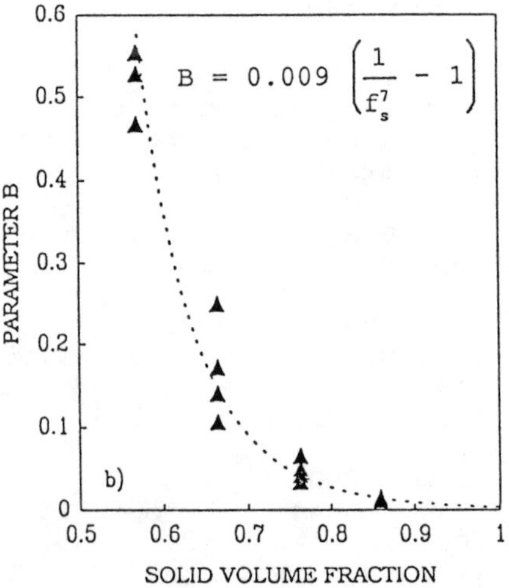

FIG.4 Parameter B versus solid volume fraction.

TRANSIENT THERMO-VISCOELASTIC BEHAVIOUR OF CRACKS IN A LAYERED STRUCTURE

A.P.S. SELVADURAI and M.C. AU
Department of Civil Engineering
Carleton University
Ottawa, Ontario, Canada K1S 5B6

ABSTRACT

The paper examines the transient thermo-viscoelastic behaviour of a cracked pavement structure. The constitutive law for a thermorheologically simple viscoelastic material is used in the theoretical and associated numerical developments. The analysis focusses on the evaluation of the time dependent stress intensity factor at a crack tip in a segmented layered pavement structure.

INTRODUCTION

Asphalt concrete materials exhibit viscoelastic behaviour even at low temperatures of the order of -9°C [1]. Consequently it is desirable to investigate the influence of viscoelastic asphalt properties on the behaviour of fractures and other defects located in layered pavement structures. This paper extends the authors' earlier studies [2] to include the combined influences of heat conduction within the pavement structure and viscoelastic behaviour of the asphalt surface layer. The finite element technique is used to study the transient behaviour of a crack located at the surface of a pavement structure, which is subjected to a sudden lowering of the surface temperature in the form of a step function. The time dependent variation in the mode 1 stress intensity factor at the crack tip is evaluated.

CONSTITUTIVE RELATIONS

The time-dependent stress-strain ($\sigma(t)$, $\epsilon(t)$) relationship for linear viscoelastic asphalt materials can be expressed in the form of the hereditary integral

$$\sigma(t) = R(0)\epsilon(t) - \int_0^t \frac{\partial R(t-\tau)}{\partial \tau}\epsilon(\tau)d\tau \tag{1}$$

where t is the time and $R(t)$ is the "Relaxation Modulus", which can take the form of either $K(t)$ or $G(t)$ depending upon the *dilatational* or *distortional* processes respectively. The complete constitutive tensor can be obtained by combining the two types of responses. The influence of temperature on asphalt behaviour is accounted for by assuming the temperature-time equivalence hypothesis. That is, the relaxation modulus at a certain temperature T can be obtained from the corresponding relaxation modulus at a given or reference temperature T_0 [3]. Consequently the viscoelastic material is considered to be *thermo-rheologically simple.*

NUMERICAL MODELLING

The finite element technique is applied to study the transient thermo-viscoelasticity problem. The temperature equation is given as in [2] and can be solved by using an implicit time integration scheme. A system of Volterra integral equations of the second-kind is obtained for the displacement field. Using a Dirichlet series representation of the relaxation modulus, a memory load vector which summarizes the effect of the deformation and the temperature up to the current time, can be obtained. A step by step time integration scheme similar to that proposed by Taylor et al. [4] and Srinatha and Lewis [5] can be considered for the displacement. Quadratic elements are used for the discretization and quarter point elements are applied to model the crack tip behaviour. Complete details of the formulation are given in [6].

EXAMPLE

The pavement structure studied consists of a viscoelastic asphalt layer resting in bonded contact with an elastic medium. The asphalt layer contains full depth cracks spaced $2\ell_0$ apart and a surface crack of length ℓ is located at the centre. The surface of the pavement is subjected to a sudden reduction in temperature (T_S) in the form of a Heaviside step function. The thermal properties for the layer and the medium are the coefficient of thermal expression $\alpha(T_0) = \alpha_0 = 5 \times 10^{-6}/^oC$; thermal conductivity $k(T_0) = k_0 = 2J/m/^oC/sec$; and volumetric heat capacity $\rho c(T_0) = (\rho c)_0 = 4 \times 10^6 J/m^3/^oC$. Examples of relaxation moduli for the viscoelastic layer are given in Table 1. The finite element discretization is given in Figure 1. The time-dependent variations in the mode I stress intensity factor are given in terms of the normalized parameters

$$\overline{K}_I^* = \frac{K_I\sqrt{2\ell}}{G_1(0)\alpha_0 T_S} \quad ; \quad \bar{t} = \frac{tk_0}{\ell^2(\rho c)_0} \tag{2}$$

Examples of numerical results are shown in Figures (2) and (3).

REFERENCES

1. Low-Temperature Properties of Paving Asphalt Cements, State of the Art Report 7, *Transportation Research Board*, National Research Council, Washington, D.C. (1988).

2. A.P.S. Selvadurai, M.C. Au and W.A. Phang, "The modelling of low

temperature behaviour of cracks in asphalt pavement structures", *Canadian Journal of Civil Engineering* (1990) (in press).

3. F. Schwartz and A.J. Staverman, "Time-dependence of linear viscoelastic behaviour", *Journal of Applied Physics*, 23, pp.838-843 (1953).

4. R.L. Taylor, K.S. Pister and G.L. Goudreau, "Thermomechanical analysis of viscoelastic solids", *Int. J. Numerical Methods Engineering*, 2, pp.45-49 (1970).

5. H.R. Srinatha and R.W. Lewis, "A finite element method for thermoviscoelastic analysis of plate problems", *Computational Methods in Applied Mechanics and Engineering*, 25, pp.21-33 (1981).

6. A.P.S. Selvadurai, M.C. Au and P. Joseph, "Low temperature fracture mechanics of viscoelastic pavement structures", (in preparation).

Table 1:

Material Type	$G(t)/G(0)$	$K(t)/K(0)$
(a)	$0.9 + 0.1\exp(-x)$	1.0
(b)	$0.5 + 0.5\exp(-x)$	1.0
(d)	$0.05 + 0.95\exp(-x)$	1.0

$(x = 8t/\{\ell^2(\rho c)_0/k_0\})$

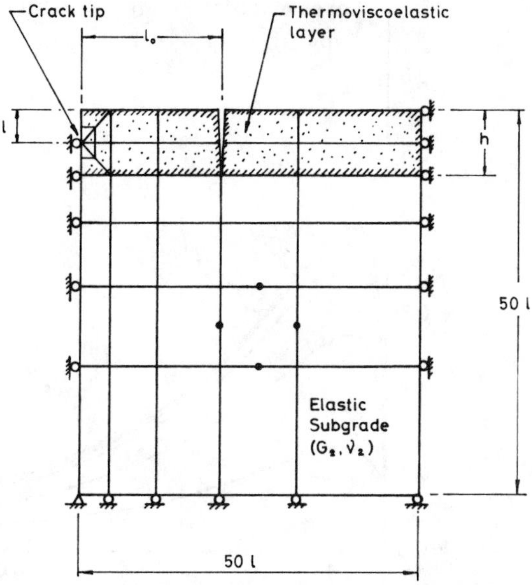

Figure 1. Finite element discretization of pavement structure.

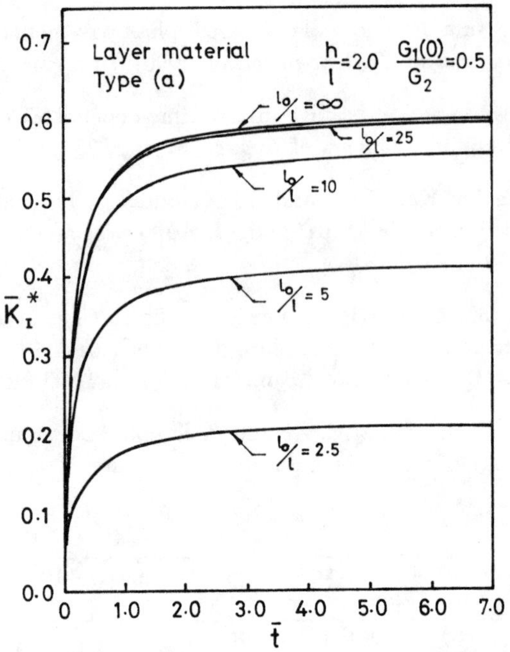

Figure 2. Time-dependent behaviour of the mode I stress intensity factor at the surface crack.

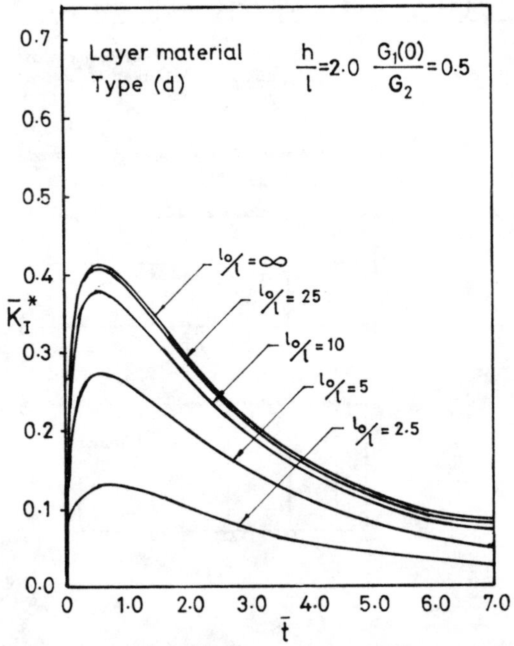

Figure 3. Time-dependent behaviour of the mode I stress intensity factor at the surface crack.

FRACTURE, DAMAGE, LOCALIZATION

ACOUSTIC MATERIAL SIGNATURE OF AN ORTHOTROPIC PLATE

By

M. A. Awal, Research Assistant, T. Kundu, Associate Professor

Department of Civil Engineering & Engineering Mechanics

University of Arizona

Tucson, Arizona 85721.

ABSTRACT

In this paper a theory is presented for the calculation of the acoustic material signature of an orthotropic plate overlain by a fluid. The analysis is carried out for low frequency acoustic waves generated by a cylindrical transducer without a lens rod and the response is measured by a line receiver. Several new features of the material signature and their possible use in fiber reinforced composite material characterization are indicated.

INTRODUCTION

Potential use of acoustic microscopy at low frequency for determining elastic properties (Weglein, 1985) and charcterizing internal defects (Miller, 1985) have been recognized in the recent years. Although conventional spherical (Lemon and Quate, 1973) and shell shaped (Liang et al., 1985) transducers have gained rapid popularity in acoustic microscopy experiments, they are not suitable for characterizing anisotropic materials (Kushibiki et al., 1981). To overcome the shortcomings associated with conventional acoustic microscopes, Kundu (1988a, b) has recently proposed a cylindrical transmitter and line receiver arrangement in an acoustic microscope for characterizing internal defects in anisotropic materials.

While the possible use of acoustic microscopy in anisotropic material characterization has been recognized, to the authors knowledge, Karim & Kundu (1989) first attempted to synthesize acoustic material signature (AMS) for such materials. In the present study, a theoretical analysis is carried out to synthesize AMS for an orthotropic plate generated by a cylindrical trunsducer and line receiver arrangement as shown in Fig. 1.

ANALYSIS

Geometry of the problem of our interest is shown in Fig. 1. An orthotropic plate overlain by a fluid is subjected to converging acoustic waves generated by a cylindrical transducer extending an angle α as shown in the figure. The acoustic beam after being reflected by the specimen surface strikes the receiver located at an angle θ (see Fig. 1) and generates a voltage V. The objective of this paper is to analytically synthesize the receiver output voltage (AMS) as a function of z, the distance between the reflecting surface of the specimen and center of curvature of the cylindrical transducer. To achieve this objective, first the potential of the incident field is computed. Then by Fourier inversion technique the reflected field is obtained (Bertoni and Tamir, 1973). For a converging beam direct Fourier transform of the incident field is very difficult. Hence stationary phase method (Jeffreys and Jeffreys, 1950) is adopted to compute the Fourier transform of the incident field in an approximate manner. Thus the final reflected field is obtained (Karim and Kundu, 1989) which can be rewritten in the following form

$$\phi_r(x,y) = \sqrt{R}e^{\frac{ik_f R}{2\pi}} \int_0^{\frac{k_f \tan\alpha}{\sqrt{1+\tan^2\alpha}}} R(k)A(k)e^{i(\eta y - kx)}dk \tag{1}$$

$R(k)$ here is different from $R(k)$ in Karim and Kundu (1989) because of the shear stress is zero at the bottom surface of the plate. Detailed steps of $R(k)$ computation is omitted here.

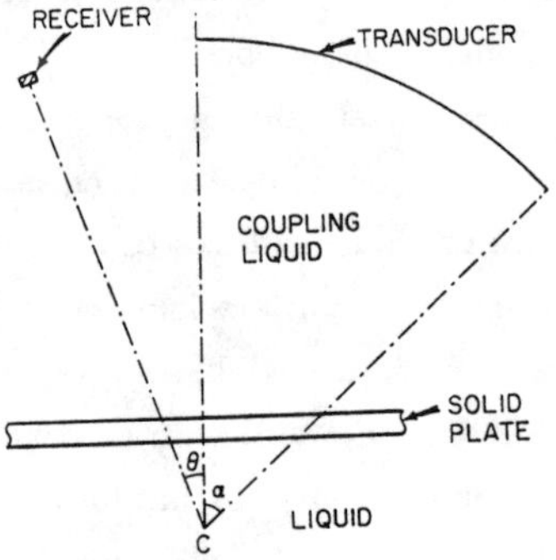

Fig. 1: Problem geometry.

RESULTS AND DISCUSSIONS

Numerical results are presented in this section for a Glass-Polyester (50% volume fraction) composite plate overlain by water. For comparison, AMS is also calculated for Glass and Polyester plate (isotropic) overlain by water. Material properties of Glass, Polyester and Glass–polyester composite are given in Table 1.

Table 1: Material Properties

Material	C_{11} GPa	C_{12} GPa	C_{22} GPa	C_{66} GPa	Density ρ (gm/cc)
1. Glass	91.20	30.40	91.20	30.40	1.35
2. Polyester	6.19	3.64	6.19	1.28	1.10
3. Composite (0^0)	36.94	3.73	11.15	4.66	1.23
4. Composite (90^0)	11.15	3.73	36.94	4.66	1.23

Water has a density of 1 gm/cc and its P-wave velocity is 1.5 mm/μsec. The transducer has a cylindrical shape, as shown in Fig. 1, with a radius of curvature $R = 170$ mm and subtended angle α, equal to 45^0 and 80^0. Frequency of the generated acoustic wave is 6 MHz. Numerical results are shown in Fig. 2. One can see from figure 2 that 45^0 and 80^0 transmitters don't produce different AMS for isotropic plates. This is because rays with incident angle greater than 45^0 cann't reach the receiver.

ACKNOWLEDGEMENT

This research was partially supported by a grant from the National Science Foundation under the contract number DMC–8807661.

REFERENCES

1. Bertoni, H. L. and Tamir, T. (1973) "Unified Theory of Rayleigh-Angle Phenomena for Acoustic Beams at Liquid Solid Interfaces", Journal of Applied Physics, 2, 157–172.

2. Jeffreys, H. and Jeffreys, B. S. (1950) *"Methods of Mathematical Physics"*, 2nd Ed., University Press, Cambridge, 505–507.

3. Karim, M. R. and Kundu, T. (1989) "Acoustic Material Signature of Fiber Reinforced Composites". Energy–source Technology Conference & Exibition Houston, Jan 22–25, 1989.

4. Kundu, T. (1988) "Acoustic Microscopy at Low Frequency", ASME Journal of Applied Mechanics, 46, 325–331.

5. Kundu, T. (1988) "A Theoretical Analysis of Acoustic Microscopy with Converging Acoustic Beams", Journal of Applied Physics B, 46, 325–331.

6. Kushibiki, J., Ohkubo, A. and Chubachi, N. (1981) "Linearly Focussed Acoustic Beams for Acoustic Microscopy", Electron letters, 17, 520–522.

7. Lemons, R. A. and Quate, C. F. (1973) "A Scanning Acoustic Microscope", Proceedings IEEE Ultrasonic Symposium, 18–20.

8. Liang, K. K., Kino, G. S. and Khuri-Yakub, B. T. (1985) "Material Characterization by the inversion of V (z)", IEEE Transaction on Sonics and Ultrasonics, SU–22, 213–224.

9. Miller, A. J. (1985) "Scanning Acoustic Microscope in Electronics Research", IEEE Transactions, SU–32, 320–3244.

10. Weglein, R. D. (1985) "Acoustic Micrometrology", IEEE Transactions on Sonics and Ultrasonics, SU–32, 225–234.

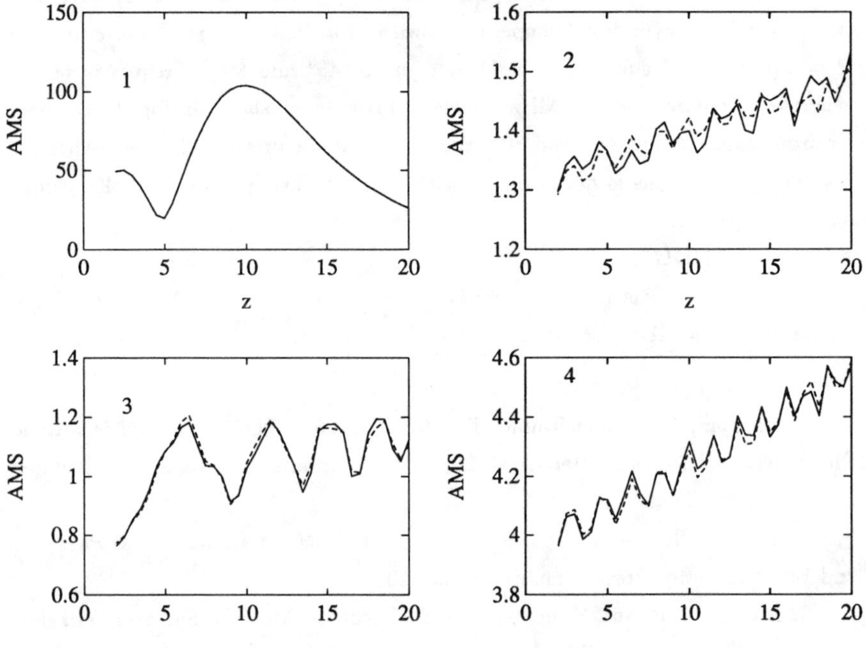

Fig. 2: AMS of four meterials (see table 1) for receiver angle $\theta = 20^0$. Material numbers are shown in the top–left corner in every plot. Solid and dotted lines corresponds to 45^0 and 80^0 transducer angles (α) respectively.

RECENT PROGRESS IN DAMAGE MODELING: NONLOCALITY AND ITS MICROSCOPIC CAUSE

Zdeněk P. Bažant
Walter P. Murphy Professor of Civil Engineering
Nortwhwestern University, Evanston, IL 60201, USA

ABSTRACT

The lecture has a twofold aim: To briefly review the current status of nonlocal modeling of damage, and to present a new justification of the nonlocality of strain-softening damage due to microcracking. The continuum damage due to microcracking is analyzed assuming a simplified micromechanics model in which interactions of adjacent microcracks are neglected and the microcracks are arranged at the centers of the cells of a cubic mesh. The release of stored energy caused by the formation of one microcrack is calculated as a function of the associated relative displacement across the cell, which corresponds to the average strain of the macroscopic continuum. After imposing the homogenization conditions of equal energy dissipation and of displacement compatibility between the micromechanics model and the macroscopic continuum, it is shown that damage is a nonlocal variable that is a function of the averaged (nonlocal) strain from a certain neighborhood of the given point. This corroborates the hypothesis of nonlocal damage, which was originally introduced for other reasons. The cause for the nonlocality of damage is that whether a microcrack will form depends on the energy stored in a finite region of nonnegligible size around the potential microcrack.

PART I. COMMENTS ON CURRENT STATUS OF NONLOCAL MODELS

A continuum with nonlocal damage has recently been shown to be an effective approach for the analysis of strain-softening structures (Bažant and Pijaudier-Cabot 1987; Pijaudier-Cabot and Bažant 1987; Bažant, Lin and Pijaudier-Cabot 1987; Bažant and Pijaudier-Cabot 1988; Bažant and Lin 1988a,b). The basic idea of the nonlocal continuum model is that only the damage is nonlocal, being a function of the strain average from a certain neighborhood of a given point, while all the other variables, especially the elastic strain, are local.

By contrast, in the original nonlocal continuum models for elastic materials (Kröner 1968; Krumhansl 1968; Kunin 1968; Levin 1971; Eringen and Edelen 1972; Eringen and Ari 1983; etc.), as well as in the first nonlocal model for strain-softening continuum (Bažant 1984; Bažant, Belytschko and Chang 1984), the elastic strain and total strain were nonlocal. This led to certain numerical difficulties (Bažant and Pijaudier-Cabot 1988), for example the existence of spurious zero-energy instability modes (which had to be suppressed artificially by overlay with local continuum), the presence of spatial integrals or higher-order instability modes (which had to be suppressed artificially by overlay with local continuum), the presence of spatial integrals or higher-order derivatives in the differential equations of equilibrium or motion and in the boundary and interface conditions, and an imbricate structure of the finite element approximation which however proved cumbersome for programming.

These difficulties were later shown to be a consequence of imposing symmetry on the integral or differential operators involved. The symmetry is lost with the nonlocal damage concept, which means that the tangential (but not the elastic) structural stiffness matrix of the finite element approximation is nonsymmetric (Bažant and Pijaudier-Cabot 1988). But the particular type of nonsymmetry obtained does not appear to cause any numerical difficulties, even for strain-softening structures with thousands of nodal displacements (Bažant and Lin 1988a).

The nonlocal concept eliminates the problems with spurious mesh sensitivity and incorrect convergence. It assures that refinements of finite element mesh cannot lead to spurious localization of strain, damage and energy dissipation into a strain-softening zone of vanishing volume. The most important physical property of nonlocal continuum damage is that for geometrically similar structures it yields a size effect that is transitional between plasticity (no size effect) and linear elastic fracture mechanics (the strongest possible size effect). This size effect is evidenced by extensive laboratory measurements on various kinds of concrete structures and fracture specimens of concrete and rock, as well as the available test data for fracture of ice (e.g. Dempsey 1990) and toughening ceramics (cf. Bažant and Kazemi 1990, with further references). The finite element codes based on local continuum cannot capture the size effect, which is a major fault when structures with damage are analyzed. Correct modeling of the experimentally observed size effect should be adopted as the basic criterion of acceptability of any finite element code for concrete structures or rock.

Physical justification by micromechanics, however, has been rather limited. In a recent study (Bažant 1987) it was suggested that the physical source of nonlocality of damage is the fact that the formation and growth of a microcrack depends on the strain energy stored in a nonzero volume of the material surrounding the microcrack, whose release drives the growth of the microcrack. Considering a quasiperiodic microcrack array and analysing the displacements due to fracture, it was shown that, under certain simplifying assumptions, the damage is a function of the of the spatially averaged fracturing strain of the macroscopic smoothing continuum, which implies damage to be nonlocal. This form of damage, however, does not seem to be the most convenient formulation, and does not quite agree with the nonlocal damage formulations used in the above-mentioned finite element models.

Aside from presenting a brief review of the advances amd problems just described, the lecture has the secondary objective of presenting the following new justification of the nonlocality of continumum damage that is due to a system of densely distributed microcracks.

PART II. JUSTIFICATION OF NONLOCALITY OF DAMAGE DUE TO MICROCRACKS

Consider an elastic material with penny-shaped microcracks of various diameters 2a. We imagine the material to be subdivided into cubical cells of side ℓ (Fig. 1a), each of which contains approximately in the middle one microcrack. For the sake of simplicity, we suppose each microcrack to be sosmall ($a \ll \ell$) that its interaction with other microcracks, as well as the energy release from the adjacent cells, is negligible.

We analyze one microcrack and allign the cell so that its one side as well as the coordinate axis x be parallel to this microcrack (Fig. 1a,b). We suppose that the microcrack plane is normal to the maximum principal stress at the center of the cell before cracking, denoted as σ, and for the sake of simplicity we assume that the normal strains in the direction parallel to the crack are constant, as illustrated by imagined sliding restraints on the sides of the cell shown in Fig. 1b. We assume the variation of σ over the cell to be sufficiently small, so that the stress intensity factor K_I of the microcrack is approximately the same as that for a penny-shaped crack in an infinite elastic solid with stress σ at infinity, which is as follows (cf. Broek 1986; Knott 1973; Tada et al. 1985; Murakami 1987)

$$K_I = 2\ \sigma \sqrt{a/\pi} \tag{1}$$

We now try to calculate the energy release due to crack formation as a function of the deformation of the cell. We begin by writing the rate of release of energy (complementary energy) W_f^* due to fracture:

$$\partial W_f^* / \partial a = 2\ \pi\ a\ K_I^2 / E' = 8\ \sigma^2 a^2 / E' \tag{2}$$

where $E' = E/(1 - \nu^2)$, E = Young's elastic modulus, ν = Poisson ratio. Since the material is elastic and thus path-independent, we may consider for the purpose of energy calculation that the crack has formed under constant stress σ. Then, by integration of Eq. 2, the total energy release caused by the microcrack is obtained as

$$W_f^* = 8\ a^3 \sigma^2 / 3\ E' \tag{3}$$

Let δ be the total relative displacement between the opposite sides of the cell ($\delta = u_2 - u_1$ where u_2, u_1 = displacements in the x-direction at the opposite sides of the cell) and δ_f the relative displacement due to crack formation, which is approximately equal to the relative displacement in an infinite solid between its opposite infinities. In the diagram of $\sigma\ell^2$ (the force acting on the sides of the cell) versus δ (Fig. 2a), W_f^* is represented by the area 0120. This triangular area is equal to $W_f^* = \delta_f \sigma \ell^2 / 2$. Setting this equal to Eq. 3, one gets

$$\delta_f = \frac{16}{3E'} \frac{a^3}{\ell^2} \sigma \tag{4}$$

The same expression can be obtained from Eq. 3 by Castigliano's theorem, which implies that $\delta_f = \partial W_f^* / \partial(\sigma\ell^2)$.

Consider now that the crack forms at constant δ rather than at constant σ. This must be equivalent to first unloading the uncracked solid from stress σ_0 to a certain stress $\sigma_0 - \sigma_f$ (path $\overline{12}$ in Fig. 2b) and, second, letting the crack grow at constant stress (path $\overline{23}$ in Fig. 2b), provided that $\sigma_f = E' u_f / \ell$ (from triangle 123 in Fig. 2b), in order to guarantee that the displacement increase δ_f due to crack formation at constant stress $\sigma_0 - \sigma_f$ restores the original total displacement δ_f (point

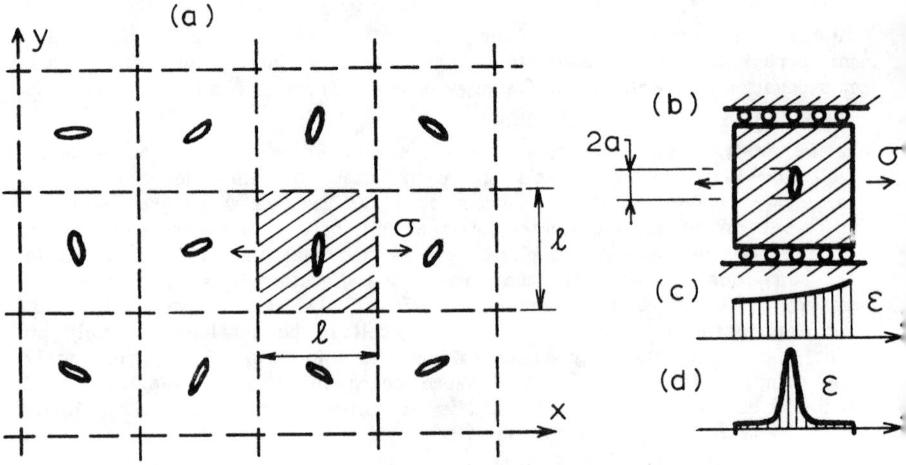

FIG. 1. (a) Array of cubical cells containing microcracks, and (b) one cell with simplified boundary conditions considered in calculations.

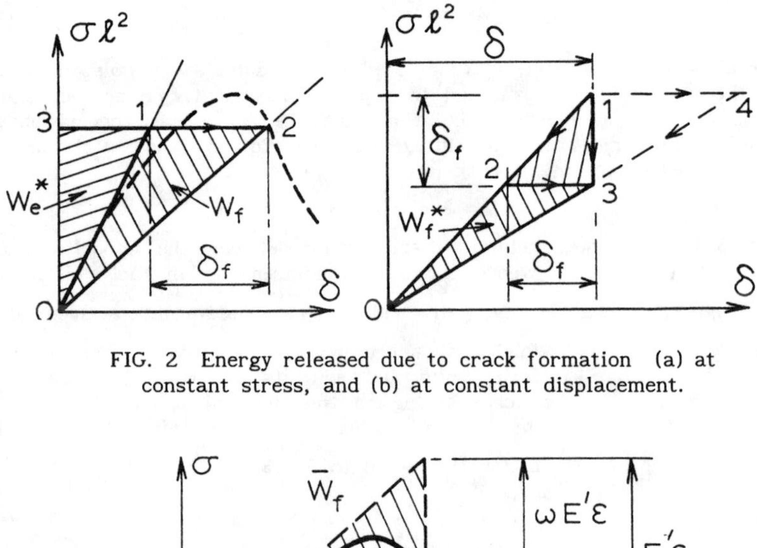

FIG. 2 Energy released due to crack formation (a) at constant stress, and (b) at constant displacement.

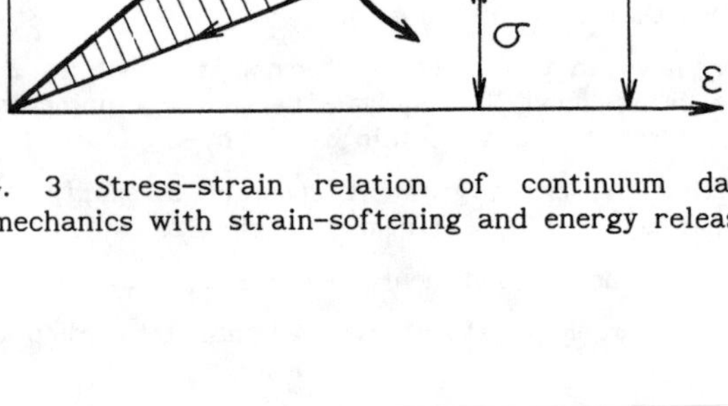

FIG. 3 Stress-strain relation of continuum damage mechanics with strain-softening and energy release.

3 in Fig. 2b). In analogy to Eq. 4, we have for this process $\delta_f/\ell = 16(a/\ell)^3(\sigma_0-\sigma_f)/3E'$ (segment $\overline{23}$ in Fig. 2c). If we set this equal to $\delta_f/\ell = \sigma_f/E'$ and solve for σ_f, we get

$$\sigma_f = \frac{\sigma_0}{1 + \frac{3}{16}\left(\frac{\ell}{a}\right)^3} \tag{5}$$

where $\sigma_0 = E'\delta/\ell$ = initial stress before cracking. The (complementary) energy released by crack formation at constant stress at σ_0-σ_f is represented by area 0230 in Fig. 2b, while the energy (not complementary energy) released by crack formation at constant displacement δ is represented by area 0130 in Fig. 2b, the value of which is

$$W_f = \sigma_f \frac{\delta}{\ell}\frac{\ell^3}{2} = \frac{E' \ell \delta^3}{2 + \frac{3}{8}\left(\frac{\ell}{a}\right)^3} \tag{6}$$

Note that since the material is elastic, that is path-independent, the same result must be obtained if one considers the path 143 (instead of 123) in Fig. 2b, for which, first, the crack is formed at constant stress σ_0 and, second, the cell is unloaded so as to restore the original relative displacement δ.

For a propagating crack we must have $K_I = K_R(a)$ = given R-curve (crack resistance curve) = critical stress intensity factor required for further crack growth, which must be determined in advance. Eq. 1 with σ replaced by $\sigma-\sigma_f$ (path $\overline{23}$ in Fig. 2b) then provides

$$a = \frac{\pi K_R^2(a)}{4(\sigma_0 - \sigma_f)^2} \tag{7}$$

Substituting $\sigma_0 = E'\delta/\ell$ and Eq. 5 for σ_f, and solving the resulting equation for δ/ℓ, we acquire the relation

$$\frac{\delta}{\ell} = \frac{1}{2E'}\sqrt{\frac{\pi}{\ell}\frac{\ell}{a}}\, K_R\left(\ell\frac{a}{\ell}\right)\left[1 + \frac{16}{3}\left(\frac{a}{\ell}\right)^3\right] = \phi\left(\frac{\ell}{a}\right) \tag{8}$$

where ϕ is a function, defined by this relation. Denoting the inverse function as ψ (and supposing function ϕ to be invertible), we may write

$$\frac{\ell}{a} = \psi\left(\frac{\delta}{\ell}\right) \tag{9}$$

Substituting this into Eq. 6 we obtain the desired result

$$W_f = \frac{E' \ell \delta^3}{2 + \frac{3}{8}\psi^3\left(\frac{\delta}{\ell}\right)} = f\left(\frac{\delta}{\ell}\right) \tag{10}$$

where f is a function.

For other geometries of microcracks and repetitive cells, one may expect similar results, but with different expressions for functions ϕ, ψ and f.

Let ε be the (local) macroscopic strain (normal strain in the direction of principal stress σ), and $\langle\varepsilon\rangle$ the average (nonlocal) macroscopic strain, defined as $\langle\varepsilon(\underset{\sim}{x})\rangle = \ell^{-3}\int_V \varepsilon(\underset{\sim}{s})dV(\underset{\sim}{s})$ where V is the volume of the cell, $\langle\ \rangle$ denotes the spatial (nonlocal) averaging operator, $\underset{\sim}{x}$ is the coordinate vector of the center of the cell, and $\underset{\sim}{s}$ are the cordinate vectors of the points of the cell. For the sake of brevity, we will delete in the following the coordinates $\underset{\sim}{x}$ and $\underset{\sim}{s}$, simply writing

$$\langle\varepsilon\rangle = \frac{1}{\ell^3} \int_V \varepsilon \, dV \tag{11}$$

As one homogenization condition, strains ε must be compatible with displacement δ due to crack, which is satisfied by

$$\delta = \ell \langle\varepsilon\rangle \tag{12}$$

The energy released by crack formation at constant δ may now be rewritten as

$$W_f = \frac{E' \ \ell^3 \langle\varepsilon\rangle^2}{2 + \frac{3}{8} \psi^3(\langle\varepsilon\rangle)} = f(\langle\varepsilon\rangle) \tag{13}$$

or more precisely $W_f(\underset{\sim}{x}) = f[\langle\varepsilon(\underset{\sim}{x})\rangle]$. This equation, which shows that W_f is a function of the average (nonlocal) strain rather than the (local) strain, is the key for the nonlocal character of continuum damage. Note that simplifying Eq. 13 as a local relation $W_f(x) = f[\varepsilon(x)]$ can make a large difference if the strain distribution approaches Dirac delta function as shown in Fig. 1d, which is known to typically happen in local continuum damage formulations. In fact, the nonlocality is what enforces smooth (nonlocalized) strain distributions.

Consider now the standard stress-strain relation of continuum damage mechanics (Kachanov 1958; Lemaitre and Chaboche 1978). Its simplest form is

$$\sigma = (1 - \omega) E' \varepsilon \tag{14}$$

where ω is called the damage, supposed here to be a scalar, for the sake of simplicity. Continuum damage mechanics assumes the unloading stiffness to be given by the secant modulus, which is equal to $(1-\omega)E'$ and corresponds to line $\overline{02}$ in Fig. 3. The energy release (dissipation) due to damage is given by the area of triangle 0120 in Fig. 3, which is $\omega E'\varepsilon^2/2$. Thus the energy release from volume V of the cell is

$$W_f = \int_V \omega \frac{E'\varepsilon^2}{2} dV = \frac{E'}{2} \ell^3 \langle \omega \varepsilon^2 \rangle \tag{15}$$

At this point, subject to further confirmation, we anticipate damage ω to be a nonlocal variable characterizing the behavior of the volume V as a whole. This permits simplifying Eq. 15 by treating ω as a constant, even though it may vary with spatial coordinates x, y, z. Accordingly, Eq. 15 may be approximately replaced by the equation

$$W_f = \frac{E'}{2} \ell^3 \omega \langle \varepsilon^2 \rangle \tag{16}$$

As a second homogenization condition, the energy releases calculated for the cell of the micromechanics model (Eq. 10) and for volume V of the macroscopic continuum having the same center (Eq. 16) must be equal. Solving this equality for ω, we acquire the following result:

$$\omega = \frac{\langle\varepsilon\rangle^2}{\langle\varepsilon^2\rangle}\left[1 + \frac{3}{16}\psi^3(\langle\varepsilon\rangle)\right]^{-1} = \frac{2 f(\langle\varepsilon\rangle)}{E' \ell^3 \langle\varepsilon\rangle^2} = \frac{\langle\varepsilon\rangle^2}{\langle\varepsilon^2\rangle} F(\langle\varepsilon\rangle) \quad (17)$$

in which F is a function. If the variation of ω within volume V is not too strong, one may use the approximation $\langle\varepsilon^2\rangle \simeq \langle\varepsilon\rangle^2$, and then

$$\omega = F(\langle\varepsilon\rangle) \quad (18)$$

This confirms that damage due to microcracking is nonlocal (a result we anticipated in our previous assumption), and must be considered as a function of the average (nonlocal) macroscopic strain rather than the local macroscopic strain.

If we did not replace Eq. 15 by Eq. 16, the result would be that

$$\langle \omega \varepsilon^2 \rangle = \hat{F}(\langle\varepsilon\rangle) \quad (19)$$

where $\hat{F}$ is a function. This still means that damage is nonlocal, but its calculation from strains in a finite element program would be implicit, and thus more involved.

CONCLUSION

Spurious localization of strain-softening damage, along with the inherent spuriousm mesh sensitivity and nonobjectivity, can be prevented by the nolocal continuum concept. An effective and easily usable formulation results when only the strain du to microcracking is considered an nonlocal while the elastic strain is local, same as in the classical material models.

A simplified micromechanics analysis shows that continuum damage due to microcracking must be nonlocal, expressed as a function of the spatially averaged (nonlocal) strain in a certain neighborhood of the given continuum point. The reason, simply stated, is that the fracturing strain due to damage is the result of the release of stored energy from the microcrack neighborhood, the size of which is not zero but finite.

ACKNOWLEDGMENT.-Financial support under National Science Foundation grant No.BCS-8818230 to Northwestern University is gratefully acknowledged. Partial support has also been obtained from NSF Center for Science and Technology of Cement-Based Materials at Northwestern University.

REFERENCES

Bažant, Z. P. (1984). "Imbricate continuum and its variational derivation," Journal of the Engineering Mechanics Division, ASCE, 110(12), 1693-1712.

Bažant, Z. P. (1985). "Mechanics of distributed cracking," Applied Mechanics Reviews, ASME 39(5), 675-705.

Bažant, Z. P. (1987). "Why continuum damage is nonlocal: justification by microcrack array," Mechanics Research Communications, 14(516), 407-419.

Bažant, Z. P., and Belytschko, T. B., and Chang, T. P. (1984). "Continuum theory for strain-softening," Journal of the Engineering Mechanics Division, ASCE, 110(12), 1666-1692.

Bažant, Z. P., and Kazemi, M. T. (1990). "Size effect in fracture of ceramics," J. of Am. Ceramic Soc., (in press).

Bažant, Z. P., and Kim, S. S. (1979). "Plastic-fracturing theory for concrete," J. of the Engng. Mech. Div., Proc. ASCE, 105, 407-428.

Broek, D. (1974). Elementary engineering fracture mechanics, Sijthoff and Noordhoff, International Publishers, Netherlands.

Dempsey, J. (1990). ASCE Materials Engineering Congress held in Denver, in press.

Eringen, A. C., and Edelen, D. G. B. (1972). "On nonlocal elasticity," International Journal of Engineering Science, 10, 233-248.

Eringen, A. C., and Ari, N. (1983). "Nonlocal stress field at Griffith crack," Crist. Latt. and Amorph. Materials, 10, 33-38.

Kachanov, L. M. (1958). "Time of rupture process under creep conditions," Izvestia Akademii Nauk, USSR, 8, 26-31 (in Russian).

Knott, J. F. (1973). "Fundamentals of fracture mechanics," Butterworth, London. forces," International Journal of Solids and Structures. 3, 731-742.

Krumhansl, J. A. (1968). "Some considerations of the relations between solid state physics and generalized continuum mechanics," Mechanics of Generalized Continua, edited by E. Kroner, Springer-Verlag, Heidelberg, Germany, 298-331.

Kunin, I. A. (1968). "The theory of elastic media with microstructure and the theory of dislocations," Mechanics of Generalized Continua, edited by E. Kröner, Springer-Verlag, Heidelberg, Germany, 321-328.

Lemaitre, J., and Chaboche, J. L. (1985). Mécanique des matériaux solides, Dunod-Bordas, Paris.

Levin, KV. M. (1971). "The relation between mathematical expectation of stress and strain tensors in elastic microheterogeneous media," Prikl. Mat. Mekh, 35, 694-701.

Mazars, J. (1984). "Application de la mecanique de l'endommagement au comportement non-lineaire et a la rupture du beton de structure," These de doctorat d'Etat es Sciences Physiques, Universite Paris VI, France.

Murakami, Y., Ed. (1987). "Stress intensity factors handbook," Pergamon Press, Oxford, New York.

Pijaudier-Cabot, G., and Bazant, Z. P. (1987). "Nonlocal damage theory," J. of Eng. Mech., ASCE 113(10), 1512-1533.

Pijaudier-Cabot, G., and Bazant, Z. P. (1990). "Propagation of interacting cracks in an elastic solid with inclusions," Preliminary Report, Northwestern University.

Tada, H., Paris, P. C., and Irwin, G. R. (1985). "The stress analysis of cracks handbook," 2nd ed., Paris Productions Inc., St. Louis (226 Woodbourne Dr.) Mo.

LOCALIZATION PHENOMENA AT THE BOUNDARIES AND INTERFACES OF SOLIDS

Ahmed BENALLAL, René BILLARDON, Giuseppe GEYMONAT
Laboratoire de Mécanique et Technologie
E.N.S. de Cachan/C.N.R.S./Université Paris 6
61, Avenue du Président Wilson, 94235-CACHAN, France

ABSTRACT

The conditions for the localization of the deformation, inside, at the boundaries or at the interfaces of rate-independent solids are studied in details, for both the linear and the non-linear case. Physical interpretations of these conditions are also given.

1- INTRODUCTION

Since the pioneering works of Hadamard[1], Hill[2], Mandel[3] and Rice[4], the localization of the deformation in rate-independent materials is treated as the bifurcation of the rate problem.

Here, we give a general view of the various bifurcation and localization phenomena for possibly heterogeneous solids made of rate-independent (including elastic-plastic, damageable) materials. Under the small strain assumption, the behaviour of these materials is described by the following piece-wise linear rate constitutive laws :

$$(1) \qquad \dot{\sigma} = \mathbb{L} : \varepsilon(\mathbf{v}) \quad \text{with} \quad \begin{cases} \mathbb{L} = \mathbb{E} \ \text{when} \ f < 0, \ \text{or} \qquad f = 0 \ \text{and} \ \mathbb{b} : \mathbb{E} : \varepsilon(\mathbf{v}) < 0 \\ \\ \mathbb{L} = \mathbb{H} \ \text{when} \qquad\qquad\qquad f = 0 \ \text{and} \ \mathbb{b} : \mathbb{E} : \varepsilon(\mathbf{v}) \geq 0 \end{cases}$$

where $\dot{\sigma}$ and $\varepsilon(\mathbf{v}) = \dot{\varepsilon}$ respectively denote the stress and strain rates, $\mathbf{v}$ the velocity, and f the yield function.

This paper constitutes a unified presentation of the results given by Benallal, Billardon & Geymonat [5-8]. It is shown that in general different types of localization phenomena may occur, depending on the failure of one of the three conditions which are described in section 2. Their physical interpretation is the following :

- *the ellipticity condition* is very classical. Its failure is the condition for localization given by Rice and linked to the appearance of *deformation modes involving discontinuities of the velocity gradient.* It has also been related to *stationary acceleration waves* ;
- *the boundary complementing condition* governs instabilities at the boundary of the solid. Its failure leads to *deformation modes localized at the boundary* and is related to *stationary surface waves* (for instance Rayleigh waves);
- *the interfacial complementing condition* governs instabilities at interfaces. Its failure leads to *deformation modes localized at each side of the interface* and is related to *stationary interfacial waves* (Stonely waves).

2- RATE PROBLEM ANALYSIS : THE LINEAR CASE

Let us consider for instance the body sketched in Fig.1 ; it may represent an inclusion within a matrix, a fibre in a composite, ... Qualitative results can be exhibited from the analysis of the rate problem for the so-called linear comparison solid (see Hill[9]). In this case, *this linear problem is well-posed if and only if* the following conditions are met :

- *the ellipticity condition* : the rate equilibrium equations must be elliptic in the closure of the body Ω, i.e.

$$\det(\mathbf{n} \,.\, \mathbb{H} \,.\, \mathbf{n}) \neq 0 \qquad \text{for any vector } \mathbf{n} \neq \mathbf{0}, \text{ and any point } M \in \overline{\Omega}.$$

- *the boundary complementing condition* : this relation between the coefficients of the field and boundary operators must be satisfied at every point P belonging to the boundary Γ where the boundary conditions are formally written as $\mathbb{B}(\mathbf{v}) = \mathbf{g}$. This condition is easily phrased in terms of an associated problem on a half space defined by $z > 0$. It requires for every vector $\mathbf{k} = (k_1, k_2, 0) \neq \mathbf{0}$, that the only solution to the rate equilibrium equations with constant coefficients (equal to those of the operator at point P), in the form

$$\mathbf{v}(x, y, z) = \mathbf{w}(z) \exp[i\,(k_1\, x + k_2\, y)]$$

with bounded $\mathbf{w}$ and satisfying the homogeneous boundary conditions $\mathbb{B}(\mathbf{v}) = \mathbf{0}$, is the identically zero solution $\mathbf{v} \equiv \mathbf{0}$.

- *the interfacial complementing condition* : this relation between the coefficients of the field operators in Ω_1 and Ω_2 must be satisfied at every point Q of the interface I between Ω_1 and Ω_2. This condition is again easily phrased in terms of an associated problem on the whole space divided by the plane interface $z = 0$. It requires for every vector $\mathbf{k} = (k_1, k_2, 0) \neq \mathbf{0}$, that the only solution to the rate equilibrium equations with constant coefficients (equal to those of the operators at point Q, in Ω_1 for $z < 0$ and in Ω_2 for $z > 0$), in the form

$$(\mathbf{v}_1(x, y, z), \mathbf{v}_2(x, y, z)) = (\mathbf{w}_1(z), \mathbf{w}_2(z)) \exp[i\,(k_1\, x + k_2\, y)]$$

with bounded $(\mathbf{w}_1,\mathbf{w}_2)$ and satisfying the continuity requirements (continuity of the velocity and the traction rates) across the interface $z = 0$, is the identically zero solution $(\mathbf{v}_1(z), \mathbf{v}_2(z)) \equiv (\mathbf{0}, \mathbf{0})$; (where $\mathbf{v}_1$ and $\mathbf{v}_2$ are the solutions, respectively for $z < 0$ and for $z > 0$).

When these three conditions are fulfilled, the rate boundary problem admits *a finite number of linearly independent solutions, which depend continuously on the data, and which constitute diffuse modes of deformation.*

Remarks :

- *these three conditions are local,* and this is particulally important when considering their numerical implementation ;
- the above-given results remain valid for an arbitrary number of non-intersecting interfaces, an interfacial condition being written for each interface ;
- the failure of these conditions can be interpreted as localization criteria as recalled in section 1. *These localization criteria can also be used as indicators of the local failure of the material* (Billardon & Doghri[10]);

- *both boundary and interfacial complementing conditions fail in the elliptic regime* of the equilibrium equations, or at the latest, when the ellipticity condition fails. Thus, localized modes of deformation at the boundary or at the interface generally occur before the onset of so-called shear banding modes.

3- THE NON-LINEAR CASE : SOME RESULTS

Although the complete analysis of the non-linear problem is not yet available, some results can be given for the possibility of emergence of deformation modes involving jumps of the velocity gradient for the bi-linear rate constitutive laws (1).

The necessary and sufficient conditions for the onset of such modes *inside the body* have been given by Borré & Maier[11], who extended the results given by Rice[4], and Rudnicki & Rice[12-13] for so-called *continuous and discontinuous localizations*. In [9], we have amplified these results by seeking *necessary and sufficient conditions* for which a discontinuity surface for the velocity gradient appears at, or reaches the boundary of the solid. These conditions are given below for the constitutive laws (1) with

$$\mathbb{H} = \mathbb{E} - \frac{(\mathbb{E} : \mathbb{a}) \otimes (\mathbb{b} : \mathbb{E})}{h}$$

where it is assumed that $h > 0$, and $\mathbb{E}$ is strictly positive definite.

At a point P of the boundary Γ where only surface traction rates $\dot{\mathbf{F}}$ are applied, *the necessary and sufficient conditions for continuous localization* [i.e. the material is in loading ($\mathbb{L} = \mathbb{H}$) on each side of the singular surface] are

$$\text{(2a)} \begin{cases} \text{i) there exists } \dot{\varepsilon}_0 \text{ such that } \mathbf{m} \,.\, \mathbb{H} : \dot{\varepsilon}_0 = \dot{\mathbf{F}} \\ \text{ii } \det(\mathbf{n} \,.\, \mathbb{H} \,.\, \mathbf{n}) = 0 \\ \text{iii) } (\mathbf{m} \,.\, \mathbb{E} \,.\, \mathbf{n}) \,.\, (\mathbf{n} \,.\, \mathbb{E} \,.\, \mathbf{n})^{-1} \,.\, (\mathbf{n} \,.\, \mathbb{E} : \mathbb{a}) = \mathbf{m} \,.\, \mathbb{E} : \mathbb{a} \end{cases}$$

At a point P of the boundary Γ where only surface traction rates $\dot{\mathbf{F}}$ are applied, *the necessary and sufficient conditions for discontinuous localization* [i.e. the material is in loading ($\mathbb{L} = \mathbb{H}$) on one side and in unloading ($\mathbb{L} = \mathbb{E}$) on the other side of the singular surface] are

$$\text{(2b)} \begin{cases} \text{i) there exists } \dot{\varepsilon}_0 \text{ such that } \mathbf{m} \,.\, \mathbb{H} : \dot{\varepsilon}_0 = \dot{\mathbf{F}} \text{ and } \mathbb{b} : \mathbb{E} : \dot{\varepsilon}_0 > 0 \\ \text{ii) } \det(\mathbf{n} \,.\, \mathbb{H} \,.\, \mathbf{n}) < 0 \\ \text{iii) } (\mathbf{m} \,.\, \mathbb{E} \,.\, \mathbf{n}) \,.\, (\mathbf{n} \,.\, \mathbb{E} \,.\, \mathbf{n})^{-1} \,.\, (\mathbf{n} \,.\, \mathbb{E} : \mathbb{a}) = \mathbf{m} \,.\, \mathbb{E} : \mathbb{a} \end{cases}$$

In these conditions, $\mathbf{m}$ denotes the unit outward normal to the boundary of the solid in P, whereas $\mathbf{n}$ denotes the unit normal to the singular surface in P.

At a point Q of an interface I, two types of singular surface may occur. The singular surface *can* either stop at, or cross the interface (see Fig.2). In the latter case, the singular surface can meet the interface at different angles on each side. Analogous conditions to conditions (2) can be exhibited.

Remarks :

- conditions (2) are to be compared with the corresponding *conditions inside the solid* given by Borré & Maier and stated as

(3) $\det(\mathbf{n} \,.\, \mathbb{H} \,.\, \mathbf{n}) \leq 0$

with *equality corresponding to continuous*, and *inequality corresponding to discontinuous localization ;*

- given conditions (2) and (3), singular surfaces of the type discussed here *generally* appear *first inside the body* ;
- similar conditions to conditions (2) can be exhibited for general boundary conditions. (Note that *for displacement boundary conditions, relation (3) applies both inside the body and at the boundary.*) ;
- conditions (2) and (3) are a priori unrelated to the boundary complementing condition.

4- REFERENCES

1. J. HADAMARD, Leçons sur la propagation des ondes et les équations de l'hydrodynamique (Paris, 1903).
2. R. HILL, J. Mech. Phys. Solids, 10, 1-16 (1962).
3. J. MANDEL, J. de Mécanique, 1,.3-30 (1962).
4. J.R. RICE in Theoretical and Applied Mechanics, 207-220, edited by W.T. Koiter (North-Holland, 1976).
5. A. BENALLAL, R. BILLARDON & G. GEYMONAT in Cracking and Damage, 247-258, edited by J. Mazars & Z.P. Bazant (Elsevier, 1989).
6. A. BENALLAL , R. BILLARDON & G. GEYMONAT, C.R. Acad. Sci. Paris, 308 (II), 893-898 (1989).
7. A. BENALLAL , R. BILLARDON & G. GEYMONAT in Actes Congrès Français de Mécanique, 1, 242-243 (AUM, Metz 1989).
8. A. BENALLAL , R. BILLARDON & G. GEYMONAT, C.R. Acad. Sci. Paris, 310 (II), 679-684 (1990).
9. R. HILL, J. Mech. Phys. Solids, 6, 236-249 (1958).
10. R. BILLARDON & I. DOGHRI, C. R. Acad. Sci. Paris, 308 (II),.347-352 (1989).
11. G. BORRE & G. MAIER, Meccanica, 24, 36-41 (1989).
12. J. RUDNICKI & J.R. RICE, J. Mech. Phys. Solids, 23, 371-394 (1975).
13. J. RUDNICKI & J.R. RICE, Int. J. Solids Struct., 16, 597-605 (1980).

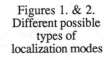

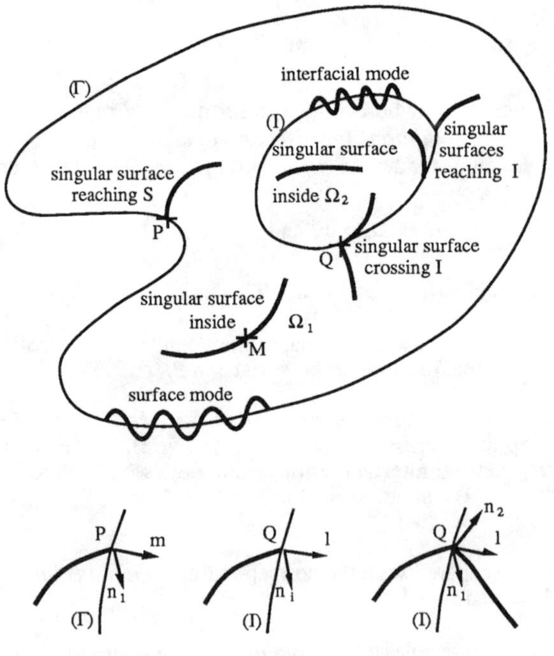

Figures 1. & 2. Different possible types of localization modes

Damage–Rheology Uncoupling for Microplane Damage Tensor, with Application to Concrete with Creep

Ignacio Carol
Assoc. Prof. of Civil Eng., Technical University of Catalonia, 08034 Barcelona, Spain

Zdeněk P. Bažant
Walter P. Murphy Prof. of Civil Eng., Northwestern University, Evanston, IL 60208

ABSTRACT

Continuum damage mechanincs with the concepts of effective stress and strain equivalence provides suitable framework for developing constitutive models for concrete–type materials. The paper presents a new model that is composed of two independent parts, one for damage and one for rheology. This has the advantage that modeling complex material behavior can be achieved by a combination of much simpler models. However, existing formulations for damage seem to be inadequate for this purpose. The present formulation for 3–D damage is based on the microplane model for concrete; a macroscopic damage tensor is obtained as an integral of the damage on microplanes over all possible orientations. In the examples presented, the new damage tensor is combined with two different models for classical continuum: linear elasticity and ageing viscoelasticity. The results obtained by the overall model show good agreement with well known experimental behavior of concrete in both intantaneous and time–dependent domains.

INTRODUCTION

The basic equation for a onedimensional continuum damage model with the concepts of effective stress and strain equivalence can be written as $\sigma=\alpha\tau$ (σ=macroscopic stress, τ=effective stress, α=damage variable of a geometric nature varying from 1 to 0). This is complemented by appropiate laws for α and τ; e.g. $\alpha=\alpha(\varepsilon)$ and $\tau=\tau(\varepsilon)$. In this approach, a clear separation of effects is present between damage and all the remaining elastic and inelastic effects (plasticity, viscosity, etc.) to which we will refer as "rheology". In 3–D, a similar approach is possible [1,2] based on the expression:

$$\sigma_{ij} = \alpha_{ijkm}\tau_{km} \tag{1}$$

where damage is represented by a fourth–order tensor (repetition of lowercase latin indices implies summation). However, while any existing theory for 3–D continuum (elasticity, plasticity, viscoplasticity, etc.), can be used for $\underset{\sim}{\tau}$, the formulation for damage (the laws for $\underset{\sim}{\alpha}$) remains the weakest point of this type of approach. Existing models for 3–D damage generally are either too simplistic or too complicated; quite often they are formulated in an abstract or phenomenological way, and good fits of test data are scarce (see a review in [3]).

MICROPLANE DAMAGE TENSOR

The new formulation for 3–D damage has been derived from the existing microplane model for concrete, which was proven to fit very closely a large number of test data under 1, 2 and 3–D loading conditions [4,5]. A microplane is any plane cutting the material with a certain orientation given by its unit normal vector with components n_i. Normal and shear stresses σ_N, σ_{T_r} and strains

ε_N, ε_{T_r} are considered on each microplane. The normal components are further split into volumetric and deviatoric parts ($\sigma_N=\sigma_V+\sigma_D$, $\varepsilon_N=\varepsilon_V+\varepsilon_D$). A kinematic constraint is assumed between the components of strain on each microplane and the macroscopic tensor: $\varepsilon_V=\delta_{km}\varepsilon_{km}/3$, $\varepsilon_D=(n_k n_m-\delta_{km}/3)\varepsilon_{km}$ and $\varepsilon_{T_r}=(\delta_{rk}n_m+\delta_{rm}n_k-2n_r n_k n_m)\varepsilon_{km}/2$, where δ_{km}=Kronecker delta. In the previous formulations of the microplane model for concrete, stress-strain relations $\sigma_V=\mathcal{F}_V(\varepsilon_V)$, $\sigma_D=\mathcal{F}_D(\varepsilon_D)$ and $\sigma_T=\mathcal{F}_T(\varepsilon_T)$ were assumed. By contrast, in the present formulation three effective stresses τ_V, τ_D and τ_{T_r} and the corresponding damage variables α_V, α_D and α_T are defined for each microplane ($\sigma_V=\alpha_V\tau_V$, $\sigma_D=\alpha_D\tau_D$ and $\sigma_{T_r}=\alpha_T\tau_{T_r}$), and the microplane laws are established for damage in the form: $\alpha_V=\mathcal{G}_V(\varepsilon_V)$, $\alpha_D=\mathcal{G}_D(\varepsilon_D)$ and $\alpha_T=\mathcal{G}_T(\varepsilon_T)$. The detailed derivation of the macroscopic damage tensor can be found in [3]. The final expression (where the integrals extend to the upper half hemisphere) is:

$$\begin{aligned}\alpha_{ijpq} &= \frac{\alpha_V}{3}\delta_{ij}\delta_{pq}+\frac{3}{2\pi}\int_\Omega \alpha_D n_i n_j\Big(n_p n_q-\frac{\delta_{pq}}{3}\Big)d\Omega+ \\ &+\frac{3}{2\pi}\int_\Omega \frac{\alpha_T}{4}(n_i n_p\delta_{jq}+n_i n_q\delta_{jp}+n_j n_p\delta_{iq}+n_j n_q\delta_{ip}-4n_i n_j n_p n_q)d\Omega\end{aligned} \tag{2}$$

NUMERICAL IMPLEMENTATION

The laws assumed for α_V, α_D and α_T are shown in Fig. 1. The envelope curves are exponential functions $\alpha=\exp[-(\varepsilon/a)^p]$ with fixed values of p, and a total of 3 parameters: a_1, a_2 and a_3. The integral over the hemisphere is computed numerically. 28 fixed ("sample") microplane orientations are considered. The evolution of damage on the 28 microplanes is evaluated, updated and stored during the computations. The remaining aspects of the numerical implementation are very similar to the last version of the microplane model described in [5].

EXAMPLES OF APPLICATION

In the first example, the damage model has been used in conjunction with linear elasticity. The example is a uniaxial test of concrete reported by van Mier in 1984 [6], in which both longitudinal stress and transverse strain were recorded. The parameters of the model are: E=2406 MPa, ν=0.18, a_1=0.0004, a_2=0.0060, a_3=0.0018. The results are represented in Fig. 5 by solid lines. The dots are the experimental data and the dashed lines are the results obtained with the previous version of the microplane model [5].

In the second example, the microplane damage tensor has been used in conjunction with ageing viscoelasticity in the form of a Maxwell chain. The parameters of the chain have been determined according to the recomendations given in [7] using a computer program which performs a least–square approximation to the values of the creep function for a concrete with compressive strength f_c'=36.8 MPa, fictitious depth of the specimen e=30 cm, and relative humidity h=90%. The chain ensures that for no damage ($\alpha_{ijkm}=\delta_{ik}\delta_{jm}$, $\sigma_{ij}=\tau_{ij}$) the ageing viscoelatic behavior is approached satisfactorily by the overall model. The parameters of the damage model have also been assumed to vary with time so that the peak values of the instantaneous onedimensional

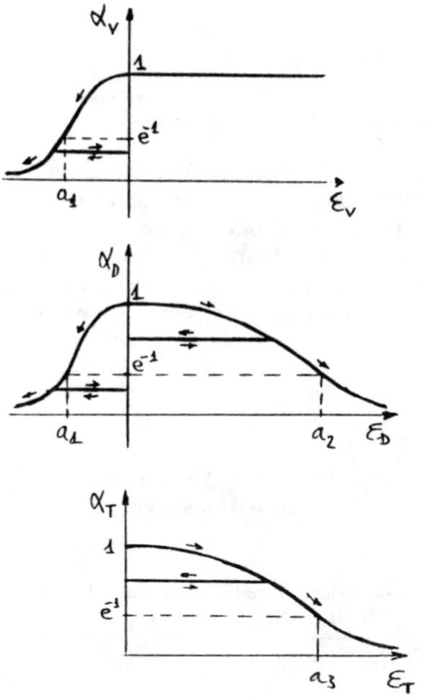

Fig.1. Laws for α_V, α_D and α_T.

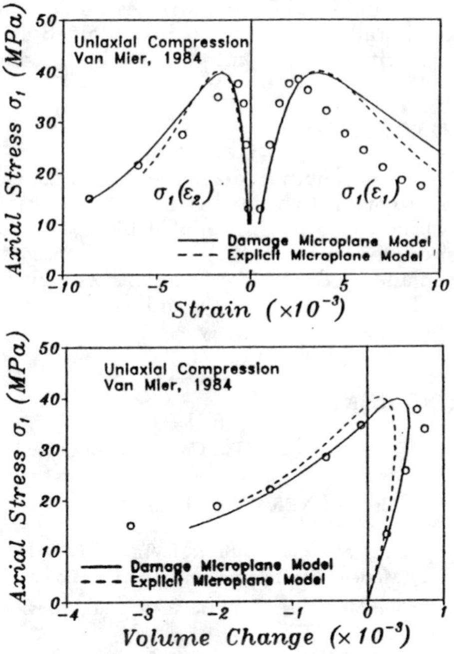

Fig.2. Results of example 1.

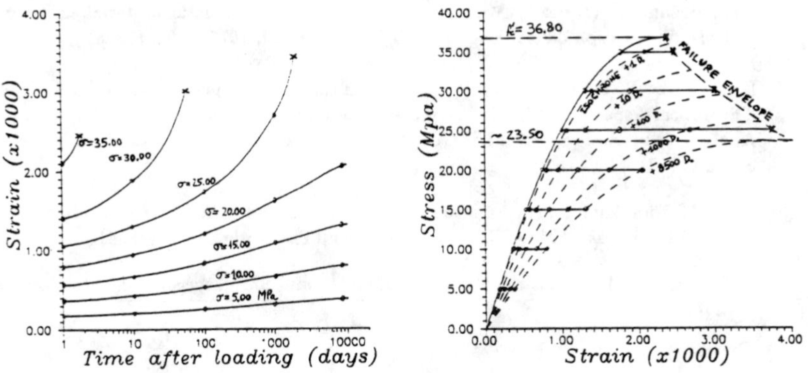

Fig. 3. Results of example 2.

σ–ε diagram at various ages coincide with the age dependence of f_c'. Various creep tests (consisting of a uniaxial step load applied at 28 days) with increasing values of the load value, have been run with the 3-D model. The stresses and strains in the axis of loading are represented in Fig. 4 showing a strain–time diagram and a stress–strain diagram with creep isochrones. Linear creep is obtained for low stresses (under about $0.4f_c'$), and nonlinear creep and failure under sustained load is obtained under high stresses. All these features agree with the well known behavior of concrete.

CONCLUSIONS

Continuum damage mechanics provides a suitable framework for developing constitutive models for concrete–type materials. Separation of the formulation into two uncoupled parts, one for damage and one for rheology, is especially convenient. A new 3-D damage tensor derived from the existing microplane model for concrete is briefly outlined. It is combined with elasticity and linear viscoelasticity, and well known experimental behavior of concrete is reproduced satisfactorily.

REFERENCES

[1] J. L. Chaboche. "The Concept of Effective Stress Applied to Elasticity and Viscoplasticity in the Presence of Anisotropic Damage". Euromech Congress 115, Grenoble (1979). Also in ONERA Report 1979/77.

[2] J. C. Simó, J. W. Ju, R. L. Taylor, K. S. Pister. "On Strain–Based Continuum Damage Models. Formulation and Computational Aspects" in Constitutive Laws for Engineering Materials, C. S. Desai et al ed., Elsevier (1987).

[3] I. Carol, Z. P. Bažant, P. C. Prat. "Geometric Damage Tensor Based on Microplane Model". Internal report, ETSECCPB, Technical University of Catalunya, Barcelona, Spain (1990).

[4] Z. P. Bažant, P. C. Prat. "Microplane model for Brittle–Plastic Material. I: Theory" and "II: Verification". J. of Eng. Mech ASCE, 114(10), 1672–1702, (1988).

[5] I. Carol, Z. P. Bažant, P. C. Prat. "New Explicit Microplane Model for Concrete: Theoretical Aspects and Unified Implementation for Constitutive Verification and F. E. Analysis". Internal report, ETSECCPB, Technical University of Catalunya, Barcelona, Spain (1990).

[6] J. G. M. van Mier. "Strain–Softening of Concrete under Multiaxial Loading Conditions". Ph. D. dissertation, Univ. of Eindhoven, The Netherlands (1984).

[7] Z. P. Bažant. Mathematical Modelling of Creep and Shrinkage of Concrete. John Wiley (1989).

Shear band bifurcation in soil modelling: a rate type constitutive law for explicit localisation analysis

CHARLIER ROBERT
Université de Liège, MSM, 6 quai Banning, 4000 Liège Belgium (in stay in IMG 10/89–08/90)

CHAMBON RENÉ, DESRUES JACQUES, HAMMAD WALID
IMG, BP 53X, 38041 Grenoble Cedex, France

Abstract

A new family of rate type constitutive laws has been developed, including consistency to a Limit Surface. Most parameters are identified on axisymmetric state, and then by an interpolation, but the shear modulus is identified on the bifurcation phenomenon. An essential property of that family is that, although it is throughly non linear, a direct shear band analysis can be performed without any linearisation assumption. The discussion of the obtained criterion is presented in relation with the question of identification of shear moduli.

1 The constitutive law

It is well known that classical elastoplasticity with smooth yield surfaces and isotropic hardening leads to unrealistic predictions of bifurcation for shear band localisation. Special features such as vertices, non normality,... are needed to improve the prediction. On the other hand, phenomenological rate type constitutive equations, which offer more flexibility, have to be carefully checked with regards to consistency requirements.

CLoE is the generic name of a class of constitutive equations designed with a special attention to consistency — in order to remain efficient in the vicinity of ultimate stress states — and localisation. The name *CLoE* is an acronym of the words *Consistency* , *Localisation* , and *Explicit.* This family has the generic form [1,2] :

$$\overset{\nabla}{\underline{\sigma}} = \underline{\underline{\boldsymbol{A}}} : \underline{\dot{\varepsilon}} + \underline{\boldsymbol{b}}\, \|\underline{\dot{\varepsilon}}\| \tag{1}$$

in which $\underline{\underline{\boldsymbol{A}}}$ is a four rank tensor and $\underline{\boldsymbol{b}}$ a second rank one. $\underline{\underline{\boldsymbol{A}}}$ and $\underline{\boldsymbol{b}}$ depends on $\underline{\sigma}$. It can be seen that , at a given state, $\underline{\boldsymbol{b}}$ controls the difference between responses to $\underline{\dot{\varepsilon}}$ and $-\underline{\dot{\varepsilon}}$. The incremental non linearity is due to the norm $\|\underline{\dot{\varepsilon}}\|$.

A conical limit surface is been assumed. Its shape in the deviatoric plane was proposed by Van Eekelen [3]; it is a smooth surface with different friction angles in triaxial compression and extension.

The general form of the constitutive tensors, when representing the stress and strain tensors in a 6 dimension space, is :

$$\underline{\underline{A}} = \begin{bmatrix} a & f' & e' & & & \\ f & b & d' & & & \\ e & d & c & & & \\ & & & g & & \\ & & & & h & \\ & & & & & j \end{bmatrix} \tag{2}$$

$$\underline{b} = \begin{bmatrix} K & L & M & 0 & 0 & 0 \end{bmatrix}^T \tag{3}$$

There are 15 coefficients that will be determined using two different ways :

- an interpolation between compression and extension axisymmetrical states for the parameters $a, b, c, d, d', e, e', f, f', K, L, M$, see [2],
- and a bifurcation analysis for the shear parameters g, h, j which are not activated during most classical experiments but only in a shear band.

It is to be noted that in plane strain and axisymmetrical analysis, g and h are not used but only j.

2 Shear band analysis

This constitutive law has been developed in order to simulate the bifurcation of the strain rate and the strain localisation. In this section we present and discuss the bifurcation criterion, following an extension of the classical RICE's development [6,7]to a non linear model .

Symbols noted ^{0} are related to the outside of the shear band, and symbols noted ^{1} are related to the inside of the shear band. The kinematical condition is

$$\underline{L}^1 = \underline{L}^0 + \vec{g} \otimes \vec{n} \tag{4}$$

where $\underline{L}$ is the velocity gradient, $\vec{n}$ is the normal to the band and $\vec{g}$ is a vector.

The statical condition is :

$$(\underline{\dot{\sigma}}^1 - \underline{\dot{\sigma}}^0) \cdot \vec{n} = 0 \tag{5}$$

where $\underline{\dot{\sigma}}$ is the derivative of the CAUCHY stress with respect to fixed axes.

Using the JAUMANN objective derivative, one can rewrite the constitutive law :

$$\dot{\sigma}_{ij} = \boldsymbol{D}_{ijkl} L_{kl} + b_{ij} \parallel \underline{\dot{\varepsilon}} \parallel \tag{6}$$

where the new constitutive tensor $\underline{\underline{D}}$ includes the objective derivative effect. Taking into account the direction of the shear band, one defines

$$\Lambda_{ik} = \boldsymbol{D}_{ijkl} n_l n_j \tag{7}$$

After some developments [5], one obtains two bifurcation criterions:
The *determinant* criterion :

$$det(\underline{\Lambda}) = 0 \tag{8}$$

And the *norm* criterion :

$$\| \frac{1}{2}(\Lambda_{ki}^{-1}\, b_{ij}\, n_j\, n_l + \Lambda_{li}^{-1}\, b_{ij}\, n_j\, n_k) \| = 1 \tag{9}$$

Bifurcation in shear bands can occur when one of these two criterions is met.

The bifurcation criterions are function of the coefficients g, h, j, of the direction θ of the shear band, and of course of the stress state. When using the criterions as an indentification tool, g, h, j are unknown quantities while θ and the critical stress state are known; on the contrary, if the criterions are used to predict localisation when integrating the law for field problems, g, h, j are material data. Consider now the plane strain state. The criterions are non linear functions of only two variables : the shear coefficient j and the direction of the normal θ. The determinant criterion is of the first order in j and second order in $\tan^2\theta$. The norm criterion is of the second order in j and fourth order in $\tan^2\theta$. The zeros of the criterion can be determined for every state of plane stress, depending always on j and $\tan^2\theta$. The figures 1 and 2 are presented here as an example. It has been observed that the norm criterion is always more severe than the determinant one.

The norm criterion has essentially negative values, but three positive areas are appearing or growing when the deviatoric stresses are increasing. The zero level contours are representing possible bifurcation states. For plane strain compression on Hostun RF sand [4], a little time before the compression, one obtains the fig. 2. The bifurcation occurs for a stress level of $\frac{\sigma_1}{\sigma_3} = 5,9$ and an angle of $\theta \simeq 60^o$ —which is in good accordance with the experiments results [8]— for a shear modulus $j \simeq 7000$.

3 Conclusion

A new rate type constitutive law has been developed. Its shear modulus is determined through a shear band bifurcation analysis. Therefore lòcalisation is explicitly included in the constitutive equation. The law and its bifurcation criterion have been implemented in a finite element code in order to evaluate the true limit state of quasi static geotechnical problems.

References

[1] CHAMBON R., 'Une classe de lois de comportement incrémentalement non-linéaires pour les sols non visqueux - résolution de quelques problèmes de cohérence',*C. R. Acad Sci. Paris*, 308 ,pp 1571-1576 ,1989

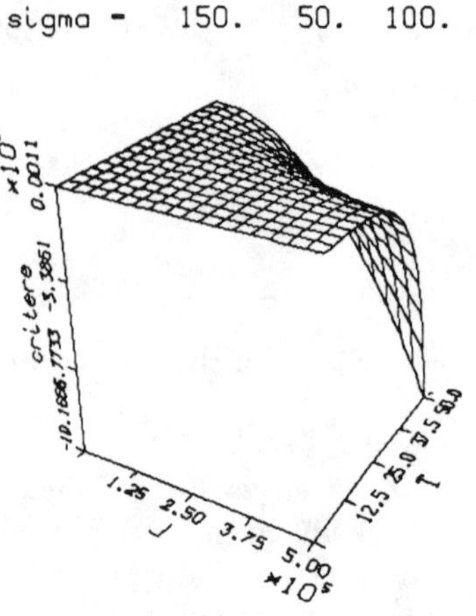

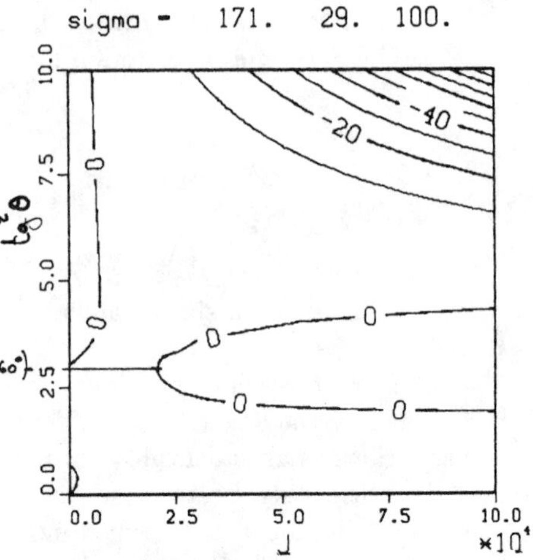

figure 1 - 3-D view figure 2- level curves
of the norm criterion for a plane strain state

[2] CHAMBON R.,DESRUES J.,HAMMAD W.,CHARLIER R. 'Soil Modelling with regard to Consistency: CLoE, a NEW Rate Type Constitutive Model',*Third int. conf. on Constitutive laws for Engng materials*, Tucson ,Ed. DESAI C.S. and KREMPL E. ,ASME Press ,1991

[3] vanEEKELEN H.A.M., 'Isotropic Yield Surface in Three Dimensions for Use in Soil Mechanics',*Int. J. for Num. and Anal. Meth. in Geom.* ,4 ,pp 89–101 ,1980

[4] CHAMBON R., DESRUES J., HAMMAD W., and CHARLIER R., "CLoE, développement d'une loi incrémentale non linéaire à Consistance et Localisation Explicite", *Rapport de recherche du groupe Geomécanique, IMG*, Grenoble, 1990.

[5] CHAMBON R. and DESRUES J., "Soil modelling with regard to consistency and bifurcation : a new approach" ,*IInd International Workshop on Numerical Methods for Localisation and Bifurcation of granular bodies*, Gdansk (Poland) september 25-29, 1989.

[6] CHAMBON R. and DESRUES J., "Bifurcation par localisation et non linéarité incrémentale : un exemple heuristique d'analyse complète", in *Plastic Instability*, Proceedings of Considere Memorial Int. Symp. on Plastic Instabilities, Presses de l'ENPC, Paris ,pp 101-119, 1985

[7] DESRUES J. and CHAMBON R., "Shear band analysis for granular materials: the question of incremental non linearity", *Ingenieur Archiv*, 59, pp 187-196, 1989.

[8] DESRUES J. "La localisation de la déformation dans les matériaux granulaires", *thèse de Doctorat es Science*, USMG & INPG, Grenoble 1984.

Soil Modelling with regard to Consistency: CLoE, a New Rate Type Constitutive Model

DESRUES Jacques, CHAMBON René, HAMMAD Walid
IMG, BP 53X, 38041 Grenoble Cedex, France

CHARLIER Robert
Université de Liège, MSM, 6 quai Banning, 4000 Liège Belgium
(in stay in IMG 10/89–08/90)

Abstract

The paper describes the main features of a new class of constitutive laws, in the framework of *Incrementally Non Linear* constitutive equations. *CLoE* is a generic name for that new class of laws, with reference to *Explicit Consistency* at the limit surface, and *Localisation* analysis. A top–down analysis of the model is presented, and illustrated by examples.

Introduction

Among the challenges that soil modellers have to face, strain localisation is one of the most exciting. *CLoE* is the generic name of a class of constitutive equations designed with a special attention to consistency — in order to remain efficient in the vicinity of ultimate stress states — and localisation. The name *CLoE* is an acronym of the words *Consistency* , *Localisation* , and *Explicit.*

1 A new class of laws

CLoE is a new development in the framework of *Incrementally Non Linear constitutive equations* for soils, which has already been promoted for years by our group (Geomechanics group, IMG Grenoble) as an alternative framework with respect to the Elastoplastic one.

1.1 Generic mathematical form

Stress rate (an objective one) and strain rate are supposed to be related by a tensorial function, depending on some state variables. Neither linearity nor piecewise linearity is assumed. No assumption is made on a decomposition of the strain rate into a plastic and an elastic part; the reason is lack of convincing experimental arguments for an objective decomposition of strain into such parts (irreversibility is acting from the beginning of loading up to the end of unloading for a given load path).

A general form for rate type constitutive equations is :

$$\overset{\nabla}{\underline{\sigma}} = \underline{f}(\underline{\sigma}, \underline{\dot{\varepsilon}}) \quad (1)$$

with $\overset{\nabla}{\underline{\sigma}}$ an objective stress rate, $\underline{\sigma}$ the stress state, and $\underline{\dot{\varepsilon}}$ the strain rate.

The new class of constitutive equations discussed here is defined by the generic form :

$$\overset{\nabla}{\underline{\sigma}} = \underline{\underline{\boldsymbol{A}}} : \underline{\dot{\varepsilon}} + \underline{\boldsymbol{b}}\, \|\underline{\dot{\varepsilon}}\| \quad (2)$$

in which $\underline{\underline{\boldsymbol{A}}}$ is a four rank tensor and $\underline{\boldsymbol{b}}$ a second rank one. $\underline{\underline{\boldsymbol{A}}}$ and $\underline{\boldsymbol{b}}$ depends on $\underline{\sigma}$. A similar formulation has been proposed by Kolymbas [6] , with explicit *a priori* formulae for $\underline{\underline{\boldsymbol{A}}}$ and $\underline{\boldsymbol{b}}$. It can be seen that , at a given state, $\underline{\boldsymbol{b}}$ controls the difference between responses to $\underline{\dot{\varepsilon}}$ and $-\underline{\dot{\varepsilon}}$. The incremental non linearity is due to the norm $\|\underline{\dot{\varepsilon}}\|$. Another useful equivalent form is :

$$\overset{\nabla}{\underline{\sigma}} = \underline{\underline{\boldsymbol{A}}} : (\underline{\dot{\varepsilon}} + \underline{\boldsymbol{b}'}\, \|\underline{\dot{\varepsilon}}\|) \quad (3)$$

1.2 Consistency at the Limit surface

We assume that a *limit surface* can be defined in stress space, which bounds the domain of the admissible stress states. The equation of this surface is :

$$\psi(\underline{\sigma}) = 0 \quad (4)$$

For sake of flexibility, we have chosen the formulation of $\psi(\underline{\sigma})$ proposed by van Eekelen [7]. When the stress state reaches the limit surface, consistency requires that no outer stress rate response can be generated by the constitutive equations. As shown in reference [2], this condition reads :

$$\boldsymbol{A}_{klmn} \frac{\partial \psi}{\partial \sigma_{kl}} = -\lambda \boldsymbol{b'}_{mn} \text{ with } \lambda > 0 \quad (5)$$

In the $v0.12$ realisation of *CLoE*, discussed here, stress state is the only state variable considered. Chambon [1] has shown that in such cases, the constitutive equations are cross-anisotropic in the principal axes of the stress. Then, the number of unknown quantities entering in $\underline{\underline{\boldsymbol{A}}}$ and $\underline{\boldsymbol{b}'}$ is reduced to 15.

1.3 Axisymmetric triaxial responses

For stress states lying on the axis σ_1, σ_2 or σ_3 in the deviatoric plane, the constitutive tensors $\underline{\underline{\boldsymbol{A}}}$ and $\underline{\boldsymbol{b}'}$ have a simpler form depending on 8 unknown quantities only. These states correspond to axisymmetric triaxial paths. The determination of the 8 unknown quantities can be made on the basis of *analytical formulations* of the axisymmetric triaxial responses in compression and in extension (loading), plus some additional information from incipient unloading responses, and so-called "pseudo isotropic" response [1]. The case of the shear

[1] more detail on the basic paths formulation can be found in ref. [5]]

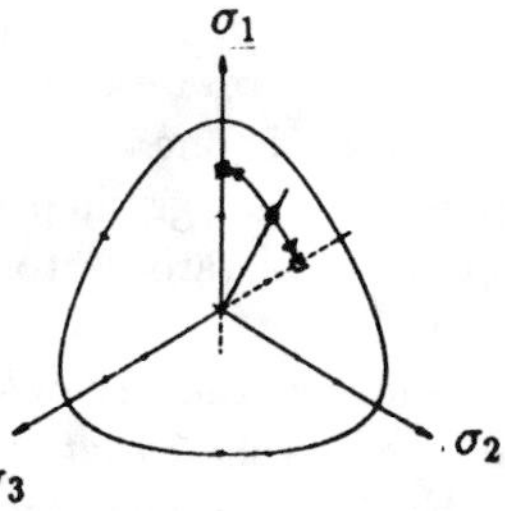

figure 1 - image-states in the deviatoric space

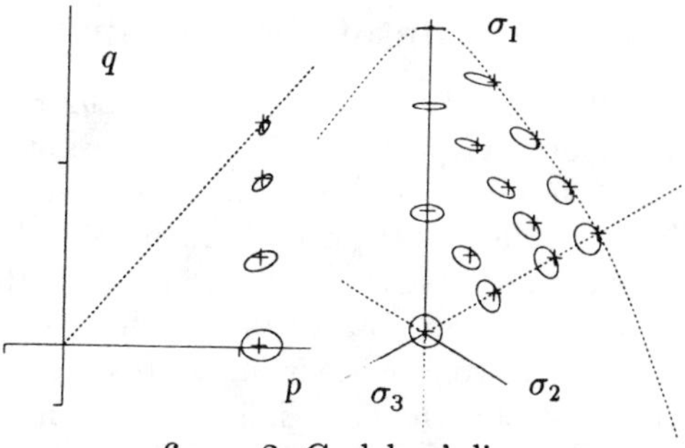

figure 2- Gudehus' diagrams at various stress states

modulus j is special; it must be discussed in relation with bifurcation analysis, see ref. [3], [5]. This is another basic feature of *CLoE*. So it appears that $\underline{\underline{A}}$ and $\underline{b}'$ can be determined for any state lying on the axisymmetric triaxial axes, leading to a complete determination of the constitutive response for these state.

1.4 Interpolation

We still have to define the response whatever the state is. The solution is an interpolation procedure between the responses for axisymmetric triaxial states. We define two *images* of the actual stress point in the 3D stress space (principal stresses) according to figure 1 : these *image-states* lie on the axisymmetric triaxial axes in the deviatoric plane of the actual state, and their deviatoric stress versus limit deviatoric stress ratio, $\bar{q}$ is the same as that for the actual state.

As discussed in section 1.3, the constitutive response for such states is known. So we get so-called *"image-laws"* associated with the *"image-states"*. The role of the interpolation procedure is to give the actual law (i.e. the $\underline{\underline{A}}$ and $\underline{b}'$ tensors for the actual state) as a function of the *image-laws* and the Lode's angle of the actual stress state. $\underline{\underline{A}}$ and $\underline{b}'$ are interpolated separately.

An essential feature of the interpolation scheme is to insure consistency at the limit surface by enforcing interpolated $\underline{\underline{A}}$ and $\underline{b}'$ to met the consistency equation (5).

2 *CLoE* for Hostun RF sand

2.1 Identification

As discussed in section 1.3, *CLoE* is based on analytical formulations of the material response on a set of basic paths.

An identification procedure has been defined, which is seen as a part of the model. The data that it takes are essentially familiar geotechnical data, like friction angle, dilatancy factor.

2.2 Illustration of the response

Most illustrative of the specificity of *CLoE* is the figure 2. Here are shown the constitutive responses at different stress states, from isotropic one to limit states. We use the representation introduced by Gudehus [4] : the egg-shaped diagrams give a section of the tridimensional envelop of stress rate vectors mapping the response to a unit strain rate vector's sphere.

First we can observe that, although highly non-linear, the response envelop remains continuous in any state. Secondly, these figures show that the consistency condition at the limit surface is met, since the stress rate envelops for limit stress states are tangent to the limit surface. This can be seen in the deviatoric plane, and in the *pq* plane.

3 Conclusion

The examples presented above show that *CLoE v0.12* fulfils the basic requirements of the project : to be a reasonable representation of sand behaviour, and to remain strictly consistent when reaching the limit admissible states. Later experience with this model, including parameter sensibility check and bifurcation tests, will give a firm basis for practical soil modelisations. First experiences of integration of that model in FEM code *LAGAMINE*, University of Liège, lead to encouraging results.

References

[1] CHAMBON R., 'Contribution à la modélisation numérique non linéaire des sols', *thèse de doctorat es sciences* ,USMG-INPG ,1981

[2] CHAMBON R., 'Une classe de lois de comportement incrémentalement non-linéaires pour les sols non visqueux - résolution de quelques problèmes de cohérence',*C. R. Acad Sci. Paris,* 308 ,pp 1571-1576 ,1989

[3] CHARLIER R.,CHAMBON R.,DESRUES J.,HAMMAD W. 'Shear band bifurcation in soil modelling: a rate type constitutive law for explicit localisation',*Third int. conf. on Constitutive laws for Engng materials*, Tucson ,Ed. DESAI C.S. and KREMPL E. ,ASME Press ,1991

[4] GUDEHUS G., 'A comparison of some constitutive laws for soils under radially symmetric loadings and unloadings',Numerical Methods in Geomechanics , Ed. W.Wittke ,Balkema ,Vol 4 ,pp 1309–1323 ,1979

[5] HAMMAD W., 'Localisation en bande de cisaillement dans les sables', *Thèse de doctorat* ,UJF - INPG ,1990

[6] KOLYMBAS D. and ROMBACH G., 'Shear band formation in generalized hypoelasticity', *Ingenieur-Archiv* ,59 ,pp 177–186 ,1989

[7] vanEEKELEN H.A.M., 'Isotropic Yield Surface in Three Dimensions for Use in Soil Mechanics',*Int. J. for Num. and Anal. Meth. in Geom.* ,4 ,pp 89–101 ,1980

THE INFLUENCE OF TEMPERATURE AND DAMAGE ON THE PLASTIC DEFORMATIONS OF CK15 STEEL

A. FELDMÜLLER, TH. LEHMANN, O. T. BRUHNS
Ruhr-Universität Bochum, Lehrstuhl für Mechanik I
4630 Bochum 1, West Germany

ABSTRACT

A constitutive model for plastic deformations concerning damage by voids is introduced. The modelling is done within a thermodynamical frame so that nonisothermal processes can be calculated. For Ck15 steel experimental data, material parameters and some calculations are presented.

INTRODUCTION

Ductile materials may be subject to large plastic deformations, like for example, in metal forming. During such processes lattice defects are produced which cause strain hardening. The dissipated mechanical energy leads to temperature changes depending on the velocity of this process and affecting the hardening of the material.
A considerable decrease in strength is caused by damage resulting from microshearbands [1] or voids. Here we deal with damage by microvoids which often appear during large ductile deformation. These voids originate, for instance, from inclusions which disturb the deformation of the surrounding material; this leads to locally increased stresses causing the damage.
The above-mentioned effects are especially important, if the deformations localize, as in the neck of a cylindrical tensile specimen.

CONSTITUTIVE EQUATIONS

A constitutive law was formulated which is thermodynamically consistent and considers the main characteristics of the material behaviour in the frame of a phenomenological continuum damage theory.
The concept of continuum requires that the voids are small. If the damage exceeds a critical value, as it is the case when macrocracks develop, a continuum approximation is no longer valid. Then it is necessary to regard a body with a different geometry, for example a body with a crack or a void inside.
In the following theory the scalar damage variable ω, which is defined as void volume fraction of the total volume, is taken into account by using the concept of effective stresses $\mathbf{S}$ [2] and by formulating an additional deformation rate $\mathbf{D}_v$ corresponding to the volume changes of the material produced by the increase in porosity.

The following decomposition of the total deformation rate $\mathbf{D}$ is applied:

$$\mathbf{D} = \mathbf{D}_r + \mathbf{D}_p + \mathbf{D}_v \tag{1}$$

The elastic part $\mathbf{D}_r$ is completely reversible and can be described by the hypoelastic law (2) where $\mathbf{T}$ is the deviatoric part of $\mathbf{S}$ and ∇ denotes the objective time derivative by Jaumann.

$$\mathbf{D}_r = \frac{1}{2G}\left(\overset{\nabla}{\mathbf{T}} + \frac{1-2\nu}{3(1+\nu)} \, \mathrm{tr}(\dot{\mathbf{S}})\mathbf{1} \right) + \alpha \dot{T}\mathbf{1} \tag{2}$$

The plastic deformation rate $\mathbf{D}_p$ obeys the normality rule (3) and is controlled by the yield condition (4) which is formulated with effective stresses (5).

$$\mathbf{D}_p = \lambda \frac{\partial F}{\partial \mathbf{S}} \tag{3}$$

$$F = \mathrm{tr}(\mathbf{T}^2) - k^2 = 0 \tag{4}$$

$$\mathbf{S} = \frac{\sigma}{1-\omega} \tag{5}$$

λ is calculated from the consistency condition $\dot{F} = 0$. The hardening function k^2 depends on the temperature T and the scalar internal variable a which is defined by evolution equation (7).

$$k^2(a,T) = a_k(T)\left[1 - \exp\left[b_k(T)\,a\right]\right] + c_k(T)\,a + d_k(T) \tag{6}$$

$$\dot{a} = \mathrm{tr}(\mathbf{T}\cdot\mathbf{D}_p) \tag{7}$$

The third part of the deformation rate is expressed by equations (8) and (9).

$$\mathbf{D}_v = \frac{\dot{\omega}}{3(1-\omega)} \tag{8}$$

$$\dot{\omega} = \left[\vartheta_k(T) + \vartheta_l(T)\,\omega\right] \sqrt{\mathbf{D}_p \cdot \mathbf{D}_p} \tag{9}$$

From the specific free enthalpie $\psi(\mathbf{S},T,a)$ which is implicitly influenced by ω because of the effective stresses, and the first law of thermodynamics we derive an evolution equation for the temperature [3]. It is important to notice that a smaller part of the external work producing plastic deformations is stored in microstress fields and is not immediately dissipated.
Fourier's law is used to calculate the heat conduction, if the temperature is inhomogeneous. The restriction of positive entropy production resulting from the second law of thermodynamics is taken into account.

EXPERIMENTAL RESULTS

The material parameters which appear in the equations (6) and (9) are given in table I for a typical low carbon steel Ck15 (DIN 17200 No. 1.1141) as coefficients of a third order polynomial of temperature T.

For the determination of k^2 isothermal tension tests at different temperatures were performed (fig. 1). It has to be mentioned that the phenomenon of strain aging which occurs during these low deformation rate tests of fig. 1 is not included in the theory presented here, so that these influences have to be treated separately. The localisation due to necking limits the validity of these global curves, so that the information for larger local strains has to be taken from photographs made at different times during the experiments (fig. 2).

To evaluate ω the specimen were cut and observed by light microscopy to determine the fraction of voids.

TABLE I. Material coefficients of Ck 15 for $25°C < T < 365°C$

Parameters	$\cdot(T/°C)^1$	$\cdot(T/°C)^2$	$\cdot(T/°C)^3$	$\cdot(T/°C)^4$
a_k [N²/mm⁴]	+1.38372·10⁵	−1.88554·10²	+1.01214·10⁰	−1.21006·10⁻³
b_k [mm²/N]	−4.41853·10⁻²	+1.11929·10⁻⁴	−7.81370·10⁻⁷	+8.28617·10⁻¹⁰
c_k [N/mm]	+5.36326·10²	−1.54249·10⁰	+6.17450·10⁻³	−8.13315·10⁻⁶
d_k [N²/mm⁴]	+3.55067·10⁴	−2.81476·10¹	+6.41417·10⁻³	−2.30519·10⁻⁴
ϑ_1 [1]	+6.41117·10⁰	−2.12246·10⁻²	+3.50033·10⁻⁴	−7.93400·10⁻⁷
ϑ_k [1]	+5.0000·10⁻⁴	+0.0	+0.0	+0.0

×	T = 364 °C	
+	T = 313 °C	
▲	T = 176 °C	$\dot{\varepsilon} \approx 10^{-4}\ s^{-1}$
○	T = 110 °C	
□	T = 25 °C	

$\dot{\varepsilon} \approx 10^{-4}\ s^{-1}$
T = 25 °C

× local data
— global data

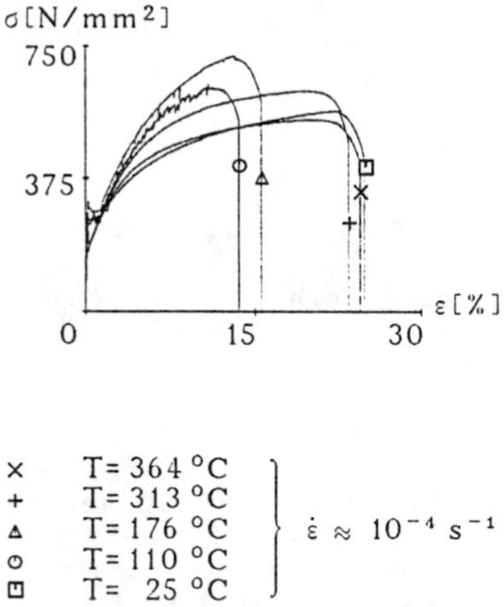

FIG. 1. Isothermal tension tests   FIG. 2. Isothermal tension test

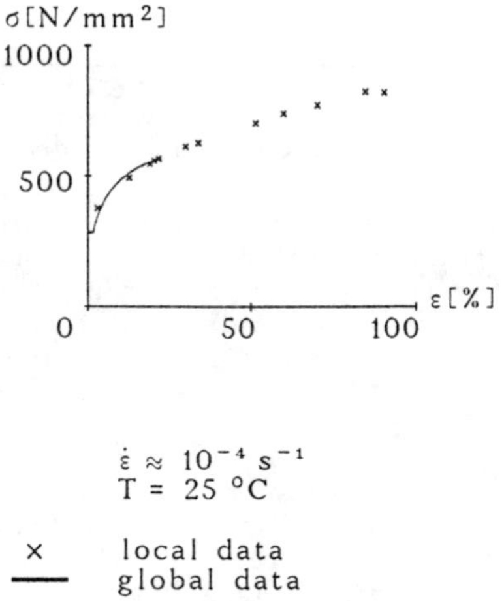

NUMERICAL RESULTS

Figure 3 shows ω as a function of the local axial strain at a constant temperature of 25°C. The experimental data are compared with the calculation of a homogeneous tension test using the parameters of table I.

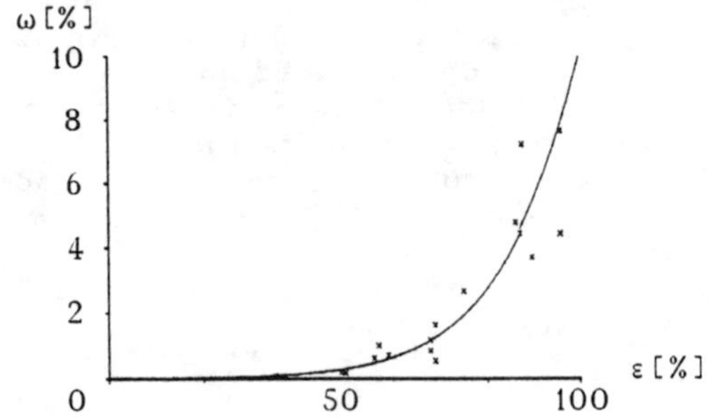

FIG. 3. Damage parameter ω vs. axial strain ε

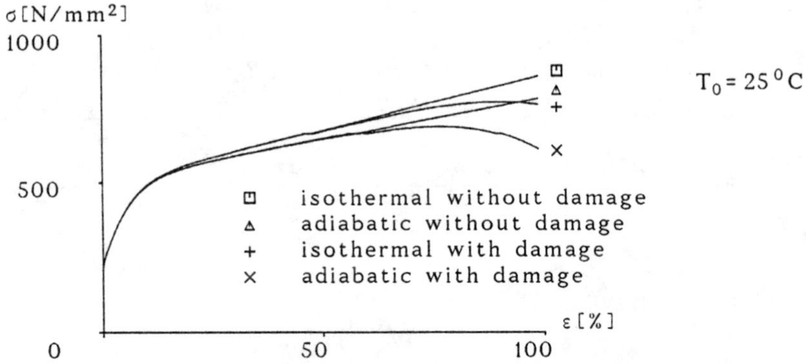

FIG. 4. Calculated homogeneous tension test

A fictitious tension test up to 100% homogeneous logarithmic strain was calculated. The stress strain curves are shown in fig. 4 for the isothermal and the adiabatic case both with or without evolution of voids. This illustrates the considerable influence of temperature and the softening effect of damage.

REFERENCES

1. O. T. Bruhns, H. Diehl, "An Internal Variable Theory of Inelastic Behaviour at High Rates of Strain", Arch. Mech. 41, 427-460 (1989)
2. J. Lemaitre, "A continuous damage mechanics model for ductile fracture", J. Engrg. Materials Techn., Trans. ASME 107, 83-89 (1985)
3. Th. Lehmann, "Some Thermodynamical Considerations on Inelastic Deformations Including Damage Processes", Acta Mech. 79, 1-24 (1989)

BOREHOLE SCALE EFFECTS AND RELATED INSTABILITIES

G. Frantziskonis, F.F. Tang and C.S. Desai

Department of Civil Engineering and Engineering Mechanics
University of Arizona
Tucson, Arizona 85721

ABSTRACT

A new mechanics based approach is proposed for scale effects and instabilities on borehole problems. In borehole type of structural systems, two types of instabilities can take place. The first is due to surface degradation growth and results into spalling of layers at the hole wall. The second is due to damage progression, and results into globally unstable response of the structure. The hole size has been found experimentally to be an important parameter in breakout instability initiation. Laboratory size holes may overestimate instability initiation properties by a large factor. At the same time, material properties such as peak stress depend largely on the size and shape of a specimen subjected to uniaxial or triaxial compression. This work attempts to incorporate size and scale effects into the instability initiation conditions. The important task of transferring information from laboratory experiments to actual large scale engineering problems is analyzed and discussed. The potential of the theory is demonstrated. The need for further experimental and theoretical work is identified.

1.0 INTRODUCTION

Borehole instabilities and breakouts are often characterized by the slabbing mode that affects a portion of the material close to the borehole wall. In addition, it is clear that the borehole size has significant effect on the initiation of breakout, Haimson and Herrick [1]. In general, "small" holes fail at higher external stresses than "large" ones. Since laboratory size boreholes are usually smaller than the ones in the field, the importance of hole size and its relation to breakouts is of a basic and quantitative nature. Thus in order to achieve a well-grounded statement about borehole stability that complies with laboratory and field observations, the dependency of borehole stability on its size (scale effect) must also be modeled.

Borehole scale effects is an example of the fact that the deformational characteristics of brittle materials depend on the size as well as on the shape of the structure (specimen). In a specimen subjected to uniaxial compressive stress, when the ratio of height to width (diameter for cylindrical specimens) of the sample is increased, the level of (macroscopic) stress at unstable failure decreases, Hudson et al [2], Desai et al [3]. In reference 3 a number of uniaxial compressive tests on cylindrical specimens are reported, from which the following conclusions are made. For the series of tests of constant sample diameter and increasing height, the peak stress reduces significantly, and a concave curve of peak strength versus height is observed, figure 1a. On the other hand, for constant height and increasing diameter the peak strength increases and a convex curve is observed, figure 1b. The above test results represent a part of the shape effects in brittle materials. The size effect calls for different response of specimens of the same shape but different size. For cylindrical specimens, size effect is observed if the height over diameter (L/D) remains constant; as the size of the sample increases, the peak strength reduces as shown in figure 1c.

Haimson and Herrick [1] studied the behavior of samples with different central hole sizes subjected to external stress. Square blocks of dry Alabama limestone having different diameters of central holes, ranging from 2 to 12 cm were subjected to uniaxial stress. All blocks had side length to borehole diameter ratio of 5:1. It was found that small diameter holes required larger stresses to induce breakouts, figure 2.

The above test results indicate that size and shape effects are significant especially for structures (specimens) of small size. The borehole tests indicate that as the hole size increases (more than 9 cm for Alabama limestone), the size effect becomes less prominent since the applied stress for breakout approaches a constant level. Similarly, the cylindrical specimen tests indicate that the peak stress tends asymptotically to a certain level. However different stress levels are reached depending on size or shape effects as shown in figure 1. Reviews, Maury [4], Guenot [5], demonstrate that classical design procedures lead to overestimation of the drilling fluid density by a factor of 2 to 8. From the test results shown in figure 2 it can be seen that breakout of small diameter holes require about 3 times the stress of large ones. Similar observations hold for tests on cylindrical specimen.

The bifurcation theory is used by Papanastasiou and Vardoulakis [6] to examine the effect of borehole radius on borehole stability. Material behavior is described by the deformation theory of plasticity, and internal length is introduced in the formulation through employment of Cosserat theory.

2.0 BACKGROUND

The model including damage proposed in [7-12] is used herein. The notation of reference 12 is adopted herein and the constitutive equations can be written as

$$\dot{\sigma}_{ij} = L_{ijkl}\,\dot{\epsilon}_{kl} - \dot{r}(\sigma^{u}_{ij} - \sigma^{d}_{ij}) \tag{2.1}$$

where

$$L_{ijkl} = (1 - r)\,C^{u}_{ijkl} + rC^{d}_{ijkl}$$

r is a scalar representing the ratio of damaged to total volume, C^{u}_{ijkl} and C^{d}_{ijkl} are the constitutive tensors for the intact and damaged fraction respectively.

Surface degradation is induced by micro-structural inhomogeneity and its growth is initially stable [13, 14]. It is important to mention that there is certain evidence that this phenomenon acts as a trigger effect on the shear band appearing in a specimen. The sudden growth of surface effects results in the occurrence and development of shear bands penetrating into the body [13, 14].

A distance ρ is from the surface is defined as

$$\rho = a\left[\int_{c} W ds - l\right] \tag{2.3}$$

where W is a weighting function (the simplest case calls for W=unity and even this case has been shown to provide satisfactory results), a is a material constant determinable from test results on different size specimens, l is the so-called surface degradation material length, and c is the path of maximum (absolute) principal compressive stress.

3.0 ANALYTICAL AND NUMERICAL SOLUTIONS

Let us consider the following problem depicted in figure 3. The borehole structure of length and width D contains a central circular hole of radius R. The ratio D/R is considered constant such that D/2R = 5. Uniaxial compressive external stress σ is applied externally. If the material is considered linear, isotropic and elastic, then the σ_{max} occurs at point A and, Roark and Young [15]

$$\sigma_{max} = \sigma_{A} = k\,\sigma_{nom} \tag{3.1}$$

where

$$\sigma_{nom} = \frac{\sigma D}{D - 2R} \tag{3.2}$$

$$k = 3.00 - 3.13\left(\frac{2R}{D}\right) + 3.66\left(\frac{2R}{D}\right)^2 - 1.53\left(\frac{2R}{D}\right)^3 \tag{3.3}$$

In order to demonstrate the capability of the theory to capture the scale effects of this problem we make (for the time being) the following simplifying assumptions. We consider that the material is linear elastic and isotropic. This, of course represents a "stiffer" material than actual rocks. In addition we assume that for the external stress levels at breakout the material in the surface degradation zone has zero stiffness. Thus at breakout initiation this problem can be analyzed by using the solution presented in equations (3.1)-(3.3) but now the radius of the hole is $R + \rho$. For this problem equation (2.3) reduces to

$$\rho = \alpha(2\pi R - l) \tag{3.4}$$

and

$$\sigma_{max} = k \frac{\sigma D}{D - 2(R+\rho)} \tag{3.5}$$

and k is given from (3.3) where $R+\rho$ is substituted instead of R. For the material discussed later it was found that α=0.21 and l was considered to be one inch [12]. If we consider that at breakout, σ_{max} in (3.5) assumes a constant value we can obtain solutions for different R where always D/2R = 5. The curve in figure 4 is obtained where the diameter (2R) is plotted on the horizontal axis and the external stress at breakout is plotted on the vertical one. Clearly, the shape and trend of this curve is similar to the experimental one given in figure 2.

The finite element method has been used, long ago, for the stress-deformation and stability analysis of borehole problems. Initial works in this area, Desai and Reese [16], considered nonlinear elastic material response and the Mohr envelope was used as the criterion for development of plastic zones. It was found that plastic zones advance close to the borehole at a certain distance depended on the geostatic loading. The extent of plastic zone may be related to the surface degradation zone discussed in this study. However, since the surface degradation zones are depended on the geometry of the structure, such a relation would be based on fixed radius of the borehole.

The problem shown in figure 3 was investigated numerically for six different values of the central hole radius. For all problems the ratio of D over R was constant such that D/2R = 5. The finite element mesh used is shown in figure 5. Eight node quadrilateral elements were used. For the six problems studied the hole radii were 0.5, 1, 2.5, 5, 7.5 and 10 cm. For each problem the surface degradation distance ρ was calculated (equation 3.4). The elements next to the central hole extend for a distance ρ as shown in figure 5. The material properties assigned for these elements are the surface degradation ones. In figure 6 the maximum (peak) stress is plotted on the vertical axis and the hole diameter is plotted on the horizontal one. The shape of this curve is similar to the shape shown in figure 2.

CONCLUSIONS

Surface degradation growth is shown to play an important role for phenomena observed in borehole problems such as scale effects and surface damage instabilities. Analytical and numerical results show the capability of the theory to predict such phenomena. At this time, only qualitative comparison of observed and predicted responses is possible. Further experimental work related to growth of damage and resulting instability close to borehole walls is needed.

Acknowledgments

A part of the research herein was supported by Grant No. AFOSR 890460 from the Air Force Office of Scientific Research, Bolling AFB.

REFERENCES

1. Haimson, B.C. and Herick, C.G., "Borehole Breakouts and In Situ Stress," Proc. Energy-Source Techn. Conf., Houston, TX, (1989).
2. Hudson, J.A., Brown, E.T. and Fairhurst, C., "Shape of the Complete Stress-Strain Curve for Rock," Proceedings 13th Symposium, Rock Mechanics University of Illinois, Urbana, (1971).
3. Desai, C.S., Kundu, T. and Wang, G., "Size effect on damage in Progressive Softening Process for Simulated Rock," International Journal for Numerical and Analytical Methods in Geomechanics, in press.
4. Maury, V., "Observations, Researches and Recent Results About Failure Mechanisms Around Single Galleries," Proceedings 6th International Congress on Rock Mechanics, Montreal Canada, 2, 1119-1128 (1987).
5. Guenot, A., "Stress and Rupture Conditions Around Oil Wellbores," Proceedings 6th International Congress on Rock Mechanics, Montreal Canada, 1, 109-118, (1987).
6. Papanastasiou, P.C. and Vardoulakis, I.G., "Bifurcation Analysis of Deep Boreholes: II. Scale Effect," Int. J. Num. Anal. Meth. in Geomechanics, in press.
7. Frantziskonis, G. and Desai, C.S., "Constitutive Model with Strain Softening," International Journal of Solids and Structures, 23, 733-750 (1987).
8. Frantziskonis, G. and Desai, C.S., "Elastoplastic Model with Damage for Strain Softening Geomaterials," Acta Mechanica, 68, 151-170, (1987).
9. Frantziskonis, G., "Damage and Free Edge Effects in Laminated Composites. Energy and Stability Propositions," Acta Mechanica, 77, 213-230, (1989).
10. Frantziskonis, G. and Desai, C.S., "Degradation Instabilities in Brittle Material Structures," Mechs Res. Comm., to appear.
11. Frantziskonis, G. and Desai, C.S., "Surface Degradation Mechanisms in Brittle Material Structural Systems," Int. J. Fracture, in press.
12. Frantziskonis, G. Tang, F.F. and Desai, C.S., "Degradation Instabilities in Brittle Material Structural Systems," Comp. Meth. Appl. Mech. & Engr., submitted or publication, (1990).
13. Kitagawa, H. and Matsushita, H., "Flow Localization in Elastic-Plastic Material Developing from the Stress-Free Surface," International Journal of Solids and Structures, 23, 351-368, (1987).
14. Yukutake, H., "Fracturing Process of Granite Inferred From Measurements of Spatial Velocities and Temporal Variations in Velocity During Triaxial Deformations," J. of Geophysical Res., 94, 15639-15651, (1989).
15. Roark, R.J. and Young, W.C., Formulas for Stress and Strain, 5th edition (1989).
16. Desai, C.S. and Reese, L.C., "Stree-Deformation and Stability Analysis of Deep Boreholes," Proc. 2nd Congr. Intl. Soc. for Rock Mechs., Belgrade, Yugoslavia, (1970).

Figure Captions

Figure 1: Schematic of Size and Shape Effects from Uniaxial Compression of Cylindrical Rock Specimens. (a) Diameter D Constant and Varying Height L, Shape Effect, (b) L constant and Varying D, Shape Effect, (c) D/L Constant, Size Effect.

Figure 2: Experimental Data for Scale Effects for Borehole Problems, After [1].

Figure 3: Geometry of a Borehole Problem Structure Under Uniaxial Compressive External Load.

Figure 4: Scale Effect Predicted through the Simplified Analysis.

Figure 5: Finite Element Mesh for the Numerical Analysis of the Borehole Problem Depicted in Figure 4.

Figure 6: Predicted Scale Effect through Examination of the Maximum External Applied Stress.

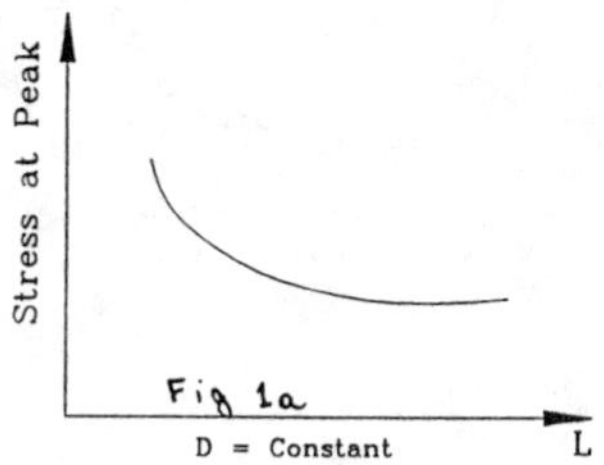
Stress at Peak
Fig 1a
D = Constant
L

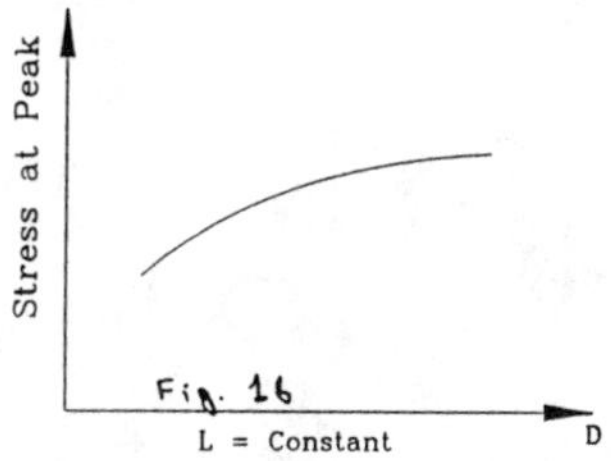
Stress at Peak
Fig. 1b
L = Constant
D

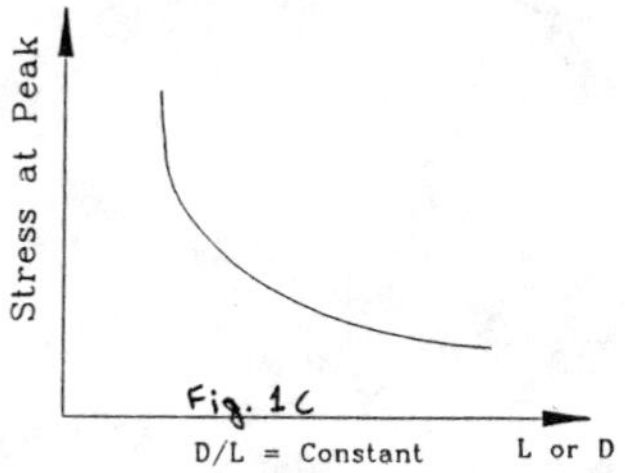
Stress at Peak
Fig. 1c
D/L = Constant
L or D

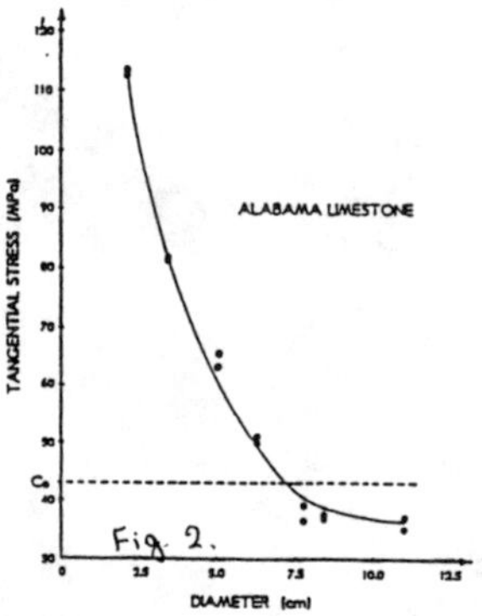
ALABAMA LIMESTONE
TANGENTIAL STRESS (MPa)
DIAMETER (cm)
Fig. 2.

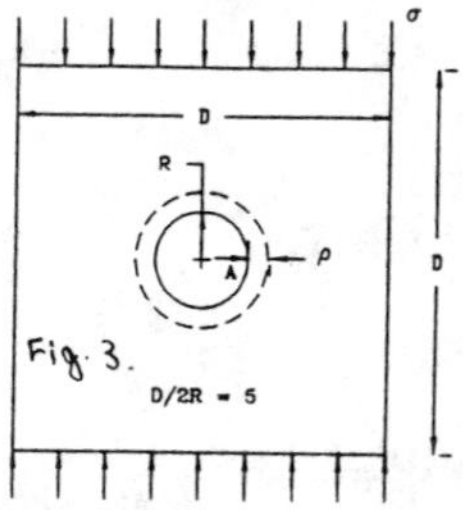
σ
D
R
ρ
A
Fig. 3.
D/2R = 5

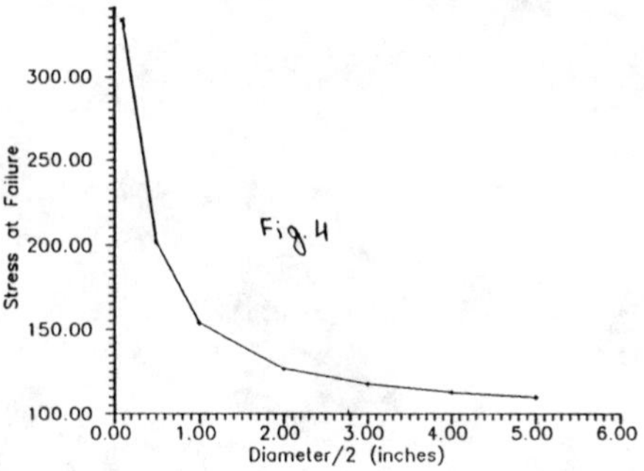
300.00
250.00
200.00
150.00
100.00
Stress at Failure
Fig. 4
0.00
1.00
2.00
3.00
4.00
5.00
6.00
Diameter/2 (inches)

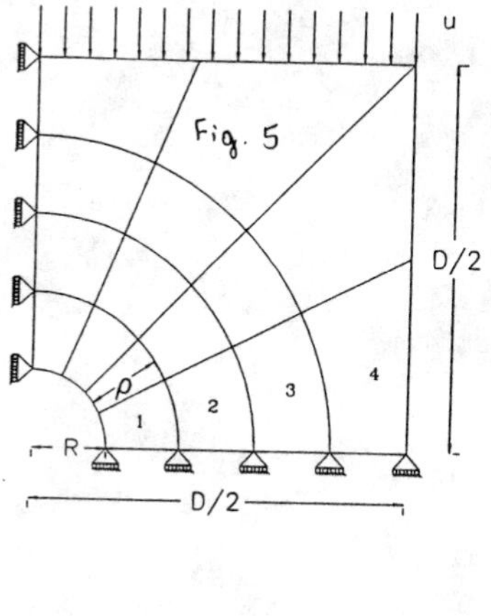
u
Fig. 5
D/2
ρ
1
2
3
4
R
D/2

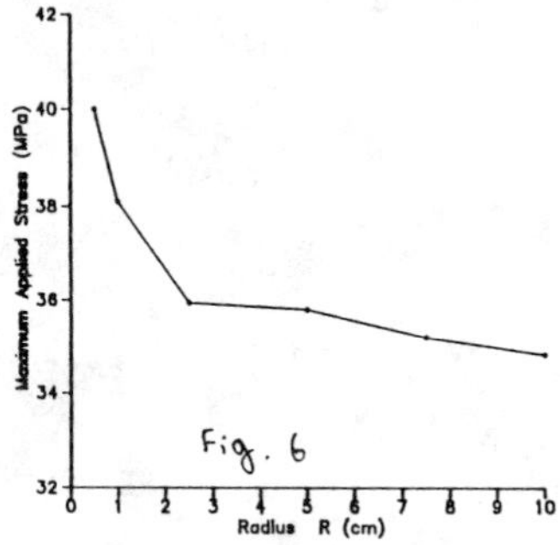
42
40
38
36
34
32
Maximum Applied Stress (MPa)
Fig. 6
0
1
2
3
4
5
6
7
8
9
10
Radius R (cm)

AN INCREMENTAL MODEL FOR INDUCED ANISOTROPIC DAMAGE IN ROCKS.

IKOGOU S., SHAO J.F., HENRY J.P.
Laboratory of Mechanics of Lille
EUDIL 59 655 Villeneuve d'Ascq - FRANCE

ABSTRACT

Experimental investigations on a sandstone allow to show induced anisotropic microcracking of rock. An incremental type damage model is developed to describe the rheological behaviour of this material. Validation of the model on proportional loading tests is presented.

INTRODUCTION

Constitutive modelling of rock materials becomes very important to evaluate the safety of structures in Petroleum engineering, nuclear and chemical waste storages and others underground works. Induced anisotropic damage is a frequently observed phenomenon for a great number of rocks susceptible to cracking under compressive stress Very important dilatancy can result from this anisotropic cracking which is difficult to modelize. In this paper, experimental results in triaxial compression tests will be presented to explain the damage mechanism for a sandstone. Then, an incremental damage model will be formulated according to the experimental investigations.

EXPERIMENTAL INVESTIGATIONS

Rock used in this study is a sandstone (99 % quartz content) whose average porosity is 10 %. Cylinder specimens (75mm in length and 37.5mm in diameter) are firstly dried and then saturated with methyl alcohol . Two strain gauges are placed in longitudinal and transverse directions respectively. The used rates are 1.2 x 10-6/s and 1.4 x 10-2 MPa/s respectively in strain and stress controlling tests [1].

A hydrostatic compression test is firstly achieved and the obtained results are presented in figure 1. We can notice that : a) the longitudinal and transverse strains are identical and consequently the inital structure of rock is isotropic ; b) the stress-strain relation is linear after a natural microcracks closure phase.

Triaxial compression tests with different confining pressures are then performed. In figure 2, longitudinal and transverse strains are presented as function of stress deviator. It can be seen that : a) longitudinal strain remains quasi linear as stress deviator increases, $d(\sigma_1-\sigma_3)>0$; b) there is a very important non linearity on transverse strain. This corresponds physically to opening of microcracks perpendicular to the minor principal stress (σ_3) ; c) the orientated microcracking leads also to an important dilatancy of rock.

In figure 3, a triaxial compression test with unloading-reloading cycles is presented. We can notice that: a) irreversible strain after completely unloading is negligible.The unloading, $d(\sigma_1-\sigma_3)<0$, corresponds to microcrack closure. Rock behaviour in unloading path can be approximately described by a linear relation using secant modulus dependent only on the last stress inversion point ; b)the reloading curve coincides pratically with the initial loading curve.

Figure 4 presents results of a reduced triaxial compression test . This stress path corresponds also to opening of microcracks, $d(\sigma_1-\sigma_3) > 0$. We obtain non linear strains as the confining pressure descreases.

From the above experimental investigations, we can conclude that the studied rock exhibits essentially two distinc behaviour domains : microcrack opening and closure domains. The rheological behaviour of rock depends completely on microcracking state of material. The induced damage is always developed in the plane perpendicular to the minor principal stress.

FORMULATION OF THE INCREMENTAL DAMAGE MODEL

Incremental rheological models have been developed by many authors [2] to describe soils and concretes behaviour. Various models have also been proposed to describe isotropic and anisotropic damage under extensive stress in concrete materials [3]. We shall use the philosophy of incremental models and the damage concept to formulate the incremental damage model.

We suppose firstly that the stress-strain relation of rock can be expressed by the following incremental form :

$$\begin{pmatrix} d\varepsilon_1 \\ d\varepsilon_2 \\ d\varepsilon_3 \end{pmatrix} = \begin{bmatrix} M_{11} & M_{12} & M_{13} \\ M_{21} & M_{22} & M_{23} \\ M_{31} & M_{32} & M_{33} \end{bmatrix} \begin{pmatrix} d\sigma_1 \\ d\sigma_2 \\ d\sigma_3 \end{pmatrix}$$

Where M is named rheological matrix whose expression depends on the microcracking state (opening or closure). In this relation, the three principal stresses are ordered as $\sigma_1 \geq \sigma_2 \geq \sigma_3$. We suppose that rock material exhibits pressure dependent linear elastic behaviour in the plane perpendicular to σ_3 (plane $0\ \sigma_1 \sigma_2$). Then the general behaviour of rock is transverse isotropic. We have consequently :

$$M_{11} = M_{22} = \frac{1}{E} \quad , \quad M_{12} = M_{21} = \frac{-\nu}{E} \quad , \quad E = E_o\left(1+a\left(\frac{\sigma_3}{P_a}\right)^m\right)$$

where E_o is the Young modulus in simple compression test ($\sigma_3 = 0$), ν the Poisson ratio, P_a the atmospheric pressure, a and m are wo material parameters.

M_{31} and M_{32} are two moduli determing the strain in the direction of microcracking induced by $d(\sigma_1-\sigma_3)$ and $d(\sigma_2-\sigma_3)$. They naturally depend on microcracking state and we have four distinc cases :

(1°) $d(\sigma_1-\sigma_3) > 0$ and $d(\sigma_2-\sigma_3) > 0$

$$M_{31} = \frac{-1}{K_{c31}} \quad , \quad M_{32} = \frac{-1}{K_{c32}}$$

(2°) $d(\sigma_1-\sigma_3) > 0$ and $d(\sigma_2-\sigma_3) \leq 0$

$$M_{31} = \frac{-1}{K_{c31}} \quad , \quad M_{32} = \frac{-1}{K_{d32}}$$

(3°) $d(\sigma_1-\sigma_3) \le 0$ and $d(\sigma_2-\sigma_3) > 0$

$$M_{31} = \frac{-1}{K_{d31}} \quad , \quad M_{32} = \frac{-1}{K_{c32}}$$

(4°) $d(\sigma_1-\sigma_3) \le 0$ and $d(\sigma_2-\sigma_3) \le 0$

$$M_{31} = \frac{-1.0}{K_{d31}} \quad , \quad M_{32} = \frac{-1.0}{K_{d32}}$$

where K_c and K_d are moduli corresponding respectively to microcracking opening and closure.

$$K_{c31} = \frac{E}{\nu}\left(\frac{\sigma_1 - \sigma_3}{Q_t}\right)^{-\beta} \quad , \quad (\sigma_1 - \sigma_3) > Q_t \qquad\qquad K_{c32} = \frac{E}{\nu}\left(\frac{\sigma_2 - \sigma_3}{Q_t}\right)^{-\beta} \quad , \quad (\sigma_2 - \sigma_3) > Q_t$$

Where Q_t represents the elastic linear limit in transverse direction, and β a material parameter. Q_t and β depend on σ_3. K_{d31} and K_{d32} are determined by the secant moduli at the last stress inversion point. M_{13}, M_{23} and M_{33} represent the strains induced par the variation of the minor principal stress σ_3. We propose:

$$M_{33} = \frac{1}{E_f} \qquad M_{13} = \frac{-\nu_{13}}{E_f} \qquad M_{23} = \frac{-\nu_{23}}{E_f} \qquad E_f = E\left(\frac{\sigma_1 - \sigma_3}{Q_t}\right)^{-\alpha}$$

where ν_{13} and ν_{23} are two pseudo Poisson ratios which would depend respectively on $(\sigma_1-\sigma_3)$ and $(\sigma_2-\sigma_3)$, α is a material parameter. The associated parameter determination procedure is presented in detail in [1] [4].

VALIDATION OF THE MODEL

In this paper, we present the validation of the model only for proportional loading tests $(\sigma_1/\sigma_3=k)$. In figure 5, comparisons between experimental results and model simulation are presented. We notice good agreements.

CONCLUSION

A microcracking state dependent damage model is proposed to describe induced anisotropic damage of rock. Validation on proportional loading tests gives good agreements. Validation on other stress paths and applications to borehole stress analysis are presented in [1] [4].

REFERENCES :

1. Ikogou S. "Etude expérimentale et modélisation du comportement d'un grès", Thèse de Doctorat Université de Lille (1990)
2. Darve F., Boulon M., Chambon R. "Loi rhéologique incrémentale des sols", Journal de Mécanique 17, n° 5, p. 679-716 (1978)
3. Mazars J. "Application de la mécanique de l'endommagement au comportement non linéaire et à la rupture du béton de structure", Thèse d'état, Université de Paris VI (1984)
4. Ikogou S., Shao J.F., Henry J.P. "Modélisation de la fissuration anisotrope d'un grès et validation", Colloque G.F.R.-A.U.G.C, Grenoble (1990)

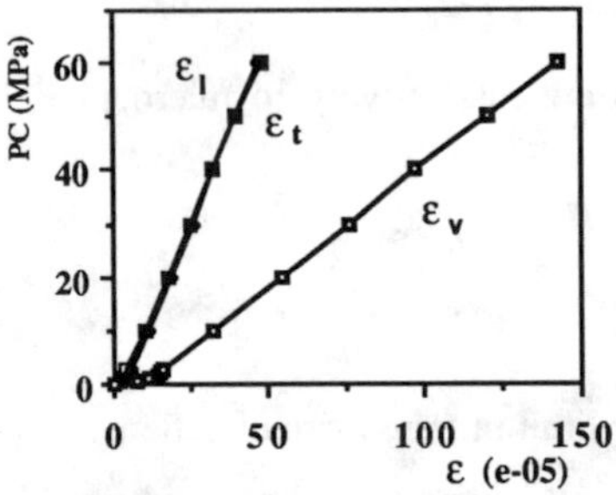

Fig.1 Hydrostatic compression test

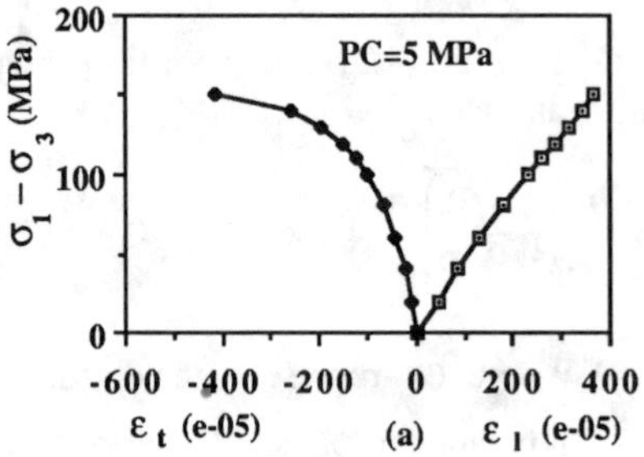

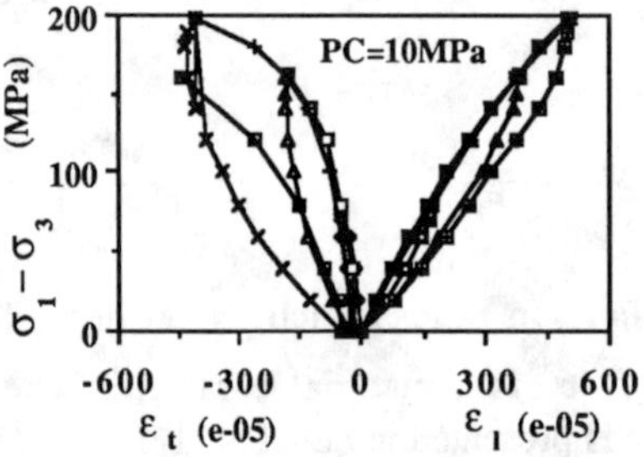

Fig.3 Triaxial compression test with unloadinid-reloading cycles

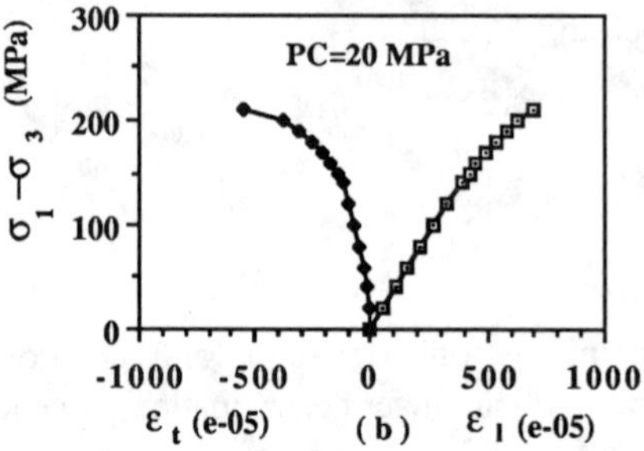

Fig.2 triaxial compression test

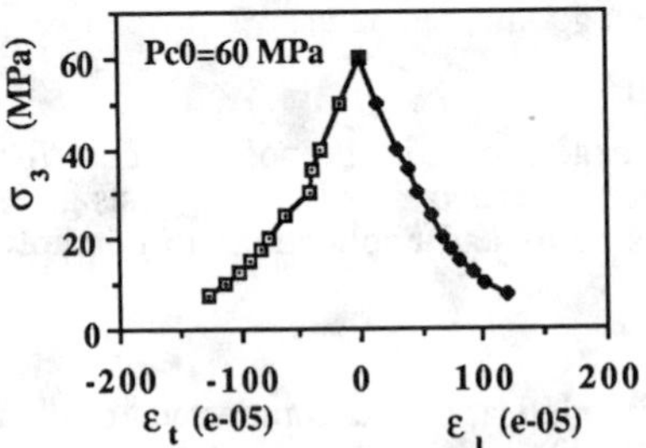

Fig.4 Reduced compression test

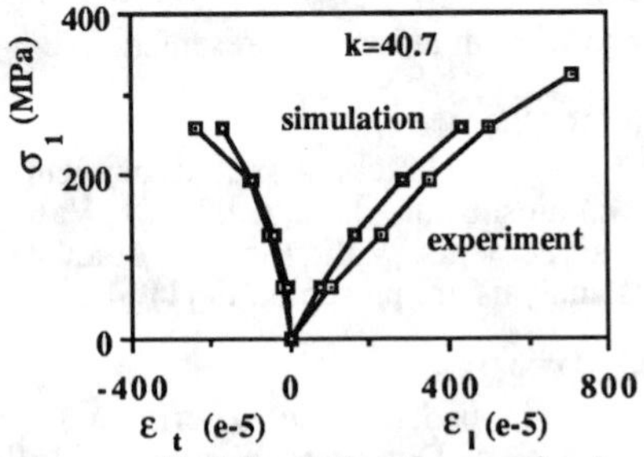

Fig.5 Simulation of proportional loading test

STATISTICAL MODELS FOR BRITTLE RESPONSE OF SOLIDS

DUSAN KRAJCINOVIC and MICHAL BASISTA
Mechanical and Aerospace Engineering
Arizona State University, Tempe, AZ 85287-6106, USA

ABSTRACT

This paper is an attempt to probe into the question of the justification and application of the deterministic models for the analysis of the so-called cooperative phenomena, i.e. brittle deformation processes dominated by the evolution of an ensemble of interacting microcracks.

INTRODUCTION

Micromechanical modeling is understood herein as an attempt to formulate an analytical relation between the microstructure and chemical composition of a solid and its macro (structural) response. For the sake of expedience and economy in applications most analysts have a strong preference for deterministic models. An often unjustified confidence in deterministic models is based on the assumption that all involved fields (such as stresses and strains) are sufficiently well defined by their averages over the so-called representative volume element.

However, even a casual glance at a micrograph reveals a certain level of disorder or randomness in the microstructure of a typical structural material. The degree of randomness ranges from large in the case of particulate composites to moderate for solids with both short- and long-range order (such as polycrystals).

Hence, it seems reasonable to inquire into the justifications, or at the very least limitations, allowing use of deterministic models for processes which are basically stochastic. Such an inquire may have a profound effect on the analytical modeling of the so-called cooperative brittle deformation processes. This rather modest, in both size and scope, paper is an attempt to discuss this important problem from a different viewpoint. The basic premise is that the question whether a deterministic approach is applicable for the considered case can be answered only on the basis of statistical analyses upon the examination of the scatter in results.

A valid statistical investigation of the influence of a random input (initial disorder of the microstructure) on the output (mechanical response of the structure) requires a statistically valid sample of results (i.e. large number of computations). Consequently, economy of computations mandates use of simplified models focusing on the qualitative qualitative aspects of the basic problem. This requirement promotes use of rather crude discretizations, such as lattices, at the expense of computationally more intensive finite element models.

DISCRETE (STATISTICAL) MODELS

For a better understanding of the problem it is, perhaps, advisable to start with a very simple model such as a loose bundle of parallel rods carrying an external tensile load F (Krajcinovic and Silva [1], Krajcinovic [2], etc.). The analysis of this parallel bar system is based on the following simplifying assumptions:

- all rods are elastic, perfectly brittle and arranged in such a manner that they equally share in carrying the external load F,
- all N rods have identical stiffness K/N and elongation,
- the strengths of the rods are different.

Once the initial (quenched) disorder is introduced assigning unequal strengths to individual bars the subsequent analysis is deterministic. It is trivial to prove that the equilibrium is satisfied when

$$F = Kx(1 - D) \tag{1}$$

where x is the elongation of the system and

$$D = (n/N) = \int_0^{Kx} p(f_R)\, df_R \tag{2}$$

is the damage variable selected as the ratio between the number of the already ruptured bars n and the total number of bars N (assumed to be very large). Also, in (2) $p(f_R)$ is the probability density function of bar strengths which in some manner defines the microstructure of the material. Assuming a uniform probability density function $p(f_R)$ = const. it is not difficult to demonstrate that: the damage is linearly proportional to the displacement x, the force is a quadratic function of displacement and that the damage at the apex of the stress-strain curve (stress rupture) is $D_c = 0.5$.

As shown in [1] the form of the stress-strain curve and the value of the damage parameter at the stress rupture depends on the adopted distribution of the bar strengths. Additionally, it should be noted that this simple model totally ignores the defect interaction assuming the force to be equally shared by all extant bars. Finally, it is important to underline that once the strengths of bars (i.e. microstructure) are prescribed it becomes unnecessary to devise the damage evolution laws.

Using expressions $F = A_o\sigma$, $\varepsilon = x/L$ and $K = EA_o/L$, where E, A_o and L are the elastic modulus, initial (undamaged) cross-sectional area and length of the macro-bar, respectively, the above expressions can be rewritten in the conventional form

$$\sigma = E(1 - D)\varepsilon \quad \text{and} \quad D = A_v/A_o \tag{3}$$

where A_v is the area initially occupied by the ruptured bars which is not any more available for the transmission of forces.

A computationally much more intensive problem of large triangular lattices was lately examined by Hansen, et al. [3]. The central idea was to examine the influence of the initial disorder and the size of the lattice on the mechanical response. The first task was accomplished assigning different strengths to different bars of the lattice and repeating computations for each distribution (initial disorder) obtaining many different physical realizations of the same initial statistics. In each case the computations of forces in the lattice members for a given displacement of the rigid end members were performed in the routine and deterministic manner. It is important to notice that the lattice model allows for the consideration of the crack interactions and stress fluctuations attributable to the incured damage. Naturally, the strength probability density function (taken as uniform as in the previous case) was identical for all cases allowing for a meaningful comparison of results. Computed quantities were then averaged over different lattices keeping the number n of ruptured bars constant ("history" averaging). The influence of the lattice size was then assessed considering lattices for which L (being the number of diagonal members along a straight line connecting the rigid end bars) was taken as 4, 8, 16 and 24.

According to the results reported in [3], modified as suggested in [4], the force-displacement relation may be written in the following form

$$F(L) = K[1 - D(L)]x \tag{4}$$

where the expression for the damage parameter is now

$$D(L) = \sqrt{3}\beta\, L^{-\beta} x \qquad (5)$$

Except for the fact that both the force and the damage parameter are functions of the lattice size these expressions are similar to those derived for the loose bundle of parallel bars. The parameter $\beta = 3/4$ was selected to allow collapsing of the pre-critical segments of the force-displacement curves for all four lattice sizes to a single curve. It is very interesting that the magnitude of the damage defined by (5) at the apex is again 0.5 [4].

It is important to emphasize the fact that the corresponding expressions for the loose bundle parallel system and the lattice are identical in form for the pre-critical regime. In other words, the response is not very sensitive to the level of sophistication in modeling. In summary, the computations reported in [3] demonstrate that in the pre-critical (hardening) regime:

(a) the average stress-strain curves for all four lattice sizes and all distributions of the initial disorder can be collapsed on a single master-curve (quadratic parabola),

(b) the relation between damage and strain is linear, and

(c) the damage at the apex of the stress-strain curve was equal to 0.5 for both models and uniform distribution of rupture strengths.

In concert with the basic premises on which the loose bundle parallel bar system is based this obviously means that in the pre-critical regime the crack interaction has little or no influence on the overall response which is, consequently, well described by simple volume averages of the involved fields. In other words, the pre-critical regime is deterministic by nature. The crack interaction and even the redistribution of stresses is not crucial for the determination of the overall response. The fractality, i.e. dependence of the overall response on the size of the lattice is, perhaps, associated with a rather small lattices and should fade away with increasing L.

In a sharp contrast, the distribution of forces in the post-peak regime is strongly multi-fractal. The inability to identify a single length scaling parameter should be related to the fact that the post-peak regime strongly depends on the distances between the adjacent interacting cracks. It seems reasonable to expect that more than one length parameter must be introduced to model that (non-local) behavior. Additionally, the post-peak response is found to depend strongly on the initial disorder. Consequently, since the initial disorder in most engineering materials is random the post-peak behavior is inherently stochastic.

PERCOLATION MODELS

The fact that the magnitude of the damage parameter at the stress rupture appears to be (according to an admittedly limited number of data [3]) an invariant, depending only on the dimensionality and the microstructure, (rupture strength distribution) provides a strong impetus to explore the same problem within the framework of the percolation theory. In contrast to the preceding case the percolation theory incorporates disorder by assuming the strength of all bars to be equal. However, even though such distribution of rupture strengths is not entirely realistic the percolation theory allows provides for useful estimates of the response in the vicinity of the percolation point in terms of universal constants and exponents.

An initial attempt to analyze lattice models within the percolation theory is due to Beale and Srolovitz [5]. The assumption that only two different rupture strengths of lattice members are possible introduced significant problems into computations and, not surprisingly, overestimates the brittleness. The subsequent analyses within the framework of the node-link-blob approximation and for-mulation of the continuum percolation models with exotic names such as swiss-cheese and blue cheese [6,7] are of some interest despite the fact that all implications have not as yet been carefully explored from the enginering point of view.

SUMMARY AND CONCLUSIONS

The present paper contains a short discussion of one of the critical issues in the analytical modeling of the so-called cooperative phenomena. In the present case, the cooperative phenomena is understood as a deformation process dominated by nucleation and growth of a large number of microcracks and their coalescence into a macrocrack of critical size.

The present analyses are based on the premise that the numerical results obtained analysing lattices with some initial (quenched) disorder may in a qualitative sense be interpreted as indicative for the continua as well. This is, naturally, not always true and in this sense some of the conclusions are tenuous at this time. Additionally, the present conclusions are based on the analyses of planar lattices. The fact that the lattices are trusses (i.e. central-force lattices) is not crucial. Qualitatively similar results are obtained when the nodes are linked together by beams having both axial and bending rigidity [8]. It is well known that the effect of the crack interaction reduces in transition to three dimensions. This aspect should have effect on the conclusions only in a quantitative sense.

Assuming, nevertheless, that the analyses of the lattices (and grids) are, indeed, in a qualitative sense an indication of the existing trends in solid materials it seems reasonable to conclude that:

(a) the pre-critical regime is deterministic, while
(b) the post-critical regime is strongly stochastic.

Consequently, the response of a structure in the pre-critical (hardening) regime can be readily modeled by the existing micromechanical and phenomenological theories assuming the solid to be homogeneous. Since the response in the post-critical (softening) regime depends on the initial disorder in an essential manner its modeling requires a stochastic theory.

ACKNOWLEDGEMENT

The authors gratefully acknowledge the financial support in form of a research grant from the U.S. Department of Energy, Office of Basic Energy Research, Division of Engineering, Materials and Geosciences which made this work possible.

REFERENCES

1. D. Krajcinovic and M.A.G. Silva, "Statistical aspects of the continuous damage mechanics", Int. J. Solids Struct., 18, 551-562, (1982).
2. D. Krajcinovic, "Damage mechanics", Mech. Materials, 8, 117-197, (1989).
3. A. Hansen, S. Roux and H.J. Herrmann, "Rupture of central-force lattices", J. Phys. France, 50, 733-744, (1989).
4. D. Krajcinovic and M. Basista, "Rupture of central-force lattices revisited", submitted for publication.
5. P.D. Beale and D.J. Srolovitz, "Elastic fracture in random materials", Phys, Rev. B, 37, 5500-5507, (1988).
6. D. Sornette, "Critical transport and failure in continuum crack percolation", J. Phys. France, 49, 1365-1377, (1988).
7. P.M. Duxbury, in Statistical Models for the Fracture of Disordered Media, edited by S. Roux and H. J. Herrmann (North Holland, Amsterdam 1990).
8. H.J. Herrmann, A. Hansen and S. Roux, "Fracture of disordered, elastic lattices in two dimensions", Phys. Rev. B, 39, 637-648, 1989.

THEORETICAL INVESTIGATIONS ON CRACK ARREST IN REINFORCED CONCRETE

Ming-Te Liang and Kuo-Chyuan Tsay
1. Associate Prof., National Taiwan Ocean University, Keelung, Taiwan, R.O.C.
2. Graduate student

ABSTRACT

The objective of the paper is theoretically to study the mechanism of crack arrest in reinforced concrete. Theoretical results show that $\sigma \propto (s)^{-1/2}$

INTRODUCTION

The lack of beneficial interaction between the steel rods and concrete matrix in reinforced concrete is not unavoidable. However, by means of equations describing compatibility between adjacent points on the reinforcement and in the concrete, it is possible to calculate these forces and, from suitable fracture mechanics applications, to interpret them in terms of a reduction in the crack extension force.

CRACK ARREST MECHANISM

Irwin [1] showed that the retionship between strain energy release rate (G) and stress intensity factor (K) is

$$G = \pi K^2 (1-\mu) E^{-1} \quad \text{(for plane strain)} \tag{1}$$

where E is Young,s modulus and is Poisson,s ratio. Romualdi and Batson [2] have shown that the crack extension force tendency is

$$G = 4(1-\mu^2)\sigma^2 a (\pi E)^{-1} \tag{2}$$

where σ is a uniform remote tensile stress and 2a is crack length. Assumes that a mass of reinforced concrete is subjected to σ. The crack is centrally located between four adjacent steel wires, and the diameter of the crack is equal to the rod spacing s (FIG.1). The section A-A of FIG.1 is drawn in FIG.2. Solutions for the interaction force distribution are obtained for discrete points along the wire. The points are spaced at a distance h and the interaction force at any point y_i is $p_i = f_i h$, where f(y) is the resulting force distribution and has the units force per unit length. Let v_i be the y-directed displacement of a discrete point y_i and let d_{ij} be the displacement of y in the concrete due to a unit force at the location of the wire at point y_j. The condition for no relative displacement between the concrete and the wire at each of the n discrete points is

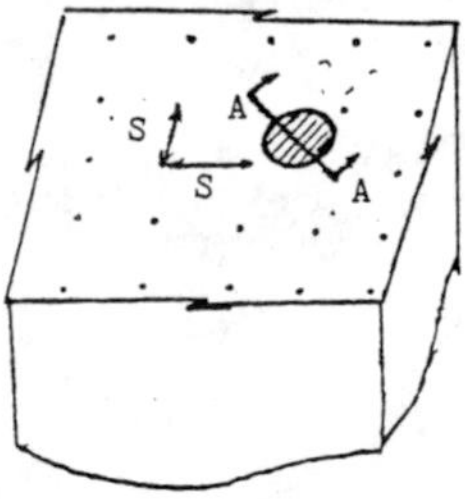

FIG.1 Cross section through wire reinforced concrete at location of flaw

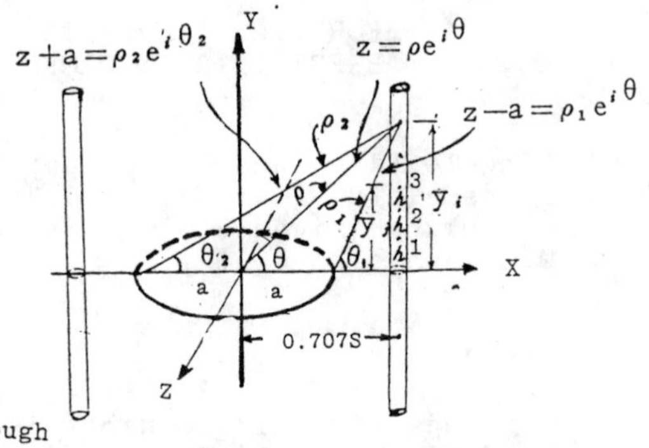

FIG.2 Section through two wires and circular crack

$$v_i - \sum_{j=1}^{n} d_{ij} P_j = 0 \tag{3}$$

Passing through Eqs.(1)and(2),one gets the remote tensile stress

$$K_\sigma = 2\sigma(a)^{1/2}\pi^{-1} \tag{4}$$

Similarly,one obtains the negative internal crack

$$K_p = 2p(a)^{1/2}\pi^{-1} \tag{5}$$

The resulting total stress intensity factor is

$$K_T = K_\sigma - K_p = 2(\sigma - p)(a)^{1/2}\pi^{-1} \tag{6}$$

Let the Airy stress function suggested by Westergaard[3] be

$$\phi = Re\bar{\bar{Z}} + y I_m \bar{Z} \tag{7}$$

where $\bar{Z}$, Z ,and Z' are successive derivative of a function $\bar{\bar{Z}}$ (z), z is the complex variable(x+iy).The appropriate stress function for the problem is

$$Z(z) = \sigma z[(z+a)(z-a)]^{-1/2} \tag{8}$$

Lame s contants Q and G for plane strain are defined as

$$Q = E\mu[(1+\mu)(1-2\mu)]^{-1} \text{ and } G = E[2(1+\mu)]^{-1} \tag{9}$$

The plain strain displacement is given by

$$2Gv = 2(1-\mu)I_m\bar{Z} - yReZ \tag{10}$$

The terms involving and can be found in terms of the coordinates shown in FIG.2.The expressions for $I_m\bar{Z}$、ReZ become

$$I_m\bar{Z} = -\sigma\rho(\rho_1\rho_2)^{-1/2}\sin[\theta-(\theta_1+\theta_2)/2] \tag{11}$$

$$ReZ = \sigma\rho(\rho_1\rho_2)^{-1/2}\cos[\theta-(\theta_1+\theta_2)/2] \tag{12}$$

Substituting Eqs.(9),(11),and(12)into Eq.(10),the displacement can be written as

$$v = -\sigma(1+\mu)\,\rho(\rho_1\rho_2)^{-1/2}E^{-1}\left\{2(1-\mu)\sin[\theta-(\theta_1+\theta_2)/2] + y\cos[\theta-(\theta_1+\theta_2)/2]\right\} \quad (13)$$

Eq.(13) includes two parts, the displacement caused by uniform stress σ .

$$v = \sigma y(1+\mu)(1-2\mu)E^{-1} \quad (14)$$

and the singularity condition obtained from that Eq.(14) must be subtracted from Eq.(13) for estimating v_i.

Dean et al[4] presented the solution for the displacement caused by two equal and oppose forces in an infinite solid medium as shown in FIG.3.Assumes that the bond forces between the concrete and stell are respectively uniform around the surface and acted along the center line of the rod.The restraining effect of the rod is neglected.In FIG.3,r_2, y-axis and D are respectively the radius, center line,and a point on the surface of the rod.The unit loads(P=1) are located at the positions $\pm y_j$ and D is

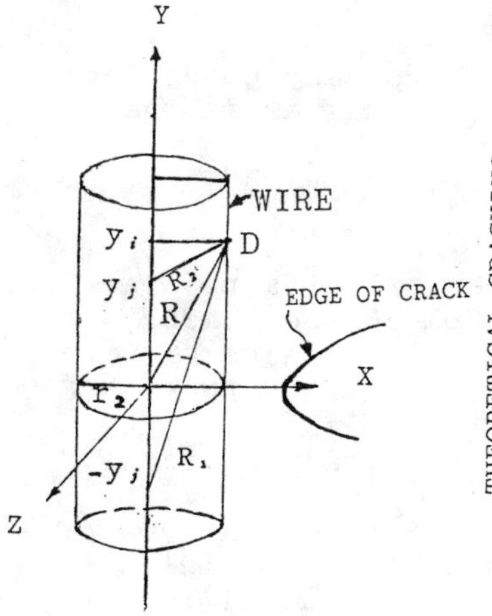

FIG.3 Geometric configuration of rod and forces

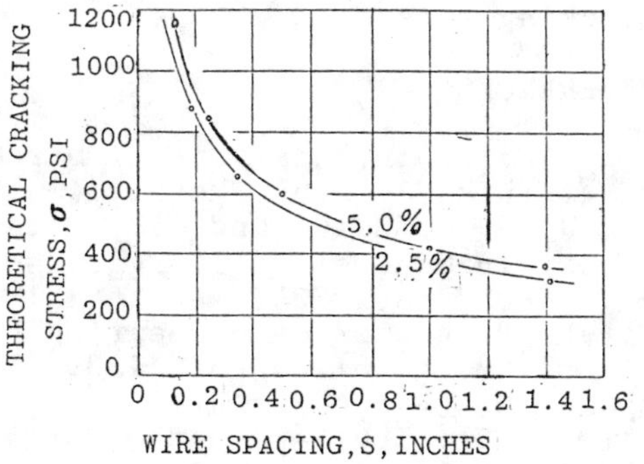

FIG.4 Cracking stress as a function of wire spacing (G_c=0.02 in.lb per sq in.,σ = 1000 psi)

the point($z=y_i$) at which the displacement is sought.The displacement at y_i caused by a unit load at y_j ,without the correction for the existence of the crack, is

$$d'_{ij} = B^{-1}[(y_i+y_j)^2R_1^{-3}-(y_i-y_j)^2R_2^{-3}+(Q+3G)(Q+G)^{-1}(R_1^{-1}+R_2^{-1})] \quad (15)$$

where $B = 8\pi G(Q+2G)(Q+G)^{-1}$ (16)

The resulting average stress $\bar{\sigma}$ is the stress to be superimposed over the crack area,with opposite sign,in order to provide the necessary displacement correction.

Eqs.(13) and(14) yield displacements in the vicinity of a crack caused by σ and also apply for the case of a crack with P in a stress-free medium.The correction to be added to Eq.(15) is then

$$d''_{ij} = -\bar{\sigma}\,(1+\mu)\rho(\rho_1\rho_2)^{-1/2}\,E^{-1}\Big\{2(1-\mu)\sin[\theta-(\theta_1+\theta_2)/2] + y_i\,\cos[\theta-(\theta_1+\theta_2)/2]\Big\} - \bar{\sigma}y_i E^{-1}(1+\mu)(1-2\mu) \tag{17}$$

Thus

$$d_{ij} = d'_{ij} + d''_{ij} \tag{18}$$

THEORETICAL RESULTS

The procedure for calculating v_i and d_{ij} was programmed for repeated solution on computer.The forces P_j were solved from Eqs.(3) and (18).The internal crack pressure P is computed and the total K_T is obtained from Eq.(6).The results of eight solutions for stell percentages (by volume) of 2.5% and 5.0% are shown in FIG.4.

CONCLUSIONS

FIG.4 is proved that $\sigma \propto (s)^{-1/2}$ .For any percentage of stell,the smaller the spacing,the higher is the cracking stress.

REFERENCES

1. G.R. Irwin,"Analysis of Stress and Strains Near End of a Crack,"J. Appl. Mech.24(3),361-364(Sep.,1957).
2. J. P. Romualdi and G.B. Batson,"Mechanics of Crack Arrest in Concrete," J.Engg. Mech. Division,ASCE,89(3), 147-168,(June,1963).
3. H. M. Westergaard,"Bearing Pressures and Cracks," J.Appl. Mech. 61,A49-A53,1939.
4. W. R. Dean,H.W. Parson,and D.W.Sneddon,"A Type of stress Distribution on the Surface of a Semi-Infinite Elastic Solid," Proc.,Cambridge Philosophical Society, 40,5(1944).

DAMAGE MODELS FOR MASONRY AS A COMPOSITE MATERIAL: A NUMERICAL AND EXPERIMENTAL ANALYSIS

G. MAIER, A. NAPPI and E. PAPA
Department of Structural Engineering
Technical Univ. (Politecnico), Milan, Italy

ABSTRACT

This paper outlines some preliminary results of a study on masonry behaviour based on the following guidelines: a damage model is adopted and experimentally calibrated for mortar, while bricks are described as brittle-elastic; homogenization procedures lead to orthotropic constitutive laws for masonry walls under monotonic and cyclic loading; resulting models for masonry are validated by experiments on miniaturized masonry panels.

INTRODUCTION

A most traditional building material such as masonry is attracting much interest in today's structural engineering especially in view of its extensive use in seismic areas for both ancient monumental and new standard structures. A survey of the abundant recent literature on masonry motivated the study outlined herein. This is meant to exploit some concepts in micromechanics of composites and continuum damage mechanics in order to achieve an enhanced description of the mechanical properties of masonry. From the above three areas to combine, only few representative publications directly related to the present purposes will be quoted here [1-3].

HOMOGENIZATION

A masonry wall (fig.1) is conceived as a two-

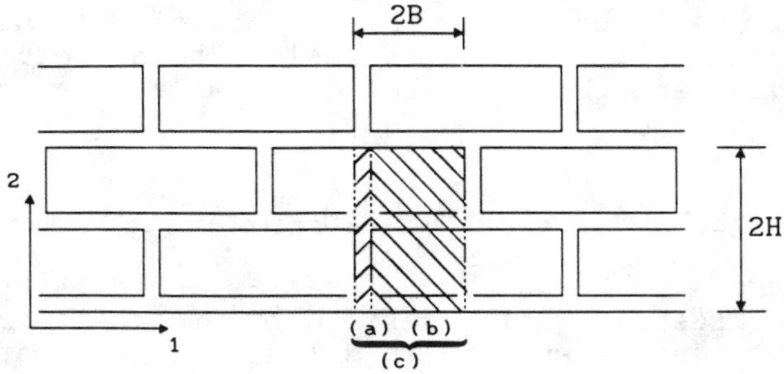

FIG. 1 - Masonry wall: homogenization procedure.

dimensional plane-stress non-homogeneous periodic continuum. Mortar and bricks play the roles of matrix and reinforcement inclusion, respectively, of a particular composite material. These phases are assumed to be isotropic and with perfect adhesive bond at the interface. Therefore, masonry as a homogenized medium turns out to be orthotropic, with an elastic stiffness matrix (order 3, in the plane) defined by four independent parameters. These are evaluated by the homogenization procedure outlined below with reference to fig.1.

First, we consider a strip in direction 2 which contains a sequence of vertical mortar layers and exhibits a periodicity defined by the representative volume (a) with period 2H. Second, a parallel strip is considered with same period and representative volume (b). Both strips are separately homogenized by means of an asymptotic expansion technique proposed in [4]. At this stage the wall can be regarded as a periodic array of homogeneous vertical layers with period 2B and representative volume (c), fig.1: a third application of the same technique, with expansion in direction 1, provides the sought stiffness matrix of masonry as homogeneous material.

In nonlinear finite element analyses the homogenization method above described for linear elasticity and encompassing three unidirectional homogenization phases, has to be carried out for each loading step at each Gauss point, using "instantaneous" moduli of mortar and bricks, still assumed as isotropic.

In view of this fact, a simplified homogenization procedure has also been implemented. This is based on an interpretation of the masonry wall as an array of individually homogeneous horizontal layers of mortar and bricks and, hence, reduces to a single unidirectional homogenization phase. The difference in the resulting moduli turns out to be a small percent for standard data.

Every analysis based upon a homogenized masonry model and, hence, performed in terms of overall (average) strains and stresses, requires to derive from these variables the local ones, primarily in order to capture failure thresholds. This "dehomogenization" process is performed here basically by means of the procedure proposed in [5], special provisions being required for wall edges and high overall variable gradients as potential sources of inaccuracy.

CONSTITUTIVE LAWS FOR BRICKS AND MORTAR

Bricks are assumed as linear-elastic-brittle, the failure threshold being defined by Grashof's criterion of maximum tensile strain. Thus four material parameters characterize the brick behaviour (Young's modulus E;

Poisson's ratio ν; limit strains ε_t and ε_c in tension and compression, respectively).

Mortar is considered as an elastic material susceptible to damage, understood as degradation of stiffness and sometimes also of strength (softening). Among various damage models proposed in the literature for concrete-like materials (e.g., Lubliner-Oñate, Martin-Resende, Ortiz) the model developed by Mazars [6] was deemed to be particularly flexible and to permit reasonable compromises between realism and simplicity in the specific present context.

The following three variants of this damage model are considered in the present study, with the common basic hypothesis of isotropic damage, described by a scalar D ($0 \leq D \leq 1$).

(α) "Basic model", constructed with the following ingredients [6]: "equivalent strain"; damage variable D expressed as a combination of two damage contributions due to suitably defined tension and compression states; evolution laws for these contributions; a threshold function that defines the current domain where damage is not produced. Besides E and ν, this model involves five material parameters, which can be identified by uniaxial tension and compression tests.

(β) "Unilateral model" [7]: as (α) but with allowance for the stiffness difference when stresses change sign, because damage-related microcracks and voids may close. There are again five parameters involved, to calibrate as above.

(γ) "Fatigue model" [8][9]: the strain domain of non-increasing damage is removed or reduced to the origin while a loading-unloading criterion is preserved. This modification of models (α) or (β) permits the simulation of fatigue damage due to cyclic stresses. There are now six parameters to be identified (on the basis of uniaxial tension/compression tests and of uniaxial cyclic tests).

(δ) "Damage-plasticity model" [9]: residual strains are accounted for as suitable functions of the damage variable (rather than by a separate conventional plasticity model). The same parameters required for model (α) are needed.

Clearly, model (α) is suitable for monotonic stresses only; variant (γ) is required in the presence of high numbers of stress cycles; model (δ) turns out to be computationally more demanding; the three variations of the basic model can be combined in different ways.

NUMERICAL AND EXPERIMENTAL RESULTS

The above outlined constitutive laws and homogenization techniques were implemented in several variants, combined with conventional finite element modelling and a predictor-corrector finite-increment

time integration scheme. Thus different variations can be compared computationally.
In order to validate the numerical analysis tools, an experimental rig has been designed for testing miniaturized masonry panels and structures. Clearly, scale effects on the complex nonlinear behaviour of masonry cannot be assessed by similarity theory. Therefore, calibration of models at small scale yield results which can be transferred only qualitatively to full scale structures. Some preliminary results obtained on these bases are summarized below.
Numerical tests carried out by model (α) for mortar and the homogenization procedures outlined earlier, substantiate the following closed-form, semiheuristical expression for the non-zero entries of the masonry stiffness matrix as functions of brick and mortar Young moduli E_b and E_m in MPa, assuming $\nu = 0.2$ for both

$$K_{11} = \frac{1}{2}\,(0.3\ E_m + 1.775\ E_b) - D\,(0.12\ E_m + 0.05\ E_b)$$

$$K_{22} = \frac{1}{2}\,(0.55\ E_m + 1.525\ E_b)\,(1 - D^n)^{1/2} \quad , \qquad K_{12} = 0.2\ K_{22}$$

$$K_{33} = 0.4\ K_{22} \qquad \text{with} \quad n = 1 + (E_m\ /\ 15{,}000)\,(E_m\ /\ E_b)^{1/2}$$

where Cartesian reference and standard geometry is assumed as in fig.1. These simple formulae, for $D \leqq 0.6$, entail differences no more than 3% with respect to the three-phase homogenization.
Failure of masonry in the present idealization means that either a limit strain in a brick is attained or the mortar strength is saturated (a "critical" value is reached by damage D, which corresponds to the limit stress). This criterion, based on model (α) with parameters identified by tests on mortar cubic specimens and three-phase homogenization, led to the failure loci (resistance domains) represented in fig.2 for two inclinations β of principal overall stress σ_1 with respect to direction 1 (β=0°, β=30°). Fig.2 also shows histograms related to our uniaxial test results (for β=0°). The resistance domains in fig.2 are qualitatively similar to the ones obtained by Page experimentally [10].
By governing displacements, uniaxial cyclic tests (loading-unloading) were performed with increasing amplitude on masonry panels. The main features observed are: at equal maximum strain attained by the last cycle the number of cycles has negligible influence; time-dependence manifestations (relaxation; creep at imposed stress) show up close to the limit stress; deformations and damage are spread (no significant strain localization) in the softening stage (so that strains preserve a meaning also along the falling-down branch of the envelope curve). Typical experimental results are

shown in figs.3-4 and compared to computer simulations based on variant (γ) of Mazars' damage model for mortar. Fig.5 shows a masonry specimen after the above mentioned tests.

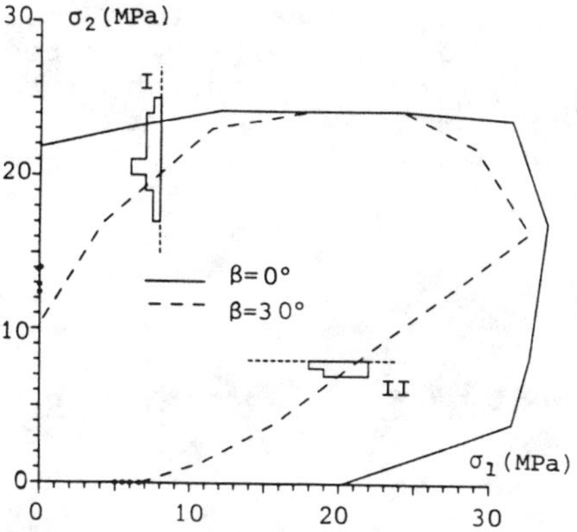

FIG. 2. Resistance domains of masonry - Dots denote experimental results with β=30°; histogram I(II) refers to 15(8) experimental tests with β=0° and σ_1=0 (σ_2=0).

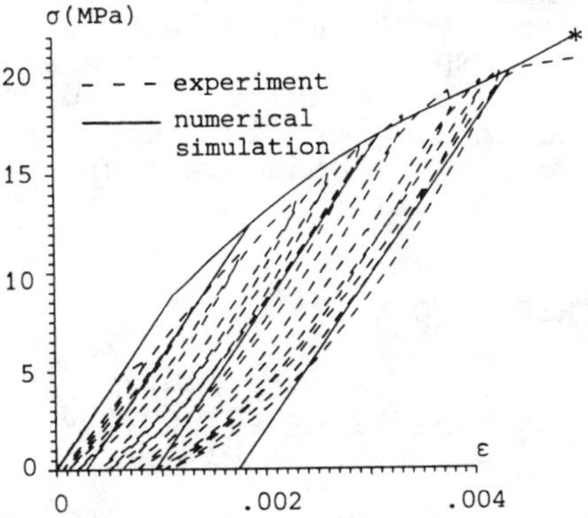

FIG. 3. Experimental vs. simulated cyclic straining process with β=0° and σ_1=0 - The asterisk refers to the ultimate strain (owing to brick failure).

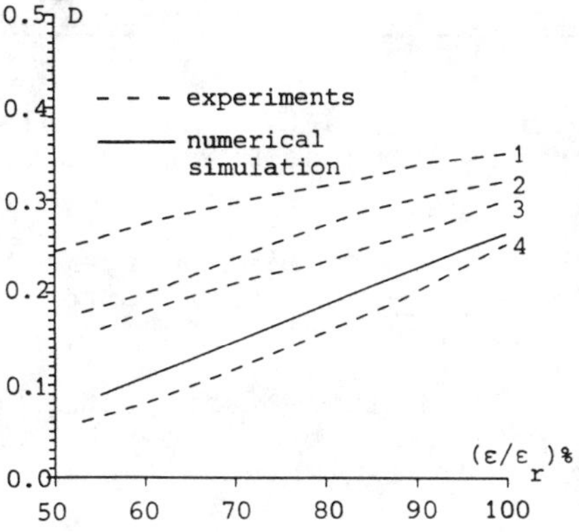

FIG. 4. Damage variable D vs. ratio between current and ultimate strain (solid line and curve 4 are related to the plots of fig. 3; curves 1-3 refer to other tests on nominally equal specimens and show their scatter).

FIG. 5. A miniaturized masonry specimen after cyclic testing.

CLOSING REMARKS

In present structural engineering, masonry as building material is characterized by growing importance in seismic zones, by unusually high scattering of experimental data and by severe limitations on experiments which are feasible in practical situations. The investigation in progress outlined herein (sponsored by CNR-GNDT) is intended to provide an insight into masonry behaviour and advanced simulation tools for the nonlinear cyclic response of masonry, by using damage models, homogenization and parameter identification techniques.

REFERENCES

[1] D.P. Adams, (Ed.), Proceedings Fifth North American Masonry Conference (Johnson Publishing Co., Boulder, Colorado 1990).

[2] R.M. Christensen, Mechanics of Composite Materials (J. Wiley, New York 1979).

[3] J. Lemaitre and J.L. Chaboche, "Mécanique des matériaux solides" (Dunod, Paris 1985).

[4] E. Sanchez-Palencia, Non Homogeneous Media and Vibration Theory (Springer-Verlag, Berlin 1980).

[5] C. Auriel, G. Boubal and P. Ladeveze, "Sur une méthode de calcul des effets locaux", in C.R. Troisièmes Journées Nationales sur le Composites (Paris, 21-23 Septembre 1982).

[6] J. Mazars, "A Description of Micro- and Macroscale Damage of Concrete Structures", Engrg. Fract. Mech., 25 (5/6), 729-737 (1986).

[7] P. Ladeveze, "Sur une théorie de l'endommagement anisotrope", Rapport Interne N.34 (Laboratoire de Mécanique et Technologie, Cachan, France 1983).

[8] J.J. Marigo, "Modelling of Brittle and Fatigue Damage for Elastic Material by Growth of Microvoids", Engineering Fracture Mechanics, 24 (4), 861-874 (1985).

[9] E. Papa, Damage Mechanics of Masonry (in Italian), Doctoral Dissertation (Dept. of Structural Engineering, Politecnico di Milano 1990).

[10] A.W. Page, "The Biaxial Compressive Strength of Brick Masonry", Proc. Inst. of Civil Engineering, 2 (71), 893-906 (1981).

A CONCRETE MODEL FOR THE ANALYSIS OF REINFORCED CONCRETE STRUCTURE SUBJECTED TO ALTERNATE LOADINGS

A.MILLARD AND G.NAHAS

ABSTRACT

This paper presents an elasto plastic fracturing model to describe the behaviour of concrete under alternate loadings. Special care has been given to crack closure aspects. An example of application to a reinforced concrete shear wall is presented.

INTRODUCTION

Among the various reinforced concrete structures which can be subjected to alternate loadings, buildings are of primary concern for safety and economical reasons. In order to predict the behaviour of a building and to appraise safety margins with reasonable accuracy, for exemple in case of seismic loadings, structural models are required, which involve detailed modeling of concrete properties, for various kinds of damages (traction, compression and shear). The aim of the present paper is to present a model of concrete under multiaxial state of stress, enabling a detailed description of local phenomena in concrete, and therefore capable of predicting the behaviour of structural reinforced concrete elements like beams and shear walls under alternate loadings. The model will be described in the next paragraph and one exemple of application will be presented in the last one.

CONSTITUTIVE MODEL FOR CONCRETE UNDER ALTERNATE LOADING

The various constitutive models which have been proposed during the last two decades for concrete fall into two main categories :
- Models based on continum damage mechanics, where the inelastic behaviour is described by means of one or several damage variables : the first models considered only one isotropic variable (2). Since then, many refinements were incorporated in order to account for anisotropy, different damage mechanisms in traction and compression, unilateral aspect of cracking and irreversible strains(3).
- Models based on an elasto plastic approach including cracking effects, as in the pionneering work of BAZANT and KIM (4). In these models, the various inelastic mechanisms can be accounted for by means of a multi-criteria approach, as in (5) for exemple. Many applications of such models have been published. However, their use is often restricted to the prediction of structural behaviour under monotonous loading. Among the problems set by the treatment of alternate loadings, a crucial one is the cracks closure. This aspect has been enlighted by some recent work, in the frame of damage (6) as well as elasto plastic models (7). One major difficulty comes from the lack of available experimental results. Some of them can be found in (8) and (9) for example, from which it can be observed (see figures 1 and 2)in case of uniaxial behaviour that :

- After cracking, when unloading in the post-peak regime, the stiffness is lower than in the elastic regime. The elastic stiffness is recovered in the compression domain after some non linear transition.
- Under zero stress, the cracks remain open.
- The transition between traction and compression is neither $\sigma = o$ nor $\varepsilon = o$

The model which is proposed here, accounts for the above mentionned features. It has been developped in the frame of a general purpose finite element code CASTEM 2000 (10). For this reason, three main failure modes have been considered :
- failure under traction loads,
- failure under shear loads,
- failure under compression loads.

Details of the three failure modes can be found in ref.(5). Here only the tensile failure will be adressed.

Tensile failure criterion :
In case of virgin concrete, a maximal principal stress criterion is assumed :

$$\underset{i=1,3}{\text{Max}}\ \sigma_i \leq R_t$$

where R_t denotes the tensile strength of concrete.
Once the criterion is reached, the principal direction is memorized and the subsequent new cracks will be orthogonal to the first one. This, the model accounts for the anisotropy induced by cracking.

Softening law :
Softening laws lead to numerous difficulties like localization and non local effects. Although this is a very important problem these aspects will not be discussed here and we will assume that the experimental strain-stress curves are intrinsic.
For tensile failure, the tensile strength decreases with a damage parameter d directly connected with the positive strain ε_i. In case of unloading, the elasticity modulus is first Es(d), up to a zero stress, then, the strain decreases up to a residual strain $\varepsilon_R(d)$ from which the initial elasticity modulus E is recovered (see figure 3). Beyond the limit strain the concrete has no more resistance in traction along this direction.

APPLICATION TO A REINFORCED CONCRETE SHEAR WALL SUBJECTED TO PERIODIC LOADING

This example deals with a shear wall which has been subjected to a periodic loading with monotonously increasing amplitude. The shear wall was fixed at the bottom and loaded at the top by an hydraulic actuator as shown on figure 5.

Details of geometry, material properties and loading can be found in ref. (11).
Different calculations were performed in order to investigate the influence of various parameters like YOUNG's modulus and tensile strength of concrete. The best agreement was found for a low tensile strength, which corresponds to the fact that the concrete was probably micro-cracked prior to the test, due to drying, handling etc. Comparison between experimental and calculated values is shown on figure 6.

CONCLUSION

An elasto-plastic fracturing model has been proposed here to describe the behaviour of concrete under alternate loadings. It is based on a multi-potential approach. Special care has been given to the treatment of concrete cracking and cracks closure which play a predominant role in the problems under consideration. One example of application has shown the capabilities of the model. However it has to be fully validated under complex loading histories. Moreover, problems associated with strain-softening, like localization, must be considered.

REFERENCES

(1) **CH.LEPRETRE,A.MILLARD,JM.REYNOUARD,O.MERABET,G.NAHAS**

"Structural analysis of reinforced concrete structures under monotonous and cyclic loadings" - SMIRT 10 - LOS ANGELES 1989

(2) **J.MAZARS**

"Application de la mécanique de l'endommagement au comportement non linéaire et à la rupture du béton de structure" -

THESIS - UNIVERSITY OF PARIS 6 - 1984

(3) **S.RAMTANI**

"Contribution à la modélisation du comportement multiaxial du béton endommagé, avec description du caractère unilatéral"

THESIS - UNIVERSITY OF PARIS 6 - 1990

(4) **Z.BAZANT, S.KIM**

"Plastic fracturing theory for concrete" -

J.ENG.MECH.ASCE VOL.105 - 1979

(5) **G.NAHAS**

"Calcul à la ruine des structures en béton armé"

THESIS - UNIVERSITY OF PARIS 6 - 1986

(6) **C.LA BORDERIE,Y.BERTHAUD,G.PIJAUDIER-CABOT**

"Crack closure effects in continum damage mechanics"Numerical implementation" - PROC."ZELL AM SEE" 1990

(7) **O.MERABET**

"Modélisation des structures planes en béton armé sous chargements monotone et cyclique - Construction et validation d'un modèle numérique" -

THESIS - INSA DE LYON - 1990

(8) **H.REINHARDT,H.CORNELISSEN,D.HORDIJK**

"Tensile tests and failure analysis of concrete"

JOURNAL STRUCT.ENG.DIV.ASCE - VOL 112 N°11 - 1986

(9) **S.RAMTANI,Y.BERTHAUD,J.MAZARS**

"A model for describing both the anisotropic and uniaxial distributed damage in concrete" PROC.SMIRT 10-LOS ANGELES 1989

(10) **P.VERPEAUX,A.MILLARD,TH.CHARRAS,A.COMBESCURE**

"A modern approach of large computer codes for structural analysis" PROC SMIRT 10 - LOS ANGELES - 1989

(11) **A.MILLARD,G.NAHAS**

"Behaviour of reinforced concrete structures under seismic loading" - To be published - 1990

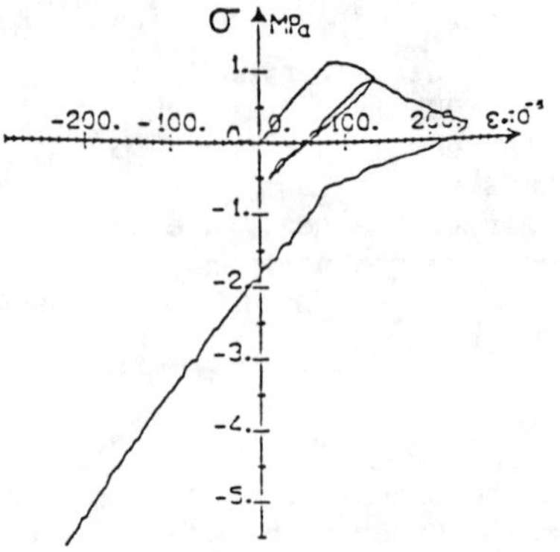

FIGURE 1 FROM REF.(6) - STRESS STRAIN CURVE FOR CONCRETE IN TENSION/COMPRESSION

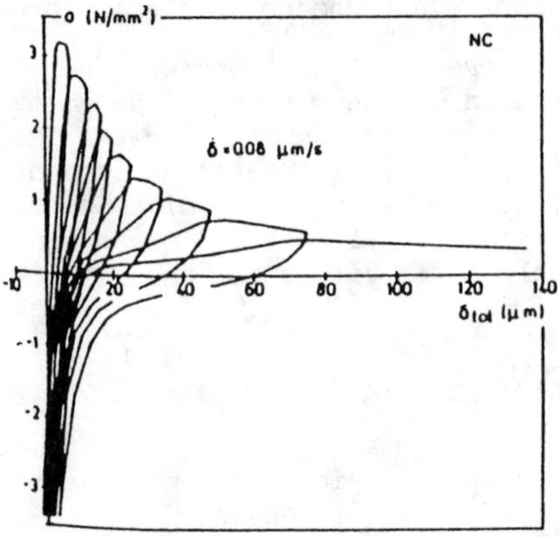

FIGURE 2 FROM REF.(8) - STRESS DEFORMATION CURVE FROM CYCLIC TESTS WITH COMPRESSIVE LOWER STRESS

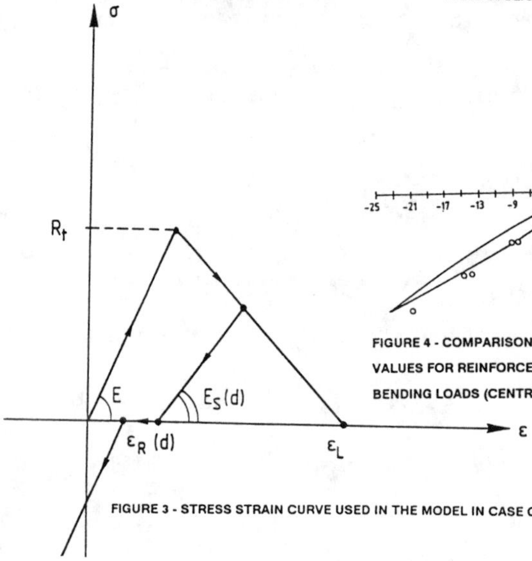

FIGURE 3 - STRESS STRAIN CURVE USED IN THE MODEL IN CASE OF UNIAXIAL TRACTION - COMPRESSION LOAD

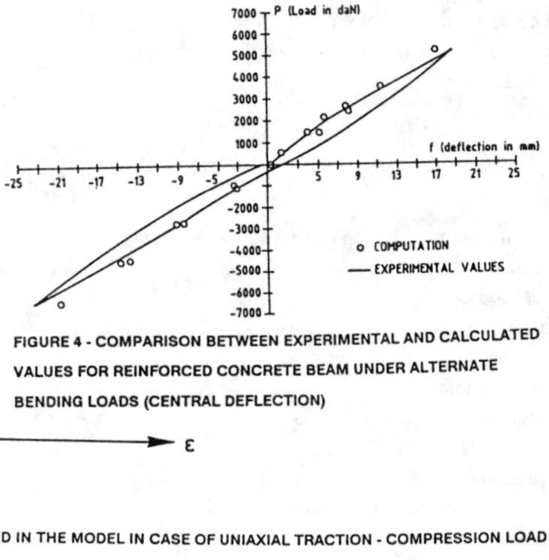

FIGURE 4 - COMPARISON BETWEEN EXPERIMENTAL AND CALCULATED VALUES FOR REINFORCED CONCRETE BEAM UNDER ALTERNATE BENDING LOADS (CENTRAL DEFLECTION)

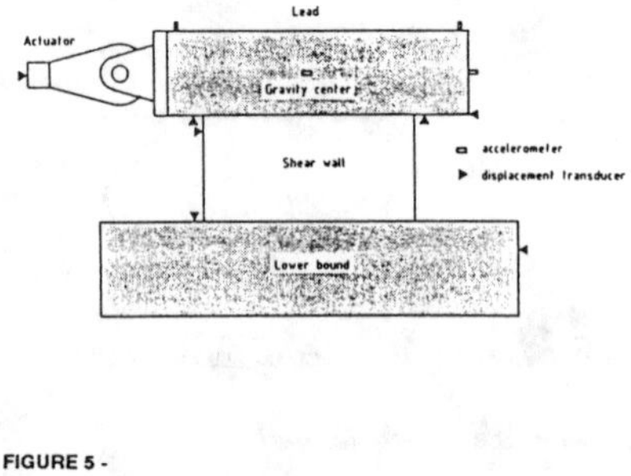

FIGURE 5 -

DYNAMICAL TESTS ON REINFORCED CONCRETE SHEAR WALL MEASUREMENTS AND TEST SETUP

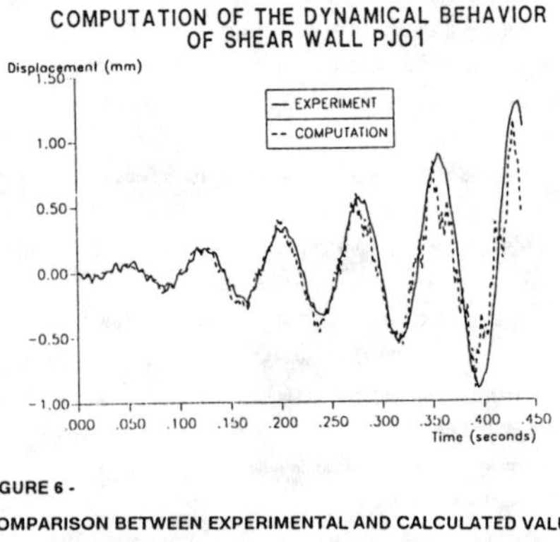

FIGURE 6 -

COMPARISON BETWEEN EXPERIMENTAL AND CALCULATED VALUES DISPLACEMENT AT THE TOP OF THE SHEAR WALL VERSUS TIME

A CONSTITUTIVE DAMAGE MODEL FOR COMPOSITE MATERIALS SUBJECTED TO HIGH-RATE LOADING*

J. A. NEMES and P. W. RANDLES
Mechanics of Materials Branch, Naval Research Laboratory, Washington, D. C., 20375, USA

ABSTRACT

A multidimensional constitutive model is developed to study the response of thick composite materials subjected to high rate loading. A vector description of damage is included in a thermodynamic formulation with internal variables for the derivation of constitutive equations for materials with transverse isotropy. Purely phenomenological evolution equations are given in rate form leading to overall rate-dependent behavior.

INTRODUCTION

The development of constitutive laws for thick composite materials subjected to high-rate loadings has received little attention to date, in part due to the much more widespread use of thin laminates and also due to the complexity of the problem. A continuum damage mechanics (CDM) model is developed for a simple class of materials and permissable damage modes. The development, based on the use of internal variables, follows the work in [1-3].

MODEL DEVELOPMENT

The material under consideration is a thick laminated composite material with no fiber reinforcement in the 1-direction (through-thickness) and a balanced arrangement of fibers in the 2,3-plane. The material will be assumed transversely isotropic, with regard to both its stiffnesses and evolving damage. The evolving modes of damage are restricted to specific types of matrix cracking, with a maximum strain criteria used for fiber breakage without evolution. The matrix damage is characterized by the vector $\underset{\sim}{V} = (V_1, V_2, V_3)$. The V_1 component of damage is intended to represent delamination cracking shown in Fig. 1a, and will be used to describe damage due to through-thickness stress waves. The remaining matrix damage, shown in Fig 1b, is characterized by a combination of V_2 and V_3 given by $V_* = \sqrt{V_2^2 + V_3^2}$ for transverse isotropic damage. Whereas the V_1 -damage evolves, perhaps catastrophically, to complete separation, the V_* -damage evolves to a spatially saturated state of damage, followed by either fiber breakage or further delamination damage.

Constitutive Equations

The development in [3] is used for the derivation of constituive equations. The assumption of infinitesimal damage is relaxed since evolution of damage to an end state is desired. Considering small strain and axisymmetry conditions leads to a free energy function

$$\rho\psi = f(I_1, I_2, \ldots, I_6) \tag{1}$$

* This work supported by DARPA Naval Technology Office.

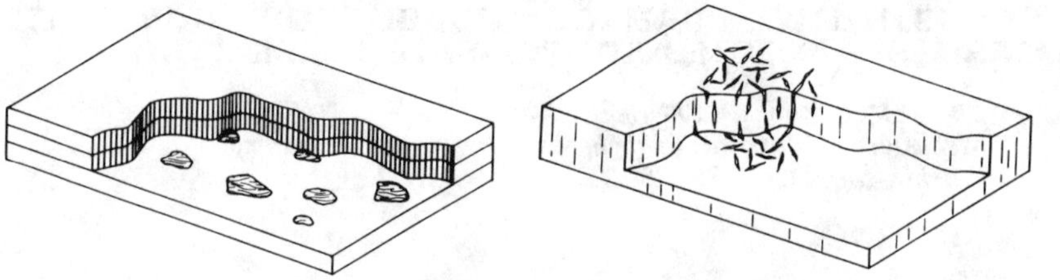

FIG. 1. Damage modes considered a) V_1-damage b) V_*-damage.

of the strain and damage invariants $I_1 = \varepsilon_{11}$, $I_2 = \varepsilon_{22} + \varepsilon_{33}$, $I_3 = \varepsilon_{22}^2 + \varepsilon_{33}^2$, $I_4 = \varepsilon_{12}^2$, $I_5 = V_1^2$, $I_6 = V_*^2$. The stress tensor components result from the derivatives of $\rho\Psi$, which is expressed for a transversely isotropic material under an axisymmetric deformation field as

$$\begin{pmatrix} \sigma_{11} \\ \sigma_{22} \\ \sigma_{33} \\ \sigma_{12} \end{pmatrix} = \begin{bmatrix} C_{11} & C_{12} & C_{12} & 0 \\ C_{12} & C_{22} & C_{23} & 0 \\ C_{12} & C_{23} & C_{22} & 0 \\ 0 & 0 & 0 & C_{44} \end{bmatrix} \begin{pmatrix} \varepsilon_{11} \\ \varepsilon_{22} \\ \varepsilon_{33} \\ \varepsilon_{12} \end{pmatrix}, \tag{2}$$

where

$$\begin{aligned} C_{11} &= (1 - \nu_{23}) E_{11} / \left(1 - \nu_{23} - 2\nu_{12}^2 E_{22}/E_{11}\right) \\ C_{12} &= \nu_{12} E_{22} / \left(1 - \nu_{23} - 2\nu_{12}^2 E_{22}/E_{11}\right) \\ C_{22} &= \frac{1}{1 + \nu_{23}} \left(1 - \nu_{12}^2 E_{22}/E_{11}\right) E_{22} / \left(1 - \nu_{23} - 2\nu_{12}^2 E_{22}/E_{11}\right) \\ C_{23} &= \frac{1}{1 + \nu_{23}} \left(\nu_{23} + \nu_{12}^2 E_{22}/E_{11}\right) E_{22} / \left(1 - \nu_{23} - 2\nu_{12}^2 E_{22}/E_{11}\right) \\ C_{44} &= 2G_{12} \quad . \end{aligned} \tag{3}$$

The damage dependence is then prescribed on the elastic parameters in (3) based on experimental evidence. For illustrative purposes the following simple damage dependences are assumed

$$\begin{aligned} &E_{11} = \left(1 - V_1^2\right) \mathring{E}_{11} \quad, E_{22} = \left(1 - \alpha_1 V_*^2\right) \mathring{E}_{22} \quad, \quad \nu_{23} = \left(1 - \alpha_4 V_*^2\right) \mathring{\nu}_{23} \\ &G_{12} = \left(1 - V_1^2\right)\left(1 - \alpha_2 V_*^2\right) \mathring{G}_{12} \quad, \quad \nu_{12} = \left(1 - V_1^2\right)\left(1 - \alpha_3 V_*^2\right) \mathring{\nu}_{12} \quad, \end{aligned} \tag{4}$$

where the superscript '0' denotes virgin properties and the parameters α_i, $i = 1,..4$, $0 < \alpha_i < 1$, control the extent of softening.

Evolution Equations

Rate dependency is introduced into the CDM model through the damage evolution equations, which are taken as a function of the current state of damage and the state of stress above a scalar threshold function F. Both the V_1 and V_* types of damage are assumed here to be governed by a threshold of the Mohr-Coulomb type

$$F = \sqrt{1 + (\tau/f_3)^2} - (f_1 - \sigma)/f_2 \quad , \tag{5}$$

where the parameters f_1, f_2, f_3 as shown in Fig. 2 are related to the tension growth threshold, σ_G, shear growth threshold, τ_G, and friction tangent, ϕ_G, which are also a function of damage. The evolution of V_1 damage is postulated to depend only on the stress components $\sigma = \sigma_{11}$ and $\tau = \sigma_{12}$ with threshold quantities defined as

$$\sigma_G = \left(1 - V_1^2\right)\sigma_{G0}, \; \tau_G = \left(1 - V_1^2\right)\tau_{G0}, \; \phi_G = \phi_{G0} + V_1^2\left(\phi_{G1} - \phi_{G0}\right) \qquad (6)$$

where the subscript '0' denotes virgin threshold properties.

The evolution or rate equation for damage is taken to be a function of the current damage and d_1, shown in Fig. 2 as the distance in σ,τ space to the surface F. The specific form considered is given by

$$\dot{V}_1 = (d_1/\sigma_{G0})^{n_1} / \left[\eta_1\left(1 - V_1^2\right)\right] , \qquad (7)$$

where n_1 and η_1 are material parameters. The $1/(1-V_1^2)$ term causes acceleration of damage as $V_1 \rightarrow 1$,simulating a brittle failure. The evolution of V_* damage follows a similar development with the exception that the damage is taken to saturate rather than to increase catastrophically as in (7). The following summarize the threshold and damage evolution equations

$$\sigma = \frac{1}{2}\left(\sigma_{22} + \sigma_{33}\right) , \quad \tau = \sqrt{\sigma_{12}^2 + \frac{1}{4}\left(\sigma_{22} - \sigma_{33}\right)^2} ,$$

$$\bar{\sigma}_G = \bar{\sigma}_{G0}/\left(1 - V_*^2\right), \; \bar{\tau}_G = \bar{\tau}_{G0}/\left(1 - V_*^2\right) \qquad (8)$$

$$\bar{\phi}_G = \bar{\phi}_{G0} + V_*^2\left(\bar{\phi}_{G1} - \bar{\phi}_{G0}\right) , \; \dot{V}_* = \left(d_*/\bar{\sigma}_{G0}\right)^{n_*} / \eta_* .$$

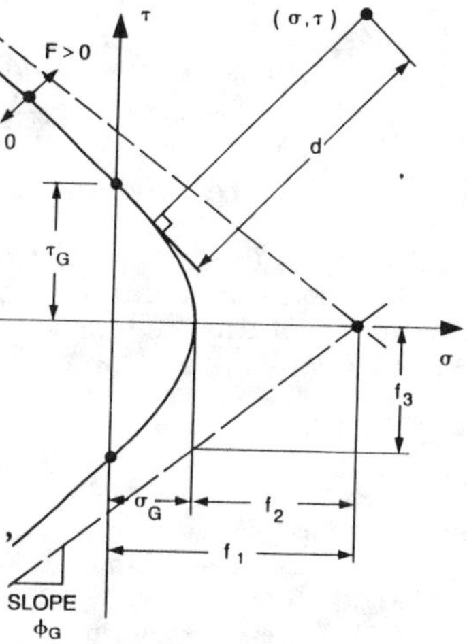

FIG. 2 Threshold surface.

Fiber breakage is the third mode of damage. Consistent with [4], a rate-independent maximum strain criteria is used.

MATERIAL RESPONSE

To illustrate the predictive capability of the model for high rate loading, calculations are performed for two loading cases. The first is for uniaxial stress in the 1-direction under specified deformation rates and the second is for radial deformation with $\sigma_{11} = 0$. For these specialized deformations a reduced set of material properties are required, which are taken here as $\nu_{12} = .04$, $E_{11} = 10000$ MPa, $\nu_{23} = .30$, $E_{22} = 50000$ MPa, $n_1 = 2$, $\eta_1 = .0001$, $\sigma_{G0} = 50$ MPa, $n_* = 1$, $\eta_* = .01$, $\alpha_1 = .75$, $\alpha_2 = .75$, $\alpha_3 = .95$, $\bar{\sigma}_{G0} = 30$ MPa and $\varepsilon_{ult} = .02$. The stress - strain response and damage evolution for the first case is shown in Fig. 3. At the lowest rate the response is indicative of a perfectly brittle material, with essentially instantaneous damage accumulation. At the higher rates, the time required for damage accumulation permits higher stress levels to be reached with a softening behavior. In the second case where loading is in the plane of the fibers, three different types of behavior result from the calculations at different rates, shown in Fig. 4. At the lowest rate the damage reaches maturity quickly, resulting in a bilinear response. At the intermediate rate, the damage

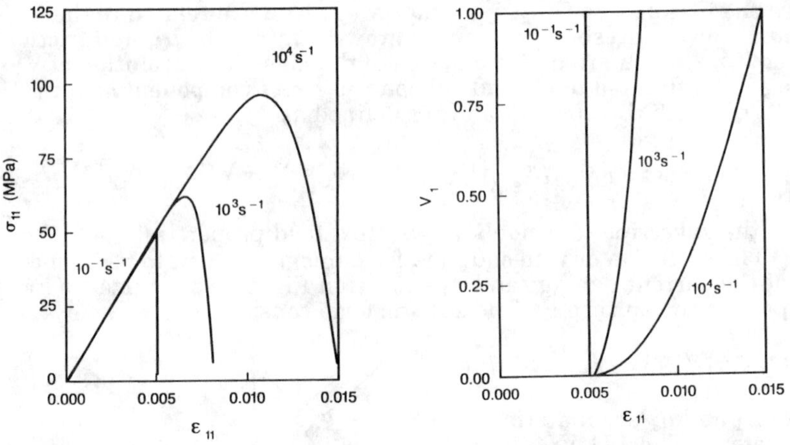

FIG. 3. Thru-thickness response a) uniaxial stress b) damage.

matures gradually, permitting some overstress to occur, followed by gradual softening. Finally at the highest rate, the time required for damage evolution is longer than that required to reach the critical strain, resulting in low amounts of damage.

REFERENCES

1. L. Davison and A. L. Stevens, "Thermomechanical constitution of spalling elastic bodies", J. Appl. Phys. 44(2), 668-674 (Feb 1973).
2. D. Krajcinovic and G.U. Foneska, "The continuous damage theory of brittle materials", J. Appl. Mech.48, 809-815 (Dec 1981).
3. R.Talreja, "A continuum mechanics characterization of damage in composite materials", Proc. R. Soc. A399 195-216,(1985).
4. J. Harding and L.M. Welsh,"A tensile testing technique for fibre-reinforced composites at impact rates of strain", J. Mat. Sci.,18, 1810-1826,(1983).

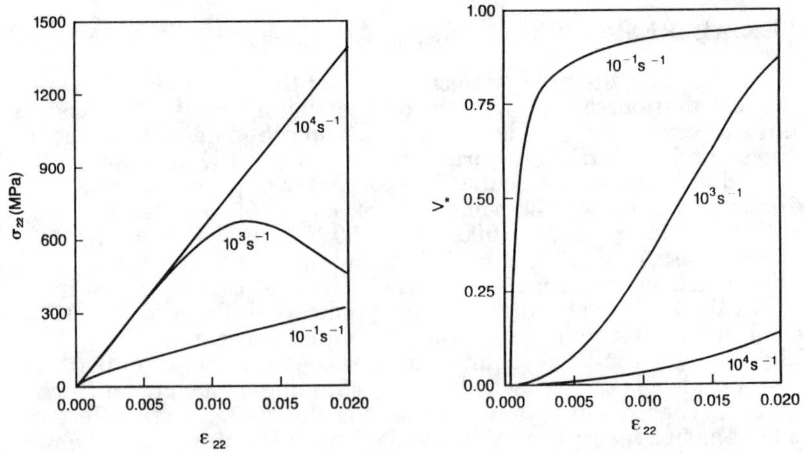

FIG. 4 In-plane response a) biaxial stress b) damage.

CONSTITUTIVE EQUATIONS FOR POLYCRYSTALLIC VOID-CONTAINING MATERIALS

J. NOVÁK
Nuclear Research Institute, Řež, Czechoslovakia

ABSTRACT

Constitutive equations for polycrystallic void-containing materials, permitting analysis of non-proportional loading, are proposed; their relation to Gurson-Tvergaard equations are defined. Properties of the equations are shown by comparison of theoretical prediction of critical strains for shear band initiation and of fracture strains.

INTRODUCTION

The effect of voids in material without strain rate sensitivity may be described by constitutive equations in the form of a relation between Jaumann derivative of stress tensor components and strain rate. Constitutive equations used most often are those suggested by Gurson [1] and modified by Tvergaard [2] by introducing 2 parameters denoted by q_1 and q_2. The matrix material behaviour is modeled by flow theory with isotropic hardening. Since modeling of crystallographic slip consequences for materials without porosity was successfully used for prediction of fracture, it is desirable to try for a synthesis of both principally different microphysical phenomena into one constitutive model.

CONSTITUTIVE EQUATIONS

The constitutive equations are constructed so that under slightly non-proportional loading the matrix behaves as a polycrystallic material: its properties are described by deformation theory. In conditions of proportional loading the plastic part of constitutive equations must coincide with commonly used flow theory equations. For zero void volume fraction it must be identical with deformation theory. The resulting equations of hypoelastic version are of the form

$$D_{ij}=D^{P}_{ij}+D^{K}_{ij}+D^{E}_{ij}=M_{ijkl}\overset{\nabla}{\sigma}_{kl}, \qquad (1)$$

where D^{P}_{ij} corresponds to Gurson-Tvergaard equations. The additive member D^{K}_{ij} may be written in the form (eq. 2)

$$D^K_{ij}=(3\sigma_M)/(2h_s\sigma_e)[\delta_{ik}\delta_{jl}-\delta_{ij}\delta_{kl}/3-(3s_{ij}s_{kl})/(2\sigma_e^2)]\overset{\nabla}{\sigma}_{kl}.$$

Factor σ_M/σ_e on the RHS corresponds to porosity effect on D^K_{ij}; $h_s=\sigma_M/\epsilon_M$ is secant modulus of matrix material. Elastic member D^E_{ij} is of the form

$$D^E_{ij}=\sigma_M/(E\sigma_e)[(1+\nu)\delta_{ik}\delta_{jl}-\nu\delta_{ij}\delta_{kl}]\overset{\nabla}{\sigma}_{kl}. \tag{3}$$

In contrast to other studies, we incorporated the effect of porosity in D^E_{ij}.

Using evolution equations for the flow stress of the matrix material σ_M and for the void volume fraction f and inverting eq. (1) we receive for hypoelastic version the relation

$$\overset{\nabla}{\sigma}_{ij}=L^{hypo}_{ijkl}D_{kl}, \tag{4}$$

$$L^{hypo}_{ijkl}=(E\sigma_e)/(A\sigma_M)\left\{\delta_{ik}\delta_{jl}+\frac{E/(2h_s)+\nu}{1-2\nu}\delta_{ij}\delta_{kl}-\right.$$

$$-\frac{[(3s_{ij})/(2\sigma_M)+AB\delta_{ij}][(3s_{kl})/(2\sigma_M)+AB\delta_{kl}]}{AD+(3\sigma_e^2)/(2\sigma_M^2)}+$$

$$\left.+\frac{A[(Ds_{ij})/\sigma_e-(B\sigma_e\delta_{ij})/\sigma_M][(Ds_{kl})/\sigma_e-(B\sigma_e\delta_{kl})/\sigma_M]}{(2h_s)/(3E)[(2(1+\nu)D)/3+\sigma_e^2/\sigma_M^2][AD+(3\sigma_e^2)/(2\sigma_M^2)]}\right\}$$

$$A=(3E)/(2h_s)+1+\nu,\ B=\alpha/(1-2\nu),\ D=(H\sigma_M)/(E\sigma_e)+(3\alpha^2)/(1-2\nu)$$

$$H=(h\delta^2)/(1-q_1f)-3\sigma_M\alpha\gamma(1-f),\ h=d\sigma_M/d\epsilon_M$$

$$\alpha=0.5fq_1q_2\sinh(0.5q_2\sigma_{kk}\sigma_M^{-1})$$

$$\gamma=q_1\cosh(0.5q_2\sigma_{kk}\sigma_M^{-1})$$

$$\delta=\sigma_e^2/\sigma_M^2+(\alpha\sigma_{kk})/\sigma_M.$$

In coordinate system with axes parallel to the stress tensor principal axes the equations (4) are simplified and it is not difficult to formulate hyperelastic version of equations - hyperelastic version is more suitable for great deformations [3].

Remark: From the analysis of model properties it followed the necessity to modify the evolution equation for σ_M, represented by the equivalent plastic work expression

$$(1-f)\sigma_M\dot{\epsilon}^P_M=\sigma_{ij}D^P_{ij}. \tag{5}$$

This expression is commonly used even if it represents only a convenient assumption . We suggest to modify this evolution equation in the following manner:

$$(1-q_1f)\sigma_M\dot{\epsilon}^P_M=\sigma_{ij}D^P_{ij}. \tag{6}$$

Eq. (6) yields an intuitively acceptable result: for

σ_{kk}=0 the critical deformation for shear band initiation does not depend on void volume fraction. Factor q_1 in eq. (6) has as a consequence a change in certain relations of Gurson-Tvergaard model - in "consistency condition" and in expression for H (for details see [3]).

PROPERTIES OF CONSTITUTIVE EQUATIONS

In accordance with the fundamental idea of Rice [4] the constitutive equations were used in bifurcation analysis of shear band initiation for prediction of shear fracture [3]. The results of calculations were compared with published experimental data obtained from tests with porous iron and structural steels for different stress triaxialities.

Fracture strain prediction for porous iron under conditions of axisymmetric deformation

Experiments of study [5] form a basis for comparing predicted critical strains with fracture strains for materials with different values of void volume fraction in initial state, in conditions of simple tension (i.e. $\sigma_{22}=\sigma_{33}=0$). Matrix material parameters are described in detail in [3]; the relation $\sigma_M=\sigma_M(\varepsilon^P{}_M)$ is linear for $\varepsilon^P{}_M>0.32$. In tab. I critical strain for shear band initiation ε_c, critical void volume fraction ϕ_c and shear band orientation ϕ are given against the initial value of void volume fraction f_O. Experimentally obtained fracture strains [5] are included in tab. I.

TABLE I. Critical parameters for strain localization and fracture strains.

f_O	ε_c	f_c	ϕ	ε_f
0.003	1.26	0.0107	1.20-1.22	1.28
0.015	1.16	0.0462	1.16-1.19	1.10
0.037	1.01	0.0943	1.10-1.14	0.75
0.062	0.883	0.135	1.05-1.09	0.50
0.111	0.684	0.193	0.98-0.99	0.25

For small values of f_O, i.e. in conditions when the strain localization is controled predominantly by a cristallographic slip, the agreement between ε_c and ε_f is satisfactory. For higher values of f_O the results confirm the significant influence of porosity inhomogeneity.

Influence of stress triaxiality on fracture strain

Suggested constitutive equations in combination with bifurcation analysis of strain localization predict in a satisfying manner the effect of stress triaxiality on the fracture strain [3]. Tab. II contains values of factors of diminishing the fracture strain at increasing value $C=\sigma_{kk}/(3\sigma_e)$ from 0.8 to 1.2 that are defined as follows: Factor according to the experimentally determined relation $\varepsilon_f = k.\exp(-1.5C)$ valid for structural steels and factor of diminishing the critical strain for shear band initiation in porous iron for different values of f_0.

TABLE II. Influence of $C=\sigma_{kk}/(3\sigma_e)$ on fracture strain

method of evaluation	$\varepsilon_{C=0.8}/\varepsilon_{C=1.2}$
empirical relation	1.82
bifurcation analysis, $f_0=0.015$	1.96
bifurcation analysis, $f_0=0.037$	1.88
bifurcation analysis, $f_0=0.062$	1.77

Effect of strain geometry on fracture strain

Constitutive equations used in strain localization bifurcation analysis predict correctly the strain geometry effect on effective deviatoric fracture strain. At sufficiently high values of stress triaxiality this critical strain is independent of strain geometry [3].

REFERENCES

1. A.L. Gurson, J.Engr.Mater.Technol. (Trans. ASME,H) 99, 2 (1977).
2. V. Tvergaard, Int.J.Fracture 17, 389 (1981).
3. J. Novák, "Constitutive equations for polycrystallic void-containing materials and shear fracture initiation", Rep. ÚJV in press.
4. J.R. Rice, "The localization of plastic deformation", in Theoretical and Applied Mechanics, ed. by Koiter (North Holland, 1976).
5. W.A. Spitzig, R.E. Smelser, D. Richmond, Acta metall. 36,1201 (1988).

AN ELASTIC-BRITTLE-DUCTILE CONSTITUTIVE MODEL FOR ROCK

G.I. OFOEGBU and J.H. CURRAN
Department of Civil Engineering, University of Toronto
Toronto, Ontario, Canada M5S 1A4

ABSTRACT

Load distribution in rock is described in terms of a transformed stress tensor, using the concept of damage to account for the fact that the imposed load is supported by a diminished fraction of the solid volume. The elastic deformability is described using a combination of the damage variable and the elastic stiffness of uncracked rock. The inelastic deformability is described in terms of the dilation ratio, using the flow rule (nonassociative) of plasticity theory.

The mechanical behaviour of *intact* rock is influenced by the occurence of microcracks, which may open or close, and along which sliding may occur, during an incremental change of load. Hence, three modes of deformation may occur:

(1) Elastic deformation of the uncracked rock matter.

(2) Brittle deformation, i.e., opening (including extension) and closing of microcracks. Because cracks reduce the effective load-bearing area of an arbitrary surface, this class of deformations affect the load-bearing capacity and elastic stiffness.

(3) Ductile deformation, i.e., sliding on interfaces. This deformation mode may affect the load-bearing capacity, because of the smoothing of interfaces, but it has no effect on the elastic stiffness.

Load distribution in rock can be described in terms of either the *apparent* stress σ_{ij}, i.e., load per unit total area, or the *latent* stress σ_{ij}^{t}, which accounts for the effect of cracks on the load-bearing area[1–3]. Latent stress is interpreted as the average stress in the uncracked rock matter[3,4]. Hence, elastic strain is related to latent stress through the elastic stiffness of uncracked rock, which is thus distinguished from the apparent elastic stiffness.

The concept of damage is applied to describe the relations (i) apparent elastic stiffness vs elastic stiffness of uncracked rock, and (ii) apparent vs latent stresses. The strength and deformability of rock is described in terms of the latent stresses.

CONCEPT OF DAMAGE

A *reduced tensor* description of damage, based on a model suggested in [1] is adopted. Damage is described as a tensor r_{ij}, which is described in terms of the two invariants, r_s (spherical damage) and r_c (deviatoric damage). The scalars r_s and r_c represent the effects of r_{ij} on the hydrostatic and deviatoric intensities, respectively, of the stress tensor. Stress-strain-damage relations are expressed in terms of r_s and r_c, so that the individual components of r_{ij} need not be known. It can be shown that[1,3,4]

$$G^* = (1 - r_c)G \tag{1}$$

$$K^* = (1 - r_s)K \tag{2}$$

where G and K are the shear and bulk moduli of uncracked rock, and G^* and K^* are the apparent shear and bulk moduli for *intact* rock. If damage is caused mainly by cracks, it can be assumed that $r_s = 0$, and hence $K^* = K$. The apparent vs latent stress relations for such situations are[1,3,4]

$$\sigma_{ij} = (1 - r_c)\sigma_{ij}^t + \frac{r_c}{3}\delta_{ij}\sigma_{kk} \tag{3}$$

The values of r_c for laboratory specimens can be determined as follows[3,4]:
(1) Evaluate G^* using the unloading or reloading curves of a cyclic compression test under constant confining pressure σ_{con} (e.g., Fig. 1). G^* initially increases to a maximum, G^*_m, and thereafter decreases towards a steady minimum. The value of G^*_m varies with σ_{con}.
(2) Evaluate G^*_m at different σ_{con} values; it increases with σ_{con}, approaching G asymptotically as σ_{con} becomes large.
(3) Using this value of G, convert each G^* history (step 1) to an r_c history, via eq. (1). As shown in [3,4], the r_c history for a specimen is a function of σ_{con},

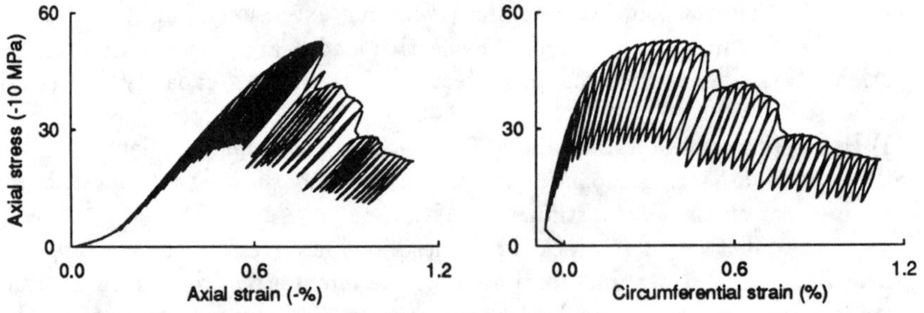

Figure 1: Stress-Strain curves for Crieghton norite at $\sigma_{con} = 30$ MPa

which can be described in terms of three parameters, r_{co} (the value of r_c under zero deviatoric stress), r_{cmin} (minimum value of r_c), and r_{cmax} (its maximum value).

The growth of damage is described as follows [3,4]:

$$\Delta r_c = 0 \quad \text{if} \quad \Delta\kappa \leq 0 \quad \text{or} \quad \kappa < \kappa_{pm} \tag{4}$$

$$\Delta r_c = g(\kappa,\ \sigma_{con}) \quad \text{if} \quad \kappa \geq \kappa_{pm} \quad \text{and} \quad \Delta\kappa > 0 \tag{5}$$

where $\kappa = -\sqrt{J_2^t}/(I_1 - 3T_o)$ is the loading parameter (tension positive), J_2^t is the second invariant of the deviatoric latent stress tensor, I_1 is the first invariant of the latent stress tensor, T_o is the tensile strength of uncracked rock (evaluation of T_o is explained in [4]), and κ_{pm} is the maximum value of κ attained in a previous loading. The function $g(\kappa,\ \sigma_{con})$ is evaluated by partial differentiation of the damage function $r_c(\kappa,\ \sigma_{con})$ obtained from laboratory data (Fig. 2).

ROCK STRENGTH: SHEAR RESISTANCE

It is argued in [4] that the driving forces which cause inelastic rock deformation are related to tensile stresses induced at the tips, and shear stresses induced on the surfaces, of microcracks; and that the magnitude and distribution of both types of stresses can be related to the loading parameter κ. Hence, rock strength is described in terms of this loading parameter, and the permissible values of κ are referred to as *shear resistance*, denoted a. It is shown in [3,4] that three characteristic values of a can be identified on a complete stress-strain curve: (1) The *initial yield* shear resistance, a_o, represents the state at which the effect of crack growth begins to exceed that of crack closure. Both material compliance and damage begin to increase at this state (i.e., $r_c = r_{cmin}$ at $\kappa = a_o$), for a specimen deformed under constant confining pressure. (2) The *maximum*

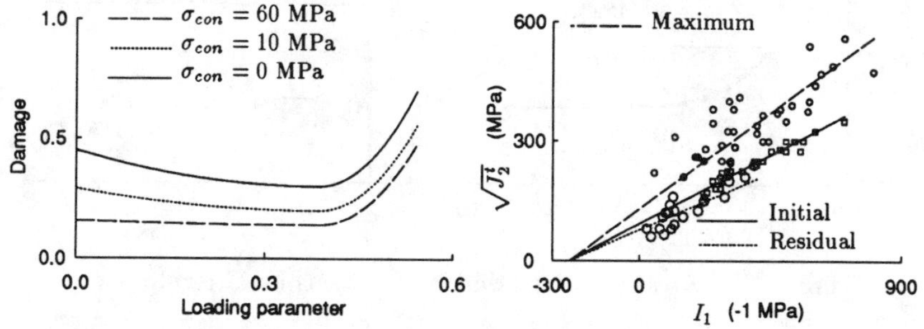

Figure 2: Damage function and yield surfaces for Crieghton norite

shear resistance, a_{max} is attained just prior to the coalescence of hitherto isolated cracks into one or more through-going interfaces. Inelastic deformation between initial yielding and this state consists of the opening and sliding of isolated cracks; subsequently, it consists of sliding on the through-going interfaces. Hence, increase in damage ceases once this state is attained (i.e., $r_c = r_{cmax}$ at $\kappa = a_{max}$), and, because of the smoothing of the sliding interfaces, the value of a decreases towards (3) a_{res}, the *residual* shear resistance. Therefore, the strength (or yield) function for *intact* rock is given by

$$F = \sqrt{J_2^t} + a(I_1 - 3T_o) = 0 \quad \text{where} \quad a = a(\Gamma^N,\ \sigma_{con}) \tag{6}$$

Γ^N being the square root of the second invariant of the deviatoric inelastic strain tensor[4]. Fig. 2 shows plots of $\sqrt{J_2^t}$ vs I_1 for Creighton norite. Three straight lines through $I_1 = 3T_o = 240$ MPa, representing a_o, a_{max} and a_{res}, are also shown. The shear resistance function $a(\Gamma^N, \sigma_{con})$ for the same rock is plotted in Fig. 3

INELASTIC STRAIN-STRESS RELATIONS

It is assumed that a function H exists, such that the inelastic strain increment Δe_{ij}^N is given by

$$\Delta e_{ij}^N = \Lambda \partial H / \partial \sigma_{ij}^t \tag{7}$$

where Λ is a positive scalar. The function H is assumed to have the same form as, but not equivalent to, the yield function F. That is,

$$H = \sqrt{J_2^t} + \beta I_1 \tag{8}$$

where β is a material property evaluated using laboratory inelastic strain data. It can be shown that $6\beta = \Delta\varepsilon^N / \Delta\Gamma^N$, where $\Delta\varepsilon^N$ is the inelastic volumetric strain increment. Hence β is referred to as *dilation ratio.* It can be described as a function $\beta(\Gamma^N, \sigma_{con})$, an example of which is shown in Fig. 3.

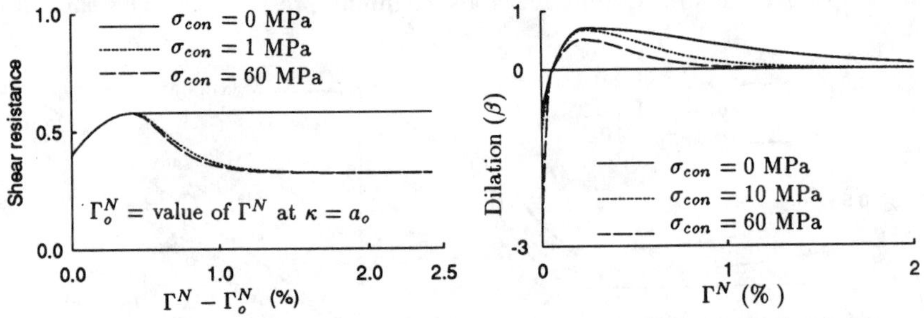

Figure 3: Shear resistance and dilation functions for Crieghton norite

REFERENCES

1. G. Frantziskonis and C.S. Desai, "Constitutive model with strain softening", Int. J. Solids Structures 23(6), 733–750 (1987).
2. J. Hult in Continuum Damage Mechanics, edited by Krajcinovic and Lemaitre (Springer-Verlag, New York 1987).
3. G.I. Ofoegbu and J.H. Curran, "Description and measurement of damage for *intact* rock", Proc. 43 Canadian Geotech. Conference, Quebec (October 1990).
4. G.I. Ofoegbu and J.H. Curran, "Yielding and damage of *intact* rock", Can. Geotech. J. (submitted).

THE EFFECTS OF SURFACE DAMAGE ON THE BEHAVIOR OF DILATANT MATERIAL DISCONTINUITIES

M. E. PLESHA*, B. C. HAIMSON#, X. QIU* and X. HUANG#
*Department of Engineering Mechanics
#Geological Engr. Program, Dept. of Materials Science
University of Wisconsin, Madison, Wisconsin 53706

ABSTRACT

An experimental and analytical investigation of surface damage mechanisms in dilatant rock joints is described. The experimental investigation was conducted on artificial joints molded in hydrostone and on natural joints in dolomite. A new wear theory based on energy was developed and incorporated into an interface constitutive model which was used to numerically simulate the laboratory tests.

INTRODUCTION

Surface damage plays an important role in the mechanical and hydraulic behavior of material discontinuities, particularly for discontinuities that are initially closely mated and display dilatancy during sliding. Dilatancy is normal deformation of an interface that accompanies sliding displacements due to asperities of one surface riding up on those of the other surface. Dilatancy can help improve the sliding resistance of an interface by generating additional compressive stresses. For rough crack surfaces that terminate at a crack tip, dilatancy can also serve to effectively toughen the crack when the mode of fracture is mode II or mode III. If the asperities had infinite strength, dilatancy would be fairly easy to characterize using a purely kinematic relationship. However, laboratory and in situ field tests of various dilatant contact problems involving rock joints and crack surfaces in concrete show that surface damage can be significant and can have an important effect on load-deformation behavior. In this presentation, we review the results of an extensive experimental and analytical investigation of the mechanics of rock joint behavior with emphasis on dilatancy, surface damage and the formulation of effective constitutive models that are amenable to numerical implementation. More thorough treatments of the topics discussed here are contained in references [1-5].

LABORATORY INVESTIGATION

We have conducted approximately 100 cyclic direct shear tests on joints molded of a model material that simulates soft natural rock (hydrostone). These samples had saw-tooth shaped asperities with orientation of 3,6 or 10 degrees with respect to the mean plane of the interface and the top and bottom surfaces of each sample were initially very closely matched. The motivation for using specimens of this type is that they permit a large number of tests, each using different load-displacement conditions, to be performed on samples which for our purposes could be considered identical. To help corroborate our observations and conclusions based on model material sample tests, we also conducted tests on natural joints in dolomite. In support of this part of the research, we designed and fabricated a precision servo-controlled direct shear testing machine. This machine is computer-driven using data acquisition and control software, has loading capacity in both the normal and shear directions of 500 kN (100 kip), accommodates joint surfaces up to 150 mm x 200 mm (6 in x 8 in) and permits a shear displacement of up to 10 cm (4 in).

Most of our tests employed a prescribed cyclic shear displacement history under a constant normal stress in the range of 0.5-10 MPa. Inspection of the resulting shear stress-normal displacement history and normal displacement-shear displacement history in conjunction with visual observations of the sample yield the following conclusions:

1. Normal deformations due to dilatancy decreased with cycling, indicating that significant asperity damage was occurring and that the interface tended to become more planar as a test progressed.
2. The rate at which the asperity surfaces decayed increased with increasing compressive stress and decreased as the number of cycles in a test progressed.
3. Under low compressive stress levels (less than 2 MPa) the mode of damage was "wear" in which destruction of the asperities was rather gradual. For high compressive stress levels (greater than about 5 MPa), the mode of damage was "asperity shearing" in which the asperity surfaces were very rapidly destroyed during the first few cycles of shearing. For moderate compressive stress levels, the mode of damage typically began as wear for several cycles, then switched to asperity shearing for several cycles, followed by wear at a very slow rate for the duration of the test.

These observations have helped guide our modeling efforts as described in the next section. Especially, we have concentrated on obtaining an understanding of and models for wear. We do not yet have an adequate constitutive model developed for asperity shearing, nor have we developed a theory for the transition between damage mechanisms observed for moderate stress levels although it appears that fatigue plays a role.

ANALYTICAL INVESTIGATION

Rock joint behavior is irreversible and history-dependent. In our approach to constitutive modeling, we employ macrostructural assumptions regarding joint deformations and stresses, and assume that the contact surface is macroscopically smooth so that normal and tangent directions can be defined; these assumptions then yield a general constitutive framework. Microstructural considerations allow us to postulate and investigate specific surface idealizations and models for rock joint frictional behavior and damage.

To help obtain realistic and effective microstructural damage models, we developed a new theory for wear based on the premise that wear is an energy related phenomena [5]. In other words, the removal of a certain volume of surface material requires a specific energy that is supplied in the form of work dissipated during sliding. We also include the possibility that damaged asperity debris can be adhesively reattached to the contacting surfaces and this also requires a specific energy per unit volume of reattached material. Through Fourier analysis of surface shape and surface shape changes due to wear, it is possible to develop a quantitative wear theory for rather general surface geometries. Under certain assumptions, it is possible to obtain the exponential wear model originally postulated in [6], however, it is clear that from this theory that the asperity degradation parameter should be a function compressive stress level; this conclusion was reached experimentally in reference [7] and is also supported by our laboratory work [1,3].

NUMERICAL SIMULATIONS OF LABORATORY TESTS

Shown in the inset of the figure is a schematic illustration of a direct shear test with constrained dilatancy. The specimen was molded of hydrostone with a joint area of 150×150 mm^2 and saw-tooth shaped asperities of 10° orientation and 2.2 mm height. The specimen was initially uncompressed and then the upper block of material was sheared by 25 mm. By servo-control, an appropriate normal stress history was applied so that throughout the test, there was no normal deformation. The resulting stress histories are shown as solid lines in the figure. This test is significant because it is characteristic of the type of behavior that joints display under in-situ conditions at depth such as in block pull-out from the wall or roof of a tunnel or cavern. The laboratory test was simulated using the constitutive law described herein; material parameters were obtained from a direct shear test under constant compressive stress for a different hydrostone sample. The results of the numerical simulation are shown by the dashed lines in the figure in which it is observed that for this rather challenging test, the agreement is good. The results of other simulations of laboratory tests are given in references [2,14].

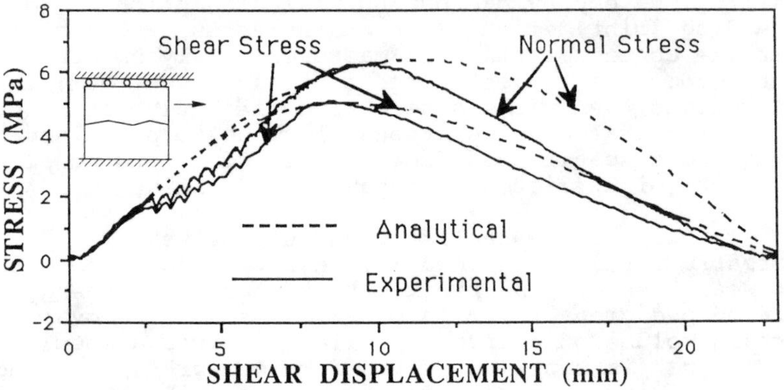

ACKNOWLEDGEMENTS

The support of the U.S. Army Research Office, Geosciences Division, through grant No. DAAL03-86-K-0134 is gratefully acknowledged.

REFERENCES

1. X. Huang, "A laboratory study of the mechanical behavior of rock joints, with particular interest to dilatancy and asperity surface damage mechanism", Ph.D. thesis, University of Wisconsin (1990).
2. X. Qiu, "Modeling Mechanical Behavior of Rock Discontinuities", Ph.D. thesis, University of Wisconsin (1990).
3. X. Huang, B.C. Haimson, X. Qiu and M.E. Plesha, "An investigation of the mechanics of rock joints-part I, laboratory investigation", in preparation.
4. X. Qiu, M.E. Plesha, X. Huang and B.C. Haimson, "An investigation of the mechanics of rock joints-part II, analytical investigation", in preparation.
5. X. Qiu and M.E. Plesha, "A theory for dry wear based on energy", J.Tribology, submitted for publication.
6. M.E. Plesha, "A constitutive model for rock discontinuities with dilatancy and surface degradation", Int'l. J. Num. Anal. Meth. in Geomech., 11, 345-362 (1987).
7. R.W. Hutson and C.H. Dowding, "Joint asperity degradation during cyclic shear", Int'l. J. Rock Mech. Min. Sci. Geomech. Abstr., 27, 109-119 (1990).

MODELING OF CONCRETE BEHAVIOR USING AN ANISOTROPIC DAMAGE APPROACH

Salah RAMTANI and Yves BERTHAUD
Laboratoire de Mécanique et Technologie
ENS Cachan / CNRS / Université Paris 6
61 avenue du Président Wilson
94235 CACHAN CEDEX
and GRECO "Rhéologie des Géomatériaux"

ABSTRACT

An attempt of modeling the behavior of concrete through a damage approach is presented. The phenomena which are studied are the elastic anisotropy and the evolution of permanent strains induced by damage and the consequences of the closure of the cracks, i.e. damage. Two damage variables are introduced in the free energy which is expanded in a Taylor series. The crack closure is described via a partition of the elastic strain tensor. Some experimental results are compared to the predictions given by our model.

INTRODUCTION

The modeling of the behavior of materials such as concrete has been studied using various theoretical approaches : homogenization [1], models with loading surfaces [2,3], damage models [4,5] and more recently micromechanics taking into account the interactions [6,7]. In the latter case, these models are not suitable for numerical implementation because of their complexity are in the other cases the damage is often considered to be isotropic. Microscopical observations reveal that microcracks and micropores exist even in the virgin material due to the chemical reaction of hydration. Under the mechanical loading the pores collapse and the crack propagate around or through the aggregates and inside the matrix. These two different defects are modeled by a set of two damage variables denoted $\mathbb{d}$ and δ. The first one is a second rank order tensor and the second one is a scalar. These variables are introduced in the free energy of the damaged material which is expanded in a Taylor series [8]. Each term of the expansion is related to a particular phenomenon. The description of the crack closure is achieved through a generalization of the crack closure parameter [9] and a partition of the elastic strain tensor. In the first part of this paper the damage model written within the framework of continuum damage mechanics is presented. Then, some predicted multiaxial loading paths are compared to experimental responses obtained in our Laboratory.

MODELLING OF THE BEHAVIOR OF CONCRETE

For an isothermal process the Helmoltz energy is written as a Taylor series :

$$\rho\psi(\varepsilon, \mathbb{d}, \delta) = \rho\psi(0, \mathbb{d}, \delta) + \rho\psi'(0, \mathbb{d}, \delta) : \varepsilon + \frac{1}{2}\rho\psi''(0, \mathbb{d}, \delta) : \varepsilon : \varepsilon \quad (1)$$

$$\rho\psi(\varepsilon, \mathbb{d}, \delta) = \rho\psi^{d} + \rho\psi^{rd} + \rho\psi^{ed} \quad (2)$$

where ρ is the mass density supposed to be constant and ψ' (resp.ψ'') denote the first (resp. second) derivative of ψ in function of ε which is the strain tensor. The quantity $\varepsilon : \varepsilon$ is equal to $Tr[\varepsilon \otimes \varepsilon]$ where $\otimes$ is the tensorial product. The first term which is equivalent to an energy has no influence in the state law and appears only in the dissipation. The third term $\rho\psi^{ed}$ is proposed as :

$$\rho\psi^{ed} = \frac{1}{2}\{2\,\mu_0 \mathrm{Tr}\,[(1-d)^{1/2} \otimes \varepsilon \otimes (1-d)^{1/2} \otimes \varepsilon)] + \lambda_0(1-\delta)\,\mathrm{Tr}^2\,[\varepsilon]\} \quad (3)$$

where λ_0 and μ_0 denote the Lamé constants of the initial material. The first part of this expression is responsible for the anisotropy of the damaged material and the second part denotes the influence of pores on the volumetric part of the elastic energy.

Considering the approach given in [ ], the second term $\rho\psi^{rd}$ is proposed as :

$$\rho\psi^{rd} = \frac{1}{2}\{\beta'\,\mathrm{tr}\Big[d \otimes \varepsilon \otimes d\Big] + \gamma'\delta\,\mathrm{tr}[\varepsilon]\,\delta\} \quad (4)$$

where β' and γ' are two coefficients which denote the sensibility of the material to microcracking. If we suppose that : $\beta' = 2\,\mu_0\,\beta$ and $\gamma' = 2\,\lambda_0\,\gamma$ we obtain :

$$\sigma = 2\mu_0(1-d)^{1/2} \otimes \varepsilon \otimes (1-d)^{1/2} + \lambda_0(1-\delta)\mathrm{tr}[\varepsilon]1 + \frac{1}{2}(2\mu_0\,\beta\,d \otimes d + \lambda_0\gamma\,\delta^2 1) \quad (5)$$

In the relation given above it can be remarked that : (i) the term σ^{rd} is only function of the damage variables d and δ, (ii) the anisotropy of this residual stress is naturally expressed by the term $d \otimes d$. Since no plasticity formalism has been introduced in our formulation, this rate of energy dissipated is only due to the damage variables.

<u>Crack closure</u>

It is possible to generalize the idea of "closure parameter" first introduced for damage mechanics [9]. This idea supposes that the state of damage i.e. microcracking is described by the variable d. That variable must be replaced by hd for "compressive" loadings where h is a material parameter. As a second rank order tensor has been used in our analysis the "closure parameter" must be a fourth rank order tensor $\mathbb{H}$. A great number of experiments would be needed to identify all the components of $\mathbb{H}$. In any case, it is not sure that these components remain constant during the test because the shape of the cracks may vary. For example Fig. 1 illustrates the evolution of the component h_{11} deduced from a compression test. So, for simplicity reasons it has been decided to replace $\mathbb{H} \otimes d$ by a new damage variable d^*. We propose heredown the new expression of the Helmoltz energy which takes into account the uniaxial phenomenon.

$$\rho\psi = \frac{1}{2}(\,2\mu_0\mathrm{tr}[1-d)^{1/2} \otimes <\varepsilon^e>_+ \otimes (1-d)^{1/2} \otimes <\varepsilon^e>_+]$$

$$+ 2\mu_0\,\mathrm{tr}[1-d)^{1/2} \otimes <\varepsilon^e>_- \otimes (1-d)^{1/2} \otimes <\varepsilon^e>_-] \; + \lambda_0\,(1-\delta)\mathrm{tr}^2[\,\varepsilon^e\,] \quad (6)$$

$<\varepsilon^e>_+$ is built with the positive eigen values of ε^e and $<\varepsilon^e>_- = \varepsilon^e - <\varepsilon^e>_+$ in the principle axes. The expression of ε^e is given by : $\varepsilon^e = \varepsilon - \varepsilon^{an}$ and ε^{an} represents :

$$\varepsilon^{an} = \frac{1}{2}\left\{\beta\;d \otimes (1-d)^{-1} \otimes d + \beta^*\;d^* \otimes (1-d^*)^{-1} \otimes d^* + \gamma\frac{\delta^2}{1-\delta}1\right\} \quad (7)$$

The state law is : $\sigma = \;2\mu_0\,(1-d)^{1/2} \otimes <\varepsilon^e>_+ \otimes (1-d)^{1/2}$

$$+ 2\mu_0\,(1-d^*)^{1/2} \otimes <\varepsilon^e>_- \otimes (1-d^*)^{1/2} + \lambda_0\,(1-\delta)\,\mathrm{tr}[\varepsilon^e]\,1 \quad (8)$$

It can be remarked that the last term is kept identical because pores cannot close. This model supposes that all the energy, excepted elastic energy which is recovered, is dissipated in heat and creation of new surfaces. This formalism is different to the plasticity that supposes that some energy can be stored into the material.

The damage growth satisfying the Clausius-Duhem inequality is generally governed by loading surfaces.In our case, we have adopted different loading surfaces for the three damage variables d, d^* and δ. The details can be found in [10].

COMPARISON BETWEEN PREDICTION AND EXPERIMENTS

Identification : this model needs only a compression test and a tensile test to be identified. The result of the identification can be seen on Fig. 2. In particular, the volumetric strain is well described.

Non radial test : it seemed to us that a new test would be a good challenge to verify the ability of our model. For this, after a radial path up to σ_2 = -25 MPa, the hydrostatic pressure has been kept constant (Tr [σ] = 112.5 MPa) while the lateral pressure was progressively removed (Fig. 3). The agreement between the prediction and the experiment is good.One can particularly remark the changing in the curvature of the volumetric strain that is well captured.

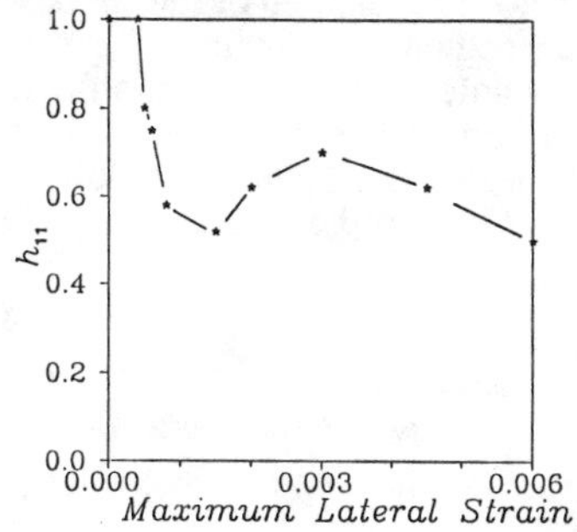

Figure 1 : Evolution of the h_{11} component in a compression test.

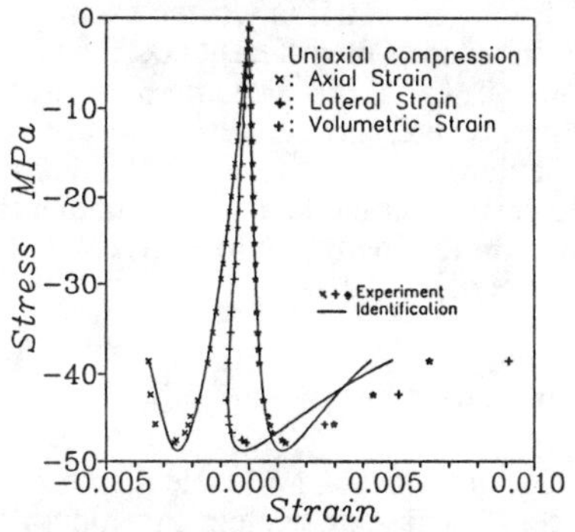

Figure 2 : Performance of the anisotropic damage model. Comparison Model Experiment for uniaxial compression.

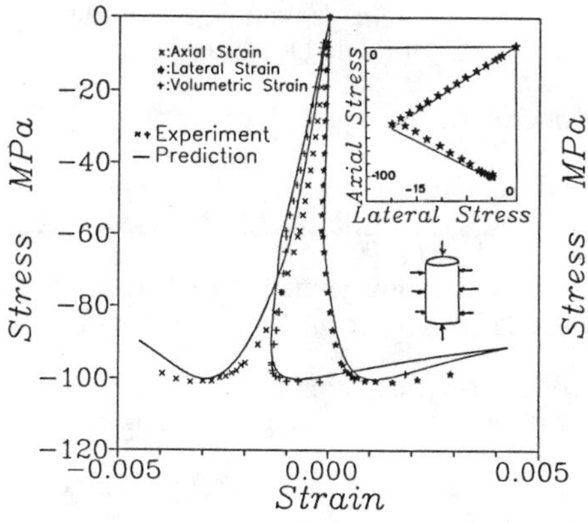

Figure 3 : Comparison Model Experiment for the non radial path : α = 2.5, Tr [σ] = 112.5 MPa.

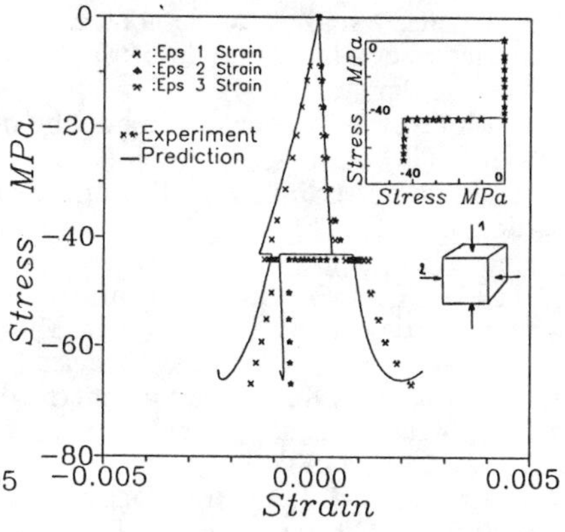

Figure 4 : Comparison Model Experiment for the biaxial test.

Biaxial test : it has been possible to predict the results of a biaxial test performed in another laboratory using a slightly different concrete [11]. The set of parameters has been kept identical excepted one of them which has been fitted on the curve (σ_1, ε_3). The two other curves have been deduced from the model (Fig. 4). It has to be pointed out that this test is difficult to perform. Nevertheless, in this case the shape of the curves are well reproduced and so we can consider that this model is relevant to describe the behavior of concrete under multiaxial loading paths.

CONCLUSION

The continuum damage theory can be applied to model the behavior of concrete through a set of appropriate damage variables. The thermodynamic potential retains three major phenomena in the behavior of damaged concrete : 1 the elastic anisotropy, 2 the evolution of permanent strains, 3 the crack closure effect. The Helmoltz potential is written as an expansion in a Taylor's series in function of the strain tensor.The different coefficients of the expansion are function of the set of the damage variables $\mathbb{d}$ $\mathbb{d}^*$ and δ. A general good agreement its obtained between experimental and theoretical results regarding the whole stress strain response and especially the variation of the volumetric strain and of the permanent strains. However, it could be pointed out that these permanent strains are sometimes lower in our predictions than those given by the experiments due to the existence of local plasticity.

REFERENCES

1. S. Andrieux, "Un modèle de matériau microfissuré, application aux roches et aux bétons", Thèse de docteur Ingénieur de l'ENPC, (1983).
2. A. Dragon, Z. Mroz , "A continuum Model for plastic-brittle behaviour of rock and concrete, Int. J. Engng Science, Vol. 17, 121-137, (1979) .
3. Z.P. Bazant, S.S. Kim, "Plastic fracturing theory for concrete", J. Engng Mech., ASCE, 105, 407-428, (1979).
4. J. Mazars, "Application de la mécanique de l'endommagement au comportement non linéaire et à la rupture du béton de structure", Thèse de Doctorat d'Etat, LMT - Université Paris 6, (1984).
5. S. Murakami, "Mechanical modeling of material damage", J. of App. Mech.,55, 280-286, (1988).
6. G. Pijaudier-Cabot, Z.P. Bazant, Y. Berthaud, "Interacting crack systems in particular of fiber reinforced composites", Proc. of the fifth Int. Conf. on Numerical Methods in Fracture Mechanics, Freiburg, 403-414, (1990).
7. Ju J.W, "On two dimensionnal micro mechanical damage models for brit-tle solids with interacting cracks", Proceedings, Micromechanics of Failure of Quasi-brittle materials, ed. by Shah S.P. et al., Elsevier Pubs., pp. 105-114, (1990).
8. G. Hermann, J. Kestin, "On the thermodynamics foundation of a damage theory in elastic solids", in Cracking and Damage, Ed. by J. Mazars and Z.P. Bazant, Elsevier Pubs, 228-232, (1988).
9. P. Ladevèze, J. Lemaitre, "Damage effective stress in quasi brittle conditions", 16th IUTAM Congress, Lyungy Denmark, (1984).
10. S. Ramtani," Contribution à la modélisation du comportement multiaxial du béton endommagé avec description du caractère unilatéral", Thèse de Doctorat de l'Université Paris 6, (1990).
11. J.M. Torrenti and B. Djebri, "Constitutive laws for concrete : an attempt of comparison", Proc. of the Second Int. Conf. on Computer Aided Analysis and Design of Concrete Structures, Zell am See, Austria, 871-882, (1990).

STOCHASTIC CONSTITUTIVE LAWS FOR GRAPH-REPRESENTABLE MATERIALS

MARTIN OSTOJA-STARZEWSKI
Department of Metallurgy, Mechanics and Materials Science
Michigan State University
East Lansing, MI 48824-1226

Telephone: (517) 355-5141
FAX (517) 353-9842

ABSTRACT

Results of an ongoing research program on effects of material microstructural randomness are reviewed. The work is set in the context of graph-representable solids. Two basic categories of problems are distinguished: "frozen-in" randomness and evolving randomness. In the first category, stochastic constitutive laws are discussed for the case of linear elastic trusses with the topology of Delaunay networks. An account is given of: relative effects of randomness in geometry and microscale physical constants, second-order random field characteristics of effective moduli, and determination of bounds on these moduli. In the second category of problems, an overview of stochastic modelling of damage formation in granular microstructures is conducted.

INTRODUCTION

Material spatial randomness may have significant effects on the macroscopic mechanical response. Hence, there arises a need to develop stochastic constitutive laws. In this paper we review and synthesize our researches on constitutive modelling of graph-representable microstructures. As pointed out in [1], one advantage of choosing a graph microstructure is the possibility of modelling several different types of materials -- granular, fibrous, cellular -- while another is the possibility to grasp both, the geometrical and the physical microscale material randomness. We distinguish two basic categories of problems:

1. determination of constitutive laws with the randomness being given a priori (or "frozen-in"),
2. determination of constitutive laws when the randomness evolves due to the applied loadings.

LINEAR ELASTIC DELAUNAY NETWORKS

We shall discuss the first category using the paradigm of linear elastic Delaunay networks [2,3], while the second one by employing the paradigm of damage formation on graphs of grain boundaries [4,5]. The common thread of both these studies is the definition of a scale dependent *representative volume element (RVE)*, also called a *window* of size $L \times L$; scale dependence is brought out through a parameter $\delta = L/d$, where d is the average microscale. The studies are carried out in two-dimensional settings.

In order to assess the effects of geometric and physical microscale randomness on the moduli of continuum approximations, four different Delaunay networks (Fig. 1a)) were investigated: a) an unmodified network dual to a standard Voronoi tessellation and having the strongest geometric disorder, b) a centroid-corrected network dual to the same standard Voronoi tessellation, c) a modified network having a weak geometric disorder, and d) a regular (equilateral) triangular network with randomness in spring constants. In cases a), b) and c) the spring constants were assigned according to the following rule: let each edge of length l be made of a linear elastic rod of the same cross section A and the same material of Young's modulus E, then the effective spring constant (of the rod) $k = EA/l$.

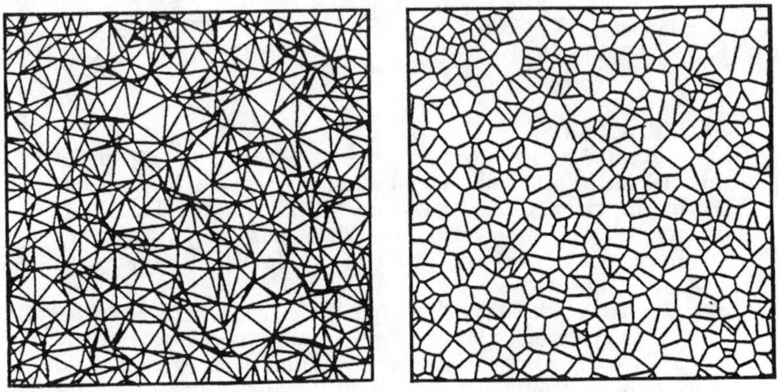

Fig. 1a) Delaunay network and dual to it b) Voronoi tessellation.

In a continuum setting the constitutive law is written as

$$\underset{\approx}{\sigma} = \underset{\approx}{C}(\underset{\sim}{x}, \delta, \omega)\underset{\approx}{\varepsilon} \tag{1}$$

where σ is the Canchy stress, ε is the infinitesimal strain controlled at window boundary, and ω is a single (deterministic) realization of microstructure in the window centered at position x. It follows that the choice of a continuum approximation -- that is the choice of an effective random tensor field $\underset{=}{C}$ - depends on the choice of δ. Although (1) is just a linear elastic law, it indicates that the scale dependence is essential in continuum formulations of other more complex (nonlinear, nonconservative, etc.) material responses.

Analysis of effective elastic moduli of the Delaunay networks was conducted by always solving, for any fixed δ, a mechanics problem of a square window under natural boundary conditions. A continuous increase of δ from ~1 through $\delta \gg 1$ results in a transition from random anisotropic properties of an RVE of statistical continuum approximation to the deterministic isotropic properties of an RVE of deterministic continuum approximation. Moreover, the following main conclusions are drawn here:

i) randomness in the network's geometry causes scatter and decrease of moduli;

ii) randomness in the network's physical properties causes scatter and increase of moduli - an effect weaker than the one in i);

iii) the autocorrelation functions of shear moduli are isotropic, while the others are amisotropic;

iv) by analogy to statistical continuum mechanics, the Voigt and Reuss bounds can be determined for Delaunay networks;

v) the uniform strain approximation results in practically identical autocorrelation functions as those obtained by the exact method.

Recently, a novel procedure, called the method of neighborhoods, for calculation of upper and lower bounds on constitutive response of random microstructures has been developed. The method is presented in [6] in the setting of a regular planar network of equilateral triangles with randomness present in the coefficients of vertex-vertex constitutive laws. Results on linear elastic networks indicate that a very good improvement on the conventional Voigt and Reuss bounds may be achieved with rather small neighborhoods, and thus with only a modest increase of the computational effort. At present, research is in progress on determination of a hierarchy of upper and lower bounds for linear viscoelastic random microstructures; here we use the correspondence principle and the generalization of dissipation function formulation to random media [7].

INTERGRANULAR DAMAGE PHENOMENA

Our studies of constitutive laws for evolving microstructures are set in the context of damage phenomena. Damage in our model is assumed to take place in the grain boundaries -- i.e. on a graph such as the Voronoi tessallation (Fig. 1b)) -- only. Since failure of any grain boundary is a random phenomenon, we employ a random a random variable describing its state -- elastic (i.e. intact) or inelastic (i.e. damaged) -- where the randomness enters through the constitutive coefficients (e.g. strength) associated with the given grain boundary. It follows that the damage states of the body are described by a binary random field. In a refinement of this model we use three damage states -- inelastic, plastic, and broken -- so that the damage states are described by a ternary random field. Our efforts are focused on the solution of problems of macroscopic response by a reduction to results in percolation theory and Markov random fields [8,9]. The following are our principal results:

i) near and at the point of transition from partial- to macro-damage, field Z is highly inhomogeneous which explains scatter in strength for any finite δ;

ii) the percolation theory approach justifies the functional form of Weibull-type statistics (except for volume dependence) and brings out the fractal character of damage phenomena;

iii) the information theoretic picture of scatter in damage geometry can be based on Gibbs probability measures and related to the thermodynamic picture;

iv) constitutive laws of a statistical continuum approximation can be derived using the thermodynamic orthogonality formalism generalized to random media.

Acknowledgement

This research is supported by the Air Force Office of Scientific Research under Grant No. AFOSR-89-0423; Lt. Col. G.K. Haritos is the program manager.

References

1. M. Ostoja-Starzewski, "Graph Approach to the Constitutive Modelling of Heterogeneous Solids," Mechanics Research Communications, Vol. 14 (4), pp. 255-262, 1987.

2. M. Ostoja-Starzewski and C. Wang, "Linear Elasticity of Planar Delaunay Networks: Random Field Characterization of Effective Moduli," Acta Mechanica, Vol. 80, pp. 61-80, 1989.

3. M. Ostoja-Starzewski and C. Wang, "Linear Elasticity of Planar Delaunay Networks - II: Voigt and Reuss Bounds, and Modification for Centroids," Acta Mechanica, 1990, in press.

4. M. Ostoja-Starzewski, "Mechanics of Damage in a Random Granular Microstructure: Percolation of Inelastic Phases," Letters in Applied and Engineering Sciences (Int. J. Engng. Sci.), Vol. 27(3), pp. 315-326, 1989.

5. M. Ostoja-Starzewski, "Damage in a Random Microstructure: Size Effects, Fractals, and Entropy Maximization," Applied Mechanics Reviews, Vol. 42(11, Part 2), pp. 202-212, 1989; also in *Mechanics Pan-America*, C.R. Steele, A.W. Leissa and M.R.M. Crespo da Silva, Editors, ASME Press, New York, pp. S202-S212, 1989.

6. M. Ostoja-Starzewski, "Bounds on Constitutive Response for a Class of Random Material Microstructures", Computers and Structures, 1990, in press.

7. M. Ostoja-Starzewski, "A Generalization of Thermodynamic Orthogonality to Random Media," Journal of Applied Mathematics and Physics (ZAMP), 1990, in press.

8. M. Ostoja-Starzewski, "Markov Random Field Methods for Heterogeneous Materials," in "Session on Material Instabilities," 11th U. S. National Congress of Applied Mechanics, Tucson, AZ, May, 1990.

9. M. Ostoja-Starzewski, "How can Random Fields be used in Micromechanics?", Workshop on "Percolation Models of Material Failure", Cornell University, Ithaca, NY, May/June, 1990.

Localization in Micropolar Elasto-Plasticity

Paul Steinmann and Kaspar Willam
Institute of Mechanics, University of Karlsruhe, D-7500 Karlsruhe 1
CEAE-Department, University of Colorado at Boulder, CO 80309-0428

Abstract

Micropolar continuum descriptions involve additional rotational degrees of freedom of the material particles. This study examines whether these additional variables remedy the well-known localization problems of classical continuum theory, namely the loss of ellipticity causing severe mesh dependence of numerical solutions.

1 Introduction

In this paper we focus on micropolar elasto-plasticity and investigate the conditions for the loss of ellipticity. To this end we will at first review the kinematic relations and the balance conditions of micropolar continua. Then linear-elastic and elasto-plastic constitutive models are introduced in a canonical form, which is convenient for studying the bifurcation problem. An 'augmented' localization tensor is derived which reveals discontinous bifurcation at the constitutive level. As an example the J_2-flow theory is adapted to micropolar plasticity to illustrate how the 'standard' singularity of the local continuum theory is influenced by the additional rotationfield.

2 Kinematics and Balance Laws

Each material particle in the micropolar continuum possesses three translational and three rotational degrees of freedom, see [2] [3] [4] [6]. As usual the translational motion is described by the displacementfield $\mathbf{u}$ whereas the rotational motion is described by the rotationfield $\boldsymbol{\omega}$. The non-symmetric strain tensor is defined as $\boldsymbol{\epsilon} = \nabla \mathbf{u} + \mathbf{e} \cdot \boldsymbol{\omega}$. In addition the curvature tensor is defined by the gradient of the rotationfield, i.e. $\boldsymbol{\kappa} = \nabla \boldsymbol{\omega}$.

The stress tensor $\boldsymbol{\sigma}$ is as usual the two-point linear mapping between the normal and the traction vector. Omitting body forces the local balance of linear momentum results in the divergence of the non-symmetric stress tensor $\operatorname{div} \boldsymbol{\sigma}^t = \mathbf{0}$. Conjugate to the curvatures additional couplestresses $\mathbf{m}$ act on the surfaces of the volume element. Omitting body couples the local balance of angular momentum results in the divergence of the non-symmetric couplestress tensor augmented by the skew-symmetric contributions of the stress tensor $\operatorname{div} \mathbf{m} + \mathbf{e} : \boldsymbol{\sigma} = \mathbf{0}$.

3 Constitutive Relations

In the case of linear elasticity the stress tensor follows the linear mapping

$$\sigma = \mathbf{E} : \epsilon \quad \text{with} \quad \mathbf{E} = 2\mu \mathbf{1}_4^{sym} + 2\mu_c \mathbf{1}_4^{skw} + \lambda \mathbf{I} \otimes \mathbf{I} \tag{1}$$

For isotropic behaviour $\mathbf{E}$ is the general fourth-order isotropic tensor above. Similarly the couplestresses are linearly related to the curvature via the isotropic material tensor $\mathbf{C}$

$$\mathbf{m} = \mathbf{C} : \kappa \quad \text{with} \quad \mathbf{C} = 2\alpha \mathbf{1}_4^{sym} + 2\beta \mathbf{1}_4^{skw} + \gamma \mathbf{I} \otimes \mathbf{I} \tag{2}$$

Under planar deformations the couplestress and curvature tensors may be represented as vectors. Then the material operator $\mathbf{C}$ reduces to the second-order isotropic tensor $\mathbf{C} = l^2 \mu \mathbf{I}$. Here l has the dimension of a length which is termed the 'characteristic length'.

The yield-condition delimits the elastic range of the micropolar-continuum by a single scalarfunction of stress and couplestress $F = F(\sigma, \mathbf{m}) = 0$, e.g. [1] [4] [5]. In analogy to classical plasticity the strain-rate and the curvature-rate are decomposed into an elastic and a plastic contribution. Assuming normality the magnitudes of the plastic strain-rate and the plastic curvature-rate depend on the same plastic multiplier $\dot{\lambda}$

$$\dot{\epsilon}_p = \dot{\lambda} \frac{\partial F}{\partial \sigma} = \dot{\lambda} \mathbf{n}^\sigma; \qquad \dot{\kappa}_p = \dot{\lambda} \frac{\partial F}{\partial \mathbf{m}} = \dot{\lambda} \mathbf{n}^m \tag{3}$$

The tangential elasto-plastic constitutive relations may be compacted into

$$\begin{pmatrix} \dot{\sigma} \\ \dot{\mathbf{m}} \end{pmatrix} = \begin{pmatrix} \mathbf{E}_{ep} & \mathbf{D}_{ep} \\ \mathbf{D}_{ep}^t & \mathbf{C}_{ep} \end{pmatrix} : \begin{pmatrix} \dot{\epsilon} \\ \dot{\kappa} \end{pmatrix} \tag{4}$$

where the suboperators are defined below

$$\mathbf{E}_{ep} = \mathbf{E} - \frac{\mathbf{E} : \mathbf{n}^\sigma \otimes \mathbf{n}^\sigma : \mathbf{E}}{\mathbf{n}^\sigma : \mathbf{E} : \mathbf{n}^\sigma + \mathbf{n}^m : \mathbf{C} : \mathbf{n}^m}; \quad \mathbf{D}_{ep} = -\frac{\mathbf{E} : \mathbf{n}^\sigma \otimes \mathbf{n}^m : \mathbf{C}}{\mathbf{n}^\sigma : \mathbf{E} : \mathbf{n}^\sigma + \mathbf{n}^m : \mathbf{C} : \mathbf{n}^m} \tag{5}$$

$$\mathbf{C}_{ep} = \mathbf{C} - \frac{\mathbf{C} : \mathbf{n}^m \otimes \mathbf{n}^m : \mathbf{C}}{\mathbf{n}^\sigma : \mathbf{E} : \mathbf{n}^\sigma + \mathbf{n}^m : \mathbf{C} : \mathbf{n}^m}$$

4 Localization as a Bifurcation Problem

Let us assume that the homogeneously deformed solid is subjected to quasi-static increments of deformation. While the displacement- and the rotationfield are assumed to be C^0-continuous i.e.

$$[[\mathbf{u}]] = \mathbf{u}^+ - \mathbf{u}^- = \mathbf{0}; \qquad [[\omega]] = \omega^+ - \omega^- = \mathbf{0} \tag{6}$$

the fields of displacementgradients and rotationgradients may exhibit a jump across the plane of discontinuity defined by the same normal $\mathbf{N}$

$$[[\nabla \mathbf{u}]] = \nabla \mathbf{u}^+ - \nabla \mathbf{u}^- \neq \mathbf{0}; \qquad [[\nabla \omega]] = \nabla \omega^+ - \nabla \omega^- \neq \mathbf{0} \tag{7}$$

Maxwells compatibility conditions require that the jumps are of the form

$$[[\nabla \mathbf{u}]] = \gamma^u \mathbf{M}^u \otimes \mathbf{N}; \qquad [[\nabla \boldsymbol{\omega}]] = \gamma^\omega \mathbf{M}^\omega \otimes \mathbf{N} \tag{8}$$

Analogous to the wave propagation argument in classical continuum theory the vector $\mathbf{M}$ denotes the polarization direction and the localization amplitude γ is related to the wave speed. The corresponding discontinuities in the strain- and curvature-rates are defined by the tensor product of the unit vectors $\mathbf{N}$ and $\mathbf{M}$

$$[[\dot{\boldsymbol{\epsilon}}]] = [[\nabla \dot{\mathbf{u}}]] + \mathbf{e} \cdot [[\dot{\boldsymbol{\omega}}]] = \gamma^u \mathbf{M}^u \otimes \mathbf{N}; \quad [[\dot{\boldsymbol{\kappa}}]] = [[\nabla \dot{\boldsymbol{\omega}}]] = \gamma^\omega \mathbf{M}^\omega \otimes \mathbf{N} \tag{9}$$

Introducing the jumps in the stress- and couplestress-rate with the aid of the elasto-plastic material operator and requiring balance of linear and angular momentum across the plane of discontinuity leads to the localization condition for micropolar continua.

$$\begin{pmatrix} \mathbf{Q}_{ee} & \mathbf{Q}_{ec} \\ \mathbf{Q}_{ce} & \mathbf{Q}_{cc} \end{pmatrix} \cdot \begin{pmatrix} \gamma^u \mathbf{M}^u \\ \gamma^\omega \mathbf{M}^\omega \end{pmatrix} = \begin{pmatrix} \mathbf{0} \\ \mathbf{0} \end{pmatrix} \tag{10}$$

Localization initiates at the first time in the deformation history when this condition is fullfilled by some directions $\mathbf{N}$ and $\mathbf{M}$. Thereby the suboperators are defined as

$$\mathbf{Q}_{ee} = \mathbf{N} \cdot \mathbf{E}_{ep} \cdot \mathbf{N}; \quad \mathbf{Q}_{ec} = \mathbf{N} \cdot \mathbf{D}_{ep} \cdot \mathbf{N}; \quad \mathbf{Q}_{cc} = \mathbf{N} \cdot \mathbf{C}_{ep} \cdot \mathbf{N} \tag{11}$$

5 J_2-Micropolar Elasto-Plasticity

For illustration the micropolar elasto-plastic J_2-model [1] is investigated for associated flow. In the absence of couplestresses this formulation degenerates to the wellknown J_2-flow theory. Yielding is defined by

$$F = F(\boldsymbol{\sigma}, \mathbf{m}) = \sqrt{3J_2} - Y = 0 \quad \text{with} \quad J_2 = \frac{1}{4}(\mathbf{s} : \mathbf{s} + \mathbf{s} : \mathbf{s}^t) + \frac{1}{2}\mathbf{m} : \mathbf{m} \tag{12}$$

Here J_2 denotes the generalized second invariant of deviatoric stress. For simplicity we restrict our attention to the stress state of pure shear and vary only the magnitude of the couplestress components.

As a starting point we consider the traditional state of pure shear for the Boltzmann continuum with $\sigma_{11} : \sigma_{22} = 1 : -1$ and $m_1 = m_2 = 0$. In two dimensions the normal vector $\mathbf{N}$ is simply $(\cos\theta, \ \sin\theta)^t$ with θ denoting the angle between the x-axis and the unit normal. For a variation of θ in the range $[0, 2\pi]$ the eigenvalues and the determinant of the augmented localization tensor are computed and normalized by the values of the elastic operator. Phase velocity plots, as presented in [7] for elasto-plastic material models in the frame of classical continuum theory, are shown in Fig. 1.1-1.3. As depicted in Fig. 1.1, minima of the localization tensor $\det \mathbf{Q}$ are found at $\theta = \pi/4$ as in the continuum model. However $\min(\det \mathbf{Q}) < 0$ due to the minor non-symmetry in the elastic constitutive operator when $0 < \mu_c < \mu$. The results for the same stress state and additional couplestresses $m_1 : m_2 = 0 : 0.2$ are shown in Fig. 1.2. Now the augmented localization tensor possesses only positive eigenvalues for all directions θ. Therefore the presence of both, the stresses and couplestresses, act like 'hardening' and stabilizes the localization operator. At last, the couplestress is further increased to $m_1 : m_2 = 0 : 2$ so that the suboperator $\mathbf{Q}^{cc}$ is getting nearly singular for $\theta = \pi/2$ while $\mathbf{Q}^{ee}$ approaches the elastic properties, see Fig. 1.3. Clearly in the limit, as $\sigma_{11} = \sigma_{22} = 0$ and $m_1 : m_2 = 0 : 1$ the full singularity will develop in the subspace of curvatures.

6 Conclusion

Introducing new field variables, which in turn bring a length scale into the constitutive description is not sufficient to eliminate localization for all conditions. In the presence of stresses as well as couplestresses the singularity of the localization tensor cannot develop for perfectly plastic conditions. Nevertheless this stabilizing effect of the micropolar description may be offset by softening and/or non-associated flow. Even though the benefits of micropolar continua have been increasingly emphasized by some authors, the problem of discontinous bifurcation at the constitutive level is not circumvented in all cases.

References

[1] R. de Borst, "A Generalisation of J_2-Theory for Polar Continua", TNO-IBBC Report BI-89-195, Delft University of Technology, Delft (1989).

[2] E. Cosserat and F. Cosserat, "Theorie des Corps Deformables", Herman et fils, Paris, (1909).

[3] W. Günther, "Zur Statik und Kinematik des Cosseratschen Kontinuums", Abh. Braunschweig. Wiss. Ges. 10, pp. 195-213, (1958).

[4] H. Lippmann, "Eine Cosserat-Theorie des plastischen Fliessens", *Acta Mechanica*, Vol. 8, pp. 255-284, (1969).

[5] H.-B. Mühlhaus, "Application of Cosserat Theory in Numerical Solutions of Limit Load Problems", *Ing.-Arch.*, Vol. 59, No. 2, pp. 124-137, (1989).

[6] H. Schaefer, "Das Cosserat-Kontinuum", *ZAMM*, Vol.47, pp. 485-498, (1967).

[7] N. Sobh, "Bifurcation Analysis of Tangential Material Operators", Ph.D. Diss., CEAE Department, University of Colorado, Boulder, (1987).

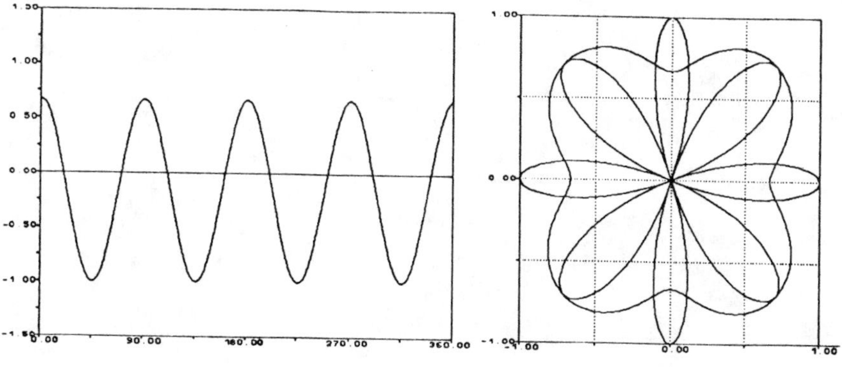

Fig. 1.1 Variation of det **Q** and phase velocity

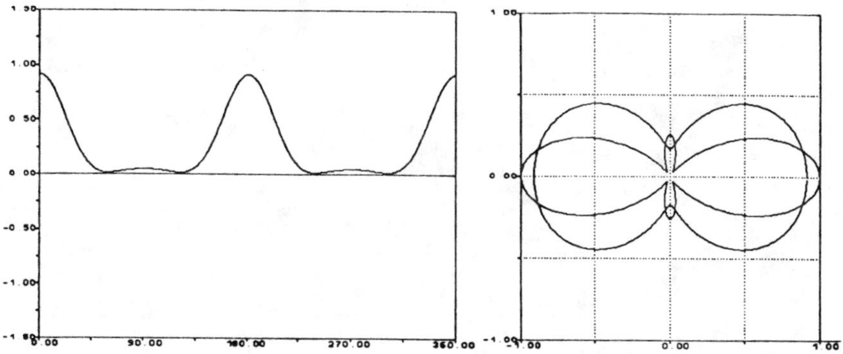

Fig. 1.2 Variation of det **Q** and phase velocity

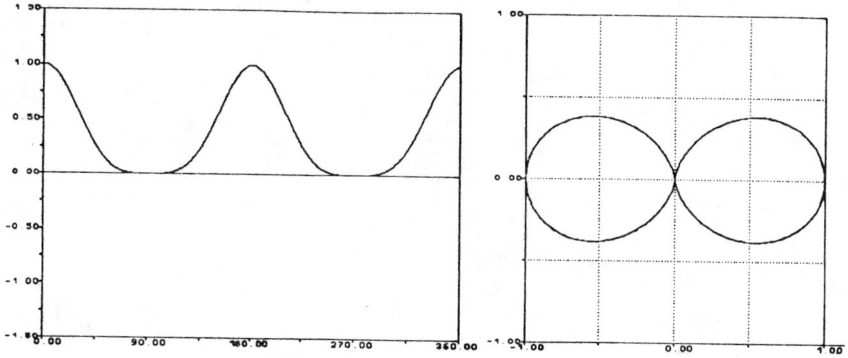

Fig. 1.3 Variation of det **Q** and phase velocity

ONE-DIMENSIONAL HYSTERESIS MODELLING OF CONCRETE AFFECTED BY ALKALI-SILICA REACTION

SU Xiaozu
Dept of Civil and Structural Engineering, Tongji University, Shanghai, People's Republic of China

Peter WALDRON and Anthony BLAKEBOROUGH
Dept of Civil Engineering, University of Bristol, Bristol, BS8 1TR, UK

ABSTRACT

A damage mechanics based uniaxial stress-strain model, which satisfies Drucker's stability postulate, is proposed for the compressive behaviour of concrete affected by Alkali-Silica Reaction. The model can describe most of the features of concrete under uniaxial compression and shows good agreement with test results.

1. INTRODUCTION

Although many constitutive relationships have been proposed for normal concrete, comparatively little modelling work has been done on concrete affected by Alkali-Silica Reaction (ASR). ASR occurs at discrete points in the matrix of certain concretes when highly alkali cement comes into contact with aggregate containing silica in the presence of water. The product of ASR is a hygroscopic gel which readily absorbs free water resulting in expansion and random cracking of the concrete. The need for modelling concrete damaged by ASR is increasing as more structures affected by ASR are being diagnosed around the world [1].

In this paper a model is proposed for the ASR-affected and normal concrete under uniaxial compression.

2. THE MODEL

The model is a series connection of a plastic element and a crack element (Fig. 1), with a damage variable to describe the deterioration of both. The plastic element models the uncracked part of the concrete, while the crack element models the stress-strain behaviour of the substances

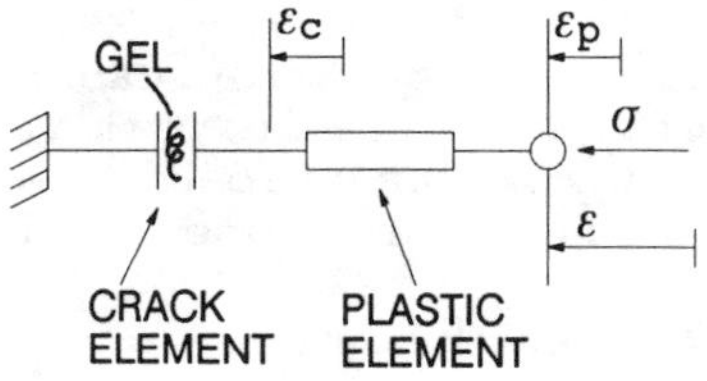

Fig. 1

(usually the hygroscopic gel) between the two surfaces of the model crack which represents possibly many cracks in the concrete.

We denote ε_p and σ_p as the strain and stress of the plastic element, and ε_c and σ_c as the strain and stress of the crack element, respectively. Hence we have $\sigma_p=\sigma_c=\sigma$, $\varepsilon_p+\varepsilon_c=\varepsilon$, where ε and σ are the strain and stress of the whole model.

In this paper, stress and strain are considered to be positive when in compression. Loading (unloading) is defined as a process in which strain increases (decreases) monotonically.

2.1. The Plastic Element

With reference to [2] the relationships

$$\sigma_p = \sigma_L + [\eta(1-\omega) - \sigma_L][1-\exp(-K(\varepsilon_p-\varepsilon_{pL})] \tag{1}$$

and

$$\sigma_p = \sigma_u - [\eta(1-\omega) + R\sigma_u][1-\exp(-K(\varepsilon_{pu}-\varepsilon_p)] \tag{2}$$

are proposed for loading and unloading respectively, where $(\varepsilon_{pL},\sigma_L)$ and $(\varepsilon_{pu},\sigma_u)$ are two points in the ε_p-σ plane from which loading and unloading begins respectively; and $R(0\leq R\leq 1)$, $K(>0)$ and $\eta(>0)$ are parameters. ω is the damage variable $(0\leq\omega\leq 1)$ for which $\omega=0$ means no damage, $\omega=1$ represents completely damaged.

2.2. The Crack Element

The stress-strain relationship is

$$\sigma_c = H_y(\varepsilon_c)H_\sigma(\varepsilon_c), \tag{3}$$

where

$$H_\sigma(\varepsilon_c) = -B(1-\xi\omega)\,\mathrm{Ln}(1-A\varepsilon_c), \tag{4}$$

and

$$H_y(\varepsilon_c) = \begin{cases} 1, & \text{for loading when } \varepsilon_c>\varepsilon_{cu} \\ 1-\alpha+\alpha(\varepsilon_c/\varepsilon_{cu})^n, \; n\geq 1,\; 0<\alpha<1, & \text{for unloading} \\ H_{yL}+(1-H_{yL})[(\varepsilon_c-\varepsilon_{cL})/(\varepsilon_{cu}-\varepsilon_{cL})]^\gamma,\; 0<\gamma<1, & \text{for loading when } \varepsilon_c\leq\varepsilon_{cu}. \end{cases} \tag{5}$$

$H_y(\varepsilon_c)$ is a function representing hysteresis behaviour. $B(>0)$ and $A(>0)$ in Eq. 4 are parameters, B representing strength and A crack width. $\xi(0\leq\xi\leq 1)$ is a damage reduction factor. When B is sufficiently small $(B\to 0)$, the crack element represents cracks containing no substances. ε_{cu} (its initial value being zero) is the historical maximum unloading strain, and $(\varepsilon_{cL},H_{yL})$ is the point in the ε_c-H_y plane at which reloading begins.

2.3. Damage Evolution

The proposed formula for evolution of damage is

$$\omega = C_1 (W_d/W_{lim})^s, \ s>0 \tag{6}$$

where W_d is the energy dissipated, and W_{lim} is the energy limit. By energy we mean energy per unit volume in this paper. C_1 in Eq. 6 will be explained later.

We define W_d as the energy that would be dissipated if complete unloading took place. This will make W_d, and hence ω, non-decreasing functionals of strain history.

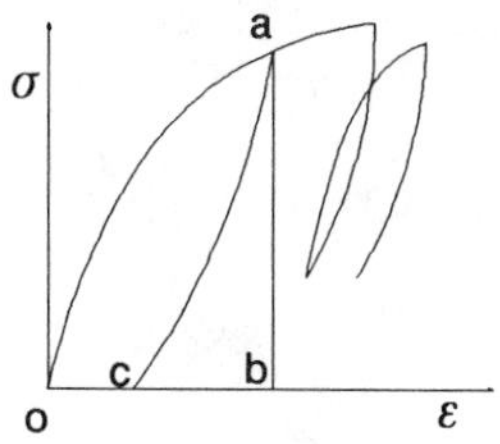

Fig. 2

Fig. 2 illustrates the definition of W_d. When the current point is at a, the would-be unloading curve is ac. Hence the value of W_d at point a is the area oac. For subsequent loading cycles, energy dissipated in each cycle is defined as the energy that would be dissipated if complete unloading to the starting load level of this cycle took place. The total dissipated energy is the sum of the energies dissipated in all cycles. It is assumed that only the energy dissipated in the plastic element will contribute to damage evolution.

For arbitrary loading cycles, intuition tells us that more damage will be incurred when the concrete experiences new values of positive strain than when the concrete deforms within the historical maximum strain. To model this phenomenon, the concepts of primary damage and secondary damage are advanced. Primary damage is defined as the damage of concrete experiencing virgin compressive strain, while secondary damage is defined as the damage of concrete deforming within the limit of historical maximum compressive strain. This is implemented by the coefficient C_1 in Eq. 6 which takes the form: $C_1=1$ for virgin compressive strain; and $C_1=h$ $(0<h<1)$ for non-virgin compressive strain.

It can be shown that the model satisfies Drucker's stability postulate.

3. COMPUTER IMPLEMENTATION AND COMPUTATION RESULTS

The implementation of incremental stress scheme is straightforward. For full range computation of the model, however, an incremental strain scheme will have to be adopted. In the latter case, the main idea of the implementation is to compare the slopes $d\sigma/d\varepsilon_p$ and $d\sigma/d\varepsilon_c$ at each increment. Then the element with the lower slope is chosen and its strain is computed by linear approximation. The strains of the two elements thus computed are then compared with the true total strain. A corresponding adjustment is made to reduce the error and is repeated until the error is acceptable.

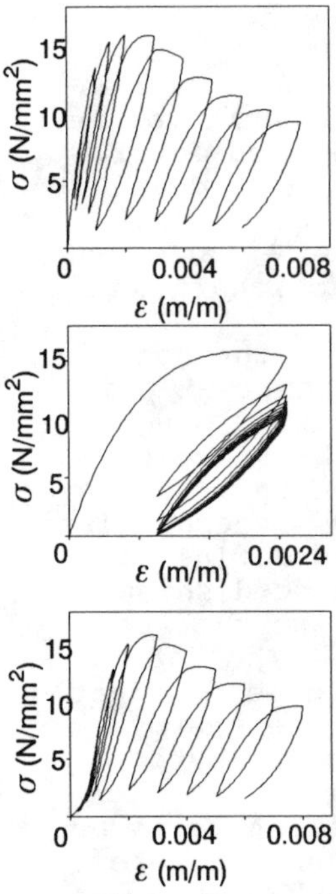

Fig. 3

A first estimate of the parameters can be made based on interpretation of their physical meaning. Refined parameter estimation can be achieved through minimisation of an error functional by using an adaptive-step-size-random-search algorithm. Fig. 3 shows computation results for demonstration of the model. Figs. 4 and 5 shows comparison of the model with available test results [3]. It can be seen that the crack-closing process and gel resistance are reflected by the concave portion of the stress-strain curve.

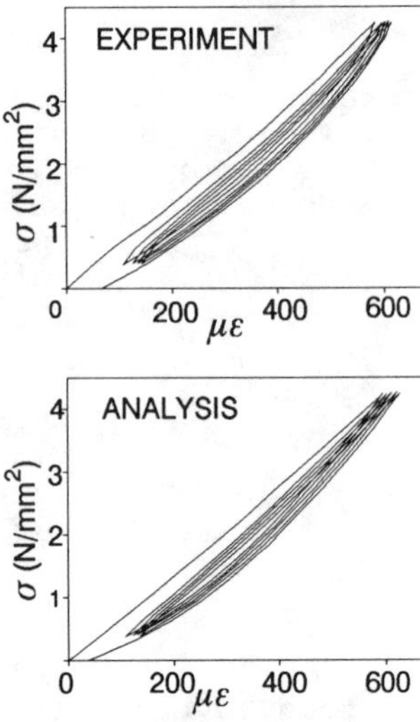

Fig. 4

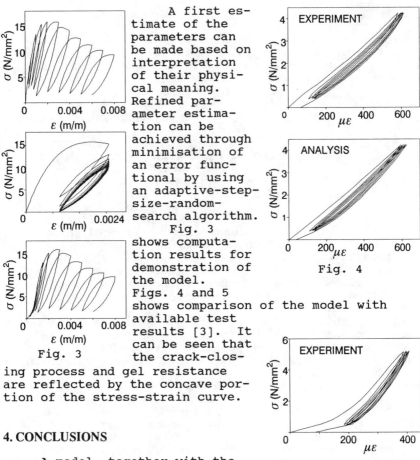

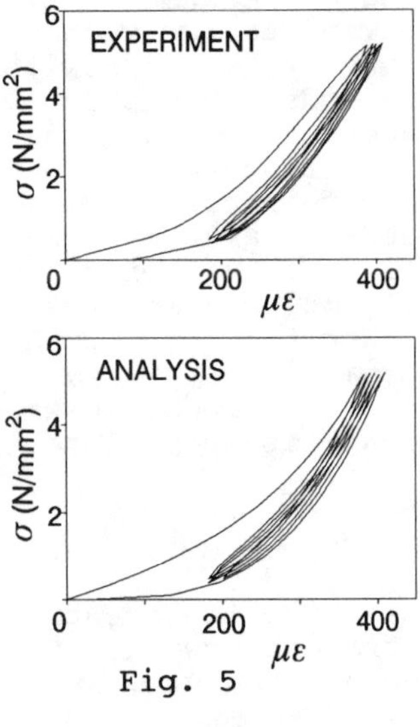

Fig. 5

4. CONCLUSIONS

A model, together with the concepts of the crack element, primary damage and secondary damage, is put forward which can describe the main features of ASR-affected and normal concretes under uniaxial compression. The model conforms well to the available test results.

5. REFERENCES

1. K. Okada et al (eds.), Proceedings of the 8th International Conference on Alkali-Aggregate Reaction, Kyoto, July 1989

2. A. Fafitis and S.P. Shah, "Rheological model for cyclic loading of concrete", ASCE J. Struct. Eng. 110(9), 2085-2102 (Sept 1984)

3. P. Waldron, "Report on stiffness damage tests for Mott Hay & Anderson Special Services Division", Univ. of Bristol, Dept of Civil Eng., Bristol, UK, Mar 1989 (Private communication)

THE EFFECT OF LOCALIZATION ON THE BEHAVIOR OF SOIL STRUCTURES AND NON-LOCAL CONTINUUM

Tadatsugu Tanaka
Faculty of Agriculture, Meiji Univ., Kawasaki, Japan

ABSTRACT

The shear band thickness is introduced as characteristic length into constitutive equation. The constitutive model for non-associated strain hardening-softening elastoplastic material including strength anisotropy is employed. A return mapping algorithm is used for integration of elastoplastic constitutive relation. In addition finite element formulation based on the concept of non-local plastic strain is proposed and analysis of plane strain compression tests of sand is presented.

INTRODUCTION

The failure of sand mass is usually progressive and linked to the development of narrow shear bands of localized deformation. The shear band thickness is introduced as characteristic length into constitutive equation using four-noded isoparametric element with reduced quadrature. This element produces psuedo-equilibrium to improve the bound of solution. An explicit dynamic relaxation method is applied to the solution of problems and probable hour-glassing is automatically suppressed by the nature of the method[1].

A constitutive model for non-associated elastoplastic material is introduced. Row's stress-dilatancy relation is applied to the constitutive model in order to determine the dilatancy angle. The internal friction of soil is modelled as close as to the actual one ; i. e. internal friction is anisotropic and pressure-level dependent and exhibits post-peak strain softening.

A return mapping algorithm proposed by Ortiz and Simo[2] is used for integration algorithm of elastoplastic constitutive relation including shear band effect. Using these numerical models, the scale effect of surface footing is analysed.

Finally finite element formulation based on the concept of non-local plastic strain is proposed and analysis of plane strain tests of sand is presented.

GENERAL RETURN MAPPING METHOD FOR ELASTOPLASTIC CONSTITUTIVE MODEL INCLUDING SHEAR BAND

In the return mapping algorithm the elastically predicted stresses σ_A are relaxed onto a suitably updated yield surface(σ_B). A change in stresses cause an associated change in the elastic strains given by

$$\dot{\varepsilon}^e = [D]^{-1}(\sigma_B - \sigma_A) \tag{1}$$

where [D] is elastic matrix. As the total strains don't change during relaxation process, the plastic strain change is balanced by an equal and opposite change in the elastic strains

$$s\dot{\varepsilon}^p = -\dot{\varepsilon}^e = -[D]^{-1}(\sigma_B - \sigma_A) \tag{2}$$

where s is the ratio of shear band area to finite element area and approximately given by $w/\sqrt{F_e}$ (w= thickness of the shear band, F_e=total area of element).

The plastic strain increments are proportional to the gradient of the plastic potential Φ

$$\dot{\varepsilon}^p = \lambda(\partial\Phi/\partial\sigma) \tag{3}$$

where λ is a positive scalar multiplier to be determined with the aid of the loading-unloading criterion. Combining equation (2) and (3) gives

$$\sigma_B = \sigma_A - s\lambda[D](\partial\Phi/\partial\sigma) \tag{4}$$

As there are changes to the plastic strains there is a change to the hardening-softening parameter κ

$$\dot{\varepsilon}_B{}^p = \dot{\varepsilon}_A{}^p + \lambda(\partial\Phi/\partial\sigma) \tag{5}$$

$$\kappa_B = \kappa_A + \dot{\kappa} \tag{6}$$

$$\kappa = \int d\bar{\varepsilon}_P,\quad (d\bar{\varepsilon}_p)^2 = 2\{(de_{XP})^2 + (de_{yp})^2 + (de_{zp})^2\} + (d\gamma_{xyp})^2 = \lambda^2 \tag{7}$$

where $de_{xp}, \cdot\cdot, d\gamma_{xyp}$ = deviatoric plastic strain increments. If Drucker-Prager type function is employed as plastic potential, κ is given by the equation

$$\dot{\kappa} = \dot{\kappa}\{\lambda(\partial\Phi/\partial\sigma)\} = \lambda \tag{8}$$

The suitably relaxed stress must satisty the yield function f

$$f(\sigma_B, \kappa_B) = f\{(\sigma_A - s\lambda[D](\partial\Phi/\partial\sigma)), (\kappa_A + \lambda)\} = 0 \tag{9}$$

The yield function is linearized around current values to obtain the state variables, the value of λ can be solved to

$$\lambda = f(\sigma_A, \kappa_A) / \{s(\partial f/\partial\sigma)[D](\partial\Phi/\partial\sigma) - (\partial f/\partial\kappa)\} \tag{10}$$

Substitution of this value of λ into equation (4) yields the updated plastic state.

The yield function is Mohr-Coulomb type and the plastic potential is Drucker-Prager type. The frictional hardening and softening functions are employed. The friction angle is anisotropic and pressure-level dependent. Refer to Tanaka & Kawamoto[3] for the details of the constitutive model.

NON-LOCAL CONTINUUM FINITE ELEMENT FORMULATION

The spacial average of the plastic strain increment at location x is given as[4]

$$\langle \Delta\varepsilon^{p}(x) \rangle = \int \bar{\alpha}(x, s)\, \Delta\varepsilon^{p}(s)\, dV \qquad (11)$$

where $\bar{\alpha}(x, s) = \alpha(s - x) / V_r(x)$

$\alpha(x)$ is weighting function, V_r is representative volume for averaging the plastic strain

s is general co-ordinate vactor

A weighting function for numerical computation is given as

$$\alpha(x) = \exp\{-(kx)^2 / l^2\} \qquad (12)$$

where l is the characteristic length, k is material constant. For each integration point of each element the averaged non-local plastic strain increment is obtained as

$$\Delta\varepsilon^{P}(x) = \Sigma \bar{\alpha}_{(i,j)}\, \Delta\varepsilon^{P}_{(j)} = m \Delta\varepsilon^{P}_{(i)} \qquad (13)$$

$\alpha_{(i,j)}$ is the values of $\bar{\alpha}(x, s)$ for integration point $x_{(j)}$ relative to point $x_{(i)}$. The subscripts in parenthes refer to the integration points of finite elements within representative volume. m denotes the coefficient of averaging and this value varies as plastic strains increase. Instead of $\Delta\varepsilon^{p}$ we can apply the averaging operation directly to λ. For a formulation consistent with the constitutive model including shear band (i. e. shear band embedded in a much larger finite element), the area ratio s is utilized in the non-local compuation (i. e. $s = 1/m$, for very small softening mudulus).

NUMERICAL RESULTS

The finite element mesh for a half area of surface strip footing is shown in Fig. 1. An incremental integration methods by both load and displacement controlling are applied along with an equilibrium iteration tolerance of force norm 10^{-2}. The calculated relations between the bearing capacity factor N_γ and normalized displacement for footing breadth B=40cm are shown in Fig. 2. Although we are using almost flat stress-strain relations for softening parts, reduction of N_γ after the peak is pronounced by displacement controlling analysis. Almost the same peak value of N_γ is obtained by load controlling but footing loads are less reliable after the peak. Fig. 3 shows the relation between peak N_γ and B.

The finite element mesh for a 1/4 area of plane

strain test is shown in Fig. 4. The material constants of constitutive model are the same as that of the footing problems. The initial stresses are given as $\sigma_3 = 0.8$, $\sigma_1 = 1.0$ kgf/cm², $\sigma_2 = (\sigma_1 + \sigma_3)/2$. The calculated relations between the stress ratio σ_1/σ_3 and principal strain ε_1 (%) are shown in Fig. 5.

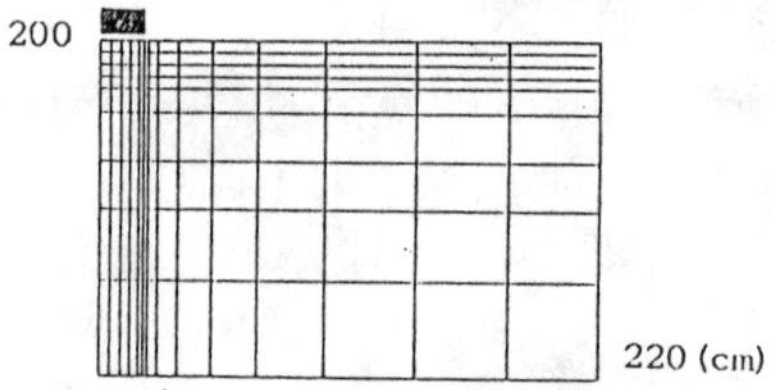

Fig. 1 Finite element model

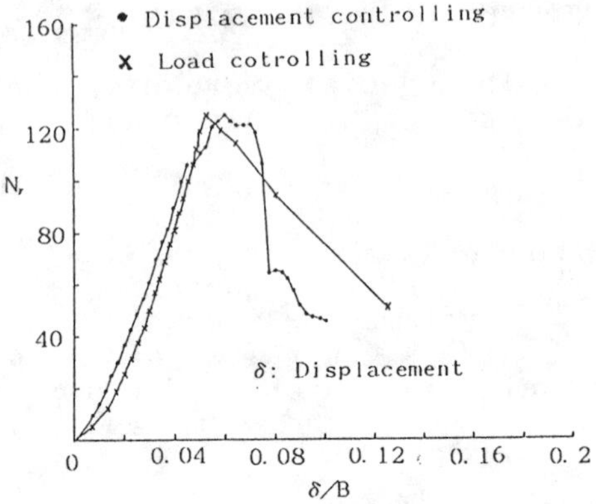

Fig. 2 Load-Displacement relations

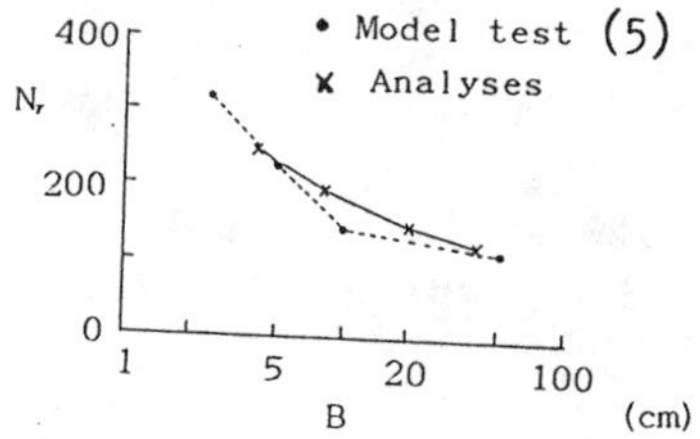

Fig. 3 Scale effect of footing

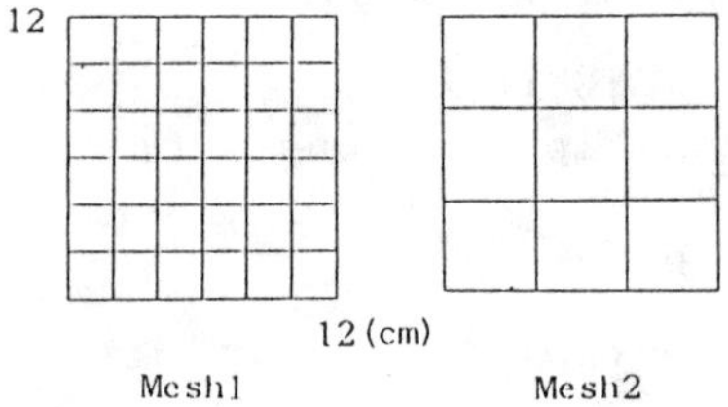

Fig. 4 Finite elements for plane strain test

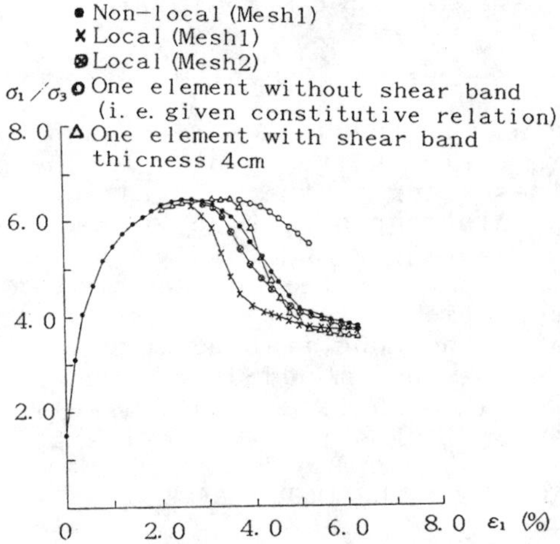

Fig. 5 Stress-strain relations

REFERENCES

1. T. Tanaka and Kawamoto O., Proc. 6th Int. Conf. Numer. Meth. Geomech., Innsbruck, 1213–1218 (1988)
2. M. Ortiz and J. C. Simo, Int. J. Numer. Meth. Engng. 23, 353–366 (1986)
3. T. Tanaka and Kawamoto O., Ing. Arch., Submitted
4. Z. P. Bazant and F. B. Lin, Int. J. Numer. Meth. Engng., 26 1805–1823 (1988)
5. F. Tatsuoka et al., 2nd Int. Workshop Numer. Meth. Localiz Bifur. Gran. Bodies, Gdansk (1989)

SIZE EFFECT AND MICROSTRUCTURE IN GRANULAR ROCKS

Ioannis VARDOULAKIS
Department of Civil and Mineral Engineering University of Minnesota USA, Department of Engineering Science, National Technical University, Athens, Greece

Jean SULEM
CERMES Ecole Nationale des Ponts et Chaussées, Paris, France

ABSTRACT

In this paper we present the constitutive equations of a deformation theory of plasticity for pressure sensitive dilatant materials incorporating second gradient into the flow rule and yield conditions. Size effects are discussed on the basis of a bifurcation analysis of triaxial compression tests on a particular sandstone.

1. INTRODUCTION

Depending on rock properties and on experimental conditions such as end constraint, confinement, shape and size of the specimen in uniaxial and triaxial compression tests various failure modes are observed; i.e. shear banding, barelling with distributed cracking and axial splitting . Under 'ideal' boundary conditions the occurence of these modes can be explained by the concept of equilibrium bifurcation. Bifurcation of the deformation process means that at some critical state the deformation process does not follow its 'straight ahead' continuation but turns to an entirely different mode. Within the framework of classical continuum theories we distinguish among two types of bifurcation modes: (a) discontinuous or localized modes and (b) continuous or diffuse bifurcation modes. As first indicated by Bazant [1] diffuse bifurcations cause secondary tensile stress which may the opening of preexisting latent cracks and lead finally to axial splitting. Localized bifurcations in form of shear bands in rocks were first addressed by Rudnicki and Rice [2]. Vardoulakis [3] on the other hand analysed diffuse and localized bifurcations in the triaxial test on granular materials. Even though classical continuum theories describe satisfactorily the onset of instabilities, they break down in the post-bifurcation regime leading to ill-posed mathematical problems. This has manifested itself quite dramatically in the numerical analysis of large scale problems where one often encounters a critical dependence of the solutions on mesh size, accompanied by stability and convergence problems. In addition, physically observed features such as shear band thickness, periodicity of shear bands, and preferred wavelengths in surface instability cannot be modeled. The missing link is the absence of an internal length in the constitutive models. Considering the microstructure of a material, the introduction of an internal length in the constitutive equations prevents the loss of ellipticity to occur in the governing equilibrium equations and consequently allows to regularize the considered bifurcation problem. Furthermore the introduction of an internal length leads to an influence of the size of the sample to the solution of the bifurcation problem (scale effect).

In a previous paper (Sulem and Vardoulakis [4]) the authors have presented the bifurcation analysis of the triaxial test using a Cosserat continuum theory which allows for independent particle displacements and particle rotations and which is suitable to model the microstructure of granular material (Mühlhaus and Vardoulakis [5]). In this paper microrotation of the grains is not considered. Following experimental and in situ observations that localization phenomena as well as slabbing or exfoliation phenomena are characterized by strong spatial variations of the volumetric deformation, the assumption is made here that plastic volumetric deformation is a function not only of the (classic) cumulative plastic shear strain γ^p but also of the second gradient $\nabla^2\gamma^p$. In a recent paper Vardoulakis and Aifantis [6] have presented a gradient flow theory of plasticity. In this paper we consider a deformation gradient theory of plasticity for pressure sensitive dilatant material. First the constitutive equations are given then the bifurcation analysis of axisymmetric triaxial test is presented. Finally, numerical results are presented and scale effects obtained within the frame of the gradient model and the Cosserat model are compared.

2. PROBLEM STATEMENT AND INCREMENTAL CONSTITUTIVE EQUATIONS

Let a homogeneous cylindrical rock specimen in an undistorted initial configuration be subjected to a monotonic axisymmetric deformation . An equilibrium bifurcation mode is said to be taking place as soon as in addition to the fundamental homogeneous axisymmetric motion of compression an other inhomogeneous perturbation solution exists that fulfils the same boundary conditions. The assumption is made here that the two end platens are perfectly lubricated so that the build up of frictional constraints is prevented.

In the prebifurcation regime, strains and stresses are homogeneous in the sample. Let (°) denote the Jaumann time derivative and (r, θ, z) be the polar coordinates with the z-axis coinciding with the specimen's axis. In the case of infinitesimal transitions emanating from a homogeneous state of axisymmetric triaxial compression we obtain the following set of rate equations for a second gradient deformation theory of elasto-plasticity:

$$
\begin{aligned}
\overset{\circ}{\sigma}_{rr} &= L_{11}\dot{\varepsilon}_{rr} + L_{12}\dot{\varepsilon}_{\theta\theta} + L_{13}\dot{\varepsilon}_{zz} - K\beta\ell^2\nabla^2\gamma^p \\
\overset{\circ}{\sigma}_{zz} &= L_{31}\dot{\varepsilon}_{rr} + L_{32}\dot{\varepsilon}_{\theta\theta} + L_{33}\dot{\varepsilon}_{zz} - K\beta\ell^2\nabla^2\gamma^p \\
\overset{\circ}{\sigma}_{rr} - \overset{\circ}{\sigma}_{\theta\theta} &= 2Gh_1(\dot{\varepsilon}_{rr} - \dot{\varepsilon}_{\theta\theta}) \\
\overset{\circ}{\sigma}_{rz} &= 2Gh_1\dot{\varepsilon}_{rz}
\end{aligned}
\qquad (1)
$$

with :

$$
\begin{aligned}
L_{11} &= G(4h_1/3+\kappa+h_2/3-h_3(1/\sqrt{3}+\kappa\beta)(1/\sqrt{3}+\kappa\mu)) \\
L_{12} &= G(-2h_1/3+\kappa+h_2/3-h_3(1/\sqrt{3}+\kappa\beta)(1/\sqrt{3}+\kappa\mu)) \\
L_{13} &= G(-2h_1/3+\kappa-2h_2/3-h_3(1/\sqrt{3}+\kappa\beta)(-2/\sqrt{3}+\kappa\mu)) \\
L_{31} &= G(-2h_1/3+\kappa-2h_2/3-h_3(-2/\sqrt{3}+\kappa\beta)(1/\sqrt{3}+\kappa\mu)) \\
L_{32} &= L_{31} \\
L_{33} &= G(4h_1/3+\kappa+4h_2/3-h_3(-2/\sqrt{3}+\kappa\beta)(-2/\sqrt{3}+\kappa\mu))
\end{aligned} \tag{2}
$$

In the above equations G and K are the elastic shear and bulk moduli respectively, κ=K/G , h_1, h_2, h_3 are given by :

$$h_1 = \frac{\chi_s}{1+\chi_s}; \quad h_2 = \frac{1}{1+\chi_s}; \quad h_3 = \frac{1}{1+\chi_t-\chi_T} \tag{3}$$

$$\chi_s = (q-p)h_s/G \; ; \chi_t = (q-p)h_t/G \; ; h_s = \mu/\gamma^p; \; h_t = d\mu/d\gamma^p(1-\alpha L^2\nabla^2\gamma^p) \tag{4}$$

$$\chi_T = -\kappa\mu(\beta-\beta'\ell^2\nabla^2\gamma^p) \tag{5}$$

and

$$K\beta\ell^2\nabla^2\gamma^p = \ell_1^2\nabla^2\dot{\varepsilon}_{rr} + \ell_2^2\nabla^2\dot{\varepsilon}_{\theta\theta} + \ell_3^2\nabla^2\dot{\varepsilon}_{zz} \tag{6}$$

with

$$
\begin{aligned}
\ell_1^2 &= \ell_2^2 = K\beta\ell^2 h_3(1/\sqrt{3} + \kappa\mu) \\
\ell_3^2 &= K\beta\ell^2 h_3(-2/\sqrt{3} + \kappa\mu)
\end{aligned} \tag{7}
$$

where ℓ^2 has the dimension of a square length and is related to the grain size of the material ($\ell^2 = r^2/10$ where r is the radius of a "representative" cell which is an assembly of mineral grains).

4. BIFURCATION ANALYSIS

To study the possibility for existence of a non-homogeneous deformation mode under the boundary conditions presented above we consider the following fields :

$$v_r = U(\rho)\cos\zeta \quad ; \quad v_z = W(\rho)\sin\zeta \tag{8}$$

where ρ and ζ are dimensionless radial and axial coordinates :

$$\rho = r/R \; , \quad \zeta = m\pi z/H \; (m = 1,2\ldots) \tag{9}$$

R and H are the current radius and height of the cylindrical specimen, m is a modal number and K_m is a shape number :

$$K_m = m\pi R/H \tag{10}$$

Using the constitutive equations (1) the equations of equilibrium lead to the following differential system in U and W:

$$a_1L_\rho(U)-a_2K_m^2U+a_3K_mW'-\ell_1^{*2}(L_\rho^2(U)-K_m^2L_\rho(U))-\ell_3^{*2}K_m(L_\rho(W')-K_m^2W') = 0$$

$$a_4K_m(U'+\frac{U}{\rho})-a_5(W''+\frac{W'}{\rho})+a_6K_m^2W-K_m\ell_1^{*2}(L_\rho'(U)+\frac{1}{\rho}L_\rho(U)-K_m^2(U'+\frac{U}{\rho})) \tag{11}$$

$$-K_m^2\ell_3^{*2}(W''+\frac{W'}{\rho}-K_m^2W) = 0$$

where a_i (i=1,6) are functions of the state of stress and of the constitutive parameters L_{ij} and G, L_ρ is the Bessel operator and $\ell_i^* = \ell_i/R$, (i=1,2).

The boundary conditions express that along the external cylindrical surface of the specimen, a constant confining pressure is acting. Furthermore as it is demonstrated in [6] the presence of higher order strain gradients requires extra boundary conditions. This extra boundary condition express that the flux of plastic strain-rate is equal to zero at the external cylindrical surface ($\partial_r \dot{\gamma}^p = 0$ for $\rho = 1$).

The bifurcation condition is derived from the requirement of non trivial solutions for the perturbation fields U and W.

<u>A numerical example:</u> We refer here to a Doddington sandstone on which an experimental program of triaxial tests has been performed by Santarelli [7]. Size effect in terms of critical plastic shear strain at barelling (m=1) vs characteristic dimension of the sample for a given value of the slenderness of the sample (H/R = 4) is shown on figure 1 for a gradient model and for a Cosserat model. We can observe that bigger specimens are less stable than smaller ones. A gradient model allows to describe larger size effect than a Cosserat model. In a near future we shall construct constitutive models which combine gradient effects and microrotation of grains.

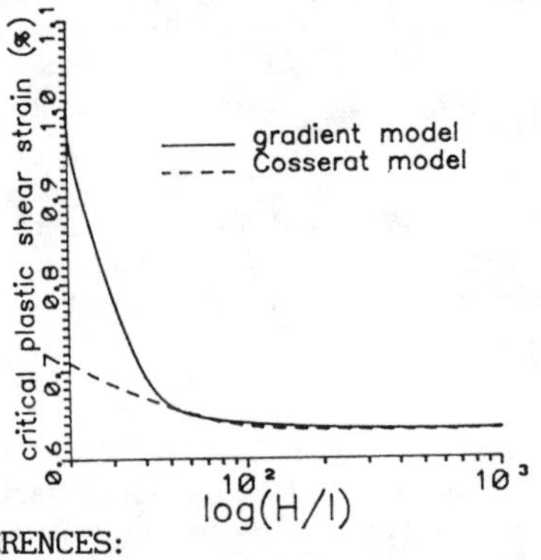

FIG. 1. Bifurcation of a sandstone specimen under triaxial compression. Size effect (m = 1, H/R = 4).

REFERENCES:

1. Bazant, Z.P., L'instabilité d'un milieu continu et la résistance en compression. Bulletin Rilem <u>35,</u> 99-112 (1967).67).).

2. Rudnicki, J.N. and Rice, J.R., Conditions for the localization of the deformation in pressure sensitive dilatant material. J. Mech. Phys. Solids <u>23</u>, 371-394 (1975).

3. Vardoulakis, I., Rigid granular plasticity model and bifurcation in the triaxial test. Acta Mechanica <u>49</u>, 57-59 (1983).

4. Sulem J., Vardoulakis I., Bifurcation Analysis of the Triaxial Test on Rock Specimens. A Theoretical Model to Shape and Size Effect. Acta Mechanica (in print),(1990).

5. Mühlhaus H.B., Vardoulakis I., The Thickness of Shear Bands in Granular Materials. Géotechnique <u>37</u>, 271-283 (1987).).

6. Vardoulakis I., Aifantis E.C., A Gradient Flow Theory of Plasticity for Granular Materials, in preparation (1990)

7. Santarelli, F.J., Theoretical and experimental investigation of the stability of the axisymmetric wellbore. PhD thesis. Imperial College, London, (1987).

FINITE SIMPLE SHEAR USING A COUPLED DAMAGE AND PLASTICITY MODEL

GEORGE Z. VOYIADJIS
Professor of Civil Engineering, Louisiana State University, Baton Rouge, LA 70803

PETER I. KATTAN
Research Associate, Dept. of Civil Engineering, Louisiana State University, Baton Rouge, LA 70803

ABSTRACT

Constitutive equations are developed for a coupled model of finite strain plasticity and continuum damage mechanics. The hypothesis of elastic energy equivalence is used to derive the necessary transformations between the damaged state and the hypothetical undamaged state. To demonstrate the use of the model, the problem of finite simple shear is investigated and the appropriate differential equations derived.

THE COUPLED PLASTICITY-DAMAGE MODEL

In a recent paper, Voyiadjis and Kattan [1] presented a thermodynamically based consistent model coupling finite strain plasticity with continuum damage mechanics. The concept of effective stress introduced earlier by Kachanov [2] is used in conjunction with the hypothesis of elastic energy equivalence [3]. Two configurations of the body are considered. The first configuration, denoted here by C, refers to the deformed damaged state, while the second configuration, denoted here by $\bar{C}$, refers to a hypothetical state of the body where it is deformed but undamaged. The relationship between the two configurations C and $\bar{C}$ is achieved through the use of the hypothesis of elastic energy equivalence. In this hypothesis, it is assumed that the elastic energy in the configuration C is the same in form as the elastic energy in the configuration $\bar{C}$ except that the stresses are replaced by the corresponding effective stresses and the damage tensor components are taken to be identically zero in $\bar{C}$.

In the formulation, the Eulerian reference system is used, i.e., all the actual quantities are referred to the configuration C, while the effective quantities referred to $\bar{C}$. A linear transformation is first introduced between the Cauchy stress tensor $\underset{\sim}{\sigma}$ and its effective counterpart $\underset{\sim}{\bar{\sigma}}$ in the form:

$$\bar{\sigma}_{ij} = M_{ijkl}\, \sigma_{kl} \tag{1}$$

where M_{ijkl} are the components of the fourth-rank damage effect tensor. Rewriting equation (1) in terms of the deviatoric stresses $\underset{\sim}{\tau}$ and $\underset{\sim}{\bar{\tau}}$, one obtains:

$$\bar{\tau}_{ij} = N_{ijkl}\,\sigma_{kl} \tag{2}$$

where N_{ijkl} are the components of a fourth-rank tensor given by

$$N_{ijkl} = M_{ijkl} - \frac{1}{3} M_{rrkl}\,\delta_{ij} \tag{3}$$

Using equation (3), one can obtain the following transformation for the second deviatoric stress invariant:

$$\bar{\tau}_{ij}\,\bar{\tau}_{ij} = H_{klmn}\,\sigma_{kl}\,\sigma_{mn} \tag{4}$$

where the fourth-rank tensor components H_{klmn} are given by

$$H_{klmn} = N_{ijkl}\,N_{ijmn} \tag{5}$$

Similar transformation equations hold for the backstress tensor $\underset{\sim}{\beta}$ and its deviatoric part $\underset{\sim}{\alpha}$, i.e., $\bar{\beta}_{ij} = M_{ijkl}\,\beta_{kl}$, $\bar{\alpha}_{ij} = N_{ijkl}\,\beta_{kl}$, $\bar{\alpha}_{ij}\,\bar{\alpha}_{ij} = H_{klmn}\,\beta_{kl}\,\beta_{mn}$, and $\bar{\tau}_{ij}\,\bar{\alpha}_{ij} = H_{klmn}\,\sigma_{kl}\,\beta_{mn}$.

Next one uses the hypothesis of elastic energy equivalence to obtain the transformation of the elastic strain tensor $\underset{\sim}{\varepsilon}'$ as

$$\bar{\varepsilon}'_{kl} = M_{klmn}^{-T}\,\varepsilon'_{mn} \tag{6}$$

It should be noted that in deriving equation (6), small (infinitesimal) elastic strains along with the existence of an elastic strain energy function, are assumed.

In a similar fashion, one uses the von Mises yield function f (with both isotropic and kinematic hardening) along with the hypothesis of elastic energy equivalence to obtain the following transformation equations for the elastic and plastic parts, $\underset{\sim}{d}'$ and $\underset{\sim}{d}''$ of the spatial strain rate tensor d:

$$\bar{d}'_{ij} = M_{ijkl}^{-T}\,d'_{kl} + \mathring{M}_{ijmn}^{-T}\,\varepsilon'_{mn} \tag{7}$$

$$\bar{d}''_{ij} = \gamma\,N_{ijkl}\,d''_{kl} - \gamma\,\dot{\Lambda}\,(\sigma_{mm} - \beta_{nn})\,N_{ijrr} \tag{8}$$

where $\dot{\Lambda}$ is the scalar multiplier used in the associated flow rule of plasticity, γ is a ratio of the two scalar factors $\dot{\Lambda}$ in the configurations C and $\bar{C}$, and a superposed "o" indicates an appropriate corotational derivative.

Using the above transformation equations, one derives the inelastic constitutive relation in the form

$$\sigma_{ij} = \bar{D}_{ijkl} \, d_{kl} \tag{9}$$

where $\bar{D}_{ijkl}$ are the components of the fourth-rank effective elasto-plastic stiffness tensor (that includes the effect of damage). The derived expression for the tensor $\bar{\underset{\sim}{D}}$ takes the form:

$$\bar{D}_{ijkl} = O_{pqij}^{-1} \, D_{pqmn} \, M_{mnkl}^{-T} \tag{10}$$

where the fourth-rank tensor components O_{pqij} are dependent on the yield and damage functions f and g, the damage variable ϕ_{ij} and the fourth-rank elasticity and elasto-plastic stiffness tensors $\underset{\sim}{E}$ and $\underset{\sim}{D}$.

FINITE SIMPLE SHEAR

The problem of finite simple shear in the x_1-direction is discussed now. For this problem, the spatial strain rate tensor $\underset{\sim}{d}$ (the symmetric part of the velocity gradient) and the spatial spin tensor $\underset{\sim}{W}$ (the antisymmetric part of the velocity gradient) are given by:

$$[d] = \frac{k}{2} \begin{bmatrix} 0 & 1 & 0 \\ 1 & 0 & 0 \\ 0 & 0 & 0 \end{bmatrix} , \quad [W] = \frac{k}{2} \begin{bmatrix} 0 & 1 & 0 \\ -1 & 0 & 0 \\ 0 & 0 & 0 \end{bmatrix} \tag{11}$$

where k is a constant representing the shearing strain rate. Using a corotational stress rate in terms of a modified spin tensor $\underset{\sim}{\Omega}$ of the form:

$$\overset{\circ}{\sigma}_{ij} = \dot{\sigma}_{ij} + \sigma_{im} \, \Omega_{mj} - \Omega_{in} \, \sigma_{nj} \tag{12}$$

and a similar equation for $\underset{\sim}{\alpha}$, along with equation (11), one obtains the following system of differential equations in $\underset{\sim}{\sigma}$ and $\underset{\sim}{\alpha}$:

$$\dot{\sigma}_{11} = k \, \bar{D}_{1112} + 2 \, \Omega_{12} \, \sigma_{12} \tag{13a}$$

$$\dot{\sigma}_{22} = k \, \bar{D}_{2212} - 2 \, \Omega_{12} \, \sigma_{12} \tag{13b}$$

$$\dot{\sigma}_{33} = k \, \bar{D}_{3312} \tag{13c}$$

$$\dot{\sigma}_{12} = k\,\bar{D}_{1212} + \Omega_{12}\,(\sigma_{22} - \sigma_{11}) \tag{13d}$$

$$\dot{\alpha}_{11} = \dot{\mu}\,(\tau_{11} - \alpha_{11}) + 2\,\Omega_{12}\,\alpha_{12} \tag{13e}$$

$$\dot{\alpha}_{22} = \dot{\mu}\,(\tau_{22} - \alpha_{22}) - 2\,\Omega_{12}\,\alpha_{12} \tag{13f}$$

$$\dot{\alpha}_{33} = \dot{\mu}\,(\tau_{33} - \alpha_{33}) \tag{13g}$$

$$\dot{\alpha}_{12} = \dot{\mu}\,(\tau_{12} - \alpha_{12}) + \Omega_{12}\,(\alpha_{22} - \alpha_{11}) \tag{13h}$$

where $\dot{\mu}$ is a scalar multiplier obtained from the consistency condition in the plasticity model. The differential equations (13) can now be solved using a Runge-Kutta-Verner type numerical technique.

REFERENCES

1. Voyiadjis, G. Z., and Kattan, P. I., "A Coupled Theory of Damage Mechanics and Finite Strain Elasto-plasticity. Part II: Damage and Finite Strain Plasticity," International Journal of Engineering Science, Vol. 28, No. 6, pp. 505-524 (1990).
2. Kachanov, L. M., "On the Creep Fracture Time," Izv Akad. Nauk USSR Otd. Tekh., Vol. 8, pp. 26-31 (1958) (in Russian).
3. Sidoroff, F., "Description of Anisotropic Damage Application to Elasticity," in IUTAM Colloquium on Physical Nonlinearities in Structural Analysis, Springer-Verlag, Berlin, pp. 237-244 (1981).

PROGRESSIVE DAMAGE ANALYSIS OF BRITTLE SOLIDS

S. YAZDANI, R. LIN, and S. KARNAWAT
Dept. of Civil Engineering and Construction
North Dakota State University
Fargo, ND 58105

ABSTRACT

The general framework of a damage mechanics model for brittle solids is developed. The theory identifies two distinct cracking patterns and covers both elastic and inelastic damage processes. The model is illustrated for mortar in uniaxial tension with loading-unloading stress path.

INTRODUCTION

The mechanical properties of brittle solids are known to be very sensitive to the kinetics of microdefects under applied stresses. Alteration of elastic properties, induced anisotropy, and transition to a ductile response under lateral pressures are a few examples of such a diverse behavior. Before deformations are localized (for example, a macrocrack formation), the microdefect growth is a fairly distributed process. For such situations, continuum damage mechanics (CDM) can be used as a viable engineering method for the material modeling. Interested readers can refer to a recent review paper by Krajcinovic [1] on the development and concepts of the CDM.

This study outlines the general formulation of a constitutive damage model for brittle solids. Both elastic and inelastic damage processes are included. The internal variable theory of thermodynamics is used and an energy dissipation inequality is established. Only small deformations are considered. It is assumed that the distributed cracking (damage) can be modeled as continuum with isothermal deformations. Following the novel idea of Oritz [2], damage is reflected through the fourth-order compliance tensor. In addition to the general formulation, specific evolutionary expressions are proposed for mode I cracking through which both dilatation and inelastic damage strains are obtained.

THEORY

In the absence of any body couple and heat source, assume that the elastic properties are affected by the process of damage. Denoting the accumulated damage as k, the current compliance tensor of the material can be decomposed into its additive parts as

$$\mathbf{C}(k) = \mathbf{C}^{o} + \mathbf{C}^{c}(k) \tag{1}$$

where the initial compliance is given by $\mathbf{C}^{o}$ and $\mathbf{C}^{c}(k)$ is the added flexibility due to damage. It is further assumed that damage is irreverssible, so that $\dot{k} \geq 0$. With the framework of the internal variable theory of thermodynamics, and its associated assumption that a neighboring equilibrium state exists, the Gibb's energy potential is given as

$$G(\sigma,C,\varepsilon^{i}) = \frac{1}{2}\sigma : C : \sigma + \sigma : \varepsilon^{i} \tag{2}$$

where σ is the Cauchy's stress tensor and ε^i is the strain due to inelastic damage processes (Yazdani and Schreyer [3]). The second law requires that dissipation, d_s, be non-negative. Mathematically, one can express this as

$$d_s = \frac{\partial G}{\partial C} :: \dot{C} + \frac{\partial G}{\partial \varepsilon^{i}} : \dot{\varepsilon}^{i} \tag{3}$$

where the superposed dots denote time derivatives and : signifies contraction operation. Equation (3) can also be interpreted as the rate of work done by the thermodynamic affinities, $\frac{\partial G}{\partial C}$ and $\frac{\partial G}{\partial \varepsilon^{i}}$, through their respective fluxes, $\dot{C}$ and $\dot{\varepsilon}^{i}$. Combining Equations (1), (2), and (3), the following form for dissipation is obtained

$$d_s = \frac{1}{2}\sigma : \dot{C}^{c} : \sigma + \sigma : \dot{\varepsilon}^{i} \tag{4}$$

To progress further, evolutionary equations are needed for $\dot{C}^c$ and $\dot{\varepsilon}^i$. With the use of damage parameter k, consider the following definitions

$$\dot{C}^{c} = \dot{k}\mathbf{R} \quad \text{and} \quad \dot{\varepsilon}^{i} = \dot{k}\mathbf{M} \tag{5}$$

where **R** is a fourth order damage response tensor which determines the direction of elastic damage processes and **M**, a second order tensor, determines the direction of $\dot{\varepsilon}^i$. Both **R** and **M** play a central role in developing a meaningful constitutive material model. Assuming penny shaped microcracks, two cracking patterns are recognized and are labeled as modes I and II. Mode I is the cleavage cracking and mode II is the combined shear sliding and crack opening (Ortiz [2], Yazdani and Schreyer [3]). To address these two modes, a conical decomposition of the stress tensor is performed (Ortiz [2])

$$\sigma = \sigma^{+} + \sigma^{-} \tag{6}$$

with σ^+ and σ^- being the positive and negative cones of σ. Further decomposition of the response tensor **R** is also made

$$\mathbf{R} = \mathbf{R}_{I} + \mathbf{R}_{II} \tag{7}$$

where subscripts I and II refer to the two fracturing modes stated above.

To characterize the induced damage, a damage criterion is needed. Define a damage potential $\Psi(\sigma,C,\varepsilon^{i})$ such that when damage occurs $\Psi=0$. This in turn defines a damage surface. The condition $\Psi>0$ is not allowed and $\Psi<0$ indicates an elastic response. Using the damage criterion together with Equations (4), (5), and (7), the general form of the damage surface is obtained

$$\Psi(\sigma,C,\varepsilon^i) = \frac{1}{2}\sigma : \mathbf{R}_I : \sigma + \frac{1}{2}\sigma : \mathbf{R}_{II} : \sigma + \sigma : \mathbf{M} - \frac{1}{2}t^2(k) = 0 \tag{8}$$

where t(k) is a damage function which characterizes damage hardening and softening.

To complete the formulation, specific forms for **R** and **M** must be provided. Such forms are proposed below for mode I cracking. Description of mode II inelasticity is beyond the space allocation for this paper. For mode I deformations, Ortiz [2] postulated the following equation

$$\mathbf{R}_I = \frac{\sigma^+ \otimes \sigma^+}{\sigma^+ : \sigma^+} \tag{9}$$

which was also used by Yazdani and Schreyer [3]. The symbol ⊗ in Equation (9) indicates the tensor product. The change in the apparent Poisson's ratio, observed experimentally for brittle solids, can not be modeled by the use of Equation (9). To improve on this perceived deficiency, the following modification is proposed

$$\mathbf{R}_I = \frac{\sigma^+ \otimes \sigma^+}{\sigma^+ : \sigma^+} + \gamma\, H(\lambda)\, (\mathbf{I} - \mathbf{i} \otimes \mathbf{i}) \tag{10}$$

where λ is the maximum eigenvalue of σ^+ and γ represents a material parameter. The Heavside function H(·) in Equation (10) is used to ensure that $\mathbf{R}_I$ is activated under tensile loading only. Brittle solids, in tension, usually display increasing volumetric strains with increase in axial deformation (Fonseka and Krajcinovic [4]). The material parameter γ is used to accommodate such deformational patterns.

The tensor **M** is also decomposed as $\mathbf{M} = \mathbf{M}_I + \mathbf{M}_{II}$ to reflect the decoupling of the two fracturing modes. A specific form for $\mathbf{M}_{II}$ is given by Yazdani and Schreyer [3] and will not be repeated here. For $\mathbf{M}_I$, experimental tests on a number of brittle solids have indicated that the cleavage cracking takes place at approximately right angles to the direction of principal stress axis. Guided by such meso-level kinetics, a simple form for $\mathbf{M}_I$ is proposed as

$$\mathbf{M}_I = \beta\, \sigma^+ \tag{11}$$

where β is a material parameter which controls the magnitude of inelastic damage strains.

To illustrate the model for a particular material, a specific form of the damage function, suitable for that material, must be used. Combining Equations (8), (10), and (11), and following the work of Ortiz [2], the following form of the damage function is established

$$t = \sqrt{1 + 2\beta}\;\; f_t\, e^{\frac{1}{1 + E_o\, k}}\; \ln(1 + E_o k) \tag{12}$$

where E_o is the initial value of Young's modulus and f_t is the uniaxial tensile strength of the material. Other forms of damage function could also be used. With E_o = 4500 ksi, Poisson's ratio of .20, f_t = .50, γ = .20, and β = 1.0, the theoretical stress-strain curve is plotted in Figure 1 for both axial and lateral deformations for mortar. The experimental results of Gopalaratnam and Shah [5], and the model prediction using Equation (9) is also plotted for comparison. It can be seen that salient features of material behavior in mode I are captured by the proposed formulation. The softening regime is represented with broken lines to indicate that localized deformations are usually observed beyond the limit state. The assumption of distributed damage used in the development of the CDM is no longer valid under such situations.

REFERENCES

1. D. Krajcinovic, "Damage mechanics", Mech. Mat., 8, 117-197, 1989.
2. M. Ortiz, "A constitutive theory for the inelastic behavior of concrete", Mech. Mat., 4(1), 67-93, 1985.
3. Yazdani and H. L. Schreyer, "Anisotropic damage model with dilatation for concrete", Mech. Mat., 7(3), 231-244, 1988.
4. G. U. Fonseka and D. Krajcinovic, "The continuous damage theory of brittle materials", part 2, J. Appl. Mech., ASME, 48, 816-824, 1981.
5. V. S. Gopalaratnam and S. P. Shah, "Softening response of plain concrete in direct tension", ACI Journal, 87, 310-323.

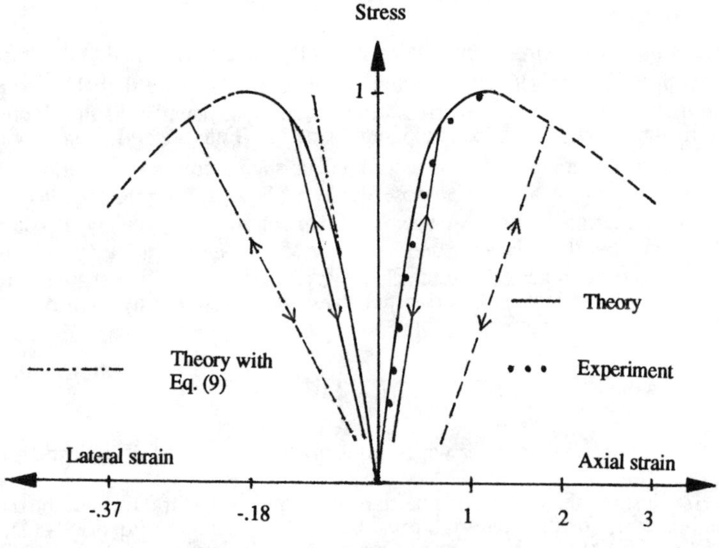

Figure 1. Theoretical stress-strain curves in uniaxial tension for mortar (normalized).

CONSTITUTIVE LAWS OF DAMAGE MATERIAL

PING ZHANG
Mechanical Engineering Department, Xiangtan University, Hunan, P.R.C.

ABSTRACT

Based on the continuum damage mechanics, a new method to establish the evolution equation of damage is provided by introducing the accompaning parameter. Problem of creep damage is investigated. The evolution equation of creep damage is obtained. Creep tests have been conducted at 700°c on GH33A alloy.

INTRODUCTION

Continuum damage mechanics (C.D.M.) has developed continuously since the early works of Kachanov and Rabotnov. It constitutes a practical tool to take into account the various damaging processes in materials and structures at a macroscopic continuum level. The establishment of an adequate damage law is unquestionably one of the important and, as yet unresolved, issues. Several possible avenues are being explored.[1,2,3,4]

From macrosscopic aspect, damage is not directly measurable as strain or plastic. However, the damaging process is an irreversible of energy dissipation and is accompanied with variations of some physical parameters. We can obtain indirectly the extent of material deterioration by measuring these physical parameters, such as density, resistivity, sound velocity and hardness etc. We call these parameters as Accompaning Parameters. Accompaning parameters play an important part in establishing the contitutive laws of damage material. From this viewpoint, we provide a new method to establish the evolution equation of damage with the aid of introducing accompaning parameters.

CONSTITUTIVE ASSUMPTIONS AND EVOLUTION EQUATION OF DAMAGE

The definition used herein is in accordance with continuum mechanics in order to solve problems of solid mechanics involving stresses, strains and failure. Those variables are defined at each mathematical point. The damage variable is considered as an internal state argument in irreversible process of progressive deterioration in describing the mechanics behavior of material from macroscopic aspect.

All the physical and chemical phenomena such as creep, fatigue, corrosion and radiation which result in damage have their own damage mechanism, and correspond respectively to different evolution laws of damage. So it is suitable and necessary that we put forward different basic damages to indicate the different characters of a variety of damage mechanism.[5,6]

To make clear, all kinds of irreversible developing processes of basic damage are called as basic damage career, i.e., damage career in short. For example, taking cumulative plastic strain when $\sigma > \sigma_w$ (damage stress threshold) as the career parameter of plastic damage; taking the time of tertiary creep as the career parameter of creep damage. In multimechanism damage, a damaging process may include several damage careers.

Consider now a single cell, Let m kinds of basic damage mechanism and p kinds of accompaning parameter exist in this cell, corresponding damage variables and accompaning parameters would be expressed respectively as W1, W2,... Wm and y1, y2,...yp. Career variables would expressed as Z1,...Zm.

At the given time t, the basic damage career would respectively be

$$\begin{cases} Z_1 = Z_1(t) \\ \cdots \\ Z_m = Z_m(t) \end{cases}$$

Signing

$$\mathbf{Z} = \{Z_1, \cdots Z_m\}$$
$$\mathbf{W} = \{W_1, \cdots W_m\}$$
$$\mathbf{Y} = \{y_1, \cdots y_p\}$$

Constitutive Assumptions

At the present state of the development, CDM is a state type of theory.

Damages are a set of internal variables which describe the current state of material deterioration. Accompanig parameters are not only related to the state of damage but also related to other state parameters. For example, the elasticity modulus and the hardness are not only related to the state of damage but also related to temperature.

The accompaning parameters are assumed to be functions of the damages and the state parameters, $\mathbf{K}$ ($\mathbf{K} = \{K_1, K_2 \cdots K_l\}$):

$$\mathbf{Y} = B(\mathbf{W}, \mathbf{K}) \tag{1}$$

Differentiating the foregoing expression, we obtain

$$d\mathbf{Y} = D_W B\, d\mathbf{W} + D_K B\, d\mathbf{K} \tag{2}$$

The Measuing Equation and Evolution Equation

In the damaging process $\mathbf{Z}$, $\mathbf{y}$ can be determined by experiment. If the response function of accompaning parameters is F ($\mathbf{Z}$) then

$$d\mathbf{y} = F(\mathbf{Z})\, d\mathbf{Z} \tag{3}$$

Equation (3) is called as the measuring equation. From equations (2) and (3), we obtain

$$d\mathbf{w} = D_w^{-1} B (F d\mathbf{Z} - D_k B d\mathbf{K}) \tag{4}$$

In one damage mechanism case, equation (4) can be writen

$$dw = \left(\frac{\partial B}{\partial w}\right)^{-1} \left[F(Z) dZ - \sum_{i=1}^{k} \frac{\partial B}{\partial K_i} dK_i\right] \tag{5}$$

Neglecting state parameter terms

$$dw = \left(\frac{\partial B}{\partial w}\right)^{-1} F(Z) dZ \tag{6}$$

In the damage process, it result in variations of several accompaning parameters even one damage mechanism case. Therefore the representation of damage is not unique.

HIGH TEMPERATURE CREEP DAMAGE

Creep damage is a term to describe the material degradation which gives rise to the acceleration of creep rate. The damaging process begins at the end of secondary creep. We choose creep rate $\dot{\varepsilon}$ as accompaning parameter and the time of tertiary creep as damage careep.

For instantaneous increases in damage the following relation is postulated

$$\frac{d\dot{\varepsilon}}{dw} = f(w) \tag{7}$$

For many materials, according to the experimentel results, the measuring equation of creep rate may be writen as

$$d\dot{\varepsilon} = h\, \dot{\varepsilon}_m^s Z^{\nu-1} dZ \tag{8}$$

Where s, ν and h are constants which are detained by material and text condition. $\dot{\varepsilon}_m$ is minimum creep rate. By (7) and (8), we have

$$f(w)\, dw = h\, \dot{\varepsilon}_m^s Z^{\nu-1} dZ \tag{9}$$

If $f(w) = A(1-w)^r$, where A and r are material and temperature dependent coefficients, then

$$A(1-w)^r dw = h\dot{\varepsilon}_m^s Z^{v-1} dZ \qquad (10)$$

Integrating the foregoing expression we obtain

$$w = 1-[1-\frac{h(r+1)}{Av}\dot{\varepsilon}_m^s Z^v]^{\frac{1}{r+1}} \qquad (11)$$

According to the critical damage value Wc=1, the critical value of damage career may be obtained

$$Z_c = [\frac{Av}{(r+1)h\dot{\varepsilon}_m^s}]^{\frac{1}{r+1}} \qquad (12)$$

From equations (11) and (12) the evolution equatio -n can be writen as

$$w = 1-[1-(\frac{Z}{Z_c})^v]^{\frac{1}{r+1}} \qquad (13)$$

Material and Experimental Results

The material used in this investigation was CH33A alloy with nomposition Ni-20Cr-0.95Al-2.8Ti-B. The heat treatment were used: 1080 c/16h A.C. . The creep tests were performed at temperature 700 c. The measuring equation of accompaning parameter is obtained as

$$d\dot{\varepsilon} = 1140\,\dot{\varepsilon}_m^3 Z dZ \qquad (14)$$

The evolution equation of creep damage is

$$w = 1-[1-(\frac{Z}{Z_c})^2]^{0.117} \qquad (15)$$

CONCLUSLONS

Establishing the evolution equation of damage depe -nds on the measurement of accompaning parameter. It is necessary to inverstigate the relationship between damage and accompaning parameter. In this paper provides a new method to establish the evolution equation of damage by introducing accompapning parameter and its application.

REFERENCES

1. L.M.kachanov, "Time of the Rupture process process Under Creep Conditions", Izv. Akad. Nauk. SSR, Otd Tekh Nauk NO. 8, 26-31 (1958).
2. Y.N.Rabotnov, Creep Problems in Structural Members (North Holland 1969)
3. J.Lemaitre, "How to Vse Damage Mechanics", Nuclear Engng. and Desinn 80, 233-245 (1984).
4. J.L. Chaboche, "Continuum Damage Mechanics", J. Appl. Mech. 55 (1), 59-64 (1988)
5. P.Zhang, "Investigation of the Behaviour of Multimechanism Damage Evolution", J. Xiangtan University 12 ((1) 27-33 (Mar. 1990)
6. P.Zhang, "Measurement for Multimechanism Damage", J. Xiangtan University 11 (4), 40-46 (Dec. 1989)

MICRO-MACRO CORRELATION

A PROPOSAL FOR A NEW UNDERSTANDING OF THE CRACK TIP: A NON-LINEAR APPROACH

Elias C. Aifantis
Department of Mechanical Engineering-Engineering Mechanics
Michigan Technological University, Houghton, MI 49931, USA

ABSTRACT

It is proposed to utilize the approach of nonlinear dynamics for self-organized microstructures in order to understand the stability and growth of the crack tip. This approach has been recently applied with some success to address certain unresolved issues in plasticity and associated deformation patterning phenomena including the velocity, width and spacing of deformation bands. No such attempt has been made, however, for the ultimate stage of deformation i.e. fracture. The purpose of the present lecture is to outline such possibility, by providing some evidence in support of the necessity of this proposal and by listing a few preliminary results in support of its feasibility.

INTRODUCTION

It is suggested to explore the nonlinear nature of the crack tip by considering its interaction with the heterogeneously evolving microstructure (e.g. point defects, dislocations, voids) ahead of it. Two previously unexplored methods to fracture mechanics may be adopted: (i) the stability of partial differential equations (SPDE) approach, and (ii) the cellular automata (CA) simulation.

In the SPDE approach a simplified system of nonlinear partial differential equations may be formulated for each type of microstructure. They will involve the applied stress field, the microstructure density, and the crack-tip opening displacement or the crack-tip velocity. Then, the stationary or dynamical stability of the crack-tip may be examined by resorting to bifurcation and nonlinear stability techniques recently advanced in the theory of nonlinear phenomena and dissipative structures. In the CA approach a specific microscopic mechanism of crack growth may be assumed for each type of microstructure. This will be implemented into a CA lattice placed ahead of the crack tip. Activation of the CA process will then simulate the macroscopic crack propagation by letting portion of the lattice elements to be "consumed" by the advancing crack. The rate of this advancement is determined by the microscopic rule assigned for the evolution and interaction of the particular microstructure under consideration with the crack tip.

It is anticipated that the presently proposed "truly" nonlinear framework to crack stability and propagation will provide new insight into some currently unresolved issues of fracture mechanics, including the determination of the crack-tip velocity and the size and stability of the dislocation cell walls ahead of it which often dominate the crack growth process. In this connection, it is pointed out that usual elastodynamic or pseudo-plastic analyses of crack propagation essentially assume a constant crack velocity given at the outset. This is in contrast to the approach proposed here where the crack velocity is to be obtained from a "travelling wave front" analysis of the governing differential equations.

Moreover, many interesting deformation patterning phenomena routinely observed at stationary crack tips including slip bands and plastic zones may be given a new possibility for interpretation. Similarly, the branching and river-like patterning phenomena often observed at travelling crack tips in an orderly periodic, fractal, or even chaotic fashion could also be addressed quantitatively.

It is thus hoped that a new fundamental understanding of fracture toughness and crack propagation phenomena will be achieved. This understanding will include both brittle and ductile fracture processes. It will be particularly useful for crack growth studies under environmental (hydrogen) or creep (high temperature) conditions, as well as for problems of interfacial fracture and fracture of composites.

EVIDENCE IN SUPPORT OF THE PROPOSED APPROACH

There is ample evidence supporting the need for a nonlinear approach to fracture with emphasis on the description of the complex deformation patterns ahead of the crack tip and the determination of the crack velocity. We briefly elaborate on such evidence here by providing a set of photographs from existing experimental work revealing crack features which cannot be described by adopting solely current elastic-like (K-field) or elastoplastic (J-field) analyses to fracture.

Figure 1 shows well known deformation patterns routinely observed ahead of crack tips. Nevertheless, there is not yet a predictive theory or even a systematic attempt for describing the topology of (a) plastic zones, (b) the dislocation cell wall sizes or (c) the number and spacing of microcracks occurring ahead of the crack tip. In the same figure we also show crack surfaces with two commonly observed spatial features, namely (d) cracklets and (e) river-like crack patterns which again cannot be explained within existing framework of fracture analysis.

Figure 2 shows another explicit nonlinear feature of crack tip behavior; namely, stick-slip fracture. Such fracture is characterized by (a) periodic-like crack-tip velocity jumps and discontinuous or step-like crack growth physically reflected by (b) texture changes on the fracture surface.

Figure 3 demonstrates the non-convex or non-monotonous character of the crack resistance curve. The negative-slope regime is a macroscopic manifestation of the microscopic deformation instabilities such as thermal softening, molecular relaxation (polymers) or negative strain rate sensitivity (metals).

The remaining half of this section is devoted to a similar discussion for plastic flow. This is done not only because of strong analogies between plastic deformation and fracture but also because more definite results are already available for plasticity. In this connection, it is noted that the present author and co-workers were among those who initiated a nonlinear dynamical approach for the heterogeneity of plastic flow and wavelength selection analysis of the associated dislocation patterns. This, again, is in contrast to the previously existing energy minimization approaches which are generally inappropriate for "far from thermodynamic equilibrium" dissipative (dislocation) structures. In the case of deformation and shear band patterning such a dynamical approach resulted in a plasticity theory with a gradient-dependent flow stress. This provided, among other things, the spacing and width of stationary shear bands, as well as the velocity of travelling Portevin-Le Chatelier bands: all of which are features that have not been captured before by classical localization of deformation analyses. Some of these results are shown in Figures 4, 5, 6 and further details can be found in references [1-4]. We include these pictures here for two reasons: The first reason is to illustrate the similarities between plastic flow and fracture in relation to instability and deformation patterning phenomena. The second reason is to provide definite evidence about the suitability of the advocated approach in predicting such nonlinear patterning phenomena. Figure 4 shows (a) the periodicity of persistent slip bands (PSB) as revealed by transmission electron microscopy, (b) the plateau of the associated cyclic stress-strain graph, and (c) the periodic solutions for the corresponding dislocation structures obtained from a gradient-dependent model of dislocation dynamics. Figure 5 shows (a) an optical micrograph revealing the width of a macroscopic shear band (MSB), (b) the softening regime in a true stress-strain graph, and (c) the solution of a nonlinear equation for the plastic strain indicating the evolution and width changes of a stationary shear band. Finally, Figure 6 shows (a) an optical micrograph and a schematic illustration revealing the width and spacing of Portevin-Le Chatelier bands, (b) the nonconvex σ - $\dot{\varepsilon}$ graph, and (c) the experimental step-like stress-strain graph for constant stress rate tests. Also in the figure we show the solutions obtained from a nonlinear gradient-dependent model of plastic flow for (d) the travelling bands and (e) the associated stress-strain graph. The agreement between theory and experiment is quite encouraging and has prompted the proposed extension of this approach to fracture.

ADVOCATED APPROACH AND PRELIMINARY RESULTS

(i) Stability of Partial Differential Equations (SPDE) Approach

This approach is a definite departure from existing crack tip analyses based on standard reversible or irreversible thermodynamic considerations. For example, the non-uniformity of the stress field has not been taken into account by Griffith or

Cherepanov in setting up their energy balance equation in the vicinity of "brittle" or "ductile" crack tips. [An exception was taken in a recent paper [5] where the Cahn-Hilliard approach for non-uniform systems was extended to describe the crack growth]. It is expected that higher-order strain gradients will play here the same role that they did in determining the width and spacing of shear bands in the localization of deformation problem. The role of gradients will also be crucial in obtaining the crack velocity from a wave-front analysis rather than assuming it at the outset. As a preliminary result we show in Figure 7 (a) the periodicity of velocity resulting from (b) an experimentally obtained non-convex crack resistance graph. This was achieved by applying the SPDE approach [6] to the problem of viscoelastic peeling whose nonlinear aspects are very similar to polymer fracture [7]. In the same figure we show the step-like peeling (cracking) as obtained (c) theoretically and as observed (d) experimentally.

Similarly, there is an abundant amount of literature on dislocation-crack tip interactions but they all are concerned with static elasticity calculations and neglect the dynamical nature and patterning behavior of dislocations. Only Hirsch and co-workers [8] have pointed out the importance of dislocation velocity and its role in determining the ductile-brittle transition (DBT). It is thus evident that one has to consider two or three dislocation populations and their interaction with themselves and the crack tip as follows:

- First, the dislocations making up the cell walls which are assumed to be immobile.
- Second, the dislocations emitted from the crack tip which are usually organized in individual or groups of slip lines.
- The interaction of these two populations lead to the shielding effect of the crack, but it may also release a third type of dislocations which tends to produce the opposite effect i.e. antishielding [9].

Thus, the suggestion is advanced here that a correct description of the dynamics and nonlinear interactions of dislocations among themselves and the crack tip is a key factor for a fundamental understanding of crack propagation. Similar arguments can be employed for other types of dominant microstructures such as hydrogen atoms (embrittlement) and voids (creep).

(ii) Cellular Automata (CA) Approach

By cellular automata we refer to the simulation of macroscopic phenomena by applying specific microscopic rules to describe the interaction and evolution of identical entities (microstructures) that govern the behavior of a properly discretized physical system (CA lattice). During the last decade CA has been successfully applied in modeling the self-organization for a wide variety of nonlinear phenomena ranging from chemistry and biology to turbulence and materials science [10]. It appears that CA is particularly suited to simulate the macroscopic crack growth as a result of the underlying microscopic mechanisms, such as hydrogen diffusion, dislocation emission, or void migration and coalescence, that occur ahead of the crack tip.

As a preliminary result [11] of this approach, we depict in Figure 8 the crack path as it evolves under the influence of external stress and the concurrent void migration and coalescence. The void velocity is assumed to be proportional to the gradient of the local hydrostatic stress provided by the elastic field ($\sigma_h \sim 1/\sqrt{r}$), while the crack growth process takes place by void coalescence at the crack tip. It is interesting to note the apparent similarity between (a) the optical micrograph and (b) the CA simulation.

REFERENCES

1. E.C. Aifantis, The physics of plastic deformation, Int. J. Plasticity 3, 211-247 (1987).

2. a) D. Walgraef and E.C. Aifantis, Plastic instabilities, dislocation patterns and nonequilibrium phenomena, Res Mechanica 23, 161-195 (1988).

b) D. Walgraef and E.C. Aifantis, Dislocation Inhomogeneity in Cyclic Deformation, in: Micromechanics and Inhomogeneity - The Toshio Mura Anniversary Volume, Eds. G.J. Weng et al., Springer Verlag, pp. 511-533, 1990.

3. H.M. Zbib and E.C. Aifantis, On the structure and width of shear bands, Scripta Metall. 22, 703-708 (1988).

4. H.M. Zbib and E.C. Aifantis, A gradient-dependent model for the Portevin-Le Chatelier effect, Scripta Metall. 22, 1331-1336 (1988).

5. J.B. Rundle and W. Klein, Nonclassical nucleation and growth of cohesive tensile cracks, Phy. Rev. Letters 63, 171-174 (1989).

6. T.W. Webb and E.C. Aifantis, Stick-slip peeling, ASME Winter Annual Meeting, San Francisco, CA, No. 89-WA/EEP-7.

7. a) G.C. Sih (Editor), Dynamic Crack Propagation, Noordhoff, Leyden (1973).

b) J.G. Williams, Fracture Mechanics of Polymers, Ellis Horwood Ltd, Chichester (1984).

c) A.J. Kinloch and R.J. Young, Fracture Behavior of Polymers, Applied Science Publishers Ltd., London (1983).

8. P.B. Hirsch, S.G. Roberts and J. Samuels, Dislocation mobility and crack tip plasticity at the ductile-brittle transition, Revue Phys. Appl. 213, 409 (1988).

9. S.M. Ohr, Antishielding dislocations at a crack tip, Scripta Metall. 21, 1681-1684 (1987).

10. D. Farmer, T. Toffoli and S. Wolfram (Editors), Cellular Automata, North-Holland, Amsterdam (1984).

11. C.C. Cusack and E.C. Aifantis, unpublished results.

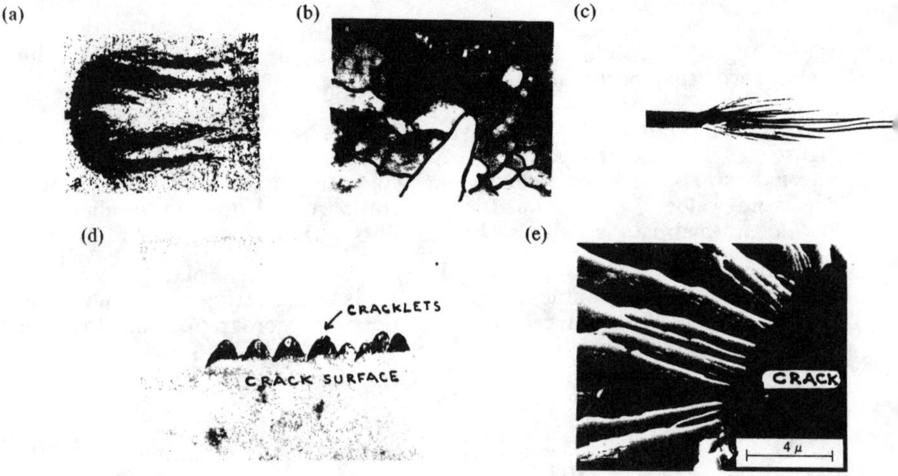

Figure 1. Deformation patterns ahead of crack tips.

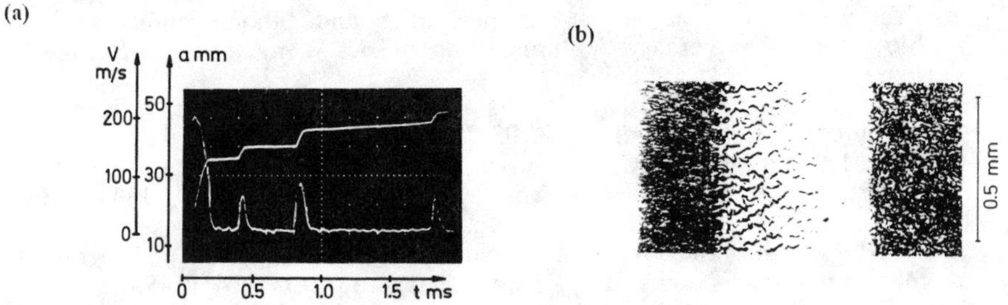

Figure 2. Stick-slip fracture.

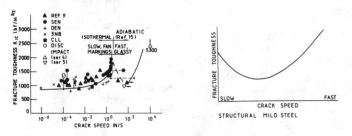

Figure 3. Non-monotonous crack resistance.

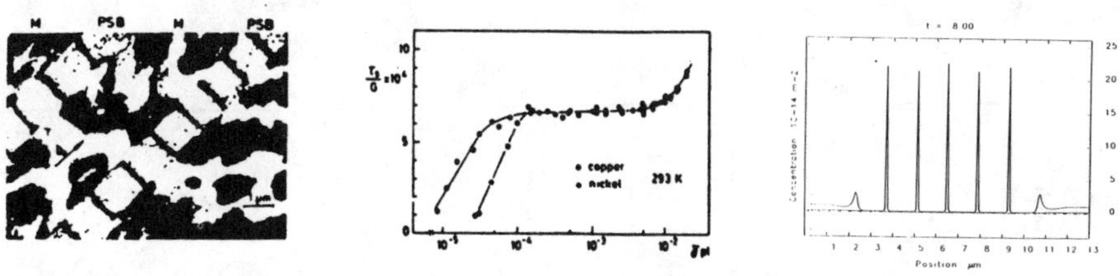

Figure 4. Persistent Slip Bands

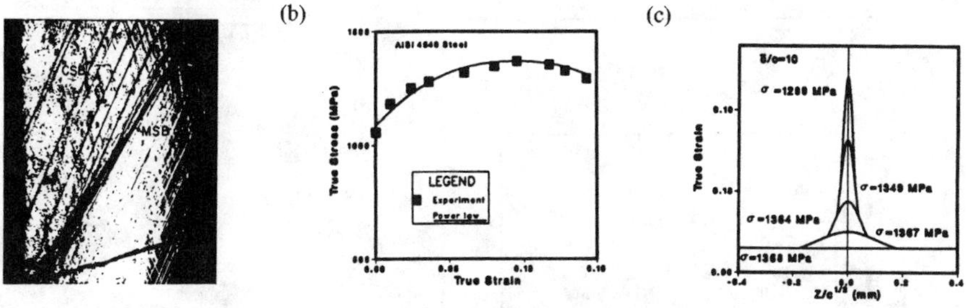

Figure 5. Stationary Shear Bands

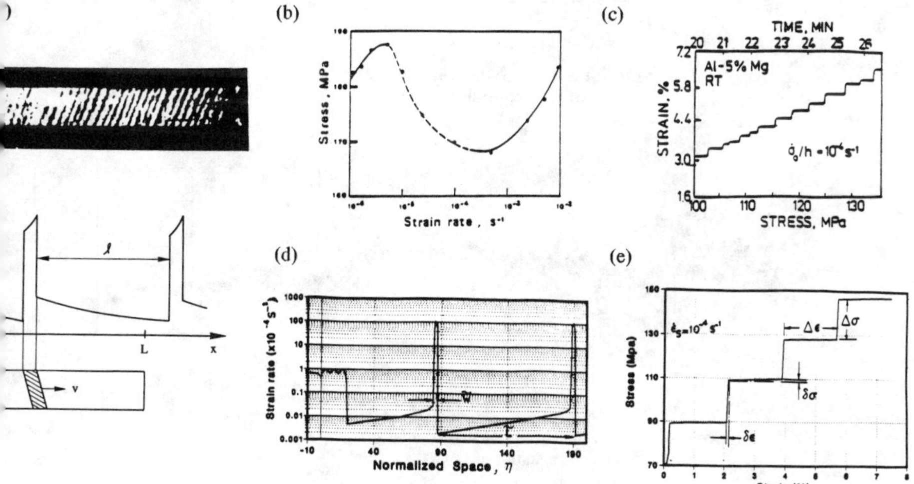

Figure 6. Portevin-Le Chatelier Bands.

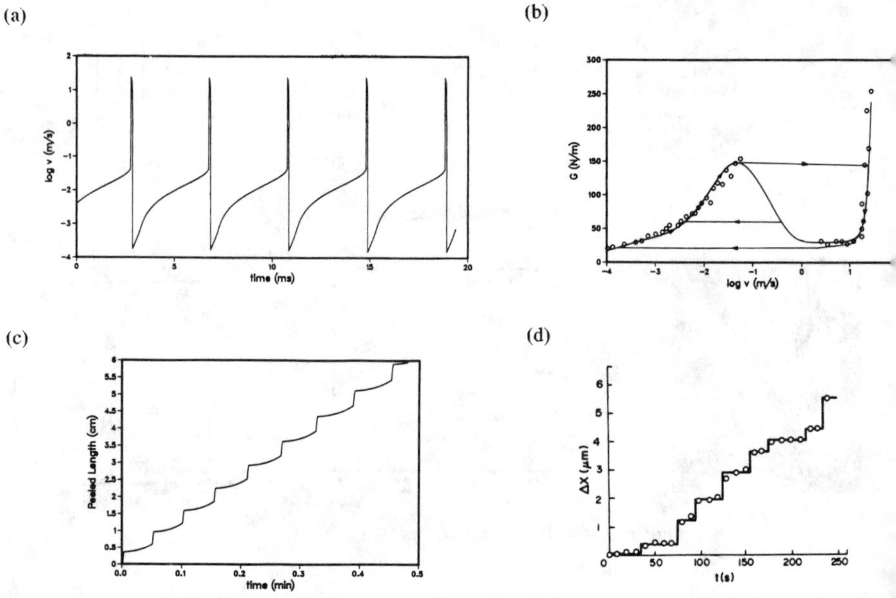

Figure 7. Preliminary results for stick-slip fracture using the SPDE approach.

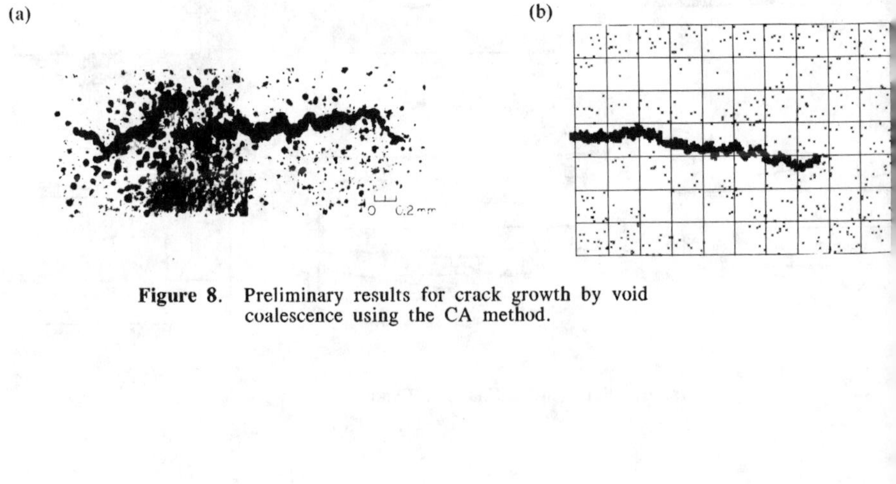

Figure 8. Preliminary results for crack growth by void coalescence using the CA method.

STRESS-STRAIN MODELLING OF HETEROGENEOUS GRANULAR SOLIDS BASED ON MICRO-MECHANICS

Ching S. Chang and Anil Misra
Department of Civil Engineering,
University of Massachusetts, Amherst, MA 01003

Abstract

Micromechanics is considered to derive a constitutive model for granular materials under finite strains with particle slidings and consequent fabric change. The stress-strain relationship is defined for an element consisting of a large number of particles. The heterogeneity in the stress and strain fields within the element are accounted by employing Hill's principle of averaging. The derived stress-strain relationship is in terms of contact properties and packing structure. Examples are shown to illustrate the capability of this model.

INTRODUCTION

Until recently most of the stress-strain modelling of granular materials was founded on a phenomenological approach using conventional theories of plasticity, elasto-plasticity, hypoelasticity etc.. Though these models have proved to be useful in the study of soil mechanics, it is desirable to develop models which rationally account for the discrete nature of the material.

Based on the micro-scale interaction between particles, constitutive models have been previously developed for granular material (1,2). However, these models do not account for heterogeneity. Previous research has shown that at the low deviatoric stress levels, the deformation field is fairly uniform (3). However, for higher deviatoric stress levels due to inter-particle rolling and sliding, the uniformity of deformation field does not hold (3).

In this paper, the overall constitutive relationship for an element of granular solid is derived by considering the overall stress and strains to be averages of local stress and strains in accordance with Hill's averaging principle (4). The derived stress-strain law is in terms of the contact stiffness, contact strength, particle size, void ratio, coordination number, and packing structure.

THEORETICAL MODEL

A granular packing is envisaged to consist of circular particles interacting with each other at inter- particle contacts through conceptual springs and sliders. Under applied loading, the springs account for the elastic deformation at a contact and the sliders account for the plastic slip.

Based on this idealized system, the interaction of a particle and its neighbors can be defined. Consequently the local constitutive law at a particle level is established. However, the stress-strain behavior of interest is for an element consisting of large number of particles. By the heterogeneous nature of granular materials, each particle is expected to behave differently. Thus an averaging process over the local behavior is required to determine the global behavior of the element. The averaging process is such that it preserves the equilibrium of each particle embedded in the element.

The theoretical model is described in the following aspects: a) local constitutive law, b) global constitutive law, and c) finite deformation with fabric change.

Local constitutive law

Based on the interaction with neighboring particles, the local constitutive law is defined at a particle level relating the local stress (due to contact forces) and local strain (due to spring/slider deformation). The local stress-strain law is given as

$$\dot{\sigma}^n_{ij} = C^n_{ijkl}\dot{\epsilon}^n_{kl} \tag{1}$$

where $\dot{\sigma}^n_{ij}$, $\dot{\epsilon}^n_{kl}$ are the stress and strain increments, respectively, and C^n_{ijkl} is the instantaneous local stiffness tensor at a particle level.

The instantaneous local stiffness tensor is derived as a function of contact properties and the position of neighboring particles including the effect of particle spin and equilibrium requirements (5).

With the consideration of particle spin and equilibrium requirements, the mechanical behavior is more accurately modelled. This definition of local constitutive law facilitates the representation of a discrete granular system as an equivalent continuum system.

Overall stress-strain relationship

Defining $\dot{\sigma}_{ij}$ and $\dot{\epsilon}_{ij}$ as the volume average of the local quantities $\dot{\sigma}^n_{ij}$ and $\dot{\epsilon}^n_{kl}$. The overall stress-strain relationship is obtained to be in the following form:

$$\dot{\sigma}_{ij} = C_{ijkl}\dot{\epsilon}_{kl} \tag{2}$$

where C_{ijkl} is the overall stiffness tensor which can be expressed as:

$$C_{ijkl} = \frac{1}{V}\sum_n C^n_{ijpq}H^n_{pqkl} \tag{3}$$

where H^n_{ijkl} is a weighting function relating local strain to the global strain

$$\dot{\epsilon}^n_{ij} = H^n_{ijkl}\dot{\epsilon}_{kl} \tag{4}$$

The weighting function is obtained by considering each particle of C^n_{ijkl} to be embedded in a material of stiffness C_{ijkl} (5).

Finite deformation and fabric change

The finite deformation and fabric change lead to the change in frame of reference during the incremental loading process. The effect of finite deformation has been incorporated in the model by using appropriate objective stress and strain rates. In contrast to the conventional finite deformation theory, the present model is unique since it incorporates the fabric changes in addition to the configuration change of the granular media.

MODEL PREDICTION

An example is shown here to illustrate model capability. The results are calculated for the packing shown in Fig. 1. The parameters required by the model are the inter-particle friction angle and contact stiffness. Fig. 2 shows the stress-strain curve obtained under biaxial compression for two values of confining stresses (20 psi and 100 psi). Under a low confining stress the stress-strain behavior is seen to be brittle while the behavior is ductile at a higher confining stress. The packing fails at different stress levels under the two confining stresses. The Mohr-Coulomb failure angle is found to be higher at low confining stress in contrast with that at higher confining stress. Fig. 3 shows the volume change behavior of the packing under the two confining stresses. The behavior tends to be more dilative at a low confining stress than that at higher confining stress. These stress-strain, strength, and volume change behavior are similar to that observed from experiments on sands.

CONCLUSION

A stress-strain model for granular material has been presented. The results show that the model can appropriately account for the increasing heterogeneity with the stress level due to inter-particle sliding. It is found that the finite deformation and fabric change significantly effects the model predictions, especially on the strength and post peak behavior.

Although the local constitutive law is characterized by a simple spring-slider system, the model captures the salient features of the global stress-strain behavior of sands, such as hardening, softening, stress dependency, dilation-contraction.

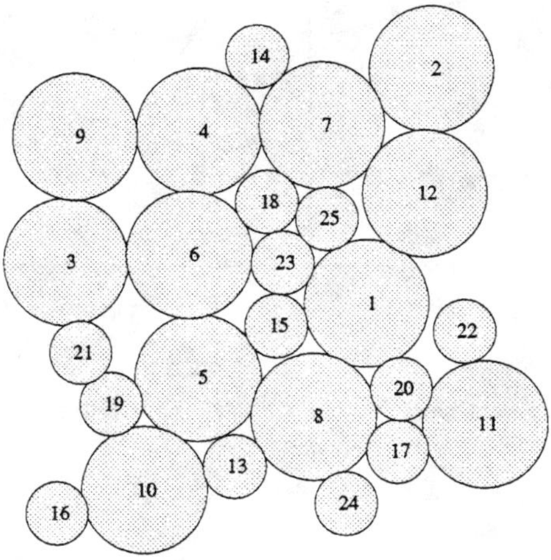

Fig. 1 Packing used in the example.

REFERENCES

1. Chang, C. S., Misra, A., and Xue, J. H.,"Incremental Stress-Strain Behavior of Packings Made of Multi-sized Particles," International Journal of Solids and Structures, Pergamon Press, Vol. 25, No. 6, 1989, pp. 665-681.

2. Chang, C. S., and Liao, C.,"Constitutive Relations for Particulate Medium with the Effect of Particle Rotation," International Journal of Solids and Structures, Vol. 26, No. 4, 1990, pp. 437-453.

3. Chang, C. S., and Misra, A., "Computer Simulation and Modelling of Mechanical Properties of Particulates," Journal of Computer and Geotechniques, Elsevier Science Publishers, Vol. 7, No. 4, 1989, pp. 269-287.

4. Hill, R., "The Essential Structure of Constitutive Laws for Metal Composites and Polycrystals," Journal of Mechanics and Physics of Solids, Vol. 15, No. 2, 1967, pp. 79-95.

5. Misra, A., "Constitutive Relationships for Granular Solids with Particle slidings and Fabric Changes," Ph.D. dissertation, University of Massachusetts, 1990.

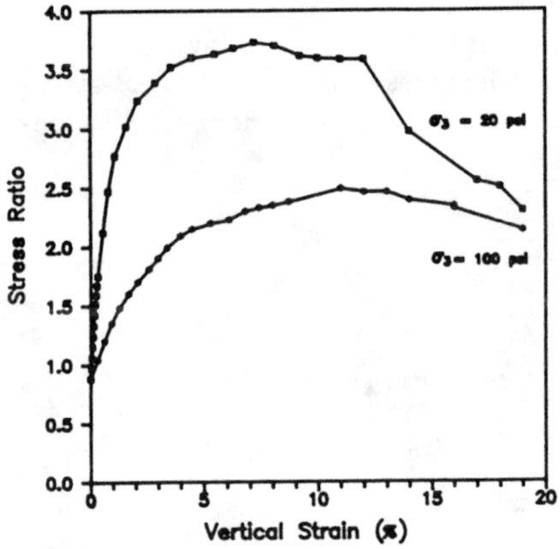

Fig. 2 Stress-strain behavior.

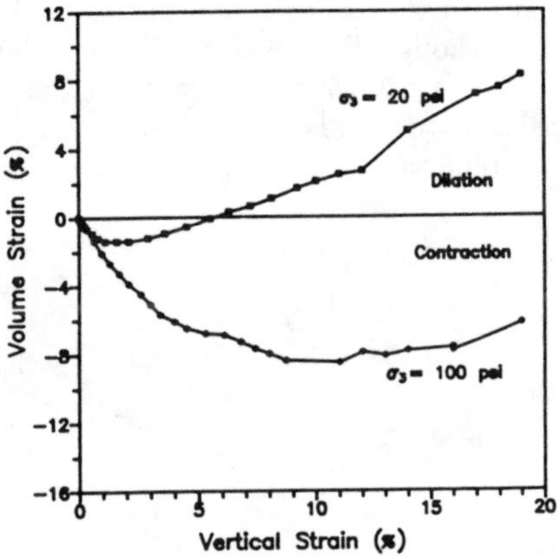

Fig. 3 Volume change behavior.

MICROMECHANICAL MODEL FOR DEFORMATION OF AGGREGATES OF FRAGMENTS, GRAINS, OR BLOCKS

D. R. CURRAN and L. SEAMAN
SRI International, Menlo Park, CA 94025

ABSTRACT

Ceramics and brittle geologic materials can deform non-elastically under compression and shear by sliding and ride up of fragments, grains, or blocks, with accompanying competition between dilatancy and pore compaction. This paper describes a micromechanical model of such deformation based on multiplane plasticity and an analogy to the dynamics of atomic dislocations. In the analogy, atomic lattice slip planes are replaced by slip on interfaces between blocks, and atoms are replaced by fragments, grains, or blocks. Atomic vacancies are replaced by spaces between the blocks.

1. INTRODUCTION

Hard, brittle materials often deform nonelastically by grain sliding, ride-up, and compaction. In this paper, we will use the terms "grain", "fragment" and "block" interchangeably to designate the sliding object.

Attempts to model inelastic sliding have usually taken one of three routes: continuum modelling (Drucker [1988] and Spencer [1982]), "discrete element" block motion modelling [Britton and Walton, 1987; Heuze' et al, 1990], or micromechanical modelling [Chang and Misra, 1989; Isida and Nemat-Nasser, 1987; Nemat-Nasser and Obata, 1988].

In this paper we take a micromechanical approach, and focus on describing the behavior of the motion of lines of voids between the sliding objects, as in atomic dislocation theory.

2. THEORETICAL MODEL

We introduce a scale size into the problem, the "material element" size, D. The value of D must be large enough to contain a statistical number of dislocations.

We assume that the fragment bed in the material element contains a finite number of fragment interface orientations or "slip planes" along which the fragments or blocks can slide. The number of such planes is the number of independent fragment face orientations. Thus, the first step in constructing a model is to specify the number and orientations of the slip planes, which is related to specifying the fragment shapes. As reviewed by Grady and Kipp [1987], there are a number of statistical procedures for generating such fragment beds; in the present paper we assume that the geometry of the bed is known.

On the ith slip plane, the slip strain rate is given by the Orowan [1940] equation:

$$\dot{\gamma}_i^p = g\, N_{di}\, b\, B\, v_{id} \tag{1a}$$

where N_{di} = number of sliding blocks (mobile dislocations) in the ith plane intersecting a unit area slice through the bed, bB is the block slip distance , g is a geometrical factor relating the active slip system to the shear strain direction, and v_{id} is the propagation velocity of the slipping through the bed on the ith plane. Thus, the variable slip distance bB replaces the constant magnitude of the Burgers vector in atomic dislocation theory.

The dislocation velocity v_{id} is equal to B divided by the time it takes for the hole to move from one side of the block to the other. That time is the time required for the block to slide the distance bB. That is,

$$v_{id} = B/(bB/v_{ib}) = v_{ib}/b$$

where v_{ib} is the block sliding velocity. Thus, (1a) becomes

$$\dot{\gamma}_i^p = gN_{di}Bv_{ib} \tag{1b}$$

This approach has a resemblance to the "splitting cracks" analyses recently summarized and extended by Nemat-Nasser and Obata[1988]. However, those analyses focus on the extension of tensile cracks driven by the sliding of shear crack faces, whereas the present analysis assumes that such tensile cracks have already coalesced to form the fragments or blocks. Our model has more in common with the dilatancy models of Britton and Walton [1987].

The amount of porosity due to all the block slippage on a plane is $aN_{bi}bB^2$ = ab, where a is a proportionality constant accounting for block shape and the degree of block interference, and $N_{bi} = 1/B^2$ is the number of blocks intersecting unit area. For a constant value of N_{bi}, the corresponding rate of increase of dilatant porosity on the ith plane $d\phi_i/dt$ is equal to av_{ib}/B, since v_{ib} = Bdb/dt. Comparison with Eq(1b) shows that

$$g\, N_{id}\, B^2\, \dot{\phi}_i = a\, \dot{\gamma}_i^p$$

However, the total rate of change of porosity will also contain a term describing pore compaction due to the confining pressure. That is,

$$\dot{\phi}_i = a\dot{\gamma}_i^p/(gN_{id}B^2) + \dot{\phi}_{ic} \tag{2a}$$

where $\dot{\phi}_{ic}$ is the ith plane component of the competitive pore compaction rate resulting from the confining pressure (to be discussed later). That is, $\dot{\phi}_{ic} < 0$. Thus, according to the model, on a given plane, increasing shear slip b at zero confining pressure would produce increasing porosity, but if significant confining pressure is present there will be competition between this dilatancy and pore compaction.

The dislocation density N_{di} is given by:

$$N_{di} = f_i/B^2 \tag{3}$$

where f_i is the fraction of blocks that have an adjacent hole large enough to allow slip into it. The dislocation density can in principle increase with time due to the nucleation of new dislocations. A nucleation rate equation could then be written with the rate expressed as a function of stress, strain, and other variables. At present, however, we treat f_i as a constant.

Thus, combining Eqs.(1b) and (3) yields:

$$\dot{\gamma}_i^p = (gf/B)v_{ib} \tag{4}$$

The next step is to determine the average block velocity v_{ib}. Our estimate comes from conservation of momentum for a single block:

$$(\tau_i - \mu\sigma_{ni})\, B^2 t_i = \rho B^3 v_{ib} \tag{5}$$

where v_{ib} is the velocity increment delivered to the block during the time t_i by the excess of the shear stress τ_i over the frictional stress on the ith plane, where μ is the (possibly rate, temperature, and pressure-dependent) coefficient of friction, and σ_{ni} is the normal compressive stress on the plane.

The time t_i is assumed to be the time for a shear wave to make a round trip in the block, $2B/C_t$. Thus, (5) becomes:

$$v_{ib} = 2(\tau_i - \mu\sigma_{ni})/\rho C_t \tag{6}$$

We now combine (4) and (6) to obtain:

$$\dot{\gamma}_i^p = (\tau_i - \mu\sigma_{ni})/rC_t^2T \tag{7}$$

where $T = B/(2gfC_t)$. Since g is of the order of unity, the characteristic time T approximates the time for the shear wave to make a round trip between dislocations. That is, smaller dislocation densities lead to larger characteristic times for slip.

The constitutive relations for the ith plane are thus Eq. (2a) for the dilatancy rate, Eq. (7) for $\dot{\gamma}_i^p$, and

$$\dot{\gamma}_i = \frac{\dot{\tau}_i}{2G} + \dot{\gamma}_i^p \tag{8}$$

where $\dot{\gamma}_i$ is the total strain rate on the ith plane, $\dot{\tau}_i/2G$ is the elastic strain rate on the plane, G is the elastic shear modulus, and the last term in (8) is the nonelastic or "plastic" strain rate on the plane due to block slipping.

Because the actual motion in a fragment bed is expected to be much more complicated than the above simple picture, we next replace the porosity and strain rate terms on the ith plane in the second part of Eq.(2) with rates of change of total porosity and equivalent plastic strain to gain in simplicity with no real loss in accuracy.

$$\dot{\phi} = a\,\dot{\gamma}^p + \dot{\phi}_c \tag{2b}$$

where ϕ is the total porosity and γ^p is the equivalent plastic strain.

Equations (2b), (7), and (8) are solved simultaneously on each slip plane with the equations of motion to obtain solutions to given boundary value problems.

We next discuss the pore compaction model that we have developed to compute the value of $\dot{\phi}_c$ in Eq.(2b).

As a block slides it will exhibit an increasing overhang to the adjacent hole, resulting in a decreasing compaction pressure. Thus, if significant confining pressure is present, the block will have an increasing tendency to be pressed into the adjacent hole. However, as slip continues the block will begin to climb over the next block and the pressure needed for compaction will increase. In short, we expect the compaction pressure to be periodic with a wavelength given by the block size.

However, in many cases this effect may be negligible compared to the decrease in compaction pressure with increasing porosity [Carroll and Holt, 1972; Carroll, 1980; Curran et al, 1987]. That is, as b approaches unity, the dilatant porosity will be large and the compaction strength for most materials will become very low without the additional softening discussed above. Therefore, in the calculations with the model to be described later, we have ignored the above periodic softening and hardening, and a simple pressure-volume pore compaction model is used [Seaman and Tokheim, 1974].

3. COMPUTATIONAL MODEL

Our computational model (named FRAGBED) partitions the slip among the active slip planes, and is composed of several sub-models:

- A multiple-plane plasticity model for the deviatoric stresses. This model describes fragment sliding in the fragmented bed. That is, the "plastic" strain is due to fragment sliding., as described above. The model has much in common with those of Batdorf and Budiansky [1949]; Peirce, Asaro, and Needleman [1981; Bazant [1985, 1987, 1988], and Zienkiewicz and Pande [1987].

- A Mie-Grüneisen relation for the pressure in the solid intact material and a pore compaction model for the treatment of the fragmented material.

- The process described above for handling the fragment sliding-induced bulking and the resulting porosity as a function of the applied shear strain.

- A tensile fracture model to allow separation of the broken material.

- A process for modifying the yield strength on each slip plane to account for softening/hardening as well as Coulomb friction; the cohesion and friction angle on each slip plane are made functions of the amount of slip.

4. USE OF THE MODEL TO INTERPRET EXPERIMENTAL DATA

Computational simulations were made to represent several laboratory experiments. The static tests of Heard and Cline [1980], the dynamic tests of Lankford [1989], and a pressure-shear impact test of Shockey and Klopp [1990] were simulated. The tests of Heard and Cline and of Lankford were simulated using the intact (unsoftened) properties of the aluminum nitride with a Coulomb yield strength and $\tan\phi = 0.33$. We used a density = 3.26 g/cm^3, the first and second terms of the bulk modulus series = 2.1 and 4.2 Mbar, the initial yield of 30 kbar, and a shear modulus of 1.26 Mbar. The simulation of these tests of the intact material resulted in only a fair match to the experimental data because we did not use the highly nonlinear pressure-dependent yield strength that would be required to fit this family of data.

A more detailed study was made of the pressure-shear impact of Shockey and Klopp. A 0.54-cm boron carbide plate carrying a 0.0577-cm aluminum nitride target plate impacted a boron carbide anvil at 672 m/s. The transverse particle velocity was measured with laser interferometry at the rear of the anvil. The impacting surfaces were slanted 20° to the axial direction to induce both a longitudinal (xx direction) and a transverse wave (yy direction) in the target. The results derived from this test were that there was a constant normal stress of 10.9 GPa and a nearly constant σ_{xy} in the aluminum nitride of 1.78 GPa.

For computationally simulating this pressure-shear impact test we used a friction angle given by $\mu = \tan\phi = 0.11$ in the granulated ceramic (after several trials with other values) and a time constant T for the shear stress relaxation of 0.1 μsec. The values of a, g, and f were set to unity. Nine slip planes were used, with normals in the 3 Cartesian directions and bisecting those coordinates in all 4 quadrants.

The calculations gave good agreement with the observed σ_{xy} history, and revealed extremely complex slip behavior, including rotation of slip planes. Most of the slip occurred on the plane whose normal bisects the x-z directions.

5. CONCLUSIONS

The model appears to be appropriate and useful for describing the highly nonlinear, anisotropic, and history-dependent deformation of fragmented beds. It can be applied to a wide range of applications from ceramic armor to jointed rock masses. It is fully 3-dimensional, and can be run in 3-D hydrocodes. We expect that it will find wide use as an interpretive and predictive analytical tool.

REFERENCES

Batdorf, S. B. and B. Budiansky, "A Mathematical Theory of Plasticity Based on the Concept of Slip," Tech. Note No. 1871 of National Advisory Committee for Aeronautics, April 1949, Washington.

Bazant, Zdenek P., and Pere C. Prat, "Creep of Anisotropic Clay: New Microplane Model," Jour. of Eng. Mech., ASCE, 113, No. 7, p 1050, July 1987.

Bazant, Zdenek P., and B. H. Oh, "Microplane Model for Progressive Fracture of Concrete and Rock," Jour. of Eng. Mech., ASCE, Vol. 111, No. 4., p. 559, April 1985.

Bazant, Zdenek P., and Pere C. Prat, "Microplane Model for Brittle-Plastic Material: I. Theory," Jour. of Eng. Mech., ASCE, Vol. 114, No. 10, p. 1672, Oct. 1988.

K. Britton and O. R. Walton, "Brittle Fracture Phenomena-an Hypothesis", in Proceedings of the Second International Symposium on Rock Fragmentation by Blasting, Keystone, Colorado, Aug. 23-28, 1987.

Carroll, M. M., and A. C. Holt, "Static and Dynamic Pore-Collapse Relations for Ductile Porous Materials," Jour. Appl. Phys., Vol 43, No. 4, p. 1626-1636, April 1972.

C. C. Chang and A. Misra, Journ. Eng. Mechanics, Vol. 115, No. 4, (April, 1989).

D. C. Drucker, Applied Mechanics Reviews, Vol. 41, No. 4, ASME Book No. AMR034, (April, 1988).

D. E. Grady and M. E. Kipp, in **Fracture Mechanics of Rocks**, Ed. by Barry Atkinson, Academic Press, (1987).

Heard,H. C. and C. F. Cline, Mechanical Behaviour of Polycrystalline BeO, Al_2O_3 and AlN at High Pressure, J. Mat. Sci., Vol 15, No. 8, p. 1889-1897, Aug. 1980.

F. E. Heuze', O. R. Walton, D. M. Maddix, R. J. Shaffer, and T. R. Butkovich, "Analysis of Explosions in Hard Rocks: The Power of Discrete Element Modeling", in Proceedings of the International Conference on Mechanics of Jointed and Faulted Rocks, Vienna, Austria, April 18-20, 1990.

M. Isida and S. Nemat-Nasser, Acta. metall., Vol. 35, No. 12, pp. 2887-2898, (1987).

Kanakia, M. D. and M. B. Peterson, Literature Review of Solid Lubrication Mechanisms, Interim Report BFLRF No. 213, for U.S. Army Belvoir Research, Development and Engineering Center, Fort Belvoir, Virginia, July 1987.

Lankford, J., private communication to R. Klopp at SRI, November 1989.

Malvern, L. E., "The Propagation of Longitudinal Waves of Plastic Deformation in a Bar of Materials Exhibiting a Strain-Rate Effect," Jour. Appl. Mech., Vol 18, p. 203-208, 1951.

S. Nemat-Nasser and M. Obata, Transactions of the ASME, 24, Vol. 55, (March, 1988).

E. Orowan, Proc. Roy. Soc. London, 52, 8, (1940).

Peirce, D., R. J. Asaro, and A. Needleman, "An Analysis of Nonuniform and Localized Deformation in Ductile Single Crystals," Report MRL E-138, Brown University for National Science Foundation, Oct. 1981.

Seaman, L., and R. E. Tokheim, "Computational Representation of Constitutive Relations for Porous Material," Final Report by SRI for Defense Nuclear Agency, DNA 3412F, May 1974.

D. A. Shockey and R. W. Klopp, Tests for Determining Failure Criteria of Ceramics under Ballistic Impact, Quarterly Progress Report No. 2, by SRI for U.S. Army Research Office, January 1990.

A. J. M. Spencer, "Deformation of Ideal Granular Materials", in **Mechanics of Solids**, Edited by H. G. Hopkins and M. J. Sewell, Pergamon Press, (1982).

Zienkiewicz, O. C., and G. N. Pande, "Time-Dependent Multilaminate Model of Rocks – A Numerical Study of Deformation and Failure of Rock Masses," Int. Jour. for Num. and Analytical Methods in Geomechanics, Vol. 1, p. 219, 1977.

ACKNOWLEDGEMENTS

This work was partially supported by the Office of Naval Research under Contract No. NOOO14-88-C-0734. The contract monitor was Dr. Roshdy Barsoum.

DAMAGE EVOLUTION AND MICRO/MACRO TRANSITION FOR LAMINATED COMPOSITES

Fan J. Zhang J. Young W.
Laboratory of Constitutive Laws,
Department of Engineering Mechanics,
Chongqing University, China, 630044

Abstract

In this paper, a noval theoretical formulation of damage evolution of laminated composites under in-situ restricted conditions, characterized by micro/macroscopic approach for an equivalent model, is proposed. An in-situ damage effective function is introduced to formulate the constitutive relation for restricted and damaged lamina. A damage evolution law, incorporating the effects of in-situ constraints, is proposed in terms of energy release rate and damage resistance curve as well as the in-situ damage effective function. A serious of experimental results are correlated and some of interesting phenomena are predicted.

INTRODUCTION

Recently intensive research activities in damage analysis can be observed from both microscopic and macroscopic points of view. It is very significant as it closely relates to degeneration evaluation and life prediction of composite structures. From microscopic point of view, various kinds of shear lag models (Balley, et al ,1978, 1979 [1]; Refsnider et al, 1979 [2]; Flaggs 1985 [3]; Laws and Dvorak, 1988 [4]; Swanson 1989 [5]), variational approach (Hashin, 1985, 1987 [6]), self-consistent approach (Dvovak, laws, Hejazi, 1983 [7]) and approximate approach of elasticity (Nuismer and Tan, 1988 [8]) were proposed to investigate damage mechanisms of composite materials. From macroscopic point of view, a thermodynamic formulation within the context of internal variables (Telreja, 1986 [9], Harris and Allen, 1987, 1988 [10]) were proposed to establish a constitutive relations which incorporate damage effects.

Reviewing the developments in the past 12 years, we can see that works to link micro/macroscopic aspects are very few. It seems that advantages from microscopic analysis haven't much taken into macroscopic analysis; this may be a reson why some macroscopic model can hardly develop damage evolution law. On the other hand the microscopic analysis, such as shear lag model, is usually limited in the application for local areas. It is difficult to connect this microscopic analysis with the overall behavior of composite material with many different layers. Taking the investigation of in-situ behavior as example, a serious of reports given by Flaggs, Mclaughlin, et al. have confirmed effects of constraints of laminates on the in-situ strength of lamina, which can be as high as 2.5 times the lamina transverse strength, and on the on

----

Financial support 9187004-6 from CNSF is fully acknowledged.

of matrix cracking. Later on Flaggs developed a 2-D shear lag model to carry on analysis for the in-situ behavior [3]. While his work is significant, the analysis are limited to very special conditions and none of effects of in-situ constraints on the damage evolution are reported. This, to our opinion, may be due to the lack of an effective model to make the connection of microscopic analysis to the whole laminates and its macroscopic behavior.

In this paper, a novel theoretical formulation of damage evolution of laminated composites under in-situ restricted conditions, characterized by micro/macroscopic approach for an equivalent model, is proposed. An in-situ damage effective function is introduced to formulate the constitutive relation for restricted and damaged lamina. A damage evolution law, incorporating the effects of in-situ constraints, is then proposed.

CONSTITUTIVE EQUATION FROR LAMINA UNDER RESTRICTIVE DAMAGE

Laminate composites are macroscopically heterogeneous materials in the sense it consistes of anisotropic layers with different orientations. However, each lamina may be considered as homogeneous material with a distribution of micro cracks or inclusions inside, and to be confined by whole material structure through shear and/or normal forces on the contact surfaces with neighbouring layers. One of the essential points of this formulation is to get the constitutive equation for this piecewise homogeneous material, which can incorporate lamina damage effects and restrictive effects.

Let's take a representative volume element (RVE) from the in-situ dameged and restricted lamina at hand, and then introduce a CDM unit cell in which the homogeneous stress and strain are macroscopic stress $\bar{\sigma}_{ij}$ and strain $\bar{\epsilon}_{ij}$. Obviously both units should have the same volume V and external surface Se and the distribution of surface traction Ti. The equivalent conditions of surface traction, strain energy and matrix property between RVE and CDM cell are used to derive the required constitutive equation.

<u>Traction Equivalence</u>: Following Hill [11], the traction effects may be expressed by a surface traction integration as

$$Ls = 1/2V\int_{s_e} (T_iX_j + T_jX_i)ds \tag{1}$$

Denoting damage volume fraction as fc, internal surface Se, matrix volume Vm, matrix average stress and strain as $\bar{\sigma}^m_{ij}$, $\bar{e}^m_{ij}$,respectively. We have

$$fc = Vm/V \quad \text{and} \quad \bar{\sigma}^m_{ij} = 1/Vm\int_{Vm}\sigma_{ij}dV \tag{2a,b}$$

Expressing Ls explicitly for both RVE and CDM units and equating each other, it results

$$\bar{\sigma}_{ij} = (1-fc)\bar{\sigma}^m_{ij} \tag{3}$$

<u>Equivalence of Strain Energy</u>. The strain energy should equal to

$$A = 1/2\int_{Se+Si} T_i\ U_i dS \tag{4}$$

Expressing A explicitly and using Hill's "macroscopic average" concept for distribution of traction, we have

$$A = 1/2\bar{\sigma}_{ij}\bar{\epsilon}_{ij}V \qquad \text{(CDM)} \tag{5}$$

$$A = \bar{\sigma}_{ik}[1/2\int_{Vm}(U_{i,k}+U_{k,i})dV - 1/2\int_{Se}(U_i N_k + U_k N_i)dS] \qquad \text{(RVE)} \tag{6}$$

with $\bar{\epsilon}^m_{ij} = 1/Vm\int_{Vm}\epsilon_{ij}dV$ (7)

Denoting α_{ij} as $\alpha_{ij} = -1/2V\int_{Si}(U_i N_j + U_j N_i)dS$ (8)

We get $\bar{\epsilon}_{ij} = (1-f_c)\bar{\epsilon}^m_{ij} + \partial_{ij}$ (9)

Equivalance of Matrix: The matrix property can be expressed by its undamaged compliance S^o_{ijkl} or stiffness tensor C^o_{ijkl} as

$$\epsilon^m_{ij} = S^o_{ijkl}\sigma^m_{kl} \quad , \quad \sigma^m_{ij} = C^o_{ijkl}\epsilon^m_{kl} \tag{10a,b}$$

It is obvious that

$$\bar{\epsilon}^m_{ij} = S^o_{ijkl}\bar{\sigma}^m_{kl} \tag{11}$$

Combining eq.s (3),(9) and (11), we obtain

$$\bar{\epsilon}_{ij} = S^o_{ijkl}\bar{\sigma}_{kl} + \alpha_{ij} \quad , \text{ or} \tag{12}$$

$$\bar{\sigma}_{ij} = C^o_{ijkl}(\bar{\epsilon}_{kl} - \alpha_{kl}) \tag{13}$$

Where

$$\alpha_{ij} = \sum_{k=1}\alpha^{(k)}_{ij} \quad , \quad \alpha^{(k)}_{ij} = -1/2V\int_{S_i}(U^{(k)}_i N_j + U^{(k)}_j N_i)dS \tag{14a,b}$$

For small deformation and linear elasticity material, α_{ij} should be a linear function of surface traction Ti or σ_{ij}. i.e

$$\alpha^{(k)}_{ij} = w^{(k)}_{ijkl}\bar{\sigma}_{kl} \quad , \quad \alpha_j = W_{ijkl}\bar{\sigma}_{kl} \tag{15a,b}$$

Substituting it into eq.s (12)(13), we get

$$\bar{\epsilon}_{ij} = S_{ijkl}\bar{\sigma}_{kl} \tag{16}$$

with $S_{ijkl} = S^o_{ijkl} + W_{ijkl}$ (17)

where $W_{ijkl} = \Sigma\, W^{(k)}_{ijkl}$ (18)

By using Vogit symbol, eq.s (16) (17) can be expressed as

$$\bar{\epsilon}_j = S_{ij}\bar{\sigma}_i \qquad S_{ij} = \overset{o}{S}_{ij} + W_{ij} \tag{19a,b}$$

MICRO/MACRO COMBINED STUDY FOR AN EQUIVALENT CONSTRAINT MODEL

1. Macroscopic Aspects

The equivalent constraint model for any layer k may consist of the upper and lower constraint layers I and II as shown in Fig.1. To make its restriction equivalent to the realistic one, from laminated theory their inplane stiffness Q^{I}_{ij} , Q^{II}_{ij} and thickness h^{k}_{I}, h^{k}_{II} can be determined, respectively, as

$$h^{k}_{I} = \sum_{r=k+1}^{n} hr \quad , \qquad h^{k}_{II} = \sum_{r=0}^{k-1} hr \tag{20a,b}$$

$$Q^{I_k}_{ij} = 1/h^{k}_{I} \sum_{r=k+1}^{n} h_r Q^{(r)}_{ij} \quad , \qquad Q^{II_k}_{ij} = 1/h_{II} \sum_{r=0}^{k-1} h_r Q^{(r)}_{ij} \tag{21a,b}$$

The stiffness matrix $[Q^{(k)}_{ij}]$ of Laminar K can be obtained by inversing the compliance equation (19a,b). We then have constitutive equation for kth lamina as

$$\{\bar{\sigma}^{(k)}\} = [Q^{(k)}] \{\bar{\epsilon}^{(k)}\} \tag{22}$$

in which Q_{ij} is a function of in-situ damage effect function, it may be rewritten this function as Λ_{ij}. For instance, we have

$$\bar{\sigma}^{(2)}_{xy} = Q_{66} (1 - \Lambda_{66}) \bar{\gamma}^{(2)}_{xy} \tag{23}$$

in which $\Lambda_{66}$are closely related to W_{66}. It is easily to show that

$$\Lambda_{22} = 1 - \frac{\bar{\sigma}^{(2)}_{x}}{\overset{0}{Q}_{22}\bar{\epsilon}^{(2)}_{x} + \overset{0}{Q}_{12}\bar{\epsilon}^{(2)}_{y}} \quad , \qquad \Lambda_{66} = 1 - \frac{\bar{\sigma}_{xy}}{\bar{\gamma}^{(2)}_{xy} \overset{0}{Q}_{66}} \tag{24}$$

If We carry on a microscopic analysis for the equivalent model and get the relations between $\bar{\sigma}^{(2)}_{x}$ and $\bar{\epsilon}^{(2)}_{x}$, $\bar{\epsilon}^{(2)}_{y}$ and the relation between $\bar{\sigma}_{xy}$ and $\bar{\gamma}_{xy}$, we then can get the expressions for Λ_{22} and Λ_{66}.

2. Microscopic Analysis to get In-situ Damage Effective Function

The equivalent layers I and II are orthotropic substructres with angles α_1 and α_2 to principal axis of layer K. For simplicity we only consider symmetry case here. The unsymmetrc results will publish elsewhere. Following Nuismer and Tan (1988), who conclude that " the solution for the in-plane normal response of the laminate decouples from the solution for the in-plane shear response due to the assumed orthotropy of laminate", each response is considered seperately. Take the overall extension of the model in X-direction as example. Following the work given by Nayfeh et al (1973, 1977), we first carry on the average for displacement across the lamina thickness $h^{(k)}$, and write down it as $u^{(k)}$, $v^{(k)}$, which are assumed as

$$u^{(k)} = u^{(k)}(x), \qquad v^{(k)} = \epsilon^{(k)}_{y}(x) y \qquad (k=1, 2) \tag{25}$$

ϵ_y denotes poisson's effect and may have a significant change due to ϵ_x changing seriously between two matrix micro crack. The equilibrium equation for laminar 1, for instance, can be expressed as

$$\frac{\partial \sigma_x^{(1)}}{\partial x} + \frac{\partial \sigma_{xy}^{(1)}}{\partial y} - \frac{\tau_x}{h_1} = 0 \tag{26}$$

$$\frac{\partial \sigma_{xy}^{(1)}}{\partial x} + \frac{\partial \sigma_y^{(1)}}{\partial y} - \frac{\tau_y}{h_1} = 0 \tag{27}$$

in which shear stresses may be expressed by

$$\tau_x = K_x(u^{(1)} - u^{(2)})$$

$$\tau_y = K_y(v^{(1)} - v^{(2)}) \tag{28a,b}$$

Since we are not interested in the distribution of stress and strain along the y-direction , so we carry on average also along width 2W for the above equation.i.e.,

$$(\tilde{\tilde{\cdot}}) = 1/2W \int_{W}^{W}(\cdot)dy \tag{29}$$

Combining the overall equilibrium equation , matrix crack boundary condition and transverse restrict conditions, we finally get the explicit relations for $\bar{\sigma}_x^{(2)}$, $\bar{\epsilon}_x^{(2)}$ $\bar{\epsilon}_y^{(2)}$, and then obtain the closed relations for Λ_{22} and Λ_{11} as

$$\Lambda_{22} = 1 - \frac{\Phi_1+\Phi_2 D\tanh(\lambda_1/D)}{\Phi_3+\Phi_4 D\text{tnh}(\lambda_1 /D)} \tag{30}$$

$$\Lambda_{66} = 1 - \frac{\Gamma_1+\Gamma_2 D\tanh(\lambda_2/D)}{\Gamma_3+\Gamma_4 D\text{tnh}(\lambda_2 /D)} \tag{31}$$

$$\chi = h_I/h_{II} \tag{32}$$

Where λ_1, λ_2, Γ_1 , Γ_2, Φ_1, Φ_2 are functions of damage variable D, defined as the ratio of thickness $h^{(k)}$ of kth lamina over the distance $S^{(k)}$ of two damage interval, i.e, $h^{(k)}/S^{(k)}$ and the equivalent compliance and thickness.

EVOLUTION LAW OF MATRIX CRACKS CONSIDERING CURING STRAIN AND IN-SITU RESTRICTION

a, Energy Release Rate

To derive a damage evolution law incorporating curing strain ϵ_i^0 and In-situ restriction, it is essential to express energy release rate $G^{(k)}$ in terms of ϵ_i^0 and the In-situ damage effective function Λ_{ij}.

The energy release rate $G^{(k)}$ have the relation with the rate of work W and strain energy U as

$$G^{(k)} = \frac{\delta W}{\delta A^{(k)}} - \frac{\delta U}{\delta A^{(k)}} \tag{33}$$

Where $A^{(k)}$ is damage area of micro-crack surface in kth laminar. Expressing W,U for the constraint equivalant model, in terms of its macroscopic compliance $\bar{S}_{ij}$, stress $\bar{\sigma}_1$ the thickness

$h^{(r)}$ and stiffness $Q_{ij}^{(r)}$ for r-th layer, and then substituting it into eq.(30), we obtain

$$G^{(k)} = -1/2 \sum_{r=0}^{n} h_r \, dQ_{ij}^{(r)}/dD^{(k)} (\bar{S}_{il}\bar{\sigma}_l + \epsilon_i^p + \epsilon_i^{o(r)})(\bar{S}_{jm}\bar{\sigma}_m + \epsilon_i^p + \epsilon_i^{o(r)}) \qquad (34)$$

in which $\overset{o}{\epsilon}_i$ and ϵ_i^p are curing strain and "plastic" strain caused by the coupling of residual strain and damage, respectively. Since the space is limited , we only list the results in eq.s(35,36), the analysis will be published elsewhere.

$$\epsilon_j^{o(\kappa)} = [\overset{o}{\bar{S}}_{im} 1/h \sum_{r=0} h^{(k)} \overset{o}{\bar{Q}}{}_{ml}^{(r)} (\alpha_l^{(r)} - \alpha_i^{(k)})](T_{serving} - T_{curing}) \qquad (35)$$

$$\epsilon_i^p = \bar{S}_{ij} 1/h \sum_{r=1}^{n} h^{(k)} Q_{jl}^{(r)} \epsilon_l^{o(r)} \qquad (36)$$

2. Damage Evolution Resistence Curve

The criterians for damage nucleation and growth are introduced as

$$G^k(\sigma) = G_C \qquad \text{------ nucleation} \qquad (37)$$

$$G^k(\sigma,\ D) = G_R(D^k) \qquad \text{------- growth} \qquad (38)$$

in which G_C is critical energy release rate and $G_R\ (D)^{(k)}$ is damage evolution resistance which is a material property, related to the energy for developing new surface and local plastic energy.

$G_r(D^k)$ can be determined by $\sigma \sim D$ demage resistance curve for speeified cross-play lamina, and calculating $G(\sigma,D)$ from eq(34), and then using equation (38) to get $G_R(D^k)$, which can be expressed as

$$G_R(D^k) = G_c + G_0(1 - e^{-RD}) \qquad (39)$$

CACULATED RESULTS

Among a serious of comparisons of calculated results with experimental data given by Wang, Flaggs, et al, Fig.2 is chosen to show the axial stiffness reduction E_x/E_x^o verses crack density h/s for [0,90,0]s, [0,90]s, $[0,90_2]$s, $[0,90_3]$s laminates. Experimental data expressed by symbols are taken from Allen, Harris (1987). Fig.3 shows the crack evolution by the curve of density N verses laminate stress σ for $[0/90_4/0]$ and $[0/90_3]$ laminates.

Experimental data expressed by symbols are taken from A.S.D.Wang from reference[4]. It can be seen these correlation is very good.

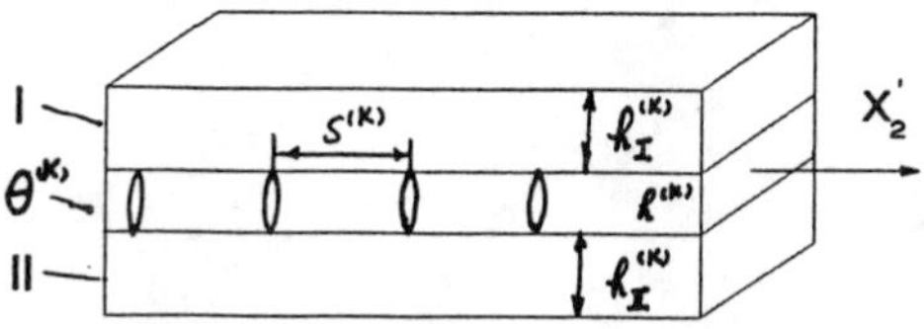

Fig.1 equivalent model

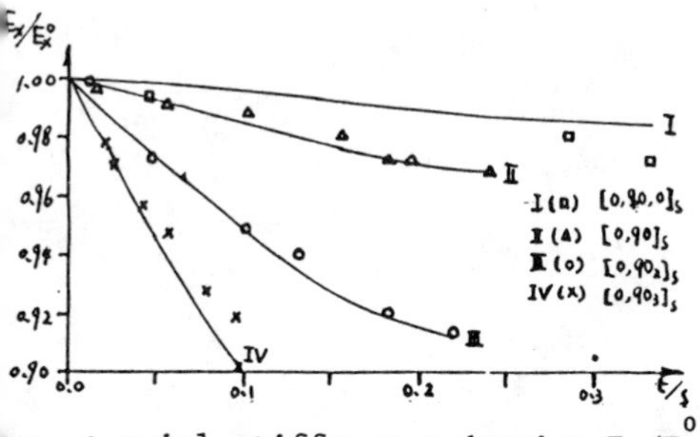

Fig.2 axial stiffness reduction E_x/E_x^0

verses crack density h/s

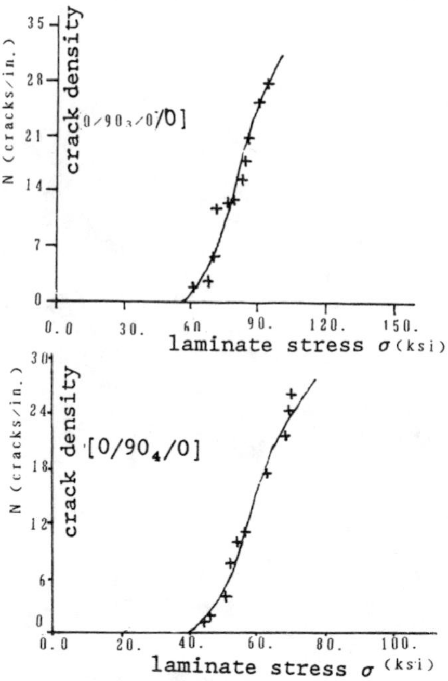

Fig.3 crack density N verses laminate stress σ

Reference

1, A.Parvizi and J.E.Bailey, On multiple transverse cracking in glass fibre epoxy cross-ply laminates, J.Material Science,V.13, 2131-2136(1978)

2, A.L.Highsmith and K.L.Reifsnider, Stiffness-reduction mechnism in composite laminates. ASTM STP775, 103-117(1982)

3, D.L.Flaggs, Prediction of tensile matrix ffailure in composite laminates, J. Composite Materials, Vol.19, 29-50(1985)

4, N.Laws and G.J.Dvorak, progressive transverse cracking in composite laminates, J.Composite Materials, V22, 900-915(1988)

5, S.R.Swanson, On the mechanics of microcracking in fibre composite laminates under combined stress, J.Engng.Materials and Technology, V01.111,145-149,(1989).

6, Z.Hashin,Analysis of arthogonally cracked laminates under tension, J.Appl.Mech.,V.54,872-879(1987).

7, N.laws,G.J.Dvorak and M.Hejazi, Stiffness changes in unidirectionel composites caused bycrack systems, Mech. of Materials,V.2'123-137(1983).

8, R.J.Nuisme and S.C.Tan, Constitutive relations of a cracked composite lamina, J.Composite Materials, V.22, 306-321(1985).

9, R.Telreja, A continuum mechanics characterization of damage in composite materials, Proc.R. Soc.Lond. 399A, 195-216(1985).

10, C.E.Harris and H.A.David, A continuum damage model of fatigue-induced damage in laminated composites, SAMPE Journal, 43-51, 1988

11, R.Hill,Elastic Properties of Reinforced Solids: some theoretical properties, J.Mech.Solids, Vol.11, 357-372(1963).

Correlations between Macro and Micro Structures in Granular Media

Jean Louis Favre & Pierre-Yves Hicher
Ecole Centrale de Paris, CNRS URA 850, Châtenay-Malabry, France.

Afif Rahma
Bureau de Recherches Géologiques et Minières, Département d'Ingénierie Géotechnique, Orléans, France.

Abstract

The logic of connection between parameters of the continuous medium and the discontinuous medium has been established for granular materials, in particular for soils. Many correlations between parameters of classic models (elasticity, plasticity) and parameters of grains are thereby consistenly proposed. This method has been extended to an elastoplastic model.

Introduction

The properties of the macroscopic medium can be explained deductively by way of the microscopic structure (almost the time very sophisticate). Even though one can easily observe these connections and their modelization statistically, one is ofter disappointed by the robustness of the models vis a vis the quality of the samples.

To avoid these problems in granular media, it is profitable to draw the logic of connections between, on the one hand, the macroscopic properties of a continuous medium *(CM)* and, on the other hand, the microscopic properties of the grains in a discontinuous medium *(DM)*.

Classification of soil parameters from the macro-micro properties connections

In observing the different mechanisms of deformability of the grains and of their assembly, one can explain the deformability and the plastic flow properties of the imaginary continuous medium *(CM)*.

This parameter classification logic in soil mechanics has alrealy been explained (Favre, Biarez, 1977) and the following adopted classes have been proposed :

1- volumetric strenght *(DM)* = volumetric weights of grains (γ_s) and fluids (γ_w, etc).

2- constitutive equation *(DM)* = constituents of grains.

3- boundary conditions *(DM)* = geometry of grains and possible number of contacts (granulometric range), considered to be parameters of interaction *(DM)*.

For convenience these three classes were called the *Nature* of the discontinuous medium *(DM)*.

This *nature* is synthesized by particular arrangements of the grains in response to normalized testing, i.e., Atterberg's limits W_l, W_p, maximum and minimum void ratios e_{max}, e_{min}, proctor density, etc.

The total assembly, as another indicator of the number of contacts, has to be characterised by the void ratio *e* and by the degree of saturation *Sr*. For convenience these parameters were called the *Compacity* of the discontinuous medium *(DM)*.
Hence, the first basic equation was :

Nature (DM) + Compacity (DM) ——> Rheology (CM)

and by combining the first two terms into *Mechanical State (DM) :* $(e_{max}-e)/(e_{max}-e_{min}) = Dr$, the relative density, we obtained the second basic equation :

Mechanical State (DM) ——> Rheology (CM).

Therefore an examination of the mechanisms at the grain dimension level provided a good guide for discussing the major correlations and for criticizing those which were not general and robust enough (Favre 1980):
- The parameters of interaction *(DM)*, which indicate the possibilities of the relative mobility of the grains (class 3) are the most explicative of the rheology *(CM)*, the meshes of the assembly being bent more than multiplied (or diminished), and bent more than the grains themselves.
- Synthetized parameters, have a good explanation potential, as for example Atterberg's Limits for clays, while for sands, in addition to e_{max} and e_{min}, the granulometric range $Uc=d_{60}/d_{10}$ must be used.

Principal correlations between the parameters of the discontinuous medium *(DM)* and those of the elasticity, consolidation and perfect plasticity models *(CM)* for sands and clays have been assembled by Favre (1980) and, subsequently, by several authors in the same way, (especially by Zervoyanis, 1982), they have now to be extended to other more sophisticated models.

Cyclade model and determination of its parameters

The constitutive model called "*Cyclade*" has been developed at the *Ecole Centrale de Paris* and the *Institut Français du petrole*. It is an elastoplastic model which employs the concept of critical state. The plastic strains are decomposed in three plane strain mechanisms. The hardening variables are the plastic volumetric strain and an interval variable linked to deviatoric plastic strains of each mechanism, whose initial value γo determines the size of the elastic domain.

We have to predict from the parameters of the discontinuous medium a set of ten parameters which can be classified as follows :
- elastic parameters : E_a, ν, n (isotropic non linear elasticity) with

$$E = E_a (p/p_a)^n \quad (1)$$

E_a, Young modulus for $p=p_a=100\ KPa$.
ν, Poisson's ratio
n, coefficient on non linearity
- size of the elastic domain : γ_o
- critical state parameters : b, p_{co}, φ with

$$p_c = p_{co} \exp b(\varepsilon_v^p) \quad (2)$$

p_{co}, initial value of the critical pressure p_c, depending on the initial density,
φ, friction angle at critical state (or perfect plasticity)
- plastic deviatoric parameters : *a*, *b*
- plastic volumetric parameter : α

These five classes can be divided into two groups:
1- experimental parameters : E_a, ν, *n*, *b*, p_{co}, φ directly determined from test results (conventional triaxial tests)
2- numerical parameters : γ_o, *a*, *b*, α, found in different equations of the model and determined only by comparing the experimental curves with the computed ones.

Nineteen sands, tested on 27 different initial compacities (150 triaxial tests) have been examined by factor analysis. 40 variables were created : among them, 6 variables of *Nature* considered to be the most explicative and at the same time the most independent of each other:
- *Nature* : d_{60}, d_{10}, $Uc=d_{60}/d_{10}$, $Cz=d_{30}^2/d_{60}.d_{10}$, *Angularity*, e_{max}.

As for clays with W_l and $I_p=W_l-W_p$, we found a good relation between e_{max} and $Ie=e_{max}-e_{min}$ but through a third variable *Uc*:

$Ie = 0.62\{e_{max} - 0.14 \log(Uc/330)\}$, coefficient of correlation $R=0.95$ (3)

due to the fact that for sand the granulometric range can vary greatly and become very influential.
- Mechanical State : $Dro = (e_{max}-e_o)/(e_{max}-e_{min})$
- Parameters of the model :

E_a, φ, $\beta_1 = \beta/(1+e_o)$ or $\beta_2 = \beta_1.Ie$
p_{co} with $\log p_{min} = \log p_{co}-Dro.\beta_2$
γ_o, *a*, *b*, α (the four numerical parameters)

Analogy with clays and considerations of the role of the granulometry range and microscopic mechanism for every parameter have given us the following results :
1- Experimental parameters
- φ is explained by the size, the shape and the mineralogy of grains
- the critical state is, approximatively, a straight line in $(Dr, \log p_c)$ plane but its slope β_2 is not constant as for clay (role of the granulometry), see in Fig. 1 :

$$\beta_2 = 4.14 + 7.5 \log Cz - 4.2\, e_{min} \qquad R = 0.93 \qquad (4)$$

the lines are classified according to the granulometry

$$\log pc_{min} = 9.9 - 8 \log Cz \qquad R = 0.98 \qquad (5)$$

2- Numerical parameters : we propose original but non unique models, representing a certain image of reality :

$$\gamma_o = 4.10\text{-}2(\log d_{60} + 1) \qquad R = 0.87 \qquad (6)$$

$$\log a = 1.15 + 0.3 \log d_{60} \qquad R = 0.86 \qquad (7)$$

$$\alpha = 1.6 + 0.35 \log d_{10} + 0.42\, Cz \qquad R = 0.79 \qquad (8)$$

$$b = 0.2 + 0.84\, Ie - 0.34\, e_{max} - 0.02 \log d_{10} \qquad R = 0.81 \qquad (9)$$

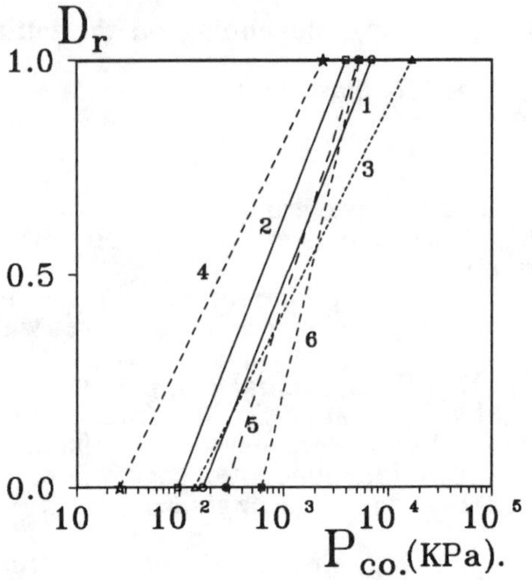

N°	Sable
1	Duncan
2	Seed
3	Been
4	Vesic
5	Mohk-Hostun
6	Bishop-Ham

Figure 1 "Lines of Critical State"

Conclusion

The logic of connection between the behaviour of the continuous medium and that of the discontinuous medium can be expressed through correlations between nature and compacity parameters of grain assemblies and rheological parameters of the equivalent continuous medium. This logic must be a guide for the determination of an initial set of parameters for sophisticated models which need an optimization procedure. We verified the similarity of the behaviour of clays and sands with the specific role of the granulometry for sands.

Acknowledgements

We thank the *B.R.G.M* which sponsored this research. The code of optimization "*Adelap*" was provided by Y. Meimon (*I.F.P*), the data by E.R. Michalski (*B.R.G.M*). This work would not have been possible without the database "*Modelisol*".

References

1- Biarez, J. & Favre, J.L. 1977. Statistical estimation and extrapolation from observations. Reports of Organisers, Spec. Session 6, IX ICSMFE, Tokyo, Vol 3:505-509.

2- Favre, J.L. 1980. Milieu continu et milieu discontinu. Mesure statistique indirecte des paramètres rhéologiques et approche probabiliste de la sécurité. Th. Doct. es Scies. Univ. Paris 6: 2T.

3- Zervoyanis, C. 1982. Etude synthétique des propriétés mécaniques des argiles et des sables sur chemin oedométrique et triaxial de révolution. Th. Doct. Ing. Ecole Centrale de Paris.

Stress Versus Temperature Dependent Activation Energies in Creep*

A. D. Freed
NASA–Lewis Research Center
Cleveland, Ohio

S. V. Raj
Cleveland State University
Cleveland, Ohio

K. P. Walker
Engineering Science Software, Inc.
Smithfield, Rhode Island

July 18, 1990

Abstract

The activation energy for creep at low stresses and elevated temperatures is that of lattice diffusion, where the rate controlling mechanism for deformation is dislocation climb. At higher stresses and intermediate temperatures, the rate controlling mechanism changes from that of dislocation climb to one of obstacle-controlled dislocation glide. Along with this change, there occurs a change in the activation energy. A temperature-dependent Gibbs free energy does a good job of correlating steady-state creep data, while a stress-dependent Gibbs free energy is less desirable. Application is made to a LiF–22mol.%CaF_2 hypereutectic salt.

1 Introduction

Choosing the free energy for activation (or Gibbs free energy), ΔG, as the thermodynamic state function implies that stress, σ, and absolute temperature, T, are the independent state variables, *i.e.* $\Delta G = \Delta G(\sigma, T)$. This state function is related to the activation enthalpy, ΔH, through the isothermal relationship

$$\Delta G = \Delta H - T\,\Delta S\ ,$$

where the activation entropy, ΔS, can be expressed as

$$\Delta S = -\left.\frac{\partial \Delta G}{\partial T}\right|_{\sigma}\ .$$

*This paper is a synopsis of *NASA TM-103192* (same title and authors as this) where application is also made to copper, and a more detailed discussion of the results can be found.

Combining these two relationships results in the expression

$$Q \equiv \Delta H = \left.\frac{\partial(\Delta G/T)}{\partial(1/T)}\right|_{\sigma}, \tag{1}$$

where Q is the activation energy. This is a useful relationship because it provides a means whereby functional forms for the free energy can be determined from isothermal experimental data.

The probability for the occurrence of an equilibrium fluctuation in energy greater than ΔG at a given absolute temperature is provided by Boltzmann's expression, such that the creep rate,

$$\dot{\epsilon} = \dot{\epsilon}_0 \exp\left(\frac{-\Delta G}{RT}\right), \tag{2}$$

is taken to be that fraction of the maximum attainable creep rate, $\dot{\epsilon}_0$, which is allocated by this probablility of fluctuation [1], where R is the universal gas constant, *i.e.* $R = 8.314\ \mathrm{Jmol^{-1}K^{-1}}$.

2 Models for Creep

Two different methods for modeling creep are presented. The first is a theory based on dislocation kinetics. The second is a phenomenological approach to creep. The capablility of each method in correlating creep data is explored.

2.1 Dislocation Theory – Based Model

At the higher temperatures and lower stresses, the well-established and prevailing mechanism for creep is diffusion-assisted dislocation climb [2], which evolves according to the relation

$$\dot{\epsilon}_c = A_c \frac{D\mu b}{kT}\left(\frac{\sigma}{\mu}\right)^n, \tag{3}$$

where

$$D = D_0 \exp\left(\frac{-Q_\ell}{RT}\right)$$

is the lattice diffusion coefficient with D_0 defining the frequency factor. Therefore from Eqns. 1 and 2, $Q = Q_\ell$ which is the activation energy for lattice diffusion. Here μ is the elastic shear modulus, A_c is the creep coefficient for climb, b is the magnitude of the Burgers vector, k is the Boltzmann constant (*i.e.* $k = 1.381 \times 10^{-23}\ \mathrm{JK^{-1}}$), and n is the power-law creep exponent. The quantities σ/μ and $kT/\mu b^3$ are normalized variables for stress and temperature.

At the intermediate temperatures and higher stresses, obstacle-controlled dislocation glide is the prominent mechanism controlling creep [1, 2], which evolves according to the relation

$$\dot{\epsilon}_g = A_g \left(\frac{\sigma}{\mu}\right)^2 \exp\left[\frac{-\Delta F}{RT}\left(1 - \frac{\sigma}{\hat{\tau}}\right)\right], \tag{4}$$

where $Q = \Delta F(1 - \sigma/\hat{\tau})$ from Eqns. 1 and 2, with ΔF being the enthalpy of activation in the absence of stress, and where $\hat{\tau}$ is the strength of the obstacle to dislocation glide, and A_g is the creep coefficient for glide.

The creep rate is taken to be given by the sum[3]

$$\dot{\epsilon} = \dot{\epsilon}_c + \dot{\epsilon}_g \tag{5}$$

which is analogous to the classical decomposition of strain-rate into creep, *i.e.* $\dot{\epsilon}_c$, and plasticity, *i.e.* $\dot{\epsilon}_g$, contributions.

The ability of Eqns. 3 and 4, *i.e.* Eqn. 5, to correlate experimental creep data for LiF–22%CaF_2 is illustrated in Fig. 1. The diffusion coefficient used here is for the diffusion of Ca^{+2} in CaF_2, *i.e.* $D_{Ca^{+2}}$, as this appears to be the diffusion process which governs the rate of dislocation climb in LiF–22%CaF_2 [4]. The exponential creep or glide response is temperature dependent in this figure. This is because therein the creep rate is normalized for dislocation climb, not dislocation glide. This is true for both the experimental data and its correlation. The ability of Eqn. 4 to correlate these data in the exponential creep domain may be considered to be satisfactory, but it is certainly not exceptional; in particular, it does not predict the correct slope for the data.

2.2 Phenomenological Model

A phenomenological model for creep which combines GAROFALO'S [5] expression for stress dependence with MILLER'S [6] expression for temperature dependence is given by

$$\dot{\epsilon} = A\vartheta \sinh^n\left(\frac{\sigma}{C}\right), \tag{6}$$

where A is the creep coefficient, C is the power-law breakdown stress, and

$$\vartheta = \begin{cases} \exp\left(\dfrac{-Q_\ell}{RT}\right) & T_m > T \geq T_t \\ \exp\left\{\dfrac{-Q_\ell}{RT_t}\left[\ln\left(\dfrac{T_t}{T}\right) + 1\right]\right\} & T_t \geq T > 0 \end{cases}$$

accounts for the thermal diffusivity, where T_t is a transition temperature, and T_m is the absolute melting temperature. Therefore from Eqns. 1 and 2, $Q = Q_\ell$ whenever $T_m > T \geq T_t$, and $Q = Q_\ell T/T_t$ whenever $T_t \geq T > 0$. At stresses less than power-law breakdown, Eqn. 6 reduces to a power-law expression like Eqn. 3; whereas, for stresses greater than power-law breakdown, Eqn. 6 becomes an exponential expression similar to Eqn. 4.

The correlative capability of this model is demonstrated in Fig. 2. Independent of model correlation, the capability of $\dot{\epsilon}/\vartheta$ *vs.* σ (found in Fig. 2) to collapse the experimental data onto a master curve with less scatter than one obtains with $\dot{\epsilon}kT/D\mu b$ *vs.* σ/μ (found in Fig. 1) distinguishes the two approaches. This is particularly true in the domain of exponential creep. By not normalizing the stress with the shear modulus, and by using ϑ instead of $D\mu b/kT$, data distributions with less scatter are observed over the entire range of stress. We know of no physical explanation for why this temperature-dependent activation energy should correlate experimental creep data better than that of a stress-dependent activation energy, but that certainly seems to be the case.

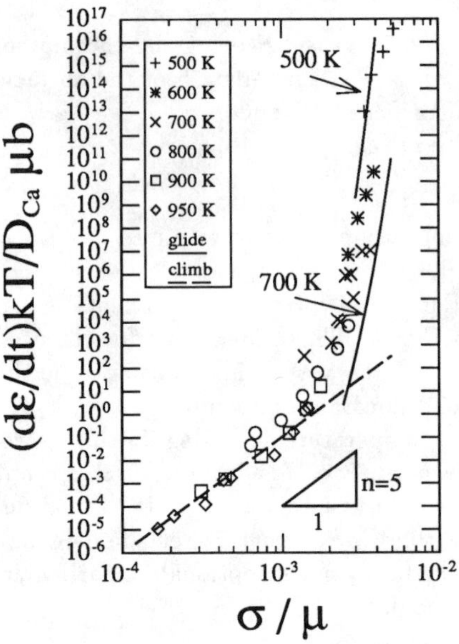

Figure 1: Theoretical creep response of LiF–22%CaF_2 hypereutectic salt. Data are from RAJ & WHITTENBERGER [4].

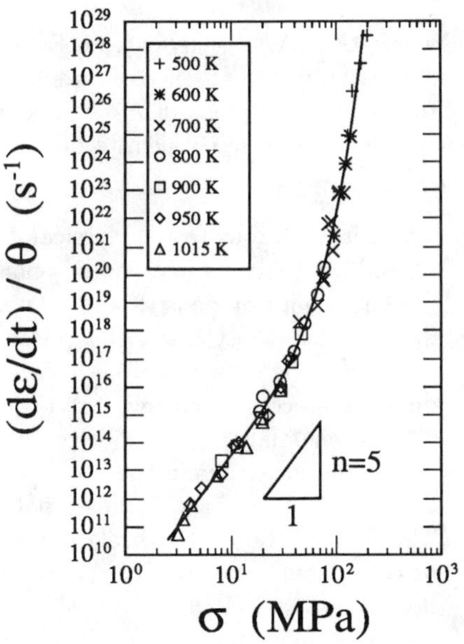

Figure 2: Phenomenological creep response of LiF–22%CaF_2 hypereutectic salt. Data are from RAJ & WHITTENBERGER [4].

References

[1] U F KOCKS, A S ARGON AND M F ASHBY, "Thermodynamics and Kinetics of Slip," *Progress in Materials Science*, **19**, eds B Chalmers, J W Christian and T B Massalski (Pergamon Press, Oxford 1975).

[2] H J FROST AND M F ASHBY, *Deformation-Mechanism Maps: The Plasticity and Creep of Metals and Ceramics* (Pergamon Press, Oxford 1982).

[3] W D NIX AND B ILSCHNER, "Mechanisms Controlling Creep of Single Phase Metals and Alloys," *Strength of Metals and Alloys*, **3**, 1503–1530, eds P Haasen, V Gerold and G Kostorz (Pergamon Press, Oxford 1980).

[4] S V RAJ AND J D WHITTENBERGER, "Deformation of As-cast LiF–22mol.%CaF_2 Hypereutectic Salt Between 500 and 1015 K," *Mater Sci Eng*, **A124**, 113–123 (1990).

[5] F GAROFALO, "An Empirical Relation Defining the Stress Dependence of Minimum Creep Rate in Metals," *Trans Metall Soc AIME*, **227**, 351–356 (1963).

[6] A MILLER, "An Inelastic Constitutive Model for Monotonic, Cyclic, and Creep Deformation: Part I—Equations Development and Analytical Procedures," *J Eng Mater Technol*, **98**, 97–105 (1976).

A MICROMECHANICAL VISCOPLASTIC MODEL FOR HASTELLOY-X

Eric H. Jordan
University of Connecticut
Storrs, Connecticut 06269

Kevin P. Walker
Engineering Software, Inc.
Smithfield, Rhode Island 02917

Shixiang Shi
University of Connecticut
Storrs, Connecticut 06269

ABSTRACT

A constitutive model for Hastelloy-X polycrystalline metals is derived from the known deformation behavior of single crystals of the same metal using a self-consistent formulation which averages the response of many randomly oriented single crystals while including the interaction effects of the differently oriented crystals. The single crystal behavior which has been used was developed by summing postulated slip behavior on crystallographic slip systems. The predictive capabilities of the single crystal model for Hastelloy-X will be presented.

Single crystal properties used for input to the model have determined experimentally for Hastelloy-X. Using these properties the behavior of polycrystal Hastelloy-X will be predicted.

In addition, methods other than the self-consistent method will be discussed. The other methods are considered in an attempt to more realistically model the heterogeneity and grain boundary sliding.

INTRODUCTION

An attempt to model the viscoplastic response of a polycrystalline metal starting from a postulated viscoplastic law for a single slip system is presented. Experiments were run on specimen made of microscopic single crystals and of polycrystal material of the same chemistry. The experiments provide a critical test of the model made. The main modeling approach was the self consistent method [1,2]. However, other approaches are also being developed.

Experiments on Single and Polycrystals

Predictions of polycrystal response from single crystal response has typically been based on assumed single crystal properties and measured polycrystal properties. In this work both single crystal and polycrystal viscoplastic behavior were determined experimentally. Nickel based alloy Hastelloy-X was chosen as the test material because it was available in large single crystal form from Pratt and Whitney Inc. The chemistry of this material is as follows:

Element	C	Cr	Co	Mo	W	Fe	Mn	Si	P	S	Ni
Percent by Weight	0.10	22.0	1.5	9.0	0.6	18.5	1.0	1.0	0.040	0.303	rem.

Single Crystal Model

The single crystal material was modeled by summing the predicted slip occurring on octahedral and cube planes of the hastelloy-X crystal. The model was previously used to model alloy PWA1480 [4]. The inelastic strain on each slip system is assumed to be governed by the following unified viscoplastic law in which the stress component assumed to be responsible for inelastic strain is the local resolved shear stress.

$$\dot{\gamma}_r = K^{-p}(\pi_r - \omega_r)\,|\pi_r - \omega_r|^{p-1} \qquad (r=1,2,\ \dots\ ,12) \qquad (1)$$

$$\dot{\omega}_r = \rho_1\dot{\gamma}_r - \rho_2\,|\dot{\gamma}_r|\,\omega_r - \rho_3\,|\omega_r|^{m-1}\omega_r \qquad (r=1,2,\ \dots\ ,12) \qquad (2)$$

γ = the inelastic shear strain rate on the rth slip direction

where K, ρ_1, ρ_2, ρ_3, p, m = material constants, the same for slip systems of the same type.

ω = respectively, a tensor valued state variable.

The overall response of the single crystal was obtained from summing the response on 12 octahedral slip systems and 6 cube systems of this FCC crystal. Octahedral constants could be determined independently from the cube constants by using tensile data from 100 oriented samples while cube constants were determined from torsion data. Because the constants are implicit, a nonlinear least square procedure was used to obtain them. A comparison of experiment and simulation is shown in Fig. 1 where it can be seen that the model does a good job of representing the experiments.

The Self-Consistent Model

One method of predicting polycrystalline properties from single crystal properties is the classical self-consistent model [1,2]. Briefly in this idealization, a spherical single crystal grain is embedded in an infinite isotropic medium. The elastic and inelastic properties of the isotropic medium are derived from the angle averaged response properties of the embedded grain.

The complete governing equations are too long to show here, but are given in [5]. The general nature of the equations can be seen from the following specialized form which applies for a problem where the spherical inclusion and the matrix are isotropic and of identical elastic properties D. For this case the usual Kroner relation is obtained:

$$<\Delta\sigma_{ij}> = D_{ijkl}(<\Delta\varepsilon_{kl}> - <\Delta\varepsilon^p_{kl}>) \qquad (3)$$

$$\Delta\sigma_{ij}(\eta,\beta,\varphi) = <\Delta\sigma_{ij}> + D_{ijkl}(S_{klmn} - I_{klmn})\,(\Delta\varepsilon^p_{mn}(\eta,\beta,\varphi) - <\Delta\varepsilon^p_{mn}>) \qquad (4)$$

In the above, I is the identity tensor and S is the Eshelby tensor known from elastic properties which are computed using Gauss integration involving various orientations in the non simplified case. The prescribed independent variable is the applied averaged strain $<\Delta\varepsilon_{kl}>$ which is therefore known. In the forward difference procedure, the inelastic strain $\Delta\varepsilon^{p}{}_{kl}$ is found from the values and the stress at the beginning of the current time step using the constitutive equation [eq. 1,2] and the volume averaged inelastic strain $<\Delta\varepsilon^{p}{}_{kl}>$ is easily obtained from the values of the individual grain inelastic strains. One may obtain $<\Delta\sigma_{ij}>$ from eq. 3, use it in eq. 4 with the other known quantities on the R.H.S. to find $\Delta\sigma_{ij}(\eta,\beta,\phi)$. The procedure is therefore explicit once the elastic properties are known. This is in contrast to the elastic problem which is implicit.

Results from the above procedure are currently under development.

Future Work

The idealization used under estimates the heterogeneity compared to the real material because of the averaging involved. In addition, the model includes no grain boundary sliding. To help improve the realism of the model, the following additional models will be investigated.

1. A three layer model involving a small volume fraction layer of viscous material will be made using an assumption of constant strain in each constituent.

2. A model with stress free grain boundaries will be made using the solution of Mura and Furuhashi [6].

3. A model which will explicitly be represented randomly oriented grains will be made. The collection of grains will be assumed to occur in an infinite repeating pattern.This approach will be solved using Fourier series in the manner described in [5].

REFERENCES

1. Budiansky, B., and Wu, T.T., Proc. 4th U.S. Nat. Congr. Appl. Mech., p. 1175.

2. Hill, R., "A Self-Consistent Mechanics of Composite Materials", J. Mech. Phys. Solids 13, pp. 213-222, 1965.

3. Jordan, E.H. and Chan, T., "A Unique Elevated Temperature Tension-Torsion Fatigue Test Rig", 3rd Annual Hostile Environment and High Temperature Measurements Conference, Cincinnati, OH, March 25-26, 1986.

4. Walker, K.P. and Jordan, E.H., "Biaxial Constitutive Modeling and Testing of a Single Crystal Superalloy at Elevated Temperature", Biaxial and Multiaxial Fatigue, EGF 3 (edited by M.W., Brown and K.J. Miller), Mechanical Engineering Publications, London, pp. 145-170, 1989.

5. Walker, K.P., and Jordan, E.H. and Freed, A.D., "A Nonlinear Mesomechanics of Composites with Periodic Microstructure: First Report", NASA Technical Memorandum 102051, 1989.

6. Mura, T., and Furuhashi, R., "The Elastic Inclusion with a Sliding Interface," Journal of Applied Mechanics, Transactions of ASME, Vol. 45, 308, 1984.

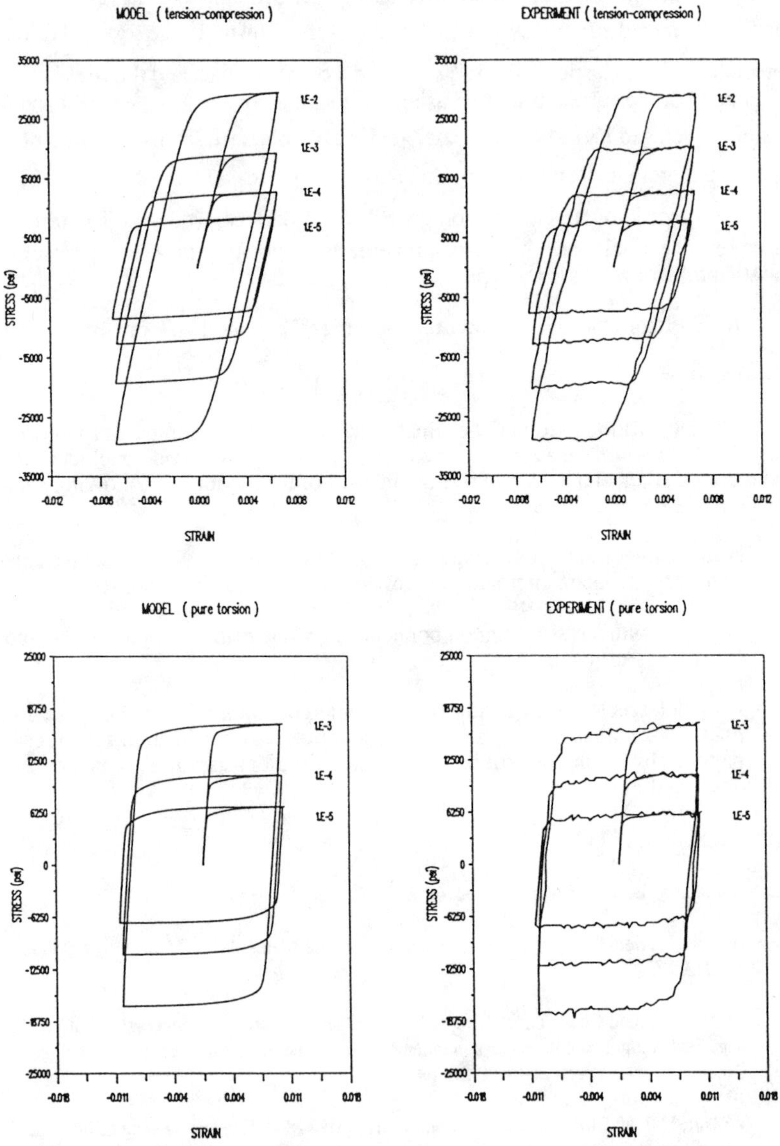

Fig. 1 Strain rate dependence of Hastelloy-X single crystal (001) at 982°C, model and experimental results

MICRO-MECHANISM OF INDUCED ANISOTROPY IN THE STRESS-STRAIN BEHAVIOR OF GRANULAR MATERIALS

Anil Misra and Ching S. Chang
Department of Civil Engineering,
University of Massachusetts, Amherst, MA 01003

Abstract

The material symmetry of granular solids undergoes alteration during a loading process. This change of material symmetry is termed as induced anisotropy. In this paper, a computer simulation technique is employed to investigate the mechanism of induced anisotropy in granular materials from the level of inter-particle interactions. Computer simulation experiments are performed on packings of circular particles under jump rotational stress paths. The evolution of directional distribution of contacts, directional distribution of contact forces, and shear forces in the packing during the loading process is investigated.

INTRODUCTION

From the experimental tests on sands it has been observed that the anisotropy of the stress-strain behavior of sands is greatly influenced by the loading history (1). The change of material anisotropy due to a loading history is termed as induced anisotropy. It is reasonable to surmise that the change of material anisotropy is due to the micro-level mechanisms of particle interactions. Due to the difficulty of observing each particle during a deformation process, the micro-mechanical aspects of the behavior cannot be explicitly investigated in most of the experimental studies of granular materials. In recent years, with availability of increased computational power, computer simulation methods have been developed to simulate the deformation of soil as an assembly of circular particles.

In this work, we employ the computer simulation technique to investigate the micro-mechanisms that transpire during a deformation process. Computer simulation experiments are performed on packings of circular particles under jump rotational stress paths consisting of a loading-unloading cycle followed by reloading in a direction inclined to the original loading axis. It is seen that the loading-unloading cycle induces anisotropy in an initially isotropic sample. The evolution of directional distribution of contacts, directional distribution of contact forces, and shear forces in the packing during the loading process is investigated.

COMPUTER SIMULATION METHOD

We employ the computer simulation method for granular materials developed by Cundall and Strack (3). In the computer simulation method, granular material is envisioned to be composed of rigid circular particles. The particles are assumed to interact at inter-particle contacts via deformable springs. Under an increment of load, the particles in the granular assembly move relative to each other. This relative movement between particles is resisted by the springs at the inter-particle contacts. The sprint constants are non-linear, obtained from the Hertz-Mindlin theory of frictional contacts (2). For each particle, the full equation of motion is considered in order to obtain the particle movement.

A typical simulation is carried out in the following steps:

Step 1. Generation of packing with a given void ratio and initial confining stress.

Step 2. Shearing the generated packing in accordance with a specified stress path. In order to follow the given stress path a servo controlled mechanism is employed (3).

LOADING CONDITION

The generated particle assembly under a given isotropic stress is loaded-unloaded followed by reloading in a direction inclined to the original loading axis. The stress path followed is illustrated in Fig. 1. For example, a $\alpha = 0°$ loading-unloading cycle will follow the path OAO. During the loading process the minor principal stress is kept constant. The loading-unloading cycle is expected to introduce anisotropy in the packing such that a sample, which is isotropic to begin with, will exhibit an anisotropic behavior on subsequent reloading. The induced anisotropy can be studied by reloading the sample along an axis inclined to the original loading-unloading direction. For example, a reloading stress path of $\alpha = 30°$ subsequent to a load-unload cycle along $\alpha = 0°$ stress path will follow the path OB. This loading-unloading-reloading stress path is termed as a jump rotational stress path since the principal stress axis is rotated suddenly during the loading process.

RESULTS

A 2-dimensional packing consisting of two-sized particles (85 of 0.21mm radius and 60 of 0.105 mm radius) is generated and compressed to an initial isotropic stress of 3 psi. In order to check for inherent anisotropy present in the sample, the sample is loaded along various directions. The stress-strain curve obtained for $\alpha = 0°$ and $90°$ loading tests are shown in Fig. 2. It can be seen that the stiffness and strength of the sample is very similar for the two cases indicating absence of any inherent anisotropy. The isotropy of the sample can also be seen from the isotropy of contact normal distribution of the packing at the initial condition (see Fig. 3).

To introduce anisotropy, the sample is loaded-unloaded along $\alpha = 0°$ (along OAO) stress path to a stress ratio of 1.9. Subsequently, to study the effect of induced anisotropy, the sample is reloaded along $\alpha = 0°, 30°, 45°, 60°$, and $90°$ directions from the same unloaded state. The stress-strain curves for reloading are given in Fig. 4. The sample displays a stiffer response for $\alpha = 0°$ test compared to tests with major principal direction more and more inclined to the

0° direction. The volume strain behavior is also different from that displayed by the original sample. The sample is more dilative for 0° reloading compared to the other stress paths. The relative softer behavior along the 90° is related to the lower number of contacts in that direction as can be observed from the contact normal distribution of the packing at unloaded stage (see Fig. 3). In addition to that, from the contact force distribution one can observe the presence of shear forces at the contact in the unloaded stage (see Fig. 3). The shear forces are negligibly small at the initial stage.

CONCLUDING REMARKS

The loading-unloading history induces anisotropy in the assembly of disks such that the material displays a softer response along the direction of minor principal stress. Similar observations have been made on sands tested in a directional shear cell under loading-unloading and reloading with principal stress axis rotated. From computer simulation results it is observed that the loading-unloading history causes the disappearance of contact in the minor principal stress direction. Furthermore it introduces shear forces at the contacts. It is seen that micro-mechanisms have a significant influence on mechanical behavior of granulates and should be accounted in the stress-strain modelling of these materials.

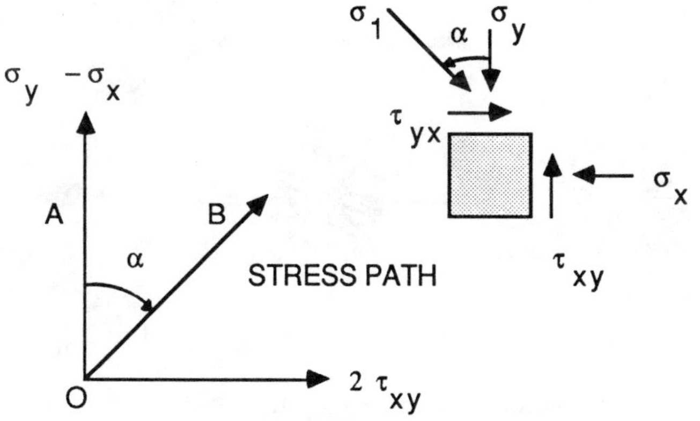

Fig. 1 Loading paths.

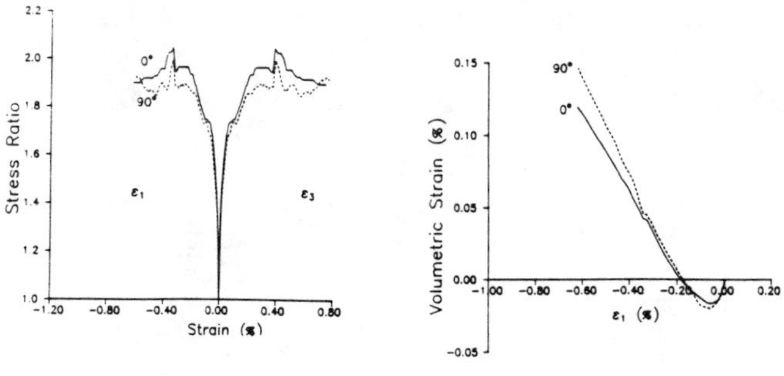

Fig. 2 Stress-strain behavior under loading along $\theta = 0°$ and 90° stress paths.

REFERENCES

1. Arthur, J.R.F., Berkenstein, S., Germaine,J.T. and Ladd, C.C. (1981). "Stress path tests with controlled rotation of principal stress directions," ASTM STP 740, Eds. R. Yong and F. Townsend, 516-540.

3. Chang, C.S. and Misra, A. (1989). "Computer simulation and modelling of mechanical properties of particulates," Computers and Geotechnics, 7(4), 269-287.

4. Cundall, P.A. and Strack, O.D.L. (1979). "A discrete numerical model for granular assemblies," Geotechnique, 1979, 29(1), 47-65.

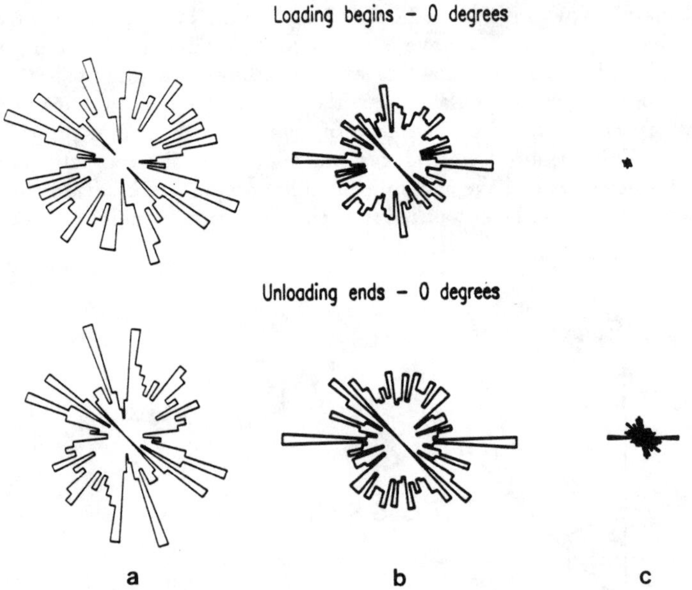

Fig. 3 (a) Contact normal distribution, (b) Normal force distribution, and (c) Shear force distribution.

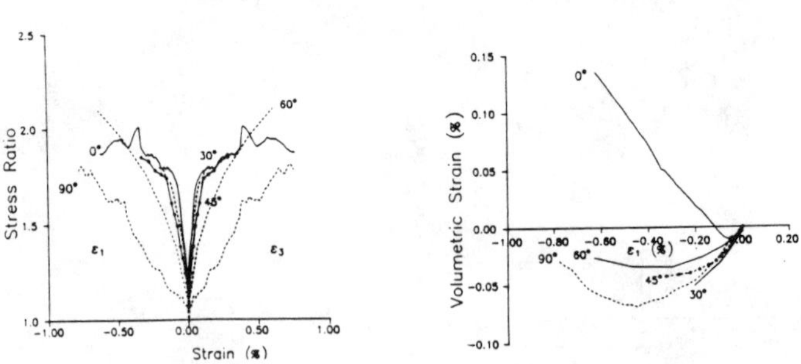

Fig. 4 Stress-strain behavior under reloading.

A GENERALIZATION OF SCHMID'S LAW IN SINGLE CRYSTAL PLASTICITY

Qing Qin and John L. Bassani
Department of Mechanical Engineering and Applied Mechanics
University of Pennsylvania, Philadelphia, PA 19104-6315, USA.

August 14, 1990

Abstract

Some single crystals are found not to obey the well-known Schmid's law for slip on individual systems. For example, in many $L1_2$ intermetallic compounds such as Ni_3Al, the critical resolved shear stress on the primary slip system at yield is a function of the of the orientation of the loading axis and the sense of load. To accommodate such behaviors, a generalization of Schmid's law is proposed, where stress components other than the Schmid stress are also incorporated. Finally, we demonstrate that the existence of non-Schmid stress effects can make it difficult to activate many systems simultaneously even in cubic crystals.

1 Introduction

Even though Schmid's law has been widely accepted as the yield criterion for single crystals, exceptional behaviors have been observed, e.g., in BCC single crystals. In recent years, a class of intermetallic compounds having the $L1_2$ structure, including Ni_3Al (Ezz et al, 1987), has been found to display strong non-Schmid yield behaviors, for example, the critical resolved shear stress (CRSS) depends both upon the orientation of the loading axis as well as the sense of the load. Paidar et al (1984) proposed a microscopic model which successfully explains various aspects of the anomalous behavior in Ni_3Al. In this paper, a yield criterion incorporating non-Schmid behaviors is developed based upon these observations as well as a general thermodynamic model. We also demonstrate that the existence of non-Schmid stresses in the yield criterion makes it more difficult to activated simultaneous slips as compared to cubic crystals obeying Schmid's law.

2 The Proposed Yield Criterion

An idealized thermodynamic model due to Rice (1971) is employed to provide a framework for the motivation of the non-Schmid yield criterion. The thermodynamic stress ($\tau^{*\alpha}$) that is conjugate to the shear (γ^{α}) on a slip system α can be identified from the model (Rice, 1971):

$$\tau^{*\alpha} = \tau_0^{\alpha} - \frac{\partial \phi}{\partial \gamma^{\alpha}} \tag{1}$$

where $\tau_0^{*\alpha}$ is the resolved shear stress on slip system α, and ϕ is the free energy per unit volume. The following generalization of Schmid's law is proposed:

A slip system α is potentially active if a generalized stress measure $\tau^{*\alpha}$ reaches a critical value, that is,

$$\tau^{*\alpha} = \tau_{cr}^{*\alpha} \tag{2}$$

Note, if the free energy of the system is independent of the shears, then equation (2) reduces to the conventional Schmid's law.

As an example, the yield behavior of $L1_2$ intermetallic compounds (e.g., Ni_3Al) is studied. A yield criterion that depends on other shear stress components as well as τ_0^{α} and is consistent with PPV theory (Paidar, Pope and Vitek, 1984), is chosen as:

$$\tau^{*\alpha} \equiv \tau_0^{\alpha} + A|\tau_{pe}^{\alpha} + k\tau_{se}^{\alpha}| + B\tau_{cb}^{\alpha} = \tau_{cr}^{*\alpha} \tag{3}$$

where A, B and k are material parameters and are taken to be the same for each slip systems due to the symmetry of the FCC crystals, and α is the slip systems index varying from 1 to 12 for FCC crystals. The planes and directions that define the shear stresses that enter in (3) are listed in Table 1.

Table 1: Orientations of shear stresses

α	τ_0^{α}	τ_{pe}^{α}	τ_{se}^{α}	τ_{cb}^{α}
1	$(111)[01\bar{1}]$	$(111)[\bar{2}11]$	$(1\bar{1}\bar{1})[2\bar{1}\bar{1}]$	$(100)[01\bar{1}]$
2	$(111)[\bar{1}01]$	$(111)[1\bar{2}1]$	$(\bar{1}1\bar{1})[\bar{1}\bar{2}\bar{1}]$	$(010)[\bar{1}01]$
3	$(111)[1\bar{1}0]$	$(111)[11\bar{2}]$	$(\bar{1}\bar{1}1)[\bar{1}\bar{1}\bar{2}]$	$(001)[1\bar{1}0]$
4	$(1\bar{1}\bar{1})[0\bar{1}1]$	$(1\bar{1}\bar{1})[\bar{2}\bar{1}\bar{1}]$	$(111)[\bar{2}11]$	$(100)[0\bar{1}1]$
5	$(1\bar{1}\bar{1})[\bar{1}0\bar{1}]$	$(1\bar{1}\bar{1})[12\bar{1}]$	$(\bar{1}\bar{1}1)[\bar{1}21]$	$(0\bar{1}0)[\bar{1}0\bar{1}]$
6	$(1\bar{1}\bar{1})[110]$	$(1\bar{1}\bar{1})[1\bar{1}2]$	$(\bar{1}1\bar{1})[\bar{1}12]$	$(00\bar{1})[110]$
7	$(\bar{1}1\bar{1})[011]$	$(\bar{1}1\bar{1})[21\bar{1}]$	$(\bar{1}\bar{1}1)[2\bar{1}1]$	$(\bar{1}00)[011]$
8	$(\bar{1}1\bar{1})[10\bar{1}]$	$(\bar{1}1\bar{1})[\bar{1}\bar{2}\bar{1}]$	$(111)[1\bar{2}1]$	$(010)[10\bar{1}]$
9	$(\bar{1}1\bar{1})[\bar{1}\bar{1}0]$	$(\bar{1}1\bar{1})[\bar{1}12]$	$(1\bar{1}\bar{1})[1\bar{1}2]$	$(00\bar{1})[\bar{1}\bar{1}0]$
10	$(\bar{1}\bar{1}1)[0\bar{1}\bar{1}]$	$(\bar{1}\bar{1}1)[2\bar{1}1]$	$(\bar{1}1\bar{1})[21\bar{1}]$	$(\bar{1}00)[0\bar{1}\bar{1}]$
11	$(\bar{1}\bar{1}1)[101]$	$(\bar{1}\bar{1}1)[\bar{1}21]$	$(1\bar{1}\bar{1})[12\bar{1}]$	$(0\bar{1}0)[101]$
12	$(\bar{1}\bar{1}1)[\bar{1}10]$	$(\bar{1}\bar{1}1)[\bar{1}\bar{1}\bar{2}]$	$(111)[11\bar{2}]$	$(001)[\bar{1}10]$

Simulations of uniaxial stressing are carried out for orientations of the loading axes in the spherical triangle as well as for the two different loading senses. The resolved shear stress on the primary slip system is plotted in Fig. 1(A) for a particular choice of A, B and k. Figure 1(B) is the experimentally observed tension/compression asymmetry for Ni_3Al (Pope and Ezz, 1984), which agrees well with the present model. It is the terms in absolute value that give rise to the tension/compression asymmetry.

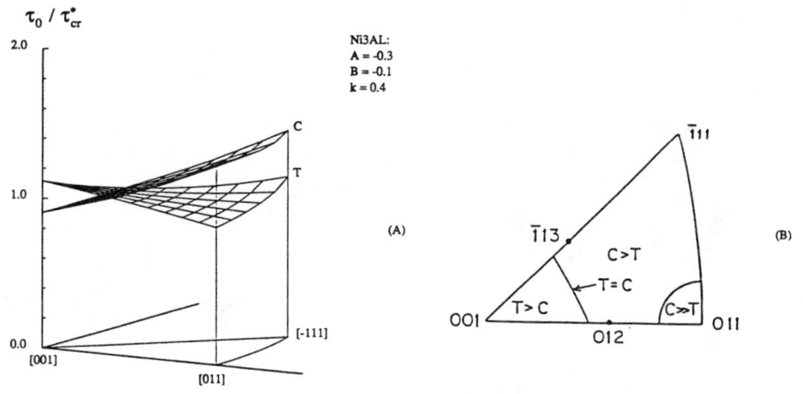

Figure 1: Resolved shear stress on the primary slip system at yield for both tension (T) and compression (C) with respect to the orientation of loading axis for Ni_3Al.

3 Restricted Slip

The orientations of slip systems associated with the FCC lattice possess a high degree of symmetry. Consequently, for Schmid's law the hyperplanes that define yield for each slip system are highly symmetric in stress space. As Bishop and Hill (1951) observed, there are many special stress states (28 for FCC crystals) that can simultaneously activate five or more (actually, six or eight) slip systems.

In contrast, non-Schmid terms in the yield criterion tend to distort the yield surfaces and thereby destroy symmetries. For example, Fig. 2 are plots of projections of yield loci on π-plane for one of the special stress states corresponding to FCC crystals obeying Schmid's law and $L1_2$ intermetallic compounds characterized by the yield criterion in (3). The numbers adjacent to each facet tell how many overlapping slip systems there are. For the latter, the 180° rotational symmetry does not exist, which is consistent with the experimentally observed tension/compression asymmetry. Also, there are not as many multi-fold vertices as in the regular FCC crystals. Therefore, non-Schmid stresses distort the yield surface and simultaneous activation of slip systems is more difficult.

4 Summary

A generalization of Schmid's law for single crystals is proposed to incorporate various non-Schmid behaviors, where shear stresses other than the resolved shear stress on the slip system contribute to the driving force for the dislocation motion. Several important consequences of this non-Schmid behavior have been studied and will be shown in the forthcoming papers. For example, with non-Schmid stresses, the plastic flow is not associated with the yield surface. This could have strong effects on the post-yield plastic behavior, such as strain localization in the form of shear bands. Also, simultaneous activation of many slip systems becomes more difficult. In polycrystals of grains with non-Schmid

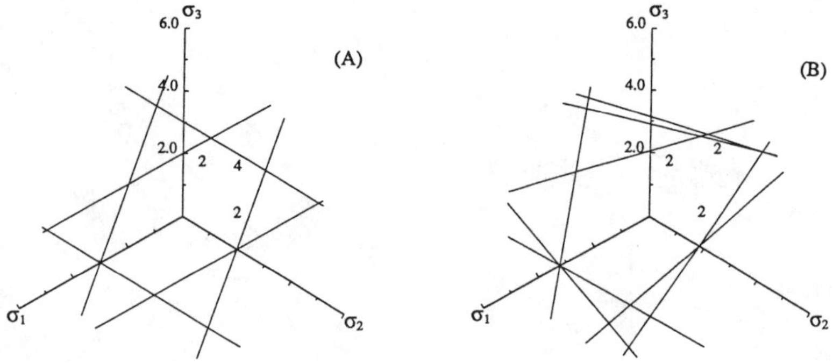

Figure 2: Comparison between yield loci of (A) FCC single crystals and (B) $L12$ intermetallic compounds for one of the special stress states (Bishop and Hill, 1951).

behaviors, this gives rise to additional plastic hardening raising the stress level and therefore, the tendency for failure.

Acknowledgements

The authors gratefully acknowledge many stimulating discussions with Profs. D.P.Pope and V.Vitek and the support of the NSF/MRL program at the University of Pennsylvania under Grant No. DMR88-19885 and the U.S. Dept. of Energy under Grant No. DE-FG02-85ER45188.

References

[1] Bishop, J.F.W. and Hill, R., 1951, *Phil Mag.*, Vol.42, p.1298.
[2] Paidar, V., Pope, D.P., and Vitek, V., 1984, *Acta metall.*, Vol.32, No.3, pp435–448.
[3] Pope, D.P. and Ezz, S.S., 1984, *International Metal Reviews*, Vol.29, No.3, pp136-167.
[4] Rice, J.R., 1971, *J. Mech. Phys. Solids*, Vol.19, pp.433–455.

Micromechanical Constitutive Behavior of Granular Media Under Dynamic Loading Conditions

M. H. Sadd, A. Shukla, Q. M. Tai and Y. Xu
Department of Mechanical Engineering and Applied Mechanics
University of Rhode Island
Kingston, RI 02881

Abstract

This paper addresses the dynamic constitutive behavior of granular materials from a micromechanical viewpoint. Such materials transmit mechanical loadings through contact mechanisms between adjacent particles. Specific contact theories were investigated and these were used in a numerical code employing the distinct element method in order to determine wave speed, amplitude attenuation and dispersion. Numerical predictions are compared with experimental strain gage results.

Introduction

The constitutive behavior of granular and particulate materials poses a formidable task due to the material's *microstructurally dependent load transfer characteristics*. Cohesionless particulate media can be described as a collection of distinct particles which can displace from one another with some degree of independence and which interact basically only through contact mechanisms. These types of materials establish discrete paths along which major load transfer occurs. The local microstructure or *fabric*, i.e. the local geometrical arrangement of particles, plays a dominant role in establishing this discrete load transfer phenomena. Our overall aim is to understand the dynamic behavior of this type of material when it is subjected to explosive loadings of short duration which produce propagating stress waves. It is found that the discrete nature of the media creates a *structured wave guide network of selective paths* for the waves to propagate along. The propagational characteristics of wave speed, amplitude attenuation, and dispersion (wave-form spreading) are thus related to the local fabric and the established wave paths. The present study is concerned with the specific contact laws which govern the dynamic constitutive behavior. Results using linear, nonlinear and nonlinear hysteretic contact laws are given. Computational and experimental studies have been conducted on specific model aggregate assemblies composed of circular disks in order to simulate granular and particulate materials.

Constitutive Laws and Numerical Implementation

The simulation of wave propagation in granular media was accomplished through the use of the *distinct element method* which was originally developed by Cundall et.al. [1]. This technique models the wave propagation process by determining at various time steps, the intergranular contact forces and movements of each disk in a large assembly. The numerical method is based on the use of an explicit scheme whereby the interaction of the granules is modeled using rigid-body dynamics assuming each particle interaction is governed by particular force-deformation and/or force-deformation rate contact laws. The movements of each of the disks is a result of the propagation through the medium of disturbances originating at the input loading points. Obviously the contact law used between particles will have major control on establishing the resulting wave propagation phenomena. Figure 1 illustrates the contact interaction between two typical disks. Each pair of disks are allowed to overlap (simulating deformation), and therefore a contact force *F* will develop. Previous work [2,3] using a simple linear force deformation law and linear velocity dependent damping produced satisfactory results for the wave speed and peak amplitude attenuation; however, unrealistic wave-form dispersion occurred. The goal of the present work is to include nonlinear contact behavior and to replace velocity dependent damping with a hysteretic energy dissipation mechanism. The contact laws under study are shown in Fig. 2.

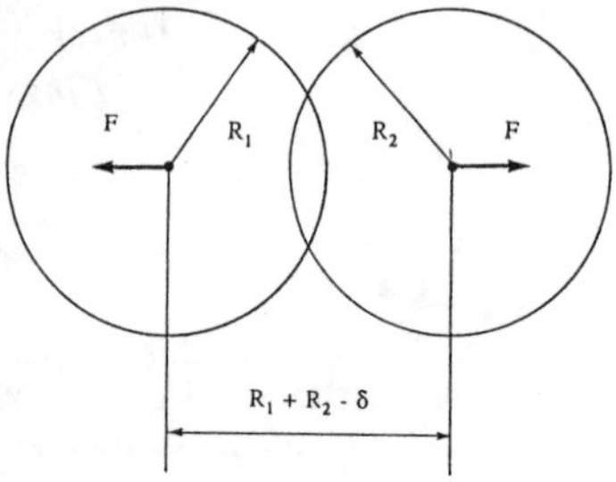

Fig. 1 Schematic of disk interaction

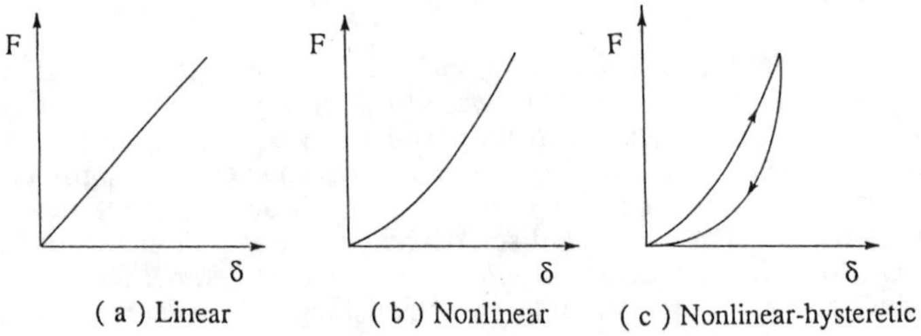

Fig. 2 Contact laws

For the nonlinear case as shown in Fig. 2b, the contact response is given by the equation

$$F = \alpha\delta^n \tag{1}$$

where δ is the relative displacement between the centers of the two adjacent disks (see Fig.1), and α and n are model parameters dependent on the disk material and geometry. The parameter n is larger than unity, which produces an increasing stiffness with increasing deformation, and this agrees with the non-Hertz contact mechanics proposed by Johnson [4]. Without damping the nonlinear contact law eliminates the large dispersion which appeared in linear case, and thus the wavelength was found to remain constant (see Fig.3). However, once viscous damping was introduced, this model also produced unacceptably large dispersion as shown in Fig. 3. Since some form of damping must be included to match experimental attenuation data, this nonlinear law had to be modified, and this lead to the hysteretic model.

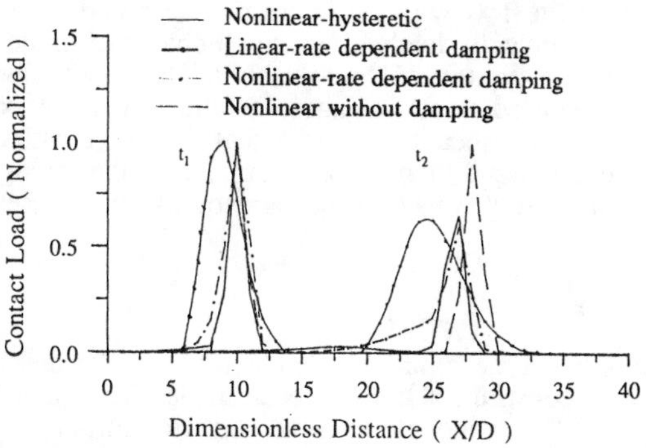

Fig. 3 Wave shapes at t_1 and t_2 (t_1 = 210μs, t_2 = 600μs)

The nonlinear-hysteretic contact law is shown in Fig. 2c and can be written as

$$F = \begin{cases} \alpha_L \delta^{\,n} & \text{... } \textit{loading} \\ \alpha_U \delta^{\,n+m} & \text{... } \textit{unloading} \end{cases} \qquad (2)$$

where α_L, α_U and n are parameters dependent on the disk material and geometry, and m is a function of the maximum deformation of the disks. Results from the distinct element model using this law are compared with the previous cases in Fig. 3. It was found that for this case both the amplitude and the dispersion characteristics were now acceptable.

Experimental Methods

Our experimental studies have employed the use of *dynamic photoelasticity* and *strain gage methods*. The photomechanics technique employed the usual photoelasticity methods in conjunction with high speed photography to collect full-field photographic data of the wave propagation process in model assemblies of birefringent circular disks. The strain gage methods involved placing strain gages on selected disks in similar assemblies. Entire dynamic strain profiles were then recorded at specific locations in the assembly, and this allowed the calculation of wave speed and attenuation along particular wave paths. As an example of the strain gage technique, an experiment with 40 disks arranged in a straight chain is shown in Fig. 4. The disks were

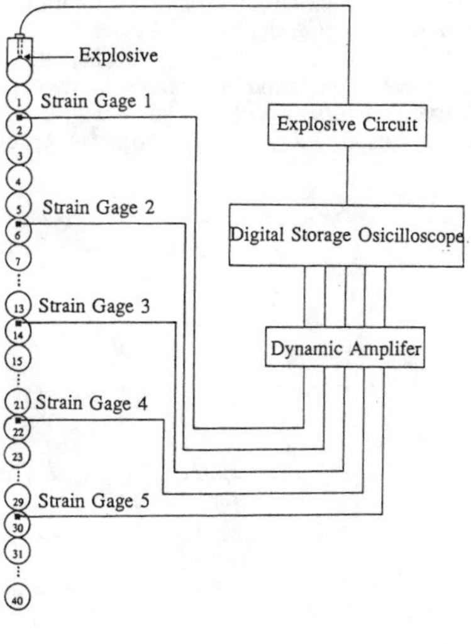

Fig. 4 Setup of single chain experiment

fabricated from Homalite-100 with 25.4 mm diameter and 6.35 mm thickness. The assembly was dynamically loaded by a small charge of explosive which was contained in a holder on the top of the assembly. The stress wave generated by the explosion was monitored by strain gages, and the normalized contact loads are shown in Fig.5. It can be seen from Fig.5 that the wave attenuation is large through the first few disks and then becomes smaller further along the chain. No significant wave dispersion (wave-form spreading) was observed from the experiments.

Conclusions

Using a simple geometry of a single chain of circular disks, a study of the effects of the contact constitutive law has been made. The numerical simulation was carried out using the distinct element method with a triangular input force of 20μs duration, and with a 2μs time step. Comparisons with experimental strain gage data indicated that the nonlinear hysteretic model provided numerical results which compared more favorably with the data than results from the linear or nonlinear contact models. The peak amplitude attenuation comparison for the nonlinear hysteretic case is shown in Fig. 6, and comparisons of the wave speed and wave dispersion also indicate good agreement.

Acknowledgement

Support from the Air Force Office of Scientific Research under contract No. F49620-89-C-0091 is sincerely appreciated

References

[1] Cundall, P. A. and Strack, O. D. L. (1979) "A Discrete Numerical Model for Granular Assemblies," *Geotechnique*, Vol.29, No.1, pp.47-65.
[2] Sadd, M. H., Shukla, A., and Mei, H., (1989) "Computational and Experimental Modeling of Wave Propagation in Granular Materials," *Proc. of the 4th International Conference on Computational Methods and Experimental Measurements*, pp.325-334, Capri, Italy
[3] Sadd, M. H., Shukla, A., Mei, H., and Zhu, C.Y., (1989) "The Effect of Voids and Inclusions on Wave Propagation in Granular Materials," **Micromechanics and Inhomogeneity** -The Toshio Mura Anniversary Volume, ed. Weng, G.L. Taya, M., and Abe, H., Springer-Verlag, New York
[4] Johnson, K.L., (1985), **Contact Mechanics**, Cambridge University Press

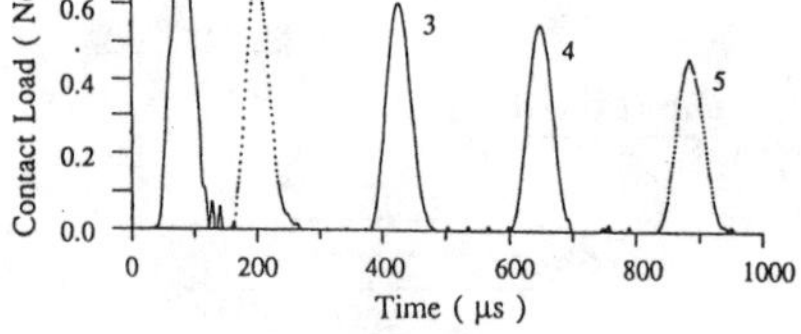

Fig. 5 Contact load versus time (experiment)

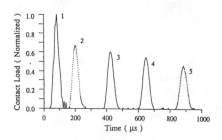

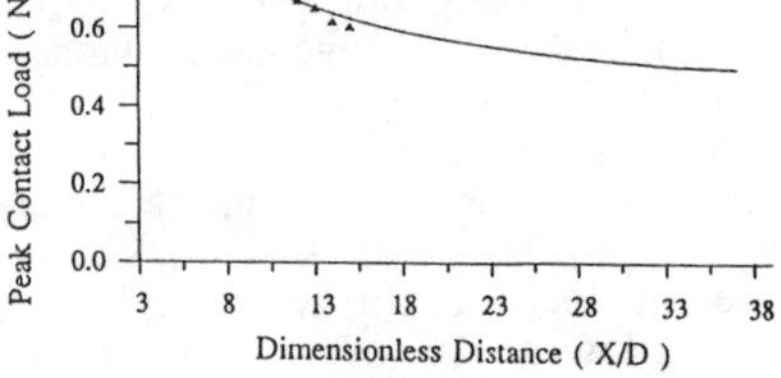

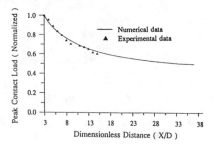

Fig. 6 Peak contact load versus distance

PREDICTIONS OF BAUSCHINGER CURVES WITH THE LIN'S POLYCRYSTAL MODEL

HIROSHI TAKAHASHI
Dept. Precision Engng.,Yamagata Univ.,Yonezawa JAPAN.

NOBORU ONO
Dept. Precision Engng.,Tohoku Univ.,Sendai JAPAN.

ABSTRACT

Bauschinger curves of polycrystal aluminium are predicted numerically with the Lin's polycrystal model. Bauschinger effect is taken into account in the constitutive law of single crystal by introducing the backlash model. This model is based on the author's experimental findings in torsional stress strain curves of polycrystal aluminium pipes that stress reversal yields transient region of nonworkhardening. The calculated results show the good agreements with the experiments for large prestrains. The comparison is made with the polarity model proposed by Hess & Sleeswyk.

INTRODUCTION

Several attempts have been made to predict Bauschinger curves of polycrystal metals. Hutchinson[1] and Ono et al.[2] calculated the Bauschinger curves with the KBW model[3] assuming the Taylor's isotropic law for single crystal workhardening. Since the Taylor's law does not take the Bauschinger effect of single crystal into consideration, their calculation evaluate the effect of internal stress due to intergranular constraint, i. e. Heyn stress. However, Bauschinger effect of polycrystal mainly depends on that of single crystal.

Budiansky & Wu[3] assumed the Ishlinsky-Prager's kinematic hardening law for the yielding function of single crystal. Weng[4] also proposed the combined hardening law of isotropic and kinematic. Kratochvil & Tokuda [5] proposed the stress memory hardening law. All of these anisotropic hardening models are intended to account for stress decreasing due to stress reversal.

Hasegawa et al.[6] showed that cell structures developed during prestraining are partially dissolved by stress reversal and new cell structures consisting of opposite sign dislocations are formed. During resolution and reformation of cell structures, the flow stress is maintained constant. The plateau at the Bauschinger curve has been observed by Takahashi et al.[7] for aluminium at large strain, by Miyauchi[8] for steel and by Christoudoul et al.[9] for copper.

The kinematic hardening model can not explain the stagnation of workhardening in the Bauschinger curves. The present study tries to reflect the above experimental findings on the constitutive law of single crystal.

The author[7] found that the stress reversal yields a transient region where workhardening pauses. After the transient region, workhardening restarts as if the material forgets the stress reversal. Since this feature resembles backlash in gear contact, the present article proposes a new model named “backlash model”. We calculate the Bauschinger curves of aluminium polycrystal with the Lin's polycrystal model[11].

BAUSCHINGER CURVES OF POLYCRYSTAL ALUMINIUM

The torsional tests of aluminium pipe specimens were carried out to find the features of Bauschinger curves especially at large strain[7]. The results are shown in Fig.1, where γ_0 is prestrain and

$$\overline{\tau} = |\tau|, \quad \overline{\gamma} = \int |d\gamma|. \tag{1}$$

The initial loading curve is approximated by the equation,

$$\overline{\tau} = F \cdot \overline{\gamma}^{n} \quad (F=59\text{Mpa},\ n=0.24). \tag{2}$$

Here we introduce the non-dimensional stress index ζ as

$$\zeta = (\overline{\tau}/F)^{1/n}. \tag{3}$$

Converting the ordinate $\overline{\tau}$ of Fig.1 into ζ, we get Fig.2.

We see the Bauschinger curves run parallel to the initial loading curve after the stagnation of workhardening. In other words, all of the curves coincide with the initial loading curve shifted parallel to the strain axis. These shift strains $\Delta\overline{\gamma}$ are shown in Fig.3, which is approximated by the following equation.

$$\Delta\overline{\gamma} = 2f(\gamma_0) \tag{4}$$

where $f(x)=x$ for $x<x_0$ and $f(x)=\beta(x-x_0)+x_0$ for $x>x_0$ with $\beta=0.4$ and $x_0=2\%$.

We define the effective strain for workhardening by

$$\gamma_{eff} = \overline{\gamma} - \Delta\overline{\gamma}. \tag{5}$$

The relation between γ_{eff} and γ is represented by the hardening process diagram as shown in Fig.4. When the loading direction is reversed, the material has the transient region, where workhardening pauses and the flow stress is decreased by

$$\Delta k = 8 \cdot \gamma_0 \quad (\text{MPa}). \tag{6}$$

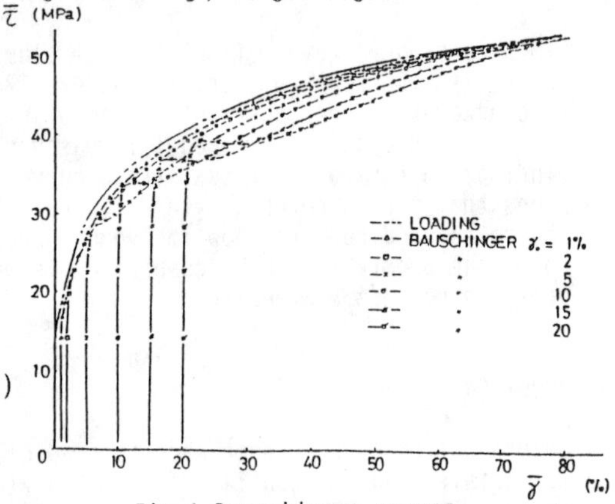

Fig.1 Bauschinger curves

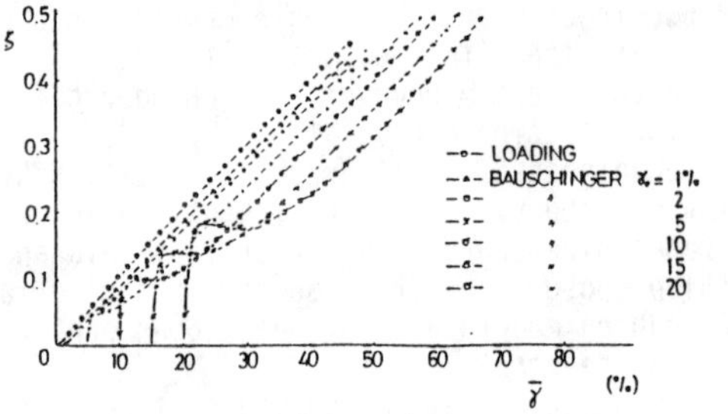

Fig.2 Linearized Bauschinger curves.

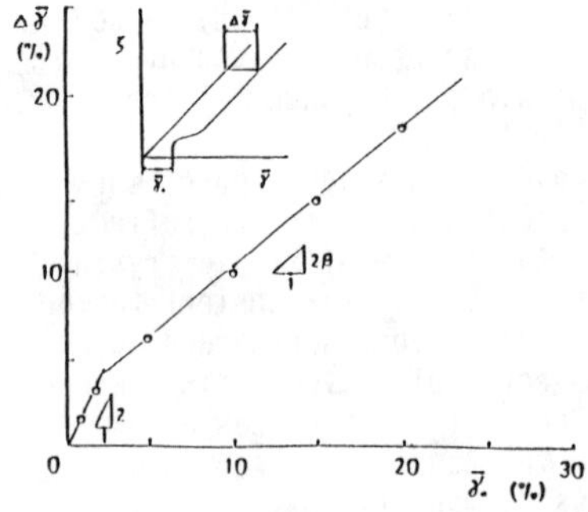

Fig.3 Shift strains

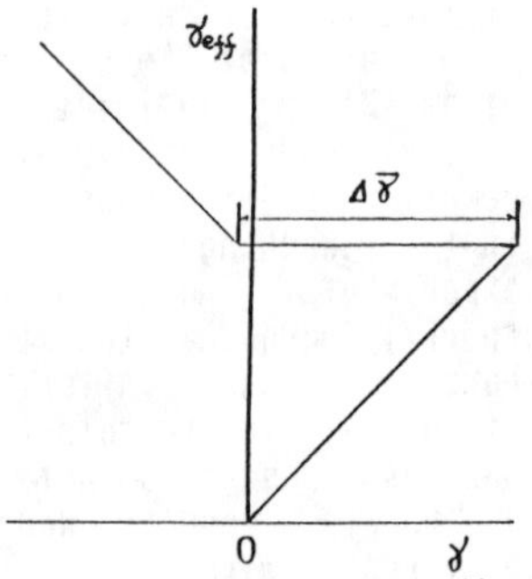

Fig.4 Hardening process diagram.

CONSTITUTIVE LAW OF SINGLE CRYSTAL

Backlash model

We propose the "backlash model" for single crystal workhardening. Suppose two kinds of strains for each slip system, reversible strain $\gamma^{(r)}_{rev}$ and effective strain for isotropic workhardening $\gamma^{(r)}_{eff}$ (≥ 0). We assume the workhardening condition for the effective strain to increase

$$z^{(r)}\gamma^{(r)}_{rev} = f(\gamma^{(r)}_{eff}) \tag{7}$$

where $z^{(r)} = \mathrm{sign}(\tau^{(r)})$. Then the strain increments are given by

$$d\gamma^{(r)}_{eff} = z^{(r)}\,d\gamma^{(r)}, \quad d\gamma^{(r)}_{rev} = f'\,d\gamma^{(r)} \tag{8}$$

During monotonic loading, all of the active slip systems satisfy the workhardening condition and the both of $\gamma^{(r)}_{eff}$ and $\gamma^{(r)}_{rev}$ increase.

Once the loading direction is reversed, the sign of stress $z^{(r)}$ is changed and we find $z^{(r)}\gamma^{(r)}_{rev} < f(\gamma^{(r)}_{eff})$. Then the strain increments are given by

$$d\gamma^{(r)}_{eff} = 0, \qquad d\gamma^{(r)}_{rev} = d\gamma^{(r)}. \tag{9}$$

The reversible strain $\gamma^{(r)}_{rev}$ of previous sign is anihilated and then the one of present sign is accumulated again up to $f(\gamma^{(r)}_{eff})$

The critical yielding stress $k^{(r)}$ is assumed as

$$k^{(r)} = F\cdot[\sum_s \gamma^{(s)}_{eff}]^n - \Delta k^{(r)}. \tag{10}$$

where $\Delta k^{(r)}$ is the stress drop just after the stress reversal given by

$$\Delta k^{(r)} = b\{f(\gamma^{(r)}_{eff}) - \langle z^{(r)}\gamma^{(r)}_{rev}\rangle\} \tag{11}$$

where $\langle x \rangle = (x+|x|)/2$ and b is the material constant. For a monotonic loading, $\Delta k^{(r)}$ is always zero. Just after the loading direction is reversed, however, we find $\langle z^{(r)}\gamma^{(r)}_{rev}\rangle = 0$.

Polarity model

Recently Hess & Sleeswyk[10] proposed a new model based on the polarity in cell structures. The dislocation density $\nu^{(r)}$ is consisting of two polarized dislocation densities, $\nu^{(r)}_{+}$ and $\nu^{(r)}_{-}$.

$$\nu^{(r)} = \nu^{(r)}_{+} + \nu^{(r)}_{-}. \tag{12}$$

The generation and annihilation of them are given by

$$\nu^{(r)}_{+} - \nu^{(r)}_{-} = \gamma^{(r)},$$

$$\dot{\nu}^{(r)} = |\dot{\gamma}^{(r)}|(1-2\delta\,\nu^{(r)}_{\mp}/\nu^{(r)}) \qquad \text{if} \quad \dot{\gamma}^{(r)} \gtrless 0 \tag{13}$$

where δ is the anihilation probability and $(\dot{\ })$ denotes rate.

The critical shear stress is assumed as

$$k^{(r)} = F\cdot[\sum_s \nu^{(s)}]^n \tag{14}$$

NUMERICAL PREDICTIONS WITH LIN'S POLYCRYSTAL MODEL

Each crystal has cubic shape and numerous blocks consisting of 27 crystals having different orientations are piled up as illustrated in Fig.5. The shear strain increment $d\gamma^{(r)}$ in each crystal are determined by the simplex method based on the Lin's polycrystal model[11].

Now we caluculate Bauschinger curves with the backlash model, where F=32(Mpa), n=0.24 and b=20(Mpa). The calculated curves are shown in Fig.6 and linearized in Fig.7. For the large prestrains the calculations simulate well the features

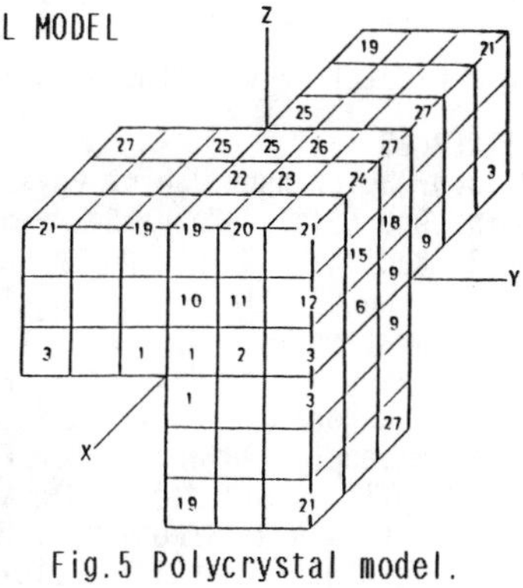

Fig.5 Polycrystal model.

of the experimental curves of Fig.1. For the small prestrains, however, the stagnation of workhardening seems to be too conspicuous.

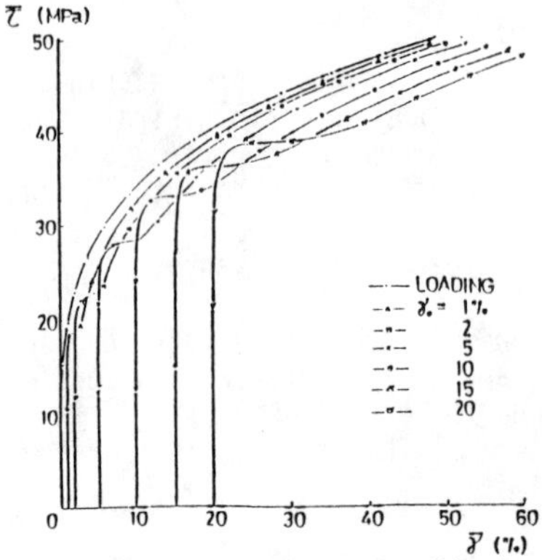

Fig.6 Prediction with the backlash model.

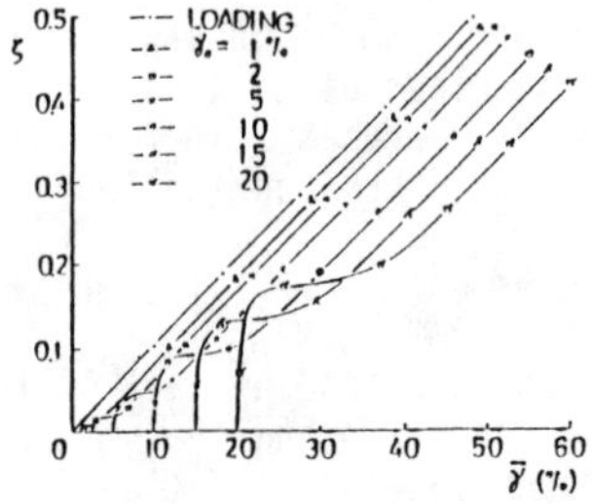

Fig.7 Linearized curves of Fig.6.

Next we caluculate the Bauschinger curves with the polarity model, where δ is 0.55 and ν_0, initial value of $\nu_+^{(r)}$ and $\nu_-^{(r)}$, is $0.75\text{x}10^{-3}$. The calculated results are shown in Fig.8 and linearized in Fig.9. For the small prestrains the curves more resemble the experiments than the backlash model. But the stress drop can not be seen and the parallelism is wrong at large prestrains.

We can conclude that the backlash model is good for large prestrains and the polarity model is good for small prestrains.

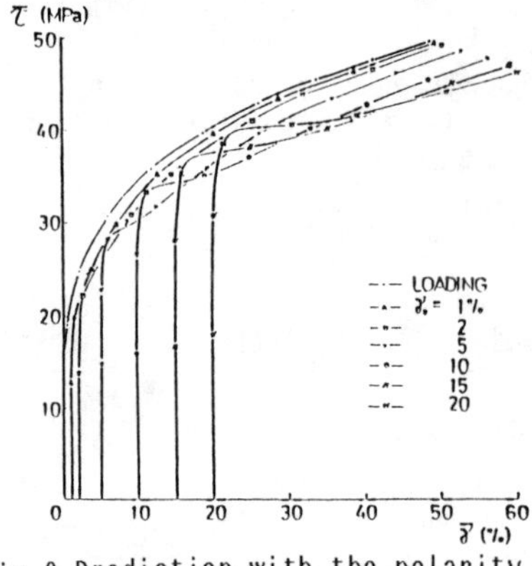

Fig.8 Prediction with the polarity model.

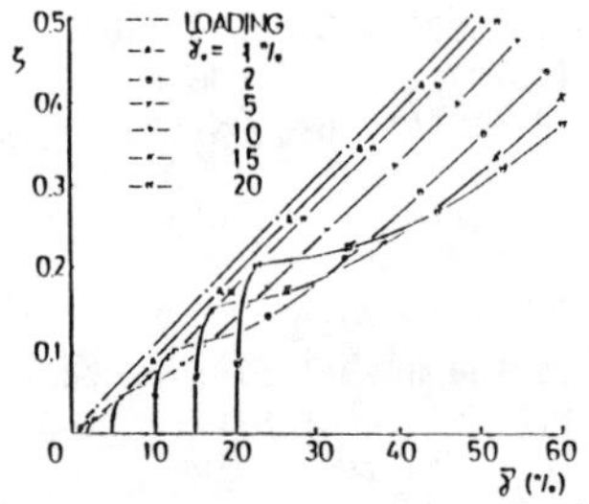

Fig.9 Linearized curves of Fig.8.

REFERENCES

1. J.W. Hutchinson, J.Mech.Phys.Solids, 12(1964), 11.
2. N. Ono et al., Mater.Sci. and Engng., 59(1983), 223.
3. B. Budiansky & T.T. Wu, Proc.4th Congr.Appl.Mech., 2(1962), 1175.
4. G.J. Weng, Acta Mecha., 37(1980), 217.
5. J. KratochviL & M. Tokuda, Trans.ASME.J.Engng.Mater.Tech., 106(1984), 299.
6. T. Hasegawa, T. Yakou & S. Karashima, Mater.Sci.Engng., 20(1975), 267.
7. H. Takahashi, I. Shiono, N. Chida & K. Endo, Bull.JSME., 27(1984), 2095.
8. K. Miyauchi, Advanced Tech.Plasticity, 1(1984), 623.
9. N. Christodoulou, O.T. Woo & S.R. MacEwen, Acta Metall., 34(1986), 1553.
10. F. Hess & A.W. Sleeswyk, J.Mech.Phys.Solids, 37(1989), 735.
11. H. Takahashi, Int.J.Plasticity, 4(1988), 231.

DISCONTINUITIES: INTERFACES, JOINTS

FRICTIONAL SLIDING IN THE PRESENCE OF THERMALLY INDUCED PORE PRESSURES

R.O. DAVIS and G. MULLENGER
Department of Civil Engineering, University of Canterbury, Christchurch, New Zealand

ABSTRACT

In this paper we consider a single degree of freedom elastic slider with a rate and pore pressure dependent friction law. Pore pressures are thermally driven by frictional dissipation. Increasing temperature induces increases in pore pressure, which results in weakening and unstable behavior.

INTRODUCTION

Single degree of freedom elastic sliders such as shown in Fig. 1 have found wide use in seismological literature as analogs for rate-dependent stick-slip behavior observed in rocks [1-4]. The load point velocity v_0 is held constant while the displacement and velocity of the block are monitored. Various forms of friction law are incorporated at the base of the block in attempts to model the typical stick-slip behavior seen in experiments on crustal rocks. The analogy to earthquake instabilities is clear [2]. State variable friction laws such as that of Ruina [3] are most widely used.

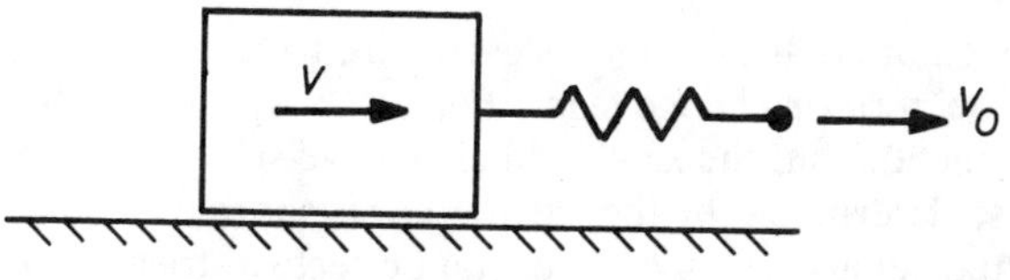

FIG. 1. The elastic slider.

Below we will incorporate a different type of frictional behavior into the elastic slider. A slightly simplified form of the pore fluid frictional heating model of Davis, et al. [5] will be used. This model was proposed to represent the behavior of creeping landslides. Frictional dissipation on the sliding surface results in pore fluid heating which in turn causes increasing pore pressures. This reduces the effective stress and hence the strength, and may, under certain conditions, lead to a complete loss of stability.

When the friction law of [5] is used together with pore fluid heating in the elastic slider, we find a rich variety of response. For some material parameters and initial conditions, changes in the load point velocity lead to stable response with the slider velocity approaching v_0 as time increases.

For other conditions, flutter type stick-slip instabilities develop. It has been observed [6] that slow moving landslides may behave in this way. The model may be useful in seismological considerations as well.

THE FRICTION LAW

The friction law used here is both rate and pore pressure dependent. It is similar to that used in [5]. We have the following relationship between the shear stress τ on the block base, the effective normal stress σ', and the block velocity v, for all $v \geq 0$.

$$\frac{\tau}{\sigma'} = \mu + \beta \sinh^{-1}\left(\frac{v}{v_R}\right) \tag{1}$$

Here μ and β are constants relating to the static strength and the velocity dependence of the slipping interface. The velocity v_R is an arbitrary reference velocity. In [5] the $\sinh^{-1}$ function was replaced by the ℓn function. In that form, (1) is similar to the soil friction law of Singh and Mitchell [7], as well as the rock friction laws used in [1-4].

In the rock friction laws, β is not constant, but instead evolves according to a state variable or variables. This permits the law to show weakening with increasing velocity. Here we hold β constant. Weakening (or hardening) may result from changes in the effective stress σ' caused by pore fluid heating effects. In some circumstances we find that the pore pressure increase due to frictional heating may be sufficient to reduce the effective stress to zero. In that event, the shear strength is also reduced to zero, and the block slides without frictional resistance until pore pressure dissipation allows a recovery of strength.

Finally we note that the $\sinh^{-1}$ is used here instead of ℓn because the velocity v may be zero or change sign. For $v >> v_R$, the $\sinh^{-1}$ function mimics the ℓn function, but the $\sinh^{-1}$ will also move smoothly through zero. Typical response determined by the friction law is shown in Figure 2 where strength is plotted versus velocity for constant effective stress. The strength response for negative velocities exactly mirrors that for positive.

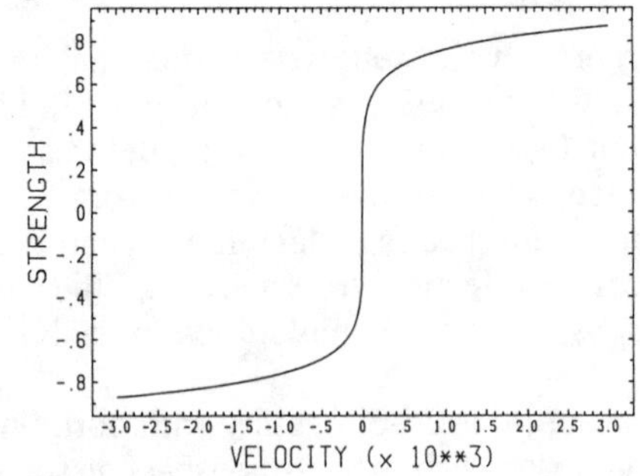

FIG. 2. Frictional resistance for constant effective stress

TYPICAL RESPONSE

We have performed numerical integrations of the slider motion incorporating inertia for a range of material parameters and initial conditions. The calculations were carried out with considerable care using a fourth order Runga-Kutta-Felhberg integration scheme. Despite these precautions, one must be aware that the calculated response can only represent the difference equations, and the cautionary remarks of Tongue [8] regarding the numerical computation of any dynamical system should be borne in mind.

In all our calculations we take a steady state sliding condition for the initial state of the block. At time t = 0, the load point velocity is increased and then held constant. The velocity of the block, the spring force, the frictional resistance, and the pore pressure are all monitored. These quantities and time are made dimensionless as follows

$$\bar{v} = \frac{v}{\ell}\sqrt{\frac{m}{k}}\,, \quad \bar{F} = \frac{F}{k\ell}\,, \quad \bar{S} = \frac{S}{k\ell}\,, \quad \bar{u} = \frac{u\ell}{k}\,, \quad \bar{t} = t\sqrt{\frac{k}{m}} \tag{2}$$

Here k is the spring constant, m the slider mass, and ℓ the thickness of the block (which is important in regard to heat transport upwards through the block). This results in a natural period of 2π for the block in the absence of frictional resistance. In all calculations a time step was used equal or less than 10^{-3} times the natural period. In the results described below, the model parameters were taken to be the same as in [5].

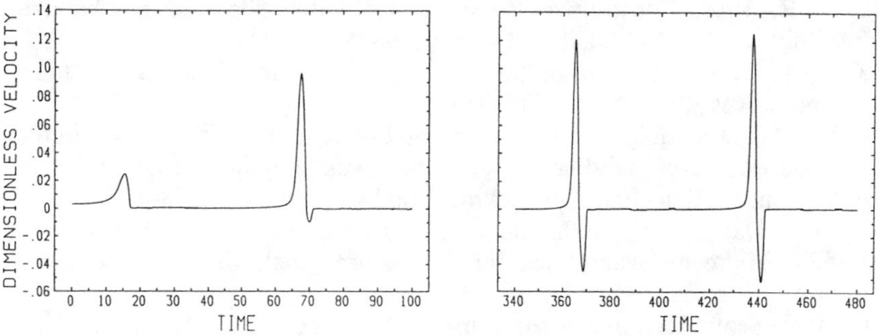

FIG.3 **Initial response (a) and limit cycle response (b) of elastic slider with v_0 = 0.003.**

Figure 3 illustrates the behavior of the slider with dimensionless load point velocity equal to 0.003. The initial steady state velocity was 0.0029. The response of the slider is dominated by long periods of zero velocity interrupted by brief slip episodes. Figure 3(a) shows the slider velocity during the first 100 time units following the change in load point velocity.

The system settles into a stable limit cycle shown in Figure 3(b). The maximum slider velocity during the slip episodes is approximately 0.12, roughly 40 times the load point velocity.

For other load point velocities, a range of different types of behavior of the slider may be found. For example, with $v_0 = 0.03$ the slider quickly finds a limit cycle in which the stick or zero-velocity time is very brief, and overall the slider flutters with nearly simple harmonic motion. Conversely, with $v_0 = 0.0003$, a limit cycle attractor does not exist and the slider very slowly approaches a steady state fixed point in which $v = v_0$.

CONCLUSION

Pore fluid heating, driven by frictional dissipation, gives rise to a wide range of response for the single degree of freedom elastic slider. Both stable steady state response, and stable limit cycles are found. For certain initial conditions the slider will exhibit stick-slip behavior. Long periods of zero velocity are interrupted by short sharp oscillations. This behavior may be exploited in the analysis of creeping landslides as well as in simulation of earthquake cycles.

REFERENCES

1. J.H. Dieterich, Time-dependent friction in rocks, Jour. Geophys. Res., 77, 3690-3697 (1972).
2. J.R. Rice, Constitutive Relations for Fault Slip and Earthquake Instabilities, PAGEOPH, 121, 443-475 (1983).
3. A.L. Ruina, Slip instability and state variable friction laws, Jour. Geophys. Res.88, 10359-10370 (1983).
4. J.C. Gu, J.R.Rice, A.L. Ruina, and S. Tse, Stability of frictional slip for a single degree of freedom elastic system with non-linear rate and state dependent friction, Jour. Mech. Phys. Solids, 32, 167-196 (1984).
5. R.O. Davis, N.R. Smith, and G. Salt, Pore fluid frictional heating and stability of creeping landslides, Int. Jour. Num. Anal. Methods Geomech., in press.
6. R.F. Scott, Incremental movement of a rockslide, in Rockslides and Avalanches, ed. B. Voight, 1, 659-668, (Elsevier, Amsterdam 1978).
7. A. Singh and J.K. Mitchell, General stress-strain-time functions for soils, Jour. Soil Mech. Found. Div. ASCE, 94, 21-46 (1968).
8. B.H. Tongue, Characterization of numerical simulations of chaotic systems, Jour. Appl. Mech., 54, 695-699 (1987).

A MICROMECHANICS MODEL FOR ROUGH INTERFACES

M.P. Divakar and A. Fafitis
Department of Civil Engineering
Arizona State University
Tempe, AZ 85287-5306

ABSTRACT

A new constitutive model for transfer of interfacial tractions along rough cracks in strain softening composites is presented. The model relates the normal and shearing stresses on the rough crack to the corresponding displacements in terms of the interface strength, contact areas, the contact angle, and the crack closing pressure. The performance of the constitutive model was evaluated by predicting experimental results. The comparison between predicted and experimental results appear to be very satisfactory.

INTRODUCTION

A constitutive model based on micromechanical concepts is presented in this paper. The mechanisms of shear transfer were isolated into sliding, interlocking, overriding and fracturing components. The deformation properties of the rough crack were modeled using the concepts of critical state soil mechanics and statistical simulation of rough surfaces. The main strategy for calibrating some parameters of the constitutive model was to use idealized test results.

THE MICROMECHANICAL MODEL

Figure 1a shows the micromechanical representation of a rough crack in concrete, subjected to shearing and normal stresses [1]. The coarse aggregates are treated as inhomogeneities of different sizes embedded randomly in a cement mortar matrix with elastic constants, μ_O, K_O.

In order to achieve a simpler mathematical model, the micromechanical model is replaced by equivalent mechanical models representing the various components. The interlocking mechanism (Fig. 1b) is represented by trapezoidal asperities which are interlocked with each other. The overriding component (Fig. 1c) is explained by a model containing triangular asperities of various sizes and internal angles, randomly distributed along the rough crack surface. The fracturing component (Fig. 1d) is modeled by rectangular asperities of various sizes.

Figure 2 shows the continuum model of a rough crack. It is assumed that the two faces of the rough crack are initially matched. The interlocking mechanism is represented by springs connecting the two halves of the sliding blocks. The triangular asperity is deformable and is the statistical equivalent of the rough crack surfaces. It models the overriding and fracturing components of the rough crack. Using the continuum model, the constitutive law is written as [1]:

$$\begin{Bmatrix} \sigma_t \\ \sigma_n \end{Bmatrix} = \sigma_I \, a_c(w,\delta_t) \begin{bmatrix} 1 & \mu \\ -\mu & 1 \end{bmatrix} \begin{Bmatrix} \sin\alpha(\delta_t,\delta_n) \\ \cos\alpha(\delta_t,\delta_n) \end{Bmatrix} - \begin{Bmatrix} 0 \\ \sigma_s(w) \end{Bmatrix} \tag{1}$$

where, σ_t and σ_n are the tangential and normal stresses acting on a unit area of the crack, σ_I is the interface strength on the active side of the triangular asperity, μ is the average frictional coefficient, σ_s is the closing pressure, a_c is the actual area of contact (function of displacements), and α is the asperity angle which varies with the displacements. The functions $a_c(w, \delta_t)$ and $\alpha(\delta_t, \delta_n)$ were determined based on experimental data.

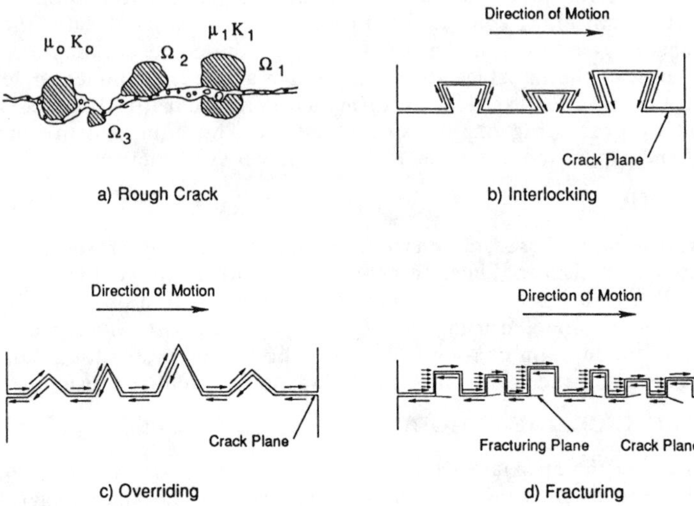

Figure 1. Micromechanical and equivalent models of the rough crack

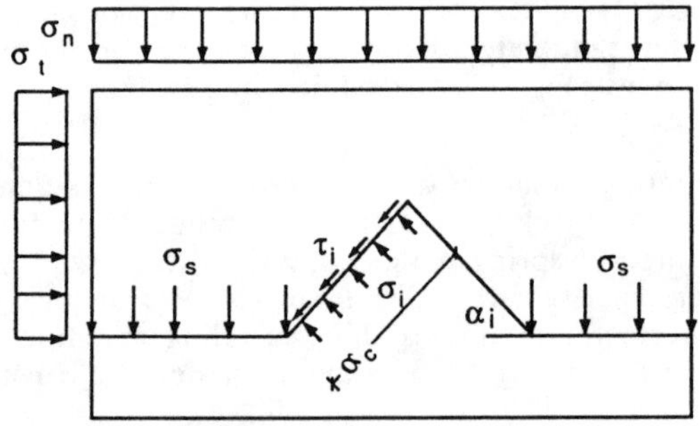

Figure 2. Continuum model of the rough crack

PARAMETERS OF THE CONSTITUTIVE MODEL

The strategy adopted for calibrating the parameters of the model was to use idealized test results. A comprehensive database of experimental results was built [1] for this purpose. The database includes the test data of Loeber [2], Walraven and Reinhardt [3], Daschner [4], Millard and Johnson [5], and Nissen [6].

Parameters such as the coefficient of friction, the crack closing pressure and the interface strength were evaluated based on the knowledge from existing literature. The friction coefficient assumed for the present model is 0.55. The interface strength, σ_I is a material constant, given by the following equation:

$$\sigma_I = G f'^{\,0.56}_c \tag{2}$$

where the constant, G, is 5.83 for normal strength concrete and 6.40 for high strength concrete. For the closing pressure, σ_S, the following formula is proposed:

$$\sigma_s = \sigma_p \exp\left(-K w^{\lambda}\right) \tag{3}$$

in which σ_p is the peak tensile strength, ($6.5\sqrt{f'_c}$ psi where f'_c is in psi), and w is the crack width in μ in. The values of the constants were determined by regression ($K = 1.544x\,10^{-3}$ and $\lambda = 1.01$).

The remaining parameters were established based on detailed analyses [1].

PERFORMANCE OF THE CONSTITUTIVE EQUATION

The performance of the constitutive equation was evaluated by predicting experimental results. The test data with constant normal stresses, constant crack widths, and variable normal stress and crack widths were used for testing the performance of the constitutive equation.

The prediction of variable normal stress and crack width experiments was attempted on 46 data sets from the database. The data of Walraven and Reinhardt [3] and Millard and Johnson [5] were included in the study. Figure 3 shows some of the results of prediction. In this type of tests, all four variables change during the test, and the responses are coupled. It was difficult to predict both shear stress, normal stress vs. slip, dilation responses accurately.

CONCLUSIONS

A new constitutive model for interface shear in strain softening composites based on micromechanics concepts is proposed. The constitutive equation relates the normal and shear stresses with the interface strength, contact area, friction coefficient, asperity angle and the crack closing pressure. The parameters of the constitutive model were determined using test results obtained under extreme boundary conditions. The responses of the rough crack under all three boundary conditions were very well predicted with the proposed constitutive model.

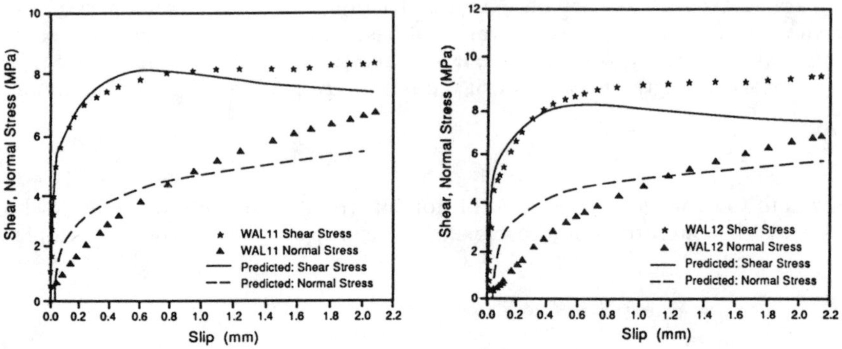

Figure 3. Prediction of variable normal stress and crack width test results

REFERENCES

1. Divakar M.P. "Constitutive Modeling and Numerical Implementation of Interface Shear in Concrete", Dissertation Presented in Partial Fulfillment of the Requirements for the Degree of Doctor of Philosophy, Arizona State University, Tempe, AZ, May 1990.

2. Loeber P.J. "Shear Transfer by Aggregate Interlock", M.Sc. Thesis, The University of Canterbury, Christchurch, NewZealand, 1970.

3. Walraven J.C. and Reinhardt H.W. "Experiments on Shear Transfer in Cracks in Concrete: Part I; Description of Results", Stevin Laboratory Report No. 5-79-3, Delft University of Technology, The Netherlands, 1979.

4. Daschner F. "Schubkraftubertragung in Rissen von Normal- und Leichtbeton", ***Report,*** Technische Universitat Munchen, Lehrstuhl fur Massivbau, West Germany, 1980.

5. Millard S.G. and Johnson R.P. "Shear Transfer across Cracks in Reinforced Concrete due to Aggregate Interlock and Dowel Action", ***Mag. Concr. Res.,*** 36(126), 9-21, 1984.

6. Nissen I. "Rissverzahnung Des Betons Gegenseitige Rissuferverschiebungen Und Uebertragene Kraefte", Dissertation Submitted in Partial Fulfillment of the Requirements for the Degree of Doctor of Engineering, Technical University of Munich, West Germany, 1987.

VERIFICATION FOR USING THIN LAYER FINITE ELEMENTS TO MODEL ROCK JOINTS VIA COMPARISON WITH RESULTS FROM ALTERNATIVE COMPUTATIONAL SCHEME AND LABORATORY MODELING

K.L. FISHMAN and SADDAM AHMAD
Assistant Professor and Graduate Student, State University of New York at Buffalo.

ABSTRACT

This study provides valuable data which contributes to verification of numerical results obtained utilizing thin layer finite elements for modeling the response of rock joints.

INTRODUCTION

Rock masses are generally not comprised of homogeneous, isotropic, continuous material and may contain joints faults and/or bedding planes. It is often necessary to account for the presence of such planes of weakness in numerical analyses. Within the context of the finite element method special elements are often employed which describe the localized response of rock joints. A thin layer element discussed by Desai et al [1] offers advantages over alternative elements. In this study the ability of thin layer elements to provide reliable results is demonstrated through comparison of results obtained via an alternative element for modelling rock joint response, and from a controlled laboratory model test.

FINITE ELEMENT ANALYSIS

For this study the case of a circular opening embedded in a jointed rock mass is investigated. Since the computer model results are to be compared with measurements from a laboratory model the dimensions and material properties for the finite element analysis are taken directly from the laboratory model. The model is 21 in. high by 21 in. wide by 12 in. thick with a 3 in. diameter hole cored through the center. Smooth horizontal joints are created in the model by partitioning the formwork for a solid block into slab sections with thin sections of sheet metal. After a 24 h curing period the slabs are separated from the dividers and placed together prior to coring the circular hole through the center. Details on the construction, instrumentation, testing and analysis of the model may be found in ref. [2].

Figure 1 depicts the mesh utilized for the finite element study which includes both solid and joint elements. Analyses are performed using thin layer elements [1] and Goodman elements [3] to model rock joint response. Both solid and thin layer elements

Table 1. Material Properties
For Finite Element Analysis

	Material	
Property	Concrete	Joints
Young's Modulus (ksi)	3000.	30.000.
Poissons Ratio	0.3	0.3
Shear Modulus (ksi)		0.1
Shear Stiffness. Ks (pci)		3700.
Normal Stiffness. Kn (pci)		1.111.111.

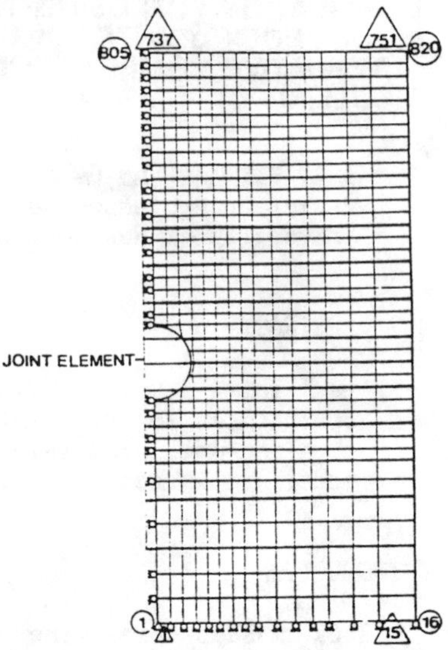

Figure 1. Finite Element Mesh

are represented by four node isoparametric elements. All materials are considered linear elastic and properties for concrete and joints are provided in Table 1.

A 10 psi surcharge is applied to the top of the model and results from the finite element analysis are presented in figures 2 and 3. The attenuation of tangential stress along the crown and springline respectively are displayed. Also shown for reference are results considering a solid model, i.e. no joints.

A separate mesh was used for results represented by the solid model. One should expect that results obtained using joint elements with high shear stiffnesses will be close to results obtained for a solid material. Figures 2 and 3 clearly indicate this to be the case for both analyses with thin layer elements (high G) and Goodman elements (high Ks).

A comparison of analyses incorporating joints with low shear stiffness indicates close agreement between results obtained with thin layer elements (low G) and Goodman elements (low Ks).

The above results illustrate that use of thin layer elements to model rock joint behavior may render results similar to those obtained using methods based on different computational principles. However, whether or not correct finite element results are realized for the case of a circular opening embedded in a jointed material should be demonstrated. In order to accomplish this a comparison shall be made between measurements made on a laboratory model and back predictions from finite element analysis.

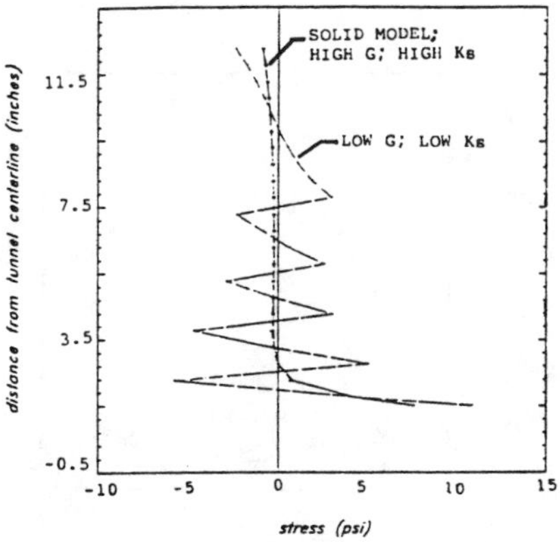

Figure 2. FEM Results Tangential Stress Along The Crown.

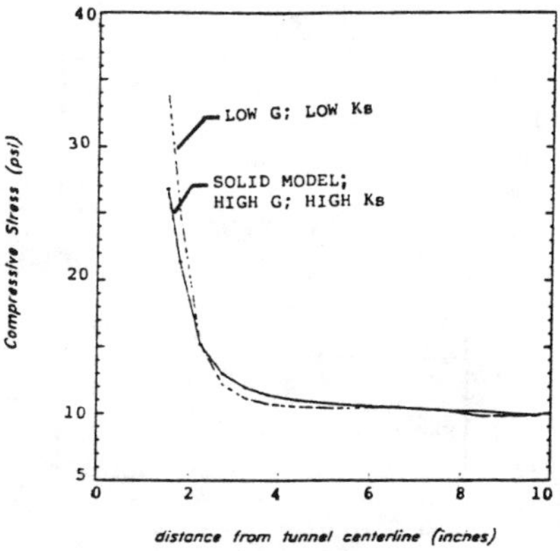

Figure 3. FEM Results Tangential Stress Along The Springline.

MEASUREMENTS

A distributed axial stress is applied to the top of the model corresponding to an average stress of 300 psi distributed for the full width of the model at a location of 2.5 in. below the top of the concrete. Measurements are made during loading of the model with an array of strain gauges placed along the springline and crown of the opening.

Results are presented in Figures 4 and 5 from the crown and springline respectively. The results compare measurements with finite element calculations for the jointed case as well as calculations performed with the Kirsch solution. The Kirsch solution serves as a reference since this analysis has been shown to represent well the stress and strain field due to surcharge loading around a circular opening embedded in an isotropic, homogeneous elastic medium without joints. The surcharge considered in the Kirsch solution is the average stress measured on the laboratory model 2.5 in. below the top of the concrete. Tangential strains are measured and are compared with results of the finite element analysis and Kirsch solution in terms of these quantities.

Figure 4 confirms the results obtained with the finite element analysis along the crown and indicates that the presence of joints does indeed have a pronounced effect on the response of the rock mass. Figure 5 shows that strains measured by gauges along the springline compare well with finite element calculations with the exception of two gauges located 0.5 in. and 1.0 in. from the periphery. This discrepencey is quite large and is likely due to instrumentation error and complications associated with the attachment of strain gauges to a

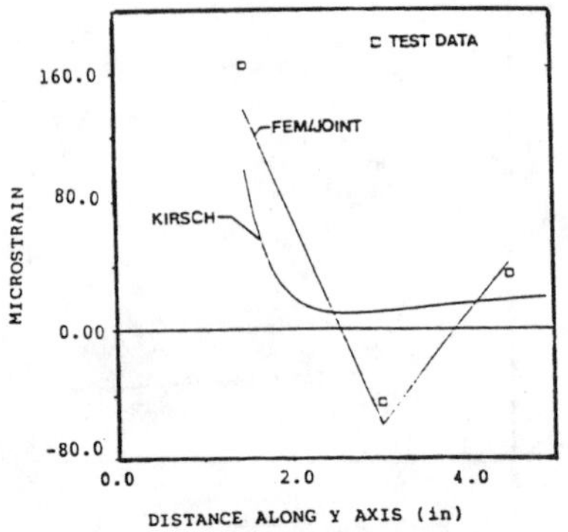

Figure 4. Measured and Computed Tangential Strain Along The Crown.

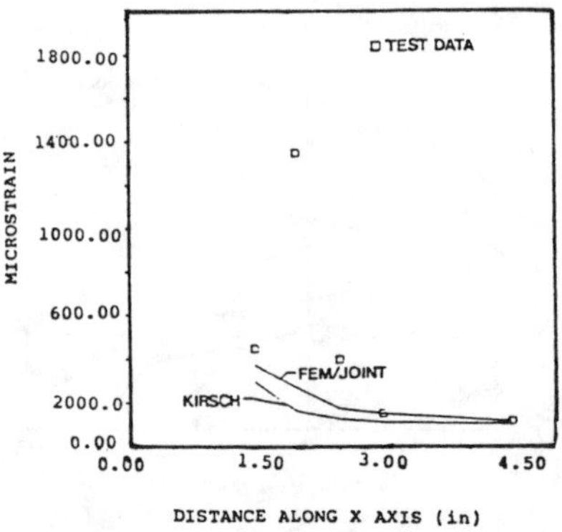

Figure 5. Measured and Computed Tangential Strain Along The Springline.

nonhomogeneous material such as concrete. Other measured strains match those calculated almost exactly. The observation that stress concentration near the tunnel periphery is higher due to the presence of rock joints is also confirmed by measurements.

CONCLUSIONS

Comparison between results from thin layer finite elements, finite element analysis with alternative elements, and the laboratory model are considered satisfactory. Therefore, for the case considered it has been demonstrated that reliable results may be obtained with thin layer finite elements.

Further verification on the use of thin layer elements is necessary. Future laboratory models should consider the effect of different boundary conditions, excavation, the presence of water, and joint characteristics such as spacing, orientation, roughness, and filler materials.

REFERENCES

1. Desai, C.S., Zaman, M.M., Lightner, J.G., and Siriwardane, H.J., "Thin Layer Element for Interfaces and Joints", Int. J. Num. & Analyt. Methods in Geomechanics, 8, 9-43, (1984).
2. Fishman, K.L., Derby, C.W., and Palmer, M.C., "Verification For Numerical Modelling of Jointed Rock Mass Using Thin Layer Elements", International Journal For Numerical and Analytical Methods in Geomechanics, 14, (1990).
3. Goodman, R.E., Tayler, R.L., and Brekke, T.L., "A Model for the Mechanics of Jointed Rock", J. of Soil Mech. and Found. Div., 94(SM 3),(ASCE 1968).

A THREE DIMENSIONAL CONSTITUTIVE MODEL FOR ROCK JOINTS WITH ANISOTROPIC FRICTION AND SURFACE DEGRADATION

LANRU JING and **ERLING NORDLUND**
Luleå University of Technology, Sweden

ABSTRACT

A 3-D constitutive model was formulated using non-associated plasticity theory. The non-linear behavior of rock joints, including anisotropic friction, surface degradation, non linear dilatancy and dependence of normal stiffness on the normal closure, can be simulated. The second law of thermodynamics was was used to restrict the values of certain parameters so that the increments of the disspative work are always non-negative. The prediction from the model agrees well with the experimental results.

INTRODUCTION

Due mainly to absence of test results under truly 3-dimensional loading conditions, most of the constitutive models for rock joints are developed in 2 dimensions. However, in many engineering problems, 3-dimensional loading conditions and the anisotropy of friction angle of joints must be considered. In this paper, test results on anisotropic friction and the development of a 3-dimensional constitutive model for rock joint behavior are presented.

EXPERIMENTAL RESULTS OF ANISOTROPIC FRICTION OF JOINTS

48 concrete replicas of one natural granite joint were tested after tilting tests. The samples were sheared in 12 directions with 30 degree interval under 4 levels of normal stress. The distribution curves are plotted on a polar diagram as shown in Fig. 1a. With the increase of the normal stress, the degree of anisotropy of friction angle tends to decrease. These curves can be fitted approximately by ellipses.

Assuming that the basic friction angle, β, of the material is isotropic, it is suggested that the total friction angle, ψ, be expressed as $\psi = \beta + \alpha$. α here is the asperity angle whose magnitude may be written as

$$\alpha = \sqrt{[C_1\cos\phi - C_2\sin\phi]^2 + [C_1\sin\phi + C_2\cos\phi]^2} \qquad (1)$$

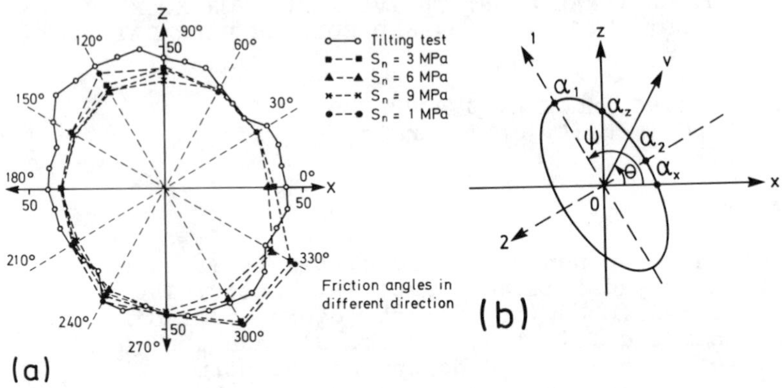

Fig.1 a) Polar diagrams of friction angle of joint samples, b) idealized asperity ellipse

where $C_1 = \alpha_1 \cos(\theta-\phi)$, $C_2 = \alpha_2 \sin(\theta-\phi)$. α_1 and α_2 are the major and minor principal values on the ellipse of α. ϕ is the direction angle of major semi-axis, and θ is the shear direction angle, as shown in Fig. 1b.

FORMULATION AND BEHAVIOR OF THE PROPOSED MODEL

The joint surface is assumed to be nominally planar with many tiny asperities. Fluid flow, thermal effects and rotational friction are not considered. For the local coordinate system defined on the joint surface, the X-axis is parallel with the strike of the surface, the Z-axis is parallel with the plunge, and the N-axis is in the normal direction. Repeated indices are summed over the range x, z, and n unless stated otherwise.

Let $d\sigma_i$ and du_j be the increments of stresses and relative displacements of the joint surfaces and assume that displacement increments can be divided into reversible and irreversible parts, $du_j = du_j^e + du_j^p$. The reversible part of the displacement can be written as [1]

$$d\sigma_i = k_{ij} du_j^e \tag{2}$$

where k_{ij} is the stiffness tensor. The normal stiffness is given as [2]

$$k_{nn} = k_{nn}^o/(1.0 - u_n/u_n^m)^2 \quad (u_n^m \le u_n \le 0) \tag{3}$$

where k_{nn}^o is the initial normal stiffness and u_n^m is the maximum closure of the joint. $k_{nn} = 0$ when $u_n > 0$. The slip function, F, and the sliding potential, Q, are written as

$$F = \sqrt{T_x^2 + T_z^2} + \sigma_n \ , \ Q = \sqrt{T_x^2 + T_z^2} + \omega \tan(\alpha) \tag{4}$$

$$T_x = \tau_x/\tan(\varphi + \alpha_x), \ T_z = \tau_z/\tan(\varphi + \alpha_z) \tag{5}$$

where α_x and α_z are the asperity angles in X- and Z-directions and ω is a material constant. The sliding rule is defined as

$$du_j^p = \begin{cases} \lambda \ \partial Q/\partial\sigma_j & (F \geq 0, \ \dot{F} = 0) \\ 0 & (F < 0) \end{cases} \tag{6}$$

where λ is a non-negative scalar. By applying the consistency condition $dF = 0$ an explicit constitutive relation can be obtained and written here as

$$d\sigma_i = (k_{ij} - \frac{k_{ir}\frac{\partial Q}{\partial\sigma_r}\frac{\partial F}{\partial\sigma_s}k_{sj}}{\frac{\partial Q}{\partial\sigma_p}k_{pq}\frac{\partial F}{\partial\sigma_q}})du_j \tag{7}$$

The surface degradation is simulated by adoption of an exponential law for α [3], written here as

$$\alpha_i = \alpha_i^o \exp[-D(\sigma_n/\sigma_c)W^p] \qquad (i=1,2) \tag{8}$$

where α_i^o is the initial value of α_i, σ_c is the uniaxial compressive strength of the rock material, D is a material constant and $W^p = \int dW^p = \int \sigma_j du_j^p$ is the accumulated plastic work. For the isothermal problems, the second law of thermodynamics requires that $dW^p \geq 0$. This restriction can be satisfied when $\alpha \leq \beta$ and $\omega \geq 0$.

The proposed model was tested against measured results of a joint sample sheared cyclically in first X- and then Z-direction under a normal stress equal to 2 MPa. The predicted and measured joint dilatancy are shown in Fig. 2. The shear stress - shear displacement curves in X- and Z-directions are shown in Fig. 3. The material parameters used were: D = 0.0001, $\beta = 36^\circ$, ω = 0.3, k_{xx} = 1.4 GPa/m, $k_{zz} = k_{nn}$ = 2 GPa/m, u_n^m = 0.1 mm, $\alpha_1^o = 14^\circ$, $\alpha_2^o = 3^\circ$, $\varphi^o = 36^\circ$, and σ_c = 52 MPa.

CONCLUSIONS AND ACKNOWLEDGEMENT

This 3-D model can simulate approximately the basic behavior of rock joints with anisotropic friction and surface degradation under both monotonic and cyclic loading conditions. The predictions from the model agree well with the test results. The model was implemented into a 3-dimensional distinct element program, 3DEC. Further development is underway.

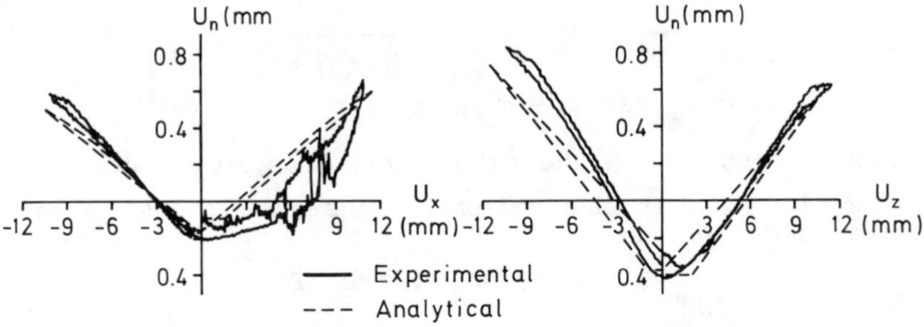

Fig. 2 Joint dilatancy in X- and Z- directions

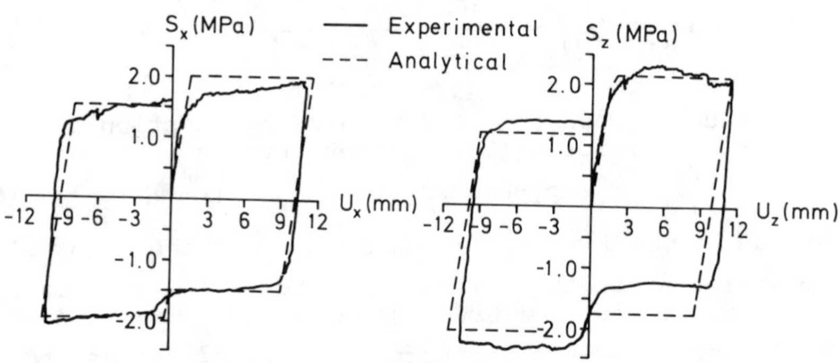

Fig. 3 Shear stress - shear displacement curves

The financial support from Swedish Natural Research Foundation and help from personals in Division of Rock Mechanics, Luleå University of Technology, Sweden, are gratefully appreciated by the authors.

REFERENCES

1. R. Michalowiski and Z. Mroz, "associated and non-associated sliding rules in contact friction problems", Achieves of Mechanics 30(3), 259-276 (1978).

2. S. C. Bandis, A. C. Lumsden, and N. Barton, "Fundamentals of rock joint deformation", Int. J. Rock Mech. Min. Sci. 20 (6), 249-268, (1983).

3. R. W. Hutson, "Preparation of duplicate rock joints and their changing dilatancy under cyclic shear", Ph.D Thesis, Northwestern University, USA, (1987).

A CONSTITUTIVE RELATIONSHIP INCORPORATING DILATANCY FOR INTERFACES AND JOINTS

PROF.(DR.) PIYUSH PARIKH
M.S. UNIVERSITY OF BARODA
BARODA, GUJARAT, INDIA

DR.U.D. DATIR
CENTRAL DESIGNS ORGANISATION
GANDHINAGAR, GUJARAT, INDIA

ABSTRACT

A new approach for the analysis of jointed rock is proposed with a constitutive relationship for the joint element which directly incorporates dilation parameter. This parameter has been identified in terms of change in principal strain ratio 'ν^*' at every instant. The new constitutive relationship is applied to typical cases for establishing its efficiency and efficacy.

INTRODUCTION

The classical laws of friction are followed in case of sliding on plane surfaces, while deviation from these laws is observed in case of sliding over joint surfaces having asperities. The deviation is as a consequence of volume change occurring during sliding. The generalized equation of sliding friction for all types of surfaces shall be the integration of the classical sliding friction equation rigorously true at every instant. The process of integration should include the consideration of distortion of plane at every instant. The resultant distortion of plane at every instant of sliding can be accounted for in terms of a geometrical parameter, consisting of ratio of principal strain(ν^*), a quantity analogous to poisson's ratio (ν) in classical mechanics.

A NEW APPROACH INCORPORATING DILATANCY

The sliding in jointed rock is associated with continuous alterations of surface characteristics. In physical terms, the process of sliding is occurring at new plane of orientation continuously. The phenomenological parameter representative of the alteration of surface characteristics has been identified in terms of ratio of principal strains (ν^*).

A new constitutive relationship for the joint element is devloped from the basic phenomenon of sliding incorporating change in ratio of principal strains (δ) at every instant. The constitutive matrix for 2D plane stress case can be given in the form:

$$[E^*] = [D] \; [\tau]$$

$$\text{Where } [D] = \frac{1}{E^*} \begin{bmatrix} 1 & 0 & 0 \\ 0 & 2D(1-\nu^*)(1+2\nu^*) & 0 \\ 0 & 0 & 1+\nu^* \end{bmatrix}$$

In the above [D] matrix, D is a parameter defined as D = iav / io

Where iav = average dilation angle under given normal stress observed during the direct shear test.

io = average angle of asperity.

It is observed that value of ν^* for the joint is below unity when the joint is in the compressional mode, it invariably becomes unity when the joint just begins the dilational mode and it increases further until a little earlier of the point of maximum normal strain, when it starts retarding towards unity.

APPLICATIONS OF NEW CONSTITUTIVE RELATIONSHIP FOR INTERFACES AND JOINTS

In order to establish the efficiency and efficacy of the new constitutive relationship, it is incorporated in the 2D plane stress/plane strain finite element programme developed by Hinton and Owne (1977). The joint elements are considered to be solid elements. The properties of joint elements are expressed in terms of ν^* and E^*, quantity analogus to poission's ratio (ν) and young's modulus [E]. These properties are considered to be variables. The programme is applied to typical cases some of which are as under:

(i) Laboratory direct shear tests conducted on samples having regular asperities by Datir (1989)
(ii) In situ direct shear tests conducted by Datir (1981).
(iii) Single joint element discussed by Desai et al(1984).
(iv) Case of thick circular cylinder, typical of a tunnel in jointed rock, discussed by Hiton and Owen(1977).

These are briefly discussed hereafter:

(i) Discretization of direct shear test sample having regular asperties is shown by fig.1.

Observed and computed values of normal displacement are shown in fig.2. It is evident from this figure that the two values are very close to each other.

(ii) Discretization of in situ shear test block is in fig.3. Comparison of observed and computed normal displacements is shown in fig.4, which indicate a close conformity.

(iii) Desai and Zaman(1984) have discussed application of a thin layer element developed by them to a single element shown in fig.5. It is claimed that the computed shear stress is very close to the applied shear stress and the computed displacements at the top nodes are close to the exact solution. The same problem is analysed by the new approach discussed above. The results computed by Desai and Zaman are shown in Table-I. The results computed by the authors are also superimposed in the same table. The figures in the table are self explanatory. This establishes the efficacy of the new approach in computing logical results.

(iv) Hinton and Owne(1977) have discussed case of a thick circular cylinder analogus to a circular opening in a jointed rock mass. Discretication used is similar to one used by Hinton and Owen except for the joint element introduced in the present case. Radial stress distribution along the joint element is plotted in fig.6 as arrived at by Hinton and Owen (without joint element) and as arrived at by the authors. The results are quite logical.

CONCLUSION

It is possible to realistrically model the sliding behaviour of jointed rock by following the new approach put forward by the authors.

REFERENCES

1 U.D. Datir, "On Mechanical Behaviour of Jointed Rocks", Ph.D. Thesis, submitted to M.S. University of Baroda, Gujarat,India.(1989).

2 U.D. Datir, "In Situ shear Tests on Rock',M.E. Dissertation, submitted to M.S. University of Baroda, Gujarat,India(1981).

3 C.S. Desai et al, "Thin layer element for interfaces and joints", Int.J.Num.and Ana. methods in Geomech.Vol.8,19-43 (1984).

4 E.Hinton et al, "Finite Element Programming" Academic Press,London(1977).

TABLE-I SINGLE JOINT ELEMENT

GAUSS POINT NO.	SHEAR STRESS REPORTED BY DESAI ET AL (1984)	SHEAR STRESS COMPUTED FOR		
		$\nu^* = 0.3$	$\nu^* = 0.99$	$\nu^* = 1.5$
1	10.00350	10.0000	10.0000	9.9985
2	9.99825	10.0000	10.0000	9.9985
3	10.00350	10.000	10.000	9.9985
4	10.00350	9.9994	10.0000	10.0020
5	10.00350	9.9994	10.0000	10.0000
6	10.00 350	10.0000	10.0000	9.9985
7	9.99835	10.0000	10.0000	9.9985
8	10.00350	10.0000	10.0000	9.9985
NODE NO	NORMAL DISPLACEMENT REPORTED BY DESAI ET AL (1984) $\times 10^{-4}$	NORMAL DISPLACEMENT COMPUTED FOR $\times 10^{-4}$		
		$\nu^* = 0.3$	$\nu^* = 0.99$	$\nu^* = 1.5$
1	0.00000	0.00000	0.00000	0.00000
2	0.00000	0.00000	0.00000	0.00000
3	0.00000	0.00000	0.00000	0.00000
4	1.68272	3.83037	0.101916	-6.84017
5	-1.68272	-3.83037	-0.101916	6.84017
6	2.27373	5.10720	0.135888	-9.12001
7	0.00000	0.00000	0.00000	0.00000
8	-2.27373	-5.10720	-0.135888	9.11998

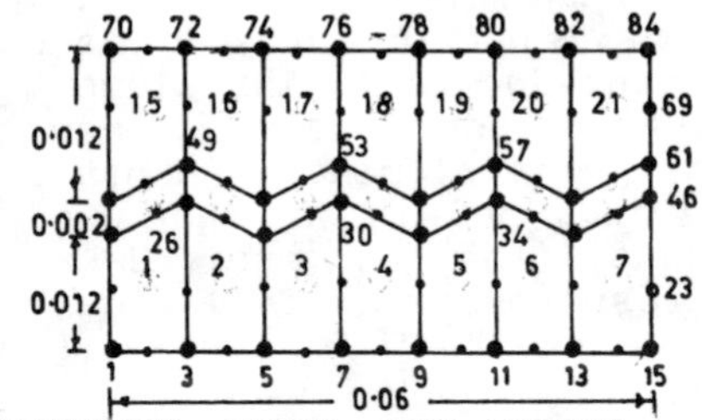

FIG.1 DIRECT SHEAR TEST SAMPLE.

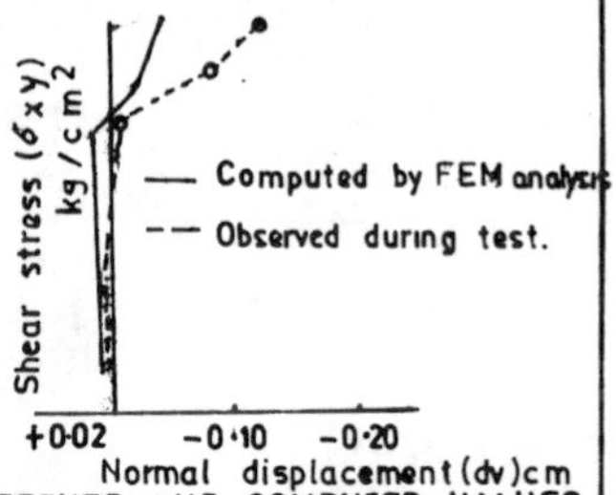

FIG. 2. OBSERVED AND COMPUTED VALUES

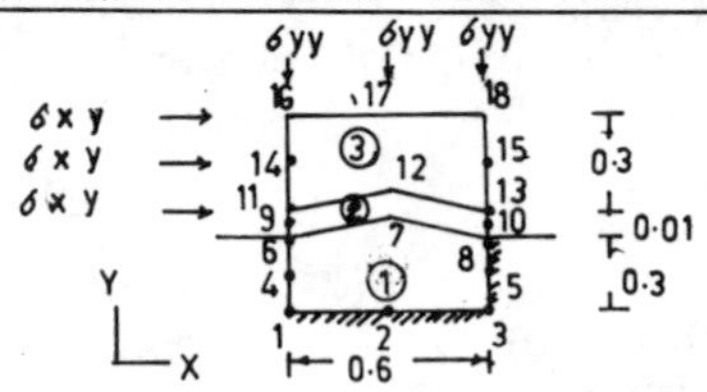

FIG 3 IN SITU SHEAR TEST BLOCK

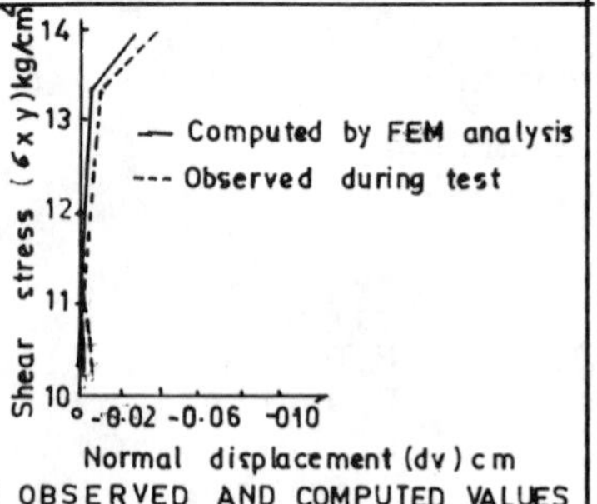

FIG·4 OBSERVED AND COMPUTED VALUES

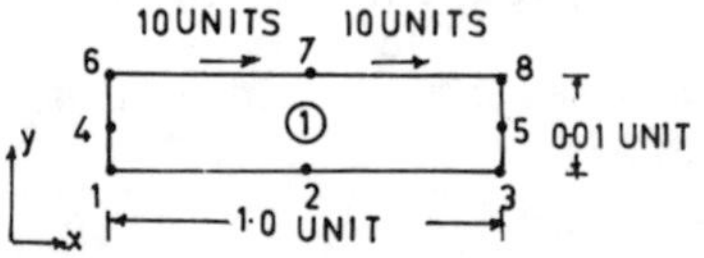

FIG·5 SINGLE ELEMENT

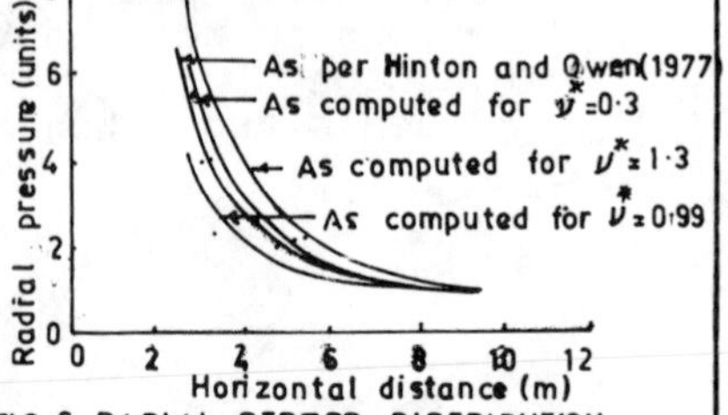

FIG 6 RADIAL STRESS DISTRIBUTION

Constitutive Equation of Anisotropic
Rock Mass with Discontinous Joints

Cao Ping Sun Zong Qi
Central South University of Technology
Changsha 410083 P.R.China

Pan Chang Liang
Central South University of Technology

ABSTRACT

Effective Young's moduli of rock mass with unfilled discontinous sets of joints have been discussed applying the energy theory of fracture mechanics at the present work. On the fundation of the effective Young's moduli, a method determining the constitutive equation of the jointed rock mass is put forward.

STRESSES ANALYSIS OF UNFILLED DISCONTINOUS JOINT

Rock mass is such a material containing joints, its properties are more complex than that of matellic materials. If there are several sets of joints in rock mass, it behaves anisotropicly. The paper attempes to derive the constitutive equation of rock mass with unfilled discontinous joints applying theory of fracture mechanics.

It is known that rock mass is usually in compression and/or compression-shear stress condition acted with structure stresses, deadweight stresses and excavation performance. In the state the two surfaces of joint contact each other. For such rock mass with unfilled discontinous joints, the following assumptions are made:

(i) The unfilled discontinous joints in rock mass are seen as cracks to handle;

(ii) Under the action of compression and/or compression-shear stresses, joints are closed, there are closing stresses in joint. The stresses applied to joint by load and the closing stresses are uniform along surfaces of joint (as Fig. 1);

(iii) There is no relative normal displacement between surfaces of joint.

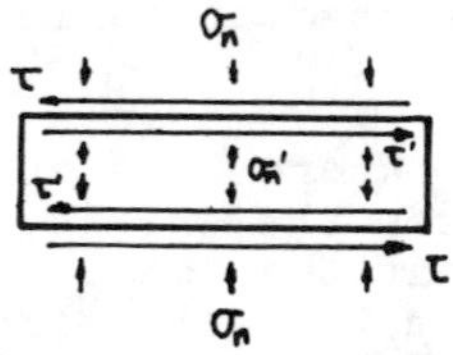

Fig.1 Stresses analysis for a joint

From fracture mechanics point of view, it is known that if the closing stresses in joint are considered, the stress intensity factors of joint tip may be reduced and are equal to the superposition of stress intensity factors caused by load and by closing stresses respectively. Thus introducing effective stress components will be expressed as following:

$$\left.\begin{aligned}\sigma_n^* &= \sigma_n - \sigma_n' \\ \tau^* &= \tau - \tau'\end{aligned}\right\} \quad (1)$$

Where σ_n and τ are normal and shear stress component respectively produced by applied load. σ_n' and τ' are closing normal and shear stress component in joint caused by the closing of joint. From assumptions (i), (ii) and (iii), it can be derived that $\sigma_n^* = 0$, and τ^* is determined with the slipping or slipping tendency between the surfaces of joint. So the effective stresses can be expressed as following for a set of joints (as Fig. 2):

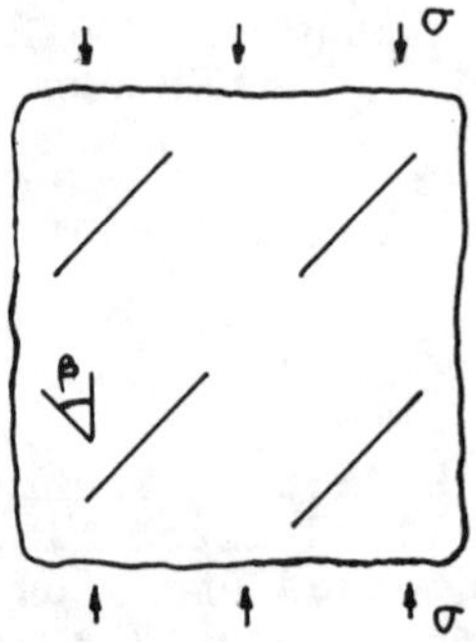

Fig. 2 The model of rock mass with a set of joints

$$\left.\begin{aligned}\sigma_n^* &= 0 \\ \tau^* &= \begin{cases} 0 & \text{if } \tau \le C + \sigma_n \mathrm{tg}\varphi \\ \sigma\cos\beta(\cos\beta - \sin\beta\,\mathrm{tg}\varphi) - C & \text{if } \tau > C + \sigma_n \mathrm{tg}\varphi \end{cases}\end{aligned}\right\} \quad (2)$$

Where C is cohesion, φ is friction angle of joint. β is the angle between the normal direction of joint and acting direction of stress σ.

EFFECTIVE YOUNG'S MODULI OF ROCK MASS WITH ONE OR SEVERAL SETS OF JOINTS

Considering two models of rock mass, first of them is abstracted from rock mass with one set of joints and the second of them is abstracted from 'ideal' rock mass without joint but containing some microcracks which length can be considered equal to zero. There are same boundary condition for the two models. Tt is assumed that the joints in the first model are formed by the stable extension of the microcracks of the second model. From the fracture mechanics point of view the energy extending the microcracks is surpplied by the releasing of strain energy of the second model. The releasing rate of energy can be expressed as following for a joint:

$$G = -\frac{\partial U}{\partial A} \quad (3)$$

Where U is strain energy, A is the area of the joint. From the energy theory of fracture mechanics, it is known that the total increase of strain energy extending a micro crack of the first model to a relative joint of the second model is:

$$U_e = -\frac{2}{E}\int_0^a \left(K_I^2 + K_{II}^2 + K_{III}^2(1+\nu)\right) da \quad (4)$$

for plane stress problem

$$U_e=-\frac{1-\nu^2}{E}\int_0^a(K_I^2+K_{II}^2+K_{III}^2/(1-\nu))2\pi a\,da \tag{5}$$

for space problem

Where K_I, K_{II}, and K_{III} are the joint tip stress intensity factors for I,II,and III modes deformation respectively.From Betti's reciprocal theorem,the relation of Young's moduli between the two models is:

$$-\frac{\sigma^2V}{2\bar{E}}=-\frac{\sigma^2V}{2E}+U_e \tag{6}$$

Where $\bar{E}$--effective Young's modulus of jointed rock mass in the direction of stress σ

E--Young's modulus of 'ideal' rock mass

U_e--the increse of strain energy due to the presence of joints

V--the volume of the body containing joints

For space problem the joints are considered as penny shaped and their radius is a.For plane problem using a expresses the half length of joint.The stress intensity factors can be calculated with the effective stresses,i.e.:

$$\left.\begin{aligned}&K_I=K_{III}=0\\&K_{II}=\alpha\tau^*\sqrt{\pi a}\end{aligned}\right\} \tag{7}$$

for plane problem

$$\left.\begin{aligned}&K_I=0\\&K_{II}=\frac{4\alpha\tau^*\cos\theta\sqrt{\pi a}}{\pi(2-\nu)}\\&K_{III}=-\frac{4\alpha\tau^*(1-\nu)\sin\theta\sqrt{\pi a}}{\pi(2-\nu)}\end{aligned}\right\} \tag{8}$$

for space problem

Where α is correcting coefficient of stress intensity factor of a set of group joints as Fig.2.From formulas (2) to (8) and taking spherical average over angle θ, the effective Young's moduli can be obtained as following:

$$\text{If } \tau\le C+\sigma_n tg\varphi \qquad \bar{E}=E \tag{9}$$

$$\text{If } \tau> C+\sigma_n tg\varphi$$

$$\bar{E}=E\{1+2[\cos\beta(\cos\beta-\sin\beta tg\varphi)-C/\sigma]^2\pi a^2\alpha^2\chi\}^{-1} \tag{10}$$

for plane stress problem

$$\bar{E}=E\left\{1+\frac{32(1-\nu^2)[\cos\beta(\cos\beta-\sin\beta tg\varphi)-C/\sigma]^2a^3\alpha^2\chi}{3(2-\nu)}\right\}^{-1} \tag{11}$$

for space problem

Where χ is the number of joints containing in rock mass with unit volume.For unfilled joints,C can be considered as equal to zero approximately,so it is known from formulas (10) and (11) that the effective Young's moduli are constant values for given jointed rock mass.

If there are M sets of joints in rock mass,the total increase of strain energy is equal to the summary of that of every set of joints which can be calculated with the introduced method befor approximately,i.e.:

$$U_e=\sum_{i=1}^{M}\overset{i}{U_e}$$

Where $\overset{i}{U_e}$ is the increase of energy due to the present of the i'th set of joints.

CONSTITUTIVE EQUATION OF ROCK MASS WITH ORIENTATION SETS OF JOINTS

In cartesian coordirate system, the constitutive equations of 'ideal' rock mass without joint and rock mass with joints can be expressed as following respectively:

$$\{\varepsilon\} = [S]\{\sigma\} \tag{12}$$

$$\{\varepsilon\} = [S^*]\{\sigma\} \tag{13}$$

Where

$$\{\varepsilon\}^T = \{\varepsilon_x, \varepsilon_y, \varepsilon_z, \gamma_{xy}, \gamma_{yz}, \gamma_{zx}\}^T = \{\varepsilon_1, \varepsilon_2, \cdots, \varepsilon_6\}^T$$

$$\{\sigma\}^T = \{\sigma_x, \sigma_y, \sigma_z, \tau_{xy}, \tau_{yz}, \tau_{zx}\}^T = \{\sigma_1, \sigma_2, \cdots, \sigma_6\}^T$$

From elastic theory, there are $S_{ij} = S_{ji}$, $S^*_{ij} = S^*_{ji}$. The components of compliance matrix $[S^*]$ can be calculated in the following way. At the first, the leading dalgoral components of the matrix $[S^*]$ are calculated by setting $\sigma_i = 1$ ($i=1,2,...,6$) succssively. Substituting the relative components of formulas (12) and (13) into (6), and setting $\sigma = \sigma_i$, there is:

$$S^*_{ii} = S_{ii} - 2(U_e)_{ii} \tag{14}$$

The subscripts here and below don't satisfy the summary arrangement. The second step is to calculate the non-diagoral components of the matrix $[S^*]$ by setting $\sigma_i = 1$, and $\sigma_j = 1$ (for $i=2,3,...,6; j=1,2,...,i-1$) succssively. From the introduced method and the formulas (12) to (14) there is:

$$S^*_{ij} = S_{ij} + (U_e)_{ii} + (U_e)_{jj} - (U_e)_{ij} \tag{15}$$

The upper rigth components of matrix $[S^*]$ can be calculated by setting $S^*_{ji} = S^*_{ij}$. If C=0, it can be proved that $[S^*]$ is a constant matrix for given jointed rock mass.

CONCLUSIONS

(i) If the cohesion of joints is not considered the effective Young's moduli is a constant value for given jointed rock mass.

(ii) The reduction of Young's moduli of jointed rock mass is caused by the effective shear stress applied to joints.

(iii) The components of compliance matrix of jointed rock mass are not equal to zero generally.

REFERENCES

1. G.C.sih, Handbook of stress intensity factors, (Lehigh University, 1973)

2. J.Kemeny and N.G.W.Cook, "Effective modeli, Non-linear Deformation and Strength of A Cracked Elastic Solid", Int. J. Mech. Min. Sci. & Geomech. Abstr. 23(2) 107-118 (April 1986)

3. Ayal de S.Jayatilaka, Fracture of Engineering Brittle Material, (Applied Science Publishers LTD, London 1979)

A FAILURE MODEL FOR FIBERS REBONDING IN FIBER REINFORCED COMPOSITE

M. REZA SALAMI AND SAMEER A. HAMOUSH

Assistant and Adjunct Professors, Department of Civil Engineering, North Carolina Agricultural and Technical State University, Greensboro, North Carolina, 27411

ABSTRACT

An analytical model is proposed to predict the ultimate tensile strength of fiber-reinforced composite when the failure is governed by fiber debonding. The analytical analysis is based on the principle of the compliance method in fracture mechanics with the presence of an interfacial crack between the fiber and the matrix interface. The model is developed based on the assumption that both the matrix and the fiber behave elastically and the matrix strain at a zone far from the matrix fiber interface equals to the composite strain. Also, it is assumed that a complete bond exists between the fiber and the matrix in the unbonded zone, the matrix transfers the loading without yielding and the crack faces are traction free. It is shown that the separation strain energy release rate for fiber reinforced composites can be obtained for cases with and without the existence of an interfacial crack. Numerical examples are presented and compared to results obtained in the literature by using finite element analyses and experimental tests. The comparison demonstrates the accuracy and the convergence of the model.

INTRODUCTION

The slip and the interface bond failure in the fiber reinforced composites have received a considerable attention in the recent research studies [1-4]. The interactions between fibers and the matrix are extremely complex. The first attempt to explain the reinforcing effect of the fibers was based entirely on elastic interactions. This was first described by Cox [2]. Piggott [3,4] studied the behavior of short fiber reinforced composites. The study includes the estimation of the strain as well as stress at the bonded interface. The matrix tensile strain was assumed to be constant and equal to the composite tensile strain. It was concluded that the shape of the stress-strain curve is strongly dependent on the fiber aspect ratio, and the adhesion between the fiber and matrix has relatively little effect on the stress-strain curve.

In this paper, the behavior of a singly reinforced composite body is considered to obtain the analytical value of the pull-out strain energy release rate, the analytical value of the strain energy release rate can be used to determine the failure of uniaxial composites.

FRACTURE MECHANICS APPROACH (COMPLIANCE METHOD)

The reciprocal of the gradient of the load-deflection curve is called a compliance, C, and governed by

$$C = \frac{u}{P} \tag{1}$$

where C is the compliance of the body, P is the applied forces and u is the boundary displacement in the P direction. The strain energy release rate , G, can be shown to satisfy

$$G = \frac{P^2}{8\pi r}\left(\frac{\partial C}{\partial a}\right) \tag{2}$$

where a is the length of the interfacial crack and r is the radius of the fiber. In equation (2), the strain energy release rate becomes defined if the function ($\partial C/\partial a$) is introduced, the function ($\partial C/\partial a$) has been obtained by Hamoush and Salami [1].

The schematic model for the analyzed problem is depicted in Figure 1. In Figure 1, when the load P is applied, the stress distribution will be in a circular symmetry, i.e. the matrix displacement w and the shear stress, τ, do not vary with orientation about the fiber axis as shown in Figure 2. Both the shear stress and the longitudinal displacement are a function of the radial distance, z, from the fiber center. The total compliance of the force P can be obtained [1] from three different cases. These cases, from Ref. [1], are (1). Compliance of the embedded portion of fiber, (2) Compliance of the free-edge portion of fiber and (3) Compliance due to matrix deformation.

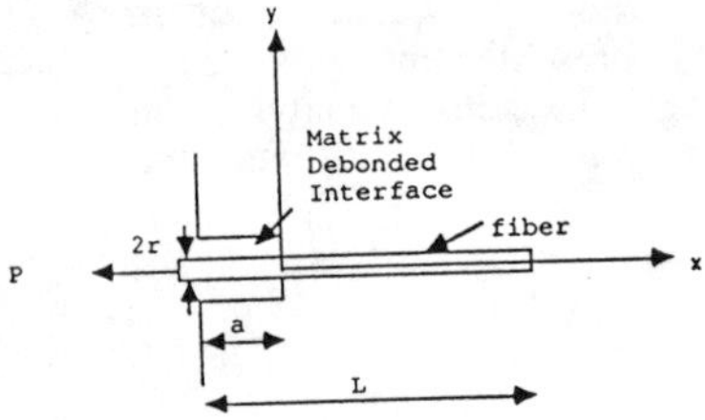

FIG. 1. Fiber with interfacial crack.

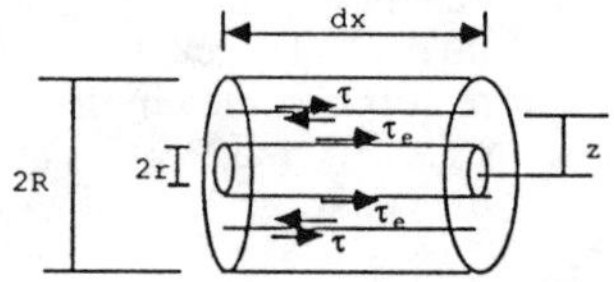

FIG. 2a. Short length of fiber and surronding matrix.

TOTAL CHANGE COMPLIANCE FOR WHOLE BODY.

The total change in compliance for the whole body is given by Hamoush and Salami [1] as

$$\frac{\partial C}{\partial a} = \frac{S-2}{SA_fE_f}\left[\frac{1-\cosh k(L-a)}{\sinh^2 k(L-a)}\right] - \frac{1}{SA_fE_f} + \frac{1}{A_fE_f} + \\ + \frac{k^2r^2}{h^2E_f}\left[\frac{S-1}{SA_f}\cosh k(L-a) + \frac{1}{SA_f}\right]\frac{1}{\sinh^2 k(L-a)} \tag{3}$$

where
$$S = \left(V_f + \frac{1-V_f}{n}\right)\left(\frac{1-V_f}{V_f} + n\right), \tag{4}$$

$$V_f = \frac{A_f}{A_f + A_m} = \text{volume fraction of fibers,} \tag{5}$$

$$n = \frac{E_f}{E_m} = \text{modular ratio,} \tag{6}$$

A_m, A_f = the areas of matrix and fiber,

E_f = young modulus of fiber,

E_m, ν_m = young modulus and Poisson's ratio of matrix,

a = length of the interfacial crack,

L = length of the interfacial crack + length of fiber,

r = radius of the fiber,

$$h^2 = \frac{E_m}{(1 + \nu_m)\, E_f \ln \frac{R}{r}} \quad , \tag{7}$$

$\frac{R}{r}$ = depends on the fiber packing, and

R = from the fiber surface to a distance (as shown in Fig.2),

$$k^2 = \frac{h^2}{r^2} \tag{8}$$

The strain energy release rate ,G, can be obtained analytically by substituting equation (3) into equation (2).

NUMERICAL INVESTIGATION

To examine the applicability of the analysis proposed the predicted tensile strength of the composite was compared with the observed data from a straight steel fiber, and matrix material in the literature [5]. The effect of the fiber stiffness (the ratio r/L) on the strain strain energy release rate can be obtained by examining of equation (3). The plots of equation (3) for different ratios of L/r is shown in Figure 3 and Table 1. In Table 1 the fiber modulus of elasticity is 30000 ksi., the modulus of elasticity for the matrix is 2000 ksi., the applied force is 354 lb. and the poisson's ratio for the matrix is 0.1. This Figure shows that the interface crack becomes significant if the ratio of L/r is decreasing. Also, it can be noted from the Fig. 3 that the interface crack has no major effect on the strain energy release rate when that interface crack is less than 0.4 of the total fiber length L. This observation matches the work done by Mandel [5] when he used a finite element model and experimental test results.

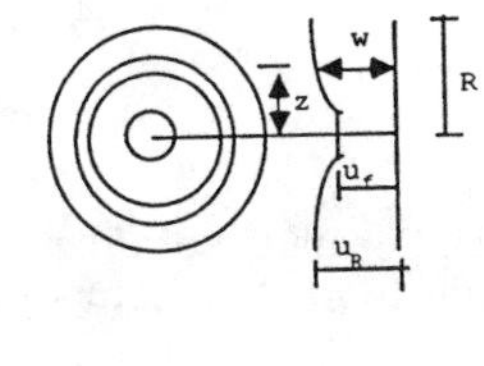

FIG. 2b. Fiber with nearest neihgbours.

FIG. 2. Stresses at the fiber and the associate displacements.

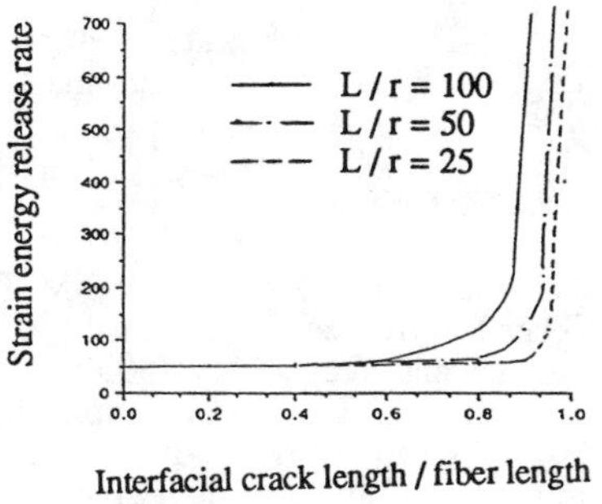

FIG. 3. Effect of fiber stiffness on the strain energy release rate.

TABLE 1. Effect of Fibre Stiffness on the Strain Energy Release Rate. (for Fiber Spacing equals to 12.5 of the Fibre Diameter)

Interface Crack/ Fiber Length	Fiber Length/ Fiber Radius (L/r)		
a/L	100	50	25
0.00	51.69	51.69	51.69
0.20	51.69	51.69	53.74
0.40	51.69	53.13	56.71
0.60	52.91	53.74	61.23
0.80	53.74	61.23	120.18
0.90	61.25	120.18	623.50
0.95	120.18	623.5	2323.32
0.98	722.93	3108.70	14210.73
1.00	∞	∞	∞

The fiber modulus of elasticity is 30000 ksi., the modulus of elasticity for the matrix is 2000 ksi., the applied force is 354 lb. and the poisson's ratio for the matrix is 0.1

CONCLUSION

An analytical model is proposed to predict tensile strength of fiber reinforced cement based composites whose failure is governed by the strength of the fiber matrix interface. The proposed model is based on the concept of fracture mechanics and theory of elasticity.

The debonding is treated as interfacial crack of two dissimilar materials, the total strain energy release rate for interfacial crack is obtained from the concept of compliance method in fracture mechanics, the proposed model is capable of obtaining the strain energy release rate when the interface crack doesn't exist. In the proposed model there is a simplifying assumption made which the interface crack is traction free and no elastic bond slip exists at the interface.

It can be concluded from this model that: (1) For elastic fiber with the ratio L/r is more than 100, the interface crack and the fiber lengths have no major effect on the pull out strain energy release rate; (2) The interface crack becomes significant if the fiber is rigid; (3) Increasing the fiber spacing increases the strain energy release rate.

ACKNOWLEDGEMENTS

The authors gratefully appreciate the support by the Department of Civil Engineering at the North Carolina A&T State University.

REFERENCES

1. Sameer A. Hamoush and M. Reza Salami, "Interfacial Separation of Fibers in Fiber Reinforced Composites", Accepted for Publication in Journal of Composite Science and Technology, (1990).
2. Cox, H. L., Brit. J. Appl. Phys., (1952) 72.
3. Piggott, M. R., "Expressions governing stress-strain curves in short fiber reinforced polymers," Journal of Material Science 13, (1978), pp. 1709-1716.
4. Piggott,M. R., "Load Bearing Fiber Composites", International Series on the Strength and Fracture of Material and Structures, Pergamon Press.
5. Mandel, J. A.,Wei, S. and Said, S., "Studies of the Properties of the fiber-matrix interface in steel fiber reinforced mortar; ACJ Material Journal, Mar-Apr. 1987, pp. 101-109.

A GENERALISED PLASTICITY MODEL WITH MULTILAMINATE FRAMEWORK

A.Varadarajan, Professor
Dept.of Civil Engineering, IIT Delhi,
New Delhi-110016, India

ABSTRACT

Multilaminate models are consceptually attractive and they easily capture the effect of anisotropy. A generalised plasticity model developed with stress-invariant framework is presented with multilaminate framework. The predictions by the two approaches are compared.

INTRODUCTION

A number of constitutive models have been proposed to characterise the behaviour geologic materials. Several kinematic hardening models have been proposed to include the anisotropic behaviour. These models use some form or the other of stress invariants. They employ 'translational rule' and it involves an extensive exercise in geometry. An alternative approach is to adopt the multilaminate framework proposed by Pande and Sharma (1). This model is conceptually very sound and attractive and captures the effect of anisotropy automatically.

The multilaminate model describes the behaviour of any material through the responses of infinite number of planes (laminates) at a point. In this approch, the key step is to evolve a constitutive model for a single plane and through integration of the contributions of infinite number of planes, the constitutive behaviour of the continuum at a point is obtained. During loading, each plane follows a stress-strain curve of its own and at any stage of loading, the response record of every plane, which is distinct, is available and this depends on the nature of loading and the direction of principal stresses used. If the principal stresses are rotated, the effect of it is clearly broughtout by the characteristics of the various planes. With the multilaminate framework, the stress induced anisotropy has been brought out using critical state model by Pande and Sharma (1) and Vardarajan and Pande (2).

Pietruszczak and Pande (3) presented plasticity formulation of the multilaminate frame work. With this approch, an attempt is made herein to present the generalised plasticity model proposed by Desai and his coworkers in the multilaminate framework.

GENERALISED PLASTICITY MODEL

Invariant Frame

The basic concept, theoretical devlopment and application of the model are given in various publications (4,5). The main attractive feature of this model is that it allows the complexity of the model to be varied consistent with the practical requirement of a given problem. The continuous yielding and ultimate yield behaviour is given as,

$$F = J_{2D} - \left(\frac{-\alpha}{\alpha_o^{n-2}} J_1^n + \gamma J_1^2\right)(1-\beta S_r)^m \qquad \ldots (1)$$

Where J_{2D} = second invariant of the deviatoric stress tensor, S_{ij}, of the total stress tensor σ_{ij}; S_r = stress ratio= $J_{3D}^{1/3}/J_{2D}^{1/2}$, J_{3D} =third invariant of S_{ij}, and α, n, γ, β and m are are the response functions. The value of the normalising constant, α_0, is equal to 1.0 MPa.

The constant related to ultimate condition (α=0) are m, γ and β. The value of m is found to be -0.50 for many geologic materials. The parameter n represents the phase change point indicating transition from compressive to dilative volume change. The hardening behaviour is given as

$$\alpha = \frac{a_1}{\xi^{\eta 1}} \qquad \ldots (2)$$

where a_1 and η_1 are constants and ξ is related to plastic strain.

The nonassociative plasticity is incorporated through plastic potential function obtained by correcting the yield function F to $\bar{F}$ by modifying α to α_Q (6) as

$$\alpha_Q = \alpha + \kappa(\alpha_1 - \alpha)\ (1 - r_v) \qquad \ldots (3)$$

where κ is a material parameter, α_1 and r_v are related plastic strain at strategic states.

The incremental stress-strain relationship is obtained with the normality rule and consistency condition as

$$d\underset{\sim}{\sigma} = \underset{\sim}{C}^{ep}\ d\underset{\sim}{\varepsilon} \qquad \ldots (4)$$

where $d\sigma$ and $d\varepsilon$ are vectors of stress and strain components, and $\underset{\sim}{C}^{ep}$ is constitutive matrix.

Multilaminate Frame

The details of model development with multilaminate frame are given in (1,3). Herein only the significant steps together with the proposed model are given. Fishman and Desai(7) specialised the model for continuum(Eq.1) to describe yielding behaviour of joints as,

$$F' = \tau^2 + \alpha' \sigma_n^{n'} - \gamma' \sigma_n^2 = 0 \quad ..(5)$$

where σ_n and τ are normal and shear stresses on the plane. The constant $\gamma'$is related to the ultimate state. The parameter n is a function of the phase change point. The function α' represents hardening and is a function of normal and shear deformations on the joint. To depict nonassociative plasticity, the plastic potential function is obtained by modifying $\alpha'$to α_Q'in Eq.(5) as given earlier.

Using the yield function for the joint, the yield function for an ith plane is given as

$$F_p = (\tau^2 + \alpha'\sigma_n^{n'} - \gamma'\sigma_n^2)\ F_s = 0 \quad ...(6)$$

where F_s has the same meaning as given in Eq.(1). With the normality rule and consistency condition, the stress-strain relationship is obtained for the plane. Using appropriate transformation and numerical integration, the stress-strain response at a point is given as

$$d\varepsilon^p = \left[\sum_{i=1}^{q} W_i\ T_i \left(\frac{\delta F_p}{\delta \underset{\sim}{\bar{\sigma}_i}}\right) \left(\frac{\delta F_p}{\delta \underset{\sim}{\bar{\sigma}_i}}\right)^T T_i \right] d\underset{\sim}{\sigma} \quad ...(7)$$

where $\bar{\sigma}_i = \{\tau, \sigma_n\}$, T_i = transformation matrix q = number of sampling planes and W_i= weighting factor.

PREDICTION

The stress-strain-volume change responses of rock salt obtained already with invariant formulation (7) and with present multilaminate formulation by suitably `guessing' various parameters is shown in Fig. (1). There is qualitative agreement between the two predictions.

CONCLUSIONS

A generalised plasticity model is presented with the multilaminate framework. The comparison of predictions between the invariant frame and multilaminate frame is comparable but is not very close to each other. Further work is necessary in the evaluation of parameters for multilaminate model.

REFERENCES

1. Pande, G.N. and K.G.Sharma, Int. J. Num. Anal. Meth, Geomech, 7, (1983).
2. Vardarajan, A. and Pande, G.N. Under Communication, (1990).

3. Pietruszczak, S. and Pande, G.N. Int. J. Num. Anal. Meth. Geom. 11, (1987)

4. Desai, C.S. et al. Int. J. Num. Anal. Math. Geom. 10, (1986)

5. Desai, C.S. and Vardarajan, A. J. Geoph. Res. Oct. (1987)

6. Frantziskonis et. al. J. of Eng. Mech. Div. ASCE, 112 (1986)

7. Fishman and Desai, In. Const. Laws for Engg. Mat. Ed. Desai et al. Elsvier, (1987)

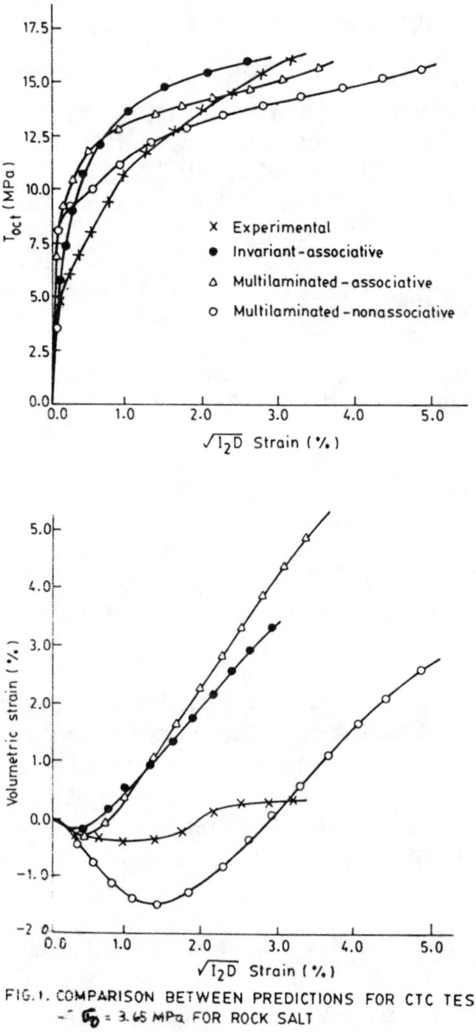

FIG. 1. COMPARISON BETWEEN PREDICTIONS FOR CTC TEST – σ_0 = 3.45 MPa FOR ROCK SALT

SPECIAL MATERIALS; METAL MATRIX COMPOSITES, CERAMICS, POLYMERS, ENGINEERED MATERIALS

Anisotropic Channel Flow of Fiber Suspensions

M. Cengiz Altan
School of Aerospace and Mechanical Engineering
University of Oklahoma, Norman OK 73019

Selçuk İ. Güçeri R. Byron Pipes
Center of Composite Materials and
Department of Mechanical Engineering
University of Delaware, Newark, DE 19716

Abstract
Anisotropic flow of fiber suspensions is studied by using a rheological model for anisotropic fluids. A fourth-order orientation tensor is used to describe the orientation field. The final form of equations presents a convenient and tractable model for spatially non-uniform flows. This constitutive model is used to analyze the flow field resulting from the introduction of randomly oriented particles in a fully developed straight channel flow. Before steady state is obtained, significant transients in the velocity field are observed up to several channel widths downstream.

Introduction
In both natural sciences and engineering applications, the investigation of the flow of particles in a viscous medium has been the focus of a number of scientific research activities. Studying the flow behavior of the anisotropic fluids and understanding the driving mechanisms for the orientation structures are particularly important. For non-homogeneous flow fields where the velocity gradient tensor is spatially nonuniform, it is necessary to investigate the effect of fibers on the flow kinematics by using the proper constitutive equations and numerical techniques.

In this study, the anisotropic flow of the fiber suspensions in a straight channel is studied. The steady state flow field is evaluated as a function of a parameter C which can be determined from the physical quantities, such as fiber volume fraction and fiber aspect ratio. Results are presented for semi-concentrated fiber suspensions, and are compared with the zero volume fraction limit ($C = 0$) calculations.

Theory
The motion of an ellipsoidal particle immersed in a Newtonian fluid is first studied by Jeffery [1]. Later, Jeffery's equation was expressed in tensorial form by Giesekus [2] and Bretherton [3]. The constitutive equation presented by Evans [4] for dilute suspensions of long, rigid cylindrical fibers is also proposed in a very similar form by Dinh and Armstrong [5] as a rheological equation of state for non-dilute fiber suspensions. The model is claimed to be valid for semi-concentrated region where the average spacing between fibers is greater than its diameter d but less than its length l. For these material systems, the extra stress due to the presence of particles can be written as:

$$\sigma_{ij}^{(p)} = \mu C u_{k,l} S_{ijkl} \tag{1}$$

$$S_{ijkl} \equiv < p_i p_j p_k p_l > \equiv \oint p_i p_j p_k p_l \psi(\vec{p}) d\vec{p} \tag{2}$$

where, p_i is the i'th component of the unit vector describing the fiber orientation, $u_{k,l}$ is the velocity gradient tensor ($\frac{\partial u_k}{\partial x_l}$), and μ is the fluid viscosity. The parameter C is proposed to be a function of average fiber spacing where its magnitude can be estimated by using fiber aspect ratio a_p ($a_p = \frac{l}{d}$) and fiber volume fraction ϕ_v as [5]:

$$C = \mathcal{O}\left(\frac{\phi_v a_p^2}{\ln(\frac{1}{a_p \phi_v})}\right) \tag{3}$$

The components of the orientation tensor S_{ijkl} have all the information needed to describe the fiber orientation. For planar orientations, S_{ijkl} has four independent components which can be found by integrating over a predetermined distribution function. The distribution function $\psi(\vec{p},t)$ describes the orientation state and is defined as the probability of having a fiber at an orientation $\vec{p}$ at a given time. In constitutive equations for very long fibers, the motion of fibers is commonly obtained by assuming the fibers as rigid line elements. The resulting expression can be written as the Jeffery's equation with infinite aspect ratio, which also is used for the motion of rigid-dumbbells of polymer kinetic theory.

$$\dot{p}_i = u_{i,q}p_q - u_{k,q}p_k p_q p_i \tag{4}$$

Instead of performing the integration and tedious calculation of the distribution function, an equivalent system of differential equations can be written as:

$$\frac{dS_{ijkl}}{dt} = u_{i,m}S_{jklm} + u_{j,m}S_{iklm} + u_{k,m}S_{ijlm} + u_{l,m}S_{ijkm} - 4u_{m,n}S_{ijklmn} \tag{5}$$

The resulting set of equations makes it possible to solve for the independent components of the tensor from the specified velocity gradients. However, it contains sixth-order orientation tensor which needs to be approximated. In this work, the sixth-order orientation tensor is approximated by using a quadratic closure approximation as,

$$S_{ijklmn} \cong S_{ijkl}S_{mnpp} \tag{6}$$

The particulate and suspension flows have usually low Reynolds numbers where the viscous effects are dominant over the inertia effects. Utilizing stream function Ψ, and after a little manipulation, the governing equation which satisfies momentum and continuity relations is obtained for a steady two-dimensional incompressible flow as:

$$\begin{aligned} 0 = {} & \frac{\partial^4\Psi}{\partial x_1^4} + \frac{\partial^4\Psi}{\partial x_2^4} + 2\frac{\partial^4\Psi}{\partial x_1^2\partial x_2^2} + C\frac{\partial^2}{\partial x_1\partial x_2}\left[(S_{1111} - 2S_{1122} + S_{2222})\frac{\partial^2\Psi}{\partial x_1\partial x_2} + \right. \\ & \left. (S_{1112} - S_{1222})\frac{\partial^2\Psi}{\partial x_2^2} + (S_{1222} - S_{1112})\frac{\partial^2\Psi}{\partial x_1^2}\right] + \\ & C\frac{\partial^2}{\partial x_2^2}\left[(S_{1112} - S_{1222})\frac{\partial^2\Psi}{\partial x_1\partial x_2} + S_{1122}\frac{\partial^2\Psi}{\partial x_2^2} - S_{1122}\frac{\partial^2\Psi}{\partial x_1^2}\right] - \\ & C\frac{\partial^2}{\partial x_1^2}\left[(S_{1112} - S_{1222})\frac{\partial^2\Psi}{\partial x_1\partial x_2} + S_{1122}\frac{\partial^2\Psi}{\partial x_2^2} - S_{1122}\frac{\partial^2\Psi}{\partial x_1^2}\right] \end{aligned} \tag{7}$$

In the present study, a constant C value is specified for the numerical experiments, and the final results are judged to correspond to a number of cases with different aspect ratio and volume fractions. For a given suspension, there are five equations, Eq. (5), and Eq. (7), and five unknowns, Ψ, and four independent S_{ijkl}, that need to be solved.

Numerical Solution Technique

The solution of the anisotropic flow equations for a straight channel is investigated. At the inlet of the domain, the flow field is taken as fully developed with random fiber orientation. A finite difference scheme will be employed in order to solve for the stream function. There are two distinctive steps of our numerical treatment of the problem:

a. The solution of Eq. (7) over the generated fixed mesh.

b. The solution of Eq. (5) along the particular streamlines for each nodal point.

Due to the channel symmetry, only half of the channel will be considered. A descriptive figure for the channel geometry, and the boundary conditions used are shown in Fig. 1a. The standard finite difference discretization techniques are used for Eq. (7), and an explicit Gauss-Seidel iteration technique is used sweeping the inner nodal points of the domain. Thirteen point discretization is used for the biharmonic operator ∇^4. The maximum relative error is specified to be 10^{-5} for the convergence and the resulting velocity gradients are used to solve for the orientation structure S_{ijkl}.

For the planar orientation solutions in the channel, Eq. (5) is solved by using the public domain software LSODE which uses a predictor-corrector method. A relative error of 10^{-10} is used in the calculations. Since the fiber orientation is coupled back to the flow field, the information about orientation structure needs to be fed back to Eq. (7). In order to overcome this difficulty, we have implemented a tracing technique for every nodal point inside the flow domain. The stream function value for each nodal point is traced back to the channel inlet individually, thus determining the streamline terminating at that nodal point. Then along this streamline, the fiber orientation equations are solved for every subinterval up to the nodal point considered.

Results and Discussions

Within the semi-concentrated region, C has an approximate range of 0 to 18 for the volume fractions from 0.04 to 0.2. Therefore, in this paper, the solutions with two nonzero C values, which are 6 and 16, will be presented. The iteration between the stream function and orientation field is carried out until the convergence in stream function is obtained with a maximum relative error of 10^{-4}. Typically, the convergence is reached between 5 to 15 iterations between the stream function and orientation field. Details of the numerical algorithm and finite difference discretization are contained in Altan [6]. Since the channel width appears to be too narrow in the figures drawn to scale (up to $16W_{ch}$), the streamlines are shown up to $4W_{ch}$. The streamlines for a fully developed channel flow are depicted in Fig. 1b. This trivial solution for stream function represents the flow field to which the $C \neq 0$ solutions will be compared.

The results for $C = 6$ is given in Fig. 1c. For $C = 6$, a small decrease in the u_1 velocity is observed near the inlet section of the symmetry line. Similarly, this causes a small increase of this velocity component at the inlet section near the bottom wall. The change in velocity field is observed in the first 1-$2W_{ch}$ downstream, and then the velocity field is slowly returned to its initial parabolic profile over the next $10W_{ch}$.

In Fig. 1d, the results for $C = 16$ are depicted. The streamlines in Fig. 1d show that with the presence of fibers, the streamline transients become important at the inlet section. Even at the $4W_{ch}$ downstream, the velocity profile is found to be different from the $C = 0$ solution.

In order to illustrate the change in the velocity profile due to the presence of fibers, the u_1 velocity component, which is parabolic for $C = 0$, is shown for three different nonzero C values. In Figs. 2a-b, the u_1 profiles are depicted for $x_1 = W_{ch}$ and for $4W_{ch}$. The biggest difference between the $C = 0$ solution and the nonzero C values are obtained for $x_1 = W_{ch}$ as shown in Fig. 2a. The velocity profile is changing towards plug flow behavior with the increase in C value. This phenomenon can be

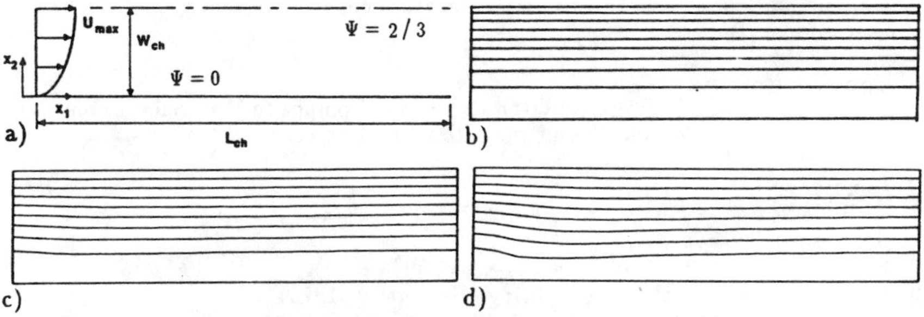

Figure 1: The channel geometry and the streamlines for: b) $C = 0$ c) $C = 6$ d) $C = 16$. The figure is drawn to scale up to $4W_{ch}$.

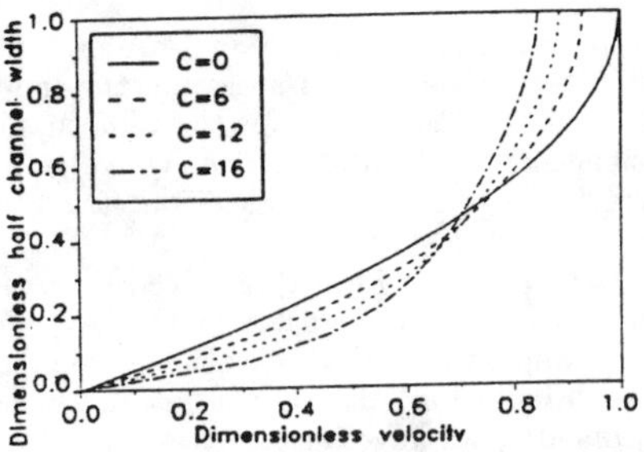

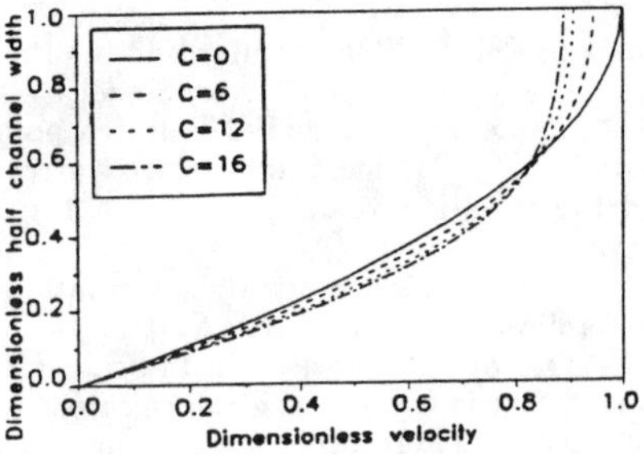

Figure 2: The u_1 velocity profiles across the channel for different C values at: a) $x_1 = W_{ch}$, b) $x_1 = 4W_{ch}$.

explained by the following mechanisms:

a) The fluid near the wall encounters less resistance since most of the fibers are relatively aligned near the wall. This orientation structure near the wall increases the magnitude of the velocity.

b) On the contrary, due to the relative randomness near the symmetry axis, the fluid experiences relatively higher resistance. This high resistance in comparison with the regions near the wall affects u_1, reducing its magnitude.

This combined effect of the orientation structure changes the velocity profile to a plug flow type behavior near the inlet, but as can be observed in Fig. 2b, the u_1 velocity recovers back slowly to its initial parabolic profile further downstream. The alignment of all fibers with the flow direction downstream decreases the difference in the orientation structure across the channel width; therefore, the fluid experiences a similar resistance across the channel, resulting in a parabolic profile.

Concluding Remarks

The governing equation for two-dimensional, steady state anisotropic flow with negligible inertia is derived. The proposed numerical scheme of determining the velocity and orientation field separately, and, then, iterating for the convergence is found to be satisfactory for the channel flow geometry. The solution of fiber orientation equations along the streamlines requires a special tracing technique for each nodal point since the accurate determination of fiber orientation is needed on each node for the stream function solution. The developed numerical scheme yields stable results within the validity of the model.

For the investigated channel geometry, the velocity and orientation fields are observed to reach a steady state slowly after the first 5 to 15 half channel widths downstream depending on the magnitude of the coefficient C which governs the effect of the fiber concentration and aspect ratio. The larger C values (i.e., a larger aspect ratio or concentration) are found to exhibit longer transients. However, most of the significant transients in the velocity and orientation fields are observed up to 4 half channel widths downstream. Near the inlet, the centerline velocity is found to be decreasing and then increasing within next few channel widths. The opposite behavior is observed near the wall.

Acknowledgments

This study was supported by the Center for Composite Materials through the National Science Foundation Engineering Research Centers program.

References

[1] G. B. Jeffery, *Proc. Royal Soc.*, A102, 161-179 (1922).
[2] H. Giesekus, *Rheol. Acta*, 2, 50 (1962).
[3] F. P. Bretherton, *J. Fluid Mech.*, 14, 284-304 (1962).
[4] J. G. Evans, Ph. D. thesis, Cambridge University (1975).
[5] S. M. Dinh and R. C. Armstong, *J. Rheology*, 28, 207-227 (1984).
[6] M. C. Altan, Ph. D. thesis, University of Delaware (1989).

CONSTITUTIVE MODELLING OF POLYMERIC MATERIALS AT LARGE DEFORMATIONS

YANNIS F. DAFALIAS
Department of Civil Engineering, University of California at Davis,
and Division of Mechanics, National Technical University of Athens

ABSTRACT

A general perspective of the constitutive modelling for polymeric materials at large deformations is presented. The constitutive framework with internal variables adopted in this study encompasses other material classes as well, but the molecular chain network structure of polymers and its affine deformation attrributes particular features to some aspects of the constitutive formulation. Emphasis is placed on the orientational part related to constitutive spins along which different approaches are discussed.

INTRODUCTION

The molecular chain-network structure of polymeric materials affects in many ways their mechanical properties. One important aspect is their capability to sustain large strains before failure, which can be plastic, elastic, viscoplastic, viscoelastic or combination of the above, depending on the glassy, rubbery or intermediate state of a polymer. This combination of material and geometrical nonlinearities, renders the constitutive modelling of polymeric materials under large deformations one of the most challenging task of continuum mechanics.

The general constitutive framework adopted here will be that of tensorial internal state variables, whose evolution is the macroscopic manifestation of microstructural changes at the molecular level. Such concepts as the relation among stress, strain, orientation and stretching of molecular chains have been the objectives of numerous papers, e.g., [1]. The results of these investigations were used qualitatively, quantitatively and/or in combination, to formulate constitutive models within the framework of continuum mechanics, e.g. in [2-7]. Certain issues pertaining to such constitutive formulations under large deformations are addressed in the present paper.

THE BASIC FORMULATION

A basic Eulerian constitutive formulation can be succinctly presented by

$$\mathbf{D} = \mathbf{D}^e + \mathbf{D}^p \; ; \qquad \mathbf{W} = \mathbf{W}^e + \mathbf{W}^p \tag{1}$$

$$\sigma = \bar{\sigma}(\lambda_i, \mathbf{n}_i, k, \alpha, T) \; ; \qquad \mathbf{D}^p = \bar{\mathbf{D}}^p(\sigma, k, \alpha, T, \mathbf{D}) \tag{2}$$

$$\dot{k} = \bar{k}(\sigma, k, \alpha, T, \mathbf{D}) \; ; \qquad \overset{\circ}{\alpha} = \dot{\alpha} - \omega\alpha + \alpha\omega = \bar{\alpha}(\sigma, k, \alpha, T, \mathbf{D}) \tag{3}$$

where $\mathbf{D}$, $\mathbf{D}^e$ and $\mathbf{D}^p$ are the material, elastic and plastic rates of deformation, respectively, $\mathbf{W}$, $\mathbf{W}^e$ and $\mathbf{W}^p$ are the material, elastic and plastic spins, respectively, σ is the Cauchy stress, λ_i the principal elastic stretches along the principal unit directions $\mathbf{n}_i$, k and α are scalar and tensor-valued internal variables, respectively, T is the temperature and ω a proper constitutive spin defining the corotational rate of α. Invariance requirements under superposed rigid body rotations renders the constitutive functions $\bar{\sigma}$, $\bar{\mathbf{D}}^p$, $\bar{k}$ and $\bar{\alpha}$ isotropic functions of their arguments. A constitutive law for $\mathbf{D}^e$ is not presented since it can be inferred from Eq. $(2)_1$ by proper time differentiation.

Postponing for later the discussion on the spin terms, one can observe that different classes of material response can be described by the above set of equations. Rubber elasticity is provided by Eq. $(2)_1$, where the presence of k and α may account in addition for damage and anisotropic characteristics. Large deformation viscoelasticity of elastomers can be described by the set of Eqs. $(2)_1$ and (3), where the last two equations provide the viscous element as well as damage and anisotropic development. Large deformation elastoplasticity and elastoviscoplasticity can be described by the full set of Eqs. (1)-(3). The way $\bar{\mathbf{D}}^p$, $\bar{k}$, and $\bar{\alpha}$ depend on $\mathbf{D}$, determines rate-dependent or rate-independent (homogeneous of order 1 in $\mathbf{D}$) response. For example in elastoplasticity the $\mathbf{D}$ appears linearly in $\bar{\mathbf{D}}^p$ via the loading index, while in viscoplasticity is absent. Also $\mathbf{D}^p$ may substitute for $\mathbf{D}$ in $\bar{k}$ and $\bar{\alpha}$, but given Eq. $(2)_2$ the $\mathbf{D}$ will appear implicitly if it enters $\bar{\mathbf{D}}^p$. While Eqs. (1)-(3) may refer to materials other than polymers, as well, what distinguishes the polymers' application is the particular forms of the constitutive functions. For example the form of σ is dictated by the kinematics of the molecular chains [2], but can also be determined on phenomenological grounds based on the particular response of rubber elasticity [3,4]. In the inelastic regime, the forms of $\bar{\mathbf{D}}^p$ and $\bar{k}$ are related, for example, to the structure of glassy polymers as in [7].

The corotational rate entering Eq. $(3)_2$ suggests that one must consider not only the "extensional" part in a rate equation of evolution, but also the "orientational" part, portrayed here by the constitutive spin ω. In some formulations the necessity to define ω is bypassed by presenting an equation corresponding to Eq. $(3)_2$ in a Lagrangian formulation [5,6]. The orientational aspects then are not explicitly considered, and often lost. In a different approach, one can define directly α in terms of a kinematical tensor in the current (or relaxed) configuration. In [8] the orthotropic axes and in [7] the eigenvectors of a back-stress tensor were determined along the Eulerian principal stretch directions, and their intensities related to the stretches. However, it is well known that the anisotropic characteristics are defined by the texture which in general cannot be directly related to the plastic strains; also the definition of a back stress in terms of a plastic strain has failed to produce realistic material response in reverse loading [9]. While the affine deformation gives credence to the assumed direct interdependence between the back-stress and principal stretch directions and values [7], it needs further validation by simulating reverse loading response in the spirit of investigation in [9]. It must also be observed that the evolution law of the back-stress introduced in [7] can be brought into the form of Eq. $(3)_2$ with ω defined as the spin Ω^E of the Eulerian strain ellipsoid and the form of $\bar{\alpha}$ properly adjusted, although such procedure is not operationally advantageous.

ON CONSTITUTIVE SPINS

It is clear from the foregoing that the definition of the constitutive spin ω, or any other equivalent orientational constutitive tool, cannot be supressed without loosing an important aspect of a constutitive relation. In this section a brief exposition of different approaches to determine ω will be made.

By differentiating the kinematics of the continuum from the kinematics of the underlying substructure, and assuming here small elastic strains for simplicity, Mandel [10] showed that $\omega = \mathbf{W}^e$ in Eq. $(1)_2$. Thus, in order to obtain ω one must postulate constitutive relations for the plastic spin $\mathbf{W}^p$ along the lines of Eq. $(2)_2$ for $\mathbf{D}^p$. Such proposition for $\mathbf{W}^p$ was also made in [11], while definite constutitive expresions for $\mathbf{W}^p$ were given first in [12,13]. The necessity to provide constitutive relations for $\mathbf{W}^p$ has been debated in a number of papers, e.g., [7,14,15], but the reason for such debate seems to be a misinterpretation of what Mandel and followers have defined as plastic spin. A brief elaboration on this point is necessary, bearing on the subsequent development.

Assuming the multiplicative decomposition of the deformation gradient $\mathbf{F} = \mathbf{F}^e\mathbf{F}^p$ into elastic and plastic parts with respect to an arbitrary relaxed configuration κ_u, and following the notation in [16], the kinematics at κ_u are given by

$$\mathbf{D}^p_u = (\overset{\Delta}{\mathbf{F}}{}^p\mathbf{F}^{p-1})_s = (\dot{\mathbf{F}}^p\mathbf{F}^{p-1})_s \; ; \qquad \mathbf{W}^p_u = (\overset{\Delta}{\mathbf{F}}{}^p\mathbf{F}^{p-1})_a = (\dot{\mathbf{F}}^p\mathbf{F}^{p-1})_a - \omega_u \tag{4}$$

where the subscript u indicates reference to κ_u (e.g., ω_u is the ω in reference to κ_u), and the subscripts s and a indicate symmetric and antisymmetric parts. The definition of $\mathbf{D}^p_u$ and $\mathbf{W}^p_u$ at κ_u is given in terms of the corotational with ω_u rate $\overset{\Delta}{\mathbf{F}}{}^p$, and not in terms of $\dot{\mathbf{F}}^p$. This is exactly the point of misinterpretation for the plastic spin encountered in different works, e.g., [7,14,15], where the plastic spin is often identified as the $(\dot{\mathbf{F}}^p\mathbf{F}^{p-1})_a$. The latter would be correct only if κ_u is chosen to be the isoclinic configuration [10,13,16] for which by definition $\omega_u \equiv 0$. In fact one can postulate any value, constant (including zero) or variable, for $(\dot{\mathbf{F}}^p\mathbf{F}^{p-1})_a$, but still he must constitutively define $\mathbf{W}^p_u$ in order to determine ω_u. This corresponds to the fact that the choice of the unstressed configuration κ_u, has no bearing on the final constitutive formulation as shown in [16]. In conclusion, the form of Eq. $(4)_2$ clearly shows the important fact that the plastic spin is not, in general, the material spin at κ_u, but the relative spin of the continuum with respect to the underlying substructure.

While the foregoing are motivated by crystalline plasticity, where the substructure is defined by the lattice, care should be exercised before corresponding concepts are applied to polymers. For polymeric materials the substructure can be defined in terms of the molecular chain network, but whether a physically meaningful relative spin between continuum and subtructure can be similarly defined, is an open question, given the affine nature of the deformation.

The definition of ω in Eq. $(3)_2$ can also be pursued in a more direct way, without having to define first a relative spin between continuum and substructure. One approach calls for the identification of characteristic unit directions $\mathbf{n}_i$ embedded in the continuum (in the current or relaxed configuration), whose spin $\mathbf{W}_i$ is given by the standard expression of continuum mechanics

$$\mathbf{W}_i = \mathbf{W} - [(\mathbf{n}_i \otimes \mathbf{n}_i)\mathbf{D} - \mathbf{D}(\mathbf{n}_i \otimes \mathbf{n}_i)] \qquad \text{(no sum over i)} \tag{5}$$

These directions can be along specific line segments or eigenvectors of kinematical or internal variable second order tensors. Use of such $\mathbf{W}_i$ in combination, can determine ω in a variety of ways [17-20]. In fact such approach may be advantageous for polymeric materials whose substructure and its evolution are characterized by molecular line segments and their orientations [1]. Another approach of direct identification of ω in Eq. $(3)_2$ can be purely kinematical. The ω can be identified as the spin of rotating frames such as the Eulerian or Lagrangian strain ellispoids, as well as the relative spin between these two (i.e., $\omega = \dot{\mathbf{R}}\mathbf{R}^T$ with $\mathbf{F} = \mathbf{R}\mathbf{U}$), as for example in [12,19,21-23], but more physical insight is necessary before any one of the above suggestions is pursued.

In view of the affine deformation of the molecular substructure of polymers, the identification of ω with the Eulerian spin Ω^E appears promising. However, the $\omega = \Omega^E$ in Eq. $(3)_2$ will not necessarily yield an internal variable α coaxial with the Eulerian ellipsoid. Such coaxiality may or may not be desirable, or at most may be an equilibrium orientation state not immediately attainable. To address this issue, a proposition made in [24] for elastomeric materials can be extended to polymers in general. It entails the introduction of the concept of Internal Spin Ω^I, such that ω at κ_u can be defined by

$$\omega_u = \Omega^E - \Omega^I \; ; \qquad \Omega^I = \eta_r (\alpha \mathbf{V}^p - \mathbf{V}^p \alpha) \tag{6}$$

with all quantities entering Eq. (6) referring to the relaxed configuration κ_u (the same applies then to Eq. $(3)_2$), and $\mathbf{V}^p$ defined from the polar decomposition $\mathbf{F}^p = \mathbf{V}^p\mathbf{R}^p$. The role of Ω^I is to "push" α towards coaxiality with $\mathbf{V}^p$ with an intensity proportional to a constitutive parameter η_r and the degree of non-coaxiality and deviatoric magnitude of α and $\mathbf{V}^p$, portrayed by the terms in parenthesis in Eq. $(6)_2$.

CONCLUSION

The molecular structure of polymeric materials has a significant effect on certain aspects of their constitutive modelling not encountered in other material classes. Particularly the affine deformation feature affects the orientational constitutive part expressed by proper spin terms in the rate equation of evolution of internal variables. Yet, many questions remain open for future research and validation by experiments. In this endeavor the rich constitutive modelling framework from other material classes can be helpful, but caution must be exercised when transferring established concepts to the case of polymeric materials.

Acknowledgements

The support by the NSF grants MSM-8619229 and MSS-8918531 is acknowledged.

REFERENCES

1. D.J. Brown and A.H. Windle, "Stress-orientation-strain relationshis on non-crystalline polymers," J. Materials Sci. 19, 2013-2038 (1984).
2. L.R.G. Treloar, The Physics of Rubber Elasticity (Oxford University Press, New York 1958).
3. K.C. Valanis and R.F. Landel, "The strain-energy function of a hyperelastic material in terms of the extension ratios," J. Appl. Phys. 38 (7), 2997-3002 (1967).
4. R.W. Ogden, "Large deformation isotropic elasticity-on the correlation of theory and experiment for incompressible rubberlike solids," Proc. R. Soc. Lond. A. 326, 565-584 (1972).

5. J. Lubliner, "A model of rubber viscoelasticity," Mech. Res. Comm. 12 (2), 93-99 (1985).
6. J.C. Simo, "On a fully three-dimensional finite strain viscoelastic damage model: Formulation and computational aspects," Comp. Meth. Appl. Mech. Eng. 60, 153-173 (1987).
7. M.C. Boyce, D.M. Parks and A.S. Argon, "Large inelastic deformation of glassy polymers. Part I: rate dependent constitutive model," Mech. of Materials, 7, 15-33 (1988).
8. R. Hill, The Mathematical Theory of Plasticity (Oxford University Press, New York 1950).
9. M.A. Eisenberg and A. Phillips, "On nonlinear kinematic hardening," Acta Mechanica 5, 1-13 (1968).
10. J. Mandel, Plasticite Classique et Viscoplasticite (Courses and Lectures No. 97, ICMS, Udine, Springer, New York 1971).
11. J. Kratochvil, "Finite-strain theory of crystalline elastic-plastic materials," J. Appl. Phys. 42, 1104-1108 (1971).
12. Y.F. Dafalias, "Corotational rates for kinematic hardening at large plastic deformations," ASME, J. Appl. Mech. 50, 561-565 (1983).
13. B. Loret, "On the effect of plastic rotation in the finite deformation of anisotropic elastoplastic materials," Mech. of Materials 2, 287-304 (1983).
14. E.T. Onat, "Representation of elastic-plastic behavior in the presence of finite deformations and anisotropy," private communication (1987), also submitted to Int. J. of Plasticity.
15. M. Obata, Y. Goto and S. Matsuura, "Micromechanical consideration on the theory of elasto-plasticity at finite deformations," Int. J. Engng. Sci 28, 241-252 (1990).
16. Y.F. Dafalias, "Issues on the constitutive formulation at large elastoplastic deformations, Part 1: Kinematics, and Part 2: Kinetics," Acta Mechanica, 69, 119-138 (1987) and 73, 121-146 (1988).
17. E.H. Lee, R.L. Mallet and T.B. Wertheimer, "Stress analysis for anisotropic hardening in finite-deformation plasticity," ASME, J. Appl. Mech. 50, 554-560 (1983).
18. Y.F. Dafalias, "A missing link in the macroscopic constitutive formulation of large plastic deformations," in Plasticity Today, edited by Sawczuk and Bianchi (Elsevier Appl. Sci., Essex, 1985).
19. H.M. Zbib and E.C. Aifantis, "The concept of relative spin and its implications to large deformation theories," Acta Mechanica 74, 15 (1988).
20. E. van der Giessen, "On a continuum representation of deformation-induced anisotropy," in Advances in Constitutive Laws for Engineering Materials, edited by Jinghong and Murakami (Int. Acad. Publ., Pergamon, New York, 1989).
21. A. Dogui and F. Sidoroff, "Quelques remarques sur la plasticite anisotrope en grandes deformations," C.R. Acad. Sci. Paris, 299, Serie II, No. 18, 1225-1228 (1984).
22. R. Sowerby and E. Chu, "Rotations, stress rates and strain measures in homogeneous deformation processes," Int. J. Solids Structures 20, 1037-1048 (1984).
23. R.N. Dubey, "Corotational rates on principal axes," S.M. Archives 10, 245-255 (1985).
24. Y.F. Dafalias, "Constitutive model for large viscoelastic deformations of elastomeric materials," to appear in Mech. Research Communications.

A MODEL FOR NODULAR GRAPHITE CAST IRON COUPLING ANISOTHERMAL ELASTO-VISCOPLASTICITY AND PHASE TRANSFORMATION

Nacera HAMATA, René BILLARDON, Didier MARQUIS
Laboratoire de Mécanique et Technologie
E.N.S. de Cachan/C.N.R.S./Université Paris 6
61, Avenue du Président Wilson, 94235-CACHAN, France

Ahmed BEN CHEIKH
RENAULT, Direction de la Recherche, Service 0954
8-10 Avenue Emile Zola, 92109-BOULOGNE-BILLANCOURT, France

ABSTRACT

In this paper, constitutive laws for nodular graphite ferritic cast iron during the post-solidification phase of the casting process are given.

1- INTRODUCTION

In the case of nodular graphite ferritic cast iron, the typical cooling rates observed during the casting process lie between 0.01 and 0.2 °Cs^{-1}. For such low cooling rates, the material can be considered as : first, at high temperature, a γ-phase (i.e. an austenitic matrix with graphite nodules) ; second, at low temperature, an α-phase (i.e. a ferritic matrix with graphite nodules) ; and third, during the transformation, an evolutive mixture of these two phases.

2. MODELLING OF EACH PHASE

The thermo-mechanical behaviour of each phase is described by an anisothermal elasto-viscoplastic model [1] based on classical continuum thermodynamics concepts [2]. *As a first approximation,* it is assumed that *the evolutive carbon fraction* of the austenitic and ferritic matrixes is modelled through *the dependence of this model upon temperature.* The thermo-mechanical state of each phase is described by the following set of independent state variables :

$$\mathbb{V} \equiv (\varepsilon, T, \varepsilon^p, \mathbf{V}) \qquad \text{with} \qquad \mathbf{V} \equiv (p, \boldsymbol{\alpha}),$$

where ε, ε^p denote the total and plastic strain tensors respectively, T the temperature, $(\varepsilon^p, \mathbf{V})$ the set of the internal variables, p the accumulated plastic strain and $\boldsymbol{\alpha}$ a tensorial variable .

The reversible behaviour is described by the Helmholtz specific free energy $\Psi = \Psi(\varepsilon, T, \varepsilon^p, \mathbf{V})$, so that the thermodynamic forces $\mathbb{A} \equiv (\boldsymbol{\sigma}, s, -\boldsymbol{\sigma}, \mathbf{A})$ are defined from the following *state laws* :

$$\boldsymbol{\sigma} = \rho \frac{\partial \Psi}{\partial \varepsilon} = -\rho \frac{\partial \Psi}{\partial \varepsilon^p}, \qquad s = -\rho \frac{\partial \Psi}{\partial T}, \qquad \mathbf{A} \equiv (R, \mathbb{X}) = \rho \frac{\partial \Psi}{\partial \mathbf{V}},$$

where ρ denotes the mass density, $\boldsymbol{\sigma}$ the stress tensor, s the entropy, R the isotropic hardening, and $\mathbb{X}$ the kinematic hardening.

The irreversible behaviour is described by a dissipation potential

$$\Phi = \Phi(\mathbb{A} ; \mathbb{V}) = \Phi(\sigma, \mathrm{A}, \overrightarrow{\mathrm{Grad}}\mathrm{T} / \mathrm{T} ; \mathbb{V}),$$

from which the following *evolution and Fourier's laws* are derived :

$$\dot{\varepsilon}^p = -\frac{\partial\Phi}{\partial\sigma}, \qquad \dot{\mathbf{V}} = -\frac{\partial\Phi}{\partial\mathbf{A}}, \qquad \vec{q} = -\frac{\partial\Phi}{\partial[\overrightarrow{\mathrm{Grad}}\mathrm{T} / \mathrm{T}]},$$

where $\vec{q}$ denotes the heat flux.

3. COMPLETE CONSTITUTIVE EQUATIONS TAKING ACCOUNT OF THE PHASE TRANSFORMATION

In the model proposed in this paper, the description of *the behaviour of a two-phase material* is based on *the theory of mixtures.* A simplified approach consists in taking *the state variables* ε, T *and* ε^p, *as variables common to the two phases.* To model the phase transformation (from γ to α), *an additional scalar internal variable* $\varkappa$ ($\varkappa \in [0,1]$) is introduced. It represents *the evolutive volumetric fraction of the α-phase* and the associated variable Z is the phase transformation "energy release rate". Therefore, the set of the thermodynamical variables becomes :

$$\mathbb{V} \equiv (\varepsilon, \mathrm{T}, \varepsilon^p, \varkappa, \mathbf{V}) \text{ with } \mathbf{V} \equiv (p, \alpha) ; \; \mathbb{A} \equiv (\sigma, s, -\sigma, Z, \mathbf{A}) \text{ with } \mathbf{A} \equiv (R, \mathbb{X})$$

Further assumptions must be made to define the *couplings* [3] between the transformation mechanism on one hand, and the mechanisms involved in the anisothermal elasto-viscoplastic behaviour of each phase on the other hand. These assumptions lead to particular forms of the potentials $\Psi(\mathbb{V})$ and $\Phi(\mathbb{A} ; \mathbb{V})$, and consequently, of the evolution and Fourier's laws ($\dot{\varepsilon}^p, \dot{\mathrm{V}}, \dot{\varkappa}, \vec{q}$).

The following *state couplings* are assumed :

$$\rho\,\Psi = [(1-\varkappa)\,\rho_\gamma \Psi_\gamma + \varkappa\,\rho_\alpha \Psi_\alpha] + \rho_\alpha \Psi_{vtr}(\varepsilon, \mathrm{T}, \varepsilon^p, \varkappa) + \rho_\alpha \Psi_{Ltr}(\mathrm{T}, \varkappa)$$

where $(\rho_\gamma \Psi_\gamma)$ and $(\rho_\alpha \Psi_\alpha)$ correspond to the γ- and α- phases respectively. The first term corresponds to a linear mixing law, and the last two terms, to the volumetric variation and to the latent heat emission associated to the phase transformation respectively.

The following *dissipation uncouplings* are assumed :

$$\Phi = \Phi_{vp}(\sigma, \mathbb{X}, \mathrm{R} ; \mathbb{V}) + \Phi_{th}(\overrightarrow{\mathrm{Grad}}\mathrm{T} / \mathrm{T} ; \mathbb{V}) + \Phi_{tr}(\mathrm{Z} ; \mathbb{V})$$

where the first term refers to viscoplasticity, the second term to thermics, and the third term to phase transformation. Following the assumption that the viscoplastic strain ε^p is split into a classical viscoplastic strain $\varepsilon^p{}_{cvp}$ and a transformation viscoplastic strain $\varepsilon^p{}_{tr\text{-}vp}$, the first term takes the following particular form :

$$\Phi_{vp} = \Phi_{cvp}(\sigma, \mathbb{X}, \mathrm{R} ; \mathbb{V}) + \Phi_{tr\text{-}vp}(\sigma ; \mathbb{V})$$

The following *evolution couplings,* modelled through the introduction of state variables as parameters in the dissipation potential, are assumed.
First, influence of the phase transformation on the viscoplastic flow :

$$\Phi_{cvp} = (1-\varkappa)\,\Phi_{\gamma cvp}(\sigma, \mathbb{X}, \mathrm{R} ; \mathrm{T}) + \varkappa\,\Phi_{\alpha cvp}(\sigma, \mathbb{X}, \mathrm{R} ; \mathrm{T})$$

where $\Phi_{\gamma cvp}$ and $\Phi_{\alpha cvp}$ correspond to the γ- and α- phases respectively.

Second, influence of the thermomechanics on the kinetics of the phase transformation :

$$\Phi_{tr} = \Phi_{tr} (Z ; \varepsilon^p, T)$$

Third, influence of the phase transformation on the coefficients of the Fourier's law :

$$\Phi_{th} = \Phi_{th}(\overrightarrow{\mathrm{Grad}}T / T ; T, \varkappa)$$

Fourth, influence of the phase transformation on the evolution of the transformation viscoplastic strain :

$$\Phi_{tr\text{-}vp} = \Phi_{tr\text{-}vp}(\sigma ; T, \varkappa)$$

4. EXPERIMENTAL IDENTIFICATION

For each phase, the plastic strain evolution is written as :

$$\dot{\varepsilon}^p = \frac{3}{2} \frac{1}{K(T)} < f >^{n(T)} \frac{(\mathbb{S} - \mathbb{X})}{J_2(\sigma - \mathbb{X})} ,$$

where $\mathbb{S}$ denotes the stress deviator and f a yield function which defines the *elastic domain* such that

$$f = J_2(\sigma - \mathbb{X}) - R - \sigma_y (T) \leq 0 ,$$

where σ_y denotes the yield stress.

The viscous behaviour [K(T), n(T)] at different temperatures of the α-phase is identified from isothermal relaxation tests subsequent to tensile tests at various strain rates. Since the hardening of the γ-phase appears practically nill, the viscous behaviour at different temperatures of this phase is identified from isothermal creep tests at different stress levels. Results of these tests are given in Fig.1.

The elasto-plastic behaviour [E(T), ν(T), σ_y(T), $\mathbb{X}$(T), and R(T)] of the α-phase is identified from isothermal tension tests with elastic unloadings performed, down to the yield stress in compression, at different values of the plastic strain. Results of these tests are given in Fig.2.

For each phase, the thermoelasticity law is written as :

$$\varepsilon^e = \varepsilon - \varepsilon^p = \frac{1 + \nu(T)}{E(T)} \sigma - \frac{\nu(T)}{E(T)} \mathrm{tr}(\sigma) \mathbb{1} + d(T) (T - T_0) \mathbb{1}$$

where T_0 denotes the room temperature, and d the thermal expansion coefficient. This coefficient d is identified from dilatometric tests without any mechanical load. The result is given in Fig.3.

Dilatometric tests with different constant mechanical loadings enable to identify the transformation viscoplastic strain.

The phase transformation kinetics [$\dot{\varkappa}$] is identified from the evolution of the electrical resistivity R_e of tension specimens subjected to different constant mechanical loads while cooled at different temperature rates. The result for a zero mechanical loading and a constant cooling rate equal to 0.1°Cs^{-1} is given in Fig.4. The electrical resistivity of the mixture is assumed such that :

$$R_e = (1 - \varkappa) R_{e\gamma} + \varkappa R_{e\alpha}$$

where $R_{e\gamma}$ and $R_{e\alpha}$ correspond to the γ- and α- phases respectively.

5. REFERENCES

1. A. BENALLAL & A. BEN CHEIKH in Constitutive laws for Engineering Materials : Theory and Applications, 667-674, edited by C.S. Desai et al. (Elsevier, 1987).
2. P. GERMAIN, Q.S. NGUYEN & P. SUQUET, J. Appl. Mech, A.S.M.E., 50, 1010-1020 (1983).
3. D. MARQUIS & J. LEMAITRE, Revue de Physique Appliquée, 23, 615-624 (1988).

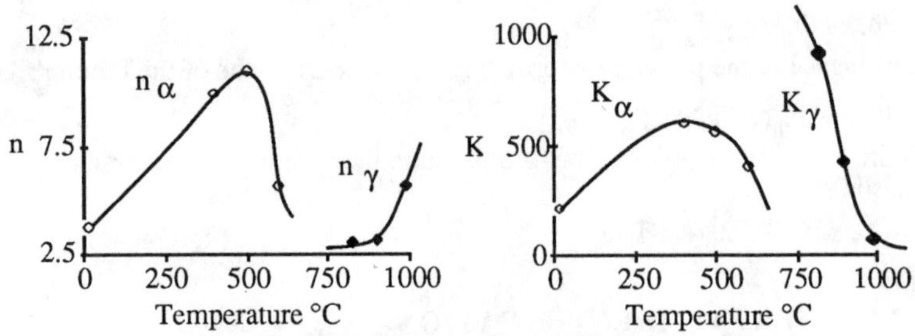

Fig.1 Evolution of viscosity coefficients (n, K)

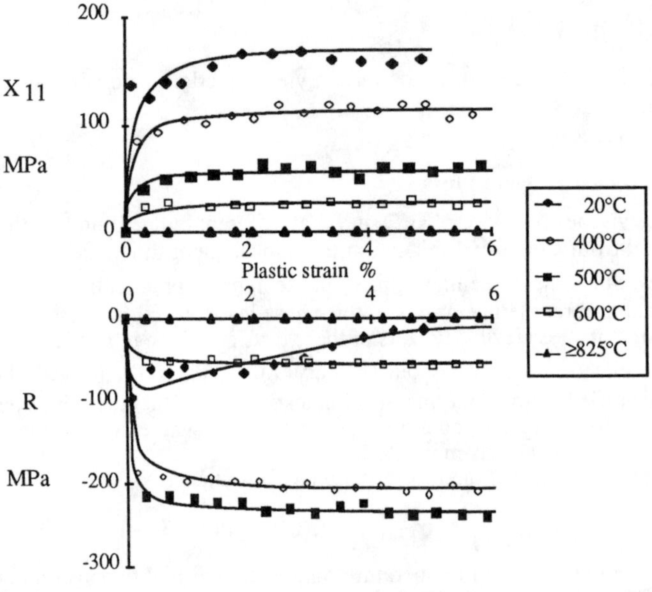

$$\sigma_{11} = X_{11} + R + \sigma_y + K\ \dot{\varepsilon}^{p\ 1/n}$$

Fig.2 Evolution of kinematic (X_{11}) and isotropic (R) hardenings

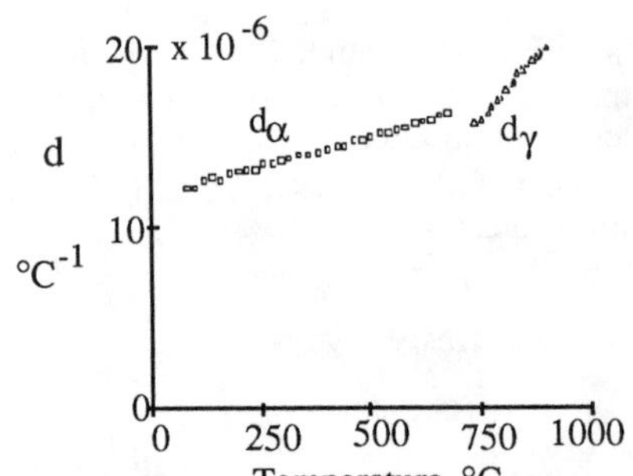

Fig.3 Evolution of thermal expansion coefficient (d)

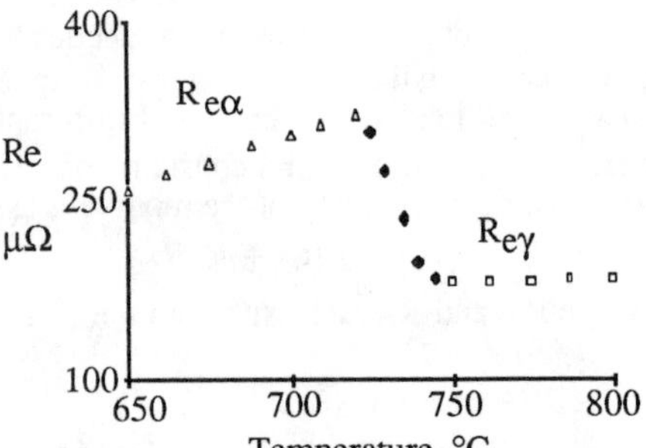

Fig.4 Evolution of electrical resistance (R_e)

ENGINEERED AND ADAPTIVE MATERIALS--CHALLENGES FOR SOLID MECHANICS

George K. Haritos, Lt Colonel, USAF
Program Manager and Acting Director, Aerospace Sciences
Air Force Office of Scientific Research
Bolling AFB, Washington DC 20332-6448
USA

Abstract - Performance requirements for emerging and projected aerospace systems have, in many cases, eliminated conventional materials as potential candidates for these applications. It is now envisioned that future aerospace vehicles will require considerable amounts of such emerging materials as metal-matrix composites, ceramics and ceramic-matrix composites, carbon-carbon composites, et.al., as well as a host of hybrid multimaterials which would combine features and/or ingredients of all of the above. All of these materials will have to be custom-tailored, "engineered", to satisfy specific performance requirements. This engineering will take place over a wide range of dimensional scales and will involve a number of disciplines; principal among them are chemistry, materials science, and mechanics. The solid mechanics community can play a critical role in this process. This paper outlines a number of basic research issues associated with families of emerging materials to underscore the need for an integrated, interdisciplinary approach for developing such materials. Research needs in ceramics, ceramic composites, and carbon-carbon composites will be addressed. The possibility of studying natural multimaterials as a source of ideas for innovation in the design of manmade materials is also briefly visited.

Future systems, most notably in the aerospace arena, are placing unprecedented demands on materials and structures. Performance requirements for emerging and projected aerospace structures and engines, such as sustained operating temperatures and temperature gradients, structural-to-gross weight ratios, engine thrust-to-weight ratios, enhanced operating life, and stricter safety criteria, have, in many cases, virtually eliminated from consideration most conventional monolithic materials. Figure 1, for example, shows the expected temperature distribution for sustained performance over the hypersonic vehicle, while Figure 2 contrasts the structural-to-gross weight ratios of existing systems, including the Space Shuttle, to those of various conceptual designs for the hypersonic vehicle. The long-term research and development emphasis for meeting such needs has thus shifted to such emerging materials as metal-, ceramic-, and carbon-matrix composites. At present, they appear to be the most promising for satisfying such requirements.

The approach for creating these new multiphase materials represents a radical departure from the classical practice of processing naturally occurring materials; we are now combining at the microstructural level constituents selected from a widely diversified array of sources to produce the desired macroscopic behavior [1]. This concept in the design of materials represents the revolutionary impact of advanced fiber-reinforced composites [2].

The resulting materials are characterized by very complex structure and highly anisotropic behavior. Thus, by their very nature, these materials do not lend themselves to classical mechanics treatment. The fundamental postulate of the classical approach that the critical volume element can be made arbitrarily small, breaks down in the face of phase interactions whose influence domain is nonlocal. The basis of such cornerstones of continuum mechnics as the gradient and divergence theorems needs to be re-examined. As pointed out first in [3], and then in greater detail in [4], these theorems are ultimately founded on the assumption that microstructural feature interactions are localized, at an arbitrarily infinitesimal range. This enables the application of the concept of limit to the interior and the boundary of the material body which, in turn, makes possible the mathematical connection of the deformation (and resulting strains and stresses) of the material to a set of externally applied loads. As our materials-synthesis capabilities became more sophisticated, as they did in recent years, the range of scales over which microstructural-features evolution and interaction influence the material's macro-behavior has expanded dramatically. The fundamental challenge of mesomechanics outlined in [4], and depicted schematically in Figure 3, is the mathematical linking of the kinematics of such micro-feature evolutions to the material's macroscopic stress-strain behavior. Ultimately, this will lead to accurate life predictions, based on the current state of damage in the material.

Traditionally, the key sciences involved in the synthesis, processing, and initial characterization of new material systems are chemistry and materials science. The mechanics community does not usually get involved until a new material has been identified as a promising candidate for a range of applications. All too often, the mechanics of materials' input is not solicited until premature material failures occur. It is the thesis of this paper that there is an urgent need for the mechanics community to participate in the process of new materials development at an as early as possible stage. The potential contributions to saving time and resources and to producing materials better suited for specific missions are very significant. Examples of mechanics contributions which have already been made are discussed in the ensuing paragraphs. At the same time, a number of issues awaiting illumination associated with the development of novel material systems are identified, as a means of underscoring the critical need for interdisciplinary cooperation as the most efficient way for their resolution.

Ceramics and ceramic-matrix composites have emerged as strong candidates for a wide array of high-temperature aerospace applications. Many products are already available and in use at low and moderate temperatures, such as automotive engine components. Ceramics offer a number of advantages over metal- and polymer-based materials. They have high melting points, good strength (and, in the case of composites, good toughness), they retain most of their strength to temperatures as high as 1200C, they are relatively light weight (good specific strength and modulus), and they cost less than most other aerospace materials. Their biggest advantage, however, is their environmental stability. As Karl Prewo put it, "Polymers decompose and burn while metals corrode and melt under conditions that have little or no effect on many ceramics." [5] Although reinforced ceramics offer a broader applications potential, understanding, describing, and predicting the behavior of monolithic ceramics is also very important; their potential high-temperature uses include bearings, heat exchangers, and low-load components. Most importantly, since they are the matrix material for ceramic-matrix composites, their behavior plays a key role in the overall composite behavior.

Research in the behavior of monolithic ceramics at low temperature has recorded good progress in the areas of toughness, R-curve behavior, role of microstructure and the interfaces. Significant progress was achieved, for instance, by a University of Cambridge, United Kingdom team of investigators who addressed the physical fracture processes underlying the R-curve behavior of Al_2O_3 ceramics [6]. Figure 4 summarizes their findings. They directly observed, identified, and modeled three distinct crack-propagation inhibitingmechanisms. They then classified the contribution of each mechanism to the R-curve behavior, and made suggestions to material microstructure designers for enhanced crack growth resistance. At low temperatures, a key remaining issue is the identification and modeling of the fatigue behavior of monolithic ceramics. Although some investigators have been considering various aspects of cyclic crack growth behavior, e.g. [7, 8], it is felt that this issue is far from being resolved.

Aspects of the behavior of monolithic ceramics at high temperatures are not as well understood. While some progress has been made in understanding overall creep behavior, creep rupture, and oxidation, research is needed in the areas of fatigue, cavitation, and microstructure creep. A prospective of the state of understanding regarding these issues can be found in [9] and [10].

A similar situation exists in the state of understanding of the behavior of ceramic-matrix composites. That is, good progress has been achieved in identifying the mechanisms of deformation and damage at low temperatures, e.g. [11, 12], while very little is known about their high-temperature behavior [13]. At room temperature, additional research is needed in fatigue crack propagation. At elevated temperatures, a host of issues remain: toughness, creep, oxidation, thermo-mechanical fatigue, failure modes, role of interfaces and microstructures. It is anticipated that the physical mechanisms controlling all aspects of the materials' behavior at high temperatures will be distinct than those at work at low temperatures. Of interest is not only a description of the behavior of selected ceramic composites, but, more importantly, the understanding of the microstructure and processing connection to the macro thermo-mechanical behavior, so that in cooperation with chemists and materials scientists the mechanics community can affect improvements in the composition and processing of these materials for desired applications. Two examples of progress in the mechanics of continuous-fiber reinforced ceramics are given in Figures 5 and 6 [14]. Figure 5 shows active mechanisms involved in suppressing fiber failure at the front of an advancing matrix crack; also plotted are the conditions which promote debonding over fiber failure. Figure 6 identifies the various failure modes which have been observed under tension and bending.

The general mechanics research goals for ceramics and ceramic-matrix composites are (1) accurate prediction of behavior and life times in service-like environments including cyclic thermomechanical loading at high average temperatures and (2) guiding the development of materials for specified performance.

Carbon/carbon composites are light weight, highly refractory, exhibit superior thermal shock and creep resistance, and retain their strength to very high temperatures, as shown in Figure 7. Their present useage is limited to either relatively low temperatures (less than 1650C) for extended periods of time (hundreds of hours), or to short times (tens of hours) at high temperatures. Future needs include applications at high temperatures for extended periods of time and under thermal cycling conditions. The key obstacle to satisfying these needs is the low resistance of carbon/carbon composites to oxidation. Carbon begins to oxidize at approximately 400C. Considerable effort has been devoted over the last several years toward devising effective methods for enhancing the oxidation resistance of these materials; see, e.g. [15] and [16]. Approaches investigated for oxidation-protection systems include matrix doping with chemical inhibitors, and fiber and/or component coatings. The influence that each of these approaches will have on the thermo-mechanical properties of the resulting material system will determine the material's suitability for structural applications. It is therefore prudent that the mechanics community become involved with the chemists and material scientists working to develop oxidation-resistant carbon/carbon materials, in order to assist in guiding their efforts, by predicting the behavior of the product of each of the approaches being considered. Mechanics modeling has proven quite effective in predicting properties of both fibers and of unidirectional carbon/carbon composites, as demonstrated in Figures 8 and 9 [17]. Thus, the ultimate mechanics research goal associated with carbon/carbon materials is to develop the required analytical capability for guiding the development of oxidation protection systems for structurally useful composites.

The performance/environment demands being placed on future systems point to a direction of multi-functional, adaptive materials and structures. There is already a strong movement toward that direction. We can no longer afford separate materials and components for each task. It is expected that adaptive, multifunctional components and materials will, in future systems, become the rule rather than the exception. In this possible future scenario, one has to marvel at the ability of nature to produce routinely such (living) multifunctional and adaptive systems. Early in 1990, a new research thrust, entitled "Biomimetics," was successfully advocated to senior Air Force management by two program managers at the Air Force Office of Scientific Research [18]. It was motivated from the realization that both inspiration and innovation can be gained in approaching the engineering of new materials by studying and imitating whenever prudent (and possible) naturally occurring materials and structures. The basic research goal of Biomimetics is to understand and describe the structure and function of naturally-evolved materials, to enable the technology goal of producing aerospace materials with superior properties by mimicking the processing and design principles mastered by nature. An example of how nature manages to combine weak ingredients and still produce materials and structures with quite respectable properties is given in Figure 10 [19].

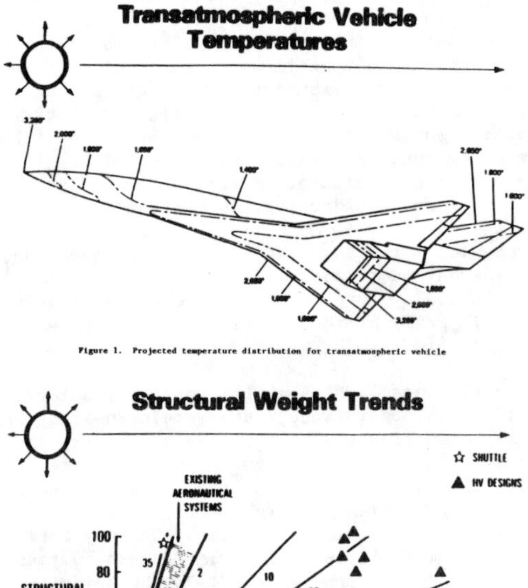

Figure 1. Projected temperature distribution for transatmospheric vehicle

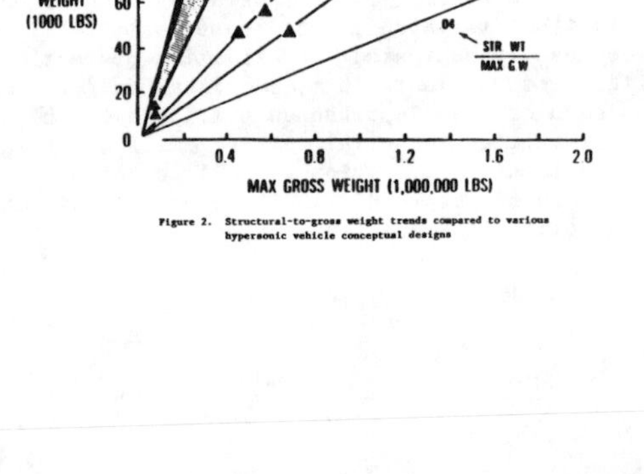

Figure 2. Structural-to-gross weight trends compared to various hypersonic vehicle conceptual designs

Mesomechanics: Constitutive Material Behavior

- **CURRENT APPROACH**
- **GOAL**

Continuum
Mechanics

Real Material

Heterogeneous
Approach

- Failure Criteria Damage-Independent
- Unable to Predict Multiple-Failure Modes
- Cannot Predict Interactions Among Damage Micromechanisms

- Damage Evolution Included in Constitutive Description
- Failure Criteria Based on Damage State

Include Time and Damage in Material Behavior

PAYOFF: Improved Durability, Improved Materials

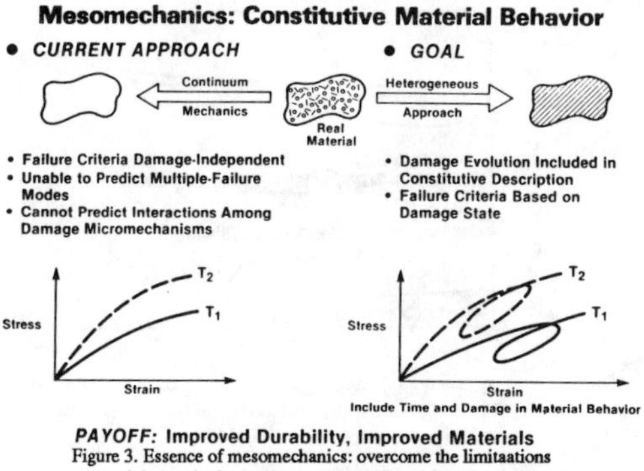

Figure 3. Essence of mesomechanics: overcome the limitations inherent in the continuum approach and face the challenges inherent in the heterogeneous medium approach

Frictional Grain Pullout found to be the most effective toughening mechanism in monolithic ceramics

- $\Delta G = c\,\mu\,f\,\sigma_{max}\,d$

  Coef. of Friction — Fraction of Crack Face Interacting — Grain Size

- $\Delta G = 0.6\,\mu\,f\,\sigma_{max}\,d_{min}$

  Transverse Grain Size

- $\Delta G = \dfrac{1}{18}\,f\,\dfrac{\sigma_{max}^2}{E}\,d$

- Microcracking Mechanism not present

For Tougher Ceramics
- Maximize friction by increasing normal stress between grains
- Include elongated grains with high fracture strength

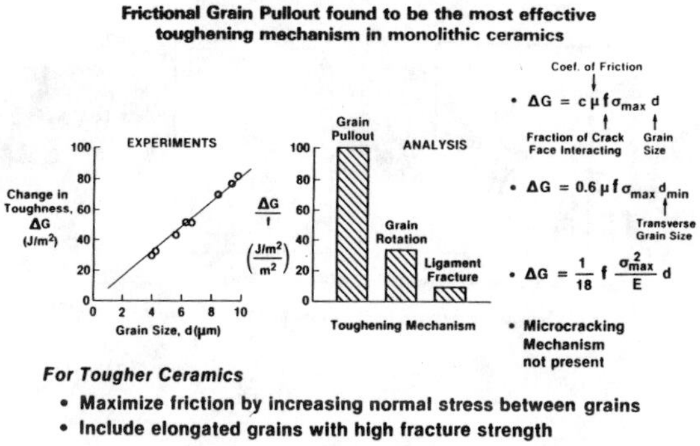

Figure 4. Toughening mechanisms in Al_2O_3 ceramics [6]

FIBER FAILURE SUPPRESSED
AT MATRIX CRACK FRONT

CRITICAL ENERGY RELEASE RATE
REQUIRED FOR CRACK DEBONDING

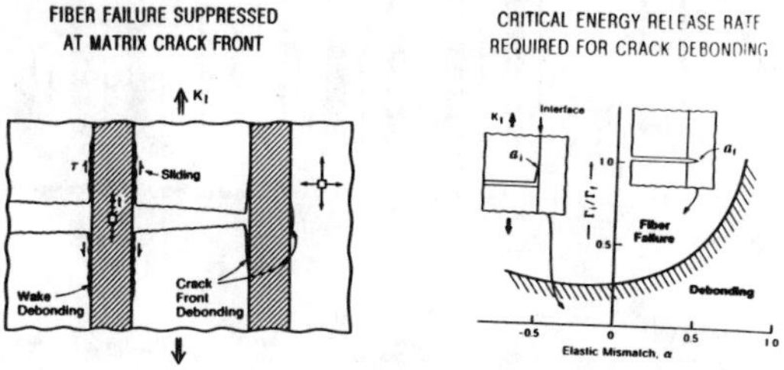

Figure 5. Mechanisms at metrix crack/fiber intersection in ceramic-matrix composites [14]

FAILURE IN TENSION

BENDING FAILURE

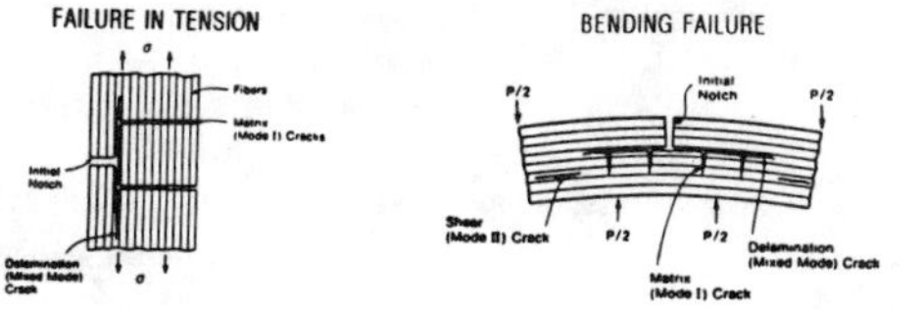

Figure 6. Failure modes in ceramic-matrix composites [14]

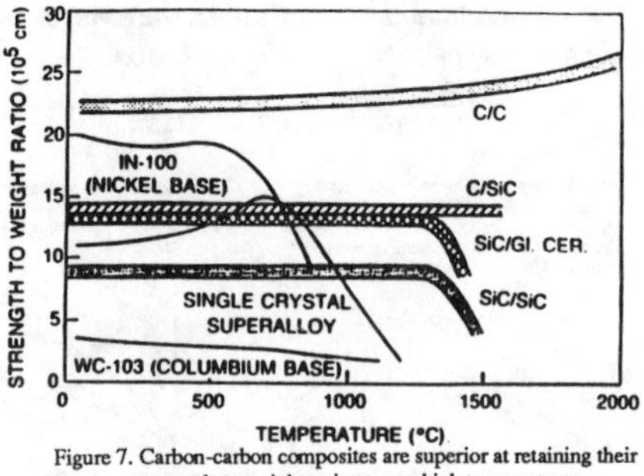

Figure 7. Carbon-carbon composites are superior at retaining their strength-to weight ratio at very high temperatures.

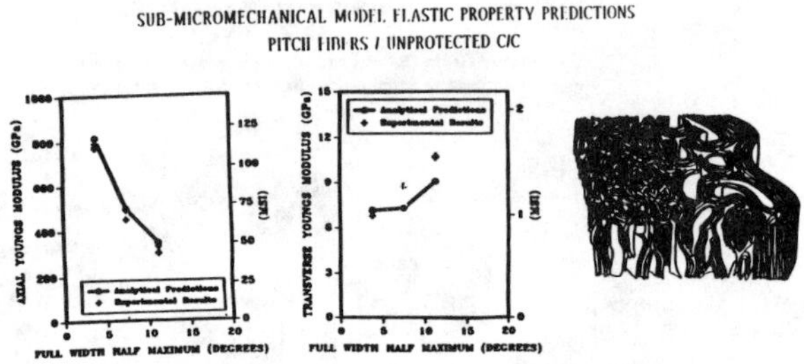

Figure 8. Micromechanics predictions for fiber properties in carbon/carbon [17]

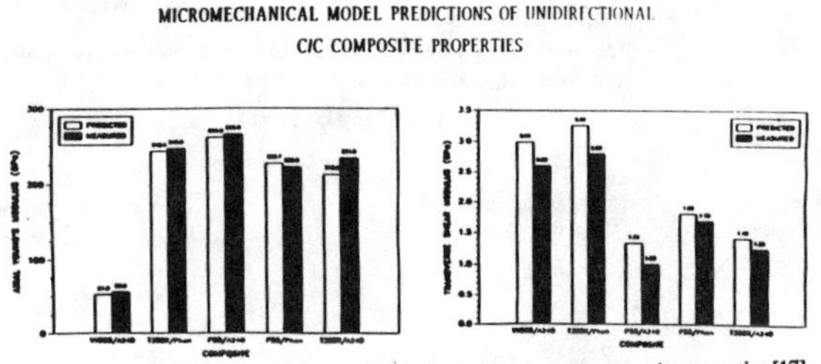

Figure 9. Micromechanics predictions for unidirectional carbon/carbon composite properties [17].

A PHYSICAL MODEL OF NACRE

CONSTITUENTS: 95% $CaCO_3$ (CHALK) AND 5% ORGANIC GLUE

STRUCTURE: "BRICK AND MORTAR"
$CaCO_3$ BRICKS (0.5 μm THICK)
ORGANIC MORTAR (20 - 30 nm)

PROPERTIES: FRACTURE STRENGTH, $\sigma_F = 185 \pm 20$ MPa
FRACTURE TOUGHNESS, $K_{IC} = 8 \pm 3$ MPa $\sqrt{m}$

COMPARES FAVORABLY WITH MOST "HIGH-TECH" CERAMICS

Figure 10. A physical model of nacre (mother-of-pearl) [18].

REFERENCES

1. Drucker, D. C., "Preliminary design on the microscale and macroscale," Advances in Aerospace Structures and Materials, AD-01, pp 1-3, ASME, New York, 1981.

2. Salkind, M. J., "Fiber Composite Structures," Proceedings of ICCM, Vol 2, pp 5-30, AIME, New York, 1976.

3. Haritos, G. K., Hager, J. W., Amos, A. K., Salkind, M. J., and Wang, A. S. D., "Mesomechanics: the microstructure-mechanics connection," Proceedings of 28th Structures, Structural Dynamics, and Materials Conference, Part 1, pp 812-818, AIAA, New York 1987.

4. Haritos, G. K., Hager, J. W., Amos, A. K., Salkind, M. J., and Wang, A. S. D., "Mesomechanics: the microstructure-mechanics connection," Int. J. Solids Structures, Vol 24, No 11, pp 1081-1096, 1988.

5. Prewo, K. M., "Fiber Reinforced Ceramics--New Opportunities for Composite Materials," 4th US-Japan Conference on Composite Materials, June 27-29, 1988, Washington DC.

6. Vekinis, G., Ashby, M. F., and Beaumont, P. W. R., "Direct Observation of Fracture and Damage Mechanics of Ceramics," Cambridge University Engineering Department Technical Report CUED/C-MATS/TR 174, March 1990.

7. Luh, E. Y., Dauskardt, R. H., and Ritchie, R. O., "Cyclic Fatigue-Crack Growth Behavior of Short Cracks in SiC-Reinforced LAS Glass-Ceramic Composite," J. of Mat. Sci. Letters, Vol 8, 1989, in press.

8. Steffen, A. A., Dauskardt, R. H., and Ritchie, R. O., "Cyclic Fatigue-Crack Propagation in Ceramics: Long and Small Crack Behavior," to appear in FATIGUE 90, Proceedings of the Fourth International Conference on Fatigue and Fatigue Thresholds, Honolulu, Hawaii, H. Kitagawa and T. Tanaka, Eds.

9. Page, Richard A., Lankford, J., and Chan, Kwai S., "Study of High Temperature Failure Mechanisms in Ceramics," Southwest Research Institute AFOSR Annual Report under Contract F49620-88-C-0081, Nov 1989.

10. Hutchinson, John W., "Mechanisms of Toughening in Ceramics," Theoretical and Applied Mechanics, P. Germain, M. Piau, and D. Caillerie, eds., Elsevier Science Publishers B. V. (North-Holland), IUTAM, 1989.

11. Budiansky, Bernard, Hutchinson, John W., and Evans, Anthony G., "Matrix Fracture in Fiber-Reinforced Ceramics," J. Mech. Phys. Solids, Vol 34, No 2, pp 167-189, 1986.

12. Brennan, John J., and Prewo, Karl M., "Silicon Carbide Fibre Reinforced Class-Ceramic Matrix Composites Exhibiting High Strength and Toughness," J. Matl. Sci., Vol 17, pp 2371-2383, 1982.

13. Schioler, Liselotte J., "Recent Progress and Research Needs in Ceramic Matrix Composites," Proceedings 3rd International New Materials Conference and Exhibition, Osaka, Japan, 1-5 October 1990.

14. Evans, A. G., and Marshall, D. B., "The Mechanical Behavior of Ceramic Matrix Composites," Acta Metall., Vol 37, No 10, pp 2567-2583, 1989.

15. Fitzer, Erich, "The Future of Carbon-Carbon Composites," Carbon, Vol 25, No 2, pp 163-190, 1987.

16. Strife, James R., and Sheehan, James E., "Ceramic Coatings for Carbon-Carbon Composites," Ceramic Bulletin, Vol 67, No 2, pp 369-374, 1990.

17. Sullivan, B. J., and Rosen, B. W., "Microstructural Evaluation of Carbon-Carbon Constituents, Bundle, and Multidirectional Composite Properties," Proceedings 14th Annual Conference on Composite Materials and Structures, Cocoa Beach, FL, Jan 1990.

18. Hedberg, Frederick L., and Haritos, George K., "Biomimetics--Advancing Manmade Materials Through Guidance from Nature," AFOSR FY92 Research Initiative Presentation, Feb 1990.

19. Currey, J. D., "Mechanical Properties of Mollusk Shell, Symp. Soc. exp. Biol., No 34, pp 75-97, 1980.

Constitutive Models for Piezoelectric Materials

Shiv P. Joshi
Aerospace Engineering Department
University of Texas at Arlington, Arlington, TX 76019

ABSTRACT

Piezoelectric material produces electric charges when mechanically deformed and an electric potential causes a mechanical deformation. This property makes it suitable for sensor and transducer applications. The understanding of the electroelastic constitutive behavior is critical to predicting the response of a structure with embedded piezoelectric material. A concise formulation of linear constitutive relations is presented in this paper. The nonlinear constituive behavior is also discussed briefly.

INTRODUCTION

The Curies first showed the presence of piezoelectricity in crystals in 1880. The first practical use of the piezoelectric effect was during World War I when Langevin's sonar emitter was effectively used to detect German submarines. Prior to World War II, researchers at MIT discovered that certain ceramics such as PZT (Lead-Zirconate-Titanate) could be polarized to yield a high piezo response [1]. Piezoceramics consists of a large number of small crytallites sintered together and polarized by an external electric field. Kawai [2] discovered that the polarized homopolymer of vinylidene fluoride (PVDF) developed far greater piezo activity than any other synthetic or natural polymer. Poled PVDF still dominates all other materials in terms of its intensity of piezo activity [1].

Although the behavior of piezoelectric materials in non-structural applications has been investigated extensively, the treatment is often simplistic and adhoc. The recent interest in "Smart Structures" has put espacial emphasis on the regorous understanding of elctroelastic behavior of piezoceramics as an integral part of a structure.

The nonlinear theory of dielectrics has been studied by Toupin [3], Nelson [4] and Tiersten[5]. The relation between the equations of linear piezoelectricity and the more general electroelastic equations is discussed by Tiersten [5]. Nelson presented a completely deductive derivation of the dynamical equations and constitutive relations for elastic, electric, and electroelastic phenomena based on the fully electrodynamic Lagrangian theory of elastic dielectrics. Penfield and Hans [6] developed a linear piezoelectric theory which does not account for gradient of polarization and electrostatic interference. Mindlin [7] derived a system of two dimensional equations for high frequency motions of crystal plates accounting for coupling of mechanical, electrical and thermal fields. Readers interested in this area may refer to books by Nye [8], Berlincourt et.al. [9], and Landau and Lifshitz [10]. A phenomenological description of the dynamic response of piezoceramics to an external electric field, including domain reorientation processes and the dynamics of dipole moment in each domain, has been developed by Chen et.al. [11,12,13,14].

A concise formulation of linear constitutive equations for piezoelectric materials is preseted here. The linear relations are espacialized for piezoceramic materials available for structural applications. Nonlinear effects are briefly discussed.

PIEZOELECTRIC FIELD EQUATIONS

The physics involved in the piezoelectric theory may be regarded as a coupling between Maxwell's equations of electromagnetism and elastic stress equations of motion. The coupling takes place through the piezoelectric constitutive equations.

Maxwell's equations in vector form are written as,

$$\nabla \times \boldsymbol{E} = -\mu_o \frac{\partial \boldsymbol{H}}{\partial t} \qquad \nabla \times \boldsymbol{H} = \frac{\partial \hat{\boldsymbol{D}}}{\partial t} \tag{EQ 1}$$

where $\boldsymbol{E}$ is electric field intensity, $\boldsymbol{H}$ is magnetic field vector and $\hat{\boldsymbol{D}}$ is electric flux density also known as displacement vector. The free-space permeability, μ_o, is used because piezoelectric materials are nonmagnetic. In the quasi-electrostatic approximation [15], which is usually adequate for the study of piezoelectric phenomena, time-derivative terms in the electromagnetic equations may be dropped. Then the electric field may be expressed as,

$$\boldsymbol{E} = -\nabla \phi \tag{EQ 2}$$

and the only electromagnetic equation which need to be considered is

$$\nabla \cdot \hat{\boldsymbol{D}} = 0 \tag{EQ 3}$$

The elastic stress equation of motion is,

$$\nabla \cdot \sigma = \rho \ddot{u} \tag{EQ 4}$$

where σ is stress tensor, ρ is mass density and $\ddot{u}$ is acceleration vector. Coupling among eqn. 2-4 is introduced by piezoelectric constitutive equations.

CONSTITUTIVE EQUATIONS

We will adopt index notations in the remainder of the paper for convinience. We will employ the thermodynamic Gibbs potential to derive constitutive equations and will consider the σ_{ij} (stress components), E_k (electric field components), and T (absolute temperature) as independent variables.

$$G = U - \sigma_{ij}\varepsilon_{ij} - E_k\hat{D}_k - TS \tag{EQ 5}$$

where G is the Gibbs potential, S is the entropy, and U is the internal energy. For adiabetically insulated reversible system, the total differential of internal energy is

$$dU = \sigma_{ij}d\varepsilon_{ij} + E_k d\hat{D}_k + TdS \tag{EQ 6}$$

and therefore the total differential of Gibbs potential is

$$dG = -\varepsilon_{ij}d\sigma_{ij} - \hat{D}_k dE_k - SdT \tag{EQ 7}$$

Expressing the Gibbs potential in Taylor series and neglecting higher order terms, we obtain,

$$dG = \left(\frac{\partial G}{\partial \sigma_{ij}}\right)_{E,T} d\sigma_{ij} + \left(\frac{\partial G}{\partial E_k}\right)_{\sigma,T} dE_k + \left(\frac{\partial G}{\partial T}\right)_{\sigma,E} dT \tag{EQ 8}$$

From eqs. 7 and 8

$$\varepsilon_{ij} = -\left(\frac{\partial G}{\partial \sigma_{ij}}\right)_{E,T} \qquad \hat{D}_k = -\left(\frac{\partial G}{\partial E_k}\right)_{\sigma,T} \qquad S = -\left(\frac{\partial G}{\partial T}\right)_{\sigma,E} \qquad \text{(EQ 9)}$$

The total differentials of dependent variables ε_{ij}, $\hat{D}_k$, and S is given as a function of independent variables as

$$\begin{aligned} d\varepsilon_{ij} &= \left(\frac{\partial \varepsilon_{ij}}{\partial \sigma_{lm}}\right)_{E,T} d\sigma_{lm} + \left(\frac{\partial \varepsilon_{ij}}{\partial E_n}\right)_{\sigma,T} dE_n + \left(\frac{\partial \varepsilon_{ij}}{\partial T}\right)_{\sigma,E} dT \\ d\hat{D}_k &= \left(\frac{\partial \hat{D}_k}{\partial \sigma_{lm}}\right)_{E,T} d\sigma_{lm} + \left(\frac{\partial \hat{D}_k}{\partial E_n}\right)_{\sigma,T} dE_n + \left(\frac{\partial \hat{D}_k}{\partial T}\right)_{\sigma,E} dT \\ dS &= \left(\frac{\partial S}{\partial \sigma_{lm}}\right)_{E,T} d\sigma_{lm} + \left(\frac{\partial S}{\partial E_n}\right)_{\sigma,T} dE_n + \left(\frac{\partial S}{\partial T}\right)_{\sigma,E} dT \end{aligned} \qquad \text{(EQ 10)}$$

where

$$s_{ijlm}^{E,T} = \left(\frac{\partial \varepsilon_{ij}}{\partial \sigma_{lm}}\right)_{E,T}, \qquad d_{ijn}^{T} = \left(\frac{\partial \varepsilon_{ij}}{\partial E_n}\right)_{\sigma,T} = \left(\frac{\partial \hat{D}_n}{\partial \sigma_{ij}}\right)_{E,T}, \qquad \alpha_{ij}^{E} = \left(\frac{\partial \varepsilon_{ij}}{\partial T}\right)_{\sigma,E} = \left(\frac{\partial S}{\partial \sigma_{ij}}\right)_{E,T},$$

$$\varepsilon_{kn}^{\sigma,T} = \left(\frac{\partial \hat{D}_k}{\partial E_n}\right)_{\sigma,T}, \qquad p_k^{\sigma} = \left(\frac{\partial \hat{D}_k}{\partial T}\right)_{\sigma,E} = \left(\frac{\partial S}{\partial E_k}\right)_{\sigma,T}, \qquad \frac{\rho c^{\sigma,E}}{T_0} = \left(\frac{\partial S}{\partial T}\right)_{\sigma,E}$$

are elastic compliance coefficients, piezoelectric strain constants, coefficients of thermal expansion, dielectric permittivities, pyroelectric coefficients, respectivly; and $c^{\sigma,E}$ is the specific heat and ρ is the mass density. Integrating eq. 10, we obtain,

$$\begin{aligned} \varepsilon_{ij} &= s_{ijlm}^{E,T}\sigma_{lm} + d_{ijn}^{T}E_n + \alpha_{ij}^{E}\Delta T \\ \hat{D}_k &= d_{klm}^{T}\sigma_{lm} + \varepsilon_{kn}^{\sigma,T}E_n + p_k^{\sigma}\Delta T \\ \Delta S &= \alpha_{lm}^{E}\sigma_{lm} + p_n^{\sigma}E_n + \frac{c^{\sigma,E}}{T_0}\Delta T \end{aligned} \qquad \text{(EQ 11)}$$

Piezoceramics are widely used, therefore we will especialize eq 11 for them. The constitutive equations of the polarized piezoceramics are equivalent to the equations for a pizocrystal of the hexagonal 6mm symmetry class. In abbreviated subscript notation these equations may be written as

$$\begin{aligned} \varepsilon_{11} &= s_{11}^{E,T}\sigma_{11} + s_{12}^{E,T}\sigma_{22} + s_{13}^{E,T}\sigma_{33} + d_{31}^{T}E_3 + \alpha_1^{E}\Delta T \\ \varepsilon_{22} &= s_{12}^{E,T}\sigma_{11} + s_{11}^{E,T}\sigma_{22} + s_{13}^{E,T}\sigma_{33} + d_{31}^{T}E_3 + \alpha_1^{E}\Delta T \\ \varepsilon_{33} &= s_{13}^{E,T}\sigma_{11} + s_{13}^{E,T}\sigma_{22} + s_{33}^{E,T}\sigma_{33} + d_{33}^{T}E_3 + \alpha_3^{E}\Delta T \end{aligned}$$

$$\varepsilon_{23} = s_{44}^{E,T}\sigma_{23} + d_{15}^{T}E_2, \qquad \varepsilon_{13} = s_{44}^{E,T}\sigma_{13} + d_{15}^{T}E_1, \qquad \varepsilon_{12} = \left(\frac{s_{11}^{E,T} - s_{12}^{E,T}}{2}\right)\sigma_{12} \qquad \text{(EQ 12)}$$

$$\hat{D}_1 = d_{15}^{T}\sigma_{13} + \varepsilon_{11}^{\sigma,T}E_1 + p_1^{\sigma}\Delta T, \qquad \hat{D}_2 = d_{15}^{T}\sigma_{23} + \varepsilon_{11}^{\sigma,T}E_2 + p_1^{\sigma}\Delta T$$

$$\hat{D}_3 = d_{31}^{T}(\sigma_{11} + \sigma_{22}) + d_{33}^{T}\sigma_{33} + \varepsilon_{33}^{\sigma,T}E_3 + p_3^{\sigma}\Delta T$$

$$\Delta S = \alpha_1^{E}(\sigma_{11} + \sigma_{22}) + \alpha_3^{E}\sigma_{33} + p_1^{\sigma}(E_1 + E_2) + p_3^{\sigma}E_3 + \frac{c^{\sigma,T}}{T_0}\Delta T$$

The strain, electric displacement and entropy are assumed to depend linearly on the stress, electric field and temperature (eq. 10) in deriving eq. 11. Some higher order effects can be brought about by including second order terms in eq. 10. Among other nonlinear coefficients, elctrostriction coefficients and electrooptical cofficients are qudratic effects which can be viewed as corrections to piezoelectric coefficients and dielectric permittivities, respectively.

ACKNOWLEDGEMENT

This work is a part of the preliminary studies on damage survivability and damage tolerance of "smart" laminated composites sponsered by the Army Research Office.

REFERENCES

1. Manual, "Kynar Piezo Film", Pennwalt Corporation.
2. Kawai, H., "The Piezoelectricity of Polyrinydene Fluoride", Japan Journal of applied physics, No. 8, 1979, pp. 975-976.
3. Toupin, R.A., "A Dynamical Theory of Elastic Dielectrics", Int. J. Eng. Sci., Vol. 1, 1983, pp 101-126.
4. Nelson, D.F., "Theory of Nonlinear Electroacoustics of Dielectric, Piezoelectric Crystals", J. Acoust. Soc. Am., Vol. 63, June 1978, pp. 1738-1748.
5. Tiersten, H.F., "Electroelastic Interactions and the Piezoelectric Equations", J. Acoust. Soc. Am., Vol. 70, December 1981, pp. 1567-1576.
6. Penfield, P. Jr and Hans, H.A., "Electrodynamics of moving media", Research Monograph, No. 40, Massachusetts Institute of Technology Press, Cambridge, Massachusetts, 1967.
7. Mindlin, R.D., "Equations of High Frequency Vibration of Thermopiezoelectric Crystal Plates", Int. J. Solids Struct., Vol. 10, No. 6, 1974, pp. 625-637.
8. Nye, J.F., "Physical Properties of Crystals", Oxford, Clarendon Press, 1964.
9. Berlincourt, D.A., Curren, D.R., Jaffe, H., "Piezoelectric and Piezomagnetic Materials and Their Function as Transducers, In: Physical Acoustics, Mason W.P. (Ed.), Vol. 1-Part A, Academic Press, New York, 1964.
10. Landau, L.D., Lifshitz, E.M., "Electrodynamics of Continuous Media", Pergamon Press, Oxford-London-New York-Paris, 1960.
11. Chen, P.J., "Characterization of the Three-Dimensional Properties of Poled PZT-65/35 in the Absence of Losses", Acts Mech., Vol. 47, 1983, pp. 95-106.
12. Chen, P.J., "Three-Dimensional Dynamic Electromechanical Constitutive Relations of Ferroelectric Materials", Int. J. Solids Struct., Vol. 16, No. 12, 1980, pp 1059-1067.
13. Chen, P.J., Peercy, P.S., "One-Dimensional Dynamic Electromechanical Constitutive Relations for Ferroelectric Materials, Acta Mech., Vol. 31, No. 3, 1979, pp. 231-241.
14. Chen, P.J., Tucker, T.J., "Determination of the Polar Equilibrium Properties of the Feroelectric Ceramic PZT-65/35", Acta Mech., Vol. 38, No. 3-4, 1981, pp. 209-218.
15. Auld, B.A., "Wave Propagation and Resonance in piezoelectric materials", J. Acoust. Soc. Am., Vol. 70, No. 6, December 1981, pp. 1577-1585.

MICROPHYSICAL MODEL TO DESCRIBE IMPACT BEHAVIOR OF CERAMICS

A.M. RAJENDRAN, M.A. DIETENBERGER, and D.J. GROVE

University of Dayton, Research Institute, Dayton, Ohio 45469

ABSTRACT

We recently reported [1] the development of an internal state variable based dynamic constitutive model. This model describes microcrack nucleation and growth processes in conjunction with viscoplastic pore collapse. Degradation of the elastic properties, under both tension and compression, due to damage growth is modeled using micromechanics based theories. Pore collapse is described by a pressure dependent yield function. This three-dimensional, continuum damage model has been incorporated into a finite element code and simulated damage in silicon nitride target plates.

INTRODUCTION

The high velocity impact behavior of ceramics is dominated by stiffness loss and inelastic deformations under impact generated compressive loading. Inelastic deformations in ceramics are assumed to be due to microcracking and plastic flow. The proposed model is based on microcrack nucleation and growth, and pore collapse mechanisms. Damage is defined in terms of average crack density and is treated as an internal state variable. Nucleation of damage is described as a statistical process, while stiffness reduction due to microcracking is modeled using analytical damaged moduli. The evolution law for the damage variable is formulated from a generalized Griffith criterion. Damage evolution under both tensile and compressive loading was modeled. The pore collapse mechanism is described using a viscoplastic formulation of a pressure dependent yield function.

IMPACT DAMAGE MODEL FOR CERAMICS

We followed the approach of Margolin et al. [2] to derive the stress-strain relationship for a damaged brittle material. The applied strain is decomposed into three parts: the average elastic strain, the inelastic strain due to crack growth, and the plastic strain. We used the pressure dependent plastic flow equations to describe the plastic strain as well as the pore collapse. The stress-strain relationship for a cracked aggregate is calculated as,

$$S_{ij} = M_{ijkl}(\gamma)\ [\ e^{e}_{kl} + (\epsilon^{e}_{v}/3)\ \delta_{kl}] \qquad (1)$$

γ is a scalar describing the crack density of randomly orientated cracks. M is the damaged moduli which generally depends on crack orientation and the stress state. The evolution law for the internal state variable γ is derived from fracture mechanics based equations. The damage growth rate is related to crack growth rate and is given by,

$$\dot{a} = \begin{cases} 0 & \text{for } G_I < G_c \\ nC_R[1 - G_c/G_I)] & \text{for } G_I > G_c \end{cases} \quad (2)$$

where G_I is the applied energy release, G_c is the critical G, and a is the current crack size. The constitutive model has been incorporated into a finite element code, EPIC-2.

IMPACT EXPERIMENTS

One of the most often used laboratory experiment to model impact response of materials is the plate impact test shown schematically in Figure 1. In this configuration, a flyer plate impacts a stationary target plate at high velocity (>50 m/sec). The diagnostic measurement involves a stress gauge embedded between the target and a back plate. This gauge records the stress history with respect to time. The deformation and damage processes that occur inside the target are usually interpreted from the measurements and general features of the gauge signal.

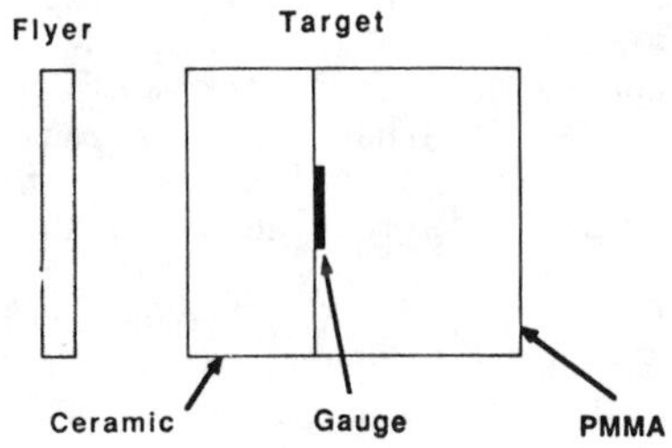

FIG. 1. Schematic of Plate Impact Experiment

We performed several plate impact tests on silicon nitride targets and determined the Hugoniot Elastic Limit (HEL) as 12 GPa. For modeling, we considered two different tests at impact velocities 175 m/s and 800 m/s. The stress amplitude (3.0 GPa) in the low velocity test is therefore within the elastic range. Damage under compression is not expected to occur at this stress level. The target fails due to the tensile waves generated due to wave reflections from the stress free back surface of the flyer and the target plates. However, above the HEL stress level, complex inelastic deformations due to microplasticity and microcracking under compressive loading usually occur in the target material. In fact the failure process is almost complete prior to the arrival of any tensile waves.

We modeled the 'below HEL' and 'above HEL' experiments using the EPIC-2 code [3].

MODEL SIMULATIONS

Figure 2a compares the model generated stress history with the low velocity experiment. The flat top (between A and B) in this figure indicates that there is no microfracturing under compression and the deformation is elastic. However, the target spalled due to the tensile waves as indicated by the signal between C and D. In the absence of spall, the stress history will follow the dotted line. In Figure 2b, the model simulated stress history inside the target is shown. The shock wave arrives at A and the material remains in compression until the unloading wave arrives at C. Tensile loading begins at D, followed by stress relaxation due to tensile damage between E and F. The dashed line in Figure 2b shows the damage evolution inside the target. Complete failure is assumed when damage reaches a value of one.

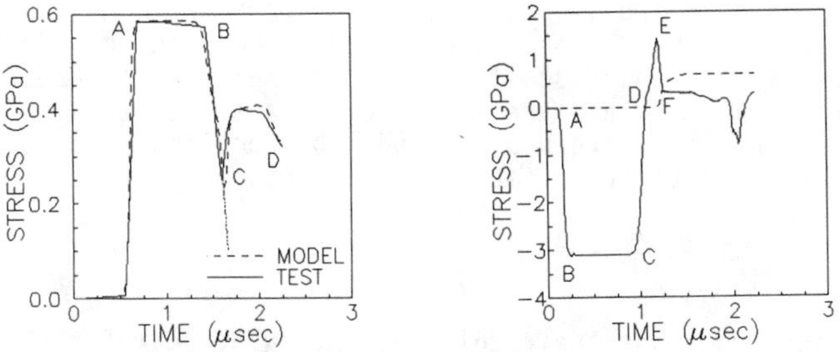

FIG. 2. a) Comparison between the model and gauge signal
b) Model generated stress history inside the target

In ceramics, it is easier to nucleate and propagate microcracks under tension than under compression. Therefore, in tension, n in equation (2) is assumed to be equal to one. However, in compression, n is usually small (<< 1) and is calibrated by matching the gauge signal.

In Figure 3a, the compressive damage can be seen between points A and B. Note that the portion between A and B was almost flat in the low velocity experiment (see Figure 2a), indicating no damage. The model generated compressive damage growth is also shown by the dashed line in Figure 3b. Damage initiation and growth leads to stress relaxation between B and C. At C, a release wave from the flyer unloads the stress. In general, the ability of the model in reproducing the experimentally measured stress histories was extremely good.

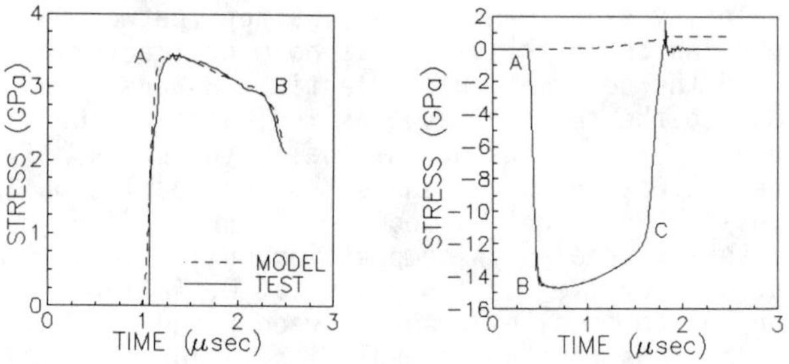

FIG. 3. a) Comparison between the model and gauge signal b) Model generated stress history inside the target

SUMMARY

We successfully modeled impact damage evolution in silicon nitride under both tensile and compressive loading conditions. The strength and stiffness degradations were modeled through an internal state variable based scalar damage parameter γ representing the crack density in the material. We demonstrated the capabilities of the microphysical model by modeling both 'below HEL' and 'above HEL' plate impact experiments.

ACKNOWLEDGMENTS

The authors thank their colleague Dr. N.S. Brar for the plate impact experimental data.

REFERENCES

1. M.A. Dietenberger, A.M. Rajendran, and D.J. Grove, in 'Shock Waves in Condensed Matters', ed. S.C. Schmidt, J.N. Johnson, and L.W. Davision (North-Holland Press, 1990).
2. L.G. Margolin, B.W. Smith, and G. P. DeVault, Shock Waves in Condensed Matter, ed. Y.M. Gupta, (Plenum Press, NY, 1986).
3. G.R. Johnson and R.A. Stryk, Air Force Report No. AFATL-TR-86-51, 1986

MODELING SECOND ORDER EFFECTS IN PIEZOELECTRIC MATERIALS

M. SALLAH
Center for Manufacturing Research

G. R. BUCHANAN
Department of Civil Engineering
Tennessee Technological University, Cookeville, TN 38505

ABSTRACT

The nonlinear equations ofelectroacoustics are solved using a perturbation method coupled to the finite element method to obtain numerical results. The physical application is for a degenerate waveguide convolver since this microelectronic device has been discussed in the literature. Linear solutions for surface waves on a piezoelectric substrate are obtained using a finite element analysis. The higher order correction is based upon the wave speed for the material and the form of the constitutive equations for that govern second order effects.

INTRODUCTION

The equations of nonlinear electroacoustics have been discussed in numerous technical documents and the text by Maguin [1] can be used as a reference for studying the development of the nonlinear relation between mechanical displacements and electrical potential. In this paper a perturbation method of analysis used by Ganguly and Davis [2] and later by Planet *et al*. [3] is combined with a finite element analysis, as suggested in [3], and used to obtain a correction to the linear analysis for displacement and potential.

The constitutive relations are nonlinear and characterized by higher order elastic constants, piezoelectric constants, dielectric and electro-strictive constants. In this research the emphasis is placed on formulating and using the finite element method to compute the nonlinear correction to the classical linear piezoelectric problem.

GOVERNING EQUATIONS

The equations of linear piezoelectricity are

$$C_{ijkl}\, S_{kl,i} - e_{kij}\, E_{k,i} = \rho \partial^2 u_j / \partial t^2, \tag{1}$$

$$e_{ikl}\, S_{kl,i} + \epsilon_{ik}\, E_{k,i} = 0. \tag{2}$$

A three dimensional solution can be obtained for a two dimensional

(x_1, x_2 space) model by assuming

$$u_1 = U_1 sin(k_3x_3) exp(-i\omega t), \qquad u_2 = U_2 sin(k_3x_3) exp(i\omega t),$$

$$u_3 = U_3 cos(k_3x_3) exp(-i\omega t), \qquad \phi = \Phi cos(k_3x_3) exp(i\omega t). \tag{3}$$

where k_3 is an assumed wave number and ω is the circular frequency. Equations (1-3) are solved using the finite element method to obtain the solution for surface waves on a substrate of $LiNbO_3$. Lithium niobate is a microelectronic material characterized by higher order effects. The nonlinear correction to the linear solution is obtained by assuming

$$u_i = u_i^o + u_i^1, \qquad \phi = \phi^o + \phi^1. \tag{4}$$

where u_i^1 and ϕ^1 are the nonlinear corrections. The correction terms are responses to the driving stresses τ_{ij}^{NL} and electrical displacements D_i^{NL} that are functions of gradients of the linear displacements and potential. The nonlinear corrections are computed as [1,2,4]

$$\tau_{ij}^{NL} = H_{ijklmn}\, u_{k,l}^o\, u_{m,n}^o + F_{lijmn}\, u_{m,n}^o\, \phi_{,l}^o - \tfrac{1}{2} l_{lmij}\, \phi_{,l}^o\, \phi_{,m}^o \tag{5}$$

$$D_i^{NL} = \tfrac{1}{2} F_{iklmn}\, u_{k,l}^o\, u_{m,n}^o - l_{ilmn}\, u_{m,n}^o\, \phi_{,l}^o + \tfrac{1}{2} b_{ilm}\, \phi_{,l}^o\, \phi_{,m}^o \tag{6}$$

where

$$H_{ijklmn} = C_{inkl}\delta_{jm} + \tfrac{1}{2} C_{ijnl}\delta_{km} + \tfrac{1}{2} C_{ijklmn}$$

$$F_{lijmn} = e_{lnj}\delta_{im} + e_{lijmn}$$

and C_{ijkl}, e_{lnj} and δ_{ij} are the linear stiffness constants, piezoelectric constants and Kronecker delta, respectively. The higher order constants are the elastic stiffness C_{ijklmn}, piezoelectric constants e_{lijmn}, dielectric constants b_{ilm} and electrostrictive constants l_{lmij}.

ANALYSIS

The governing equations are solved for a model of a degenerate waveguide convolver. A substrate of $LiNbO_3$ has two transducers engraved on its surface and surface acoustic waves with frequency ω are generated by the transducers and meet at the center, between the transducers. The waves have the same frequency for a degenerate convolver and the interaction of the two waves gives a nonpropagating frequency of 2ω with wave number k. The linear problem is solved a second time with a forcing function $e^{-2i\omega t}$*exp*. The forcing function at each node is computed using (5) and (6) as an elasticity driving force F_t^{NL} and an electric charge driving force F_e^{NL} computed as follows

$$F_t^{NL} = \partial\tau_{ij}^{NL}/\partial x_j, \qquad F_e^{NL} = \partial D_i^{NL}/\partial x_i. \tag{7}$$

The linear solution is obtained using assumptions similar to Eqs. (3) and the gradients appearing in Eqs. (5) and (6) were computed using finite difference approximations.

A nine node isoparametric finite element was used to model a two-dimensional space. A ten element model was sufficient to obtain results. The model was fixed at the base to represent a section of the substrate. The surface wave velocity was computed for a semi-infinite half-space of $LiNbO_3$ and was found to be 3487.6 m/s with wave number k_3 = 0.31 for a shorted surface. Similarly, for a nonshorted surface the wave velocity was 3397.0 m/s with k_3 equal 0.45. The initial displacements and electric potentials for this problem are plotted in Fig. 1.

The nonlinear corrections were computed using the theory outlined above and for a shorted surface are shown in Fig. 2. A detailed discussion of the finite element formulation and subsequent computations is given in [5]. Similar computations were carried out for a nonshorted surface and thoes results are shown in Fig.3.

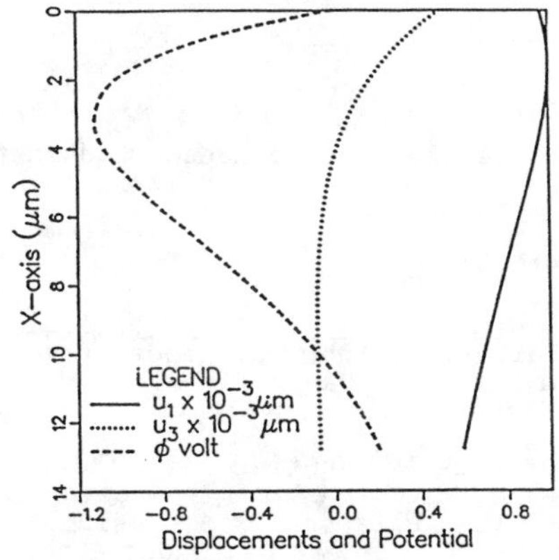

Figure 1. Displacements and potential for $LiNbO_3$

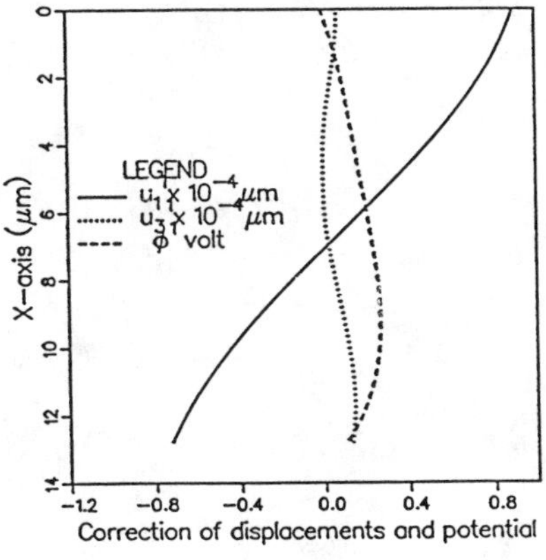

Figure 2. Corrections for a shorted surface

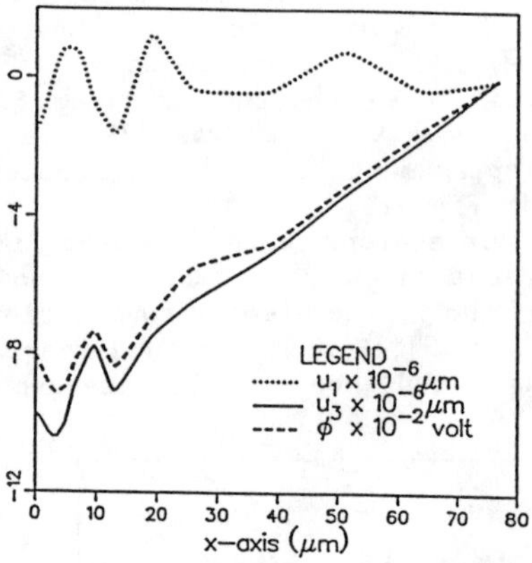

Figure 3. Corrections for a nonshorted surface

REFERENCES

1. G. A. Maugin, Nonlinear Electromechanical Effects and Applications, World Scientific Pub., Singapore (1985)
2. A. K. Ganguly and K. L. Davis, "Nonlinear interaction in degenerate surface acoustic wave convolvers", J. Appl. Phys., 51, 920-926 (1980).
3. M. Planet, G. Vanderborck, M. Gautier and C. Maerfeld, Linear and nonlinear of a piezoelectric waveguide convolver using a finite element method", Proc. IEEE Ultrasonics Symposium, 166-171 (1982).
4. M. Planet, Propagation Non Linear des Ondes Acoustiques dans les Solides,These, Universite de Franche-Comte, Besancon, France (1984)
5. M. Sallah, Finite Element Analysis of Microelectronic Materials with Application to Field Effect Transistors and Waveguide Convolvers, PhD Dissertation, Tennessee Technological University, Cookeville, TN 38505 (1989).

Transverse Constitutive Response of Titanium-Aluminide, Metal-Matrix Composites

M. Sohi and J. Adams
Allied-Signal Aerospace Company
Garrett Engine Division
Phoenix, Arizona

Rabin Mahapatra
Naval Air Development Center
Warminster, Pennsylvania

ABSTRACT

Titanium-aluminide, metal-matrix composites (MMCs) offer excellent room and elevated temperature specific strength and stiffness in the fiber direction. However, in the transverse direction the current level of strength is lower than that offered by the monolithic matrix material. This review describes an analytical study of the transverse constitutive response of SCS-6/Ti-25Al-10Nb-3V-1Mo (atomic percent) MMC using the finite-element (FE) method. Since the residual stresses induced during consolidation have a strong influence on mechanical response of MMCs, they were included in the analysis. Calculation of the residual stresses was performed by nonlinear thermoelastic treatment of the cool-down process from the consolidation temperature. The effect of matrix relaxation on the reduction of residual stresses was evaluated. The analysis of the transverse constitutive response included matrix plasticity and fiber-matrix debonding. Debonding was the more significant contributor to the composite nonlinear behavior.

INTRODUCTION

Titanium-aluminide MMCs have been identified as candidate materials for high-temperature applications of up to 1400F. Intensive research on the processing and mechanical behavior of these materials initiated in the early 1980s. A concise (open literature) summary of the experimental data has been generated on these materials.[1]

The existing data indicates that titanium-aluminide MMCs provide excellent strength and stiffness properties in the longitudinal direction. However, in the transverse direction the strength is lower than the monolithic matrix material. This restricts the benefits derived from their applications in biaxial stress fields and necessitates understanding of transverse constitutive response.

The analytical efforts to date have mostly concentrated on modeling of MMC longitudinal behavior and calculation of residual stresses generated during cool-down from processing temperatures of about 1700 to 1800F.[2,3] These analytical efforts along with the recent studies on MMC transverse behavior are intended to result in processing guidelines and constituent materials requirements to eliminate composite cracking during consolidation and to define and improve design limits.

The transverse constitutive response of MMCs is affected by several factors; such as the fiber and matrix properties, the cool-down residual stresses, and the fiber-matrix interfacial strength. The current analysis includes the effects of these factors. A unit cell of the MMC was modeled using the finite-element (FE) method. Nonlinear thermoelastic analyses were performed to estimate the cool-down residual stresses induced in the constituents during cool-down from the hot-isostatic-pressing (HIPing) temperature.

These residual stresses play a significant role on mechanical behavior of titanium-aluminide MMCs. Due to low room temperature ductility of the matrix material, the residual stresses can induce cracks in the composite prior to any mechanical loading. Furthermore, these residual stresses can hasten the onset of plasticity in the matrix material upon application of the mechanical loading. Additional analyses were also performed to determine the effect of hold times and matrix relaxation on the reduction of residual stresses.

Matrix plasticity is one of the sources of nonlinearity in the composite transverse stress-strain response. Another source is fiber-matrix debonding. Interface elements were used to model the fiber-matrix interface for predicting the onset of fiber-matrix debonding and the post-debonding transverse constitutive response. The onset of nonlinearity was related to the fiber-matrix chemical bond strength and the residual radial compressive stresses at the interface. The composite transverse modulus was related to the tangential shear friction between the fiber and matrix.

The following paragraphs describe the fiber and matrix material properties used in the analysis, the finite-element model with the associated boundary conditions, and the analysis results.

Constituent Material Properties

The properties of the SCS-6 fiber and the Ti-25Al-10Nb-3V-1Mo matrix were used in this study. SCS-6 is a 5.6 mil fiber fabricated by chemical vapor deposition of silicon carbide on a carbon monofilament. This fiber contains a carbon rich layer on the surface to reduce its reactivity with titanium alloys. Figures 1 through 3 show the fiber and matrix properties used in the analysis. The alloy used in the measurement of these properties was subjected to a heat treatment that nearly simulated that experienced by the matrix material during MMC consolidation.

The temperature-dependent thermoelastic properties of Figure 1 were used in the calculation of the residual stresses. The creep curves of Figure 3 were used to develop primary and secondary hyperbolic sine creep models that were input to the FE code and used to estimate the reduction of residual stresses due to hold times during cool-down from the maximum HIP consolidation temperature. The room-temperature, elasto-plastic curve of Figure 2 was used to determine the room-temperature composite transverse constitutive response under mechanical loading.

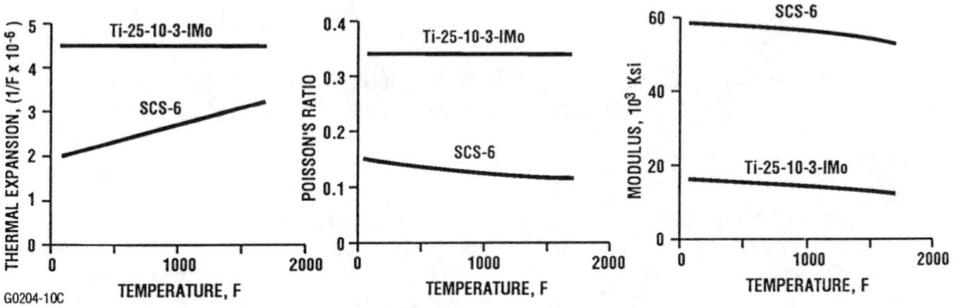

Figure 1. SCS-6 Fiber and Ti-25-10-3-1Mo Matrix Properties Used in the Analyses.

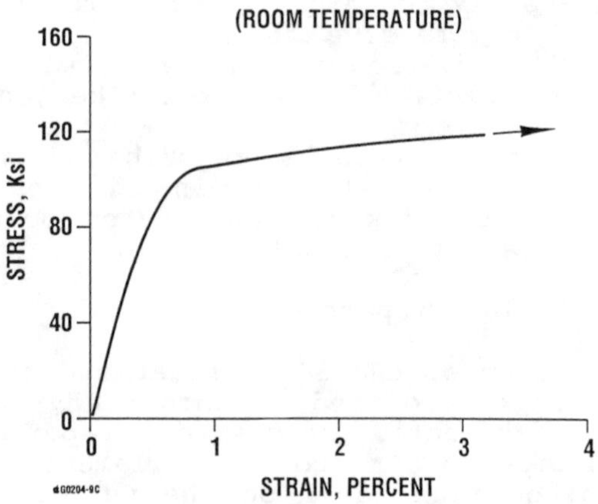

Figure 2. Ti-25-10-3-1Mo Properties Used in the Analyses.

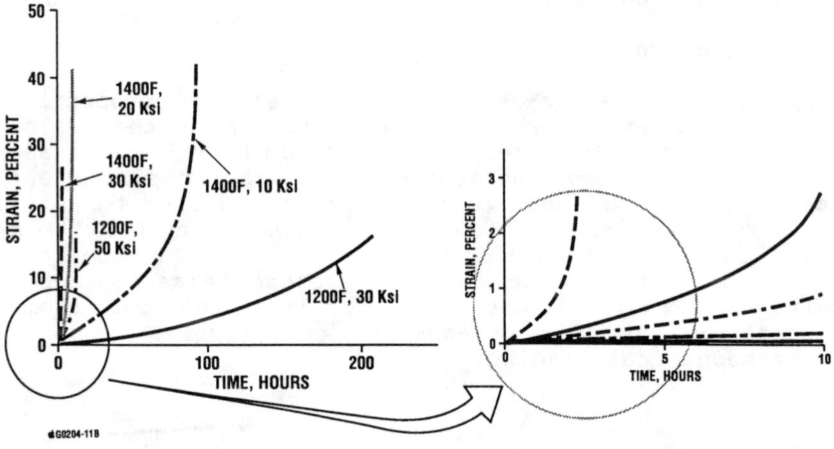

Figure 3. Ti-25-10-3-1Mo Primary and Secondary Creep Properties Used in the Analyses.

Finite-Element Modeling

Figure 4 depicts the approach used in the finite-element idealization of the MMC unit cell. A square packing configuration and a fiber volume fraction of 35 percent were assumed in the analyses. Due to symmetry considerations, only a quarter of a unit cell was required for modeling. Figure 4 also indicates the displacement boundary conditions imposed on the quarter unit cell model. The fiber and matrix areas were meshed using 3-D, isoparametric eight-noded solid elements

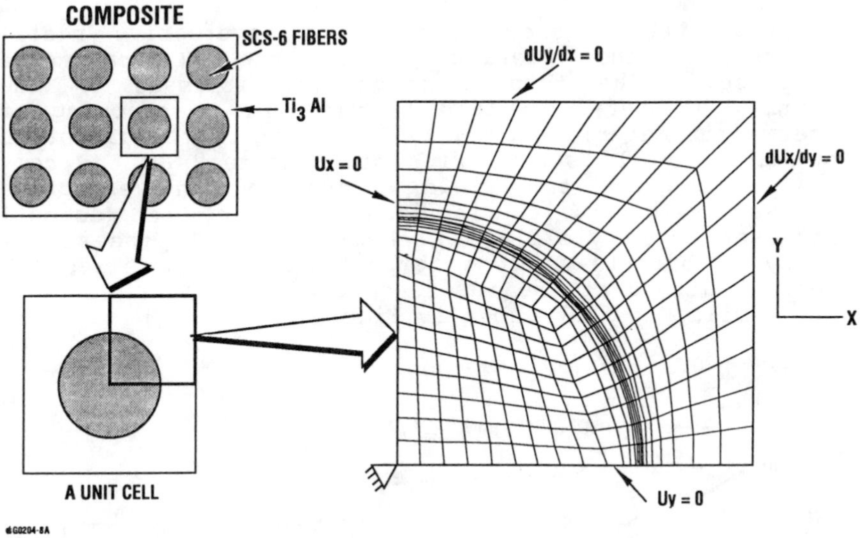

Figure 4. Transverse Constitutive Response Modeled.

plane strain conditions. The fiber and matrix were joined by interface elements that could support compressive load and any user-prescribed tensile load. The interface elements also had a stick-slip feature with an associated user-defined coefficient of friction. The analysis was performed using the ANSYS* general purpose finite-element code.

RESULTS AND DISCUSSIONS

Residual Thermal Stresses

Titanium alloy MMCs are usually consolidated at HIP temperatures of about 1700 to 1800F. Upon cool-down from this temperature, residual stresses develop in the fiber and matrix material due to the mismatch in their coefficient of thermal expansion (CTE). Figure 5 shows the maximum matrix residual stresses calculated for the SCS-6/Ti-25-10-3-1Mo system during cool-down from 1700F. Also shown are the locations of the maximum stresses in the unit cell. These locations are at the fiber-matrix interface. Since the CTE of the matrix material is higher than that of the fiber, the tangential stress is tensile and the radial stress is compressive.

---

*ANSYS is a registered trademark of the Swanson Analysis Systems Inc.

As will be shown later, the compressive radial stresses at the interface improve the MMC transverse response. The tensile tangential stresses in the matrix, however, can lead to yielding or cracking of the matrix material. Figure 5 also shows the variation of matrix yield stress with temperature.[4] As the comparison between the yield stress and the effective stress indicates, it is feasible that the residual stresses cause yielding of the matrix. If the matrix alloy is brittle at room temperature, the tensile tangential stress at the interface can cause matrix cracking similar to that shown in Figure 6.

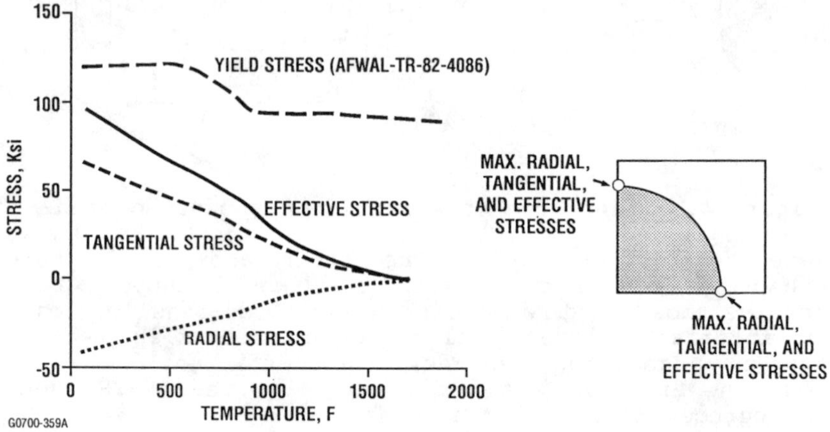

Figure 5. Residual Stresses Result During Cool-Down.

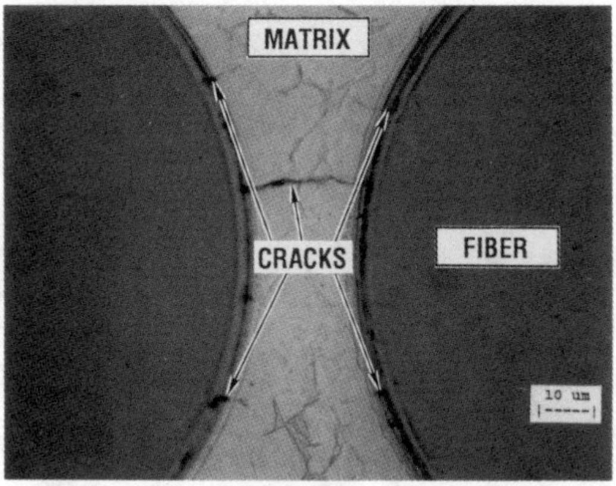

Figure 6. Excessive Residual Thermal Stresses Can Cause Cracking of Matrix and/or Interface.

The residual stresses may be reduced by holding the MMC at an appropriate intermediate temperature, causing matrix relaxation. Figure 7 compares the effects of 10-hour holds at 1100 and 1300F. For the matrix alloy considered in this analysis, only the 1300F hold was effective in causing a reduction of the maximum residual effective stress (over 20 percent). As the hold time and/or temperature increase, the reaction between the fiber and matrix increases, leading to degradation of the fiber and thus the composite properties.

Transverse Mechanical Loading

Elastic-plastic analyses were performed to predict the MMC response under transverse loading. The residual stresses (no hold) were included in the analyses. Figure 8 provides the contrast between the contributions by matrix plasticity and fiber-matrix debonding to composite nonlinearity. In the top curve, the fiber-matrix debonding is prevented and the composite nonlinearity is only due to matrix plasticity. In the bottom curve, the fiber and matrix are allowed to debond; that is, when the compressive residual stress at the interface is overcome, separation occurs (i.e., no chemical bond between the fiber and the matrix).

As can be seen, the debonding contribution to composite nonlinear behavior is much greater than that of the matrix plasticity. Furthermore, since the onset of gross material nonlinearity such as that shown in the bottom curve sets the design limit (in the transverse direction), it is desirable to increase this limit by

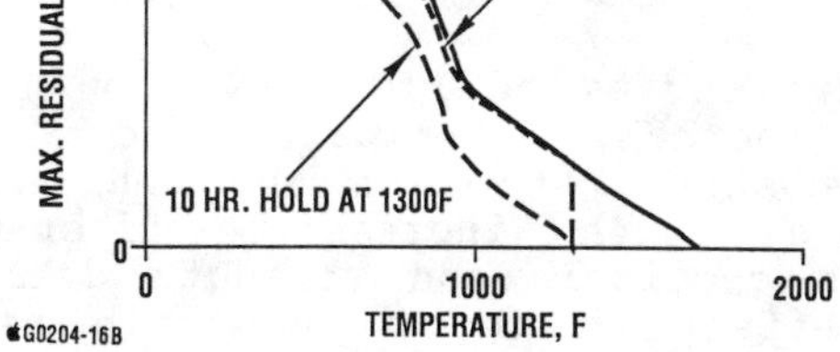

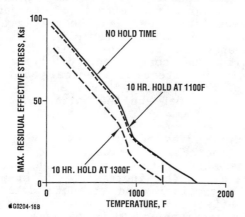

Figure 7. Effect of Hold Time on Residual Effective Stress During Cool-Down Predicted.

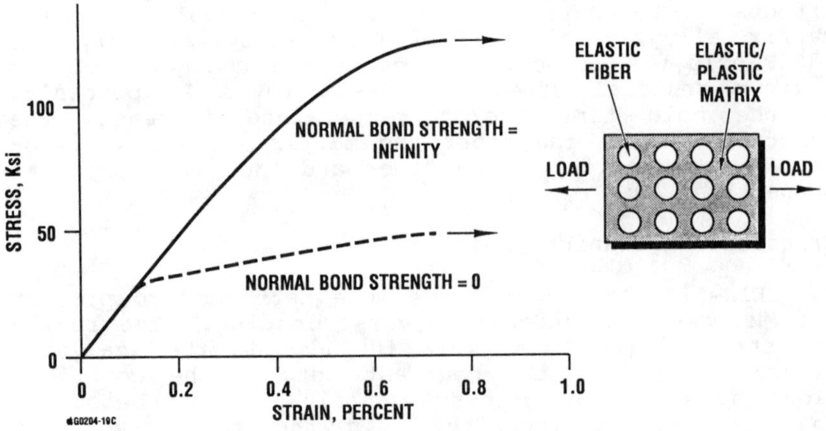

Figure 8. Nonlinearities Result From Matrix Plasticity and Fiber-Matrix Debonding.

either increasing the residual radial stress or by enhancing the chemical bond between the fiber and matrix. Increasing the residual radial stress by increasing the fiber-matrix CTE mismatch, however, has the adverse effect of increasing the residual tangential stresses that can lead to matrix cracking or yielding.

Figure 9 shows the effect of tangential shear friction between the fiber and the matrix and also the influence of fiber-matrix chemical bond strength on composite transverse response. The initial modulus increases with the coefficient of friction. In the evaluation of the effect of chemical bond strength, a coefficient of friction of $\mu=1$ was assumed to ensure no slip between the fiber and the matrix prior to debonding.

The top two curves of Figure 9 show the effect of interfacial (normal) chemical bond strength. As expected, the stress at which the nonlinearity initiates increases with increasing fiber-matrix bond strength. The chemical bond strength does not have a significant effect on the high strain region of the stress-strain response. This is because in cases where there is a chemical bond between the fiber and the matrix, once fiber-matrix debonding occurs, it propagates very rapidly along the interface.

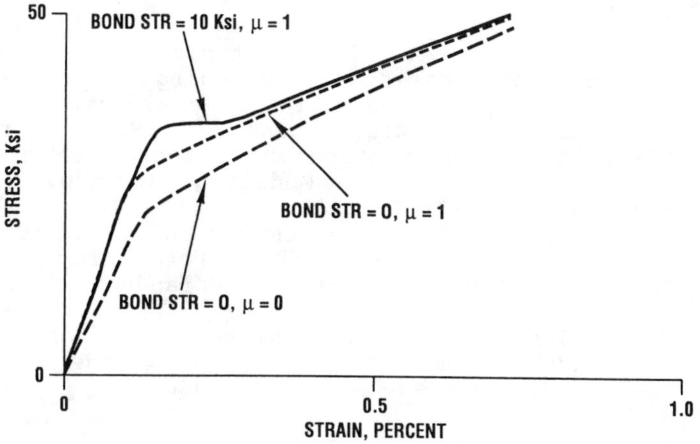

G0204-22D

Figure 9. Interfacial Normal Strength and Coulomb Friction Affect Composite Transverse Response.

CONCLUSIONS

1. The residual stresses induced during cool-down can be large enough to cause yielding or cracking of the matrix material.
2. The residual radial stresses are beneficial to MMC transverse response; the residual tangential stresses are detrimental to MMC microstructure and transverse response.
3. Matrix relaxation due to hold at appropriate intermediate temperatures can be used to reduce the residual stresses.
4. The initial MMC modulus increases with increased tangential friction between the fiber and the matrix.
5. Fiber-matrix debonding is a more significant contributor to nonlinearity in the MMC transverse response than matrix plasticity.
6. The onset of gross nonlinearity in the MMC increases in stress with an increase in fiber-matrix chemical bond (interfacial normal strength).

ACKNOWLEDGMENT

This work was conducted under Task III of the Naval Air Development Center Contract N62269-86-C-0248.

REFERENCES

1. J. M. Larson, et al., "Titanium Aluminides for Aerospace applications," Proceedings of the 1989 TMS Fall Meeting, Indianapolis, Indiana, Oct. 1-5, 1989.
2. D. Buchanan, "Micromechanical Study of MMC Fiber/Matrix Interactions," Presented at the USAF MMC Mechanical Behavior Modeling Workshop, Orlando, Florida, Nov. 17-18, 1989.
3. W. S. Johnson, "Fatigue Testing and Damage Development in Continuous Fiber Reinforced Metal Matrix Composites," NASA Technical Memorandum 100628, June 1988.
4. M. J. Blackburn and M. P. Smith, "R & D on Composition and Processing of Titanium Aluminide Alloys for Turbine Engines," AFWAL-TR-82-4086.

PROGRESSIVE FAILURE ANALYSIS OF CEMTITIOUS COMPOSITES

T. STANKOWSKI and S. STURE
Dept. of Civil, Environmental, and Architectural Engrg.
University of Colorado, Boulder, CO 80309-0428

K. RUNESSON
Dept. Structural Mechanics, Chalmers University of Technology,
Göteborg, Sweden

ABSTRACT

Analysis of progressive failure in cementitious particle composites such as concrete requires a realistic description of the topology of the composite structure and the modeling of the interaction of the constituents at their interface. The interface model has to account for fundamental mechanisms such as adhesion, debonding, dilatancy and mobilized friction. In conjunction with a decohesive softening material model for the cementitious matrix, macroscopical response characteristics of the composite material can be simulated. This paper addresses issues at both the constitutive and the macro system levels.

INTRODUCTION

The failure characteristics of cementitious particle composites such as concrete depend strongly on the magnitude and the combination of stresses, spanning the full range from brittle to ductile post-peak response. Progressive failure is accompanied by the formation of distinct transverse cracks in tension, axial splitting in uniaxial compression, and formation of shear bands or diffuse microcracking as the confining stresses increase. These macroscopic observations are closely tied to degradation mechanisms in the internal structure of the heterogeneous material. In order to analyze these mechanisms, the cementitious material is considered as a two-phase composite consisting of individual aggregate particles embedded in a mortar matrix.

The interaction between these two constituents is modeled by means of an interface, which is considered as a very important component of the composite structure. It has been modeled in the past often by a simplified Coulomb friction models with uncoupled tensile and shear failure mechanisms. The interface model employed here in the analysis of the response behavior of composite materials accounts for coupled normal and shear failure mechanisms due to adhesion between the composite phases.

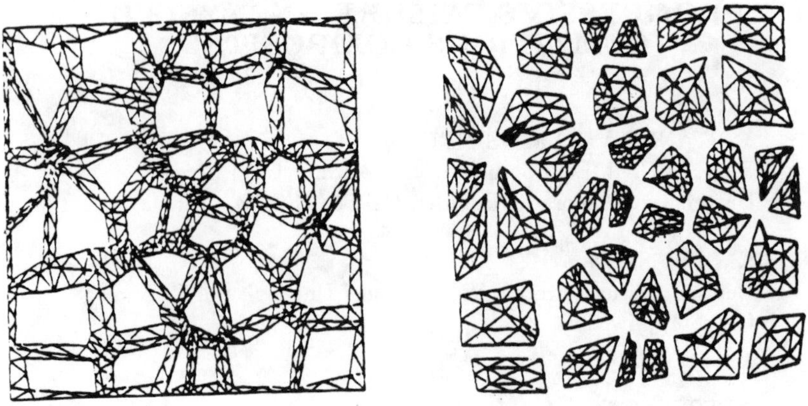

Figure 1: Triangulation (CST) of Mortar and Aggregate

DESCRIPTION OF COMPOSITE

Topology of the Composite

The composite material considered herein is comprised of aggregates, cement paste and voids. Since it is infeasible and perhaps inconceivable that each separate particle can be considered explicitly in the topological description, the description is restricted to a two-phase matrix particle composite in which only the larger inclusions are accounted for. The aggregate phase is idealized by polygonal approximations of crushed angular particles, which are embedded in the mortar matrix, [1,2]. The distribution pattern of the aggregates particles as well as their shapes are randomly generated, however, the aggregate area fraction and the size distribution are controlled at the same time.

The polygonal description of particles has been adopted earlier by [3] for the idealization of granular solids with rigid particles and sliding interfaces, where the interfaces have the lumped properties of the physical interfaces and the mortar matrix. This method is also known as the *Distinct Element Method.* In the present case, the constituents of the composite are considered individually. A typically generated composite along with the finite element mesh is shown in Fig.(1).

Constitutive Properties of Aggregate and Mortar

The mechanical properties of the aggregate phase is generally far superior to those of the mortar phase and the interfaces for normal strength concrete in terms of strength and stiffness. It is therefore assumed that the (isotropic)

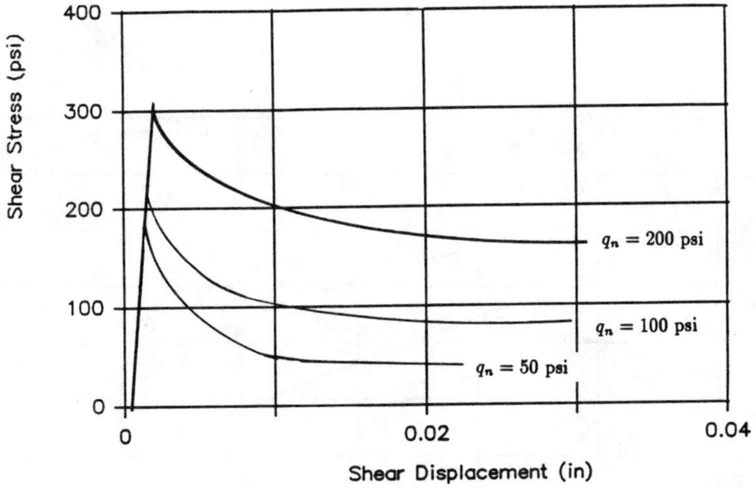

Figure 2: Deformation of Mortar Granite Interfaces under Constant Compressive Load

aggregate particles remain in the linear elastic regime throughout the loading history.

The mortar matrix is modeled by an elementary linear elastic/linear softening *Drucker Prager* formulation with decohesive softening. Associated flow is assumed during softening, thus the friction angle Φ is constant and identical with the dilatancy angle of Ψ during the entire deformation process. The model is implemented within the framework of invariant response models, [4], and does not contain any feature that reduces mesh size dependency of the softening computation such as an internal length or volume measure, [5,6].

The adoption of these rather basic constitutive assumptions allows a qualitative evaluation of the contribution of the constituents to the macroscopic properties of the composite material.

Constitutive Model for the Interfaces

Fig.(2) displays the experimentally obtained load-deformation behavior of mortar-granite interfaces in unconstrained dilatation direct shear experiments for different levels of compressive normal load. The strength characteristics are shown in Fig.(3). A constitutive model for the phase interfaces which is able to model this interface behavior has been developed recently. It is formulated in analogy to the theory of incremental plasticity for a continuum. This model describes fracture and slip of the interface for an arbitrary combination of the normal and tangential traction. The fracture criterion that defines the onset of tensile and/or shear debonding is of a nonlinear, smooth Mohr-Coulomb type. Degradation of the adhesive bond due to fracturing occurs for an arbitrary mode of loading comprising shear and compression or tension whenever the fracture criterion is satisfied. The debonding process is

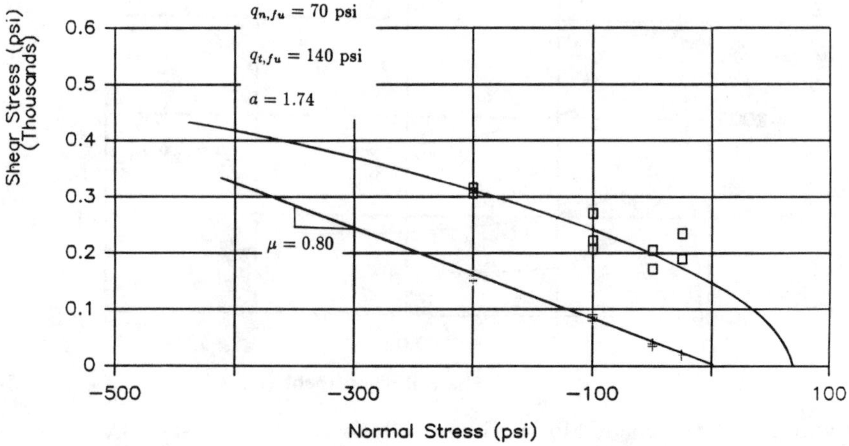

Figure 3: Peak and Residual Strength of Mortar Granite Interfaces

described via a fracture energy release measure analogous to a work softening hypothesis. In the residual state, frictional resistance is only available in the presence of normal compressive stresses. Interface dilatancy, i.e. relative normal displacement caused by the application of pure shear, is monitored by a non-associated slip rule. The constitutive relations are regularized by the introduction of a recoverable adhesion component in the tangential as well as the normal direction.

Of particular interest with regard to the progressive failure analysis of quasi brittle composites is that the interface model decribed above exhibits ultimately brittle failure in mixed control in the presence of any tensile normal stresses and increasing shear displacements, while the response for compression shear experiments is always ductile. Details of this model phenomenon are discussed in [1].

3 COMPOSITE FAILURE SIMULATION

Fig.(4) illustrates the extent of failure of the composite specimen shown in Fig.(1) when subjected to uniaxial tension at three stages during the progressive failure analysis: (a) close to peak load, (b) at peak load and, (c) at residual strength. The load is applied under displacement control with full lateral restraint at top and bottom faces simulating rigid loading platens. The extent of failure in the matrix is visualized in terms of active zones of plastic work dissipation.

The response predictions reproduce realistically the dominant failure modes, which can be observed in laboratory experiments. For the tensile load history, as shown in Fig. (4), distributed failure is confined to interface cracking with

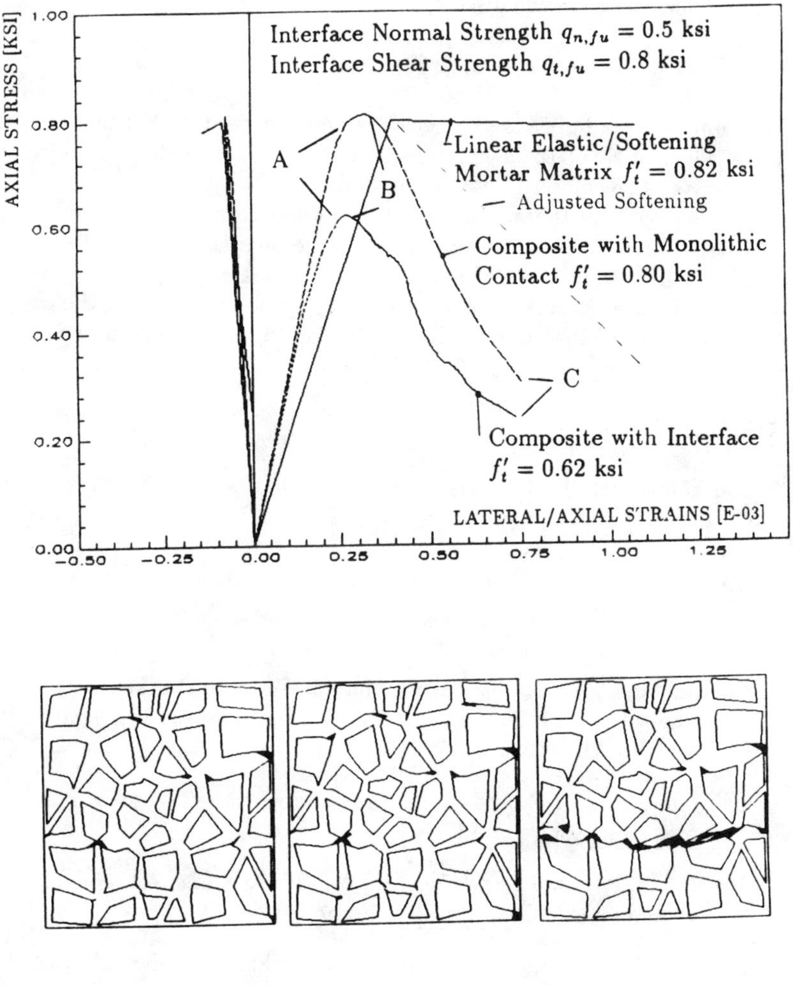

Figure 4: Nominal Response in Uniaxial Tension and Tensile Crack Pattern at Stages A, B, C

limited spread into the mortar matrix. Only in the final stages of failure emerges a single discontinuity through the cross section. Stages of crack arrest and crack bridging between the aggregate particles across the mortar matrix can be identified from the discontinuous global response.

4 CONCLUDING REMARKS

The composite failure analysis provides new insight into the failure mechanisms which take place in heterogeneous matrix aggregate composites such as concrete. The computational results indicate that initial failure is very distributed throughout the specimen, either in terms of interface failure in tension or matrix failure in compression, and only the final stages of failure are truly localized. The simulation of tensile load histories confirm the experimental observations of a surface dominated failure process. In contrast, the compressive load history simulation exhibits shear failure modes, which appear to be volume dominated.

Acknowledgements

The authors gratefully acknowledges the collaboration with Dr. Kaspar Willam and the financial assistance provided by Grant AFOSR-89-0289 and the support of Dr. Spencer Wu.

REFERENCES

1. Stankowski, T.,"Numerical Simulation of Progressive Failure in Particle Composites " Ph.D. Thesis, C.E.A.E. Department, University of Colorado, Boulder, 1990

2. Finney, J.L.,"A Procedure for the Construction of Voronoi Polyhedra", Comp. Phys., 32, pp. 137-143

3. Cundall, P.A. and Strack, O.D.L., A Discrete Numerical Model for Granular Assemblies, Geotechnique, 29, pp. 47-65

4. Ottosen, N.S.,"Theoretical Framework for Modeling the Behavior of Frictional Materials", Int. J. Solids and Structures, 22, No. 11, p. 1325

5. Pietruszczak, S. amd Mroz, Z.,"Finite Element Analysis of Deformation of Strain Softening Materials", Int. J. Num. Meth. Eng., 1981, 17, pp. 327-334

6. Willam, K., Montgomery, K.,"Fracture Energy Based Softening Plasticity for Shear Failure", Proc. Int. Symp. on Interaction of Non-Nuclear Munition with Structures, II, Mannheim, 1987, pp. 679-691

Local and Homogenized Elastic Response in a Periodic Tungsten/Copper Composite

Kevin P. Walker
Engineering Science Software, Inc.
Smithfield, Rhode Island 02917

Alan D. Freed
NASA-Lewis Research Center
Cleveland, Ohio 44135

Eric H. Jordan
University of Connecticut
Storrs, Connecticut 06268

July 12, 1990

Abstract

Local elastic fields in the unit cell of a periodic composite are examined numerically with an integral equation approach. Solutions are obtained using a Fourier series method with rectangular subvolume elements. Specific results are presented for a tungsten/copper composite which is being considered as a candidate material for improved rocket nozzle liners.

1 Introduction

The combustion chamber in the main engine of the space shuttle has a liner material which is fabricated from a copper alloy. Temperature gradients are generated within this liner material during the space shuttle's launch which are large enough to cause substantial thermally induced stresses. A tungsten/copper composite is being considered as a substitute to increase the strength and improve the durability of the combustion liner. The candidate replacement material is comprised of tungsten filament wires embedded in an oxygen-free, high conductivity copper matrix and may be characterized as a ductile-ductile type material.

2 Local and Homogenized Response

In order to perform a structural analysis of a fibrous composite component, it is necessary to divide the structure into finite elements. Each finite element is assumed to be comprised of a doubly periodic row of unit cells containing a tungsten fiber surrounded by the copper matrix. As far as the Gaussian integration point is concerned, the surface of the finite element is many unit cells away, so that the problem of determining the local fields within the unit cell reduces to determining the response within a periodic cell of an infinite lattice when the strain increment $\Delta\varepsilon^0_{kl}$ given by the finite element code is applied at infinity.

The local response at any point $\mathbf{r}$ within the unit cell is obtained from the relation $\Delta\varepsilon^T_{kl}(\mathbf{r}) = M_{klrs}(\mathbf{r})\Delta\varepsilon^0_{rs}$, where $M_{klrs}(\mathbf{r})$ represents the magnification or strain concentration factor which magnifies the strain increment applied at the surface of the finite element—*i.e.* at its nodes—and gives the strain increment at any point $\mathbf{r}$ in the unit periodic cell. The tensor magnification factor $M_{klrs}(\mathbf{r})$ is a complicated function of the geometry and constitutive properties of the constituent materials comprising the unit periodic cell. Once the total strain increment $\Delta\varepsilon^T_{kl}(\mathbf{r})$ at any point $\mathbf{r}$ is known, the stress increment can be computed *via* Hooke's law in the form $\Delta\sigma_{ij}(\mathbf{r}) = D_{ijkl}(\mathbf{r})\Delta\varepsilon^T_{kl}(\mathbf{r})$, where $D_{ijkl}(\mathbf{r})$ is the elasticity tensor at the point $\mathbf{r}$. The overall homogenized stress increment, $\Delta\sigma^0_{ij}$, required for calculating the stiffness of the finite element, can then be obtained by volume averaging over the unit cell in the form, $\Delta\sigma^0_{ij} = \dfrac{1}{V_c}\iiint\limits_{V_c} \Delta\sigma_{ij}(\mathbf{r})\, dV(\mathbf{r})$, where V_c denotes the volume of the unit cell.

Once the homogenized stress increment, $\Delta\sigma^0_{ij}$, is calculated at each Gaussian integration point in each finite element in the composite structure, the finite element analysis will yield the strain-temperature histories at the damage critical locations. These strain-temperature histories can then be applied to the finite element containing the damage critical Gauss point and the Fourier series method will yield the local variation of the total strain increment from the relation, $\Delta\varepsilon^T_{kl}(\mathbf{r}) = M_{klrs}(\mathbf{r})\Delta\varepsilon^0_{rs}$.

It may thus be seen that the methods are used in a complementary fashion. First to homogenize and obtain the overall macroscopic response of the composite and then to "zoom in" and calculate the local response in and around the fibers in a unit periodic cell. In obtaining the overall homogenized response it is necessary to use rapid methods for estimating the magnification tensor $M_{ijkl}(\mathbf{r})$ since this is used at each Gauss point of the structure for each strain increment of the loading history. A much more accurate value of the magnification tensor, $M_{ijkl}(\mathbf{r})$, can be used in postprocessing the finite element results to look at the local stress-strain variations throughout the unit cell.

3 Integral Equation

The determination of the stress and strain increments throughout the fibrous composite material under isothermal elastic conditions requires the solution of the integral equation [1]–[5],

$$\begin{aligned}\Delta\varepsilon^T_{kl}(\mathbf{r}) &= \Delta\varepsilon^0_{kl} - \frac{1}{A_c}\sum_{n_p=0}^{\pm\infty}\sum{}' g_{klmn}(\boldsymbol{\zeta})\iint\limits_{A_c} e^{i\boldsymbol{\xi}.(\mathbf{r}-\mathbf{r}')}\times \\ &\quad \times \vartheta(\mathbf{r}')\left(D^f_{mnrs} - D^m_{mnrs}\right)\Delta\varepsilon^T_{rs}(\mathbf{r}')\, dS(\mathbf{r}'), \qquad (1)\end{aligned}$$

in which $\vartheta(\mathbf{r}') = 1$ in the fiber and $\vartheta(\mathbf{r}') = 0$ in the matrix and where

$$g_{klij}(\boldsymbol{\zeta}) = \frac{1}{2}\left(\zeta_j\zeta_l\left(D^m_{pikq}\zeta_p\zeta_q\right)^{-1} + \zeta_j\zeta_k\left(D^m_{pilq}\zeta_p\zeta_q\right)^{-1}\right), \tag{2}$$

with $\zeta_p = \xi_p/\sqrt{\xi_m\xi_m} = \xi_p/\xi$ being a unit vector in the direction of the Fourier wave vector $\boldsymbol{\xi}$, and $\xi = \sqrt{\xi_m\xi_m}$ denoting the magnitude of the vector $\boldsymbol{\xi}$. In equation 1 the sum is taken over integer values in which

$$\xi_1 = \frac{2\pi n_1}{L_1}, \quad \xi_2 = \frac{2\pi n_2}{L_2}, \tag{3}$$

and L_1, L_2 are the dimensions of the unit periodic cell in the x_1, x_2 directions, so that $A_c = L_1L_2$. The values of n_1, n_2 are given by

$$n_p = 0, \pm1, \pm2, \pm3, \ldots, etc., \quad \text{for } p = 1, 2 \tag{4}$$

and the prime on the double summation signs indicates that the term with $n_1 = n_2 = 0$ is excluded from the sum. The superscripts f and m on the elasticity tensor denote the values corresponding to the fiber and matrix, respectively.

The solution of the preceding integral equation yields the local strain increment in terms of the applied strain increment, from which the magnification tensor can easily be obtained.

4 Numerical Example

Fig. 3 shows the transverse stress concentration factors, σ_{11}, within the unit periodic cell, when a tungsten/copper fiber composite is loaded in the transverse direction with an overall stress of $\sigma^0_{11} = 1000$ KPa. The tungsten fiber occupies a volume fraction $f = 9/49 = .184$ in the unit cell of the composite, with $E_{\text{Tungsten}} = 395$ GPa, $\nu_{\text{Tungsten}} = 0.28$, $E_{\text{Copper}} = 127$ GPa and $\nu_{\text{Copper}} = 0.34$.

A Fourier series approach is used to calculate both the stress concentration throughout the unit cell and the homogenized transverse elastic modulus. Within the unit cell nine (9) subvolumes are used to calculate the stress variation throughout the tungsten fiber, whilst forty (40) subvolumes are used in the copper matrix. The stress concentration in the tungsten fiber varies from 1175 KPa to 1392 KPa, with a volume average of 1297 KPa. In the copper matrix the stress concentration varies from a minimum of 682 KPa to a maximum of 1210 KPa. With forty nine (49) subvolumes in the unit cell the overall homogenized transverse elastic modulus is calculated to be $E_{\text{homogenized}} = 156.3$ GPa. It can be seen that the transverse stress in the square planform fiber forms a ridge valley in the direction of the transverse stress, the average stresses in the ridges being 1340, 1392, 1340 KPa and the average stresses in the valley being 1179, 1175, 1179 KPa. This behavior can be noticed in a similar problem where a cuboidal inclusion in an infinite matrix suffers a uniform eigenstrain or transformation strain. The problem is outlined on page 91 of Mura's book "*Micromechanics of Defects in Solids*", [3]. The ridge valleys are present even when the fiber is isolated, and may be contrasted to the case of an isolated fiber with circular (or ellipsoidal) cross-section which, by Eshelby's analysis [6], would possess a uniform stress distribution within the fiber.

top of unit cell

	826	819	835	837	835	819	826	
	939	879	739	682	739	879	939	
	1131	1191	1340	1392	1340	1191	1131	
$\sigma_{11}^0 \Longleftarrow$	1206	1210	1179	1175	1179	1210	1206	$\Longrightarrow \sigma_{11}^0$
	1131	1191	1340	1392	1340	1191	1131	
	939	879	739	682	739	879	939	
	826	819	835	837	835	819	826	

bottom of unit cell

Figure 3: *Transverse stress concentration, σ_{11}, for an applied transverse stress of $\sigma_{11}^0 = 1000$ KPa. The 9 central subvolumes in the unit cell represent the tungsten fiber embedded in 40 surrounding subvolumes representing the copper matrix. Each unit cell, with its 49 subvolumes, is embedded in a doubly periodic array of identical cells. The homogenized transverse Young's modulus is $E = 156.3$ GPa. Each number represents the volume averaged stress over the square subvolume.*

5 Acknowledgment

This work is being supported by the NASA-Lewis Research Center under Grant Number NAG3-882 and by the Department of Energy under Contract DE-AC02-88ER13895.

References

[1] Walker, K.P., Jordan, E.H., and Freed, A.D., 1989, *Nonlinear Mesomechanics of Composites With Periodic Microstructure: First Report*, NASA TM-102051.

[2] Walker, K.P., Jordan, E.H., and Freed, A.D., 1989, *Equivalence of Green's Function and the Fourier Series Representation of Composites With Periodic Microstructure, Micromechanics and Inhomogeneity—The Toshio Mura Anniversary Volume*, Springer–Verlag, pp. 535–558.

[3] Mura, T., 1982, *Micromechanics of Defects in Solids*, Martinus-Nijhoff Publishers, The Hague/Boston/London.

[4] Nemat-Nasser, S., Iwakuma, T. and Hejazi, M., 1982, *On Composites with Periodic Structure*, Mechanics of Materials, **1**, pp. 239–267.

[5] Iwakuma, T. and Nemat-Nasser, S., 1983, *Composites with Periodic Microstructure*, Computers and Structures, **16**, Nos. 1–4, pp. 13–19.

[6] Eshelby, J.D., 1957, *The Determination of the Elastic Field of an Ellipsoidal Inclusion, and Related Problems*, Proc. Roy. Soc., **A241**, pp. 376–396.

A CONSTITUTIVE THEORY FOR THE INELASTIC BEHAVIOUR OF FIBROUS COMPOSITES

R. VAZIRI
Dept. of Metals & Materials Engineering, University of British Columbia, Vancouver, B.C., Canada, V6T 1W5

M.D. OLSON and D.L. ANDERSON
Dept. of Civil Engineering, University of British Columbia, Vancouver, B.C., Canada, V6T 1W5

INTRODUCTION

It is well known that micromechanical effects such as matrix cracking, fibre rupture, or debonding between fibre and matrix may take place in fibre-reinforced composites (FRC) at early stages of loading. These phenomena, together with the inherent nonlinearities in the deformational response of the usually soft matrix phase, are largely responsible for the observed inelastic behaviour of FRC laminates. It is the objective of this paper to present a descriptive outline of a recently developed constitutive model [1,2] simulating the inelastic response of FRCs from a phenomenological point of view. A brief description of the finite element code which incorporates the model will also be given. Finally, an example application showing the accuracy of the theoretical results will be presented.

CONSTITUTIVE MODEL

The new material model is essentially representative of the mechanical behaviour of a single layer (or ply) of the laminate, under plane stress conditions. The methodology is developed within the framework of orthotropic plasticity. In this development, only the rate-independent behaviour for an isothermal process is considered. Moreover, in order to focus on the material nonlinearities, the formulation is limited to considerations of small strains and small displacements only.

The proposed model may best be divided into three distinct regimes: the elastic regime, the plastic regime and the post-failure regime. The linear elastic orthotropic stress-strain relation is used first until the combined state of stress satisfies a quadratic yield function, which marks the incipience of plastic flow. In particular the Tsai-Hill and Puppo-Evensen criteria are adopted to represent the onset of nonlinearity (or yielding) in unidirectional and bidirectional (e.g. woven) fibre composites, respectively. These criteria are generalizations of Mises-type plastic flow theories for isotropic materials, and use 3 yield parameters to account for the orthotropy. Failure criteria, which are assumed to have the same functional forms in stress space as the yield criteria, are used to represent the upper limit of plastic flow. Once failure is reached according to these criteria, it is ascribed to either matrix or fibre failure depending on the relative magnitudes of the various stresses. To simulate the post-failure behaviour, two types of failure modes are defined, namely, brittle and ductile. For the brittle failure mode, the composite layer is simply assumed to lose its entire stiffness and strength in the offending stress direction. For the ductile failure mode, the layer retains its strength while losing all of its stiffness in the failure direction. These two extreme modes were chosen to provide effective bounds on the real behaviour.

This new formulation requires that the ply properties be known. In particular uniaxial tensile tests along the two principal axes of orthotropy, and an in-plane shear test are needed to determine the material functions and constants of the theory.

FINITE ELEMENT CODE

The foregoing analytical constitutive model has been incorporated into a nonlinear finite element program designed to perform progressive yielding and/or failure analyses of laminated FRC structure under quasi-static in-plane loading. The assumptions of simple laminate theory are employed wherein the usual linear orthotropic elasticity law is replaced by the orthotropic elastic-plastic-failure theory described above. The finite element program is based on the conventional displacement method using two-dimensional 8-node isoparametric elements. The nonlinearities in the equilibrium equations are handled by a mixed incremental and Newton-Raphson iterative procedure. All stress and strain quantities are monitored at each Gaussian integration point in each layer of every element. For every iteration within a given load increment, all stresses are checked for yield or failure and the appropriate constitutive matrix is used.

NUMERICAL RESULTS

Example applications contained in [1-3] considered uniform laminates with and without a central hole loaded to failure, with comparisons to experimental data if available. As an example of the results here we show the cyclic response of a uniform six layer laminate, made up of unidirectional Boron/Aluminum plies, loaded in uniaxial tension. The fibres are orientated in the 0°, +45°, -45° directions with respect to the loading direction, such that the laminate is symmetric about the mid plane. The experimental data is taken from a report by Sova and Poe [4]. They included tests for the longitudinal and transverse stress-strain response of the individual plies and these were approximated by bilinear curves for the present numerical work. Since no experimental data was available for the shear response, the latter was derived from the uniaxial tensile stress-strain curve of a [±45°] laminate following the procedure outlined by Rosen [5]. The material parameters used in the calculations are listed below:

Elastic moduli	GPa	$E_1 = 210$	$E_2 = 107$	$G = 32$
Poisson's ratio		$\nu_{12} = 0.2$		
Tangent moduli	GPa	$E_{T1} = 202.7$	$E_{T2} = 24.3$	$G_T = 1.5$
Yield stress	GPa	$X_o = 1.2$	$Y_o = 0.09$	$S_o = 0.045$
Failure stress	GPa	$X_u = 1.7$	$Y_u = 0.12$	$S_u = 0.11$

Figure 1 shows the average stress (in the six layers) in the loaded direction vs the strain in the same direction for three cycles of uniaxial loading progressing to ultimate failure. The present model results are shown as solid lines, while the experimental ones are dotted. As may be seen the model captures the relevant features of the response associated with such cyclic loading. Thus, for instance, the residual strains are correctly predicted and the hysteresis loops predicted by the model follow the same trend as the experimental ones. The kink in the third unloading curve was due to the $\pm 45°$ plies yielding in compression. Calculations were also performed for a pure monotonic loading case and essentially the same failure point was reached. This agrees with the experimental findings of Sova and Poe [4], who showed that the load cycles did not affect the ultimate stress and strain level of the laminate.

CONCLUSIONS

An orthotropic plasticity failure model has been developed to describe the mechanical properties of fibre reinforced composite layers (or plies) in their elastic, plastic and post-failure regions. A plane finite element program has been developed incorporating this new constitutive model so as to calculate the response of laminates comprised of several layers or plies. Application of the program to uniform laminate problems in [1,2] shows very promising comparisons to experimental results, both in following a deformation path and in predicting ultimate failure. Solutions to non-uniform laminate problems reported in [2,3] also show good agreement, although few experimental results are available for comparison. The calculations demonstrate that the ultimate strength and ultimate deformation of a laminate comprised of many layers having different ply directions is very strongly influenced by the post-failure nature of the individual plies.

The practical significance of the present results is that plasticity theory, which is a rather simple model in comparison with those based on fracture mechanics or continuum damage models, can be used to model the nonlinearities in the stress-strain behaviour of fibre reinforced composites.

REFERENCES

1. R. Vaziri, M.D. Olson and D.L. Anderson, "A Plasticity-Based Constitutive Model for Fibre-Reinforced Composite Laminates", Accepted for publication in the Journal of Composite Materials, (January 1991).

2. R. Vaziri, "On Constitutive Modelling of Fibre-Reinforced Composite Materials", Ph.D. Thesis, Department of Civil Engineering, University of British Columbia, Vancouver, (1989).

3. R. Vaziri, M.D. Olson and D.L. Anderson, "Finite Element Analysis of Fibrous Composite Structures: A Plasticity Approach", submitted to Int. J. Num. Meth. Engng., (1990).
4. J.A. Sova and C.C. Poe, "Tensile Stress-Strain Behaviour of Boron/Aluminum Laminates", NASA-TP-1117 (1978).
5. B.W. Rosen, "A Simple Procedure for Experimental Determination of Longitudinal Shear Modulus of Unidirectional Composites", J. Comp. Mats., 6, pp. 555-557 (1972).

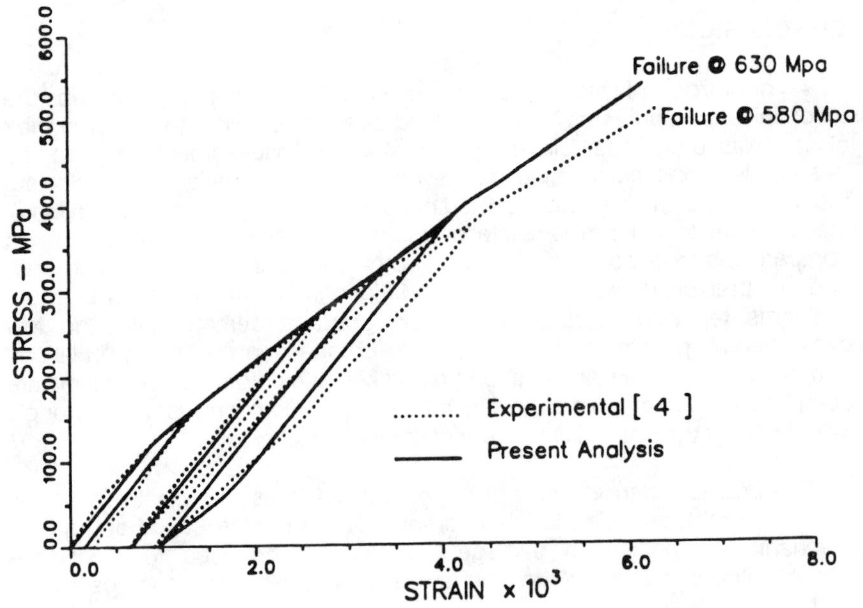

Fig. 1. Tensile Stress-Strain Response for a [0°/±45°] Symmetric Boron/Aluminum Laminate Subject to Three Load Cycles.

Constitutive Modelling of Particle-Reinforced Metal Matrix Composites at High Strain Rates

R. Vaziri, A. Poursartip, G. Pageau and R. Bennett
Composites Group, Department of Metals & Materials Engineering,
The University of British Columbia, Vancouver, B.C., Canada V6T 1W5

Abstract

The behaviour of selected particle-reinforced metal matrix composites (MMC) under impact conditions is being studied. Some important issues that need to be considered in constitutive modelling of MMC are outlined. Some results from Taylor and penetration tests are presented. A plasticity-based model is proposed that accounts, in a unified sense, for many of the characteristic features of MMC.

Introduction

There is a large body of literature covering the analytical and experimental approaches to the constitutive modelling of MMC at low, conventional, strain rates. Similarly, there is an even larger body of literature on the constitutive modelling of the high strain rate behaviour of isotropic, homogeneous materials. This paper introduces the issues, both analytical and experimental, that must be addressed to develop constitutive models for the high strain rate behaviour of MMC.

In this first instance, we confine ourselves to particle-reinforced MMC, as the degree of anisotropy is minimized, even though the material is still heterogeneous. We focus on the impact behaviour at intermediate velocity range (≤ 1.5 km/s). In this range, much of the motion occurs at strain rates of 10^4 1/s and pressures generated are low enough (~10 GPa) so that material strength, as opposed to thermodynamic effects, become important [1]. We propose that the ability to predict what happens during the penetration of MMC targets constitutes the ultimate test of our modelling capability.

Analytical Issues

There are several recent review articles that describe the current state of the numerical simulation of impact and penetration [2,3]. Computer programs that accommodate the physics of impacting solids, provide a convenient numerical tool for handling the complex calculations involved in penetration problems. These programs, generally known as *hydrocodes*, also serve as a numerical testbed for developing material models. A review of hydrocodes with potential application to deep penetration problems can be found in [3].

A viable approach would be to use a production hydrocode with the existing high strain rate constitutive models already developed for homogeneous isotropic materials. However, a recent review of literature [4] reveals that no model currently covers, in a unified manner, the following features of MMC behaviour:

1. A significant difference in tensile and compressive strengths [5].
2. Enhancement of strength and ductility under increasing confining pressure [6].
3. Increasing brittleness of the material (higher strength and lower strain to failure) and increasing stiffness with increasing volume fraction of particles .
4. Strain-rate dependency of both elastic and plastic properties under dynamic loading conditions [7].

In addition, the effects of localized yielding, thermal softening, particle/matrix interface decohesion, and damage accumulation prior to fracture must also be considered. As very little experimental data are available in the literature, these must be generated to guide, calibrate, and validate the constitutive models.

Experimental Issues

In order to perform parametric studies, we have chosen to consider two material systems both based on a 6061-T6 aluminum matrix. The first, available from DURALCAN is an alumina particle (~30 micron) reinforced system made by rheo-casting with 10, 15, and 20% reinforcement volume fractions, respectively. The second, available from DWA Composites, is a silicon carbide particle (~10 micron) reinforced system made by a powder metallurgy route, with 15, 30, and 55% reinforcement volume fractions, respectively. The materials were extruded for optimal properties. Relevant material properties are summarized in Table I, where unreinforced 6061-T6 Al and 7075-T6 Al are included for comparison.

To study the high strain rate behaviour of MMC, deep penetration tests were conducted by firing relatively rigid tungsten rods (4.5 mm diameter, 27 mm long, 7.1 gram mass) at cylinders of the material of interest with dimensions of 50.8 mm diameter and 152.4 mm length. The incident velocity was 850 m/s. Penetration results for some of the materials tested are provided in Table I.

After the test, the specimens are X-rayed, sectioned, and subjected to traditional metallurgical examinations. A typical example of the performance of a 30% SiC_p MMC is shown in Figure 1. Inspection of the penetrated MMC specimens indicates that bulging and elastic rebound of the tungsten penetrator can occur. Examination of the particle distribution around the crater nose indicates that there is clustering and increased local volume fraction of the particles. A thin coating of aluminum has been detected on the crater surface indicating possible local melting of the matrix phase. Residual stresses are created, as shown by the macro-cracks that form during sectioning of the tested specimen. There has been no evidence of shear banding.

In order to decouple the many complicated effects (large plastic deformation, failure and post-failure behaviour of material) observed in the penetration tests, we resort to simple high strain rate uniaxial tests to study the plastic deformation separately. This test, known as the Taylor anvil test, involves impacting a short flat-ended cylinder of the material of interest on a flat rigid target. The plastic deformation conditions created in such tests are comparable to those existing in a typical penetration test. The average dynamic flow stress can be determined from measurements of the deformed shape, and knowledge of the original kinetic energy [8]. Typical examples for MMC are shown in Figure 2 where radial cracking of the 30% SiC_p material can be observed. This is not unexpected, as Table I shows that the strain to failure and hence the ductility of this material is comparatively low. Such low ductility is obviously not dominant in the deep penetration tests, as witnessed by the improved behaviour of the MMC. Furthermore, as shown in Table I, the unreinforced 7075-T6 Al, despite its slightly superior strength properties, exhibits a slightly deeper penetration than the 30% SiC_p MMC. This suggests that other factors such as pressure dependency, increased local volume fraction of particles, temperature and so forth must be considered in strength calculations for MMC.

Proposed Constitutive Model

The description most commonly applied to metals, under the range of conditions of interest, is to divide the deformation behaviour into volumetric and deviatoric components. It is generally held that the details of the volumetric model, described by an equation of state, are relatively unimportant at velocities considered here (~ 1 km/s), as the pressures generated are modest [1]. In this regime, an accurate description of the deviatoric behaviour of the material assumes the primary role.

We propose a phenomenological plasticity-based constitutive model that contains the terms necessary to describe the behaviour outlined in the previous sections and lends itself to an efficient implementation in hydrocodes. Since there is virtually no information available on the combined effects of pressure (p), temperature (T), strain-rate ($\dot{\varepsilon}$), strain-hardening (ε), particulate volume fraction (V_f) and damage (D) on the yield strength (Y), a simple approach is to consider these effects separately as

$$Y = f_1(\varepsilon, \dot{\varepsilon}, T)\, f_2(p)\, f_3(D)\, f_4(V_f) \tag{1}$$

The function f_1 is taken to be of the same form as that suggested for FCC metals in [9]. The use of this particular function is supported by its mechanistic nature as well as its close agreement with the high strain rate experimental data for aluminum alloys. In order to model the transition from brittle to ductile deformation with increasing hydrostatic pressure, we assume that the function f_2 follows the exponential form given in [10,11]. To account for the reduction of strength due to damage (micro-crack) formation, the yield strength is relaxed by a damage-dependent function [12], $f_3 = 1 - D$, such that the material loses its strength completely when a suitable measure of D reaches one. The effect of V_f can either be implicit in the previous functions, or be explicitly specified by f_4. A power law ($f_4 = V_f^{2/3}$) has been suggested in [13].

Conclusions

The behaviour of MMC at high strain rates is sufficiently different from that of isotropic homogeneous metals that current high strain rate consititutive models are not applicable. There are little experimental data available currently, though tests to date show very promising results. On the basis of work in the literature and our own results, a candidate phenomenological constitutive model for the high strain rate behaviour of MMC has been proposed.

Acknowledgements

This work is supported by the Defence Research Establishment Valcartier (DREV), Québec, Canada, with Mr. René Larose as Scientific Authority. The assistance and encouragement of the staff at DREV is gratefully acknowledged.

References

1. NMAB Committee, "Materials Response to Ultra-High Loading Rates", National Materials Advisory Board, NMAB-356, Washington, D.C. (1980).
2. Walter, W.P. and Zukas, J.A., Fundamentals of Shaped Charges (John Wiley, New York 1989) pp. 266-308.
3. Vaziri, R., Pageau, G. and Poursartip, A., "Review of Computer Codes for Deep Penetration Modelling of Metal Matrix Composite Materials", Report, Dept. Metals & Materials Eng., U.B.C. (December 1989).
4. Vaziri, R., Pageau, G. and Poursartip, A, "Constitutive Modelling of Materials at High Strain Rates - A Comprehensive Review", Report, Dept. Metals & Materials Eng., U.B.C. (July 1990).
5. Arsenault, R.J. and Wu, S.B., "The Strength Differential and Bauschinger Effects in SiC-Al Composites", Mater. Sci. Eng. 96, 77-88 (1987).
6. Vasudevan, A.K., *et al.*, "The Influence of Hydrostatic Pressure on the Ductility of Al-SiC Composites", Mater. Sci. Eng. A107, 63-69 (1989).
7. Marchand, A., *et al.*, " An Experimental Study of the Dynamic Mechanical Properties of an Al-SiC_w Composite", Eng. Frac. Mech. 30, 295-315 (1988).
8. Jones, S.E., *et al.*, "A Rate-Dependent Interpretation of the Taylor Impact Test", J. Energy Res. Tech. 111, 254-257 (1989).
9. Zerilli, F.J. and Armstrong, R.W., "Dislocation-Mechanics-Based Constitutive Relations for Material Dynamics Calculations", J. Appl. Phys. 61(5), 1816-1825 (1987).
10. Desai, C.S. and Siriwardane, H.J., Constitutive Laws for Engineering Materials (Prentice Hall, NJ 1984) p. 296 and 314.
11. Karr, D.G., *et al.*, "Asymptotic and Quadratic Failure Criteria for Anisotropic Materials", Int. J. of Plasticity 5, 303-336 (1989).
12. Rajendran, A.M. and Kroupa, J.L., "Impact Damage Model for Ceramic Materials", J. Appl. Phys. 66(8), 3560-3565 (1989).
13. Mahoney, M.W. and Ghosh, A.K., "Superplasticity in a High Strength Powder Aluminum Alloy with and without SiC Reinforcement", Metall. Trans. A 18A, 653-661 (1987).

Table I - Material data, penetration and Taylor test results

Material	ρ (kg/m^3)	E (GPa)	ν	Y_s (MPa)	UTS (MPa)	Y_d (MPa)	ε_f (%)	P_{exp} (mm)
6061-T6 Al	2710	69.0	0.34	275	310	313	15.2	79
7075-T6 Al	2805	71.5	0.34	495	570	597	11.3	50
30% SiC_p MMC	2844	120.7	0.27	434	552	575	3.0	45
15% Al_2O_{3p} MMC	2869	87.6	0.32	317	359	368	5.4	63

Y_s =static tensile yield strength at 0.2% strain; Y_d =dynamic yield strength calculated using Eqs (11) to (17) of [8]; ε_f =failure strain; P =experimentally measured penetration depth.

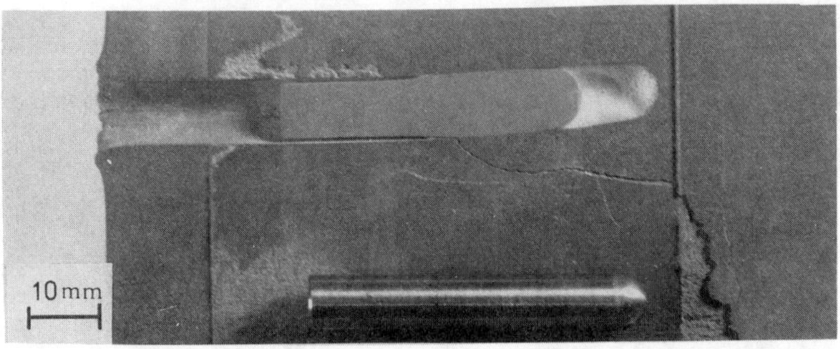

Fig. 1 - Section through the penetrated 30% SiC_p specimen

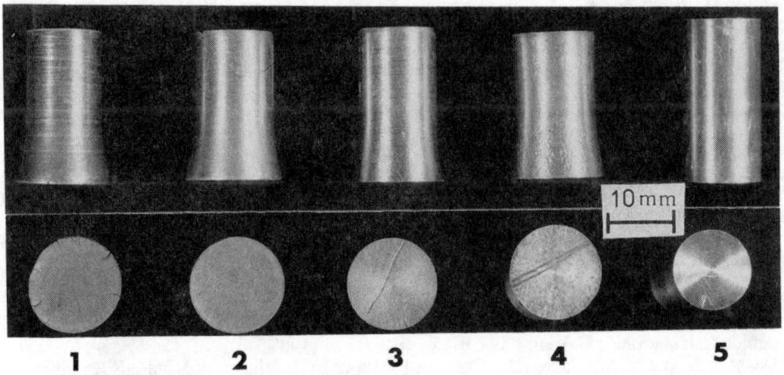

Fig. 2 - Side views (upper) and impact faces (lower) of Taylor cylinders: (1) 30% SiC_p MMC (2) 15% Al_2O_{3p} MMC (3) 7075-T6 Al (4) 6061-T6 Al (5) undeformed 6061-T6 Al

AN ELASTO-PLASTIC CONSTITUTIVE MODEL FOR THE BEHAVIOUR OF CHEMICAL POWDERS AT LOW STRESSES

L. LANCELOT & I. SHAHROUR

Laboratoire de Mécanique de Lille
Département Sols-Structures
IDN - B.P. 48
59651 Villeneuve d'Ascq Cedex - FRANCE

ABSTRACT

An elasto-plastic model has been developped using experimental data obtained with a special triaxial cell. The model includes a deviatoric mechanism, involving a Mohr-Coulomb failure surface, an isotropic strain-hardening and strain-softening rule, a non-associated plastic potential, and an associated volumic mechanism, involving a purely volumic strain-hardening rule. The model has then been validated on triaxial and oedometric tests.

INTRODUCTION

Current draftcodes for the design of silos are based on a simplified analysis of the state of stress in silos derived from the theories of Janssen [2] and Jenike [3]. A rigid-plastic constitutive law is thus assumed for bulk solids.

However, deformations play a major part in the behaviour of bulk materials [5]. A few elasto-plastic constitutive models, allowing a better quantification of deformation, have recently been proposed for bulk solids [1]. However, sophisticated models require that experimental data be available, first for parameters determination, then for validation on various stress paths.

The study undertaken in Lille involves two aspects :

• the development of a new specific experimental equipment adapted for low stresses, and

• the development and validation of a constitutive model based on experimental data obtained with the new testing equipment on two kinds of chemical bulk solids.

The experimental part of the study has been fully described elsewhere [4]. Two powders were tested : a crystallized organic acid (powder A) and a very fine mineral powder (powder B). Experimental results have been interpreted using common concepts in soils modelling (critical state, characteristic threshold, strain hardening...), and a constitutive model has been developped according to these concepts. This model is briefly described here, and a few simulations of triaxial and oedometric tests are presented on powder A.

A CONSTITUTIVE MODEL FOR POWDERS

The model is based on elasto-plasticity theory, where a strain increment $\dot{\varepsilon}$ is the sum of a recoverable part $\dot{\varepsilon}^e$ (elastic) and an irrecoverable part $\dot{\varepsilon}^p$ (plastic) :

$$\dot{\varepsilon} = \dot{\varepsilon}^{e} + \dot{\varepsilon}^{p} \quad (1)$$

For an elasto-plastic model, expressions have to be derived for :

- elastic behaviour,
- failure criterion and yield surface,
- hardening and/or softening rule,
- plastic potential (flow rule).

Elasticity

In our model, non-linear elastic behaviour was expressed using the following equations :

$$E = E_A \cdot \frac{p}{P_A} \quad (2)$$

$$\nu = \nu_o \quad (3)$$

where E and ν are Young modulus and Poisson's ratio respectively, p is the mean stress, E_A and ν_0 are material constants, and P_A is atmospheric pressure.

Shear yielding mechanism

A Mohr-Coulomb failure criterion has been chosen. A hardening function R_s has then been introduced to account for the hardening behaviour of powder A (figure 1):

$$f_s(p,q,\varepsilon_d^p) = q - M_f\,(p + C)\,R_s(\varepsilon_d^p) \quad (4)$$

where M_f and C are material constants and R_s ranges from R_0 (initial elastic radius) to unity at failure.

The following expression was shown to give satisfactory results for the evolution of elastic radius R_s with plastic shear strain ε_d^p [4] :

$$R_s(\varepsilon_d^p) = 1 - \exp(-a\,\varepsilon_d^p) \quad (5)$$

where a is a material constant.

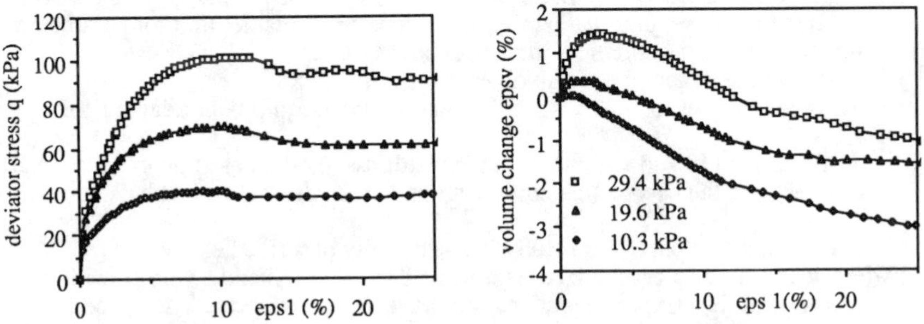

Figure 1 : Typical triaxial test results for 3 confining pressures

Flow rule

Plastic strain increment is supposed to derive from a plastic potential g. To account for the characteristic threshold concept, g can be chosen such that :

$$\frac{\partial g}{\partial p}(p, q) = M_g - \frac{q}{p} \quad (6)$$

where M_g is a material constant. In the present model, stabilization of $\varepsilon_v{}^p$ for large shear strain, as observed on laboratory triaxial tests (figure 1), was obtained by modifying the above equation, so that :

$$\frac{\partial g}{\partial p} = A_g\left(M_g - \frac{q}{p}\right) \tag{7}$$

where :

$$A_g = e^{-\alpha_g . \varepsilon_d^p} \tag{8}$$

This expression for $(\partial g/\partial p)$ introduces another parameter α_g.

Strain softening can also be observed on experimental curves for powder A (figure 1). The expression of R_s will then be modified so that R_s passes through a maximum and then decreases until it stabilizes for large shear strain. To achieve this, a dependency of R_s with volumic strain $\varepsilon_v{}^p$ is introduced :

$$R_s(\varepsilon_d{}^p, \varepsilon_v{}^p) = (1 - \exp(- a\, \varepsilon_d{}^p))\exp(\mu\, \varepsilon_v{}^p) \tag{9}$$

μ is a new parameter setting the stress peak magnitude.

<u>Consolidation yielding mechanism</u>

Stress paths involving low deviatoric stress levels q/p, such as in oedometric or isotropic compression tests, are not satisfactorily reproduced by a single shear yielding mechanism model. Along such paths, the shear yielding surface expands very little, and the plastic strains generated are too low [4]. In this model, a second, purely volumic yielding mechanism defining a plane yield surface $f_c(p, \varepsilon_v{}^p) = 0$, with an associated flow rule, is introduced.

The most suitable expression for f_c was found to be [4] :

$$f_c\,(p, \varepsilon_v^p) = \left(\frac{p}{P_{co}}\right)^{0.5} - R_c \tag{10}$$

$$R_c = \exp\,(\beta\, \varepsilon_v^p) \tag{11}$$

where P_{c0} and β are material constants.

TYPICAL RESULTS

Oedometric tests were carried out for the determination of the compressibility coefficient β, and for the validation of the model. As shown on figure 2 for a typical simulation, oedometric tests are satisfactorily reproduced with the model, although plastic strains are slightly underestimated.

Triaxial tests involve three loading steps : oedometric loading for sample preparation, isotropic loading at a given confining pressure, and triaxial shear loading until sample failure [4].

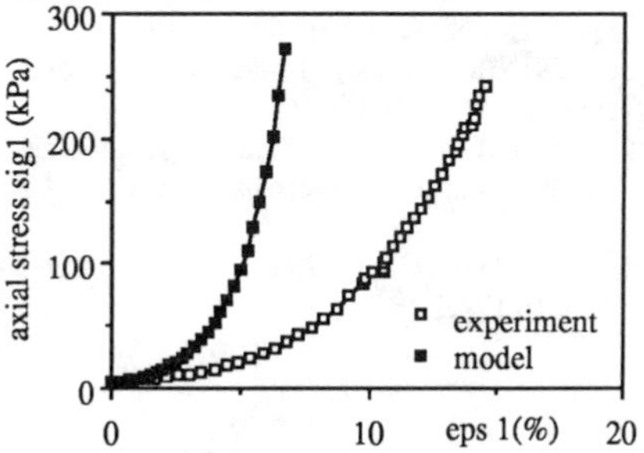

Figure 2 : Simulation of oedometric test

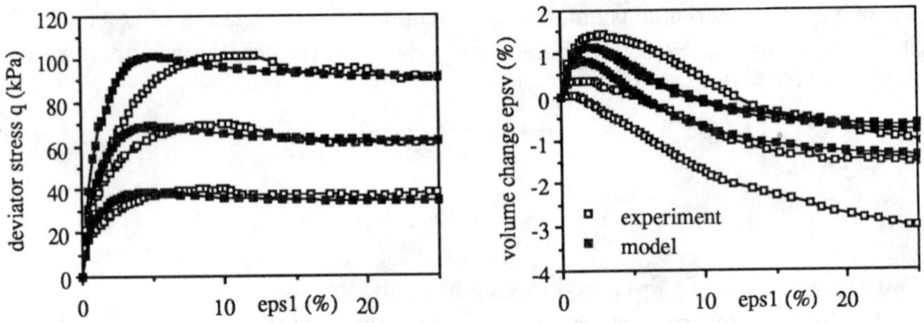

Figure 3 : Simulations of triaxial tests

Figure 3 presents typical simulations for the triaxial compression tests shown on figure 1. All the experimentally observed features are satisfactorily reproduced : progressive plastification (hardening), stress peak, softening, and eventual stabilization of deviator stress and volume change for large shear strains.
However, the deviatoric stress level after oedometric loading as computed by the model is overestimated and results in a R_S value of 0.9. This large value for the elastic radius induces an almost elastic-perfectly plastic simulated behaviour for the subsequent triaxial shear loading. A kinematic hardening rule would thus improve the simulations on such stress paths.

CONCLUSION

In this paper, a nine constant constitutive model is presented and used to simulate the behaviour of chemical bulk solids. The modelled response of powders for oedometric and triaxial tests is satisfactory, although isotropic strain hardening was found to be insufficient to reproduce stress paths involving large stress inversions (such as oedometric loading-unloading followed by triaxial reloading). Further developments of the model will include the implementation of a kinematic hardening rule and the introduction of viscosity in order to reproduce real flow in silos.

REFERENCES

1. EIBL J. & ROMBACH G., "Numerical investigations on discharging silos", Int. Conf. on Numerical Meth. in Geomechanics, Innsbrück, edited by Swoboda, 317-320, (1988).

2. JANSSEN H. A., "On the pressure of grains in silos", Proceedings Institution of Civil Engineers, 553, (London 1896).

3. JENIKE A. W., "Storage and flow of bulk solids", Utah Eng. Exper. Station, Bulletin 123, University of Utah (1964).

4. LANCELOT L., "Etude experimentale et modélisation du comportement des poudres de l'industrie chimique", Thèse de Génie Civil, Université des Sciences et Techniques de Lille (1990, in French).

5. SCHWEDES J., "Entwicklung der Schüttguttechnik seit 1974", Aufberietungs Technik, n° 8, 403-410 (1982).

IMPLEMENTATION, EVALUATION

COMPARISON OF VARIOUS PLASTICITY/VISCOPLASTICITY MODELS

K.Aazizou, H.Burlet, G.Cailletaud
Centre des Matériaux de l'Ecole des Mines de Paris,
UA CNRS 866, BP 87, 91003 EVRY CEDEX, FRANCE

ABSTRACT

This paper shows constitutive equations including plasticity and viscoplasticity with a coupling between them. The implementation in a finite element code is discussed. An evaluation of the performance of the numerical treatment is made, by comparison with more classical models.

INTRODUCTION

Non linear kinematic hardening is going to become an international standard for the representation of cyclic behaviour of materials [1]. It can be used either in the framework of time independent plasticity (TIP), or in viscoplasticity (VIP). Using the same concept, the present paper shows a model with plasticity, viscoplasticity, and a coupling between the hardening modes (VPC) : in this model, the inelastic strain is the sum of plastic and viscoplastic strain, which are simultaneously integrated for any kind of loading; the coupling is produced by linear relations between the plastic and viscoplastic variables used for kinematic hardening. The main interest of this model is its hierarchical character : it includes as special cases pure time independent plasticity and pure viscoplasticity, so that the identification of the material coefficients is modular [2]. The coefficients of the plastic part of the model are found from monotonic tensile and from fatigue tests, these concerning the viscoplastic part from creep or relaxation tests, or from cyclic tests with hold periods. The coupling term (if existing) is defined by tests which sequentially involve plasticity and viscoplasticity. This model has successfully be applied in the past for stainless steels [2] or aluminium alloys [3]. In the following, we give some elements concerning its numerical implementation in a finite element code, and the computational cost which is attached with.

FORMULATION OF THE MODEL.

We describe now a basic version of the model, which is expressed in a completely symmetrical form. The kinematic variables are the tensor $\mathbf{X_p}$ for the plastic part, and the tensor $\mathbf{X_v}$ for the viscoplastic part, the isotropic variables are the scalar R_p for plasticity and the scalar R_v for viscoplasticity. For both plasticity and viscoplasticity, a von Mises criterium is used :

$f_p = J(\sigma-\mathbf{X_p})-R_p$, with $J(\sigma-\mathbf{X_p}) = \left(1.5(\sigma'-\mathbf{X'_p}):(\sigma'-\mathbf{X'_p})\right)^{0.5}$, for plasticity,

$f_v = J(\sigma\text{-}\mathbf{X_v})\text{-}R_v$, with $J(\sigma\text{-}\mathbf{X_v}) = \left(1.5(\sigma'\text{-}\mathbf{X_v'}):(\sigma'\text{-}\mathbf{X_v'})\right)^{0.5}$, for viscoplasticity, ($\sigma'$, $\mathbf{X_p'}$, $\mathbf{X_v'}$ represent the deviatoric part of σ, $\mathbf{X_p}$ and $\mathbf{X_v}$).

The expression of the hardening variables versus state variables is shown in Table I, and the state variables evolution in Table II. In these Tables and in the following of the text, the material coefficients are denoted by *italic* characters.

TABLE I. Expression of the hardening variables.

$\mathbf{X_p} = \frac{2}{3}\, c_p . a_p . \alpha_\mathbf{p} + \frac{2}{3} c_{vp} . \alpha_\mathbf{v}$
$\mathbf{X_v} = \frac{2}{3}\, c_{vp}\ \alpha_\mathbf{p} + \frac{2}{3} c_v\, a_v\, \alpha_\mathbf{v}$
$\mathbf{R_p} = R_{ps} + (R_{po} - R_{ps})\exp(-b_p .p)$
$\mathbf{R_v} = R_{vs} + (R_{vo} - R_{vs})\exp(-b_v .v)$

TABLE II. Evolution of the variables.

$\dot{\alpha}_\mathbf{p} = \dot{\varepsilon}_\mathbf{p} - c_p\ \alpha_\mathbf{p}\ \dot{p}$
$\dot{\alpha}_\mathbf{v} = \dot{\varepsilon}_\mathbf{v} - c_v\ \alpha_\mathbf{v}\ \dot{v}$
$\dot{p} = (\frac{2}{3}\ \dot{\varepsilon}_\mathbf{p} : \dot{\varepsilon}_\mathbf{p})^{0.5}$
$\dot{v} = (\frac{2}{3}\ \dot{\varepsilon}_\mathbf{v} : \dot{\varepsilon}_\mathbf{v})^{0.5}$

The plastic and viscoplastic strain rates are then respectively deduced from the normality rule :

$$\dot{\varepsilon}_\mathbf{p} = \dot{p}\, \mathbf{n_p}\ ;\quad \dot{\varepsilon}_\mathbf{v} = \dot{v}\, \mathbf{n_v}, \qquad \text{with } \mathbf{n_p} = \frac{\partial f_p}{\partial \sigma}, \text{ and } \mathbf{n_v} = \frac{\partial f_v}{\partial \sigma}$$

The viscoplastic strain rate $\dot{v}$ is determined by the equipotential surface defined by f_v function, and the plastic one $\dot{p}$ by the consistency condition :

$$\dot{v} = \left(\frac{J(\sigma - \mathbf{X_v}) - R_v}{K}\right)^n, \text{ and } \dot{p} = (\ \mathbf{n_p} : \dot{\sigma} - \frac{2}{3}\, c_{vp}\ \ \dot{\alpha}_\mathbf{v} : \mathbf{n_p}\ ) / H,$$

with $H = \frac{2}{3} c_p\ a_p\ \mathbf{n_p} : (\mathbf{n_p} - c_p\ \alpha_\mathbf{p}) + b_p\ (R_{ps} - R_{po})\exp(-b_p .p)$

It is worth noting that the plastic strain rate depends on the viscoplastic hardening rate, so that it is indirectly time dependent. In fact, the plastic-viscoplastic partition of inelastic strain will change according to the strain rate. The general scheme of inelastic flow is illustrated in Figure 1.

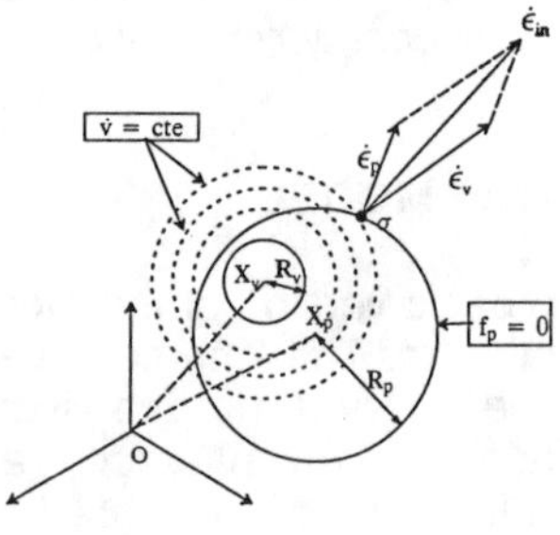

FIG.1. General scheme of plastic and viscoplastic flow for VPC model.

NUMERICAL IMPLEMENTATION AND TEST.

The most sensitive part of the computational procedure is the local integration of the constitutive equations, which may be time consuming for models with many internal variables. Note that for the real materials, we use a version of the VPC model with one isotropic but two kinematic variables for each straining mode, and so that the local system present 38 variables in 3D cases, including plastic and viscoplastic strains. The classical methods have been implemented for TIP, VIP and the VPC models, radial return [3], mid-point rule [4] or θ–method associated with a local resolution by a Newton method. The radial return method is classically known for being equivalent to a pure implicit integration and the mid-point rule to a semi-implicit one. This is still valid for linear kinematic hardening, but this is no longer true for non linear kinematic hardening. That can be studied on error maps, like in [5]. The general conclusion is not surprising : the pure implicit method gives the best results, specially for plane stress conditions and in the case of large increments, if a "consistent stiffness" matrix [6] is used on the global level. The semi-implicit integration is more precise for small increments, and is very efficient when associated with a substepping technique. These conclusions should be tempered for very large problems. For models expressed here, if the mesh size exceeds 10000 dof , the calculation and inversion of the stiffness matrix becomes too heavy, and alternative methods like BFGS [7] are more attractive. These results are detailed elsewhere [8].

Two examples are now shown in order to quantify the performances obtained with the various formulations. The first one simply concerns a single element. The coefficients are chosen for VIP model so that the behaviour is practically time independent. The VPC model presents the same type of identification. Figure 2 shows the curves obtained for a 2% tension under strain control, with strain rates varying between 10^{+1} and 10^{-8} s^{-1}. The time needed to integrate the three models for this loading is given in Fig.3. The TIP model produces the lower bound; for VIP model, the CPU time slowly increases when strain rate decreases, that is due to the fact that the viscoplastic strain increases in the same time. Two regions can be distinguished for the VPC model : for strain rates faster than 10^{-6} to 10^{-7} s^{-1}, the two straining modes are active, and all the equations are present in the local system. For lower strain rates, the "elastic limit" in sense of the time independent plasticity is not reached, and only viscoplastic flow is present. In these conditions, the model becomes faster than VIP, because the identification of VPC which gives the same viscous effect of VIP involves a well conditioned coefficient set. The same exercice has been made on a small structure, made with an axisymmetric notched specimen (see a characteristic result in Fig.4). In this test, the three models are still identified to give similar results during a tension test. The amount of inelastic strain obtained for a 0.1 mm tension is about the same in each case, as shown in Table III. The time spent for VPC model is about three times higher than for TIP and VIP models. The price to pay for using the VPC model is then quite reasonnable.

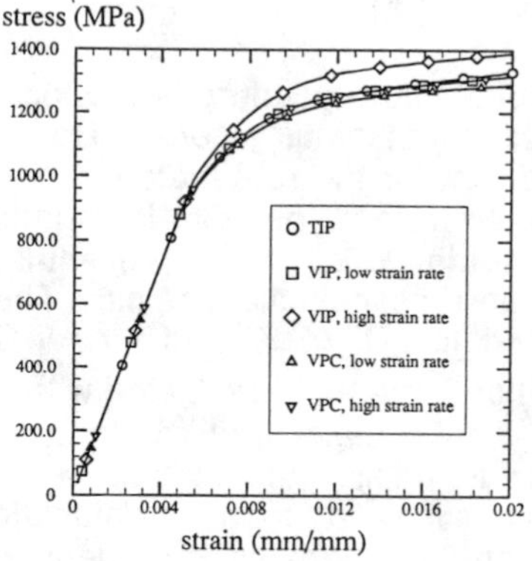

FIG.2 Stress response for TIP, VIP and VPC model under strain control.

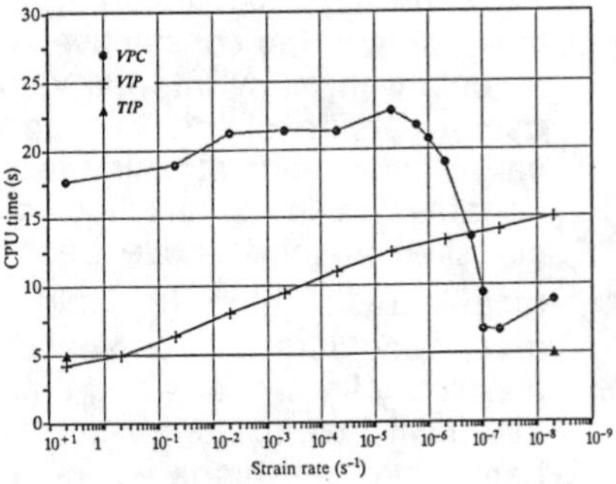

FIG.3. CPU time for the 3 models vs. strain rate.

TABLE III. Max inelastic strain and CPU time for the 3 models.

	ε_{max}	CPU time[a)]
TIP	0.115	528 s
VIP	0.095	676 s
VPC		1971 s

a) on a Sun4/330

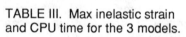

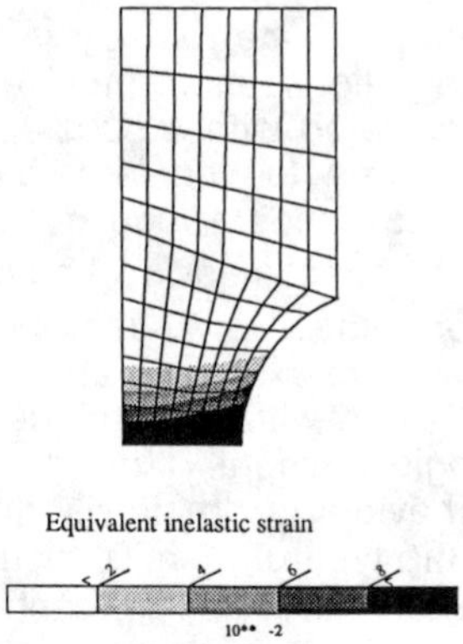

FIG.4 Test on axisymmetrical notched specimen.

REFERENCES.

1. J.L.Chaboche, " Constitutive equations for cyclic plasticity and cyclic viscoplasticity", Int.J. of Plasticity , 5, 247-302 (1989).
2. E.Contesti and G.Cailletaud, "Description of Creep-Plasticity Interaction with non-unified Constitutive Equations : Application to an austenitic stainless steel", Nucl.Eng.Design, 116, 265-280 (1989).
3. M.L. Wilkins, "Calculations of elastic-plastic flow", in Methods of computational Physics, edited by B.Adler et al., Vol.3 (Academic Press, New York 1964).
4. J.R.Rice and D.M. Tracey, "Computational Fracture Mechanics", in Numerical and Computer Methods in Structural Mechanics, edited by S.J.Fenves (Academic Press, New York 1973).
5. R.D.Krieg and D.B.Krieg, "Accuracies of numerical solution for the elastic-perfectly plastic model", J.Press.Vessel Tech.510-515 (1977).
6. J.C.Simo and R.L.Taylor, "Consistent Tangent Operators for Rate-Independent Elastoplasticity", in Computers Methods in Applied Mechanics and Engeneering, 48, 101-118, (North Holland 1985).
7. H.Matthies and G.Strang, "The solution of non linear Finite Element equations", Int.J. Numer.Methods Engeneering, 14, 1613-1626 (1979).
8. K.Aazizou, H.Burlet and G.Cailletaud, submitted at Engineering computations.

A FEM ANALYSIS OF STARIN LOCALIZATION USING A NON-LOCAL STARIN-SOFTENING PLASTICITY

Toshihisa ADACHI and Feng XHANG
Dept. of Transportation Engr., Kyoto Univ., Japan

Fusao OKA and Atsushi YASHIMA
Dept. of Civil Engr., Gifu Univ., Japan

ABSTRACT

In the present study, the behavior of a sedimentary soft rock under the drained triaxial compression and the vertical pressure on lowering panel were analyzed using a constitutive relation with strain hardening and strain softening.

CONSTITUTIVE MODEL

Adachi-Oka's model

Adachi and Oka [1] developed an elasto-plastic constitutive model for overconsolidated clay based on the non-associated flow rule, considering a special plastic potential function in the overconsolidated region. They extended this model by use of the stress history tensor in order to express both strain hardening and strain softening characteristics. They proposed that the stress history tensor is a functional of the stress history with respect to the strain measure. This stress history tensor, σ^*_{ij}, is given by

$$\sigma^*_{ij} = \frac{1}{\tau}\int_0^z exp((z-z')/\tau)\sigma_{ij}(z')dz', \tag{1}$$

$$dz = (de_{ij}de_{ij})^{1/2}, \tag{2}$$

in which z is the strain measure, τ is the material parameter which controls the strain hardening and strain softening phenomena and e_{ij} is the deviatoric strain tensor. Although the deviatoric stress increases in the early stage of loading, reaches the peak strength and then decreases and finally reaches the residual strength, the stress history tensor monotonously increases during the whole deformation process and leads to the residual strength state. At large strain, the difference between stress and stress history tensor becomes small. Therefore, it can be said that the difference between the stress and the stress history tensor corresponds to the cohesive strength due to cementation or bonding. The cohesive strength can be destructured due to deformation. Strain measure z is similar to the intrinsic time used in the endochronic theory advocated by Valanis [2].

We summarize the constitutive model as follows:

Yield function:

$$f_y = \bar{\eta}^* - k_s = 0, \qquad \text{where} \quad \bar{\eta}^* = (S^*_{ij}S^*_{ij}/\sigma^{*2}_m)^{1/2}, \tag{3}$$

in which S^*_{ij} is the deviatoric component of σ^*_{ij} and σ^*_m is the spherical component of σ^*_{ij}.

Hardening-softening parameter:

$$k_s = \frac{M_f^* G' \gamma^p}{M_f^* + G' \gamma^p}, \tag{4}$$

where γ^p is the second invariant of the deviatoric plastic shear strain and M_f^* is the value of $\bar{\eta}^*$ at failure.

Plastic potential:

$$f_p = \bar{\eta} + \tilde{M} ln((\sigma'_m + b)/(\sigma_{mb} + b)) = 0, \tag{5}$$

where $\tilde{M}$ and $\bar{\eta}$ are defined by

$$\tilde{M} = -\frac{\bar{\eta}}{ln((\sigma'_m + b)/(\sigma_{mb} + b))} \quad \text{and} \quad \bar{\eta} = \left(\frac{S_{ij}}{\sigma'_m + b}\frac{S_{ij}}{\sigma'_m + b}\right)^{1/2}. \tag{6}$$

Nonlocal generalization of the Adachi-Oka's model

Bazant [3] clarified that the main shortcoming of the continuum model for strain softening is that the energy dissipated at the failure state is incorrectly predicted to be zero, and that the finite element solutions converge to these incorrect solutions as the mesh is refined. Bazant et al.[4] developed a nonlocal continuum model with local strains to describe the softening behavior on a continuum level. According to their theory, only the damaging parameter is considered nonlocal. A nonlocal model such as Bazant's may be one reasonable way to describe strain softening.

We can generalize the Adachi-Oka's model based on Bazant's concept of the non-localization of the material model. The nonlocal stress history tensor, $\bar{\sigma}^*_{ij}$ in Adachi-Oka's model is obtained by using the following equation:

$$\bar{\sigma}^*_{ij} = \frac{1}{V_{r(x)}} \int_v \alpha(\mathbf{x} - \mathbf{x}')\sigma^*_{ij}(\mathbf{x}')dv, \tag{7}$$

where

$$V_{r(x)} = \int_v \alpha(\mathbf{x} - \mathbf{x}')dv, \tag{8}$$

and $\alpha(\mathbf{x} - \mathbf{x}')$ is a given weighting function of the distance $|\mathbf{x} - \mathbf{x}'|$. For the sake of simplicity, we have used the normal distribution function for $\alpha(\mathbf{x} - \mathbf{x}')$,

$$\alpha(\mathbf{x} - \mathbf{x}') = e^{-(k|\mathbf{x}-\mathbf{x}'|/L)^2}. \tag{9}$$

In this study, two different values are used for material parameter L, i.e., $L = 1.0cm$(case-1) and $L = 2.0cm$(case-2). In the case of $L = 2.0cm$(case-2), the effect of weighting function is cut off when the distance of the center of the element is longer than the critical value ($|\mathbf{x} - \mathbf{x}'| = 1.3cm$).

NUMERICAL EXAMPLES BY FEM

Triaxial compression test

We have applied the Adachi-Oka's model for the behavior of sedimentary soft rock under drained triaxial compression tests. For this simulation, we have used 4 noded linear element. The calculation was carried out under an axisymmetric condition. Material properties and initial condition are listed in Table 1. A comparison between the computed and experimental results for the stress-strain relationships are shown in Fig.1.

In order to examine the mesh size sensitivity, different numbers of elements (8, 50, or 200) were used to analyze a quarter of the cylindrical specimens. The average stress-strain relations computed are shown in Fig.2. We have also prepared Fig.3 to examine the convergence of the solution with increase of the number of finite elements. In Fig.3, stress difference denotes the difference of deviatoric stress at the deviatoric strain = 0.96 % between the case in which the number of finite element mesh is one and the others with 8 to 200 finite element meshes. In this figure, results by nonlocal model are also shown. It appears that the solution converges very well even for the local case. It is found that the convergence is a little bit better for the nonlocal case than for the local case. This may be related to the uniqueness of the solution in the sense of Valanis[5].

Vertical pressure on lowering panel

Experiments with lowering panels are often performed in order to clarify the earth pressure on a tunnel. In this study, the lowering panel experiment was simulated using the local Adachi-Oka's model. For this simulation, 8 noded isoparametric elements were used. To avoid the excessive stress concentration around the edge of the lowering panel due to the discontinuous descent of the panel, the declined area between the lowering panel and the base of the box was introduced. Finite element mesh, boundary conditions, and material properties are shown in Fig.4.

The acting load on the panel between point A and B while the panel is being lowered is plotted in Fig.5. That is to say, the load decreases rapidly due to the descent of the panel, after passing through the minimum value, it again increases. This result correpsonds to the empirical fact that "if a ground is loosened excessively, the acting load to a tunnel increases".

Contours of octahedral strain at the stage at which the panel was lowered down to 8mm is shown in Fig.6. It is clear that the strain localization occurs around the edge of the declined panel.

REFERENCES

1. F.Oka and T.Adachi, Proc. 5th ICNMIG, Nagoya, 1985, pp.293-300
2. K.C.Valanis, Part I, Arch. of Mech., Vol.23, No.4, 1971, pp.517-533
3. Z.P.Bazant, J. Appl. Mech., ASME, Vol.55, 1988, pp.287-293
4. Z.P.Bazant et al., Proc. Int. Conf. on Com. Plas., 1987, pp.1757-1779
5. K.C.Valanis, J. Applied Mechanics, Vol.52, 1985, pp.649-653

Table 1. Material properties.

E	kgf/cm^2	13,500
K	kgf/cm^2	3,700
G'		1,000
b	kgf/cm^2	40
σ_{m0}	kgf/cm^2	1.0
σ_{mb}	kgf/cm^2	150
M_f		1.97
τ		0.09

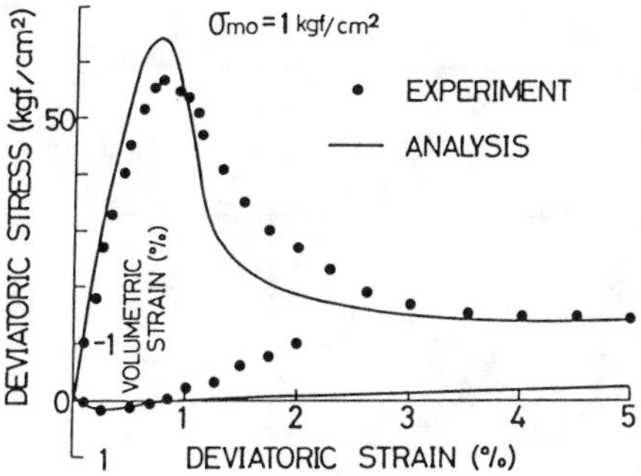

Fig.1. Stress-strain relation.

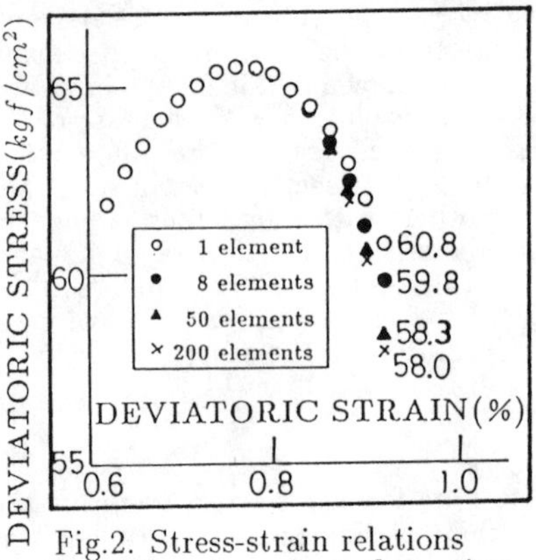

Fig.2. Stress-strain relations for different mesh configurations.

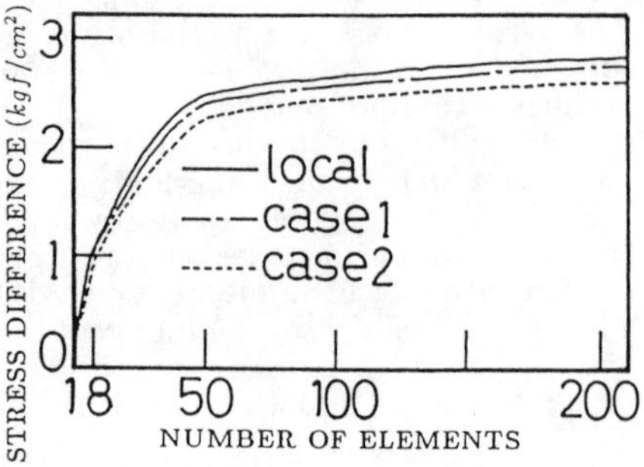

Fig.3. Convergence of solution.

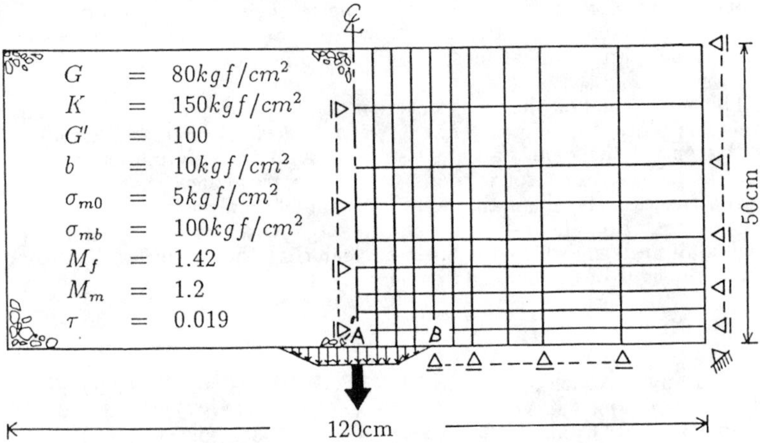

Fig.4. Numerical conditions for lowering panel problem.

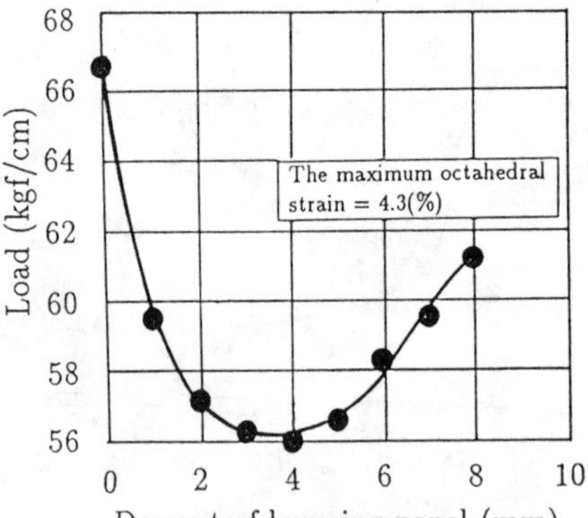

Fig.5. Relation between acting load and descent of the panel.

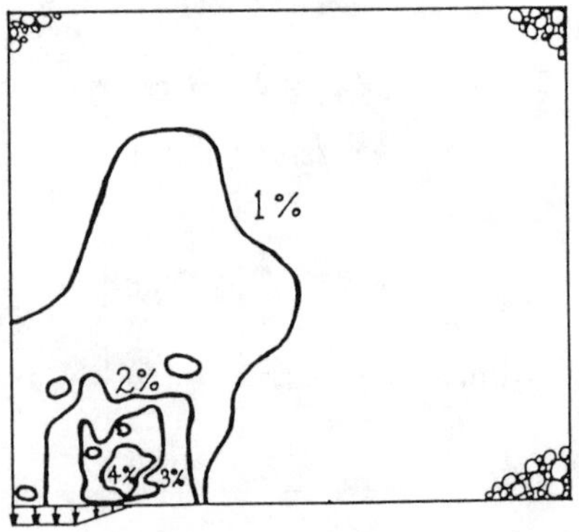

Fig.6. Contours of octahedral strain at descent of the panel of 8mm.

On the Stability and Well–Posedness of Nonmonotonic Material Models

T. Belytschko and M. Kulkarni

Department of Civil Engineering, Robert R. McCormick School of Engineering and Applied Science, The Technological Institute, Northwestern University, Evanston, IL 60208–3109.

ABSTRACT

The stability and well-posedness of material models which violate Drucker's stability postulate are examined. Both rate-dependent and rate-independent materials are considered. It is shown that certain norms of engineering interest are insensitive to the detailed character of localization fields.

INTRODUCTION

The analysis of nonmonotonic material models has received considerable attention since the thought provoking publication of Sandler and Wright [1]. Nonmonotonicity or strain softening is a mathematical representation of failure of a material in which the load carrying capacity of the material decreases with increasing deformation. Materials that exhibit a descending branch in their stress–strain curves may then be thought of as unstable materials. Sandler and Wright argued that mathematical models of nonmonotonic materials are ill-posed in that small changes in the loading conditions resulted in large changes in the response. They did not however clearly distinguish this behavior from similar behavior such as dynamic buckling or snapthrough. In buckling or snapthrough, large changes in response also occur with small changes in the initial data. If the perfect model of a beam is subjected to a load substantially greater than it's buckling load, then time integration of the discrete equations of a perfectly straight beam will not exhibit buckling. However, if the initial data is changed by introducing an imperfection, the physically correct buckling response can be simulated. Similar behavior is observed in fluid mechanics: a laminar flow simulation will remain laminar beyond the critical Reynolds number unless instabilities are triggered by imperfections in the initial data or round off errors. These problems are therefore sensitive to initial data, yet the models governing these phenomena are not considered ill-posed.

The dilemma concerning well-posedness perhaps stems in part from interpretations of the definitions of Ritchmeyer and Morton, which can be paraphrased as follows; a problem is well-posed if small changes in the data make only small changes in the solution. However, this definition was probably not intended to be applied to perfect models of unstable physical problems: it is unavoidable in such models that small changes in data lead to large changes in response. Such large changes in response are the very reason these models can describe unstable physical phenomena correctly, a striking example are the viscoplastic constitutive equations. Instability of these equations lead to close predictions of the Hopkinson torsional test. It may be emphasized that in such models, only an assumption of initial imperfections leads to physically meaningful results. Therefore it seems that the notion of well-posedness as defined in Ritchmeyer and Morton is not directly applicable to problems with instabilities: either the character of changes in initial data or the character of the changes in response must be restricted appropriately. Some proposals for such restrictions are given herein.

Some of the misunderstanding results from a failure to distinguish physical instability and stability of a solution. The solution to a physically unstable problem should be a stable solution if the correct instability is captured. Generally however, a solution is unstable in the regime where the physical response is unstable.

A striking difference between, for example, analysis of beam buckling and that of unstable materials is that strain softening is almost always accompanied by localization of deformation in a narrow region. In rate independent material models this region of intense straining localizes to a set of measure zero. Many regularization schemes have been proposed to circumvent this problem. Aifantis [2], Mulhaus and Aifantis [3] have proposed gradient regularization schemes wherein the effects of strain gradients are incorporated in the constitutive law; Bazant, Belytschko and Chang [4] used nonlocal constitutive equations. Several authors subscribe to the school of thought that treats rate independent models as the limit of a rate dependent model: Needleman [5], Loret and Prevost [6]. Batra and Kim [7] have studied the effects of strain–softening in the context of nonpolar constitutive laws. A combination of viscous and gradient regularization has been studied by Belytschko and Kulkarni [8]. However, all of these regularization schemes introduce substantial numerical difficulties. Therefore, if the regularization schemes do not affect the stability of the problem , they may in some cases be superfluous.

Even in the case of regularized models, such as rate dependent models, imperfections play a crucial role in the evolution of the problem; in the absence of an imperfection, the solution is strikingly different from that of a perfect model. The central issue is that the effect of initial imperfections must be considered to obtain a clear understanding of the behavior of nonmonotonic models. In this paper, a framework is presented within which the stability of material models can be examined and is applied to a one dimensional model problem. This approach, which is based on dynamical systems theory, may be used for both rate dependent and rate independent materials. It is shown that for certain norms, which are useful for engineering purposes, rate dependent and rate independent models exhibit similar behavior when viewed in this framework.

STABILITY OF EVOLUTION PROBLEMS

A useful and rigorous framework for an examination of stability of a non–linear system of differential equations is available in the theory of dynamical systems. In practice these systems of equations are tensor equations, however for simplicity we consider the following one-dimensional equation, which governs the evolution of a scalar variable in time

$$\frac{d}{dt} u(x,t) = g(x) \tag{1}$$

Stability of u can be examined as follows: if a solution $u'(x, t)$ is obtained with any perturbation $a'(x)$ on the initial conditions which are small in the sense that a norm

$$\| a'(x) \| < \varepsilon \tag{2}$$

where ε is an arbitrarily small parameter, and if the norm of the solution $u'(x, t)$ is in a finite hypersphere about the solution to the unperturbed problem $u(x, t)$, i.e. if

$$\|u(x, t) - u'(x, t)\| < C\,\varepsilon, \ \forall t \tag{3}$$

where C is a finite constant, then the solution is stable. Here $\| \cdot \|$ is a norm. The particular norm is a matter of choice by the analyst. One example of a suitable norm for a one-dimensional problem such as simple shearing of a slab of length L, is the L_2 norm, given as the following integral

$$\|u(x,t) - u'(x,t)\|^2 = \int_0^L [u(x,t)-u'(x,t)]^2 \, dx \tag{4}$$

Note that the solution u(x,t) may represent a *physical instability*. Nevertheless, to be a meaningful solution to the unstable problem, the solution itself must be *stable*.

Simple shearing of a slab of material has been accepted as an appropriate model for describing the Hopkinson torsional experiment. This one-dimensional problem may be analyzed using the framework described above. The equations governing the response of a slab subject to simple shear for a rate–independent model are the following

the momentum equation

$$\sigma_{,x} = \rho v_{,t} \tag{5a}$$

the constitutive equation

$$\dot{\sigma} = E^{tan} \dot{\varepsilon} \tag{5b}$$

the strain displacement equation

$$\varepsilon = u_{,x} \tag{5c}$$

Here σ represents the shear stress, ε, the shear strain, a superposed dot represents the material time derivative and a comma indicates differentiation with respect to the variable which follows. E^{tan} represents a tangent modulus obtained from the yield condition, consistency condition and the plastic flow rule. For rate dependent materials the constitutive equations are

$$\dot{\sigma} = E(\dot{\varepsilon} - \dot{\gamma} \operatorname{sgn}\sigma) \tag{5d}$$

and the evolution for the internal variable γ

$$\dot{\gamma} = f(\sigma,\gamma) \tag{5e}$$

We present here an analysis for rate independent materials, for a detailed analysis of rate dependent models the reader is referred to Belytschko, Moran and Kulkarni [9]. Stability is examined by first solving the problem with homogeneous initial conditions, i.e. with a(x)=0. The solution is then

$$\text{for } r(t) < \frac{\sigma_Y}{E} L: \; F = A\,E\frac{r(t)}{L},\; u(x,t) = \frac{r\,x}{L},\; \varepsilon(x,t) = \frac{r}{L} \tag{6}$$

$$\text{for } r(t) \geq \frac{\sigma_Y}{E} L:$$

$$F = A\;E^{tan}\frac{r(t)}{L} + \sigma_Y\,(1 - \frac{E^{tan}}{E}),\; u(x,t) = \frac{r\,x}{L},\; \varepsilon(x,t) = \frac{r}{L} \tag{7}$$

where $E^{tan} = \frac{EH}{E+H}$ and H is the plastic modulus (H < 0 implies softening). This is the homogeneous solution to the problem, which is shown in Fig. 1a.

We now examine its stability. Consider an arbitrary imperfection such as random C^1 noise in the cross-sectional area of the bar, i.e. let

$$A = A_0\,(1 - a'(x)) \tag{8}$$

Such perturbations are physically unavoidable and thus must be considered when they have a significant effect. Softening will first occur at the point where a(x) is a maximum, denoted by x_m, and will only occur at that point.

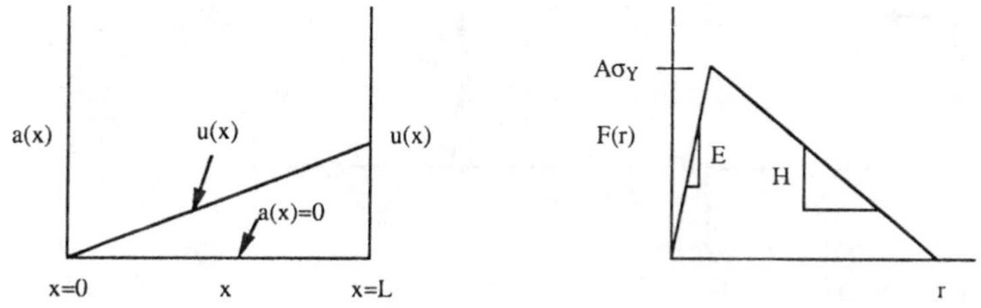

Fig. 1a

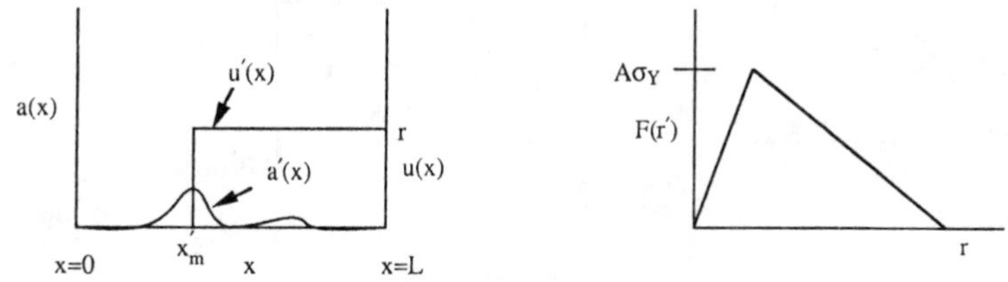

Fig. 1b

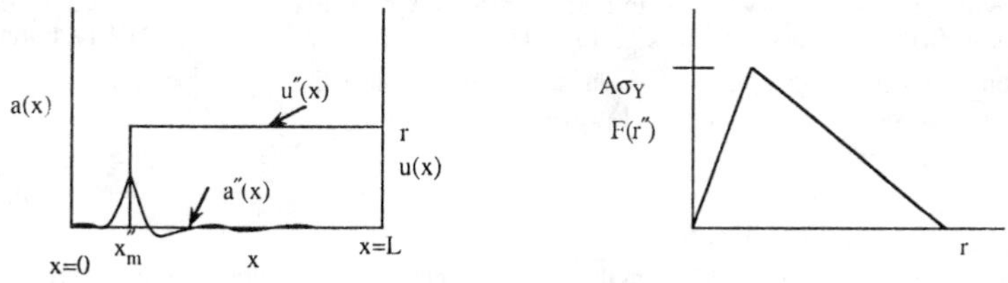

Fig. 1c

Thus for the function $a'(x)$ given in Fig. 1b the solution for $u'(x)$ is that shown in Fig. 2 with infinite strain at x_m'. The solution, for $r \geq \frac{\sigma_Y}{E} L$, is

$$u'(x,t) = r(x,t)\, H\,(x-x_m), \quad \varepsilon(x,t) = r(x,t)\, \delta(x-x_m) \tag{9}$$

This solution is unique but it depends on the location x_m. This solution is not close to the solution (7), in the sense of (3), which is obvious from comparing the plots of $u(x)$ and $u'(x)$, so the homogeneous solution is unstable. Strain-softening thus localizes to a point in a rate-independent material. Note that this conclusion is reached without recourse to thermodynamical arguments, in contrast to Ottosen [10] or Bazant [11].

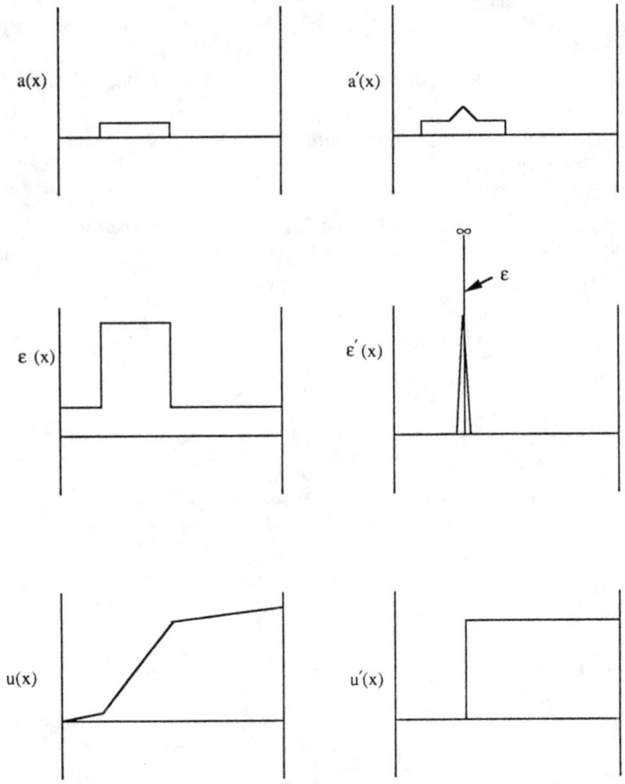

Fig. 2a Fig. 2b

One peculiarity in strain–softening problems is that a perturbation with a different maximum will have a solution which is in a sense strikingly different. That is, if we consider a perturbation $a''(x)$ with a maximum at x_m'', then the solution is as shown in Fig. 2. The norm of this solution will not be close to u', i.e. $\|u'(x, t) - u''(x, t)\| > C\,\varepsilon$. This peculiarity is also observed in rate dependent materials [9]! However, if

$$\|a'(x) - a'''(x)\| < \varepsilon^2 \tag{10}$$

then

$$\|u'(x, t) - u'''(x, t)\| < C\varepsilon \tag{11}$$

i.e. if a''' is a small perturbation of a', then the solution is changed very little, so u' can in this sense be considered a stable solution. Another way to consider stability is to consider another norm. Note that

$$\|F(r') - F(r'')\| < C\,\varepsilon \tag{12}$$

and in fact the two response functions are equal. Another way of expressing this is in terms of the L_∞ norm defined by

$$\max_{x \in \Omega} |u'(x, t) - u'''(x, t)| < C\varepsilon$$

The solution to the physically unstable problem is stable in this norm even when it is not stable in the L_2 norm. The L_∞ norm appears to be quite insensitive to the exact structure of the imperfection. From an engineering viewpoint, the maximum displacement is of principal interest, so stability of the solution in this norm may be adequate.

One difficulty of these solutions is that the strain ε is not in the H^1 or even L_2 function spaces, which makes convergence of numerical solutions quite problematical. The solution is physically unreasonable in that it associates no dissipation with the discontinuity at x_m, as noted for transient solutions of this problem by Bazant and Belytschko [12].

A common fallacy in the strain-softening literature has been to tacitly take the initial perturbation to be a step-function, i.e.

$$a(x) = \varepsilon\ (H(x)-H(\alpha L)) \tag{13}$$

where H(x) is the Heaviside function. This perturbation in A(x) has a unique solution given in the paper and shown in Fig.1. However, this solution is not a stable solution, in the sense of equations (3) or (12). A small perturbation of (13) leads to a strikingly different solution. To illustrate this, consider the perturbation shown in Fig.2. The solution is now that given by equation (9), with x_m replaced by x_m''. Thus even a small perturbation of these initial conditions leads to a strikingly different solution. This solution is unstable even in the L_∞ norm.

In the viscoplastic material model, which is considered well–posed, the solution exhibits a similar sensitivity to initial imperfections! The localization zone does not reduce to a set of measure zero. However, as shown by Belytschko, Moran and Kulkarni [9], the structure of the strain field depends strongly on the Fourier spectrum of the imperfection. The L_2 norm exhibits a similar sensitivity to the character of the imperfection: the localization zone always occurs at the weakest point. However, the L_∞ norm in rate-independent material in insensitive to the exact structure of the imperfection as in the rate dependent model. Furthermore, as time dependent effects are decreased, the L_∞ norm of the rate–dependent model converges to that of the rate–independent model.

CONCLUSIONS

In summary, well–posedness can not be defined as insensitivity of a perfect model to changes in the initial data and the stability of solutions cannot be properly examined without considering the effect of perturbations of initial conditions, and these initial perturbations must be quite arbitrary. In particular, step-function perturbations such as are tacitly assumed in many finite element solutions are not suitable candidates for an examination of stability for non–monotonic materials.

The model which is chosen to solve a problem with non–monotonocity should depend on the purpose at hand. If the details of the strain field in the localization zone are of interest, then models which do not localize to a set of measure zero must be used. However if the goal is to obtain the global response, which is characterized by the L_∞ norm, the rate–dependent models may be satisfactory. In both rate–independent and rate–dependent models, a careful study of the effect of the various imperfections must be made, for they can critically affect response; in rate–independent models the morphology of the strain field and the global response also depend on imperfections.

The behavior of structures with nonmonotonic materials present special difficulties in defining the stability of solutions because are extremely sensitive to initial data in certain regimes. This has been exemplified by the large changes in the L_2 norm which are brought about by the imperfection in Fig. However, this also appears to be a characteristic of the actual physical problem which is modelled: in the absence of a notch, the location of a shear band can be quite random. How to deal with this difficulty and characterize it requires further study, and we hope that this paper has provided some guidance in this effort.

ACKNOWLEDGEMENT

The support of the Office of Naval Research under award N00014-89-J-3066 to Northwestern University is gratefully acknowledged.

REFERENCES

[1] I. Sandler, and J. Wright, "Summary of Strain-Softening", in Theoretical Foundations for Large Scale Computations of Nonlinear Material Behavior, DARPA-NSF Workshop, edited by S. Nemat-Nasser (1984).
[2] E. C. Aifantis, "On the Microstructural Origin of Certain Inelastic Models", J.Engrg. Mat. Tech., 106, 326-330.
[3] H. B. Mulhaus, and E. C. Aifantis, "The Influence of Microstructure-Induced Gradients on the Localization of Deformation in Viscoplastic Materials." M. M Report No. 171, Dept. of Mech. Engrg., Michigan Technological University, (December. 1989).
[4] Z. P. Bazant, T. Belytschko, and T. P Chang, "Continuum Theory for Strain Softening", J. Engrg. Mech, ASCE, 110 (3), 1666-1692, (1984).
[5] A. Needleman,. "Material rate Dependence and Mesh Sensitivity in Localization Problems", CMAME, 67, 69-85, (1988).
[6] J. H. Prevost, and B. Loret,. (1990). "Dynamic Strain Localization in Elasto-(Visco-)-Plastic Solids, Part 2: Plane Strain Examples", CMAME., to appear (1990).
[7] R. C. Batra, and C. H. Kim, "Adiabatic Shear Banding in Elasto-Viscoplastic Nonpolar and Dipolar Materials.", Int. J. Plasticity, 6, 127-141, (1990).
[8] T. Belytschko, and M. Kulkarni, " On Imperfections and Spatial Gradient Regularization in Strain-Softening Viscoplasticity", in Proc. Symp. on Failure Criteria and Analysis in Dynamic Response, edited by H. E Lindberg, to appear, (1990).
[9] T. Belytschko, B. Moran and M. Kulkarni, "Stability and Imperfections in Quasistatic Viscoplastic Solutions", App. Mech. Reviews, 43, (5), 251-256, (1990)
[10] N. S. Ottosen, "Thermodynamical Consequences of Strain-Softening in Tension." J. Engrg. Mech., ASCE, 112(11), 1152-1164, (1986).
[11] Z. P. Bazant, "Instability, Ductility and Size Effect in Strain-Softening Concrete." J. Engrg. Mech., ASCE, 102(2), 331-334, (1976).
[12] Z. P. Bazant, and T. Belytschko, "Wave Propagation in Strain-Softening Bar: Exact Solution.", J. Engrg. Mech., ASCE, 111, 381-389, (1985).

Numerical Integration of Incrementally Non-Linear Constitutive Relations with Adaptive Steps

BOULON MARC, DARVE FÉLIX, EL GAMALI HAMID
IMG BP 53X 38041 Grenoble Cedex France

TOURET JEAN PIERRE
EDF SEPTEN 12,14 av. Dutriévoz 69628 Villeurbanne Cedex France

Abstract

The utilization of incrementally non-linear constitutive relations in the finite element method is worth considering carefully the choice of integration steps length. In this paper we shall first consider the effect of increment size on the integration results with constant step. We shall then present the principle of integration with adaptive steps. Loading conditions with severe curvatures in stress diagrams will be studied as good demonstrative examples.

Introduction

The behaviour of Geomaterials is severely non-linear after global and incremental points of view. For this reason constant step integration cannot actually lead to both a suitable integration accuracy and reasonable computation duration with the finite element method.

1 The incrementally non-linear constitutive relation

The general incremental formulation of the constitutive relations for non-viscous materials can be written on the following form (Darve [1]:

$$d\epsilon_\alpha = M_{\alpha\beta}(u_\gamma)d\sigma_\beta \quad ; \quad (\alpha, \beta, \gamma = 1, 6) \qquad (1)$$

with :

$$d\underline{\epsilon} = \left[d\epsilon_{11}, d\epsilon_{22}, d\epsilon_{33}, \sqrt{2}d\epsilon_{23}, \sqrt{2}d\epsilon_{31}, \sqrt{2}d\epsilon_{12}\right] \text{ : increment stain vector}$$

$$d\underline{\sigma} = \left[d\sigma_{11}, d\sigma_{22}, d\sigma_{33}, \sqrt{2}d\sigma_{23}, \sqrt{2}d\sigma_{31}, \sqrt{2}d\sigma_{12}\right] \text{ : increment sress vector}$$

$$\underline{u} = \frac{d\underline{\sigma}}{||d\underline{\sigma}||_2} \text{ : direction of } d\underline{\sigma} \ (||d\underline{\sigma}||_2 = \sqrt{d\sigma_{ij}d\sigma_{ij}} = \sqrt{d\sigma_\beta d\sigma_\beta})$$

$$\underline{\underline{M}} \ (= \underline{\underline{D}}^{-1}) \text{ : constitutive matrix}$$

The description of the incremental non-linearity, wich is also the one of the plastic irreversibilities, depends on the variation of the matrix $\underline{\underline{M}}$ with the direction of the incremental stress $d\underline{\sigma}$. In fact, for physical reasons that seems reasonable to consider that $\underline{\underline{M}}$ varies continuously with $\underline{u}$.

The polynomial serial development of functions $M_{\alpha\beta}$ allows us to obtain the general form of the incremental non-linear of second order laws: (Darve [2]):

$$d\epsilon_\alpha = M^1_{\alpha\beta} d\sigma_\beta + \frac{1}{||d\underline{\sigma}||_2} M^2_{\alpha\beta\gamma} d\sigma_\beta d\sigma_\gamma \tag{2}$$

Three basic assumptions allow to simplify the expressions of tensors $\underline{\underline{M}}^1$ and $\underline{\underline{\underline{M}}}^2$:

- orthotropy of incremental behaviour
- quasi-elastic shear moduli ($\forall \;\; \beta \geq 4 \;\; \gamma \geq 4 \;:\; M^2_{\alpha\beta\gamma} = 0$)
- no "crossed terms" ($\forall \;\; \beta \neq \gamma \;:\; M^2_{\alpha\beta\gamma} = 0$)

Equation 2 becomes in orthotropy axes:

$$d\underline{\epsilon} = \begin{bmatrix} \underline{\underline{\mathrm{N}}} & & & \\ & 2G_1 & & \\ & & 2G_2 & \\ & & & 2G_3 \end{bmatrix} d\underline{\sigma} \tag{3}$$

with:

$$\underline{\underline{N}} = \frac{\underline{\underline{N}}^+ + \underline{\underline{N}}^-}{2} + \frac{\underline{\underline{N}}^+ - \underline{\underline{N}}^-}{2} \begin{bmatrix} \frac{d\sigma_{11}}{||d\underline{\sigma}||_2} & 0 & 0 \\ 0 & \frac{d\sigma_{22}}{||d\underline{\sigma}||_2} & 0 \\ 0 & 0 & \frac{d\sigma_{33}}{||d\underline{\sigma}||_2} \end{bmatrix} \tag{4}$$

The matrices $\underline{\underline{N}}^+$ and $\underline{\underline{N}}^-$ are computed from the material behaviour on "generalized triaxials" paths, wich is completly described on an analytic way (Darve & Dendani [3]) with discrete and continuous internal variables.

2 Adaptive step integration of the constitutive equation

2.1 principle

The serial development, around the instant τ of $\underline{\sigma}$ can be written:

$$\sigma_{\tau+d\tau} = \sigma_\tau + \dot{\underline{\sigma}}_\tau \frac{d\tau}{1!} + \ddot{\underline{\sigma}}_\tau \frac{d\tau^2}{2!} + ...$$

$$\sigma_{\tau+d\tau} = \sigma_\tau + [\underline{\underline{D}}_\tau + \frac{1}{2} d\underline{\underline{D}}_\tau + ...].d\underline{\epsilon} \tag{5}$$

In order to limit ourselves to an approximation of the first order in an explicit mode, it will be necessary to satisfy the following relationship (Boulon [4], Chau [5]):

$$\frac{1}{2}\|d\underline{\underline{D}}_r\|_\infty \ll \|\underline{\underline{D}}_r\|_\infty \tag{6}$$

assuming the terms of superior orders to be negligible.

By calculating the precision index $p = \frac{2\|\underline{\underline{D}}_r\|_\infty}{\|d\underline{\underline{D}}_r\|_\infty}$ throughout the integration and by fixing an interval of precision $[p_1, p_2]$, it is possible to regulate the integration step.

A supplementary control of this step may be realized in order that the stress state remains within the limit surface.

2.2 application

The considered integration path is an undrained one with severe curvatures in stress diagram.

On figure 1 we notice that integration results with 5.10^{-3} and 10^{-3} steps have no physical sense since some stress states are out of the limit surface.

Figure 2 shows that the adaptive step integration and the integration with 10^{-5} constant step give similar results; the adaptive step requives 129 increments instead of 2000 (16 times less!)

Figure 3 presents the successive choosen increment sizes used, fonction of the prescribed precision index p.

References

[1] F. Darve, "Une formulation incrémentale des lois rhéologiques; application aux sols", *Thèse de doctorat es sciences*, INPG, 1978

[2] F. Darve, "Une loi rhéologique incrémentale non-linéaire pour les solides", *Mech. Res. Comm.*, 7(4),1980

[3] F. Darve, H. Dendani, "An incrementally non-linear constitutive relation and its predictions", *In Constitutive Equations for Granular Non-Cohesive Soils*, Cleveland, Ed. SAADA A.S. and BIANCHINI G., BALKEMA, Rotterdam, 1989

[4] M. Boulon, "Contribution à la mécanique des interfaces sols-structures. Application au frottement latéral des pieux", *Habilitation de recherche*, Grenoble, 1989

[5] B. Chau, "Simulation numérique du comportement des ouvrages en terre par la méthode des éléments finis", *Thèse de docteur ingénieur*, 1989, Grenoble.

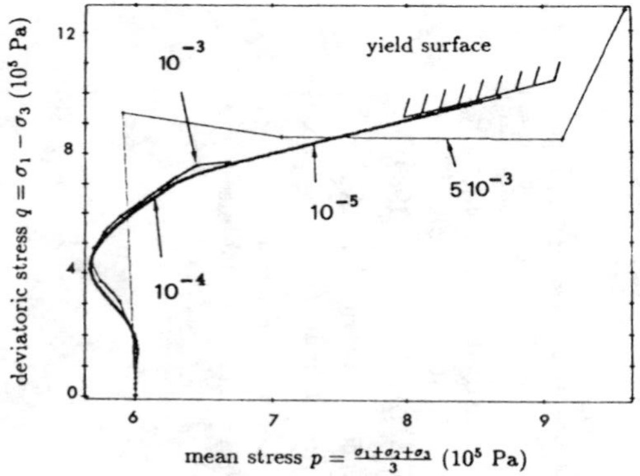

Figure 1: Integration by explicit method with a constant step: increment size effect.

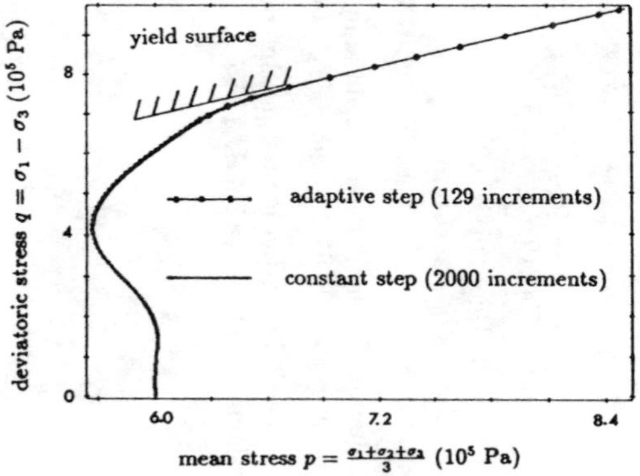

Figure 2: Integration by explicit method with constant step and adaptive step: comparison.

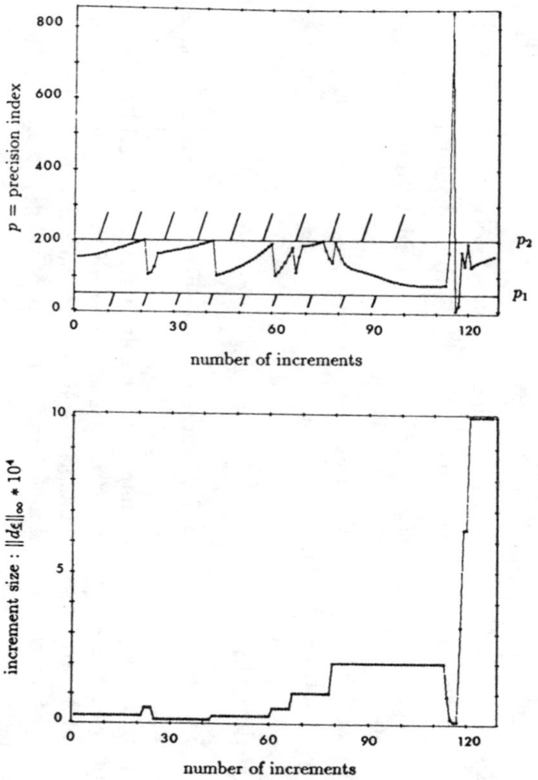

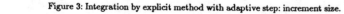

Figure 3: Integration by explicit method with adaptive step: increment size.

FINITE ELEMENT AND FINITE BLOCK METHODS IN GEOMECHANICS

W. F. CHEN and AUSTIN D. PAN
School of Civil Engineering
Purdue University

ABSTRACT

The limitations of the finite element method are discussed. A new procedure - the finite block method - is introduced. The finite block method extends the range of applicability of limit analysis to frictional materials. A bearing capacity example shows the procedure gives realistic results.

INTRODUCTION

Much work has been conducted in applying the finite element method for problems in geomechanics. Although much success can be credited to the finite element method, most researchers would agree that the method has many limitations and difficulties. For example, finite element applications typically require the idealization of geomaterials as elastic-plastic materials and a relatively fine mesh size to capture limit loads. The finite element method is more suited to elastic, continuous materials that exhibit equally high tensile and compressive strength. Indeed, the method was originally developed in the aerospace industry and it is in the analysis metal structures where the most successful applications have been made.

Geomaterials are in the most part nonlinear discontinuous materials typically strong in compression, but weak in tension. Many of the basic assumptions implicit in the finite element method, such as compatibility and continuity, do not apply to geomaterials. Only through significant modifications has the method been adapted for problems in geomechanics.

This paper discusses the applications of a new method -- the finite *block* method -- which has shown great potential in solving problems in geomechanics without many of the limitations and difficulties encountered by the finite element method. The paper also shows that the finite block method is in essence a method of limit analysis, but contains many significant improvements and advantages over existing limit analysis techniques.

THE FINITE BLOCK METHOD

Shi and Goodman [1] are credited with the original development of the finite block method as discussed in this paper. Some of the basic concepts of the finite block method can be found in a number of earlier methods, notably the discrete element method by Cundall [2] and the discrete structural models by Kawai [3].

In the finite block method a discontinuous rock or soil medium is modelled as an assemblage of blocks. Unlike the finite element method in which compatibility is imposed on common nodal points, the finite block method allows individual blocks to separate or slide away from each other, satisfying the condition of no tension between blocks. A system of equilibrium equations for the block assemblage is derived through the minimization of the potential energy. These equations are then solved iteratively in each load or time increment until the constraint of no penetration between blocks is fulfilled. When blocks are in contact, the Coulomb friction law is applied on the contact surfaces to model friction. The major advancement made by Shi and Goodman is the development of a complete kinematic theory that enables a large assemblage of blocks to move and deform without the penetration of one block into another.

LIMIT ANALYSIS BY THE FINITE BLOCK METHOD

Limit analysis is recognized as a most powerful procedure for computing the collapse load of numerous problems in geomechanics in a direct manner. In engineering practice where determining the collapse load or stability of a system is usually the primary goal, limit analysis has proven itself to be more practical and realistic than the finite element method. The bearing capacity of footings, the earth pressure problems and the stability of slopes are examples for which limit analysis provides the most efficient and direct method of engineering solution. Even though the method is an approximate method, solutions obtained are realistic and compare closely with those substantiated by theoretical and experimental results. Improving the accuracy and capability of limit analysis would therefore be an important contribution. Herein, it will be shown that the finite block method is in essence a method of limit analysis, but the existing computer program of finite block method has already offered many significant improvements over existing limit analysis techniques and provide useful answers in geotechnical engineering.

In existing techniques of limit analysis [4], a lower-bound solution is obtained when the load, determined from a distribution of stress alone, satisfy: (a) the equilibrium equations; (b) stress boundary conditions; and (c) nowhere violates the yield criterion. Or simply stated, the lower-bound technique considers only equilibrium and yield, and gives no consideration to kinematics. Conversely, a upper-bound solution is obtained when the load, determined by equating the external rate of work to the internal rate of dissipation of energy in an assumed deformation mode (or velocity field), satisfy (a) velocity boundary condition; and (b) strain and compatibility conditions. The upper-bound technique is concerned mainly with kinematics and considers only the failure or velocity modes and energy dissipation, the stress distribution need not be in equilibrium. By suitable choice of stress and velocity fields, the collapse load can be bracketed by the lower-bound and upper-bound solutions. The proof of the limit theorems is based on the fundamental assumption of perfect plasticity with associated flow rule for the material. If plasticity is not appropriate for the soil, then the answer obtained will be not very significant. Proper alternative theorem and method need be developed. The finite block method may provide such a reasonable alternative for the soil idealized as a frictional material. This is described in the forthcoming.

The finite block method contains characteristics from both the upper-bound and lower-bound techniques. The method is similar to the upper-bound technique in that a failure mode must be predetermined, block boundaries must be defined by the user, and the solution is based on energy dissipation. The frictional dissipation requires the normal stress on the plane of sliding. The finite block method satisfies equilibrium and provides this information. The following upper bound theorem would provide a reasonable design on analysis procedure for a frictional soil. Collapse is assumed to occur for an assumed failure mode, if the work done by the applied loads exceeds the frictional dissipation which is computed from an equilibrium distribution of normal stresses on the assumed surface of slidings [6].

The capability of the finite block method to model blocks as frictional materials is very significant. Coulomb's law is applied to the contact surfaces when blocks are in contact. Existing upper-bound techniques are largely based on materials assuming perfect plasticity. The difficulty of determining normal contact forces between blocks have previously impeded the modelling of frictional materials. Real geomaterials are quite complex and are neither truly frictional in behavior, nor are they plastic. They are probably somewhere in between. While upper-bound techniques have always provided the ideal plastic solution on the one extreme, the finite block method can now give us the friction solution on the other extreme.

In short, the finite block method extends the existing limit analysis techniques to frictional materials and reduces the range in which the true solution is bracketed by the upper and lower bounds. Overall, the finite block method more resembles the upper-bound technique since a failure mechanism must be initially defined by the user. As the method is based on the assumption of frictional material and equilibrium is satisfied, the collapse load would thus be lower than the solution obtained by an upper-bound technique based on perfect plasticity. By trying out different configurations for the failure mechanism, the solution that provides the lowest collapse load would be closest to the exact solution. It has been shown in many instances [4] that if reasonable failure mechanisms are assumed, the solutions obtained from the various mechanisms do not differ greatly from each other, and the exact solution is very close to the limit analysis solution.

AN ILLUSTRATIVE EXAMPLE

The bearing capacity of a strip footing based on the Hill mechanism is shown in Fig. 1 and will now be used to illustrate the finite block method. The results will also be compared with the upper-bound limit analysis solution [4].

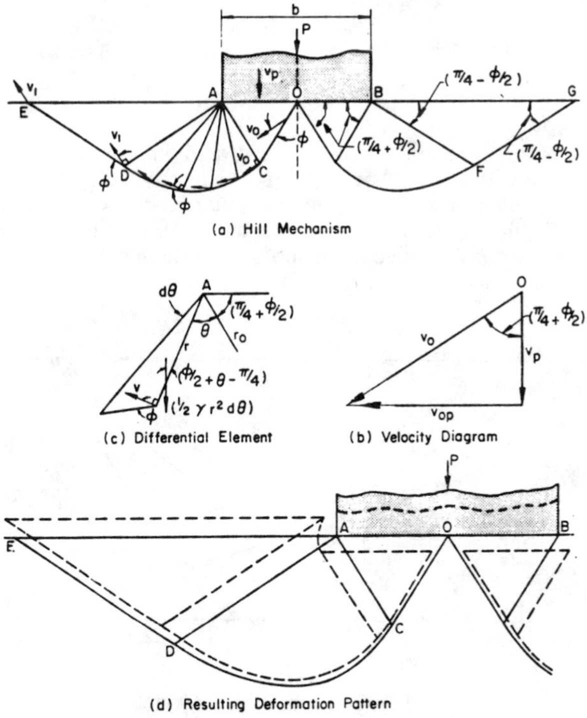

Fig. 1. Bearing Capacity Calculation Based on Hill Mechanism

Since the Hill mechanism is symmetric about the axis of the footing it is only necessary to consider the left half of the failure mechanism as shown in Fig. 1(d). The Hill mechanism consists of three elements: a triangular wedge under the footing (AOC), a triangular wedge under the free surface (ADE), and a logspiral shear zone (ACD). The geometry of the mechanism is dependent on the angle of internal friction ϕ of the soil medium. Complete details of the upper-bound solution for the bearing capacity of the footing is contained in Reference [4]. Briefly, the solution considers the weight of the soil and assumes a smooth surface footing resting on a cohesionless soil (c=0). With the aid of the velocity diagram of Fig. 1(b) and the differential element of the logspiral of Fig. 1(c) the work done by the load P and the work done by the soil elements are calculated. The collapse load, P_u, for the limit analysis is then given as:

$$P_u = \gamma \frac{1}{2} b^2 N_\gamma \tag{1}$$

where b is the width of the footing, γ is the unit weight of the soil, and the dimensionless bearing capacity coefficient, N_γ, is defined as:

$$N_\gamma = \frac{1}{4}\tan\left(\frac{1}{4}\pi+\frac{1}{2}\phi\right)\left[\tan\left(\frac{1}{4}\pi+\frac{1}{2}\phi\right)e^{\left(\frac{1}{2}3\pi\right)\tan\phi}-1\right] \tag{2}$$

$$+\frac{3\sin\phi}{1+8\sin^2\phi}\left\{\left[\tan\left(\frac{1}{4}\pi+\frac{1}{2}\phi\right)-\frac{\cot\phi}{3}\right]e^{\left(\frac{1}{2}3\pi\right)\tan\phi}+\tan\left(\frac{1}{4}\pi+\frac{1}{2}\phi\right)\frac{\cot\phi}{3}+1\right\}$$

Taking the same example of Fig. 1 and applying the finite block method, a block model is developed as shown in Fig. 2. In the block model, the logspiral shear zone is modelled by a series of ten triangular blocks with reduced corners. In the limit analysis, the logspiral shear zone assumes an infinite number of triangular blocks. The reason for the reducing the corners of the blocks is important and will be discussed later.

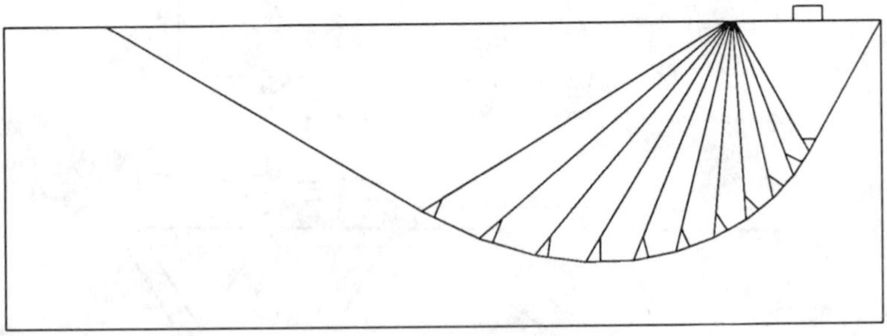

Fig. 2. Finite Block Model

The analysis result of the finite block method is plotted in Fig. 3. The limit analysis by Equations (1) and (2) is also shown in Fig. 3 as a horizontal line at the limit load. The results shown in Fig. 3 are based on friction angle: $\phi = 30$ degrees, unit weight of soil $\gamma = 120$ pcf, and $b = 20$ ft. The finite block analysis is based on ten load increments with a time factor of 1.5 seconds. The analysis was performed on a 386 microcomputer using the program DDA developed by Shi [5]. An example of the block displacements produced by the finite block method is shown in Fig. 4 for a model with five triangular blocks modelling the shear zone.

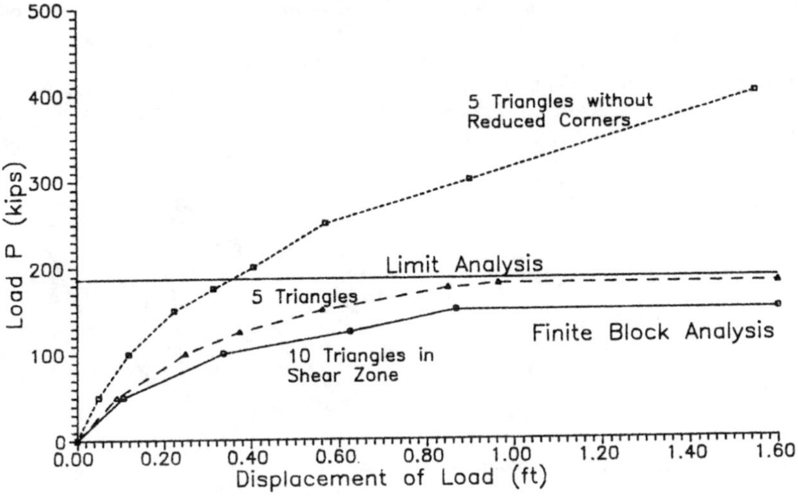

Fig. 3. Analysis of Bearing Capacity Based on Hill Mechanism

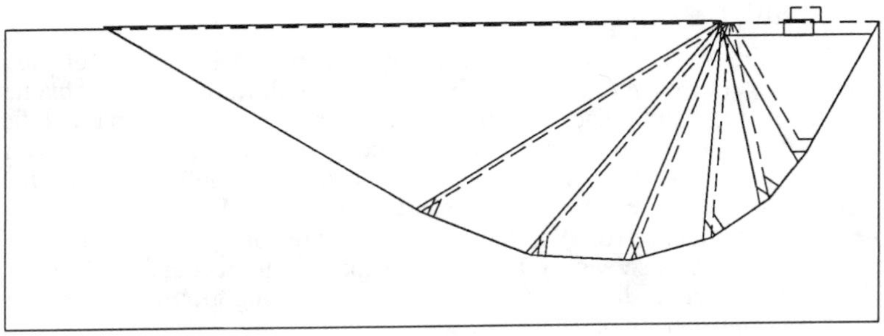

Fig. 4. Deformation Pattern Resulting from Finite Block Analysis

Comparing the two solid lines in Fig. 3, the finite block method results in a collapse load that is lower than the the upper-bound limit analysis solution. Since the finite block method is based on a frictional material, the collapse load is expected to be lower than the limit analysis solution which is based on a perfectly plastic material. The true solution would lie between these two solutions as the soil medium in neither a perfectly plastic nor frictional material.

It should be noted that the finite block solution is sensitive to how well the logspiral shear zone is approximated by a finite number of triangular bocks. If only five instead of ten blocks were used as shown in Fig. 4, the model becomes excessively stiff and approaches the limit analysis solution as shown by the dashed line in Fig. 3.

As previously mentioned, it is important that the finite block model incorporate reduced corners for some of the blocks (see Fig. 2). Reduced corners are typically necessary for corner-to-corner contacts such as those in the logspiral shear zone of Fig. 2. If corners are not reduced as shown in Fig. 5, the block model becomes excessively stiff and 'locking' of the blocks occurs. Extra energy is then required to lift and rotate blocks, resulting in a higher collapse load as shown by the dotted lined plotted in Fig. 3. Reducing the corners of blocks is not without theoretical basis. In reality, blocks with perfect sharp corners do not exist and on contact high stress concentrations develop to reduce the corners.

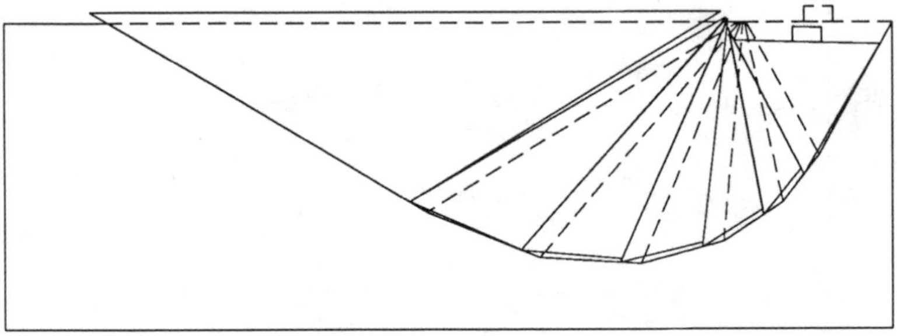

Fig. 5. Deformation Pattern of Finite Block Model without Reduced Corners

CONCLUSION

The finite block method shows great promise in providing solutions to many problems in geomechanics. By satisfying both equilibrium and kinematics and with its ability to handle frictional materials, the finite block method significantly extends the range of applicability of the existing limit analysis techniques to geomaterials. Further research is needed to study the implications of this new method as applied to realistic geomechanics problems. The problem of reduced corners has been identified. The proposed procedure for frictional materials leads to realistic answers as shown by the bearing capacity application. It is evident that the finite block method holds great potential of becoming a powerful new tool for research and engineering practice.

REFERENCES

1. G. H. Shi and R. E. Goodman, "Generalization of Two-Dimensional Discontinuous Deformation Analysis for Forward Modelling", International Journal for Numerical and Analytical Methods in Geomechanics, Vol. 13 (4), pp. 359-380 (August 1989).

2. P. A. Cundall, "A Computer Model for Simulating Progressive, Large-Scale Movements in Blocky Rock Systems", Proceedings of the International Symposium on Rock Fracture, II-8, Nancy, France (1971).

3. T. Kawai, "New Discrete Structural Models and Generalization of the Method of Limit Analysis," Finite Elements in Nonlinear Mechanics, Vol. 2, Norwegian Institute of Technology, Tronheim, pp. 885-906 (1977).

4. W. F. Chen, Limit Analysis and Soil Plasticity, Elsevier, Amsterdam, The Netherlands (1975).

5. G.H. Shi,"Discontinuous Deformation Analysis: A New Numerical Model for the Static and Dynamics of Block Systems", Ph.D. Dissertation, Department of Civil Engineering, University of California at Berkeley (1988).

6. D. C. Drucker, "On Stress-Strain Relations for Soils and Load Carrying Capacity, " Proc. of the 1st Int. Conf. on the Mech. of Soil-Vehicle Systems, Torino-Saint Vincent 12-16, Giugno, pp. 15-27, (1961).

A SECANT STRUCTURAL SOLUTION SCHEME FOR INCREMENTAL DAMAGE AND PLASTICITY

Z. CHEN
New Mexico Engineering Research Institute, Univ. of New Mexico

M. K. NEILSEN
Applied Mechanics I, Sandia National Laboratories

H.L. SCHREYER
Dept. of Mechanical Engineering, University of New Mexico

ABSTRACT

There exist two major difficulties for nonlinear structural analysis involving softening. One consists of the ill-conditioning of the tangent stiffness matrix near critical points, and the other is the choice of a suitable constraint to obtain the solution path. In an attempt to make failure simulation a routine procedure, a new solution procedure is proposed in which a secant stiffness matrix is used in conjunction with a local constraint condition. The procedure is formulated in terms of continuum damage mechanics and, with the definition of an artificial secant matrix, the method is applied to plasticity softening as well. Numerical solutions have been obtained for both plane strain and plane stress to show that snap-back and snap-through associated with shear band formation can be efficiently predicted.

INTRODUCTION

The existence of structural softening has been demonstrated conclusively by a number of experimentalists using displacement controlled devices. Furthermore, if the displacement measurement used for load control is made across a softening zone, a reversal or "snap-back" in the space of load versus structural displacement can be exhibited. To predict such phenomena it is necessary to introduce constitutive models that exhibit softening with a nonlocal feature to limit the size of the failure zone or to use a cracking model in which the softening is given directly in terms of traction versus crack opening. With either approach, there are formidable computational problems in that the governing tangent matrix becomes singular at a bifurcation or limit point, and multiple solution paths exist beyond these critical points. The successful existing solution procedures are very inefficent so only a handful of numerical solutions exist.

Most techniques involve the use of the tangent stiffness matrix with iterations imposed to satisfy equilibrium to within a suitable tolerance. Several techniques are used to circumvent the difficulties associated with critical points. One method is to monitor the negative pivots obtained from the decomposition of the tangent stiffness matrix and to add a matrix to shift the critical eigenvalue away from zero [1]. A second method suppresses iterations around the critical point [2]. Because it is almost impossible to obtain converged solutions around critical points by simply prescribing external load increments, it has been suggested that the load level become a function of another variable through a suitable constraint. Direct or indirect displacement control [3, 4], arc-length control [5, 6] and self-adaptive "hyperelastic" constraints [7] provide a means for changing the load level from one iteration to the next. In order to simulate more efficiently the localization that accompanies softening, the constraint condition must be sensitive to the state variables inside the localization zone. One approach is to use the eigenvector associated with the lowest eigenvalue of the tangent stiffness matrix as a guide to obtain the appropriate solution [8, 9].

Even with these advances in understanding, lack of convergence is still sometimes a problem after the limit or bifurcation point has been traversed. The reason might be due to the fact that in these solution strategies the load increment for the first iteration must be prescribed according to some criterion. The result is that at a later iteration there may be a change in the sign of the load increment and, consequently, changes in the domains that are hardening, softening and unloading. A new approach [10] to help circumvent this problem is to identify the point in the body which is undergoing the most severe deformation and to prescribe the increment of one of the strain components at that point. Since such an increment can be assumed to be monotonically increasing, the load increment is determined indirectly and does not change sign. The procedure is analogous (and equally successful) to that used for obtaining snap-back and snap-through from tests on experimental specimens. Although the approach has proven to be robust for a class of plasticity softening problems, the rate of convergence is quite slow, especially if a smooth form (continuous derivative) is assumed for the hardening-softening function . The reason is that the tangent stiffness matrix may be poorly conditioned even though the smallest eigenvalue is not zero (critical point).

Recently, continuum damage mechanics [11] has evolved as an important field for describing the response of materials which develop microcracks and microvoids. One of the distinguishing features of the theory is that the elasticity tensor changes and, in fact, the elasticity tensor can be used to reflect the state of damage. The approach is reflected in structural analysis as a governing matrix equation which contains the secant rather than the tangent stiffness matrix. The result is that equilibrium is governed by a matrix which does not become singular even though bifurcations and limit points continue to exist as defined by the tangent stiffness matrix. A description of the approach is outlined here based on preliminary calculations which are very promising [12].

Of course, some materials may be governed by plasticity rather than continuum damage mechanics so there remains the issue of how to efficiently solve plasticity softening problems. Also given here is a possible approach in which a pseudo secant tensor is constructed in an incremental fashion based on the equations of plasticity. In effect, continuum damage mechanics is used as the basis for constructing a computational algorithm for nonlinear problems in which the governing tangent matrix may become singular.

ELEMENTARY EQUATIONS FOR DAMAGE

For purely mechanical processes, assume the internal energy, U, is a function of strain, **e**, and internal variables which are defined to be the components of the elasticity tensor, **E**, for the case of "ideal" damage. For linear elasticity, the internal energy is quadratic in strain:

$$U = \frac{1}{2}\mathbf{e}:\mathbf{E}:\mathbf{e} \tag{1}$$

The Clausius- Duhem inequality yields the constitutive equation for stress, **s**, and the dissipation inequality:

$$\mathbf{s} = \frac{\partial U}{\partial \mathbf{e}} = \mathbf{E}:\mathbf{e} \qquad D \equiv -\frac{\partial U}{\partial \mathbf{E}}::\dot{\mathbf{E}} = -\frac{1}{2}(\mathbf{e}\otimes\mathbf{e})::\dot{\mathbf{E}} \geq 0 \tag{2}$$

Suppose the damage process is described through a monotonically increasing parameter, ω, by the evolution equation

$$\dot{\mathbf{E}} = -\dot{\omega}\mathbf{R}(\mathbf{E},\mathbf{e}) \tag{3}$$

in which the function **R** is also a function of strain and elasticity. Then the dissipation inequality is satisfied if **R** is positive semi-definite. The dissipation requirement is used to define a damage function as follows:

$$\Phi = \mathbf{e}:\mathbf{R}:\mathbf{e} - H^2(\mathbf{E},\mathbf{e}) \tag{4}$$

If $\Phi < 0$, no damage is occurring; $\Phi > 0$ is not permitted and, if $\Phi = 0$, damage is occurring with the dissipation inequality automatically satisfied. The damage hardening-softening function, H, must also be given as a scalar function of the strain and elasticity.

The functions **R** and H are normally constructed from microstructural arguments. Several forms exist. Frequently, experimental data provide information in the form of a surface in stress space as to when damage is occurring. If the surface can be cast in the form of (4), then an expression for **R** can be deduced directly. For example consider a Mises-type surface:

$$\Psi \equiv \mathbf{s}^d:\mathbf{s}^d - h^2 = 0 \tag{5}$$

in which $\mathbf{s}^d$ denotes the stress deviator and h is an experimentally determined hardening and softening function. If $\mathbf{P}^d$ denotes the deviatoric projection, then

$$\mathbf{s}^d = \mathbf{P}^d:\mathbf{s} = \mathbf{P}^d:\mathbf{E}:\mathbf{e} \tag{6}$$

Since $\mathbf{P}^d:\mathbf{P}^d = \mathbf{P}^d$, it follows that terms in (5) can be equated to corresponding terms in (4) if $h = H$ and if

$$\mathbf{R} = \mathbf{E}:\mathbf{P}^d:\mathbf{E} \tag{7}$$

Therefore, not only does "ideal" damage lead to a fairly simple constitutive formulation but experimental data in the usual form of an inelastic surface in stress space and a hardening-softening function can be used directly to provide the necesary functions.

THE COMPUTATIONAL ALGORITHM

Consider a static problem in which the spatial discretization process yields the secant stiffness [K] based on the current value of the components of the elasticity tensor, and the load vector {q} which is assumed to be applied proportionally, i.e., if μ is the magnitude of the load, then

$$\{q\} = \mu \{q^*\} \qquad\qquad \{q^*\}^T\{q^*\} = 1 \tag{8}$$

The problem is to solve the nonlinear matrix equation

$$[K]\{u\} = \{q\} \qquad\qquad [K] = [K\{u\}] \tag{9}$$

in which the dependence of the secant stiffness on the displacement vector {u} is emphasized.

Suppose the load parameter is to be incremented. To begin the iterative process, let $[K]_I$ denote the secant stiffness from the previous iteration. Determine the new displacement field $\{u^*\}_{I+1}$ associated with the unit force, $\{q^*\}$:

$$[K]_I\{u^*\}_{I+1} = \{q^*\} \tag{10}$$

Then if the magnitude of the force, μ, is known, the actual displacement field will be $\{u\} = \mu\{u^*\}$. However, instead of prescribing μ, suppose the magnitude, η, of a constraint condition is specified:

$$\eta = \{c\}^T\{u\} = \mu\{c\}^T\{u^*\} \tag{11}$$

where {c} is a vector of the type $\{c\}^T = \langle 0,..., 0, 1, 0, 1, 0,..., 0\rangle$ which is used merely to relate two or more of the degrees of freedom of the problem. Then the magnitude of the force vector is determined indirectly from the relation

$$\mu = \frac{\eta}{\langle c\rangle^T\{u^*\}} \tag{12}$$

The displacement field is the product of μ and $\{u^*\}$, both of which are known. The equilibrium relation (10) is satisfied unless additional damage is predicted from the level of damage computed from the previous iteration in which case the secant matrix must be updated and the sequence of equations (10), (11) and (12) solved again. The process is continued until the change in damage from one iteration to the next meets a convergence criterion:

$$\left|\{\omega\}_{I+1}\right| - \left\|\{\omega\}_I\right| \le \varepsilon \tag{13}$$

There are two significant advantages to the approach. First, the constraint criterion, which is chosen to be consistent with the condition used in the damage function to describe when damage is occurring, can be identified with that point in the body which is undergoing the most severe damage. That point usually changes during the loading process. Second, the secant stiffness is not singular at critical points as defined through the use of the tangent stiffness matrix.

POSSIBLE APPLICATION TO PLASTICITY

The procedure outlined above is given in terms of a constitutive equation based on continuum damage mechanics. The amount of computer time dropped by an order of magnitude when this model was used instead of the tangent stiffness approach associated with plasticity. If a secant stiffness could be derived for plasticity, then there is a possibility that the same numerical algorithm could be used for plasticity problems. Although the feasibility of the method has not been demonstrated through the use of model problems, the following derivation shows the theoretical basis for such an approach.

As before, let **E** denote the current elasticity (secant) tensor and let **T** be the tangent tensor as obtained using the conventional equations of plasticity. Instead of the use of damage mechanics for providing the appropriate expression for **E**, the condition is invoked that the evolution of an artificial **E** must satisfy the rate equation expressed in terms of the secant tensor and the tangent tensor:

$$\dot{\mathbf{s}} = \dot{\mathbf{E}}:\mathbf{e} + \mathbf{E}:\dot{\mathbf{e}} = \mathbf{T}:\dot{\mathbf{e}} \tag{14}$$

With a rearrangement of terms, this equation yields

$$\dot{\mathbf{E}}:\mathbf{e} = (\mathbf{T} - \mathbf{E}):\dot{\mathbf{e}} = (\mathbf{T} - \mathbf{E}):\frac{(\dot{\mathbf{e}} \otimes \mathbf{e})}{\mathbf{e}:\mathbf{e}}:\mathbf{e} \tag{15}$$

Therefore, a possible form for a pseudo secant tensor is to construct one incrementally from the initial elasticity tensor from the relation suggested by (15):

$$\Delta\mathbf{E} = (\mathbf{T} - \mathbf{E}):\frac{(\Delta\mathbf{e} \otimes \mathbf{e})}{\mathbf{e}:\mathbf{e}} \tag{16}$$

Unfortunately, this equation yields an unsymmetric increment even if the tangent and initial elasticity tensors are symmetric. An alternative form, obtained by symmetrizing (17), is the following:

$$\Delta\mathbf{E} = (\mathbf{T} - \mathbf{E}):\Delta\mathbf{e} \otimes \frac{\Delta\mathbf{e}:(\mathbf{T} - \mathbf{E})}{\Delta\mathbf{e}:(\mathbf{T} - \mathbf{E}):\mathbf{e}} \tag{17}$$

It is easily seen that (14) is satisfied and that the increment in the secant tensor is symmetric if both the secant and tangent tensors are symmetric. There is the possibility that the denominator will be zero but with the use of (15) the denominator can be written

$$\Delta\mathbf{e}:(\mathbf{T} - \mathbf{E}):\mathbf{e} = \mathbf{e}:\Delta\mathbf{E}\ \mathbf{e} \tag{18}$$

The result is strictly less than zero if $\Delta\mathbf{E}$ is negative definite which is plausible based on physical considerations.

CONCLUSIONS

An efficient computational algorithm has been described for a constitutive equation based on idealized damage which is defined to be the situation where the internal variables in the internal energy are chosen to be the components of the elasticity tensor. Singularities are avoided and the use of a localized constraint permits solutions to be obtained when there is a reversal in the load-deflection curve. The procedure is analogous to the one used to obtain load reversals in experiments. Although the process of developing a psuedo secant tensor for plasticity is not unique, there is the potential for using the same computational structure for plasticity problems.

REFERENCES

1. T. Belytschko and D. Lasry, "Loacalization limiters and numerical strategies for strain-softening materials," Cracking and Damge: Strain Localization and Size Effect, Edited by J. Mazars and Z.P. Bazant, p. 349 (1989).
2. E. Ramm, "Strategies for tracing non-linear responses near limit points," Non-Linear Finite Element Analysis in Structural Mechanics, Edited by W. Wuderlich, E. Stein and K.J. Bathe, p. 68 (Springer-Verlag, New York 1981)
3. J.H. Argyris, "Continua and discontinua," Proceedings of the 1st Conference in Matrix Methods in Structural Mechanics, Wright-Patterson Air Force Base, Ohio, p. 11 (1965).
4. R. de Borst, Non-Linear Analysis of Frictional Materials, Ph.D. Dissertation, Delft University of Technology, The Netherlands (1986).
5. M.A. Crisfield, "A fast incremental/iterative solution procedure that handles 'snap through'," Computers and Structures , 13, p. 55 (1981).
6. E. Ramm, "The Riks/Wempner approach - an extension of the displacement control method in nonlinear analysis," Recent Advances in Non-Linear Computational Mechanics, Edited by E. Hinton, D.R.J. Owen and C. Taylor, p. 63 (1982).
7. J. Padovan and S. Tovichakchaikul, "Self-adaptive predictor-corrector algorithms for static nonlinear structural analysis," Computers and Structures, 15, p. 365 (1982).
8. R. de Borst, "Computation of post-bifurcation and post-failure behavior of strain-softening solids," Computers and Structures, 25, p. 211, (1987).
9. M.A. Crisfield and J. Wills, "Solution strategies and softening materials," Computer Methods in Applied Mechanics and Engineering, 19, p. 1269 (1988).
10. Z. Chen and H.L. Schreyer, "A numerical solution scheme for softening problems involving total strain control," To appear, Computers and Structures (1990).
11. D. Krajcinovic, "Damage mechanics," Mechanics of Materials, 8, p. 117 (1989).
12. Z. Chen and H.L. Schreyer, "Failure-controlled solution strategies for damage softening with localization," Micromechanics of Failure of Quasi-Brittle Materials, Edited by S.P. Shah, S.E. Swartz and M.L. Wang, (Elsevier Applied Science, New York 1990).

ANALYSIS OF FLOW DEFORMATION IN SOIL STRUCTURES WITH LIQUEFIED ZONES

W.D. Liam FINN
Department of Civil Engineering
University of British Columbia
Vancouver, B.C., Canada

R.H. LEDBETTER
US Army Corps of Engineers
Waterways Experiment Station
Vicksburg, MS, U.S.A.

M. YOGENDRAKUMAR
Royal Military College
Kingston, Ont., Canada

ABSTRACT

A method is presented for the dynamic response analysis of earth structures containing zones of liquefiable material. An important element of the analysis is the tracking of potential flow deformations after liquefaction has been triggered. Application of the method to estimating potential post-liquefaction deformations in Sardis Dam, Mississippi is described.

ANALYSIS OF POST-LIQUEFACTION BEHAVIOUR

An important consideration in evaluating the seismic stability of earth structures with potentially liquefiable materials is whether a flow failure will occur after liquefaction. If the driving shear stresses due to gravity on a potential slip surface through liquefied materials in an embankment are greater than the post-liquefaction (residual) strength, deformations will occur until the driving stresses are reduced to values compatible with static equilibrium [1]. The more the driving stresses exceed the residual strength, the greater the deformations to achieve equilibrium.

The deformations of earth structures containing potentially liquefiable soils may be estimated using the computer program, TARA-3FL [2], which is a specialized derivative of the general program TARA-3 [3].

Structure of Program TARA-3FL

The basic theory of the finite element program TARA-3 has been reported by Finn [4,5]. Only procedures specific to TARA-3FL will be described here.

In a particular element in the soil structure the shear stress-shear strain state which reflects pre-earthquake conditions is specified by a point P_0 on the stress-strain curve shown in Fig. 1. When liquefaction is triggered, the strength will drop to the residual value. The post-liquefaction stress-strain curve cannot now sustain the pre-earthquake condition and the unbalanced shear stresses are redistributed throughout the dam. In the liquefied elements, the stresses are adjusted according to the following equation,

$$\partial\tau = \frac{\partial f}{\partial \sigma'_m} d\sigma'_m + \frac{\partial f}{\partial \gamma} d\gamma \tag{1}$$

where $\tau = f(\sigma'_m, \gamma)$. This process leads to progressive deformation until equilbirium is reached at a state represented by P_2.

The residual strength will be triggered in all elements that will liquefy according to the criteria developed by Seed [6] and Wang [7].

Since the deformations may become large, it is necessary to update progressively the finite element mesh. Each calculation of incremental deformation is based on the current shape of the dam, not the initial shape as in conventional finite element analysis.

CASE HISTORY: SARDIS DAM, MISSISSIPPI

The general configuration of Sardis Dam is shown in Fig. 2. During the design earthquake, liquefaction is predicted to occur in the core and in thin discontinuous seams of silty clay in the top stratum clay in the foundation. The thin layer may be seen clearly in Fig. 3.

The residual strength, S_r, in the core was estimated to be 5 kPa (100 psf) based on Seed's correlation between corrected standard penetration resistance $(N_1)_{60}$ and residual strength [8]. From a variety of studies, the residual strength in the thin layer of silty clay was assumed to be 0.075 p' where p' is the initial vertical effective stress [9]. The original strength of the thin layer was taken as 100 kPa (2000 psf).

The large differences between the initial and post-liquefaction strengths resulted in major load shedding from affected elements to stronger sections of the dam. This put heavy demands on the ability of the program to track accurately what was happening and on the stability of the algorithms. In such cases it is imperative to

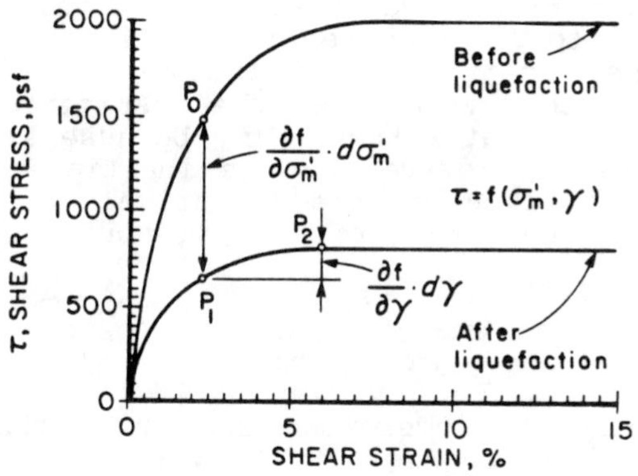

Figure 1. Adjusting Stress-Strain State to Post-Liquefaction Conditions.

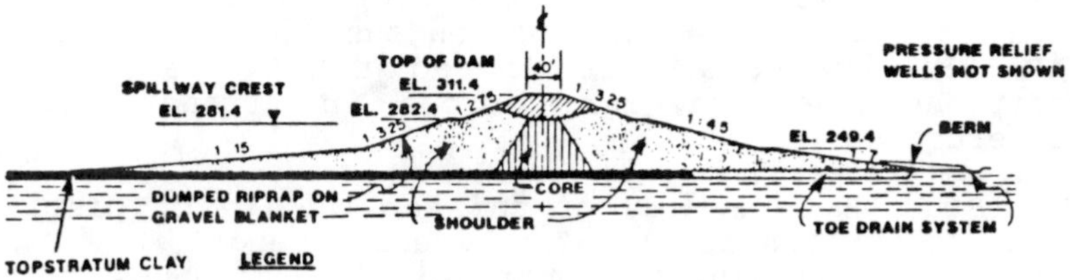

Figure 2. Typical Section of Sardis Dam in Mississippi.

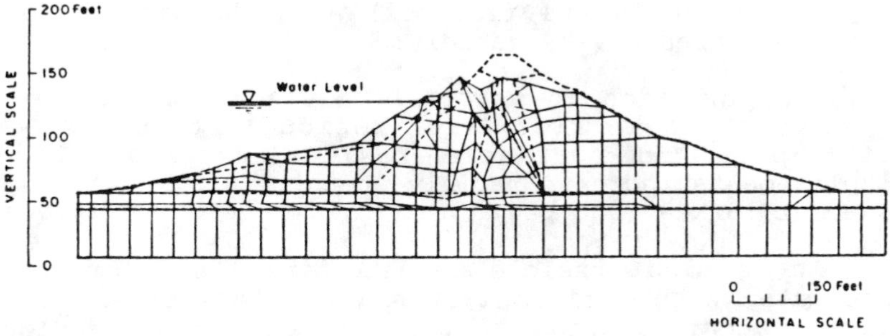

Figure 3. Initial and Post-Liquefadtion Configurations of Sardis Dam.

have an independent check that the final deformed position is indeed an equilibrium position. The most direct check is to run a conventional stability analysis on the deformed position. If the major deformations occurred during the earthquake, the resulting factor of safety should be greater than unity because some of the deformation field is driven by the inertia forces. If the major deformations occurred relatively slowly after the earthquake, the factor of safety should be close to unity.

Sardis Dam was analyzed for the residual strength values specified above with a minimum residual strength in the thin layers of silty clay of 17.5 kPa (350 psf). The initial and final deformed shapes of the dam for this case are shown in Fig. 3. Very substantial vertical and horizontal deformations may be noted, together with intense shear straining in the weak thin layer. The factor of safety of the deformed shape was found to be 0.97. It is also interesting to note that the critical slip surface exited the slope near the location suggested by the finite element analysis.

Many results of this type, for different assumptions about the residual strengths, suggest that the TARA-3FL analysis does indeed achieve equilibrium positions even for large drops in strength due to liquefaction.

Studies were made of the sensitivity of displacements to various levels of residual strength. In Fig. 4 the variations in the vertical displacements at the upstream edge of the crest (curve 1) and in the horizontal deformations at the midpoint of the upstream slope (curve 2) are shown for various levels of constant residual strength in the thin liquefied layer in the foundation. The increase in displacement is gradual with decrease in residual strength until the strength drops to about 20 kPa (400 psf) when the displacements begin to increase very rapidly.

The variation in vertical displacements (curve 3) is also shown in Fig. 4 for residual strengths S_r = 0.075 p'. For variable residual strengths, the displacements increase rapidly when the minimum strength is about 15 kPa (300 psf).

It is also possible to determine the deformations associated with various factors of safety based on the original configuration of the dam for various residual strengths. The variation of vertical crest displacement with factors of safety of the undeformed dam for various values of residual strength are shown in Fig. 5. For the first time a designer has available the deformation fields associated with different factors of safety for a

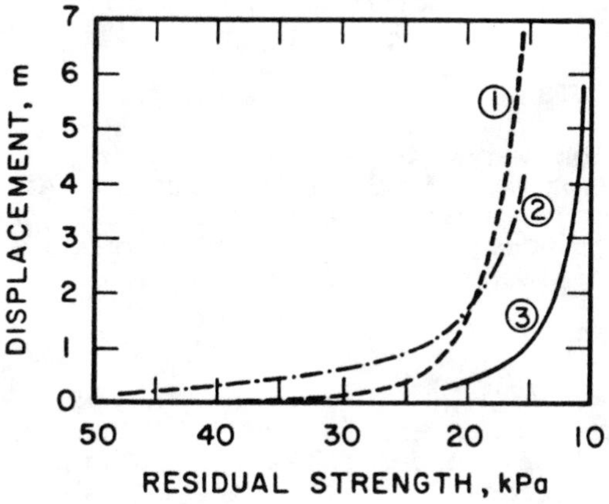

Figure 4. Variation of Dam Displacements with Residual Strength.

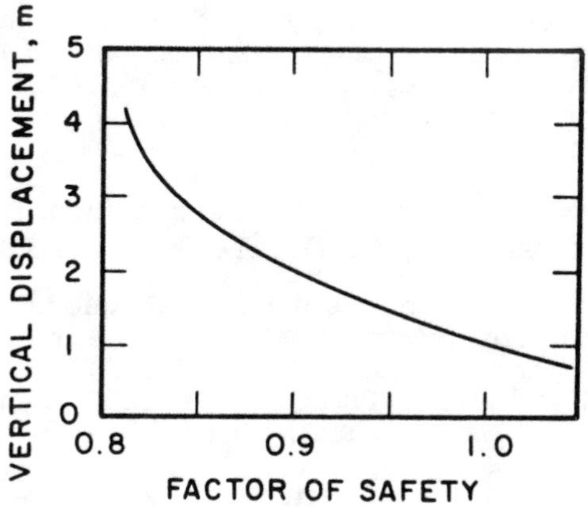

Figure 5. Variation in Vertical Displacements with Factor of Safety of the Undeformed Dam.

particular dam. This information is helpful in deciding on an appropriate factor of safety.

ACKNOWLEDGEMENTS

Permission was granted by the Chief of Engineers, U.S. Army Corps of Engineers, to publish data from the Sardis Dam study. Development of TARA-3FL was supported by Sato-Kogyo Co., Japan, and National Science and Engineering Research Council of Canada.

REFERENCES

1. G. Castro, S.J. Poulos and F.D. Leathers, "A Re-examination of the slide of the Lower San Fernando Dam", Journal of Geotechnical Engineering Division, ASCE, Vol. 111, GT9 (1985).
2. W.D. Liam Finn, and M. Yogendrakumar, "TARA-3FL - Program for Analysis of Liquefaction Induced Flow Deformations", Dept. of Civil Eng., University of British Columbia, Vancouver, B.C., Canada (1989).
3. W.D. Liam Finn, M. Yogendrakumar, N. Yoshida, and H. Yoshida, "TARA-3: A Program for Nonlinear Static and Dynamic Effective Stress Analysis", Soil Dynamics Group, University of B.C., Vancouver, B.C. (1986).
4. W.D. Liam Finn, "Dynamic Effective Stress Response of Soil Structures: Theory and Centrifugal Model Studies", Proc. 5th Int. Conf. on Num. Methods in Geomechanics, Nagoya, Japan, Vol. 1, 35-36 (1985).
5. W.D. Liam Finn. Seismic Analysis of Embankment Dams, Dam Engineering, Vol. 1, Issue 1, pp. 59-75 (1990).
6. H.B. Seed, "Earthquake-Resistant Design of Earth Dams", in Seismic Design of Embankments and Caverns, Terry R. Howards, Editor, ASCE, pp. 41-64 (1983).
7. W. Wang, "Some Findings in Soil Liquefaction, Water Conservancy and Hydroelectric Power Scientific Research Institute", Beijing, China, August (1979).
8. H.B. Seed, "Design Problems in Soil Liquefaction", Journal of Geot. Eng., ASCE, Vol. 113, No. 7, August, pp. 827-845 (1987).
9. Woodward Clyde Consultants Ltd, Private Communication (1989).

CONTINUUM MODEL AND FINITE ELEMENT ANALYSIS OF CORROSION IN REINFORCED CONCRETE

K. F. FONG
CORRPRO Company, Inc., Medina, OH 44225

G. R. BUCHANAN
Tennessee Technological University, Cookeville, TN 38505

ABSTRACT

A pit corrosion model for corrosion of reinforcing steel is formulated and numerical solutions are obtained using a coupled time dependent finite element method of analysis. Typical results for the diffusing and migrating chemical constituents are presented graphically.

INTRODUCTION

Corrosion of reinforcing steel in reinforced concrete can be considered to be a two-fold process. The corrosive reagent, the chloride anion from the saline solution, diffuses through the concrete cover by way of capillary pore action and will be absorbed by the cement gel paste. When the chloride concentration near the reinforcing bar reaches a critical threshold value it activates the electropotential and electric field across the protective oxide layer through polarization. Additionally, iron dissolution occurs and triggers the flow of free electrons on the reinforcing bar surface from anodic to cathodic areas. The electrochemical cell for redox reaction is complete and the result is pit corrosion.

In this paper a generalized anaerobic pit corrosion finite element model is analyzed based on the assumption that the metal ions react through hydrolysis in the corroding cavity or pit. The products from the electrochemical reactions are transported by means of diffusion and electrochemical migrationwithin the pit because the ions are suspended in a mixture of saline electrolyte solution.

FUNDAMENTAL EQUATIONS

The chemical equations are based upon physical assumptions that limit metal dissociation to the bottom of the pit described as

$$Fe^{2+} + H_2O \underset{K_{1b}}{\overset{K_{1f}}{\rightleftharpoons}} FeOH^+ + H^+ \qquad (1)$$

followed by hydrolysis equilibria defined by

$$H^+ + OH^- \underset{K_{2b}}{\overset{K_{2f}}{\rightleftharpoons}} H_2O \tag{2}$$

The fundamental balance equation from the theory of mixtures is the balance of mass for the "ath" constituent [1]

$$\rho \partial c_a/\partial t = m_a + A_a - h_{ak,k} \tag{3}$$

where c_a is the concentration, m_a is an external mass supply, assumed zero in this study, A_a is mass generation caused by chemical reactions and h_{ak} is the relative mass flux. Equation (3) has the specific form

$$\rho \partial c_a/\partial t = [D_a c_{a,k} - z_a D_a (c_a \phi_{,k})/RT]_{,k} + A_a \tag{4}$$

where the term in brackets is h_{ak}, z_a is the valence number, D_a is the diffusion coefficient, ϕ is the electrochemical potential, R is the ideal gas constant and T is the absolute temperature. Additional discussion concerning the derivation and assumptions that lead to Eq. (4) can be found in [2]. The six chemical constituents are defined as c_1=Fe^{2+}, c_2=$FeOH^+$, c_3=Cl^-, c_4=Na^+, c_5=H^+ and c_6=OH^-. The governing mass transport equations are derived using (4) as a model.

$$\partial c_1/\partial t = D_1 c_{1,kk} - 2D_1(c_1\phi_{,k})_{,k}/RT - K_{1f}c_1 + K_{1b}c_2c_5 \tag{5}$$

$$\partial c_2/\partial t = D_2 c_{2,kk} - D_2(c_2\phi_{,k})_{,k}/RT - K_{1b}c_2c_5 + K_{1f}c_1 \tag{6}$$

$$\partial c_3/\partial t = D_3 c_{3,kk} + D_3(c_3\phi_{,k})_{,k}/RT \tag{7}$$

$$\partial c_4/\partial t = D_4 c_{4,kk} - D_4(c_4\phi_{,k})_{,k}/RT \tag{8}$$

$$\partial c_5/\partial t = D_5 c_{5,kk} - D_5(c_5\phi_{,k})_{,k}/RT + K_{2b}c_5c_6 + K_{1b}c_2c_5 - K_{2f} - K_{1f}c_1 \tag{9}$$

$$\partial c_6/\partial t = D_6 c_{6,kk} + D_6(c_6\phi_{,k})_{,k}/RT + K_{2b}c_5c_6 - K_{2f} \tag{10}$$

It is assumed that there is no accumulation or loss of ions in the bulk electrolyte and the distribution of electrode potential within the electrolyte should satisfy the Laplace equation,

$$\phi_{,kk} = 0 \tag{11}$$

Similarly, it is assumed that charge is conserved within the pit area and in terms of corrosion cell current density [3,4]

$$i_{c,k} = 0 \tag{12}$$

Since i_c is a function of ϕ it is possible to compute electric field distributions. Equations (5-11) are solved using a coupled finite element formulation. Four node two-dimensional isoparametric elements with seven degrees of freedom per node are used to model the interior of circular shaped pits.

NUMERICAL RESULTS AND DISCUSSION

Various physical parameters are obtained from [5-7]. The diffusivity coefficients for Fe^{2+}, $FeOH^{+}$, Na^{+} and Cl^{-} are $1.0(10^{-9})m^2/sec$ and for H^{+} and OH^{-} are $9.3(10^{-9})$ and $5.3(10^{-9})$, respectively. At equilibrium the forward and backward rate constants of Eqs. (1-2) are equal, $K_{1f} = K_{1b} = 1.0(10^{-14})$ sec^{-1} and $K_{2f} = K_{2b} = 1.5(10^{-14})$.

A finite element model of two adjacent pits is shown in Fig. 1 with the various boundary conditions.

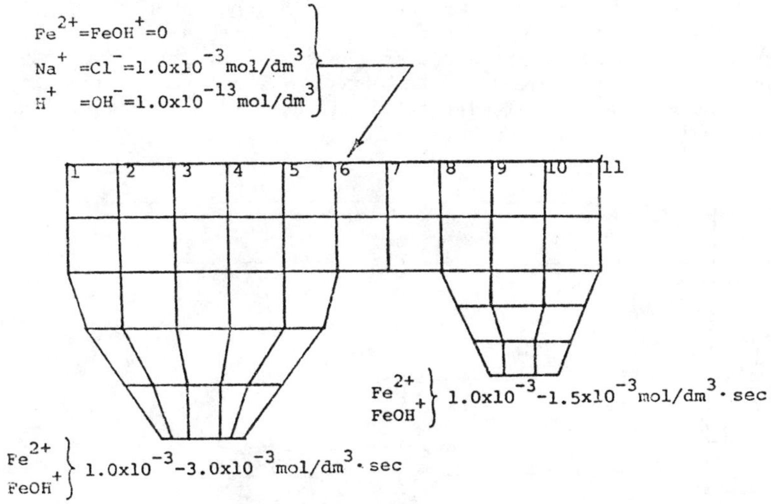

Figure. 1. Adjacent corrosion pits on a steel surface

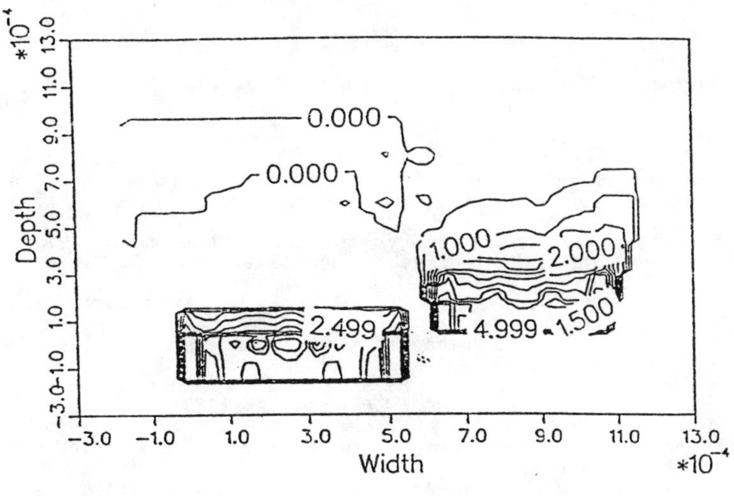

Figure 2. Fe^{++} concentration ($10^{-3}mol/dm^3$)

Typical results are illustrated in Fig. 2. as a contour plot of Fe^{++} and indicates that, at a given time, Fe^{++} at the bottom of the deeper pit is unable to migrate away even under the influence of a relatively strong electropotential. However, the smaller pit pattern indicates that Fe^{++} is able to leave the area near the bottom of the pit. This implies that the smaller pit is corroding more rapidly than the larger pit. A typical contour plot of the computed electric field in the vertical direction is shown in Fig. 3. Similar results indicate that the pH of the smaller pit is less than that of the larger pit and, again, that supports the idea that the smaller pit has a higher corrosion rate.

The actual pit corrosion problem is undoubtedly far more complicated than this analysis can explain. In this finite element model various chemical pressures in the capillary pores and actual concrete conditions have been neglected. However, these idealized results do give a general picture of pit corrosion propagation and possible electrochemical reactions.

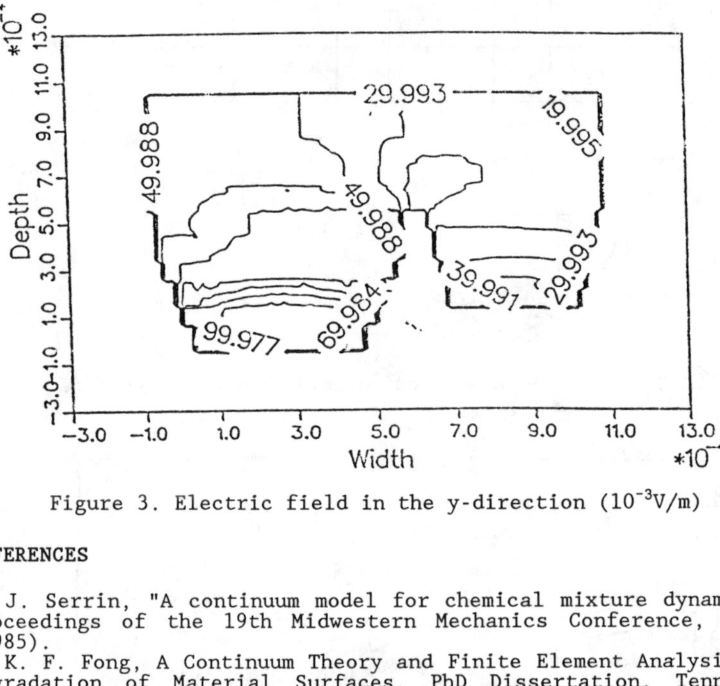

Figure 3. Electric field in the y-direction (10^{-3}V/m)

REFERENCES

1. J. Serrin, "A continuum model for chemical mixture dynamics", Proceedings of the 19th Midwestern Mechanics Conference, 59-64 (1985).
2. K. F. Fong, A Continuum Theory and Finite Element Analysis for Degradation of Material Surfaces, PhD Dissertation, Tennessee Technological University, Cookeville, TN (1989).
3. J. W. Fu, "A finite element analysis of corrosion cell", Corrosion, 38, (1982).
4. J. W. Fu, "A finite element method for modeling localized corrosion", Corrosion, 40, (1984).
5. J. R. Gralvele, "Transport processes and mechanisim of pitting of metal", J. Electrochem. Soc., 123, 464-474 (1976).
6. S. M. Southerlan and P. W. Tasker, "A mathematical model of crevice and pitting corrosion-I, the physical model", Corrosion Science, 28, 603-620 (1988).
7. W. Strumm and J. J. Morgan, Aquatic Chemistry, Wiley Interscience (1970)

Mapping Shear Failure Using Velocity Discontinuities

C.J. Flockhart, Y.C. Lam,

DSTO, Materials Research Laboratory
Department of Mechanical Engineering, Monash University

R.G. O'Donnell and R.L. Woodward

DSTO, Materials Research Laboratory

Abstract

The link between velocity discontinuities and adiabatic shear failure has been established. Using the DYNA 2D finite element code, problems of compression and penetrator break-up were simulated successfully. The velocity discontinuities could be identified by maxima in shear strain-rate from the finite element solution and corresponded closely to the observed failure sites. The possibility of writing a shear failure algorithm suitable for either adiabatic shear or shear fracture is indicated.

Introduction

The susceptibility of metals to adiabatic shear failure depends on their thermomechanical properties. However, even for materials susceptible to adiabatic shear, failure by this mechanism may not occur as a shear fracture may intervene. Indeed, the band of intense shear strain is only observed when a shear velocity gradient exists normal to the plane of the band. Backman and Finnegan [1] suggest a correlation of adiabatic shear bands with the slip lines associated with slip line field solutions to plastic deformation problems, and in particular with those slip lines which are also velocity discontinuities. Such an hypothesis allows nucleation associated with the characteristic slip surfaces, failure on those velocity discontinuities which persist, and an explanation of the velocity requirements for adiabatic shear failure.

Velocity Discontinuities

Consider an arbitrary surface within a deforming material. On one side of the surface velocity components are v and u normal and tangential to the surface respectively, and on the other side these are v' and u'. Continuity requires that $v = v'$ but there is no such requirement on the tangential components for a rigid/plastic material so that a discontinuity of amount $u' - u$ may exist. If $u' - u$ is finite, then on crossing the surface, an element will experience a finite shear strain $(u' - u)/v$ in an infinitessimal time interval, implying an infinite shear strain rate. Such a velocity discontinuity must correspond with a slip line, as the Levy-Mises rules require a correspondence between stresses and strain rates, and the maximum shear stresses define the slip lines. Nevertheless all slip lines are not necessarily velocity discontinuities as the tangential velocity may vary continuously across a slip line. It is thus necessary to be able to identify velocity discontinuities separately from maximum shear stress directions. Two options for doing this are (a) by examination of the velocity distribution in a deforming body, looking for changes in direction of the velocity vectors and (b) by looking for maxima in the shear strain rate.

Figure 1 illustrates several slip line field solutions for two dimensional problems with velocity discontinuities indicated by a distinctive heavy outline. For all cases in Figure 1 failure sites analogous to the regions outlined by the velocity discontinuities can be identified in dynamic axisymmetric problems.

FIG. 1 Plane strain slip line fields for a range of problems which mirror the analogous axisymmetric cases. Slip lines which are also velocity discontinuities are indicated by heavy lines.

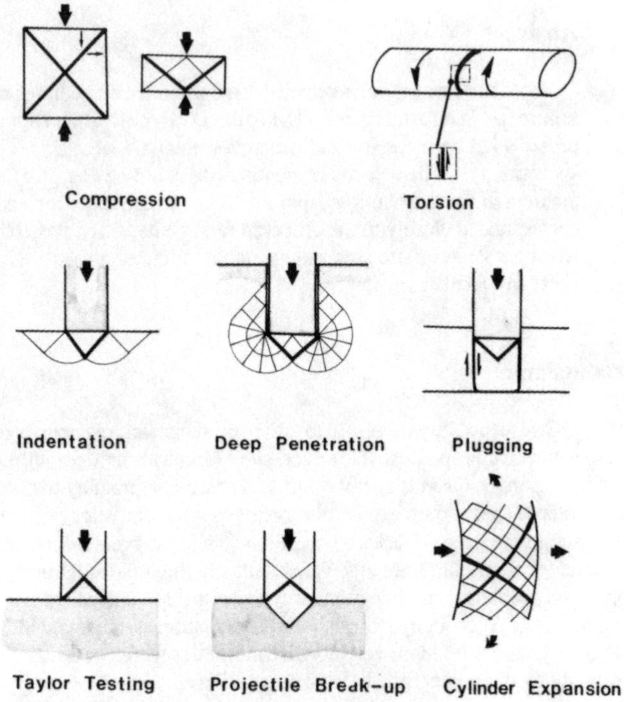

Finite Element Simulation of Impact Events

The DYNA 2D finite element wave propagation computer code was used to simulate impact deformations for simple compression and the Taylor test. The purpose was to demonstrate that velocity discontinuities can be related to observed shear failures for both axisymmetric and plane strain problems, while this may not be so for other parameters such as strains.

Figure 2 shows the strain rate, strain and velocity distribution in the frictionless plane strain compression of aluminium. Velocity discontinuities should correspond with a maximum in the change in velocity vector in moving through the solid. This can be qualitatively seen in the velocity vectors but is more evident in the distribution of shear strain rate. The pattern is as expected from the slip line field solution to this problem and in this case the strain distribution also mirrors these velocity discontinuities.

The Taylor test involves impact of a flat ended projectile onto a semi-infinite hard target upon which the impact end of the cylinder mushrooms. Figure 3 shows a typical example of the separation of a conical surface of failure at the impact end which is one characteristic failure mode. Figure 4 shows plots of effective strain and shear strain rate for a Taylor test at 10 μs after impact. The distribution of effective plastic strain is remarkably similar to that assumed in simple one dimensional analytical models of the Taylor test, and is consistent with their ability to predict approximately the final profiles, however it in no way relates to observed failures of the type shown in Figure 3. The contours of shear strain rate on the other hand indicate both a maximum and a velocity discontinuity in such a geometry as to separate a conical section from the impact end.

Discussion and Conclusion

For dynamic problems, inertial effects and wave propagation are important and, under impact conditions, the geometry changes rapidly with time. It is therefore surprising that simulations of such problems produce distinct strain rate maxima which mirror closely slip line field velocity discontinuities, a feature of steady state deformation problems. A requirement for such velocity discontinuities to result in failure is that they persist for a finite time. That the present approach is useful is exemplified by the observation of simple failures even in very high velocity fragment impact problems.

Although velocity vectors in principle could be used to identify the sites of shear failure, it is difficult to achieve by inspection. The detection of the magnitude of velocity changes may be incorporated into the finite element code. However, the same requirement could be met in the present program, by the plotting of the shear strain rate distribution. Writing of a shear failure algorithm should be relatively straight forward, having identified its location by a persistant maximum in shear strain rate. The strain at which softening initiates having been established from the simple adiabatic stress/strain calculations indicated above, then material which both exceeds that strain and is within a band or zone of shear strain rate maximum could be assigned to fail and carry no further shear or tensile stress, in the next computational step. Alternatively, if the ductility of the material in shear is known to be less than the strain for the onset of softening then a failure may be inserted representing a shear fracture.

The present study has thus established a correspondence between shear failure and the appearance of velocity discontinuities in high rate deformation problems. It is shown that such velocity discontinuities can be identified in finite element solutions by examination of shear strain rate maxima and that such maxima identify closely with observed failures in compression and the Taylor test. Velocity discontinuities may be used for constructing a shear failure algorithm.

Reference

1. M.E. Backman and S.A. Finnegan, "The propagation of adiabatic shear" in Metallurgical Effects at High Strain Rates, edited by Rohde, Butcher, Holland and Karnes (Plenum N.Y. & London, 531-543, 1973).

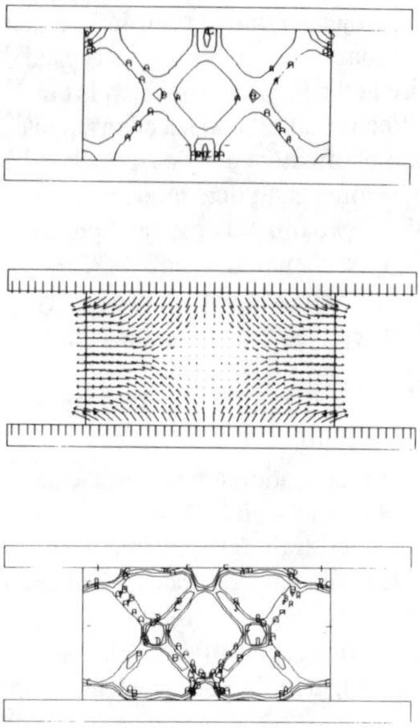

FIG. 2 Plane strain slip line fields for an aluminium alloy using frictionless contact and a strain rate of 1000 s^{-1}. Top to bottom: Effective plastic strain, velocity distribution and shear strain rate.

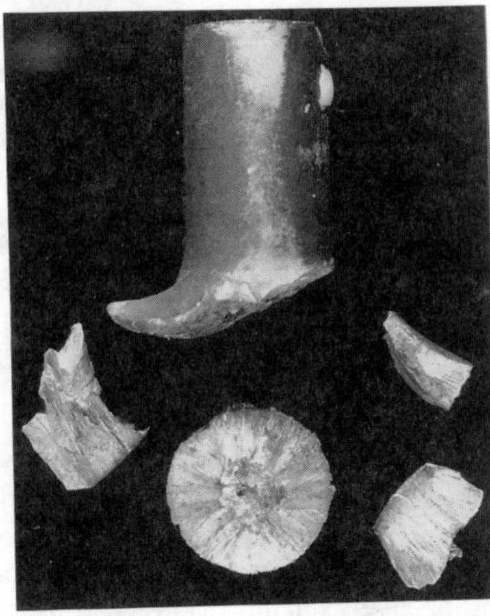

FIG. 3 Failure of a high strength steel Taylor test cylinder showing the separation of a conical section at the impact end.

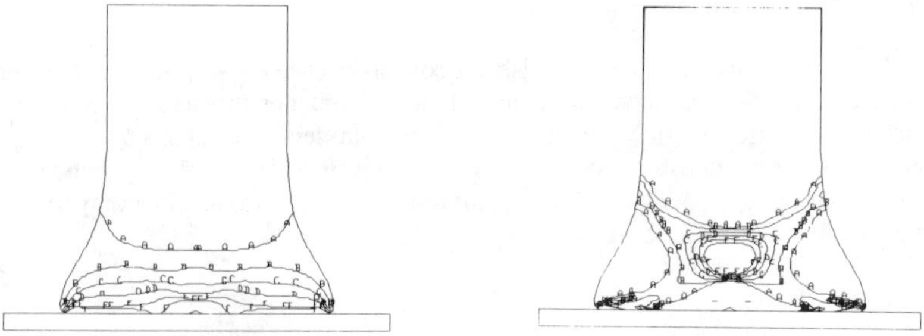

(a) (b)

FIG. 4 Taylor test impact of steel cylinder showing (a) contours of effective plastic strain, and (b) shear strain rate. Time 10 μs after impact at velocity of 200 ms^{-1}.

BASIC IDEAS FOR INTEGRATING ELASTIC-ZIEGLER KINEMATIC HARDENING PLASTICITY FORMULATIONS WITH AN "A-PRIORI" ERROR CONTROL

ALBERTO FRANCHI
Professor, Dep. of Civil Eng., Universitá di Brescia, Italy

FRANCESCO GENNA
Assistant Professor, Dep. of Structural Eng., Politecnico di Milano, Italy

PAOLO RIVA
Research Fellow, Dep. of Civil Eng., Universitá di Brescia, Italy

ABSTRACT

The incremental elastic-plastic with Ziegler's kinematic hardening constitutive law is presented as a Linear Complementarity Problem (LCP) with reference to Maier's work. The integration scheme is thought of as a two steps algorithm: a "predictor" step described by the "traditional" LCP mentioned before, and a "corrector" step, interpreted as a "neutral" equilibrium phase which is described again as a strictly convex problem. Both steps assume the stress path to be always inside a prescribed error on the value of the uniaxial yield limit. An application is presented to compare the proposed scheme with other more traditional adopted by standard plasticity finite elements codes.

INTRODUCTION

The question of accuracy of the incremental elastic-plastic problem numerical integration was rediscovered and imposed to the public only recently [1,2]; the attention was moved from the integration of the structural problem to the integration of the constitutive law. Within this framework, Hodge Jr. presented a technique for the "a posteriori" linearization of the yield surface with a pre-assigned error control [3].

This paper is developed within this framework and intends to first recall the formulations of the incremental elastic-plastic constitutive law with Ziegler kinematic strain-hardening in terms of mathematical programming, then to present a general integration algorithm. The results are compared with those obtained using some of the most reliable algorithms implemented in common computer programs [4].

MATERIAL CONSTITUTIVE LAW WITH ZIEGLER KINEMATIC HARDENING

Let σ_{ij} and ϵ_{ij} denote the stress and strain symmetric tensors in orthogonal Carthesian coordinates x_i, $i = 1,2,3$, in a given configuration Ω. The stress tensor σ_{ij} must satisfy the yield condition: $\varphi = f(\sigma_{ij} - \alpha_{ij}) - \sigma_0 \leq 0$, where $\dot{\alpha}_{ij} = h_k \frac{\dot{\epsilon}^{pl}}{\sigma_0}(\sigma_{ij} - \alpha_{ij})$ is the Ziegler kinematic hardening evolution law, in which parameters h_k and σ_0 are determined from a uniaxial test, and the equivalent plastic strain rate $\dot{\epsilon}^{pl}$ is defined through the relation: $\dot{\epsilon}^{pl} = \frac{1}{\sigma_0}(\sigma_{ij} - \alpha_{ij})\dot{\epsilon}^p_{ij}$.

The stable non-holonomic material response to a given strain rate tensor $\dot{\epsilon}^0_{ij}$

Strain rate tensor $\dot{\epsilon}^0_{ij}$ is conceived as the sum of an elastic $\dot{\epsilon}^e_{ij}$ and a plastic part $\dot{\epsilon}^p_{ij}$, i.e. $\dot{\epsilon}^0_{ij} = \dot{\epsilon}^e_{ij} + \dot{\epsilon}^p_{ij}$. The elastic stress rate-strain rate relationship reads: $\dot{\sigma}^e_{ij} = D_{ijhk}\dot{\epsilon}^0_{ij}$ and $\dot{\sigma}_{ij} = D_{ijhk}\dot{\epsilon}^e_{ij}$, where D_{ijhk} is the symmetric positive definite elastic tensor.

The flow rule can be described as follows [5]:

$$\begin{aligned} &if \quad \varphi(\bar{\sigma}_{ij}, \bar{\alpha}_{ij}, \sigma_0) < 0 \;\; then \;\; \dot{\epsilon}^p_{ij} = 0 \;\; else \\ &\dot{\varphi} = \frac{\partial f}{\partial \sigma_{ij}}\dot{\sigma_{ij}} + \frac{\partial f}{\partial \alpha_{ij}}\dot{\alpha_{ij}} \leq 0 \;\; ; \;\; \dot{\epsilon}^p_{ij} = \frac{\partial \varphi}{\partial \sigma_{ij}}\dot{\lambda} \, , \; \dot{\lambda} \geq 0 \;\; ; \;\; \dot{\varphi}\dot{\lambda} = 0 \end{aligned} \quad (1)$$

For a given strain-rate tensor $\dot{\epsilon}^0_{ij}$ exists a unique plastic multiplier rate $\dot{\lambda}$.

The neutral equilibrium holonomic material response due to a negative rate of the uniaxial strength and zero rate of deformations

The idea of the "return" step is that of keeping the strain constant ($\dot{\epsilon}^0_{ij} = 0$) and prescribing a decrease of the uniaxial strength $\dot{\eta}\sigma_0$ ($\dot{\eta} < 0$). This step is named "neutral equilibrium holonomic" because: (i) an infinite set of stress solutions σ_{ij} is found at the same "load" level ϵ^0_{ij} (defining a neutral equilibrium configuration); (ii) holonomic because the stress path is of a purely mathematical nature and has no physical basis.

In analytical terms:

$$D^{-1}_{ijhk}\dot{\sigma}_{hk} + n_{ij}\dot{\lambda} = 0 \;\text{ and }\; \dot{\varphi} = \mathrm{n_{ij}}\dot{\sigma}_{\mathrm{ij}} - \frac{\mathrm{h_k}}{\sigma_0}\dot{\lambda}\mathrm{n_{ij}}(\sigma_{\mathrm{ij}} - \alpha_{\mathrm{ij}}) - \dot{\epsilon}\sigma_0 = 0 \quad (2)$$

INTEGRATION SCHEME

The integration scheme strategy is described with the help of Fig. 1 for the special case of perfect plasticity. Two surfaces are represented in the stress space: the first, $\varphi(\sigma_{ij},\ \sigma_0) = 0$, is the original yield surface, while the second, $\varphi(\sigma_{ij},\ \sigma_0(1+\eta)) = 0$ is the same yield surface with a yield stress value increased by the quantity $\eta\sigma_0$. The sequence of "predictor" and "corrector" steps is such that the stress point trajectory always lies between the two surfaces.

More specifically, point A is the elastic limit; for a given increment of the deformation vector $\Delta\epsilon^0_{ij} = \dot{\epsilon}^0_{ij}\Delta t^0$, the behaviour is considered elastic up to point 1; then the increment of total deformation is taken equal to zero and the rates of the stress tensor ${}_1\dot{\sigma}_{ij}$ and ${}_1\dot{\lambda}$ are found by solving problem 2, where ${}_1n_{ij}$ is the gradient to the yield surface in ${}_1\sigma_{ij}$. If n_{ij} is constant in the finite step, $\dot{\sigma}_{ij}$ and $\dot{\lambda}$ are constant as well and therefore explicit integration rules can apply:

$${}_1\Delta\lambda = {}_1\dot{\lambda}\,{}_1\Delta t \;\text{ and }\; {}_1\Delta\sigma_{\mathrm{ij}} = {}_1\dot{\sigma}_{\mathrm{ij}}\,{}_1\Delta \mathrm{t} \quad (3)$$

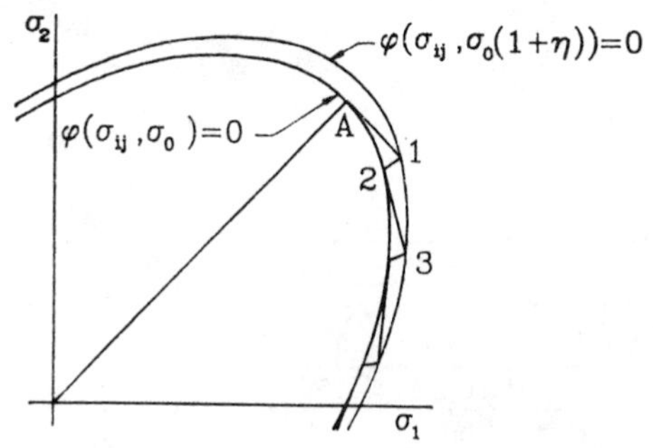

Figure 1: Integration scheme.

The stepsize ${}_1\Delta t$ is determined by the condition that the final stress point must stay on a plane tangent to the actual yield surface.

At point 2, problem 1 is solved for a given rate of total deformation $\dot{\epsilon}^0_{ij}$ where $n_{ij} = {}_1n_{ij}$, i.e. the gradient to the yield surface at point 1. The rate problem 1 gives ${}_2\dot{\sigma}_{ij}$ and ${}_2\dot{\lambda}$ which can be integrated according to Eqs. 3 where the index 1 is changed with 2.

The step-size ${}_2\Delta t$ is determined by finding the intersection of the stress rate vector ${}_2\dot{\sigma}_{ij}$ with the outer yield surface $\varphi(\sigma_{ij},\ \sigma_0(1+\eta)) = 0$. Having found point 3 on the outer yield surface, the new gradient ${}_2n_{ij}$ is computed and the procedure continues as at point 1.

NUMERICAL EXAMPLE

Example 1 from ABAQUS Manual [4] has been tested. Data load history and results of the example are given in Fig. 2. Comparison of the results shows the effect of different prescribed errors η. A considerable saving in the number of steps and iterations per step is observed using the proposed approach.

CONCLUSIONS

1. An algorithm is presented for integrating the rate plasticity problem by describing both the "tangent" predictor and the "return" steps in term of strictly convex formulations;

2. the explicit integration scheme allows the stress path to satisfy an "a priori" error control;

3. numerical experiments have been performed with reference to standard finite element algorithms.

REFERENCES

1 Krieg, R.D., Krieg, D.B., "Accuracies of Numerical Solution Methods for the Elastic-Perfectly Plastic Model," *Transaction of the ASME*, Vol. 99, No. 4, November 1977, pp. 510-515.

2 Schreyer, H.L., Kulag, R.F., and Kramer, J.K., "Accurate Numerical Solutions for Elastic Plastic Models," *Transaction of the ASME*, Vol. 101, August 1979, pp. 226-235.

3 Hodge, P.J.Jr., "Automatic Piecewise Linearization in Ideal Plasticity," *Computer Methods in Applied Mechanics and Engineering*, November 1977, pp. 249-272.

4 ABAQUS, *"Theoretical Manual,"* Version 4.6, Hibbitt, Karlsson & Sorensen, Inc., Providence, R.I., 1987.

5 Maier, G., "A Matrix Structural Theory of Piecewise Linear Elastoplasticity with Interacting Yield Planes," *Meccanica*, March 1970, pp. 54-66.

Load Step	ABAQUS Number of Iterations	ABAQUS σ_x (ksi)	ABAQUS ε_x^{pl} (10^{-3})	STRUPL-2 Number of Iterations	$\eta = 0.05$ σ_x (ksi)	$\eta = 0.05$ ε_x^{pl} (10^{-3})	$\eta = 0.01$ σ_x (ksi)	$\eta = 0.01$ ε_x^{pl} (10^{-3})
1	1	10.00	0.000	2	10.50	0.050	10.10	0.010
2	3	15.00	0.500	1	25.00	1.500	25.00	1.500
3	3	20.00	1.000	2	4.50	1.450	4.90	1.490
4	3	25.00	1.500	1	0.10	1.010	0.10	1.010
5	1	12.55	1.500	2	20.59	1.060	20.20	1.020
6	3	0.10	1.010	1	30.00	2.000	30.00	2.000
7	1	15.05	1.010	2	9.50	1.950	9.90	1.990
8	3	30.00	2.000	1	0.10	1.010	0.10	1.010
9	1	15.05	2.000					
10	3	0.10	1.010					

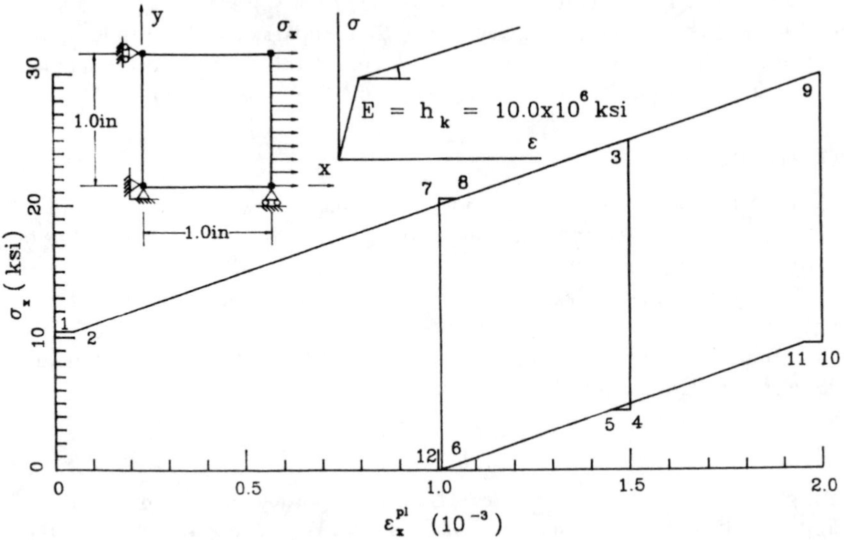

Figure 2: Application example [4].

ON THE SOLUTION OF VISCOPLASTIC & CREEP PROBLEMS

M. S. Gadala* and M. Hamid

EMRC, 1607 East Big Beaver Road, Troy, Michigan 48083

*(*Also Adjunct. Assoc. Prof. at University of Michigan-Dearborn)*

ABSTRACT

An integrated procedure for the solution of viscoplastic and creep problems is presented. The procedure involves local iterative stress loop for enhancing the convergence. The integration of the viscoplastic constitutive law is performed by a mixed implicit/explicit scheme.

INTRODUCTION

The finite element solution of viscoplastic and creep problems requires an efficient and stable numerical algorithm. Among the difficulties faced are the proper calculation of the tangent stress-strain tensor, for use in stiffness and pseudo force computations, and the accurate integration of the inelastic constitutive relations. In this paper, we present an integrated algorithm for the solution of viscoplastic and creep problems discussing the above two points.

In viscoplastic problems, we assume that the total strain, ε is the sum of the elastic and viscoplastic components, ε^e and ε^{vp}, respectively. The viscoplastic component of strain is governed by a viscoplastic flow rule in the form

$$\dot{\varepsilon}^{vp} = d\lambda < \Psi(F) > \frac{\partial F}{\partial \sigma} \tag{1}$$

where $F = 0$ represents a yield condition, and the function $< \Psi(F) >$ is equal to zero for $F < 0$, and is equal to F for $F \geq 0$ and $d\lambda$ is a time variable fluidity coefficient [1]. The incremental constitutive relation is given by

$$\Delta\sigma = \boldsymbol{D}\left(\Delta\varepsilon - \Delta\varepsilon^{vp}\right) \tag{2}$$

where $\boldsymbol{D}$ is the matrix of elastic constants. Using the α–method [2], the incremental viscoplastic strains may be written as

$$\Delta\varepsilon^{vp} = \Delta t\left[(1-\alpha)\ {}^{t}\dot{\varepsilon}^{vp} + \alpha\ {}^{t+\Delta t}\dot{\varepsilon}^{vp}\right] \tag{3}$$

The viscoplastic strains at a time $t+\Delta t$ are calculated by utilizing a limited Taylor series expansion

$${}^{t+\Delta t}\dot{\varepsilon}^{vp} = {}^{t}\dot{\varepsilon}^{vp} + {}^{t}\left(\partial\dot{\varepsilon}^{vp}/\partial\sigma\right)\ {}^{t}(\Delta\sigma) \tag{4}$$

which when combined with Eqns. (2, 3) gives

$$\Delta\sigma = {}^t\hat{D}\left(\Delta\varepsilon - \Delta t \cdot {}^t\dot{\varepsilon}^{vp}\right) \tag{5}$$

where

$${}^t\hat{D} = \left[I + \Delta t \cdot \alpha \cdot {}^t\left(\partial\dot{\varepsilon}^{vp}/\partial\sigma\right) D\right]^{-1} D \tag{6}$$

and I is the identity matrix.

SOLUTION ALGORITHM

In the solution procedures, the following steps are followed:

i. Select a time step Δt and calculate the effective stress-strain matrix $\hat{D}$, the corresponding stiffness matrix and pseudo force vector, i.e.;

$${}^tK = \int_v B^T\, {}^t\hat{D}\, B\, dV \quad \& \quad \Delta^t f^{vp} = \int_V B^T \cdot {}^t\hat{D}\left(\Delta t \cdot \dot{\varepsilon}^{vp}\right) dV$$

ii. after the solution, perform the following iterative loop in the stress calculation routine:

- calculate $\Delta\sigma$ using original $\hat{D}$ at start of the increment
- update stresses and hence calculate new $\dot{\varepsilon}^{vp}$
- update $\hat{D}$ based on new stress and strain values
- continue until an appropriate measure for the stress (e.g., effective stress converges)

iii. compute global residual and perform the global iterative loop.

In the integration of the viscoplastic constitutive relations we use a mixed implicit/explicit scheme. In the implicit scheme [3], the stresses are updated in two steps, first, an elastic predictor moves the stress state from ${}^t\sigma$ to ${}^{t+\Delta t}\sigma^*$, then the stresses are subsequently mapped onto a suitable yield surface by projecting along the initial plastic flow direction ${}^t\boldsymbol{d}$ to obtain ${}^{t+\Delta t}\sigma^{(i-1/2)}$, which will be finally mapped onto the updated flow direction ${}^{t+\Delta t}\boldsymbol{d}$ to obtain ${}^{t+\Delta t}\sigma^{(i)}$. Figure (1) shows a graphical representation of the scheme and the following equations formulate the procedure:

$${}^{t+\Delta t}\sigma^* = {}^t\sigma\, \hat{D}\left({}^{t+\Delta t}\Delta\varepsilon - {}^{t+\Delta t}\Delta\varepsilon^{vp}\right) \tag{7}$$

$${}^{t+\Delta t}\sigma = {}^{t+\Delta t}\sigma^* - (d\lambda)\, \hat{D}\left[(1-\alpha)\, {}^t\boldsymbol{d} + \alpha\, {}^{t+\Delta t}\boldsymbol{d}\right] \tag{8}$$

In the second part of the procedure, the explicit scheme, a simple Euler-forward integration is used:

$${}^{t+\Delta t}\sigma^{(i)} = {}^t\sigma + \int_{{}^t\varepsilon}^{{}^{t+\Delta t}\varepsilon^{(i)}} \hat{D}^{vp}\, d\,\varepsilon \tag{9}$$

where $\hat{D}^{vp}$ is the viscoplastic constitutive matrix at the start of the iteration, and the solution is obtained by integrating over a sufficient number of sub-increments.

If the Gauss point is elastic and loading the implicit part of the procedure is used whereas if the Gauss point is already plastic and loading, the explicit part of the procedure is used. This is shown to reduce the number of iterations required by eliminating the overestimated stress state by the elastic predictor. The reduction in the number of iteration depends on the spread of plastic zone in the structure.

EXAMPLES & DISCUSSION

The above procedures have been implemented in the NISA II finite element package [4] and in the following section, we present some numerical examples from the package [5].

Uniaxial Tension of a Square Plate

A square plate of sides 1 in and thickness 0.1 in is subjected to a uniform tension of 1200 psi and a uniform temperature of 100 °F. The plate material has modulus of elasticity, $E = 24.07 \times 10^6$ psi, hardening modulus, $E_t = 7.3 \times 10^5$ psi, yield strength, $Y = 8990.0$ psi, and coefficient of thermal expansion, $\alpha = 11.18 \times 10^{-6}(1/°F)$. The problem is designed to test the efficiency of the proposed scheme for the integration of constitutive relations. The loading is divided into two steps, one step to the elastic limit and one step in the plastic zone. Displacement results and number of iterations are shown in Table I together with the exact hand-calculated values. It is noted that when the Gauss points are plastic and loading (which is the case specially designed in this example) the proposed scheme is superior to the generalized implicit scheme. In practical situations, the improved number of iterations will depend on the spread of the plastic zone in the structure and the status of the Gauss points.

TABLE I. Results for uniaxial example.

Method	Results		
	UX in ($\times 10^3$) (4.621906)*	No. of iterat.	
		Step 1	Step 2
Proposed Scheme	4.62183	1	2
General. Implicit	4.63133	1	38

* Exact solution

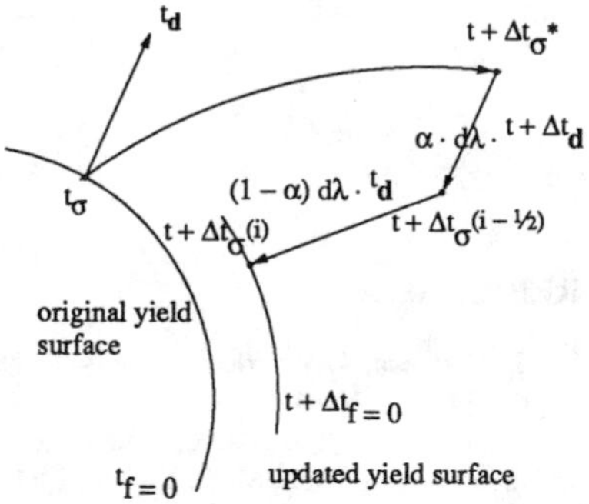

FIG. 2. Generalized trapezoidal scheme

Creep Analysis of a Rotating Disk

A thin disk of 12 in outside diameter and 2.5 in inside diameter is rotating at 15,000 rpm. The material of the disk is assumed to have the following creep law: $\varepsilon^c = 4.41 \times 10^{-32} \sigma_e^{6.2} t$, where t is the time, and σ_e is the effective stress. The material also has a modulus of elasticity of 18×10^6 psi, yield strength of 58×10^3 psi and density of 7.3177×10^{-4} 16 sec^2 / in^4. Ten 8-noded axisymmetric elements are used to model one half of the disk. The creep analysis of the disk utilized an α–method implicit time integration scheme with $\alpha = 1/2$ with automatic time stepping based on strain criterion and a maximum limit of 3 hours for the step. As indicated in Fig. (2), the NISA results compare favorably with the analytical and finite element results of Ref. [6,7], respectively. The current finite element analysis cannot predict the analytical rising stress because the thickness and the centrifugal forces are kept constant with time. More specific comparisons are not available for this example and more detailed cases are under investigation.

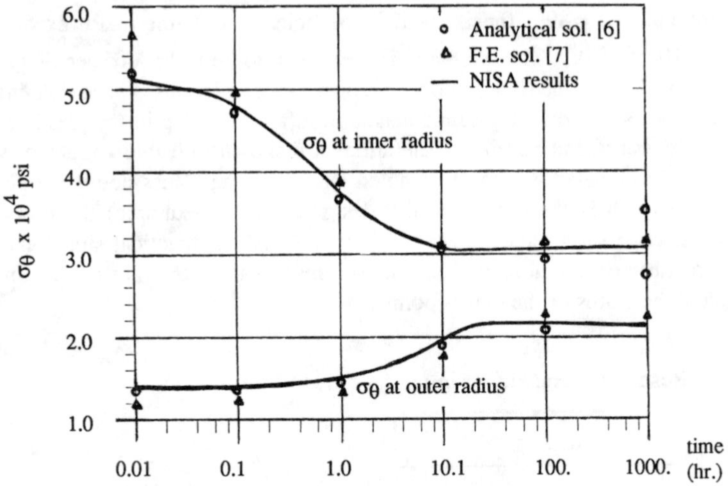

FIG. 2. Stress relaxation for rotating disk

REFERENCES

1. I. Cormeau, "Numerical stability is quasi-static elasto-visco-plasticity," Int. J. Num. Meth. Engng., 9(1975) 109-127.
2. M. Kojic and K. J. Bathe, "The effective stress-function algorithm for thermo-elasto-plasticity and creep," Int. J. Num. Meth. Engng., 24(1987) 1509-1532.
3. M. Ortiz and E. P. Popov, "Accuracy and stability of inegration algorithms for elastoplastic constitutive relations," Int. J. Num. Meth. Engng., 21(1985) 1561-1576.
4. NISA II User's Manual, EMRC, Troy - Michigan, Version 90.0 (1990).
5. NISA II Verification Problem Manual, EMRC, Troy - Michigan, Version 90.0 (1990).
6. A. Mendelson, Plasticity: Theory and Applications, McMillan, New York, (1968).
7. T. W. Yang, C. K. Chen and H. K. Shee, "Endochronic viscoelastic creep analysis of 2-D structures, Computers & Structures," 26(1987) 425-429.

A NEW VOID EVOLUTION LAW AND ITS FEM FORMULATION

Gao.Z Fan.J* Fung.s Zheng.X
Laboratory of Constitutive Laws,
Department of Engineering Mechanics,
Chongqing University, China, 630044

Abstract

An analytical analysis of a cylinderical model by directly using isotropic-kinematic hardening law for matrix is given. A new evolution law of voids is then proposed and embeded into a constitutive framework and, in turn, incomplemented into a nonlinear finite element analysis.

INTRODUCTION

Since the growth, nucleation and interaction of voids are main sources of degeneration and failure of ductile materials, a lot of works in this subject have been done in the past twenty years. Some of the original contributions may be owed to McClintock [1], Rice and Tracey [2], Gurson [3], who carried on the analysis of void growth in assuming that the matrix behaves as perfect rigid plasticity. One main direction to develop their works is to apply more realistic constitutive law in the analysis. Yamamoto, Duva and Hutchson introduced average flow stress or average effective stress and strain ,respectively, to indirectly account for matrix hardening [4] [5]. Perzyna used the linear hardening law for matrix of viscoplastic materials [6].

Following the approach given by Carroll and Halt [7], Fan and Zheng developed a closed solution for void growth of a spherical model under external mean stress when matrix material behaves as mixed isotropic-kinermatic hardening [8]. TO consider the effect of the deviatoric deformation on the void growth, an effective deviatoric function K is introduced. While the direct application of mixed hardening law for the matrix seems to be a progress. The fact that K is determined by large deformation FEM analysis make its application quite limied. In this paper the mentioned difficulties have been circumvented based on a cylinderical model by directly using mixed isotropic-kinematic hardening for matrix, in which the closed relations for both macroscopic mean stress and effective deviatoric stress, in terms of initial and current void volume fraction and material constants, have been obtained. The evolution law of voids is then proposed and, in turn, embed into a constitutive framework. Further, a nonlinear finite element approach incorporating the developed void evolution law has been implemented and applied.

VOID GROWTH IN MATERIALS WITH ISOTROPIC-KINEMATIC HARDENING

A cylinder with internal and exteral radius a and b, respectively, is used as unit cell of materials. While the common

----

Financial support 9187004-6 from CNSF is fully acknowledged.

assumptions of incompressibility and rigid plasticity of matrix are adopted, the matrix materials are taken as isotropic-kinematic mixed hardening type. The macroscopic logrithmatic strain and cauahy atress will be denoted by E_{ij} and Σ_{ij} and the corresponding microscopic strain and stress denoted by e_{ij} and σ_{ij}, respectively.

The cylinderical model is uniformly extended along radius direction with no presure inside. The bounary condition at r=b and matrix incompressibility condition are expressed in (2), (3), respectively,

$$\dot{V}_i = E_{ij} X_j \tag{2}$$

$$\dot{\epsilon}_{11} + \dot{\epsilon}_{22} + \dot{\epsilon}_{33} = 0 , \qquad \text{with } \dot{\epsilon}_{ij} = \dot{\epsilon}^p_{ij} \tag{3}$$

assuming x1 ,x2 ,x3 are arthogonal principle axis, we have

$$dE_{11} = dE_{22} \qquad , \; dE_{ij} = 0 \; , \text{ if } i \neq j \tag{4}$$

The contraction in x3-direction or the value of dE_{33}/dE_{11} is very important, since it determines the deformation in z direction and relates the effect of stress triaxity on void growth. A parameter ω is therefore introduced as

$$dE_{33} = -\omega dE_{11} \tag{5}$$

and the volume strain can be expressed as

$$\epsilon = dv/v = dE_{kk} = B(\omega) dE_{11} \geq 0 \; , \quad \text{with } B(\omega) = 2 - \omega \tag{6}$$

$B(\omega)$ may be called strain restriction function which determines the effects of triexity on voids evolution. In fact, if $\omega = 0$ the material is in plane strain state, the deformation restriction in z-direction is very strong and, b(w) takes the maximum value 2, then the volume change (or void growth) takes also the maximum value under the same applied strain dE . However, if the deformation restriction in z-direction is very weak ω will takes large value. It results in that $B(\omega)$ and T are smaller, then the void growth must be also smaller. When $\omega = 2$, then B(w) eqeal zero; no volume dilatation at all. The range of $B(\omega)$ may be varied in the interval $0 < \omega < 2$. Denoting fo. f as initial and current void volume fraction respectively, we have the following matrix conti- nuity condition

$$df = (1-f)\, dE_{\kappa\kappa} \tag{7}$$

The macroscopic and microscopic intrinsic time (or Odquist parameter) are denoted by $\bar{\zeta}$ and ζ, respectively, which are defined as Eucledean norm of deviatoric plastic strain E^{ρ}_{ij} and $\dot{\epsilon}^{\rho}_{ij}$ as

$$d\bar{\zeta} = (dE^{\cdot p}_{ij}\, dE^{\cdot p}_{ij})^{1/2} \qquad , d\zeta = (\epsilon^{\cdot p}_{ij}\, d\epsilon^{\cdot p}_{ij})^{1/2} \tag{8a,b}$$

Due to the assumptions of rigid plasticity and incompressibility for matrix, we have

$$d\epsilon^{\cdot p}_{ij} = d\epsilon_{ij} \qquad , \; d\zeta = (d\epsilon_{ij} d\epsilon_{ij})^{1/2} \tag{9a,b}$$

$$dE^{\cdot p}_{ij} = dE^{\cdot}_{ij} = dE_{ij} - (dE_{kk}\delta_{ij})/3, \; d\zeta = (1+\omega)(2/3)^{1/2} dE_{11} \tag{10a,b}$$

Combining eq.(7) with (6b) and (10b) and solving the mentioned bounary value problem, the rate of void growth and the increment of microscopic strain can be expressed as, respectively,

$df/d\bar{\xi} = (3/2)^{1/2}(1-f)B/(3-B), \qquad d\epsilon_{ij} = q_{ij}d\zeta$ (11a, b)

where q_{ij} are explicit functions of ω and r.

The constitutive equation of rigid plasticity matrix with mixed isotropic-kinematic hardening can be written as [11]:

$S_{ij} = S_y^o f(\zeta)d\epsilon_{ij}/d\zeta + \alpha_{ij}$ (12)

where the first term of right hand describe isotropic hardening through introduction of hardening function $f(\zeta)$ and initial deviatoric stress S_y^o. If we neglect the coupling effect of isotropic hardening on kinematic hardening, the second term α_{ij}which denotes kinematic hardening can be expressed as

$\partial_{ij} = \int_0^3 \rho(\zeta\_\zeta^\bullet)\partial\epsilon_{ij}/\partial\zeta^\bullet d\zeta^\bullet$ (13)

Substituting eq.(11b) into eq.s (12) and (13), we obtain

$S_{ij} = q_{ij}F(\zeta) \qquad , \ F(\zeta) = S_y^o f(\zeta) + \int_0^\zeta \rho(\zeta\_\zeta^\bullet)d\zeta^\bullet$ (14a, b)

If we take kernel function ρ and hardening function f as

$\rho(\zeta) = \sum_{i=1}^{3} c_i \mathrm{Exp}(\partial_i\zeta)$ and $F(\zeta) = 1+k\zeta$ (15a, b)

we may approximately express $F(\zeta)$ as

$F(\zeta) = A_1 + A_2\zeta$ (16)

where $A_1$and A_2 are combined material constants.

Siminar to the work given by Duva and Huchinson in 1984 [6], take Φ, ϕ as macroscopic and microscopic potential function, respectively, the macroscopic stress can be expressed as

$\Sigma_{ij} = \partial\Phi/\partial E_{ij} = 1/v \int_{Vm}\partial\phi/\partial E_{ij}dv = 1/v\int_{Vm}S_{kl}\partial\epsilon_{kl}/\partial E_{ij}dv$ (17)

Substituting the related relations of $S_{kl}$and ϵ_{ij}into eq(17) we can express all macroscopic stress in terms of Ω_i, which are functions of f_0 ,f, $B(\omega)$ and material parameters A_1 and A_2 . Specifically, macroscopic mean stress $\Sigma_m$and effective deviatoric stress $\Sigma_e$can be expressed as

$\Sigma_m = A_1\Omega_1 + A_2\Omega_3\_(A_1\Omega_2+A_2\Omega_4), \qquad \Sigma_e = 3(A_1\Omega_2+A_2\Omega_4)$ (18a, b)

It is well known that triaxity T , which has important effect on void growth, is defined as the ratio of Σ_m over Σ_e. Therefore, It can be expressed analytically by using eq.s(18a,b). Furthermore, numerical results show that T is mainly determined by strain restriction function $B(\omega)$ and the relation B =f(T) can be expressed as

$B/(3-B) = \alpha_0 \mathrm{Exp}[\alpha_1 T/(1+T)]$ (19)

Substituting this relation into eq.(11), we obtain the growth law of voids as:

$df/d\bar{\xi} = c(1-f)\mathrm{Exp}[\alpha_1\ T/(1+T)]$ (20)

for 1018 steel, our experimental data shows c= 0.001 and = 3.253.

The relation between T and $df/d\bar{\xi}$ denotes finite maximum rate of void growth which is different with the one, proportional to Sinh T and therefore with infinite rate if T being infinite [9]. The law of void nucleation developed in [8] together with eq.(18) results in the evolution law of voids as:

$$\dot{f} = C(1-f)Exp[\alpha_1 T/(1+T)]+[H(f)(\phi/\Phi_0\_^1)^n]/(1-f) \tag{21}$$

for 1018 steel, $H(f)= 0.000032 + 0.025f$ and $n=5$.

FEM STIFFNESS MATRIX FOR LARGE PLASTICITY INCORPORATING VOID EVOLUTION

Since the plastic deformation in microelement with voids of ductile materials may be very large. A new formulation of large deformation plasticity and nonlinear finite element code have been developed. The predominant characteristics are following. The framework of current configuration with corresponding multiplicative decomposition related to fictitious unloaded reference is adopted. Hencky strain and its conjugated stress Jaumann rate with a body-fixed coordinate system are used in the formulation of FEM.

Denoting d^i_k and $\overset{\nabla}{S}{}^i_{\cdot\kappa}$ as deformation rate and deviatoric Jaumann rate, respectively, We have the following constitutive equation

$$(\overset{\nabla}{S})^i_{\cdot k} + \bar{Q}^i_k = Ad^i_{k(p)} \quad , \qquad \text{with} \qquad \bar{Q}^i_k = \Sigma\alpha^{(r)}\, Q_k^{i(r)} \tag{22a,b}$$

where A and $\alpha^{(r)}$ are material parameters and $Q_k^{i(r)}$ is generized frictional force for α - dissipative mechanism. From eq.22a we get

$$\dot{\sigma}^i_{\ \kappa} = \dot{\sigma}^m_m \delta^\iota_\kappa (1/3-A\bar{\mu}/9k) + \bar{\mu}Ad^i_k - \bar{\mu}\bar{Q}^i_k + S^i_r d^r_k - d^i_r S^r_k \tag{23}$$

Using Hook's law and eq.(7), we get $\dot{\delta}^m_m = 3k[\dot{d}^\kappa_\kappa - \dot{f}_g/(1-f)]$ (24)

Combining eq.s(23) and (24) results in the stiffness matrix as

$$[\dot{\sigma}^i_{\ k}] = [Dep][d^i_k] - \bar{\mu}[\bar{Q}^i_k] - \bar{\mu}[F^i_k] \tag{25}$$

Where [Dep] is elastic-plastic matrix, $[F^i_k]$ is the matrix envolving void evolution and can be expresssd as

$$F^i_k = [(k-A\bar{\mu}/3)f_g \delta^i_k] \,/\, [\bar{\mu}\ (1-f)] \tag{26}$$

Eq.s (25)(26) are used in the FEM formulation. Since the space is limited, the numerical algorism will be pubulished elsewhere.

REFERENCES

1. F.M.McClintock, A Criterion for ductile fracture by the grouth of Voids, J.Appl Mech. 35. 363 (1968).
2. J.R.Rice and D.M.Traccy, On the Ductile Enlargement of Voids in triaxial stress fields, J.Meth.Phys.Solids, 17. 201 (1969).
3. A.L.Garson, Continum Theory of Dactile Rupture by Voids Nucleation and Growth: Part I-Yield Criteria and Flow Rules for Porous Ductile Media, ASME, J.of Eng.Mat. and Tech., 2-15,(1977).
4. H.Yamamoto. Int.J.Fracture, 14(1978).
5. J.M.Dava, Constitutive Potentials For Dilutily Voided Nonlinear Materials, Mechanice of Materials, 3.41-54 (1984).
6. P.Perzyna, Internal State Variable Description of Dynamic Fracture of Ductile Solids, Int.J. Solids Structure. 1986.
7. M.M.Carroll and A.C.Holt, J.Appl.Phys.43,(1972).
8. Fan J.and Zheng x., Micro/Macro Combined Study for Constitutive Law Incorporating Damage Evolution with Application. Proc. of Int. Conf.of Constitutive Laws for Engineering Materials 65-73, 1989.

An Endochronic-Plasticity Computational Model for Undrained Behavior of Soils

M. E. Al-Gassimi [1] and S. Sture [2]
[1] Graduate student, Dept. CE, Colorado State University, Ft. Collins, CO 80523.
[2] Professor, Dept. CEAE, University of Colorado at Boulder, CO 80309-0428.

ABSTRACT

Incremental solution strategies for endochronic plasticity model behavior under undrained conditions are developed. Two categories of model control at the constitutive level are considered. Typically, a fully strain-controlled model is used in finite element applications, however, it is often of interest to conduct analysis under mixed control where equilibrium and incremental incompressibility conditions are iteratively enforced at each loading step. Solution strategies are developed for stress and mixed controls which are categories often used in laboratory tests.

1. INTRODUCTION

Recent developments in the endochronic theory have made it one of the most promising analytical tools available today. The coupling term introduced by Valanis [1] into the hydrostatic rate equations gave rise to a coupling between shear and hydrostatic components that can produce full dilatency. The recent development [2,3] accounts for anisotropic features and simulates soil behavior under nonproportional and cyclic loading quite realistically. In this paper we extend the range of applicability of the recent endochronic model by devising two incremental solution strategies for the model behavior under undrained conditions.

2. INCREMENTAL SOLUTION STRATEGIES

Computational procedures for the calculation of the response of soil under drained conditions have been developed for stress- and strain-controls [2]. In order to apply the same procedures to undrained conditions we still use the effective stresses to calculate and incrementally update the endochronic equations. Undrained conditions are iteratively enforced at each loading step according to the model control desired, which will then permit the calculation of the pore pressure increment [4]. The model only uses principal stresses and strains and assumes that all state variables are fully known before the application of any new load increment. An asterisk is used to identify those variables that are either given or known from a previous solution.

2.1. STRESS CONTROL

In a stress-controlled model we initially calculate the pore pressure increment by assuming it to be equal to the increment in mean total stress, where the increment in total principal stress, ΔS_i^*, is given.

$$\Delta u^0 = \frac{1}{3} \sum_{i=1}^{3} \Delta S_i^* \tag{1}$$

Effective stresses can then be calculated which permits the use of the endochronic model equations to calculate strain response $\Delta\epsilon_i$ from which the volumetric strain $\Delta\epsilon$ is calculated.

Since the pore-fluid is considered to be completely incompressible, then $\Delta\epsilon$ is supposed to be zero, otherwise the initial assumption with regard to the pore pressure is revised based on the calculated $\Delta\epsilon$ and the bulk modulus of the soil skeleton, K:

$$\Delta u^{m+1} = \Delta u^m + \frac{1}{3} K \,\Delta\epsilon \quad , \quad m = 0, 1, 2, \cdots \tag{2}$$

Iterations are repeated until the solution converges within a specified tolerence.

2.2. MIXED CONTROL

The general mixed form of control is represented by an appropriate tangent modulus relation, whose formulation is accomplished by part-inversion of the incremental elastic relationship:

$$\begin{bmatrix} \Delta\sigma_1 \\ \Delta\sigma_2 \\ \Delta\sigma_3 \end{bmatrix} = \begin{bmatrix} K_{11} & K_{12} & K_{13} \\ K_{21} & K_{22} & K_{23} \\ K_{31} & K_{32} & K_{33} \end{bmatrix} \begin{bmatrix} \Delta\epsilon_1^e \\ \Delta\epsilon_2^e \\ \Delta\epsilon_3^e \end{bmatrix} \tag{3}$$

where superscript e is used to indicate the elastic part of an increment. Later, superscript p will be used to denote the plastic part.

For 3-dimensional problems two cases of mixed control analysis are possible. The first is the one where two total principal stress increments and one principal strain increment are known, and the other comprises of the case where two principal strain increments and one total principal stress increment are known. In this paper we will focus only on the first case, where $\Delta\epsilon_1^*$, ΔS_2^* and ΔS_3^* are known, because of the similarities in the equations governing the two cases.

Equation (3) can be written as:

$$\begin{bmatrix} \Delta\sigma_1 \\ \Delta\sigma_2 \\ \Delta\sigma_3 \end{bmatrix} = \begin{bmatrix} K_{11} & K_{12} & K_{13} \\ K_{21} & K_{22} & K_{23} \\ K_{31} & K_{32} & K_{33} \end{bmatrix} \begin{bmatrix} \Delta\epsilon_1^* - \Delta\epsilon_1^p \\ \Delta\epsilon_2 - \Delta\epsilon_2^p \\ \Delta\epsilon_3 - \Delta\epsilon_3^p \end{bmatrix} \tag{4}$$

and total stresses are given by:

$$\begin{bmatrix} \Delta S_1 \\ \Delta S_2^* \\ \Delta S_3^* \end{bmatrix} = \begin{bmatrix} \Delta\sigma_1 \\ \Delta\sigma_2 \\ \Delta\sigma_3 \end{bmatrix} + \Delta u \begin{bmatrix} 1 \\ 1 \\ 1 \end{bmatrix} \tag{5}$$

The response variables $\Delta\epsilon_2$, $\Delta\epsilon_3$ and pore pressure increment Δu are initially unknown. If we use an entirely strain-controlled endochronic model, and if we are given the nth increments $\Delta\epsilon_1^*$, $\Delta\epsilon_2$, $\Delta\epsilon_3$ and Δu, we may choose any integration operator to update the response variables ${}^n\sigma_1, {}^n\sigma_2, {}^n\sigma_3, {}^nS_1, {}^nS_2, {}^nS_3$ and $\Delta\epsilon$. As the updated variables ${}^nS_2^*, {}^nS_3^*$ and $\Delta\epsilon = 0$ are known, we may determine $\Delta\epsilon_2$, $\Delta\epsilon_3$ and Δu from the incremental equilibrium and continuity equations:

$$\begin{aligned} g_2 &= {}^nS_2 - {}^nS_2^* = 0 \\ g_3 &= {}^nS_3 - {}^nS_3^* = 0 \\ \Delta\epsilon &= \Delta\epsilon_1^* + \Delta\epsilon_2 + \Delta\epsilon_3 = 0 \end{aligned} \tag{6}$$

where g_2 and g_3 are the nth out-of-balance total stresses and ${}^nS_2$ and ${}^nS_3$ can be decomposed as:

$$\begin{aligned} {}^nS_2 &= {}^{n-1}\sigma_2^* + K_{21}\Delta\epsilon_1^* + K_{22}\Delta\epsilon_2 + K_{23}\Delta\epsilon_3 + {}^n\sigma_2^p + {}^{n-1}u^* + \Delta u \\ {}^nS_3 &= {}^{n-1}\sigma_3^* + K_{31}\Delta\epsilon_1^* + K_{32}\Delta\epsilon_2 + K_{33}\Delta\epsilon_3 + {}^n\sigma_3^p + {}^{n-1}u^* + \Delta u \end{aligned} \tag{7}$$

The unknown variables $\Delta\epsilon_2$, $\Delta\epsilon_3$ and Δu can be solved iteratively from eq. (6) via scaled gradient iterations:

$$\begin{bmatrix} \Delta\epsilon_2^{m+1} \\ \Delta\epsilon_3^{m+1} \\ \Delta u^{m+1} \end{bmatrix} = \begin{bmatrix} \Delta\epsilon_2^m \\ \Delta\epsilon_3^m \\ \Delta u^m \end{bmatrix} - \begin{bmatrix} K_{22} & K_{23} & 1 \\ K_{32} & K_{33} & 1 \\ 1 & 1 & 0 \end{bmatrix}^{-1} \begin{bmatrix} g_2^m \\ g_3^m \\ \Delta\epsilon^m \end{bmatrix} \quad , \quad m = 0, 1, 2, \cdots \tag{8}$$

where K_{ij} is the elastic matrix which is considred an approximation of the Hessian matrix. The initial values $\Delta\epsilon_2^0$, $\Delta\epsilon_3^0$ and Δu^0 are conveniently obtained as the elastic solution from the conditions $g_2 = 0, g_3 = 0, \Delta\epsilon = 0, {}^n\sigma_2^p = 0$ and ${}^n\sigma_3^p = 0$:

$$\begin{bmatrix} \Delta\epsilon_2^0 \\ \Delta\epsilon_3^0 \\ \Delta u^0 \end{bmatrix} = \begin{bmatrix} K_{22} & K_{23} & 1 \\ K_{32} & K_{33} & 1 \\ 1 & 1 & 0 \end{bmatrix}^{-1} \begin{bmatrix} \Delta S_2^* - K_{21}\Delta\epsilon_1^* \\ \Delta S_3^* - K_{31}\Delta\epsilon_1^* \\ -\Delta\epsilon_1^* \end{bmatrix} \tag{9}$$

3. PROGRAMMING CONSIDERATIONS

Each numerical algorithm was coded as a separate subroutine that includes the basic endochronic equations, and it performs explicit numerical integration over the given load increment. The load increment is subdivided into smaller subincrements within each subroutine to avoid dependency of the solution on the number of subdivisions used. Special care is taken to interface the different subroutines to the main driver program so that a user can switch between subroutines freely whenever needed. This flexibility makes it possible to study soil response due to loading along complex stress paths, where the control variables are different for each segment in the path. For example, a typical triaxial undrained test for sand starts from an isotropic consolidation state. The consolidation part of the test usually involves three principal stresses as control variables. On the other hand, the undrained deviatoric part of the loading history involves a mixed form of control, where two confining stresses and one axial strain are known.

4. ASSESSMENT OF UNDRAINED MODEL BEHAVIOR

Model parameters reported by Valanis and Read [2] for ISST soils are used to demonstrate the ability of the model to predict undrained soil behavior. Even though the parameters were evaluated from simple tests on drained sand, the model gave good results when used to predict undrained soil behavior due to cyclic triaxial compression–extension loading. The simulation test starts from an isotropic consolidation state where an all-around effective stress of 3.45 MPa (500 psi) is applied. Axial stress is then cycled between 4.45 MPa (645 psi) and 2.45 MPa (355 psi) to simulate triaxial compression–extension load history. Fig. (1) shows the response of σ_1, σ_2, and ϵ_1 with the number of load cycles applied. It is evident that as the number of cycles increased, the undrained strength of soil dropped in a manner consistent with test experience. In Fig. (2), the cyclic behavior is shown in terms of σ_3, $\sigma_1 - \sigma_3$, and the pore pressure u. Figs. (3) and (4) show p–q diagram and the octahedral shear behavior respectively.

REFERENCES

[1] Valanis, K. C., 1987, "Endochronic theory of Soils and Concrete", Proceedings, Second International Conference on Constitutive Laws for Engineering Materials, Tucson, Arizona, Vol. 1, pp 247–248.

[2] Valanis, K. C. and Read, H. E., 1988, "An Endochronic Plasticity Model for ISST Soils", U.S. Army Waterways Experiment Station, Report No. SL–88–3.

[3] Valanis, K. C. and Peters, J. F., 1988, "Thermodynamics of Frictional Materials, Report–I: Constitutive Theory of Soils with Dilatant Capability", U.S. Army Waterways Experiment Station, Report No. GL–88–20.

[4] Runesson, K., Axelsson, K. and Sture, S., 1989, "Aspects on Undrained Behavior in Soil Plasticity", University of Colorado, Dept. CEAE Report.

[5] Valanis, K. C. and Peters, J. F., 1989, "An Endochronic Plasticity Theory with Shear–Volumetric Coupling", Private Communication.

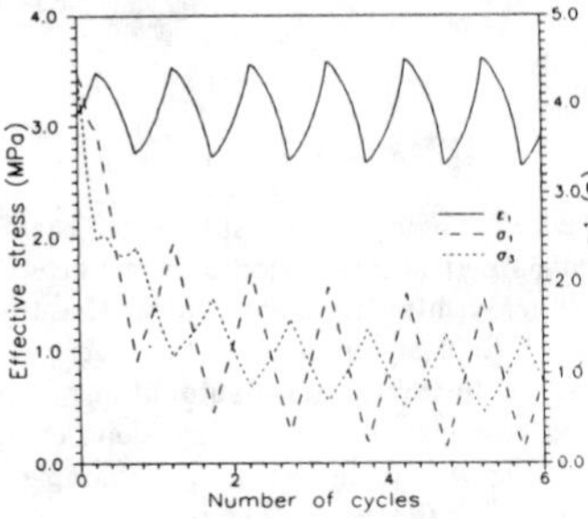

Fig. 1. Response of effective stresses and axial strain with number of load cycles.

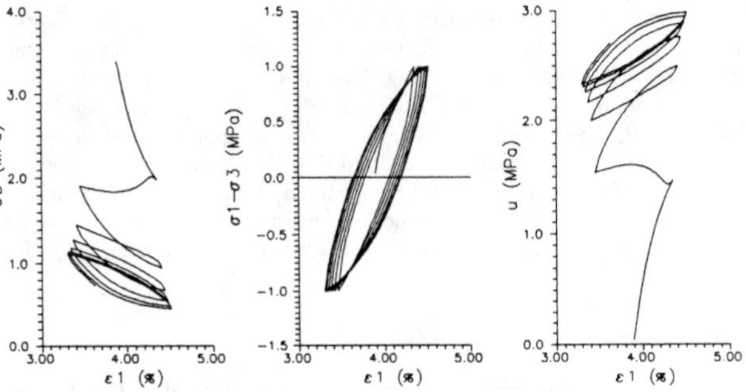

Fig. 2. Cyclic model behavior.

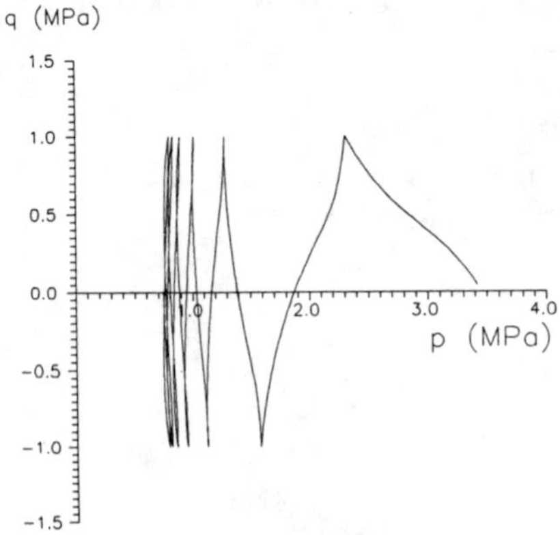

Fig. 3. p–q diagram.

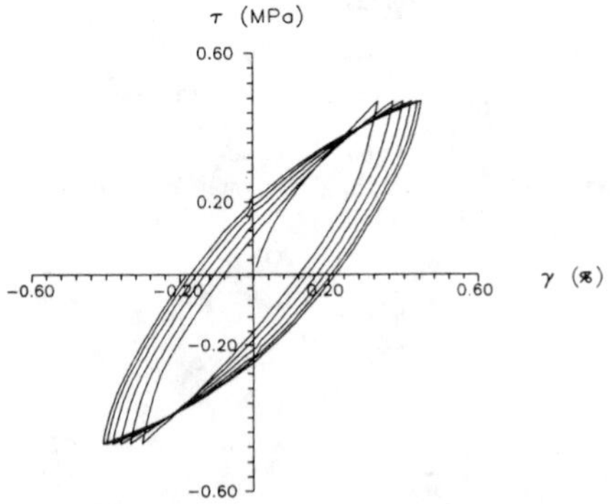

Fig. 4. Octahedral shear behavior.

Finite element analysis of failure mechanisms in elasto- plastic soil slopes

D. V. Griffiths and D. J. Kidger
Department of Engineering
University of Manchester
Manchester M13 9PL, U.K.

Abstract

Finite element methods in conjunction with elasto-plastic constitutive laws are able to give accurate estimates of the factor of safety of soil slopes. The mechanisms of failure produced by these analyses however, are often diffuse and poorly defined as compared with those assumed in classical limit analysis techniques. This paper describes two methods for improving their visualisation. It is shown that by introducing a small degree of post-peak material softening into the constitutive law, a considerably more localised mechanism is achieved. In contrast to previous attempts to 'trigger' mechanisms whereby select elements were weakened, all elements are given the same constitutive law in the present approach. A second method involves a regridding approach in which the displacement field is enhanced by interpolating between the nodal values.

INTRODUCTION

The finite element method is capable of giving good predictions of collapse loads in many geomechanics applications. Researchers will continue to seek more efficient algorithms for tackling these elasto-plastic problems and debate will continue over the *best* type of element or integration strategy to employ in the stress redistribution processes. In 2-d analyses, the 8-node *serendipity* element with reduced integration has received the most exposure (see e.g. Zienkiewicz *et al* 1975, Griffiths 1982, Potts *et al* 1990), and has been shown to be robust in a wide range of collapse load problems. The examples in this paper will therefore use this element.

Although the finite element method can give good collapse load predictions, this is done without reproducing in any detail the mechanisms assumed by the corresponding limit analyses. Admittedly, conventional finite element meshes are constrained to remain continuous, so it would be unrealistic to expect highly localised mechanisms to be reproduced without gross element distortions.

A method used in the past to produce more localised failure mechanisms has been to introduce weak elements at strategic locations within the mesh

(Prevost 1984). This has the effect of *triggering* a mechanism which inevitably starts to form at the weakest point which has been artificially introduced.

This paper describes two methods of enhancing the failure mechanism within a finite element mesh in problems of slope stability. The first method relates to the constitutive law incorporated in the material behavior and represents a form of *triggering* with the important difference that all elements are given the same properties throughout the analysis. The second method is purely interpretive, and involves extracting as much information as possible from the finite element analysis by using the shape functions to interpolate the displacement field between the nodal values.

REVIEW OF SLOPE STABILITY ANALYSES

The field of slope stability represents a rather well-posed collapse problems in geomechanics for analysis by finite elements. One of the reasons for this is that the problem is relatively unconfined, enabling a mechanism to form easily. A consequence of this is that the flow rule is not of great importance in drained analyses, so a non-associated flow rule with no plastic volume change is usually sufficient to give adequate results.

The present work concentrates on '$\phi_u = 0$' soils, whose strength is suitably modeled by a von Mises or Tresca failure criterion. Figure 1 shows a finite element mesh of a slope of undrained clay which was brought to failure by gradually reducing the shear strength c_u through the stability number ($N = c_u/\gamma H$) while monitoring the crest displacement. It is seen that close agreement with Taylor's (1937) solution ($N \approx 0.19$) is obtained (Figure 2), although a rather diffuse failure mechanism is indicated by the deformed mesh in the elastic-perfectly plastic case.

Introduction of softening

Greater localisation of the failure mechanism can be achieved by introducing a small degree of post-peak softening into the constitutive law. This is a form of *triggering*, but differs from other methods in that every element in the mesh is given the same softening property. The degree of softening is measured by the parameter M which represents the ratio of the negative modulus of the material post-peak to its elastic value pre-peak. Figure 3 shows the effect of a modest amount of softening ($M = -0.25$) on the failure mechanism in the same slope. It is seen that a well-defined mechanism is now visible, although the softening has had the effect of weakening the slope an indicated by a higher stability number at failure (Figure 2).

It has been observed that the degree of softening as measured through the parameter M, influences the precise location of the zone of soil at failure, and this is currently a topic of further investigation.

Regridding method

Kidger (1990) described this method in which a regular rectangular grid is superimposed on the finite element mesh. The location of each grid point is defined by the local coordinates of the parent element within which each is positioned. After the analysis has been performed and the slope has reached failure, the deformed grid is redrawn. This is achieved by retrieving the local coordinates of each grid point and plotting its position as defined by the new nodal coordinates.

Figures 4 and 5 show examples of the method applied to embankments with varying foundation layer depth. The examples were chosen to demonstrate the different types of mechanisms that would be predicted by Taylor's limit analysis approaches (see e.g. Terzaghi and Peck 1967)

The first case (Figures 4a and b) shows the deformed mesh and grid, the latter clearly indicating a *toe circle* which is tangent to the lower stratum and outcrops at the toe of the slope. The second case (Figures 5a and b) considers the same slope on a deeper foundation layer. In this case, the grid indicates a *midpoint circle* which is tangent to the lower stratum and whose center lies on the centerline of the slope face. The grid plots are further enhanced by plotting the difference in displacements generated by two trial factors of safety just prior to failure (see Smith and Griffiths 1988). In each case, the finite element solutions gave good agreement with the computed stability number from limit analysis. The regridding approach leads to an enhanced visualisation of the failure mechanism which would not be readily apparent from the nodal displacements of the relatively crude finite element discretisation.

CONCLUDING REMARKS

Two methods of enhancing the visualisation of slope stability mechanisms have been described. Firstly, it is shown that by introducing a small degree of post-peak material softening into the constitutive law, a considerably more localised mechanism is achieved. All elements are given the same material properties which is in contrast to previous attempts to 'trigger' mechanisms whereby select elements were weakened. Secondly, a regridding approach is used which shows that interpolation of the displacement field within elements leads to mechanism visualisation which would not have been readily noticeable from the nodal displacements alone.

References

[1] D.V. Griffiths. Computation of bearing capacity factors using finite elements. *Géotechnique*, 32(3):195–202, 1982.

[2] D.J Kidger. *Visualization of finite element eigenvalues and three dimensional plasticity*. PhD thesis, Department of Engineering, University of Manchester, 1990.

[3] D.M. Potts, G.T. Dounias, and P.R. Vaughan. Finite element analysis of progressive failure of carsington embankment. *Géotechnique*, 40(1):79–102, 1990.

[4] J.H. Prevost. Localisation and deformation in elastic-plastic solids. *International Journal for Numerical and Analytical Methods in Geomechanics*, 8:187–196, 1984.

[5] I.M. Smith and D.V. Griffiths. *Programming the Finite Element Method.* John Wiley and Sons, Chichester, New York, 2nd edition, 1988.

[6] D.W. Taylor. Stability of earth slopes. *J. Boston Soc. Civ. Eng.*, 24:197–246, 1937.

[7] K. Terzaghi and R.B. Peck. *Soil Mechanics in Engineering Practice*, page 237. John Wiley and Sons, New York, 1967.

[8] O.C. Zienkiewicz, C. Humpheson, and R.W. Lewis. Associated and non-associated viscoplasticity and plasticity in soil mechanics. *Géotechnique*, 25:671–689, 1975

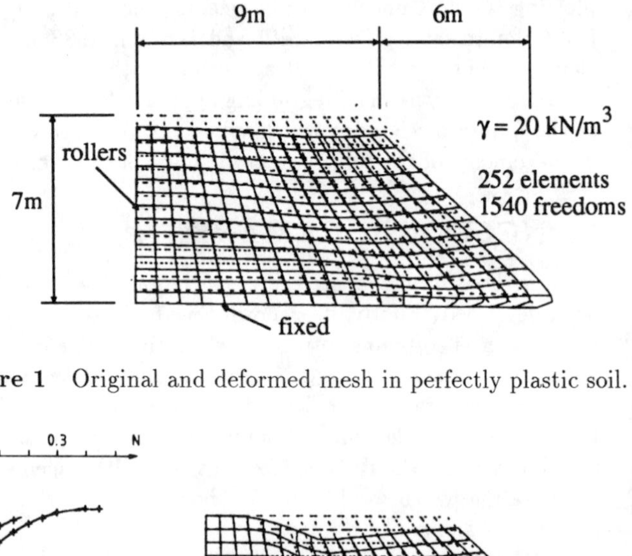

Figure 1 Original and deformed mesh in perfectly plastic soil.

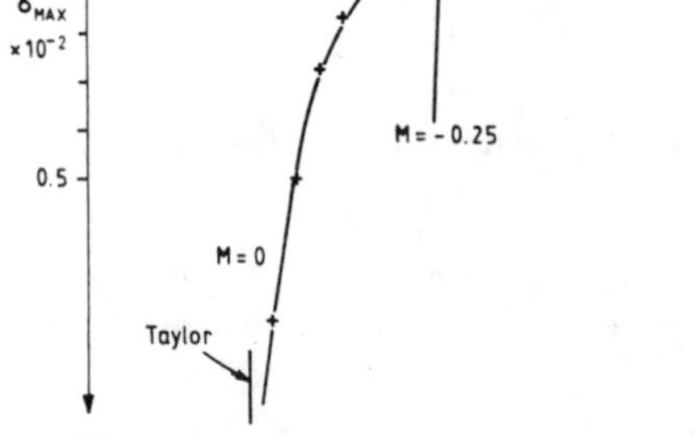

Figure 2 Computed stability number at failure.

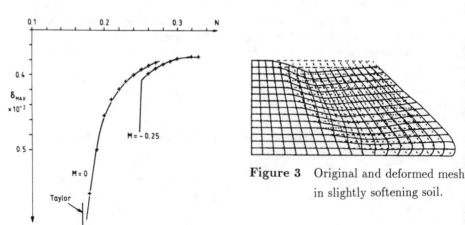

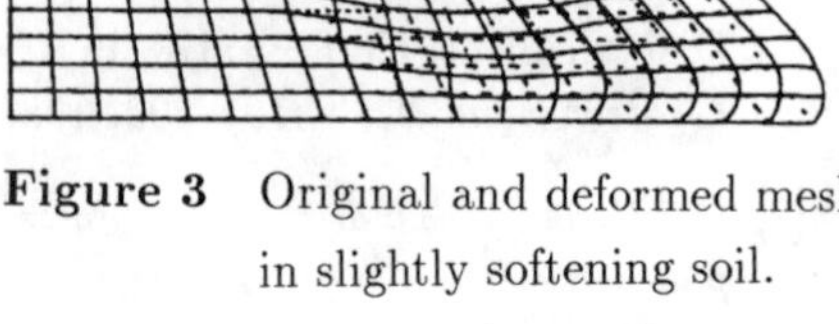

Figure 3 Original and deformed mesh in slightly softening soil.

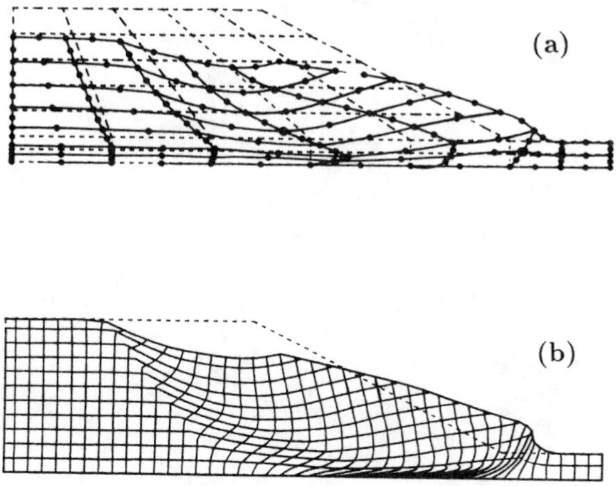

Figure 4 Toe circle: original mesh (a), after regridding (b).

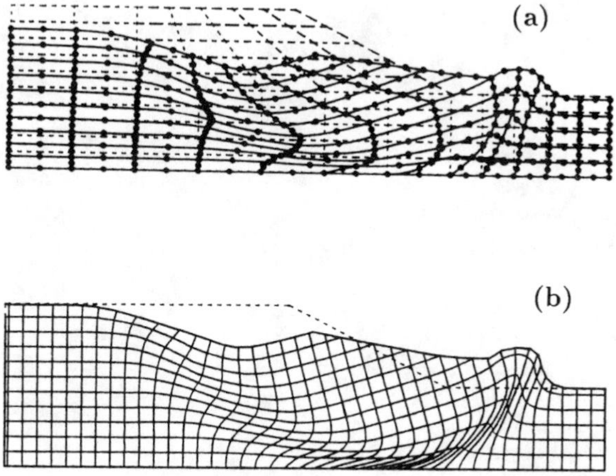

Figure 5 Slope circle: original mesh (a), after regridding (b).

Thermo-elastoplastic deformations and stresses in a cylinder head made of grey cast iron

H. E. Hjelm
Division of Solid Mechanics, Chalmers University of Technology
S–412 96 Göteborg, SWEDEN

ABSTRACT

A fully three-dimensional numerical study of thermo-elastoplastic deformations and stresses in a cylinder head made of grey cast iron for a Diesel truck engine is presented. The temperature field as well as the strain and stress fields in the cylinder head under normal working conditions are calculated by use of the FE-code ABAQUS. A modified von Mises yield criterion is used to model the elasto plastic behaviour of grey cast iron. This criterion employs two yield stresses, one in compression and the other in tension. Large in-plane membrane stresses are found in the bottom plate of the cylinder head.

INTRODUCTION

Grey cast iron is used in many modern industrial applications, for example, in engine blocks and cylinder heads. Good forming capability, low manufacturing cost, and high internal damping are major advantages of this material.

The design loads of a cylinder head made of grey cast iron are high enough to cause plastic yielding at several locations. However, since a very limited amount of experimental data is available (especially biaxial data) about the elasto-plastic behaviour of grey cast iron, numerical studies of the mechanical behaviour of grey cast iron components are normally performed using an ordinary thermo-elastic material model as implemented in numerous FE-codes.

PRESENT STUDY

In the present paper, a constitutive model for the elasto-plastic behaviour of grey cast iron is proposed. It is based on the modified von Mises yield condition with the yield function:

$$f = \frac{1}{2} s_{ij} s_{ij} + \frac{1}{3} \sigma_{kk} (\sigma_{\mathrm{C}} - \sigma_{\mathrm{T}}) - \frac{1}{3} \sigma_{\mathrm{C}} \sigma_{\mathrm{T}} \tag{1}$$

in combination with an associated flow rule and nonlinear isotropic hardening. With this constitutive model, which uses two yield limits, one in tension (σ_{T})

and one in compression (σ_C) it is possible to model, the strong difference in yield limits in compression and in tension, typically $\sigma_C \simeq 3\,\sigma_T$, and a stronger, more nonlinear, hardening in tension than in compression. These characteristics of grey cast iron have been observed in uni-axial cyclic tests performed at working temperatures on a grey cast iron used in cylinder heads for Diesel engines.

The above constitutive model has been implemented, using a scheme presented by [1], into a commercial FE-code for nonlinear analysis, ABAQUS [2]. The different hardening behaviour in tension and in compression is modelled by use of different effective plastic strains in tension and compression (a tensile stress state is characterized by a positive value for the first invariant of the stress tensor, σ_{kk}). To improve the convergence behaviour in the global equlibrium iterations, a consistent tangent modulus matrix has been employed, as shown in [3].

The material model has been used to analyse the three-dimensional deformation – and stress field in a cylinder head. A symmetrical portion of one quarter of the part of the cylinder head that covers one cylinder is analysed. The numerical FE-model is shown in Figure 1 and it is assumed to be representative enough for this investigation. It includes a bottom plate, a top plate, one wall facing outwards, one and a half bolt centering sleeve with bolts, one exhaust fume channel and a quarter of a fuel inlet.

During normal operation of a Diesel engine there is a long-term variation due to start and stop of the engine and a short-term variation under the combustion cycle. This paper focusses on the long-term variation, thus the stress field in a cold cylinder head and that in a cylinder head at operating temperatures are calculated.

The stationary heat flow and corresponding temperature field in the cylinder head is calculated numerically by use of ABAQUS considering for example a spatial variation of the heat transfer coefficient at the top of the combustion chamber, see [4] for details. The mechanical loading was divided into the following five steps:

1 The forces from prestressed boltings (tightening the cylinder head to the engine block) and the fixed displacement of the lower edge of the outer wall (simulating partial contact with the engine block) are applied.
2 The temperature field of a full working condition is added.
3 The temperature field in step 2 is removed.
4 The temperature field in step 2 is applied again.
5 The gas pressure from the combustion chamber is added to step 4.

CALCULATED RESULTS

The maximum calculated temperature, 290° C, in the cylinder head is found at the bottom of the cylinder head. Generally, high temperatures, above 200° C are only seen at the part facing the combustion chamber.

The material in the bottom plate is severly constrained especially in the 3-direction by the fixed boundary condition below and the stiff exhaust fume channel and the bolt centring sleeve above. As expected, very large compressive hydrostatic stresses ($\frac{1}{3}\sigma_{kk}$), up to 850 MPa, are calculated in this region with the thermo-elastoplastic material model. The corresponding value calculated for a thermo elastic material model is somewhat lower, about 270 MPa. The pressure-dependent yield function chosen here, see (1), in combination with the (necessary) thermo elastic predictor in the constitutive algorithm employed inevitably leads to a strong hydrostatic stress state in this region. It is believed that at least a part of this stress state is artificial, and originates from the numerical method. The best way to restrict such effects would probably be to use a more realistic boundary condition which permits a vertical translation of the lower part of the cylinder head and contact with a flexible underlying material. The deformed mesh, calculated after the fifth load step, is shown in Figure 1. Figure 2 shows in-plane stresses for the upper and the lower part of the bottom plate at line **A** which is marked in Figure 1. The temperature load results in high compressive, mainly elastic, membrane stresses. The additional bending stresses caused by the gas pressure leads to yielding at line **A**.

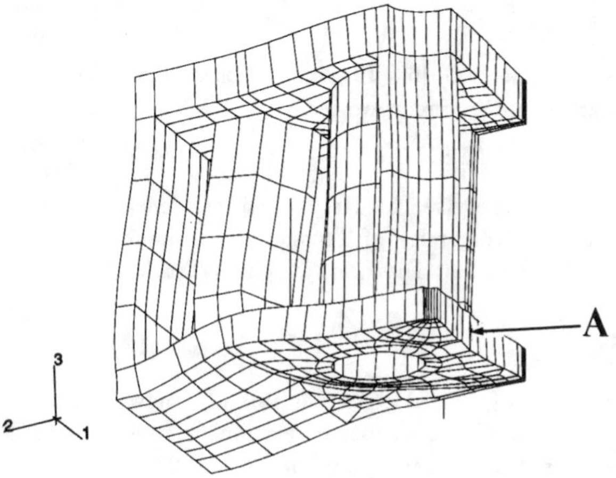

Figure 1: Calculated deformed mesh after fifth load step for thermo-elastio plastic material model.

DISCUSSION

As noted previously, and also in [5], the background information on the mechanical behaviour is very limited and comes primarily from uni-axial experiments. The yield function chosen here, the modified von Mises yield function (1), can display, in one single expression, the different yield strengths observed in tension and compression. However, there may be discrepancies between predictions made by this constitutive model and the (scarce) bi-axial experimental results (almost entirely for fracture surfaces, not yield surfaces) in particular for cases with bi-axial stress states. The yield function proposed contains a σ_{kk}-term, thus making yielding dependent on the first stress invariant. With an associated flow rule this leads to a plastic volume increase, which has been

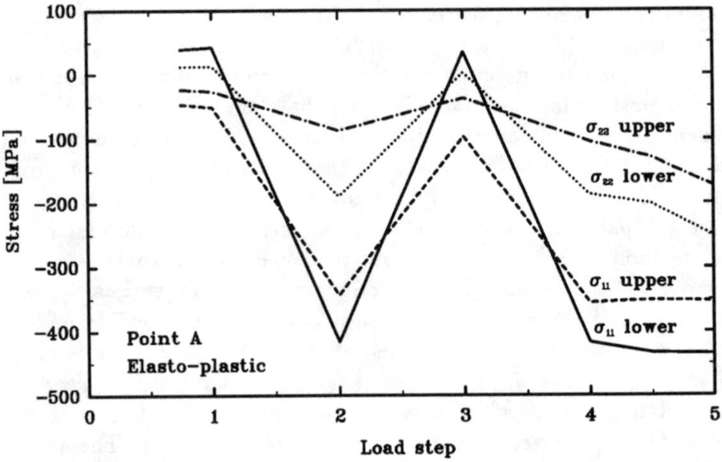

Figure 2: Calculated in plane stresses at line **A** (see Figure 1) for elasto plastic material model.

found in experiments, see [6]. In a more complex yielding function, the third stress invariant could be incorporated.

ACKNOWLEDGEMENTS

This work was supervised by Dr. B. Lennart Josefson, Division of Solid Mechanics, Chalmers University of Technology. Financial support was given by the Product Development Division, at Volvo Truck Corporation in Göteborg, Sweden.

REFERENCES

1. M. Ortiz, and J.C. Simo, "An analysis of a new class of integration algorithms for elastoplastic constitutive relations", International Journal for Numerical Methods in Engineering 23, 353–366 (1986).
2. ABAQUS User's Manual, version 4.8, (Hibbitt, Karlsson & Sørensen, Inc., 100 Medway Street, Providence, RI 02906, USA 1989).
3. G.P. Mitchell, and D.R.J. Owen, "Numerical solutions for elasto-plastic problems", Engineering Computations 5, 274–284. (1988).
4. H.E. Hjelm, "Thermo-elastoplastic deformations and stresses in a cylinder head made of grey cast iron", Research report F132 (Division of Solid Mechanics, Chalmers University of Technology, Göteborg, Sweden, 1990).
5. B.L. Josefson, and H.E. Hjelm, "Elasto-plastic deformations in grey cast iron", Research report F131 (Division of Solid Mechanics, Chalmers University of Technology, Göteborg, Sweden, 1990).
6. L.F. Coffin Jr, "The flow and fracture of a brittle material", Journal of Applied Mechanics, Transactions of the ASME 72, 233–248 (1950).

EFFICIENT ALGORITHMS FOR THE FE-ANALYSIS OF PRESSURE DEPENDENT CONSTITUTIVE EQUATIONS INCLUDING SWELLING

K. HORNBERGER, E. DIEGELE, B. SCHINKE

Kernforschungszentrum Karlsruhe - Institut für Material- und Festkörperforschung - P.O. Box 3640 - D-7500 Karlsruhe Federal Republic of Germany

ABSTRACT

The FE-implementation of a broad class of material models is applied to lifetime assessments of the First Wall of a fusion reactor.

INTRODUCTION

Over the past two decades, a large number of constitutive models characterizing the inelastic deformation behavior of metals have been introduced in the literature. The need for reliable FE-implementations of material models subject to rapid changes has caused the present work to be conducted, in which efficient solution procedures for a broad class of constitutive models with internal variables are described. In the models considered, the inelastic strain rate embodies a deviatoric part as well as a volumetric part. The latter is used to model phenomena such as irradiation induced swelling.

IMPLEMENTATION OF THE GENERAL MATERIAL MODEL

Under the assumptions of small strains and small rotations, the total strain rate ($\dot{\varepsilon}$) is the sum of the elastic ($\dot{\varepsilon}^e$), inelastic ($\dot{\varepsilon}^p$), and thermal ($\dot{\varepsilon}^{th}$) strain rates: $\dot{\varepsilon} = \dot{\varepsilon}^e + \dot{\varepsilon}^p + \dot{\varepsilon}^{th}$. The elastic strain rate and the stress rate $\dot{\boldsymbol{\sigma}}$ are related by Hooke's law, the inelastic strain rate is governed by a flow law,

$$\dot{\varepsilon}^p = f_1(\,.\,)\,\boldsymbol{\Sigma}' + f_2(\,.\,)\,\mathbf{1} \tag{1}$$

where $\mathbf{1}$ is the second-order unit-tensor, and the overstress $\boldsymbol{\Sigma} = \boldsymbol{\sigma} - \boldsymbol{\alpha}$ is the difference between the applied stress $\boldsymbol{\sigma}$ and the backstress $\boldsymbol{\alpha}$. A prime mark (′) on second-order tensors is used to indicate the deviatoric part. The model is completed by kinetic equations for the internal variables, the tensorial backstress $\boldsymbol{\alpha}$,

$$\dot{\boldsymbol{\alpha}} = h_1(.)\,\dot{\boldsymbol{\varepsilon}}'^{p} - r_1(.)\,\boldsymbol{\alpha}' + g_2(.)\,\mathbf{1} \tag{2}$$

and a set of an arbitrary number of isotropic hardening variables K_A labeled by the subscript (A)

$$\dot{K}_A = \Gamma_A(.)\,\dot{\eta} + \Theta_A(.)\,\dot{T} - \Lambda_A(.) \tag{3}$$

A subscript (1) applied to functions refers to the viscoplastic part of the material model, whereas a subscript (2) refers to the volumetric portion, i.e. the viscoplastic model is recovered by discarding all functions with the subscript (2).
The non-linear functions f_1, f_2, h_1, r_1, g_2, Γ_A, Θ_A and Λ_A depend on K_A, the temperature T, and possibly additional external variables, and on the invariants of $\boldsymbol{\sigma}$ and $\boldsymbol{\alpha}$ up to degree 2:

$$J_1(\boldsymbol{\sigma}) = \frac{1}{3}\,\mathrm{tr}(\boldsymbol{\sigma}) \qquad J_1(\boldsymbol{\alpha}) = \frac{1}{3}\,\mathrm{tr}(\boldsymbol{\alpha})$$

$$J_2(\boldsymbol{\sigma}') = \frac{1}{2}\,\boldsymbol{\sigma}':\boldsymbol{\sigma}' \qquad J_2(\boldsymbol{\alpha}') = \frac{1}{2}\,\boldsymbol{\alpha}':\boldsymbol{\alpha}' \qquad J_2(\boldsymbol{\Sigma}') = \frac{1}{2}\,\boldsymbol{\Sigma}':\boldsymbol{\Sigma}'$$

and on the history-dependent variables η_1, η_2 which may be chosen from either

$$\eta^s_1 = \int (d\boldsymbol{\varepsilon}'^p : d\boldsymbol{\varepsilon}'^p)^{1/2}\,, \quad \eta^s_2 = (1/3)\int |\mathrm{tr}(\dot{\boldsymbol{\varepsilon}}^p)|\,dt\,, \qquad \text{strain hardening, or}$$
$$\eta^w_1 = \int \boldsymbol{\sigma}' : d\boldsymbol{\varepsilon}'^p\,, \quad \eta^w_2 = (1/3)\int |\mathrm{tr}(\boldsymbol{\sigma})\,\mathrm{tr}(\dot{\boldsymbol{\varepsilon}}^p)|\,dt\,, \quad \text{work hardening.}$$

Equations (1)-(3) provide a general framework in which many of the constitutive models commonly used for metals are contained as special cases.

The system of rate equations (1)-(3) is integrated numerically using the generalized midpoint rule (GMR). A detailed discussion of the GMR along with numerical examples is given in [1]. The nonlinear system of algebraic equations arising is projected to a subspace of the full stress space. Due to the presence of volumetric terms, the subspace used in viscoplasticity must be extended by an additional "base vector" parallel to **1** [2].
Owing to its excellent convergence properties, Newton's method in combination with a line search algorithm is used to solve the projected system. For an initially isotropic material, the subspace iteration furnishes the exact solution, and the consistent Jacobian can be calculated in closed form immediately in the subspace once the iteration has converged. A detailed description of the numerical procedures used is given elsewhere [1].

APPLICATION TO THE FIRST WALL OF A FUSION REACTOR

Material Model

The general material model given in Equations (1) - (3) has been implemented in the FE code ABAQUS [3]. In a special form it has been applied to lifetime assessment of First Wall (FW) components in a fusion reactor. In normal operation, the FW is periodically loaded by a surface heat flux and internal heat generation as well as high neutron fluence.

Thus, the function f_1 is split into a function f_1^{cb}, which is given by Chaboche's model and a function f_1^{n}, which takes into account neutron induced creep ($f_1 = f_1^{cb} + f_1^{n}$). The function f_2 reflects the neutron swelling rate induced by the neutron dose $D = \int \Phi \, dt$ of a fluence Φ .

$$
\begin{aligned}
f_1^{cb} &= \frac{3}{2} \left\langle \frac{\sqrt{J_2(\Sigma')} - R - k}{L} \right\rangle^{n} \\
f_1^{n} &= \frac{3}{2} \Phi \frac{dc}{dD}, \quad f_2 = \Phi \frac{dS}{dD} (1 + B\, tr(\sigma)) \\
h_1 &= H, \quad \Gamma_A = b(Q - K_A), \quad r_1 = \gamma f_1^{cb}
\end{aligned}
\tag{4}
$$

The parameters of Chaboche's model (L,R,k,n,H,b,Q.γ) are given in [4], and h_2, g_2, Θ_A and Λ_A are 0.
For the functions c and S the relations by Horie [5] have been used.

FE-Analysis and Results

A cross section of a typical FW realization and its cyclic loading history are shown in Fig.1 and Fig.2, respectively. The thermo-mechanical cycles are characterized by a burn time t_b with a surface heat flux (maximum 40 W/cm^2) and volumetric heat deposition (maximum 14 W/cm^3) and a off-burn time t_d. The fatigue behavior of the structure is characterized by the cyclic variation in stress and strain. These fields depend on the neutron dose and flux.
In steady state operation of the reactor, in a first part of the lifetime the stresses relax due to thermal and irradiation induced creep. Once an incubation dose is exceeded swelling becomes the dominating process leading to a pronounced increase in stress.
For a periodically operated (demonstration) reactor these conditions will hold for the mean stress instead of the stress. As it is impossible to calculate stress and strain for several 10,000 cycles, the calculations are split up into several steps: First, a small number of cycles (n) are calculated with periodic loading until a saturation of peak stress and strain levels is reached. Next, the component is subjected to maximum thermal and neutron loading during a hold time t_H, thus simulating $N = t_H/t_b$ cycles. Subsequently, both steps are executed alternately.
Thermal and mechanical analyses have been performed with the FE code ABAQUS. The FE-mesh as shown in Fig. 1 comprises 228 2-D elements with 773 nodes. In mechanical analyses, generalized plane strain (normal to the plane of the figure) has been considered to be the most suitable boundary condition.
The change in stress for a reactor operating in the steady state at exposed locations (F1, F2 at the FW front part, and C1, C2 at the coolant tube) is shown in Fig. 3. As another example, contour plots of axial stresses at the end of burn of a cyclically operating reactor are given in Fig. 4. After an irradiation dose of 50 dpa, redistribution and a minor stress relaxation is observed. In Fig. 5, the axial stresses at the end of the cycles (i.e., at end of the dwell time) are plotted. The redistribution of the stress fields after a neutron dose of 70 dpa shows a small increase.

REFERENCES

1. K. Hornberger, H. Stamm, "An implicit integration algorithm with a projection method for viscoplastic constitutive equations", Int. J. Num. Meth. Eng. 28 , 2397 - 2421 (1989).
2. H. D. Hibbitt, "Some issues in numerical simulation of nonlinear structural response", Workshop on computational methods for structural dynamics, NASA, Langley Research Center, 1 - 13, 1985.
3. Hibbitt, Karlsson, Sorensen, Inc., 'ABAQUS User's Manual', Providence, RI, (1988).
4. J. Lemaitre, J.-L. Chaboche, Mecanique des materiaux solides , (Paris 1985).
5. T. Horie et. al., "An analytical and experimental study on lifetime predictions for fusion reactor first walls and divertor plates", IAEA Report TEDCDOC-393, Lifetime Predictions for First Wall and Blanket Structures of Fusion Reactors, 107 - 133 (1986).

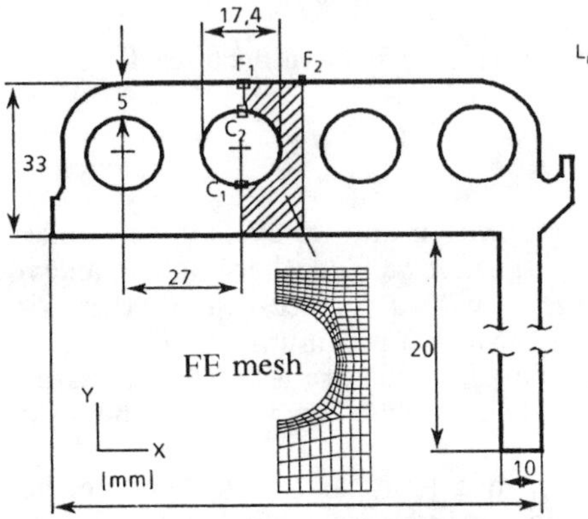

Fig. 1. Cross section of FW structure

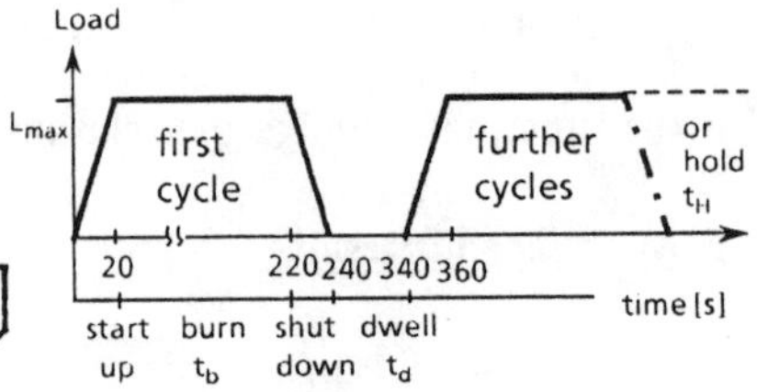

Fig. 2. Load history

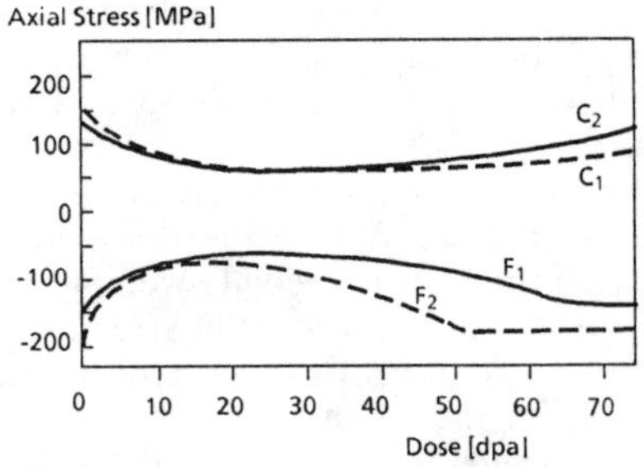

Fig. 3. Stress history at exposed locations under steady state conditions

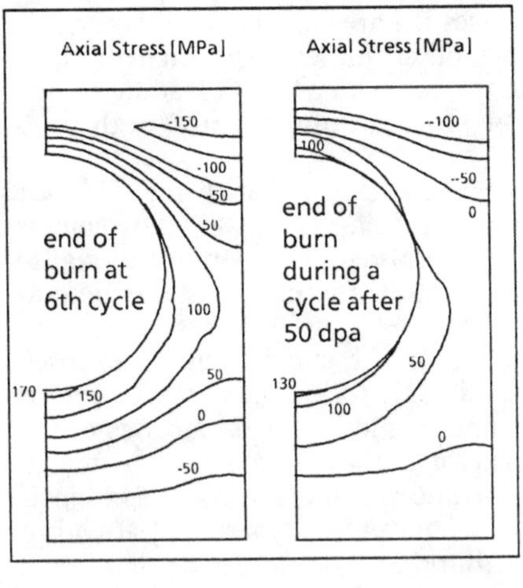

Fig. 4. Stress in FW at the end of burn at zero dose compared to 50 dpa.

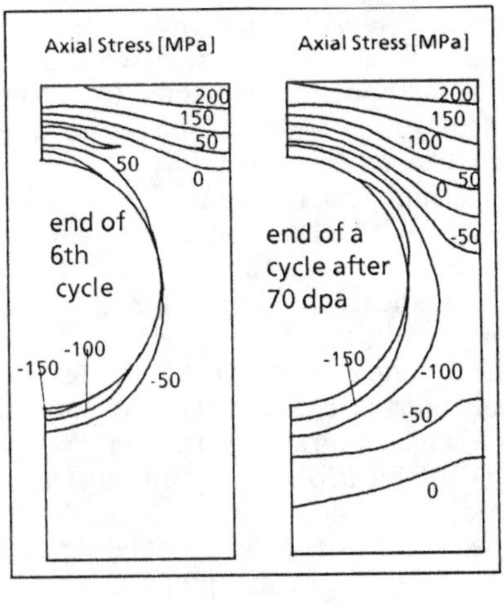

Fig. 5. Stress in FW at the end of dwell at zero dose compared to 70 dpa.

FINITE ELEMENT IMPLEMENTATION OF CONSTITUTIVE EQUATIONS FOR PERLITE

by
Camille A. Issa
Department of Aerospace Engineering,
Mississippi State University, MS 39762
and
John F. Peters
US Army Waterways Experiment Station
Vicksburg,-Mississippi, 39180

ABSTRACT

An endochronic plasticity model for perlite insulation was implemented into a finite element stress analysis code for double-wall cryogenic storage vessels subjected to static, dynamic, and thermal loads. The paper discusses the problems created by the highly crushable nature of the perlite and development of a stable constitutive driver for the endochronic model.

Introduction

A finite element analysis was required for a double-wall cryogenic storage vessels used to store liquid H_2 for test firing of NASA rocket engines. The focus of the analysis was the load transfered between the vessel walls by highly crushable perlite insulation. Laboratory experiments demonstrated that the perlite was similar to a crushable sand that, at low stress with suficiently high density, would even dilate. An endochronic model based on the coupled theory of Valanis and Peters [1] was well suited to model the complex volumemetric response of the material.

Finite Element Iteration Scheme

The modified Newton method was used to compute the non-linear response of the perlite to a load increment $\Delta\mathbf{Q}$. The equilibrium relationship is written in incremental form for the ith iteration as:

$$\mathbf{K}\delta\mathbf{q}^i = \Delta\mathbf{Q} + \mathbf{R}(\Delta\mathbf{q})^{i-1} \tag{1}$$

where

$$\mathbf{R}(\Delta\mathbf{q})^{i-1} = \Delta\mathbf{Q} - \int_V \mathbf{B}^T \Delta\sigma^{i-1} dV \tag{2}$$

and

$$\Delta\mathbf{q}^i = \Delta\mathbf{q}^{i-1} + \delta\mathbf{q}^i \tag{3}$$

with $\Delta\sigma$ being the stress generated by the strain $\Delta\epsilon = \mathbf{B}^T\Delta\mathbf{q}$. The relationship between $\Delta\sigma$ and $\Delta\epsilon$ is obtained from integrating the endochronic constitutive relationships described in the section that follows. The iterations were continued until $||\mathbf{R}^i||/||\Delta\mathbf{Q}|| \leq 10^{-4}$. This interative scheme was incorporated into a time integration based on Newmark's method.

Constitutive Equations

The volumetric response of perlite is quite complex [2] and to obtain accurate agreement between the model and experiment, k and F_h in Equations (9) and (10) below had to be made functions of volumetric strain and hydrostatic stress. However, from a computational standpoint, the shear response was the most troublesome and discussion in this brief paper will be limited to the shear response. The differential relationships for the case where strain increments are given are as follows:

$$de^p_{ij} = \frac{1}{2G+A}(2G\,de_{ij} + Q_{ij}\,dz_s) \tag{4}$$

$$dQ^{(r)}_{ij} = A_r\,de^p_{ij} - \alpha_r Q^{(r)}_{ij}\,dz_s \tag{5}$$

$$s_{ij} = \sum_{r=1}^{N} Q^{(r)}_{ij} \tag{6}$$

$$Q_{ij} = \sum_{r=1}^{N} \alpha_r Q^{(r)}_{ij} \tag{7}$$

$$A = \sum_{r=1}^{N} A_r \tag{8}$$

$$dz^2 = de^p_{ij}de^p_{ij} + k^2 d\epsilon^{p2} \tag{9}$$

$$dz_s = \frac{dz}{F_s}. \tag{10}$$

The constants k, A_r, α_r and the function F_s are experimentally determined. s_{ij} is the shear stress and de_{ij} increment of shear strain. The superscript p indicates that the strain is plastic wherein:

$$de^p_{ij} = de_{ij} - \frac{1}{2G}ds_{ij} \tag{11}$$

where G is the elastic shear modulus. There are similar equations for the hydrostatic components which are coupled to the shear to produce shear-induced volume change (see References [3,4,5]). The tensor $Q^{(r)}_{ij}$ represents the internal "force" associated with each internal variable and is defined in the computation by its initial value and its evolution as given by (5). Thus, given the current state, defined by s_{ij}, and $Q^{(r)}_{ij}$, the increments of strain, de_{ij}, and the endochronic time increment, dz, the plastic strain rates can be determined from Equation (4). The stress increment can then be determined from Equation (11).

Explicit Scheme for Stress Computation

The increment of endochronic time is needed at each step to "drive" the integration scheme described above. By substituting Equation (4) and and its hydrostatic counterpart into (9), dz can be computed from the current state and the strain increments. However, the relationship is only valid for infinitismal strain quantities. Further, Peters [3] and Sengupta [4] found this explicit stepping scheme prone to numerical instability for large plastic strains. For the constitutive relationships developed for perlite the problem was particularly severe because essentially all strains were plastic.

Improved Stress Computation

The difficulty posed by the explicit scheme described above is created by the exponential nature of the solution to the differential equation. As with most explicit schemes when applied to differential equations of this type, significant errors arise unless exceedingly small step sizes are used. The evolution of the plastic state is described by N such equations and the error is different for each evolution equation; as integration proceeds, the error in $Q_{ij}^{(r)}$ becomes greatest for the equation having the largest α_r.

A more accurate update formula, similar to that used by Valanis and Fan [6], can be derived form the exact solution to Equations (4) to (10):

$$Q_{ij}^{(r)} = Q_{ij}^{(r_o)} e^{-\alpha_r \Delta z_s} + \int_0^{\Delta z_s} A_r e^{-\alpha_r(\Delta z_s - z')} \frac{de_{ij}^p}{dz'} dz' \tag{12}$$

where $Q_{ij}^{(r_o)}$ is $Q_{ij}^{(r)}$ at the start of the load step. Equation (12) is approximated by Equation (13).

$$Q_{ij}^{(r)} \approx Q_{ij}^{(r_o)} e^{-\alpha_r \Delta z_s} + \frac{A_r}{\alpha_r}(1 - e^{-\alpha_r(\Delta z_s - z')}) \frac{de_{ij}^p}{dz_s}. \tag{13}$$

Equation (13) is accurate for finite increments of strain and can be used, in place of Equation (4), to obtain a non-linear relationship that can be solved iteratively for Δz. Altough the computation for Δz is more expensive than that of the explicit scheme, the resulting stress computation is considerably more accurate and is stable even for large plastic strains.

To test the ability of the model to reproduce the response of perlite, a single axisymmetric element was subjected to loads sufficiently great to induce predominately plastic strains. As shown in Figure 1 the analysis accurately simulates the non-linear behavior. It is especially important to note that the unloading is reproduced well even though the model does not employ yield surfaces or special unloading criteria. Also reproduced well is the shear-induced dilation of the dense perlite versus contraction of the crushable loose perlite.

Concluding Remarks

The model described in this paper was used successfully for analysis of a liquid H_2 tank under thermally-induced and dynamic loads. The convergence was found to be slow but this is a more a results of using the modified Newton method and the large plastic strains than a property of the endochronic formulation. From experience with integration of endochronic constitutive models we conclude that explicit relationships are not suitable for general analysis.

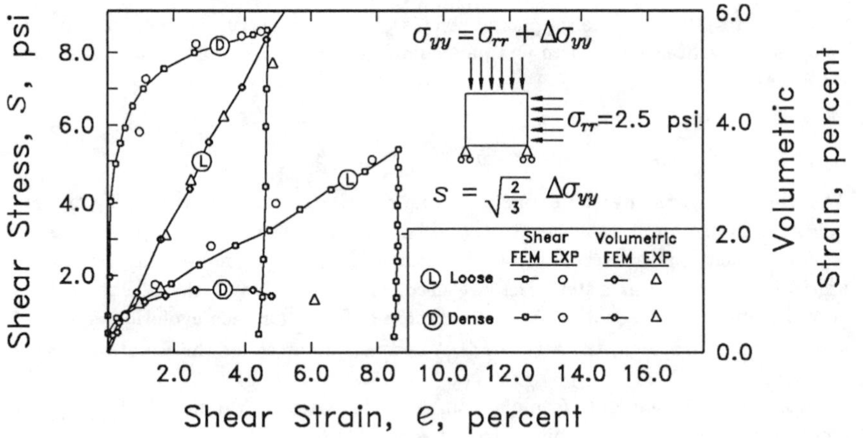

Figure 1: One element simulations of perlite behavior

Acknowledgement

This paper is based on a joint project by the Waterways Experiment Station and Mississippi State University supported by Sverdrup Technology, Inc. under contract with the National Aeronautics and Space Administration (NASA). Appreciation is extended to Greg Garic, Contract Supervisor for Sverdrup Technology, Inc., and to Warren Swanson, NASA contact for this project, for their guidance and support. Permission was granted by the Chief of Engineers to publish this information.

References

1. K. C. Valanis and J. F. Peters, "An Endochronic Plasticity Theory with Shear-Volumetric Coupling," to appear in the *International Journal for Numerical and Analytical Methods in Geomechanics*, 1990.

2. K. C. Valanis, "Mechanical Properties of Perlite", Final report by Endochronics Inc. to Waterways Experiment Station, Report No. WES-88-1, 1988.

3. J. F. Peters, "Internal Variable Model for Frictional Materials", *Constitutive Equations for Granular Non-Cohesive Soils*, edited by A. Saada and G. Bianchini (A. A. Balkema Publishers), 551-569, 1989.

4. A. Sengupta, "Application of a Thermodynamic Model to Soils and Concrete", PhD Dissertation, Illinois Institute of Technology, Chicago Illinois, (also IIT Report No. IIT-CE-89-01) 1988.

5. K. C. Valanis and H. E. Reed, "Endochronic Plasticity: Basic Equations and Applications," Notes for Short Course on Constitutive Laws for Engineering Materials, Tucson, Arizona, 1987.

6. K. C. Valanis and J. Fan, "A numerical Algorithm for Endochronic Plasticity and Comparison with Experiment," *Computers and Structures* **19**, (5/6), 717-724, 1984.

IMPLEMENTATION ALGORITHM FOR A NEW ELASTO-PLASTIC CONSTITUTIVE MODEL

JAY A. ISSA AND EDWARD C. CLUKEY
Exxon Production Research Company
P.O. Box 2189, Houston TX 77252-2189

POUL V. LADE
School of Engineering and Applied Science
University of California, Los Angeles

ABSTRACT

This paper presents a numerical method for implementing a new elasto-plastic material model into a finite element (FE) computer program (ABAQUS). The techniques used are based on the displacement method for the solution of 3-D nonlinear boundary value problems. The incremental form of the model is presented showing the procedures relating stress increment to strain increment. The implementation is based on the solution of the stress increment from the tangent stiffness matrix (Jacobian) at the the start of the step and a corresponding stress increment correction to assure compatibility between the updated yield surface and the computed plastic work. The procedures used were verified by comparing the results computed on a 3-D cube.

INTRODUCTION

The tracking of a nonlinear stress-strain relation in a FE analysis involves a proper procedure for satisfying the required constitutive relations. The stress increment is obtained by integration of the strain increment calculated from the FE solution using the displacement method for the field equations. Numerically it is convenient to consider small strain increments so that linear approximations may be used.

In this paper, a mathematical algorithm is presented for generating incremental elasto-plastic constitutive relations for materials modelled by a single work hardening/softening non-associated plastic flow model. An important feature of this model is the application of a single isotropic yield surface shaped as an asymmetric tear-drop that describes contours of equal plastic work. This concept produces mathematical and computational efficiency in FE programs. The non-associated flow rule is derived from a potential function which describes a 3-D surface shaped as a cigar with asymmetric cross-section in the stress space. The model is derived such that the transition from hardening to softening occurs abruptly at the peak failure point.

COMPONENTS OF CONSTITUTIVE MODEL

Elastic Behavior. The elastic strain increments, which are recoverable upon unloading, are calculated from Hooke's law as:

$$E_{ur} = K_{ur} \cdot p_{atm} \cdot [\sigma_{min}/p_{atm}]^n \tag{1}$$

The dimensionless modulus number, K_{ur}, and the exponent n are determined from triaxial compression tests.

Failure Criterion. A general 3-D failure criterion has been developed for granular materials [1]. The criterion is expressed in terms of the first and the third stress invariants as:

$$f_n = [I_1^3/I_3 - 27] \cdot [I_1/p_{atm}]^m \leq \eta_1 \quad (2)$$

Flow Rate. The plastic strain increments are calculated from:

$$d\epsilon^p_{ij} = d\lambda \, \partial g_p/\partial\sigma_{ij} \quad (3)$$

in which g_p is a plastic potential function and $d\lambda$ is a proportionality constant. The plastic potential function is written in terms of the three stress invariants [2]:

$$g_p = [\Psi_1 \cdot I_1^3/I_3 - I_1^2/I_2 + \Psi_2] \cdot [I_1/p_{atm}]^\mu \quad (4)$$

The material parameters Ψ_2 and μ are dimensionless constants. The parameter Ψ_1 is related to the curvature parameter m for the failure criterion and it acts as a weighting factor between the triangular shape and the circular shape.

Yield Criterion and Work Hardening/Softening Law. The yield surfaces are derived from surfaces of constant plastic work, i.e.

$$f_p = f_p'(\sigma) - f_p''(W_p) \quad (5)$$

in which $$f_p' = [\Psi_1 \; I_1^3/I_3 - I_1^2/I_2] \cdot [I_1/p_{atm}]^h \cdot e^q \quad (6)$$

The parameters Ψ_1 and h are constant for a given material. The parameter q varies with the stress level f_n/η_1 from zero at the hydrostatic axis to unity at the failure surface according to:

$$q = \alpha(f_n/\eta_1)/(1-(1-\alpha)(f_n/\eta_1)) \quad (7)$$

where α is a material constant. During strain hardening the yield surface becomes larger due to increasing plastic work, i.e.,

$$f_p'' = [1/D]^{1/\rho} \cdot [W_p/p_{atm}]^{1/\rho} \quad (8)$$

where D and ρ are material constants. At failure this relationship produces strain-softening, decreasing yield surface, i.e.,

$$f_p'' = A \cdot e^{-B\left(\frac{W_p}{p_{atm}}\right)} \quad (9)$$

In Eq.(9), A and B are positive constants set so Eq. 8 and 9 agree at the point of failure.

INCREMENTAL FORM OF THE CONSTITUTIVE MODEL

To use the constitutive model in ABAQUS, it is necessary to develop an incremental form of the model,

$$\{d\sigma\} = [C^{EP}] \, \{d\epsilon\} \quad (10)$$

where $[C^{EP}]$ is the elasto-plastic tangent stiffness matrix.

The total strain increments are assumed to be divisible into elastic components $\{d\epsilon^e\}$ and plastic components $\{d\epsilon^p\}$. During plastic flow, the yield function f_p is zero and it remains zero during any increments in stress or strain, thus

$$f_p\ (\sigma_{ij},\ W_p) = f_p'\ (\sigma_{ij}) - f_p''\ (W_p) = 0 \tag{11}$$

and

$$df_p = \partial f_p/\partial\sigma_{ij} \bullet d\sigma_{ij} + \partial f_p/\partial W_p \bullet dW_p = 0 \tag{12}$$

$$d\sigma_{ij} = C^E_{ijkl}\ (d\epsilon_{kl} - d\epsilon^p{}_{kl}) \tag{13}$$

$$dW_p = \sigma_{ij}\ d\epsilon^p_{ij} = \sigma_{ij}\ d\lambda \bullet \partial g_p/\partial\sigma_{ij} \tag{14}$$

Substitution of (13) and (14) into (12) gives

$$\frac{\partial f_p}{\partial\sigma_{ij}} C^E_{ijkl}\ (d\epsilon_{kl} - d\epsilon^p_{kl}) + \frac{\partial f_p}{\partial W_p}\sigma_{ij}\ d\lambda\ \frac{\partial g_p}{\partial\sigma_{ij}} = 0 \tag{15}$$

or

$$d\lambda = \frac{\partial f_p/\partial\sigma_{ij} \bullet C^E_{ijkl}\ d\epsilon_{kl}}{\partial f_p/\partial\sigma_{ij} \bullet C^E_{ijkl} \bullet \partial g_p/\partial\sigma_{kl} - \partial f_p/\partial W_p \bullet \sigma_{ij} \bullet \partial g_p/\partial\sigma_{ij}} \tag{16}$$

Therefore

$$d\epsilon^p_{ij} = \frac{\partial f_p/\partial\sigma_{ij} \bullet C^E_{ijkl}\ d\epsilon_{kl}}{\partial f/\partial\sigma_{ij} \bullet C^E_{ijkl} \bullet \partial g_p/\partial\sigma_{kl} - \partial f_p/\partial W_p \bullet \sigma_{ij} \bullet \partial g_p/\partial\sigma_{ij}}\ \frac{\partial g_p}{\partial\sigma_{ij}} \tag{17}$$

and

$$d\sigma_{ij} = C^E{}_{ijkl} - \frac{C^E_{ijmn} \bullet \partial g_p/\partial\sigma_{mn} \bullet \partial f_p/\partial\sigma_{rs} \bullet C^E_{rskl}}{\partial f_p/\partial\sigma_{ij} \bullet C^E_{ijkl} \bullet \partial g_p/\partial\sigma_{kl} - \partial f_p/\partial W_p \bullet \sigma_{ij} \bullet \partial g_p/\partial\sigma_{ij}}\ d\epsilon_{kl} \tag{18}$$

According to the yield criterion in (11), the derivatives of f_p with regard to stresses and plastic work become:

$$\partial f_p/\partial\sigma_{ij} = \partial f_p'/\partial\sigma_{ij} \text{ and } \partial f_p/\partial W_p = \partial f_p''/\partial W_p \tag{19}$$

Since $g_p\ (\sigma_{ij})$ is being homogeneous of degree μ, using Euler's theorem

$$\sigma_{ij}\ \ \partial g_p/\partial\sigma_{ij} = \mu\ g_p \tag{20}$$

The incremental stiffness matrix for the elasto-plastic material model then takes the form

$$[C^{EP}] = [C^E] - \frac{[C^E]\ \{\partial g_p/\partial\sigma\}\ \{\partial f_p'/\partial\sigma\}^T\ [C^E]}{\{\partial f_p'/\partial\sigma\}^T\ [C^E]\ \{\partial g_p/\partial\sigma\} + \mu\ \partial f_p''/\partial W_p \bullet g_p} \tag{21}$$

NUMERICAL IMPLEMENTATION IN FINITE ELEMENT PROGRAM

In finite element applications the displacement formulation is commonly used. The numerical algorithm developed makes use of equation (21), i.e., the input quantities are the stress components $\{\sigma^t\}$ and the plastic work W_p^t obtained at time t and the components of the new strain increments $\{\Delta\epsilon^{t+\Delta t}\}$ at time $t+\Delta t$ calculated using the displacement formulation for the field equations. The output quantities are the new values of the stress components $\{\sigma^{t+\Delta t}\}$ and updated model parameters such as plastic work $W_p^{t+\Delta t}$, as well as the tangent stiffness matrix, $[C^{EPt+\Delta t}]$.

Numerical Algorithm. The first step of the numerical algorithm is to determine the possible stress path resulting from a given strain increment $\{\Delta\epsilon^{t+\Delta t}\}$. To accomplish this, a set of elastic trial stress increments $\{\Delta\sigma^{Et+\Delta t}\}$ are computed by the following

$$\{\Delta\sigma^{Et+\Delta t}\} = [C^{Et}]\ \{\Delta\epsilon^{t+\Delta t}\} \tag{22}$$

The algorithm was formulated such that negative principal stresses do not develop. The new stress point components, $\{\sigma_N\} = \{\sigma^t\} + \{\Delta\sigma^{Et+\Delta t}\}$, and the invariants of $\{\sigma_N\}$ and $\{\sigma^t\}$ are then calculated. The yield function is then calculated by

$$F_1 = f_p(\sigma_N) = f_p'(\sigma_N) - f_p''(W_p^t). \tag{23}$$

If $F_1 \le 0$, then the stress point is inside the yield surface, the material behavior is entirely elastic, and the new stress at $t+\Delta t$ is $\{\sigma^{t+\Delta t}\} = \{\sigma_N\}$. Young's modulus is then calculated to form the Jacobian matrix, i.e., $[J^t] = [C^{Et}]$. The new stress state and Jacobian are then returned to ABAQUS.

If $F_1 > 0$ and $F_2 = f_p(\sigma^t) < 0$, where the initial stress point $\{\sigma^t\}$ is inside the yield surface. The portion of the stress path that is purely elastic must be determined. This is done by finding R such that $f_p(\sigma_t + R\Delta\sigma^{Et+\Delta t}) = 0$. An initial adjustment using linear interpolation gives

$$R_1 = -f_p(\sigma^t)/(f_p(\sigma_N) - f_p(\sigma^t)) \tag{24}$$

A second, fine correction is also included using Taylor expanpansion. If $F_2 = 0$, R is set equal to 0.

The stress state at yield, $\{\sigma_1\} = \{\sigma^t\} + R\ \{\Delta\sigma^{Et+\Delta t}\}$, and corresponding invariants are then determined. The remaining strain increment (1-R) $\{\Delta\epsilon^{Et+\Delta t}\}$ is divided into M equal steps, $\{\Delta\epsilon_i\} = (1-R)/M\ \{\Delta\epsilon^{Et+\Delta t}\}$ and the process repeated M times. The value of M depends on the magnitude of the time step used in ABAQUS.

Next the elastic parameters corresponding to the current stress state and the values of $\{\partial f_p'/\partial\sigma\}$ and $\{\partial g_p/\partial\sigma\}$ are calculated. The stress, $\{\Delta\sigma_i\} = [C^{EP}]\ \{\Delta\epsilon_i\}$, and plastic strain increments, $\{\Delta\epsilon_i^p\} = d\lambda\ \{\partial g_p/\partial\sigma\}$, can then be determined. A stress correction is then applied so that $f_p = f_p - f_p'' = 0$.

The final stress, $\{\sigma^{t+\Delta t}\} = \{\sigma^t\} + \{\Delta\sigma_i\} + \{\delta\sigma_i\}$, and updated plastic work, $W_p^{t+\Delta t} = W_p^t + \{\sigma^t\}^T\ \{\Delta\epsilon_i^p\}$, are then determined. The solution is then checked to see if the peak failure point is reached, $f_n(\sigma) = \eta$. At this point the failure parameters A and B are calculated.

$$A = (f_p')_{peak}\ e^{BW_p}, \quad \text{and} \quad B = df_p''/dW_p\ \big|_{peak} \cdot 1/(f_p')_{peak} \tag{25}$$

The direct Jacobian matrix is then calculated for the time step,

$$\{\Delta\sigma^{t+\Delta t}\} = [J]\ \{\Delta\epsilon^{t+\Delta t}\} \tag{26}$$

REFERENCES

1. Lade, P.V., "Elasto-Plastic Stress-Strain Theory for Cohesionless Soil with Curved Yield Surfaces," Int. J. Solids Struct., Vol. 13, pp. 1019-1035, 1977.
2. Lade, P.V. and Kim, M.K., "Single Hardening Constitutive Model for Frictional Materials, III, Comparisons with Experimental Data," Computers and Geotechnics, Vol. 6, pp. 31-47, 1988.

A FINITE ELEMENT WITH EMBEDDED CRACK BASED ON TENSILE AND SHEAR FRACTURE

M Klisinski and K. Runesson
Dept. of Structural Mechanics, Chalmers University of Technology, Gothenburg, Sweden

S. Sture
Dept. of Civil, Environmental and Architectural Engineering, University of Colorado, Boulder, USA

INTRODUCTION

A new technique for modeling localized deformation within a softening band is described, where softening is attributed to a displacement discontinuity within an element. The fictitious crack model concept by Hillerborg is used to describe fracture behavior. The shape functions of the finite element provide the necessary information about discontinuity. In this manner objectivity with regard to the element configuration is automatically satisfied. Results of numerical simulations show independence on mesh design.

FICTITIOUS CRACK CONCEPT

The fictitious crack concept by Hillerborg et al. [1] is based on the experimental evidence that, in the softening regime, the stress–relative displacement (not strain) relations are unique representations of material behavior. The area under stress–relative displacement curve describes the fracture energy required to form a unit crack area. In this approach there is no need to convert measurements to strains (other than for possible formal unification with strain–based models). Therefore, there is no need for introducing an internal length, that is widely used in smeared crack models [2], [3], [4].

FINITE ELEMENT WITH EMBEDDED CRACK

In our approach a crack forms within the element. Therefore, the finite element consists of two parts: Continuum and Discontinuity.

Nodal displacements are decomposed in the corresponding parts

$$\bar{\mathbf{u}} = \bar{\mathbf{u}}^c + \bar{\mathbf{u}}^d \tag{1}$$

We note that this decomposition holds independently on the magnitude of displacements, which may correspond to small or large deformations. The description of an uncracked continuum is standard. The nodal displacements $\bar{\mathbf{u}}^d$, that correspond to a discontinuity, are related to crack opening and shear slip via special shape functions. For a discrete crack the most proper choice is a jump function with a unit jump at the location of discontinuity. The alternative is a smeared version with a linear shape function distributing the unit jump within the entire element. Both types are indicated in Fig. 1.

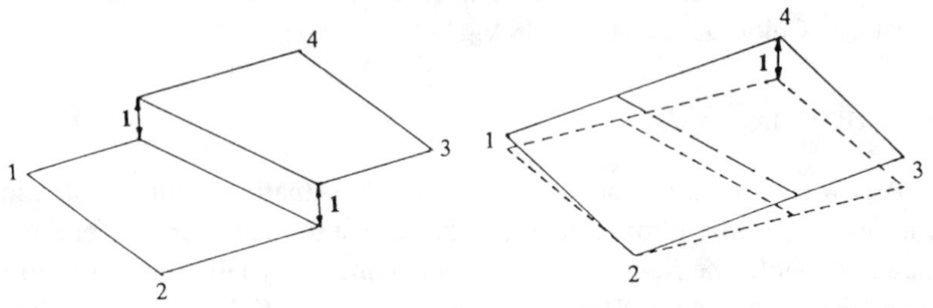

FIG. 1. Shape functions for (a) single band, (b) smeared bands.

A formal representation of the shape function is defined via the redistribution matrix $\mathbf{A}$ which in the case of a unit jump has the form

$$\mathbf{A}^T = \begin{bmatrix} -\beta & 0 & -\beta & 0 & 1-\beta & 0 & 1-\beta & 0 \\ 0 & -\beta & 0 & -\beta & 0 & 1-\beta & 0 & 1-\beta \end{bmatrix} \quad 0 \le \beta \le 1$$

It should be noted that no additional degrees of freedom have been introduced, and since the softening layer is initially represented by a line the description is independent of the element size.

EXAMPLE OF MATERIAL DESCRIPTION

For illustration purposes the simplest material description has been chosen in this paper. Two types of cracks are considered: For tensile cracks Rankine's failure criterion is used, whereas for shear cracks Tresca's criterion is chosen. These failure criteria determine at the same time the presence and direction of the crack discontinuity. It is possible to introduce more general failure criteria such as Mohr–Coulomb's criterion. The intact material is assumed to be linear elastic, and the post-crack softening is also described by linear stress–relative displacement laws. (More general stress–displacement relations can, of course, easily be adopted).

Two different concepts of unloading can be envisioned. One is consistent with linear fracture mechanics and assumes that the crack closes completely during unloading. The second one is rather in the spirit of plasticity that each crack opening is irreversible. In practice (and in our model) these two concepts can be combined.

ELEMENT STIFFNESS MATRIX

Based on the assumption that the stress state at the discontinuity must be consistent with (equilibrate) the stress state in the surrounding intact continuum, the following tangent stiffness matrix can be derived

$$\mathbf{D}^* = (\mathbf{I} + \mathbf{D}^e\mathbf{BAC})^{-1}\mathbf{D}^e \qquad \dot{\sigma} = \mathbf{D}^*\mathbf{B}\dot{\bar{\mathbf{u}}} \tag{2}$$

where $\mathbf{D}^e$ is the elastic stiffness matrix, $\mathbf{C}$ is the compliance matrix describing the behavior of the softening layer, $\mathbf{A}$ is the redistribution matrix, $\mathbf{B}$ the element strain transformation matrix and, finally, $\mathbf{I}$ is a unit matrix. We note that it is possible to define a fictitious strain measure $\dot{\epsilon} = \mathbf{B}\dot{\bar{\mathbf{u}}}$ for the whole element without introducing any extra constitutive assumption about an internal length. The necessary "length" information is provided entirely by the element geometry and the shape functions.

FINITE ELEMENT PROCEDURE

The procedure for solving softening problems is quite straight–forward, since the softening layers are accommodated one at a time. When the stress state within an element signals failure, stress redistribution takes place in the surrounding elements with possible new formation of softening layers. In the two extreme cases the softening bands align themselves in form of serial or parallel chains. In the former case a single discontinuity is sufficient to cause structural failure, while in the later case softening layers must develop in all the parallel elements in order to produce such overall failure.

NUMERICAL EXAMPLE

A plane stress panel is subjected to tension, while the ends are restrained horizontally. Three different uniform meshes were used for the computations in order to demonstrate the mesh objectivity. Figures 2a and 2b depict the localization patterns for the half of the panel and the force versus displacement for all meshes.

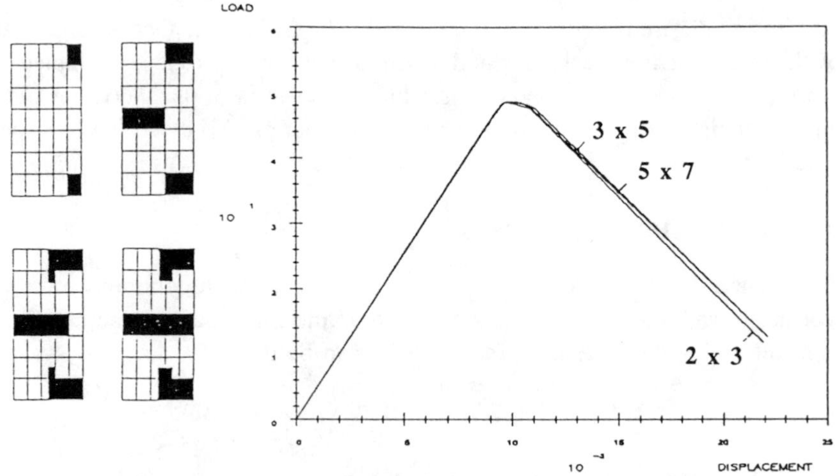

FIG.2. (a) Developement of cracked zones, (b) Applied load versus vertical displacement.

CONCLUSIONS

The method assures objectivity with respect to mesh design (magnitude of elements) as shown by the numerical example. This follows from the basic assumption inherent in the fictitious crack model. The advantages of the presented method include simplicity in formulation and numerical implementation. Moreover, the same number of degrees of freedom as in conventional formulations is employed, and there is no need for any mesh manipulations. The extension to large displacement description is straight–forward, and the concept of nodal displacement decomposition creates a solid theoretical base for such modification.

REFERENCES

[1] A. Hillerborg, M. Modeer, P.L. Peterson, "Analysis of crack formation and crack growth in concrete by means of fracture mechanics and finite elements", Cement and Concrete Research, 6, 773–782, (1976).

[2] T. Belytschko, J. Fish, B.E. Engelmann, "A finite element with embedded localization zones", J. Comp. Methods Appl. Mech. Engng., 70, 59–89, (1988).

[3] O. Dahlbom, N. Ottosen, "Smeared crack analysis using a generalized fictitious crack model", ASCE, J. Engng. Mech., 116, 55–76, (1990).

[4] S. Pietruszczak, Z. Mroz, "Finite element analysis of deformation of strain–softening materials", Int. J. Num. Meth. Engng., 17, 327–334, (1981).

A new approach of an incremental non-linear constitutive law

SAFWAN LABANIEH

IMG, BP 53X, 38041 Grenoble Cedex, France

Abstract

We propose in this paper a new approach of an incremental non-linear constitutive law of the second ordre. The new approach is based on the direct description of the current derivatives (tangent pseudo-modulus and tangent pseudo poisson's ratio) of the calibration paths (conventional compression and extension triaxial tests) rather than an analytic description of these paths. This approach leads to an important simplification and reduction of constitutive parameters. The introduction of a state parameter $\psi = e_c - e_i$ permits to easily take into consideration the influence of the void ratio as well as the confining pressure on the constitutive characteristics of the material.

Introduction

The incremental non-linear constitutive law of the second ordre as given by Darve [1] has the following form:

$$d\epsilon_\alpha = M^1_{\alpha\beta} d\sigma_\beta + \frac{1}{\| \underline{d\sigma} \|} M^2_{\alpha\beta\gamma}\, d\sigma_\beta d\sigma_\gamma \tag{1}$$

The constitutive parameters of this law are obtained from the material behaviour on "generalized triaxial paths" given by the functions:

$$\begin{aligned} \sigma_k &= f(\epsilon_k, \sigma_j, \sigma_l) & \quad (\sigma_j, \sigma_l\ constants) \\ \epsilon_j &= g(\epsilon_k, \sigma_j, \sigma_l) & and \\ \epsilon_l &= h(\epsilon_k, \sigma_j, \sigma_l) & (k, j, l = 1, 2, 3\ and\ k \neq j \neq l) \end{aligned}$$

These functions are obtained from a stress-strain and a void ratio-strain relationships of conventional compression and extension triaxial tests (calibration stress paths) and two hypothesis concerning the influence of the intermediate stress.

The complexity of soil behaviour leads to rather complex expressions of these relationships specially for the void ratio-strain one.

1 Incremental description of calibration stress paths

1.1 Stress-Strain relationship

A stress-strain relationship is characterized by an initial modulus (E), a stress asymptotic value (σ_p) and a given law for the evolution of tangent pseudo-modulus ($d\sigma_1/d\epsilon_1$) (Labanieh[4]). Triaxial tests on the Hostun sand ($d_{50} = 0.7mm$) (Darve & Labanieh[2]) showed that for equal strain distant experimental points (in the domaine of homogeneous strains) the ratio of two successif secant pseudo-modulus ($\Delta\sigma_1/\Delta\epsilon_1$) is rather constant (Labanieh [4]). This implies that the evolution of secant pseudo-modulus follows a geometric series. We can then note:

$$\sigma_p - \sigma_i = \sum_1^{\infty} E_n \Delta\epsilon \tag{2}$$

where the current secant pseudo-modulus is given by:

$$E_n = E_i R_r^{n-1} \tag{3}$$

where σ_i and E_i represent the confining pressure and the initial modulus respectively and R_r^{n-1} represents a reference reason of the geometric series for a reference small strain increment $(\Delta\epsilon_1)_r$.

The current deviator is obtained by summing (2) up to the increment n:

$$\sigma_a - \sigma_i = \sum_1^{n} E_n \Delta\epsilon \tag{4}$$

1.2 void ratio-axial strain relationship

The void ratio-axial strain relationship is characterized by the initial void ratio value, an initial poisson's ratio (ν), a maximum value of strain rates ratio (tangent pseudo poisson's ratio) ($\nu_{max} =\mid \dot{\epsilon}_3/\dot{\epsilon}_1 \mid_{max}$) and a given law of evolution of ν between its two extreme values. Experimental results permitted to adopt, for the tested sand a geometric series also. In the case of compression, the series is divergent ($R > 1$), where in the case of extension the series is divergent only up to the inflexion point of the void ratio-axial strain curve and then convergent. experimental results permitted to propose the following relation between the reason of the stress-strain series and the new divergent one:

$$R_r^{'} = 1 + \frac{1 - R_r}{7} \tag{5}$$

In the case of extension the reason of the convergent series is given by:

$$R_r^{''} = \frac{1}{R_r^{'}} \tag{6}$$

2 Constitutive parameters

Constitutive parameters for the Hostun sand ($d_{50} = 0.7mm$) were determined from conventional compression and extension triaxial tests realized for different densities and four confining pressures on a conventional triaxial apparatus with local measurements of density by Gamma rays and lateral strain by photo-electric cells (Labanieh [4]). The introduction of a state parameter (ψ) representing the relative position of the initial void ratio from the critical void ratio ($e_c - e_i$) permitted to consider a single variable to describe the influence of the initial void ratio as well as the confining pressure (Darve & Labanieh [2],[3]).

3 Results

We give in figure [1] a comparison betwwen theoretical and experimental results for a compression and extension calibration tests on the dense Hostun sand under a 100 KPa confining pressure. In figure [2] we give a comparison between theoretical and experimental results for compression and extension calibration tests on the loose Hostun sand under 800 KPa confining pressure. These conditions correspond to the extreme conditions tested for $e_c - e_i$ (from $\approx 0.21to \approx 0.$). In figure [3] we give an example of the validation of this approach on a constant mean pressure test ($\sigma_m = 300KPa$) on the dense Hostun sand realized on the true triaxial apparatus of the IMG. We note on these figures the very good concordance between theoretical and experimental results. A complete validation of this approach is given in Labanieh [4]. This approach is beeing now applied to the RF hostun sand.

References

[1] F. Darve "Une loi rheologique incremental non-lineaire pour les solides", Mech. Res. Comm., 7 (4), 1980.

[2] F. Darve, S. Labanieh "Incremental constitutive law and for sands and clays. Simulations of monotonic and cyclic tests", Int. J. Num. and Anal. Meth. in Geomechanics, 6, 1982

[3] F. Darve, S. Labanieh "An incremental non-linear constitutive law and cyclic behaviour of sands", Int. Symp. on Num. Mod. in Geomechanics, Zurich, 13 – 17 Sept. 1982.

[4] S. Labanieh "Modélisations non-linéaires de la rhéologie des sables et applications", Thèse de doctorat es siences, Univ. J. Fourrier, Grenoble, 1984.

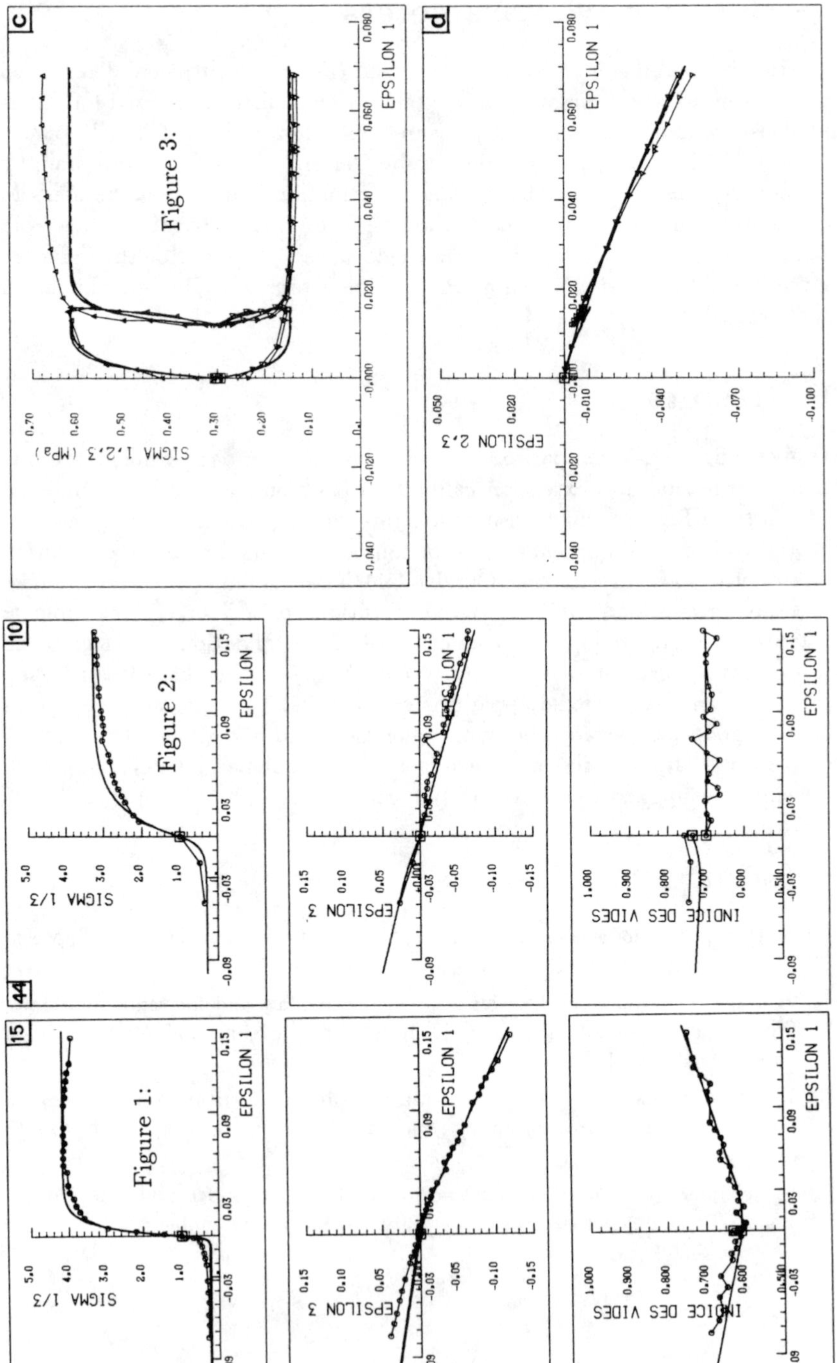

Figure 1:

Figure 2:

Figure 3:

NUMERICAL VALIDATION OF A NONLOCAL DAMAGE THEORY

H. Murakami and D. M. Kendall
Department of Applied Mechanics and Engineering Sciences, R011
University of California at San Diego
La Jolla, California 92093

K. C. Valanis
ENDOCHRONICS Inc.
Vancouver, Washington 98665

ABSTRACT

A nonlocal damage theory recently proposed by Valanis [1] is implemented into a finite element wave code. The model retains hyperbolicity even under strain softening thus eliminating mesh-sensitivity. Convergence and mesh insensitivity are shown numerically by solving the problem considered by Bazant and Belytschko [2].

INTRODUCTION

Strain softening behavior has been observed in heterogeneous materials, such as concrete and rocks. Sandler and Wright [3] reported uncomfortable mesh sensitivity in finite element analyses of initial value problems with rate independent strain softening materials. This mesh sensitivity is due to deformation trapping [4] which is caused by the loss of the hyperbolicity of the constitutive model. Thus, in order to avoid mesh sensitivity in finite element analyses, a rate-independent strain softening constitutive model must retain hyperbolicity. An excellent review on strain softening was compiled by Read and Hegemier [5].

A nonlocal damage theory recently proposed by Valanis [1] retains hyperbolicity even under strain softening. The objective of the present report is to investigate numerically the mesh insensitivity of the model by implementing the model into a finite element wave code. Convergence and mesh insensitivity are demonstrated numerically by solving the problem considered by Bazant and Belytschko [2]. A finite rod is subjected to constant end velocities, which generate tensile waves propagating inward.

BASIC EQUATIONS OF NONLOCAL DAMAGE THEORY

Consider one dimensional wave propagation in an elastic-fracturing solid in the axial direction denoted by coordinate x. The axial deformation is described by a map, $x=x(X, t)$, between the material coordinate, X, and the spatial coordinate, x, at time t. Axial Cauchy stress, strain rate, and velocity are denoted by σ, $\dot{\varepsilon}$, and v ($=\dot{x}$), respectively. In what follows, overdot implies material time derivative. The basic equations of the damage theory proposed by Valanis [1] become:
(a) Equation of motion

$$\frac{\partial \sigma}{\partial x} = \rho \dot{v} \tag{1}$$

where ρ is mass density;

(b) Constitutive Relation

$$\overset{\circ}{\sigma} = \dot{\sigma} = E\,\dot{\varepsilon} + \dot{E}\,\varepsilon, \quad E = E_0\,\varphi^2, \quad \dot{\varepsilon} = \frac{\partial v}{\partial x} \tag{2a,b,c}$$

where $\overset{\circ}{\sigma}$ is the Jaumann rate of the Cauchy stress, E_0 and E are the intact and current elastic moduli, respectively, ε is the logarithmic strain, and φ is the integrity parameter which describes the stiffness degradation and lies in the range: $0 \le \varphi \le 1$; the undamaged state is represented by $\varphi(0)=1$;

(c) Damage Evolution Law

$$\dot{\varphi} = -\,E_0\,\varphi\,\varepsilon^2\,\dot{\xi} \tag{3}$$

where ξ (> 0) is the damage coordinate and evolves according to:

$$\dot{\xi}\,(x, t) = \max\left(0.0, \int_\Omega k\,(x-x')\,\dot{\varepsilon}\,(x',t)\,dx'\right) \quad \text{for } \varepsilon(x,t) > 0 \text{ and } \dot{\varepsilon}\,(x,t) > 0 \tag{4a}$$

$$= 0 \qquad \text{otherwise} \tag{4b}$$

in which Ω is the one dimensional domain occupied by the solid and the kernel function, k(x), is compact and defined as:

$$k\,(x) = k_0\sqrt{\frac{b}{\pi}}\,\exp\,(-\,b\,x^2) \tag{5}$$

where k_0 and b are material constants; especially, the constant b represents the range of nonlocal interaction for the evolution of the damage coordinate;

(d) Appropriate boundary data on $\partial\Omega$ and initial conditions at time t=0.

A corresponding local damage theory may be obtained by setting the kernel function equal to the Dirac delta function:

$$k\,(x) = k_0\,\delta\,(x) \tag{6}$$

FINITE ELEMENT IMPLEMENTATION

The governing equations of motion, (1), and stress strain relations, (2), are in local form. Therefore, updated Lagrangian formulation based upon the principle of virtual work holds. Consequently, the new damage model can be easily incorporated into existing finite element codes by simply adding a subroutine for the nonlocal damage evolution, (3) and (4).

An explicit updated Lagrangian code was developed by using two-node elements. Since the implementation of updated Lagrangian codes is well documented, only the implementation of the damage evolution law, (4), is briefly explained. For a typical integration point at x, whose ε and $\dot{\varepsilon}$ are both positive, the increment of the damage coordinate $\Delta\xi$ under time increment Δt is obtained from (4a) by using two-point Gauss quadrature over the elements which cover the support of k(x-x'):

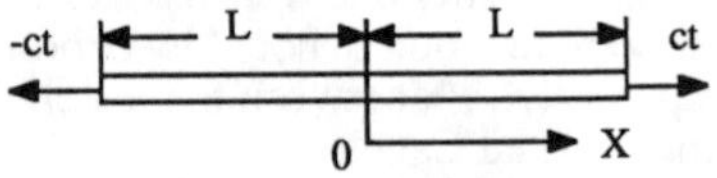

FIG. 1. Strain Softening Bar

$$\Delta\xi\,(x) \approx \sum_e \left\{ \sum_{p=1}^{2} k\,(x - x'(\eta_p))\,\Delta\varepsilon(x'(\eta_p))\,\frac{L_e}{2} \right\}, \quad \eta_2 = -\,\eta_1 = \frac{1}{\sqrt{3}} \tag{7}$$

where the first summation is over the elements on the support, the second summation is over the Gauss integration points, and L_e is the element length.

NUMERICAL RESULTS

Consider a quiescent bar of length 2L, shown in Fig. 1, which is suddenly loaded by end velocities, c and -c, at time t=0. The boundary conditions are

$$v(x(L,t),t) = c\,, \qquad v(x(-L,t),t) = -c \tag{8}$$

The exact solution for softening materials with linearly elastic response up to softening strain ε_c was obtained by Bazant and Belytschko [2]. When the tensile waves meet at x=0, a strain exceeding ε_c is generated, softening response is triggered, and the rod splits into two pieces. The softening zone does not spread out: it is confined at x=0 due to deformation trapping [4]. Bazant, Belytschko and Chang [6] showed mesh sensitive, yet convergent finite element results for a local theory.

The above problem was solved by employing the nonlocal damage theory [1] defined by (1)-(5). For comparisons, the local model defined by (1)-(4) and (6) was also employed. The parameter k_0 in (5) is related to the softening strain by considering monotonic loading:

$$\sigma = E_0\,\varepsilon\,\exp\left(-\frac{2}{3}k_0 E_0 \varepsilon^3\right)\,, \qquad k_0 = \frac{1}{2\,E_0\,\varepsilon_c^3} \tag{9}$$

A constant time step defined by the Courant number computed by the intact elastic modulus is used. In the calculation, the half length of the rod, L, is 15 inch, and the material parameters are: E_0=30 x 10^6 psi, ρ=7.5 x 10^{-4} lb-sec^2/in^4, b=4/9 in^{-2}, ε_c=0.002. To investigate mesh sensitivity, three meshes with 60 elements (Mesh A), 120 elements (Mesh B), and 240 elements (Mesh C) are employed. Figure 2 shows the evolution of stress at X=L/2 obtained by using the local damage model. Due to the arrival of the traction free wave, which is induced by the splitting of the rod, a tensile pulse of finite duration is observed when softening occurs. The results exhibit similar mesh-sensitivity as reported by Bazant, Belytschko and Chang [6]. It was observed that only one or two elements at the center experience strain softening. Figures 3 show the evolutions of stress and displacement for the nonlocal damage theory. The evolutions of strain and stress, and the stress-strain response at the center element (x≈ 0) are also illustrated in Figs. 4. The above results demonstrate the mesh insensitivity of initial value problems of softening materials described by the Valanis nonlocal damage theory [1].

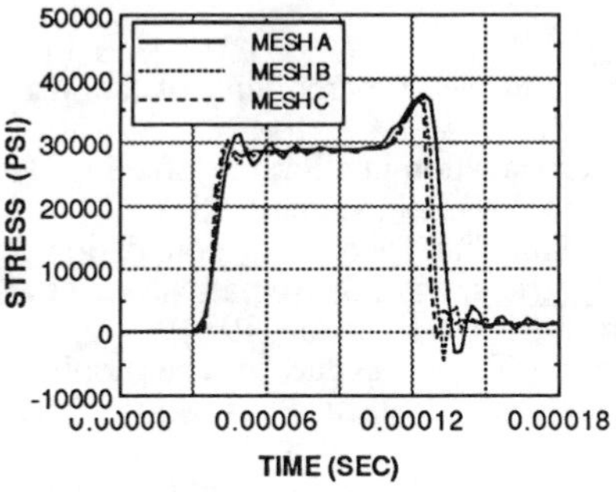

FIG. 2. Evolution of Stress at X=L/2
(Local Theory)

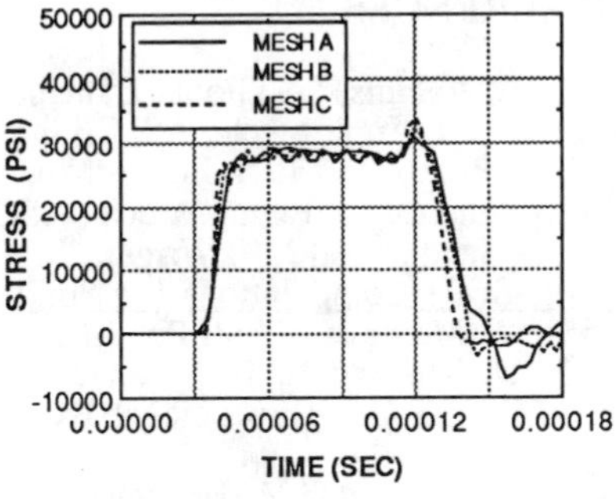

FIG. 3a. Evolution of Stress at X=L/2
(Nonlocal Theory)

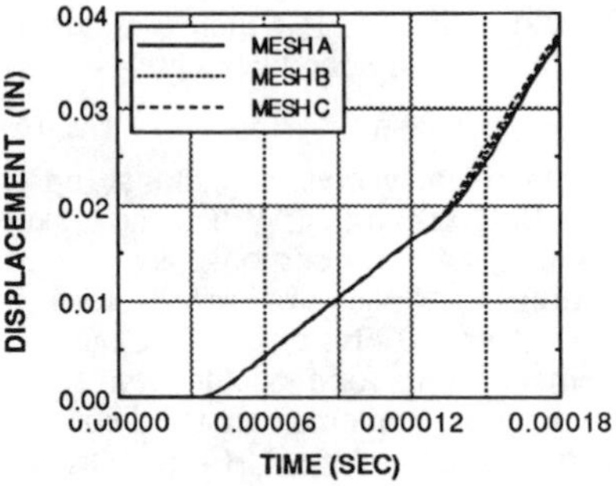

FIG. 3b. Evolution of Displacement at X=L/2 (Nonlocal Theory)

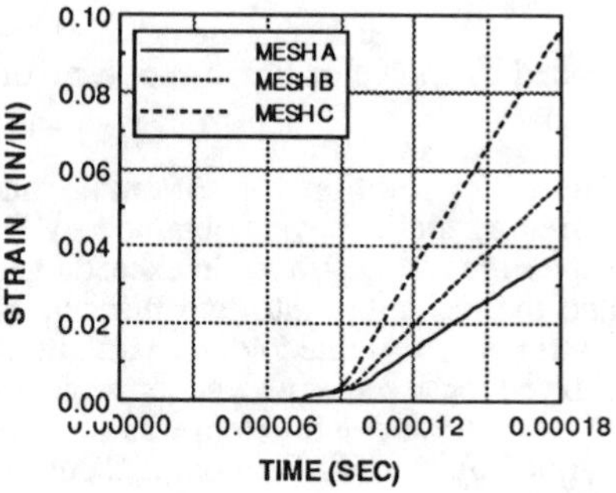

FIG. 4a. Evolution of Strain at x=0 (Nonlocal Theory)

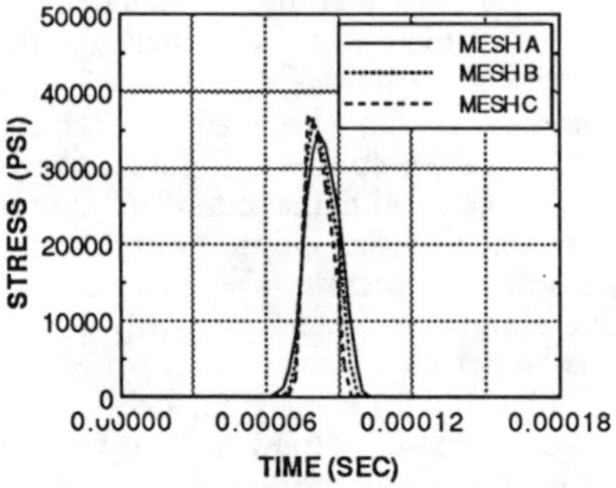

FIG. 4b. Evolution of Stress at x=0 (Nonlocal Theory)

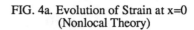

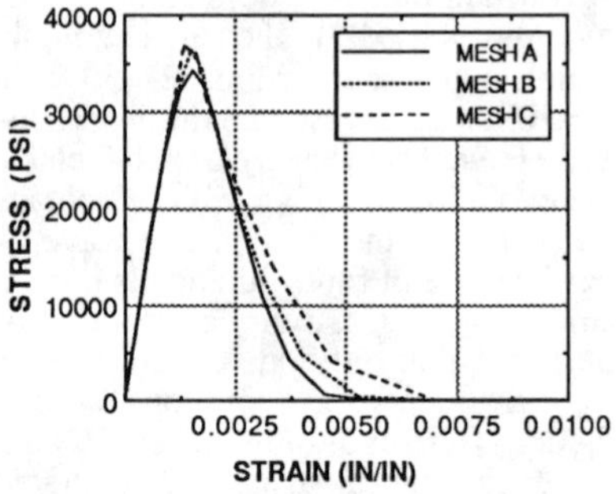

FIG. 4c. Stress-Strain Response at x=0 (Nonlocal Theory)

REFERENCES

1. K. C. Valanis, "A Global Damage Theory and the Hyperbolicity of the Wave Problem," J. Appl. Mech., ASME, in press (1989).
2. Z. P. Bazant and T. B. Belytschko, "Wave Propagation in a Strain-Softening Bar: Exact Solution," J. Engng. Mech., ASCE 111(3), 381-398 (March, 1984).
3. I. S. Sandler and J. P. Wright, "Strain Softening," in Theoretical Foundation for Large-Scale Computations for Nonlinear Materials, eds., S. Nemat-Nasser et al., 285-296 (Martinus Nijhoff Publishers, Dordrecht, The Netherland, 1984).
4. F. H. Wu and L. B. Freund, "Deformation Trapping due to Thermoplastic Instability in One-Dimensional Wave Propagation," J. Mech. Phys. Solids 32(2), 119-132.
5. H. E. Read and G. A. Hegemier, "Strain Softening of Rock, Soil and Concrete- A Review Article," Mech. Materials 3, 271-294 (1984).
6. Z. P. Bazant, T. B. Belytschko and T.-P. Chang, "Continuum Theory for Strain-Softening," J. Engng. Mech., ASCE 110(12), 1666-1692 (1984).

STABILITY ANALYSIS OF ROCK-TUNNEL LINING SYSTEM

Kazem Najm
Department of Resourses Engineering,
Faculty of Engineering,
Hokkaido University, Japan.

ABSTRACT:

A simple but very practical concept of " passive resistance quality of rock " is used as a basis for the stability analysis of a tunnel, through the rock mass, and the tunnel's deformations are investigated using Finite Element modeling. The method is described and its application is proposed.

INTRODUCTION

The proposed method is an example of a compromise between the structural and practical conceptions of rock-tunnel lining system. For a selected problem,a relatively simple, but dependable, solution has been obtained,which enables more attention to be paid to the proper choice and determination of the fundamental system parameters affecting the result.

It is assumed that the wall rock is sotropic, homogeneous and its contact surface in the passive portion is modeled linearly elastic by springs normal to the contact surface. The spring's constant(stiffness) is assumed to be Kn,the coefficient of normal passive resistance of the rock mass. The solution can be readily applied to nonlinear problems,with slight modification.

CONCEPT

The concept of passive ressistance was first introduced in Soil Mechanics,[1]. Driving of a tunnel through a stressed rock mass,by any method,produces in the mass a local destressing effect, which is instantaneously followed by the displacement of the mass towards the tunnel cavity. Depending on the relative rigidity of the tunnel lining and its surrounding rock mass, and also its real or apparent movement relative to the rock around it, (figure-1-a), the extreme pressures to which a lining structure may be subjected are:

1- The **effective field pressure Po**,if no movement of the wall rock occurs.

2- The **active pressure P**, which occurs if the wall rock moves, even slightly, towards the lining.

3- If the situation is such that the deformed

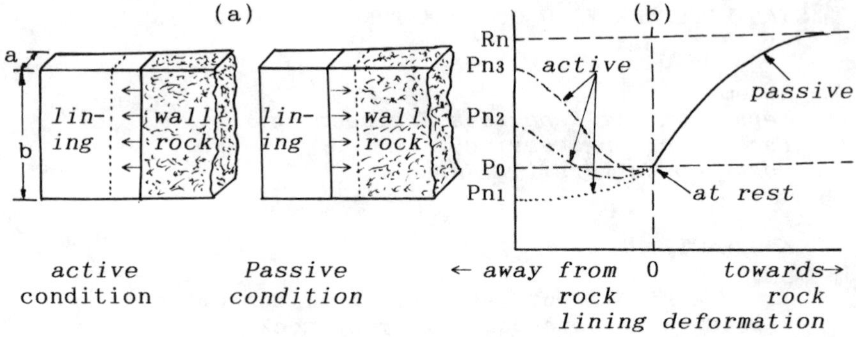

Figure(1)-Active and passive stress formation due to interaction between lining and its surrounding rock.

lining leans against the rock mass at some points, tending to displace it, this is opposed by virtue of rock's **passive resistance R**, trying to keep the lining in place. In figures-1-b, a typical charachtristic curve of passive state and three of active case is presented.

Detailed study of the value of **Kn** for the tunnels of different geometry revealed the fact that it is a charactistic property of the rock mass, and do not depend ,largely, on the size and shape of tunnel. An empiricl formula for the evaluation of **Kn**, for a rock mass free from major geological disturbances,has been deduced[2]; that is:

$$K_n = \frac{c.E}{m.(1 - \nu^2)}$$

where,

E = Young's Modulus;
ν = Poisin's Ratio;
m = b/a

Table-1 variation of constant "c" with the extent of disturbed zone S· ***r****, around a tunnel of radius* ***r****.*

S	3	4	5	6	7	8
c	4.1	4.5	4.9	5.2	5.4	5.6

MATHEMATICAL MODEL

Attention is given to the case of an already excavated tunnel with concrete lining which is deformed. The conatct surface between the lining and its surrounding rock is divided into active and passive parts, according to whether its deformation is away from rock mass or towards it. Both parts are then divided into equal number of steps of lengths b_i and b_j, respectively. The steps width a along the tunnel axis is taken to be unity.

For an elastic rock mass,normal passive resistance **Rn** is supposed to be in proportion to the induced displacement Un which activates it[3]; that is,

$$Rn_j = - Kn \cdot Un_j \qquad (1)$$

Where, **Kn** is the coefficient of normal passive resistance of the rock, (kg/cm^2). The reason for considering only the normal component, represented by subscript $_n$, is due to the fact that the normal loads on the lining are mainly responsible for its deformation.

Suppose that normal active loads $\{Pn_i\}$, acting on the lining of a tunnel, deform the lining whose convergence D_j along, at least, NPS(total Number of Passive Steps) lines are measured in the field. The concept is that if we could, somehow, find the value of corresponding convergences $d_{j,i}$ caused by unit normal active load applied on the active steps "i",then:

$$\{D_j\} = [d_{j,i}] \cdot \{P_i\} \qquad (2)$$

Instead of convergences D_j, one could measure stresses or strains at, at least, NPS points in any part of the lining, and proceed in the same way. Generally applicable computer programs are made for each particular case.

NUMERICAL FORMULATION OF THE MODEL

Let the normal active forces $\{Pn_i\}$ and normal passive resistances $\{Rn_j\}$ acting on their respective sides, be represented by their X and Y components; then:

$$\begin{Bmatrix} Px_i \\ Py_i \\ Rx_j \\ Ry_j \end{Bmatrix} = \begin{bmatrix} ST_{i,n} \\ ST_{j,n} \end{bmatrix} \cdot \begin{Bmatrix} Ux_n \\ Uy_n \end{Bmatrix} \qquad (3)$$

[ST],being the stiffness matrix of size n n for the lining. By allowing the passive resistances $\{Rn_j\}$ of the rock to act on each of the passive steps,in response to the deformations brought about by the active loads $\{Pn_i\}$, Equation (4) may be written as:

$$\begin{Bmatrix} Px_i \\ Py_i \\ 0 \\ 0 \end{Bmatrix} = \begin{bmatrix} ST_{i,n} \\ ST^*_{j,n} \end{bmatrix} \cdot \begin{Bmatrix} Ux_n \\ Uy_n \end{Bmatrix} \qquad (4)$$

With [ST*] being the modified [ST] matrix. Applying a unit normal active load P_{ni} on the active step "i" and solving the equation (4), we get the corresponding displacements $\{U_n\}$, from which the resulting convergence between any two points on the inner surface of the lining, namely $[d_{j,i}]$ could be found out. Now, having calculated $[d_{j,i}]$ and measured $\{D_j\}$, the active loads $\{P_{ni}\}$ could be readily established by solving the equation (2).

CASE SUDY

Deformation of Okurige tunnel, in Hokkaido, Japan, was studied using the proposed techn-ique. The evaluated and observed deformations of the lining are represented in (figure-2).

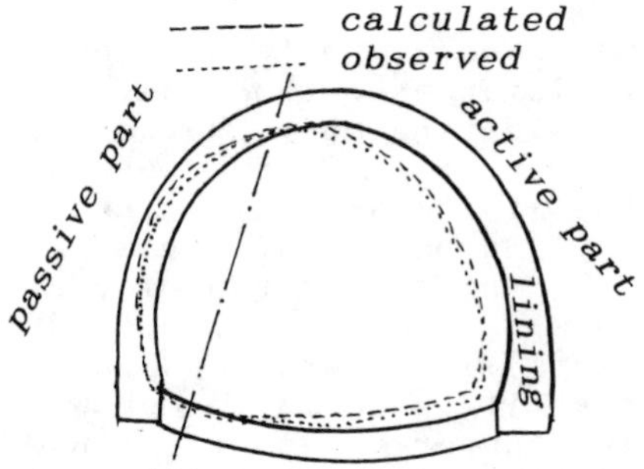

Figure(2)-deformations of Okurige tunnel's lining.

CONCLUSION

The stability evaluation procedure based on the passive resistance conceptual model is applicable to a wide range of tunneling conditions, specially those through weak rock[4]. When this method is used to evaluate the loads responsible for the deformation of a lining-rock system, valuable informations can be gained about the response of the rock mass to the excavation and supporting system. It has the following advantages:

1- Variety of field deformation data may be used as the input to the solution process, and therefore in every situation the most easy to measure and reliable data may be safely selected.

2- Boundary conditions imposed are structurally realistic and would guarantee a reliable result.

3- Coeffiecient Kn can be used as a standard Finite Element Code for a particular site, allowing speedy and reliable further implementation.

References:

1-Terzaghi Karl; "Theoretical soil mechanics"; Chapman and Hall; pp 100-118; (1954).

2-Najm K.; "Passive Resistance Method for the stability analysis of tunnels"; not yet published; (1989).

3-Aldorf J. and Exner K; "Mine openings: stability and support"; Elsevier; pp 145-168; (1986).

4-Jaeger and Cook; "Fundamentals of rock mechanics"; Chapman and Hall; pp 177-183; (1984).

A Structural Constitutive Algorithm Based on Continuum Damage Mechanics for Softening with Snapback

M.K. Neilsen
Applied Mechanics I, Sandia National Laboratories

Z. Chen
New Mexico Engrg. Research Institute, Univ. of New Mexico

H. L. Schreyer
Department of Mechanical Engineering, Univ. of New Mexico

Abstract

A structural constitutive algorithm for elements with localized deformation zones is developed. The algorithm is based on the assumption that the material is described by an elastic-damage model. One- and two-dimensional problems are solved to demonstrate the effectiveness of the approach for generating solutions to strain-softening material problems that are not mesh dependent.

Introduction

When structural members constructed from strain-softening materials are loaded into the softening regime they exhibit deformations that are localized into small regions. If no precautions are taken, finite element solutions to problems with softening materials will predict a localization band that has a width the same size as a typical element. This mesh dependence can lead to highly inaccurate solutions; thus, techniques that eliminate the mesh dependence in softening problems are needed.

The size of localization zones observed in experiments involving various materials are generally much smaller than the structure being analyzed. Thus for finite element analyses of complete structures, techniques that can capture localized deformations that are smaller than the elements being used are needed. Recent work in this area by Pietruszczak and Mroz [1] and Belytschko et al. [2] has produced encouraging results. In this paper, a new structural constitutive algorithm that describes the behavior of an element with an embedded localized deformation zone is developed. The primary difference between this work and that of previous workers is that this new approach is based on the assumption that the material is elastic-damaging whereas the previous work was developed in incremental or rate form which is appropriate for elastic-plastic materials.

Also, Schreyer and Chen [3] and Schreyer [4] have shown that snapback in the load-displacement space will occur when the softening zone is small compared with the size of the structure. Therefore, the ability to capture snapback within a single element is needed if large elements are to be used. Such a capability is demonstrated.

One-dimensional Problem

Consider a bar of length a that is subjected to a uniaxial stress σ (Fig.1) and constructed from a material with the constitutive relation shown in Fig. 2. When the bar is loaded into the softening regime, softening is assumed to occur uniformly in a region of length s (s is a material parameter). Material outside the softening zone elastically unloads. If the bar is modeled with finite elements and the elements are chosen such that the softening zone is modeled with one element, then the constitutive relation shown in Fig. 2 is appropriate. The difficulty arises when the entire bar is modeled with a single element. The constitutive relation in Fig. 2 describes the material behavior, but cannot describe the behavior of an element in which some of the material is strain-softening and the remaining material is elastically unloading.

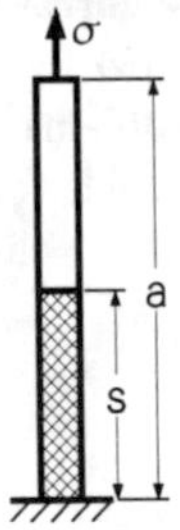

Fig. 1. 1-D Model Problem

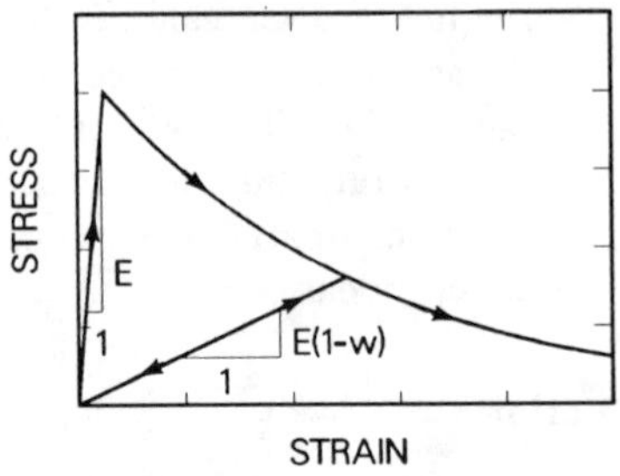

Fig. 2. Constitutive Relation

If the material that is elastically unloading is not damaged, then strain in the unloading region, ϵ_u, is given by

$$\epsilon_u = \frac{\sigma}{E} \tag{1}$$

where E is the elastic modulus. Strain in the softening zone, ϵ_s, is given by

$$\epsilon_s = \frac{\sigma}{E(1-w)} \tag{2}$$

where w is damage in the softening zone. The following expression for average element strain, ϵ, can be derived using Eqs. 1 and 2

$$\epsilon = [1 - w + \frac{s}{a}w]\epsilon_s \tag{3}$$

If the element stress, σ, and strain, ϵ, are related by

$$\sigma = E(1-\overline{w})\epsilon \tag{4}$$

where $\overline{w}$ is the element damage, then it can be shown that the element damage is related to damage in the softening zone, w, by the following equation.

$$\overline{w} = \frac{sw}{[a - aw + sw]} \tag{5}$$

The above equations are used in the new structural constitutive algorithm.

The procedure for solving problems with the new structural constitutive algorithm includes the following steps. First, trial nodal displacements are computed using the secant stiffness matrix corresponding to the current damaged state. The average element strains are then computed from the nodal displacements using a conventional approach. Next, softening zone strains are computed from the average element strains and the current damage using Eq. 3. The constitutive relation for the material (Fig. 2) is then used to compute damage in the softening zone. Element stress and damage are then computed using Eqs. 4 and 5. Finally, the process is repeated as needed to satisfy equilibrium and the constitutive relations.

The static, one-dimensional model problem shown in Fig. 1 was analyzed using a finite element code with the new structural constitutive algorithm. Numerical solutions were obtained using a procedure developed by Chen and Schreyer [5] in which a constraint that limits the amount of damage generated during each step is introduced. Plots of applied load vs. end displacement for bar models with one, five, and twenty elements are shown in Figure 3. These plots show that the load-displacement response is not mesh dependent and that snapback can be captured with a single element.

Extension to Two-dimensions

For the new algorithm to be practical, it must be generalized to two- and three-dimensions. A three-dimensional damage evolution equation that defines the material behavior has been derived which reduces in two-dimensions to a form developed by Bazant and Oh [6]. The equation is based on the assumption that damage increases the compliance of the material only in a direction perpendicular to the softening zone. Two-dimensional problems were solved using a procedure similar to the one-dimensional approach. The softening zone strain was computed using Eq. 3 and the maximum principal element strain. The problem analyzed in this initial study was that of a block loaded in uniaxial tension; thus, the damage zone orientation was fixed. Results obtained from analyses using three different finite element meshes (Fig. 4) show that the applied load vs. end displacement curve is not mesh dependent.

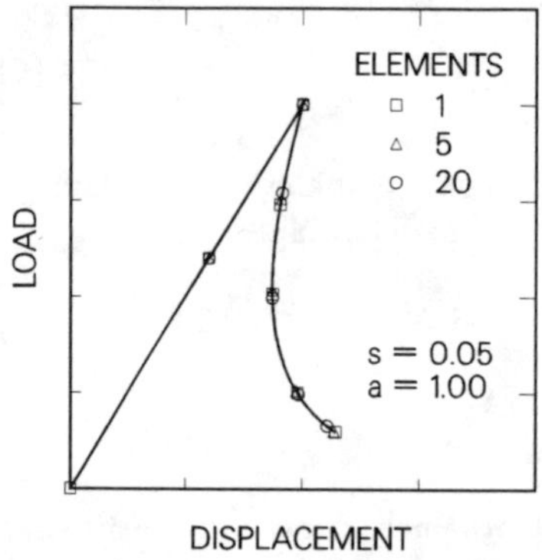

Fig. 3. Results from 1-D Finite Element Analyses

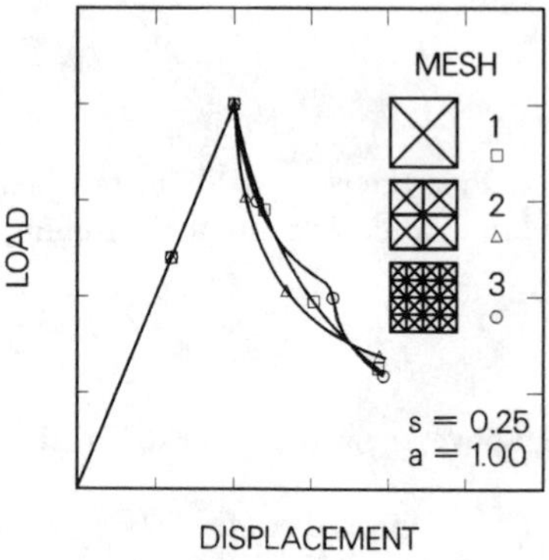

Fig. 4. Results from 2-D Finite Element Analyses

Conclusions and Future Work

Softening, localization and snapback are behaviors that are difficult to model. The response of a structure depends on not only the material response but also the size and orientation of localized deformation zones. A structural constitutive relation that captures the response of an element with a localized deformation zone was developed. The new constitutive relation was used in the solution of problems that display localized failure and snapback. The capability of snapback within a single finite element is important to show the feasability of using large elements. However, several questions remain concerning the effects of having softening zones with various orientations, locations and sizes within an element.

References

1. St. Pietruszczak and Z. Mroz, *Finite Element Analysis of Deformation of Strain-Softening Materials*, **Int. J. Num. Meth. Engr.**, 17, 327-334 (1981)
2. T. Belytschko, J. Fish and E. Engelmann, *A Finite Element with Embedded Localization Zones*, **Comp. Meth. Appl. Mech. Engr.**, 70, 59-89 (1988)
3. H.L. Schreyer and Z. Chen, *One-Dimensional Softening with Localization*, **J. Appl. Mech.**, 53, 791-797 (1986)
4. H.L. Schreyer, *The Need for Snapback in Constitutive Algorithms*, in **Computational Plasticity - Models, Software and Applications**, edited by D.R.J. Owen, E. Hinton and E. Onate (Pineridge Press, 1987)
5. Z. Chen and H.L. Schreyer, *Failure-Controlled Solution Strategies for Damage Softening with Localization*, in **Micro-mechanics of Failure of Quasi-Brittle Materials**, edited by S.P. Shah, S.E. Swartz and M.L. Wang, 135-145 (1990)
6. Z.P. Bazant and B.H. Oh, *Rock Fracture via Strain-Softening Finite Elements*, **J. Engr. Mech.**, 110(7), 1015-1035 (1984)

A Comparison of the Predictive Capabilities of Two Unified Constitutive Models at Elevated Temperatures

J. Olschewski, R. Sievert and A. Bertram
Bundesanstalt für Materialforschung und -prüfung (BAM), Berlin, FRG

ABSTRACT

The viscoplastic constitutive models proposed by Chaboche and Bodner-Partom have been compared in their predictive capabilities of the material behaviour of the cast nickel base alloy IN 738 LC at the service temperature of 850 °C. In particular, the ability to describe the material response under multiaxial strain-controlled loading conditions is demonstrated in comparing the experimental data obtained from a 90° out-of-phase test with hold times at each stress peak with those obtained by model predictions.

INTRODUCTION

A first step toward an advanced life time prediction concept for gas turbine engines is the selection of appropriate viscoplastic constitutive models which are capable to model the high temperature deformation behaviour of NI-based superalloys unter multiaxial loading conditions at service temperatures. For that purpose, two different unified models based on the proposals of Chaboche [1] and Bodner-Partom [2] have been selected and calibrated on the basis of uniaxial test data solely. Various combined tension/torsion cyclic tests performed under nonproportional strain paths have shown that IN 738 LC exhibits no additional hardening. This is in agreement with results obtained by Chan et al. [3] on B1900 + Hf. In the present paper, the 90° out-of-phase test with strain holds at each stress peak will be used to demonstrate the applicability of both unified constitutive models for modelling the high temperature multiaxial deformation behaviour in the absence of additional hardening effects.

CONSTITUTIVE MODELS

The complete set of constitutive equations which has been used in this study is given in Table I below. The material constants presented in Table II have been calculated by using an optimization procedure based on the Levenberg-Marquardt algorithm. Details concerning the identification process can be found in [4]. It is worth to note, that only uniaxial tension, LCF and creep tests have been taken into account. To evaluate the validity of unified models an extensive experimental program covering a broad variety of loadings at different temperatures has been performed. Due to the limited space only some results obtained at the service temperature of 850 °C will be given below.

TABLE I. Summary of the Chaboche and Bodner-Partom model

Chaboche model	Bodner-Partom model
A1: strain decomposition	
$\mathbf{E}=\mathbf{E}_e+\mathbf{E}_i$	
A2: Hooke' s law	
$\mathbf{S}=\overset{<4>}{\mathbf{C}}_e\cdot\cdot\mathbf{E}_e$	
A3: flow rule	
$\dot{\mathbf{E}}_i=\sqrt{3/2}\left\langle\frac{J_2(\mathbf{S}'-\mathbf{X})-R}{K}\right\rangle^n\frac{\mathbf{S}'-\mathbf{X}}{\lVert\mathbf{S}'-\mathbf{X}\rVert}$	$\dot{\mathbf{E}}_i=\sqrt{2}\,D_o\exp\left[-\frac{1}{2}\left(\frac{Z}{J_2(\mathbf{S}')}\right)^{2n}\right]\frac{\mathbf{S}'}{\lVert\mathbf{S}'\rVert}$
$J_2(-)=\sqrt{3/2}\lVert(-)\rVert\,,\quad \dot{p}=\sqrt{2/3}\lVert\dot{\mathbf{E}}_i\rVert$	
A4: hardening rules	$Z=Z^I+Z^D$
▪ isotropic hardening	
$\dot{R}=(Q_o-R)\dot{p}\,,\ R(p=0)=k$	$\dot{Z}^I=m_1(K_1-Z^I)\dot{W}_i-A\left(\frac{Z^I-K_2}{K_1}\right)^r$, $Z^I(t=0)=K_o$
▪ kinematical/directional hardening	
$\dot{\mathbf{X}}=c\left(a\frac{\mathbf{S}'-\mathbf{X}}{J_2}-\Phi(p)\mathbf{X}\right)\dot{p}-d\left(\frac{J_2(\mathbf{X})}{a}\right)^r\frac{\mathbf{X}}{J_2(\mathbf{X})}$	$\dot{\mathbf{B}}=m_2\left(K_3\frac{\mathbf{S}}{\lVert\mathbf{S}\rVert}-\mathbf{B}\right)\dot{W}_i-A\left(\frac{\lVert\mathbf{B}\rVert}{K_1}\right)^r\frac{\mathbf{B}}{\lVert\mathbf{B}\rVert}$
$\Phi(p)=\Phi_\infty-(\Phi_\infty-1)e^{-\omega p}$	$Z^D=\mathbf{B}\cdot\cdot\frac{\mathbf{S}}{\lVert\mathbf{S}\rVert}$

TABLE II Material constants for IN 738 LC tested at 850 °C

Chaboche model		Bodner-Partom model	
$E = 149\,650$ MPa, $\nu_e = 0.33$			
K	= 1150	D_o	= 2.45 10^6 s^{-1}
n	= 7.7	n	= 0.289
k	= 153 MPa	K_o	= 4.18 10^5 MPa
Q_o	= 0.0	K_1	= 3.76 10^5 MPa
a	= 311 MPa	K_2	= 3.07 10^5 MPa
b	= 317 MPa	K_3	= 1.54 10^5 MPa
c	= 201	m_1	= 0.581
Φ_∞	= 1.1	m_2	= 0.344
ω	= 0.04	A	= 4.59 10^3 MPa/s
d	= 2.27 10^{-2} MPa/s	r	= 5.4
r	= 4.8		

COMPARISON OF MODEL PREDICTIONS AND EXPERIMENT

In Figs. 1-2 the predictive capabilities of Chaboche model (CH) and Bodner-Partom model (BP) to describe the uniaxial deformation behaviour of IN 738 LC at 850 °C in different loading situations are shown in comparing the model predictions of creep tests and a tension test with strain rate jumps with experimental data (EX).

The simulation of monotonic creep tests for three different stress levels at 850 °C is compared with experimental data in Fig. 1. As it is indicated in Fig.1 the model calculations of tests which have been used to calibrate the models can be considered in general as good. In order to verify the unified models used in this investigation a comparison is made between experimental data and model predictions for a monotonic strain controlled tension test with varying strain rates at 850 °C. The results for a test with strain rate jumps from $\dot{\varepsilon} = 10^{-6}\,s^{-1}$ to $\dot{\varepsilon} = 10^{-2}\,s^{-1}$ and back to $\dot{\varepsilon} = 10^{-6}\,s^{-1}$ are given in Fig. 2. Both the qualitative and quantitative agreement of the model simulations with the experimental result are excellent.

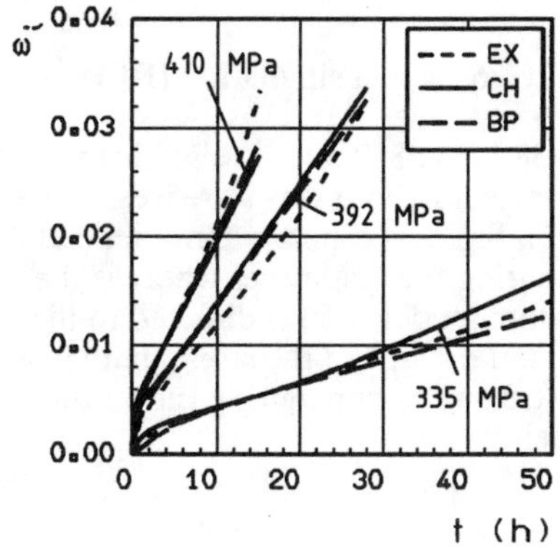

FIG. 1. Comparison of the experimental and simulated data for monotonic creep tests

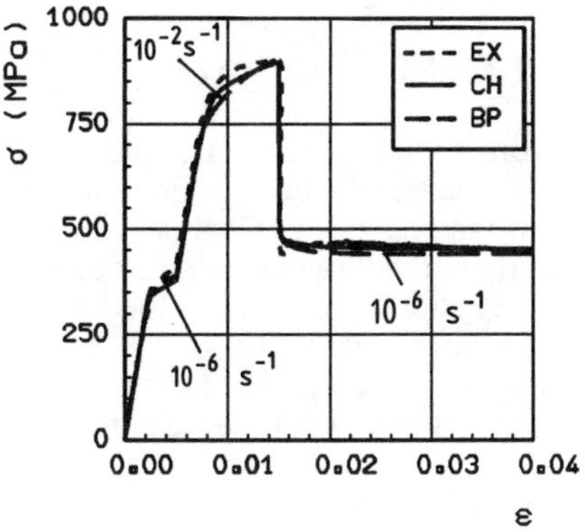

FIG. 2. Comparison of the experimental and simulated data of a tension test with strain rate jumps

Figure 3 shows the strain path of a 90° out-of-phase cycling ($\varepsilon_{eq} = 0.006$, $\dot{\varepsilon}_{eq} = 10^{-3}\,s^{-1}$) with strain holds of $t_H = 120$ s imposed at each of the stress peaks.

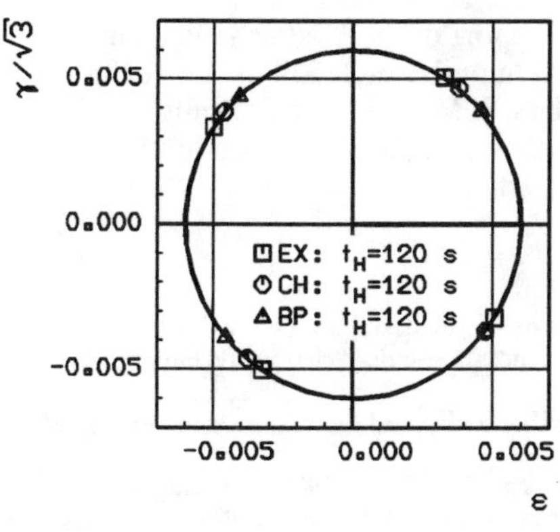

FIG. 3. 90° out-of-phase with strain hold

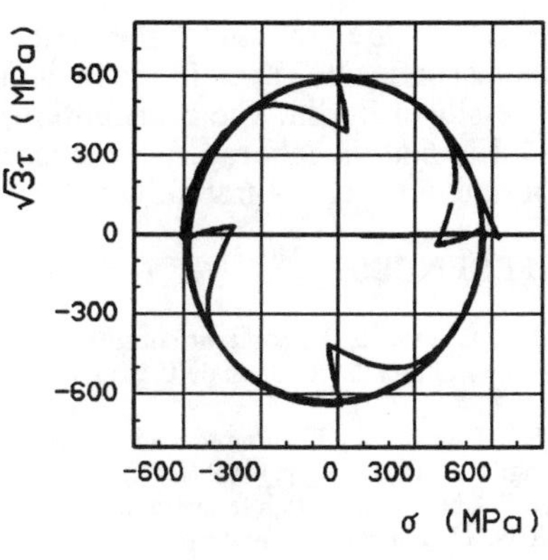

FIG. 4. Material response (EX)

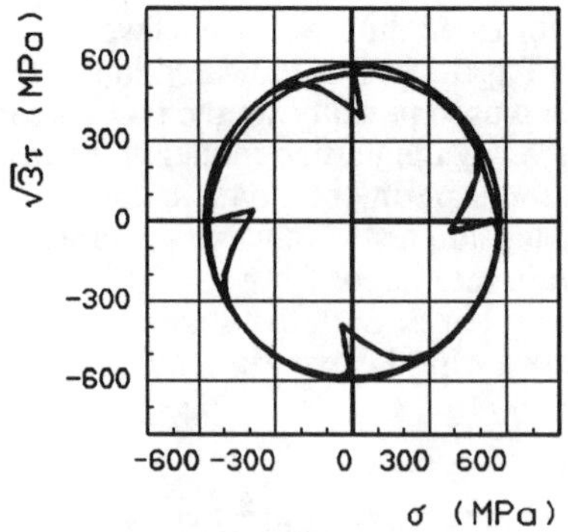

FIG. 5. Model simulation (CH)

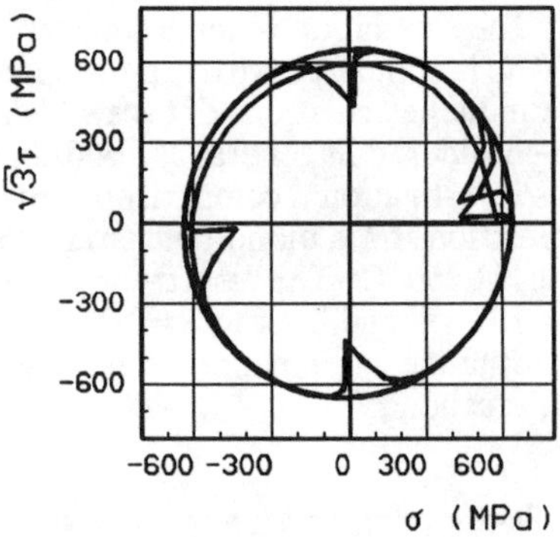

FIG. 6. Model simulation (BP)

The corresponding material response of IN 738 LC at 850 °C with significant stress relaxation following a nonlinear stress trajectory in the σ- $\sqrt{3}\tau$ stress space is shown in Fig. 4. The Chaboche model (CH), Fig. 5, is capable to model the variation of the direction of stress reduction during the hold times whereas the stress relaxation predicted by the Bodner-Partom model (BP) is directed to the origin in stress space, Fig. 6. But, nevertheless, the shape of the stress path in stress space as well as the magnitude of the stress reduction during the strain holds were predicted accurately by both models.

CONCLUSIONS

It has been observed that the cyclic hardening behaviour of IN 738 LC under in-phase and out-of-phase loading was essentially the same. Therefore, unified constitutive models adjusted to the material response on a uniaxial data base only, as the Chaboche model and the Bodner-Partom model in the present investigation, are capable to describe the multiaxial out-of-phase deformation behaviour accurately.

ACKNOWLEDGEMENTS

The authors are grateful for the financial support of the Deutsche Forschungsgemeinschaft (DFG) within the Sonderforschungsbereich 339 of the Technical University of Berlin. The contribution of Drs. J. Meersmann, J. Ziebs and Mr. H.-J. Kühn at the laboratory for mechanical testing (BAM) in conducting the experimental work is gratefully acknowledged.

REFERENCES

1. J.-L. Chaboche, "Viscoplastic constitutive equations for the description of cyclic and anisotropic behaviour of metals", Bulletin de L'Academie des Sciences, Serie des Sciences Techniques, Vol. XXV, 33-42 (1977).
2. K.S. Chan et al., "Phenomenological modeling of hardening and thermal recovery in metals", J. Eng. Mat. Techn. 110(1), 1-8 (1988).
3. K.S. Chan et al., " High temperature inelastic deformation of the B1900 + Hf alloy under multiaxial loading: theory and experiment", J. Eng. Mat. Techn. 112(1), 7-14 (1990).
4. R. Sievert, " Ein Beitrag zur Bestimmung der Materialparameterwerte viskoplastischer Stoffgesetze mittels eines Optimierungsverfahrens. Bundesanstalt für Materialforschung und -prüfung (BAM), Berlin (BAM-1.01 90/3, 1990).

FINITE ELEMENT ANALYSIS OF BOUNDARY VALUE PROBLEMS INVOLVING STRAIN LOCALIZATION

D. PERIĆ and S. STURE
Dept. of Civil, Enviromental, and Architectural Engrg.
University of Colorado, Boulder, CO 80309-0428

K. RUNESSON
Dept. Structural Mechanics, Chalmers University of Technology,
Göteborg, Sweden

ABSTRACT

Numerical examples are presented that illustrate significant differences between load-displacement response curves for soil specimens loaded in conventional triaxial compression (CTC) and plane strain passive (PSP) tests. The analyses were performed by using a nonlinear finite element code, which incorporates closed form bifurcation solutions for a three-invariant plasticity model. Results show excellent qualitative agreement with experimental results, which also suggest that softening observed under plane strain conditions is a geometrical effect (i.e. global or structural softening) which cannot be attributed to the constitutive behavior of the material.

INTRODUCTION

A modified three-invariant Lade's model (MRS Lade model) [Sture et al., 6], which includes cohesion was used to generate analytical bifurcation solutions based on general expressions presented by Ottosen and Runesson [4]. These solutions provide information about critical stress levels and corresponding bifurcation directions. The bifurcation solutions were also used as a diagnostic tool for the mesh design and alignment within a nonlinear finite element code which utilizes the initial load method. These analyses have been described by Perić [5].

In these analyses a medium dense and dry sand was used [Gemperline, 2]. It contains some silt and has maximum particle size of 2.38 mm, while the mean particle size is 0.60 mm. Calibration of the MRS Lade model for the sand was performed by using optimization techniques described by Abifadel [1]. A series of CTC experiments which were performed at confining stresses corresponding to 172, 344 and 689 kPa, were used as input tests for the calibration procedures and subsequent analyses.

NUMERICAL ANALYSES

Samples with height to width (diameter) ratio 2:1 were analyzed under PSP and CTC conditions. The entire sample was modeled for the PSP tests, while only one half of the sample was discretized for the CTC tests by assuming axisymmetry. Simplified boundary conditions which allow development of full friction between the specimen and end platens were used. Thus localization was triggered by the presence of frictional boundaries rather than by using the concept of a "weak element." Consequently, the specimen was assumed to be perfectly homogeneous.

Extensive bifurcation studies at the constitutive and global levels have shown that the plane strain state is extremely favorable for early bifurcation, while CTC test conditions are not [Peric, 5]. Four meshes denoted by R0, R1, R2 and R3 which are shown in Fig. 1 were selected in order to explore the influence of mesh design on the load-displacement response of the specimen. Mesh R0 consists of nine-node quadrilateral elements, while meshes R1, R2 and R3 consist of six-node triangular elements. It is noted that mesh R0 is totally unbiased, while mesh R3 is totally biased. The influence of the shear band width on the response behavior was not studied.

Loading was performed in two phases. First phase consisted of isotropic compression (CTC tests) or anisotropic compression under plane strain condition (PSP tests). In the second phase vertical displacements were applied at the top of the specimen. Load displacement response curves for all four meshes are shown in Fig. 2 in terms of nominal vertical stress versus nominal vertical strain.

It is evident that mesh R3 gives minimum peak load and the most severe post-peak softening. Fig. 3 shows the deformed R3 mesh at nominal vertical strain of 20%. Only this mesh was capable of reproducing this familiar failure mechanism due to the proper mesh alignment which is based on a diagnostic bifurcation analysis.

Finite element analyses of specimens loaded under CTC conditions have indicated that the possibility for discontinuous bifurcation exists only far beyond peak strength. Even at that stage a shear band does not develop throughout the specimen. Consequently, the mesh design is not important in this case. Moreover, the response at the constitutive level gives a satisfactory solution, which differs insignificantly from the finite element solution. Fig. 4 shows the comparison of CTC and PSP load-displacement response curves for samples subjected to a confining pressure of 344 kPa. It is interesting to note that there exists significant resemblance between these numerical solutions and experimental results presented by Marachi et al. [3].

CONCLUSIONS

The analyses have shown that stress states or loading configurations have a significant influence on load-displacement response of specimens. The plane strain state was found to be extremely susceptible to early bifurcation which results in severe softening even as the specimen steadily hardens in the local sense. Capture of the kinematics of the failure mechanism was possibly due to the mesh alignment which was performed based on the bifurcatoin solutions. Since CTC test conditions were found to be extremely resistant to bifurcation, the consequences are of major importance for determination of strength parameters and design of soil structures. Pertinent PS strength properties can be found only by application of the modified finite element analysis which is tailored especially for the strain localization problems or by conducting the actual experiments.

Acknowledgements

The authors gratefully acknowledge support from NASA Grant NAGW - 1388 to the *Center for Space Construction*, University of Colorado at Boulder.

REFERENCES

1. Abifadel, N., Klisinski, M., Sture, S., and Ko, H.-Y. (1988), "Unconstrained Optimization for the Calibration of an Elasto-Plastic Constitutive Model", Internal Technical Report, Dept. of Civil, Environmental, and Architectural Engrg., University of Colorado at Boulder.

2. Gemperline, M. C. (1988), "Centrifugal modeling of shallow foundations", in Soil Properties Evaluation from Centrifugal Models and Field Performance, Proc. ASCE, Eds. Towsend and Norris, pp. 45-70

3. Marachi, N. D., Duncan, J. M., Chan, C. K., and Seed, H. B., (1981), "Plane Strain Testing of Sand", in Laboratory Shear Strength of Soil, ASTM STP 740, Eds. Young and Towsend, American Society for Testing and Materials, pp. 294-302

4. Ottosen, N. S., and Runesson, K. (1990), "Properties of Discontinuous Bifurcation Solutions in Elasto-Plasticity", to appear in Int. J. Solids and Structures

5. Perić, D. (1990), "Localized Deformation and Failure Analysis of Pressure Sensitive Granular Materials", Ph. D. Dissertation, University of Colorado at Boulder.

6. Sture, S. Runesson, K., and Macari-Pasqualino, E. (1989), "Analysis and Integration of a Three-Invariant Plasticity Model for Granular Matrials", Ingenieur-Archiv, 59, pp. 253-266

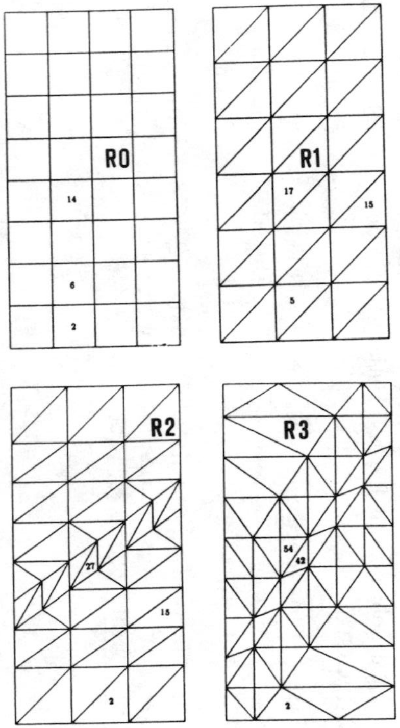

Figure 1: FEM Meshes R0, R1, R2, and R3

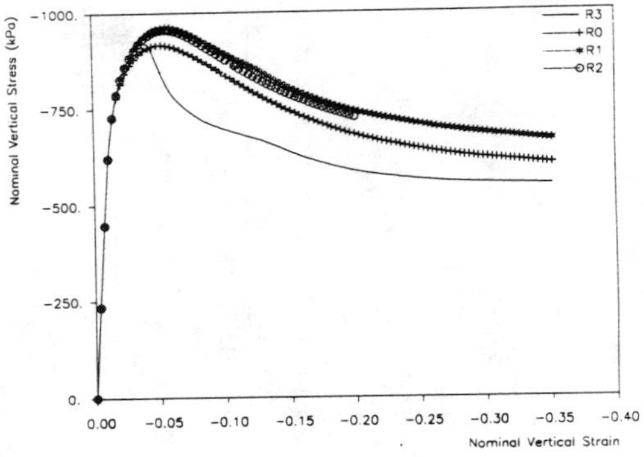

Figure 2: Stress-Strain Responses for Meshes R0, R1, R2, and R3 (172 kPa)

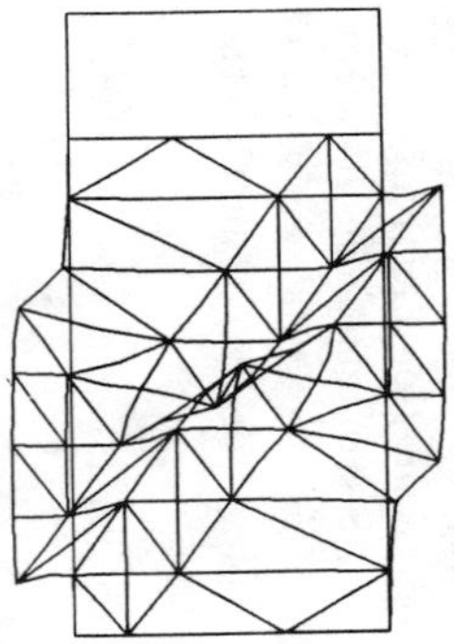

Figure 3: Deformed Meshes R3 at 20% Nominal Verical Strain (172 kPa)

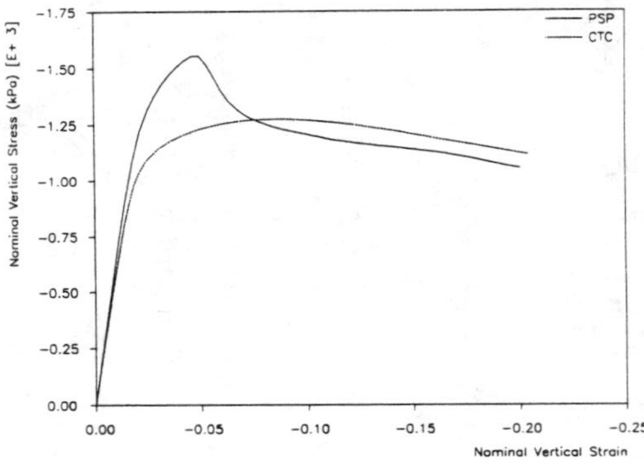

Figure 4: Stress-Strain Responses for PSP and CTC Tests

PENETRATION RATES OF A RIGID CORE IN AN ICE PLATE

U.G.A. PUSWEWALA and R.K.N.D. RAJAPAKSE
Dept. of Civil Engrg., University of Manitoba, Canada R3T 2N2.

ABSTRACT

Penetration of a circular rigid core in an ice plate is simulated by plane strain finite element analysis using the steady-state power law creep model for ice. The core-ice system represents a simplified model for a laterally loaded pile in ice. Numerical results are compared with available test data, and a non-linear dashpot model for the core-ice system is suggested. Effects of seperation between core and ice, and performance of circular vs. square core are discussed.

INTRODUCTION

Laterally loaded piles are encountered in many engineering applications in the cold regions of the world, where pile foundations are used to support oil rigs, transmission towers, pipe-lines, and many other structures. Foundations in frozen media undergo continuing deformation with time due to the creep behaviour of the embedding material. Such pile-ice interaction problems can be studied by selecting a suitable constitutive model for ice, and a convenient solution scheme like the Finite Element Method (FEM) to solve the relevant non-linear boundary value problem.

Experimental evidence shows that laterally loaded steel rods in ice attain a steady rate of lateral displacement [1]. A suitable constitutive model to describe this behaviour is the steady-state power law creep model [2]. The 3-dimensional problem of a pile in ice under a lateral load can be simplified by using a model of a beam on a non-linear dashpot mechanism. The dashpot relationship for displacement rate can be developed by assuming a horizontal plane section of the pile-ice system and analyzing the penetration of a rigid core in an ice plate under plane strain conditions. In the present study, the above core-ice interaction problem is analyzed by using a finite element code developed earlier [3], and the steady state power law creep model for ice [2]. The parameters needed for the creep model are evaluated by an appropriate method. Simulations using a circular core under different loads are compared with test results [1], and an equivalent dashpot model for the core-ice system is developed. Simulations are also performed for a square core, and effects of the seperation between ice and back of the core are discussed.

CREEP MODEL AND NUMERICAL SCHEME

Minimum creep strain rate developed in ice is related to applied stress as [2]:

$$\dot{\epsilon}^c = A\sigma^B , \tag{1}$$

where ϵ^c= creep strain, σ= applied stress, A, B= creep parameters, and superposed dot denotes the time derivative. The uniaxial relationship of eq. (1) is generalized to multi-axial situations by assuming isotropy and incompressibility of the material, and written as [3]:

$$\dot{\epsilon}_e^c = A \sigma_e^B , \tag{2}$$

where σ_e and $\dot{\epsilon}_e^c$ are equivalent stress and equivalent creep strain rate, respectively, given by,

$$\sigma_e = \sqrt{\frac{3}{2} s_{ij} s_{ij}} \quad \text{and} \quad \dot{\epsilon}_e^c = \sqrt{\frac{2}{3} \dot{\epsilon}_{ij}^c \dot{\epsilon}_{ij}^c} . \tag{3}$$

Indicial notation is used in eq. (3), ϵ_{ij}^c denotes components of the creep strain tensor, and s_{ij} denotes deviatoric stress tensor. Using the Prandtl-Reuss relations to relate the components of the creep strain rate tensor and stress tensor, the following expressions are finally obtained as the constitutive model for the numerical study [3]:

$$\dot{\epsilon}_{ij}^c = \frac{3}{2} A \left[\frac{3}{2} s_{kl} s_{kl}\right]^{\frac{B-1}{2}} s_{ij} ; \qquad \sigma_{ij} = D_{ijkl}(\epsilon_{kl} - \epsilon_{kl}^c) , \tag{4}$$

where σ_{ij} denotes stress tensor, D_{ijkl} denotes Hooke's tensor for isotropic linear elasticity, and ϵ_{kl} denotes total strain tensor. A finite element computer code was developed to analyze structure-frozen media interacion problems using an iterative, time-incrementing, unconditionally stable numerical algorithm. Details of the basic formulation, the numerical algorithm, selection of appropriate time intervals, and verification of the code were provided elsewhere [3].

NUMERICAL ANALYSIS AND DISCUSSION:

Previously, tests had been conducted [1] on steel rods of diameter 76 mm embedded to a length of 610 mm in ice (at -2^oC) and loaded by two equal lateral loads applied on the protruding ends of the rods at the top and bottom of the ice mass. However, it was not possible to directly evaluate the creep parameters (A, B) or elasticity constants of ice using those test results. Two qualitative properties of ice evident from the results [1] were: (*i*). the rods attained a constant rate of penetration, and (*ii*). an appreciable instantaneous deflection (about 2 mm) occurred. Observation (*i*) above indicates that ice seemed to had deformed at an almost constant strain rate during the time span considered, a phenomenon that can be conveniently described by an equation of the type (1), and (*ii*) indicates that ice probably possessed a low value of Young's modulus (E). Thus E = 500 MPa was selected for the numerical simulations along with a Poison's ratio (ν) of 0.47 to represent the almost total incompressibility of ice.

Values suggested in the literature for A and B of ice vary over a considerable range, depending on pressure, temperature, texture, fabric, grain boundary structure, etc. [2,4,5]. Thus determination of an exact value of B for ice used in the tests [1] was impossible. However, since a value of B in the vicinity of 3 seemed reasonable, B = 3.17 proposed by Glen [2] was selected, and A was evaluated by an indirect (numerical) method using some of the test results [1] and the selected value of B, as described below.

A rigid core of diameter 76 mm was considered embedded in an ice plate under plane strain conditions (unit thickness in third dimension). Half the plane domain was considered for the numerical simulations, due to symmetry. 8-node isoparametric finite elements were used for discretization, and the core was assumed to be fully bonded to the ice. A lateral load (P) of 114 N was applied on the core, and its penetration into ice was simulated by the computer code for a set of values of A ranging from 0.0035 to 0.0250 $(mm^2/N)^B$ and $B = 3.17$. The applied load was equivalent to one of the four load cases used in the tests [1] (34.8 kN at each end of rod), and the initial value of A (0.0035) was obtained at $-2°C$ by graphically interpolating the values of A given by Glen [2] for different temperatures. Each simulation resulted in a constant penetration rate of the core after a breif period of primary creep, similar to the behaviour seen in tests [1]. By plotting penetration-rate vs. parameter A, which resulted in a straight line, the value of A corresponding to the experimental rate [1] of 0.18 mm/hour was estimated as 0.0214 $(mm^2/N)^B$.

With the above A and B values, simulations were performed using the same mesh for the remaining three loading cases considered in the tests, these being 40.6, 46.5, and 52.3 kN at each end of the rod [1], or equivalently, 133.12, 152.46, and 171.48 N for plane strain simulations. Numerical results from these simulations are shown in Fig. 1. Penetration rates evaluated from fig. 1 (by computing the slopes of the lines) are shown in fig. 2, along with the test results [1] for comparison. For $P = 114$ N, the numerical and test results coincide due to the above evaluation procedure, but the remaining loading cases provide a measure of the accuracy of prediction. From fig. 2, a fairly good comparison is seen for $P = 133.12$ and 152.46 N, where the numerical values underestimate the test results by 10% and 15%, respectively. This indicates that the constitutive assumptions can be regarded valid for the relevant stress range and time span. But for $P = 171.48$ N, the deviation is 29%. These comparisons can be affected by the following factors: (a). Different A and B values may prevail at higher loads (stresses). (b). Possible seperation between ice and back of the core. (c). Inaccuracy of the value of E. (d). Use of small displacement formulation to model comparatively large penetrations.

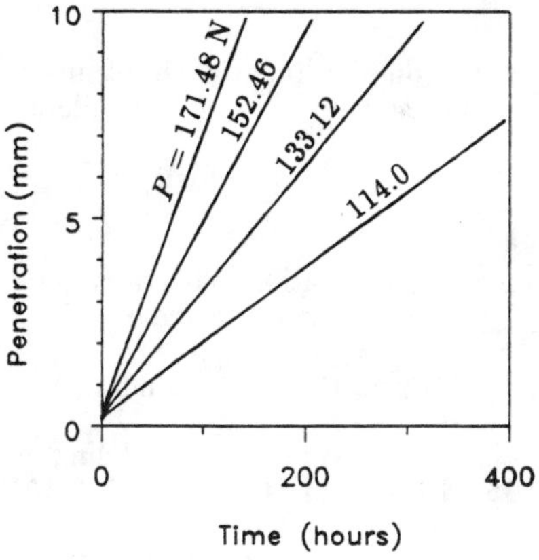

Figure 1

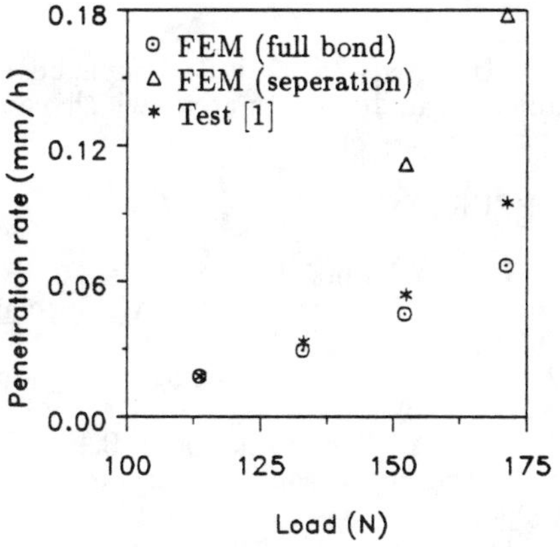

Figure 2

Due to (d), the response curves may be inaccurate as time increases (fig. 1). However, the effect on rates would not be great since the latter were computed from the initial parts of the response curves. Influence of (c) was found to be minimal, as simulations performed for different E values (150–9000 MPa) with all other conditions identical yielded almost identical penetration rates. Effect of (b) was investigated by using a mesh where half the circumferential length of the core at its back was completely detached from ice. Penetration rates predicted for this mesh with same parameters as earlier are also shown on fig. 2, for P = 152.46 and 171.48 N, and these overestimate the test values by about 100%. Thus the penetration rate for a seperated core is more than twice that of a fully bonded core, and the test values for higher loads are between these two extremes. It is indeed possible that seperation may occur over an area less than half the circumferential length, and a thorough study of this effect requires a more elaborate model involving adfreeze/bond elements to simulate the bond between the core and ice. The possibility of (a) is very real [5] but lack of experimental data and space limitations prevent a full presentation here.

By using the least squres method to fit the numerical results for the fully bonded core (fig. 2) to a power relationship, the following equivalent non-linear dashpot model can be proposed for the circular core-ice system with full bond:

$$\dot{d} = (0.51737 \times 10^{-8}) P^{3.1835} , \tag{5}$$

where $\dot{d}$= penetration rate and P = applied load. A similar set of simulations using a square-shaped core with circumferential length equal to that of the circular core (i.e. each side 29.845 mm) indicated that a square core would show a lesser penetration rate than a circular core. The equivalent dashpot model for the square core-ice system with full bond is:

$$\dot{d} = (0.3457 \times 10^{-8}) P^{3.2256} , \tag{6}$$

Using eqs. (5) and (6), simple beam elements on non-linear dashpot mechanisms can be formulated, which can then be used for analyses of pile-ice systems.

REFERENCES

1. L. Domaschuk, D.H. Shields, and R. Kenyon, 'Penetration rates of spheres and cylinders into ice', POAC Speciality Conference, Lulea, Sweden, (June 1989).
2. J.W. Glen, 'The creep of polycrystalline ice', Proc. Roy. Soc. London, Ser. A, 228, 519-538 (1955).
3. U.G.A. Puswewala, and R.K.N.D. Rajapakse, 'Numerical modeling of structure-frozen soil/ ice interaction', ASCE, J. of Cold Reg. 4(3), 133-151 (1990).
4. D.C. Sego, and N.R. Morgenstern, 'Deformation of ice under low stresses', Can. Geotech. J., 20, 587-602 (1983).
5. R. Leb. Hooke, 'Flow law for polycrystalline ice in glaciers: comparison of theoretical predictions, laboratory data, and field measurements', Rev. Geophy. and Space Phy., 19(4), 664-672 (1981).

COMPUTATIONAL ASPECTS OF DYNAMIC FINITE ELEMENT ANALYSIS OF IMPACT IN SAND

JORGE A. RODRIGUEZ, and CHAIM J. PORAN
Civil Engineering Dept. Univ. Of North Carolina At Charlotte, USA

ABSTRACT

A numerical model for sand is presented in this paper. The model, based on the multiple yield surface theory for deviatoric and volumetric plasticity with large deformations capability, is used for the numerical analysis of the response of sand to successive impacts. In addition to material nonlinearities, the sand response in the near field of the impact zone is further complicated by very large deformations and change in material properties resulting from each impact. The paper discusses several of the computational aspects involved in such analysis including remeshing and reassignment of material properties after each impact. Computed impact induced deformations in the sand show good correlation to measured data available from laboratory experiments.

INTRODUCTION

The paper describes computational aspects of the implementation of a numerical model to analyze the response of sand subjected to repeated impact loading by a rigid pounder. The main computational problems are related to very large stresses and deformations induced in the zone near the impact resulting in the development of a crater and significant densification of the sand mass. A dynamic nonlinear finite element analysis was used to implement the model. The axisymmetric formulation uses two dimensional bilinear isoparametric elements for large deformations and rotations. The material model is based on the multiple yield surfaces theory for deviatoric and volumetric plasticity with kinematic hardening, and an implicit direct integration is used in the time domain.

NUMERICAL MODEL

The constitutive equation used for rate form plasticity is based on the stress-strain relationship of the Hookian elastic material and the Prandtl-Ruess definition of the total strain rate tensor. The radial return mapping algorithm and the concept of virtual yield surfaces developed by Salah-Mars [4] are used during plastic yielding. The numerical solution is advanced in incremental iterative form, where the stresses are calculated from the strain history using an initial tangent operator for the computation of the trial stresses. A direct time integration is used to solve the equation of motion. The average acceleration method from the Newmark family of solutions is used by means of an explicit predictor, implicit-multi-corrector algorithm [2]. This is an unconditionally stable formulation with second order accuracy.

A multiple yield surface model based on kinematic hardening plasticity theory was used to represent the behavior of sand. The model is formed by selecting a sufficient number of yield surfaces, each characterized for a given stress level and tangent modulus, such that almost any particular stress-strain curve can be modeled. A detailed description of the multiple yield surfaces formulation can be found in [3] and [4]. The Drucker-Prager yield criteria is used for the yield surfaces of deviatoric plasticity as well as for the failure or limiting surface. A non associative flow rule was used. The plastic potential corresponds to a cylinder shape in the stress space, similar to the Von-Mises yield criteria. To account for volumetric plasticity the constitutive model includes an isotropic hardening volumetric plasticity component. The yield surfaces for volumetric plasticity are planes perpendicular to the hydrostatic axis with an associative flow rule.

COMPUTATIONAL ASPECTS AND RESULTS

Selection of material parameters and code verification were conducted based on comparison with available experimental data [1], which included triaxial compression tests, and static plate bearing tests carried out to large plate deformations [3]. The calibrated model was used to analyze sand response to repeated surface impacts applied by a rigid pounder. For the analysis, impact loading was applied as surface acceleration of a rigid mass based on experimental impact acceleration data from Heh [1].

Analytical results indicate that the extents of the affected soil volume, yielding, and failure depend on the stress level induced during the impact by the acceleration pulse which is used to simulate the loading. Furthermore, after a single pulse is over the system continues to deform until a final stable state is reached. The final density distribution determines the soil behavior for the subsequent drop. To properly model successive impacts, a reassignment of properties is performed after each drop based on the current density, using correlations derived from available experimental data [1].

During the analysis of each loading pulse, elements near the impact area undergo extremely large deformation. The initial mesh may be distorted to such extent that element degeneration may prevent mapping and the computation is aborted. A special mesh design and a remeshing procedure are necessary in order to overcome this problem. The remeshing is coupled with the reassignment procedure for material properties. It is performed automatically at the end of each drop in order to produce a restart file for the analysis of the subsequent impact with the new geometry and material properties. Figure 1 shows the initial mesh and the mesh at the end of the fifth impact.

Numerical problems of convergency and accuracy were encountered in the solution procedure of the incremental form of the constitutive model. One of the problems was caused by the closeness of the initial stresses to the apex of the limit surface and the large magnitude of stress changes imposed by the loading. In the incremental procedure, trial stresses laying outside the limiting surface produced stress updates leading back to the apex. As a result, some elements near the punching edge of the rigid loading plate were no longer contributing to the residual forces vector and thus preventing local equilibrium with the applied load. Significant errors, especially in the computed stresses, were observed. In addition, the punching of the pounder into the sand produces a strong geometric singularity which affects the quality of the finite element approximation. The reassignment of material properties after each drop may also induce singularities of element properties which affects the accuracy of the finite element solution.

To overcome these problems the mesh was refined and designed to improve accuracy and facilitate remeshing and reassignment of model properties without geometric or material singularities. An elastic static case of a rigid circular plate pushed into a softer material was considered for mesh selection. The strain energy of the elements around the plate was computed and the mesh was refined so that all elements had relatively uniform strain energy. This procedure lead to a finer mesh near the plate, with a distribution that was approximately proportional to the magnitude of the stress. In the nonlinear problem the stress magnitude is limited by material strength and the stress concentration is not as high, while the strains are increased due to plastic deformation. The highly graded mesh obtained for the elastic solution was adapted for the nonlinear dynamic cases to accommodate the plastic flow so that small elements were not excessively distorted during the impact analysis, as shown in Figure 1. The elements distribution was adopted so that the reassignment of properties after each drop could be done for successive layers of elements with a distribution similar to the density variation. Analysis was performed using short time steps to enable very small strain increments required for proper solution procedures. Computed results of 12 successive impacts correlate well with experimental data available from Heh [1], as shown in Figure 2.

CONCLUSIONS

The highly nonlinear character of this problem, due to both large deformations and the stress-strain relationships of the material model, impose special conditions affecting the stability and accuracy of the finite element solution. The main computational aspects of this type of analysis are the effects of meshing and assignment of model properties on the computed results. Coupled procedures for remeshing and reassignment of material properties are necessary to accommodate the large changes in geometry and stress state following each impact. These restart procedures must take the changes of sand density into account to enable accurate modeling of successive impacts.

REFERENCES

1. Heh, K.S. (1990). "Dynamic Compaction of Sand", PhD Dissertation, Polytechnic University, Brooklyn, New York.
2. Hughes, T.J.R. (1987). The Finite Element Method: Linear Elastic Static and Dynamic Finite Element Analysis, Prentice Hall Inc., Englewood Cliffs, New Jersey.
3. Poran C.J., and Rodriguez J.A. (1990). "Finite Element Analysis of Impact Response of Sand,", *Intl. J. for Numerical and Analytical Methods in Geomechanics*. (Submitted for publication).
4. Salah-Mars, S. (1988). "A Multiple Yield Surface Plasticity Model for Response of Dry Soil to Impact Loading", PhD Dissertation, Stanford University.

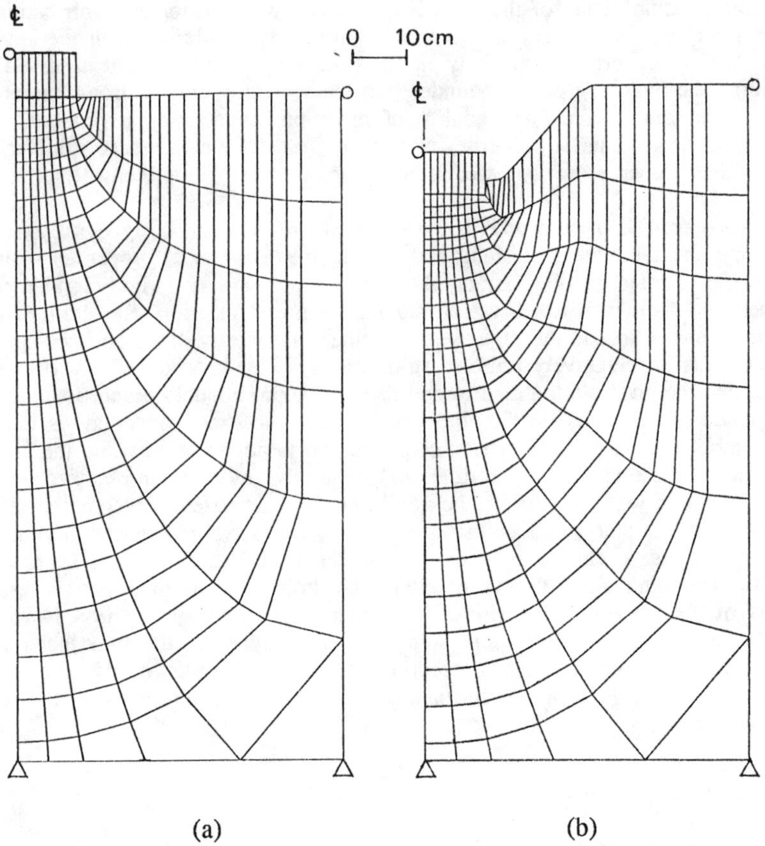

FIGURE 1. Finite Element Mesh: a) Initial, b) After The Fifth Impact.

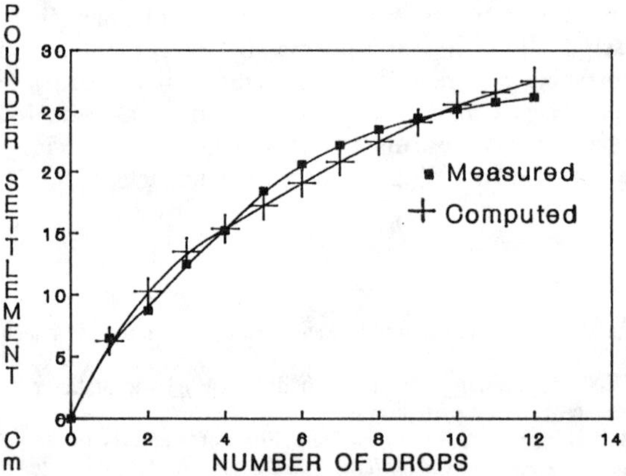

FIGURE 2. Pounder Settlement vs Number of Impacts

MODELLING AND COMPUTER ANALYSIS OF LONG-TERM STABILITY OF A SALT DOME IN RELATION TO COMPRESSED AIR ENERGY STORAGE (CASE) CAVERN DEVELOPMENT

M. REZA SALAMI

Assistant Professor, Department of Civil Engineering, North Carolina Agricultural and Technical State University, Greensboro, North Carolina, 27411

ABSTRACT

This paper contains the results of stability assessment of CAES caverns to be excavated in the McIntosh salt dome, Alabama. The study consists of two major parts: the stability assessment of the whole dome under the operation of Compressed Air Energy Storage (CAES) caverns and the operation of Olin chemical corporation, and the determination of safe boundary distance for CAES caverns using a single cavern. In this paper the whole dome was analyzed

INTRODUCTION

For many years, Compressed Air Energy Storage (CAES) has been the subject of research among electric utility companies, in order to determine the feasibility of storing off-peak excess electricity in the form of compressed air. Conceptually, the compressed air would be stored in underground geological formations which have pressure integrity and are stable under CAES operating pressure and temperature ranges. This compressed air would later be converted back into electricity to meet peak hour demand by running the turbans. It has been realized for many years that underground salt domes are very suitable for this purpose, since the first CAES cavern, excavated in a salt dome in Huntorf, West Germany, has been operating successfully since 1977 [1,2].

SCIENTIFIC BASIC OF STUDY

The development of high speed digital computers for the use of engineering analysis has made the finite element method (FEM) a powerful tool for obtaining solutions to complex geomechanical boundary value/initial value problems. The early developments of FEM were incorporated with linear elastic theory for constitutive behavior of materials, which were at later stages extended to include the theory of plasticity. These theories based on classical plasticity were inadequate in representing the long-term behavior of caverns surrounded by complex brittle-ductile media. for this reason, [1,3,4] has developed a proprietary finite element program (FEM) based on the rheological characterization of material behavior. The present FEM program is a finite element program specially designed and field-validated to interrelate real earth material properties with the corresponding behavior of the ground opening.

Within the scope of this study, the effects of the initial stress state and the initial material properties on the stability of the cavern field were investigated.

ANALYSIS

Due to the uncertainty and unavailability of the full data required for analysis, numerical finite element analysis of solution mining and cavern storage networks has been difficult. The data needed for computer simulation include the geology of the area, the geometry of the cavern network, the properties of the materials involved, in situ stress level, and cavern operating conditions. The existing data in these categories should be carefully evaluated to obtain the necessary information related to the particular analysis. The following subsections describe the data which were used to analyze of the McIntosh salt dome cavern field.

MATERIAL PROPERTIES AND IN-SITU STRESS

The results of the finite element analysis depend greatly on the material property parameters which are used to characterize the ground material. The reliable methods of obtaining such parameters are (1) laboratory testing on cores obtained from the particular sites; and (2) in situ measurements.

In situ stress is the primary cause of cavern deformation and perhaps the most important variable in the analysis of deep geological structures.

CAVERN PRESSURE

All the caverns in the McIntash dome, except the CASE caverns, were assumed to be filled with brine, which has a density of 1.2 g/cm^3. In all our analysis, a pressure of 875 psi was used for CAES caverns.

METHOD OF STUDY

The method of study involves construction of finite element models(FEM) of the McIntosh salt dome, representing the actual excavation geometries of existing brine wells and the geometries of the proposed CAES caverns [1].

Site-Specific Meshes

In order to simulate the 3-dimensional nature of the deformation, two 2-dimensional plane strain meshes of the whole dome, as shown in Figs 3 and 4, were constructed:

1. A horizontal section through the entire salt dome, including all brine wells and CAES caverns.
2. A vertical section passing through the maximum possible number of brine wells and CAES caverns.

The exact dimensions of Olin's cavern were used to construct the meshes. The following "norm" dimensions were used, allowing the possibility of changing them at later stages:

Diameter of CAES cavern, D = 150 ft ; Height of CAES cavern, L = 622 ft ; Depth from the top of the CAES cavern to ground surface, H = 1500 ft ; Distance from center of CAES cavern to dome boundary, C = 1000 ft ; and Center-to center distance between CAES caverns, S = 625 ft.

The vertical section of the finite element mesh consists of 2205 4-node elements and 2337 nodes. The horizontal section mesh consists of 2751 4-node elements and 2782 nodes.

RESULTS OF VERTICAL AND HORIZONTAL PLAIN STRAIN ANALYSIS

Results obtained for the vertical and horizontal sections using plane strain analysis are shown in Figs 1-3.

CONCLUSIONS

The following conclusions are drawn from this study:

1. Stability of Cavern Field.
Finite element analysis shows a stable condition around the cavern field.

2. Stability of CAES Caverns
The whole dome analysis shows that the site selected for the study is stable, where the sensitivity analysis strengthens this conclusion. The cavern location at more than 200 ft. from the dome boundary is satisfactory.

3 Stress Distributions
A large plastic region is developing at the bottom of the solution caverns, but not around the CAES caverns.

4. Material Property Parameters
Correct estimation of the material property parameters plays a major role in this analysis.

It is worth while to mention here that the above analysis is important for the future use in analyzing and designing all the compressed under-ground storage used for storing oil, chemical or air.

ACKNOWLEDGMENT

The authors acknowledge the support that was provided by the serata Geomechanics, Inc. and Electrical Power and Research Institute(EPRI). Also the authors acknowledge the support provided by Department of Civil Engineering of North Carolina A and T State University,

REFERENCES

1. M. Reza Salami, Serata Geomechanics Inc., "Preliminary Computer Analysis on Long-Term Stability of McIntosh Salt Dome in Relation to Compressed Air Energy Storage (CAES) Cavern Development," Preliminary report for EPRI, Presented at Solution Mining Research Institute Spring Meeting, Tulsa, Oklahoma, (1987).
2. Electric Power Research Inst. ,"General Electric Company, Conceptual Design for a Pilot/Demonstration Compressed Air Storage Facility Employing a Solution-Mined Salt Cavern," section 7, (1977).
3. Serata Geomechanics, Inc.,"Theoretical Definition and Determination Procedures A Property Coefficient Values of FEM. Constitutive Model, Laboratory Method of Material Testing," Rock Mechanics School - Solution Mining, Houston Texas, (1987).
4. S. Serata, "Prerequisites for Application of Finite Element Method to Solution Cavities and Conventional Mines", Proc. 3rd Symposium on Salt, The Northern Ohio Geological Society, Inc., Cleveland, Ohio, (1973).

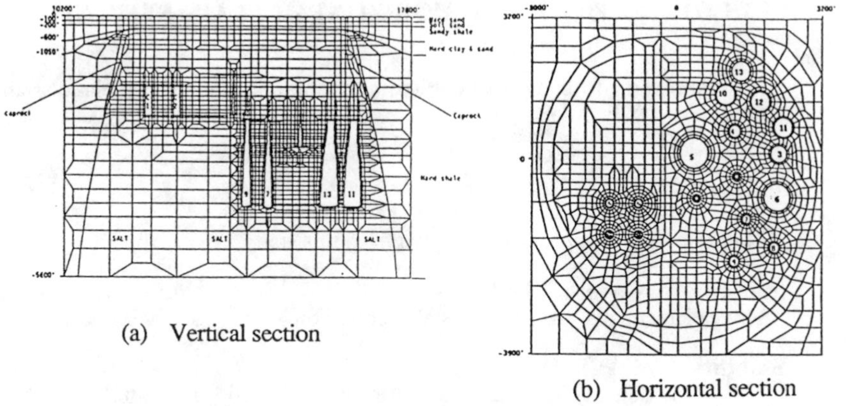

FIG. 1. Close-up of finite element network of 15 - cavern system illustrating the cavern geometry and the geological formations for the vertical and horizontal sections.

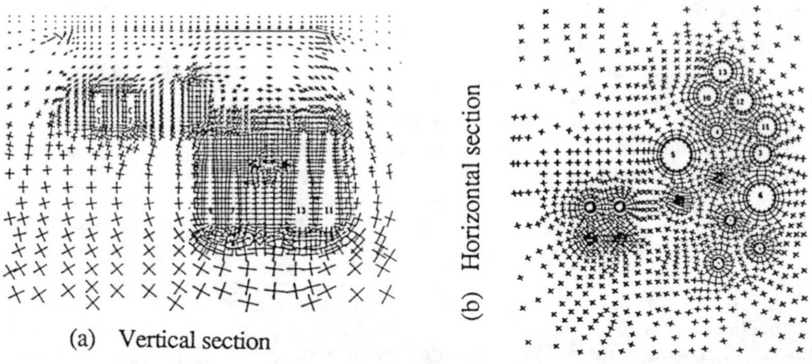

FIG. 2. The principal stress distribution patterns around caverns for vertical and horizontal sections (scales: 1' = 15,000 psi for vertical section and 1" = 20,000 psi for horizontal section at the depth of 2,400 ft).

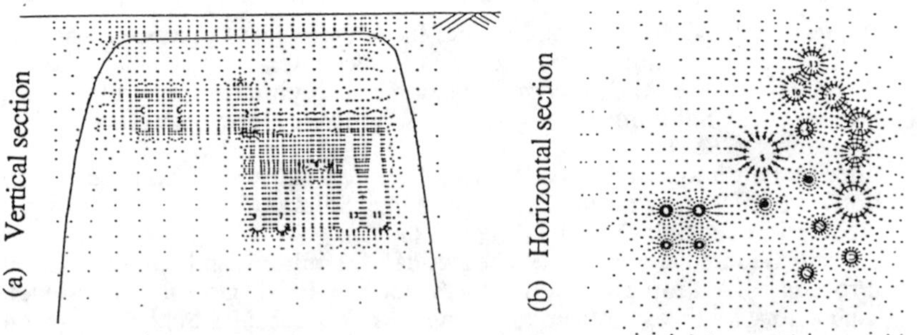

FIG. 3. The principal strain distribution patterns around caverns for vertical and horizontal sections (scales: 1' = 25 % for vertical section and 1" = 2.5 % for horizontal section at the depth of 2,400 ft).

A DIRECT METHOD OF DETERMINING SOIL CONSTITUTIVE EQUATION

Lu Tinghao
Dept. of Agricultural, Hohai University, Nanjing, China

ABSTRACT

In this paper, a strain component yield criterion is illustrated and a new elasto-plastic constitutive model of soil is proposed which are independent of classical yield surface theory. Constant stress ratio path test and verification show that the model developed in this paper can fairly good simulate the actual performance.

INTRODUCTION

Up to now, more than one hundred of soil constitutive models have been developed already. Although a considerable number of models are available, it seems that still more effective model should be developed. This roport proposes strain component yield criterion and a direct method of determining soil constitutive equation in accordance with stress strain functions which are obtained through triaxial isotropic stress test and conventional triaxial drained test. It is not based on classical yield surface theory and no longer has so many assumptions and lengthy deductions as in classical yield surface theory.

STRIN COMPONENT YIELD CRITERION

Under stress increment $\{d\sigma\}$, strain increment $d\varepsilon_i$ of any component of soil element can be represented by

$$d\varepsilon_i = d\varepsilon_i^e + \alpha_i d\varepsilon_i^P \tag{1}$$

Here, $d\varepsilon_i^e$ is elastic strain increment, $d\varepsilon_i^P$ is plastic strain increment when $\alpha_i=1$ and $\alpha_i d\varepsilon_i^P$ is actual plaslic straim increment.

$$0 \leqslant \alpha_i \leqslant 1 \tag{2}$$

α_i is called as loading coefficient which will be discussed in following.

Let ε_{it} and $d\varepsilon_{i\Delta t/2}$ express total strain at t moment and strain increment under half load increment (during $\Delta t/2$ period) respectively. The strain ε'_i at $t+\Delta t/2$ moment is

$$\varepsilon'_i = \varepsilon_{it} + d\varepsilon_{i\Delta t/2} \tag{3}$$

Let ε_{im} records the maximum of strain component ε_i which occurs in its strain history up to moment t, but the original value $\varepsilon_{im}=0$ when $t=0$. Let $d\varepsilon'_i=\varepsilon'_i-\varepsilon_{im}$ and $\alpha_i=d\varepsilon'_i/d\varepsilon_{i\Delta t/2}$.

When $|\varepsilon'_i|>|\varepsilon_{im}|$, plastic strain increment will occur, $0<d\varepsilon'_i\leqslant d\varepsilon_{i\Delta t/2}$. Here, actually there are two cases (1) continuous keeping loaded, as Fig. 1 (a), $d\varepsilon'_i=d\varepsilon_{i\Delta t/2}$, thus $\alpha_i=1$; and (2) reloaded after loaded,

as Fig. 1 (b), $d\varepsilon'_i \leqslant d\varepsilon_{i\Delta t/2}$, then $0 < \alpha_i \leqslant 1$.

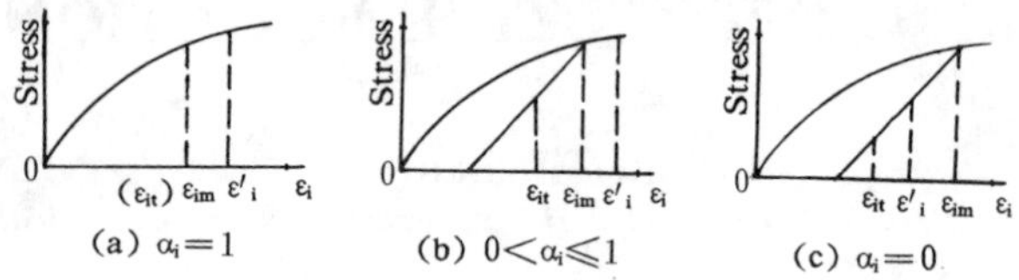

FIG. 1. Loading coefficient α_i

When $|\varepsilon'_i| \leqslant |\varepsilon_{im}|$, as Fig. 1 (c), plastic increment is not developed, we press $d\varepsilon'_i = 0$ and take $\alpha_i = 0$.

As mentioned above, the loading coefficient discussed in strain component yield criterion can simply be summarizied as follows

$$\left.\begin{aligned} d\varepsilon'_i &= \begin{cases} \varepsilon'_i - \varepsilon_{im} & when\ |\varepsilon'_i| > |\varepsilon_{im}| \\ 0 & when\ |\varepsilon'_i| \leqslant |\varepsilon_{im}| \end{cases} \\ \alpha_i &= d\varepsilon'_i / d\varepsilon_{i\Delta t/2} \end{aligned}\right\} \tag{4}$$

CONSTITUTIVE MODEL

In principal stress space, constitutive equation is expressed in followings

$$\{d\varepsilon\} = [C]\{d\sigma\} \tag{5}$$

or $$\{d\varepsilon\} = \{d\varepsilon^e\} + \{\alpha\}\{d\varepsilon^P\} = [C^e]\{d\sigma\} + [\alpha][C^P]\{d\sigma\} \tag{6}$$

In Eq (5) and (6), $[\alpha]$ is called as loading coefficient matrix. It is diagonal matrix and its diagonal element α_i is determined from Eq (4). $[C]$、$[C^e]$ and $[C^P]$ are respectively called as elasto-plastic、elastic and initial plastic seftness matrix. On the assumption that principal strain increment vector corresponds to poincipal stress vector. The constitutive equation can be extended to general coordinate system from principal stress space. That is

$$\{d\varepsilon\} = ([R]^T[C][R])\{d\sigma\} \tag{7}$$

or $$\{d\varepsilon\} = ([R]^T[C^e][R])\{d\sigma\} + ([\alpha][R]^T[C^P][R])\{d\sigma\} \tag{8}$$

Here $$[R] = \begin{bmatrix} l_1^2 & m_1^2 & n_1^2 & 2l_1m_1 & 2m_1n_1 & 2n_1l_1 \\ l_2^2 & m_2^2 & n_2^2 & 2l_2m_2 & 2m_2n_2 & 2n_2l_2 \\ l_3^2 & m_3^2 & n_3^2 & 2l_3m_3 & 2m_3n_3 & 2n_3l_3 \end{bmatrix} \tag{9}$$

Among, l_j、m_j and n_j are the angle cosine between principal stress σ_j vector and coordinate axials x、y and z.

Our target is to determine all elements c_{ij} and c^e_{ij} (i, j=1, 2, 3) in the $[C]$ and $[C^e]$ in principal stress space. Then the matrix $[C^P]$ will be written as

$$[C^P] = [C] - [C^e] \tag{10}$$

The stress increment $d\sigma_1$、$d\sigma_2$ and $d\sigma_3$ exerting on soil element can be devided into (1) $d\sigma_3$ which is equal in all directions, and (2) two deviator stress increments ($d\sigma_1$-$d\sigma_3$) and ($d\sigma_2$-$d\sigma_3$). Then total strain increment can be written by

$$\{d\varepsilon\} = \{d\varepsilon\}_{(1)} + \{d\varepsilon\}_{(2)} \tag{11}$$

Here, $\{d\varepsilon\}_{(1)}$ are $\{d\varepsilon\}_{(2)}$ are respectively obtained by means of triaxial isotropic stress test (stress ratio $\eta=dq/dp=0$) and conventional triaxial drained test ($\eta=3$).

In the triaxial isotropic stress test, the stress strain function is

$$\varepsilon_i = \frac{1}{3}\varepsilon_v = f(\sigma_3) \qquad (i = 1,2,3) \qquad (12)$$

differentiating Eq(12), we can obtain $\{d\varepsilon\}_{(1)}$. In $\{d\varepsilon\}_{(1)}$, the element

$$d\varepsilon_i = \frac{df}{d\sigma_3}d\sigma_3 \qquad (i = 1,2,3) \qquad (13)$$

In conventional triaxial drained test, the stress strain function can be represented as follows

$$\varepsilon_1 = f_1(\sigma_1 - \sigma_3, \sigma_3) \qquad (14)$$

$$\varepsilon_2 = \varepsilon_3 = f_2(\sigma_1 - \sigma_3, \sigma_3) \qquad (15)$$

$\{d\varepsilon\}_{(2)}$ developed by two deviator stress increments $(d\sigma_1-d\sigma_3)$ and $(d\sigma_2-d\sigma_3)$ can be written as

$$\{d\varepsilon\}_{(1)} = [B]\{d\sigma\} \qquad (16)$$

The elements b_{ij} (i, j=1, 2, 3) in [B] are determined as follows

$$b_{11} = \frac{\partial f_1}{\partial(\sigma_1 - \sigma_3)} \qquad (17)$$

$$b_{21} = b_{31} = \frac{\partial f_2}{\partial(\sigma_1 - \sigma_3)} \qquad (18)$$

According to the symmetrical principle of symbols, we have

$$b_{22} = \frac{\partial f_1}{\partial(\sigma_2 - \sigma_3)} \qquad (19)$$

$$b_{12} = b_{32} = \frac{\partial f_2}{\partial(\sigma_2 - \sigma_3)} \qquad (20)$$

Since $d\sigma_3-d\sigma_3=0$ and $\sigma_3-\sigma_3=0$ then

$$b_{13} = b_{23} = b_{33} = 0 \qquad (21)$$

Submitting Eq (17) ~ (21) into Eq (16) and combining Eq (16) and Eq (13), we have

$$\{d\varepsilon\} = [C]\{d\sigma\} \qquad (22)$$

here

$$[C] = \begin{bmatrix} \frac{\partial f_1}{\partial(\sigma_1 - \sigma_3)} & \frac{\partial f_2}{\partial(\sigma_2 - \sigma_3)} & \frac{df}{d\sigma_3} - (\frac{\partial f_1}{\partial(\sigma_1 - \sigma_3)} + \frac{\partial f_2}{\partial(\sigma_2 - \sigma_3)}) \\ \frac{\partial f_2}{\partial(\sigma_1 - \sigma_3)} & \frac{\partial f_1}{\partial(\sigma_2 - \sigma_3)} & \frac{df}{d\sigma_3} - (\frac{\partial f_2}{\partial(\sigma_1 - \sigma_3)} + \frac{\partial f_1}{\partial(\sigma_2 - \sigma_3)}) \\ \frac{\partial f_2}{\partial(\sigma_1 - \sigma_3)} & \frac{\partial f_2}{\partial(\sigma_2 - \sigma_3)} & \frac{df}{d\sigma_3} - (\frac{\partial f_2}{\partial(\sigma_1 - \sigma_3)} + \frac{\partial f_2}{\partial(\sigma_2 - \sigma_3)}) \end{bmatrix} \qquad (23)$$

If unloading and reloading tests are carried, we can also obtain elastic stress strain functions f^e、f_1^e *and* f_2^e and elastic softness matrix $[C^e]$ which are similar to Eq (12), (14) and (15) and Eq (23)

TEST AND VERIFICATION

In order to verify the model, constant stress ratio path tests are carried with compacted clay.

Fig. 2 shows deviator stress q vs. axial stram ε_a and volume strain ε_v vs. axial strain ε_a curves. Dotted curve T is test curve. Curves D、SJ、J and H are respectively the computated curves for Duncan's model[1]、Modified Combridge model[2]、Yin's two yielding surface model[3] and the model developed in this report. It is obvious that the model developed in this paper can better simulate the actual performance.

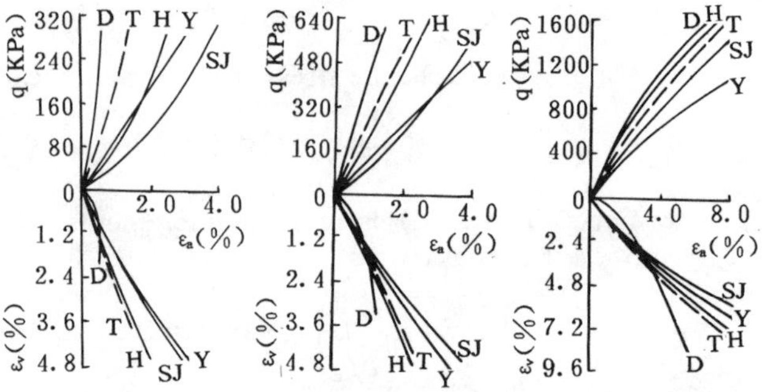

(a) $\eta=0.375$ $P_c=147.1$KPa (b) $\eta=0.75$ $P_c=294.3$KPa (c) $\eta=1.5$ $P_c=441.4$KPa

FIG. 2. Test and verification

CONCLUSION

The loading coefficient α_i is taken as $0 \leqslant \alpha_i \leqslant 1$. It will overcome shortcomings about either $\alpha_i = 1$ or $\alpha_i = 0$. Our method avoids the difficult in classical yield surfuce theory such as yield surface function、plastic potential function and hardening parameters are reasonably determined. Test and verification show that the constitutive model developed in this paper can better simulate the actual performance. The expression of the model does not change with material type、stress path、stress strain function, etc. Its physical and geometry meaning is clear. Although strain component yield criterion and constitutive model of soil are discussed in this paper, yet perhaps they are useful to some other materials.

REFERENCES

1. J. M. Duncan, and C. Y. chang, Nonlinear Analysis of stress and strain in soil, proc. ASCE, Vol. 96, No, SM 5 (1970).
2. K. H. Roscoe, and J. B. Burland, on the Generalized stress-strain Behavior of Wet Clay Engineering Plasticity, ed. J. Heyman and F. A. Leckie, Cambridge University 539-609 (1968).
3. Z. Z. Yin, A Stress-Strain Model with Two yield Surface for Soil, Chinese Journal of Geotechnical Engineering, Vol. 10, No. 4, 64-71 (July 1988).

MODELLING OF ACOUSTIC RESPONSE OF FLUIDS CONTAINING AIR DOUBLE AND SINGLE WEDGE COLUMNS-Shear Relaxation Considered

Bernd Wendlandt
Department of Defence, Materials Research
Laboratory, DSTO, PO Box 50, Ascot Vale,
Victoria 3032, Australia

ABSTRACT

An Eulerian numerical analogue of the wave equation governing the propagation and scatter of dilatation waves in discontinuous media is used to study scatter of acoustic dilatation waves by composites. Visco-elastic relaxation mechanisms are considered. The analogue is used to compute the time evolution of scatter of acoustic dilatation waves in elastic and visco-elastic materials which contain air filled wedges. Echo reduction properties of a composite are calculated.

INTRODUCTION

A second order, implicit and alternating direction Eulerian numerical approximation to the wave equation is used to compute scatter of dilatations from air columns in elastic fluids and from a row of double and single wedge shaped air columns in a visco-elastic medium backed by a layer of air. The scheme considers visco-elastic relaxation due to compression and shear.

MODEL

The propagation of acoustic waves is governed by the laws of conservation of momentum and mass and equation of state which relates the stress in a medium to the associated strain through material properties. The Kelvin model [1] provides the simplest relationship between strain and stress, which, when applied to compression and shear separately, yields a stress-strain relationship which reflects observed relaxation in dilatation and shear. This relationship can be written as $\sigma_{ij} = \lambda\left(1+\beta_\lambda \frac{\partial}{\partial t}\right)\epsilon_{mm}\delta_{ij} + 2\mu\left(1+\beta_\mu \frac{\partial}{\partial t}\right)\epsilon_{ij}$ where σ_{ij} is the stress tensor and ϵ_{ij} is the strain tensor. Material properties are described by λ,the first Lamẽ constant, and μ, the modulus of rigidity. The visco-elastic relaxation times of the material for direct and shear strains are β_λ and β_μ respectively.

When $\lambda \gg \mu$ the propagation of an acoustic excitation in the medium may be described by the dilatation , Θ . Dilatation is defined by the divergence of the incremental displacement of the strain. In regions where the medium is uniform, the dilatation Θ is related to the pressure by $p = -\rho c^2\Theta$, where

ρ is the density and c is the speed of sound characterizing the medium. The propagation of the dilatation can be described by a wave equation

$$\rho \frac{\partial^2 \Theta}{\partial t^2} = \nabla^2 \{\lambda^* + < \mu^* >\} \Theta + < \mu^* > \nabla^2 \Theta \, + Q_s \tag{1}$$

where $\lambda^* = \lambda(1 + \beta_\lambda \frac{\partial}{\partial t})$ and $\mu^* = \mu(1 + \beta_\mu \frac{\partial}{\partial t})$.

The Q_s describes the scatter of dilatation waves by material inhomogeneities and can be written as $Q_s = (\nabla < \mu^* >) \cdot \nabla \Theta - \frac{\nabla \rho}{\rho} \cdot [\nabla(\lambda^* + < \mu^* >) \Theta + < \mu^* > \nabla \Theta]$. The $< \mu^* >$ represents the average value of μ^* over a small volume of computational cell of the numerical analogue of Equation 1. This approximation is justified through the integral form of the divergence theorem [1]. The Q_s is a source term for the generation of dilatation wavelets at material discontinuities [1]. Shear losses dominate when $\mu \beta_\mu \frac{\partial}{\partial t} \gg \lambda \beta_\lambda \frac{\partial}{\partial t}$.

NUMERICAL ANALOGUE OF WAVE EQUATION

Equation 1 assumes that the material parameters are differentiable with respect to spatial coordinates. Their use in discontinuous media may require special consideration of boundary conditions to avoid any smearing of material interfaces. However, very acceptable solution schemes for wave equations have been obtained, [2] and [3], which do not require special treatment of boundary conditions and are able to trace the propagation of acoustic Gaussian pulses and wave-packets through water-elastomer-air-elastomer-steel-water sandwiches to an accuracy of 95 to 98 %.

Using second order, centered differences for the derivatives of Equation 1 a numerical analogue was written in an implicit formulation, which can readily be integrated using the unconditionally stable alternating direction technique, ADI [4]. The ADI time steps the analogue equation to Equation 1 implicitly, first along one space dimension , leaving all terms involving differentiation along the other space dimension at the previous time level temporarily to provide an interim approximation to Θ. This process is repeated along the other space dimension. An updated value for Θ is considered to have been obtained after the completion of this second step [4].

CASE STUDIES

The scheme was used to calculate the acoustic response of rectangular, cylindrical and double wedge air columns in water to Gaussian and five wavelength sinusoidal wave-packets. To indicate a practical use of this scheme, the acoustic echo of an elastomer layer, filled with double and single air columns and backed by air, was calculated, assuming $\lambda \gg \mu$.

The acoustic responses of the cavities to a Gaussian shaped wave-packet, are shown in scalar, or modulus, form of the dilatation in Figures 1 - 3. The Gaussian pulse is incident from the left in the figures and the time snapshots show the column responses when the pulse has passed the row of air filled cavities, C. The figures show a reflected component moving to the left and the part of the pulse not affected by the cavity to the right. A component which has been slowed at the surface of the cavities through surface relaxation is trailing the main portion of the pulse as expected from surface wave theory [1]. The

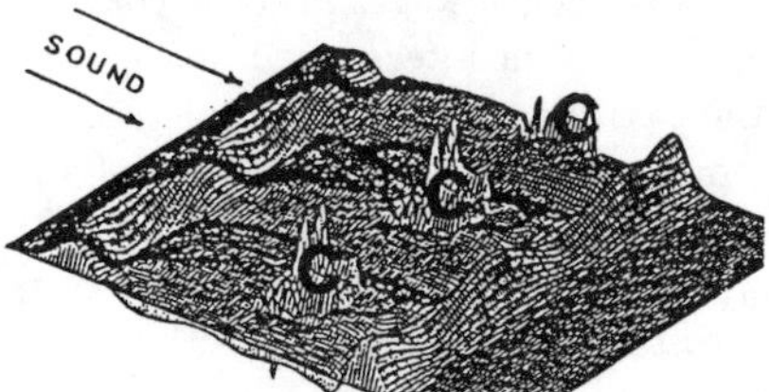

Figure 1: Scatter of acoustic pulse by square air columns in water

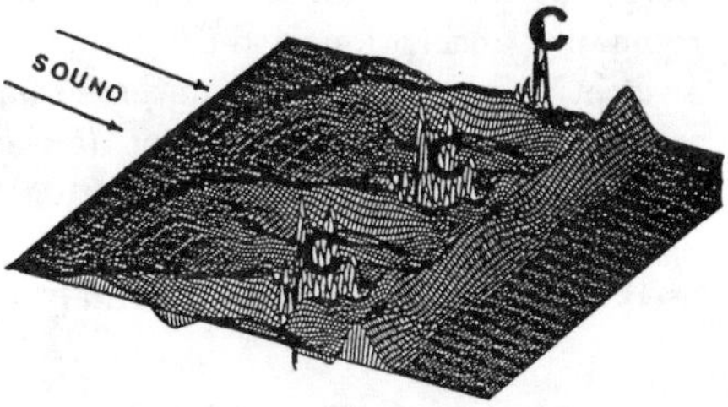

Figure 3: Scatter of acoustic pulse by air double wedge columns in water

response of a single double wedge shaped air column in water to a Gaussian pulse is shown in Figure 4. The pulse is travelling from lower right to upper left and has passed the double wedge. The primary reflection has moved away from

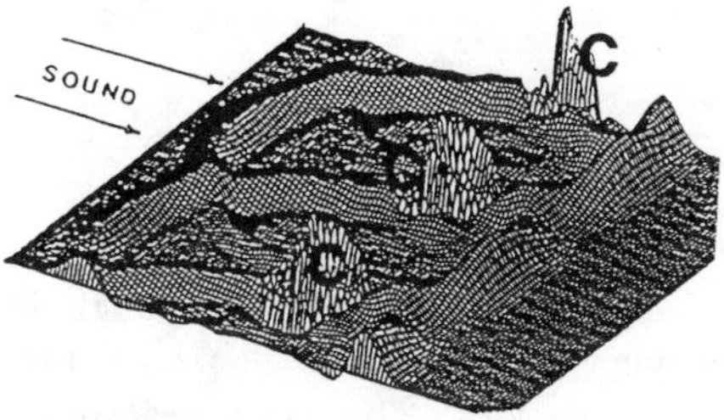

Figure 2: Scatter of acoustic pulse by air cylinders in water

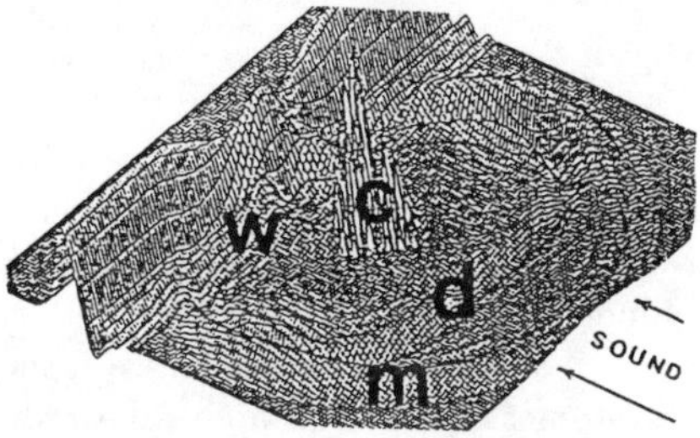

Figure 4: Monopole, **m**, dipole, **d**, and wavelet , **w**, emission by air double wedge

the air column and is followed by the dipole reflection. The corner at the back of the wedge generates a wavelet as is expected from Q_s and theory of surface waves [1].

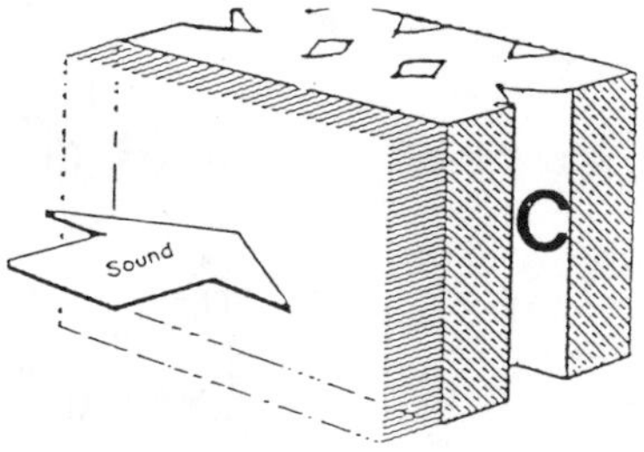

Figure 5: Elastomer/air-wedge layer

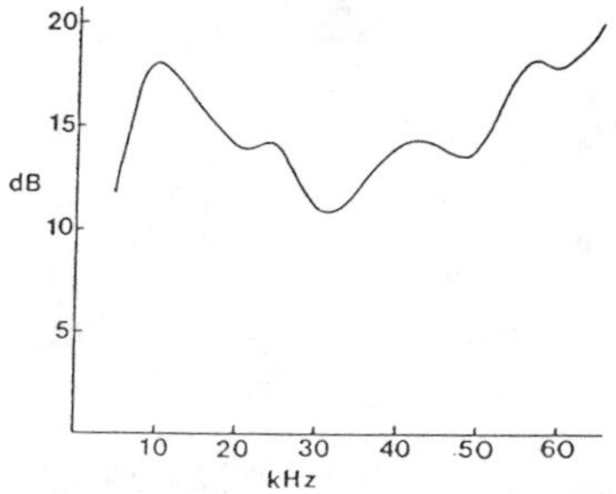

Figure 6: Echo of elastomer/air-wedge layer

Acoustic echo calculations for a selected frequency band of a row of air filled wedges immersed in an elastomer layer loaded with water, Figure 5, are shown in Figure 6. The shape, size and spacing of the double and single wedges were chosen to maximize the scatter of short wavelength waves into the coating, resonance absorption of medium wavelength waves in inter-wedge spaces and absorption of longer wavelengths around the cavities.

The reflection coefficient R was computed from the amplitude of the standing wave Θ_t generated in front of the layer and the amplitude of the incident wave Θ_i through $R = -20\log_{10}\{(\Theta_t - \Theta_i)/\Theta_i\}$. Layer parameters used are shown in Table I. Analysis indicates that the echo reduction peak

Table I. MATERIAL PROPERTIES AND GEOMETRY OF ELASTOMER LAYER

Material	Density	Lamé constant	Loss Tangent
	kg/m^3	kg/sm^2	$\omega(\lambda\beta_\lambda + \mu\beta_\mu)/(\lambda+\mu)$
water	1000	2.25×10^6	0.0
air	1.29×10^{-3}	14.0	0.0
elastomer	1130	2.54×10^6	0.650
Thickness of layer	$0.050m$	Height of double wedge	$0.020m$
Width of wedge	$0.015m$	Separation of wedges	$0.015m$
ω	: angular	excitation frequency	

at $10kHz$ may be due to monopole resonance, or resonance of circumferential waves around the air column, and the small peak at $22kHz$ may be due to a dipole resonance of the double wedge air column. The peak at $42kHz$ might be due to a resonance set up between the double wedge and the single wedge of the elastomer layer. The high echo reduction properties of the generic elastomer layer for frequencies greater than $50kHz$ may be attributed to scattering and absorption within the elastomer.

The wave-equation describing the propagation and scatter of a dilatation through piecewise continuous elastic and visco-elastic media has been presented. The numerical analogue is able to model wave/material interface and discontinuity interactions, and address problems of practical interest in noise engineering.

References

[1] A. Sommerfeld, Mechanics of Deformable Bodies,(Academic Press, New York, 1964).

[2] D. L. Brown, " A Note on the Numerical Solution of the Wave Equation with Piecewise Smooth Coefficients", Math. Comp.,42, (166), 369-391 (April 1984).

[3] B. C. H. Wendlandt "The Acoustic Properties of Layered Coatings",MRL Report, MRL-R-1034,(DSTO, March 1988).

[4] P. Roache Computational Fluid Dynamics, (Hermosa Publishers, Alberquerque,1982).

A TANGENT STIFFNESS APPROACH TO LARGE DEFORMATION INELASTIC PROBLEMS WITH A GENERALIZED MID-POINT RULE

NICHOLAS ZABARAS and ABUL FAZAL M. ARIF
Department of Mechanical Engineering
University of Minnesota,
111 Church Street S.E.
Minneapolis, MN 55455

ABSTRACT

A two parameter family of incrementally objective schemes for the integration of elasto-viscoplastic equations in large deformation analysis is given. Full linearization of the principle of virtual work in an updated Lagrangian framework and a calculation of the consistent material moduli are performed. The method is demonstrated with three examples for integration parameters which lead to the radial return method and the logarithmic strain increment.

INTEGRATION SCHEME AND VARIATIONAL FORMULATION

The main task associated with an updated Lagrangian analysis of large deformation inelastic problems consists of the constitutive part and the computation of the deformation gradient through the solution of equilibrium equations. In the constitutive part, one calculates the material state (stress and state variables) at the end of a time step with the material state and the relative deformation gradient given at the beginning of the step. The second part computes the deformation field with a proper development of a Newton-Raphson scheme from the linearization of the principle of virtual work and includes the calculation of the consistent material moduli.

In this work rate dependent hypoelastic-viscoplastic models with an internal scalar state variable s are assumed. Following familiar notation these unified viscoplastic models are briefly described as

$$\overset{\Delta}{\mathbf{T}} = \mathcal{L}^{e}[\,\mathbf{D} - \mathbf{D}^{p}\,] \tag{1}$$

$$\mathbf{D}^{p} = \sqrt{\frac{3}{2}}\,\dot{\tilde{\varepsilon}}^{p}\,\mathbf{N}^{p}(\mathbf{T}',\tilde{\sigma}) \text{ where } \mathbf{N}^{p}(\mathbf{T}',\tilde{\sigma}) = \sqrt{\frac{3}{2}}\,\frac{\mathbf{T}'}{\tilde{\sigma}} \tag{2}$$

$$\dot{\tilde{\varepsilon}}^{p} = f(\tilde{\sigma}, s) \qquad \dot{s} = g(\tilde{\sigma}, s) = h(s)\,\dot{\tilde{\varepsilon}}^{p} \tag{3}$$

where the Jaumann rate in (1) is defined using as material spin $\mathbf{W}$ the antisymmetric part of the velocity gradient.

To derive the generalized mid-point integration rule, one can write the following objective equations (from now on bar quantities denote rotational-neutralized quantities with rotation

$\mathbf{Q}(t)$ defined as the solution of $\dot{\mathbf{Q}}(t)\,\mathbf{Q}(t)^T = \mathbf{W}(t)$ with $t_n \le t \le t_{n+1}$ and $\mathbf{Q}(t_n) = \mathbf{I}$)

$$\bar{\mathbf{T}}_{n+1} = \mathbf{T}_n + \mathcal{L}^e[\,\Delta\bar{\mathbf{E}} - \Delta\bar{\mathbf{E}}^p\,] \tag{4}$$

$$s_{n+1} = s_n + g(\tilde{\sigma}_{n+\beta}, s_{n+\beta})\;\Delta t \tag{5}$$

where the strain increments $\Delta\bar{\mathbf{E}}$ and $\Delta\bar{\mathbf{E}}^p$ are defined as

$$\Delta\bar{\mathbf{E}} = \bar{\mathbf{D}}_{n+\alpha}\Delta t \quad \text{and} \quad \Delta\bar{\mathbf{E}}^p = \bar{\mathbf{D}}^p_{n+\beta}\Delta t \tag{6}$$

with $0 \le (\alpha,\beta) \le 1$, and, $\bar{\mathbf{D}}_{n+\alpha}$ and $\bar{\mathbf{D}}^p_{n+\beta}$ denoting the total and plastic rates of deformation at some intermediate (to B_n and B_{n+1}) configurations $B_{n+\alpha}$ and $B_{n+\beta}$, respectively.

The interpolation scheme for the calculation of $\bar{\mathbf{D}}_{n+\alpha}$ is derived with the following approximations

$$\bar{\mathbf{x}}_{n+\alpha} = (1-\alpha)\bar{\mathbf{x}}_n + \alpha\bar{\mathbf{x}}_{n+1} \text{ and } \bar{\mathbf{x}}_{n+1} = \mathbf{x}_n + \Delta t\,\dot{\bar{\mathbf{x}}}_{n+\alpha} \tag{7}$$

One can then show that $\bar{\mathbf{F}}_{n+1} = \mathbf{U}_{n+1}$, $\mathbf{Q}_{n+1} = \mathbf{R}_{n+1}$, and that

$$\Delta\bar{\mathbf{E}} = \left(\mathbf{U}_{n+1} - \mathbf{I}\right)\bar{\mathbf{F}}^{-1}_{n+\alpha} \tag{8}$$

The above approximation gives a family of incrementally objective integration schemes [1]. Also note that for $\alpha = 0.5$, $\Delta\bar{\mathbf{E}}$ becomes a Padé approximation of the logarithmic strain increment $\ln(\mathbf{U}_{n+1})$.

Full linearization of the virtual work principle leads to the following incremental equation [1]

$$dG = \int_{B_n} \Big\{ \mathbf{R}_{n+1}\bar{\mathcal{M}}^{ep}[d\Delta\bar{\mathbf{E}}]\mathbf{R}^T_{n+1} + (d\mathbf{R}_{n+1}\mathbf{R}^T_{n+1})\mathbf{T}_{n+1} + \mathbf{T}_{n+1}(\mathbf{R}_{n+1}d\mathbf{R}^T_{n+1})$$

$$- \mathbf{T}_{n+1}\mathbf{F}^{-T}_{n+1}d\mathbf{F}^T_{n+1} + \mathrm{tr}(d\mathbf{F}_{n+1}\mathbf{F}^{-1}_{n+1})\mathbf{T}_{n+1}\Big\} \cdot \left\{\frac{\partial\tilde{\mathbf{u}}}{\partial\mathbf{x}_n}\mathbf{F}^{-1}_{n+1}(\det\mathbf{F}_{n+1})\right\} dV$$

$$- \; d(\text{forcing terms}) \; = \; 0 \tag{9}$$

where $\bar{\mathcal{M}}^{ep}$ are the consistent linearized moduli. For $\beta = 1$, these moduli take the form corresponding to the radial return method. In the following examples $\alpha=0.5$ and $\beta=1.0$ are used in the integration scheme. Problems with other α and β values are investigated in [1].

NUMERICAL RESULTS

The incrementally objective time-integration scheme and the global jacobian described in this paper are implemented and tested via several examples using two different types of rate-dependent constitutive models. The first model used was suggested by Perzyna and has the first-order power law form, whereas the second one is a single scalar internal state variable type of model due to Anand. The model parameters and the other material properies used in this work are the same as those in [1].

The objectivity of the proposed integration scheme with large rotation effects is first verified by solving a simple shear problem with superimposed rigid rotation. A total shear strain of 0.2 and 360 degrees of rotation are applied in 1, 4 and 10 equal increments using Perzyna's model. The results shown in Fig.1 are in excellent agreement with the analytical solution of a rate-independent elastic-perfectly plastic material. In Fig.1, the normalized stress is defined as $(\bar{T}_{12}/(s_o/\sqrt{3}))$

The next example is the 60% upsetting of an axisymmetric billet of 20mm in diameter and 30mm high. Perzyna's model is used and complete sticking between the die and the workpiece is assumed. The loading is applied by prescribing the displacement of the rigid die. The present algorithm is used with very large incremental steps and the results are in good agreement with those reported in [2]. The undeformed and deformed mesh with nine increments are shown in Fig.2(a) and the load versus deflection curves with step sizes of 6.77%, 10%, 15% and 30% reduction in height per step are given in Fig.2(b).

Finally, extrusion of a cylindrical bar through a smooth curved quintic shaped die with a reduction of cross-sectional area of 25% is solved using Anand's model. The total initial length of the billet was 4.2 times the initial radius R_o. The reduction in area is achieved over a length of $1.2R_o$. An automatic time stepping algorithm is employed such that the maximum increment in equivalent plastic strain is not greater than 0.05. The undeformed and deformed meshes at steady-state are shown in Fig.3(a). Fig.3(b) shows the load versus applied end displacement curve and the normalized longitudinal stress distribution is given in Fig.3(c).

CONCLUSIONS

In this work, a tangent operator has been proposed with the resulting iteration procedure always referring to a converged state. Numerical examples demonstrate the capability of this approach to handle very large incremental steps for simple problems where step size is not restricted by curved boundaries or very stiff constitutive equations. Further research for this type of constraint problems is recommended.

REFERENCES

1. N. Zabaras and A.F.M. Arif, "A family of integration algorithms for constitutive equations in finite deformation elasto-viscoplasticity", Int. J. Numer. Meths. Engrg., Submitted for publication (1990), (Also University of Minnesota Supercomputer Institute Research Report, UMSI 90/56, April (1990)).

2. L. M. Taylor and E. B. Becker, "Some computational aspects of large deformation rate-dependent plasticity problems", Comp. Meths. Appl. Mechanics and Engrg., 41, 251-277 (1983).

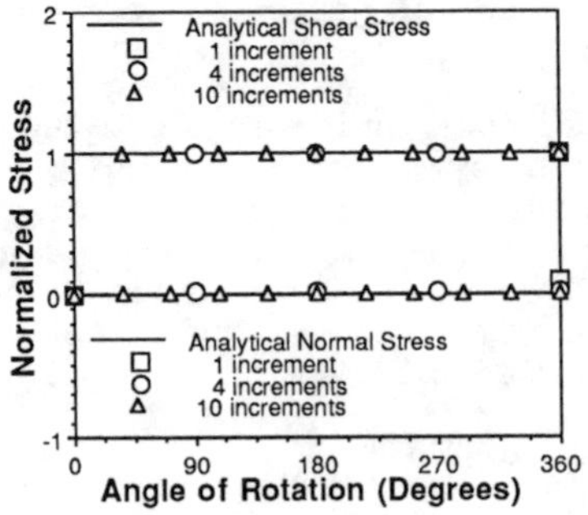

Fig.1. Normalized stress vs angle of rotation.

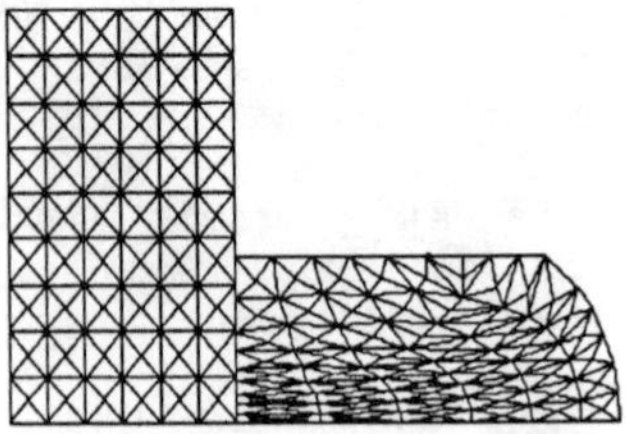

Fig.2(a). Undeformed and deformed shape.

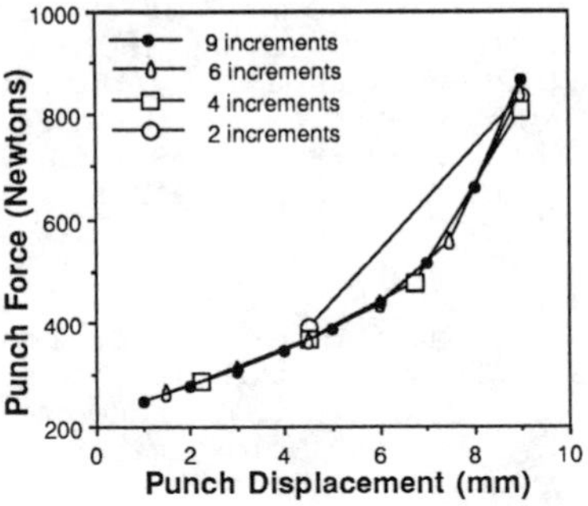

Fig.2(b). Force vs displacement for upsetting.

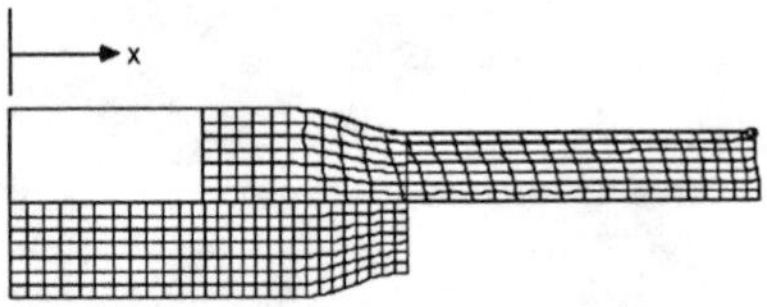

Fig.3(a). Undeformed and deformed configuration.

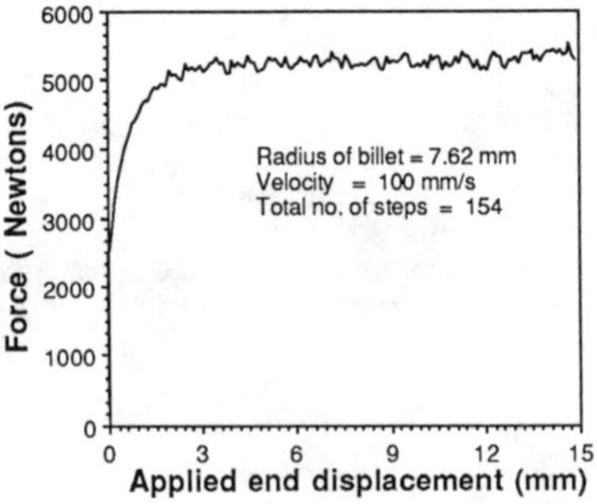

Fig.3(b). Force vs displacement curve for extrusion.

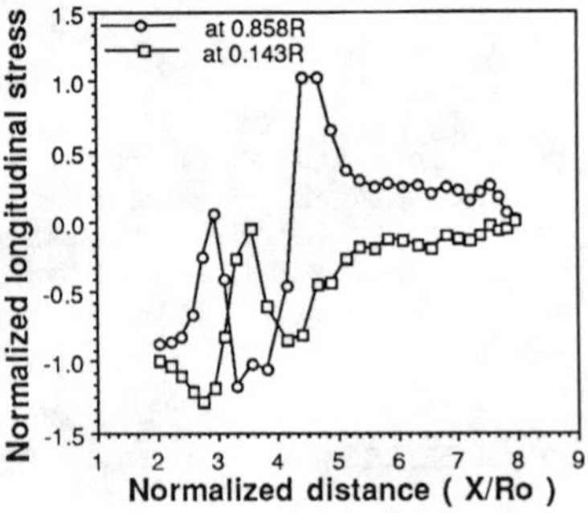

Fig.3(c). T_{xx}/s_o vs x/R_o.

EVALUATION OF HIERARCHICAL MODELS USING PUEBLO SAND DATA

M. ZAMAN, M.O. FARUQUE AND R. TABBAA
The University of Oklahoma, Norman, Oklahoma

ABSTRACT

The single surface hierarchical models (δ_0 and δ_1) developed by Desai and co-workers are employed to characterize the stress-strain and volumetric response of Pueblo sand under various stress paths. Microcomputer-based codes are used to evaluate the associated material constants and to predict stress-strain response. Relative strengths and weaknesses of each model are discussed.

INTRODUCTION

Single surface hierarchical models have been developed by Desai and co-workers [1-5]. Evidently these models are powerful in characterizing the stress-strain response of geomaterials and plain concrete. This paper addresses an application of the δ_0 (based on associated plasticity theory) and the δ_1 (based on non-associated plasticity theory) models in predicting stress-strain and volumetric responses of a granular soil obtained from Pueblo, Colorado. The experimental data used in this study were obtained by Munster [5] from a series of laboratory tests using a cubical device. A microcomputer aided approach is utilized to evaluate the associated material constants. Four categories of tests are considered in the back-prediction exercise: (i) hydrostatic test, (ii) shear tests without loading and unloading response, (iii) shear tests including loading and unloading response, and (iv) shear tests not included in the evaluation of material constants. Pertinent observations are discussed based on the comparison of predicted response and experimental data.

EVALUATION OF MATERIAL CONSTANTS AND BACK-PREDICTION

Details of the δ_0 and δ_1 models are given by Desai et. al. [1,3,5]. The δ_0 model includes two elastic constants, three constants (γ,β and m) associated with the failure state of a material, two constants (a_1 and η_1) associated with the hardening or history of inelastic deformation and one constant (n) related to transition between the compressive and dilative states of a material. For the δ_1 model, an additional constant (κ) appears that accounts for the non-associative response.

To facilitate evaluation of these material constants, Desai et. al. developed computer programs for mainframe systems. In this study, these codes were modified and implemented on microcomputers (IBM PC-XT or equivalent). The code (HIPAR) has provision to accept some or all of the material constants calculated externally by the user. For example, the user may decide to evaluate the ultimate constants (γ, β and m) by hand, use these as input and let the code evaluate the other constants. Also, the code provides warning when a given parameter exceeds certain limiting values. A general purpose computer program, LOTUS 123, is used for plotting and graphical representation of input data and predicted response. This capability greatly facilitates checking the accuracy of input data by obtaining an echo of the input data in terms of stress-strain and volumetric plots. The computer program HIPRE back-predicts the stress-strain and volumetric responses of a material using the δ_0 and δ_1 models. The code generates plots by overlayering predicted response on experimental data for the purpose of comparison. The δ_0 or δ_1 model is accessed by specifying the appropriate value of the control parameter (JNON). Detail flow charts, description of variables and program documentation are given by Tabbaa [6].

COMPARISON OF PREDICTED RESPONSE AND EXPERIMENTAL DATA

The test data used in this study were obtained from the work of Munster [4]. The tests were conducted on a granular soil using a fluid-cushion cubical device under the supervision of Prof. Desai, University of Arizona. The granular soil has a uniformity coefficient of 3.2, specific gravity of 2.59, optimum moisture content of 9% at a maximum density of 139.2 pcf and was obtained from the Urban Mass Transportation Authority (UMTA) Test Section, Colorado. Test data for some selected stress paths (CTC, CTE, HC, TC, TE and SS) [4] are used here.

Using the test data for CTC (10, 15, 20, 25), CTE (10, 20), RTE (10), RTC (20), TC (25), and TE (20) tests, where numbers in the parantheses indicate the confining pressures in psi at which the tests were conducted, and the computer code HIPAR, the following material constants were obtained: Elastic constants (E,)ν: $E = 1.5 \times 10^4$ psi, $\nu = 0.4$; Ultimate Behavior Constants (γ, β, m): $\gamma = 0.05$, $\beta = 0.9$, $m = -0.5$; Hardening Constants: $a_1 = 3.3$, $\eta_1 = 0.5$; Phase Change Constant: n = 3.3; Nonassociative Constant: $\kappa = 0.19$.

A comparison between the predicted and experimental response for the CTC (10) test is presented in Fig. (1a). The predicted stress-strain response for both models closely approximates the experimental values up to a deviatoric stress of approximately 14 psi. At higher stress levels, the predicted strains by the δ_1 model show better correlation with experimental data in the compression (vertical) direction and underestimate the experimental values in the extension (transverse) directions. An opposite trend is observed for the δ_0 model. The predicted failure stress is larger than the experimental value. A similar comparison of the volumetric-axial strain response is presented in Fig. (1b). Unlike the stress-strain response in Fig. (1a), the volumetric response is predicted well by the δ_1 model. The onset of dilatancy, as predicted by the δ_0 model occurs much earlier than experimentally observed. Also, the magnitude and the rate of dilation for a given axial strain are much larger than the corresponding measured values. Figs. (2a) and (2b) show a similar comparison for the CTC (15) test. The correlation with experimental data is significantly lower in this case, particularly for the volumetric-axial strain response. The experimental data for volumetric response appear to be inconsistent because the specimen did not exhibit any dilation although the confining pressure (15 psi) was higher than in the previous case (10 psi). Also, the test was terminated at a very low strain level (less than 1% axial strain) for some reason. Therefore, failure stresses determined from extrapolation of experimental data may be subject to error and personal judgment. The model predicted volumetric strains are much larger than the experimental values.

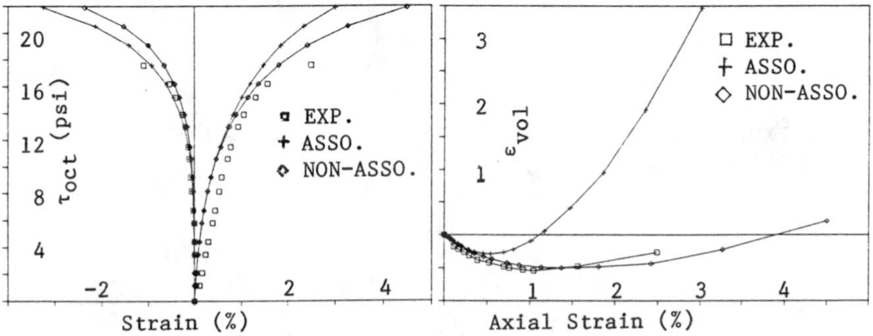

(a) Stress-strain response (b) Volumetric response

FIG. 1 Conventional triaxial compression (CTC 10) test

Comparison of stress-strain and volumetric-axial strain responses (δ_0 model) for an extension test (conventional triaxial extension) at 10 psi confining pressure, including unloading-reloading sequences, is shown in Figs. (3a) and (3b), respectively. A comparison of the δ_0 and the δ_1 models for volumetric response is presented in Fig.4. From these figures it is evident that the stress-strain responses are predicted well, particularly at low stress levels. Unloading-reloading features are captured reasonably well. Fig. (4) shows that both models predict dilatancy, but the experimental data do not exhibit such behavior.

A comparison between the predicted and experimental mean pressure-volumetric strain values is shown in Fig. (5). The experimental data display a much stiffer response than the model predictions. This phenomenon is a possible result of the hyperbolic form of the hardening function (α) used [1-3].

Also, only two parameters (a_1 and η_1) are used to define the hardening function, which may not be enough to control all of the important features related to hardening.

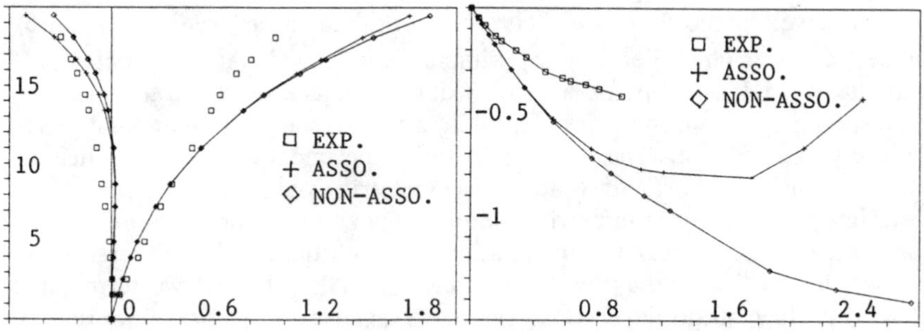

(a) Stress-strain response (b) Volumetric response

FIG. 2 Conventional triaxial compression (CTC 15) test

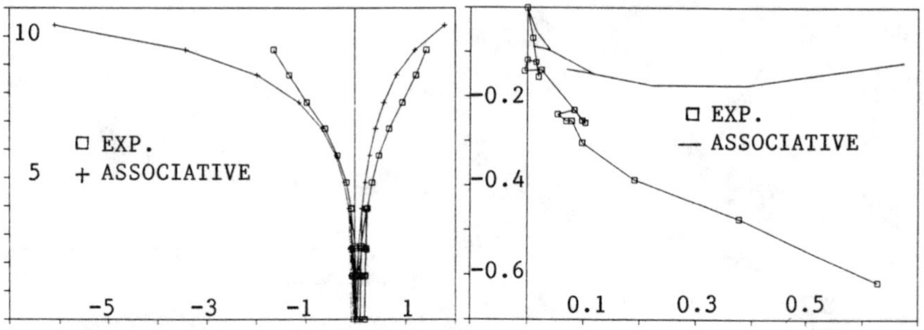

(a) Stress-strain response (b) Volumetric response

FIG. 3 Conventional triaxial extension (CTE 10) test by δ_0 model

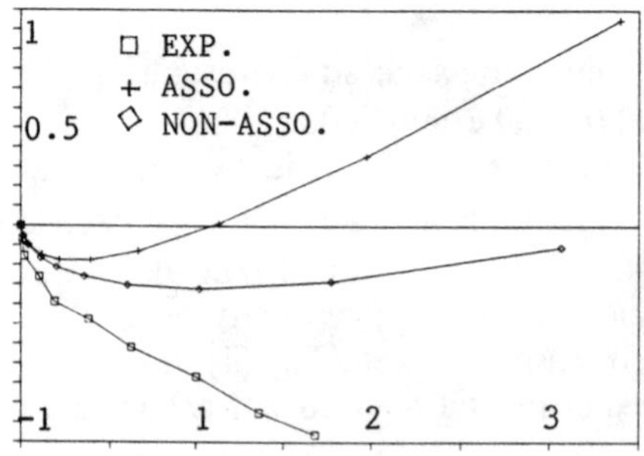

FIG. 4 CTE (10): volumetric response by δ_0 and δ_1 models

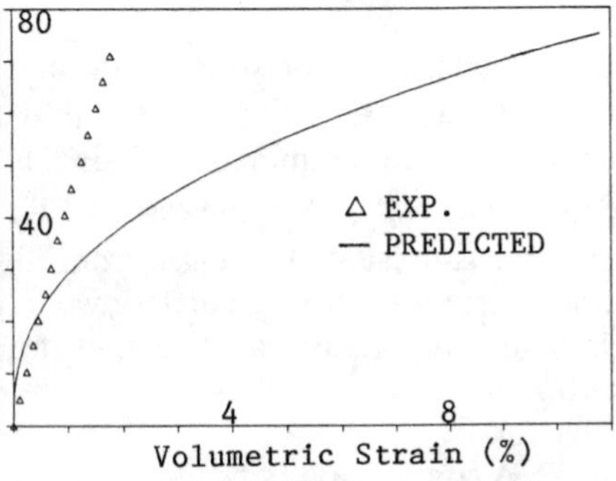

FIG. 5 Hydrostatic compression (HC) test

Similar comparisons of predicted response and experimental data were made [6] for other stress paths, but the results could not be presented here due to space limitation.

CONCLUDING REMARKS

An attempt was made in this study to employ the well known single surface hierarchical (δ_0 and δ_1) models developed by Desai and co-workers in describing the stress-strain and volumetric behavior of a granular soil that was obtained from Pueblo, Colorado. Microcomputer-based codes were used to evaluate the associated material constants and to back-predict the stress-strain and volumetric responses. A general purpose program, LOTUS 123, was employed to facilitate checking of the input data and graphical representation of results.

From comparison of the predicted response and experimental data it is evident that both δ_0 and δ_1 models predicted the stress-strain responses fairly well. The δ_1 model predicted the volumetric response in the dilative regime much better than the δ_0 model, an indication that the Pueblo sand exhibits nonassociative behavior. Substantial difference was observed in predicting the stress-strain response for the hydrostatic compression stress path. Overall, the hierarchical models provide a useful means of characterizing stress-deformation and volumetric behavior of granular materials.

ACKNOWLEDGEMENTS

The authors are thankful to Prof. C.S. Desai, University of Arizona for providing the original computer codes. Also, his permission to use the experimental data is gratefully acknowledge.

REFERENCES

1. C.S. Desai and M.O. Faruque, "Constitutive model for (geological) materials", J. Eng. Mech., ASCE, 110 (9), 1391-1408 (1984).
2. C.S. Desai, H.M. Galagoda and G.W. Wathugala, "Hierarchical modelling for geologic materials and discontinuities - joints, interfaces", Proc. 2nd Intl. Conf. on Constitutive Laws for Eng. Materials, (I) 81-94 (1987).
3. C.S. Desai and Q.S.E. Hashmi, "Analysis, evaluation, and implementation of a nonassociative model for geologic materials", J. of Plasticity, 5, 397-420 (1989).
4. C.L. Munster, "Constitutive modelling of a granular soil under three-dimensional states of stress", M.S. thesis, Virginia Tech., Blacksburg (1981).
5. G. Frantziskonis, C.S. Desai and S. Somasundaram, "Constitutive model for nonassociative behavior", J. Eng. Mech., ASCE (1986).
6. R. Tabbaa, "A comparative study of some selected plasticity models for granular soil", M.S. thesis, Univ. of Oklahoma, Norman (1990).

TESTING, PARAMETER IDENTIFICATION

EFFECT OF DENSITY ON DAMPING AND SHEAR MODULUI OF COHESIONLESS SOILS USING TWO NEW TRANSFER FUNCTION ESTIMATORS

FARSHAD AMINI
Assistant Professor, Civil Engineering Department
University of the District of Columbia, Washington, D.C. 20008

ABSTRACT

In this research, random torsional loading, in addition to the conventional sinusoidal loading were used to study dynamic soil properties using a resonant column device and an FFT analyzer. Two improved estimators of the transfer function (H3 and H4) were utilized. The purpose of the additional work contained in this paper is to compare the soil properties obtained by H3 and H4 with those evaluated using the conventional estimator, H1, at variable densities.

INTRODUCTION

The two primary dynamic soil properties, i.e. damping and shear modulus, for response calculations are usually evaluated using sinusoidal loading. In order to obtain results that are more representative of actual earthquake conditions, random loading was used in this study. The improved transfer function estimator as well as the conventional estimator, were utilized during random vibration analysis.

Measured transfer functions are usually used for frequency domain identification of the dynamic characteristics of structural systems under random loading conditions. With the advent of the Fast Fourier Transform (FFT) Analyzer, several techniques have emerged for the estimation of the transfer functions. By converting many samples of the measured data from the time domain into the frequency domain using appropriate windowing functions, the transfer function is obtained. Many researchers have indicated that the sole selection of the conventional estimator (H1) be seriously reconsidered as the source of the transfer function estimator ([1] through [3]). As a result, new techniques for transfer function estimation have been developed. This paper evaluates dynamic soil properties using H3 and H4 as well as H1 at different soil densities and variable strain levels.

THEORY

In addition to the conventional estimator of the transfer function, H1, three other estimators (H2, H3, and H4) have been recently developed. H1 is defined as the ratio of the cross spectral density of the input to the output. This is the method presently being used by the most manufactures of FFT machines. H2 is the ratio of the autospectral density of the output to the cross spectral density of the input, and this estimator is preferable at the resonance. H3 is the arithmetic mean of H1 and H2, and though it is more uniformly biased than H1 and H2, it suffers from double noise contamination [1].

Fabunmi [2], making use of the fact that H1 and H2 are the least square solutions of the true transfer function which minimizes the output and the input noise respectively, combined both estimators through a pseudoinverse transform to yield an estimator H4 which closely approximates the behavior of H1 at antiresonance and H2 at resonance. A theoretical study of the nature of the weighing function for this estimator has been shown by Fabunmi and Tasker [3].

EXPERIMENTAL PROCEDURE AND COMPUTER ANALYSIS

In the sinusoidal loading test, the input, output, and the resonant frequency of the system were recorded for different excitation levels. In the random test, both the input and the output signals were connected to an FFT analyzer for analysis. The number averages, resolution, sensitivity and the proper weighing function were then selected. In this research, a microcomputer was used for the purpose of obtaining the new estimators of the transfer function (H3 and H4). Two types of cohesionless soils, Ottawa 20-30 and Ottawa 30-50 sand, were utilized. Soil densities used in this study ranged from approximately 1650 to 1170 kg/cum. A schematic of the equipment used in this study is shown in Figure 1. Methods and procedures for testing and analysis, as well as concept of root mean square (rms) strain are described in references [4] through [9].

RESULTS AND CONCLUSION

The purpose of this research was to study the effect of soil density on dynamic soil properties using improved transfer function estimators. The behavior of the improved estimators at different densities for damping and shear modulus are shown in Figures 2 and 3, respectively. As soil densities decreased, the difference in damping values obtained by the different estimators increased. This is because the effect of the input noise is more important at lower densities and therefore the various estimators

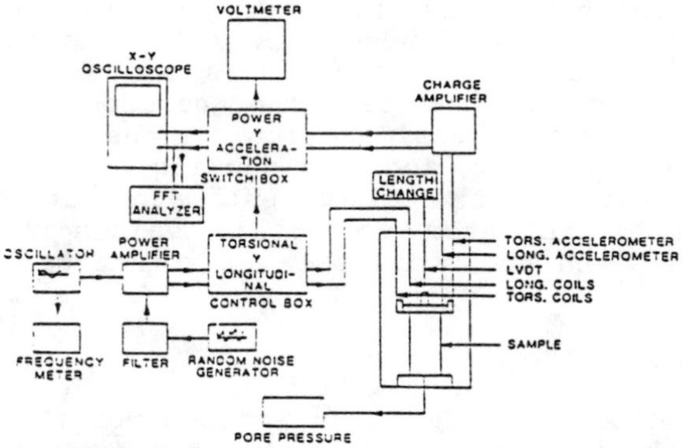

FIG. 1. Schematic of the equipment used.

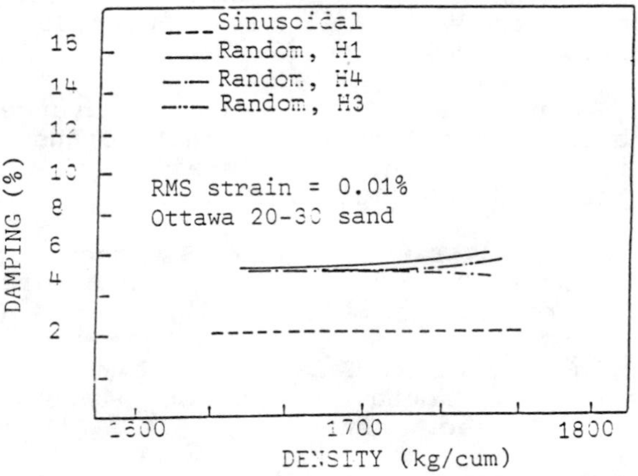

FIG. 2. Effect of soil density on damping using different estimators

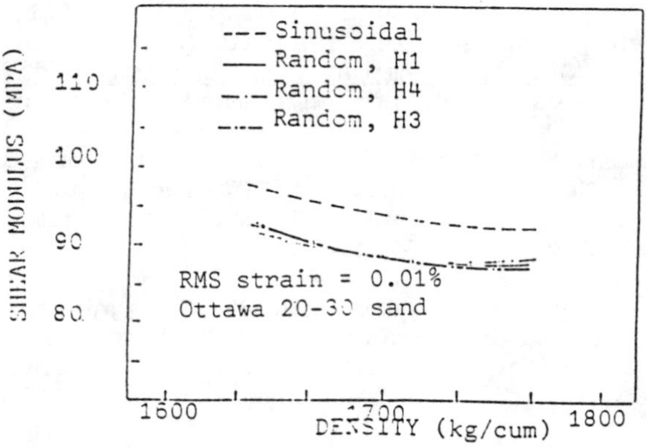

FIG. 3. Effect of soil density on shear modulus using different estimators.

behaved somewhat differently. In addition, the change in densities did not influence the relationship between the natural frequencies and thus the shear moduli obtained by the different estimators. The results of this study indicated that the consideration for the variation in soil density is important, and that damping values during random loading using the conventional estimator, at low densities, may need to adjusted to obtain values that are more consistant with the modal parameter estimation schemes.

REFERENCES

1. L. D. Mitchell, "Improved methods for the Fast Fourier Transform (FFT) calculation of the frequency response function", Transaction of the ASME, J. of Mech. Design, 104, 277-290 (April 1982).

2. J. A. Fabunmi, "Developments in helicopter ground vibration testing", J. Amer. Helicopter Soc., 31(3), 54-59 (March 1986).

3. J. A. Fabunmi, and F. A. Tasker, "Advanced techniques for measuring structural mobilities", Trans. of the ASME, J. of Vibration, Acoustics, stress, and reliability in Design, 110, 345-349 (July 1988).

4. F. Amini, K. S. Tawfiq, and M. S. Aggour, "Cohesionless soil behavior under random excitation conditions", J. of Geotech. Engrg. Div., ASCE, 114(8), 896-914 (Aug. 1988).

5. M. S. Aggour, M. R. Taha, K. S. Tawfiq, and F. Amini, "Cohesive soil behavior under random excitation conditions", Geotec. testing J., ASTM, 12(2), (June 1989).

6. F. Amini, "Dynamic soil properties using improved transfer function methods," Internat. J. of Soil Dynamics and Earthquake Engineering, in press, (Sept. 1990).

7. M. S. Aggour, K. S., Tawfiq, and F. Amini, F., "Effect of frequency content on dynamic properties of cohesive soils", in Soil Dynamics and Liquefaction, edited by Cakmak, (Elsevier Publishing company, New York, 1987).

8. F. Amini, "Effect of void ratio on Dynamic Soil Properties Using an Improved FFT Technique", Proceedings of Eighth International Modal Analysis Conference, Kissimmee, Florida, 474-478 (Feb. 1990).

9. F. Amini, K. S. Tawfiq, and M. S. Aggour, "Damping of sandy soils using autocorrelation function," Proc. 8th Sympo. on Earthquake Eng., Roorkee, India, I, 181-18[illegible] (Dec. 1986).

APPLICATION OF DOUBLE-SCALE PHYSICAL MODELLING IN GEOMECHANICS

KHOSROW BAKHTAR and FEREIDOON BAKHTAR
Bakhtar Associates
Geomechanics, Structures and Mechanics Consultants
6695 E. Pacific Coast Highway
Long Beach, California 90803
U. S. A.

ABSTRACT

Application of scale model testing, based on material scaling at normal G, is extended to a double-scale modelling taking advantage of similitude relationships at normal and elevated gravities between model and prototype. The method is used to demonstrate how a crustal block of the earth, with overall dimensions 6.25 miles (10 km) x 6.25 miles (10 km) x 5 miles (8 km), can be constructed to investigate issues of major concerns to the storage of high-level radioactive waste in geologic repositories.

INTRODUCTION

The growing demand for safe design of high-level radioactive waste repositories has directed attention to rock mass deformation and subsequent impacts on hydraulic conductivity and possible changes of ground water configuration as a result of seismic activities. The candidate site for storage of civilian high-level radioactive waste generated by the nuclear power plants is the Yucca Mountain site in the State of Nevada. Licensing of the site for construction has been delayed because of inability to predict the stability of hydraulic conductivity of the jointed rock mass. The stability of hydraulic conductivity is the most important issue needed to be accounted for in order to assess the long-term performance of the site comprising of tuff as the main rock formation to host a high-level radioactive waste repository. Any event that could cause radical changes in flow velocities are of potential concern and should be thoroughly investigated in order to assess the suitability of the site. The available numerical codes used for performance assessment studies can not be accepted in their existing forms because of difficulties involved to verify them. This paper describes how a conceptual model of the site can be constructed based on the "double-scale" physical modeling technique to achieve a general understanding of the rock mass deformation. The main requirement for scale model testing is the ability to develop low strength rock-like materials to satisfy the similitude conditions between the model and prototype.

ROCK MASS DEFORMATION

For the geologic formation at the proposed repository site area the randomly oriented joints or fractures provide most of the weakness, deformability and conductivity of the carbonates, volcanics, and basin fill (sedimentary) rock masses. The stiffness of these rocks are very much dependent on the in-situ stress conditions. The parameter which varies the most as a function of change in the in-situ stress field is the joint aperture or conductivity.

The complex deformation phenomena observed in fractured rock masses can be more readily understood [1] when the individual components of deformation are separated. In most rock masses the intact materials separating the joints can be considered as elastic or pseudoelastic. Relative to the fractures, intact rock is generally stiff and its high modulus is complemented by a low value of Poisson's ratio. Lateral expansions are limited, at least at moderate stress levels.

The second component of deformation is the normal stress-closure behavior of the joints. When the deformation of the intact rock is subtracted from the stress closure curve for the whole jointed block, a highly nonlinear, hysteric stress-closure curve is obtained for the individual joint.

The third and potentially largest component of deformation of the rock mass is the joint shear component. The rock mass at the Yucca Mountain site are highly jointed and can be thought of a deformable system. A deformable system is particularly susceptible to earthquake damage because of restructuring of in-situ stress field and resulting conductivity enhancement.

GROUND WATER TABLE

Reports on earthquake effects on mines, [2] and [3], provide comments such as "mine was flooded,"existing fractures were opened wider causing increase in water influx and almost flooding mine," are indicative of seismic pumping of the ground water. The seismic pumping effects and subsequent changes of water table configuration at the Yucca Mountain is another important issue needs to be addressed prior to licensing of the site.

APPROACH

Application of physical modelling techniques at

elevated and normal gravities can be used to investigate the phenomenological aspects of seismic pumping, changes in water table configuration, and stability of hydraulic conductivity. The most difficult task in physical modelling of tectonic processes is the development of low strength brittle material models with scaled mechanical properties similar to rocks. A discussion on choice of material models has been presented in a recent paper [4] and in general depends on:

- nature of investigation
- limitation of the testing facility
- economic constraints

For modelling a crustal block, an isolated body of the system can be considered, Figure 1, which is subjected to external forces $\vec{t}$ per unit area and $\vec{f}$ per unit mass. The local stress equation of motion is:

$$\sigma_{ij,j} + \rho f_i = \rho a_i \qquad (1)$$

where:

σ_{ij} = total stress field

ρ = density of rock

f_i = components of gravitational force

a_i = components of acceleration

and boundary conditions can be represented by:

$$t_i = \sigma_{ij} n_j \quad (n_j = \text{normal vectors}) \qquad (2)$$

For modelling crustal block, the geologic features should also be accounted for, and the problem becomes mechanically determinate if local stresses, boundary conditions, geologic features, and material properties can be defined. The derivations of scaling relationships between model and prototype has been shown [4] and can be represented by the general expression

$$\sigma^* = \ell^* \rho^* g^* \qquad (3)$$

where:

σ^* = stress scale factor $= \sigma_p/\sigma_m$

ℓ^* = geometric scale factor $= \ell_p/\ell_m$

ρ^* = density scale factor $= \rho_p/\rho_m$

g^* = gravity and acceleration scale factor $= g_p/g_m$

subscripts m and p refer to model and prototype, respectively.

For modelling at normal gravity, acceleration scale factor becomes unity [4], and at elevated gravity, Equation (3) should be satisfied between model and prototype. In practice, ratio between densities of model and prototype is very closed to unity and the density scale factor (ρ^*) can be assumed to be one. Therefore, Equation (3) can be used to determine the strength of the material model required to construct a conceptual model test specimen to study phenomena associated with the crustal tectonic processes by incorporating the major geologic features, such as faults, by increasing the gravity. The same equation can also be used at normal gravity, $g^* = 1$, for component testing of a representative volume of rock mass with the secondary faults or joints incorporated.

EXAMPLE

A typical geometric scale factor of 10^4, with $\rho^* \cong 1$, will require the stress scale factor to be reduced by 10^4. For tuff rock with unconfined compressive strength of about 10,000 psi, the required strength of the model material, based on Equation (3), will become about 1 psi which is difficult to achieve. However, if the gravity is increased using a centrifuge by 100, i.e. $g^* = 1/100$, then the required strength of the model material becomes about 100 psi. Such modelling materials have been developed [4] and are easily obtainable using a mixture of cement based grout, bentonite, barite, air entraining agents, glass beads, and water.

Using a centrifuge, with 35 cubic feet (1 m^3) bucket and 100 g capacity, a cubic test specimen with the associated fault can be tested. A layer of saturated porous material can be incorporated at the bottom portion of the centrifuge bucket to account for the water table. The boundary conditions can be simulated using a series of rubberized flat jacks. For testing, the specimen is loaded to 100 g, by controlling the boundary conditions, the simulated fault is subjected to monotonically load-increase under biaxial conditions until the rupture along the plane of discontinuity and subsequent release of strain energy occurs. Post-test sectioning of the test specimen can then provide information on the effects or extent of seismic pumping. Dye can be introduced into the water to facilitate the ease of investigation.

For component testing, a similar block can be constructed with associated joints and loaded on top of a shaker simulating the earthquake events. The changes in the mechanical aperture of the joints can be monitored using a high resolution camera and subsequently converted into hydraulic or conducting aperture [5].

REMARKS

From the foregoing discussion it is apparent that physical modelling under normal and elevated gravities can be applied to solve problems associated with long-term performance assessment of structures designed in rock mass. Physical modelling should be performed to supplement the field observations and numerical modelling. It is particularly useful for investigation of problems associated with seismic response of fractured system and studies related to changes in joint aperture or conductivity as a result of coupled shear displacement-dilation of fractured system [6]. The choice of material models and requirements for large scale test equipments make the associated cost for physical modelling above the ordinary laboratory testing. However, proper choice of material model [7] and utilization of the available testing facilities at universities, national laboratories, and private sector, bring the cost for large scale physical modelling to an affordable range.

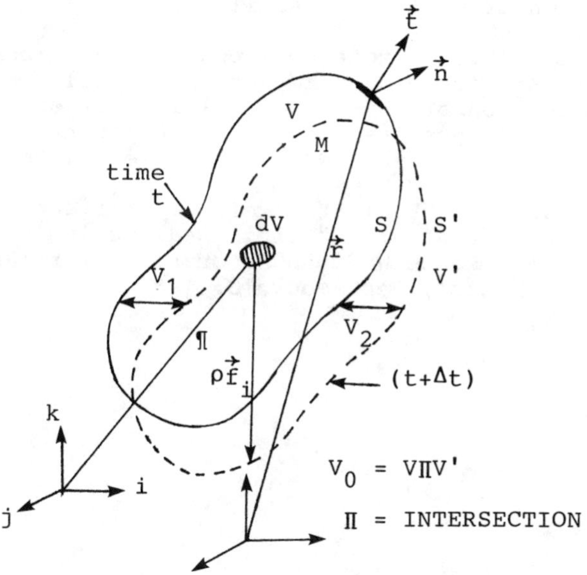

FIG. 1. Representation of gravitational and external forces on an isolated crustal block.

REFERENCES

1. N. Barton, "Effects of rock mass deformations in tunnel performance in seismic regions," Int. Rock Mech. Min. Sci. Geomech. Abstr. Vol. 20, 249-268, 1984

2. C. R. McClure, "Damage to underground structures during earthquakes," Proc. Workshop Seismioc Performance of Underground Facilities. 75-106, Augusta 1981.

3. P. R. Stevens, "A review of the effects of earthquake on underground mines," Open File Report 77-313, U. S. Geological Survey, 1977.

4. K. Bakhtar, "Physical modelling at constant G," 2nd International Conference on Constitutive Laws for Engineering Materials; Theory and Applications," 2, 1435-1441, University of Arizona, Tucson, Arizona, January 5 - 8, 1987.

5. N. Barton and K. Bakhtar, "Rock joint description and modelling for hydrothermomechanical design of nuclear waste repository," CANMET Mining Research Laboratories, Ottawa, Ontario, Canada, April 1983.

6. K. Bakhtar, "A note on Yucca Mountain rock mass deformation and subsequent impact on hydraulic conductivity - application of physical modelling for site performance assessment," Presented at U. S. Department of Energy, Yucca Mountain Project Office, Las Vegas, Nevada, 1990.

7. K. Bakhtar, "Evaluation of blast effects on model underground structures," Civil Engineering Research Seminar, Air Force Weapons Laboratory, Kirtland Air Force Base, New Mexico, September 12, 1986.

Undrained Triaxial Compression Behavior of Saturated Indiana Limestone

Daniel E. Chitty, Scott E. Blouin, Alan F. Rauch
Applied Research Associates, Inc., South Royalton, VT 05068

Kwang J. Kim
Comtec Research, 6416 Stonehaven Ct., Clifton, VA 22024

Douglas H. Merkle
Applied Research Associates, Inc., P.O. Box 40128, Tyndall AFB, FL 32403

ABSTRACT

In undrained triaxial compression tests on saturated porous limestone, conventional instrumentation and analysis techniques indicated more than 5% volumetric expansion with pore fluid pressure remaining essentially constant, a result that was not considered credible. Specially instrumented tests and two-phase finite element simulations of the tests, including fluid transport, were conducted. Both the tests and simulations lead to the conclusion that while the material undergoes volumetric expansion at the center of the triaxial specimen, the ends of the specimen, under the additional effective confinement resulting from endcap friction, experience volumetric compaction. This nonuniform strain field induces flow of pore fluid from the ends of the specimen toward its center, resulting in pore pressure that is consistent with the net volume change of the entire specimen. The strain rate and permeability of the limestone permit pore fluid migration with negligible pore pressure gradients.

INTRODUCTION

This work is an outgrowth of an experimental and analytical investigation of material properties and constitutive modeling techniques applicable to saturated porous rocks. The material under study was limestone with porosity of 0.135 from the Salem formation near Bedford, Indiana, USA. Included in the test effort was a series of undrained conventional triaxial compression tests on saturated specimens at constant confining pressures. When interpreted in the conventional manner, as illustrated in Figure 1 and described below, those tests exhibit several percent of volumetric expansion in the post-yield regime while the pore fluid pressure remains essentially constant. This result implies that net pore volume remains nearly constant during shear instead of dilating as indicated by conventional interpretation of the test data. These contradictory results prompted further research to resolve the apparent discrepancies.

CONVENTIONAL TRIAXIAL COMPRESSION TESTS

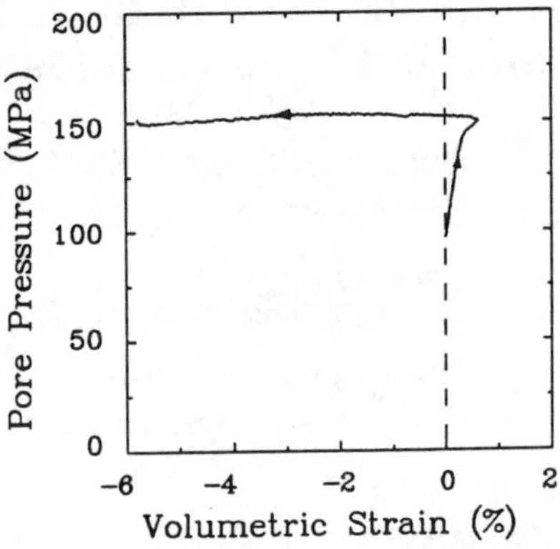

Figure 1 Relationship between pore pressure and volume strain from a conventional triaxial compression test at 200 MPa total confining pressure.

In a conventional triaxial compression test, a cylindrical specimen is prepared with hardened steel caps on its ends and an impermeable jacket sealed to the endcaps at both ends to separate the confining fluid from the pore spaces of the specimen. It is loaded hydrostatically to a predetermined level which is held constant while compressive axial strain is imposed by loading on the endcaps. As the specimen and endcaps deform the difference in moduli between them results in friction which effectively imposes an incremental radial confinement on the ends of the specimen in addition to that applied by the confining fluid. Consequently, radial expansion is greater at mid-height of the specimen than at its ends and the specimen takes on a barrel shape which is observed post-test. The volumetric strains presented in Figure 1 were derived in the conventional manner from radial deformation measurements at the center of the specimen and axial deformation measurements over the full length of the specimen. In a conventional interpretation of such test data, it is assumed that endcap friction has negligible influence on the material at the center of the specimen and that the deformational response computed from measurements at the center is representative of the material response. As discrepancies in the undrained test data show, however, this interpretation is clearly not applicable to the analysis of undrained test data.

SPECIALLY INSTRUMENTED TRIAXIAL COMPRESSION TESTS

When the discrepancy between volume strain and pore fluid pressure was discovered, it was hypothesized that the problem was related to the nonuniform strain field induced by endcap friction. To investigate, measurements were made to determine the actual volume change of the entire specimen under undrained triaxial compression loading [1]. Two radial deformation measurements at different elevations were added to the usual gage array. Figure 2 presents axial strain data along with radial strains at elevations corresponding to 17%, 32%, and 50% of specimen height, and designated 1, 2, and 3, respectively. The observed variation in radial deformation along the axis of the specimen is clearly reflected in the test data. In addition to the three active radial deformation gages, a more detailed passive set of measurements was used to define the post-test shape of the specimen. Based on the passive measurements, a shape function was derived to compute total current volumetric strain from the three active measurements, where total volumetric strain is defined as the difference between the initial and current volumes of the entire specimen divided by its initial volume. The results of that computation are presented in Figure 3 along with the volumetric strain determined in the conventional manner. Analysis of the radial deformation measurements indicates that the material near the ends of the specimen experiences net volumetric compaction, resulting in a total volume change in the specimen that is much smaller than would be concluded from the conventional measurements.

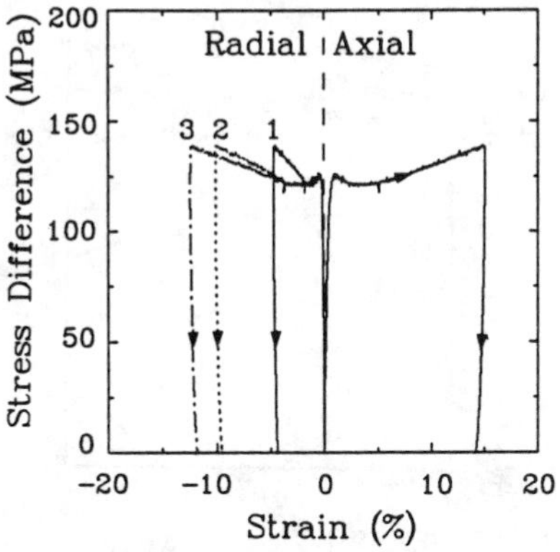

Figure 2 Measured deformations, axial and three radial locations, from 200 MPa undrained triaxial compression test.

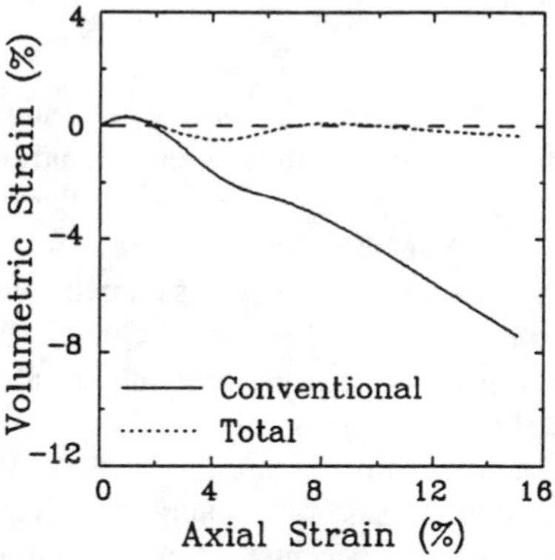

Figure 3 Comparison of conventional volumetric strain and total volumetric strain derived from three radial deformations; 200 MPa triaxial compression test.

An additional finding from the passive pre- and post-test geometry measurements is that the nonuniformity in radial deformations is accompanied by non-uniform axial deformations. In the center where radial expansion is greatest, the axial shortening is also greater than average and at the ends where smaller radial deformations occur there is correspondingly less axial deformation.

TWO-PHASE FINITE ELEMENT ANALYSIS WITH FLUID TRANSPORT

A finite element analysis of the test specimen was undertaken with the double objective of validating the analysis technique and investigating the distribution of pore fluid pressure in the specimen. This was accomplished using MEM [2], a special purpose finite element code that uses a two-phase formulation, i.e. separate degrees of freedom represent rock skeleton deformation and pore pressure. Another essential feature of MEM is the ability to model fluid flow between elements. Taking advantage of the horizontal plane of symmetry, half of the triaxial test specimen was modeled with a 2-D axisymmetric 10 x 10 element grid. A three-invariant constitutive model [3] was used to model the rock skeleton with parameter values derived from triaxial tests on the porous limestone in a dry state. The volumetric behavior of the skeleton under shear loading was derived from conventionally instrumented tests which rely on a single radial deformation measurement at mid-height. Permeability was also determined by laboratory tests.

The confining effect of endcap friction was approximated by fixing the radial degrees of freedom along the end of the specimen. Axial loads were applied quasi-statically using a displacement boundary condition at a rate approximating the laboratory loading rates.

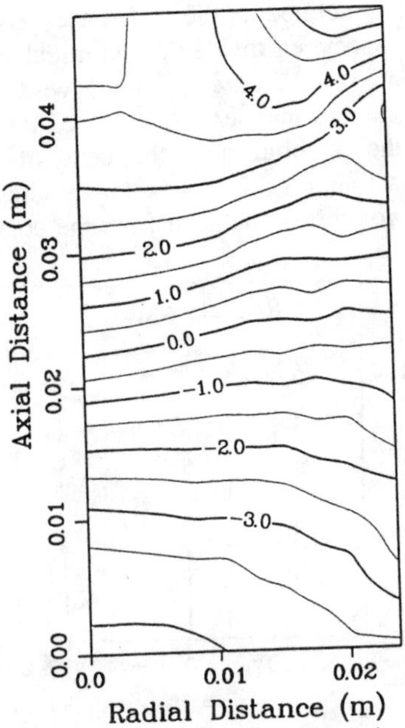

Figure 4 Contours of volumetric strain computed by MEM analysis at 15% average axial strain with 200 MPa confining pressure.

Contours of volumetric strain derived from the finite element analysis are shown in Figure 4. As in the lab tests, there is volumetric expansion at the center of the specimen and compaction at the ends. Overall the computation indicates very nonuniform volume strains throughout the specimen, raising questions as to the appropriateness of assuming uniform strains at the center of the specimen and the adequacy of the test and measurement techniques. The nonuniformity in axial deformations that was detected by passive dimensional measurements is also present in the finite element results.

Two different volumetric strain computations were described above, the conventional approach which assumes that radial deformation is everywhere equal to that measured at specimen mid-height, and the total approach which considers the actual volume change of the entire specimen. From the lab test data, both quantities were estimated and presented in Figure 3. The same two quantities were computed from the finite element calculation results. Figure 5 presents comparisons between lab test and calculation results for both quantities as functions of axial strain, the variable which is controlled during a test. The agreement between test data and calculation in Figure 5 shows that, when volume strains are computed

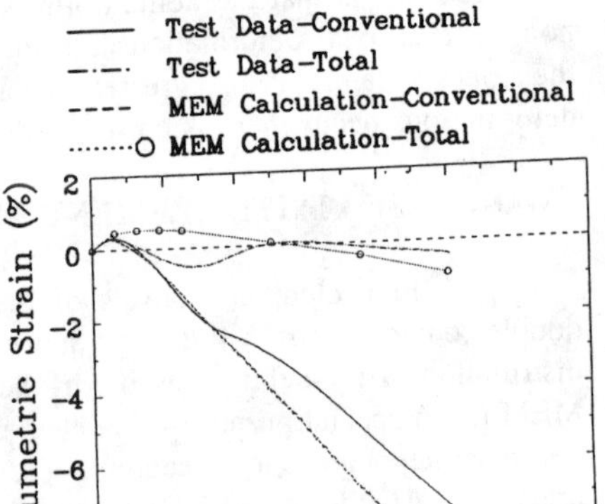

Figure 5 Comparison of experimental and finite element results for volume strains computed by two approaches for a 200 MPa triaxial compression test.

consistently, there is good agreement between the test data and the finite element simulation.

The finite element results also indicate that there was virtually no spatial variation in pore pressure at any time in the triaxial compression loading. This indicates that the loading took place slowly enough that pore fluid was able to flow from one region of the specimen to another thereby minimizing any pore pressure gradients that tend to develop as a result of nonuniform volumetric strain.

VOLUME STRAIN MEASUREMENT CONSIDERATIONS

Due to nonuniform strain fields that can be induced in both drained and undrained triaxial compression specimens as a result of endcap friction, a volume strain computation based on a single radial deformation measurement at specimen mid-height can significantly overestimate the volume change of the specimen as a whole. In undrained testing, this approximation is clearly not acceptable, and the authors [1] have developed an approximate procedure to obtain a measure of total volumetric strain. The conventional volume strain approach might be acceptable if the test is drained and the result is considered applicable to the central third of the specimen. However, for the porous limestone used in this research, the nonuniform radial strains are accompanied by nonuniform axial strains. At the center of the specimen where the largest radial expansion occurs, the axial compression is also larger than the average over the entire length of the specimen. Thus, if the axial deformation is estimated by measuring the deformation of the entire specimen, it underestimates the magnitude of the local compressive axial strain. When average axial deformation is used in combination with the radial deformation at mid-height, the computed volume change overestimates the volumetric expansion, or dilatancy, even at the midpoint.

CONCLUSIONS

The laboratory test results and numerical simulations are both consistent with the fundamental view that pore fluid pressure response is a function of the total volume change of the material within the test specimen. In a typical undrained triaxial compression test performed on a specimen of saturated porous limestone, there was flow of pore fluid from the regions of net volumetric compaction at the ends of the specimens to the region of net volumetric expansion at the center of the specimen. The permeability of the Salem limestone is sufficient that for the strain rates used in the laboratory, approximately 10^{-5}/s, this flow takes place at pore pressure gradients that are insignificant in comparison with the average stress in the specimen. Thus, there is no significant spatial variation in pore pressure in the specimen.

The two-phase finite element program, MEM, along with the three-invariant elasto-plastic material model can correctly predict the results of a saturated undrained experiment using rock skeleton properties derived from dry material property tests.

For the material tested, axial compressive strain is smaller than average in the constrained material near the specimen ends and larger than average in the less constrained material at the center of the specimen. Thus, volume strain response computed at the center of the specimen based on radial deformation at the center and average axial strain tends to significantly overestimate the volumetric expansion at the midpoint of the specimen. If used for modeling, this could introduce significant errors in model parameters, causing the models to predict more dilatancy than actually occurs.

Improvements in laboratory testing procedures at high pressures are needed to more accurately and completely characterize the strain conditions within the test specimens and to minimize endcap friction to that strains are more uniform throughout the test specimen.

ACKNOWLEDGEMENTS

The research described in this paper was partially sponsored by the U.S. Air Force Office of Scientific Research under Contract F49620-89-C-0037. The United States Government is authorized to reproduce and distribute reprints for government purposes notwithstanding any copyright notation hereon.

The laboratory testing was conducted under U.S. Army Engineer Waterways Experiment Station Contract DACA39-89-C-0059.

REFERENCES

1. Chitty, D.E. and S.E. Blouin, "Special-Purpose Mechanical Property Tests of Salem Limestone," Applied Research Associates, Inc. report to U.S. Army Engineer Waterways Experiment Station, Vicksburg, MS, February 1990.
2. Blouin, S.E., D.E. Chitty, A.F. Rauch, and K.J. Kim, "Dynamic Response of Multiphase Porous Media, Annual Technical Report 1," Applied Research Associates, Inc. report to U.S. Air Force Office of Scientific Research, Washington, DC, March 1990.
3. Merkle, D.H., W.C. Dass, and J.L. Bratton, "A Three Invariant Constitutive Model for Soil Dynamics," in *Proceedings of the Second International Conference on Constitutive Laws for Engineering Materials, Theory and Applications*, Vol. 2, Tucson, AZ, January 1987.

THE STEADY FLOW OF POLYCRYSTALLINE ICE UNDER COMBINED ACTION OF COMPRESSION AND TORSION

DR. JIH-JIANG CHYU, P.E., MASCE
Consulting Engineer, New York

ABSTRACT

The steady flow of polycrystalline ice behaves nonlinearly in general and the present study confirmed this in the case of combined compression and torsion. Moreover, this work also determined the limits of validity of a linear flow law. The experimental results indicated that this linear law holds true for a loading region in which the ratio of load intensity between torsion and compression is less than unity.

It is suggested that the experimental study be extended in several directions. This is a pilot program in the exploration of the mechanics of ice by using the simplest possible apparatus. More sophisticated laboratory equipment is surely to facilitate future research along many lines in this area, and it is hoped that such endeavors will soon be realized.

INTRODUCTION

The present work serves as an exploratory study striving for shedding some light on the theme problem by using only simple apparatus and basic theoretical treatment. Relevant literature includes the following: Sunder and Wu [1] proposed a differential model for the flow of ice on the basis of the concept of state variables, with experimental verification of uniaxial case. Harper [2] investigated several possibilities for a relationship between uniaxial creep tests and constant strain rate tests at the failure of ice and found that no such correspondence could be expected in general. Fethi [3] analyzed experimental results obtained by others with the assumption that ice creeps according to a power law during the primary creep stage. On the basis of data from unconfined compression tests, he was able to identify the starting point of tertiary point on the strain time curve.

In glaciology, ice is considered as essentially incompressible. From this and the assumption of isotropy, a general three dimensional steady flow law model was recommended. Using summation convention of indices, we have the strain rate and stress relation as

$$e_{ij} = Ad_{ij} + Bf_{ij} + Cf_{ik}f_{kj} \quad (1)$$

where d_{ij} is Kronecker's delta, e_{ij} is a strain rate component, f_{ij} is the corresponding stress component, and A,B,C are functions of the stress invariants S_1, S_2, S_3.

Consider the stress deviator defined by

$$g_{ij} = f_{ij} - (1/3)S_1d_{ij} \quad (2)$$

Then the three invariants of this tensor are

$$R_1 = 0$$

$$R_2 = 1/2 g_{ik} g_{kj} \tag{3}$$

$$R_3 = (1/3) g_{ij} g_{jk} g_{ki}$$

Thus, we have finally

$$e_{ij} = Dd_{ij} + Eg_{ij} + Fg_{ik}g_{kj} \tag{4}$$

where D, E, F are functions of R_2, R_3.

Expression (4) is the most general steady flow law for polycrystalline ice under the stipulations indicated earlier. As a consequence of linear law assumption, this reduces to

$$e_{ij} = Gg_{ij} \tag{5}$$

where G is a function of R_2 only.

The experimental setup was to perform combined compression and torsion tests on polycrystalline ice by holding R_2 constant while varying R_3. Thus, the second invariant of the stress deviator is given by

$$P^2 + 3T^2 = K^2 \tag{6}$$

where K is a constant, P, T, are the compressive and torsional shear stresses, respectively. Let U be the uniaxial compressive strain rate, and V be the shear strain rate. It follows that the linear flow law for the case of combined compression and torsion of polycrystalline ice is

$$U = (2/3)HP$$

$$V = HT \tag{7}$$

where H is a constant for constant K.

EXPERIMENTS

The specimens were hollow ice cylinders with an inner diameter of 8cm and an outer diameter of 10cm, and a length of 15cm. Loads were applied by dead weights. Pure torsional loads were applied diametrically, opposite to each other, to eliminate lateral loading on the specimen. Tests were carried out to satisfy the condition of loading expressed by (6), with K=2 bars, and 4 bars. The temperature in the test room was kept at -5C° on the average and steady flow of ice specimen commenced at the end of 24 hours approximately for both displacement components. The duration of the sustained loading of the specimens ranged from 100 to 120 hours. No specimens were crushed during the tests. Note that tests with smooth contact surfaces between metal and ice in an effort to avoid end constraint could not be performed. Therefore, these metal surfaces which are in contact with ice were serrated. Test results were based on this. Typical displacement time graphs were shown in Fig. 1, 2.

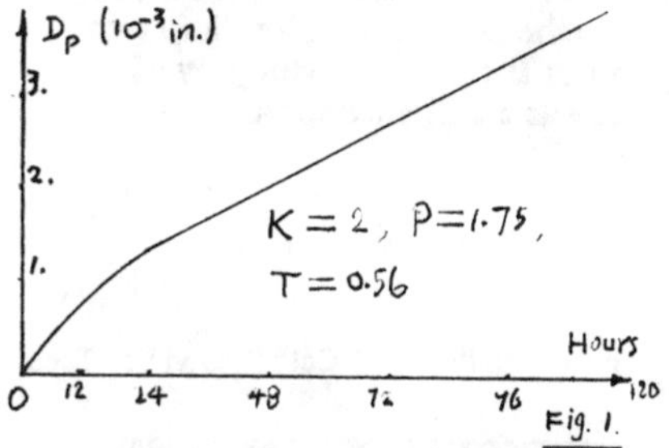

Fig. 1.

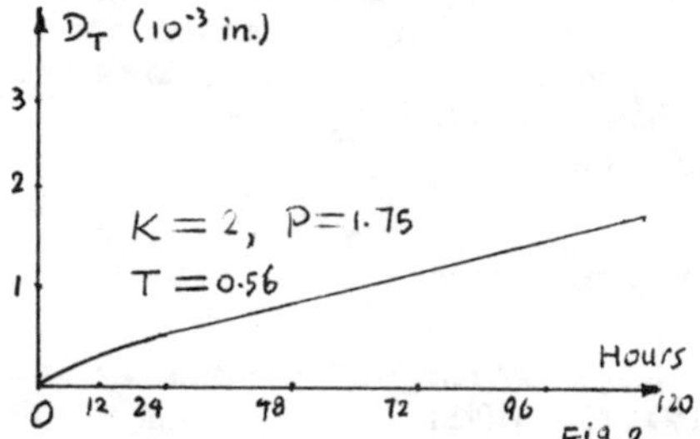

Fig. 2

CONCLUSIONS AND RECOMMENDATIONS

Based on the test results in this study, the steady flow law of polycrystalline ice under combined compression and torsion is that a component of the strain rate is a nonlinear function of the corresponding stress deviator component in general. The criteria regarding linearity of the flow law can be explained by the graph of the loading schedule condition

$$P^2/K^2 + 3T^2/K^2 = 1 \tag{8}$$

which is an ellipse with major axis K and minor axis $K/\sqrt{3}$. (Fig. 3)

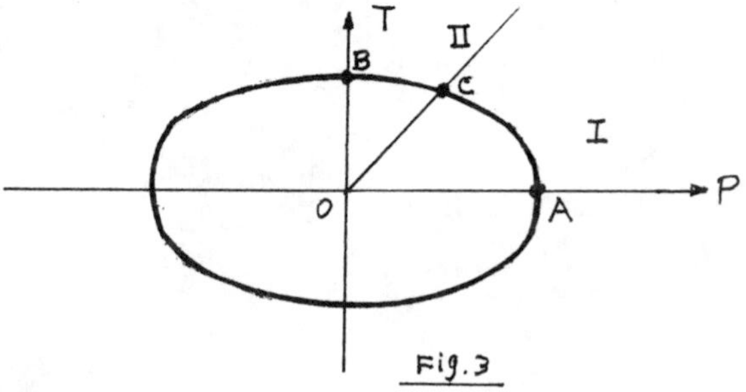

Fig. 3

The flow is linear if T/P, which is equal to the ratio of T/K to P/K, is less than unity and this is represented by the arc AC. The arc BC is associated with the nonlinear counterpart. Point C is critical point.

Different K values give rise to different loading levels. For each K value, there is a T-P ellipse associated with it. By connecting the critical points on several T-P ellipses, we obtain a straight line, passing the origin, that divides the first quadrant which represents the loading zone for the present problem, into regions I, II. Region I corresponds to the sector formed by the major axis OP and the critical line OC, whereas region II is formed by the minor axis OT and the critical line OC. It is expected that there is a closure curve for each region. The characteristics of these closure curves can be determined from further tests. Moreover, it is interesting and practically useful as well to note that correlations among different loading levels for the same T/P ratio, or vice versa, can be established to facilitate the implementation of a big scale research project. Some of the directions for future work are: (1) along the same line as the present study, but with more K values, (2) similar to the first but at several different temperatures, (3) holding R_3 constant while varying R_2, with temperature as a parameter. It is interesting to note, in passing, that the maximum T for region I corresponds to about 86.6% of the minor axis of the T-P ellipse. Therefore, we can divide the original problem into two categories, each corresponding to either region I or II for further study by subdividing each into subregions using T/P value as a guidance for such refined divisions.

REFERENCES

1. S.S. Sunder, M.S. Wu, "Differential flow model for polycrystalline ice," Cold Region Sci. Tech. 16(1), 45-62 (1989).

2. B.D. Harper. "Some implications of a nonlinear viscoelastic constitutive theory regarding interrelationships between creep and strength behavior of ice at failure," Journ. Offshore Mech. Arc. Eng. 3(2), 144-148 (1989).

3. A. Fethi. "Primary creep of polycrystalline ice under constant stress," Cold Region Sci. Tech., 159-165 (1989).

AN AUTOMATED PROCEDURE FOR CONSTITUTIVE MODEL PARAMETER SELECTION

JOHN E. CRAWFORD and NICHOLAS J. CARPENTER
Members of the Technical Staff, TRW Inc.,
One Space Park, Redondo Beach, Ca 90278

ABSTRACT

Nonlinear constitutive laws employ parameters that are often established through a trial and error curve-fitting process. This paper describes a general approach to the model-fitting problem and presents a computer architecture that fully automates the procedure and can be applied in most situations independent of the specific material model.

PROGRAM DESCRIPTION

The degree of sophistication and accuracy of constitutive laws is often a critical consideration in nonlinear finite element analyses. Nonlinear constitutive laws, such as the "cap" model, employ parameters that are usually established through a trial and error curve-fitting process. Consequently, parameter selection is a highly subjective task and a source of substantial uncertainty. To minimize this uncertainty, various computer procedures, which have been reported in the literature [1,2], are available to aid in the fitting for a particular material model. This paper describes a general approach to the model-fitting problem and presents a computer architecture that fully automates the procedure and can be applied in most situations independent of the specific material model.

The Automated Parameter Identification Procedure (APIP) computer program was developed by integrating the three primary components of a general fitting procedure: (1) experimental data, (2) constitutive models, and (3) optimization procedures. A schematic illustration of the complete APIP program is shown in Figure 1. The dashed line boxes in the illustration depict the four primary modes of operation, which range from a simple data base input/output mode to the full exercise and integration of the model, data, and optimization routines. A menu of program options allows control of the identification procedures.

Depicted within the dashed boxes are the various coding blocks of APIP. These coding blocks include "data base" which provides for the creation, storage, retrieval, and modification of material data. The "material model" block contains the coding for the material models. The architecture of the APIP code allows the material model coding to be taken directly, without modification, from finite element programs. This "non-modification" strategy provides a high degree of confidence in the material model coding and facilitates material model development because identical coding is used in both the APIP and finite element programs.

The "material test simulator" coding block provides the means to exercise the various material models in a manner that simulates load paths and responses that correspond to the selected material test data. During the optimization phase, this coding block accesses the data base to determine what type of test is to be

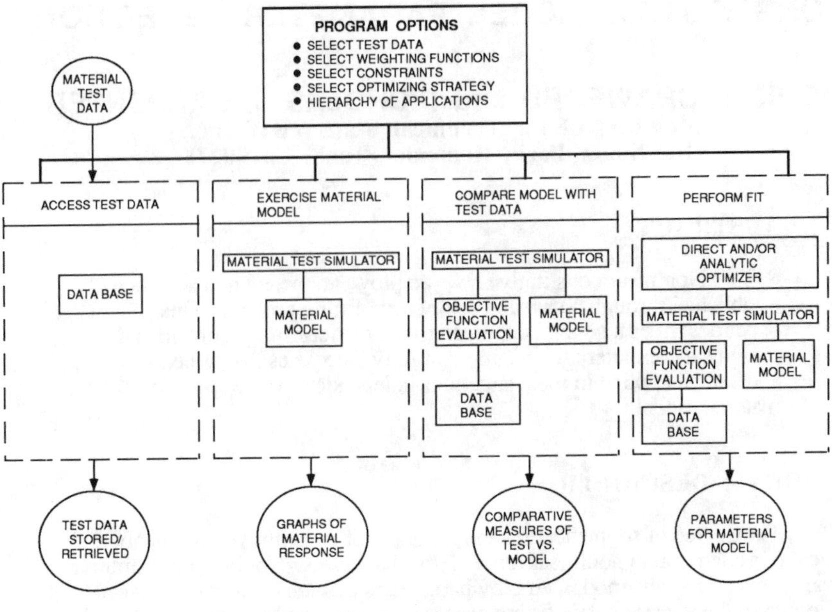

Fig. 1. Schematic depiction of the APIP Program

simulated, determines the stress or strain loads needed to recreate the test, modifies the units of the test data to match those of the material model, adjusts the sign convention of the test to match the simulation, defines scaling parameters for each region of a test (to insure a uniform evaluation of the objective function), and sets up the interface to the material model in such a way that the material model "thinks" it is being called by the finite element program from which it was taken.

The "objective function evaluation" block computes the differences between the actual and simulated test data utilizing different strategies for curve interpolation (i.e., linear or cubic splines) and difference measuring. The definition of the dependent variable (i.e., the one that is differenced) and the independent variable (i.e., the one by which the load is applied) is set as a function of test type. For example, axial stress is the dependent variable and axial strain is the independent variable in a unaixial test. The selection of the independent variable is essential to successful simulation because for each test there are possibly inadmissible states of stress that the optimizer may "stumble" into, which can be avoided by appropriate selection of the independent variable.

Direct and analytic optimization techniques are available in APIP within the "optimizer" block. Both techniques utilize an objective function that is based on a weighted sum of the square of differences between constitutive law and experimental data.

Direct optimization employes a random or pattern search technique for sampling the objective function. For the random search technique, model parameter values are randomly selected within specified parameter bounds; for the pattern search, they are obtained by subdividing a bounded parameter interval. For each set of parameters, the objective function is evaluated and recorded in such a way that "good" parameter values may be identified. Direct optimization is also useful for obtaining the model parameters which are needed to initiate the analytic optimization procedure.

The analytic optimization routine utilized in APIP is an iterative scheme based on unconstrained optimization techniques. For parameters which are subject to constraints, trigonometric transformations are employed, mapping the constrained parameter domains into uniformly scaled unconstrained domains [1]. The APIP optimization routine features full-Newton and Gauss-Newton Hessian options [4]; both options include line search with initial quadratic and successive cubic backtracking, and a descent bias factor. The minimum eigenvalue of the Hessian is evaluated for determining the descent bias factor, a procedure which was found to perform better than a model-trust approach. Also, the full-Newton Hessian option was found to converge much more rapidly that the Gauss-Newton method in most cases.

RESULTS OF DIRECT OPTIMIZATION

One advantage of direct optimization is its ability to provide a direct measure of the effect of varying model parameter values. Results using McCormick Ranch sand [3] are depicted in Figure 2. The "data" used to measure the fit were analytically (i.e., artificially) created using the parameters which are shown in the figure with the label "exact" and those given in [3]. Four parameters (D, W, K, and R, see [3] for definition) of the McCormick Ranch cap model were varied in a systematic way to determine their effect on the "fit" as measured by the magnitude of the objective function. The histogram of Figure 2 depicts the direct optimizer results in a statistical way by determining for each parameter variation the associated minimum value of the objective function. In this case variation 1 of parameter 2 (i.e., D = 0.38) is shown to have a marked deleterious effect on the objective function. Examination of these results indicates that a "good" set of material parameters would be variations 3, 5, 2-3, and 4-5, corresponding to parameters W, D, K, and R, respectively. These agree favorably with the "exact" parameter values, which are as shown in the figure.

RESULTS OF ANALYTIC OPTIMIZATION

To illustrate the effectiveness of the analytic optimizer, a soil known as flume sand was used. For this example, the objective function was chosen to

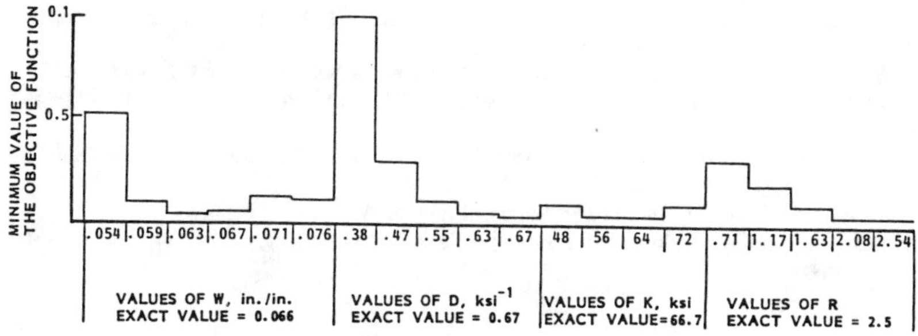

Fig. 2. Results of direct optimization

include both uniaxial and triaxial tests, with equal weighting for each test. Test data for flume sand [5] are shown in Figures 3 and 4 as well as a fit determined by trial and error (T&E fit). Also shown in these figures is the response of the "optimal" fit obtained from APIP. There are two major differences between the parameters of the T&E and "optimal" fits: a 35 percent decrease in K (the bulk modulus) and a 25 percent decrease in B (B effects the shape of the failure envelope [3]). Neither of these changes is obviously related to the data; however, they aptly demonstrate the utility of the analytic optimizer. As depicted by the figures and the values of the objective function (i.e., 2.3 for T&E fit and 1.1 for the optimal fit), it is apparent that an overall better fit to the two tests is achieved by relaxing (worsening) the fit to the uniaxial test. Thus, while a 60 percent worsening of the fit occurs in the uniaxial test (i.e., the value of the objective function for just the uniaxial test is increased by 60%), the overall fit is improved by more than a factor of 2 (i.e., the objective function considering both tests is halved).

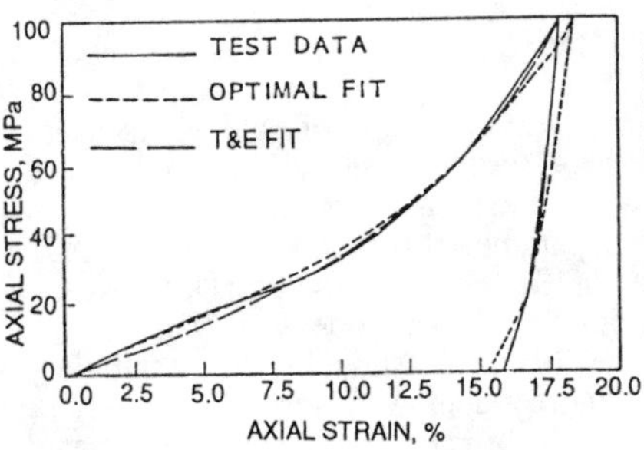

Fig. 3 Uniaxial response

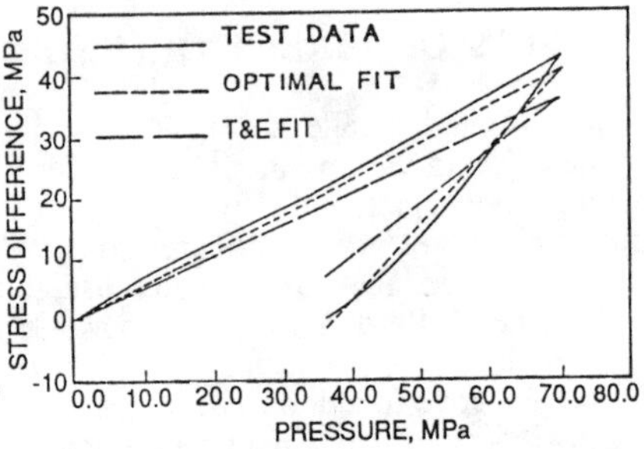

Fig.4. Triaxial response

CONCLUSION

The APIP code is straightforward and simple to operate, and offers a general purpose tool to replace the time consuming and subjective trial and error method of fitting material model parameters. It also provides a convenient format to store material data.

REFERENCES

1. J. W. DeNatale, "On the Calibration of Constitutive Models by Multivariate Optimization," Dissertation University of California, Davis, 1982.
2. J. W. Ju, J. C. Simo, K. S. Pister, and R. L. Taylor, "A Parameter Estimation Algorithm for Inelastic Material Models," Constitutive Laws for Engineering Materials, Elsevier Science, 1987.
3. I. S. Sandler and D. Rubin, "An Algorithm and a Modular Subroutine for the Cap Model," International Journal for Numerical and Analytical Methods in Geomechanics, Vol. 3, pp 173-184, 1979.
4. J. E. Dennis and R. B. Schnabel, Numerical Methods for Unconstrained Optimization and Nonlinear Equations, Prentice-Hall, 1983.
5. J. E. Windham, "Shear Friction Test Support Program: Mechanical Property Recommendations for WES Flume Sand Backfill," Structures Laboratory, WES, Vicksburg, MS, December 1984.

RATCHETTING OF A 316 STAINLESS STEEL AT ROOM TEMPERATURE

J. DANKS and S. J. HARVEY
Coventry Polytechnic, Coventry, England

ABSTRACT

The results of cycling tubular specimens of 316 stainless steel at room temperature under a primary constant axial load and a secondary cyclic torsional strain are described. Rapid cyclic hardening took place together with ratchetting in the primary load direction. The ratchet ting did not shake down but continued at a reducing cyclic rate throughout the test duration. An enhancement in tensile strength was observed in post cycling tensile tests.

INTRODUCTION

The specimens were taken from a sheet of hot rolled steel, 48 inches wide, 96 inches long and 1 inch thick. The longitudinal axis of the specimens was parallel with the direction of rolling. An analysis of the material [1] gave the following composition.

C .052%; Mn 1.36%; Si 0.71%; Ch 16%; Ni 10.9%;
B .0023%; N 539 ppm; remainder iron.

The samples tested were tubular with an outer diameter of 16 mm and a wall thickness of 1 mm. To ensure uniformity with tests being carried out in other laboratories, all specimens were solution treated at 1050°C for 1 hour and cooled naturally in air to room temperature.

TEST RIG AND TEST DETAILS

The test rig cycles the specimens under alternating torsional strain, whilst simultaneously placing the specimen in axial tension. The torsional strain used was ±1% total strain and the axial tensile stress was either 0%, 25%, 50% or 75% of the stress at the limit of proportionality in tension. Measurements were taken of twist and axial extension, which were then fed into a microprocessor controller which ensured that the cyclic strain remained at ±1%, compensation being made to the twist to allow for ratchetting in the axial direction. Tests were continued up to a maximum of 300 cycles. Some preliminary tensile tests were performed on the material to determine any anisotropy, from measurements of strain and of strength initial anisotropy was minimal.

CYCLIC TEST RESULTS

The first and most apparent result was that for all axial load conditions the material rapidly hardened, and then softened at a more gradual rate until the peak shear stress stabilised, Fig. 1. Several tests were repeated and a degree of scatter was observed. The scatter was more predominant in the hardening rather than the softening phase. The scatter was less for higher axial stresses. The stress axis in Fig. 1 is plotted as Shear Stress Range. In every test the backward peak shear stresses were higher than their forward equivalent. The peak shear stresses correspond with considerable plastic flow, and observation of the hysteresis loops show that a Bauschinger effect was evident. This behaviour is in agreement with tests published by Jaske [2].

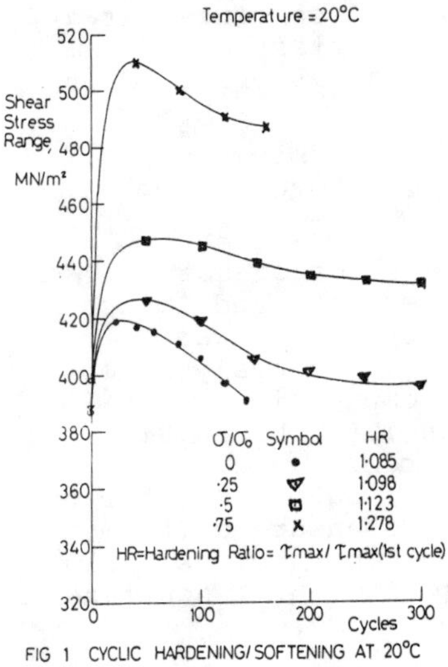

FIG 1 CYCLIC HARDENING/SOFTENING AT 20°C

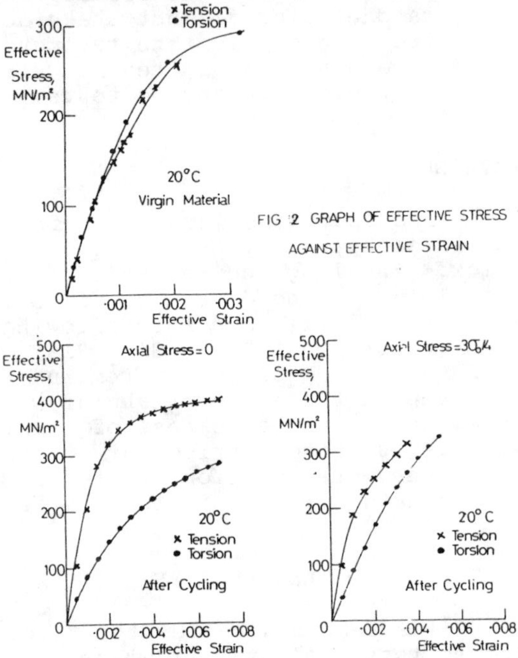

FIG 2 GRAPH OF EFFECTIVE STRESS AGAINST EFFECTIVE STRAIN

There is evidence of the development of anisotropy due to cycling. Figure 2 shows a comparison between the tensile and torsion behaviour, plotted in Von Mises effective stress and strain space, for uncycled material and cycled material with two different axial stresses. The mechanisms have been identified [3] which are thought to be responsible for the behaviour of 316 under cyclic loading. No dependence on rolling direction was found as crack planes could occur both parallel and normal to the rolling direction. As the cyclic straining proceeds there is a growth n the direction of the axial stress, Fig. 3. This is a plastic ratchetting strain and accumulates at a reducing cyclic rate, the rate eventually becoming almost constant. This cyclic effect is different from the ratchetting effects [4] [5] which show linear ratchetting strain accumulation at one extreme, or rapid shakedown at the other. The results reported here fall between these two extremes and it is suggested that the difference in behaviour is a function of the magnitude of the secondary stress range and secondary to primary stress ratio, the secondary stress being an elastic stress determined from the secondary strain range.

The pattern of in-cycle ratchet strain behaviour is shown in Fig. 4. The period of zero axial growth, AB, is seen to coincide with the elastic part of the hysteresis loop. The region of relaxation, BC, corresponds with a low or zero value of shear stress. This relaxation at low stress may be associated with strain relaxation due to low stresses interacting with dislocation pile-ups. As the shear strain increases and begins to generate higher shear stresses, the axial extension rises once more, CD. The modelling of the axial growth by an anisotropic hardening plastic potential [6] predicts the pattern of these variations.

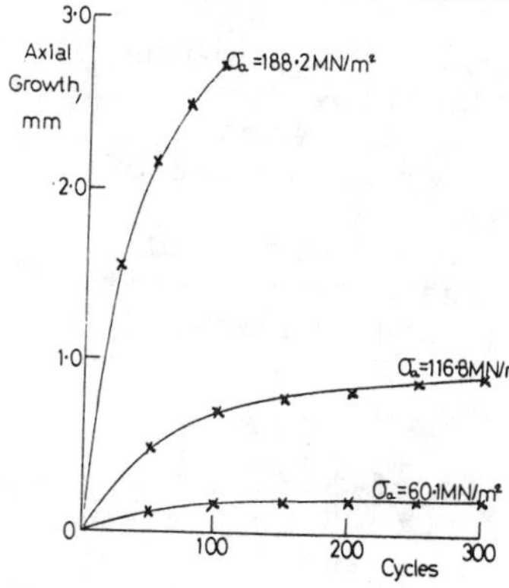

FIG 3 AXIAL GROWTH DUE TO CYCLING AT 20°C

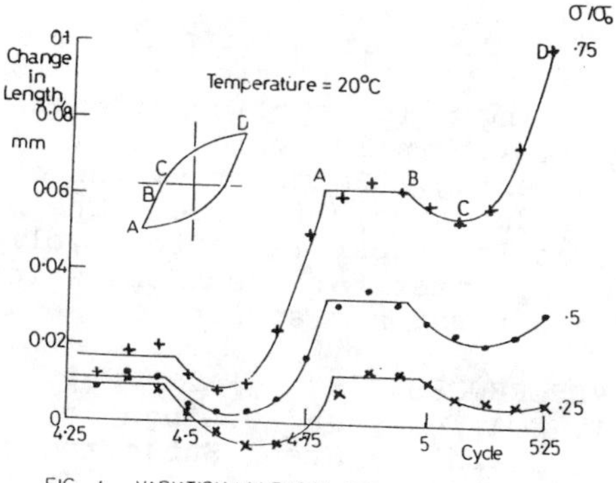

FIG 4 VARIATION IN LENGTH DURING THE 5th CYCLE

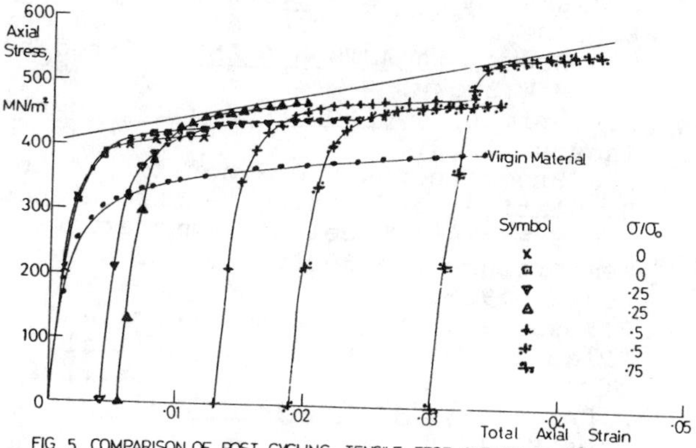

FIG 5 COMPARISON OF POST CYCLING TENSILE TEST CURVES AT 20°C

POST CYCLING TENSILE PROPERTIES

The test specimens were tensile tested after the cyclic tests, a strengthening is reported which seems to be dependant upon the total axial strain in the specimen and the primary axial stress, Fig. 5. The strain axis on which these results are plotted is one of total strain, that is the sum of accumulated plastic ratchetting strain and the strain added during the tensile test. The first and most significant point is the general rise in strength over the virgin material curve. The second point, also of importance is that the strength appears to increase as the total strain increases. Confirmation of these results were made by taking hardness readings. This observed phenomena of post cycling tensile strengthening has been used [7] to develop an expansion/translation yield surface model to estimate the hardening and tensile strengthening of this material due to cyclic straining.

CONCLUSION

The test material, 316 stainless steel, which was initially isotropic developed anistropy during the cycling. The cyclic hardening was rapid, reaching a peak within 20 cycles, followed by a small amount of softening. Ratchetting was observed, accumulating at a reducing cyclic rate. The material had enhanced tensile properties after cycling. The enhance- ment was greater for specimens which had experienced a larger ratchet strain.

REFERENCES

1. D.A.Kelly and G.Singh, "Creep Life of AISI 316 Stainless Steel Subjected at 550°C to Step Changes in Tensile Load", Fat. of Eng. Mats. and Structures, 7, No. 1, 1984.
2. C.Jaske, **Batelle University Report No. 23834**, 1980.
3. E.R. De Los Rios, F.A.Kandil, K.J.Miller and M.W.Brown, in **ASTM STP 853**, edited by K.J.Miller and M.W.Brown.
4. M.R.Bright, Cyclic Creep under Complex Stress, PhD Thesis, 1979.
5. D.J.Brookfield and D.N.Moreton, "The use of a Bree Simulation to Investigate Strain Accumulation in 316 Stainless Steel at Temperatures Between Room Temperature and 500°C", J. of Strain Analysis, 24, No. 2, 1989.
6. F.Yoshida, N.Tajima, K.Ikegami and E.Shiratori, "Plastic Theory of the Mechanical Ratchetting", B. of the Japanese S. of Mech. Eng., 21, No.153, 1978.
7. J.Danks, Cyclic Creep of T.316 Stainless Steel, PhD Thesis, 1989.

Incremental non-linear constitutive equations and probe stress paths on sand

Doanh T., Royis P.
Ecole Nationale des Travaux Publics de l'Etat. France.
Gréco "Géomatériaux"

ABSTRACT

Three series of special probe tests conducted with Hostun RF sand are performed. Most of the tests consist of a primary triaxial loading, and a directional loading with different angles of the stress increment. The purpose of these tests is to investigate the incremental response of this sand under stress probe tests.

This paper extends the previous works during the last years, and aimes at showing that the non-linear incremental constitutive equations can describe the behaviour of saturated sand subjected to various stress loadings in drained conditions.

CONSTITUTIVE EQUATIONS

The constitutive equations, restricted to the generalized triaxial paths, can be written in the following form in the orthotropic axes of the material:

$$d\varepsilon = M(h,d)\ d\sigma$$

where d is the direction of the stress increment tensor $d\sigma$, $d\varepsilon$ the strain rate tensor, and M a tensor of second order depending on the whole stress - strain history h and on the direction d. The model has been used in specialized workshops (Grenoble 1982, Cleveland 1987).

For any direction d of $d\sigma$, and for a given history, M is interpolated from M^+ and M^-, whose columns correspond to the values obtained in its orthotropical directions. M^+ and M^- are generally expressed by using the Pseudo-Young's moduli E_i and the Pseudo-Poisson's ratio υ_{ij}. The fundamental assumptions of constitutive equations are tested, mainly the directional dependency and the interpolation process of the elementary stress paths. The response envelope displayed by the material has a great importance in the evaluation of the interpolation functions.

EXPERIMENTAL DESCRIPTION

Three initial states of stress level are used, (isotropic, intermediate, close to failure). In each series, the stress paths are chosen to cover the whole space (σ_1, $\sqrt{2}\ \sigma_3$), and to obtain accurately the elliptical response generated by the strain increments.

The amplitude of the stress increment is small enough to represent the incremental behaviour of the material.

From each initial state, the results of each probe test is associated with the direction $\alpha\delta\sigma$ of the stress increment, defined as $\alpha\delta\sigma = \arctan(\delta\sigma_1, \sqrt{2}\ \delta\sigma_3)$. Figure 1 represents the results of 11 probe tests starting from the isotropic point ($\sigma_1 = 100$ kPa, $\sigma_3 = 100$ kPa), and in figure 3 we have 8 tests from the closest point to the failure surface ($\sigma_1 = 400$ kPa, $\sigma_3 = 100$ kPa). All of these tests are bilinear, or trilinear stress paths with two sharp bends. After getting the initial state with a compression, the properly probe test is conducted, and we terminate with another straigth stress path toward failure, if the path length exceeds 600 kPa.

A new interpolation function is developped (Royis 1990), and to clarify the history aspect of constitutive equations, the numerical integration of the complete probe tests, ie the proportional stress paths until failure from these initial stress states, is presented in this paper.

The parameters of the models are identified with only two tests of the first series: the classical compression $\alpha\delta\sigma$ =90° and extension $\alpha\delta\sigma$ =-90° triaxial tests. The experimental response curves of the isotropic state are well simulated (figure 2), except for the probe test of $\alpha\delta\sigma$ =0°; but the prediction curves match partially the observed aspects of the last probe tests (figure 4).

CONCLUSIONS

Experimental results of three series of stress probes test under drained triaxial compression condition on Hostun's dense sand designed to explore the behaviour of granular material are presented.

These multi-linear tests are selected to check the prediction capability of a non linear incremental constitution model of interpolation type. Several aspects of observed responses are simulated by the model.

REFERENCES

Robinet J.C., Mohkam M., Deffayet. M., Doanh T."A non linear constitutive law for soils". Constitutive relations for soils. Int. Workshop. 1982. pp 405-418. Gudehus, Darve, Vardulakis Eds. Balkema Publi.

Doanh T., Di Benedetto H., Y. Golcheh, M. Kharchafi."Non linear incremental constitutive equation: application to sands". Constitutive Equations for Granular Non Cohesive Soils. 1987. pp 255-273. Saada, Bianchini Eds. Balkema Publi.

Doanh T. "Réponses incrémentales: synthèse d'expériences et étude comparative des modèles".1989. Report of Greco Géomatériaux. pp 220-227.

Royis P. "A new familly of incrementally non-linear constitutive laws". 1990. Proc. of 2nd World Congress on Computational Mechanics, Stuttgard.

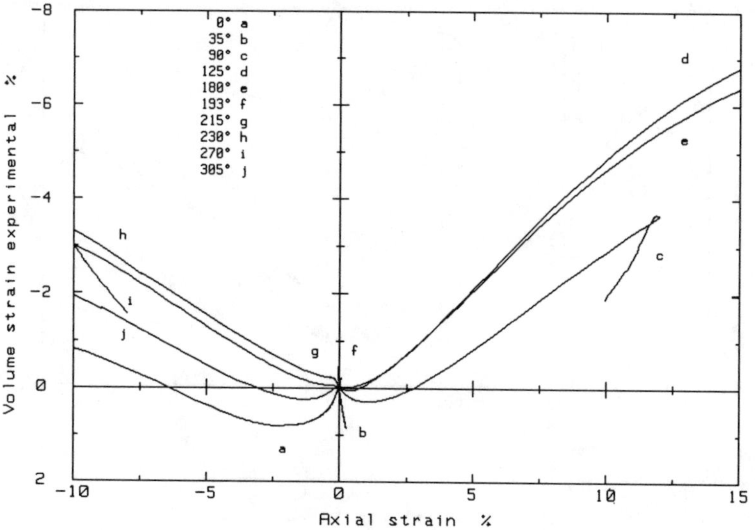

Figure 1: Probe tests from Point 100,100

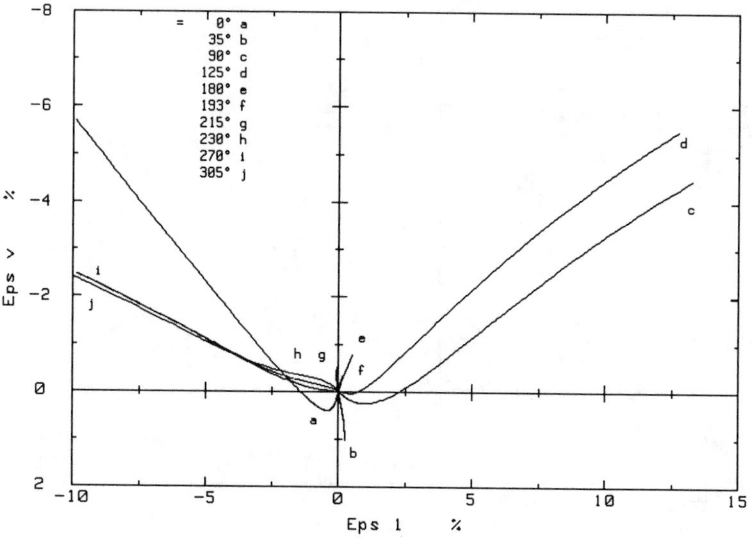

Figure 2: Numerical results of Point 100,100

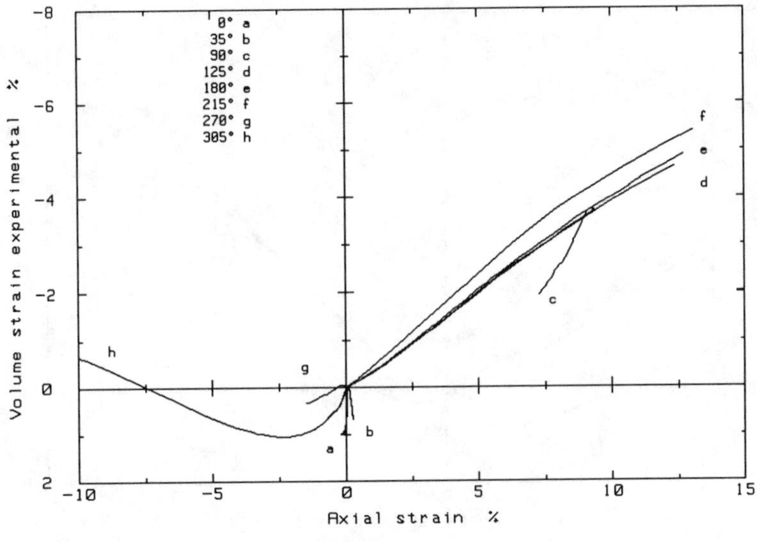

Figure 3: Probe tests from Point 400,100

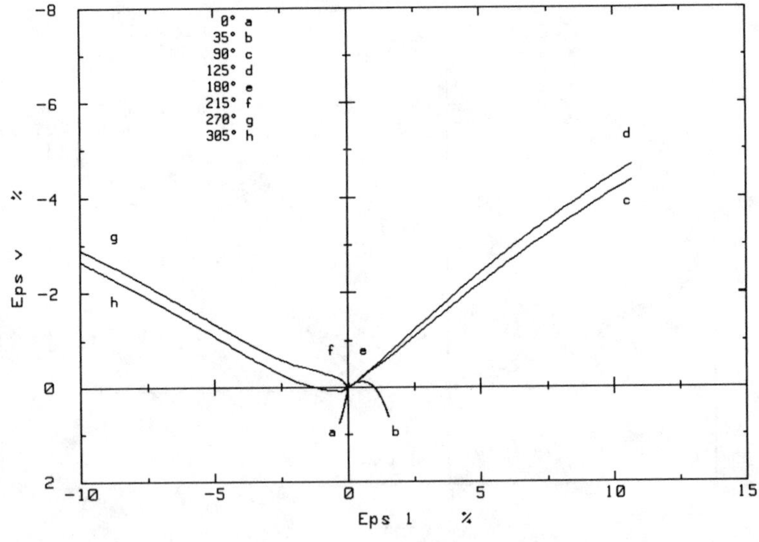

Figure 4: Numerical results of Point 400,100

MECHANICAL PROPERTIES OF MATERIALS ON THE NANOMETER SCALE

B. D. Fabes
Department of Materials Science and Engineering
University of Arizona
Tucson, AZ 85721

W. C. Oliver
Metals and Ceramics Division
Oak Ridge National Laboratory
Oak Ridge, TN 37831

ABSTRACT

Recent advances in instrumentation have lead to the development of hardness testers which can probe extremely small depths (< 1 nm) of material. This equipment promises to provide new insight into the structure and properties of material surfaces and interfaces. Thorough understanding, however, awaits development of new models for interpreting the data from these instruments. Here we describe a sub-nanometer indentation tester, show how it has been used to obtain information on the structure of an alkoxide-derived coating, and demonstrate the need to develop models of the deformation processes which take place in coatings during small-scale indentations.

INTRODUCTION

The mechanical properties of material surfaces are crucial in any application where brittle components are subjected to mechanical loading. Often, thin coatings (< 100 nm) are used to modify the surface properties of such components. Understanding the mechanical properties of these coatings is clearely important, but obtaining information without interference from the underlying material is extremely difficult, if not impossible, using standard hardness tests.

Recently, the development of mechanical property microprobes (such as the Nanoindenter, manufactured by Nanoinstruments, Knoxville, TN) have enabled the mechanical properties of extremely small depths of materials to be probed in a routine fashion. In this paper we will describe the Nanoindenter – a hardness tester with a load resolution of less than 0.3 μN. We will also present experimental results from alkoxide coatings on sapphire substrates, and show how these data can be used to obtain information on the structure of thin films. Finally, we will demonstrate the need to isolate the effects of the coating, interface, and substrate from the measured, composite hardness of this system, and describe our initial efforts in this area.

INSTRUMENTATION

A schematic diagram of the Nanoindenter is shown in Figure 1. The indenter is similar to a conventional hardness tester in so far as hardness is deterimined by indenting the sample with a pyramid-shaped diamond and measuring the deformed area per unit of applied force. Instead of dead weights, the nanoindenter uses a dc-driven loading coal to drive the indenter into the sample. This allows extremely small loads (< 0.3 μN) to be applied. However, Because the resulting indents are so small, their size cannot be determined by direct observation, as in conventional indentation testing. Instead, the displacement of the diamond is monitored continuously during the indentation using a capacitance displacement gage, and, with a knowledge of the geometry of the diamond, the contact area is determined as a function of indenter depth. The hardness is then determined from the load/displacement curve by dividing the load at each displacement with the corresponding diamond area.

A complication arises in that not all of the deflection of the indenter is due to plastic deformation within the indent area: elastic strains also contribute to the total displacement. (In conventional indentation experiments, where the indent sizes are measured optically, the elastic deformation is automatically removed when the indenter is lifted.) Therefore, to account for the elastic deformation, an AC signal is input to the loading coil on top of the dc signal and, using a lock-in amplifier, the resulting in-phase displacements are measured. Thus, a value of the elastic modulus is determined at each depth, and this is used to subtract the elastic displacements from the permanent deformation. (See Reference 1 for more detail.) The result is usually a plot of hardness vs. indentation depth, as shown in Figure 2.

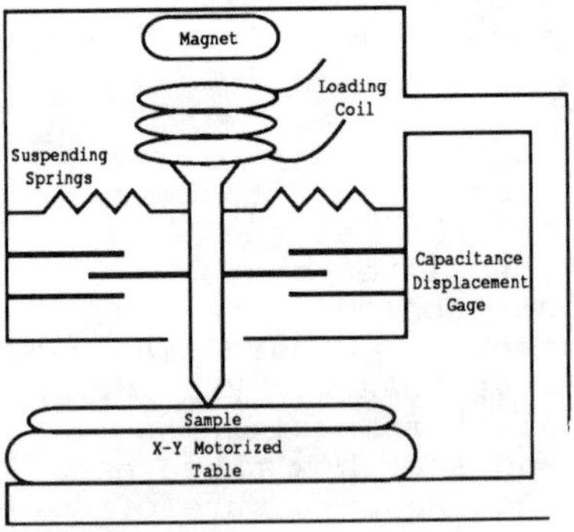

Figure 1: Schematic ofthe Nanoindenter. (after Reference 1).

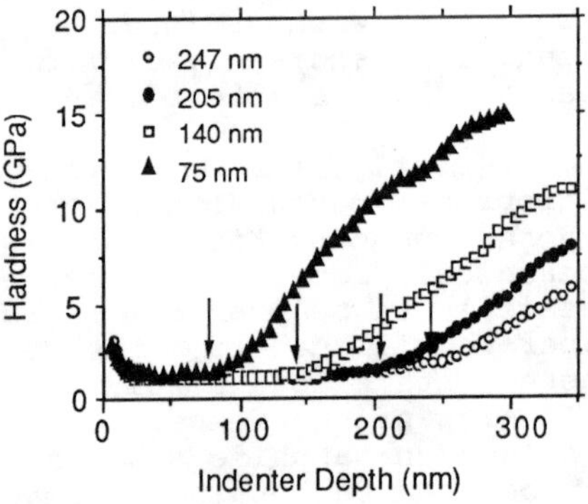

Figure 2: Hardness of TEOS coating on sapphire. Arrows indicate coating thickness.

INDENTATION OF THIN ALKOXIDE FILMS

Tetraethoxysilate (TEOS) films were coated onto sapphire substrates using sol-gel processing techniques, as described in Reference 2. The coatings were deposited in a wedge shape so that a range of thickness (from 50 to 300 nm) could be examined on one sample. The hardness of an unfired coating, as measured by the nanoindenter at four positions along the wedge is shown as a function of indentation depth in Figure 2. The top 30 nm appears harder than the interior for each of the coating thicknesses. The presence of a more dense surface "crust," previously undocumented, has been confirmed using etch rate and TEM studies [3].

At larger indentation depths, the same hardness (1 GPa) is measured for each thickness. This value appears to correspond to the intrinsic, substrate-independent hardness of the film. However, when the coating is deposited on a gold-coated sapphire substrate (an extremely weak interface) this value changes slightly (Figure 3). Hence, even at depths less than 10% of the coating thickness, the interface affects the measured properties of the coating.

Finally, at indentation depths corresponding to the thickness of the coating, the effect of the hard sapphire substrate appears.

MODELING

To understand how the hardness of the coating, the coating/substrate interface, and the hardness of the substrate all affect the measured hardness of a thin film, we are developing two models. The first model is based on Sargent's idea [4] that the net (composite) hardness of a thin coating should be a weighted average of the plastically deformed volumes in the film and in the substrate. Using this model, the indentation process can be separated into three stages: In the first stage, the plastic zone does not propagate into the substrate, so that only the properties of the coating are probed. In the second stage, as the plastic field associated with the indenter extends beyond the coating into the substrate, the properties of the substrate begin to appear; and in the third stage, where the indenter penetrates into the substrate, the composite hardness is dominated by that of the substrate. These three stages are labeled I, II, and III in Figure 3, and drawn schematically in Figure 4.

The second model is based on the definition of hardness as the force per deformed area, so that the net (composite) hardness is modeled as a weighted average of the plastically deformed *areas* instead of volumes. Here, the form of the equation describing the

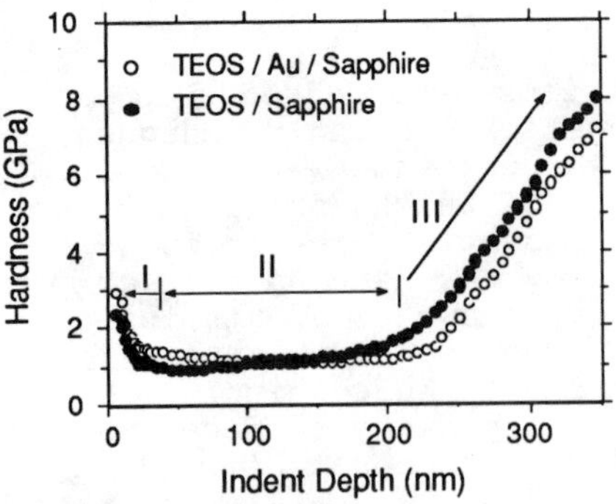

Figure 3 (above): Hardness vs. indent depth for 205 nm TEOS coatings.

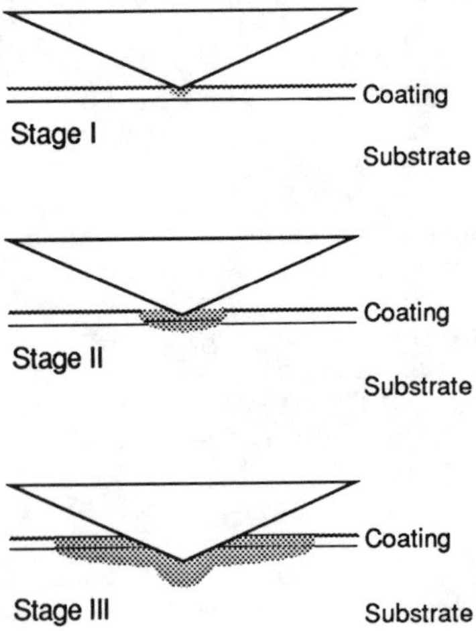

Figure 4 (right): Extents of plastic zones (shaded regions) in volume fraction model.

net hardness is similar to that for the volume fraction model. Only the fitting parameters are different.

By carefully examining these fitting paramteres, our goal is to determine which model more accurately fits the experimental data. Once the better model is determined, these parameters will provide us with clues to the important deformation mechanisms (e.g., dislocation glide, or viscous flow) under small indents which should help poinjt the way to tailoring the mechanical properties of surfaces.

REFERENCES

1. M.F. Doerner and W.D. Nix, *J. Mater. Res.*, 1 (1986) 601-609.
2. B.D. Fabes and W.C. Oliver, *J. Non-Cryst. Solids*, 121 (1990) 348-356.
3. B.D. Fabes, D.L. Klein, and L.J. Raymond, to be published in Better Ceramics Through Chemistry IV, B.J.J Zelinski et al eds, Mat. Res. Soc., (1990).
4. P.M. Sargent, in Microindentation Techniques in Materials Science and Engineering, ASTM STP 899, P.J. Blau and B.R. Lawn, eds., ASTM, Philadelphia (1986) 160-174.

ACKNOWLEDGEMENT

This research was sponsored in part by the Division of Materials Science, U.S. Department of Energy, under contract DE-AC05-84OR2100 with Martin Marietta Energy Systems, Inc.

THREE DIMENSIONAL DEM ANALYSIS FOR EFFECT OF BOUNDARY TO SHEAR DEFORMATION TEST OF GRANULAR MATERIAL

MAKOTO KAWAMURA
Associate professor, Toyohashi University of Technology, Toyohashi 440 JAPAN

SIGENORI MORIMOTO AND KEIZOU YOKOI
Graduates, Toyohashi University of Technology

ABSTRACT

This paper describe the results of the three dimensional DEM analysis for the effect of the boundary to the shear deformation test of granular material. Vertical compressive force are applied on the boundary of the specimen at first. Then the lower boundary is moved to induce the shear deformation of the specimen like a simple shear test while the vertical force is kept constant. As the results of the analysis it is made clear that the distribution of contact forces in the horizontal plane of the specimen is not necessarily uniform even the boundary moves like a simple shear test.

INTRODUCTION

Distinct element method which is developed by Cundall, is suitable to analize the motion of the granular material. DEM is based on the application of Newton's law for the motion of the particles and relation between the contacting force and differential displacement at the contact points. The displacement of the each particles is calculated from the contact force applying the law of the motion. The contact force can be estimated from the displacements of the particle using the constituitve law. The schematic diagram of the flow of the calculation is shown in Fig.1. In this paper an example of three dimensional analysis for simple shear test is introduced and some results due to the movement of the boundary are described.

ANALYTICAL METHOD

The dimension of the test specimen, which is the object of the study, is 16 cm long ,8 cm wide and 4 cm high as shown in Fig.2. The specimen consists of 512 spheres. The diameter of the spheres is uniform and

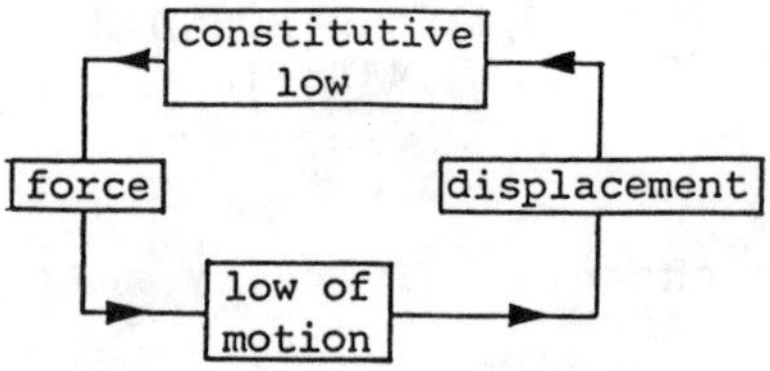

Fig.1 Schematic diagram of the calculation

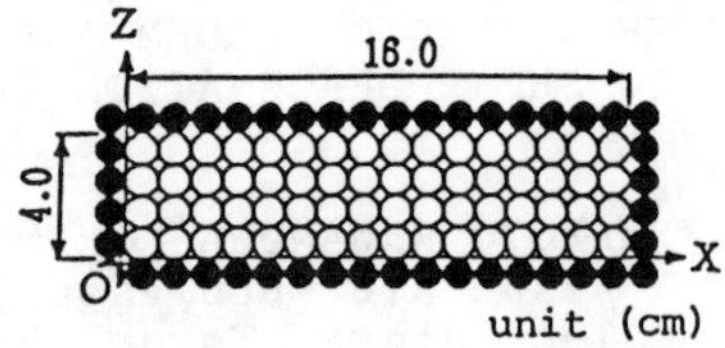

Fig.3 Arangement of the particles

Table 1 Value of the parameters for the analysis

radius	0.5 (cm)
density	2.5 (g/cm^3)
shear rigidity	250.0 (kgf/cm)
normal rigidity	500.0 (kgf/cm)
coeficient of friction	0.0 (kgf/cm^2)
vertival stress	0.5 (kgf/cm^2)
time step	$1.62x10^{-4}$ (sec)
shearing velocity	0.2 (cm/sec)

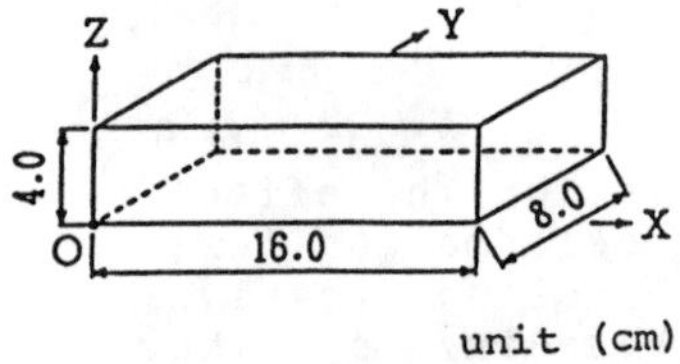

Fig.2 Model of the shear test for the analysis

1 cm. The arangement of the particles in the Y plane is 4 spheres in the z direction and 16 in the x direction, as shown in Fig.3. 8 spheres are in the y direction. In the series of the analyses boundary is taken into account by the dummy particles. The size of the dummy particle is as lärge as the spheres which are the specimen. The black circle in Fig.3 represents the dummy particle.

Values of several parameter such as density of the particle, rigidity at the contact points, are listed on the Table 1. Shear velocity is the speed of the displacement of the lower boundary toward the x direction. Time step denotes the interval of numerical integration in the time domain. An example of the calculation process are explained in Table 2. One cycle correspond one time step mentioned above. At the first stage the particles are generated inside the spacemen and the aranged regularly in this case. At this moment the particles do not move. At the second stage the space is strained in the z direction until the normal stress to the Z plane, σzz reaches 0.5 kgf/cm^2. At the third stage the lower boundary are

Table 2 Process in the analysis

process	number of cycles	shearing velocity (cm/sec)	vertical stress (kgf/cm^2)
generation of particles	0	0.0	0.0
compression	600	0.0	0.5
shearing	14600	0.2	0.5

moved toward the x direction to induce shear deformation of the assebly of the particles with the speed of 0.2 cm/sec. The lateral boundary which corresponds with the X plane at the initial condition, is moved with the speed that changes lenearly from the top to the bottom.

RESULTS OF THE ANALYSIS

The changes of the stress induced in the specimen are plotted against number of cycles in Fig.4. The forces against the all particles which are normal to the plane are sumed up and the stress are calculated as the ratio of the force to the area of the cross setion. When the shear is applied to the specimen, σ_{zz} is kept constant. σ_{xx} is increasing according to the shear deformation.

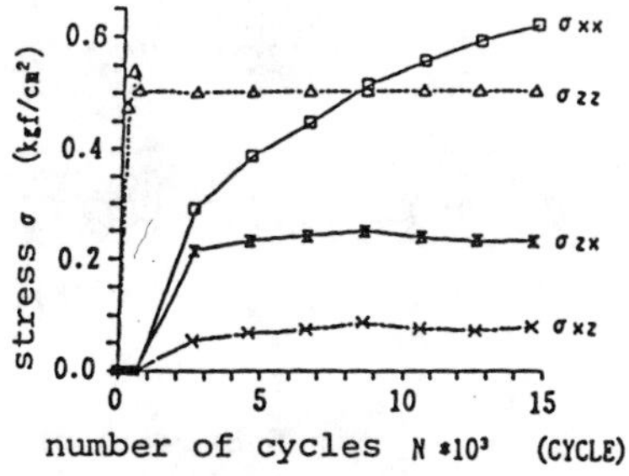

Fig.4 Changes of stress for the number of cycles

This means that the contact forces are affected by the force to displace boundary dummy particles. σ_{zx} and σ_{xz} represents the shear stress induced in the vertical and horizontal plane individualy. The value of shear stress becomes a peak value when the number of cycyles are 8600. This peak is not so sharp.

Graphic display of the analytical results , such as the displacement and the contact force of each particle, makes us easy to understand and judge the apropriety of the results. Fig.5 is an example of the graphic display. The acting force of the particles, the velocity and the contact forces in the Y plane at Y=3.5 cm are drawn as vectors. The vectors in the figure represents the projection to the Y plane. Acting force are estimated as the sumation of the forces acting each particles. The vertical component of acting force are large in the upper layer because of the influence of the stress control in the Y direction. The horizontal component of acting force is large in the lower layer because of the movement of the lower boundary.

Velocity vector of particles has same direction with acting force vector. The velocity in the y direction is equally zero in this case. Expression of the distribution of the contact force vector shows the change of the shear pattern in the spacemen. The portion enclosed by the solid line in Fig.5 represents the area where shear component of contact force is

number of cycles

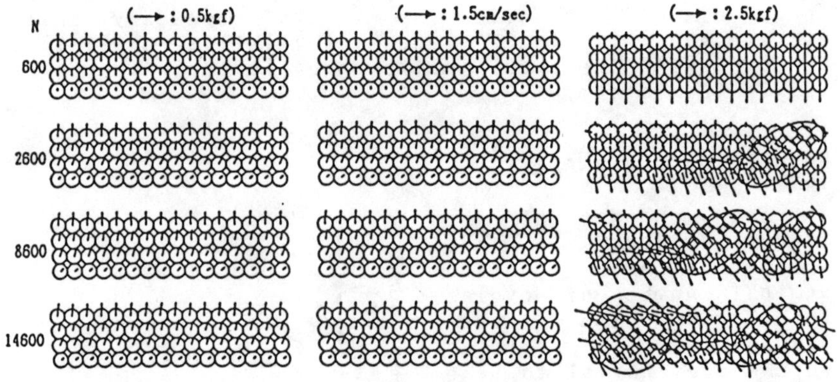

(a) acting force vector (b) velocity vector (c) contacting force vector

Fig.5 Example of graphic display of results of analysis

dominant. When the number of cycle is 2600, the shear zone is seen at the right hand side. Central portion is added as shear zone when the cycle number is 8600 and shear stress becomes peak. When the number is 14600 the shear zone in left hand side appears instead of the central portion. Comparing the contact forces, the distribution of the acting force is regular.

CONCLUSION

As the results of the analysis followings are made clear.

(1) The contact force in the horizontal plane of the specimen is not necessarily uniform even if the boundary moves like a simple shear est.

(2) Due to the effect of the boundary displacement and shear band the distribution of the contact force vector changes much during the shear deformation of the specimen.

REFERENCES

1. P.A. cundall and O.D. Strack, " The distinct element method as a tool for research in granular media, Part 1", Report to National Science Foundation , (1978).

2. P.A. cundall and O.D. Strack, " The distinct element method as a tool for research in granular media, Part 2", Report to National Science Foundation , (1979).

Critical State
And
Constitutive Parameters Identification

SAFWAN LABANIEH

IMG, BP 53X, 38041 Grenoble Cedex, France

Abstract

The paper gives a new definition of "Critical State" permitting to define it from conventional triaxial tests in their range of homogeneous strains (up to pic of stresses). The critical void ratio (e_c) ,or density, is defined as the initial value of void ratio (or density) leading to a maximum value of strain rates ratio ($\mid \dot{\epsilon}_3/\dot{\epsilon}_1 \mid$) of 0.5. On the other hand, the paper gives correlations of constitutive characteristics of soil behaviour with a state parameter (ψ) defined as the relative position of the initial value of void ratio from the critical void ratio ($e_c - e_i$).

Introduction

Critical density is usually defined as the value of the soil density corresponding to no volumetric strains. It is obtained from tests data in the range of large axial strains. Dense sand samples or samples tested under small values of confining pressure exhibit often diffused or total localisation. This falsifies naturally the value of the critical void ratio that the material is supposed to attain under the given confining pressure.

1 Critical density

It is clear then, that if the notion of critical density is valid (volumetric strains can not be undefinite for dense as well as loose sands), its determination from large strain data is not. The definition of critical void ratio (or density) should then be done in the range of homgeneous strains. We define the critical density as the initial density for which, under a defined confining pressure, the maximum strain rates ratio $\mid \dot{\epsilon}_3/\dot{\epsilon}_1 \mid$ is equal to 0.5 (Darve & Labanieh [1], Labanieh [2]). This critical density represents, in the range of large strains, an upper or lower limit, depending on whether the sample is dense or loose, that the material can not exceed.

2 Critical state diagram

The maximum of strain rates ratio ($| \dot{\epsilon}_3/\dot{\epsilon}_1 |$) which is reached by the sample, with a good approximation at the "pic" strain, is almost the same whatever the type of test is ($H/D = 1, H/D/ = 2$, lubrificated end plates or not) (Figure 1).This means that strains are still rather homogeneous. The evolution of the maximum of strain rates ratio with the initial void ratio is linear which permits us easily to determine, for each confining pressure, the initial void ratio for which $| \dot{\epsilon}_3/\dot{\epsilon}_1 |$ is equal to 0.5 (no volumetric strains). The corresponding friction angle (φ_c) is then determined (Figure 2) and cosequently the mean pressure value for the given confining pressure. Folowing the same procedure for different values of the confining pressure we can then plot the critical state diagram ($e_c - \psi$). The sand we studied is the Hostun sand (RF) using the tests data of J-L. Dangus (Colliat-Dangus [3]). We dispose of 83 drained triaxial tests for different densities and covering a large range of stresses going from 50 KPa to 5MPa. From these tests, 39 are with lubrificated end plates with $H/D = 1$ (AF 1), 19 with lubrificated end plates with $H/D = 2$ (AF 2) and 25 with unlubrificated end plates with $H/D = 2$ (F 2). The only data used to determine the critical state diagram is that of lubrificated end plates with $H/D = 1$, the others are presented for comparison. Figure 3 gives the obtained critical state diagram which illustrates the very good coherence of the results and gives a good validtion of this definition of the critical state for the given large range of stresses.

3 Constitutive parameters identification

The state parameter $\psi = e_c - e_i$ (Darve & Labanieh [1], Labanieh [2]) permits to take into consideration the influence of the density as well as the confining pressure on the soil behaviour. We present on figures 5, 6 and 7 the variation of the friction angle (φ), the maximum of strain rates ratio ($| \dot{\epsilon}_3/\dot{\epsilon}_1 |$) and initial modulus/confining pressure (E/σ_0) with the state parameter (ψ). We remark here the good correlations betwween the constitutive characteristics and the state parameter, that is with the void ratio as well as the confining pressure. The proposed correlations fit with experimental observations (Colliat-Dangus [3]) in the range of high level confining pressures.

References

[1] F. Darve, S. Labanieh "Incremental constitutive law and for sands and clays. Simulations of monotonic and cyclic tests, Int. J. Num. and Anal. Meth. in Geomechanics", 6, 1982

[2] S. Labanieh "Modélisations non-linéaires de la rhéologie des sables et applications", Thèse de doctorat es siences, Univ. J. Fourrier, Grenoble, 1984.

[3] J-L. Colliat-Dangus "Comportement des matériaux granulaires sous fortes contraintes-Influeence de la nature minéralogique du matériau étudié", Thèse de Docteur de Spécialité, Univ. J. Fourrier, Grenoble, 1986.

Figure 1:

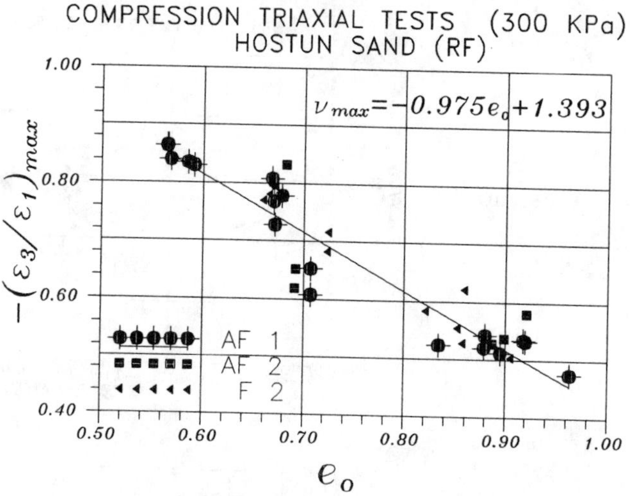

Figure 2:

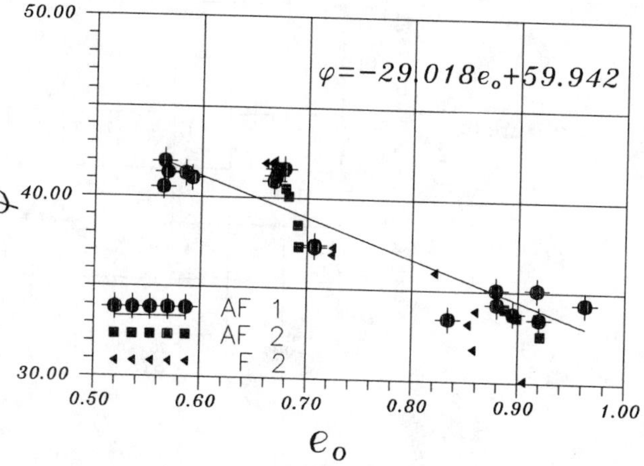

Figure 3:

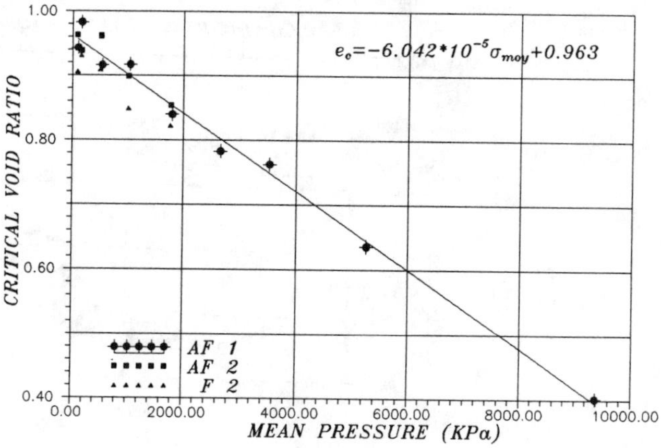

Figure 4:

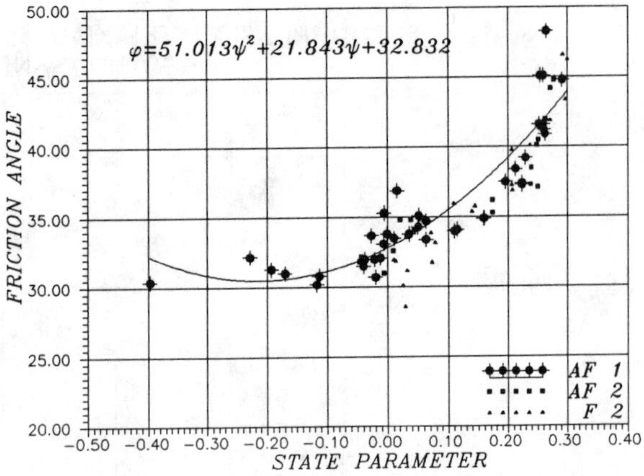

Figure 5:

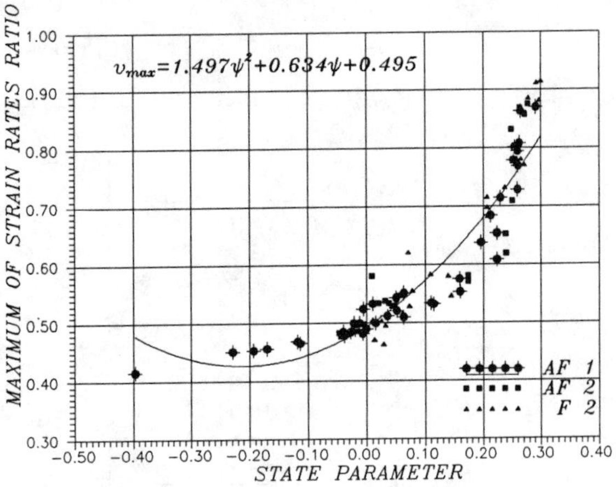

Figure 6:

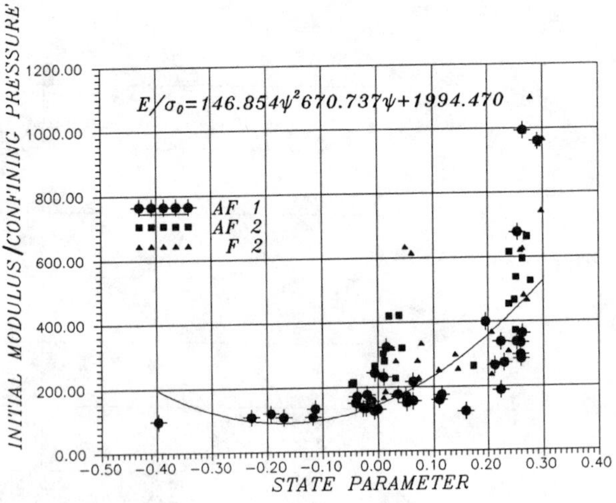

DESCRIPTION OF STRAIN HISTORY AND STRAIN RATE HISTORY IN PURE TITANIUM AND PURE COPPER

Koji MIMURA
Dept. of Mechanical Engineering, Kobe Univ., Kobe, Japan

Akio SHINDO
Dept. of Electro-Mechanical Engineering, Osaka Electro-Communication Univ., Osaka, Japan

ABSTRACT

In order to clarify the influence of strain history and strain rate history on the subsequent loading , the equi-equivalent plastic strain surfaces of pure titanium and pure copper after the quasistatic pre-tensile straining were carefully examined over the wide strain rate range, 10^{-5}/sec ∿ 600/sec. The obtained results are discussed on the basis of the corresponding numerical simulations using the Perzyna type overstress theory.

INTRODUCTION

In many experimental studies, the effects of the strain history and the strain rate history on the subsequent loading were investigated separately because of the difficulties in the measurement techniques. For example, the former was usually examined by means of the stress probe method at a quasi-static strain rate and the jump test from a certain strain rate to another under the uniaxial loading condition was used to examine the latter. The purpose of this paper is to clarify both effects on the subsequent loading simultaneously and to describe them in the constitutive relations. For this purpose, the jump tests under the multi-axial loading condition from the quasistatic strain rate to the elevated ones were carried out on pure titanium and pure copper and their subsequent loading surfaces were carefully examined.

Experimental Procedures : The apparatus used to perform the jump tests were the hydraulic proof system (HPS) and the clamp type split Hopkinson bar (CSHB) which allow the axial-torsional combined loading of the specimen at the quasistatic and/or the dynamic strain rates. In the low strain rate range of 10^{-5}/s to 10/s, HPS was mainly used and CSHB was employed at more higher strain rates up to 600/s. In the experiment, a tubular specimen was used and the total displacements (axial and torsional) were measured directly by using the strain gauges. The jump tests under the multi-axial condition were performed by combining the quasistatic loading and the dynamic loading. The quasistatic loading in the tensile direction was first applied to the specimen until the axial plastic strain had attained to 0.015, then the dynamic combined loading at the various strain rates followed without unloading.

Experimental Results : The initial yield surfaces (the quasistatic and the dynamic) of pure titanium and copper were almost the Mises type. The difference in the equivalent stresses between the axial and the torsional direction was very small. Figure 1 shows the strain rate sensitivities for both metals obtained by the tension test at the constant strain rates. The typical equi-equivalent plastic strain surfaces (or the loading surfaces) obtained by the jump test under the multi-axial loading condition were shown in Fig.2 (for titanium) and Fig.3 (for copper). In the figure, the experimental results are shown by some symbols and the solid lines. The quasistatic loading surfaces of both metals (Fig.2.a and Fig.3.a) show the remarkable distortion due to the Bauschinger effect at the earlier part of the subsequent loading. This distortion in the loading surface gradually disappears as the subsequent strain increases. In the case of copper, the cross effect is also observed. Here, our interest is not only the strain rate history itself but also its dependence on the subsequent loading directions (on the strain history), namely, whether the subsequent dynamic loading surface is the simple expansion of the corresponding quasistatic loading surface or not. With respect to this point, pure titanium and copper show the different behaviors (Fig.2.b and Fig.3.b). In the case of copper, we can suppose that the subsequent dynamic loading surfaces are the isotropic expansion of the corresponding quasistatic loading surfaces. On the contrary, the dynamic loading surfaces of pure titanium are not the simple expansion of the quasistatic ones. The distortion in the loading surface at an elevated strain rate disappears more rapidly than that at the quasistatic strain rate.

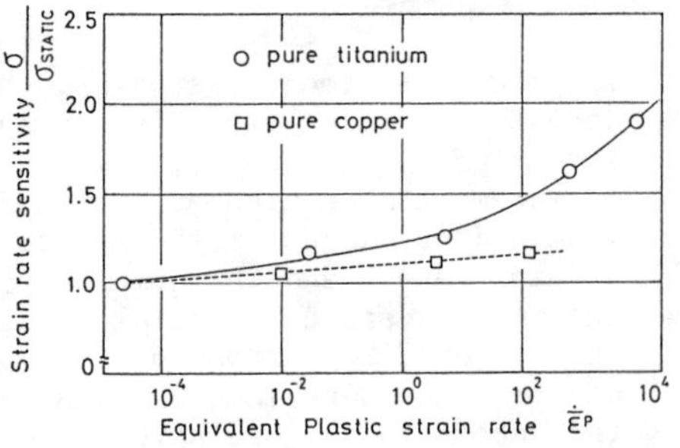

Fig. 1. Strain rate sensitivities of pure titanium and pure copper.

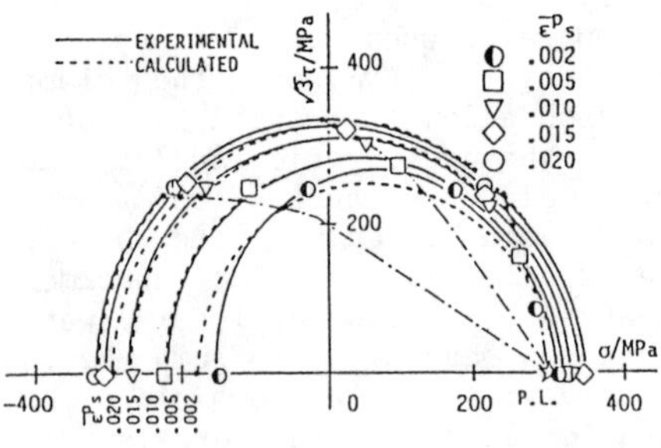

Fig. 2a. Subsequent loading surface of pure titanium. $\dot{\varepsilon}p=3\times10^{-5}$

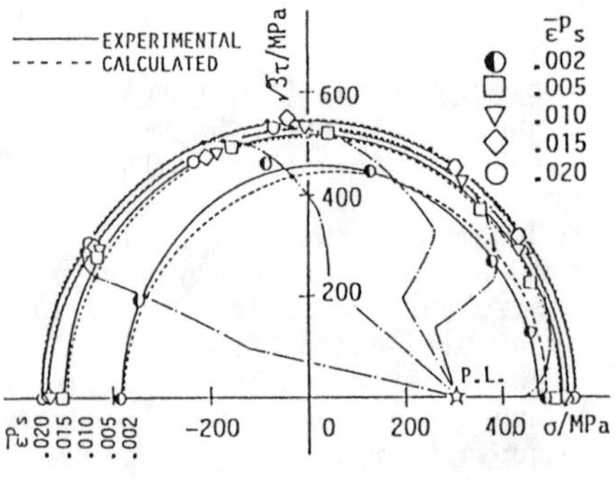

Fig. 2b. Subsequent loading surface of pure titanium. $\dot{\varepsilon}P=600/s$

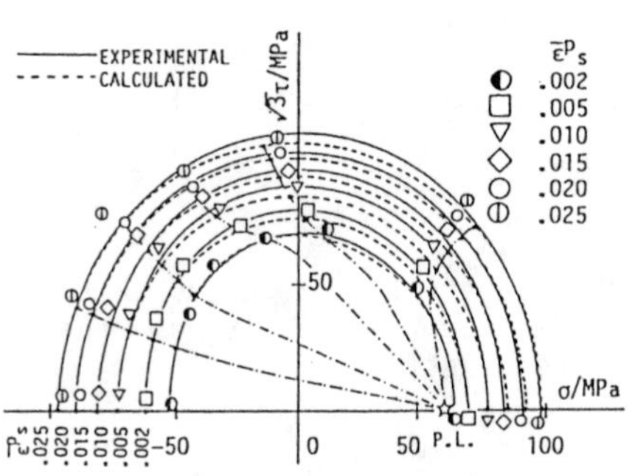

Fig. 3a. Subsequent loading surface of pure copper. $\dot{\varepsilon}^P=3\times10^{-5}$/s

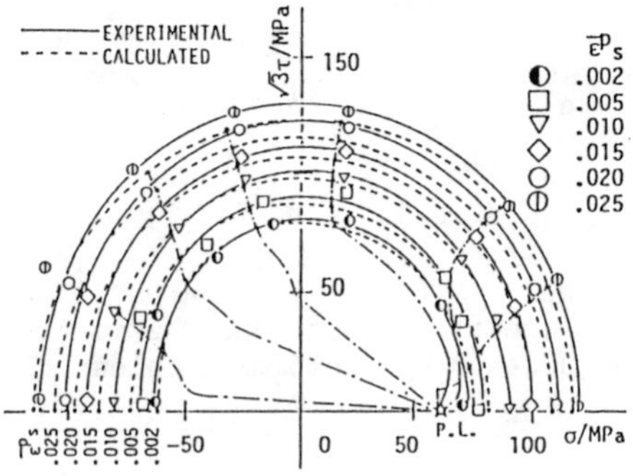

Fig. 3b. Subsequent loading surface of pure copper. $\dot{\varepsilon}^P$=130/s

Constitutive Relation and Simulation : For the description of the subsequent loading surfaces, the constitutive equations based on the internal state variables [1] and the Perzyna's overstress theory [2] are formulated. The quasistatic yield function f is supposed as

$$f = (3/2)\, \tilde{S}_{ij}\tilde{S}_{ij} - 1 = 0 \;, \quad \tilde{S}_{ij} = (\, S_{ij} - \alpha'_{ij}\,) \,/\, \bar{S}y \tag{1}$$

where $\bar{S}y$ denotes the quasistatic yield radius, S_{ij} and $\alpha'ij$ are the deviators of the stress and the back stress, respectively. Based on the Perzyna's overstress theory, the relation between the plastic strain rate and the overstress is given by

$$\dot{\varepsilon}^p{}_{ij} = \eta \,\langle\Phi(F)\rangle \,(\, \partial f \,/\, \partial \tilde{S}_{ij}\,) \;, \quad F = \bar{S}y^* \,/\, \bar{S}y - 1 \tag{2}$$

Where F and $\bar{S}y^*$ denote the overstress and the dynamic yield radius, respectively. Here, the overstress function Φ is supposed to take the form [3] :

$$\eta \,\langle\Phi(F)\rangle = K_1 \,/\, (\, \Gamma_I + \Gamma_{II}\,) \tag{3}$$

and

$$\Gamma_I = \exp\{K_2 - K_3(F+Fo)\} \quad (\text{ for thermal activation })$$
$$\Gamma_{II} = \exp\{K_4/(F+Fo)\} \quad (\text{ for viscous drag }) \tag{4}$$

$K_1 \sim K_4$ are the material constants and Fo is a proper offset so that Φ(F) is nearly equal to zero at the quasistatic strain rate. In order to describe the influence of the strain history on the subsequent loading, the internal state variable of the fourth-order tensor **ρ** and the non-proportional loading function ζ are introduced. They are used in the evolutional equations of the back stresses. Further details were presented in ref.[1].

$$\dot{\rho}_{ijkl} = D\rho(e_{ij}e_{kl} - \rho_{ijkl})\dot{\bar{\varepsilon}}^p \;, \quad e_{ij} = (2/3)\dot{\varepsilon}^p{}_{ij}/\dot{\bar{\varepsilon}}^p \tag{5}$$

$$\zeta = \{\, 1 - (\, 3\, e_{ij}\rho_{ijkl}\rho_{klmn}e_{mn}\,)/(\, 2\, \rho_{pqrs}\rho_{pqrs}\,)\,\}^m \tag{6}$$

where Dρ and m are the material constants. $\dot{\bar{\varepsilon}}^p$ denotes the equivalent plastic strain rate. To describe the strain rate history under the multi-axial loading condition, we suppose that the back stress is also the function of F. Thus,

$$\dot{\alpha}'_{ij} = \sum_{k=1}^{3} \dot{\alpha}'^{(k)}_{ij} \quad , \quad \dot{\alpha}'^{(k)}_{ij} = (1 + B^{(k)}\zeta)Da^{(k)}\{ 1 + \lambda\Delta F^{\ell} \} \times \{ Ca^{(k)}e_{ij} - \alpha'^{(k)}_{ij}\} \dot{\bar{\varepsilon}}^{p} \tag{7}$$

and

$$\Delta F = F - (K_1 / \Gamma_I) \tag{8}$$

where $B^{(k)}$, $Ca^{(k)}$ and $Da^{(k)}$ are the material constants. With respect to the quasistatic yield radius, we define it as the function of the dimensionless effective plastic work :

$$\bar{S}y = \bar{S}y^{o}(1 + R\, W^{p})^{n}, \quad \dot{W}^{p} = (\tilde{S}_{ij}/\bar{S}y^{o}) \dot{\varepsilon}^{p}_{ij} \tag{9}$$

where $\bar{S}y^{o}$ denotes the initial yield radius. The numerical simulations based on the proposed constitutive equations are performed to compare with the experimental results. The material constants used are shown in Table 1. The simulated loading surfaces and the stress paths are shown in Fig.1 and Fig.2 by broken lines and dash-dot-dash lines, respectively. The small difference between the simulated surface and the experimental surface is observed in the range of $\bar{\varepsilon}^{p}s$ (subsequent equivalent plastic strain) ≦ 0.002. However, it is improved if the deformation induced anisotropy in the subsequent yield surfaces is also taken into the consideration [1]. In the range of $\bar{\varepsilon}^{p}s$ ≧ 0.005, the simulated results well express the experimental results. Especially, in the case of pure titanium, the rapid recovery of the distortion in the loading surface similar to the experimental one can be observed at the high strain rates.

Table 1. Material constants.

Rate sensitivity			Strain History		
	Ti	Cu		Ti	Cu
K_1	5×10^5	1×10^7	$D\rho$	25.0	25.0
K_2	39.5	80.0	m	1.0	1.8
K_3	26.0	70.0	Ca^1	0.5	0.8
K_4	10.0	4.0	Da^1	100.0	200.0
Fo	0.5	0.75	B^1	0.5	25.0
			Ca^2	0.0	0.8
λ	1400	0.0	Da^2	0.0	10.0
ℓ	2.6	0.0	B^2	1.0	25.0
			R	1.0	13.5
			n	0.5	0.45

<u>**Conclusions**</u> : The strain rate jump tests under the multi-axial loading condition were carried out on pure titanium and pure copper to investigate the influence of the strain history and the strain rate history. The obtained dynamic loading surfaces are not always the simple expansion of the corresponding quasistatic loading surfaces. In the case of pure titanium, the distortion in the subsequent loading surfaces due to the Bauschinger effect more easily disappears at the high strain rates. The numerical simulations based on the proposed constitutive equations successfully express these experimental results.

REFERENCES

1. Shindo,A., Mimura,K. "On the Formulation of a Constitutive Equation Using a Modified 4th-Rank Anisotropic Moduli Tensor" Proc. of Advances in Plasticity 1989, p.107, 1989
2. Perzyna,P. "The Constitutive Equation for Rate Sensitive Plastic Materials ", Q.Appl.Math. 20,p.321,1963
3. Hojo,A., Chatani,A. "A Constitutive Equation for Carbon Steels in Wide Range of Strain Rates", J. of the Soc. of Material Science, Japan. 34-387, p.1400, 1985

STUDY OF TENSILE FRACTURE IN POLYPROPYLENE FIBER REINFORCED CONCRETE USING LASER HOLOGRAPHIC INTERFEROMETRY

B. Mobasher
Member of Technical Staff, USG Corporation Research Center, Libertyville, IL.

A. Castro-Montero
Graduate Research Assistant, Dept. of Civil Eng., Northwestern University, Evanston, IL.

S.P. Shah
Professor of Civil Engineering and Director of NSF Science and Technology Center for Advanced Cement-Based Materials, Northwestern University, Evanston, IL.

ABSTRACT

Toughening of cement-based materials containing relatively high fiber volume are studied. Crack propagation and damage distribution were examined by laser holographic interferometry. Based on fracture mechanism observed during experimental studies, a R-curve approach proposed earlier was used to predict toughening of matrices due to fiber reinforcement. The theoretical predictions show a good agreement with the experimental results.

INTRODUCTION

Fiber reinforced cement-based materials are increasingly being used for building products. Using conventional concrete construction practice, the amount of fibers normally incorporated is about less than one percent by volume of concrete. However, with use of superplasticizers and specially developed fabrication techniques, it is possible to incorporate larger volume fraction of fibers, i.e. up to 15% by volume. Fibers in such large quantities seem to fundamentally alter the nature of cementitious matrices. Tensile stress-strain relationships of cement-based matrix reinforced with two volume fraction of polypropylene fibers is shown in Fig. 1. The ultimate strength of the matrix is referred to as the Bend Over Point (BOP) which occurs at approximately 300-1000 $\mu\epsilon$ depending on the fiber volume fraction. Inset of the figure represents the contribution of the matrix phase, and indicates that the strength of the matrix can reach values as high as 15 MPa. Also at 2% strain the Matrix is carrying up to 8 MPa. This indicates that the tensile strength and strain capacity of the matrix itself is fundamentally enhanced.

HOLOGRAPHIC INTERFEROMETRY

Earlier works [1] had indicated existence of microcracking in the initial loading stages. In order to characterize the initiation and propagation of microcracks in cement based composites, reflection holographic Interferometry was developed. Holographic interferometry offers a field of view orders of magnitude larger than that of optical microscopy, and yet the displacement sensitivity remains in the order of one micron. The non destructive nature of holographic interferometric measurements allows for the characterization of the response at various strain levels for the same specimen.

Fibrillated continuous uniaxial polypropylene fiber specimens were manufactured by means of a pulltrusion process as described in [2]. Volume fraction of the fibers used was in the ranges of 8-12% by volume. Uniaxial tensile dog bone shaped specimens of dimensions 10x15x110 mm (0.394x0.594x4.33 in) were used. The Elongations were measured using a LVDT, and the load-elongation response was continuously recorded using a chart recorder.

Single beam reflection holograms (Lippmann-Denisyuk type [3]) were used. Figure 2 shows the optical arrangement for the recording of the holograms. The laser light which passes through the plate is the reference beam. As this light is reflected from the object, it interferes with the oncoming reference beam, the interference is hence recorded on the photographic emulsion of the plate. The hologram is thus recorded. By applying an incremental displacement to the specimen between two exposures recorded on the same plate, interferometric fringes are obtained. Since the holographic plate is rigidly attached to one point on the specimen, the only relative displacement of the plate with respect to the specimen is that due to straining. After reconstruction, an image analysis system was used for the enhancement and quantitative measurement. Experimental details for image reçonstruction and image analysis are provided elsewhere [4],[5].

As shown in figure 3.a the first cracks are apparent in the specimen at a stress level of about 8 MPa ($\epsilon = 290\mu\epsilon$). Cracking is verified by the localized variations in an otherwise smooth fringe patterns. The sensitivity of the measurement is in terms of displacement discontinuity of the order of 0.25 microns. By using a lettering system as the guide, cracks which were apparent are referred to as cracks "A", "B", and "C". The average spacing of these cracks for composites with 12% volume fraction is 13 mm.

During the following loading stages the cracks at location "A", "B", and "C" seem to propagate. The dotted lines indicate a superposition of the cracks at the previous stages onto the present state. At the third loading stage ,(figure 3.b, $\epsilon = 630\mu$ str), there is not a significant activity observed at the cracks "A", and "C", while crack "B" seems to be the only crack which propagates in this loading increment. Simultaneous propagation of microcrack bands are observed to take place. At $\epsilon = 0.1\%$, the crack "A" is 14 mm long. The bend over point coincides with crack "A" reaching the specimen boundary at $\epsilon = 0.12\%$. After the BOP cracks at all previously mentioned locations propagate across the specimen, while the fringes at the vicinity of these cracks indicate additional cracking. At around 0.3% strain the cracking is so closely distributed that it was not possible to identify individual cracks. As the cracking saturates in the specimen at the strain level of 2.7%, as shown in Fig 3.c, no discontinuity in fringe pattern is observed indicating that the specimen is homogeneously strained.

R-CURVE APPROACH

A primary character of fracture of quasi brittle materials is the existence of stable crack growth prior to the crack reaching its critical length. Such behavior is often referred to as R-curve behavior. For brittle materials, there is no stable crack growth. Whenever the strain energy release rate is equal to the fracture toughness of a material, the material fails (horizontal line in Fig. 4). For concrete, however, crack growth is heterogeneous and tortuous, and is accompanied by grain boundary sliding. The existence of fracture process zone results in the stable crack growth prior to the peak load. This increased energy requirement is represented as a rising R-curve in Fig. 4. For fiber reinforced concrete, the presence of fibers introduce additional toughening, which requires more energy dissipation during crack growth. This is indicated by a second rising R-curve in Fig.4. An R-curve approach has been recently proposed [6][7]. By assuming that the critical crack length, a_c, is proportional to the initial crack, a_0 as shown in eq. 1, R-curve can be derived as a function of crack extension:

$$a_c = a_0 + \Delta a_c = \alpha a_0 \quad (1)$$

$$R = \beta\,(a - a_0)^d\,;\quad d = \frac{1}{2} + \frac{\alpha-1}{\alpha} - \left[\frac{1}{4} + \frac{\alpha-1}{\alpha} - \left(\frac{\alpha-1}{\alpha}\right)^2\right]^{\frac{1}{2}} \quad (2)$$

where Δa_c is precritical stable crack growth, and α, and β are constant which reflect the material properties and geometrical loading conditions. Evaluation of these parameters can be achieved by using a fracture model proposed by Jenq, and Shah [8], as the failure criterion. Their model is based on an effective elastic crack approach and the role of fibers can be modelled by means of a closing pressure. Both the stress intensity factor and crack tip opening displacement consist of contributions from far-field stresses and closing pressure of fibers. The crack stability conditions for the matrix can be expressed as,

$$K_{IC}^{s}=K_{I}^{m}(\sigma_{c},\frac{a_{c}}{b})-\int_{0}^{a_{c}}p(\frac{x}{a_{c}})K_{I}^{F}(\frac{x}{a_{c}})dx \tag{3}$$

$$CTOD_{c}=CTOD_{m}(\sigma_{c},\frac{a_{c}}{b},\frac{a_{0}}{a_{c}})-\int_{0}^{a_{c}}p(\frac{x}{a_{c}})Q(\frac{a_{c}}{b},\frac{a_{0}}{a_{c}},\frac{x}{a_{c}})dx \tag{4}$$

Based on holographic observation, a single-notch specimen subjected to uniaxial tension as shown in Fig. 5 was considered. K_I^m and $CTOD_m$ are the stress intensity and crack tip opening displacement at the tip of an effective traction free crack due to far-field stress. σ_c is the critical far-field stress; a_c is the critical crack length; *p(x/a)* is the closure force at an arbitrary point x along the crack face, Q is the Green's function for the crack closure at point a_0 (initial crack mouth) due to a unit force applied at point x along the crack length; b represents the specimen width. $K_I^F(x/a_c)$ is the stress intensity due to a unit load applied at point x along the crack surface. K_I^m, $CTOD^m$, K_I^F and Q are obtained based on LEFM [9].

A bilinear formulation of closing pressure was assumed and is shown in Fig. 5. The closing pressure increases linearly up to a length characterized by C_0a_c, where the magnitude of stress reaches a constant fraction of the far field stress, σ_∞, ($\lambda\sigma_\infty$, where λ is defined as the stress transmission factor, $0 < \lambda < 1$),

$$p(x)=\lambda\sigma_{\infty} \qquad a_{0}<x<C_{0}a_{c} \tag{5}$$

$$p(x)=\lambda\sigma_{\infty}\left[1-\frac{C_{0}a_{c}-x}{C_{0}a_{c}}\right] \qquad C_{0}a_{c}<x<a$$

$\lambda = 0$ represents the case of no closing pressure, (i.e. plain matrix).
$\lambda = 1$ represents the case where all the force is being transmitted across the crack.

λ represents the effectiveness of type, and volume fraction of fibers in transmitting the far field stress across the crack faces. As λ approaches unity, the length of $(1-C_0)a_c$ does not contribute significantly to the stress intensification. The closing pressure yields the form used by Marshall et al. [10].

The two coupled non-linear integral equations of 3 and 4 can be solved for the two unknowns, a_c and σ_c. Once a_c is known, α and d can be calculated from equations 1, and 2. Assuming that the response of matrix and closing pressures can be both combined using the R-curves as a monolithic material with the closing pressure effect of fibers included as a matrix property, one can write:

$$R_{IC}^{s} = \frac{K_{IC}^{s\,2}}{E_m} = \frac{\left[K_I^m(\sigma_c,a_c)+K_I^f(a_c)\right]^2}{E_m} = \beta(a_c-a_0)^d \tag{6}$$

Equation 6 reflects the condition that the critical point is reached when the difference between these applied and closing pressure stress intensity factors reaches K_{IC}^{S}. One can use the obtained R-curve to develop a theoretical load-deformation response. [5]

Comparison with Polypropylene Fiber Reinforced Composites

Comparison of theoretical and experimental results for stress at the BOP for polypropylene fiber composites is indicated in Fig. 6. The matrix properties used are based on experimental results reported by Jenq and Shah [8]. The experimental results tested by Krenchel and Stang [2], and Mobasher, Stang, and Shah [11] were used. A constant value of $C_0 = 0.1$ was used. This study assumes that increasing λ can simulate the fiber volume fraction, and hence the closing pressure. Furthermore it is assumed that λ is constant throughout crack propagation. In figure 6 the theoretical results are shown for various values of λ, The experimental results are shown for the actual volume fraction of fibers used. One can observe for the theoretical results that as λ increases both the strength and the strain increase. Note that the same behavior is exhibited by increasing the volume fraction of fibers in the experimental results. The present choice of pressure predicts lower values of both stress at BOP as compared to the experimental data. Although not shown, one can observe a prediction of increased strain capacity.

To enable comparison with the present theory with the ACK theory [12], a value of interfacial shear strength, τ, is required. Depending on the fiber volume fraction used, τ/r as estimated by the crack spacings measured in reference [1, 11] results in a range of 100-200 Mpa/mm. These two values were used in the ACK model to compute the tensile strength and are also presented in figure 6. Note that the ACK model provides a good correlation with experimental data if the value of τ/r = 200 Mpa/mm is used, however this value seems to be significantly higher than values reported in the literature. [13]

CONCLUSIONS

The dependance of matrix ultimate strength on the fiber type and volume has been verified by both experimental and theoretical means. Holographic interferometry was used to quantitatively measure microcracking in polypropylene fiber reinforced cement based composites under tensile loading. Formation of microcracks have been verified to occur prior to the bend-over point. An R-curve approach was used to simulate the toughening of cement-based matrices due to fiber reinforcement. Toughening mechanism of fibers was modelled by means of applying closing pressure on crack surfaces.

ACKNOWLEDGEMENT

The support of National Science Foundation Center for Science and Technology of Advanced Cement-Based Materials (ACBM) Grant #DMR-8808432, and NSF Grant MSM-8906937 is gratefully appreciated.

1. Mobasher, B., Stang, H., and Shah, S. P., "Microcracking in Fiber Reinforced Concrete," Cement and Concrete Research, Vol.20, 1990, in press.

2. Krenchel H., and Stang, H. "Stable Microcracking in Cementitious Materials", Proceedings-2nd International Symposium on Brittle Matrix Composites Cedzyna, Poland, (Sept. 1988).

3. Denisyuk, Y.,"Photographic Reconstruction of the Optical Properties of an Object in its own Scattered Radiation Field," Sov. Phys-Dokl. 7, 543 (1962).

4. Mobasher, B., Castro-Montero, A, and Shah, S. P., "A study of Fracture in Fiber Reinforced Cement-Based Composites Using Laser Holographic Interferometry," Experimental Mechanics, Vol. 30, 1990, in press.

5. Mobasher, B., "Reinforcing Mechanism of Fibers in Cement Based Composites," Ph.D. Dissertation, Department of Civil Engineering, Northwestern University, Evanston, IL. (June, 1990).

6. Ouyang, C., Mobasher, B., and Shah, S. P., "An R-Curve Approach for Fracture of Quasi-Brittle Materials," Engineering Fracture Mechanics, Vol. 37, 1990, pp. 1-13.

7. Mobasher, B., Ouyang, C., and Shah, S. P., "Modeling of Fiber Toughening in Cementitious Materials Using an R-Curve Approach," International Journal of Fracture, 1990, in press.

8. Jenq Y. S., and Shah S. P.,"Two Parameter Fracture Model for Concrete", Journal of Engineering Mechanics, ASCE, 111, 1985, 1227-1241.

9. Tada, H., Paris, P. C., and Irwin, G. R., The Stress Analysis of Cracks Handbook, 2nd Ed., Paris Production Inc., St. Louis, Missouri, 1985.

10. Marshall, D. B., Cox, B. N., and Evans, A. G.,"The Mechanics of Matrix Cracking in Brittle-Matrix Fiber Composites," Acta Metal., Vol.33, No. 11, 1985, pp. 2013-202.

11. Stang, H., Mobasher, B., and Shah, S. P., "Quantitative Damage Characterization in Polypropylene Fiber Reinforced Concrete," Cement and Concrete Research, Vol. 20, 1990, in press.

12. Aveston, J., Cooper, G. A., and Kelly, "Single and Multiple Fracture," *The Properties of Fiber Composites, National Physical Laboratory*, Guildford, Surrey: IPC Science and Technology Press Ltd. (1971).

13. Beaudoin, J. J.,HANDBOOK OF FIBER-REINFORCED CONCRETE: Principles, Properties, Developments and Applications, Noyes Publications, N.J., 1990.

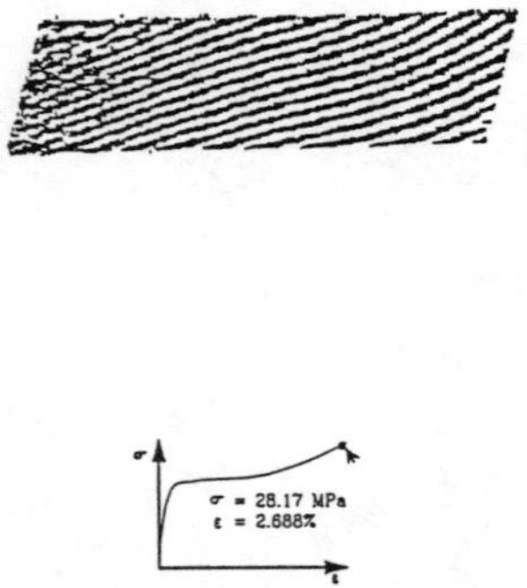

Fig. 3.c Hologram of Microcracks at $\epsilon = 2.7\%$, $\sigma = 28.2$ MPa

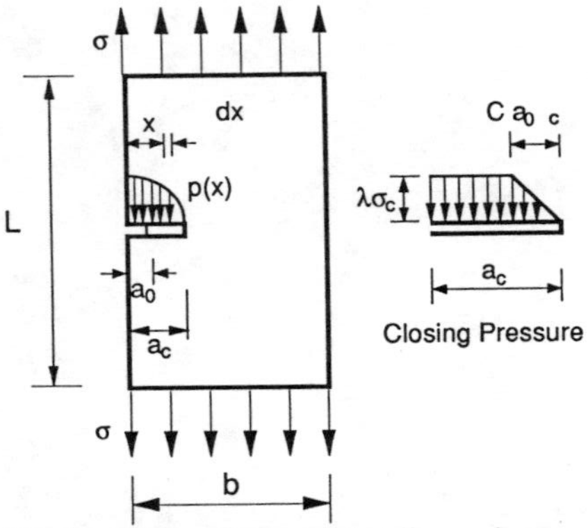

Fig 5. Single edge notch tensile specimen and the closing pressure parameters.

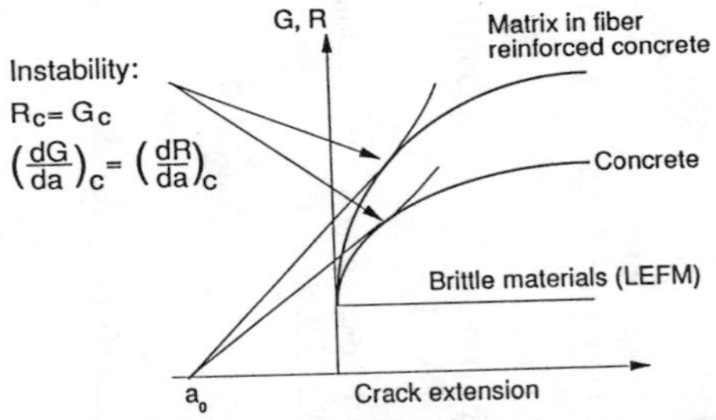

Fig 4. R-Curve Behavior of Quasi-brittle materials

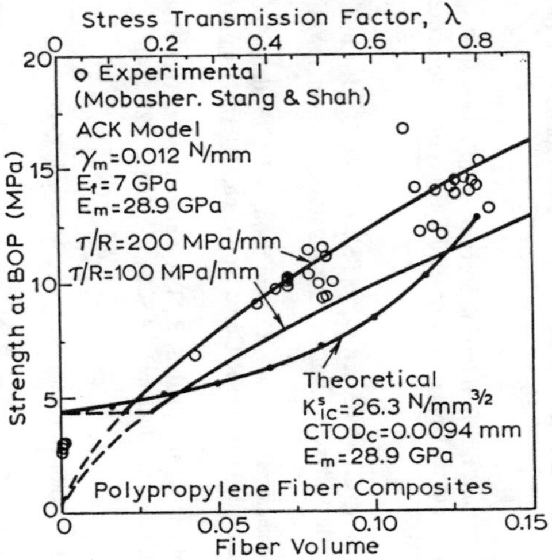

Fig 6. Experimental and theoretical comparison for polypropylene fiber reinforced composites.

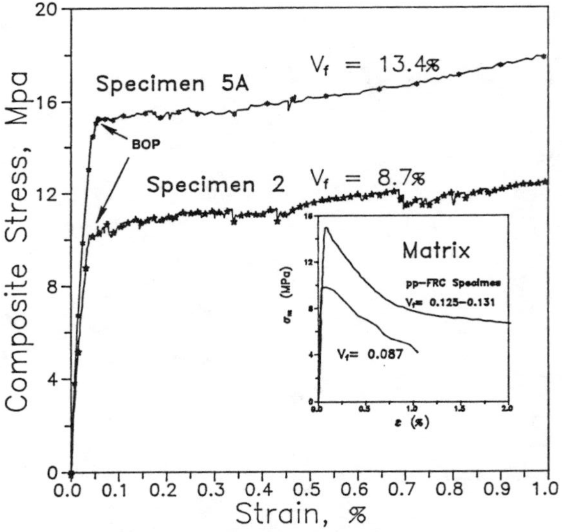

Fig 1. Tensile stress strain response of polypropylene fiber reinforced composites.

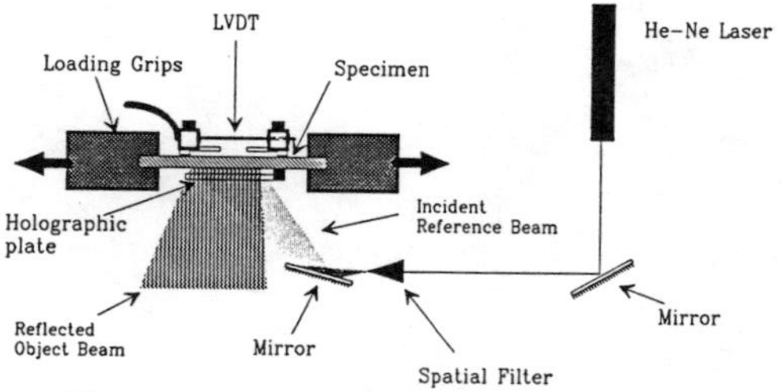

Fig. 2 Reflection holography optical set-up

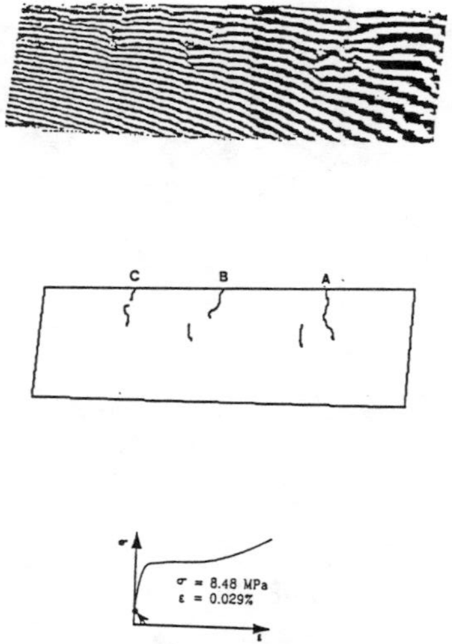

Fig. 3.a Hologram of Microcracks at $\epsilon = 290\,\mu$str, $\sigma = 8.5$ MPa

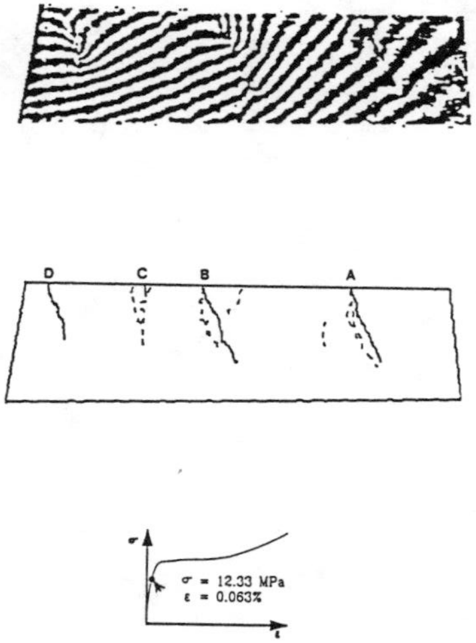

Fig. 3.b Hologram of Microcracks at $\epsilon = 630\,\mu$str, $\sigma = 12.3$ MPa

AN EXPERIMENTAL INVESTIGATION OF YIELD SURFACES IN GRANULAR MEDIA

EMMANUEL PETRAKIS, RICARDO DOBRY, PAUL VAN LAAK,
LI LIU, AND PANOS KOTSANOPOULOS
Rensselaer Polytechnic Institute, Civil Eng. Dept.
Troy, NY 12180–3590, USA

ABSTRACT

The yield surfaces of granular media are investigated by constant mean stress experiments on glass beads specimens, and by numerical simulations. In both tests and simulations the yield surfaces distort and form a smooth apex in the direction of loading, similar to observations in metals.

INTRODUCTION

A number of constitutive laws have been proposed for soil materials based on the incremental theory of plasticity. For dry granular soil, the kinematic strain hardening hypothesis is often used, which assumes that the yield surface(s) translate in stress space without changing size or shape during plastic flow. Also, multiple nested yield surfaces are introduced to represent the observed nonlinear stress–strain response [1,2].

There is need for further experimental verification of these laws, as well as for relating the stress–strain behavior to the underlying micromechanical phenomena. Similar studies in metals [3–5] have generally shown a distortion of the yield surface in the direction of loading, Fig. 1. This paper summarizes results of such an experimental investigation on a granular medium.

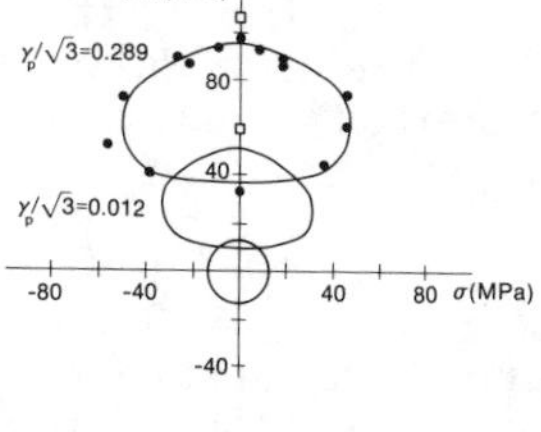

Fig. 1. Aluminum: experimental yield loci [5].

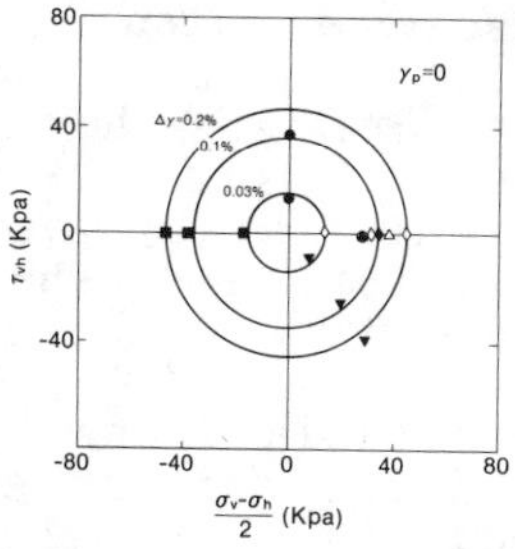

Fig. 2. Granular medium: experimental loci.

EXPERIMENTAL STUDY

The tests were designed to determine the hardening characteristics of the yield surfaces (yield loci in 2-D) of a granular medium before and after shear loading. Hollow cylindrical glass bead specimens were isotropically loaded to an identical initial condition, and then were sheared at constant mean stress along a variety of stress paths. These paths typically consisted of monotonic and cyclic parts, similar to those used by Phillips, et al. [4].

To define a "yield point" corresponding to the "initial yield surface" of a granular medium or soil is not trivial. A possibility is to use as yield criterion the engineering threshold octahedral strain, $\gamma_t \approx 10^{-2}\%$, at which irreversible volumetric strains start occurring in sand [6]. However, this is difficult to implement. The alternative adopted herein is to define the yield surface as the locus of all points in stress space having a certain value of the octahedral shear strain, larger than γ_t but close to it. This is similar to the total strain approach previously used by Peters [7].

The hollow cylindrical specimens (D_0 = 7.11 cm, D_i = 5.08 cm, H = 14 cm) were composed of a mixture of glass beads of two sizes: sieve 40–50 and 60–80, mixed in a 1:2 ratio in weight. Each specimen was prepared by tamping to the desired void ratio (e ≈ 0.68). Then it was consolidated isotropically to σ_c = 140 KPa and sheared. During shearing the mean stress was kept constant and equal to 140 KPa by changing the cell pressure accordingly. Therefore, all experiments are contained in the same π-plane in stress space.

Eight tests were conducted: i) four in compression ($\sigma > 0$), ii) one in compression–unloading, also called extension or axial extension in soil mechanics parlance ($\sigma < 0$), iii) one in torsion, and iv) one in combined compression–torsion. The repeatability of the tests, initial specimen isotropy and accuracy of the measurements at the small strains of interest were verified by plots such as Figs. 2–3. These figures show the circular yield loci from the monotonic part of the loading, for an octahedral strain, $\Delta\gamma$, ranging between 0.03% and 0.2%.

The yield locus corresponding to $\Delta\gamma$ = 0.03% was selected for further study. Figure 3 illustrates the measured change of this yield locus after loading the specimen in compression to a maximum octahedral strain γ_p = 0.25%, corresponding to point 1 on the plot. The rest of the yield locus was obtained by identifying

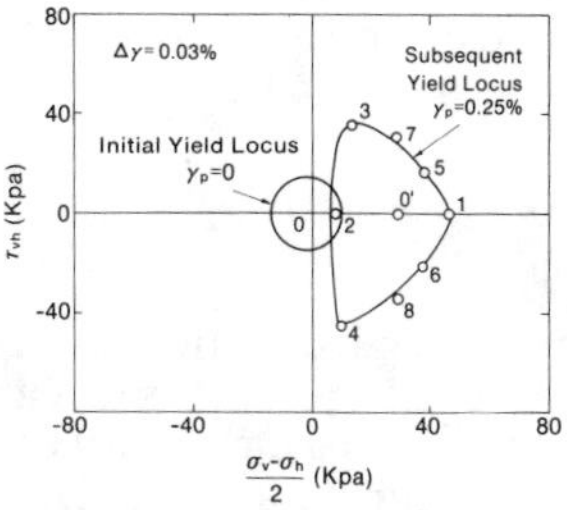

Fig. 3. Granular medium: experimental loci

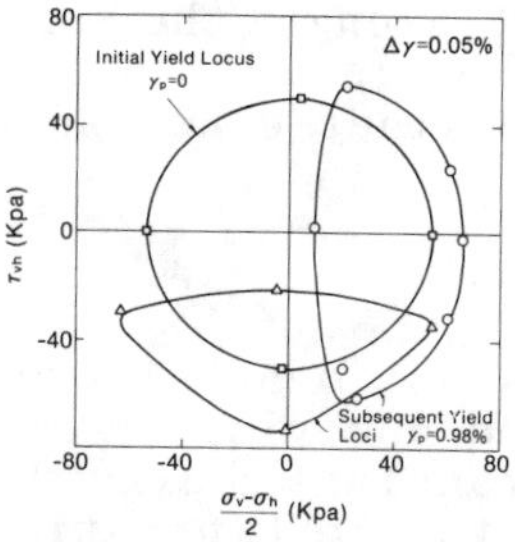

Fig. 4. Granular medium: numerical simulations.

first the displaced position of the new center (point 0'), and then by implementing probes of amplitude $\Delta\gamma$ = 0.03% in different directions from 0'. Therefore, the sequence of loading during this test was 0–1–0'–2–0'–3–0'–4–0'–5–0'–6–0'–7–0'–8–0'–2.

Figure 3 shows clearly the increase in size, development of an apex in the direction of loading and flattening of the yield locus in the opposite direction. This is very similar to the results of the experiment in aluminum shown in Fig. 1, and contradicts the hypothesis of pure kinematic strain hardening.

NUMERICAL SIMULATIONS AND MICROMECHANICS

Numerical simulations were also conducted along similar stress paths to those used in the experiments, to provide independent verification and additional insight on the micromechanics of the observed behavior. Distinct element program CONTACT [8] was used. A two-dimensional, two-size random array of 531 spheres having the properties of quartz was generated in the computer, loaded isotropically to σ_c = 132 KPa, and then prestrained monotonically at constant mean stress in compression, extension or pure shear to various levels of γ_p, with probes of $\Delta\gamma$ = 0.05% used to determine the initial and subsequent yield surfaces.

Figure 4 presents some of the results. Except for the increase in size of the yield locus, which is present in the experiments but not in the simulations, Fig. 4 predicts the same type of changes in the shape of the yield loci shown by the experimental plot of Fig. 3. An investigation of the micromechanical statistics of the simulations of Fig. 4 suggests that the formation of the apex and flattening of the yield surface is the result of the preferential allignment of the slip systems of the individual particles and particle clusters along the direction of loading. This explanation is similar to that used by researchers for metals considered as random polycrystals [3,5].

CONCLUSION

Constant mean stress experiments on glass beads and associated numerical simulations show that yield surfaces of granular media change in shape during loading, contrary to the kinematic strain hardening assumption.

ACKNOWLEDGMENTS

This investigation was supported by the U.S. Air Force Office of Scientific Research through research grant AFOSR-89-0350, and DOD instrumentation grant AFOSR-89-0172, with Major Steve C. Boyce as Technical Monitor. Mr. Bidjan Baziar assisted in setting up the equipment for the tests. Program CONBAL was developed by Prof. Tang-Tat Ng, who provided helpful advice. All this support is gratefully acknowledged.

REFERENCES

1. Z. Mroz, "On the Description of Anisotropic Workhardening," J. Mech. Phys. Solids 15, 163-175 (1967).
2. J. H. Prevost, "A Simple Plasticity Theory for Frictional Cohesionless Soils," Soil Dynam. & Earthquake Eng. 4(1), 9-17 (1985).
3. T. H. Lin, and M. Ito, "Theoretical Plastic Distortion of Polycrystalline Aggregate Under Combined and Reversed Stresses," J. Mech. Phys. Solids 13, 103-115 (1965).
4. A. Phillips, J. L. Tang, and M. Ricciutti, "Some New Observations on Yield Surfaces," Acta Mechanica 20, 23-29 (1974).
5. D. E. Helling, A. K. Miller, and M. G. Stout, "An Experimental Investigation of the Yield Loci of 1100-0 Aluminum, 70:30 Brass and an Overaged 2024 Aluminum Alloy after Various Prestrains," J. of Eng. Materials & Tech. 108, 313-320 (1986).
6. R. Dobry, R. Ladd, F. Y. Yokel, R. M. Chung, and D. Powell, "Prediction of Pore Water Pressure Buildup and Liquefaction of Sands During Earthquakes by the Cyclic Strain Method," Building Science Series 138, National Bureau of Standards (1982).
7. J. F. Peters, "Kinematic Hardening Under Jump Rotations," Proc. 7th Eng. Mechanics Conf., Blacksburg, VA, 207 (1988).
8. T.-T. Ng, and R. Dobry, "A Nonlinear Numerical Model for Soil Mechanics," submitted for publication (1990).

A PENNY-SHAPED FLAW FRACTURE TEST FOR BRITTLE SOLIDS

A.P.S. SELVADURAI
Department of Civil Engineering
Carleton University
Ottawa, Ontario, Canada K1S 5B6

ABSTRACT

The paper deals with the development of a fracture test for the evaluation of Mode I fracture toughness of concrete and other brittle elastic solids. The test focusses on the uniaxial loading of a cylindrical specimen which contains a penny-shaped crack. The boundary element technique is used to evaluate the Mode I stress intensity factor at the crack tip. The numerical estimates for the stress intensity factor can be used to compute the fracture toughness of the material. The relative dimensions of the penny-shaped flaw is varied to establish experimentally the accuracy of the fracture toughness evaluation. The paper focusses on the boundary element modelling of the problem.

INTRODUCTION

The evaluation of the fracture toughness properties of concrete and other brittle elastic solids is important to the accurate modelling of damage and degradation in such materials. Many investigators have proposed techniques for the evaluation of the Mode I fracture toughness of concrete [1]. These include flexural testing of beams with notches and direct tension testing specimens with cracks. Owing to the particulate nature of materials such as concrete, the specimens to be used in fracture toughness testing needs to be large and consequently large loads are necessary to initiate fracture. This paper presents a novel fracture test in which a penny-shaped crack is formed at the interior of a cylindrical test specimen. The specimen is subjected to tensile loads through reinforcing elements. The symmetry of the loading induces a flaw opening mode stress intensity factor at the crack tip. The numerical modelling of this fracture test is carried out by using a boundary element scheme and the Mode I stress intensity factor is evaluated for different dimensions of the penny-shaped flaw in relation to the geometry of the specimen.

BOUNDARY ELEMENT MODELLING

The boundary integral equation applicable to axisymmetric deformations of the elastic medium can be written as (Brebbia et al. [2])

$$c_{\ell k}u_k^{(\alpha)} + \int_{\Gamma_{(\alpha)}} \left\{P_{\ell k}^{*(\alpha)}u_k^{(\alpha)} - u_{\ell k}^{*(\alpha)}P_k\right\} \frac{r}{r_i} d\Gamma = 0 \tag{1}$$

where $P_{\ell k}^{*(\alpha)}$ and $u_{\ell k}^{*(\alpha)}$ are, respectively, the traction and displacement fundamental solutions [2], $c_{\ell k}$ are constants and $P_k^{(\alpha)}$ and $u_k^{(\alpha)}$ are the tractions and displacements. The superscripts (α) refer to the regions associated with the elastic medium containing the cracks $(\alpha = m)$ and the elastic bar which applies the axial loads $(\alpha = b)$ (see inset in Figure 1). The discretized version of (1) can be written as

$$\begin{aligned} c_{\ell k}u_k^{(\alpha)} \quad &+ \quad \sum_e \int_{-1}^{1} P_{\ell k}^{*(\alpha)} \left[N(\xi)\right] |J| \frac{r}{r_i} d\xi \, \{u_k\}^e \\ &= \quad \sum_e \int_{-1}^{1} u_{\ell k}^{*(\alpha)} \left[N(\xi)\right] |J| \frac{r}{r_i} d\xi \, \{P_k\}^e \end{aligned} \tag{2}$$

where $N(\xi)$ are shape functions, e is the element number and $|J|$ is the boundary Jacobian, which, for the axisymmetric as is given by

$$|J| = \left[\left(\frac{\partial r}{\partial \xi}\right)^2 + \left(\frac{\partial z}{\partial \xi}\right)^2\right]^{1/2} \tag{3}$$

Upon completion of the integrations and summations, the equation (2) can be written in the matrix form

$$\left[\mathbf{H}^{(\alpha)} \;\; \mathbf{H}_I^{(\alpha)}\right] \begin{bmatrix} \mathbf{u}^{(\alpha)} \\ \mathbf{u}_I^{(\alpha)} \end{bmatrix} = \left[\mathbf{M}^{(\alpha)} \;\; \mathbf{M}_I^{(\alpha)}\right] \begin{bmatrix} \mathbf{P}^{(\alpha)} \\ \mathbf{P}_I^{(\alpha)} \end{bmatrix} \tag{4}$$

where $\mathbf{u}_I^{(\alpha)}$ and $\mathbf{P}_I^{(\alpha)}$ are respectively the displacement and tractions at the interface between the cylindrical bar and the elastic medium. In (4) $\mathbf{H}^{(\alpha)}$ and $\mathbf{M}^{(\alpha)}$ refer to the influence coefficients matrices derived from the integration of the fundamental solutions $P_{\ell k}^{*(\alpha)}$ and $u_{\ell k}^{*(\alpha)}$ respectively. For complete bonding at the bar-elastic medium interface

$$\mathbf{u}_I^{(b)} = \mathbf{u}_I^{(m)} = \mathbf{u}_I \quad ; \quad \mathbf{P}_I^{(b)} = -\mathbf{P}_I^{(m)} = \mathbf{P}_I \tag{5}$$

Using the above constraints, the boundary element matrix equation (4) can be written as

$$\begin{bmatrix} \mathbf{H}^{(b)} & \mathbf{H}_I^{(b)} & 0 \\ 0 & \mathbf{H}_I^{(m)} & \mathbf{H}^{(m)} \end{bmatrix} \begin{bmatrix} \mathbf{u}^{(b)} \\ \mathbf{u}_I \\ \mathbf{u}^{(m)} \end{bmatrix} = \begin{bmatrix} \mathbf{M}^{(b)} & \mathbf{M}_I^{(b)} & 0 \\ & -\mathbf{M}_I^{(m)} & \mathbf{M}^{(m)} \end{bmatrix} \begin{bmatrix} \mathbf{P}^{(b)} \\ \mathbf{P}_I \\ \mathbf{P}^{(m)} \end{bmatrix} \tag{6}$$

In the boundary element discretizations, quadratic elements are used to model the boundaries of the matrix and elastic bar regions, i.e.,

$$\left.\begin{array}{l} u_i^{(\alpha)} \\ P_i^{(\alpha)} \end{array}\right\} = a_0 + a_1\zeta + a_2\zeta^2 \tag{7}$$

where ζ is the local coordinate of the element and a_r $(r = 0, 1, 2)$ are interpolation constants. To account for the singular behaviour at the crack tip we employ the singular traction quarter point boundary element proposed by Cruse and Wilson [3] where $u_i^{(\alpha)}$ and $P_i^{(\alpha)}$ take the forms

$$u_i^{(\alpha)} = b_0 + b_1\sqrt{r} + b_2 r \quad ; \quad P_i^{(\alpha)} = \frac{c_0}{\sqrt{r}} + c_1 + c_2\sqrt{r} \tag{8}$$

where b_i and c_i $(i = 0, 1, 2)$ are constants. Owing to the symmetry at the crack plane only the Mode I stress intensity factor is present and it can be evaluated by applying a displacement correlation method which utilizes the nodal displacements at four locations A, B, E, D and the crack tip i.e.,

$$K_I^{(\alpha)} = \frac{G_\alpha}{4(1-\nu_\alpha)}\sqrt{\frac{2\pi}{\ell_0}}\left\{4\left[u_z(B) - u_z(D)\right] + u_z(E) - u_Z(A)\right\} \tag{9}$$

where ℓ_0 is the length of the crack tip element.

NUMERICAL RESULTS

The boundary element scheme is applied to evaluate K_I at the boundary of the penny-shaped crack. There are a number of variables associated with this problem including (i) the length/diameter ratio of the cylindrical specimen (H/D); (ii) the radius of the crack in relation to the radius of the bar (b/a); (iii) the elastic bar-elastic matrix modular ratio (E_b/E_m) and Poisson's ratios (ν_m, ν_b) and (iv) the grip length-bar radius ratio (L/D). For purposes of presentation of the numerical results some of these non-dimensional parameters are assigned specific values consistent with certain experimental configurations. The Figure 1 illustrates the manner in which the flaw opening mode stress intensity factor at the tip of the penny-shaped crack is influenced by the bar-matrix modular ratio and the crack-bar radii ratio.

REFERENCES

[1] S.P. Shah, S.E. Swartz and B. Barr (Eds.) *Fracture of Concrete and Rock: Recent Developments*, Elsevier Appl. Sci., New York (1989).

[2] C.A. Brebbia, J.C.F. Telles and W.C. Wrobel, *Boundary Element Techniques, Theory and Applications in Engineering*, Springer Verlag, Berlin (1984).

[3] T.A. Cruse (Ed.), *Advanced Boundary Element Methods, Proc. IUTAM Symposium*, San Antonio, Texas, Springer Verlag, Berlin (1987).

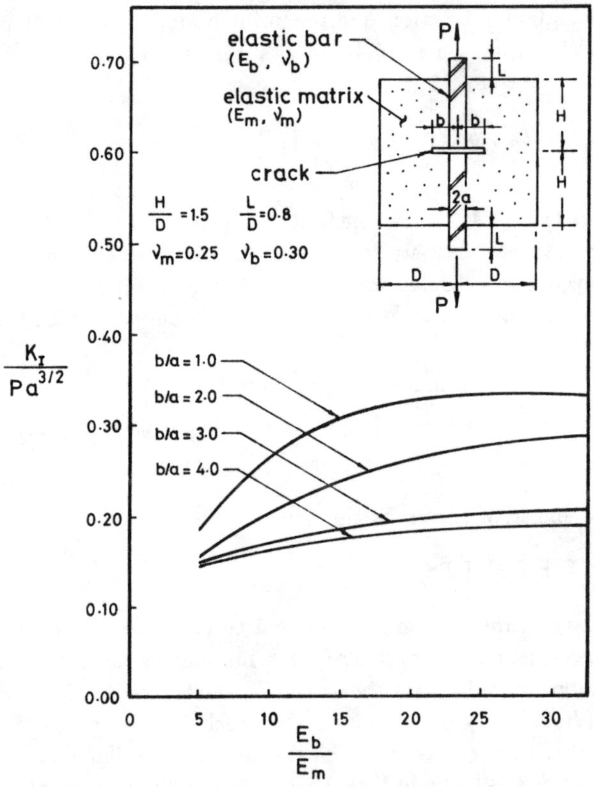

Figure 1. The stress intensity factor at the penny-shaped crack.

PAVEMENT MATERIAL PROPERTIES USING NON-DESTRUCTIVE TESTING

Raj Siddharthan, Gary M. Norris and Zia Zafir
University of Nevada, Reno

ABSTRACT

Current conventional procedures use a number of simplifying assumptions to backcalculate pavement layer moduli from the falling weight deflectometer data. The study reported here presents a relatively simple case study in which the validity of some of the assumptions is investigated. All pavement layer moduli values are underpredicted by the conventional procedures. The underprediction in some cases can be more than 35% in the granular base and the subgrade.

1.0 INTRODUCTION

One of the important input parameters used in pavement condition evaluation is the structural condition/strength of the existing pavement layers. Pavement condition/strength is often characterized by the stiffness (modulus) of the pavement layers. Laboratory and non-destructive testing (NDT) are often used to estimate the modulus of pavement layers. The laboratory evaluation is destructive, expensive, and, often, time consuming. The latest and more widely accepted NDT device is the Falling Weight Deflectometer (FWD) which imparts an impulse load similar to that of a moving traffic load on the pavement.

In the case of the FWD, the pavement layer moduli are computed using a "backcalculation analysis" in which the deflections measured along the pavement surface serve as input. Almost all of the methods that are available to carry out the backcalculation procedure assume static loading conditions for the FWD (conventional procedure). The impulse load caused by the FWD is dynamic in nature, and studies have shown that the inertia effects are important for a realistic estimation of pavement layer stiffnesses [1,2]. Two additional considerations that are important are the stress dependent pavement material properties and the development of failure conditions in the unbound materials of pavement [3].

The paper presented here, investigates the validity of the conventional procedure relative to the backcalculation analysis. The pavement sections selected in the validation study are two thin asphalt concrete pavement with thin and thick subgrade layers.

2.0 PROPOSED METHOD OF ANALYSIS

The FWD loading on the pavement represents an axisymmetric problem. Though the pavement layers are horizontally layered, the stiffness characteristics of these layers vary in lateral as well as the vertical direction because of the stress-dependant nature of the unbound materials. The finite element method is best suited for such circumstances.

In this paper, the Structural Analysis Program IV (SAPIV), which is a multi-purpose finite element program [4] having the capability of computing the displacement response of a linear elastic medium under static and dynamic loading conditions, has been used.

Since a granular base cannot carry tension, the effective stiffness of the elements under tension will be substantially lower. The failure of the granular and the coarse-grained materials was assumed to be given by the Mohr-Coulomb criterion defined in terms of an angle of the internal friction with zero cohesion. Under these circumstances, it was necessary to modify the SAPIV program to incorporate nonlinear, stress-dependent, material properties and developing failure conditions. An iterative procedure was required for this purpose. The iterative scheme adopted with SAPIV is similar to that of ILLI-PAVE [3] in which the analysis is repeated until the assumed and computed (stress-dependent) stiffnesses are within acceptable limits.

3.0 BACKCALCULATION

The FWD loading corresponds to a circular loaded area (uniform pressure) with 150 mm radius with the load equal to 40 kN. When the dynamic loading conditions are used, the pressure-time variation is assumed to be given by a Haversine function with a durations of 30 msecs. The bottom boundary is assumed to be rigid, and the lateral boundaries are assumed to be on rollers that permit only vertical motions.

A backcalculation procedure is necessary to validate the conventional method (Static layered-elastic). A recently developed finite element based backcalculation program FEDPAN is used for this purpose [5]. This program is also based on SAPIV and uses backcalculation algorithms including regression equations from the programs BISDEF and EVERCALC [6].

4.0 ANALYSIS OF SIMULATED FIELD CONDITIONS

Two representative pavement sections are used in the study. (Sections I and II). These sections represent two thin asphalt concrete (AC) pavements with two different subgrade thicknesses (Fig. 1). In Section I, the subgrade thickness is 1.5 m (shallow rigid boundary) and in Section II, it is 6 m (deep rigid boundary). The AC and granular base layer thicknesses are set at 0.076 and 0.3 m respectively. Representative resilient modulus values for the AC, base and subgrade are selected to be 2.07 Gpa, 207 MPa and 82.7 MPa respectively. The case of the nonlinear stress-dependent characterization of material properties, these values apply at the mid height in the respective layers at a large (far field) distance from the FWD load. The modulus values vary from point to point in close proximity to the load. The variation in modulus in both granular base and subgrade was achieved by assuming that the modulus is proportional to the first invariant stress to the power 0.6.

4.1 BACKCALCULATION RESULTS

In actuality, FWD displacement measured in the field should be closely matched by the results computed using the dynamic stress-dependent soil behavior model ("realistic displacement"). Therefore, the effectiveness of a backcalculation procedure can be investigated on a theoretical basis by using such "realistic displacements" as input in a backcalculation exercise from which the uniform (static) layered elastic moduli are obtained. The backcalculated moduli can be compared with the reference or far field moduli used to produce the displacements.

The moduli obtained using conventional backcalculation procedures (static, layered elastic) relative to "realistic displacement" (generated from dynamic, stress-dependent modeling) are shown in Table 1. The numbers in parentheses reflect the percent difference from the far field moduli values in various layers.

TABLE 1. RESULTS FROM THE FINITE ELEMENT BACKCALCULATION

Pavement Layer	BACKCALCULATED MODULI (MPa)	
	Subgrade Thickness = 1.5 m (Section I)	Subgrade Thickness = 6.0 m (Section I)
Asphalt Concrete	1,644 (-20.5)*	1,948 (-5.8)
Granular Base	137.5 (-33.5)	124.1 (-40.0)
Subgrade	67.3 (-18.6)	51.5 (-37.8)

* Percent difference from the far field values

5.0 CONCLUDING REMARKS

The paper presents a relatively straightforward case study in which applicability of current backcalculation procedures are investigated. The thin asphalt concrete pavement sections with deep and shallow rigid base locations were considered in the case study.

It can be noted that backcalculated AC modulus for the shallow rigid boundary case is as much as 20.5% lower than the assumed value. The granular base course and subgrade modulus are substantially underpredicted. The largest underprediction can be as much as 40% for the base modulus in the case of deep rigid boundary. The underprediction for the subgrade can be somewhat lower, about 19% for the case of shallow rigid boundary.

Additional studies relative to thick asphalt concrete pavement sections and also implication of predicted lower pavement moduli values in relation to long term pavement performance are underway.

ACKNOWLEDGEMENTS

This study was funded by a National Science Foundation Grant (CES-

8713508). The support is gratefully acknowledged.

REFERENCES

1. B. Sebaaly, T.G. Davies, and M.S. Mamlouk, "Dynamics of Falling Weight Deflectometer," J. of Transportation Engr., ASCE, Vol. 111 (November 1985).

2. M.S. Mamlouk and T.G. Davies, "Elasto-Dynamic Analysis of Pavement Deflections," J. of Transportation Engr., ASCE, Vol. 110 (November 1984).

3. M.R. Thompson, "ILLI-PAVE User's Manual," University of Illinois, Dept. of Civil Engr. (November 1982).

4. K.J. Bathe, E.L. Wilson, and F.E. Peterson, "SAPIV: A Structural Analysis Program for Static and Dynamic Response of Linear Systems," Earthquake Engr. Research Center, Report No. EERC 73-11 (April 1974).

5. C.L. Ong, "Finite Element Dynamic Pavement Analysis: Backcalculation Program for a Three-Layer Pavement," M.S. Thesis, University of Nevada, Reno (May 1990).

6. A.J. Bush III, "Nondestructive Testing of Light Aircraft Pavements, Phase II," Final Report No. FAA-RD-80-9-II, Federal Aviation Administration (November 1980).

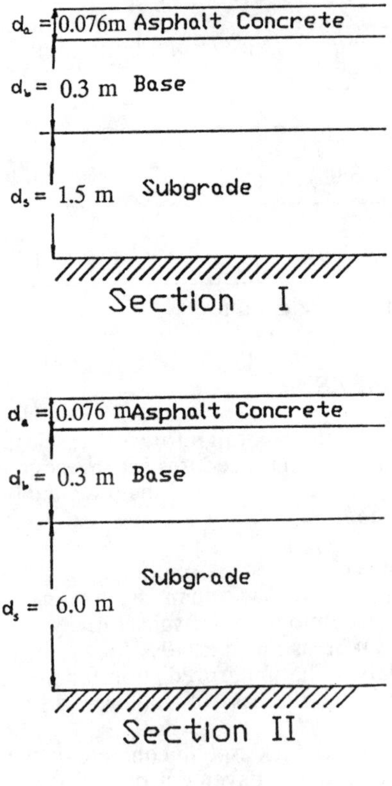

Fig. 1. Two Field Sections Used in the Study

DETERMINATION OF MATERIAL CONSTANTS FOR A HIERARCHICAL CONSTITUTIVE MODEL

A. Varadarajan, Professor
Dept. of Civil Engineering, IIT Delhi
New Delhi-110016, INDIA

C.S. Desai, Regents' Professor and Head
Dept. of Civil Engg. and Engg. Mech.
The University of Arizona, Tucson, AZ, USA

ABSTRACT

Procedure and evaluation for determination of material constants for a generalized plasticity model and the minimum tests required for are presented. Drained triaxial tests of a dense sand are used to find the constants. It is shown that even a single triaxial test result is sufficient to evaluate the approximate values of the constants for the model considered.

INTRODUCTION

A number of constitutive models have been proposed in recent times to characterize the behaviour of soils. One of the important steps in using the constitutive models is to determine material constants. This paper examines evaluation of the material constants for a constitutive model and the type and the number of tests required for the evaluation. The generalized plasticity models proposed by Desai and co-workers (1,2) using hierarchical single surface approach is considered herein. Triaxial test results for Badarpur sand, found near Delhi, India, have been used.

CONSTITUTIVE MODEL AND DETERMINATION OF MATERIAL CONSTANTS

The basic concept, theoretical development and application of the model are given previously (1). The continuous yielding and ultimate field behaviour is given by a compact and specialized form of the general polynomial representation as

$$F = J_{2D} - \left(\frac{-\alpha}{\alpha_o^{n-2}} - J_1^n + \gamma J_1^2 \right) (1-\beta S_r)^m = J_{2D} - F_b F_s = 0 \tag{1}$$

where J_{2D} = second invariant of the deviatoric stresses tensor, S_{ij}, of the total stress tensor σ_{ij}; S_r = stress ratio = J_{3D}/J_{2D}, J_{3D} = third invariant of S_{ij}, and α, n, γ, β and m are the response functions. The value of the normalizing constant, α_o, is equal to 1.0 KPa. Different S_r values designate different stress-paths.

The functions F at ultimate state (α = 0) is given by

$$F = J_{2D} - \gamma J_1^2 \, (1-\beta S_r)^{\,m} = 0 \tag{2}$$

With m, γ and β as constants, Eq. (2) gives a straight line envelope of the ultimate state. For a curved enveloped,

$$F_s = \left[\left(\exp \frac{\beta_1}{\beta_o}\right) - \beta S_r\right]^m \tag{3}$$

where β and β_1 are constants and β_o = 1.0 kPa, ormalizing constant. The value of m for many geologic materials is found to be -0.50. To find the constants γ and β in Eq. (2), at least two tests with two different stress paths (i.e., with two S_r values) are essential. However, if the extension is same, even one test is sufficient. To find β and β_1 values from Eq. (3), the minimum tests required are four with two stress-paths (compression and extension in this case) at two J_1 values. Parameter n represents the phase change point indicating transition from compressive to dilation volume change. At the phase change first, $(\delta F/\delta J_1) = 0$. This leads to

$$1 - \frac{2}{n} = \frac{J_{2D}/J_1^2}{F_s\,\gamma} \tag{4}$$

Equation (4) allows computation of n. For finding the value of n, at least one test is required.

The hardening behaviour included in the model is given as

$$\alpha = \frac{a_1}{\zeta^{\eta}} \tag{5}$$

where a_1 and η_1 are hardening constants and $\xi = \int (d\epsilon_{ij} d\epsilon_{ij})$ = trajectory of total plastic strains. Approximate values of a_1 and η_1 can be determined even from a single test result.

To allow for realistic prediction of volume change at higher shear stresses, a nonassociative flow rule is used. The plastic potential function for this case is developed by correcting the yield function F to F (3). F is formulated by correcting the growth function α in Eq. (1) as α_Q and is expressed as

$$\alpha_Q = \alpha + \kappa\ (\alpha_I - \alpha)\ (1 - r_v) \tag{6}$$

where $r_v = \xi_v/\xi$ and ξ_v is the volumetric part of ξ, κ is the material parameter and α_I is the value of at the initiation of nonassociativeness. In Eq. (6), the value of κ is assumed to apply to all stress paths. To allow for the influence of stress-path on nonassociative behavior, Eq. (6) is modified in this paper as

$$\alpha_Q = \alpha\ (\kappa_I + \kappa_2\ S_I)\ (\alpha_I - \alpha)\ (1 - r_v) \tag{7}$$

where κ_1 and κ_2 are material constants. These constants are found from the observed volume behavior near the ultimate using Eq. (7) and the relationship

$$\frac{d\epsilon_v^p}{d\epsilon_{11}^p} = \frac{3\ \partial Q/\partial J_1}{\partial Q/\partial \sigma_{11}} \tag{8}$$

To find the values of the constants κ_1 and κ_2 at least two tests with two different stress-paths are required. For finding stress path independent κ value in Eq. (6), one test is sufficient.

The Young's modulus, E, and the Poisson's ratio, ν, are evaluated using unloading-reloading response.

LABORATORY TESTS

For the determination of the various material constants, the experimental results of a locally available Badarpur sand near Delhi were used. A number of drained triaxial tests were conducted using standard triaxial device under six different stress-paths: one hydrostatic compression (HC); three conventional triaxial compression (CTC) tests and three triaxial compression (TC) tests with three initial confining pressures, one test; and one test called triaxial extension modified (TEM), in which $(\sigma_1 + \sigma_3)/2$ remained constant.

MATERIAL CONSTANTS

Table 1 shows the constants found for the Badarpur sand. The material constants were determined treating the ultimate envelope as curved (Case A), and straight line (Case B) and by using an average nonassociative constant κ and stress-path dependent nonassociative constants κ_1 and κ_2. Also, the effect of number of tests on the constants was investigated by using (i) two tests, one CTC and one TE test (Case C) and (ii) one CTC test only (Case D). For both these two cases, straight envelope was adopted. The parameters for these cases are also shown in Table 1. It can be seen that most of the constants for Cases C and D are not much different from those in Case B which uses a great number of tests; only in the case of κ, there is about 30 percent difference between Cases D and B.

VERIFICATION

The constants for different cases were used to back predict the laboratory test data. It was found that the back predictions for constants found from only one (CTC) test, although less accurate than others, provided satisfactory correlation between predictions and observations.

CONCLUSIONS

The determination of material constants for a generalized plasticity model from standard triaxial test results is presented. The minimum number of tests required for finding constants is discussed. It is shown that even a single triaxial test result is adequate to determine the approximate values of the constants for the model considered.

REFERENCES

1. Desai, C.S., Somasundaram, S. and Frantziskonis, G., "A Hierarchical Approach for Constitutive Modelling of Geologic Materials," Int. J. Num. Analyt. Meth. Geomech., Vol. 10, No. 5, 1986.

2. Desai, C.S. and Varadarajan, A., "A Constitutive Model for Short Term Behaviour of Rock Salt," J. of Geophysical Research, Oct. 1987.

3. Frantziskonis, G., Desai, C.S. and Somasundaram, S., "Constitutive Model for Nonassociative Behaviour," J. of Eng. Mech. Div., ASCE, Vol. 112, No. 9, 1986.

Table 1. Material Constants for Badarpur Sand

		Curved Envelope Case A	Straight Envelope Case B	Straight Envelope Case C	Case D
Tests Used		Nine Tests for Ultimate and Five Tests for Other Plasticity Parameters		Two Tests One CTC and One TE Test	One CTC Test ($\phi_c=\phi_E$)
Elasticity	E, ν	150,000 kPa, 0.30			
Ultimate	γ	0.07020	0.07022	0.06981	0.06190
(m=-0.50	β	-0.6900	-0.70169	-0.66408	-0.73586
β_o=1.0kPa)	β_1	0.00003	-	-	-
Phase Change	n	3.00	3.00	2.8	2.9
Hardening	a_1	0.003166	0.003850	0.003664	0.003131
(α_o=1.0kPa)	η_1	0.49317	0.45000	0.50480	0.50400
Nonassociative	κ_1	0.402	0.400	0.420	*
	κ_2	0.100	0.101	0.110	*
	κ	-	0.400		0.271

* Insufficient test to find two constants

MANUFACTURING ASPECTS; WORKSHOP

High Homologous Temperature Constitutive Behavior of FCC Metals

Stuart Brown, Pratyush Kumar, and Vivek Dave
Department of Materials Science and Engineering
Massachusetts Institute of Technology

ABSTRACT

New materials processing technologies such as continuous casting and semi-solid processing involves the deformation of metals at very high homologous temperatures. The flow behavior and constitutive response of metals at these temperatures, however, is very poorly characterized. This paper presents constitutive data for a model system of very high purity lead (99.9999%). Experiments include constant strain rate and strain rate change experiments. The resulting constant structure strain rate dependence is found to have a much higher rate dependence than is normally used in current constitutive modeling. Also, the data does not necessarily suggest a power law dependence between strain rate and stress.

INTRODUCTION

Although rate-dependent constitutive behavior has been fairly well surveyed for slow strain rate, creep applications [5,3], the constitutive behavior of high strain rate, high homologous temperature behavior is not well characterized. This article addresses a portion of high temperature characterization, the appropriate rate equation for high temperature deformation within an internal variable framework, where the flow and microstructure evolution processes are explicitly represented by a set of coupled, first order differential equations:

$$\dot{\epsilon}_{ij}^{vp} = \frac{d\epsilon_{ij}^{vp}}{dt} = \hat{f}_{ij}(\sigma_{kl}, T, s_1, \ldots, s_m) \tag{1}$$

$$\dot{s}_n = \frac{ds_n}{dt} = \hat{g}_n(\sigma_{kl}, T, s_1, \ldots, s_m), \quad 1 \le n \le m \tag{2}$$

where

ϵ_{ij}^{vp}	=	viscoplastic strain,
$\hat{f}_{ij}(\ldots)$	=	flow function,
σ_{kl}	=	state of stress,
T	=	temperature,
$s_1, \ldots, s_m$	=	m internal variables, and
$\hat{g}_n(\ldots)$	=	evolution equation for internal variable n

One experimental practice of decoupling the flow equation, $\hat{f}_{ij}(\ldots)$, from the evolution equations $\hat{g}_n(\ldots)$ involves the class of constant structure experiments, where stress, strain rate, or temperature are changed suddenly and the instantaneous change in stress or strain rate recorded. If the change is sufficiently rapid, given testing machine dynamic responses and rates of material evolution, the material transient response represents constant structure behavior ($s_n = \text{constant}_n$).

EXPERIMENTAL PROCEDURE

Compression specimens with a height-to-diameter ratio of 1 to 1 were machined from very high purity lead (99.9999%)[1]. The compression specimens were machined with concentric, shallow end grooves that held lubricant to permit large homogeneous deformations. The lubrication was sufficient to give virtually homogeneous deformation for strains exceeding 50 percent true strain.

Figures 1 and 2 present digitally smoothed constant true strain rate data for the high purity lead for two elevated temperatures over a range of strain rates. Figure 3 presents the results of a set of strain rate change experiments, at a homologous temperature of 0.95 and a jump strain of 0.4.

DATA ANALYSIS

Stress/strain data was obtained from the constant strain rate tests by first subtracting the effect of machine compliance and then digitally smoothing the data. A locally weighted regression scatter plot smoothing (lowess) technique was used for smoothing [2].

Data analysis was performed to remove extraneous experimental factors distorting the transient response of the jump tests. First, the load/displacement compression data was converted to true stress and true strain while removing the elastic compliance of the test machine[2]. The constant structure rate dependence was then determined under the assumption that first order constitutive behavior could be characterized by equations (1) and (2), where equation (2) involves only one scalar internal variable and is of the form:

$$\dot{s} = H\dot{\epsilon}^{vp} - R_d\dot{\epsilon}^{vp} - R_s - R_x \tag{3}$$

where

H = athermal hardening rate,
R_d = dynamic recovery rate,
R_s = static recovery rate,
R_x = net softening rate due to recrystallization

[1] Johnson Matthey, AESAR, Inc.
[2] The compliance of the specimen was negligible relative to that of the machine.

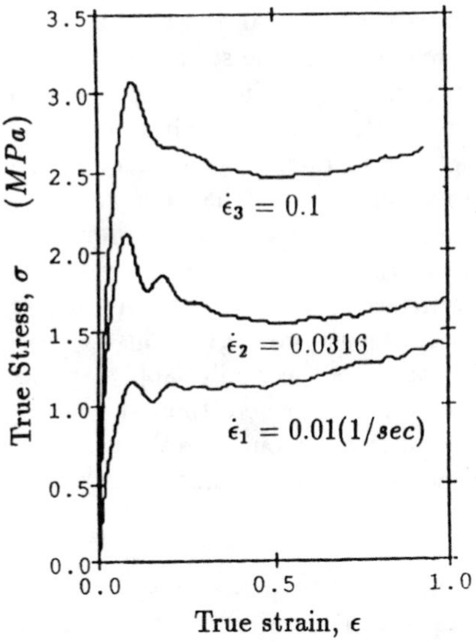

FIG. 1. Constant true strain rate compression tests (99.9999 % lead, $297°C = 0.95\ T_m$)

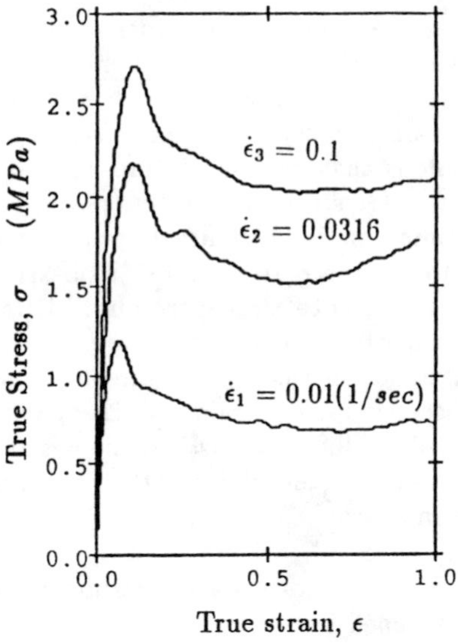

FIG. 2. Constant true strain rate compression tests (99.9999 % lead, $321°C = 0.99\ T_m$)

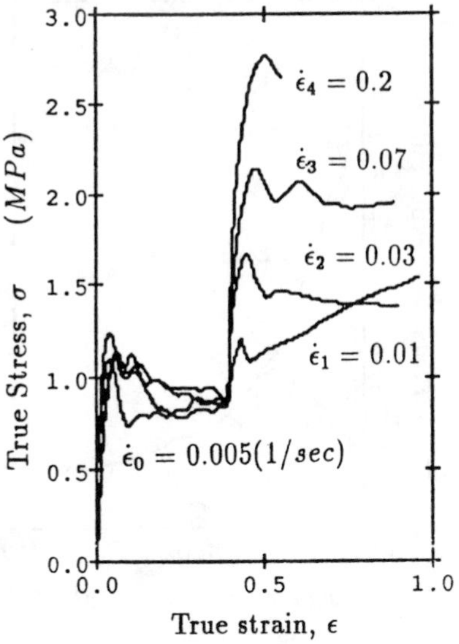

FIG. 3. True strain rate compression constant structure tests (99.9999 % lead, $297°C = 0.95\ T_m$)

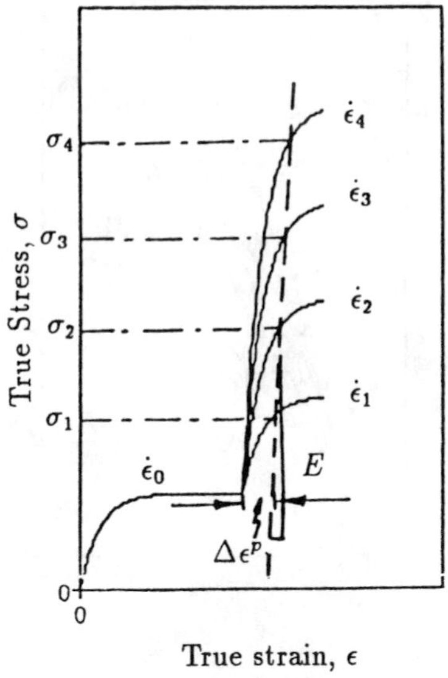

FIG. 4. Schematic of strain rate change data, illustrating plastic strain offset criterion for obtaining constant structure strain rate dependence.

Previous investigations indicate that static recovery rates at higher strain rates are a much lower than dynamic recovery rates, and can usually be neglected[1]. Similarly, the period of the oscillations of the stress/strain data at constant true strain rates indicate that the change in structure due to "dynamic" recrystallization is significantly less than the hardening rate. The result is that over short transients at high strain rates, the primary mechanisms for structural evolution are dynamic recovery and hardening. Since both mechanisms require dislocation motion, we assume they are proportional to the imposed strain rate. **The stress responses to different changes in strain rate therefore should be compared at an identical plastic strain offset from the point of the strain rate change.** This is illustrated schematically in Figure 4. Figure 5 presents the results applying this procedure to the strain rate change data of Figure 3, where the data is presented for different plastic strain offsets. These data demonstrate the variation in strain rate dependence that could be measured using a particular plastic strain offset.

Figure 6 presents the variation of constant structure **power law** exponent determined from the strain rate change data as a function of plastic strain offset. Note that the power law rate dependence decreases from a value of 9 to the value of 4 associated with steady state behavior.

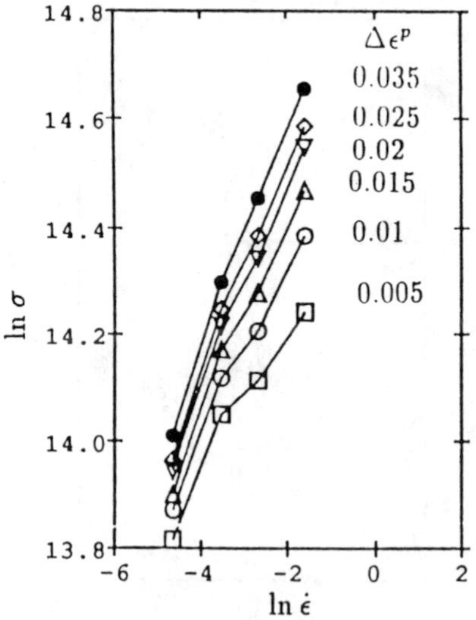

FIG. 5. Constant structure rate dependence as a function of plastic strain offset. (99.9999 % lead, $297°C = 0.95\ T_m$)

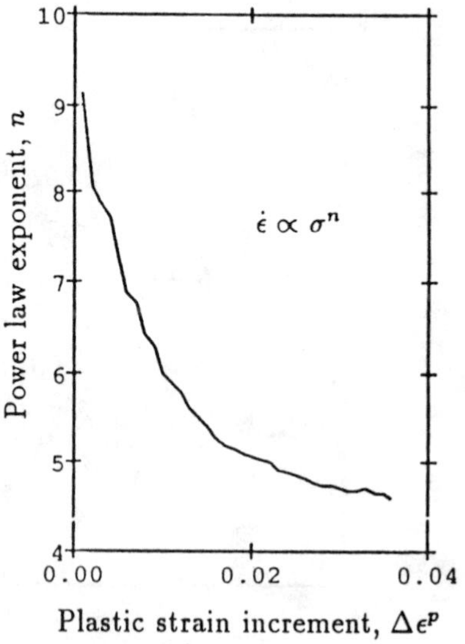

FIG. 6. Variation of constant structure power law ($\dot{\epsilon} \propto \sigma^n$) exponent n as a function of plastic strain offset $\Delta\epsilon^p$. (99.9999 % lead, $297°C = 0.95\ T_m$)

CONCLUSIONS

Internal variable constitutive models require greater attention to the appropriate flow equation that represents rate dependence at a given structure. Given the larger power law exponent, certainly a power law dependence with an exponent of 4 is incorrect. The steady state exponent of 4 observed during steady state creep is a consequence of the coupling of flow behavior and steady state structure.

Moreover, the absence of a power law functional form is also most likely incorrect. Current models of high strain rate dislocation motion involve a stress-modified, thermal activation that yields an exponential dependence of strain rate on stress [4]. Power law models provide certain analytical conveniences, but do not correlate to any fundamental physical representation of dislocation dynamics.

ACKNOWLEDGEMENT

Support for this work was provided by the U.S. National Science Foundation (Grant No. DMR-8806901)

REFERENCES

[1] S B Brown, K H Kim, and L Anand. An internal variable constitutive model for hot working of metals. *International Journal of Plasticity*, 5:95–130, 1989.

[2] J M Chambers, W S Cleaveland, Kleiner B, and Tukey P A. *Graphical Methods for Data Analysis.* Wadsworth International Group and Duxbury Press, 1983.

[3] K S Chan, U S Lindholm, S R Bodner, and K P Walker. A survey of unified constitutive theories. In *2nd Symposium on Nonlinear Constitutive Relations for High Temperature Applications*, NASA Lewis Research Center, June 1984.

[4] U F Kocks, A S Argon, and M F Ashby. *Thermodynamics and Kinetics of Slip.* Pergamon Press, 1975.

[5] 1985 ASM Materials Science Seminar. *Mechanisms of Time-Dependent Flow and Fracture of Metals*, October 1985.

ANISOTROPIC CONSTITUTIVE LAWS FOR SHEET METAL FORMING ANALYSIS BY THE FINITE ELEMENT METHOD

M. BRUNET
Laboratoire de Mécanique des Solides
I.N.S.A. 304 Villeurbanne 69621 FRANCE

G. MONFORT and J. DEFOURNY
Centre de Recherches Métallurgiques
11 rue Solvay B. 4000 Liège BELGIQUE

ABSTRACT

For sheet metal forming analysis, it is shown that a physical description of strain hardening could be obtained by mean of three slips which appear to be a set of new state variables .The development of an elasto-plastic model is proposed for implementation in F.E. codes where a non-associated flow rule is assumed . Numerical and experimental illustrative examples are presented .

THE PHYSICAL MODEL "3G":

The physical model assumes in the local orthotropic frame that any plastic deformation of the sheet results from superimposition of plastic shears, of greater or lesser extend, which occur in 6 planes at $\pi/4$ to 1,2,3 directions . A given state of stress in the orthotropic frame will therefore results in three slips G12,G23,G31 which are not identified with the plastic shear strain components .If we consider for example one of the two planes oriented at $\pi/4$ to 1 and 2 directions, a certain tangential stress τ exists in this plane and if the stress τ is great enough, a very rapid rate of slip dG12/dt is observed. As the stress τ may be positive or negative, it is presumed that the rate dG12/dt has the same sign as τ and a possible relationship may take the form :

$$2|\tau_{12}| = A_{12} + W \ln|dG_{12}/dt| \qquad (1)$$

That has been done in respect of stress τ_{12} can be done for the other 2 stresses τ_{23} and τ_{31} ,which leads to 2 relations of the same type as (1).

It is well known that plastic deformation has the effect of strain-hardening the material and this effect can be take account by introducing at any moment the plastic slip accumulated in an additive relationship more complete than eq.(1):

$$2|\tau_{12}| = A_{12} + W \ln|dG_{12}/dt| + B_{12} |G_{12}|^{n_{12}} \qquad (2)$$

where the strain-hardening law is characterised both by the constant B_{12} and by the exponent n_{12}. When writing eq.(2), it is assumed that the rate of slip is influenced only by the value of the corresponding slip. In reality, the phenomena is more complex and there is also indirect or latent strain-hardening. In these conditions, the complete set of relationships is :

$$2|\tau_{12}|^s = A_{12} + B_{12}\,[\,|G_{12}| + a\,|G_{23}| + a\,|G_{31}|\,]^{n_{12}}$$

$$2|\tau_{23}|^s = A_{13} + B_{13}\,[\,b\,|G_{12}| + |G_{23}| + c\,|G_{31}|\,]^{n_{13}} \qquad (3)$$

$$2|\tau_{31}|^s = A_{13} + B_{13}\,[\,b\,|G_{12}| + c\,|G_{23}| + |G_{31}|\,]^{n_{13}}$$

and

$$|dG_{ij}/dt| = g_o \exp[(|\sigma_i - \sigma_j| - 2|\tau_{ij}|^s)/\,W\,] \qquad (4)$$

This set of equations correspond to the case of a sheet which may present a plastic anisotropy outside its plane but has a lower plane anisotropy. If the main stresses in the orthotropic frame 1,2,3 are known at a given moment,the set of eq.(3) and (4) enables the rate of slips to be calculated and the rate of plastic strains in the same axes are given by :

$$\dot{\varepsilon}_1 = \dot{G}_{12} - \dot{G}_{31} \quad \dot{\varepsilon}_2 = \dot{G}_{23} - \dot{G}_{12} \quad \dot{\varepsilon}_3 = \dot{G}_{31} - \dot{G}_{23} \qquad (5)$$

The physical model introduces 10 constants which have to be determined experimentally by combining various tests . There are series of uniaxial tensile tests carried out at different speeds and tensile tests on very wide specimens and also a series of measurements of yield strength carried out on tensile micro-specimens cut from a carefully selected area of a drawn product . However , study of the laws of strain hardening and tensile tests show that certain constants of the material (W,c,a,b) do not change in value from one steel to another , as long as the structure remains unchanged , cubic centred for example . The other constants are obtained by means of a simple tensile test , only the result of which are utilised differently from the conventional method .

FINITE-ELEMENT IMPLEMENTATION :

From a computational point of view, they are many ways to introduce the physical model but all make use of the rotating frame formalism and since elastic deformations are assumed to be small, it is convenient to ensure the objectivity of the hypoelastic law by writing it in the same rotating frame.

The elasto-viscoplastic approach involves directly the set of eq.(3),(4),(5) by expanding the strain rate into a truncated Taylor's series opens the possibility to obtain a symmetric tangent stiffness or a non-symmetric one if higher order terms are not neglected. However, if good results are obtained, numerical difficulties can appear in some cases.

Also, we present here an non-associated elastoplastic formulation which is obtained by the combination of the Hill's criterion for the yield surface and the previous model slightly modified as the plastic potential surface. Then, in the rotating frame we have the following relations :

Hill's criterion for loading and unloading :

$$f(\{\sigma\}) - \rho(k) = 0 \quad \text{and} \quad \{a\} = \partial f/\partial\{\sigma\} \qquad (6)$$

The plastic flow equations can be written in the form:

$$\{d\varepsilon\}^p = d\lambda\ \partial g/\partial\{\sigma\} = d\lambda\ \{b\} \tag{7}$$

where it is assumed a plastic potential g such that :

$$\exp(F_{12}) + \exp(F_{23}) + \exp(F_{31}) - 3 = 0 \tag{8}$$

$$\text{with } F_{ij} = (|\sigma_i-\sigma_j|-2|\tau_{ij}|^s)/\ A_{ij} \qquad (A_{23} = A_{31} = A_{13}) \tag{9}$$

depending of the new state variables G12,G23,G31 by eq. (3),(4) where $d\lambda$ replace the product $g_o dt$.

Since the 1-direction is choosen to be the direction of reference, the hardening rule may be :

$$\rho(k) = A_{12} + B_{12}\ [\ k(1+a)/\ 2\ ]^{n_{12}} \tag{10}$$

where it is readily shown that the hardening parameter k is :

$$dk = d\lambda\ \{\sigma\}^T\ \{b\}\ /\ \rho(k) \tag{11}$$

for a work-hardening material.

This formulation has been implemented in our specific finite-element code and in the 2D. elements of the general purpose non-linear incremental finite-element code "ABAQUS" version 4.7 by mean of the user subroutine "UMAT" . The non-symmetric tangent stiffness matrix and the stresses are calculated by an explicit scheme using subincrements with an iterative end-step correction.

NUMERICAL AND EXPERIMENTAL EXAMPLES :

The first example shown by fig.(1) is the strains calculated (curves) and measured (points) on an axisymmetric blank in biaxial stretching by a spherical headed punch with a friction coefficient μ=0.03 . The material parameters are as follows:

thickness = 0.7mm

Lankford's r value : r = 1.51

A12 = 162 MPa

A13 = 166 MPa

B12 = 529 MPa

B13 = 535 MPa

a = 0.88

b = 0.88

c = 1.0

n12 = 0.45

n13 = 0.45

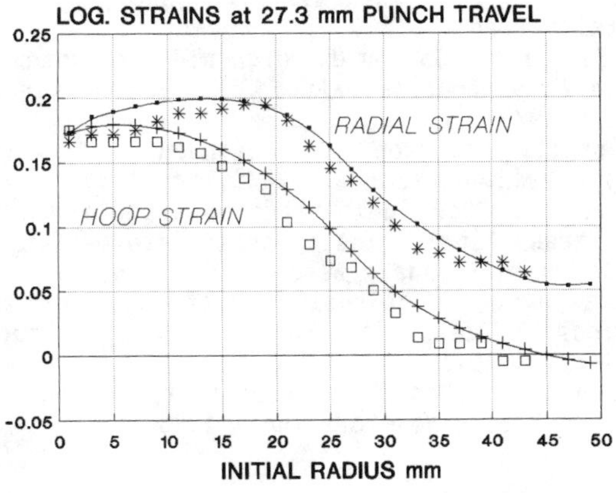

FIG. 1

The second example shown by fig.(2) is the strains calculated (curves) and measured (points) on an axisymmetric blank in deep drawing by a flat headed punch where the coefficient of friction is μ=0.10 against the punch and μ=0.16 against both the die-part and the holder-part of the tools. The material parameters are as follows:

Thickness = 0.68mm

Lankford's r value : r = 0.77

A12 = 53 MPa

A13 = 50 MPa

B12 = 757 MPa

B13 = 750 MPa

a = 0.67

b = 0.67

c = 1.0

n12 = 0.174

n13 = 0.174

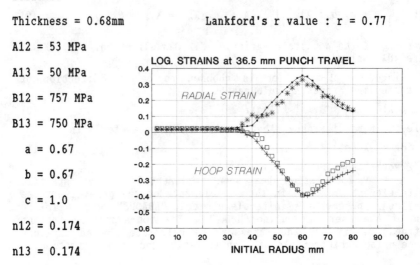

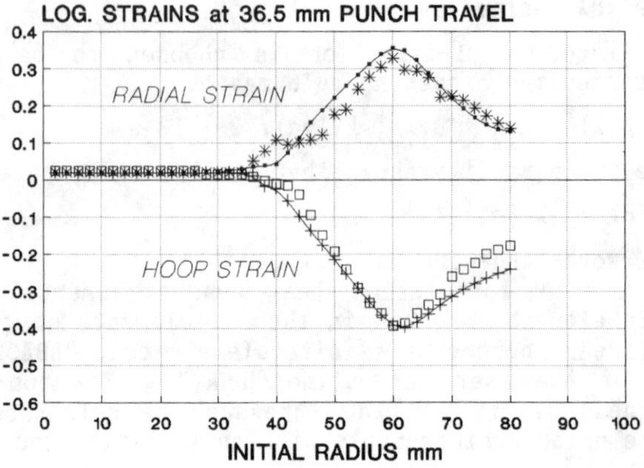

FIG. 2

The calibration of our model with experimental results has been established. The development of the constitutive model and its implementation in the 3D. shell element of the general purpose non-linear code is currently being made .

REFERENCES :

1. Hill,R. *The mathematical theory of plasticity.* (Oxford University Press), 1960.
2. D'Haeyer,R. Gouzou,J. Mignon,J. Franssen,R. Vanosmael,A. and Bragard,A. *Emboutissabilité des toles minces.* C.R.M. Internal Report S 49/77, 1977.
3. Hibbitt, Karlsson and Sorensen . "ABAQUS" *Users'Manual and theory Manual.* Version 4.7, HKS,Inc. 1988.
4. Brunet,M. "Some computational aspects in three-dimensional and plane stress finite elastoplastic deformation problem". *Engineering Analysis with Boundary Elements* ,Vol. 6, nº 2, pp.78-83, 1989.
5. Kocks,U.F. "Constitutive relations for slip". in *Constitutive Equations in Plasticity.* Edited by Ali S. Argon (The M.I.T. Press) pp. 81-115 , 1975.
6. Chen,W.F. Han,D.J. *Plasticity for Structural Engineers.* (Springer-Verlag New-York Inc.) 1988.

Acknowledgement:
This work is performed under the financial support of RENAULT (France) and COCKERILL-SAMBRE (Belgium).

A CONSTITUTIVE MODEL FOR REINFORCED GLULAM BEAMS

Julio F. Davalos and Ever J. Barbero
Constructed Facilities Center
College of Engineering
West Virginia University

ABSTRACT

The development and commercial production of hardwood glulam beams reinforced with pultruded composite materials is discussed. A linear constitutive relation to predict the response of these composite beams is presented and validated through an experimental program.

INTRODUCTION

The economical and efficient application of timber to engineered structures (bridges, 3-d frames, lattice domes) is achieved primarily through the use of glued laminated timber (glulam). The advantages of glulam over other materials are: economical production of tapered and curved members, excellent energy absorption characteristics (seismic, damping, and acoustical responses), high chemical and corrosion resistance, and better fire resistance than steel. One disadvantage of glulam is its relatively low modulus of elasticity, which is about 2 x 10^6 psi for softwoods (Southern pine, Douglas fir) and 1.5 x 10^6 psi for hardwoods (Yellow poplar). For this reason, the construction of large-span timber roofs and bridges usually require glulam members of large depths (36 to 48 inches). To significantly increase the stiffness of glulam, the members can be reinforced with composite materials (fibers embedded in a matrix) at top and bottom flanges. The reinforcement reduces the required depth of the member, which in turn reduces its weight, and the necessity of bracing the member to prevent lateral buckling.

The concept of reinforcing glulam beams with fiber reinforced composites (FRC) has been explored by other investigators [1]. However, a composite glulam product has not been manufactured commercially, probably due to the following reasons: (1) the FRC composites used in previous research were produced by costly hand lay-up procedures and utilized expensive fibers (keblar, graphite); (2) the formulation of the composite material for compatibility with wood adhesives has not been addressed; and (3) the manufacturing process of the FRC-glulam beams in current glulam plants, without significant changes to existing processing operations, has not been addressed. The research reported herein aims at the commercial implementation of reinforced glulam products by resolving the aforementioned issues.

In this study, four Yellow poplar glulam beams are reinforced with glass fiber reinforced composites produced by pultrusion. The process of pultrusion is probably the most cost-competitive method for the mass-production of composites for infrastructural applications. In this process, with an average output of two feet per minute, fibers are continuously pulled through a bath of resin and then through a heated die that defines the cross-sectional shape. The resin system can be tailored to provide corrosion resistance, fire resistance, ultraviolet resistance, bond adherence to other materials, etc.

The industrial production of FRC-Glulam beams requires research efforts to resolve various issues, such as: compatibility of the adhesive at the interface, manufacturing procedures for this new product compatible with current equipment and practices, durability and long-term performance, development of design procedures for incorporation in codes of practice, etc. However, to make FRC- Glulam a structural material of choice and to expand its potential use in structural engineering, constitutive models are needed that can be efficiently used in numerical methods of analysis, such as the finite element method. Such models must be practical, computationally feasible, and sufficiently accurate. Thus, the objective of this paper is to present a constitutive relation that can be used to predict the linear response of FRC-Glulam beams. The theoretical predictions are validated by test results of Yellow poplar glulam beams reinforced with a composite, which consists of glass fibers embedded in a polyester matrix.

MODELING OF GLULAM BEAMS

Based on the cylindrical symmetry of the growth rings of the cross section of a tree, wood is often modeled as an orthotropic material. The orthotropic assumption is too complex and not readily applicable to glulam beams, because their cross sections usually exhibit a random orientation of the growth rings (Fig. 1). A more practical and valid approach is to model the beams as transversely isotropic [2]. Based on Timoshenko's beam theory, which includes shear deformation, the transverse isotropy assumption permits representing the constitutive matrix of a 3-d glulam beam in terms of two elastic constants: the longitudinal modulus E, and the shear modulus G.

MODELING OF THE PULTRUDED REINFORCEMENT

Pultruded composite members of rectangular cross sections (e. g., 6" x 1"), consisting of uniformly distributed unidirectional fibers, are transversely isotropic. Thus, once again, the constitutive relation can be expressed in terms of E and G.

CONSTITUTIVE RELATION FOR FRC-GLULAM BEAMS

Based on lamination beam theory (LBT), which includes shear deformation [3], the expressions for the axial, bending, and shear stiffnesses of transversely isotropic laminated composite beams (Fig. 2) are:

$$A=b\sum_{i=1}^{n} E_i t_i \text{ , axial stiffness} \tag{1}$$

$$D=b\sum_{i=1}^{n} E_i \left(t_i \bar{y}_i^2+\frac{t_i^3}{12}\right) \text{ , bending stiffness} \tag{2}$$

$$F=\frac{1}{\kappa} b\sum_{i=1}^{n} G_i t_i \text{ , shear stiffness} \tag{3}$$

where, E_i and G_i are the elastic and shear moduli of the laminae; b, t_i, and $\bar{y}_i$ are the geometric constants shown in Fig. 2; and κ is the shear correction factor [4]. Using expressions (1) to (3), the normal stress at a layer i^{th} can be expressed as

$$\sigma_i=\frac{MyE_i}{D}\pm\frac{NE_i}{A} \tag{4}$$

Similarly, the shear stress at an interface k^{th} can be expressed as

$$\tau^k=\frac{V}{D}\sum_{i=k}^{n} t_i \bar{y}_i E_i \tag{5}$$

where, M is the bending moment, N is the axial force, and V is the shear force at the section of interest. The transverse deflections of the beam can be computed using expressions (2) and (3).

EXPERIMENTAL AND ANALYTICAL RESULTS

The experimental program includes four Yellow poplar glulam beams reinforced with glass fiber composites. Before assembling into beams, the longitudinal and shear moduli of the constituent laminae were obtained from bending and torsion tests. The first phase of the study is concerned with the bending behavior of the composite beams. As an example the response of one of the beams is described next.

The stacking sequence of the beam is described in Table 1. The beam was tested in bending with and without the glass fiber composite, and the experimental response is shown in Fig. 3. It is significant that the addition of 1/8 inch thick composite increases the bending stiffness by 29%. The laminated beam theory (LBT), including shear effects, predicts quite well the bending response of the test sample. The LBT analysis also predicts increases of 14% shear stiffness and 8% axial stiffness. These predictions will be checked by conducting torsion and axial tests. The development, manufacturing, and structural response of these beams is currently being investigated by the researchers and further results will be presented at the conference.

REFERENCES

1. D.A. Tingley, "Reinforced Glued-Laminated Wood Beams," Wood Science and Technology Institute, Report, (Fredericton, New Brunswick, Sept. 1980).
2. J.F. Davalos, J.R. Loferski, S.M. Holzer, and V. Yadama, "Transverse Isotropy Modeling of 3-D Glulam Beams," ASCE Journal of Materials in Civil Engineering (to Appear, 1990).
3. E.J. Barbero, "Laminated Beam Theory for Complex Structural Shapes," CFC Report (West Virginia University, Morgantown, West Virginia, July, 1990).
4. H. L. Langhaar, Energy Methods in Applied Mechanics (Wiley, New York, 1962).

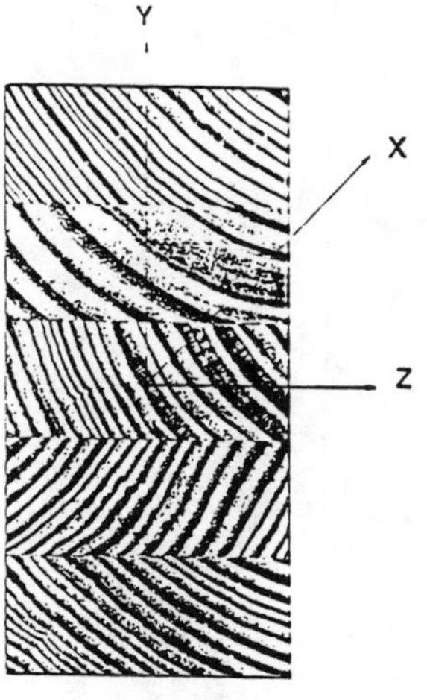

FIG. 1. Cross section of a glulam beam.

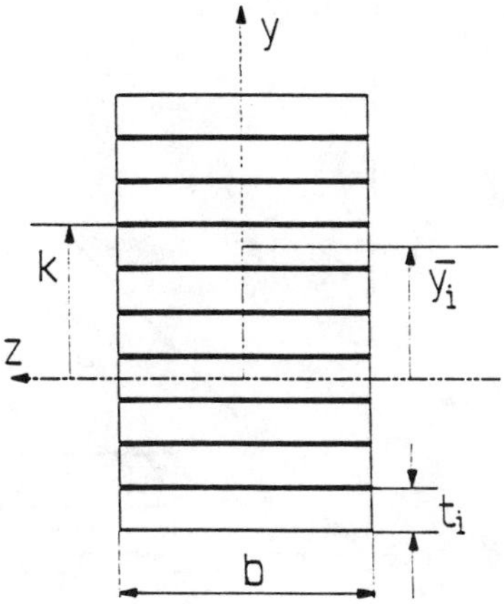

FIG. 2. Laminated beam: geometric parameters used in Eqs. (1) through (5)

Table 1. Reinforced Yellow poplar glulam beam: Dimensions and Properties.

Layer	Material	M.C. (%)	Depth (in)	E (10^6 psi)	G (10^6 psi)
1	wood	10.4	0.763	2.554	0.1406
2	wood	10.0	0.762	1.715	0.1285
3	wood	10.1	0.761	1.636	0.0923
4	wood	10.4	0.761	1.911	0.0821
5	wood	10.0	0.762	1.888	0.1062
6	wood	10.4	0.762	2.226	0.0849
7	FRC		0.125	5.800	0.5400

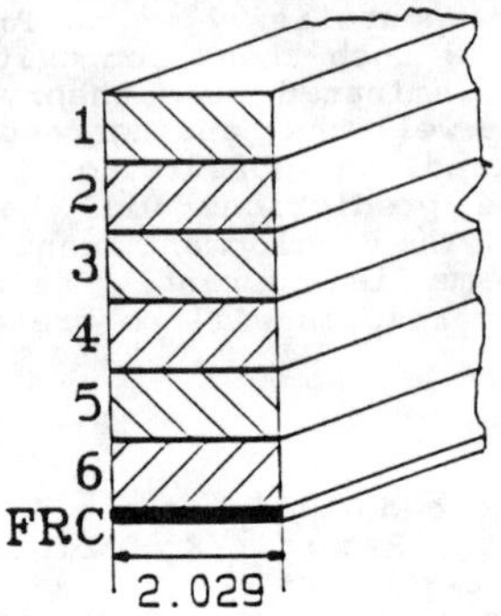

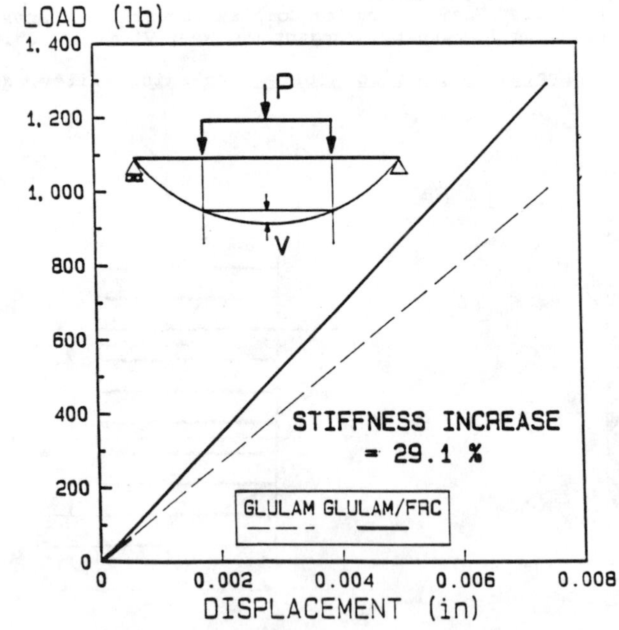

FIG. 3. Experimental comparison of bending stiffness: Glulam alone vs. FRC-Glulam.

MECHANICAL PROBLEMS IN PACKAGING OF SEMICONDUCTOR DEVICES

Kozo IKEGAMI

Tokyo Institute of Technology
Nagatsuta, Midori-ku,
Yokohama, Japan

ABSTRACT

Mechanical problems of semiconductor devices are reviewed on the basis of current investigations in Japanese research organizations. The emphasis of this article is placed on the problems relating to deformation and strength of the interconnection and plastic packaging in the devices. Thermal loading is a main factor for the deformation and strength of the device elements. The examples describe for the thermal stress distributions and the bonding strength in the interconnection and plastic package.

INTRODUCTION

Advanced technology is effectively supported by developing semiconductor devices. The high reliability of the device is demanded because of their wide applications. The device is fabricated in complicated manner by different kinds of materials with minute dimensions. The mechanical problems on deformation and strength of the minute parts in the device is concerned with the reliable performance. This paper reviews some mechanical problems in chip interconnection and plastic packaging of the semiconductor device. The author reports already a reviewed paper on the mechanical problems in the production process of semiconductor devices in another journal [1].

CHIP INTERCONNECTION

The bumps of a chip are connected to the outer elctronodes and the chips become an electron component. The connection is achieved between different materials. The thermal mismatch exists between the connected materials and causes the thermal stress and strain. The connecting strength is important for the reliability of devices.

(1) Thermal stress and strain

The flip chip is a direct method connecting between integrated circuit and external circuit of a substrate.

This is achieved by bonding the device to the external circuit through connection bumps plated on to the pads of the silicon circuit or the substrate circuitry. The number of intermediary bonds is reduced and all device connections are made simultaneously. The thermal expansion of the chip and the substrate differs according to temperature change. This causes the stresses in the soldered joints. The thermal cycles produce the stress cycles to the joints and the fracture occurs in the joints. Figure 1 shows the strain distributions in soldered joints with different solder shape under temperature change [2]. The strains are indicated by equivalent values. The maximum strain is produced at the joint root for both cases of different solder shapes. The maximum strain value depends on the solder shape. Convex soldering is more effective in reducing the strain than concave soldering.

The interconnection in tape automated bonding is achieved by an intermediary tape with a suitable pattern for an electric circuit etched from a clad metal film. The etched pattern is matched to the bonding pads on the chip surface and the substrate circuitry. The tape is attached at both ends by solder or hot pressure bonding.

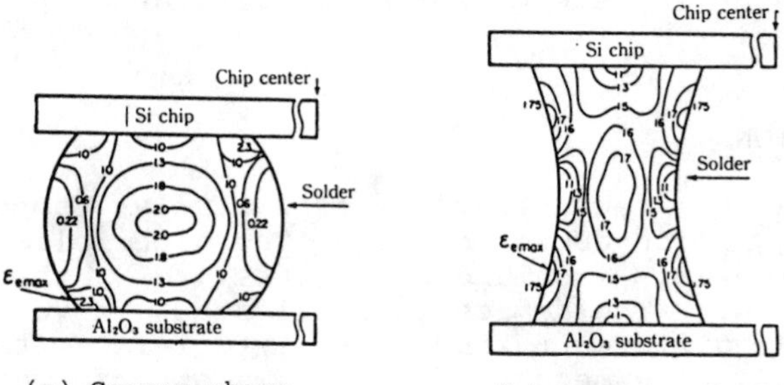

(a) Convex shape (b) Concave shape

Fig. 1 Distributions of effective strain in soldered joint

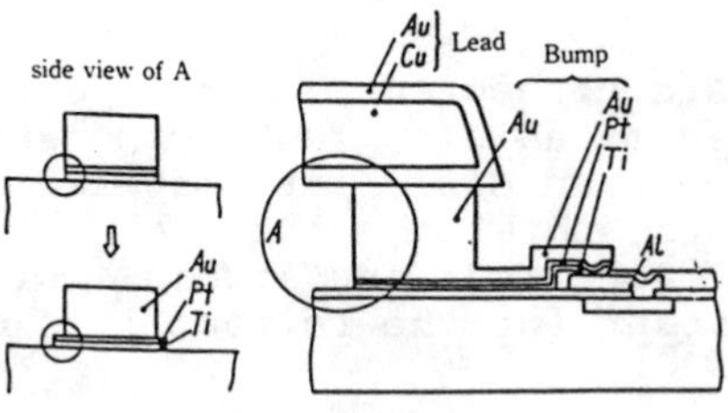

Fig. 2 Schematic figure of material constitution in bump

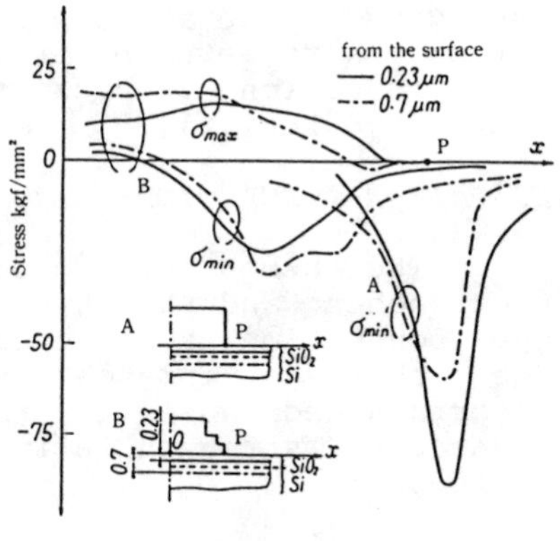

Fig. 3 Thermal stresses in bumps with different shapes

Connection pillars are used to ensure bonding at all sites around the chip. The hot pressure for bonding produces mechanical stresses as well as thermal stresses in the connected bumps. The stress can fracture the bumps and the substrate. The bump dimension and material constitution affect the stress distribution in the bump and substrate. Figure 2 is a schematic figure of an interconnection by the tape automated bonding [3]. The thermal stresses for the two bumps of different shapes are illustrated in Fig. 3. The results show the distribution of the maximum and minimum principal stresses after 10μs when the bump is heated to 350 °C. The large stress concentration is produced at the bump root of the type A. In the type B, the stress concentration is reduced by the stepped configuration bump.

(2) Bonding strength

The metallized pads on the surface of the chip are connected to the circuit terminal on a substrate by thin aluminum or gold wire. The connections are made by ultrasonic thermo-compression or thermo-sonic welding. Ball bonding method is a technique for wire bonding. In the method, a ball forms on the end of the wire by a hydrogen flame or by a spark discharge. This ball is then ultrasonically or hot pressure welded to the pad on the silicone chip using a capillary bonding tool. The connecting strength is dominated by the delaminating strength or the breaking strength at the welded part.

The shear strength against the ball deformation is shown for different wires in Fig. 4 [4]. The balls are ultrasonically bonded to the aluminum electrodes by varying the ball strain defined by the value in the inserted figure. The shear strength is examined by the shear test using a knife edge. The bonding strength increases for every wire and changes significantly with wire materials.

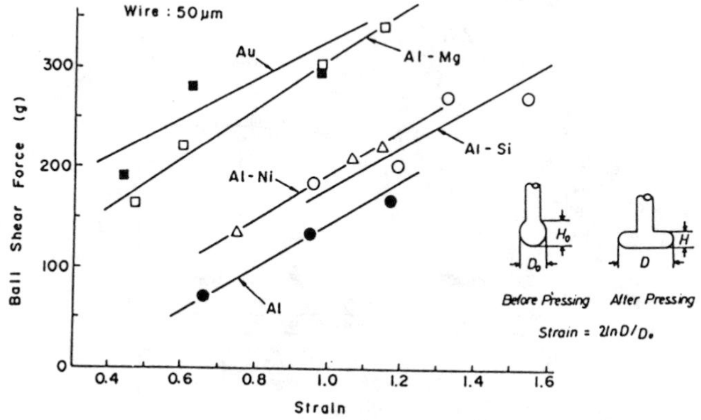

Fig. 4 Effects of ball strain on bonding shear strength

PLASTIC PACKAGING

The chip bonded with a substrate is sealed in a package for ease of handling and to afford protection from external force or thermal shock. The packaging process is achieved by packing a device in a small container or by coating a device with resin. Encapsulation of a chip in resin is widely used for plastic packaging due to high productivity. With increasing circuit density, small packages are in great demand. The encapsulated resin suffers heat cycles both in molding and operating processes. The temperature change in the process creates internal stresses in an encapsulated device due to different thermal expansion between resin and chip. Shrinkage of packaging resin also produces internal stresses. The stress causes cracking in molded resin, delamination in the interface between resin and chip, and breakage of the interconnected wire bonding.

(1) Internal stress

A silicon chip is mounted on a lead frame of copper with Ag-paste bonding and encapsulated by epoxy resin. Figure 5 shows the calculated thermal stress distribution in the resin over the chip under thermal shock by cooling from 150°C to -55° C. The stress components are indicated by the principal stress and the maximum shear stress. The stresses have the maximum values near the edge of the chip; resin cracking can be expected at this position.

The stress concentration in the resin near the chip corner is analyzed by the boundary element method as shown in Fig. 6 [6]. The stress is assumed to be induced by heating from operating chip. The analyzed stress distributions are illustrated by nondimensional values of their radial and circumferential components in a polar coordinate. The radial stress is concentrated in the

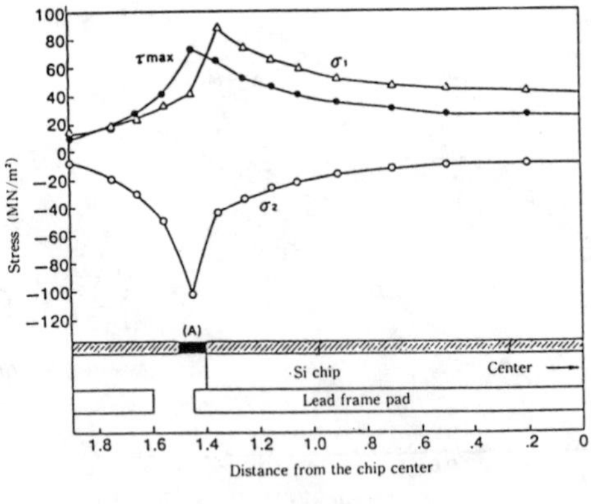

Fig. 5 Thermal stress distribution in plastic package

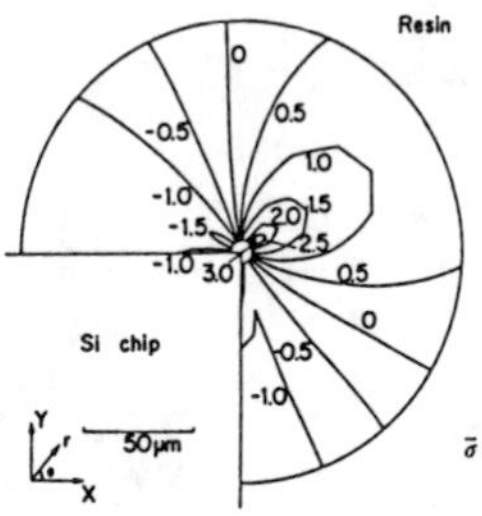

(a) Radial stress

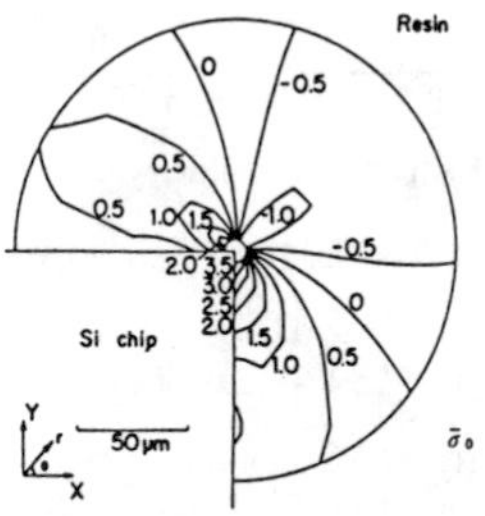

(b) Circumferential stress

Fig. 6 Thermal stress concentration at chip edge

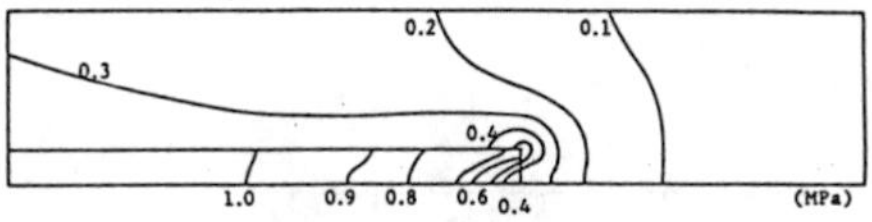

(a) 170'C

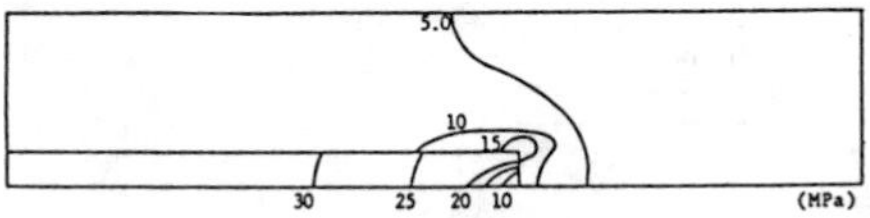

(b) 100'C

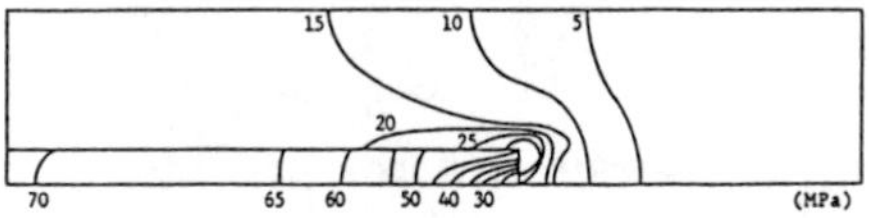

(c) 20'C

Fig. 7 Variations of internal stress during curing process

direction of 35° from the chip edge, while the circumferential stress is large in the vertical interface of the chip and also in the direction of 140° from the chip edge.

The internal stress in plastic packaging is induced by shrinkage of resin as well as thermal mismatch between chip and resin. The encapsulated resin in curing process changes the mechanical properties and the volume. The internal stress in the resin encapsulated steel plate is analyzed by using the finite element method including the mechanical and vlolumetric changes of the resin [7]. Figure 6 shows the stress variations of the internal stress during the curing process. The stress values are represented by the equivalent stress magnitude. The distribution is dependent on the temperature condition and the stress value increase with falling temperature.

(2) Packaging strength

The surface mounting method is adopted for the plastic package. This is achieved by reflow soldering an encapsulated chip on printed circuit boards. The packaged chip is heated by the reflow soldering. Moisture absorbed in the encapsulated resin is evaporated by the heating process and the vapor pressure causes the fracture in the interface between chip pad and resin. Figure 8 shows the deformed state of plastic package by vapored pressure [8]. The thick lines indicate the shape

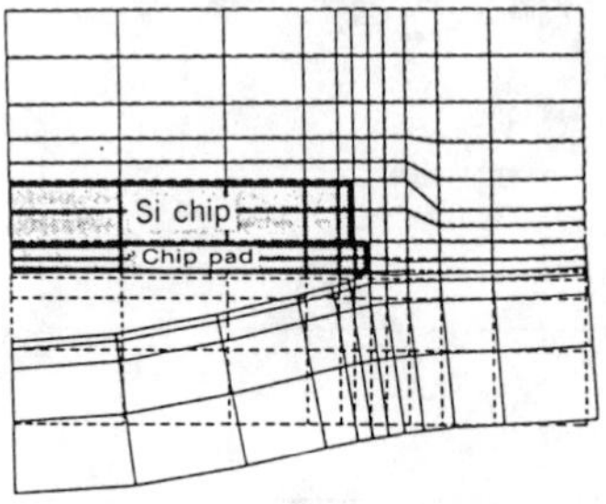

Fig. 8 Deformation of plastic package by vapor pressure

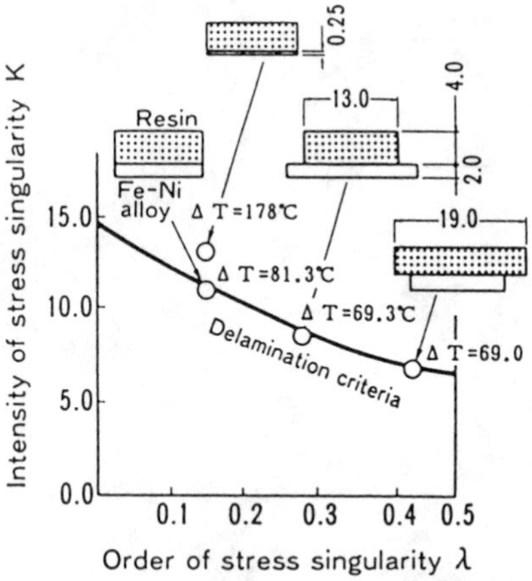

Fig. 9 Delamination criterion based on stress singularity

after deformation. The magnitude of the deformation is enlarged ten times compared with the model dimension. Large strain concentration is found near the bottom edge of the chip pad. Cracking of resin is expected to be initiated from this position.

The prediction of crack initiation and propagation in the interface between encapsulated resin and chip pad is important for the estimation of device reliability. The criterion for the initial cracking in the interface between dissimilar materials is suggested by using the stress singularity at the interface [9]. The intensity and order of the stress singularity are main parameters to determine the criterion. Figure 9 is the delamination condition based on the stress singularity criterion for the interface of models which consist of epoxy resin and Fe-Ni alloy with different dimensions.

CONCLUSIONS

The semiconductor device has minute parts and thin layers of different materials. There are many interfacial layers in the device. Thermal loading during the production process as well as the operating process induces the internal stress inside of the device due to the thermal mismatch among the constituent materials. The stress gives the damage to the device materials and their interfaces. The material characterization for minute material and thin multi-layered materials is important for solving the mechanical problems relating with the stress analysis and strength estimation of semiconductor devices.

REFERENCES

1. K. Ikegami, "Mechanical problems in the production process of semiconductor devices", Int. J. Japan Soc. Mech. Engrs., Ser. I, 33(1), 1-12(Jan. 1990).
2. R. Satoh, M. Ohshima, K. Hirota, and I. Ishi, "Micro-solder bonding technology for IC-LSI", Bull. Japan. Inst. Metals, (in Japanese), 23(12), 1004-1013 (Dec 1984).
3. T. Yamaguchi, M. Hashimoto, and T. Mizutani," Thermal stress analysis for IC bump in thermocompression", (in Japanese), J. Japan Soc. Mech. Engrs, 84(755), 1070-1076(Nov. 1983).
4. J. Onuki, M. Suwa, M. Koizumi and T. Iizuka, "Investigation of aluminum ball bonding mechanism", IEEE Trans. Comp., Hybr., and Manuf. Tech., CHMT-10(2), 242-247(June 1987).
5. M. Miyake, H. Suzuki and S. Yamamoto," Heat transfer and thermal stress analysis of plastic-encapsulated ICs", IEEE Trans. Reliability, R-34(5), 402-409(Dec. 1985).
6. M. Sato, R. Yuuki and S. Yoshioka," Boundary element analysis of steady-state heat conduction and thermal stresses in the LSI package", (in Japanese), Trans. Japan Soc. Mech. Engrs., Vol. 55A(514), 1437-1444(June 1989).
7. Y. Kawata and K. Ikegami, "Viscoelastic properties and residual stress of the resin for plastic packaging", (in Japanese), Trans. Japan Soc. Mech. Engr., (to be published)
8. M. Kitano, S. Kawai, T. Nishimura and K. Nishi, "A study of package cracking during the reflow soldering process", (in Japanese), Trans. Japan Soc. Mech. Engr., 55A(510), 356-363(Feb. 1989).
9. T. Hattori, S. Sakata, T. Hatsuda and C. Murakami, "A Stress singularity parameters approach for evaluating adhesive strength", (in Japanese), Trans. Japan Soc. Mech. Engrs., 54A(499), 597-603 (March 1988).

EVALUATION OF INELASTIC CONSTITUTIVE MODELS: FEM ANALYSIS OF NOTCHED BARS OF 2·1/4Cr-1Mo STEEL AT 600°C

- The Third Report of Benchmark Project -

Subcommittee on Inelastic Analysis of High Temperature Materials, the Society of Materials Science, Japan
Yoshida-Izumidono-cho 1-101, Sakyo-ku, Kyoto 606, Japan

ABSTRACT

Inelastic constitutive models are evaluated by implementing them in finite element analyses of notched bars of 2·1/4Cr-1Mo steel. This work is a part of the third phase of the benchmark project by the Subcommittee on Inelastic Analysis and Life Prediction of High Temperature Materials, JSMS. Eight types of constitutive models are utilized to compare analytical predictions with experiments for two types of notched bars subjected to four kinds of loads at 600°C.

INTRODUCTION

The Subcommittee on Inelastic Analysis and Life Prediction of High Temperature Materials, the Society of Materials Science, Japan, whose members are listed in the footnote, has been performing a cooperative project consisting of the following two tasks: (A) review and evaluation of inelastic constitutive models relevant to material response under plasticity-creep interaction condition, and (B) examination of prediction methods of material life in fatigue-creep regime in consideration of the effect of plasticity--creep interaction on stress-strain response.

Members of the Subcommittee are T. Inoue (Kyoto Univ., Chairman); T. Igari (Mitsubishi Heavy Ind., Secretary); N. Ohno (Nagoya Univ., ibid), S. Imatani (Kyoto Univ., ibid), M. Kawai (Univ. Tsukuba, ibid), S. Kishi (Toshiba Co., ibid), Y. Niitsu (Tokyo Inst. Tech., ibid), M. Okazaki (Nagaoka Univ. Tech., ibid), M. Sakane (Ritsumeikan Univ., ibid), F. Yoshida (Hiroshima Univ., ibid); Y. Asada (Tokyo Univ.), T. Hiroe (Kumamoto Univ.), S. Kubo (Osaka Univ.), K. Motoie (Hiroshima Denki Inst. Tech.), S. Murakami (Nagoya Univ.), S. Nagaki (Okayama Univ.), E. Tanaka (Nagoya Univ.), K. Tanaka (Tokyo Metro. Inst. Tech.), T. Yokobori (Tohoku Univ.), K. Fujiyama (Toshiba Co.), Y. Fukuta (Hitachi Co.), K. Kanazawa (National Res. Inst. Metals), H. Koto (Mitsubishi Heavy Ind.), S. Koue (Kawasaki Heavy Ind.), M. Mizumura (Nippon Steel Co.), K. Nagato (Kawasaki Heavy Ind.), T. Shimizu (Hitachi Co.), A. Suzuki (Ishikawajima Harima Heavy Ind.), Y. Takahashi (Central Res. Inst. Elect. Power Ind.), K. Tamura (Kawasaki Heavy Ind.), K. Tokimasa (Sumitomo Metal Ind.), T. Uno (Toshiba Co.), Y. Wada (Power Reactor and Nucl. Fuel Dev. Co.)

As the first and second phases toward the accomplishment of the above two tasks, benchmark studies for uniaxial and multiaxial inelastic behavior in uniform fields of stress and strain were completed to evaluate ten kinds of constitutive models and to examine eight types of fatigue-creep life prediction methods, by use of 2·1/4Cr-1Mo steel at 600°C [1-6].

In the third phase of the project, stress/strain fields near notch root of bars are analyzed by implementing inelastic constitutive models in finite element methods, followed by the life prediction which will be reported somewhere. It allows us further evaluation of inelastic constitutive models and examination of life prediction methods in practical problems. Eight types of constitutive models are utilized to compare analytical predictions with experiments on two types of notched bars subjected to four kinds of loads at 600°C.

MATERIAL AND SPECIMENS

The material used is a normalized and tempered 2·1/4Cr-1Mo steel, the same as employed in the first and second phases of the project.

Two types of notched specimens are used (Fig. 1); i.e., round notch of strain concentration factor K_t = 1.5, and V-shape notch of K_t = 3.0. Axial strain near notch root is measured by means of a high-temperature capacitance-type extensometer marketed under the trade name of Strain Pecker. It is spot-welded to the notch root of each specimen, so that average strain at the notch root for the gauge length 0.5 mm, ε_{sp}, is measured. Average strain for the gauge length 30 mm, ε_{GL}, is also measured.

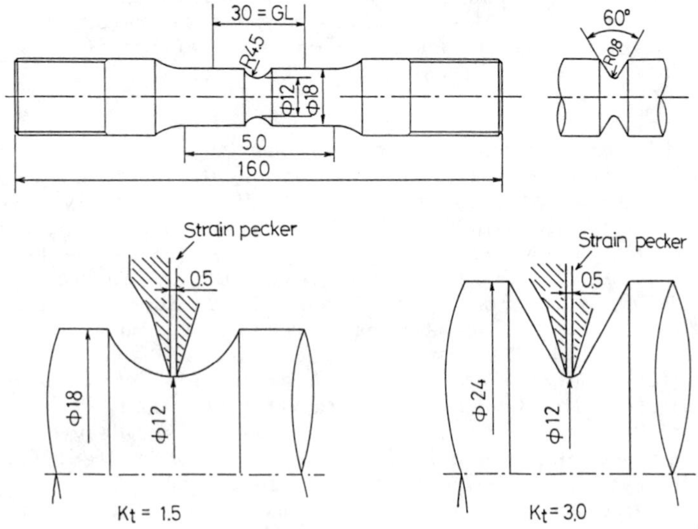

Fig. 1 Shape and dimension of notched bar specimens; (a) elastic strain concentration factor K_t = 1.5, and (b) K_t = 3.0.

BENCHMARK PROBLEMS

The two types of specimens are tested under the following four loading conditions (Table 1):

I. Creep under σ_{nom} = 120 MPa, where σ_{nom} indicates the nominal stress on the minimum cross sections of the specimens.

II. Fast-fast cyclic straining under $\dot{\varepsilon}_{GL}$ = 0.025 %/s for strain range $\Delta\varepsilon_{GL}$ = 0.2 %.

III. Slow-slow cyclic straining under $\dot{\varepsilon}_{GL}$ = 0.001 %/s for strain range $\Delta\varepsilon_{GL}$ = 0.2 %.

IV. Fast-fast cyclic straining with tension hold time t_h; $\Delta\varepsilon_{GL}$ = 0.2 %, $\dot{\varepsilon}_{GL}$ = 0.025 %/s, and t_h = 600 s.

CONSTITUTIVE MODELS EXAMINED

Eight types of constitutive models are expected to be used for the analysis in this third phase of the project. Currently the following four types of constitutive models have been implemented in finite element methods to analyze the benchmark problems; i.e., the Superposition model in which inelastic strain is decomposed independently into plastic and creep components, the Chaboche model based on the nonlinear kinematic hardening rule and the overstress concept, the Bodner model in which an exponential type creep equation is extended by means of two kinds of internal variables, and the Ohno-Murakami model based on the cyclic nonhardening strain region substantiated in a unified manner.

Table 1 Benchmark problems.

Problem	Loading Pattern	Condition	Experiment	Analysis
I	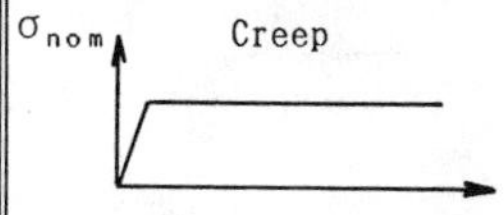	σ_{nom} = 120 MPa	~ 200 h	~ 100 h
II	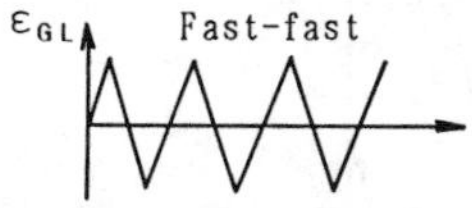	$\Delta\varepsilon_{GL}$ = 0.2 % $\dot{\varepsilon}_{GL}$ = 0.025 %/s	to rupture	~ N=10
III	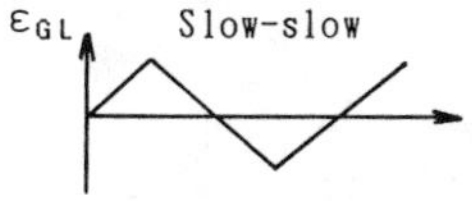	$\Delta\varepsilon_{GL}$ = 0.2 % $\dot{\varepsilon}_{GL}$ = 0.001 %/s	to rupture	~ N=10
IV	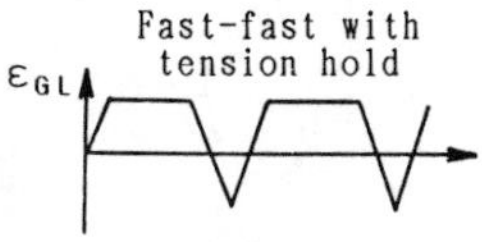	$\Delta\varepsilon_{GL}$ = 0.2 % $\dot{\varepsilon}_{GL}$ = 0.025 %/s t_h = 600 s	to rupture	~ N=10

RESULTS OF ANALYSES AND EXPERIMENTS

Experimental results and some predictions in early stages for the benchmark problems II to IV are shown in Figs. 2-7.

Table 2 deals with comparison between the experimental and predicted results for the fast-fast cyclic straining test of the specimen with K_t = 1.5. With respect to $\Delta\varepsilon_z$ at notch root, the predictions are fairly close to each other and do not deviate from the experiment so much. This is mainly because not load but dis-

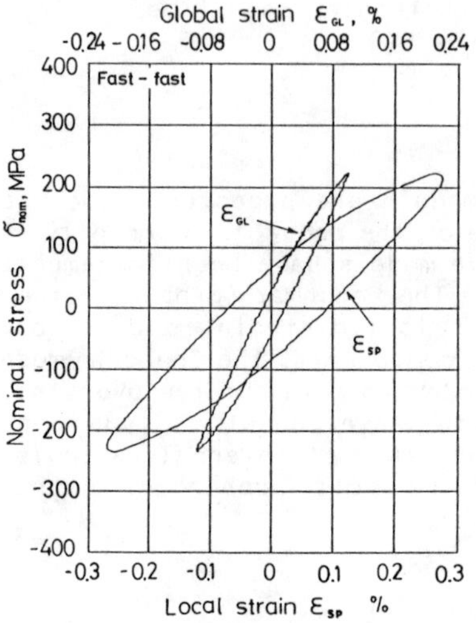

Fig. 2 Experimental result - Problem II, K_t = 1.5 -.

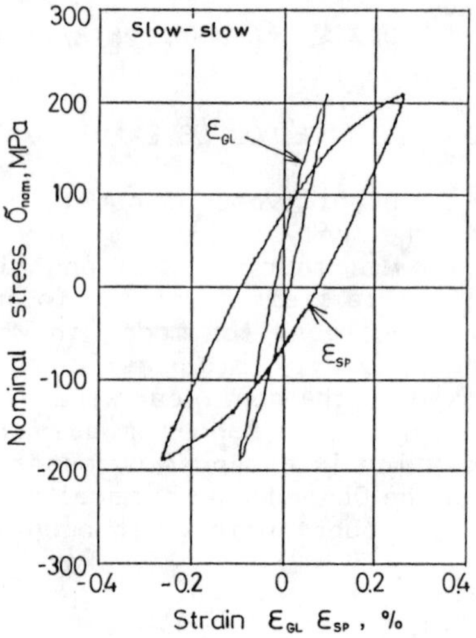

Fig. 3 Experimental result - Problem III, K_t = 1.5 -.

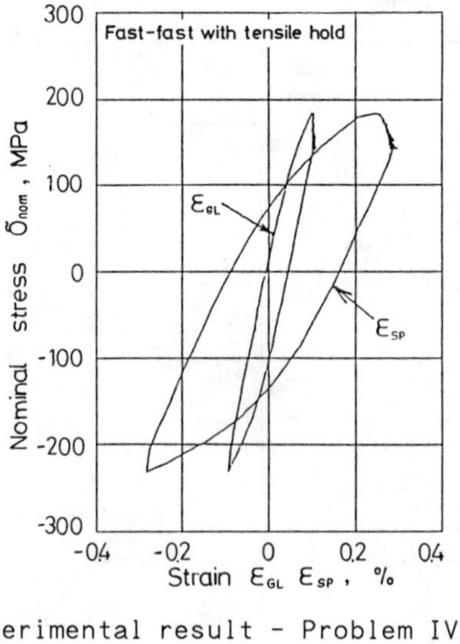

Fig. 4 Experimental result - Problem IV, K_t = 1.5 -.

splacement was controlled in the test. With respect to $\Delta\sigma_z$ at notch root, on the other hand, the Bodner model gives a considerably large value in comparison with the other model. It may be ascribed to the identification of the material parameters in the model.

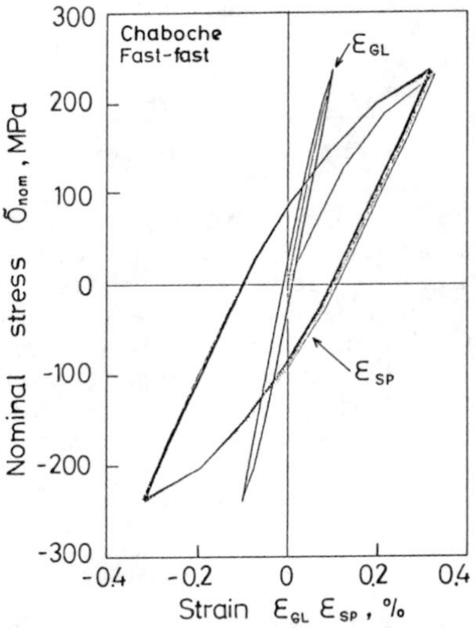

Fig. 5 Prediction by Chaboche Model - Problem II, K_t = 1.5 -.

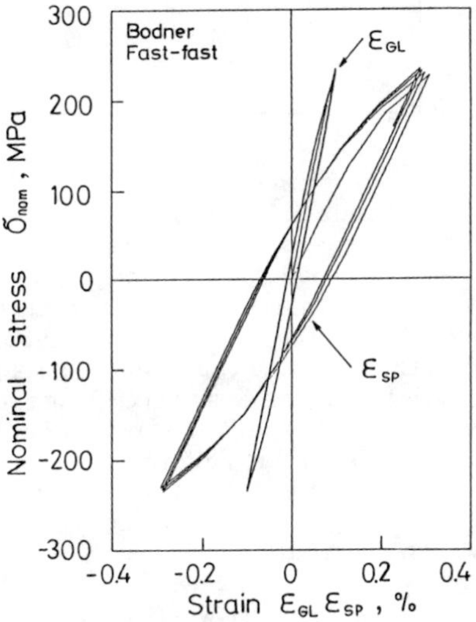

Fig. 6 Prediction by Bodner Model - Problem II, K_t = 1.5 -.

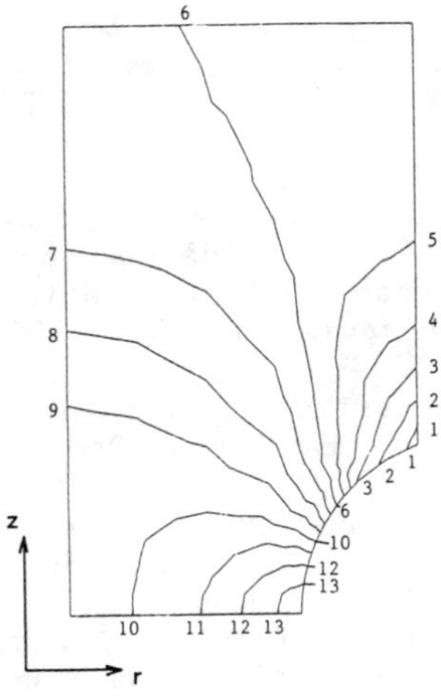

(a)

σ_z (MPa)

No.	Value
1	0
2	20
3	40
4	60
5	80
6	100
7	120
8	140
9	160
10	180
11	200
12	220
13	240

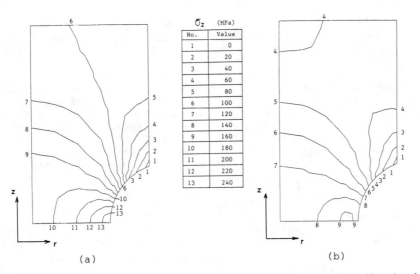

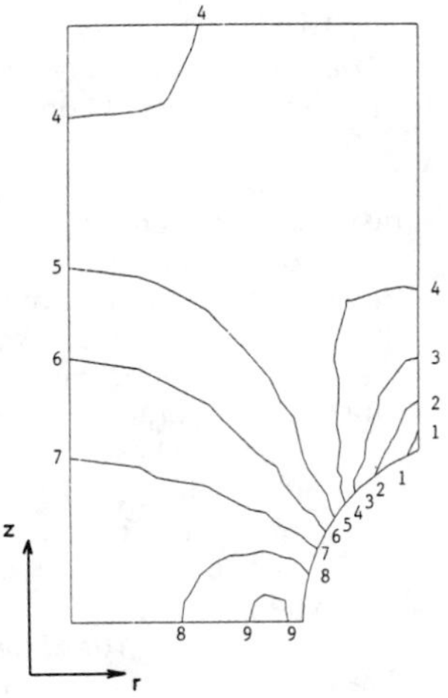

(b)

Fig. 7 Distribution of axial stress at N = 3 by Ohno-Murakami Model - Problem IV, K_t = 1.5 -; (a) just before tension hold, and (b) just after tension hold.

Table 2 Comparison between experiment and predictions with respect to the axial stress and strain ranges at notch root, $\Delta\varepsilon_{z\ notch}$ and $\Delta\sigma_{z\ notch}$, - Problem II, K_t = 1.5 -.

	Experiment	Analysis			
		Superp.	Chaboche	Bodner	Ohno-Murakami
$\Delta\varepsilon_{z\ notch}$	0.55 (=$\Delta\varepsilon_{sp}$)	0.59	0.63	0.57	0.61
$\Delta\sigma_{z\ notch}$	———	498	483	558	507

strain (%), stress (MPa)

Lives of the notched specimens can be predicted, if notch root stress-strain hysteresis loops obtained by the FEM analyses are combined with appropriate fatigue-creep life prediction methods. This is another task performed in the third phase of the project. Currently four types of life prediction methods are utilized, i.e., the linear damage rule, the strain range partitioning method, the Lemaitre-Plumtree-Chaboche method, and the Majumdar method. Comparison of such predicted lives and experimental results will be reported somewhere.

ACKNOWLEDGMENT

The present work is supported by the Ministry of Education as Grant-in-Aid for Co-operative Research A (No. 01302026).

REFERENCES

1. T. Inoue, T. Igari, F. Yoshida, A. Suzuki, and S. Murakami, "Inelastic behaviour of 2·1/4Cr-1Mo steel under plasticity--creep interaction condition", Nucl. Eng. Des. **90**, 287-297 (1985).
2. T. Inoue et al., "Report of the benchmark project on inelastic deformation and life prediction of 2·1/4Cr-1Mo steel at 600°C", in Proc. 2nd Int. Conf. on Constitutive Laws for Engineering Materials: Theory and Application, Tucson, Arizona, edited by C.S. Desai et al., Vol.II, 959-966 (Elsevier, New York 1987).
3. T. Inoue, N. Ohno, A. Suzuki, and T. Igari, "Evaluation of inelastic constitutive models under plasticity-creep interaction for 2·1/4Cr-1Mo steel at 600°C", Nucl. Eng. Des. **114**, 295-309 (1989).
4. T. Inoue, T. Igari, M. Okazaki, M. Sakane, and K. Tokimasa, "Fatigue-creep life prediction of 2·1/4Cr-1Mo steel by inelastic analysis", Nucl. Eng. Des. **114**, 311-321 (1989).
5. T. Inoue, F. Yoshida, N. Ohno, M. Kawai, Y. Niitsu, and S. Imatani, "Evaluation of inelastic constitutive models under plasticity-creep interaction in multiaxial stress state", Nucl. Eng. Des., (submitted).
6. T. Inoue, M. Okazaki, T. Igari, M. Sakane, and S. Kishi, "Fatigue-creep life prediction of 2·1/4Cr-1Mo steel at 600°C in multiaxial stress state", Nucl. Eng. Des., (submitted).

Formulation of Constitutive Law For Two-Phase Aggregates of SUS 304 Subjected to Strain-Induced Phase Transformation By Rigid Plastic Finite Element Analysis of Cold Forming.
- Energy-Based Formulation By Using Phase Strains -

CORTES JORGE, TSUTA TOSHIO, SHIRAISHI MITSUNOBU
Faculty of Engineering, Hiroshima University,
Higashi Hiroshima City, Japan.

YAMAUCHI KEISUKE
Kawasaki Heavy Industries Ltd.

OSAKADA KOZO
Faculty of Engineering Science, Osaka University.

ABSTRACT

A constitutive law for SUS 304 austenitic stainless steel under cold forming conditions has been formulated by using strain and temperature functions. Volume fraction, flow stress and deformation of each phase are included explicitly into the analysis. To define the martensitic flow stress, the criterion that the sum of energies consumed to deform each phase is the energy required to deform the two-phase alloy was propounded. By means of numerical simulations, using the rigid plastic finite element method, RP-FEM, strain of each phase was determined according to several volume fractions of martensite.

INTRODUCTION

In unstable austenitic stainless steels, such as the SUS 304, the development of a strain-induced phase transformation, austenite to martensite, cause that the mechanical properties prediction become difficult to carry out. The individual mechanical properties and the thermal conditions modify mainly the plastic properties of the material. The approximation of mechanical properties from knowledge of properties of the individual phases, has been carried out in the past by using the rule of mixtures and experimental data. Furthermore, an equation to determine the volume fraction of martensite generated was identified as a function of alloy strain [1]. The present investigation reports attempts to improve such approximation. Accounting with a quantitative expression for phase transformation, a constitutive formulation is determined by means of theoretical and numerical analyses, in which the mechanical behaviour of each phase is determined explicitly.

QUANTIFICATION OF STRAIN INDUCED MARTENSITE

An expression to determine the volume fraction of martensite, V_{fm}, has been determined by using the magnetic technique and compression tests at several constant temperatures [1]. Such expression is

$$V_{fm} = [\,1 + (\varepsilon/\varepsilon_c)^{-B}\,]^{-1} \qquad \ldots (1)$$

in which

$$\varepsilon_c = C_1 \; e^{(C_2/T)} \qquad \ldots (2)$$

where ε, T and ε_c are total strain, temperature in°K and characteristic strain, at which 50% total volume changes to martensite. For SUS 304, B= 2.46, C_1=22930, C_2= 3158.

AUSTENITIC FLOW STRESS

Since it is possible to inhibit the martensitic phase transformation, under experimental conditions, the flow stress of austenitic phase can be experimentally evaluated through differential compression tests, in which pre-strains are first added to each specimen under higher temperature conditions, where the martensitic phase is not generated, and then another compression tests with different pre-strains, are continued after unloading in order to find the yield point at each pre-strain level [1]. The austenitic flow stress as a function of temperature is expressed by the following equation

$$\sigma_a = \sigma_o + A_1\,[\,1 - e^{-(A_2\,\varepsilon_a)}\,]^{A_3} \qquad \ldots (3)$$

in which

$$\sigma_o = D_1 \; e^{(D_2/T)} \qquad \ldots (4)$$

where ε_a, σ_a and σ_o are austenite strain, austenite flow stress in kg/mm^2. and yield stress, respectively. The constants have values of , A_1= 130, A_2= 1.3, A_3= 0.861, D_1= 2.84, and D_2=620.5 .

MARTENSITE FLOW STRESS

Experimentally, the inhibition of austenitic phase is not as easier as the martensitic phase inhibition, because 100% martensitic phase can not be reached under conventional conditions. To approximate the mechanical behaviour of this phase a numerical method is developed. This method consists in the formulation of a balance of deformation energies to each phase, in which the martensite deformation energy is the unknown fraction. The formulation is based on a energetic criterion, which establish that the sum of deformation energies, consumed separately by the austenite and martensite, is equivalent to that consumed by the alloy during its deformation. This principle applied to a RP-FEM scheme was used to simulate the compression test of the alloy with high martensite content, such as 74%, constant in the overall strain range and whose mechanical behaviour at 100°C is known experimentally. The martensite flow stress, accompanied with several martensite stress-strain pairs of intermediate strain stages, are determined using the Ludwik stress-strain relation. By least square technique, the strain hardening exponent and the strength coefficient in such relation are identified, forming the data of the next simulation. Hence, iteratively the martensitic deformation energy converges to a stationary condition which defines the martensite flow stress behaviour [2]. The martensite flow stress, at 100 C, can then expressed by

$$\sigma_m = \kappa \, \varepsilon_m{}^n \qquad \ldots (5)$$

where σ_m is martensite flow stress in kg/mm^2 and ε_m is martensite strain. κ = 170 and n = 0.201 .

PHASE INTERACTIONS DURING DEFORMATION

Plastic strain of each phase during the alloy deformation is analyzed by means of numerical simulations using the RP-FEM. The analysis consists in to carry out simulations of homogeneous compression by using randomly distributed two-materials models, where the austenite and martensite mechanical properties are well defined and in which the number of elements that correspond to the martensitic phase, is constant during the simulation. The development of several simulations with distinct V_{fm} contents lead to correlate both the total deformation and the volume fraction of each phase to the plastic deformation of austenitic and martensitic phase. Such correlation is shown in the figure 1 and in the following expressions. Considering that V_{fa}, volume fraction of austenite, is $1-V_{fm}$.

$$\varepsilon_a = \varepsilon (V_{fa})^{-K} \quad , \quad \varepsilon_m = \varepsilon V_{fm}{}^{-L} \qquad \ldots (6)$$

$$K = E_1 (E_2 + \varepsilon)^{-E_3} \quad , \quad L = F_1 (F_2 + \varepsilon)^{-F_3} \qquad \ldots (7)$$

E_1 = 0.451, E_2 = 2.161, E_3 = 2.128, F_1 = 0.164, F_2 = 0.439 and F_3 = 0.882 . Since the phase strains are in function of

$$\varepsilon_a = \varepsilon_a(\varepsilon, V_{fa}) \quad , \quad \varepsilon_m = \varepsilon_m(\varepsilon, V_{fm}) \qquad \ldots (8)$$

the increment of austenite and martensite strains can be defined as

$$d\varepsilon_a = \frac{\partial \varepsilon_a}{\partial \varepsilon} d\varepsilon + \frac{\partial \varepsilon_a}{\partial V_{fa}} dV_{fa} \ , \ d\varepsilon_m = \frac{\partial \varepsilon_m}{\partial \varepsilon} d\varepsilon + \frac{\partial \varepsilon_m}{\partial V_{fm}} dV_{fm} \qquad \ldots (9)$$

Since

$$dV_{fm} = 0 \qquad \ldots (10)$$

The relationship between the strains is rewritten as

$$\frac{d\varepsilon_a}{d\varepsilon} = \Gamma_a \quad , \quad \frac{d\varepsilon_m}{d\varepsilon} = \Gamma_m \qquad \ldots (11)$$

where functions Γ_a and Γ_m can be determined from the results shown in fig. 1.

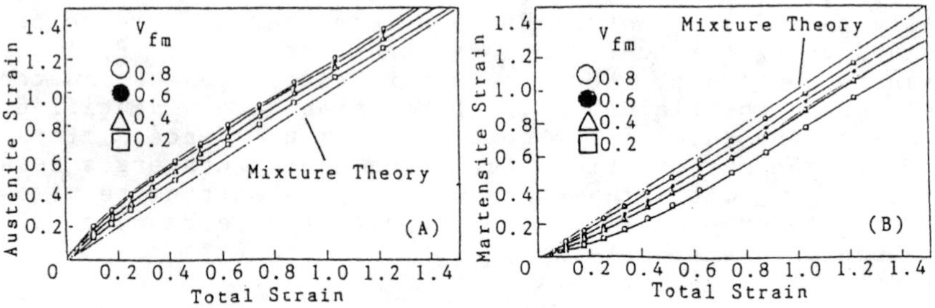

Fig. 1. Relationship between the phase strains and total strain. (A) Austenitic Phase. (B) Martensitic Phase.

CONSTITUTIVE EQUATION

Based on the deformation energies of each component of the system, under uniform compression, the stress for the aggregated, σ_T, can be derived.

$$\sigma_T \, d\varepsilon = V_{fa} \sigma_a \, d\varepsilon_a + V_{fm} \sigma_m \, d\varepsilon_m \qquad \text{... (13)}$$

By eq. (11)

$$\sigma_T = V_{fa} \, \sigma_a \, \Gamma_a + V_{fm} \, \sigma_m \, \Gamma_m \qquad \text{... (14)}$$

which is the constitutive model to describe and to predict the stress created by the strain-induced martensitic and austenitic phases.

CONCLUSIONS

The use of a energetic criterion and the RP-FEM, to relate mechanical and metallurgical problems, allowed to determine a constitutive law, leading to evaluate the mechanical properties of SUS 304 during its cold forming, and whose application would permit to analyze a great variety of cold forming process where the 304 austenitic stainless steel were used.

REFERENCES

1. Cortes, J., et al., "Simulation of Cold Forging Process of 18-8 Stainless Steel Considering Metallurgical Factors", Proc. of 3rd. Int. Conf. on Tech. of Plasticity. 1, 65-70 (Kyoto, July 1990).

2. Cortes J., et al., "Formulation of Constitutive Model For SUS 304 And Rigid Plastic Finite Element Analysis of Cold Forming Considering The Temperature Rise Due to Deformation and Non-Homogeneity of Austenite and Martensite Phases", Proc. of the 68 th National Conference of Japan Soc. of Mech. Eng. (September 1990).

ROCKET SHOCK ISOLATION WITH DEADSPACE ISOLATORS A NONLINEAR ANALYSIS WITH STOCHASTIC PARAMETERS

ZOLTAN A. KEMENY
Vice President- LORANT GROUP
1617 E. Highland Avenue, Phoenix, Arizona 85016

ABSTRACT

This numerical case study shows that internal bracings in rocket fuselage with deadspace type dissipative isolators make glued-on dampers obsolete and enhance impact and fatigue life of the fuselage.

INTRODUCTION

Metal - viscoelastomer composite universal joints[1] add delayed strength and service vibration suppression to vehicle fuselage subjected to stochastic impact load like rockets.

Experiments show[2] that structural damping can attain 60% of the critical value and that the impact load induced fuselage stresses will be well shifted in phase and natural period and reduced in amplitude when using such isolators (Fig. 1).

NUMERICAL SIMULATIONS

A single degree-of-freedom (SDF) model has been evaluated for simplicity as shown in Figure 2 to generate rocket shock and to get the system response, in DESIRE[3].

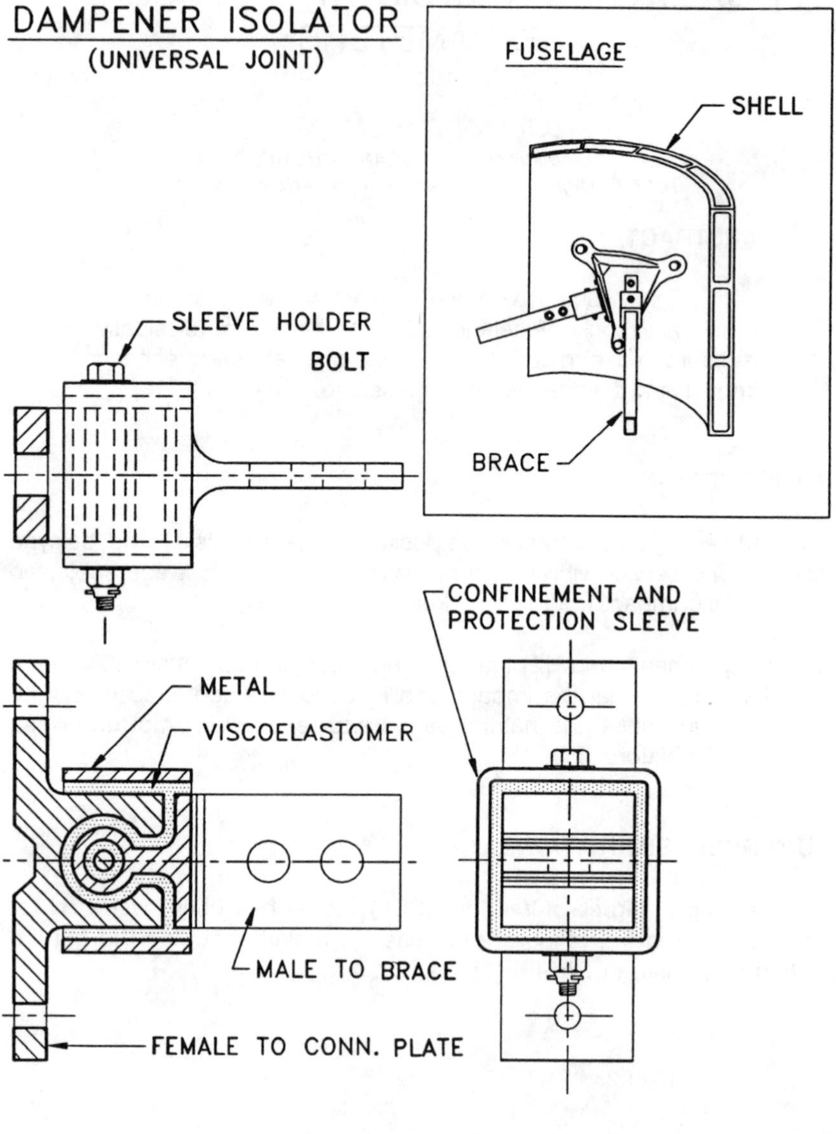

Fig. 1. Deadspace Isolator for Vehicles

STOCHASTIC NONLINEAR KELVIN MODEL

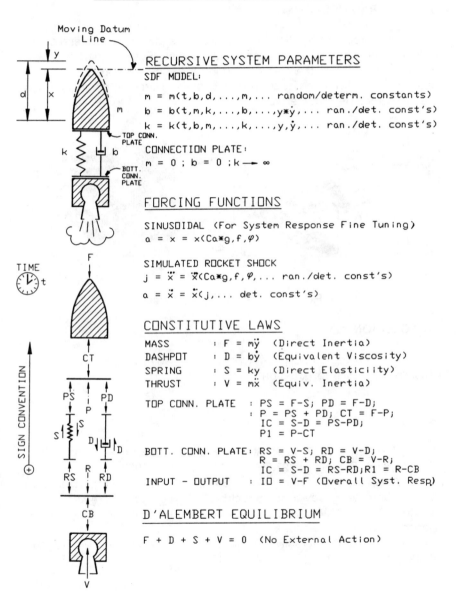

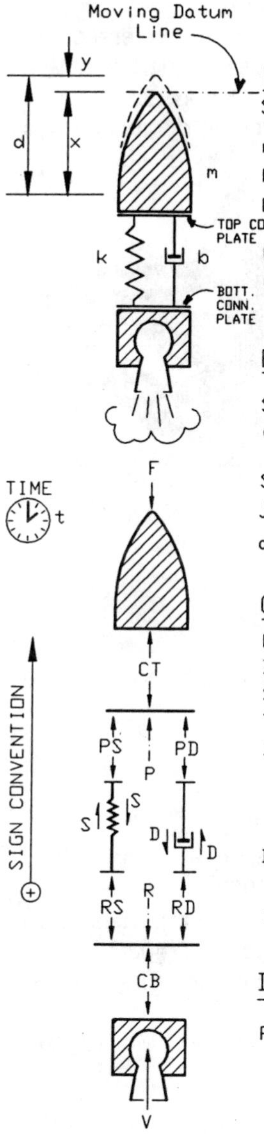

RECURSIVE SYSTEM PARAMETERS

SDF MODEL:

m = m(t, b, d, . . . , m, . . . random/determ. constants)

b = b(t, m, k, . . . , b, . . . , y*$\dot{y}$, . . . ran./det. const's)

k = k(t, b, m, . . . , k, . . . , y, $\dot{y}$, . . . ran./det. const's)

CONNECTION PLATE:

$m = 0$; $b = 0$; $k \to \infty$

FORCING FUNCTIONS

SINUSOIDAL (For System Response Fine Tuning)

$a = \ddot{x} = \ddot{x}(Ca*g, f, \varphi)$

SIMULATED ROCKET SHOCK

$j = \dddot{x} = \dddot{x}(Ca*g, f, \varphi,$ . . . ran./det. const's)

$a = \ddot{x} = \ddot{x}(j,$ . . . det. const's)

CONSTITUTIVE LAWS

MASS : $F = m\ddot{y}$ (Direct Inertia)

DASHPOT : $D = b\dot{y}$ (Equivalent Viscosity)

SPRING : $S = ky$ (Direct Elasticiity)

THRUST : $V = m\ddot{x}$ (Equiv. Inertia)

TOP CONN. PLATE : PS = F−S; PD = F−D;
: P = PS + PD; CT = F−P;
IC = S−D = PS−PD;
P1 = P−CT

BOTT. CONN. PLATE: RS = V−S; RD = V−D;
R = RS + RD; CB = V−R;
IC = S−D = RS−RD; R1 = R−CB

INPUT − OUTPUT : IO = V−F (Overall Syst. Resp)

D'ALEMBERT EQUILIBRIUM

F + D + S + V = 0 (No External Action)

Fig. 2. Rocket Shock Simulation Model

Payload pounding and structural - nonstructural movement interaction has been accounted for stochasticly, while the deadload degradation was contemplated as stochasto-deterministic.

STRUCTURAL RESPONSE

The three forces different in nature are shown on a hysteretic plot in Figure 3a, where D is the viscose dissipative, F is the inert and V is the rocket thrust v.s. the elastic restoring force (S). Figure 3b is a plot of the conservative forces v.s. dissipative ones. Here, the diagonal wings are typical for deadspace type isolators.

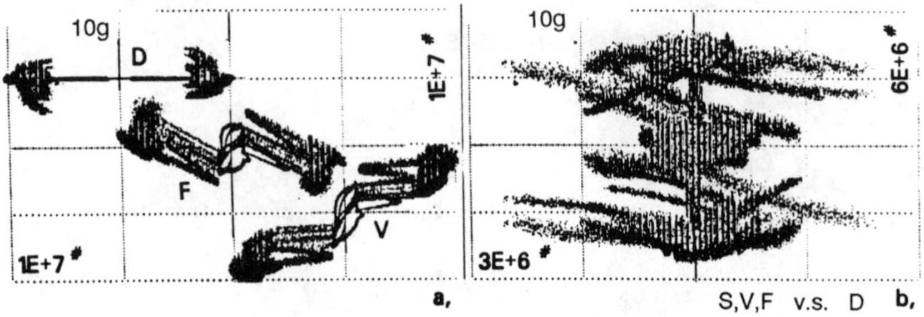

Fig. 3. Hysteretic rocket forces

CONCLUSION

The studied deadspace type internal isolation proved to be a feasible alternative to glued-on dampers.

REFERENCES

1. Z.A. Kemeny, "Building Structure Shock Isolation System", Official Gazette, U.S. Pat. No. 4,727,695, p. 42, March 1, 1988.

2. A. Fafitis, at al., "Study of the Static and Dynamic Properties of an Energy Dissipating Device for Structures", Project No. XCT 4028, Vol. 2-9 and XCT 4282, Vol. 2-4, Arizona State University, Civil Engineering Department, Tempe, Arizona, 1986-90.

3. G. A. Korn, "Interactive Dynamic System Simulation", McGraw-Hill Book Co., New York, ..., Toronto, 1988.

BUILDING SEISMIC RETROFIT WITH DISSIPATIVE ISOLATED BRACES A NONLINEAR ANALYSIS WITH STOCHASTIC PARAMETERS

ZOLTAN A. KEMENY, Vice President
LORANT GROUP, 1617 E. Highland Avenue
Phoenix, Arizona 85016

ABSTRACT

This numerical case study shows that added steel braces with dissipative isolated moment connections enhance the strength and ductility of masonry buildings in localities of high seismicity.

INTRODUCTION

Steel-polyurethane composite moment connection plate assemblies[1] redundantly add strength, ductility and damping to under reinforced masonry buildings (Fig. 1). The added strength is delayed until the elastomer gets compacted (interstory isolation). The added stiffness builds up smoothly. The interlocking keys in the isolator are designed to yield before the yielding of the brace or the reinforcing bars in the building.

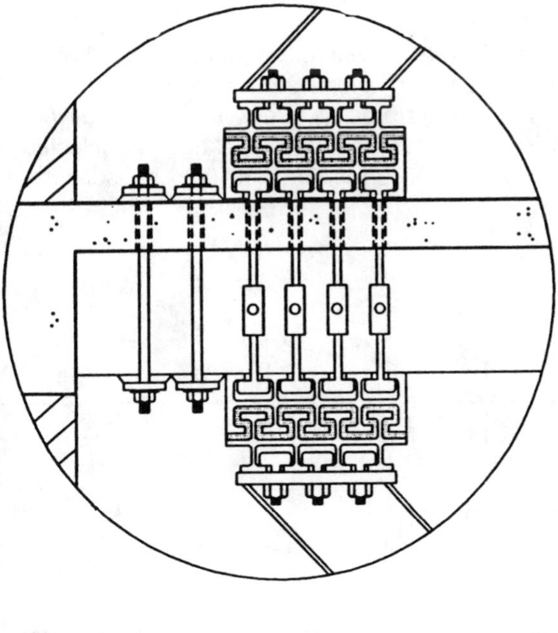

Fig. 1. Isolated moment connection

NUMERICAL SIMULATIONS

A five degree-of-freedom (5DF) lumped mass dynamic model response has been compared with an equivalent single degree-of-freedom (1DF) model response. The effect of rate dependency, elasto-plasticity, visco-elasticity, stress or strain hardening, confinements, strength and strain limits are accounted for in a deterministic fashion.

Stochastic system parameters intended to model live load pounding and sliding, cumulative structural and nonstructural damage, deadload shake-down and the impulse-damping of the incipient cracking. The 1DF model has been found useful to select the most critical simulated seismic ground motion out of numerous random simulations and some historical data. The 5DF model was effective to detect local instabilities, failures and to verify the equivalent 1DF model parameters.

Finally, four soil layers have been added and on a 9DF model assumptions about the soil interaction effect on the previous models had been verified and confirmed. The numerical simulations had used fourth-order variable-step Runge-Kutta-Nisse integration rules in DESIRE[2].

ISOLATOR AND BUILDING RESPONSE

In Figure 2 hysteretic plots indicate that the isolators develop at 0.5g peak spectral acceleration stress bouncing under 0.5 Hz frequency and impose high bracing stresses (S) on relatively low base shear (V) at 0.25 Hz when the elastomer thickness is selected to be a critical value of 0.5 in., which corresponds to 0.25 in. isolator deadspace.

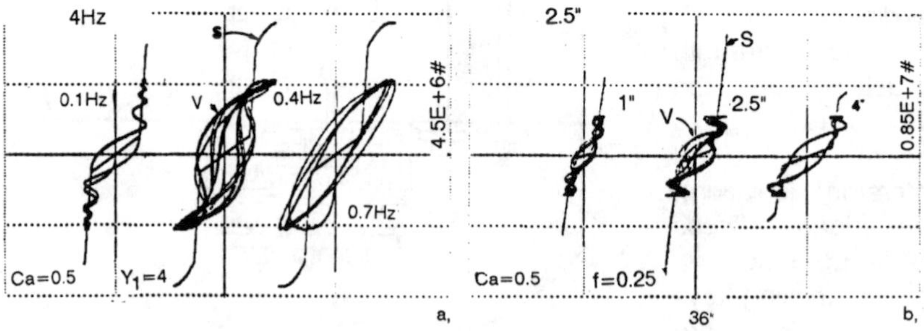

Fig. 2. Stress-bouncing and isolator deadspace opening studies

The five-floor displacement history in Figure 3 shows that both isolation and damping is achieved effectively by using interstory dissipative isolators.

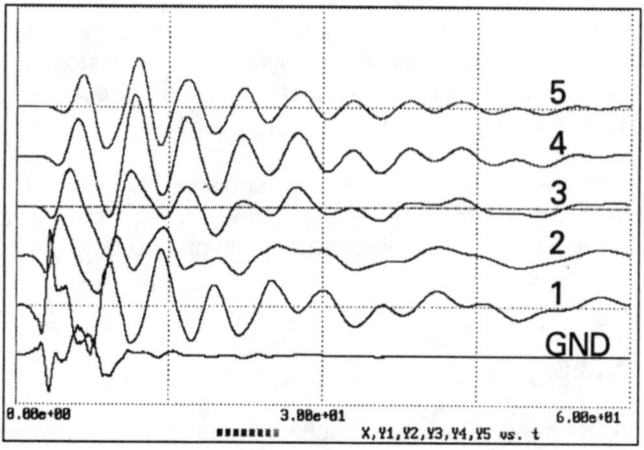

Fig. 3. Seismic floor displacement histories

The design of the base anchorage would be acceptable according to the current design code[3] although in Figure 4 it is clearly seen that the anchorage forces on the reinforcing (CB) falls in great part, outside of the strength envelope, even though the base shear (V) falls mainly within, while both having maxima less than the strength maxima itself.

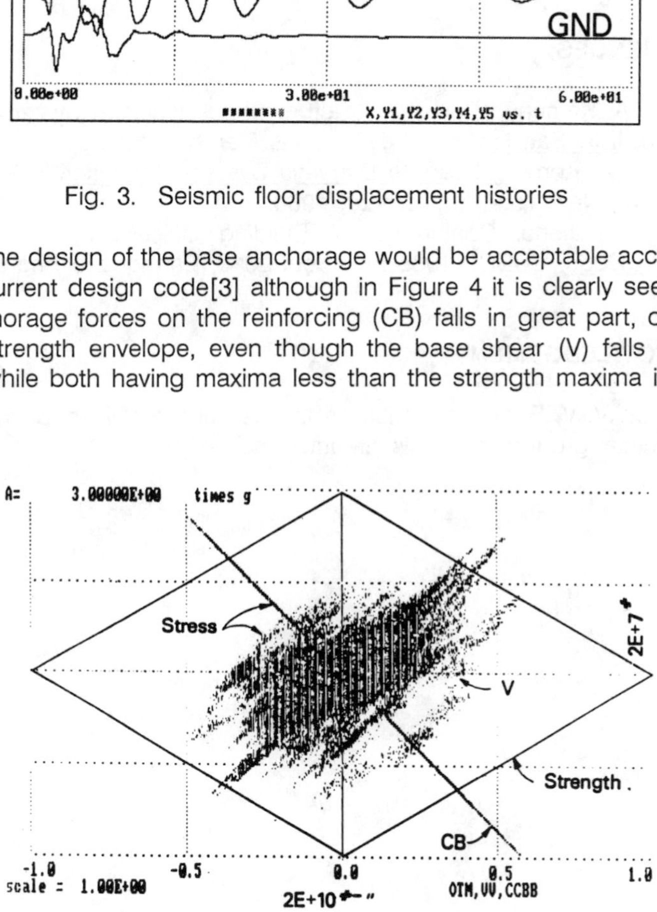

Fig. 4. Base shear-overturning moment interaction

CONCLUSION

The generalized lumped mass method proved to be efficient to simulate the response of highly nonlinear engineered building components, such as interlocking keyed steel - elastomer composite moment connections and to simulate building response considering stochastic parameters as well.

Dissipative isolated moment connections predicted to be economical and effective in adding delayed strength, stiffness and ductility to a five-story understrength masonry building in the highest seismic risk zone.

REFERENCES

1. Z.A. Kemeny, "Building Structure Shock Isolation System", Official Gazette, U.S. Pat. No. 4,727,695, p. 42, March 1, 1988.
2. G. A. Korn, "Interactive Dynamic System Simulation", McGraw-Hill Book Co., New York, ..., Toronto, 1988.
3. International Conference of Building Officials (ICBO), "Uniform Building Code", (UBC), 1988 Ed., Sec. 2312, Earthquake Regulations.

ACKNOWLEDGEMENT

Dr. V. V. Bertero's advise on the use of the 1DF models to select only critical ground motion is greatfully acknowledged.

State of the development and application of inelastic constitutive equations for metals in Germany

F.G. Kollmann
Technische Hochschule Darmstadt
Darmstadt, Germany

Abstract
In the first section of this paper issues of the development of inelastic constitutive models are addressed. Not only recently developed models are presented but also methods to identify material parameters from experimental data are discussed. Further numerical comparisons of different existing models are presented. In the second section numerical procedures for the solution of engineering problems are presented. A time integration routine for stiff constitutive equations is discussed. Further, inelastic shell analysis with mixed and hybrid strain finite elements is addressed. Finally, items for future research are presented.

1 Development of constitutive equations

In this paper by "inelastic" rate dependent as well as rate independent deformations are understood. Since both kinds of deformations are caused by micromechanical processes which can not be separated, they usually are observed jointly. Therefore, in the last two decades worldwide strong efforts have been made to develop and test so called "unified" constitutive equations to model inelastic deformations of metals. In this paper a report on the development of inelastic constitutive equations and their applications to practical problems in Germany is given.

In principle two approaches are known to formulate inelastic constitutive equations. Both start with an additive decomposition of either the strain rate tensor (small strain formulation) or the rate of deformation tensor (finite strain theory). We consider in the following only small strain theories. Then the additive decomposition is

$$\dot{\epsilon} = \dot{\epsilon}^e + \dot{\epsilon}^n . \tag{1}$$

Here $\dot{\epsilon}$, $\dot{\epsilon}^e$ and $\dot{\epsilon}^n$ are the total, the elastic and the inelastic strain rates, respectively. For the elastic part $\dot{\epsilon}^e$ of the strain rate tensor Hooke's law in rate form is supposed to hold.

In the first approach a purely phenomenological theory is formulated, which can be based on thermodynamical considerations. One such theory is given by

Bruhns and coworkers [1,2,3]. It is assumed that up to a certain limit the material behavior is purely elastic. Beyond this limit the material shows rate dependence. And finally, long term creep and relaxation phenomena can be reproduced by separate considerations. Then the evolution of the inelastic strain tensor can be described by

$$\dot{\epsilon}^n = \boldsymbol{f}(\boldsymbol{\sigma}, \dot{\boldsymbol{\sigma}}, \boldsymbol{q}_{\rm i}) \,, \tag{2}$$

where $\boldsymbol{\sigma}$ denotes the stress tensor and $\boldsymbol{q}_{\rm i}$ is a set of N internal variables (i = 1, 2,···,N). For the internal variables - which in the case of Bruhn's model are a scalar and a tensorial quantity - also evolution laws exist

$$\dot{\boldsymbol{q}}_{\rm i} = \boldsymbol{h}(\boldsymbol{\sigma}, \dot{\boldsymbol{\sigma}}, \boldsymbol{q}_{\rm i}) \,. \tag{3}$$

The quantities $\boldsymbol{f}$ and $\boldsymbol{h}$ in eq.(2) and eq.(3) are tensorvalued functions of their arguments. The tensorial internal variable is identified as an internal stress. Further it is assumed that a v.Mises yield condition holds for the difference of the external and this internal stress. Finally, the normality rule is adopted. Bruhns and Hübel [3] give also methods to identify the material parameters and functions (from uniaxial experiments) of the model. Several uniaxial homogenous deformation histories have been traced with the model and a satisfactory correspondence with experimental results has been achieved.

A completely different approach has been choosen by Steck [4,5,6]. He also uses the additive decomposition (1) and the form (2) of the evolution equation for the inelastic strain rate. However, his approach is basically not phenomenological but micromechanical. It is assumed that flow units as e.g. dislocations or dislocation packages are situated in the crystal lattice in front of obstacles as e.g. grain boundaries or dispersions. Each obstacle is characterized by a potential barrier of some height. The externally acting stresses reduce the obstacle height in the direction of their action and increase it in the opposite direction by the same amount. This model was extended by assuming that there exist not only barrier heights for the transition of the flow mechanisms but there is a phase space defined for these barriers over which the flow units (whose number is supposed to be constant) are distributed according to the past history of the material. Additionally, recovery is introduced, where the rate of recovery increases with increased hardening. During the slip processes, which are responsible for the inelastic deformations, the flow units have to overcome the barriers, which are represented by internal stresses of different magnitude. A given distribution of the flow units changes its position in phase space with transition probabilities depending on these internal stresses. The state vector $\boldsymbol{z}(t + \Delta t)$ representing the distribution of active glide systems at time $t + \Delta t$ is related to the same vector at time t by a stochastic matrix $\boldsymbol{SM}$

$$\boldsymbol{z}(t + \Delta t) = \boldsymbol{SM}\, \boldsymbol{z}(t) \,. \tag{4}$$

It can be shown that eq.(4) is a Markov chain. Exploiting known results on Markov chains Steck obtained macroscopic equations for stationary creep and transients. It is specifically remarkable, that the evolution equations for the inelastic strain rates are not mathematically stiff. The work of Steck was extended by Schlums and Steck [7] to cyclic deformation processes.

Rohleder and Kollmann [8] have performed extensive experiments with the powder metallurgical alloy AlMn 10 to test Steck's stochastic model. They have identified the material parameters of the model from uniaxial creep experiments. The steady state creep strain rate depends only on stress and temperature and is independent on the prior loading history as predicted by the model. Furthermore, positive transients after instantaneous changes of stress or temperature could be described well by the model. However, it was not possible to identify parameter sets uniquely for modelling negative transients.

In [8] a methodology for identification of material parameters of an inelastic constitutive model was used, which imitates principles of biological evolution [9]. The parameter identification represents an optimization problem. All material parameters to be identified are combined to a parameter vector $\boldsymbol{G}$. Further a scalar quality function $q(\boldsymbol{G})$ is defined. Then the optimization problem is

$$q(\boldsymbol{G}) \rightarrow min \tag{5}$$

From a start vector $\boldsymbol{G}_0$ a new parameter vector $\boldsymbol{G}_1$ is derived by a random number generator. In the same manner a population of n parameter vectors is built up. All parameter vectors are tested for the quality function $q(\boldsymbol{G})$ and the one which gives the maximal value is disposed. Then again a new parameter vector is generated and all parameter vectors are tested for the quality function. The process is truncated, once one parameter vector has been found for which the value of the quality function is smaller than a prescribed tolerance. This method has been applied successfully for parameter identification for different inelastic constitutive models as e.g. Hart's model [10].

Of significant interest are comparative studies of different constitutive models. Hartmann and Kollmann [11] have performed uniaxial numerical experiments on the inelastic models by Hart [10] and Miller [12,13]. Hartmann [14] has extended this work to a comparison of the models of Miller and Walker [15]. In both studies the numerical results were compared with experimental data whenever available. It turned out, that each set of constitutive equations can model some specific experiments quite well while for other experiments significant discrepancies between simulations and experiments were observed. At the present time it seems not to be possible to take an inelastic model from the literature and to implement into an existing finite element code in the hope to produce reasonable and reliable results for real engineering problems.

2 Application of inelastic constitutive models

In this section a brief description of applications of inelastic constitutive equations on technical problems is given. Mathematically the balance equations of continuum mechanics and the inelastic constitutive equations pose an initial-boundary-value problem for all but homogeneous deformations. Since all known inelastic constitutive equations are highly nonlinear, only numerical techniques can be applied. The main difficulty normally stems from the initial, value problem which is posed by the inelastic constitutive model. Almost all known inelastic models are mathematically stiff. Therefore unconditionally stable time integration schemes have to be applied. Cordts and Kollmann [16] have used the implicit Euler rule which requires the iterative solution of a highly nonlinear algebraic problem. In this process not only the inelastic strain rates $\dot{\epsilon}^n$, the rates of the internal variables $\dot{\boldsymbol{q}}$ but also the stress rates $\dot{\boldsymbol{\sigma}}$ have to be integrated. However, only for the inelastic strain rates and the internal variables analytic representations (from the inelastic constitutive model) are available. The values of the stress rates are only available at discrete times as solutions of the boundary value problem, e.g. by the finite element method (FEM). Therefore, Newton's method can not be applied for the iteration of the stresses, since the related part of the Jacobian matrix can only be computed by numerical differentiation which requires repeated solutions of the FEM-system and is therefore infeasible. This problem has been solved by applying Newton's method only to the inelastic strain rates and the internal variables. For the determination of the stresses the accelerated Jacobian iteration has been applied. With this method excellent numerical results for different constitutive models have been obtained.

Of significant importance in mechanical engineering are thinwalled shell strucures. Kollmann and Mukherjee [17] have developed a general, geometrically linear theory of inelastic shells. They start from a variational principle originally published for elastic three-dimensional bodies by Oden and Reddy [18], which contains only velocities and strain rates as primary quantities to be varied independently. By a projection a strictly two-dimensional shell theory was obtained. Kollmann and Bergmann [19] have derived from this shell theory a hybrid strain finite element model for axisymmetric shells using Hart's constitutive model. Kollmann, Cordts and Hackenberg [20] have developed from [17] a family of mixed and hybrid strain finite elements. They have performed intensive numerical studies for elastic shells. It turned out, that stresses converged with the same order with mesh refinements as displacements. Further most of the elements are free from membrane and shear locking. Therefore, this new family offers an interesting potential for large scale computations of inelastic axisymmetric shells. In a next step this work has to be generalized for arbitrarily shaped shells.

The following items deserve special attention for future research:

- Improvement of existing geometrically linear and nonlinear inelastic constitutive models.

- Improvement of methods to identify the set of internal variables in inelastic constitutive models from micromechanical considerations.

- Development of general mixed finite shell elements for inelastic analysis.

- Improvement of time integration algorithms for inelastic analysis.

References

[1] O. T. Bruhns, B. Boecke, and F. Link. The constitutive relations of elastic-inelastic materials at small strains. *Nuc. Engng. Des.*, 83:325 – 331, 1984.

[2] O. T. Bruhns. On the constitutive relation of austenitic stainless steels. In C.S. Desai, editor, *Constitutive relations for engineering materials: Theory and application*, pages 675 – 682, Elsevier, Amsterdam, 1987.

[3] O. T. Bruhns and H. Hübel. Constitutive equations for the inelastic behaviour of stainless steels. In *Recent advances in design procedures for high temperature plants*, pages 17 – 22, IMechE, 1988. Mechanical Engineering Publication.

[4] E. A. Steck. A stochastic model for the high-temperature plasticity of metals. *Int. Journ. Plasticity*, 1:243 – 258, 1985.

[5] E. A. Steck. A stochastic model for the interaction of plasticity and creep in metals. *Nuc. Engng. Des.*, 114:285 – 294, 1989.

[6] E. A. Steck. The description of the high-temperature plasticity of metals by stochastic processes. *Res Mechan.*, 29:1 – 19, 1990.

[7] H. Schlums and E. A. Steck. Description of cyclic deformation processes with a stochastic model for inelastic behaviour of metals. *Int. Journ. Plasticity*, 1990. To appear.

[8] N. Rohleder and F.G. Kollmann. Creep experiments for parameter identificationof the stochastic model of steck with the powder metallurgical alloy AlMn 10. *Int. Journ. Plasticity*, 6:109 –122, 1990.

[9] D. Müller and G. Hartmann. Identification of materials parameters for inelastic constitutive models using principles of biologic evolution. *Trans. ASME, Journ. Engng. Technol.*, 111:299 – 305, 1989.

[10] E.W. Hart. Constitutive relations for the non-elastic deformation of metals. *Trans. ASME, Journ. Eng. Mat. Tech.*, 98:97–105, 1976.

[11] G. Hartmann and F.G. Kollmann. A computational comparison of the inelastic constitutive models of hart and miller. *Acta Mechan.*, 69:139 – 165, 1987.

[12] A.K. Miller. An inelastic constitutive model for monotonic, cyclic, and creep deformation. Part 1: Equations develpoment and analytical procedures. *Trans. ASME, Journ. Eng. Mat. Tech.*, 98:97–105, 1976.

[13] A.K. Miller. An inelastic constitutive model for monotonic, cyclic, and creep deformation. Part 2: Applications to type 304 stainless steel. *Trans. ASME, Journ. Eng. Mat. Tech.*, 98:106–113, 1976.

[14] G. Hartmann. Comparison of the uniaxial behavior of the inelastic constitutive models of miller and walker by numerical experiments. *Int. Journ. Plasticity*, 6:189 – 206, 1990.

[15] K.P. Walker. *Research and development program for nonlinear structural modeling with advanced time-dependent constitutive relationships.* Technical Report CR-165533, NASA, Lewisville, 1980.

[16] D. Cordts and F.G. Kollmann. An implicit time integration scheme for inelastic constitutive equations with internal state variables. *Int. Journ. Num. Meth. Eng.*, 23:533–554, 1986.

[17] F.G. Kollmann and S. Mukherjee. A general, geometrically linear theory of inelastic thin shells. *Acta Mechan.*, 57:41–67, 1985.

[18] J.T. Oden and J.N. Reddy. On dual complementary variational principles in mathematical physics. *Int.Journ.Engng.Sci.*, 12:1–29, 1974.

[19] F.G. Kollmann and V. Bergmann. Numerical analysis of viscoplastic axisymmetric shells based on a hybrid strain finite element. *Comp.Mech.*, 1990. To appear.

[20] F.G. Kollmann, D. Cordts, and H-P. Hackenberg. *Implementation and numerical tests of a family of mixed finite elements for the computation of axisymmetric viscoplastic shells.* MuM-Report 90/1 (unpublished), Technische Hochschule Darmstadt, Darmstadt, Federal Republic of Germany, 1990.

AN EULERIAN ELASTO-VISCOPLASTIC FORMULATION FOR DETERMINING RESIDUAL STRESSES

A. M. MANIATTY, P. R. DAWSON, and G. G. WEBER
Sibley School of Mechanical and Aerospace Engineering, Cornell University, Ithaca, NY 14853

ABSTRACT

An Eulerian finite element formulation for modeling isotropic, elasto-viscoplastic, large deformations is discussed herein. The formulation is applied to the analysis of flat rolling to compute residual stresses following deformation.

INTRODUCTION

Residual stresses inevitably arise when metals undergo large inhomogeneous deformations as is the case in common industrial forming processes. These residual stresses have an important impact on the quality of the resulting products. This work focuses on accurately modeling steady-state deformation processes in order to determine the resulting residual stresses. The use of appropriate constitutive models based on microstructural phenomenology in conjunction with an appropriate structure of kinematics for finite elastic-plastic deformations is of primary importance. An Eulerian formulation is employed because it is especially well-suited for modeling steady-state processes such as rolling and extrusion. Most previous Eulerian analyses of forming processes neglect the elastic part of the deformations since it is usually relatively small, for example [1]. Although much information can be gained from such an analysis, no information is provided on the residual stresses. Other Eulerian analyses which do include elasticity either neglect work hardening [2] or do not incorporate rate-dependence [3]. The analysis here is novel in that it incorporates elasticity in an Eulerian framework with accurate viscoplastic constitutive models.

DEFINITION OF THE PROBLEM

Let B be the two-dimensional spatial domain under consideration with boundary ∂B, and let velocities, $\hat{v}_i$, be specified on the part of the boundary ∂B_{1i}, and tractions, $\hat{t}_i$, be specified on the part of the boundary ∂B_{2i}, where $\partial B_{1i} \cup \partial B_{2i} \equiv \partial B$ and $\partial B_{1i} \cap \partial B_{2i} \equiv \emptyset$, i=1,2. Then the following boundary value problem results from the equilibrium equation and the boundary conditions

$$\mathrm{div}\mathbf{T} + \mathbf{b} = \mathbf{0} \qquad \text{on } B \tag{1}$$

$$\mathbf{v}\cdot\mathbf{e}_i = \hat{v}_i \qquad \text{on } \partial B_{1i} \qquad i=1,2 \tag{2}$$

$$\mathbf{Tn}\cdot\mathbf{e}_i = \hat{t}_i \qquad \text{on } \partial B_{2i} \qquad i=1,2 \tag{3}$$

where $\mathbf{T}$ is the Cauchy stress tensor, $\mathbf{b}$ is the body force, $\mathbf{e}_i$, i=1,2 forms an orthonormal basis on the two-dimensional space, and $\mathbf{n}$ is a unit normal on ∂B.

A multiplicative decomposition of the deformation gradient into elastic and plastic components is assumed here for the structure of the kinematics as first proposed by Lee [4]. The plastic part of the deformation is assumed to be isochoric and the plastic spin tensor is taken to be zero [5].

The materials to be considered here are assumed to be isotropic and homogeneous and to obey an elasto-viscoplastic constitutive law. The elastic part of the deformation is taken to be linear and of the form

$$\overline{\mathbf{T}} = 2G\overline{\mathbf{E}}^e + \lambda\,(\mathrm{tr}\,\overline{\mathbf{E}}^e)\,\mathbf{I} \tag{4}$$

where G and λ are the Lamé parameters, $\overline{\mathbf{E}}^e$ is the logarithmic elastic strain and $\overline{\mathbf{T}}$ is the corresponding work conjugate stress. From the normality rule, the plastic rate of deformation may be expressed by

$$\overline{\mathbf{D}}^p = \sqrt{\frac{3}{2}}\, d^p\, \overline{\mathbf{N}} \tag{5a}$$

where

$$\overline{\mathbf{N}} = \sqrt{\frac{3}{2}}\,\frac{\overline{\mathbf{T}}'}{\overline{\sigma}} \quad , \qquad \overline{\sigma} = \sqrt{\frac{3}{2}\,\overline{\mathbf{T}}'\cdot\overline{\mathbf{T}}'} \tag{5b}$$

$$\overline{\mathbf{T}}' = \overline{\mathbf{T}} + \overline{p}\,\mathbf{I} \quad , \qquad \overline{p} = -\frac{1}{3}(\mathrm{tr}\,\overline{\mathbf{T}}) \quad . \tag{5c}$$

The viscoplastic part of the deformation is modeled by a scalar internal variable constitutive law which is comprised of a yield condition together with an evolution equation for the state variable, s, which is of the form

$$\overline{\sigma} = \hat{\overline{\sigma}}(d^p,s,\theta) \quad , \qquad \dot{s} = \hat{s}(d^p,s,\theta) \tag{6}$$

where θ is the temperature and $(\dot{\ })$ denotes a material time derivative.

The problem then is to solve equations (1) - (6) for the velocities, stresses, and the state variable on the domain of interest. It is also possible to incorporate thermal coupling as shown in [1].

SOLUTION PROCEDURE

The solution of the problem is divided into two parts; the solution of the boundary value problem (equations (1) - (3)) and the integration of the constitutive equations (4) - (6). The boundary value problem is solved here using the finite element method. Expressing the variational form of the equilibrium equation one obtains the following familiar statement

$$\int_B \mathbf{T}\cdot\delta\mathbf{D}\; dV = \int_B \mathbf{b}\cdot\delta\mathbf{v}\; dV + \int_{\partial B} \mathbf{t}\cdot\delta\mathbf{v}\; dA \quad . \tag{7}$$

Using equations (4) - (6), it can be shown that

$$\mathbf{T} = 2\mu^*(\mathbf{D} - \mathbf{D}^e) - p\mathbf{I} \ , \qquad \mu^* = \frac{\mu}{\det\mathbf{F}} \quad , \quad p = \frac{\overline{p}}{\det\mathbf{F}} \tag{8a}$$

where

$$\mu = \hat{\mu}(d^p,s) \quad , \qquad \overline{\mathbf{T}}' = 2\mu\,\overline{\mathbf{D}}^p \quad . \tag{8b}$$

Substituting the equation for **T** from (8a) into equation (7) gives an equation for the velocities and pressures. Still one more equation is needed relating the velocities and pressures. This comes in the form of the following volumetric constraint

$$\int_B (\operatorname{div} \mathbf{v} + \frac{\dot{\bar{p}}}{K}) \, \delta p \, dV = 0 \tag{9}$$

where K is the bulk modulus. Applying the finite element method to equations (7) and (9), one can solve for the velocities and the pressures.

The resulting system of equations is nonlinear and depends on $\mathbf{d}^p$, s, det**F**, and $\mathbf{D}^e$ which can be obtained by integrating the constitutive equations along the streamlines. By first assuming values for theses variables, the aforementioned boundary value problem can be solved for the velocities and the pressures everywhere in the domain. Using these computed velocities, the streamlines and the velocity gradients everywhere on the streamlines can be determined . In addition, if initial conditions are known at the locations where the material enters the domain, the constitutive equations then can be integrated along the streamlines, and new values of $\mathbf{d}^p$, s, det**F**, and $\mathbf{D}^e$ can be computed. These new values can then be used in the solution of the boundary value problem. The alternate solution of the boundary value problem and the streamline integration continues until convergence. The integration of the constitutive equations is performed using an algorithm described in [5] and [6].

NUMERICAL EXAMPLE

A rolling problem is analyzed and the resulting residual stresses are presented. In this example, eight-noded isoparametric elements are used and plane strain conditions are assumed. The constitutive model used is a hyperbolic sine law presented in [7] for 1100-Aluminum where a constant temperature of 400°C is assumed. In addition, the initial material is assumed to be stress free. The mesh for the rolling problem is shown in Figure 1 where a=2 cm and b=1.8 cm, and the roll radius is r=10 cm. The initial value of the state variable is taken as s_0=40 MPa.

In this example, the effect of non-symmetric boundary conditions will be examined so symmetry is not assumed about the x-axis. In most simulations of rolling, symmetry is assumed about the x-axis, but this is not generally the case in an actual rolling process. In addition, from a theoretical point of view, allowing for non-symmetry will give information about the uniqueness and stability of the solution. If a small change in the boundary conditions has a large effect on the results, then the solution is not stable and may not be unique. It would be of interest both practically and theoretically to see the effect of not assuming symmetry and allowing for non-symmetric boundary conditions. In practice, the friction coefficient on the top roll is often lower than that on the bottom roll. In this case, a simple sliding friction law is assumed for the top roll, while sticking friction is assumed for the bottom roll. The sliding friction law is of the form

$$t_t = \beta\,(v_o - v_t)$$

where t_t is the tangential traction, v_o is the surface roll velocity, v_t is the surface velocity of the workpiece, and the constant β is taken as β = .8 GPa/(m/s). In addition, so that the material does not bend down as it exits, the two nodes on the top and bottom immediately after the exit are fixed in the y-direction while all other surfaces are free. The results for this case are plotted in Figure 2 along with two other cases, one where both rolls are assumed to have sliding friction like the top roll in the previous case and one where both rolls are assumed to have sticking friction like the bottom roll. The Txx component of the residual stresses normalized with respect to s_o is plotted through the thickness in the exit region. It is interesting to note that on the side with the sticking friction, the curve for the case with the mixed boundary conditions follows the curve for the case with the sticking friction on both rolls, and on the side with the sliding friction, the curve jumps up to the curve for the case with sliding friction on both rolls. The difference in the boundary conditions on the top and bottom roll is significant in this case. It was found in other examples not shown here, that varying the friction conditions on the top and bottom roll only slightly had almost no noticeable effect on the residual stresses. This and the above results seem to indicate that the solution is stable. Finally, the overall trend of these results, i.e. tensile residual stresses on the surface and compressive stresses in the center, are consistent with what has been observed in experiments and is predict in theory for this geometry. For a more detailed discussion see [6].

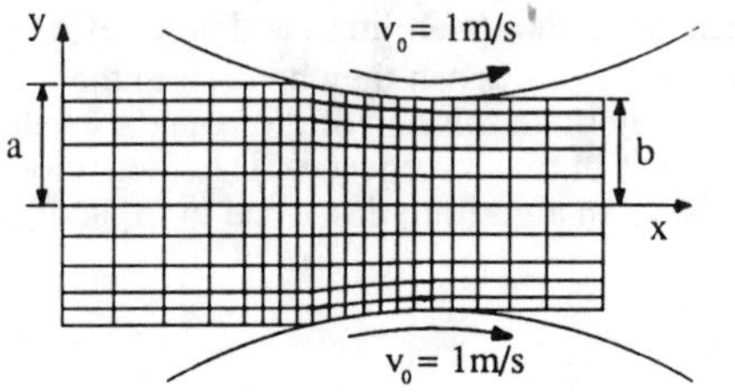

Figure 1.

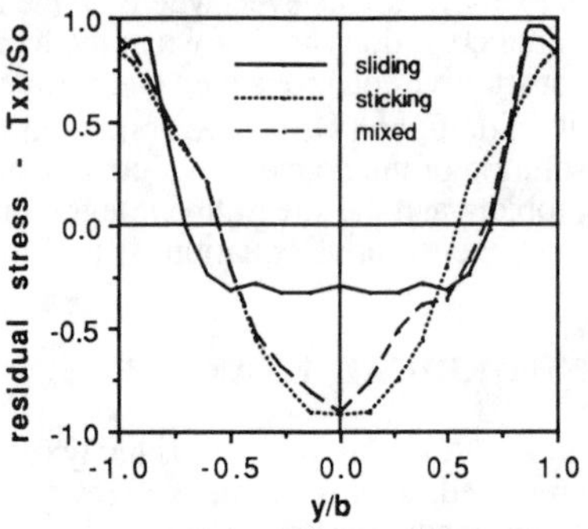

Figure 2.

REFERENCES

1. P. R. Dawson, "On modeling of mechanical property changes during flat rolling of aluminum", Int. J. Solids Structures 23 (7), 947-968 (1987).

2. E. G. Thompson and H. M. Berman, "Steady-state analysis of elasto-viscoplastic flow during rolling", in Numerical Analysis of Forming Processes, edited by J. F. T. Pittman et al., 269-283 (John Wiley, 1984).

3. S. Yu and E. G. Thompson, "A direct Eulerian finite element method for steady-state elastic plastic flow", in Numiform 89, edited by Thompson et al., 95-103 (Balkema, Rotterdam 1989).

4. E. H. Lee, "Elastic-plastic deformations at finite strains", J. Appl. Mech. 69 , 1-6 (1969).

5. G. Weber and L. Anand, "Finite deformation constitutive equations and a time integration procedure for isotropic, hyperelastic-viscoplastic solids", Comput. Meth. Appl. Mech. Engrg. 79, 173-202 (1990).

6. A. Maniatty, P. Dawson, and G. Weber, "An Eulerian elasto-viscoplastic formulation for steady-state forming processes", submitted to Int. J. Mech. Sci. (1990).

7. S. Brown, K. Kim, and L. Anand, "An internal variable constitutive model for hot working of metals", Int. J. Plast. 5, 95-130 (1989).

EQUIVALENT HOMOGENEOUS MODEL FOR SOIL STRATUM WITH WICK DRAINS

Leonard R. Herrmann & Amir M. Amirebrahimi
Dept. of Civil Engineering, Univ. of Calif., Davis, CA, 95616

ABSTRACT

A mathematical model for describing the effects of discrete wick (or sand or stone column) drains in soft saturated soil deposits is reported. The model represents the composite (soil plus discrete drains) system as an equivalent homogeneous material.

INTRODUCTION

As prime sites become evermore scarce, it is has become increasingly necessary to construct roads, and buildings on soft, unconsolidated soil deposits. In such a situation, to avoid excessive and/or differential settlement of the structure, it is often necessary that the site be quickly drained and consolidated. Such requirements have led to the development of various schemes to drain large areas of soft soils through the installation of vertical drains, such as wick drains, e.g., see [1,2].

In many situations it is necessary to analyze the soil deposit containing the drains and any completed or partially completed structure built on or within it in order to determine their behavior during the consolidation process. Such an analysis is often required in order to verify the integrity of the structure. Analytical and empirical solutions are available for the flow field around a single drain (assuming all drains behave identically), e.g., see [1,2]. However, when the configuration is such that these conditions are not met then the problem is quite complicated (for the purposes of this paper such conditions will be referred to as nonuniform conditions).

Nonuniform conditions exist near the edge of a drained area, and/or a superimposed structure, when the drained layer is of nonuniform thickness and/or spatially inhomogeneous, and when the water table level has a spatial gradient. A cross-section of such a configuration is shown in Figure 1. In order to obtain

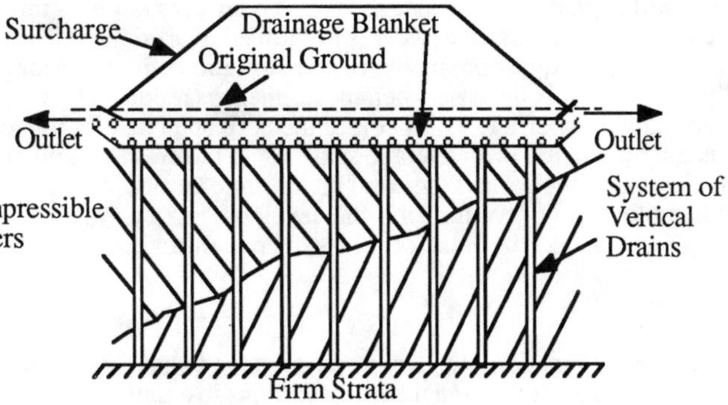

Fig 1. Example Problem Configuration

realistic predictions for such nonuniform conditions it may be necessary to preform a large scale finite element analysis. The problem is how to model the wick drains.

One approach, at least in theory, is to include each and every wick drain in the numerical model by means of series of vertical discrete one-dimensional line elements (a discrete model). For most situations such an approach is entirely impractical because of the large numbers of drains involved (thus, necessitating a prohibitively fine three-dimensional mesh). The alternative is to model the composite system as an equivalent homogeneous mass.

To date, the approach that has been used to develop a homogeneous model for wick drains is to attempt to represent the drainage from the wicks by using a modified (increased) vertical permeability coefficient for the soil, e.g. see [4]. The shortcoming of such an approach is that it really does not represent the reality of the drainage process. In such a model, for water to be drained from a volume of soil at some depth, the water must pass through all of the soil between the location in question and the drainage layer (the layer into which the wick drains discharge). Hence, the calculated vertical distribution of pore pressure can not be correct .

EQUIVALENT HOMOGENEOUS MODEL

For a soil stratum that contains an array of drains, three modes of possible water movement must be accounted for. Water will of course flow from the soil to the drains and then up to the surface drainage layer. In addition, some water will flow vertically directly from the soil to the drainage layer and finally there will usually be global (as opposed to the local flow to the drains) horizontal flow of water. Global horizontal flow of water occurs whenever there are nonuniform conditions in one or both of the horizontal directions.

The vertical and global horizontal flows of water in the soil are modeled by assigning to the soil the actual values (vertical and horizontal) of soil permeabilities. The flow of water to and out of the drains is modelled by means of a *distributed sink term* that is added to the continuity equation for the pore water. The distributed sink term is written as $q_d(x,y,z) = \alpha\,[u(x,y,z) - u_w(x,y,z)]$, where q_d (per unit volume of drained soil) is the flow of the water out of the drains (from the soil at the point in question), α is a model parameter (its calibration will be subsequently discussed), u is the excess pore water head in the soil, and u_w is the excess water pressure in the drain (due to well resistance [2]). The dependence of u_w on z is obvious; a dependence on x and y occurs for nonuniform conditions in the x,y plane. Because an equivalent homogeneous model is being developed, u is the area average of the excess pore pressure in a unit cell of soil containing one drain (see Figure 2).

The distributed sink term represents a transport of water directly from the point in the soil deposit to the drainage layer (i.e., it does not pass through any other part of the deposit). Because the water is not forced to artificially flow through the soil above the point in question, the vertical distribution of pore water pressure is not artificially modified as is the case when the drainage is modeled by using an effective vertical permeability coefficient. Of course the horizontal distribution of pore water pressure is only predicted in an average sense (over the cross section of the unit cell) by any model that homogenizes (smears) the discrete vertical drains. The key to the successful employment of the model is the development of a systematic procedure for determining (calibrating) the value of the model parameter α.

MODEL CALIBRATION

The determination of the appropriate value to use for the model parameter α involves consideration of the actual flow conditions in a unit cell of soil that surrounds a single drain. It is assumed that any global horizontal flow in the soil and any vertical flow directly to the drainage layer, are properly modeled as a consequence of using the correct soil permeabilities in the flow equation. What remains is the modeling of the flow to the drains. For this purpose a *fundamental problem* of identical flow to all drains, for the case of a step function increase in pore water pressure at time zero, is considered. The fundamental problem further assumes that only horizontal flow takes place. The selection of α involves attempting to make (for the fundamental problem) the predicted histories of the amount of drainage and the average soil pore water pressure to be correct.

The fundamental problem is analyzed i) with the proposed homogenized model and ii) using the actual geometry of the unit cell; the value of α is selected so as to make these two solutions approximately equal. Because of the brevity of this paper details of this calibration process are not given. However, Figure 3 gives a graph of α as a function of the ratio (N) of the effective diameters of the unit cell drainage area (D_e) and of the drain (D_w). The unit weight of water is γ_w, the horizontal soil permeability is k_h and r_e=D_e/2.

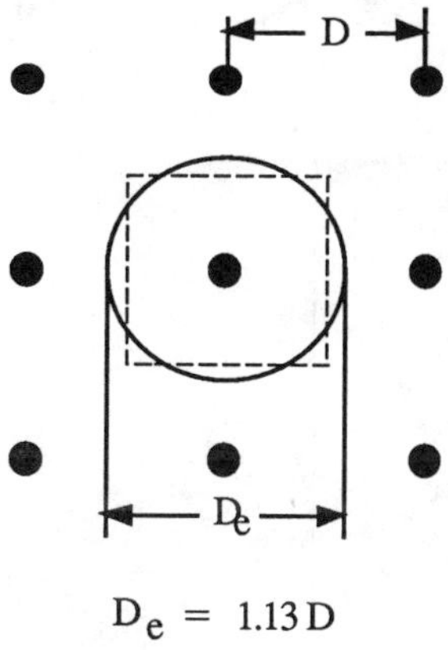

Fig 2. Example Unit Cell

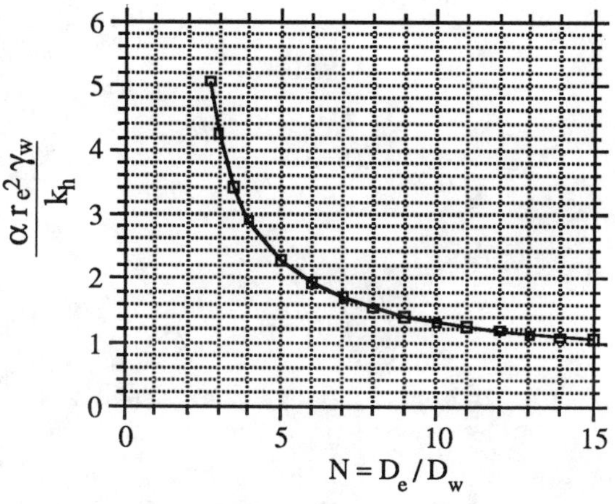

Fig. 3. Calibration Curve for α

NUMERICAL IMPLEMENTATION AND MODEL VALIDATION

The numerical implementation of the model is relatively straight forward and the details will not be presented here. The total validation study will involve comparisons of model predictions to available analytical, numerical and experimental results.

The single comparison presented here is for the case of uniform conditions in the x-y plane, hence, the flow to each drain is identical. For such ideal conditions and a homogeneous soil deposit, an analytical solution exists for the discrete problem [2]. These results are compared in Figure 4 to the predictions found using the equivalent homogeneous model. The time factor T_h and the measure of well resistance, L, are defined in [2]. The comparisons are excellent for both the average consolidation for the total soil layer and the depth distribution of the average consolidation in a horizontal plane. This comparison demonstrates the ability of the model to simultaneously predict flow to drains and vertical flow directly to the drainage layer.

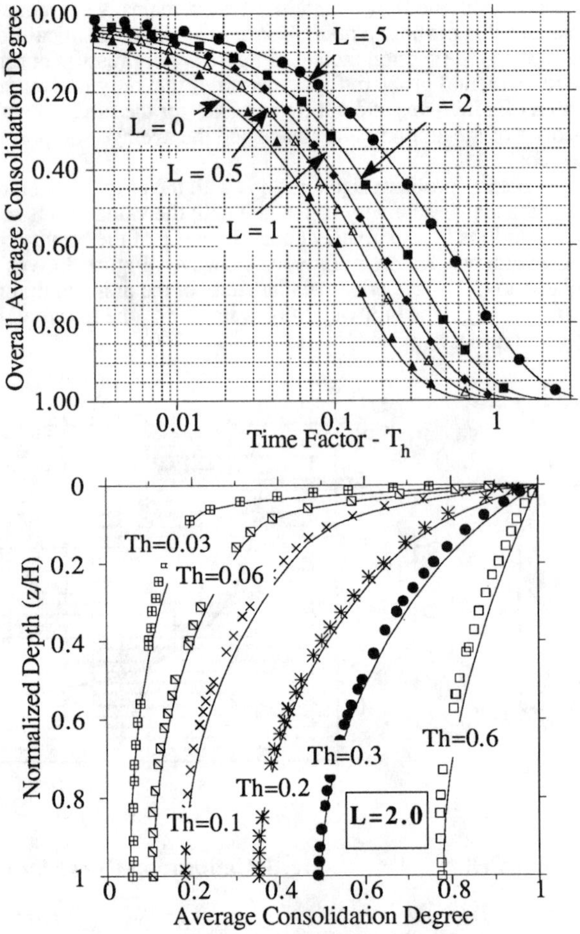

Fig. 4. Comparison of Model Predictions (Lines) to Results (Points) from Yoshikuni[3]

REFERENCES

1. S. Hansbo, "Consolidation of clay by band-shaped prefabricated drains", Ground Engineering, 12 (5), 18-35 (July 1979).
2. H. Yoshikuni and H. Nakanodo, "Consolidation of soils by vertical drain wells with finite permeability", Soils and Foundations, 14 (2), 35-46 (June 1974).
3. C. J. Poran, V.N. Kaliakin, et al, "Prediction of trial embankment behavior in Hertfordshire County Councils-Stanstead Abbots", Proc. of the Prediction Symp. on a Reinforced Embankment on Soft Ground, King's College, Strand London, eds. R. H Bassett and K. C. Veo (1988).
4. L. R. Herrmann, V. N. Kaliakin, et al, "Numerical Implementation of a plasticity model for cohesive soils", J. of Eng. Mech. Div., ASCE, 113 (4), 500-519 (April 1987).

REINFORCEMENT OF SOIL UNDER FOUNDATION USING METALLIC AND NON-METALLIC MATERIALS

R.S.KARMARKAR, Prof. in Civil Engg.,
College of Engg.,Ponda-goa, 403401, India.
D.S.SHRIDHAR, Research Scholar, Civil Engg. Dept.
I.I.T., Powai,Bombay, 400076,India.

ABSTRACT

Modern Civilization has given rise to the necessity of developing and using metallic and non-metallic materials to reinforce soil, the available natural base material on which buildings are to be supported. In order to explore the profitability of their usage for Ground Improvement, the present work undertakes an analytical study, employing FEM. Parameters pertaining to reinforcement have a considerable effect on strength and settlement characteristics of soil and these are mainly covered in this study, to obtain guidelines for optimum utility. Elastic-plastic response of soil has been accounted for. The results presented may form useful guidelines for design of reinforced soil foundation.

INTRODUCTION

The technique of providing reinforcement to soil for Ground Improvement is being developed quite recently, with respect to use in foundations. The improvement in strength and settlement characteristics is mainly attributed to resisting frictional forces developed at soil-reinforcement interfaces. The load on the foundation helps directly in increasing this resistance. It becomes necessary to try out preferred directions and dimensions of reinforcement. Treating the material as composite does not serve the purpose. FEM with provision to account for an elastic-plastic idealisation of soil response is found to be quite useful for the parameteric study undertaken.

Binquet and Lee [1,2] investigated the mechanism and the potential benefits of using reinforced earth slabs to improve bearing capacity of granular earth. The results of model tests with strip footing are presented. Akinmusuru [3] used rope fibers as reinforcement to carry out laboratory study of improvement in bearing capacity of square footings. While work on analysis and model studies is found in literature, the progress of research in development of cheaper and suitable materials and exploring use of waste materials appears not to have received the required attention.

ANALYSIS

The problem under consideration consists of a relatively rigid strip footing resting on reinforced soil. (Fig.1). Quadrilateral finite elements with condensed internal node (4 CSTs) have been used for discretizing the footing, soil and reinforcement. An available computer program [5] has been used after modifications required to simulate elastic-plastic response of soil. Following are the assumptions made in the analysis.

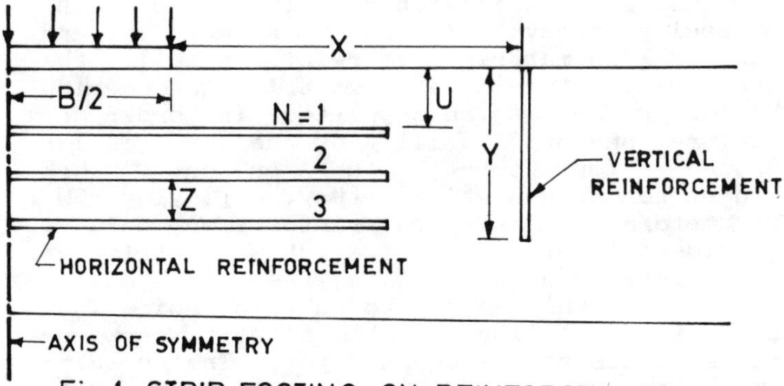

Fig.1 STRIP FOOTING ON REINFORCED SOIL

i) Soil is homogeneous, isotropic and elastic-plastic. In this study, it is also assumed as cohesionless.

ii) Uniform load is applied on the strip footing in stages until ultimate load is reached.

iii) Uniform load is applied on the strip footing in stages until ultimate load is reached.

iv) Slip between soil and reinforcement is manifested through distortion of yielded interface soil elements only.

The Drucker-Prager yield criterion has been used to indicate plasticity of soil :

$$f = \alpha\,(\sigma_x+\sigma_y+\sigma_z) + \{1/6[(\sigma_x-\sigma_y)^2+(\sigma_y-\sigma_z)^2+(\sigma_z-\sigma_x)^2] + \tau_{xy}^2 + \tau_{yz}^2 + \tau_{zx}^2\}^{1/2} - k = 0 \qquad ......(1)$$

where, $\alpha = \sin\phi/(3(3+\sin^2\phi))^{1/2}$, $k = \sqrt{3}\ c\ \cos\phi/(3+\sin^2\phi)^{1/2}$ and for the present case, $k = 0$, $\sigma_z=\mu(\sigma_x+\sigma_y)$, $\tau_{yz}=\tau_{zx}=0$.

For each load increment, the computed stresses at element centroids are checked against the above yield criterion. For particular elements for which Eqn.1 is satisfied, indicating yield, the element stiffness matrix is modified using reduced modulus of rigidity and the whole process is repeated and continued unless and until no additional elements indicate yield under a given loading. The number of iterations involved was found to be three to four.

NUMERICAL RESULTS AND DISCUSSION

The following observations are based on the study made, however, the presentation is limited to typical results pertaining to reinforcement only. The various parameters are shown in Fig.1.

The results presented refer to reinforcement thickness = .01B and the following material properties :

Sr.No.	Material	Modulus of Elasticity Kg/cm^2	Poisson's Ratio μ
1.	Soil	750,000	0.3
2.	Footing	1.16×10^9	0.3
3.	Gavanised steel(ST)	2.1×10^{10}	0.28
4.	Aluminium(AL)	7.14×10^9	0.3
5.	Nylon(NY)	2.04×10^9	0.25

The following definitions of the aspects presented have been adopted.

Q_o = Ultimate load capacity of soil without reinforcement

Q_n = As above for N layers of horizontal reinforcement
BCR = Q_n/Q_o

qn = Increase in Q_o per layer of reinforcement expressed as percentage of Q_o

pn = Increase in load, per layer in a system with N layers, expressed as percentage of load without reinforcemet, for a given value of settlement, S.

Geometrical and Configurational parameters

Horizontal layers of reinforcement : From Table I, it is seen that as N increases from 0 to 6, BC (Bearing Capacity) increases 2.625 times for steel; however the increase per layer beyond fourth to fifth layer drops considerably (Fig. 2). Similar trend exists for settlement response. Also it is seen that provision of less than 3 layers is not advantageous from BC point of view. Hence it may be justifiable to provide 3 to 5 layers.

It has been observed that upto Z/B = 0.25, BCR increases, whereas thereafter it reduces (Table I). It is advisable to adopt spacing of Z/B = 0.25. For N =4 and Z/B =0.25, this amounts to location of reinforcement within a depth of only 1.25 times the width of footing. This has a practical significance.

TABLE I : TYPICAL BEARING CAPACITY RATIOS FOR HORIZONTAL LAYERS OF REINFORCEMENT

L/B=3, U/B=0.5, Z/B= 0.25			
N	ST	AL	NY
1	1.100	1.075	1.050
2	1.275	1.225	1.075
3	1.625	1.500	1.200
4	2.075	1.900	1.450
5	2.450	2.325	1.825
6	2.625	2.475	1.950

L/B=3, N=5, Z/B=0.25			
U/B	ST	AL	NY
0.25	2.400	2.250	1.750
0.50	2.450	2.325	1.825
0.75	2.000	1.800	1.350
1.00	1.500	1.400	1.088
1.50	1.450	1.325	1.050

N=5, U/B=0.5, Z/B=0.25			
L/B	ST	AL	NY
1	1.725	1.600	1.125
2	2.125	2.000	1.475
3	2.450	2.325	1.825

L/B = 3, N = 5, U/B = 0.50			
Z/B	ST	AL	NY
0.25	2.450	2.325	1.825
0.50	2.425	2.275	1.750
1.00	2.375	2.200	1.675
1.50	2.225	2.000	1.550

BCR is found to expectedly increase as length of reinforcement increases (Table I). However it has been observed that the gain beyond L/B =3 goes on dropping. It may be economically appropriate to restrict L/B to 3.

It is categorically observed that for all the cases, U/B =0.5 is the optimum value.

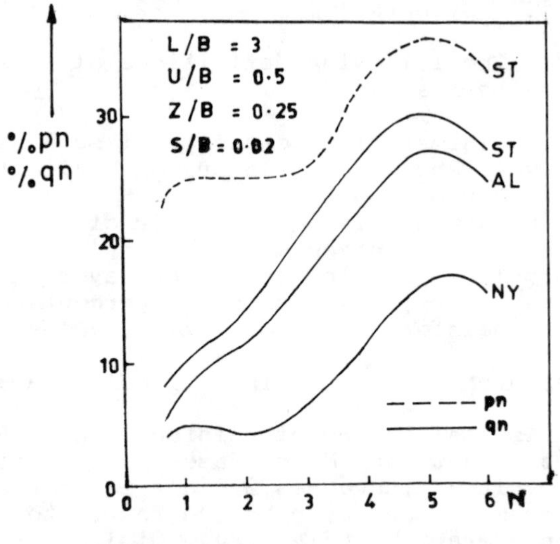

Fig.2 EFFECT OF No. OF LAYERS

<u>Combination of vertical reinforcement with horizontal reinforcement</u>:

Table II shows this effect. It is seen that there is a substantial increase in BC due to provision of vertical reinforcement, attributable to partial lateral confinement similar to skirting. The distance of vertical reinforcement from edge of footing equal to width of footing is found to give maximum gain in BCR, to the tune of 600%.

TABLE II : EFFECT OF PROVISION AND LOCATION OF VERTICAL REINFORCEMENT

X/B	BCR		REMARKS
	PRESENT ANALYSIS	MODEL STUDY [4]	
0	7.4	6.0	N = 4, Y/B = 2
1	8.3	7.5	T/B = 0.0167
2	7.5	7.0	BCR WITHOUT VERTICAL REINFORCEMENT = 2.25

Material Parameters

The reinforcement material parameters are represented by the values of elastic constants E_R and μ_R. Table I and Fig.2 indicate the effects for galvanized steel, aluminium and nylon type of materials. For N = 3 to 5, it is seen that compared to steel, qn, the increase in BCR per layer for aluminium with E_R roughly one third, is 2 to 3% lower and for nylon, with E_R roughly one tenth, is 13 to 14% lower. The ratio of E_R to E_s, E for soil, is also important.

It could be predicted that materials with relatively low values of E_R will not serve the purpose of reinforcing soil effectively.

Effect of Reinforcement on Settlement

Table III shows typical load improvement factors.

TABLE III : TYPICAL LOAD IMPROVEMENT FACTORS

MATERIAL	LIF		REMARKS
	S/B= 0.02	S/B= 0.04	
ST	2.70	2.60	N = 5, V/B = 0.5
AL	2.57	2.35	L/B = 3, Z/B = 0.25
NY	2.21	2.05	

It is seen that reinforcement is quite effective in reducing settlements, more so at lower loads, say near about working loads.

CONCLUSIONS

The study of geometrical, configurational and material parameters pertaining to reinforcement made by employing FEM to the problem of reinforced elastic-plastic foundation soil, provides useful quantitative results, It is found that for a strip footing of width B, optimum and economical values of length (L = 3B), number (N = 3 to 5), spacing (U = 0.5B) for horizontal reinforcement and distance from edge of footing of vertical reinforcement (X = B) can be adopted as guidelines.

By using a moderate amount of reinforcement upto a depth of about 1.5 times width of footing, it is possible to achieve considerable improvement in bearing capacity, to an extent of about 250%, 230% and 180% with steel, aluminium and nylon type of reinforcements. Provision of additional vertical reinforcement increases the bearing capacity enormously and it could be in the form of micropiles.

Reinforcement improves settlement characteristics also.

Comparison of analytical results with some available model test results show fair agreement.

REFERENCES

1. Binquet J & Lee. K. L., "Bearing capacity analysis of reinforced earth slabs". Jl. of Geotech. divn., A.S.C.E., 101 (4712), 1246-1255 (1975)
2. Binquet J. & lee K. L., "Bearing capacity tests on reinforced earth slabs". Jl. of Geotech. divn., A.S.C.E., 101, (4712) 1257-1276. (1975)
3. Joe. O. Akinmusuru and Jones A. Akinbolade, "Stability of loaded footings on reinforced soil". Jl. of Geotech. divn., A.S.C.E. 107 (GT6), 819-827 (1986)
4. Shivakumar , Model study of reinforced soil foundation. M.E. thesis presented to Goa University India (1989)
5. Desai and Abel., Introduction to Finite Element Method. (Litton Education Publishers, 1972)
6. Shridhar D.S., Study of Reinforced Soil Foundation using FEM, M.E. Thesis presented to Goa University, India (1989)

CONSTITUTIVE LAWS OF DAMAGED WOVEN FABRIC CERAMIC COMPOSITES

B. Lebon, P-M. Lesne and J. Renard

Office National d'Etudes et de Recherches Aérospatiales
BP 72. F -92322- CHATILLON Cedex (FRANCE)

ABSTRACT

We propose a model of degradation which takes into account physical degradation in the material. This first model is limited to orthogonal weave fabrics which provides bidirectional reinforcement. A numerical finite element method is used to determine stress distribution in the material for different configuration of the weaveness pattern. The local damage is analyzed by an homogenization method which leads to model the onset of damage and its evolution. The macroscopic behavior of the material containing degradations is described by the constitutive homogenized relations obtained through effective calculated modulus. At a certain level of loading, these results can be included in an iterative process to describe the non linear behavior of the damaged material. Moreover macroscopic damage growth equations are proposed.

1 INTRODUCTION

The behavior of woven fabric ceramic composite materials is a subject of considerable interest of recent years due to the increasing use of these materials at high temperature. Our purpose is to develop a model of degradation which takes into account local degradations of the fabric. We first describe the mechanisms of failure in ceramic composite materials resulting from in situ observations during tensile tests. Our first model is limited to orthogonally weaved fabrics which provide bidirectional reinforcement with an equal repeated pattern in both directions.

2 NUMERICAL MICRO APPROACH OR LOCALIZATION

The local damage is analyzed by an homogenization method which leads to model the onset of damage and its evolution [1]. The structure is periodic by zone and we assume that it is possible to describe it by means of two cells. Each of them takes into account a particular kind of damage and its growth. The influence of each cell upon the description of the global behavior of the material is proportional to the probability of apparition of each defect. Althougth a three dimensional cell is necessary to correctly treat the problem, in the case of our plain weave, we first used a pseudo three dimensional finite element analysis based on a beam theory. Away from the ends, the field displacements in any x=constant plane is assumed to be : $u_i(x, y, z) = \varepsilon^G_{xi}x + u(y, z)$, where ε^G_{xi} are uniform axial and shear strains and $u(y, z)$ are functions of the coordinates y and z alone. We used three-node elements.

3 STUDY OF VARIOUS CELLS

Two cells are studied as indicated on figure 1. The out of phase cell, where longitudinal cracks are modelled and the in-phase cell where the transverse cracks are studied. To model both kind of degradation, we simulate in each cell a zone where the corresponding cracks initiate and propagate. At each step, we calculate homogenized effective moduli of the differents cells [2].

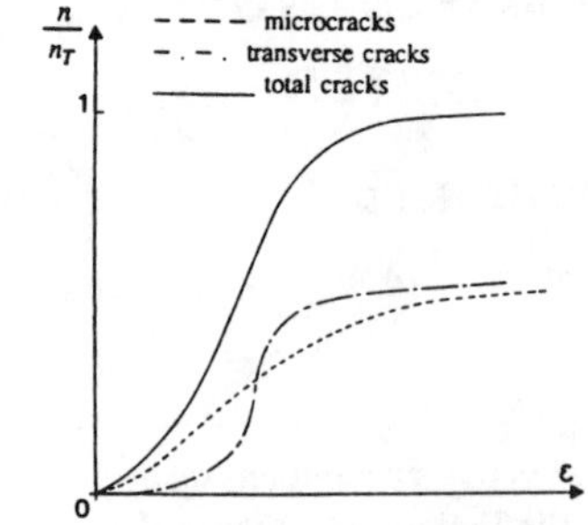

Figure 3 : Experimental crack density curves

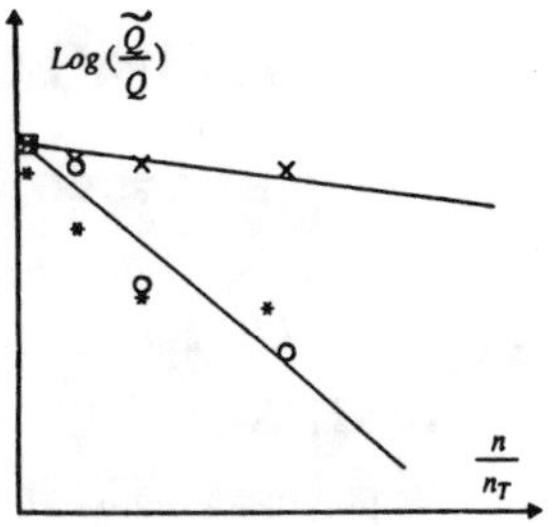

Figure 4 : Stiffness reduction

4 STIFFNESS REDUCTION LAW $\tilde{Q}/Q$ FOR IN PHASE CELL.

The conclusions of the calculation are that stiffnesses decrease with an exponential law when the damage parameter increases, i. e the distance between cracks becomes smaller.- Possible laws can be written as : $\tilde{Q}_{ij} = Q_{ij}e^{-k_i(n/n_T)}$ if σ_i>0' ; $\tilde{Q}_{ij} = Q_{ij}$ if σ_i<0 where Q_{ij} are the stiffnesses of the undamaged laminate, n and n_T are the number of cracks during the degradation and at saturation. That means that crack spacing cannot become smaller.The figure 4 gives a comparison between numerical results and exponential laws.

5 DAMAGE GROWTH

Concerning longitudinal cracks, we decreased stiffnesses by a global reduction after transverse cracks saturation. Concerning transverse cracking, we used a thermodynamical formulation to describe its evolution with an associated damage variable $\alpha = n/n_T$. The free energy ψ, taken as the thermodynamic potential, gives the law of thermoelasticity coupled with damage [4] :

$$\psi = \tilde{Q}_{ij}(\alpha)\,\varepsilon_i e_j \qquad ; \qquad A(\varepsilon, \alpha) = \frac{\partial \psi}{\partial \alpha}$$

The thermodynamic force A is the damage energy release rate, conjugate to the damage variable α. Let us note that our variable α does not correspond to the classical concept of damage variable which represents experimental stiffness decrease. In our case α includes all the damaging effects and is defined as a percentage of transverse cracking.

5.1 DISSIPATION

The second law of thermodynamic imposes that the intrinsic dissipation has to be positive : $-A\dot{\alpha} > 0$. At the initiation, we obtain the quadratic criterion A^0_c for $\alpha = 0$. We propose to employ the energy function A to characterize the damage loading/unloading conditions. The state of damage in the material is then characterized by means of a damage criterion $f(A;\alpha) = A - A^t_c(\alpha) \le 0$ where A^t_c is the value of the damage threshold at a given time (i. e the radius of the damage surface). If $A^0_c(\alpha)$ denotes the initial damage threshold, we must have $A^t_c \ge A^0_c$. This condition states that damage in the material is initiated when the damage energy release rate A exceeds the initial damage threshold A^0_c. This energy-based damage criterion is linked to the history of elastic behavior of the material. Damage grows if we have $f = 0$ and $\dot{f} = 0$ so that A^t_c is given by $A^t_c = max|A|$. If we plot the number of cracks versus the applied loading, we can express the strain ε_j as a function of the damage parameter α, by inversing the stress-strain relation. Then we get $A(\varepsilon, \alpha) = A(\alpha)$ for different crack density and the expression of $A^t_c(\alpha)$.

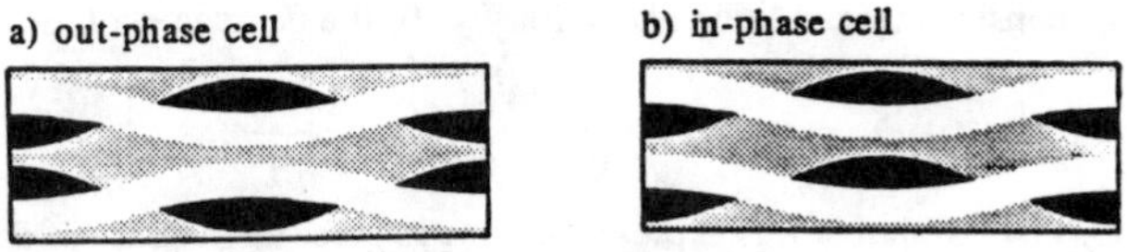

Figure 1 : Periodic cells

3.1 OUT OF PHASE CELL

It is noticeable that this kind of material has a very high porosity (about 15 %). The biggest pores (500 microns) exhibit sharp angles that can act as stress concentrations.

In the region where the 0° bundles from neighbouring clothes are very close, longitudinal cracks occur parallel to the direction of the applied load, because of stress concentration due to the narrowest bundles. In most cases, longitudinal cracks appear after transverse cracking saturation. Figure 2 shows the elementary cell and the stress concentration σ_{yy} when we apply the macro-strain $\overline{E_{yy}}$. We get a linear stiffness decrease of about 20 %.

Figure 2 : σ_{yy} isostress

3.2 IN PHASE CELL

The matrix rigidity is more than 1. 5 higher than fiber's rigidity $E_m = 350$ GPa, $E_f = 200$ GPa which is not the case for other ceramic matrix composites such as C/SiC or SiC/LAS. Consequently, when the non damaged composite is loaded, the matrix is subjected to very high stresses, compared to the fibers. Further, the ultimate strain of the matrix is lower than the ultimate strain in the reinforcement, so that matrix cracking occurs prior to fiber fracture. First microcracks initiate from the biggest pores. Further, coalescence of fiber-matrix debonding in the 90° bundles and micro-cracks in the matrix give initiation of macro-cracks which propagate through the matrix and 90° weave patterns. They increase until their saturation, characterized by a crack density and a crack spacing. This curve is plotted on figure 3. As transverse cracking is the main cause of degradation, we choose a damage parameter which describe it. Assuming the crack density homogeneous in the damaged material, we use the concept of damage mechanics to model this phenomenon [3]. We substitute an equivalent homogeneous material without crack to the cracked woven fabric composite. Cracks are taken into account through the "effective modulus" concept. A characteristic parameter is introduced $\alpha = n/n_T$: n is the crack number for a certain loading level and n_T is the crack number at saturation. We get $\alpha = 0$ and $n = 0$ when no crack exists and $\alpha = 0$ and $n = n_T$ at saturation. To simulate this evolution of cracks until saturation, we introduce several cracks in the elementary cell. At each step, stiffnesses are computed by homogenization.

The damage consistency condition gives for $\dot{f} = 0$, the damage evolution law by :

$$d\alpha = \left(\frac{\partial^2\psi}{\partial\varepsilon_{ij}\partial\alpha}d\varepsilon_{ij}\right) / \left(-\frac{\partial A^t_c(\alpha)}{\partial\alpha} - \frac{1}{2}\frac{\partial^2\psi}{\partial\alpha^2}\varepsilon_{ij}\varepsilon_{kl}\right)$$

We verify that the sign of this expresssion always gives an increase of damage as a function of $d\varepsilon_i$.

6 EXPERIMENTS AND NUMERICAL ANALYSIS COMPARISONS

Our numerical simulation of a non linear coupled damage problem, uses an iterative process where stresses and damage are calculated at each step.The $\sigma - \varepsilon$ curve obtained is compared with experimental curve (figure 5). We obtain a good correlation for the first part of the curve until transverse crack saturation. Out of phase cell modelling gives a 20% decrease for a 1 mm crack length. We plotted it on our numerical $\sigma - \varepsilon$. Our simulation is too rigid because of other types of damage which are not taken into account but lead the material to rupture as fiber breakage, pull out of fibers, . . .

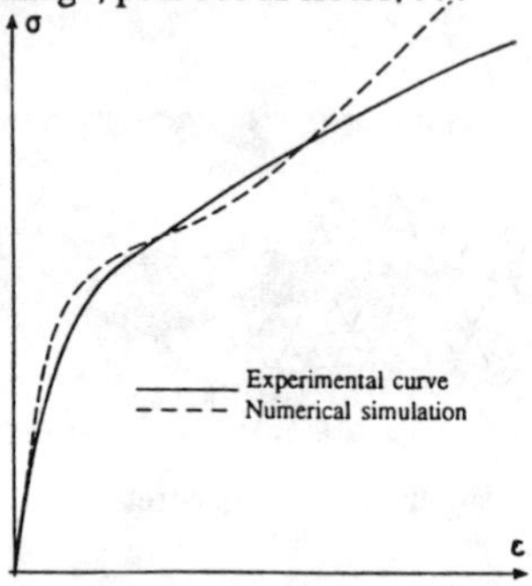

Figure 5 : Comparison between experimental and calculated tension curves

7 CONCLUSION

We propose a model of degradation for woven fabric SiC/SiC ceramic which takes into account local degradation in the material. We perform tests and the observed defects are simulated by an homogenization method. Two elementary cells are studied. A damage parameter is introduced which allows us to formulate damage evolution laws based on a thermodynamic theory of irreversible process. We simulate the non linearity of the material behavior by an iterative numerical process. We find a good agreement between experimental and numerical curves except for the final state, when an interaction between several types of damage occurs. To complete this modelling a three dimensional cell will be investigated to simulate a more realistic growth of defects and to confirm this first step approach.

Part of this work has been realized with the support of S. E. P Bordeaux wich is gratefully acknowledged.

REFERENCES

1. LEBON, B., LESNE, P.M., RENARD., ARNAULT, V., , Non linear behavior of damaged woven fabric ceramic composites 2nd Int. Conf. CADCOMP Brussels 1990
2. RENARD J. , "Study of matrix behavior in a composite material by a homogenization method". ICCM "Advancing with composites", Milan(Italy), May 10-12, 1988.
3. RABOTNOV Y. N. , "Creep problems in structural members", North-Holland (1969).
4. LEMAITRE J. , CHABOCHE J. L. ,Mechanics of Solids Materials, Cambridge University Press, 1989.

CONSTITUTIVE LAW FOR GREEN SHOTCRETE - A STEP TOWARDS REALITY

Rudolf Pöttler
ILF Consulting Engineers, Innsbruck/Austria

ABSTRACT

In this paper, a new constitutive model for green shotcrete in tunnelling is being presented. It can be claimed that the new shotcrete model simulates reality accurately and thus contributes considerably to a greater acceptance of numerical investigations among engineers working in practice and, what is most important, attaches greater importance to the results of numerical analyses in tunnel design. It has been successfully used in the NATM sections on the U.K. side of the Channel Tunnel.

INTRODUCTION

Shotcrete is an essential support measure for tunnels driven according to the principles of the New Austrian Tunnelling Method (NATM).

In the past, the use of numerical models was criticized mainly for the unrealistically high shotcrete stresses achieved in the calculations. These values were in stark contrast to practical experience, according to which the shotcrete stresses measured had always been nearly constant and relatively low.

NUMERICAL MODEL

In the meantime, experts have succeeded in proving the pronounced relaxation of green shotcrete in laboratory tests and expressing this phenomenon in a form applicable in numerical calculations [1] by expressing the creep rate as a function of stress. The formula developed in [1] was based on a modified Burger Model, which has the disadvantage that with high stresses (above 10 MPa) a trend reversion in the calculation of the creep rate is observed, i.e. the higher the stresses the lower the creep rate calculated. So in [2] a formula was developed, in which this shortcoming was overcome (Fig.1).

$$\dot{\underline{\epsilon}}^{in}(t) = (\alpha(t)\ \sigma^2{}_v + \beta(t)\ \sigma_v{}^3) * \frac{3}{2} * \underline{M}_2 * \underline{\sigma} \qquad (1a)$$

$$\sigma_v{}^2 = 0.5\ (\sigma_I - \sigma_{II})^2 + (\sigma_{II} - \sigma_{III})^2 + (\sigma_{III} - \sigma_I)^2)$$

$\sigma_I, \sigma_{II}, \sigma_{III}$ ..... principal stresses
$\underline{M}_2$ .... condensation matrix

$$\alpha(t) = (0.02302 - 0.01803\ t + 0.00501\ t^2) * 10^{-3} \qquad (1b)$$

$$\beta(t) = (0.03729 - 0.06656\ t + 0.02396\ t^2) * 10^{-4} \qquad (1c)$$

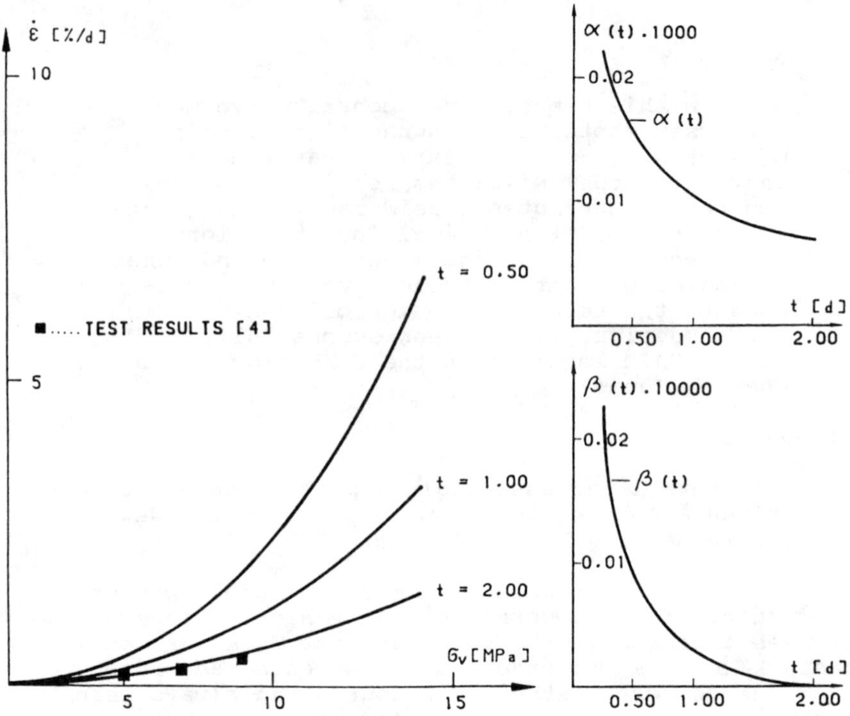

Fig.1 Strain rate as a function of stress σ and time t

LOADING OF SHOTCRETE LINING DURING EXCAVATION PROCESS

Using this formula, a realistic curve of the development of shotcrete stresses and deformations over time during tunnel excavation can be calculated. An example of this is shown in Fig.2. The uneven loading is clearly visible in the figure. During each round, the shotcrete stresses suddenly increase. Afterwards, these stresses relax and forces are again transferred from the shotcrete into the rock mass. The higher the stress the higher the relaxation. The farther the cross section considered is located off the working face, the smaller the stress increase during each round will be. Beyond a distance of 1 - 2 diameters from the working face, no more loading of the shotcrete is observed at all. Relaxation prevails. As extensive parameter studies have shown, shotcrete stresses relax almost independently of the loading history and the boundary conditions to a value of σ = 4 MPa [2]. This is in accordance with the above mentioned measurements

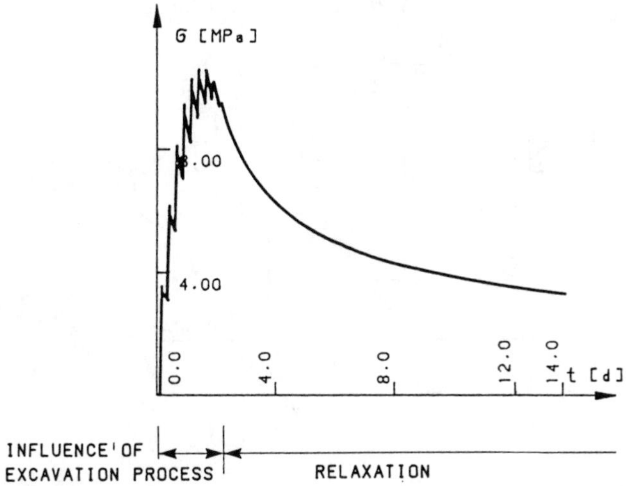

Fig.2 Stresses in a shotcrete lining within the first 14 days after excavation [2]

made on site. The higher stress values occurring in the initial phase usually are not measured by the engineers on site, which is due to constructional conditions.

CROSS SECTIONAL FORCES: AXIAL FORCE - BENDING MOMENT

Another significant feature of shotcrete can now be understood: It is true that in the shotcrete lining mainly axial forces occur, whereas bending moments are minor. Numerous relaxation tests [3] using various initial strain distributions revealed that moments decrease more strongly than axial forces. Relaxation tests were chosen as this behaviour is most specific to shotcrete loading/unloading in tunnelling (Fig.2, [2]).

A 250 mm thick shotcrete specimen was subjected to an initial compressive strain. The stresses occurring after 1 day and after 10 days were calculated (Fig.3, Table I).

TABLE I. Stresses after 0, 1 and 10 days

Strains		day 0		1 day		10 days	
ε_o	ε_u	M	N	M	N	M	N
[‰]	[‰]	[MNm]	[MN]	[MNm]	[MN]	[MNm]	[MN]
-2.50	0	-0.033	-1.020	-0.007	-0.369	-0.001	-0.063
-2.50	-1.25	-0.017	-1.531	0.001	-0.480	-0.000	-0.078
-2.50	-2.50	+0	-2.041	0	-0.510	0	-0.081

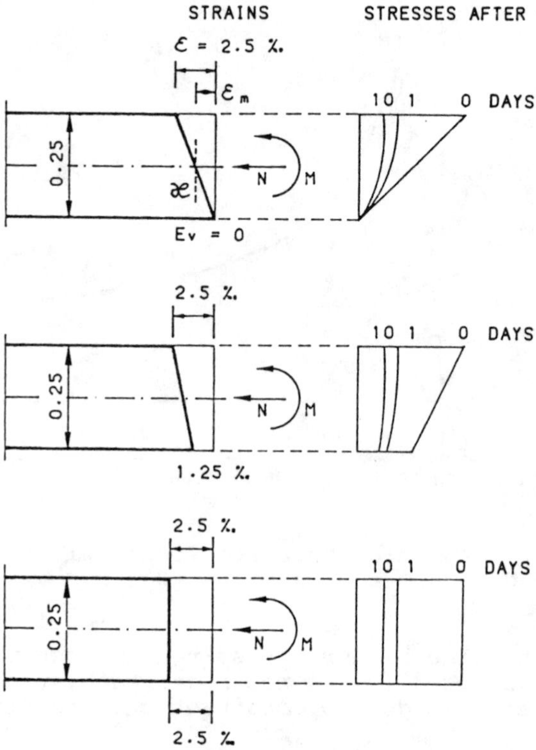

Fig.3 Time-dependent stresses and strains in the numerical relaxation test

REFERENCES

1. P. Petersen, "Geostatische Untersuchungen für tiefliegende Regionalbahnen am Beispiel Hannover" (Geostatic Analyses for Deep Tunnels of Regional Railways on the Basis of the Hannover Railway), Forschungsergebnisse aus dem Tunnel- und Kavernenbau 12, (Technical University of Hannover 1989).

2. R. Pöttler, "Time-Dependent Rock-Shotcrete Interaction. A Numerical Shortcut." Accepted for publication in Computers and Geotechnics.

3. R. Pöttler, "Konsequenzen für die Tunnelstatik aufgrund des nichtlinearen Materialverhaltens von Spritzbeton" (Impacts of Non-Linear Material Behaviour of Shotcrete on Tunnel Stability Analyses), Felsbau 8 (3), (1990).

STRESS ANALYSIS AND DESIGN GUIDELINES FOR REINFORCED EARTH SLABS IN COHESIVE SOILS

Naresh C. Samtani
Graduate Student, Dept. of Civil Engrg. and Engrg. Mech.,
Univ. of Arizona, Tucson, AZ 85721, U.S.A

ABSTRACT

Based upon results and observations of model tests involving strip footing on compacted cohesive soil, a stress analysis has been performed and guidelines for design of Reinforced Earth slabs in cohesive soils are suggested. An equivalent cross-anisotropic material is considered to represent the reinforced soil mass.

INTRODUCTION

A Reinforced Earth slab consists of a bed of soil strengthened by horizontal layers of reinforcements with relatively high tensile strength and capable of developing good bond with the soil. Studies on laboratory models [1],[2],[4],[5] have shown the benefits of incorporating reinforcements in soil. Reinforced Earth slabs are very useful and economical where loads are heavy and the subgrade is poor [6]. Based on the above studies (mainly by [4],[5]) a stress analysis of the Reinforced Earth Slab in cohesive soil has been performed and guidelines for design of Reinforced Earth slabs is presented.

DEFINITION OF PARAMETERS

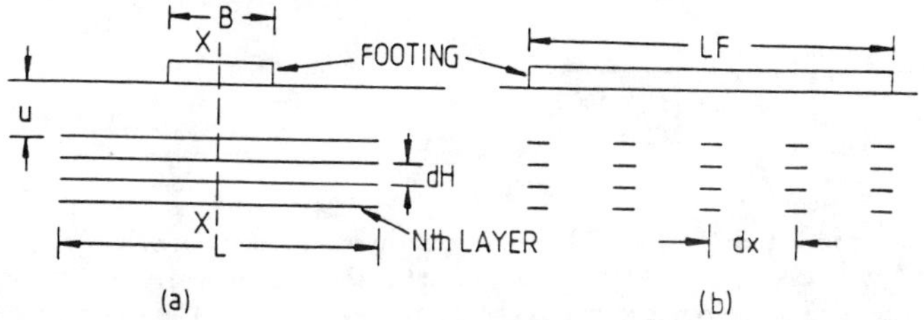

FIG. 1. Geometry of System : (a) Transverse Cross Section; and (b) Longitudinal Cross Section (Section X-X)

The geometry of the system alongwith the possible parameters for study is shown in Figure 1. Let n represent the number of reinforcing ties per layer below the foundation (in Figure 1, n=5) and w represent the width of each reinforcing tie. Let LR ($=n.w$) represent the length of the footing covered by the reinforcing strips. The density of reinforcements can now be conveniently represented by a *Linear Density Ratio, LDR* as follows

$$LDR(\%) = \frac{100.LR}{LF} = \frac{100.n.w}{LF} \tag{1}$$

The parameter dx, (Fig 1-b), can be expressed in terms of LF, n and w as follows

$$dx = \frac{LF - w}{n - 1} \tag{2}$$

For comparing strengths of reinforced and unreinforced soils, a ratio of bearing capacities for both the cases is considered [2],[5] as follows

$$BCR = \frac{q_r}{q_u} \; ; \; BCR^* = \frac{q_r}{q_u^*} \tag{3}$$

where q represents the bearing capacity, the subscripts u and r denote unreinforced and reinforced soils respectively. The definitions of q_r, q_u and q_u^* are shown in Figure 2.

OBSERVATIONS FROM MODEL STUDY

A detailed model study for the investigation of bearing capacity aspects of Reinforced Earth slabs in cohesive soils was carried out. The details of the experimental set-up, a summary of test results, bearing capacity analysis and mechanism of failure have been discussed elsewhere [4],[5]. Observations from this model study program which are useful while formulating design guidelines for reinforced earth slabs in cohesive soils are : a) The failure surface in reinforced mass is circular as in the case of the unreinforced soil, but the radius of the failure arc in reinforced soil is much larger than in unreinforced soil. However, the plan length of the failure surface (L_{plan}) remains approximately the same for unreinforced soil and soil reinforced with different *LDR*. b) Longer reinforcing strips for a given *LDR* do not significantly increase the bearing capacity. c) The load transfer between the soil and the ties takes place according to the free-body diagram shown in Figure 3. In this mechanism, the strips are subjected to tension equal to the load on the strips. d) The relation between *LDR* and *BCR* or *BCR is not linear as shown in Figure 4.**

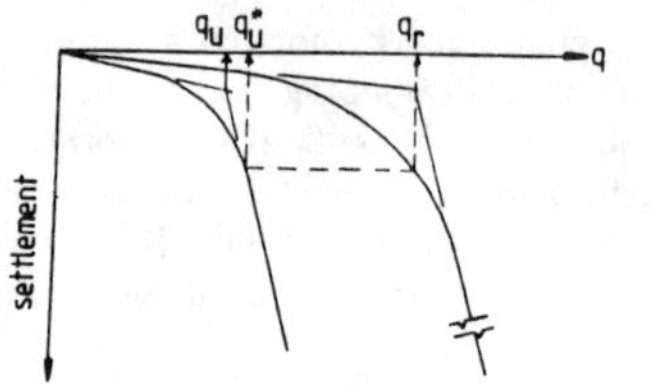

FIG. 2.

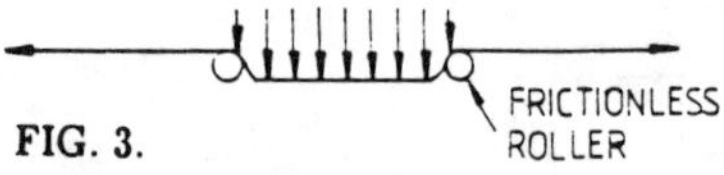

FIG. 3.

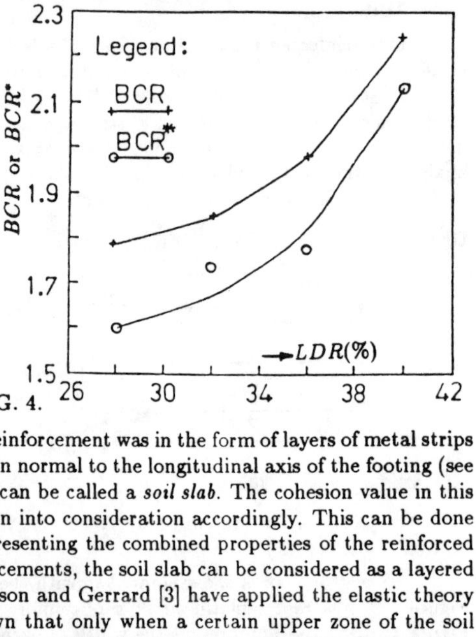

FIG. 4.

In the model study program, the reinforcement was in the form of layers of metal strips (also called *ties*) placed only in the direction normal to the longitudinal axis of the footing (see Figure 1). The zone of the reinforcements can be called a *soil slab*. The cohesion value in this zone is an enhanced one and must be taken into consideration accordingly. This can be done by considering an equivalent material representing the combined properties of the reinforced earth mass. Due to the provision of reinforcements, the soil slab can be considered as a layered mass like the Westergaard material. Harrison and Gerrard [3] have applied the elastic theory to Reinforced Earth media and have shown that only when a certain upper zone of the soil is reinforced there occurs a decrease in the vertical stress with depth. They have further shown that in the unreinforced underlying layer the vertical stress is considerably decreased and the radial stress considerably increased thereby producing a far more stable condition. This statement finds support in the increased bearing capacity in the case of reinforced soil as compared to unreinforced cohesive soil at the same settlement (this is given by BCR^*). The depth of reinforcing can safely be taken as the depth to which the failure surface would develop in the case of unreinforced soil under the load corresponding to the ultimate bearing capacity of unreinforced soil. Below this depth the effect of the imposed stresses is negligible in the case of unreinforced soil.

ANALYSIS OF TIE FORCES

When a load is applied to soil, it is resisted upto a certain limit by the soil. This limit is called the ultimate bearing capacity of the soil. The stresses within the soil mass are, in practice, limited to the safe bearing capacity of the soil. In the case of reinforced earth slabs, loads of a much higher magnitude (can be) are applied. In such a case the difference between the imposed stresses, σ_{zr} and the safe bearing capacity of the soil, σ_{zs}, plus the tension in the soil is carried by the reinforcements. The load is transferred to the ties by the mechanism shown in Figure 3. The stress distribution at each level of the reinforcement is required for estimation of the tensile force at that level. As the earth slabs represent a layered soil mass in which relatively soft and very stiff layers alternate with each other Westergaard's stress distribution theory appears to be more suited. The evaluation of the tie forces proceeds by assuming that the forces are evaluated for the same size of the footing, B, and the same settlement, ρ, for a footing on unreinforced and reinforced soil. For Westergaard material [7], assuming an equivalent cross-anisotropic material, Harrison and Gerrard [3] have obtained solutions for a wide range of circular and strip loads by applying elastic theory. For the case of a strip load of uniform vertical displacement, Harrison and Gerrard, 1972, have given the following solution for vertical stress at depth z

$$\sigma_z = \frac{T_z}{\pi x_o} \frac{1}{(\eta^2 z^2 + 1)^{\frac{1}{2}}} \quad \text{where} \quad \eta = \sqrt{\frac{1 - 2\nu_s}{2(1 - \nu_s)}} \tag{4}$$

where T_z is the total vertical force per unit length of strip, x_o is the half loaded width of the strip ($=B/2$), B is the loaded width of the strip, z is the distance measured downward from the bottom of the loaded strip in the direction of the central load axis, and η represents the characteristics of the Westergaard medium (ν_s is the value of Poisson's ratio for the soil).

The strip load intensity (at depth z) along a strip B units wide will be $\sigma_z B$. The tensile force, T, carried by a particular layer of reinforcements per unit length of the footing is given by

$$T = (\sigma_{zr} - \sigma_{zs})B \quad \text{or} \quad T = \frac{2(T_{zr} - T_{zs})}{\pi(\eta^2 z^2 + 1)^{\frac{1}{2}}} \tag{5}$$

where T_{zr} and T_{zs} are total vertical forces per unit length of foundation corresponding to σ_{zr} and σ_{zs} respectively.

However, from model test results presented by Samtani and Sonpal (1989) it was found that the Eq (5) overestimates the stress (probably because the Westergaard material assumes continuous sheets of stiff material). Using back calculations from model test results, Eq (5) can be modified to obtain a similar equation suiting the conditions existing in reinforced earth slabs. It was found that an equation similar to Eq (5) having a numerical coefficient of 0.28 (when the units of kN and meter are used) instead of $2/\pi$ predicts the ties forces correctly. The modified equation is as follows

$$T = 0.28 \frac{(T_{zr} - T_{zs})}{(\eta^2 z^2 + 1)^{\frac{1}{2}}} \tag{6}$$

The above equation has the units of force per unit length of foundation. Note that the coefficient has to be appropriately converted while using units other than kN and meter.

Using Boussinesq's stress distribution theory (which is more applicable to homogeneous soils) Binquet and Lee (1975b) have also presented equations for estimation of tie forces. But, these equations predict that the bottom layer of reinforcing strips will break first. This is in contrast to Binquet and Lee's findings [2] that breakage of reinforcing strips occured in the uppermost layers first. This finding is also supported by this study [4],[5].

DESIGN GUIDELINES

For designing stable foundations on cohesive soil using Reinforced Earth technique, following guidelines are suggested :

1) Calculate the total load, T_{zr}, to be supported per unit length on the Reinforced Earth slab. Also obtain the following quantities : (a) Cohesion of soil compacted at OMC, c, (b) Factor of safety against bearing capacity failure in unreinforced soil, FS, (c) Allowable settlement in unreinforced soil, ρ, (d) Tensile strength of ties, f_y, (e) Factor of safety against tie break, FS_B, and (f) Depth of foundation, D.

2) Assume a width of the foundation, B. Calculate the ultimate bearing capacity of unreinforced soil, q_o at depth D. Based on the assumption of circular failure surfaces in cohesive soils, G. Wilson (1941) proposed a simple expression [8], for long footings below the surface of highly cohesive soils as follows

$$q_o = 5.5c\left(1 + 0.38\frac{D}{B}\right) \tag{7}$$

3) By knowing q_o and c obtain the failure surface profile by using Fellenius approach ([7]) of best equilibrium method. Calculate the depth of the critical failure surface, d, and the length of the critical failure surface in plan, L_{plan}. For homogeneous cohesive soils, Wilson (1941) has given a set of curves, Figure 5-b, by which the coordinates of the centre of the critical slip circle may be found out. Let x and y be coordinates of O' with respect to the origin O given in Figure 5-a. For any given value of D/B, the value of y/B and x/B may be found from Figure 5-b. From these values the center of the critical circle O' may be located (Note that Eq 7 complements Figure 5-b). Alternatively, Wilson's coordinates can be used for the first trial center, and other trial circles be drawn with centers near the first. In this case by taking the moments about O' (Figure 5-a) we can write

$$q_o.B.l_o = Wl_{o'} + cR \tag{8}$$

where R is the radius of the slip circle. This process is repeated for several other trial surfaces and the failure surface with the minimum value of R is taken as the critical failure surface. The depth d and plan length L_{plan} of the this critical failure shall be determined with respect to the horizontal plane passing through the base of the footing (at depth D). For Wilson's method, $d = R - y = \sqrt{x^2 + y^2} - y$ and $L_{plan} = 2x$ where x and y are as shown in Figure 5.

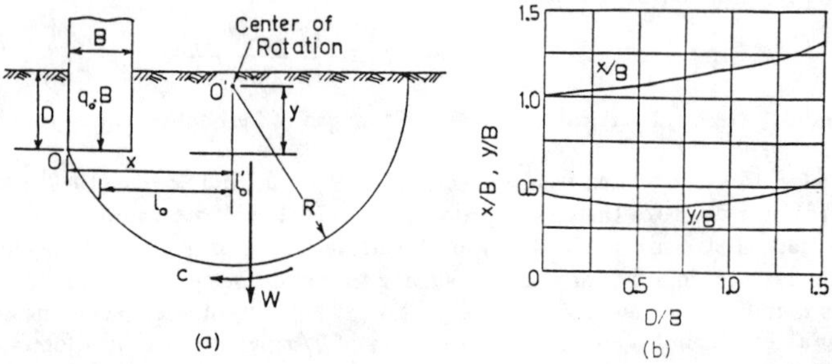

FIG. 5. (a) Fellenius best equilibrium approach, (b) Coordinates of center of slip surface (after Wilson 1941).

4) Since longer reinforcing strips (for a given LDR) do not significantly increase the bearing capacity and the plan length (L_{plan}) of the failure surface remains approximately the same for reinforced and unreinforced soil, we can calculate the minimum length of the reinforcement, L_{min} as follows

$$L_{min} = 2L_{plan} - B \tag{9}$$

Equation (9) would provide a length of reinforcement which would extend just beyond the failure surface which would have developed in the case of unreinforced soil. The length of the ties beyond the critical failure surface can be considered as anchorage length. Some extra length (e.g. 10%) may be provided; this may be considered as factor of safety against pullout. The length of reinforcements in all the layers can be of the same length.

5) Assume suitable values of u, dH, and N. Recommended values are $u \leq 0.5B$, $dH = u$ and $N = d/dH$. The values of u, dH and N can be rounded off.

6) Calculate the allowable load on unreinforced soil, q_{all}. The allowable load is calculated from shear (q_o/FS) as well as settlement consideration. Adopt the lower of the two values as the allowable load (q_{all}). Convert this load intensity into line load intensity, T_{zs}.

7) Calculate the tie force, T, in each layer of reinforcement using Eq (6) by assuming that the load T_{zr} is shared equally by all the layers of reinforcements. If some other stress distribution theory is adopted then obtain an equation similar to Eq (6) and use it to calculate the tie force T.

8) Assume a value of LDR. Figure 4 may be used as a guide (Note that Eq (3) can also be written as $BCR = T_{zr}/T_{zs}$). Assume a width w of reinforcing ties and calculate the number of strips n per unit length from Eq 1. The center to center distance, dx, can be calculated from Eq (2).

9) **Calculate the tie thickness to resist tie break as follows, for each layer**

$$t = \frac{(FS_B)(T)}{(LDR)(f_y)} \tag{10}$$

10) Adopt maximum tie thickness and make allowance for corrosion. Regional design codes must be refered for such guidelines. Let the adopted thickness (after corrosion allowance) be t_1. If the available thickness in the market is t_2 then modify LDR accordingly in direct proportion. If the thickness t_1 is too large then repeat the procedure assuming more layers of reinforcements.

The above design guidelines are based on results and observations from the model study program. The guidelines presented are simple and can easily be programmed. An illustrative example is worked out below.

Example

Design a continuous foundation in cohesive soil that will carry a load of 600 KN/m. Given: *Soil:* c= 65 KPa, ν_s=0.35; Assume D=1, FS=3. *Ties:* f_y=2.5E+6 KPa, FS_B=3
Solution From the data given we have, T_{zr}=600 KN/m. Assume B=1 m. From Eq 7, we calculate q_o=493.35 KPa. Using Wilson's method and Figure 5 we obtain, x=1.17 m; y=0.4125 m; $R = \sqrt{x^2 + y^2}$=1.24 m; $d = R - y$= 0.828m; $L = 4x - B$=3.68m. Assume u=0.3m=dH which would mean $N \simeq 3$. Since FS=3, we get $q_{all} = q_o/FS$=164.45 KPa which means T_{zs}= 164.45 KN/m (since B=1). Dividing the imposed load T_{zr} into 3 parts and using the resulting value in Eq 6 we obtain the tie force in each layer of reinforcements as $T_{z=0.3m}$= 9.85 KN/m, $T_{z=0.6m}$= 9.56 KN/m, $T_{z=0.9m}$= 9.14 KN/m, Assuming LDR=0.65 we obtain the thicknesses for the three layers of reinforcing ties as $t_{z=0.3m}$= 1.81 mm, $t_{z=0.6m}$= 1.76 mm, $t_{z=0.9m}$= 1.68 mm. Select t= 1.8 mm. Assuming the width of each reinforcing strip x=75 mm we obtain (from Eq 1) n= 8.67 $\simeq$ 9 ties per meter length of the foundation. The center to center distance , dx, between the strips works out to be (using Eq (2)) 102.78 mm (with two strips at the end of the strip footing as shown in Figure 1-b).

In the above design appropriate allowance for corrosion can be made according to the regional design codes. The above example illustrates the simplicity of the design procedure.

CONCLUSIONS

Observations and results from the model study program on reinforced earth slabs in cohesive soil show that the mechanism involved provides a simple and elaborate basis for design guidelines to be framed. With more study similar guidelines can be evolved for different shape of footings as well as for different reinforcing materials.

REFERENCES

1. Akinmusuru, J. O., and Akinbolade, J. A.,"Stability of loaded footing on reinforced soil." J. Geotech. Engrg., ASCE, 107(6), 810-827 (June 1981).
2. Binquet, J. and Lee, K. L., "Bearing capacity analysis on reinforced earth slabs." J. Geotech. Engrg., ASCE, 101(2), 1257-1276 (Feb. 1975b).
3. Harrison, W. J., and Gerrard, G. M.,"Elastic theory applied to reinforced earth." J. Soil Mech. and Found. Div., ASCE, 98(12), 1325- 1345 (December 1972).
4. Samtani, N. C., Reinforced earth slab foundations in cohesive soils, Master's thesis presented to L. D. College of Engineering, Gujarat University, Ahmedabad, India (1986).
5. Samtani, N. C., and Sonpal, R. C., "Laboratory tests of strip footing on reinforced cohesive soil." J. Geotech. Engrg, ASCE, 115(9), 1326-1330 (September 1989).
6. Steiner, R. S., "Reinforced earth bridges highway sinkhole." Civil Engrg. ASCE, 45(7), 54-56 (July 1975).
7. Taylor, D. W., Fundamentals of soil mechanics, 573-575 (Asian Student Edition 1948).
8. Wilson, G., "The calculation of bearing capacity of footings on clay." London: J. Inst. of Civil Eng., (November 1941).

SUBJECT INDEX

D

E

F

H

I

K

L

M

N

P

R

S

AUTHOR INDEX

E

F

G

H

I

J

K

L

M

N

O

P

Q

R

S

T

W

X

Y

Z